Proceedings of the 2015 International Conference on

Sustainable Development

(ICSD2015)

Proceedings of the 2015 International Conference on

Sustainable Development

(ICSD2015)

Liandong Zhu
Vaasa University, Finland

Ahmed Ouadha
USTO-MB Algeria

World Scientific

NEW JERSEY • LONDON • SINGAPORE • BEIJING • SHANGHAI • HONG KONG • TAIPEI • CHENNAI • TOKYO

Published by

World Scientific Publishing Co. Pte. Ltd.
5 Toh Tuck Link, Singapore 596224
USA office: 27 Warren Street, Suite 401-402, Hackensack, NJ 07601
UK office: 57 Shelton Street, Covent Garden, London WC2H 9HE

British Library Cataloguing-in-Publication Data
A catalogue record for this book is available from the British Library.

SUSTAINABLE DEVELOPMENT (ICSD2015)
Proceedings of the 2015 International Conference on Sustainable Development

ISBN 978-981-4749-90-9

Printed in Singapore

Preface

Since the emergence of climate and global warming onto the international agenda, research in sustainability has been underpinned by the development in energy and environmental science. Highlighted 30 years ago by the Brundtland Commission, "sustainable development" was defined as: *meeting the needs of the present without compromising the ability of future generations to meet their own needs*. This has very much defined the scope and aims of this conference.

This conference proceedings book contains the selected papers presented in the 2015 International Conference on Sustainable Development (ICSD2015) held in September 25-27, 2015 in Wuhan, Hubei, China. The conference positions itself as an international forum for researchers all over the world to come together to share and discuss their findings and contributions in all aspects of sustainability; including theory, methodology and applications covering wide spectrum of topics and issues. The conference proceedings put together a total of 119 papers in sustainable development, covering issues in environmental, energy, and economical aspects of the subjects.

Invited by the committee to edit, we are both honored and excited to have this opportunity to work for and compile this proceedings. First of all, we would like to express our gratitude to the dedication of the reviewers for the effort and time devoted during the peer review process. Their attentive comments and critiques have undoubtedly improved the quality of the selected papers. Secondly, on behalf of the ICSD2015 conference committee, we would like to thank the sponsors who have contributed so much to the success of the conference. In particular, Advanced Science, Technology and Industry Research Center, Xi'an University of Science and Technology, Anhui University of Science Technology, Guangdong University of Technology and many other research institutes have supported us in many ways.

Finally, we look forward to seeing you again at the next conference on sustainable development, possibly in one or two years' time.

Professor Liandong Zhu, The University of Vaasa, Finland
Guest Editor of ICSD2015

Professor Ahmed Ouadha, USTO-MB, Algeria
Guest Editor of ICSD2015

2015 International Conference on Sustainable Development (ICSD2015)

Table of Contents

Chapter 1
Sustainable Agriculture

Chinese cabbage infected by *Plasmodiophora brassicae* was first identified in Changde, China

J. Du, X.W. Luo, J. Peng and L.F. Wu
College of Plant Protection, Hunan Agricultural University, Changsha 410128, China

S.B. Zhang, D.Y. Zhang and Y. Liu
Key Laboratory of Integrate Diseases and Pests Management in Horticultural Crops in Hunan, Hunan Plant Protection Institute, Changsha 410125, China
E-mail: zsongb2008@163.com

Z.B. Zhao
Shaoyang University, Shaoyang 422000, China

The pathogen of Chinese cabbage from Changde city, Hunan province, China with swell root was identified as Plasmodiophora brassicae based on symptoms, characteristics of hypnosporangium, and specific PCR detection. The homologous analysis of partial 18S rRNA gene hinted this population may be a novel physiological race of P. brassicae. The clubroot is a serious threaten to many import crops of Cruciferae, and it is difficult to prevent highly infectious, quick spreading, diverse dissemination ways. So the concern on monitoring, taking precaution and preventing should be increasing.

Keywords: Chinese cabbage; Clubroot; Identification; Changde.

1. Introduction

Clubroot, caused by the soil borne obligate parasite *Plasmodiophora brassicae* Woronin, is a destructive disease of brassica crops, such as Chinese cabbage in many regions of China. Clubroot is general results in a 20%-30% reduction in yields, and over 60% up to whole reduction when it is outbreaking in China (Dixon 2009; Howard et al. 2010). Clubroot was spread by multiple ways, such as soil, infected plant debris. In addition, the hypnosporangium could survive over twenty years and bound in soil and plant debris (Kageyama & Asano 2009; Castlebury & Domier 1998), so it is extremely difficult to prevent (Yang et al. 2009).

Clubroot was first observed in Europe, and till now, was wide spreading over Europe, North America, Japan and several provinces of China, partly because of variety of Cruciferae increasing, growth area enlarging and commercial seeds exchanging (Bulman et al. 2001; Sharma et al. 2011) There was only one publication of certainly damage of Chinese cabbage by clubroot in Changsha city, Hunan province, China in 1990s (Yang et al. 1992). After that, no publications of infection of plant in the Brassicaceae family in Hunan, China were issued.

Suspected clubroot of Chinese cabbage was found in Changde city, Hunan province, China in Aug 2012, and following identification tests were carried out.

2. Materials and Methods

2.1. *Sampling*

Infected Chinese cabbage was sampling in Aug 2012 in Changde city, Hunan province, China. Two infected and two healthy samples were sampled and stored at 4 ^{o}C.

2.2. *Morphology of hypnosporangium*

Transverse section of swelling root of infected Chinese cabbage was observed in optical microscope (BX13, Olympus, Japan) (Kobelt et al. 2000 & Sharma et al. 2011). Briefly, the root of sampled chinese cabbage was washed and placed in a fixative solution (1:1, 95% acetic acid: 95% ethanol). Samples were cut in transverse sections of 0.5-cm-long root segments from the top 0 to 1cm of the main taproot. The tissue was dehydrated through through an ethanol series and then embedded in paraffin. Cross-sections (6-μm thick) were cut with a microtome, then was observed in optical microscope.

2.3. *Specific PCR detection and sequence analysis*

Total DNA of root of Chinese cabbage was extracted by TRIzolR reagent according to manufacturer's inductions (Life Technologies, CA, USA). Specific PCR primers (TC1F: 5'-GTGGTC GAACTTCATTAAATTTGGGCTCTT-3', TC1R: 5'-TTCACCTACGGAACGTATATGTGCATG TGA-3' and TC2F: 5'-AAACAACGAGTCAGCT TGAATGCTAGTGTG-3', TC2R: 5'-CTTTAGT TGTGTTTCGGCTAGGATGGTTCG-3') was synthesized by BGI Incorporated (Beijing, China) according to previous publication (Rennie et al. 2007). All PCR amplifications were conducted using a StepOne rapid PCR System (Applied Biosystems) in a 20 μL volume containing 0.5 μL template DNA solution, 0.5 μL of each of a 10 mM TC1F/TC1R or TC2F/TC2R, and 10 μL 2×Reaction mix, 1 μL Enzyme mix. Reaction conditions consisted of an initial heat denaturation step at 95 ^{o}C for 2 min, followed by 35 cycles of 95 ^{o}C for 15 s and

60 °C for 60 s and 72 °C for 1 min, finally with 72 °C for 10 min (Schnerr et al., 2001; Etebu & Osborn, 2010).

The target bands of PCR were sequencing by BGI Incorporated (Beijing, China), and the homology was analyzed by nblast in GenBank (http://blast.ncbi.nlm.nih.gov/).

3. Results and Discussion

3.1. *Occurrence area and symptoms*

The infected Chinese cabbages were collected from Changde city, Hunan Province, China. There was about 1.5 ha was infected, and no infected samples was found in Chinese cabbage and others crops of brassica in adjacent city with following survey.

The symptoms of infected Chinese cabbage were shown as Figure 1. There are typical symptoms of clubroot with infected Chinese cabbage, including club-shaped root, and wilting, yellowing of whole plant.

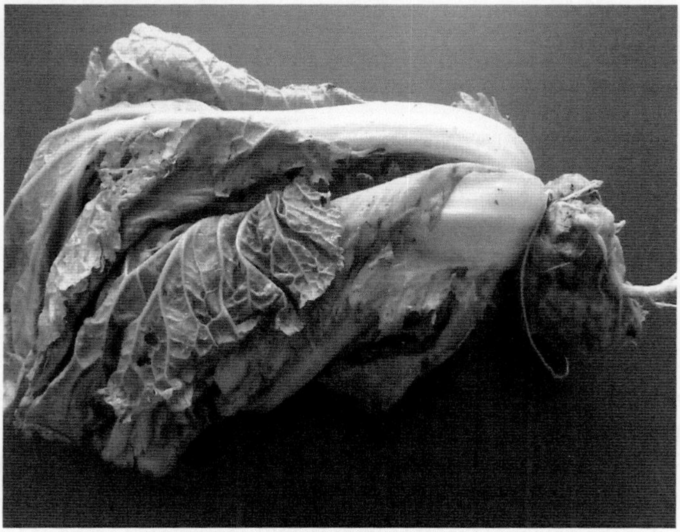

Figure 1. Symptom of infected Chinese cabbage.

3.2. *Morphology of hypnosporangium*

As shown in Figure 2, there are aggregating hypnosporangium in the infected cells of root of Chinese cabbage. And hypnosporangium was circular or approximately circular. It is typical characteristics of hypnosporangium of *Plasmodiophora brassicae* Woronin (Hwang et al. 2012; Williams 1996).

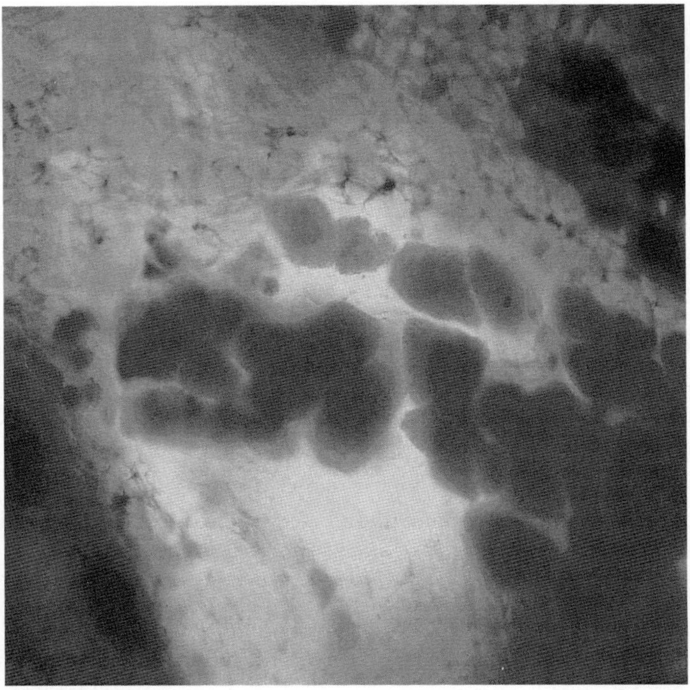

Figure 2. Characteristics of hypnosporangium in swelling root of infected Chinese cabbage (×100)

3.3. *Specific PCR detection*

From the infected Chinese cabbage samples, there are special target bands in line with expectations by both specific PCR primers in agarose gel, and not from healthy Chinese cabbage sample (Figure 3). Together with the symptoms (Figure 1) and microscopic hypnosporangium (Figure 2), the pathogen of infected Chinese cabbage sampled from Changde city, Hunan province, China, was identified as *Plasmodiophora brassicae* Woronin (Cavalier-Smith & Chao 2003). This pathogen could infect most of crops of Cruciferae, and hardly to preventing by conventional fungicides (Cao et al. 2009, Diederichsen et al. 2009), so it is the potential threat to importantly commercial crops of Cruciferae, such as rape, field mustard and Chinese flowering cabbage in Hunan province, China.

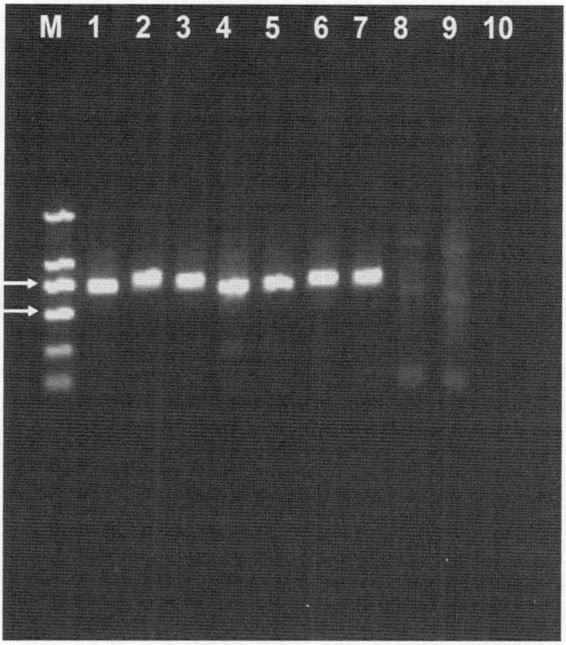

Figure 3. Specific PCR detection of infected Chinese cabbage. (M: DL2000 Marker; 1-3: PCR products of infected samples by primer TC1F/TC1R; 4-7: PCR products of infected samples by primer TC2F/TC2R; 8: PCR of healthy Chinese cabbage by primer TC1F/TC1R; 9: PCR of healthy Chinese cabbage by primer TC2F/TC2R; 10: CK)

3.4. *Sequence homology*

The partial 18S rRNA gene of *Plasmodiophora brassicae* Woronin of infected Chinese cabbage in Changde city (named as strain CD-1), which amplified by PCR with primer TC1F/TC1R was sequenced (GenBank accessing number: KF129413), and only 92% and 93% homologous to corresponding sequences of *P. brassicae* isolates from United Kingdom (Ward & Adams 1998) and Switzerland, respectively, 93% homologous to corresponding sequences of isolates from Asia, including Japan (GenBank accessing number: AB0949983 and AB094977) and Korea (GenBank accessing number: AF353998). All of these suggested that the *P. brassicae* population (strain CD-1) from Changde city, Hunan province, China may be from the different evolutionary origin as those from above districts, and also hints it is possible a novel physiological race. It was well-understanding that there is significant difference of pathogenicity to different crops infected by different physiological races of *P. brassicae* (Cao et al. 2007, Hwang et al. 2012). Hence, it is of great concern to identify the physiological race and distribution of *P. brassicae* strain CD-1.

Acknowledgement

This work was supported by the Agriculture Research System of China (CARS-25-B-05).

References

1. Bulman, S. R. Kuhn, S.F. Marshall, J.W. & Schnepf, E. 2001. A phylogenetic analysis of the SSU rRNA from members of the *Plasmodiophorida* and *Phagomyxida*. *Protist* 152: 43-51.
2. Cao, T. Manolii, V.P. Hwang, S.F. Howard, R.J. & Strelkov, S.E. 2009. Virulence and spread of *Plasmodiophora brassicae* [clubroot] in Alberta, Canada. *Canadian Journal of Plant Pathology* 31: 321-329.
3. Cao, T.S. Tewari, J & Strelkov, S.E. 2007. Molecular detection of *Plasmodiophora brassicae*, causal agent of Clubroot of Crucifers, in plant and soil. *Plant Disease* 91: 80-87.
4. Castlebury, L.A. & Domier, L.L. 1998. Small subunit ribosomal RNA phylogeny of *Plasmodiophora brassicae*. *Mycologia* 90: 102-107.
5. Cavalier-Smith, T & Chao, E.E. 2003. Phylogeny and classification of *Phylum Cercozoa* (Protozoa). *Protist* 154: 341-358.
6. Diederichsen, E. Frauen, M. E. Linders, G.A. Hatakeyama, K. & Hirai, M. 2009. Status and perspectives of clubroot resistance breeding in crucifer crops. *Journal of Plant Growth and Regulation* 28: 265-281.
7. Dixon, G.R. 2009. The occurrence and economic impact of *Plasmodiophora brassicae* and clubroot disease. *Journal of Plant Growth and Regulation* 28: 194-202.
8. Etebu, E. Osborn, A.M. 2010. Molecular quantification of the pea footrot disease pathogen in (*Nectria haematococca*) in agricultural soils. *Phytoparasitica* 38: 447-454.
9. Howard, R.J. Strelkov, S.E. & Harding, M.W. 2010. Clubroot of cruciferous crops - new perspectives on an old disease. *Canadian Jounal of Plant Pathology* 32: 43-57
10. Hwang, S.F. Strelkov, S.E. Feng, J. Gossen, B. & Howard, R.J. 2012. *Plasmodiophora brassicae*: a review of an emerging pathogen of the Canadian canola (*Brassica napus*) crop. *Molecular Plant Pathology* 13: 105-113.
11. Kageyama, K & Asano, T. 2009. Life cycle of *Plasmodiophora brassicae*. *Journal of Plant Growth and Regulation* 28: 203-211.
12. Kobelt, P. Siemens, J & Sacristan, M. D. 2000. Histological characterization of the incompatible interaction between *Arabidopsis*

thaliana and the obligate biotrophic pathogen *Plasmodiophora brassie*. *Mycological Research* 2: 220-225.

13. Rennie, D.C. Manolii, V.P. Cao, T. Hwang, S.F. Howard, R.J. & Strelkov, S.E. 2011. Direct evidence of surface infestation of seeds and tubers by *Plasmodiophora brassicae* and quantification of spore loads. *Plant Pathology* 60: 811-819.
14. Schnerr, H. Niessen, L & Vogel, R.F. 2001. Real time detection of the tri5 gene in Fusarium species by LightCycler^{TM}-PCR using SYBRR Green I for continuous fluorescence monitoring. *International Journal of Food Microbiology* 71: 53-61.
15. Sharma, K. Gossen, B.D & McDonald, M.R. 2011. Effect of temperature on cortical infection by *Plasmodiophora brassicae* and clubroot severity. *Phytopathology* 101: 1424-1432.
16. Ward, E. & Adams, M.J. 1998. Analysis of ribosomal DNA sequences of Polymyxa species and related fungi and the development of genus- and species-specific PCR primers. *Mycology Research* 102: 965-974.
17. Williams, P.H. 1966. A system for the determination of races of *Plasmodiophora brassicae* that infect cabbage and rutabaga. *Phytopathology* 56: 624-626
18. Yang, P.W. Shang, H. Dong, L.Y. Liu, S.Y & Li, J.R. 2009. Study on occurrence factors of *Plasmodiophora brassicae* Wor. and controlling strategies. *Southwest China Journal of Agricultural Science* 22 : 663-666.
19. Yang, Y.H. Zhang, X.F & Zou, L.J. 1992. Studies on the occurrence of Chinese cabbage clubroot in Changsha. *Journal of Hunan Agricultural College* 18: 854-858.

An analysis on periodicity of international grain prices fluctuation

S.P. Wang* and Y.Y. Zhu†
School of Economics and Management, North China University of Technology, Beijing, P. R. China.
*Email: *dwangshuping@163.com; †yanyun_zhu@sina.com*

Grain is integral to people's lives. International grain prices caused widespread concern because of frequent and sharp fluctuation. This paper studied the periodicity of the grain prices fluctuation. Firstly, it analyzed maize, wheat and rice prices' fluctuation visually, and on this basis, it analyzed price's volatility further by using spectrum analysis and wavelet analysis. The empirical result shows that there is a cycle about 2-4 years in which the grain price's changes are related to food production cycle. The superposition of these different periods makes grain prices' fluctuation present multiple cycle type.

Keywords: International grain prices; Periodicity; Spectral analysis; Wavelet analysis.

1. Introduction

Grain is not only the important strategic materials about the people's livelihood and national economic security, but also is the essential means of subsistence of people. The grain prices as a base of others prices, they fluctuate would affect the people's lives, as well as has more influence on the country's macroeconomic. In recent years, grain prices fluctuate sharply attracted the national attention. Although China is a major importer of grain, but lack of power in the market. Sharply higher grain prices also caused a great impact on China's economy. To decrease the volatility impact, we need to have more accurate forecasting about prices. This paper is researching on periodicity of grain prices to find out the law of food price volatility.

Since the 19th century, scholars studied and found that the economic development cycle, such as the French economist Juglar put forward the 9-10 years economic fluctuations in "The French, British and American business crisis and its cycle" which published in 1862; Kichin proposed 2-4 years of economic cycle in 1923, the main concern is the change of the inventory; Kuznets made 15 to 25 years long economic cycle in1930 (K. Andrey, T. Sergey, 2010).The international grain market is an important part of economic system, it should also exists certain periodicity. C. K. Harley (1980) conducted a periodic wheat research, and study object was volume of wheat. The result showed that volume appeared Kuznets cycle. C. Labys, Kouassi & Terraza (2000) analyzed the 21 kinds of major commodities' monthly price of the NBER statistical database by using STS models, the results showed that 21 categories fluctuations in 1960-1995 show two species cycle, the long period reflects the basic law of economic operation, and the short cycle is under attack of the short-term factors. Pustilnik & Din (2003) studied relationship between the solar activity and the wheat market, found that changes in basically the same. It indicated the cycle of solar activity is similar to wheat. CSC Sekhar (2003) analyzed monthly prices in international market as well as domestic market by the GARCH model, the results showed that fluctuations in the international and domestic agricultural prices cycle is not consistent, the length of the fluctuant periodicity, the international grain prices had to be shorter than the domestic prices. H. Cao (2007) used the monthly price of wheat mutation rate divided cycle. Study showed that the average cycle length of wheat was 29 months, and supply and demand determine the long-term trend of the price of wheat, at the same time the national policy had control effect on volatility of wheat. F. X. Meng et al (2008) used the domestic food retail price index to evaluate its fluctuation and make a simple experience judgment. The conclusion is that there are three major cycles of grain prices since 1990s, the rapid rise in 1990-1997, the depth adjustment in 1997-2002, the moderate rise in 2003-2007, and thus determine that China's grain prices fluctuations are mainly a basic cycle length for six years. X. Fu, B. J. Niu (2009) through the descriptive statistics analysis of agricultural price volatility characteristics showed that international agricultural prices had experienced six major phases. Erten & Ocampo (2012) studied the deviation from the long-term trend of 20% - 40% in grain prices which called super cycle fluctuation, the empirical results showed that the international grain prices had super cycle, and super cycle amplitude is bigger in tropical agriculture. W. Fang et al (2013) analyzed the market price of rice based on X-12 and HP filter, the results showed that the rice price has obvious seasonal characteristics and the average period was 47 months. Fluctuation in the price level (CPI) was the main factor affecting the rice price fluctuations. At present, spectral analysis and wavelet analysis are used in research on signal data processing, this two methods are used in the study of commodity prices periodicity (L. Ding, 2010; L. Zunino et al, 2011).

In general, some literatures mentioned the periodicity of grain prices volatility, but it was a kind of qualitative and subjective analysis primarily and rarely established mathematical model for in-depth analysis. At the same time, it lacked in-depth theoretical explanation for the cause of grain prices cyclical fluctuations. This paper selects maize, wheat and rice as the research object, using spectral analysis and wavelet analysis method to study the periodic fluctuation and giving specific cycle length, while explains the cause of cyclical fluctuations in grain prices by relevant economic theory.

This paper is organized as follows. Section 2 analyzes the grain prices periodicity intuitively, it judges the periodic existence and length of period by prices' fluctuations. Section 3 respectively uses spectral analysis and wavelet analysis to study the grain prices period by specific mathematical analysis way. Section 5 concludes the paper.

2. An Intuitive Analysis in Grain Prices Periodicity

Maize, wheat and rice are the three major grain crops in the world, as well as they are staple crops in international import and export trade. Their volatilities represent the grains' fluctuations. The United States is a major exporter of maize, while CBOT maize futures are subject to 2nd yellow maize for delivery. In the global market, the top two wheat futures exchange are Chicago board of trade (CBOT) and Kansas City board of trade (KCBT), this two exchanges are choosing hard red winter wheat as the delivery variety and hard red winter wheat production in the United States is the largest species. Rice exporter mainly concentrated in seven countries. Thailand is the largest exporter, and the Thailand rice classification is numerous of which 100% of the grade Brice has the most exports. This paper selects the USA 2nd yellow maize, American hard red winter wheat and 100% of the grade B Thailand rice as representatives to research the grain price periodic volatility, which time is from January 1994 to October 2014, including 250 data totally. Data sources: maize and rice, Food and Agriculture Organization of the United Nations-FAO; wheat, the United States department of agriculture.

Figure 1 shows the three varieties of grain prices.

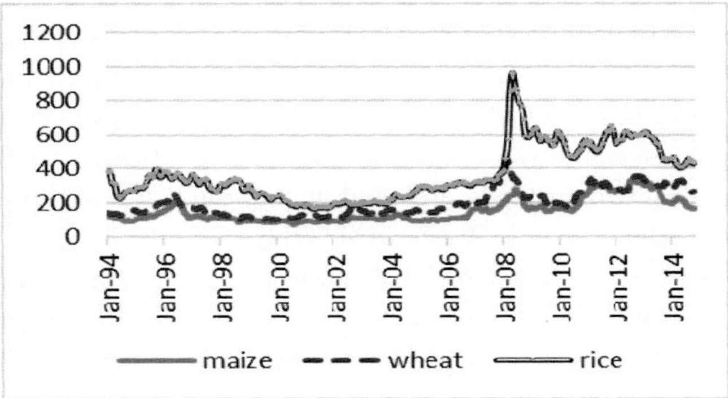

Figure 1. The prices of corn, wheat and rice.

It is clear that three grains have same trend in during this time, and prices movements can be roughly divided into three periods.

The first cycle: 1994-2001. At this stage, the grain prices were low growth and reached the stage of vertices in beginning of 1996, mainly due to Uruguay Round Agreement on Agriculture affect the rising price. Various countries adjust the agricultural policy, especially reduce price support and export subsidies, followed by the United States, the European Union's economic recovery and the rapid economic development in East Asia as well as Latin America pulled the rising grain prices. Grain prices began to decline in 1996 until 2001, which is mainly caused by oversupply. On the supply side, the agricultural economy of the developed countries implemented grain subsidies, resulting in export surplus. The OECD pointed out that agricultural subsidy in some countries account for 25% of total revenue, while Iceland and some countries accounted for 60% in the report "agricultural outlook 2003-2008". The corresponding demand side, the former Soviet Union economy continued to decline, the change of the demand for agricultural products in developing countries, as well as the existence of environmental standards, sanitary and other non-tariff barriers makes grain less demand.

The second cycle: 2002-July 2010. In 2002 to June 2007, grain prices rose in shock state. The main reason was reducing grain production and increasing consumption, at the same time currency liquidity surplus. Plenty of money into the agricultural futures market, which promoted the rising food prices. From July 2007 to May 2008, grain prices were rising rapidly, mainly due to imbalance between supply and demand. Global warming, major grain-producing countries such as Australia, the European Union, the United States suffered drought, which was the direct cause of rising grain prices. India, Russia and other countries limited the export of food in order to guarantee the domestic supply, reducing global food reserves. Followed by oil prices rising and development of biomass liquid fuels, which was performance in two aspects, one was the increasing energy prices increased grain production cost, the second was improving the grain demand to produce more biofuel. The final reason was international capital investment, which played an important role in rising prices. Since 2007, the risk of securities, bonds and other investment products added for impact of the US subprime crisis. "Hot money" flows into the grain market. In June 2008 to July 2010, grain prices were falling fast, partly because prices rose too fast on last stage, followed by the world grain production. And the most important reason was that the United States strengthened to the regulation of the commodity futures markets, "hot money" out flowed and grain prices fell rapidly.

The third period: August 2010 up to now. From August 2010 to June 2012, the grain prices rose in fluctuation. According to economic indicators from the United States and the European Union, the economic recovery is not stable during that time. The situation was unrest in the Middle East and North Africa, which promoted the rise in the price of corn and other related agricultural products. At the same time, United States and Australia and other major grain producing areas suffered droughts and floods, it affected the growth of wheat, maize etc. In June 2012 to now, grain prices showed downward trend, mainly due to high record in grain production and global grain inventories remain high.

Grain prices intuitively analyze period can be seen, the trend of grain prices was consistent with the fluctuations in economic development. It showed that grain prices periodicity associated with economic periodicity. But intuitive analysis can see cycle approximately, it is a visual and rough judgment and not reflecting prices cycle truly. The following paper will use spectral analysis and wavelet analysis to research by specific mathematical way.

3. An Empirical Analysis in Grain Prices Periodicity

3.1. *Spectral analysis*

Spectral analysis is a useful tool to study time series periodic change, its core idea is a limited Fourier transformation about time series, the data sequence is decomposed into a combination of different amplitudes and wavelengths of sine and cosine wave. x_t as sample time series, where $\mathrm{t}=1,\cdots,\mathrm{n}$. x_t is be decomposed by Fourier transform into:

$$x_t = \mu + \sum_{k=1}^{m}\left[a_k \cos(\omega_k t) + b_k \sin(\omega_k t)\right] \tag{1}$$

Where, a_k, b_k are Fourier coefficients, ω_k is Fourier frequency, $\omega_k = 2k\pi / n$. ω_k which corresponding period is $\mathrm{T} = n / k$. m is the number of frequency: if n is an even number, then $m = n/2$, if n is an odd number, then $m = (n-1)/2$. μ is a constant term.

The scatter diagrams of Fourier coefficients a_k, b_k about frequency or wavelength can form periodograms. Amplitude periodogram $J_k = \frac{n}{2}\left(a_k^2 + b_k^2\right)$ can be viewed as the k-th harmonic frequency ω_k contribution to the total sum of squares.

Before spectral analysis, the three grain varieties have a preliminary processing to remove trend term, then the SAS9.3 Proc Spectra is used to complete spectral analysis.

Here only to analyze the sample periodogram of maize. Figure 2 and figure 3 are frequency - sample periodogram and cycle - sample periodogram.

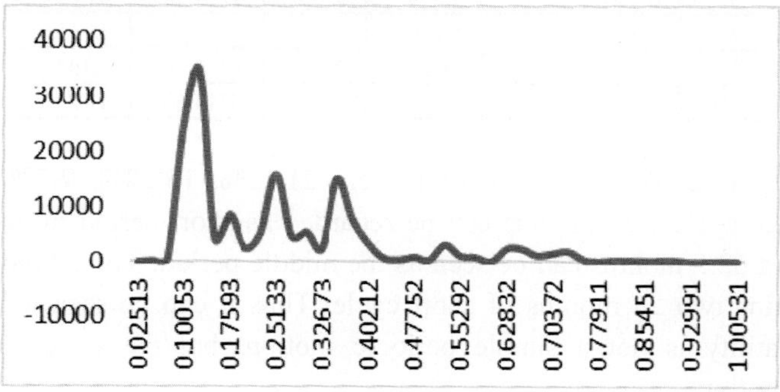

Figure 2. The price of corn: frequency - sample periodogram.

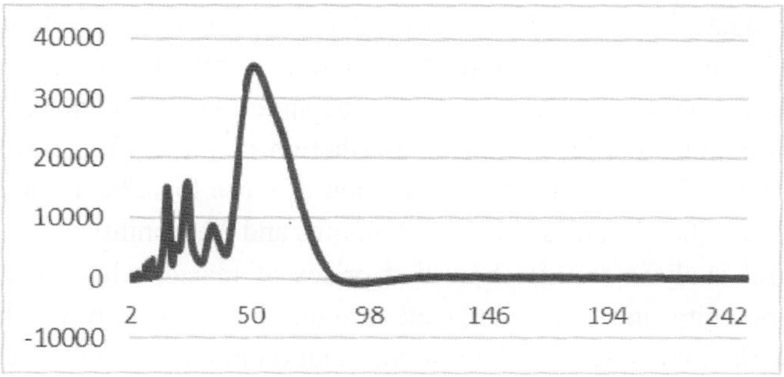

Figure 3. The price of corn: cycle - sample periodogram.

As can be seen from the graphs, during the investigation period, there is a strong cycle of 50 months, which frequency of 0.126 in Figure 2 and cycle of 50 months in Figure 3, sample periodograms lie in the peak position. At the same time, the removal of the main peak of 50 months, there are still some minor peak in Figure 3. Table 1 shows the sample periodogram maximum data and corresponding Fourier coefficient estimation.

In Table 1, the total contribution of ten cycles to the sample variance reaches 77%, in addition to the largest contribution for the 50 months, there are also two important cycle: 62.5 months and 25 months.

Table 1. Maize price spectral analysis and Fourier coefficient.

Obs	Freq	Period	Cos	Sin	P	Degree
1	0.100	62.5	-12.206	-7.174	25058.29	15.281
2	0.125	50.0	-13.615	9.576	34636.95	21.123
3	0.150	41.7	5.565	1.926	4335.23	2.643

4	0.175	35.7	3.071	-7.687	8567.48	5.224
5	0.226	27.8	0.922	6.351	5149.8	3.140
6	0.251	25.0	9.395	6.267	15944.96	9.724
7	0.276	22.7	4.859	3.744	4704.66	2.869
8	0.301	20.8	5.195	4.055	5429.84	3.311
9	0.351	17.9	-5.040	-9.723	14993.98	9.144
10	0.376	16.7	-7.052	3.496	7744.66	4.723

Three cycles of the contribution were 21.12%, 15.28%, 9.72%. For the above three cycles, 25 months can be regarded as short-period in maize price, and 50 and 62.5 months can be seen as the middle period. The 50 months cycle just contains two 25 months of short cycle. Thus it can be seen that the corn price volatility is not a simple periodic motion, but the complex multiple periodic motion.

Now, let's analyze the sample period of wheat and rice in the chart the 10 maximum data as well as corresponding Fourier coefficient estimation. The wheat total variance contribution of ten cycles nearly to 80%, of which 50 months, 35.7 months and 62.5 months for the largest contribution and more than 10%.The rice total sample variance contribution of these 10 cycles more than 68%, 50 and 41.7 months cycle contribution of close to 12%, at the same time there are some short cycle, such as 19.2 months and 25 months.

Spectral analysis results show that prices of maize, wheat and rice have certain periodicity, and it is a composite of multiple cycles. Through the analysis found that 50 months have the largest contribution in three grain, as well as have some other same period. The three crops cycle is in 2-4 years, and result is not in accord with intuitive analysis. The main reason is that intuitive analysis is just a rough result, and spectral analysis is from the mathematical point of view, which results have accuracy. Simultaneously, the results keep the consistent with the Kitchin cycle. Kitchin cycle theory is aimed at the analysis to the change of quantity of the inventory manufacturers would reduce the later production when they have too much inventory. Crops production is a similar situation. When inventory rose, the international grain prices produced a downward pressure, followed by two or three years, farmers would choose cut production to avoid prices losses. On the contrary, farmers would add production to increase incomes when inventory decline and grain market supply and demand contradictory intensifies, international food prices have upward pressure. Because grain production would take a period of time, the presentation of food price is 2-4 years period.

However, spectral analysis can only analyze some fixed cycle and not think of others. So we use the wavelet analysis to study the periodicity of grain prices.

3.2. *Wavelet analysis*

Wavelet transform is to use a set of band-pass filter to select variables to adaptive filter, namely can choose appropriate scale and translation parameters. Compared with the Fourier decomposition, the wavelet transform can from two aspects of time domain and frequency domain analysis of time series at the same time.

There are several kinds of wavelet, different wavelet functions applicable to different signal characteristics, so the key to the wavelet analysis is choosing the right wavelet function. Generally, coifN, symN and dbN, the three wavelet clusters are more suitable for economic time series. Therefore, it choose the three wavelet clusters to decompose and reconstitute grain prices, then compare error between reconstituted sequence and original sequence. The wavelet function which has the minimum error is be chosen in this paper.

Table 2. Wavelet base wavelet decomposition reconstruction error table.

wavelet basis	wavelet basis1	wavelet basis2	wavelet basis3	wavelet basis4	wavelet basis5	wavelet basis6
coif	2.55E-09	3.16E-08	1.58E-09	7.32E-08	1.66E-05	
sym	2.61E-12	1.68E-09	2.04E-08	1.41E-09	4.93E-10	2.26E-09
db	2.61E-12	1.68E-09	2.04E-08	3.83E-09	5.84E-09	3.76E-09

By comparison, symN and dbN wavelet clusters have small error after reconstruction, especially sym1, db1 wavelet, but these two have lower vanishing moments compare with others. They are suitable for strong singularity series. Grain prices volatility is relatively flat, singularity is not obvious. In conclusion, it selects sym5 as wavelet function. Maize price make one dimensional continuous wavelet transform get wavelet amplitude diagram (Figure 4).

In wavelet amplitude diagram, the vertical axis represents the scale change of cycles. When the scale is less than 5, the diagram is in black area, which can't indicate any periodic sign. When the scale is greater than 5, amplitude diagram gradually shows the characteristics of black and white, this represents the periodicity. When the scale is more than 15, it has obvious characteristic that regional black and white alternate, which indicates that maize price has obvious cycle on this scale. Amplitude figure compared with price charts, it can be seen that the white area in amplitude figure corresponds to the high points in price chart. The following is further studying periodicity by wavelet coefficient contour map.

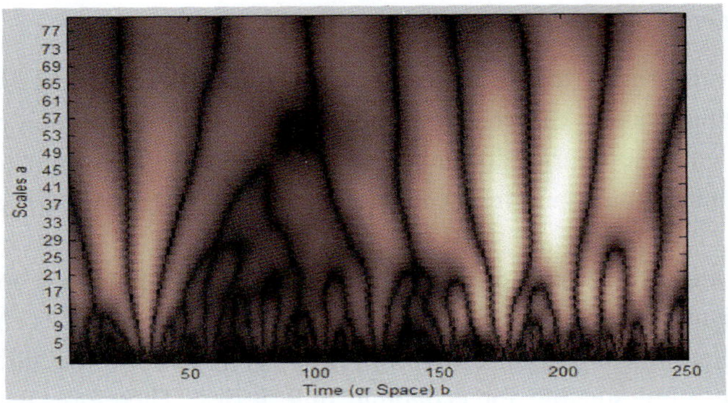

Figure 4. Maize price amplitude diagram.

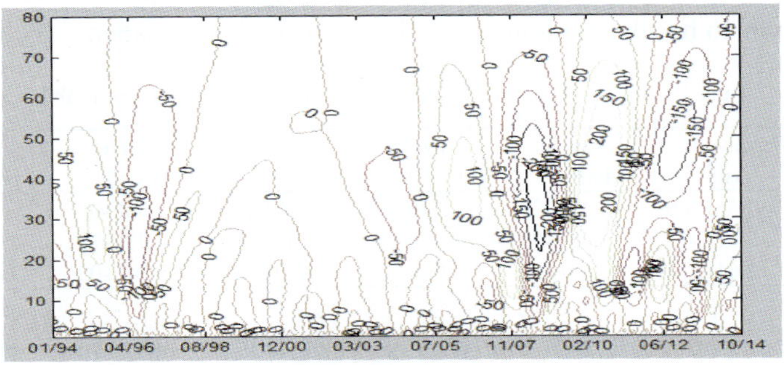

Figure 5. Maize price contour map.

In maize price contour map, the positive represents high price of maize, the negative represents low price. The contour from positive to negative shows positive and negative shocks center which corresponds length of price fluctuation in frequency domain. When scale between 25 and 50, there are several intensity centers. The high value centers are in second half of 1995, the early 1997, the second half of 2006 and the first half of 2010, the low value centers are in the second half of 1996, the middle of 2004, the early 2008 and the middle of 2012. When scale around 15, there are several weak value centers which in the early 2007, the early 2011, the middle of 2012 and 2013. The time distance between two low or high values is the length of cycle. There is a strong cycle of 40 months and a week period of 15 months about maize price. In order to prove result of the research further, it calculates wavelet variance and plots the wavelet variance diagram.

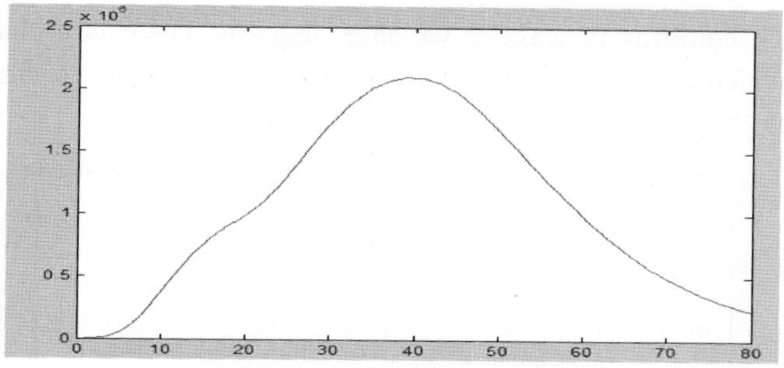

Figure 6. Maize price wavelet variance diagram.

There is a clear peak in wavelet variance diagram, which corresponds 40 months in horizontal axis. It shows strong fluctuation in the 40-month period scale, which is the main cycle. Figure has a slight bulge in 15 months, but not an obvious peak value, so it should be ignored.

Now, we present the contour map and wavelet variance diagram of wheat and rice prices.

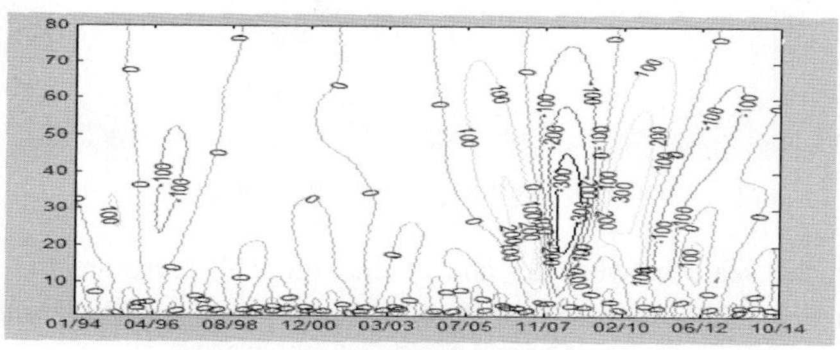

Figure 7. Wheat price contour map.

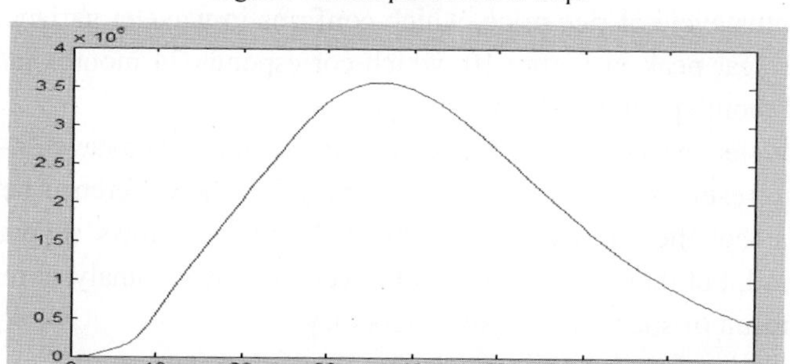

Figure 8. Wheat price wavelet variance diagram.

In Figure 7, it appears the low value center in the early 1996 and 1998 as well as the middle of 2011, it also has the high value center in the middle of 2006 and the early of 2010. It suggests that there is a 38-month cycle of wheat price, which confirms in wavelet variance diagram. There is a clear peak in Figure 8, which corresponds 36 months in horizontal axis, so 36-month period scale is main cycle. It is about 38 months in contour map.

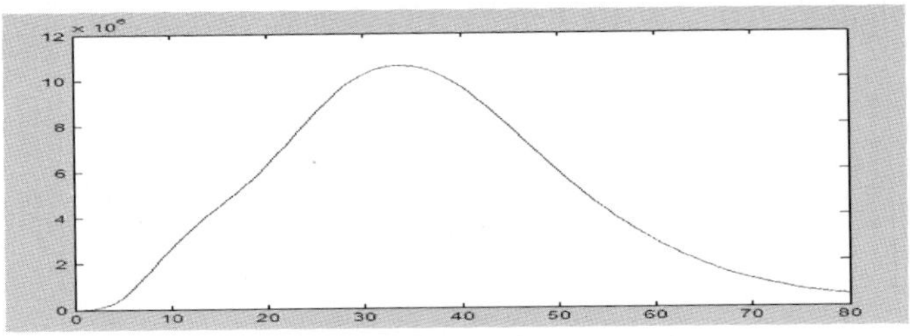

Figure 9. Rice price contour map.

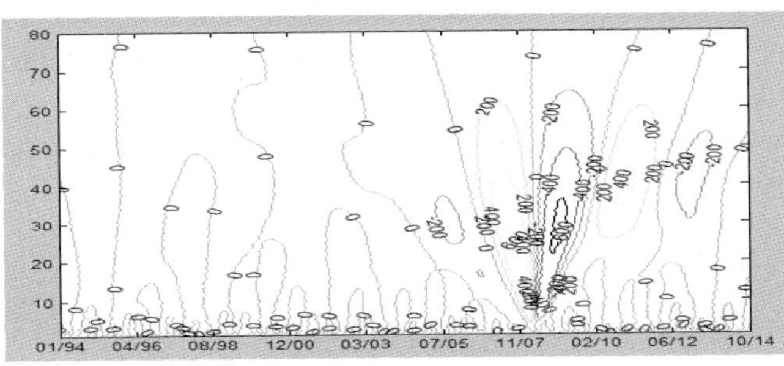

Figure 10. Rice price wavelet variance diagram.

When the scale between 25 and 50 in Figure 9, there are the low value centers in Sep. 2005, the early of 2008, the second half of 2012 and the high centers in the second half of 2007, the first half of 2010. It suggests that there are 34 months cycle of rice price, which confirms in wavelet variance diagram. There is a clear peak in Figure 10, which corresponds 34 months in horizontal axis, so 34-month period scale is main cycle.

By wavelet analysis, it can be seen that three kinds grain have 34-40 months period. The result is close to of spectral analysis. The difference between two analyses is that the consequence of spectral analysis shows several different length period, but there is only one main cycle in wavelet analysis result. It can prove the result of spectral analysis is accuracy.

4. Conclusions

Grain as a commodity, the transaction follows the law of value, thus prices presents certain periodic fluctuations. Meanwhile, grain prices have been influenced by non-market factors, including policy, especially the change about export, futures speculation, and climate change and so on. These factors would cause fluctuations in grain prices. The formation of the grain prices can be seen as the result of comprehensive effect of various factors. Therefore, the volatility of grain price is not a simple periodic motion, but it combines various aspects and compound multiple cycles.

In this paper, the research results show that there is a 2-4 years cycle in grain prices. It related that farmers adjust the production accordance with the price. Under the influence of compounding different factors and periods, grain prices have multiple periodicity. With the continuous development of grain financialization, the instability of prices is becoming more forceful. The periodicity of grain prices has the shorter trend in the future.

Acknowledgements

This research is supported by Beijing Municipal Natural Science Foundation (9152007).

References

1. K. Andrey, T. Sergey. (2010) "A Spectral Analysis of World GDP Dynamics: Kondratieff Waves, Kuznets Swings, Juglar and Kitch in Cycles in Global Economic Development, and the 2008–2009 Economic Crisis". Structure and Dynamics, 4 (1): 3–57.
2. C. K. Harley. (1980) "Transportation, the World Wheat Trade, and the Kuznets Cycle, 1895-1913". Explorations in Economic History, 17: 218-250.
3. C. Labys, Kouassi, Terraza. (2000) "Short-term Cycles in Primary Commodity Prices". The Developing Economies, 09: 330-342.
4. L. A. Pustilnik, G. Y. Din. (2003) "Influence of solar activity on state of wheat market in medieval England". http://arxiv.org/pdf/astro-ph/0312244.
5. CSC Sekhar. (2003) "Volatility of Agricultural Prices An Analysis of Major International and Domestic Markets". Research on international economic relations, Working Paper, No. 103.
6. H. Cao. (2007) "China's Wheat Prices Cycle Variation Characteristics and Reason Analysis". World Agriculture, 4: 29-32. (In Chinese)
7. F. X. Meng, P. T. Dong, X. L. Lin. (2008) "Grain Market Price Analysis about Fluctuation Law, Cause and Countermeasures". Price: Theory & Practice, 09: 20-28. (In Chinese)
8. X. Fu, J. B. Niu. (2009) "International Agricultural Price Volatility Characteristics, Rules and Trends", Chinese Rural Economy, 05: 36-45. (In Chinese)

9. B. Erten, J. A. Ocampo. (2012) "Super-cycles of Commodity Prices Since the Mid-nineteenth Century". DESA Working Paper, No. 110.
10. W. Fang, W. J. Lin, Z. Wan. (2013) "Rice Market Price Fluctuations Abnormalities Exception of Economics Evaluation: Estimation Based on the Method of X-12 and H-P". Journal of China Agricultural University, 02: 211-219. (In Chinese)
11. L Ding. (2010) "Fluctuation and Cycle of Pork Retail Price in China Based on Wavelet Analysis". Information Science and Engineering (ICISE), 2010 2nd International Conference: 2854 – 2858.
12. L. Zumino, BM. Tabak, F. Serinaldi, M. Zanin, DG. Perez, OA. Rosso. (2011) "Commodity Predictability Analysis with a Permutation Information Theory Approach". Physica A: Statistical Mechanics and its Applications 390, 876-890.

The function of soil and water conservation in agricultural development and ecological environment construction

L.L. Kang[a,*], G.J. Wei[b], Y.Yuan[c], F.F. Dong[a], Z.Y. Wang[c] and J. Sun[a]
[a]*Institute of Hydraulic Research, Zhengzhou 450003*
[b]*Qinghai University, Xining 810016*
[c]*North China University of Water Conservancy and Electric Power, Zhengzhou 450011*
[*]*E-mail: kanglingling1234@163.com*

Water and soil are the basic resources of agricultural production. Due to long-term unreasonably use and waste, as well as increasing destruction of surface vegetation, large amount of waste slag and excavation caused by the production and construction projects, which not only exacerbated the water and soil erosion, wasted the water and soil resources sharply, but also deteriorated the agricultural ecological system, all of these would endanger the sustainable development of the national economic society seriously. This paper mainly analyze the impacts and disadvantages of water and soil erosion for agricultural production and ecological environment on the basis of the reason of the water and soil erosion and discuss the function of soil and water conservation in the agricultural production development and ecological environment construction form several aspects which are soil and water conservation is conducive to agricultural production development and structure adjustment, to alleviate the contradiction between supply and demand of water resources, accelerate the development of agriculture and the ecological environment construction, reduce natural disasters and to steady development of agriculture and ecological environment improvement, soil and water conservation is the premise and foundation of the ecological environment construction, construction, is the support of rural sustainable development; finally, the paper put forward several suggestions on strengthening soil and water conservation of agriculture development project.

Keywords: Soil and water conservation; Agriculture development; Ecological environment construction.

Water is the source of life and soil is the basic of survival, they are not only the basic conditions for survival and development but also the basic resources which irreplaceable. However, the lose of soil and water has become one of the most serious environmental problems that our human have to face, which has caused the deterioration of ecological environment, restricted the development of modern agriculture, plagued the construction of new rural, farmer's production, and the improvement of living conditions, and also seriously affected the development of rural economic and ecological environment.

1. Soil and Water Loss and Its Causes

Water loss and soil erosion is that water and soil resources and the productive capacity of the land were damaged and lost under the action of natural force and human activities, such as hydraulic, wind, gravity and freeze thawing and so on. In fact, there are many reasons in view of soil and water loss, most important factors are nature and human. Natural factors are the basis and objective conditions, and human factors are the dominant factors and root causes, in my following paper, we shall elaborate the performance of human factors:

1.1. *Irrational uses of land resources*

Deforestation, steep slopes reclaimed grassland cultivation and over-grazing, disordered soil and dispatching, etc. not only destroyed the original topography, reduced the area of vegetation cover, deteriorated the natural environment, but also caused soil and water loss directly or indirectly.

1.2. *Irrational uses of water resources*

Over-exploitation of water resources leads to the ecological water can not be guaranteed and river water dramatically reduced even dry up. Adding to the exploiting groundwater that outpaced its recharge capacity and large-scale drainage cause by mining, water level drops continuously surface soil becomes drought, water and soil loss is also intensified.

1.3. *The effective control measures of development and production activities is lacked*

Along with the industrialization and urbanization in our country, there are many production and construction activities besides, with the rapid development of mining, transportation (including roads, railways, airports, ports, docks, shipping etc.), building, exploration, water, power engineering (including water conservancy project, hydropower station, nuclear power, wind power, solar power, transmission and distribution etc.), long-distance transport pipes used for conveyance of oil, gas and water, metallurgy and chemical industrial, exploitation on agriculture and forest, rural enterprise, construction of city and district etc, dumping waste soil (stone, slag) and tailings at will, occupation of farmland and other phenomena often happens in life. If we handled with it improperly or lack of effective protection measures that landslide, debris flow and other disasters will be produced, which will bring about severe soil loss. According to the investigation of soil and water loss in development projects of "Scientific Investigation of The Chinese Soil Erosion and Ecological Security" that the total spoil and abandoned dreg of various development and construction project during the "10th Five-Year Plan" was 9.21 billion tons, the total amount of soil erosion was 9.5 billion tons. Therefore, the large earth transportation work, especially dumping abandoned stone (soil, slag) has become an important factor in soil erosion.

2. The Impact of Soil and Water Loss on Agricultural Production and Ecological Environment

As it is known to all of us, the soil and water loss has not only caused the reduce of arable land, which seriously affected the development of modern agriculture, but also the sedimentation on river, lake, pond, libraries and other deposition caused by massive silt release which will weaken the sstorage features of water conservancy facilities and flood carrying capacity, aggravate the probability of floods, droughts and other natural disasters' occur, make irreversible loss of soil and water resources and greatly deteriorate the ecological environment.

2.1. *Land resources destruction and farmland erosion*

(1) Arable land reduction. Soil and water lose has caused the land resources and surface integrity serious damaged, made it fragmented and covered with ravines, because of which, cultivated and available land resources and land productivity have also reduced with the speeding up of land erosion, what's more, production conditions have been seriously deteriorate, resulting in the reduction of water resource. Large amounts of soil erosion have made the land productivity decline inevitably. Contradictory among people, places and water has been aggravated; farmland mainly distribute above the mountain ridge along ditch, especially in the hilly areas. With the storm runoff erosion, area of channel cut and terrain broken are growing larger, slope and farmland smaller. Statistically, 50 billion tons of soil lost every year in China, southern land continuously becomes more "soil", while the northern land continuously "desertification", the loss of arable land has reached nearly 67,000 hm^2 [Y]. Since 1950s, lose of arable land in entire country is about 2.667 million hm^2, which due to soil erosion. The result would have a serious impact on agricultural production, especially in the substantially decline of crop yields, increased poverty, which leading to a vicious cycle of ecological and economic systems [T, P].

(2) Land desertification. Serious soil erosion causes a huge loss of natural precipitation that we couldn't make full use of because of which, water conserva-

tion capacity and soil fertility reduces, soil structure deteriorates, land degenerated gradually, especially northern land performed constantly desertification.

2.2. *Soil fertility degeneration and land productivity reduction*

(1) Soil fertility degeneration. Soil erosion makes the hilly land into fragmented, resulting in water storage capacity of soil reduced, permeability decreased, resulted the soil organic matter and nutrient content reduced significantly, which have made the soil thin and fertility decline. Research has shown that soil erosion will not only reduce main nutrients includes nitrogen, phosphorus and potassium of soil, but also reduce boron, zinc, copper, manganese, iron and other trace elements of soil. Statistically, the amount of soil loss has reached about 50 billion tons every year in China, the loss of N, P, K elements about more than 4000 million [T].

(2) Land productivity reduction. In severe soil erosion areas, lots of surface soil lose that soil becomes increasingly thin, many sloping farmland becomes water running, soil running and fertilizer running field, which called “three-run field”, available nutrients rapidly declines, soil texture becomes coarse, ventilation deteriorates, structural damages, physical and chemical properties deteriorates, water carrying capacity of soil attenuates, land becomes barren increasingly, making the land productivity declined, food production low and unstable, even never produced. According to related statistics, lose surface soil per 1mm; cereal production reduces 10 kg/hm^2. Therefore, the productive of eroded soil usually decline about 80 percent than which doesn’t [Q].

2.3. *Water balance destruction, and floods and droughts aggravation*

(1) Water balance destruction. Water loss and soil erosion not only destroy the soil itself, but also lead the land desertificated, grassland degraded, ecological imbalanced, vegetation recovery hindered, water redistribution affected the directly. Especially with the function of the terrain recycle, vegetation destruction, rainfall interception, water retaining stagnant runoff gradually weakened, or even lost, which makes surface runoff a substantial increase, subsurface runoff a gradually reduce, thereby resulting in the reduce of soil water storage capacity, the relationship change between soil moisture and water requirements of crop, or even the balance destruction on original water.

(2) Floods and droughts aggravation. Water loss and soil erosion makes the soil properties deteriorate, permeability and carrying capacity reduce, and runoff direction change. Once rainstorm come, floods will wreak havoc, form landslides and mudslides easily and create mountain floods and other natural disasters; at

the same time, the large amount of sediment discharge caused by water and soil lose will silt up river, lakes, reservoirs and other water conservancy facilities, severely restricts the comprehensive development and efficient use of water resources, almost can be described as "rain brings floods, no rain brings drought", which leads to lower output of industrial crops and gain crops, seriously hamper the healthy development of the local agricultural economy, affect local farmers' normal production and life [X].

Meanwhile, water loss and soil erosion also cause a lot of rain resource consumption by soil erosion and ineffective evaporation, spur droughts (particularly agricultural drought) and increase the contradiction between water supply and demand. Statistically, one year average area of drought is about 20 million hm^2, disaster area is about 7 million hm^2, disaster rate reaches to 35 %, most of which are in serious soil erosion area.

2.4. *Rivers, lakes, ponds and libraries siltation, utilization function reduction*

Soil erosion causes amounts of sediment deposit in rivers, lakes, ponds and libraries, on one hand, sediment deposition reduces the storage capacity, weaken the storage capability of the reservoir and flood control capacity, and reduce the service life of reservoir, when serious, it could lead to overtopping, collapse and other disasters easily; on the other hand, the riverbed of canals and rivers raised, seriously affecting the capacity of flood passage, resulting in flood flow hard, water level rise, and floods create. In addition, the sediment, nitrogen, phosphorus, chemical pesticides and other organic pollutants loss generated by soil erosion, can easily lead reservoirs, lakes and other water bodies to eutrophication, threat to water conservancy facilities and its benefits exert seriously, hinder the sustainable development of water conservation.

According to the estimates preliminary, the capacity lose of reservoirs and ponds due to the soil erosion throughout the country has cumulated up to 200 million cubic meters, which equals to more than 200 large reservoirs capacity of 100 million cubic meters waste silt, calculated as 0.5 Yuan cost per cubic meter capacity, directly economic loss has reached about 10 billion Yuan. The lose of irrigated areas and generated electrical energy due to the flow reduction and environmental deterioration will make the lost more difficult to estimate.

2.5. *Acceleration of soil erosion and the vicious cycle of poverty, Aggravate the deterioration of the ecological environment*

(1) Acceleration of soil erosion and the vicious cycle of poverty. According to statistics from Chinese soil erosion and the scientific investigation of ecological security, ranking as China's provincial soil erosion area proportion, the northwest and the Loss Plateau regions provinces are the first, where 57% of them are national poverty counties, 44% are ethnic minority counties, 32% of the old county district belongs to the serious soil erosion ones. In addition to the topography in natural conditions, soil erosion is more closely related to the local agricultural production and economic development. In fact, most of China's soil erosion is caused by steep slope cultivation and destruction of vegetation. Therefore, soil erosion is the root of poverty. Land productivity and the obvious contradiction between people and land accelerates the vicious cycle which is called "The poorer the more cultivated, the more cultivated the poorer", especially in serious soil erosion.

(2) Aggravate the deterioration of the ecological environment. Soil erosion not only damages the agriculture, declines the land productivity and accelerates the vicious circle of poverty area, but also brings a series of environmental problems, which are mainly in the following aspects:

The first one is that it lead to the serious deposition of river, lake, pond, pool and other water conservancy facilities, not only weakened its comprehensive use function, but also caused the deterioration of water environment; the second one is that it may make large amounts of nutrients in soil and residual pesticides and fertilizers material into the water with soil together, which would decline the self purification and dilution capacity of water, as well as reduce the water environmental capacity, soil erosion would lead to the water eutrophication of river, lake, pond and reservoir of lake, pond, accelerate the speed of water pollution; the third one is that it would make the plant growth conditions worse, lead to the community degradation, severe damage of cropland andgrassland vegetation, decrease of stock volume of forest and grass vegetation, as well as the ground coverage reduction, it also weak the regulation of vegetation and soil moisture on local or regional climate; in addition, the soil erosion will increase the dust, dust storms and other inclement weather, what would influence people's living condition and production environment seriously.

In short, soil erosion is the main characteristic of ecological environment and the origin of poverty, at the same time it is also the important reason of the deterioration of ecological environment and poverty, which has become one of the most serious environmental problems humanity faced.

3. The Role of Soil and Water Conservation in Agriculture and Ecological Environment Construction

The important purpose of soil and water conservation is to promote the development of agricultural and the improvement of ecological environment. The practice shows that, soil and water conservation is to minimize the water loss and soil erosion of the original landscape and man-made soil and water loss, thus to control the further deterioration of the ecological environment, and protect natural resources and improve the ecological environment, and promote the development of agriculture and regional economic, and lead the harmony between human and environment.

3.1. *Soil and water conservation conducive to the development and adjustment of agricultural production*

In addition to the soil and water conservation in production and construction projects, which more important is to improve the ecological environment and promote economic development in rural areas, and its main purpose is to conserve soil and water, to improve soil ecological environment, and to reverse the situation of countryside, especially the vicious cycle of poverty, promoting rural economic developed to the virtuous cycle.

Practice shows that, soil and water conservation is the only effective way to control the region soil erosion, promoting the rural economic and establish diversified economic structure. The topography and natural economic characteristics must be the basis of these measures, and these measures should make full use of the land and resources, and local manpower, and the financial conditions. Then these measures would play a huge resource potential.

3.2. *Soil and water conservation ease the contradiction of water resources, accelerate the development of agriculture ecological environment*

China is a water-scarced country, per capita consumption of water resources is only 1/2 of the world average, and the distribution is uneven. According to statistics, there are more than 20 provinces were arid areas in China, the area is about 5,000,000 km^2, the amount of water shortage in agricultural was over 30 billion m^3 in these region every year[Y], and these regions are all with serious soil and water loss.

As we can know that soil and water conservation measures can effectively intercept surface runoff. Soil and water conservation measures in arid, semi-arid area can greatly increase the storage of rainfall, and ease the contradiction of water resource, providing the water for the plant, and promote the agricultural development.

3.3. *Soil and water conservation reduce natural disaster and protected the development of agriculture and ecological environment*

Practice shows that, after the comprehensive control of soil and water conservation, the ability of intercept runoff and sediment by the green reservoir basin, soil reservoirs and soil and water conservation engineering have been greatly enhanced, the amount of sediments from upstream reduced effectively. Moreover, with the improvement of comprehensive management, the effects of flood and sediment reduction have been strengthened significantly, and also the threat of river flooding reduced gradually, and the incidence of drought, floods, winds and other natural disasters reduced significantly. The measures are conducive to the agricultural and good for the ecological environment.

3.4. *Soil and water conservation is the premise and foundation of ecological environment construction*

Both of soil and water conservation and ecological environment construction are supplements for each other; they are the means of human to protect the environment and natural resources. Soil and water conservation is also the premise and foundation of the ecological environment construction.

The practice shows that, we must bring up effective measures to improve the regional ecological environment that through engineering, grass and farming. It and to improve the ecological ability of self-healing capabilities, to raise the vegetation coverage, protect soil and water resources, improve river flow conditions, to prevent water and soil loss and land desertification effectively, reducing landslides, mudslides, flash floods and other disasters, improve the capacity of river flood disaster, preventing the occurrence of dust storms and other inclement weather and finally to improve the ecological environment.

3.5. *Soil and water conservation is the support for rural development*

As an important part of agriculture and rural economy, soil and water conservation is the foundation of new socialist countryside and the carrier of sustainable development, and uncouthly plays an important role in the new socialist countryside construction.

Practice shows that, with the in-depth and comprehensive promotion of soil and water conservation, not only can it greatly improve the agricultural production capacity and promote agricultural production and income, but also adjust the industrial structure in rural areas; improving farmers' income from secondary and tertiary industries and helping farmers become rich. Meanwhile, the soil and water conservation should be combined with the construction of new countryside closely, implying the overall planning of the mountains, water, electricity, field, forest, road, grass, room toilet, pool, pipelines, towns and villages, residential, flood control, drainage, sewage and so on, and using the technical standards of soil and water conservation to guide the construction of new rural. There is no doubt that the combination could accelerate the development of rural economic and improve farmers' living standards and improve the rural ecological environment.

4. Suggestions of Strengthening Soil and Water Conservation for Agriculture Development Project

(1) Raise awareness for relations of agricultural development and ecological security

To improve the understandings for the importance and specificity of agricultural water and soil conservation through the study and research of laws and regulations and the relevant policies of soil and water conservation, as well as clarify the relationship between agricultural development and ecological security. Carrying out the work of soil and water conservation actively, selecting conservation tillage measures under the corresponding conditions at the proper time, enhancing the development momentum of modern agriculture and inhibiting the trend of ecological environment deterioration, together with accelerating the construction of agricultural development projects, in order to get the way of harmonious development which the development of agriculture would be realized, rural economic would be prospered and ecological safety would be ensured.

(2) Give comprehensive arrangements for the work of water and soil conservation for agriculture projects by the scientific development concept

Scientific development concept has overall influence on the promotion and practice of soil and water conservation for agricultural development projects, and can determine the direction and road of development. In the specific work, we should summarize the experiences and lessons of soil and water conservation work, both from domestic and foreign agricultural development project, and absorb the international advanced experience and achievements in the height of the history and the age, further defined the problem of soil and water conservation in agricultural development project that we are faced with in the new century and new stage.

(3) Set up the conception of giving priority to prevention and protection to protect water and soil resources

Water and soil, resources is the material foundation to the survival of humans, and is non-renewable or extremely difficult to renewable resources. Experience of history and reality tells us that the loss of "water" will make humans lose their foundation. Once soil resources were damaged, it is also very difficult to fix up. Even though there is a variety of existing technology of soil remediation, it needs a high cost and a long repair cycle. There must be a severely affect on society, economy, rural construction and farmers survival, etc.

Therefore, in the practical work, we should firmly give priority to protection priority, and make the prevention of soil and water conservation work as the basic principles for economic and social activities. Reverse the conception of "pay attention to the economic growth and development of agricultural production, and ignore the soil and water conservation".

(4) Raise vegetation coverage rate and rainfall infiltration, as well as increase dry and reduce flood effectively

It is one of the important contents for soil and water conservation to strengthen the construction of forest and grass vegetation in the soil erosion area (including the areas which have man-made soil erosion by the construction process of projects in agricultural development). Taking forest protection system which is combined by tree, bush and grass for soil and water and raising vegetation coverage rate, in order to increase the canopy interception and retention of grassland, reduce surface runoff, slow flow, increase rainfall infiltration, improve regional water conservation and soil drought resistance ability, and achieve the purpose of increasing dry and reducing flood effectively and promoting the development modern agriculture.

(5) Reinforce experimental study on the measures for soil and water conservation in modern agriculture

There are still varying degrees of soil erosion in most of the existing farmland (especially in dry land of orchard hills district), due to soil and water conservation measures of agricultural development is not perfect, and the technical content of agricultural water and soil conservation is relatively low, even the agricultural tillage also has limitations to adapt to regions and related conditions. So it is imminent to improve technical measures of agricultural water and soil conservation. But considering it is unlikely to completely rely on the application studied maturely by the relevant departments of soil and water conservation. Therefore we suggest that to strengthen the experimental study which is suitable for the development of agriculture structure of soil and water conservation in new skill, new methods and new technology during the construction of agricultural development projects. At the same, to do a good job in demonstration and extension work of agricultural soil and water conservation, and actively guide farmers improve soil and water conservation in the process of agricultural production, in order to promote the unity of ecological and economic benefits.

5. Conclusions

Soil erosion has become one of the China's topest environmental issues, which naturally become a constraining factor of the development of sustainable agricultural and the improvement of ecological environment. Therefore, unless we can do a good job of water and soil conservation, and maximize the prevention of soil and water erosion and its damage, maintaining the sustainable use of land and water resources, can we make the each factor such as soil, vegetation, climate, hydrology and human activities of the ecosystem which we survival stay mutual coordination, and can we ensure and promote the maximum circulation and normal development of ecological environment.

The author believes that the only way to promote the steady development of agricultural economy and gradually improved of ecological environment is to promote the work of soil and water conservation actively.

Acknowledgement

This article is done by Lingling Kang & Guanju Wei & Yue Yuan & Feifei Dong & Zeyuan Wang & Juan Sun.

References

1. PeiYuan Wang, 2011, the relationship between sustainable development and soil and water conservation, *Market Modernization (early the month periodical), 1(634):218.*
2. Lingui Gao & Hanming Deng, 2004, A survey of the relationship between agricultural water conservation and drought in Jianzhen City, *Soil and Water Conservation in Fujian Province, 3,16(1):34-36.*
3. Tianzhi Qin, 2009, Study on the Technical Assessment of Soil and Water Conservation Facilities Acceptance. *Ecological Economy [J], (10):163-169.*
4. Yong Wang, 2011, China's Soil Erosion Status and control techniques, *Energy and Environment [J], (4): 124.*
5. Yonghong Li & Zhaoliang Gao, 2011, The characteristics, harm and governance of soil erosion in the loess plateau, *Ecological Economy [J], (8):242,148-153.*

Study on effect of soil and crops on Hg Pollution in river basin, in Xiao Qinling Gold belt, China

R.P. Liu*, Y.N. Xu, J.H. Zhang, H.Q. Chen, G. Qiao, H.L. Ke and F. He
Xi'an Geology investigation center, 710054, Xi'an, China
**E-mail: lrp1331@sohu.com*

In order to study Hg enrichment and anti-pollution intensity in soil and crops in different soil parent material, we investigated field in Xiao Qinling Gold Belt. Firstly, there are more enrich Hg in soils or corps growing in the Diluvial and alluvial soil than that in Eolian loess. Secondly, the main pollution source are caused by mining activities. Corn may be one safe crop which is not obvious effected by Hg pollution. While wheat is one high risk crop, the wheat is easily rich Hg in all the type soil. Lastly, the high concentration soil could promoted Hg migrated to the crops.

Keywords: Crop; Farmland; Hg Pollution; Xiao Qinling Gold belt.

1. Introduction

Once Hg enter the soil, come into accumulation pollution, not easy decomposition process in biological material cycle and energy exchange (He Yan, 2008), not only affects the physicochemical properties of soil itself, but also affects the ecological environment and the health of human beings (Xu You-ning, 2007).

The development of metal mineral resources often lead to mine soil environmental pollution, this paper mainly studies Xiao Qinling gold mine development leads to the Hg elements lead to the soil environment pollution. The study area is the second gold producing area in China, located in Shanxi Henan border, and in the north slope of the west of Xiao Qinling Mountains, open quarry quartz vein type gold deposits, gold large-scale deposit began in the 70's of the last century. smelting activities are mainly located in the Loess Gully plateau fluvial terraces and alluvial fans area, the area of about 1169.6km^2, mainly grow wheat, corn and other crops. Study area has a large gold mining and smelting enterprises, the majority technology is amalgamation - flotation of gold. The effect of soil heavy metal pollution especially mercury pollution has become one of the main environmental problems of mine. Heavy metal elements bringing into the river, the atmosphere, soil, crop, serious harm to human health.

Farmland mainly distributed in front of bedrock mountain. Distribution alluvial slope tableland, gullied Loess Plateau, rivers alluvial terrace of three types of landform. Mainly grow wheat, corn and other crops.

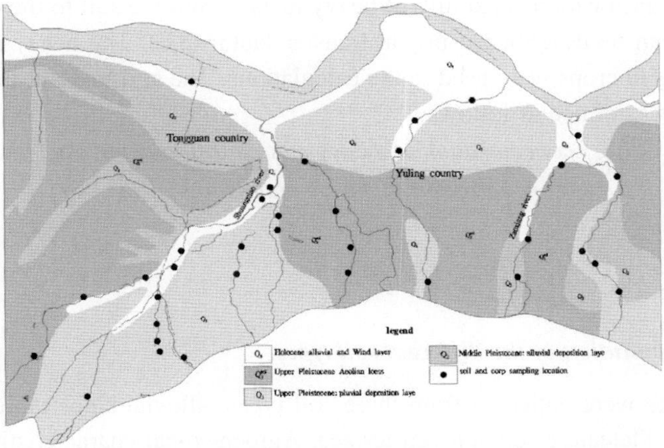

Figure 1. Distribution sampling.about corps and soils of river.

2. Research Method

2.1. *Sample collection*

Food crops grown in large areas of the reclaimed farmland were investigated in river basin of Xiao Qinling gold belt. The crops and the corresponding soil at a depth of 0–20 cm were collected in accordance with the standard 1sample/1km. When the crops were suitable for harvest. Each crop and soil sample consisted of one-to-one and weighed between 500 and 1000 g. The crop types included wheat and corns. 34 group samples of crop and soils were collected. The study areas were divided into two kinds: mining and farming, such as wastewater irrigation. Tail slag et al according to the soil forming conditions.

2.2. *Analytical method*

Hg in soil were determined by AFS-2202E double channel atomic fluorescence spectrometer(Beijing KCHG Instrument Co., Ltd), the 0.5000g sample in the 50ml colorimetric tube, to join the newly prepared 1:1 Aqua 10ml, shake. Boiling in the boiling water bath for 1 hours (every 20 minutes. Every color tube), removed after cooling in 1% hydrochloric acid diluted to the scale, shake the standby. The supernatant from the sample solution of 10ml prepared by 20ml colorimetric tube, the addition of 5ml 5% thiourea - ascorbic acid solution was diluted to the scale, shake for 30 minutes. Measurement curve on the instrument, to measure the As sample solution by reduction of 5% thiourea ascorbic acid. While absorb the Hg liquid sample solution prepared direct determination.

The translocation capability of heavy metals from the soil to the edible part of crops can be described using an transfer factor (TF). The heavy metal concentrations in crops were used in the calculations. The soil-to-plant transfer factor (TF) was defined as follows:

$$\text{TF}=\frac{concentration\ of\ heavy\ matals\ in\ plant(mg/kg)}{concentration\ of\ heavy\ matals\ in\ soil(mg/kg)} \tag{1}$$

3. Results

3.1. *The pollution of the different soil types*

All samples were collected from three soil types: alluvial slope tableland, gullied Loess Plateau, rivers alluvial terrace. Agrochemical characteristics of soils in the experimental sites are given in Table 1. From Table 1, we know that, the average content of heavy metal Hg in soil is the enrichest in Diluvial stratum soil parent material, and lowest in Eolian loess stratum. Compared with the second level of soil environmental quality standard value, all kinds of the soil are mainly polluted by Hg elements, in Xiao Qinling river.

Table 1. The contents of Hg in the the different types of soil. Unit: mg/kg.

Formation lithology	Statistics characteristic	Hg
Alluvial stratum	av.	1.96
	max.	17.43
	min.	0.09
	stdev.	3.7
Eolian loess stratum	av.	1.244
	max.	7.374
	min.	0.066
	stdev.	1.8
Diluvial stratum	av.	2.83
	max.	29.57
	min.	0.12
	stdev.	5.9
The second level of soil environmental quality standard value		≤1.0

Tables 2, 3 showed different soil parent material, the average content of Hg in food grains. In the Diluvial stratum and Alluvial stratum is more easily enrichment of Hg elements than Eolian loess. Luckly compared with standards of contaminants in foods (GB 27622005) the Hg element in the study area, the grain may pose a little risk to human.

In the different soil parent material, comparative study the maximum and average contents of Hg in wheat and maize, the value, Hg content in wheat is far greater than the corn, except lead.

Table 2. The contents of Hg in wheats on the the different types of soil. Unit: mg/kg.

Formation lithology	Statistics characteristic	Hg
Alluvial stratum	av.	0.01
	max.	0.08
	min.	0
	stdev.	0.0
Eolian loess stratum	av.	0.008
	max.	0.049
	min.	0
	stdev.	0.0
Diluvial stratum	av.	0.014
	max.	0.115
	min.	0
	stdev.	0.0
The standards of contaminants in foods		0.02

Table 3. The contents of Hg in wheats on the the different types of soil. Unit: mg/kg.

Formation lithology	Statistics characteristic	Hg
Alluvial stratum	av.	0.002
	max.	0.006
	min.	0.00
	stdev.	0.0
Eolian loess stratum	av.	0.002
	max.	0.005
	min.	0.00
	stdev.	0.0
Diluvial stratum	av.	0.002
	max.	0.006
	min.	0.000
	stdev.	0.0
The standards of contaminants in foods		0.02

In order to more clear statistical the regularities of Hg content. in different soil type. These figures were computed by log (the contents of heavy metals of alluvial soil) in Figs. 2–4.

As can be seen from Figs. 2–4, the contents of Hg of three kinds soils, wheat, corn, Hg pollution in three kinds soil, wheat, corn, the median maximum content is much higher than that of no pollution of Hg in wheat, corn, heavy metal grain planting the corresponding mining activities in the alluvial soil pollution Hg. Heavy metal content in maize grain planting Diluvial soil contaminated soil in the Hg average content is lower than the non polluted area, the remaining elements had no obvious change. As can be seen from Figs. 2–4, the median, maximum content of Hg in soil is higher than that in wheat, corn.

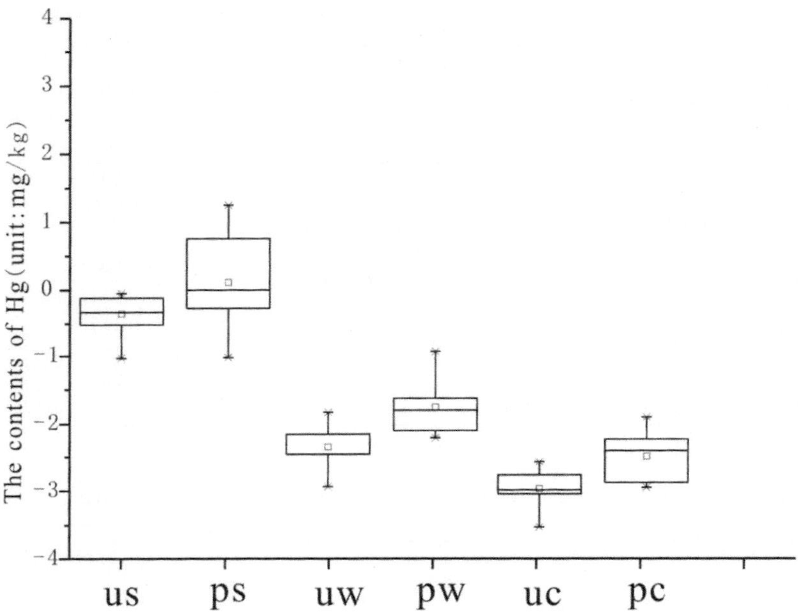

The unpolluted soil (for short us) The polluted soil (for short ps)
The unpolluted wheat (for short uw) The polluted wheat (for short pw)
The unpolluted corn (for short uc) The polluted corn (for short pc)

Figure 2. The contents of heavy metals in soils and crops on Alluvial stratum.

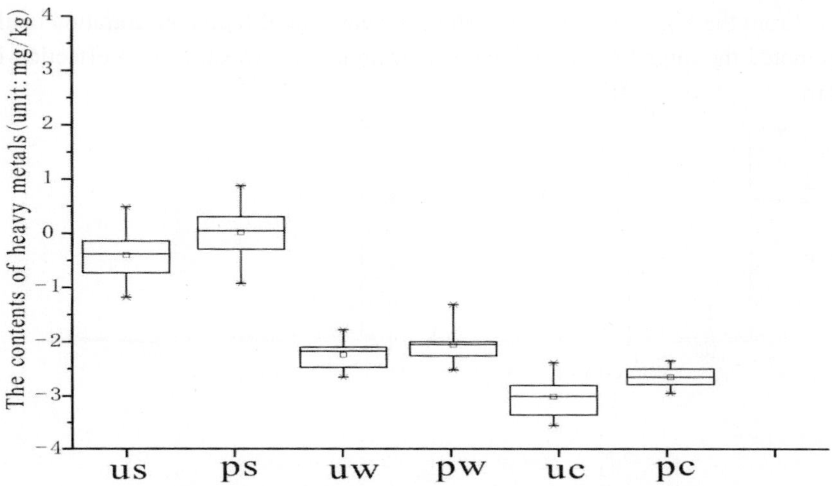

Figure 3. The contents of Hg in soils and crops on Eolian loess stratum.

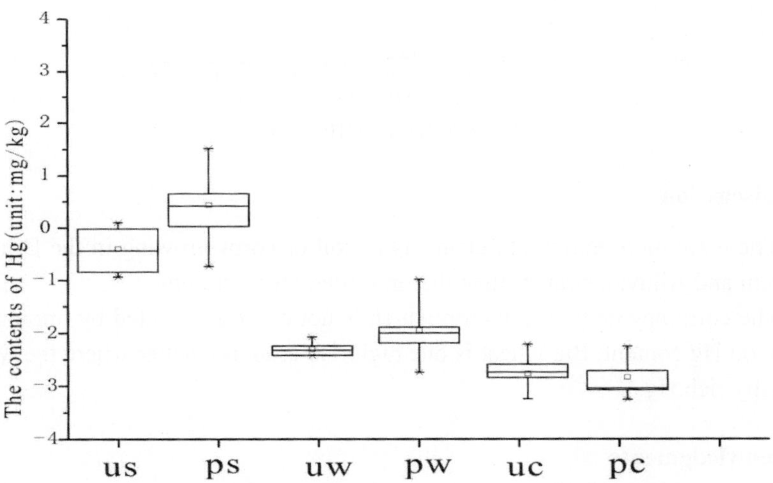

Figure 4. The contents of Hg in soils and crops on Diluvial stratum.

4. Conclusions

The main source of pollution caused by mining activities for Hg elements. Under the influence of particle size, particle more fine more easy adsorption of heavy metals,that is the reason why more easily enrich Hg elements in the Diluvial stratum and Alluvial stratum types of landform types than in Eolian loess stratum.

From the Fig. 5 of the results, the soil substrate of high concentration could promoted the migration of Hg elements from soil to wheat. But no effection in corn.

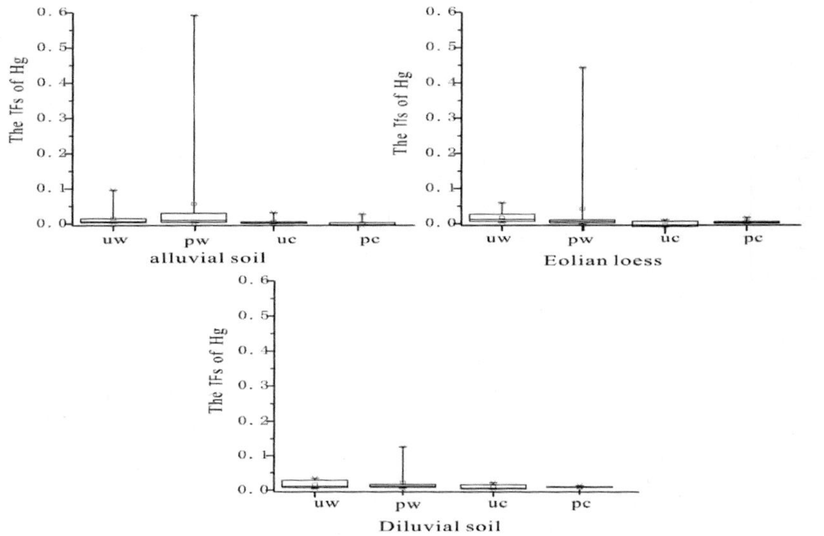

Figure 5. The TF of Hg of soil.

5. Discussion

(1) There are more enrich Hg elements in soil or corps growing in the Diluvial stratum and Alluvial stratum than that in Eolian loess stratum.

(2) The corn may be one safe crop which is not obvious effected by mining activity on Hg content. But wheat is one high risk crop, no matter where the wheat is easily rich Hg.

Acknowledgments

The study on migration transformation rule of mercury inriver under the action of water-gas exchange, gold mining area (2015JM4129). The authors would like to thank the support provided by the special research of risk of farmland soil heavy metal pollution evaluation and blockade of the Ministry of Land and Resources (2008334061), investigation and evaluation of environmental geological problems of small gold deposit in Qinling Mountains district land resources survey project (1212010741003), the Ministry of Land and Resources of Public Welfare Scientific Research (201111020), the Qinghai Tibet Plateau of mineral resources development geological environmental impact survey and evaluation

(1212011220224) and geological environment of mine – Shaanxi Tongguan field base joint funding. We are thankful to the Northwest Geological Survey and test center for statistical analysis of data.

References

1. He Yan. 2008, The Research on Heavy Metals Pollution of Soil and Cropper in Typical Area of Plain in The North of Zhe Jiang Province. Chang 'an University Xi', an, China, 15.
2. Xu You-ning, Ke Hai-ling, Zhao An-ing, et al. 2007, Assessment of Heavy Metals Contamination of Farmland Soils in Some Gold Mining Area of Xiao Qinling. Chinese Journal of Soil Science, 732-737.

Improvements of arable land pressure index model and their comparison: A case of Yunnan province*

Q. Yang*, Z.S. Yang, Y.M. He and M.J. Xu
Institute of Land & Resources and Sustainable Development, Yunnan University of Finance and Economics, Kunming 650221, China
**E-mail: yangzisheng@126.com*

Arable lands are a nation's and a region's very base of food security. It is therefore hardly surprising that the model for computing arable land pressure index has drawn increasing attentions in recent years. This paper elucidates the two major issues associated with the existing arable pressure index model, i.e. the difficulty in the computation of multiple-cropping index and issue of the steep sloping cropland either greater than or equal to 25° in the arable land base in a given mountainous area, and based on which, presents and makes a general analysis of the two improved models suggested by some Chinese scholars in recent years. An empirical comparative analysis is conducted by taking Yunnan, a mountainous border province in southwest China, as an illustrative example, and finally, concludes that the Improved Model I has successfully solved the difficulty in the computation of multiple-cropping index with the existing model. The Improved Model II is able to provide a real solution that is more suitable, feasible, and almost universal for most mountainous areas, as slopes steeper than 25° are completely excluded from the arable land base of a given mountainous area in this model. The computations show that the Improved Model II can produce a much more objective result, especially for mountainous areas where slopes steeper than 25° account for a large percentage of arable lands therein.

Keywords: Mountainous area; Arable land pressure; Model; Comparison; Yunnan province.

1. Introduction

The rapid urbanization and industrialization in recent years throughout China has led to a serious decline in the country's arable lands and threat to the country's food security, which has drawn increasing attention of experts, scholars, and governments at all levels as well. To study the relationship

---

* Supported by National Natural Science Foundation of China (41261018); Scientific Research Foundation of Yunnan Education Committee (2015YUFEC006).
Corresponding author: YANG Zi-sheng.
E-mail: yangzisheng@126.com.

between food security and availability of arable lands, the famous Chinese scholar Prof. Cai Yunlong (2002) suggested the use of the Minimum Per Capita Arable Land Threshold and of the Arable Land Pressure Index Model to measure if or not arable lands available in a given region can meet the region's need for food and to what extent (known as the degree of tension). Though widely adopted and applied in the study of arable land pressure, this model is not without weakness, in fact, it is hardly applicable directly to mountainous provinces. Considering that China is a mountainous country and as large as more than two-thirds of the country's territory is mountainous areas, highlands and hills inclusive. This paper starts by presenting the major issues associated with the existing Arable Land Pressure Model, then proceeds to make a general analysis of the two improved models suggested by some Chinese scholars base on the Cai Yunlong's Model, and finally, an empirical comparative analysis of results obtained by applying the two improved models respectively to compute the arable land pressure through Yunnan of 2009, a mountainous border province in southwest China. This paper not only provides a solid base for selecting a reasonable arable land pressure model for mountainous areas, but is of great importance and significance to enrich the research of arable land pressure, as well as to the widespread application of the arable land pressure index model, in mountainous areas.

2. Connotation and Issues of the Existing Arable Land Pressure Index Model

2.1. *Development and connotation of arable land pressure index model*

The spatial-temporal distribution of arable lands, in both quantitative and qualitative senses, as well as changes therein, is going more and more complicated. Today, the balance between food demand and supply of arable lands can no longer be fully and correctly reflected by changes in total arable area and in per capita arable area only. In this context, Prof. Cai Yunlong (2002), from the food security perspective, created the following Minimum Per Capita Arable Area Threshold and Arable Land Pressure Index Model, expressed as:

$$S_{min} = \beta \frac{G_r}{P \times q \times k} \qquad K = \frac{S_{min}}{S_a}$$

Where: S_{min} is the alert value and minimum of per capita arable area (ha), β is the food self-sufficiency rate (%), G_r is the grain consumption per capita (kg/man), p is the per-unit yield of grain, q is the percentage of total cropped grain area to total cropped area (%), k is the multiple cropping index, S_a is the actual per capita arable area (ha), and K is the arable land pressure index.

The basic implications of this model, as expressed using the above formula, are that: the balance between food demand and supply of arable lands is represented by the ratio of total arable lands demanded, which is calculated based on the actual food demand, to total arable lands available. The Arable Land Pressure Index is not only a threshold for arable land protection, but an important indicator for the regulation thereof.

2.2. ***Major issues with the existing Arable land pressure model***

2.2.1. *Issues with computation of multiple-cropping index*

The selection of correct parameters, as well as the availability of data thereof, is very crucial to having an objective, valid, and truth-reflecting results during the computation of arable land pressure index. Particularly, the grain consumption per capita and the food self-sufficiency rate vary from place to place as a result of the differences in the level of economic and social development as seen in these places. As a rule of thumb, empirical values are usually picked instead depending on a specific place. And the percentage of cropped grain area, on the contrary, is calculated using the cropped grain area and the total cropped area indicated in the Statistic Yearbook of a region.

However, computing multiple-cropping index has been challenged seriously since 1996, when “The First Detailed Land Survey of China” was completed. Take Yunnan Province as an example, the Province’s total area of arable lands and total cropped area by the end of 2013 were 6,219,800 hectares and 7,101,800 hectares respectively, and the Province's multiple-cropping index, as computed using the previous multiple-cropping index formula, was around 114.18, which was plainly discordant with the Province's farming system, which often sees two crops per year in most the Province, even three crops per year in some places. The root cause for this is that the two vital items of data used in the computation, i.e. the total area of arable lands and the total cropped area, were picked directly from the Land Survey carried out by Ministry of Land and Resources and the sample survey by Ministry of Agriculture, and the two surveys, however, differed considerably in methodology and statistical standard to be followed, which eventually resulted in a multiple-cropping index that was considerably discordant with the Province’s actualities and contributed to a distorted assessment of the Province’s arable land pressure.

2.2.2. *Issues with Arable land base*

China is in fact a mountainous country, as shown by the published statistics from the National Land Survey, carried out by Ministry of Land and Resources, that: the total area of the arable lands nationwide in 2009 was 135.385 million hectares, out of which 5.649 million hectares were either in forest regions, grasslands or in controlled areas below the highest flood level of a lake or a river in Northeast and Northwest China, and 4.314 million hectares were arable slopes steeper than 25°. In Yunnan's case, the Province's total area of arable lands as of the end of 2013 was 6.2198 million hectares, out of which slopes of 833,200 hectares were steeper than 25°, without 73,800 hectares terraces of steeper than 25°. As known that slopes are extremely prone to soil erosion as a direct consequence of violently-disturbed topsoil during the whole process of crop management and planting (through, from soil plowing, land levelling, sowing, loosening and weeding, fertilizing to harvesting), this is especially true for slopes steeper than 25°, after crops have been harvested and plowed during flood season, when they are scoured repeatedly by heavy rains, storms and surface runoff (Yang, et al., 2009). For this reason, the Article 25 of the Law of the People's Republic of China on Soil and Water Conservation explicitly stipulates that "Do not reclaim or crop on slopes steeper than 25 degrees". It is clearly foreseeable that currently-cropped slopes having a slope steeper than 25° will be given up and turned into woodlands in the near future as planned by the state.

For a mountainous area, it is commonly seen to count slopes steeper than 25° in its total area of arable lands, accordingly and naturally, many scholars count them in the Arable Land Base, too, when computing the arable land pressure index for the area. As said that these slopes will be given up and turned into woodlands in the near future for ecological protection purpose, so the final arable land pressure index was therefore nothing but a significantly-distorted value.

3. Improved Arable Land Pressure Index Model

Improvement has been made continuously to the arable land pressure model as it is widely adopted and applied. To cope with the two major issues, as mentioned above, with the existing arable land pressure index model, some scholars have suggested and put forth the improved counterparts in recent years after exploration and research.

3.1. *Improved Arable land pressure index model I*

To solve the issue with the computation of multiple-cropping index, Prof. Yang Zisheng (2008) has made improvement to the model developed by Professor Cai Yunlong, expressed as:

$$S_{min} = \frac{F_c \times D_s}{Y_a \times R_s} \qquad Y_a = \frac{Y_t}{A_t \times R_s}$$

Where: S_{min} is the alert value and minimum of per capita arable area (ha), D_s is the food self-sufficiency rate (%), F_c is the grain consumption per capita (kg/man), Y_a is the per-unit yield of grain (kg/ $hm^2 \cdot a$), R_s is the percentage of total cropped grain area to total cropped area, Y_t is the total grain yield (t), and A_t is the total area of arable lands (ha).

This model is a significant improvement and translated complement to the Prof. Cai Yunlong's classic model and use "the per-unit yield of land per year" to compute the minimum per-capita arable area, which cleverly avoid the issue with the computation of multiple-cropping index.

3.2. *Improved Arable and pressure index model II*

The Arable Land Pressure Index Model is exclusively used to measure the balance between objective demand for and actual supply of arable lands, and as already said in the preceding sections that slopes steeper than 25° in a given mountainous province will be eventually given up and turned into woodlands as required by law and for the maintenance of economic viability and environmental sustainability, so only those below the steepness threshold of 25° should be counted as available supply of arable lands. It is clear here that "the actual per capita arable area" is unable to faithfully reflect the true and effective land supply any more, especially for a mountainous province like Yunnan, where natural conditions are harsh and steep slopes make up a large percentage of its territory, even in some counties, slopes steeper than 25° comprise the majority of the county's total arable area, this explains why the arable land pressure index obtained using the existing model considerably diverged from the actual situation.

To cope with this issue, Yang Qing and Yang Zisheng (2015) have made further improvement to the Improved Model I during the computation and analysis of arable land pressure in Yunnan. With this further improved model, known as the Improved Model II, only arable lands below the steepness threshold of 25° were counted in the Arable Land Base, that is, the previous "actual per capita arable area" was replaced by a new "actual per capita suitable arable area", with slopes steeper than 25° completely excluded, to ensure a true reflection of the actual available supply of arable land in the Province.

4. Empirical Study

Yunnan is a typical mountainous province in western China, with fragile ecology, unfortunately. Statistics shows that mountain lands, including highlands and hills, having a slope either greater than or equal to 8°, make up as high as 93.6% of the Province's territory and that flat lands below 8° only 6.4% (Yang, et al., 2014). Therefore, Yunnan was selected as the object of the empirical study herein, on which the arable land pressure in 2013 was computed using the Improved Model I and II respectively, and an empirical analysis and comparison thereof were carried out subsequently to determine if or not the Province was in the shortage of arable lands and to what degree.

4.1. *Sources of basic data*

The total area of arable lands as well as the total area of slopes steeper than 25°, of the Province and of its 16 prefectures and cities, as of the end of 2013, were picked from the 2013 Land Changes Survey Data Set of Yunnan Province; the total population, food production and other socio-economic data were picked from the *2014 Statistical Yearbook of Yunnan Province* (the 1st Edition, China Statistics Press, November 2014); and the per capita food demand was assumed to be 400 kg/man, which was slightly higher than the subsistence level and close to the so-called "well-being" level (He, et al., 2008).

4.2. *Results computation and analysis*

The computation results obtained using the Improved Model I and II respectively, i.e. the arable land pressure indexes of the Province and of its 16 prefectures and cities, as of 2013, are presented in Table 1, in two groups corresponding to the Improved Model I and II respectively.

Table 1 Comparison of arable land pressure indexes of Yunnan as obtained using the Improved Model I and II respectively

Administrative Division	Arable Land Pressure Index Obtained Using the Improved Model I	Arable Land Pressure Index Obtained Using the Improved Model II	Differ by (%)
Yunnan Province	0.9879	1.1407	15.47
Kunming	2.1393	2.2945	7.25
Qujing	0.7510	0.7922	5.49
Yuxi	1.5525	1.7149	10.46
Baoshan	0.7396	0.8080	9.24

Zhaotong	1.0011	1.5469	54.53
Lijiang	1.0038	1.0669	6.29
Pu'er	0.9064	1.0224	12.79
Lincang	1.0060	1.2041	19.69
Chuxiong	0.9057	0.9996	10.37
Honghe	1.0329	1.2036	16.52
Wenshan	0.9266	1.1110	19.89
Xishuangbanna	0.9802	1.0271	4.79
Dali	0.8320	0.9365	12.56
Dehong	0.6616	0.6718	1.54
Nujiang	1.0710	2.1176	97.71
Diqing	0.9344	1.2136	29.88

It can be seen from Table 1 that the Improved Model II, to a greater or less extent, produced larger values than the Improved Model I. For example, the Province's arable land pressure index is 1.1407 with the Model II, rather than 0.9879 as with the Model I, differing by 15.47%; However, on Dehong Prefecture, Xishuangbanna Prefecture, Lijiang City, Baoshan City, Qujing City, Kunming City, and Baoshan City, the two models differ slightly, less than 10%. It is noteworthy here that the two models differ by as little as 0.01 on Dehong Prefecture, because the prefecture has very favorable natural conditions, where flat lands and gentle slopes make up the overwhelming majority of its total arable lands, and only as low as 1% are slopes steeper than 25°; and considerably on Nujiang and Diqing Prefecture, located in the alpine-valley region northwest of Yunnan, and on Zhaotong City, located in the medium-altitude mountainous region northeast of Yunnan, because the terrain in those places is harsh, and steep slopes not suitable for tilling make up a large percentage of their respective territory, especially on Nujiang Prefecture, where slopes steeper than 25° occupy 51.19% of the prefecture's total area of arable lands, the two models differ by as surprisingly high as 97.71%, indicating a "high pressure" (Yang, et al., 2015). On the other 7 prefectures and cities, however, the two models differ by a percent between 10% and 20%.

As pointed out in the preceding sections that slopes steeper than 25° are prone to serious soil erosion, increasingly thinned-out soil layer, and low and unstable per-unit yield, the Article 25 of the Law of the People's Republic of China on Soil and Water Conservation therefore explicitly stipulates that "slopes steeper than 25 degrees should be turned into woodlands or grasslands". A city or prefecture where slopes steeper than 25° make up a large percentage of its total arable lands will naturally see a significant decline in its arable lands and grain supply capacity, and therefore will face a relatively strong pressure.

In summary, the Model II can produce a more accurate reflection of the actual use and supply of arable lands in a region where steep slopes make up a large percentage of its total area of arable lands. And the two models differ little on a region where flat lands or gentle-slope mountains and hills make up a majority of its total area of arable lands.

5. Conclusions

This paper provides an objective analysis of the two major issues with the traditional arable land pressure model, namely, the difficulty in the computation of multiple-cropping index and the issue of steep sloping cropland either greater than or equal to 25° in the arable land base in a given mountainous area the counting of steep slopes. This paper further concludes that the Improved Model I can cleverly avoid the difficulty in the computation of multiple-cropping index with the existing model, and is feasible, operable, and applicable to most plain areas or areas where gentle-slope mountains and hills are prevailing, and that the Improved Model II, with ecological protection and sustainable development in mind, is more suitable for mountainous provinces where steep slopes occupies a relatively large proportion of the province's total arable lands, because with this model, slopes either greater than or equal to 25° are completely excluded from the total area of arable lands.

Yunnan Province was selected as the object of the empirical study in the paper, on which the arable land pressure was measured and computed by using the Improved Model I and II respectively, and an empirical comparative analysis of the results thereof was carried out subsequently, showing that the Improved Model II, to a greater or less extent, produced larger values than the Model I, and that the two models differ little on plain areas or areas where gentle-slope mountains and hills are prevailing, which means that the Improved Model II shall be used instead for regions where steep slopes make up a large percentage of the region's total area of arable lands in order to obtain a more accurate and objective result.

Acknowledgements

This article is the result of project (No. 41261018) supported by the National Natural Sciences Foundation of China. We thank National Natural Science Foundation Committee of China and Department of Land & Resources of Yunnan Province for their friendly help.

References

1. Cai Yun-long, Fu Ze-qiang, Dai Er-fu. The Minimum Area Per Capita of Cultivated Land and Its Implication for the Optimization of Land Resource Allocation. *Acta Geographica Sinica,* 2002, 57(2): 127-134 (in Chinese).
2. Yang Zi-sheng, He Yi-mei. Soil erosion of cultivated land in mountainous areas at China's southwest border: A case study in Yunnan Province. *Research of Soil and Water Conservation*, 2009, 16(1): 1-7 (in Chinese).
3. Yang Zi-sheng, Liu Yan-sui, Zhao Qiao-gui, He Yi-mei. *The Principles and Methods of Evaluating Regional Food Security Based on Cultivable Land Use and Its Case Study in Yunnan Province.* Beijing: Science and Technology Press of China, 2008 (in Chinese).
4. Yang Qing, Yang Zi-sheng. Calculation and Classification of the Cultivated land Pressure in Yunnan Province [J]. *Agricultural Science & Technology*, 2015, 16(7)1562-1564.
5. Yang Zi-sheng, Zhao Qiao-gui. Study on dividing flatland county, semi-mountainous & semi-flatland County and mountainous county in Yunnan Province based on the Second National Land Survey. *Journal of Natural Resources*, 2014, 29(4): 564-574 (in Chinese).
6. He Yi-mei, Yang Zi-sheng. Analysis on the standards of regional food demand per capita based on food security. *China Business (Jingji Lilun Yanjiu)*, 2008, (7): 6-8 (in Chinese).

Study on the changes of endogenous hormones in *Taxus cuspidata* female plant during the development of branch and leaf

Q.G. Qi[1], C.Y. Guo[1] and Q.C. Zhang[1,*]
[1]Forestry College of Beihua University, Jilin City, 132013, China
[*]E-mail: zqc1212@sina.com

The content variation of ABA , IAA, GA and ZR was measured with ELISA in *Taxus cuspidata* female plant during the development of branch and leaf. The results showed that among the four kinds of endogenous hormones the contents of ABA and GA were higher than those of IAA and ZR, and the contents of IAA and ZR increased during the growing period. The contents of ABA and GA in dormancy stage were more than in the growing period. The content of endogenous hormones in leaf was more than in branch.

Keywords: *Taxus cuspidata*; Female plant; Branch; Leaf; Endogenous hormones.

Abbreviations: ELISA–enzyme linked immunosorbent assay; ABA–abscisic acid; IAA–indole-3-acetic acid; GA–gibberellic acid; ZR–zeatin riboside.

1. Introduction

Taxus cuspidata is the precious relict species from Quaternary Period, recognized as the world's endangered and rare natural anti-cancer plant. It is called "national treasure" by Taiwan scholars [1]; it has been listed as an A-level rare and endangered plant in China; logging has been banned by the United Nations. Its root, stem, leaves, and bark can be used as medicine, the leaves being diuretic, effective in treating hypertension, diabetes and heart disease [2]. The bark, root, stem, leaves and other parts contain Taxol, effective in the treatment of cancer [3,4], and so called the last "defensive line for treating cancer" [5].

Taxus cuspidata features straight texture and compact structure, being flexible, with high mechanical strength, glossy, fragrant, decay-resistant, excellent coloring and bonding properties, suitable for making musical instruments, sculpture, furniture, and artistic decoration. Having a beautiful shape, red fruits, and dense green foliage in four seasons, *Taxus cuspidata* can be planted in gardens as ornamental trees.

[*] Corresponding Author

Many *Taxus cuspidata* scholars have conduced researches on the developmental physiology of seed [6], seedling photosynthesis and physiological and biochemical characteristics [7,8], polyploid mutation [9], endangered mechanism [10], forest cultivation techniques [11], and population ecology [12]. However, there have been no studies on the regularity of endogenous hormones' variation in the growth process of *Taxus cuspidata* branches and leaves. This paper expounds an exploratory study on the dynamic regularity of the contents of ABA, IAA, GA and ZR in *Taxus cuspidata* female plant during the growing period of branch and leaf by means of enzyme-linked immunosorbent assay (ELISA). The study will provide a theoretical basis for further researches on the mechanisms of *Taxus cuspidata* development.

2. Materials and methods

2.1. *Plant material*

Three staminate trees and three pistillate trees of 47-year-old *Taxus cuspidata* grown in an arboretum of the Forestry Institute of Beihua University in Jilin, China were used in the experiment.

The experiment was carried out in 2011 on the current year branch, one-year branch, two-year branch, three-year branch, fresh sample of leaf weighing 1.0g, stored in cryogenic refrigerator for the determination of endogenous hormones.

2.2. *Extraction and detection of phytohormones*

ZR, IAA, GA and ABA were detected by means of ELISA. The extracted samples were further purified with Sep-Park C18 column (Waters). Hormone levels were determined according to Wang [13]. The reagent boxes (China Agricultural University) were cooled to -20°C. The rabbit antibodies were specific to IAA, GA, ABA or ZR according to the description of the kits.

2.3. *Statistical analysis*

All experiments were repeated three times. Triplicate assays for each ELISA analysis were performed. Average values were treated with the analysis of variance; the means were presented with their standard errors.

3. Results and Analyses

3.1. *Changes of abscisic acid (ABA) content*

3.1.1. *Changes of ABA content in the branch of Taxus cuspidata female plant*

As can be seen from Figure 1, the dynamic changes of ABA content showed remarkable fluctuations from February to June, and a smaller range of variation after June. The two-year branch experienced a steady change from February to April, but a sudden increase occurred in the sprouting period in June, ABA content reaching the maximum value of 1169.07ng • g-1 (Figure 1) in the entire development of branch.

After reaching the maximum during the sprouting period, ABA content in the current branch gradually decreased. The three-year branch showed no big change in the rest of the months except the minimum value of 157.89ng • g-1 (Figure 1) in April. The one-year branch was opposite to three-year branch in content change, with the maximum value occurring in April.

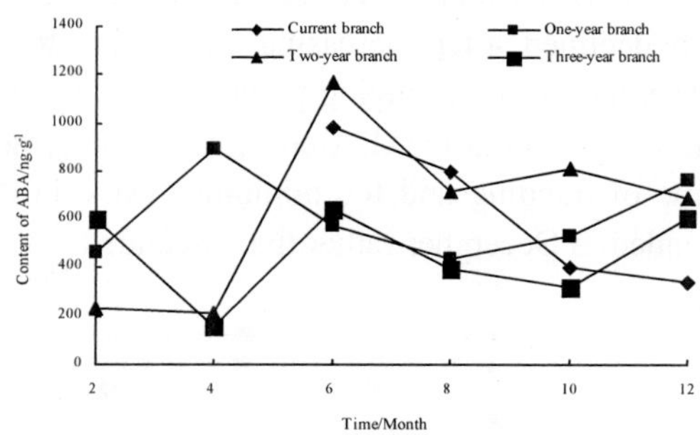

Figure 1. Changes of content of ABA in *Taxus cuspidata* during the female branch development.

3.1.2. *Changes of ABA content in the leaf of Taxus cuspidata female plant*

As can be seen from Figure 2, the older the leaf age was during the development of *Taxus cuspidata* female plant, the higher the ABA content. ABA content in the current year leaf stayed at the minimum level, presenting a gradual rise from June to October, and a decline afterwards. All the one-year, two-year, and three-year leaves had minimum values in April, and expressed a relatively parallel descent from June to August.

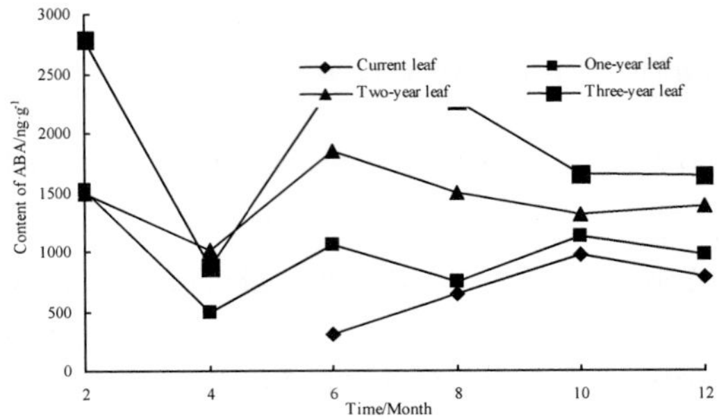

Figure 2. Changes of content of ABA in *Taxus cuspidata* during the female leaf development.

3.2. *Changes of indole acetic acid (IAA) content*

3.2.1. *Changes of IAA content in the branch of* Taxus cuspidata *female plant*

As can be seen from Figure 3, the dynamic changes of IAA content among the four kinds of endogenous hormones were relatively significant. IAA content in the current branch declined after increasing from June to August, with the changes to a lesser extent in the changing process. IAA content in the one-year branch changed in a way similar to ABA content, reaching the maximum value in April at the time of budding and the minimum value in August. However, IAA content descended in December rather than ascended.

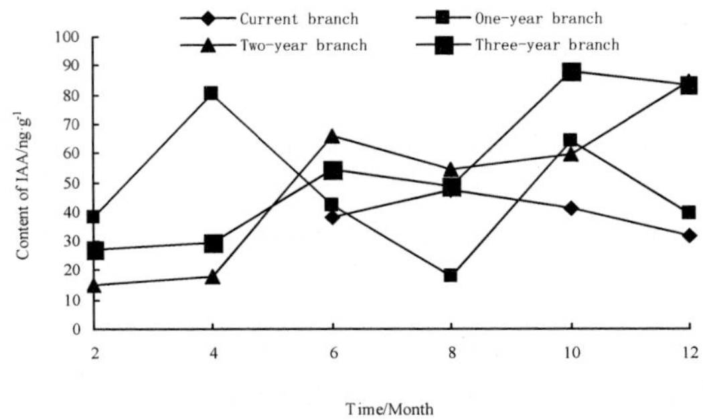

Figure 3. Changes of content of IAA in *Taxus cuspidata* during the female branch development.

3.2.2. *Changes of indole acetic acid (IAA) content in the leaf of Taxus cuspidata female plant*

As can be seen from Figure 4, the change of IAA content in the growing leaf on *Taxus cuspidata* female plant showed certain complexity. A rise in the current year leaf was followed by a steady fall and then another rise. A steady rise in the one-year leaf was followed by a sharp decline and then a steady rise, reaching the maximum value of 859.14 ng • g-1 (Fig. 4) in June, and the minimum value of 301.37 ng • g-1 (Fig. 4) in August.

A sharp rise in the two-year leaf occurred from February to April, reaching the maximum in April, and was followed by a decline. And a continued decline occurred after a slight increase in August. The change of IAA content in the three-year leaf showed a curve, with a slight decline first, followed by a rise, another descent after reaching the maximum in June, and another ascent from October.

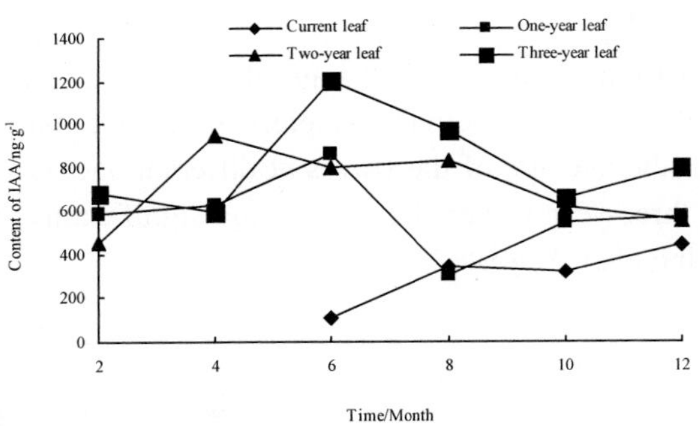

Figure 4. Changes of content of IAA in *Taxus cuspidata* during the female leaf development.

3.3. *Changes of gibberellin (GA) content*

3.3.1. *Changes of GA (GA) content in the branch of Taxus cuspidata female plant*

As can be seen from Figure 5, GA content in the current year branch fluctuated significantly during the development of *Taxus cuspidata* female plant. GA in the one-year, two-year, three-year branches expressed no major rises or falls and the changes were similar. The change of GA content in the current year branch was different from that of ABA and IAA, with an increase from June to August and a sharp decline in August, reaching the minimum value of 260.43 ng • g-1 (Fig. 5) in October, and rising afterwards. GA content in the one-year and two-year branches peaked in June during the sprouting period. GA in the three-year branch peaked in December.

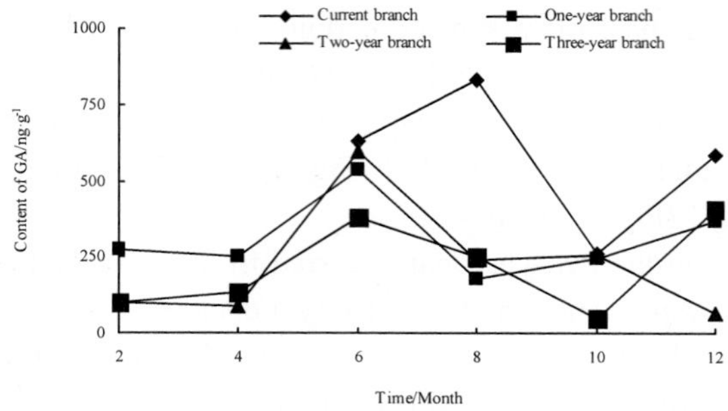

Figure 5. Changes of content of GA in *Taxus cuspidata* during the female branch development.

3.3.2. *Changes of gibberellin (GA) content in the leaf of Taxus cuspidata female plant*

As can be seen from Figure 6, the change of GA content in the leaf of *Taxus cuspidata* female plant was relatively simple. GA content in the current year leaf was lower than the average of the leaves at different age. GA in the one-year, two-year, and three-year leaves did not change significantly from February to August, yet rising after August.

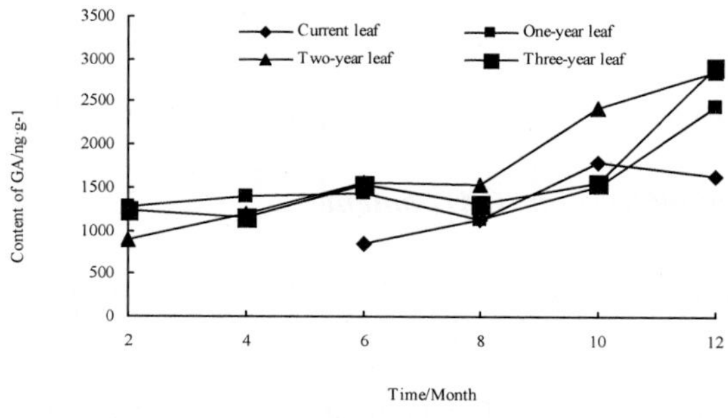

Figure 6. Changes of content of GA in *Taxus cuspidata* during the female leaf development.

3.4. *Changes of zeatin riboside (ZR) content*

3.4.1. *Changes of zeatin riboside (ZR) content in the branch of Taxus cuspidata female plant*

As can be seen from Figure 7, ZR content in the branch of *Taxus cuspidata* female plant in the development process generally showed a significant increase, followed by a decrease. The changes of GA in the current year branch were similar to the one-year branch, with an increase from June to August. The difference was the subsequent decline.

The changes of ZR content in the one-year branch were the same as ABA, IAA, with the maximum in April, then descending to the minimum value of 0.29 ng • g-1 (Fig. 7) in August, beginning to rise slowly to a value of 1.25 ng • g-1 (Fig. 7) in December. ZR content in the two-year, three-year branches reached the maximum during the sprouting period in June, and then decreased slowly.

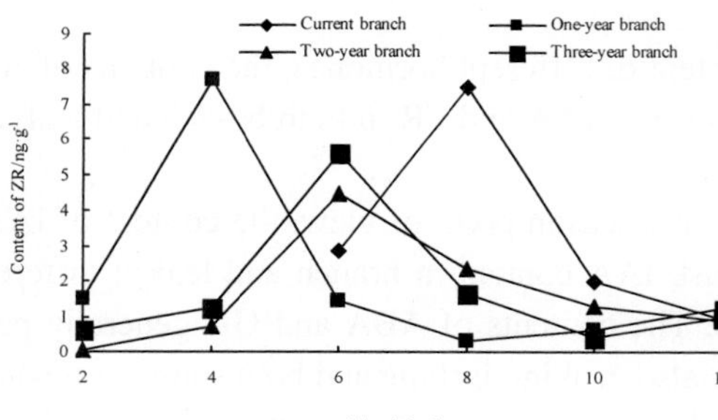

Figure 7. Changes of content of ZR in *Taxus cuspidata* during the female branch development.

3.4.2. *Changes of zeatin riboside (ZR) content in the leaf of Taxus cuspidata female plant*

As can be seen from Figure 8, the changes of ZR content in the leaf of *Taxus cuspidata* female plant was similar to ABA, namely the average ZR content in the current year leaf was lower than that in the one-year, two-year, three-year leaves, being a gradual increase. ZR content in the two-year and three-year leaves showed a relatively parallel decline from June to August. ZR content in the two-year and three-year leaves presented the same change from August to October.

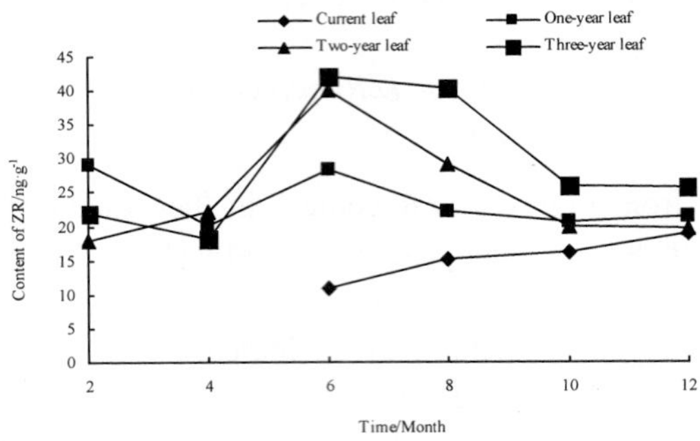

Figure 8. Changes of content of ZR in *Taxus cuspidata* during the female leaf development.

4. Conclusion and Discussion

(1) Overall, the content of four kinds of hormones in the leaf was significantly higher than that in the branch during the development of *Taxus cuspidata* female plant.

(2) As for the content of different hormones, the contents of ABA and GA were significantly higher than IAA and ZR in both branch and leaf, and the content of ZR was the lowest.

(3) From the growing season point of view, the content of IAA and ZR peaked in June and August, IAA content in branch and leaf at different age presenting complex changes. The contents of ABA and GA generally peaked in February or December, GA also peaking in June and both showing obvious regularity.

(4) Each of the hormones plays an essential and irreplaceable role in the development of plant. A variety of hormones co-exist in the same plant, and their emergence is different in time and content. As a result, the roles they play show complexity and are influenced by external conditions. Various hormones restricting and complementing each other results in certain equilibrium. And the cause of the equilibrium remains to be studied in the future.

Acknowledgements

The research was financed by the department of education of Jilin Province science foundation projects (No. 2011133), National Natural Science Foundation Project (No.30271101) and the Jilin Province science foundation projects (No. 20060553). Authors: Qige Qi, Chunyan Guo, Qichang Zhang.

References

1. Bai, G.X., Wu, B.H. and Zhang, Q.C. Research in *Taxus cuspidata*, Beijing: China Forestry Publishing House, 2002, 1~9.
2. Lili Ge, Qige Qi, Qichang Zhang, Chunyan Guo. Content Changes of Endogenous Hormones in *Taxus cuspidata* during the Flower Development. 2013 1st International Conference on Biomedical Engineering and Environmental Engineering, 2013, 683-690.
3. Wei Qiang, Zhang Guosheng, Liu Jinqi, et al. Research Progress of Chemical Components in Taxus Spices. Guangzhou Chemical Industry, 2014, 42(6): 17-23.
4. Wu Di, Li Ya-nan, Wu Li-jun, Gao Hui-yuan. Isolation and Identification of Chemical Constituents from *Taxus cuspidata* [J] Pharmaceutical Journal, 2010, 45(11): 1398-1401.
5. Li Ying, Lu Ye, Lin Li-ping. Research Progress of Antioxidative Components from Chinese yew. Strait Pharmaceutical Journal, 2013, 25(9) 2013: 13-16.
6. Qin Yiting, Li Xue, Zhai Zhixi, Dong Xuehui. Study on Physiological Property in *Taxus cuspidata* During Seed Development. Journal of Northeast Agricultural University, 2014, 45(11): 8-13.
7. Liu Tong, Cui Hai-jiao, Wu Shu-jie, Zhu Jia-yuan, Zhou Zhi-qiang. Response of Photosynthetic and Fluorescence Characteristics of Japanese yew Seedlings to Different Light Conditions. Journal of Beijing Forestry University, 2013, 35(3): 65-70.
8. Liu Tong, Zhu Jia-yuan, Li Peng, Cui Hai-jiao, Zhou Zhi-qiang. Physiological and Biochemical Characteristics of Japanese yew Seedlings as Natural Temperature Falling in Autumn and Winter. Journal of Beijing Forestry University, 2013, 35(2) 51-56.
9. Yang Na, Wang Zhe, Zhao Xiaoming, Wu Yuxiang. Preliminary Study on Polyploid Induction of Northeast yew (*Taxus cuspidata*). J. Shanxi Agric. Uni., 2012, 32(2):124-127.
10. Zhou Zhiqiang, Liu Tong, Hu Linlin, Wang Xiaochun. Tree Ring-climate Response and Endangered Mechanism Deliberation of Japanese yew(*Taxus cuspidata*) in Muling Nature Reserve. Acta Ecologica Sinica, 2010, 30(9): 2304-2310.
11. Teng Xiangjin, Sun Yugang, Qu Xian. *Taxus cuspidata* Research of Breeding and Conservation Technology. Chinese Agricultural Science Bulletin, 2009, 25(12): 82-84.

12. Liu Tong, Hu Lin-Lin, Zheng Hong, Zhou Zhi-Qiang. Researchs on soil seed bank of Japanese yew. Acta Ecologica Sinica, 2009, 29(4): 1869-1876.
13. Wang, Q.M. and Zeng, G.W. The Effect of Phytohormones and Polyamineson Sexual Differentiation of Momordicacharantia, Acta Horticulturae Sinica, 1997, Vol.24(1), pp.48-52.

The photosynthetic physiological responses of leaves from different shoots of *Lonicera edulis* to elevated CO_2

Q.C. Zhang[1,*], X.T. Zheng[1], Y.Y. Qi[1] and Y.T. Li[2]
[1]*Forestry College of Beihua University, Jilin City, 132013, China*
[2]*Baihe Forestry Bureau of Jilin Province, Baihe, 133613, China*
[*]*E-mail: zqc1212@sina.com*

Net photosynthetic rate (PN), transpiration rate (E), stomatal conductance (Cond), vapor pressure deficit at the leaf surface (VPD) and water use efficiency (WUE = PN/E) of perennial *Lonicera edulis* leaves were measured with the Li-COR 6400 portable photosynthesis system at different CO_2 concentration. For the leaves from the vegetative shoot, the reproductive shoot and the coppice shoot, PN, WUE, VPD and E increased with the increase in CO_2 concentration, the increase in PN was larger than that in E. At the CO_2 concentration elevated to $1.0*10^{-3}$, the maximal PN of the reproductive shoot, 11.8 $\mu mol \cdot m^{-2} \cdot s^{-1}$, was similar to that of the coppice shoot, 11.7 $\mu mol \cdot m^{-2} \cdot s^{-1}$. But the maximal PN of vegetative shoot was 11.0 $\mu mol \cdot m^{-2} \cdot s^{-1}$ at the CO_2 concentration of $1.6*10^{-3}$, with the maximal PN being similar to the former two, yet the CO_2 concentration appearing quite different. In the fruiting season the reproductive shoot and the coppice shoot were similar in the physiological indexes, and different from the vegetative shoot.

Keywords: *Lonicera edulis*; Vegetative shoot; Reproductive shoot; Coppice shoot; CO_2 concentration.

1. Introduction

Lonicera edulis is the perennial deciduous shrub of Caprifoliaceae, Lonicera Linn. The fruit is rich in nutrients and vitamin content. It also contains eight amino acids needed by humans from food, and a variety of trace elements. Its berries can be used as medicine, and the fruit is rich in Anthocyanin $C_{21}H_{21}O_{11}C_l$, helpful to deal with cardiovascular disease, stabilize blood pressure, treat anemia in the elderly, supplement nutrients for children, prevent broken capillaries, improve liver detoxification, perform anti-tumor function, increase white blood cell, and enhance anti-inflammatory and anti-virus capabilities [1]. In addition to being eaten raw as food, the fruit is good raw material for brewing wine, being processed into a variety of foods, and extracting natural pigment. Soft and juicy, it is suitable for being processed into a variety of products; in the meantime, rich in physiologically active substances, it contains potential nutritional value, high medicinal and commercial value. Therefore, being one of the world's rare, natural, green wild edible berries, it is known as the "third-generation fruit", and has very broad domestic and international markets [2]. Many scholars have studied *Lonicera edulis* in classification and distribution [3], cultivation techniques [4,5], and the chemical composition [6], but researches on photosynthetic physiological ecology of *Lonicera edulis* remain to be conducted. By measuring different responses of physiological indices to the elevating CO_2 concentrations and comparing the foliage of the perennial vegetative shoot, reproductive shoot and coppice shoot, their differences in response to environmental CO_2 concentration were obtained. This paper was intended for providing theoretical support and baseline data to explore the mechanism for carbon balance and photosynthesis of *Lonicera edulis*.

2. Study Area and Methods

2.1. *Study area*

The experiment was conducted on June 10, 2005 at the *Lonicera edulis* experimental base of Beihua University in Jilin City, Jilin Province. The region has a temperate continental monsoon climate, with four distinct seasons. It is lacking in rain and dry in spring, warm and rainy in summer, cool and sunny in autumn, and cold in the long winter. The annual average temperature is 3-5 °C, and the lowest average temperature is in January, generally between minus 18-20 °C, while the highest average temperature is in July, generally between 21-23 °C, with the extreme annual maximal high of 36.6°C. The frost-free period is up to 130-140d, annual rainfall 650-750mm, sunshine duration 2400-2600h.

2.2. *Methods*

The leaf chamber in a portable LI-6400 photosynthesis infrared gas analyzer (LI-6400P, LI-COR Inc.) made by LI-COR and the LI6400-01 CO_2 Injector were used to measure net photosynthetic rate (PN), transpiration rate (E), stomatal conductance (Cond), foliar saturation, instantaneous water use efficiency (WUE) and vapor pressure deficit (VPD) in the foliage of the perennial vegetative shoot, reproductive shoot and coppice shoot of *Lonicera edulis* at different CO_2 concentration. CO_2 concentration gradient was set to be (50, 100, 200, 400, 600, 800, 1000, 1200, 1400, 1600) $*10^{-6}$. With each change in CO_2 concentration, the minimal stable time was set to be 120s. The variation rate, when measured to be less than 0.05, was to be automatically recorded by the infrared gas analyzer, and repeated three times. Graph paper was used to measure any leaf area less than 6 cm^2, with the measurements entered while measuring. As for any leaf area more than 6 cm^2 it was recorded as 6 cm^2, the area of the leaf chamber. And the results got analyzed in Excel [7–9].

3. Results

3.1. *Stomatal conductance*

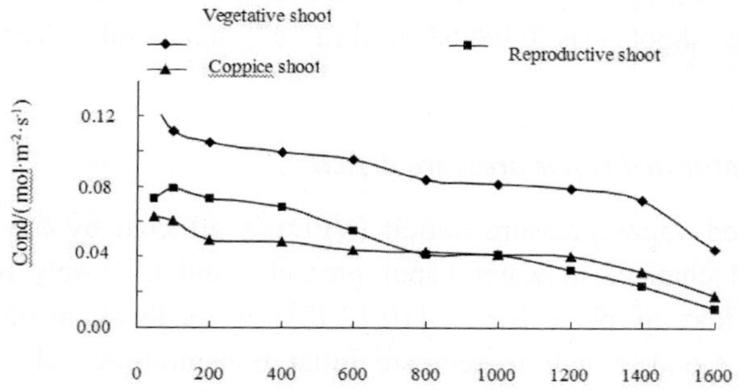

Figure 1. The stomatal conductance responses of leaves from different shoots of *Lonicera edulis* to elevated CO_2.

Table 1. Response equations of physiological parameters of different shoots of perennial *Lonicera edulis* to elevated CO_2.

Shoot type	Y	X	Equations	R
Vegetative shoot	Cond /($mol{\cdot}m^{-2}{\cdot}s^{-1}$)	CO_2	$y = 0.1234e^{-0.0005x}$	R = -0.9220
Reproductive shoot	Cond/($mol{\cdot}m^{-2}{\cdot}s^{-1}$)	CO_2	$y = 0.0951e^{-0.0011x}$	R = -0.9412
Coppice shoot	Cond/($mol{\cdot}m^{-2}{\cdot}s^{-1}$)	CO_2	$y = 0.0636e^{-0.0006x}$	R = -0.8969
Vegetative shoot	VPD/(10^{-6})	CO_2	y = 0.0074Ln(x) + 1.2715	R = +0.9428
Reproductive shoot	VPD/(10^{-6})	CO_2	y = 0.0312Ln(x) + 1.0019	R = +0.9585
Coppice shoot	VPD/(10^{-6})	CO_2	y = 0.0002x + 1.0816	R = +0.9930
Vegetative shoot	E(H_2O)/ ($mol{\cdot}m^{-2}{\cdot}s^{-1}$)	CO_2	y = 0.9964e0.0003x	R = +0.9331
Reproductive shoot	E(H_2O)/ ($mol{\cdot}m^{-2}{\cdot}s^{-1}$)	CO_2	y = 0.2479Ln(x) - 0.9036	R = +0.9579
Coppice shoot	E(H_2O)/ ($mol{\cdot}m^{-2}{\cdot}s^{-1}$)	CO_2	y = 0.1482Ln(x) - 0.349	R = +0.8934
Vegetative shoot	PN(CO_2)/ ($\mu mol{\cdot}m^{-2}{\cdot}s^{-1}$)	CO_2	$y = 3.2259e^{0.0006x}$	R = +0.9421
Reproductive shoot	PN(CO_2)/ ($\mu mol{\cdot}m^{-2}{\cdot}s^{-1}$)	CO_2	y = 0.441Ln(x) + 8.3499	R = +0.5910
Coppice shoot	PN(CO_2)/ ($\mu mol{\cdot}m^{-2}{\cdot}s^{-1}$)	CO_2	y = 0.3675Ln(x) + 8.0961	R = +0.7897
Vegetative shoot	WUE/ (10^{-3})	CO_2	$y = 2.5607e^{0.0008x}$	R = +0.9131
Reproductive shoot	WUE/ (10^{-3})	CO_2	$y = 10.11e^{0.0009x}$	R = +0.9845
Coppice shoot	WUE/ (10^{-3})	CO_2	$y = 13.351e^{0.0005x}$	R = +0.8869

Stomata serve as the channel for leaf to exchange matter with the air, and the degree of stomatal closure will significantly affect plants' fixed CO_2 transpiration rate and water loss [10–11]. As CO_2 concentration increased, stomatal conductance (Cond) in the foliage of the vegetative shoot, reproductive shoot and coppice shoot of *Lonicera edulis* showed a downward trend. Cond and CO_2 concentration were negatively correlated; Cond in the foliage of the vegetable shoot and reproductive shoot reached a significant negative correlation, while Cond in the foliage of the coppice shoot did not (Table 1). Cond values of the reproductive shoot and the coppice shoot were similar (Fig. 2), the average being 0.04954 and 0.04372 $mol \cdot m^{-2} \cdot s^{-1}$, while Cond average of the vegetative shoot was 0.08967 $mol \cdot m^{-2} \cdot s^{-1}$, apparently greater than the former.

3.2. *Foliar saturated vapor pressure deficit*

Foliar saturated vapor pressure deficit (VPD) is affected by the foliar minor environmental changes in water vapor pressure, and is closely related to the transpiration loss of plant leaves [10,12,13]. A feedback mechanism exists among VPD, Cond, and E to regulate foliar transpiration and moisture loss. VPD serves as the driving force resulting in transpiration, and when the water loss exceeds the allowable range, foliage limit it by regulating stomatal conductance, so as to achieve the effective use of limited moisture. VPD in the vegetative shoot, the reproductive shoot and the coppice shoot increased with elevated CO_2 concentration, but the rate of increase was not significant (Fig. 2).

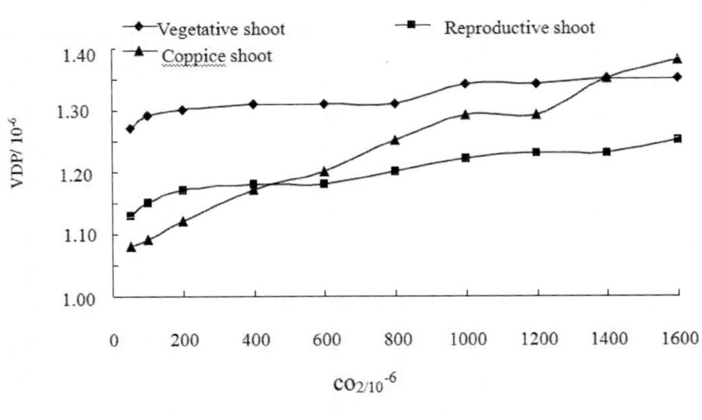

Figure 2. The VDP responses of leaves from different shoots of *Lonicera edulis* to elevated CO_2.

3.3. *Responses of transpiration rate and net photosynthetic rate to the changes in CO_2 concentration*

Transpiration rate (E) in the reproductive shoot and the coppice shoot of *Lonicera edulis* gradually increased with the increase of CO_2 concentration in the leaf chamber, and within the range of measurement of CO_2 concentration, E in the above two were not quite different, and E in the vegetative shoot was significantly larger (Fig. 4). E in the reproductive shoot and the coppice shoot averaged 0.6029, 0.5627 mol·m^{-2}·s^{-1}. E in the vegetative shoot averaged 1.2495 mol·m^{-2}·s^{-1}. Both the reproductive shoot and coppice shoot's response equations to the elevating CO_2 concentration were logarithmic equations, while the vegetative shoot's was an exponential equation. The three types of shoots were positively correlated with CO_2 concentration, and the reproductive shoot reached a significant positive correlation (Table 1).

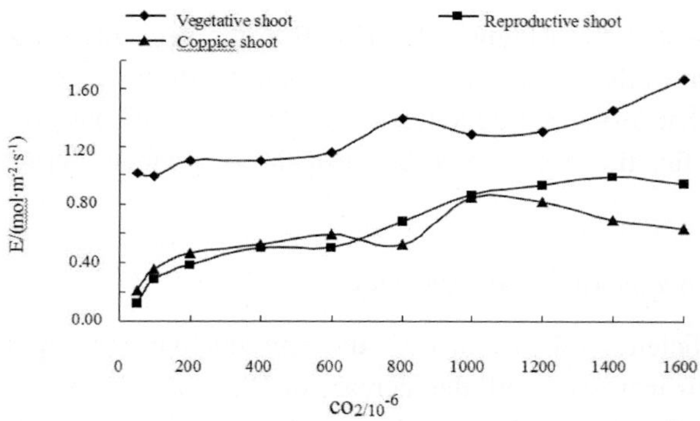

Figure 3. The transpiration rate responses of leaves from different shoots of *Lonicera edulis* to elevated CO_2.

CO_2 is one of the important raw materials for photosynthesis, and the atmospheric CO2 concentration will significantly affect net photosynthetic rate of plant leaves (PN) [14, 15]. As the CO_2 concentration increased, PN in the vegetative shoot, the reproductive shoot and the coppice shoot showed an upward trend (Fig. 5), but didn't reach a significant level. Both the reproductive shoot and coppice shoot's response curves to the elevating CO_2 concentration were logarithmic curves, while the vegetative shoot's was an exponential curve (Table 1). PN in the reproductive shoot and the coppice shoot averaged 11.1722 μmol·m^{-2}·s^{-1}, 10.4478 μmol·m^{-2}·s^{-1}. PN in the vegetative shoot averaged 5.8133 μmol·m^{-2}·s^{-1}, apparently smaller than the former.

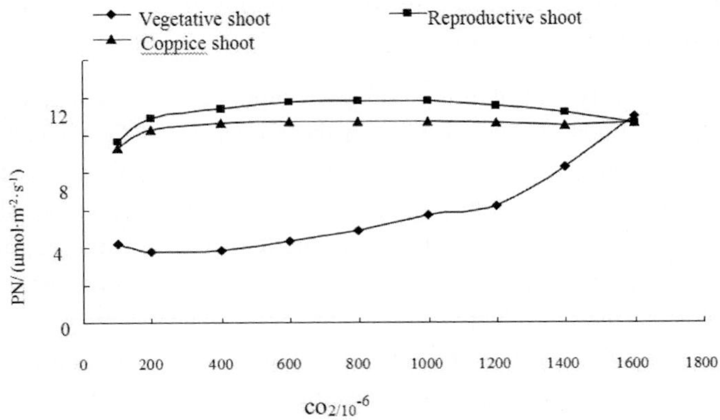

Figure 4. The net photosynthetic rate responses of leaves from different shoots of *Lonicera edulis* to elevated CO_2.

As can be seen from Figures 3 and 4, E in the vegetative shoot was much larger than that in the reproductive shoot, while PN in the vegetative shoot was smaller than that in the reproductive shoot. And a significant difference in the physiological function was shown between the two shoots during the fruiting season.

3.4. *Instantaneous water use efficiency*

Water use efficiency (WUE) in both the reproductive and coppice shoots of *Lonicera edulis* increased with the increase of CO_2 concentration (Fig. 5), being positively correlated, with the reproductive shoot reaching a significant positive correlation (Table 1). The three shoots' response equations to the elevating CO_2 concentration were exponential curve equations. WUE in the reproductive and coppice shoots averaged 22.67, 19.85*10-3(n(CO_2):n(H_2O)), while WUE in the vegetative shoot averaged 5.00*10-3(n(CO_2):n(H_2O)), apparently larger than the former. WUE was determined by the ratio of the PN and E; thus, the change of PN and E affected WUE.

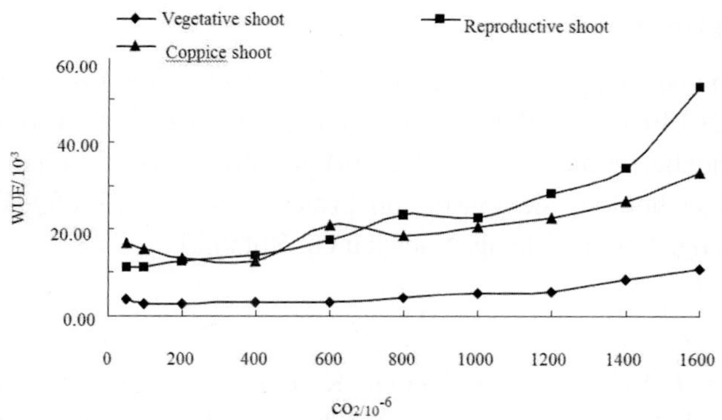

Figure 5. The water use efficiency responses of leaves from different shoots of *Lonicera edulis* to elevated CO_2.

4. Discussion

The results of this experiment showed that the response of the vegetative shoot, the reproductive shoot and the coppice shoot of *Lonicera edulis* to the elevating CO_2 concentration shared the same trend. CO_2 concentration was positively correlated with foliar saturated vapor pressure deficit (VPD), transpiration rate (E), the net photosynthetic rate (PN), and instantaneous water use efficiency (WUE), and was negatively correlated with stomatal conductance (Cond). The reproductive shoot and the coppice shoot had similar photosynthetic physiology index values, but were different from the vegetative shoot. Stomatal conductance (Cond) and transpiration rate (E) in the vegetative shoot were apparently larger than in the reproductive and coppice shoots, while the net photosynthetic rate (PN) and instantaneous water use efficiency (WUE) in the vegetative shoot were apparently smaller than in the reproductive and coppice shoots. This is mainly because *Lonicera edulis* is in its fruiting season in mid-June, when growth of the vegetative shoot is already slow, and the reproductive shoot and the coppice shoot thrive.

Many studies have concluded that as the ambient CO_2 concentration increases CO_2 can enter the mesophyll cells more easily. To reduce the loss of the limited water, pores are usually closed to some extent, lowering Cond to reduce E [13,16,17]. This study showed that the average E in three types of shoots rose, and that under the condition of high CO_2 concentration the reduction of Cond did not lead to lower E. This may be the result of the impact of larger VPD as VPD is the driving force [18] for water to evaporate from the leaf into the air.

Acknowledgements

The research was financed by the Jilin Province science foundation projects (No. 20120269, 20140307025NY), the department of education of Jilin Province science foundation projects (No. 2012130) and the central government forestry science and technology demonstration projects (No. [2013]TJQ04). Authors: Qichang Zhang, Xueting Zheng, Yueyuan Qi, Yutao Li.

References

1. Heinrich J, Svarcova I, Valentova K. *Lonicera caerulea*: A prospective functional food and a source of biologically active compounds [J]. Chemicke Listy, 2008, 102 (4):245-254.
2. Zhang Qichang, Li Liang, Zhang Yitao, Li Yutao, Sun Chengyu. Research Progress on the Utility Value of Lonicera caerulea. Journal of Beihua University (Natural Science), 2014, 15(5):661-664.
3. Plekhanova M N, Streltsyna S A, Paal T. Fruit chemical composition of Lonicera subsect. Caerulea (Caprifoliaceae) species. International conference on "Wild berry culture: an exchange of western and eastern experiences" [J]. Tartu, Estonia Metsanduslikud Uurimused, 1998, 30: 143-146.
4. Szot I, Wieniarska J. Effect of Foliar Applications of Goemar (R) BM 86 and Soil Applied Calcium Nitrate on Yield and Berry Quality of Two Blue Honeysuckle Cultivars. Acta Scientiarum Polonorum-Hortorum Cultus, 2012, 11(1): 133-144.
5. Zhang Qi-chang, Liang Qi-lan, Xia Xin-li, Yin Wei-lun. In vitro culture and plant regeneration from sprouting buds of *Lonicera edulis* Turcz. Journal of Beijing Forestry University, 2010, 32(4): 126-130.
6. Rop O, Reznicek V, Mlcek J. Antioxidant and radical oxygen species scavenging activities of 12 cultivars of blue honeysuckle fruit. Horticultural Science, 2011, 38(2): 63-70.
7. Jiang Gao-Ming and Qu Chun-Mei. Photosynthetic response of six woody species to elevated CO_2 in Quercus liaotungensis forest in the Beijing mountainous areas. Acta Phytoecologica sinsca. 2000, 24(2): 204-208.
8. Feng Yulong, Wang Lihua, Ao Hong, Nie Jiangli, Wang Wenzhang. CO_2 responses of physiological ecology characteristics in larix olgensis henry and its significance. Bulletin of Botanical Research, 1999, 19(1):53-59.
9. Wang Miao, Hao Zhanqing, Ji Lanzhu, Zhou Guangsheng. Effect of elevated atmospheric CO_2 concentrations on photosynthesis light response

characteristics of three coniferous tree species seedlings. Chinese Journal of Applied Ecology, 2002, 13(6): 646-650.

10. Sun Wei, Wang Deli. Wang Li, Yang YunFei. Responses in the transpiration properties and water use efficiency of leaves from different shoots of St ipa bacailens is to instantaneous changes in the CO_2 concentrations and photosynthetic radiation. Acta Ecologica Sinica, 2004, 24(11): 2437-2443.
11. Sun Wei, Wang Deli, Wang Li, Yang YunFei. A comparison of transpiration characteristics and water use efficiency between poaceae plants and chenopodiaceae plants under the simulated light source. Acta Ecologica Sinica, 2003, 23(4): 814-819.
12. Sun Wei, Wang Deli, Wang Li, Yang YunFei. Response of transpiration characteristics and water use efficiency of Setaria viridis to the enhancement of simulated photosynthetic radiation and CO_2 enrichment. Acta Phytoecologica sinsca. 2003, 27(4): 448-453.
13. Wang Li, Yang YunFei, Sun Wei, Wang Deli. Photosynthetic physiological response of two ecotypes of leymus chinensis to the double concentration of CO_2. Acta Agrestia Sinica, 2003, 11(1): 52-57.
14. Catherine Potvin, Liette Vasseur. Long-term CO_2 Enrichment of a Pasture Community Species Richness, Dominance and Succession [J]. Ecology, 1997, 78(3):666-677.
15. Wang RZ, Gao Q. Photosynthesis,Transpiration, and Water Use Efficiency in Two Divergent Leymus Chinensis Populations from Northeast China [J]. Photosynthetica, 2001, 39(1): 123-126.
16. Zheng FengYing, Peng Shaolin. Meta analysis of the Response of Plant Ecophysiological Variables to Doubled Atmospheric CO_2 Concentrations. Acta Botanica Sinica, 2001, 43(11): 1101-1109.
17. Tian Dalun, Luo Yong, Xiang Wenhua, Yan Wende. Photosynthetic Characteristics of Cinnamomum camphoraand Its Response to Elevation of CO_2 and Temperature. Scientia Silvae Sinica, 2004, 40(5):88-92.
18. Han Shijie, Zhou Yumei, Wang Chenrui, Zhang Junhui, Zou Chunjing. Ecophysiological response of Pinus koraiensis seedlings to elevated CO_2. Chinese Journal Of Applied Ecology, 2001, 12(1): 27-30.

The study on irrigation decision system of substrates cultivation in greenhouse

Z.G. Liu

College of Engineering, Huazhong Agriculture University, Wuhan, 430070, China
Email: liuzg@mail.hzau.edu.cn

The objective of this study is to improve agricultural water use efficiency and promote water-efficient substrate cultivation in greenhouse rapid development. A precision irrigation decision system was designed using single - chip microcomputer and sensor technology based on the crop biological principles, it includes a central control system, temperature and humidity sensors, PAR (photosynthetically active radiation) sensor, substrate moisture sensors, pump and solenoid valves, this system realize the automatic irrigation control based on the coupling action of the crop root growth, the substrate water transport and a optimum range of substrate water content. The experiment results in greenhouse show that the system has the ability to develop a scientific and rational irrigation decision and make full use for irrigation water.

Keywords: Substrate cultivation; Irrigation decision system; Greenhouse.

1. Introduction

The great changes of the word agriculture have taken place in the past few decades. The substrate cultivation in greenhouse has the advantages of avoiding the soil continuous cropping obstacles, soil pathogens and pests on crop damage, and develop rapidly at home and abroad [1–3].

Agriculture is the major water consumer in china, the water accounts for about 70% of the total water resources. The total water consumption of the word increased year by year with the development of industrialization and urbanization, and the proportion of the water used in agriculture is shrinking and facing serious challenges [4]. Water-saving agriculture should be greatly promoted, which is an effective way to improve the utilization of agriculture water and alleviate the shortage of water resources [5].

The development of the water-saving irrigation control system started early, high degree of automation, and a relatively complete control system has formed in foreign countries [6], however, the level of agriculture management in our country is relatively backward, and the irrigation equipment and control technology compared with foreign countries are still far [7–8], especially the research on substrate cultivation irrigation decision system in facilities is rarely reported, therefore, a water-saving irrigation decision system was designed to suit the substrate cultivation in greenhouse, which achieving online auto monitoring and control of substrate moisture content base on the matching degree of crop root and substrate wetting body. The irrigation decision system provides a reliable platform for studying the scientific irrigation of substrate cultivation in facilities, it is important to improve the water use efficiency and promotes the rapid development of water-saving substrate cultivation in greenhouse.

2. Systems Theory

The main crops cultivated in greenhouse are leaf vegetables, melon vegetables and solanaceous vegetables, the research shows that the crop can survive only when the moisture content in substrate above a certain threshold value [9], and the substrate water content also has a impact on the crop growth, yield and quality [10–11], therefore, it not only ensure the crop yield and quality, but also save the irrigation water if we can make a reasonable irrigation according to the crop water requirement and control the substrate water content in a suitable range for crop growth.

The goal of saving irrigation water of the irrigation decision system can be achieved by analyzing the relationship of the crop root growth, the substrate water transport and monitoring the lower limit of substrate water content. In the case of lettuce, the range of the optimum substrate moisture content can be obtained by literature [12–13], the substrate water content was controlled within the optimum range to study the dynamic changes of the lettuce roots and water transport in substrate and establish mathematical models, and integrate the models into the control algorithm.

3. System Design

3.1. *System components*

Figure 1 illustrates the irrigation decision system structure, it consists of a central control system, temperature and humidity sensors, PAR sensor, substrate moisture sensors, pump and solenoid valves.

The temperature and humidity sensors are used to monitor air temperature, humidity and substrate temperature, the PAR sensor is used to monitor canopy photosynthetically active radiation, the substrate moisture sensors are used to monitor substrate water content, the pump and solenoid valves are controlled by controller to ensure the irrigation pressure constant and achieve the solenoid valves switch, the central control system is mainly used for analysis the data detected by the sensors and decision whether irrigation based on the matching degree of crop root zoon and substrate wetting body, and a certain threshold value of substrate water content.

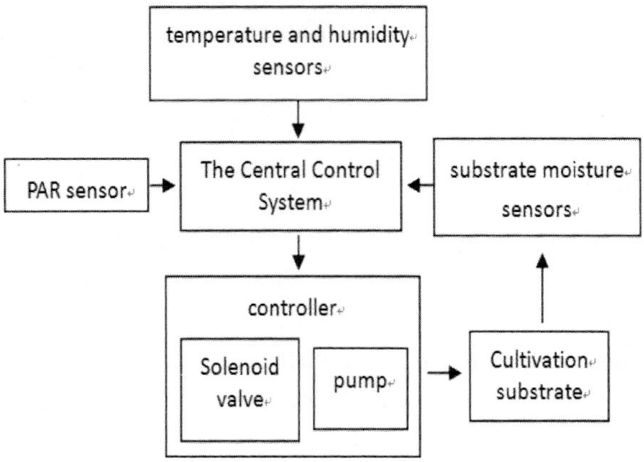

Figure 1. Structure of irrigation decision system.

3.2. *System control process*

The central control system sends the data acquisition commands to substrate moisture sensors regularly, PAR sensor, temperature and humidity sensor, calculates the root distribution and determines the moisture monitoring sensor based on the root growth model, and analysis the data returned by moisture sensors according the control algorithms to make decision and sends commands to the controller, the controller controls the pump and valve sate based on the control commands (start irrigation: first open the solenoid valve and hen open the pump;stop irrigation: first stop the pump and then off the solenoid valve).

3.3. *Substrate moisture sensor selection*

The detection accuracy and reasonableness of substrate moisture sensor have important impact on the performance of the irrigation control system. Four factors were considered to select the substrate moisture sensor: measurement range, detection accuracy and stability, cost effectiveness. The EC-5 moisture sensor developed by the company of Decagon was selected according to the work environment of the irrigation decision system and consisted of 70 MHz signal source, coaxial transmission line and foot-shaped probe, it calculates the volumetric water content by measuring the medium dielectric constant with high-frequency capacitance method [14]. The EC-5 has a low cost, low power consumption, suitable for long-term moisture monitoring under different depths, and the maximum error is less than 13% after calibration [15], it meets the system requirements.

3.4. *The hardware design of the central control system*

The hardware of the central control system plays an important role in the irrigation decision system, shown Figure 2. The single-chip microcomputer SPCE061A with small and high integrated, high reliability, low power as the control center. The SPCE061A was used as a acquisition unit of the central control system with good synchronization and real-time. The controller connect with the substrate sensors, temperature and humidity sensor, PAR sensor by 3.5mm interfaces, and control the pump and valves by relays. In addition, it also extends the SPLC501 circuit, PCF8583 clock circuit, keyboard circuit and a memory circuit to ensure the decision system is able to operate independently.

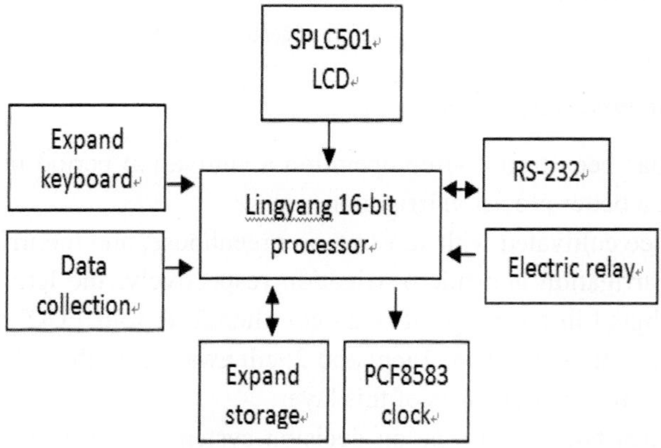

Figure 2. Hardware structure of central control system.

3.5. *The software design of the central control system*

The software of the central control system is the core of the irrigation decision system, shown Figure 3. The parameter setting function is mainly used for correcting the type of moisture sensor, substrate, crop and the optimum moisture threshold; The collection and control function is responsible for sending the collection commends of temperature and humidity, PAR and substrate moisture content, and according to a preset control algorithm to control the valve and pump action, this function has three kinds of collection methods: real-time acquisition, timing acquisition and manual acquisition; The management function is mainly used for data storage, query, display and output. In order to avoid misuse of the control device, the software has a lockout feature and a function to check the status of operation of the equipment switch after power.

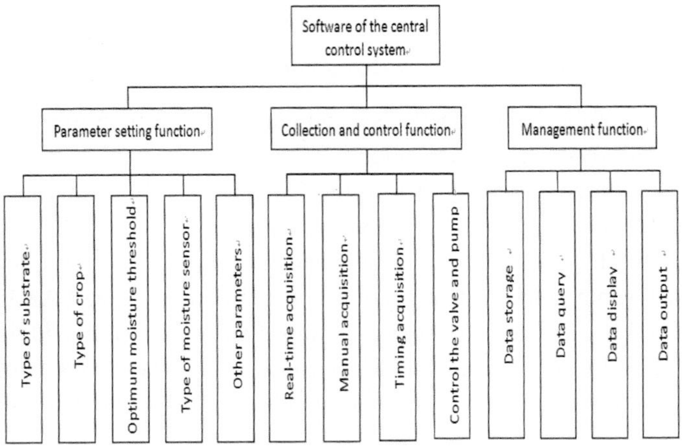

Figure 3. Software structure of central control system.

3.6. *Experimental verification*

The system has been trouble-free operation a cultivation period in greenhouse, and achieved a better precision irrigation.

The lettuce cultivated with substrate in greenhouse, and the irrigation methods are drip irrigation and micro-irrigation respectively, the lettuce root zone mainly distributed in the range of 0~25 cm, therefore, four EC-5 moisture sensors are planted at 5cm, 10cm, 15cm and 25cm away from the substrate surface to monitor the moisture changes of this layer.

The control target of irrigation decision system is to ensure the substrate moisture content in the crop root zone between the range of a certain threshold value, the system selects different substrate moisture sensors in different crop growth season based on the crop root growth model, the moisture sensor near the selected one is used to monitor the fluctuation of deeper substrate moisture content during irrigation. The irrigation threshold value is set between 50%~75% of the field capacity according the optimum moisture content of lettuce to test the performance of the irrigation decision system [12–13], and the system monitor the substrate moisture content constantly (when irrigation, sampling interval: 1min; without irrigation, sampling interval: 10min).

4. Results and Analysis

The moisture sensor planted at 10cm and 15cm away from the substrate surface are selected to monitor the fluctuation of lettuce root zone based on the lettuce root growth model, and the moisture sensor planted at 25cm is selected to monitor the fluctuation of deeper substrate moisture content. The irrigation threshold value is set between 18.5%-28% after conversion between the field capacity and volumetric water content. The substrate water content variation and trend in four-layer substrate from May 15th o 17th in 2013 is shown in Figure 4. The four curves from top to bottom represent the substrate water content variation of 5cm, 10cm and 15cm away from the substrate surface, the substrate water content of the lettuce root zone reached the irrigation threshold twice on the 15th and 16th, and the solenoid valve was opened for irrigation, the solenoid valve was closed when the substrate water content of the lettuce root zone in the best range and the substrate water content of 25cm away from the substrate surface fluctuates small, this suggests that the system can ensure the lettuce growing well and realize water-saving irrigation at the same time.

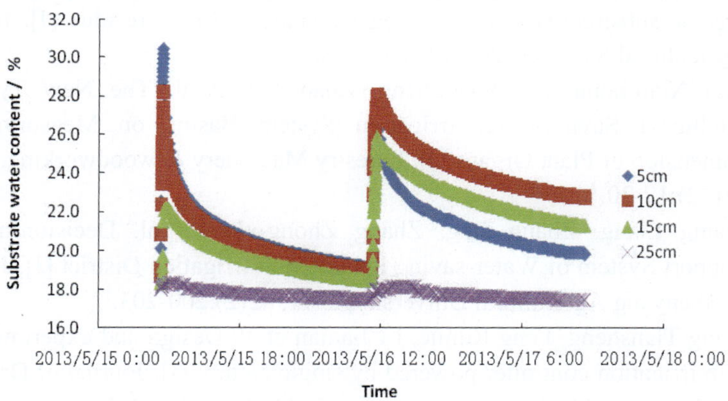

Figure 4. Substrate water content variation and trend in four-layer substrate from May 15th to 17th in 2013.

5. Conclusion

The agriculture irrigation accounts for nearly 70% of the total water consumption, therefore, research on the precision irrigation technology and improve the water use efficiency have become the focus of scientists and engineers. A irrigation decision system was designed to achieve precision irrigation and improve the water use efficiency for substrate cultivation in greenhouse based on the coupling action of the crop root growth, the substrate water transport and a optimum range of substrate water content.

The system is important for promoting the rapid development of substrate cultivation and deepening the agricultural water-saving irrigation strategies, and it is a good tool for precision irrigation of substrate cultivation in facilities.

Acknowledgments

This research was supported by the Project supported by the National Natural Science Foundation of China (Grant No.51509098) and the Fundamental Research Funds for the Central Universities (Program No.52902-0900206303)

References

1. Michael Raviv, Heinrich Lieth. Soilless Culture Theory and Practice [J]. Elsevier, 2008:6-7.
2. Guo Shirong. Soilless cultivation [M]. China Agriculture Press, 2003.
3. Li Pingping, Hu Yongguang, Wang Jizhang et al. Chemical properties of organic substrate containing vinegar residue and reed residue [J]. Jiangsu Agricultural Sciences, 2010(2):207-208.
4. Gao Xiao-hong, Li Dong-sheng, Zhao Jun et al. The New Type of Intelligent Saving-water Irrigation System Basing on Measuring the Dimension of Plant Organs [J]. Forestry Machinery & woodworking, 2004, 11(32):17-20.
5. Zhang Liang, Zhang Jing, Zhang Zhong-xiao et al. Decision-making Support System of Water-saving Irrigation in Irrigation District [J]. Journal of Shenyang Agricultural University, 2011, 42(2):200-203.
6. Hong Tiansheng, Feng Ruijue, Li Jianian et al. Design and experiment for drip irrigation controller powered by single battery [J]. Journal of Drainage and Irrigation Machinery Engineering, 2011, 29(5):455-460.
7. Yang Wan-long, Zhang Xian-rui, Liu Chun-lai et al. Research and Manufacture on Intelligent Fertilization Control System of Drip Irrigation in Greenhouse [J]. Water Saving Irrigation, 2004, 5:27-30.
8. Xu Fei-peng, Li Yun-kai, YANG Pei-ling et al. Research Status and Development Trend on Automatic Control System of Water-Saving Irrigation [J]. Innovation and Development of Urban Agricultural Engineering, 2005, 43-47.
9. Kuang Qiuming, Zhao Yandong, Bai Chenxiang. Automatic monitor and control system of water saving irrigation [J]. Transactions of the CSAE, 2007, 23(6):136-139.

10. Zou Zhirong, Li Qingming, He Zhongqun. Effects of different irrigation maximums on growth dynamics, yield and quality of cucumber during fruit-bearing stage in greenhouse [J]. Transactions of the CSAE, 2005, 21:77-81.
11. Sefer Bozkurt, Gülsüm Sayilikan Mansuroglu. The effects of drip line depths and irrigation levels on yield, quality and water use characteristics of lettuce under greenhouse condition [J]. African Journal of Biotechnology, 2011, 10(17):3370-3379.
12. Pei Yun, Bie Zhilong. Effect of different irrigation minima on the growth and physiological characteristics of lettuce under plastic greenhouse [J]. Transactions of the CSAE, 2008, 24(9):207-211.
13. Pei Yun, Bie Zhilong. Effect of different irrigation maxima on the growth , quality and physiological characteristics of lettuce in plastic greenhouse [J]. Transactions of the CSAE, 2007, 23(9):176-180.
14. Wang Xinzhong, Gu Kaixin, Liu Fei. Hill orchard irrigation control system based on wireless sensor network [J]. Journal of Drainage and Irrigation Machinery Engineering, 2011, 29(4):364-368.
15. Liu Zhigang, Li Pingping, Hu Yongguang et al. Calibration method for detection of matrix water content with dielectric-type moisture sensor [J]. Transactions of the CSAE, 2011, 27(2):199-202.

Proteomic analysis of salt-responsive proteins in the leaves of semi-mangroves *Heritiera littoralis* seedlings under experimental salt stresses

X.L. Wang

College of Environment and Ecology, Xiamen University, Xiamen, Fujian 361102, China

W.D. Han*

Guangdong Ocean University, Zhanjiang, Guangdong 524088, China
**Email: 859730249@qq.com*

C.Y. Lu

Department of Environmental Science and Engineering, Tan Kah Kee College, Xiamen University, Zhangzhou, Fujian 363105, China

Heritiera littoralis Dryand were used for studying the molecular mechanisms of semi-mangroves on salt tolerance. We used two-dimensional gel electrophoresis to measure changes in the proteome in leaves of *H. littoralis* (the total salt content in soil of 1 g/kg (control), 10 g/kg, 19 g/kg) for 50 days. The salt treatment resulted in reductions of the seedling growth in height, Proline content, and SOD activity. There were 76 protein spots were up-regulated and 61 protein spots were down-regulated. Based on MALDI-TOF-TOF analysis and NCBI database in searching the identified proteins of *H. littoralis* seedlings to high-salinity due to (1) Isoflavone reductase play a significant role in binding substrate; (2) The up-regulation of light-harvesting chlorophyll a-b binding protein likely helps maintain the integrity of the chloroplast protein complexes and the efficiency of photosynthesis; (3) The up-regulated of Cu-Zn SOD enzyme indicating that the antioxidative system may protect cells from oxidative damage.

Keywords: *Heritiera littoralis* Dryand; Salt stress; Two-dimensional gel electrophoresis; MALDI-TOF-TOF; Proteomics.

1. Introduction

In many areas of the world, salinity is one of the major environmental factors which limit plant growth and yield (Allakhverdiev et al. 2000, Munns 2002). Currently, salinity stresses to plant growth is increasingly widespread around the world. More than 50% of all arable lands may suffer serious salinization by 2050 (Wang et al. 2003). NaCl is the predominant salt in most saline environments (Turkan & Demiral 2009). In the last few years, significant progress has been made in understanding the responses at the whole plant and cellular levels when plants are subjected to high salinity (Estrella et al 2014). High salinity is detrimental to plant growth for various reasons, including a reduction in photosynthesis, ion imbalance, oxidative damage and nutrient deficiency, all of which lead to growth inhibition and even plant death (Rus et al. 2001, Ozturk et al. 2002, Tester & Davenport 2003, Munns 2005, Miller et al 2010, Ashraf & Harris 2004). Plant acclimatization to stress is associated with the changes in proteome composition (Kosova et al. 2011, Kumar et al. 2000, Veeranagamallaiah et al. 2008). Because of the directly involved in plant stress responses, proteomics-based studies contribute to the relationships between protein abundance and plant stress acclimation (Kosova et al. 2011). Proteomics is a powerful molecular to compare how different physiological conditions affect the proteome (Sobhanian et al. 2010). As yet, proteomic methods have been widely used to investigate the different stress responses in different plants, including tobacco, rice, potato, and Arabidopsis (Abbasi & Komatsu 2004, Amme et al. 2006, Aghaei et al. 2008, Labra et al. 2006, Liu et al. 2014). Two-dimensional polyacrylamide gel electrophoresis (2-DE) is one of the most sensitive and powerful techniques for separating hundreds of proteins (Weiland et al. 2014).

Mangroves, with the characteristic of intertidal plant formations in tropical and subtropical coastlines, are among the most salt-tolerant woody plants (Berlyn 1986). Mangroves can tolerate salt up to seawater concentrations (~500 mM NaCl) (Takemura et al. 2000). In spite of sharing the common responses to salt stress, different plants exhibit diverse level of tolerance toward salinity. A series of researches have been carried out to elucidate the mechanism of salt tolerance in mangrove plants, including physiological, ecological, molecular biological (Twilley et al. 1998, Hibino et al. 2001, Chen et al. 2010, Chen et al. 2011). Also, there have some reports about the proteomics of mangrove plants in response to salinity (Krishnamurthy et al. 2014, Wang et al. 2014, Zhu et al. 2012, Tada & Kashimura 2009). However, to date, still only limited information is available about salt-response proteins in *H. littoralis*, which occurs from eastern Africa to southern Asia, Australia, and Melanesia (Berlyn 1986). *H. littoralis* has an attractive characteristic with plank roots, leaves that are white to silvery on their undersides, and woody fruit (Jian et al. 2010). There are some reports about the biochemistry, genetic diversity of *H. littoralis* (Tewtrakul et al. 2010, Jian et al. 2004). Study had reported that, *H. littoralis* had little sodium and chloride in the leaves under salt stress (Paliyavuth et al. 2004)

In this study, to characterize the role of leaf blade on salt tolerance of *H. littoralis* plants, we utilized a proteomics-based approach to identify differentially expressed proteins affected by leaf blade and salt stress using 2-DE. This analysis provided a view of the proteome level changes elicited by NaCl exposure. The investigation of salt stress-responsive proteins, with the aim of gaining a better knowledge about the salt stress tolerance in *H. littoralis*, is of both fundamental and economic importance. The results give insights into the molecular mechanism in responses to salinity stress in *H. littoralis* seedlings.

2. Materials and Methods

2.1. *Plant material and experimental treatment*

Mature seeds of *H. littoralis* were collected from inter tidal zone near the seaside of Jilong Mount Isle (21°26′N, 110°23′E), Zhanjiang City, Guangdong province, China. And the seeds were cultivated in the nursery of Guangdong Ocean University (21°08′N, 110°18′E), Zhanjiang City, Guangdong province, China. After 5 months later, 5-month-old seedlings of similar size were chosen for red soil, river sand and compound fertilizer 200:200:1 medium culture under greenhouse conditions: the temperature is of 28-35 °C (day/night), relatively humidity 50%-70%. The 5-month-old seedlings were cultivated in plastic pots (high: 15 cm; Caliber: 21 cm), with amount of fresh water every 3 days. After three months, the seedlings in plastic pots grew stably. Subsequently, the seedlings were treated with salt (NaCl) stresses as showed at Tab. 1. The salt treatments were designed according to previous studies, which showed that *H. littoralis* grew well when salt gradient in the soil ≤11 g/kg, but when the salt gradient in the soil more than 14 g/kg, its growth were inhibited (Fengying 2009). The leaves were harvested after 50 days of NaCl treatments for proteomic analysis.

Table 1. Saltizing method.

Saltizing gradient Salt weight/matrix weight	Matrix weight (kg)	Total saltizing weight (g)	Concentration solution of salt (g)	Number of saltizing (times)	Salt Concentration (g/mL)	Every saltizing volume (mL)
0%	5	0	0	5	0‰	450
0.9%	5	45	9	5	20‰	450
1.8%	5	90	18	5	40‰	450

2.2. *The test methods of the total salt content*

Soil samples were collected and measured for the total salt content after the test by using mass method.

2.3. *Growth measurements and physiological indicators assays*

After the salt treatment for 30 days, seedling growth was recorded. Proline (Pro) content was estimated using the acid-ninhydrin method (Wang et al. 2008), and superoxide dismutase (SOD) activity was estimated using the inhibited NBT photochemical reduction methods (McCord et al. 2008).

2.4. *Protein extraction for 2-DE*

The frozen seedling material was homogenized with liquid nitrogen, the extract added 10mL Tris saturated phenol, the proteins were oscillated 6 time every 5 min and centrifuged at 5000 r/min for 30 min at 4°C in a refrigerated high-speed centrifuge (Hema medical instrument Co., LTD Hema TGL-18R), repeated once-through operation. Precipitated proteins with methanol solution of ammonium acetate 5 volumes as proteins, oscillated 6 time every 5 min and precipitated overnight at -20°C and centrifuged at 5000 r/min for 30 min at 4°C. The proteins were washed 5 mL acetone, and then centrifuged at 5000 r/min for 30 min at 4°C, repeated once-through operation. Then washed twice, added 4 mL acetone, centrifuged at 12000 r/min for 20 min at 4°C. The pellet was dried at room temperature, save for backup at -80°C. The concentration of proteins was determined by using BSA as a standard.

2.5. *2-DE*

The first dimension (IEF) was performed according to Duncan R., et al. (Duncan et al. 1984), and modified, using drystrip (24 cm, pH 4-7). The 120μg of protein amounts were loaded on to the gels. For running the sample voltage applied was kept at 300 V for 30 min, 700 V for 30 min, 1500 V for 1 h and 30 min, 9000 V for 3 h, and finally running at 9000 V for 4 h. The second dimension (SDS-PAGE) was performed on 30% T, 2.6% C acrylamide, 4× resolving gel buffer, 10% SDS, 10% ammonium persulfate, and TEMED. After electrophoresis, the gels were silver stained. Images were captured and spots were quantitatively analyzed by using analysis software Image Master 2D platinum 5.0 (GE).

2.6. *In-gel digestion*

Protein spots were excised from the preparative gels to 0.6 mL EP tubes, each tube was washed twice with 120 μL MilliQ water for washing, then each tube were added 120 μL 50% methanol, shaking shock 30 min at 37°C for decolorizing, and then each tube were added 100 μL ACN, shaking shock 2 min for dehydrating, and then each tube were added 50 μL 100 mM NH_4HCO_3 in 50% ACN 100 μL for imbibition, alkylated with combined and lyophilized. The resulting lyophilized tryptic peptides were dissolved in 90% ACN + 2.5% TFA, and then reconstituted in 30% ACN +0.1% TFA, then added in 5 mg/mL HCCA 0.5 uL. Until the sample completely dry, it can be fed into the mass spectrometer.

2.7. *MALDI-TOF/TOF analysis and database searching*

Mass spectra were acquired on a (Bruker Dalton) Autoflex speed™ MALDI-TOF-TOF tandem mass spectrometer, UV wavelength is 355 nm, repetition rate is 200 HZ, acceleration voltage is 20000 V, best quality resolution is 1500 Da. Scan quality range of 700-3200 Da, and then collect signals. Using software flex Analysis (Bruker Dalton) filtered the baseline peak and identified signal peak. Using Bio Tools (Bruker Dalton) software to search the NCBI database, and found matching related proteins. Meanwhile, inquiring about their capabilities to unambiguous identification of proteins for which the protein belongs to.

3. Results

3.1. *Growth and physiological responses to high-salinity stress in H. littoralis*

Salt treatments caused a decrease of the growth of *H. littoralis* seedlings. The results showed that the seedling growth decrease 75.51% at 19 g/kg salt gradient when compared to the control (Fig. 1A). Exposure to high salinity results in osmotic stress, wherefore, it is significant that plants are able to osmo-regulate to provide protection to the cells. Pro can help regulate cellular osmolarity during stress in plants. The results showed that, the SOD activity increased by 20.27% and 61.86% (with salt gradient at 10 g/kg, 19 g/kg, respectively) when compared to the controls (with salt gradient at 1 g/kg) (Fig. 1B). Similarly, the Pro content increased by 10.53% and 102.63%, respectively (Fig. 1C). The SOD activity and Pro content increased with the salt gradient increasing, through this way, *H. littoralis* seedlings protect the cell membrane system damaged by the reactive oxygen or other radicals.

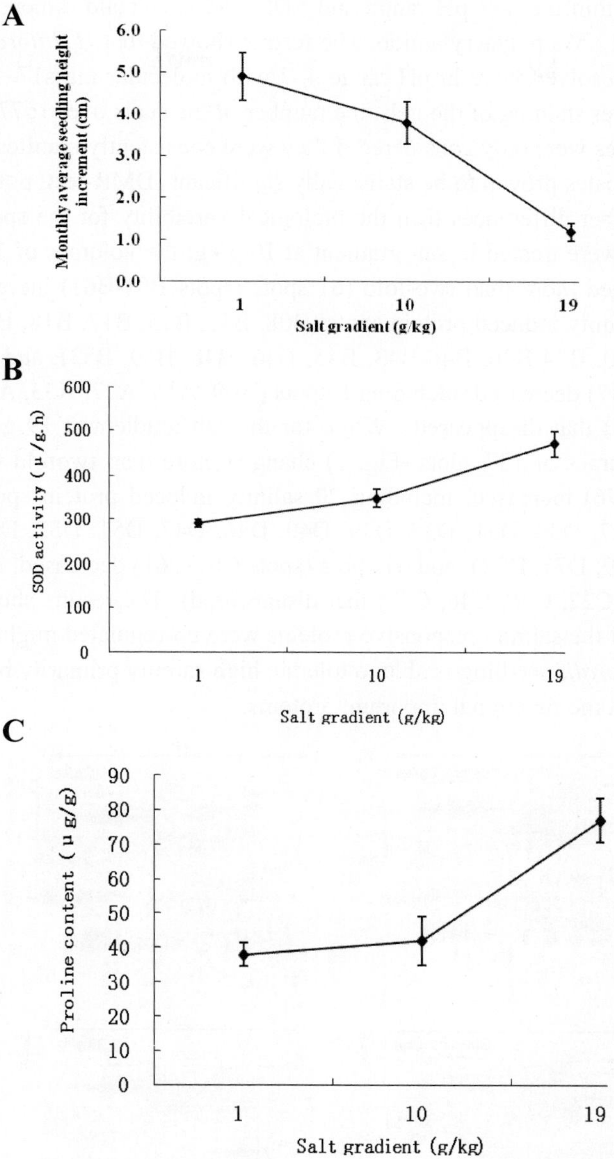

Figure 1. Changes in monthly average seedling height increment (A), SOD activity (B), Proline content (C), *H. littoralis* seedlings grown at different salt gradient (±S.D.).

3.2. *2-DE protein profile in response to salt stress*

The protein profile of seedlings was analyzed by 2-DE. In this study, IEF was carried out within the 4–7 pH range and SDS-PAGE (second dimension) was done using a 12.5% polyacrylamide. The results showed that *H. littoralis* seedling proteins resolved were in pH range 4–7 with molecular mass17–170 kDa. Following silver staining of the gels, the number of the spots over 1677 (Fig. 1). Spot differences were only considered if they were consistently manifested in all the three replicates proven to be statistically significant (DMR-test, $p<0.05$) and displaying higher differences than the biological variability for the spot. When the seedlings were treated in salt gradient at 10 g/kg, the volumes of 108 spots (Fig. 1) changed more than two-fold (61 spots (spots B01-B61) increased, including 24 salinity induced protein spots (B08, B11, B15, B17, B18, B22, B23, B26, B27, B30, B34-B36, B40-B43, B45, B46, B48, B50, B53); and 47 spots (spots A01-A47) decreased, including 8 spots (A09, A18, A27, A33, A34, A35, A39, and A45) that disappeared). While for the salt gradient at 19 g/kg treatment, the volumes of 137 spots (Fig. 1) changed more than twofold (76 spots (spots D01-D76) increased, including 20 salinity induced protein spots (D12, D19, D20, D27, D30, D31, D37, D39, D40, D46, D47, D51, D52, D57, D63, D64, D66, D69, D71, D73); and 61 spots (spots C01-C61) decreased, including 5 spots (C08, C22, C36, C46, C52) that disappeared). The results showed that the majority of the salinity responsive proteins were up-regulated might indicate that the *H. littoralis* seedling is able to tolerate high salinity primarily by the up-regulation of some functional important proteins.

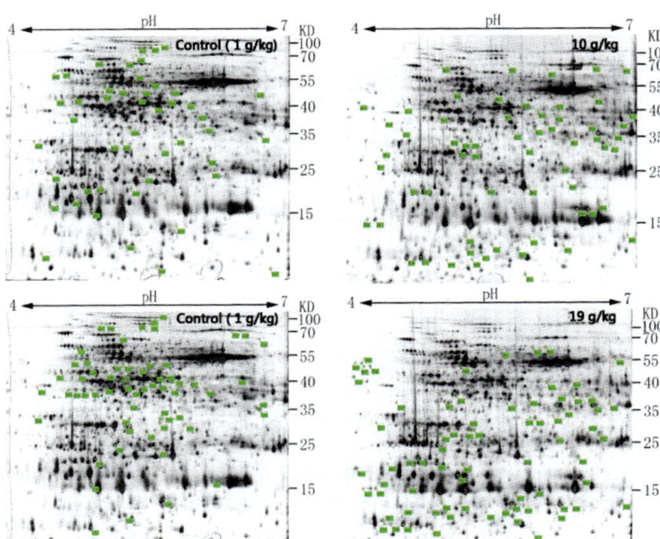

Figure 2. Representative 2-DE gel analysis of proteins extracted from the leaves of *H. littoralis* seedlings under the treatment of total salt content in soil of control (1 g/kg) and 10 g/kg, 19 g/kg for

50 d. First dimension was performed using 120 μg total soluble proteins on linear gradient IPG rods with pH 4 –7. In the second dimension, 12.5% SDS-PAGE gels were used and proteins were visualized using silver nitrate. Differentially regulated protein spots are indicated by green slashes (A, B, C, and D). Under the treatment of 10 g/kg salt gradient and 19 g/kg salt gradient, down-regulated proteins (spots A01–A47 and spots C01–C61) were respectively indicated in image of 1 g/kg salt gradient treatment, up-regulated proteins and newly induced proteins (spots B01–B61 and spots D01–D76) were respectively indicated in the image of 10 g/kg salt gradient and 19 g/kg salt gradient (Fig. 2).

We found that there were 23 spots increased and 17 spots decreased both with the treatment of salt gradient at 10 g/kg and 19 g/kg (Fig. 4). For the 17 down-regulated protein spots, we found that 8 spots (Fig. 3I) were decreased with the increase of salt gradient (spots A01/C02, A12/C15, A15/C17, A20/C22, A26/C42, A28/C45, A29/C46, A30/C47). While for the 23 up-regulated protein spots, the addition of salt gradient caused 7 spots (Fig. 3II) increased (spots B10/D10, B21/D16, B24/D18, B25/D21, B41/D37, B55/D65, B49/D54).

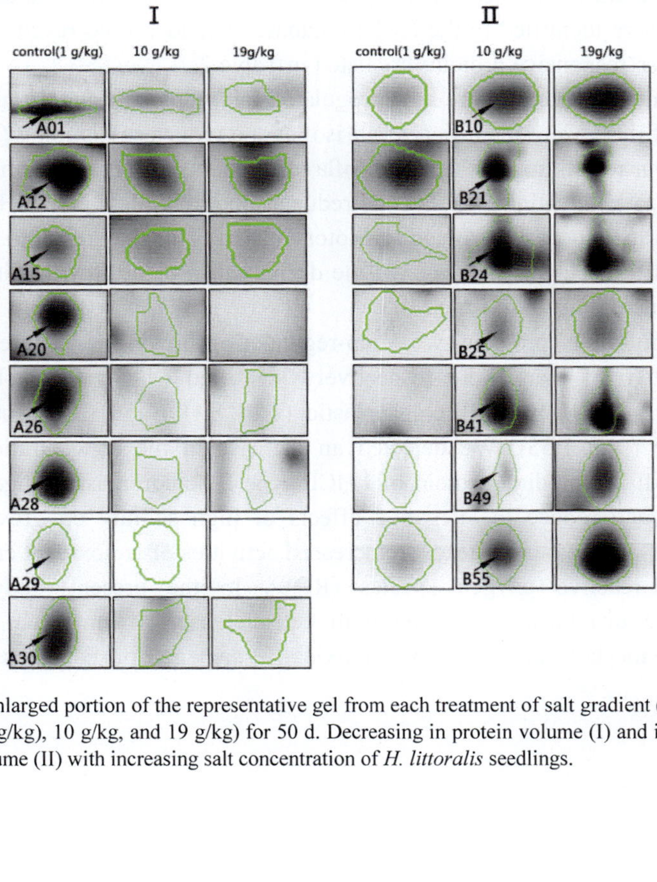

Figure 3. Enlarged portion of the representative gel from each treatment of salt gradient (left to right: (control (1 g/kg), 10 g/kg, and 19 g/kg) for 50 d. Decreasing in protein volume (I) and increasing in protein volume (II) with increasing salt concentration of *H. littoralis* seedlings.

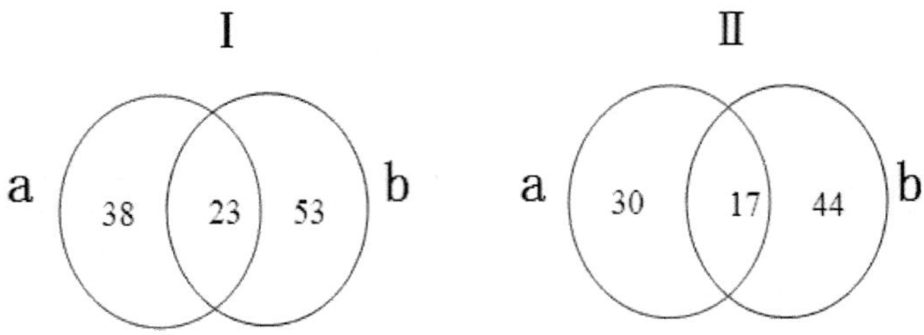

Figure 4. The Venn diagram analysis of differentially expressed protein spots of the leaves of *H. littoralis* seedlings under different concentration of salt treatment. The figure shows differentially expressed in different regions of each point between treatment and control groups; a: the salt gradient of 10 g/kg processing 50 d; b: the salt gradient of 19 g/kg processing 50 d. I: the up-regulated expression spot, II: the down-regulated expression spot.

3.3. *Biological functions of identified proteins*

The 40 spots (23 spots increased and 17 spots decreased both with the treatment of salt gradient at 10 g/kg and 19 g/kg) were excised, digested and analyzed. 3 proteins were identified in the NCBI database. The identified results of the salinity stress-responsive proteins are listed in Table 2.

Seventeen proteins were down-regulated following exposure to high salinity, one of them were identified as an isoflavone reductase homolog (spot A28; Fig. 5). Our results suggest that the isoflavone reductase homolog is inhibited by exposure to high salinity. Isoflavone reductase homolog is a NADPH- dependent aromatic reductase, involved in photosynthesis. This may result to decreased growth of the cells, as we detected the decreased in plant growth with the salt gradient increasing (Fig. 1A).

Twenty-three proteins were up-regulated following exposure to high salinity, two of them were respectively identified ChlorophyII a-b binding protein of LHCII type III chloroplastic (spot B41), and Cu-Zn superoxide dismutase (spot B55). We detected an increase in the concentration of the chlorophyll a-b binding protein of LHCII type III chloroplastic (Fig. 5). Such function may reduce the negative effects of high salinity on growth in *H. littoralis.* Salt stress often leads to increased activities of antioxidant enzymes to eliminate reactive oxygen species (ROS). In the present study, Cu-Zn superoxide dismutase associated with cell defense, which is an important protective mechanism for minimizing oxidative damage from salt stress.

Table 2. Identification of salt stress-responsive protein of the leaves of *H. littoralis* seedlings.

Spot NO.	Protein name	MV (kDa) /pl	Sequence coverage (%)	Species	Mascot score	ID
A28	Isoflavone reductase homolog	33831 / 6.16	3	Solanum tuberosm	25	IFRH_SOLTU
B41	ChlorophyII a-b binding protein of LHCII type III chloroplastic	28798 / 4.99	2	Hordeum valgare	34	CB23_HORVU
B55	Superoxide dismutase [Cu, Zn]	15219 / 5.65	19	Zea mays	79	SODC4_MAIZE

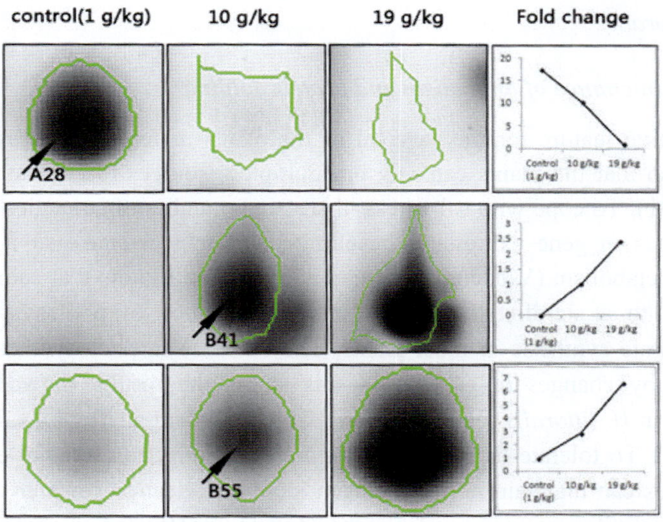

Figure 5. Enlarged portion of the representative gel from each treatment of salt gradient (left to right: control (1 g/kg), 10 g/kg, 19 g/kg) and relative abundance in terms of fold change of each spot in the gel (line plots: x-axis—stress level; y-axis—fold change).

4. Discussion

Salinity is a major abiotic stress that affects plant growth and development. Due to its high degree of salinity tolerance (Mazzucotelli et al. 2008), *H. littoralis* is a valuable model for determining the mechanisms of semi-mangroves involved in responding to salinity stress. In this study, we conducted proteomic analyses in leaves of *H. littoralis* to discuss the possible response mechanisms of *H. littoralis* to high-salinity stress.

4.1. *Biological functions of identified proteins*

The increasing of the concentration of osmolytes is regarded as a common response to osmotic stresses such as salinity or drought in the plant kingdom (Bajji et al. 2001). Pro can provide many ways for protection, such as osmotic regulation, protein structures, etc. (Hare & Cress 1997, Nanjo et al 1999). In this study, the concentration of Pro increased significantly in *H. littoralis* seedlings grown in total salt content in soil of 19 g/kg when compared to the control (Fig. 1C). Meanwhile, the concentration of 21.05% (16 in 76) of the up-regulated proteins increased significantly in *H. littoralis* seedlings grown in 19 g/kg salt gradient. These data indicate that the 19 g/kg salt gradient cause serious damage to *H. littoralis.*

4.2. *Maintenance of anabolism and protein folding*

An adaptive feature for the survival of the plants under stress is by growing slowly, so that the plants can rely on multiple resources to combat the stress (Zhu 2001). To cope with salt stress, there are a number of strategies for plants by changes in gene expression, resulting to protein changes, so it can alert certain metabolism (Vinocur & Altman 2005). We all know that salinity stress can lead to a significant decrease in the efficiency of photosynthesis and carbohydrate synthesis in most plants. The decrease in photosynthesis can be detected by changes in growth. In this study, the results showed that the decrease in *H. littoralis* seedling growth was significant at 19 g/kg salt gradient (Fig. 1A). To tolerate high salinity, by increasing electron transfer efficiency, halophytes can maintain normal or even increased electron transfer efficiency, with the activities of enzymes for carbon fixation unaffected or even stimulated (Stepien & Johnson 2009, Gomez et al. 2010). With the increase of salt gradient treatment, we detected an increase in the concentration of the chlorophyll a-b binding protein of LHCII type III chloroplastic (spot B41), the result is similar to the previous study of P. patens (Wang et al. 2008). The chlorophyll a-b binding protein of LHCII type III chloroplastic is a component of PS I in plants and facilitates light absorption and transfer of the excitation energy to reaction centers for reducing NADP+ to NADPH (Wan & Liu, 2008). We suppose that its increase guarantees *H. littoralis* to maintain a constant level of photosynthesis during the stress, the up-regulation of chlorophyll a-b binding protein of LHCII type III chloroplastic likely helps maintain the integrity of the chloroplast protein complexes and the efficiency of photosynthesis.

Isoflavone reductase homolog have a very similar polypeptide-chain fold, which is consisting of an N-terminal Ross-man-fold domain that binds NADPH, and of a C-terminal segment which largely contributes to substrate binding (Kajikawa et al. 2009). Isoflavone reductase homolog participated in biotic and abiotic stress responses of the plants (Brandalise et al. 2009). In this study, we detected a decrease in the concentration of isoflavone reductase homolog (spot A28), which indicated that some protein molecular structure of *H. littoralis* were damaged at 19 g/kg salt gradient.

4.3. *Removal of reactive oxygen species (ROS)*

Salt stress is often accompanied by a generation of reactive oxygen species (ROS) (Alvarez & Storey 1992, Watson 2000, Chatterjee & Gagnon 2001, Vaidyanathan et al. 2003), which disrupt normal metabolism via oxidative damage to lipids and proteins. Many plant species have mechanisms for ROS detoxification. Superoxide dismutase is a first line component of defense against ROS, and Cu-Zn superoxide dismutase is one of its forms, which presents a key function in the response to salt stress (Alscher et al. 2002). In this study, our results showed that the superoxide dismutase activity rose slowly with the increasing salt gradient (Fig. 1B), and also we detected an increase in the concentration of Cu-Zn superoxide dismutase enzyme (spot B55), which can enhance salinity tolerance in plants (Fig. 5). The result is similar to the previous study of salt stress of cucumber (Du et al. 2010) and sugar cane (Pacheco et al. 2013). We suggest that the Cu-Zn superoxide dismutase enzyme enable *H. littoralis* to remove ROS and protect *H. littoralis* from oxidative damage during exposure to high saline environments.

5. Conclusions

Our current study was aimed to characterizing the responses of *H. littoralis* to salt stress at proteome level. In *H. littoralis* leaves, using two-dimensional electrophoresis, we identified 3 proteins, among them, chlorophyll a-b binding protein of LHCII type III chloroplastic and Cu-Zn superoxide dismutase both were significantly up-regulated, and Isoflavone reductase homolog were down-regulated because of salt stress. This study showed that these proteins are involved in different biochemical reactions imparting tolerance to salt stress in *H. littoralis*. Our findings reveal the changes at the protein level that may provide insight into stress tolerance nature of *H. littoralis.*

Acknowledgements

The work described in this paper is supported by the grants from National Natural Science Foundation of China (Project No. 31170511, Project No. 41376115, and Project No. 41176092).

References

1. Abbasi, F.M., Komatsu, S. 2004. A proteomic approach to analyze salt-responsive proteins in rice leaf sheath, *Proteomics* 4:2072-2081.
2. Aghaei, K., Ehsanpour, A.A., Komatsu, S. 2008. Proteome Analysis of Potato under Salt Stress, *J Proteome Res* 7: 4858-4868.
3. Allakhverdiev, S.I., Sakamoto, A., Nishiyama, Y., Inaba, M., Murata, N. 2000. Ionic and osmotic effects of NaCl-induced inactivation of photosystems I and II in Synechococcus sp., *Plant Physiol* 123: 1047-1056.
4. Alscher, R.G., Erturk, N., Heath, L.S. 2002. Role of superoxide dismutases (SODs) in controlling oxidative stress in plants, *Journal of Experimental* Botany 53:1331-1341.
5. Alvarez, J.G., Storey, B.T. 1992. Evidence for Increased Lipid Peroxidative Damage and Loss of Superoxide-Dismutase Activity as a Mode of Sublethal Cryodamage to Human Sperm during Cryopreservation, *J Androl* 13: 232-241.
6. Amme, S., Matros, A., Schlesier, B., Mock, H.P. 2006. Proteome analysis of cold stress response in Arabidopsis thaliana using DIGE-technology, *J Exp Bot* 57:1537-1546.
7. Ashraf, M., Harris, P.J.C. 2004. Potential biochemical indicators of salinity tolerance in plants, *Plant Sci* 166: 3-16.
8. Bajji, M., Lutts, S., Kinet, J.M. 2001. Water deficit effects on solute contribution to osmotic adjustment as a function of leaf ageing in three durum wheat (Triticum durum Desf.) cultivars performing differently in arid conditions, *Plant Sci* 160:669-681.
9. Berlyn, G.P. 1986. The Botany of mangroves - tomlinson, PB, *Science* 234: 373-373.
10. Brandalise, M., Severino, F.E., Maluf, M.P., Maia, I.G. 2009. The promoter of a gene encoding an isoflavone reductase-like protein in coffee (Coffea arabica) drives a stress-responsive expression in leaves, *Plant cell reports* 28:1699-1708.
11. Chatterjee, S., Gagnon, C. 2001. Production of reactive oxygen species by spermatozoa undergoing cooling, freezing, and thawing, *Mol Reprod Dev* 59: 451-458.

12. Chen, J.A., Xiao, Q.A., Wu, F.H., Dong, X.J., He, J.X., Pei, Z.M., Zheng, H.L. 2010. Nitric oxide enhances salt secretion and Na+ sequestration in a mangrove plant, Avicennia marina, through increasing the expression of H+-ATPase and Na+/H+ antiporter under high salinity, *Tree Physiol* 30:1570-1585.
13. Chen, S.F., Zhou, R.C., Huang, Y.L., Zhang, M., Zhong, G.L., Shi, S.H. 2011. Transcriptome sequencing of a highly salt tolerant mangrove species Sonneratia alba using Illumina platform, *Mar Genom* 4:129-136.
14. Du, C.X., Fan, H.F., Guo, S.R., Tezuka, T., Li, J.A. 2010. Proteomic analysis of cucumber seedling roots subjected to salt stress, Phytochemistry, 71:1450-1459.
15. Duncan, R., Hershey, J.W.B. 1984. Evaluation of isoelctric-focusing running conditions during two-dimensional during two-dimensional isoelectric –focusing sodium dodecyl sulfate-polyacrylamide gel-electrophoresis-variation of gel patten with changing conditions and optimized isoelectric-focusing conditions, *Analytical Biochemistry* 138: 144-155.
16. Estrella, R. V., Barkla, B.J., Pantoja, O. 2014. Comparative 2D-DIGE analysis of salinity responsive microsomal proteins from leaves of salt-sensitive Arabidopsis thaliana and salt-tolerant Thellungiella salsuginea, *J Proteomics*, 111:113-127.
17. Gomez, S. R., Naranjo, E. M., Figueroa, M.E., Davy, A.J. 2010. Salt stimulation of growth and photosynthesis in an extreme halophyte, Arthrocnemum macrostachyum, *Plant Biology* 12: 79-87.
18. Hare, P.D., Cress, W.A. 1997. Metabolic implications of stress-induced proline accumulation in plants, *Plant Growth Regulation* 2: 79-102.
19. Hibino, T., Meng, Y.L., Kawamitsu, Y., Uehara, N., Matsuda, N., Tanaka, Y., Ishikawa, H., Baba, S., Takabe, T., Wada, K., Ishii, T., Takabe, T. 2001. Molecular cloning and functional characterization of two kinds of betaine-aldehyde dehydrogenase in betaine-accumulating mangrove Avicennia marina (Forsk.) Vierh., *Plant Mol Biol* 45: 353-363.
20. Jian, S.G., Tang, T., Zhong, Y., Shi, S.B. 2004. Variation in inter-simple sequence repeat (ISSR) in mangrove and non-mangrove populations of Heritiera littoralis (Sterculiaceae) from China and Australia, *Aquat Bot* 79: 75-86.
21. Jian, S.G., Tang, T.A., Zhong, Y., Shi, S.H. 2010. Conservation genetics of Heritiera littoralis (Sterculiaceae), a threatened mangrove in China, based on AFLP and ISSR markers, *Biochem Syst Ecol* 38: 924-930.

22. Kajikawa, M., Hirai, N., Hashimoto, T. 2009. A PIP-family protein is required for biosynthesis of tobacco alkaloids, Plant *Molecular Biology* 69:287-298.
23. Kosova, K., Vitamvas, P., Prasil, I.T., Renaut, J. 2011. Plant proteome changes under abiotic stress - Contribution of proteomics studies to understanding plant stress response, *J Proteomics* 74:1301-1322.
24. Krishnamurthy, P., Tan, X.F., Lim, T.K., Lim, T.M., Kumar, P.P., Loh, C.S., Lin, Q.S. 2014. Proteomic analysis of plasma membrane and tonoplast from the leaves of mangrove plant Avicennia officinalis, *Proteomics* 14: 2545-2557.
25. Kumar, R.G., Shah, K., Dubey, R.S. 2000. Salinity induced behavioural changes in malate dehydrogenase and glutamate dehydrogenase activities in rice seedlings of differing salt tolerance, *Plant Sci* 156: 23-34.
26. Labra, M., Gianazza, E., Waitt, R., Eberini, I., Sozzi, A., Regondi, S., Grassi, F., Agradi, E. 2006. Zea mays L. protein changes in response to potassium dichromate treatments, *Chemosphere* 62: 1234-1244.
27. Liu, C.W., Chang, T.S., Hsu, Y.K., Wang, A.Z., Yen, H.C., Wu, Y.P., Wang, C.S., Lai, C.C. 2014. Comparative proteomic analysis of early salt stress responsive proteins in roots and leaves of rice, *Proteomics* 14:1759-1775.
28. Mazzucotelli, E., Mastrangelo, A.A., Crosatti, C., Guerra, D., Stanca, A.M., Cattivelli, L. 2008. Abiotic stress response in plants: When post-transcriptional and post-translational regulations control transcription, *Plant Sci* 174: 420-431.
29. McCord, J.M., Wong, K., Stokes, S.H., Petrone, W.F., English, D. 1980. Superoxide and inflammation — a mechanism for the anti-inflammatory activity ofsuperoxide-dismutase, *Acta Physiologica Scandinavica* 25-30.
30. Miller, G., Suzuki, N., Yilmaz, S. C., Mittler, R. 2010. Reactive oxygen species homeostasis and signalling during drought and salinity stresses, *Plant Cell Environ* 33: 453-467.
31. Munns, R. 2002. Comparative physiology of salt and water stress, *Plant Cell Environ* 25: 239-250.
32. Munns, R. 2005. Genes and salt tolerance: bringing them together, *New Phytol* 167: 645-663.
33. Nanjo, T., Kobayashi, M., Yoshiba, Y., Sanada, Y., Wada, K., Tsukaya, H., Kakubari, Y., Yamaguchi-Shinozaki, K., Shinozaki, K. 1999. Biological functions of proline in morphogenesis and osmotolerance revealed in antisense transgenic Arabidopsis thaliana, *Plant Journal* 18:185-193.

34. Ozturk, Z.N., Talame, V., Deyholos, M., Michalowski, C.B., Galbraith, D.W., Gozukirmizi, N., Tuberosa, R., Bohnert, H.J. 2002. Monitoring large-scale changes in transcript abundance in drought- and salt-stressed barley, *Plant Mol Biol* 48: 551-573.
35. Pacheco, C.M., Pestana-Calsa, M.C., Gozzo, F.C., R., Nogueira, Menossi, M., Calsa, T. 2013. Differentially Delayed Root Proteome Responses to Salt Stress in Sugar Cane Varieties, *Journal of Proteome Research* 12:5681-5695.
36. Paliyavuth, C., Clough, B., Patanaponpaiboon, P. 2004. Salt uptake and shoot water relations in mangroves, *Aquat Bot* 78:349-360.
37. Qiu. F.Y. 2009. Research on Biological Characteristics, Salt-resistance, Flooding-resistance and Afforest Experiment of Several Species of Semi-mangroves in, Central South University of Forestry and Technology.
38. Rus, A., Yokoi, S., Sharkhuu, A., Reddy, M., Lee, B.H., Matsumoto, T.K., Koiwa, H., Zhu, J.K., Bressan, R.A., Hasegawa, P.M. 2001. AtHKT1 is a salt tolerance determinant that controls Na+ entry into plant roots, *P Natl Acad Sci* USA, 98:14150-14155.
39. Sobhanian, H., Razavizadeh, R., Nanjo, Y., Ehsanpour, A.A., Jazii, F.R., Motamed, N., Komatsu, S. 2010. Proteome analysis of soybean leaves, hypocotyls and roots under salt stress, *Proteome Sci* 8.
40. Stepien, P., Johnson, G.N. 2009. Contrasting Responses of Photosynthesis to Salt Stress in the Glycophyte Arabidopsis and the Halophyte Thellungiella: Role of the Plastid Terminal Oxidase as an Alternative Electron Sink, *Plant Physiology* 149:1154-1165.
41. Tada, Y., Kashimura, T. 2009. Proteomic Analysis of Salt-Responsive Proteins in the Mangrove Plant, Bruguiera gymnorhiza, *Plant Cell Physiol* 50:439-446.
42. Takemura, T., Hanagata, N., Sugihara, K., Baba, S., I., Karube, Dubinsky, Z. 2000. Physiological and biochemical responses to salt stress in the mangrove, Bruguiera gymnorrhiza, *Aquat Bot* 68:15-28.
43. Tester, M., Davenport, R. 2003. Na+ tolerance and Na+ transport in higher plants, *Ann Bot-London* 91:503-527.
44. Tewtrakul, S., Tansakul, P., Daengrot, C., Ponglimanont, C., Karalai, C. 2010. Anti-inflammatory principles from Heritiera littoralis bark, *Phytomedicine* 17: 851-855.
45. Turkan, I., Demiral, T. 2009. Recent developments in understanding salinity tolerance, *Environ Exp Bot* 67: 2-9.

46. Twilley, R.R., Rivera-Monroy, V.H., Chen, R.H., Botero, L. 1998. Adapting an ecological mangrove model to simulate trajectories in restoration ecology, *Mar Pollut Bull* 37: 404-419.
47. Vaidyanathan, H., Sivakumar, P., R., Chakrabarty, G. 2003. Thomas, Scavenging of reactive oxygen species in NaCl-stressed rice (Oryza sativa L.) - differential response in salt-tolerant and sensitive varieties, *Plant Sci* 165:1411-1418.
48. Veeranagamallaiah, G., Jyothsnakumari, G., Thippeswamy, M., Reddy, P.C.O., Surabhi, G.K., Sriranganayakulu, G., Mahesh, Y., Rajasekhar, B., Madhurarekha, C., Sudhakar, C. 2008. Proteomic analysis of salt stress responses in foxtail millet (Setaria italica L. cv. Prasad) seedlings, *Plant Sci* 175:631-641.
49. Vinocur, B., Altman, A. 2005. Recent advances in engineering plant tolerance to abiotic stress: achievements and limitations, *Current Opinion in Biotechnology* 16: 123-132.
50. Wan, X.Y., Liu, J. Y. 2008. Comparative proteomics analysis reveals an intimate protein network provoked by hydrogen peroxide stress in rice seedling leaves, *Mol Cell Proteomics* 7:1469-1488.
51. Wang, W.X., Vinocur, B., Altman, A. 2003. Plant responses to drought, salinity and extreme temperatures: towards genetic engineering for stress tolerance, *Planta* 218:1-14.
52. Wang, X.Q., Yang, P.F., Gao, Q., Liu, X.L., Kuang, T.Y., Shen,S.H., He, Y.K. 2008. Proteomic analysis of the response to high-salinity stress in Physcomitrella patens, *Planta* 228: 167-177.
53. Wang,Y., Zhu, H.W., Tam, N.F.Y. 2014. Polyphenols, tannins and antioxidant activities of eight true mangrove plant species in South China, *Plant Soil* 374: 549-563.
54. Watson, P.F. 2000. The causes of reduced fertility with cryopreserved semen, *Anim Reprod Sci* 60: 481-492.
55. Weiland, F., Zammit, C.M., Reith, F., Hoffmann, P. 2014. High resolution two-dimensiona, electrophoresis of native proteins, *Electrophoresis* 35:1893-1902.
56. Zhu, J.K. 2001. Plant salt tolerance, *Trends Plant Sci* 6:66-71.
57. Zhu, Z., Chen, J., Zheng, H.L. 2012. Physiological and proteomic characterization of salt tolerance in a mangrove plant, Bruguiera gymnorrhiza (L.) Lam, *Tree Physiol* 32:1378-1388.

The research of temporal evolution of Jinan southern mountain forest vegetation based on RS and GIS

F.Q. Bu*, B. Guo, G.Q. Zheng and X.X. Ren
Shandong Jianzhu University, Art School, Shandong, Jinan, 250101
**Email: bfq-1967@163.com*

The author studied the spatial and temporal evolution of forest vegetation in the southern mountain area of Jinan from 2003 to 2012 with the technology of RS and GIS. This investigation is based on the remote sensing images of southern mountainous area by the year of 2003 and 2012.The result shows that the distribution of forest vegetation in the southern mountain area of Jinan is very complex, and the economic forest area is the largest part and it is the most widely distributed. The conclusion includes three aspect: firstly, the total area of forest vegetation in the southern mountain area is steadily increasing; secondly, the forest resources are developing in the direction of the large patches; thirdly, the general evolution of forest vegetation tends to stay in a stable form.

Keywords: RS and GIS; Southern mountainous area; Forest vegetation; Temporal and spatial evolution.

Jinan is famous as "Spring City", which has the well-known reputation all over the world. Many springs of different forms constitute a unique to "spring" as the city's cultural center and the "Spring City" is known for its urban style, spring has become the soul and lifeline of Jinan. Jinan spring sources is in the southern mountain, which is the major urban water conservation area and a natural ecological barrier of Jinan. The region has an important role in maintaining regional ecological balance, reducing floods and other natural disasters, ensuring the city's ecological environment security. Therefore, to strengthen the development and protection of the southern mountains of the important ecological function areas, especially forest vegetation resources research to ensure the Jinan sustainable development in the future, keeping the style of “spring”plays an important role [1].

1. The General Situation of the Study Area

Southern mountains of Jinan range from north to the south boundary of Jinan. The most southern part reaches to Taian community while the most western part reaches to the boundary of Licheng and Changqing and reaches to Zhangqiu community at east, including the Licheng district Zhonggong town, Liubu town, Xiying town, and part of the land of Ganggou town and Caishi town, and part of the land of Shiliulihe offices and Dangjia offices, the area is 839.1km2. The main water supply is located in southern mountainous area. The resource conservation, the green agriculture and the tourism and leisure is the leading function of important ecological protection zone (the Green Industry Development Zone), which is considered to be the scenic spots and characteristics of cultural tourism area [2].

The area belongs to warm semi humid area of temperate zone continental monsoon climate zone, where is dry and windy in spring and hot and rainy in summer while cool and moist in autumn and cold and dry in winter. The annual average temperature is 10 ~ 20 °C and for the latest 30 years the average annual rainfall is 662.8mm. The rain and heat is at the same period and the large variability of precipitation is mostly in June to September while most of the area has less than 200 frost-free days. Cinnamon distribution area accounted 80% for the total area , but due to the topographic effect, the cinnamon soil is distributed in the mountain, which is easy to produce soil erosion. The soil pH value is 7.O ~ 8.5, with a rich calcium and easy to water leakage characteristics, which is not conducive to the development of vegetation. Regional low profile of the mainland south than in the north, ecological environment quality difference. Part of the regional forest coverage rate is high and rich in plant resources, some of them have become the stable ecological system.

2. Data Sources and Research Methods

The original data of the study come from 2003 and 2012 southern mountain region of Jinan remote sensing image, using ISO-DATA algorithm to obtain the prior knowledge, training sample selection, then use the maximum likelihood method to classify, combined with visual interpretation to modify, finally get the results. With the support of ArcGIS10.0 the geographic registration and geometric precision correction of the regional landscape plan drawn two periods and extract relevant data to establish the geographic information database; by FRAGSTATS 3.3 software computing landscape index to analyze the changes of forest vegetation of the southern mountains of Jinan and reveal the Jinan southern mountain forest vegetation distribution and temporal patterns of variation [3–7].

Research will be divided into the land use types in the southern mountains of Jinan, which includes platycladus orientalis, platycladus orientalis, cotinus coggygria, pinus thunbergii, shrub grassland, economic forest, farmland, grassland, construction land, unused land, waters and broad-leaved forest (broad-leaved forest dominated by other woodland) (11 categories). For the need of space research, the soil conditions in the southern mountain areas of Jinan were divided into two regions according to the study area.

3. Results and Analysis

3.1. *Distribution of land use types in southern mountain area of Jinan*

Application of GIS technology to the analysis of the southern mountainous geographical information database, according to 2003 and 2012 with a type of classification map and attribute table of statistical information analysis in mountainous area of southern type of detailed information to.

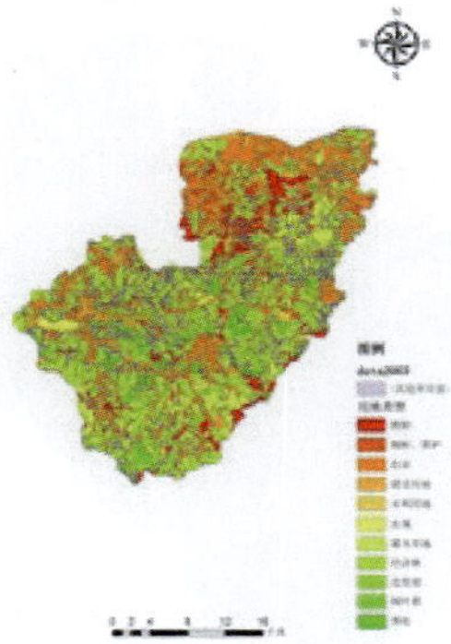

Figure 1. 2003 land use type.

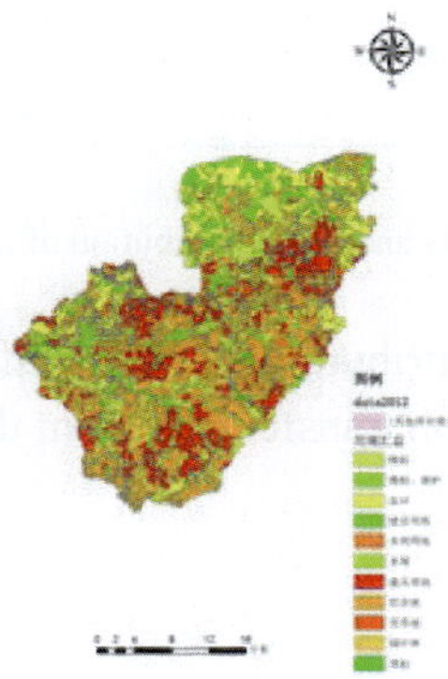

Figure 2. 2012 land use type.

Seen from the picture that the change of land use in the south mountain area of 2003 and 2012 is very limited. Combined with the land use types in the southern mountains, the spatial distribution of forest vegetation in the southern mountains is in accordance with the terrain fluctuation: in the area of flat terrain, affected by human activities, the farmland, economic forest is major; in the higher ground to the top of the mountain is platycladus orientalis, pinus thunbergii, cotinus coggygria, pine and other coniferous forest; in the middle of the buffer zone, is the shrub meadow and grassland and shrub community; broad-leaved forest is mostly preserved in the forest, the general distribution of the economic forest near.

3.2. *Analysis on the overall evolution of forest vegetation*

In 2003 and 2012, the southern mountain forest vegetation space two image processing, as the Figure 3.

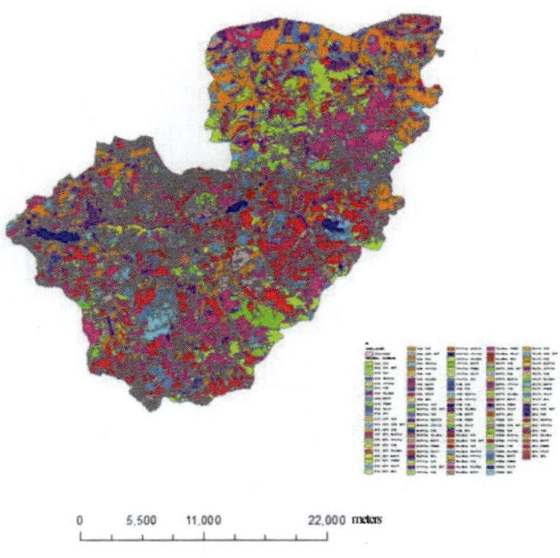

Figure 3. 2003 and 2012 distribution of forest vegetation overlay.

According to the attribute data table information of Figure 3, the south mountain forest vegetation transfer data from the perspective of the unit (km2).

Table 1. Perspective on the transfer of forest vegetation in the southern mountains (Row: 2003; column: 2012).

Line label	Platycladus orientalis	Platycladus orientalis, Cotinus coggygria	Shrub grassland	Pinus thunbergii Parl	Grassland slope	Construction land
Platycladus orientalis	66.545	4.994	5.936	0.676	1.665	0.631
Platycladus orientalis, Cotinus coggygria	1.398	23.304	1.085	0.015	0.178	0.167
Shrub grassland	3.406	0.882	95.930	0.218	3.349	2.548
Pinus thunbergii Parl	0.133	0.003	0.378	5.579	0.056	0.044
Grassland slope	0.718	0.862	4.582	0.032	10.928	0.128
Construction land	1.467	6.578	2.415	0.042	0.276	76.860
Economic forest	4.712	1.340	17.796	0.463	3.274	7.995
Broad leaved forest	1.839	0.172	3.674	0.350	0.652	0.955
Farmland	0.440	0.260	3.204	0.009	3.019	5.666
Waters	0.020	0.000	0.132	0.003	0.005	0.334
Unused land	0.090	0.010	0.438	0.006	0.011	0.135
Total	80.768	38.404	135.572	7.392	23.412	95.462

Total	Unused land	Waters	Farmland	Broad leaved forest	Economic forest
91.251	0.242	0.032	1.325	2.334	6.872
28.565	0.027	0.000	0.033	1.955	0.404
130.996	0.643	0.535	4.740	5.826	12.917
6.486	0.026	0.001	0.024	0.171	0.071
22.031	0.847	0.001	0.378	0.513	3.042
103.202	0.362	0.222	7.866	0.537	6.578
213.618	0.739	0.192	10.379	7.167	159.560
44.285	0.309	0.021	2.828	24.068	9.416
155.837	0.549	0.226	119.867	1.152	21.443
11.131	0.036	7.695	2.711	0.118	0.077
6.481	4.362	0.000	0.181	0.060	1.188
813.882	8.143	8.926	150.333	43.901	221.568

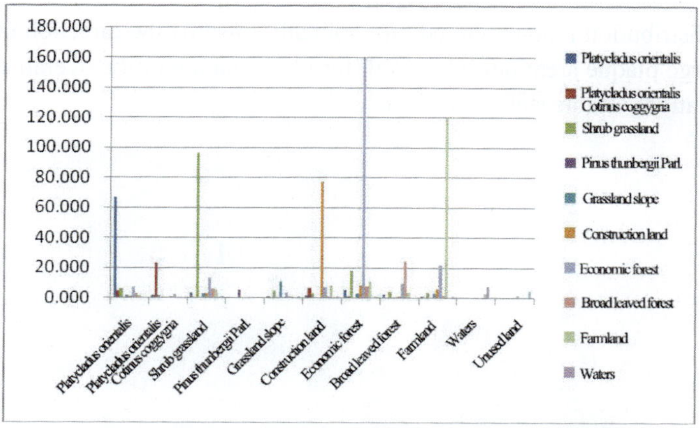

Figure 4. Statistics of forest vegetation in southern mountains.

From Table 1 and Figure 3 can be seen in 2003 to 2012 years Platycladus orientalis circulation area for 14.2km^2 accounted for 17.6% of the area of Platycladus orientalis in 2003, the main flow is 4.7km^2 of economic forest, shrub grass land 3km^2; Platycladus orientalis, Cotinus coggygria circulation area for 15.1km^2, accounted for 39.3% of the 2003 Platycladus orientalis, Cotinus coggygria area, the main flow is 5.0 km^2 of Platycladus orientalis, construction land 6.6 km^2; Shrub grassland area of circulation for 39.6km^2, accounting for 29.2% of 2003 shrub grassland area, the main flow is economic forest 17.8km^2 and barren Grassland Slope 4.6km^2, Platycladus orientalis 5.9km^2; The circulation area of 1.8km^2 pine, Pinus thunbergii in 2003 accounted for 24.5% of the area, the main flow is 0.7km^2 economic forest of Platycladus orientalis, 0.5km^2; Huangcaopo circulation area is 12.5km^2, accounting for 53.3% of the grassland area in 2003, the main flow is 3.3km^2 shrub grassland, economic forest 3.2km^2; Economic forest flow to 62.0km^2, accounting for 28% of the 2003 economic forest area, the main direction of farmland is 21.4km^2, shrub grassland 13.0km^2; The flow of the broad-leaved forest is 19.8km^2, which accounts for 45.2% of the total forest area in 2003, and the main direction of the forest is 7.2km^2, and the shrub grassland is 5.9km^2; The circulation area of farmland is 30.5km^2, accounting for 20.3% of the grassland area in 2003, the main flow of economic forest is 10.4km^2, with the construction of 7.9km^2.

According to the transfer matrix, the land use pattern of the south mountain area can be changed greatly, but the total area of forest vegetation in the southern mountain area of the southern mountain area was only 1.2%. Combined with land using landscape pattern analysis, can know the southern mountain forest vegetation showed a trend toward convergence, in the overall spatial distribution remains based on, gradually towards the direction of change in the large plaque area; but in general, forest vegetation of the overall evolution trend relatively stable state is realized.

Figure 5. Study area division of forest vegetation in southern mountains.

By Figure 5, the southern mountain area is divided into four research areas from the right to the left and the corresponding study area to four. Data analysis for each small range:

3.2.1. *Study area one*

Table 2. Forest vegetation information statistics of area one (km^2).

	2003	2012
Platycladus orientalis	21.98	26.65
Platycladus orientalis, Cotinus coggygria	24.74	19.02
Shrub grassland	19.15	18.68
Pinus thunbergii Parl	0.78	0.78
Grassland slope	1.11	1.32
Construction land	31.72	38.88
Economic forest	41.74	34.56
Broad leaved forest	1.95	1.95
Farmland	70.91	72.22
Waters	0.62	0.62
Unused land	0.33	0.33

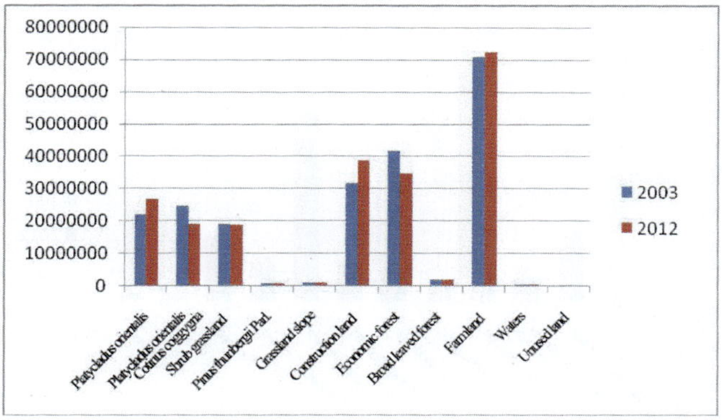

Figure 6. Forest vegetation area of area one (km^2).

The research scope of the area one is $215.01km^2$, the soil is mostly middle Ordovician limestone. In 2003 and 2012, the change of forest vegetation in this area is not large, mainly in the farmland, economic forest, other forest vegetation scattered in which. As the area near the urban area, the overall level of urbanization is higher, and the type of land use is significantly affected. Agriculture is the main economic source of this area, and the farmland is always at the highest level of land use landscape; Economic forest is affected by the process of urbanization, especially the development and construction, and the area is reduced, but it is still an important source of the region's economy; The increase in construction land is the largest, which is due to the acceleration of the process of urbanization, which is the second major landscape types in this area; The remaining forest vegetation has been in a steady state due to the change of policy protection.

3.2.2. *Study area two*

Table 3. Forest vegetation information statistics of area two (km^2).

	2003	2012
Platycladus orientalis	24.12	22.92
Platycladus orientalis, Cotinus coggygria	8.45	10.43
Shrub grassland	41.79	38.33
Pinus thunbergii Parl	4.91	4.73
Grassland slope	10.44	12
Construction land	13.56	13.55
Economic forest	63.12	66.65
Broad leaved forest	14.93	13.01
Farmland	22.69	22.24
Waters	1.76	2.17
Unused land	3.48	3.16

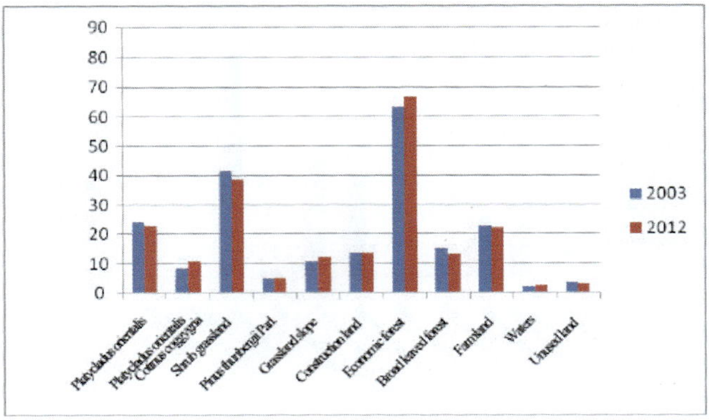

Figure 7. Forest vegetation area of area two (km^2).

The acreage of the area two is 209.19km², the soil is lower Ordovician dolomitic limestone. In addition to the economic forest and shrub grassland showed a large area of the distribution, the rest of the forest vegetation types have a lot of distribution in the area two. And in addition to farmland dependency residents distribution outside, broad-leaved forest, grassland, Platycladus orientalis, Platycladus orientalis, Platycladus orientalis, Cotinus coggygria, in the region distributed relatively uniformly dispersed. The area is mainly hilly and mountainous, with natural vegetation as the main land type, artificial influence is relatively weak. The main source of this area is economic forest, economic forest is always at the highest level of land use landscape; The geographical conditions of the affected areas of the shrub grassland accounted for a large proportion; The remaining forest vegetation is due to the smaller, stable development of the state.

3.2.3. *Study area three*

Table 4. Forest vegetation information statistics of area three (km^2).

	2003	2012
Platycladus orientalis	20.7	21.12
Platycladus orientalis, Cotinus coggygria	2.85	2.14
Shrub grassland	51.71	53.52
Pinus thunbergii Parl	1.03	0.22
Grassland slop	6.66	2.23
Construction land	24.11	19.07
Economic forest	65.89	78.57
Broad leaved forest	19	21.59
Farmland	27.5	22.19
Waters	1.22	0.91
Unused land	1.22	0.34

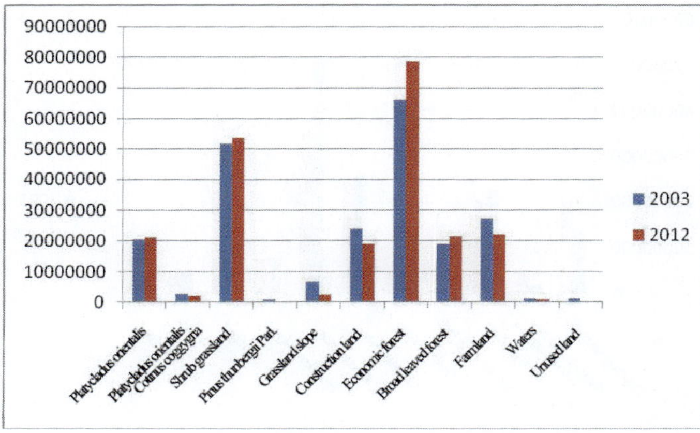

Figure 8. Forest vegetation area of area three (km^2).

The acreage of study area three is 221.90km², the soil type is Cambrian limestone, shale.In the area of economic forest and shrub grassland, economic forest is distributed in the north and south sides of the region. On the south side, the rural areas are far from Jinan, the economy is relatively backward, and the economic forest is the main source of economy; And the north side of the tourism industry, mostly subsidiary. Because of the tourism industry in the region, the regulation of ecological conservation policy, the proportion of forest vegetation resources is relatively large, the overall growth trend; And the phenomenon of returning farmland to forest and grassland, which is the main reason for the reduction of construction land.

3.2.4. *Study area four*

Table 5. Forest vegetation information statistics of area four (km^2).

	2003	2012
Platycladus orientalis	13.96	20.56
Platycladus orientalis, Cotinus coggygria	2.34	4.7
Shrub grassland	22.93	20.46
Pinus thunbergii Parl	0.67	0.75
Grassland slop	5.2	6.48
Grassland slop	26.07	23.97
Economic forest	50.82	36.2
Broad leaved forest	8.01	7.66
Farmland	29.25	39.18
Waters	5.33	5.07
Unused land	3.11	2.64

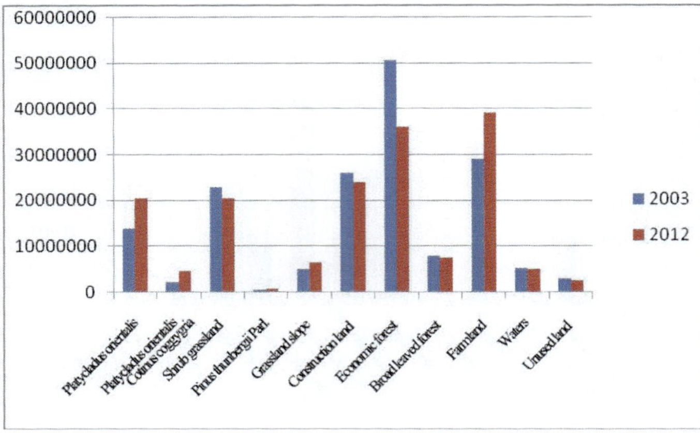

Figure 9. Forest vegetation area of area four (km^2).

The acreage of study area four is 167.68 km^2, the soil is Cambrian limestone, shale. In the region, the economic forest, farmland, construction land and other land use types, in 2003 to 2012 ten years of economic forest reduced by 14.62km^2, farmland increased by 9.93km^2, because of this place is located in the southern mountains of the southern tip of Taishan, more than the mountains, the economy is relatively backward. Platycladus orientalis, sumac, Pinus thunbergii and so vegetation found in the mountainous region, less affected by human, tends to be stable; As the economy is mainly agriculture, the reduction of economic forest is obviously influenced by the economic activities of human beings, while the farmland area is increasing.

4. Results

4.1. *Distribution of forest vegetation in the southern mountain area*

Economic forest occupies the largest area and is widely distributed; Platycladus orientalis is evenly distributed among; The distribution of Platycladus orientalis, mainly concentrated in the northern region of c.coggygria; Due to the local many small hills and mountains, in the gentle slope are mostly distributed in shrubland, so shrub grassland also in the region has a more widely distributed; Grassland is mainly concentrated in the central mountains; Broad leaved forest in the southern part of the study area has a certain distribution; Farmland rely on the construction land distribution, but there is a big change in the south of the study area.

Planning and development policy of the southern mountain area is: Afforestation, ecological conservation. Economic development needs make the area of economic forests continue to grow; With the popularization of ecological consciousness, Platycladus orientalis, sumac, Pinus thunbergii and so natural vegetation is a certain degree of protection, the area is steadily increasing; Grassland and shrubland under a certain degree of damage due to various development and construction area in reducing and need to protect.

4.2. *The evolution of forest vegetation resources in the southern mountain area*

Land use types varied greatly, but the total area of forest vegetation in the southern mountain area of the southern part of the ten years increased by 1.2%; Forest vegetation types present a trend of aggregation, which is based on the overall spatial distribution, and gradually move towards the direction of large patches. Overall, the overall evolution of forest vegetation tends to be stable.

Under the influence of the development and protection policies of the southern mountain area of Jinan, people's ecological consciousness is improved, and the evolution of forest vegetation is steadily occurring and developing; But due to improper development and construction of the southern mountain area, the forest resources subject to a certain degree of damage, especially in shrub land and grassland slope.

References

1. Li Wei. 2005. Reflections on the Environment in Southern Mountain Ecosystems of Jinan City. Chinese Hydraulic Engineering Society.
2. Meng Wei. 2009. Evaluation of Suburb Forest Ecosystem Health—A Case Study of the Southern Mountain Region in Jinan. Shandong: Shandong Normal University.
3. Geometric Correction of Remote Sensing of Baidu Encyclopedia[EB/OL] http://baike.baidu.com/view/1314421.htm.
4. Ni Jingsheng & Luan Qi & Zhe Xuejun. 2004. Basic Theory and Practice of Remote Sensing and GIS. Beijing: Publishing House of Electronics Industry.
5. Xu Xipan & Cao Weibin & Liu Jiaodi & Li Hua. 2010. Design of the Land Use Classification System Based on Remote Sensing Image—A Case Study of Shihhotze Reclamation Area. Xinjiang Agricultural Sciences, 1:204-210.
6. Jiang Chenchun. 2010. Based on Remote Sensing of Chongqing City Urban Form Evolution Research. Chongqing:Chongqing Jiaotong University.
7. Song Gangxian. 2008. The Research of Supervising the Violating Construction Land from Remote Sensing Image. Nanjing: Nanjing Agricultural University.

Distribution of soil nutrients in arable land of tropical low mountain and hill region

H.J. Huang and D.F. Wang*

Tropical Crops Genetic Resources Institute, CATAS, Danzhou 571737, China
**E-mail: dfwang@vip.163.com*

L. Zhao and D. Zeng

College of Applied Science and Technology, Hainan University, Danzhou 571737, China

B.S. Wu

Ruber Research Institute, CATAS, Danzhou 571737, China

T.Z. Wang

Hainan Province Agricultural Technology Extension and Service Center, Haikou 571100, China

To identify the distribution and variation of soil nutrients in arable land is significant to Hainan Island, which is the base of tropical crop products in China. Qiongzhong Li Autonomous County (Qiongzhong County) was selected as study area for the purpose of researching the distribution of soil pH, Organic matters, available nutrients (N, P, K, Ca, Mg, Zn). The arable land soil has acid in the research area, and the average pH was 4.5. The contents of organic matters and available N were relative rich in Qiongzhong County, while the available P and K were deficient. The contents of available Ca and Mg are 508.9 and 82.8 mg kg-1 in soil, which are deficient and unevenly distributed. The content of available Zn is as high as 4.72 mg kg-1, however, potentiality of Zn pollution exists because of high acid degree in arable land soils of Qiongzhong.

Keywords: Tropical arable land; Soil; Nutrients; Qiongzhong Li Autonomous County.

1. Introduction

Chinese arable land resources are scarce, and the arable land area per person is less than 50% of the world average, which causes a sharp contradiction with the rigid requirements of economic development (Sun et al., 2005). The arable land area per person in Hainan is only 0.086 hm^2. The arable land shortage combined with over-cultivation has led to a reduced soil quality in the area, and furthermore, excessive and unreasonable fertilization also accelerates the degradation of soil quality (Qi et al., 2007; Wang et al., 2012). Therefore, to enhance the soil fertility and improve the soil quality, it is necessary to investigate the soil nutrients and the spatial distributions of the overall quality. However, it has been more than 30 years since the last soil survey, and during the past 30 years, much has changed in Hainan, including the agricultural production, agricultural industry layout, and fertilizer input type (Wei et al., 2013). Therefore, the relevant information found in the second national soil survey differs greatly from the present-day situation, and thus, it is no longer a suitable guide for the current agricultural production. As such, an updated investigation of arable soil resources is essential.

Qiongzhong Li Autonomous County is one of the winter vegetables and tropical agricultural bases in Hainan. Investigating arable soil quality is not only important for agricultural production and layout, but it also provides references for farmers' incomes in the ethnic minority areas. By examining the arable soil in Qiongzhong County, the present study accomplished the following: it revealed the spatial distribution of soil nutrients and its major causes, served as a reference for the effective control of soil nutrient loss and the improvement of soil quality, and provided ideas for the development of precision agriculture in order to increase agricultural production and income.

2. Materials and Methods

2.1. *Study area and sampling*

Qiongzhong County sits in the middle of Hainan, and it is located at 18°43'45" - 19°25'20" N, 109°31'37" - 110°09'08" E. The county consists of ten towns with a total area of 2693.1 km^2 and a total arable area of 8673.7 hm^2. For this study, a total of 421 survey points were set up in the arable area (Figure 1), and the soil samples were collected from 0-15 cm of the topsoil.

2.2. *Analytical work*

The indoor air-dried soil samples were filtered through a 2 mm-sieve and were then ready for determining the soil pH and the available nitrogen, phosphorus, and potassium contents. Furthermore, the soil samples were filtered through a 0.25 mm-sieve. At which point, the samples underwent organic matter and microelement contents determination.

For the soil pH, the soil samples filtered through a 2 mm-sieve were used to determine the soil pH with a potentiometric method at a 5:1 ratio of water: soil. For the soil organic matter, the soil samples filtered through a 0.25 mm-sieve were measured and used to determine the soil organic matter with a potassium dichromate - sulfuric acid solution - Mo-Sb colorimetric method. In addition, the soil available phosphorus content was extracted with hydrochloric acid - ammonium fluoride and was determined with the colorimetric method. Finally, the soil available potassium content was extracted with ammonium acetate and was determined with the flame photometric method. The exchangeable calcium and magnesium were extracted with ammonium acetate, and the available zinc was extracted with 0.1M hydroxide chloride. All of the above were determined with atomic absorption spectrometry (Lu, 1999).

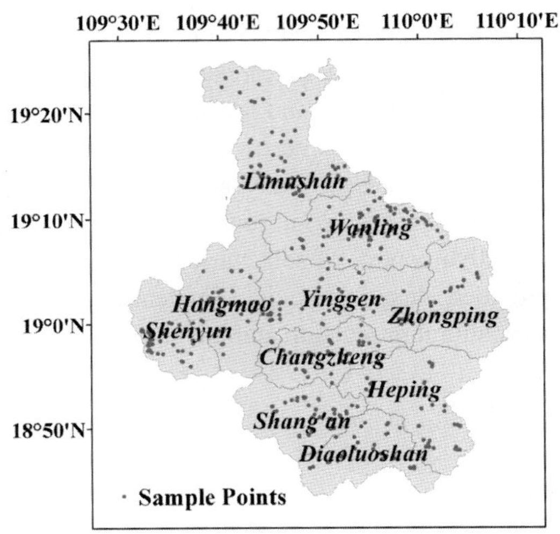

Figure 1. Study area and sample points.

2.3. *Statistical analysis*

All statistical analyses were performed with SAS 9.0.

3. Results

3.1. *Organic matters and soil pH*

The overall organic matter contents in the arable soil in Qiongzhong were rich, and the average content is 24.1 g kg^{-1} (Table 1). However, there was certain spatial variability. The soil organic matter content was highest in Wangling Town at 30.5 g kg^{-1}. In contrast, it was lowest in Shenyun Village, with a value of 18.5 g kg^{-1}. Furthermore, the soil organic matter contents in more than 90% of the arable soil in Qiongzhong County are greater than 20 g kg^{-1}, which is considered rich. In fact, only less than 10% of the arable soil had less soil organic matter than the deficit standard. The area with a lower organic matter content was mainly distributed to the north of Limu Town, west of Shenyun village, and south of Heping Town, which are primarily hilly regions.

Table 1. Distribution of organic matters and soil pH in study area.

Town	Organic matters (g kg^{-1})				Soil pH			
	Min	Max	Mean	ASD	Min	Max	Mean	ASD
Changzheng	3.4	45.0	27.6	8.6	3.9	5.5	4.8	0.4
Diaoluoshan	12.9	33.9	22.9	5.5	3.2	4.7	4.0	0.4
Heping	8.4	30.6	19.3	5.4	3.5	4.7	4.0	0.3
Hongmao	10.2	34.9	21.3	5.7	3.7	4.7	4.2	0.3
Limushan	1.1	36.2	20.1	8.2	3.4	5.1	4.3	0.4
Shang'an	3.0	39.0	24.9	6.5	3.5	4.8	4.3	0.3
Wanling	11.4	121.4	30.5	17.1	4.3	5.3	4.8	0.2
Yinggen	9.4	37.2	26.3	5.9	4.3	5.6	5.1	0.2
Shenyun	7.6	97.0	18.5	13.2	4.5	5.8	5.1	0.4
Zhongping	14.7	42.1	29.8	7.9	4.7	5.2	4.9	0.1
Total	1.1	121.4	24.1	10.9	3.2	5.8	4.5	0.5

Overall, the arable soil in Qiongzhong was acidic, and the soil pH was 4.5 with a range of 3.2-5.8. The pH range for different towns and villages was between 4.0 and 5.1. Furthermore, it was high in the northern and southern regions but low in the middle region. According to the soil pH Grading Standards from the Second Soil Survey, 90.3% of the soil in Qiongzhong is grade 5 ($pH<4.6$), thus indicating soil in the region is acidic.

3.2. *Available N, P and K*

The nitrogen content of the arable soil in Qiongzhong was 4.4 to 371.5 mg kg^{-1}, with an average value of 133.6 mg kg^{-1}, which is grade 2 according to the Nutrient Content Grading and indicates that the area is rich in nitrogen (Table 2). However, the nitrogen content was not evenly distributed. Shang'an Village had the highest, and Shenyun Village had the lowest nitrogen content, with a value of 164.8 and 96.7 mg kg^{-1}, respectively. The areas with relatively rich nitrogen contents were distributed to the north at Limu Town, the middle at Changzheng Town, and the south at Shang'an Village and Diaoluo Village. Partially arable lands located at Heping Town and Hongmao Town also had relatively high nitrogen contents. Overall, Qiongzhong is rich in nitrogen.

The phosphorus content of the arable soil in Qiongzhong was 0.1 to 456.2 mg kg^{-1}, with an average value of 7.4 mg kg^{-1}. The northern Limu Town had the highest phosphorus content at 13.1 mg kg^{-1}. While Shang'an Village, Changzheng Town, Yinggen Town, and Wailing Town had relatively lower phosphorus contents that were lower than the average value. Only scattered areas in the southern Diaoluo Village and the northern Lumian Town and Wanling Town have arable lands with phosphorus contents higher than 20 mg kg^{-1}. The different grades of phosphorus were found in different sized areas across the arable lands. For example, the 4th grade encompassed the largest area, accounting for 86.0% of the total arable land in Qiongzhong, thus indicating the scarcity of phosphorus.

Table 2. Distribution of available N, P and K in arable land soil of study area.

Town	Available N (mg kg^{-1})			
	Min	Max	Mean	ASD
Changzheng	61.4	371.5	154.9	47.3
Diaoluoshan	80.7	238.3	147.4	35.0
Heping	52.7	239.1	135.9	35.6
Hongmao	5.6	246.3	139.1	32.5
Limushan	4.4	294.8	147.6	45.5
Shang'an	65.0	294.8	164.8	36.7
Wanling	4.6	293.2	105.8	49.9
Yinggen	46.1	331.4	124.7	33.9
Shenyun	38.5	160.3	96.7	26.2
Zhongping	53.8	270.3	120.9	43.8
Total	4.4	371.5	133.6	45.7
Town	Available P (mg kg^{-1})			
	Min	Max	Mean	ASD
Changzheng	0.2	72.7	4.5	7.9
Diaoluoshan	0.3	137.2	9.1	13.0
Heping	0.4	46.4	5.1	5.6
Hongmao	0.3	168.4	9.7	14.0
Limushan	0.1	456.2	13.1	31.3
Shang'an	0.1	62.7	5.7	7.9
Wanling	0.1	275.4	5.3	16.7
Yinggen	0.1	106.7	6.1	14.2
Shenyun	0.3	54.2	4.8	7.4
Zhongping	0.5	36.1	9.4	8.9
Total	0.1	456.2	7.4	17.1
Town	Available K (mg kg^{-1})			
	Min	Max	Mean	ASD
Changzheng	5	297	37.6	34.9
Diaoluoshan	12	804	46.8	68.0
Heping	4	80	34.2	16.1
Hongmao	4	202	46.7	28.6
Limushan	5	332	28.5	25.1
Shang'an	3	271	35.3	24.5
Wanling	8	1165	42.5	63.7
Yinggen	11	182	34.7	21.3
Shenyun	14	308	64.5	50.4
Zhongping	9	253	41.5	35.2
Total	3	1165	39.7	42.1

The potassium content of the arable soil in Qiongzhong was 39.7 mg kg^{-1}, with large spatial variation. The range of the potassium content is 3 to 1165 mg kg^{-1}. Different towns and villages had very different potassium contents. For instance, Shenyun Town had the highest potassium content at 64.5 mg kg^{-1}, and all of the other towns and villages were lower than 5th grade, 50 mg kg^{-1}. From among them, Limu Town had a very low potassium content, with a value of 28.5 mg kg^{-1}. The 4th grade of potassium content encompassed the most area, which accounts for 88.0% of the total arable land in Qiongzhong, thus indicating the low amount of potassium in the arable soil.

3.3. *Exchangeable Ca, Mg and Available Zn*

The calcium content of the arable soil in Qiongzhong was 41.6 to 2055.9 mg kg^{-1}, with an average of 508.9 mg kg^{-1} (Table 3). The exchangeable calcium contents were quite different for different towns. More specifically, Hongmao Town had the highest calcium content, while Diaoluo Village and Heping Town had the lowest, which are both lower than 350 mg kg^{-1}. According to the Soil Nutrient Grading Standard, the distributions from the 3rd grade to the 5th grade had higher ratios, with values of 21.43%, 39.51%, and 22.77%, respectively. The exchangeable calcium contents in Qiongzhong were limited, which might be related to the acidic nature of the soil.

Table 3. Distribution of exchangeable Ca, Mg and available Zn in arable land soil of study area.

Town	Exchangeble Ca (mg kg^{-1})			
	Min	Max	Mean	ASD
Changzheng	41.8	1221	570.5	235.6
Diaoluoshan	54.2	1102.6	330.4	223.2
Heping	62.9	906.8	330.7	218.9
Hongmao	180.3	2055.9	758.9	437.0
Limushan	58.4	854.9	394.3	146.2
Shang'an	52.6	1230.3	474.9	313.7
Wanling	88.2	1570.9	558.1	273.1
Yinggen	41.6	903.3	401.3	203.2
Shenyun	207.3	1809.4	755.4	444.0
Zhongping	43.2	760.8	368.8	137.3
Total	41.6	2055.9	508.9	318.8
Town	Exchangeable Mg (mg kg^{-1})			
	Min	Max	Mean	ASD
Changzheng	16.4	253	110.4	56.6
Diaoluoshan	15.5	180.6	60.8	36.1
Heping	15.7	174.0	61.6	37.4
Hongmao	44.6	324.3	132.3	80.6
Limushan	14.5	114.6	54.0	21.0
Shang'an	14.7	315.5	90.0	72.4
Wanling	1.3	403.9	111.7	77.7

Town	Min	Max	Mean	ASD
Yinggen	8.0	206.4	65.8	41.8
Shenyun	14.2	185.1	60.0	40.5
Zhongping	0.1	179.8	52.0	37.9
Total	0.1	403.9	82.8	61.8

Town	Available Zn (mg kg^{-1})			
	Min	Max	Mean	ASD
Changzheng	1.73	16.28	5.00	3.12
Diaoluoshan	1.54	9.49	3.91	2.01
Heping	1.31	49.18	5.28	8.28
Hongmao	0.97	12.95	4.40	2.79
Limushan	0.71	18.41	4.02	3.36
Shang'an	1.07	11.40	4.42	2.36
Wanling	1.69	40.80	8.22	5.81
Yinggen	0.06	9.40	3.27	1.94
Shenyun	0.50	6.10	2.30	1.44
Zhongping	0.02	8.43	3.90	2.34
Total	0.02	49.18	4.72	4.32

The exchangeable magnesium content of the arable soil in Qiongzhong was 82.8 mg kg^{-1}, which is grade 4 and indicates a low magnesium content (Table 3). The exchangeable magnesium contents differed from town to town, and Zhongping Town had the lowest magnesium content. More than 90% of the arable soils in Qiongzhong County had less than 200 mg kg^{-1} of exchangeable magnesium content. The magnesium contents at the southern Diaoluo Village and Heping Town as well as the northern Limu Town were low across the entire town/village, and most of these contents were lower than 75 mg kg^{-1}.

The arable soil in Qiongzhong was rich in zinc, with an average value of 4.72 mg kg^{-1}. All of the towns/villages had a higher value than the grade 1 standard (1 mg kg^{-1}), which indicated that the arable soils in Qiongzhong possess an excellent zinc content. However, though the zinc was rich, it was not evenly distributed spatially. This was especially true for Wanling Town, which has a gentle terrain, and the zinc contents of most of the arable soils in this area were higher than 5 mg kg^{-1}. The spatial differences of the zinc content in Qiongzhong may be related to its parent material and to topographic factors. For instance, Wanling Town is surrounded by hills, and its location easily accumulates the deposition from the surrounding hills. Soil trace elements easily accumulate from the river alluvium, and thus, this could be why Wanling Town had a much higher zinc content than the other towns/villages.

4. Discussion

The soil nutrients in Qiongzhong were distributed unevenly, and there were differences among the soil nutrients from the ten towns/villages. This might be related to the soil parent material, topography, and imbalanced fertilization (Qi et al., 2009; Wang et al., 2013). Since soil trace elements primarily derived from the soil parent material (Kabata-Pendias & Pendias, 2001), the spatial distributions of calcium, magnesium, and zinc are all related to the soil parent materials. In this study of the zinc distribution in West Hainan, Wang Dengfeng et al., (2013) also demonstrated that the zinc in the soil surface and its profile distribution are both significantly influenced by the soil parent material. Moreover, the annual temperature in Hainan is high, and the cropping index is also high. Thus, the soil organic matter is likely to be decomposed. The towns/villages with low soil organic matter were mainly hilly regions. Because of the soil erosion and the unfavorable farming conditions, the soil organic matter in those areas does not easily accumulate, so there is a low content. While organic fertilizer inputs and straw returning are both important sources of soil organic matter (Yu et al., 2010), some farmers did not pay sufficient enough attention to the organic fertilizer, and they burned the straw, so using straw returning to increase the soil organic matter was not an option. Therefore, we could guide farmers to value organic fertilizer and straw returning in order to improve the soil organic matter in Qiongzhong.

Overall, Qiongzhong had weakly acidic soil and a hot and rainy climate. In addition, hills and mountains are widely distributed throughout the area. The soil is mainly brick red soil and lateritic red soil with strong allitic; thus, the soil is primarily weakly acidic. Moreover, a greater amount of chemical fertilizer is used annually, and in Hainan, the total amount of fertilizer used in 1995 was 4.04×10^5 tons, 8.63×10^5 tons in 2006 (Qi et al., 2009). The increased amount of nitrogen fertilizer is one cause for the low soil pH in this region (Guo et al., 2010).

Although the recent trend has been an annual increase in the total amount of chemical fertilizers used (Qi et al., 2009), due to the high cropping index and the mostly hilly terrain, it is not conductive to soil and water conservation. Furthermore, due to the rainy climate and the extremely concentrated rainfall during the rainy season, the climate and terrain lead to significant first flush storms in hilly areas (Gao et al., 2005; Liu et al., 2009), and soil nutrients can be lost due to the erosion of soil particles. Therefore, the annually increased amount of fertilizer did not significantly enhance the soil phosphorus and potassium contents. In addition, the nitrogen contents were low in both Wailing Town and Yinggen Town, which have relatively flat terrains. This might be due to the ease with which flatlands can be farmed, and consequently, crop growth can be achieved without nitrogen fertilizers; thus, there are relatively low nitrogen contents in the soil in those areas. Overall, the soil phosphorus and potassium deficiencies were primarily due to a lack of fertilizer application, the climate, and the topography. The soil nitrogen, phosphorus, and potassium contents can be improved through the use of a combination of multi-methods, and under the premise of rational fertilization, the prevention and control of soil erosion can be improved, together with the usage of organic fertilizer to increase the soil fertilizer capacity.

Recently, chemical fertilizers are primarily urea-, diammonium phosphate-, and potassium-based. Combined with the inadequate use of fertilizers with calcium and magnesium, these practices resulted in a low content of exchangeable calcium and magnesium in the soil. Fortunately, it is possible to enhance the calcium and magnesium contents of soil by changing the habits of farmers, such as the substitution of calcium phosphate with calcium magnesium phosphate fertilizer. Using a calcium magnesium phosphate fertilizer can increase the calcium and magnesium contents of the soil, and importantly, it can prevent soil acidification and improve the soil environment.

Finally, the soil was rich in zinc. However, excessive zinc is detrimental to human health, especially in an acidic environment. Thus, more attention should be paid to the zinc hazard in strongly acidic soil areas. By doing so, the excessive accumulation of zinc by crops would be prevented, and thus, the threat that excess zinc poses to human health would be eliminated.

Acknowledgments

Anonymous reviewers are appreciated for their comments on this study. This research was supported by National Nonprofit Institute Research Grant of CATAS-TCGRI (1630032014035) and Special Fund for Agro-scientific Research in the Public Interest (201503123-10).

References

1. Gao, C., Zhu, J.Y., Zhu, J.G., et al. 2005. Effects of extreme rainfall on the export of nutrients from agricultural land. *Acta Geographica Sinica* 60(6): 991-997.
2. Guo, J. H., Liu X. J., Zhang Y., et al. 2010. Significant acidification in major Chinese cropland. *Science* 327: 1008-1010.
3. Kabata-Pendias, A. & Pendias, H. 2001. *Trace elements in soils and plants.* Boca Raton, Florida: CRC Press Inc.
4. Liu, Y.H., Gao, C., Wang, D.F., et al. 2009. First flush effect of urban storm water runoff pollution on implications for BMPs selection. *Water Resources Protection* 25(6): 29-32.

5. Lu, R.K. 2000. *Analysis Methods of Soil Agricultural Chemistry*. Beijing: China Agricultural Science Press.
6. Qi, Z.P., Zhang, L.M. Sang, A.Y., et al. 2007. Spatial-temporal variance of paddy soil organic matter in Hainan Province. *Chinese Agricultural Science Bulletin* 23(7): 547-551.
7. Qi, Z.P., Wei, Z.Y., Li, F.Y., et al. 2009. Spatial-temporal variability of paddy soil nutrient in Hainan Province. *Chinese Journal of Soil Science* 40(6): 1292-1296.
8. Sun, X., Lin, Z.S., Sun,Y. 2005. Dynamic prediction and suggestion of total farmland in China. *Journal of Natural Resources* 20(2): 200-205.
9. Wang, D.F., Wei, Z.Y., Lv, L.W., et al. 2012. Soil productivity assessment of arable land by FMA-AHP in Chengmai Cunty of Hainan Province, China. *Chinese Journal of Tropical Crops* 33(11): 1966-1970.
10. Wang, D.F., Wei, Z.Y., Han, M.J., et al. 2013. Distribution characteristics of zinc in arable land soil of northwestern Hainan. *Journal of Southern Agriculture* 44(3): 447-451.
11. Wei, J.Z., Wang, D.F., Wei, Z.Y., et al. 2013. Temporal and spatial distribution of soil pH in arable land of Changjiang, Hainan. *Chinese Journal of Tropical Crops* 34(3): 413-417.
12. Yu, J.G., Chang, Z.Z., Huang, H.Y., et al. 2010. Effect of microbial inoculants for straw decomposing on soil microorganisms and the nutrients. *Journal of Agro Environment Science* 29(3): 563-570.

Nitrogen and phosphorus stoichiometric analyses of maize grain based on the multi-year tillage experiment in Northeast China

C.M. Jiang, S.B. Fan and W.T. Yu*

Institute of Applied Ecology, Chinese Academy of Sciences, Shenyang 110016, PR China
**E-mail: jchmjchm@163.com*

Multi-year tillage experiment was conducted in Northeast China (located at Changtu). Based on data set collected from this experiment, we found the mean value of N/P ratio in grain of maize was 5.02 with the CV of 10.6% and the relationship of grain yield vs. N yield was more robust than that of grain yield vs. P yield. The analysis using standardized major axis regression revealed that N yield varied isometrically with grain yield, while the scale exponents of P yield vs. N yield and P yield vs. yield were both significant greater than 1. These results implied the growth of maize grain was in accord with growth rate hypothesis. Conclusively, the growth of maize grain needed disproportionately more P.

Keywords: Ecological stoichiometry; Maize grain; Tillage experiment; Growth rate hypothesis.

1. Introduction

The aim of ecological stoichiometry was to track the ratio and balance of different chemical elements (mainly including carbon (C), nitrogen (N) and phosphorus (P)) within ecological processes, and then to uncover the underlying mechanism (He and Han, 2010). There were two specific theories in this discipline: homeostasis theory and growth rate hypothesis (Zeng and Chen, 2005). The former believed that the organism had the ability to keep the composition of different chemical elements relatively stable when the element composition of external environment was dramatically changed. This is the theoretical foundation of ecological stoichiometry. Growth rate hypothesis thought the ratio of C:N:P in organism was adjusted if the growth rate of organism was altered. It had been proposed that the growth rate of organism was positively correlated with its P content, and negatively correlated with N:P or C:P. Previous study had proved that the growth of microbes, insects, and crustaceans all followed the growth rate hypothesis. For plant ecology, the research about growth rate hypothesis was mainly concentrated on C:N:P in leaf and their relationships with environmental factors (Elser et al., 2000; Wright et al., 2004; Han et al., 2005; Niklas et al., 2005). It had been argued that the growth of different plant organ should be accorded with this rule (Niklas and Cobb, 2005). To the best of our knowledge, analysis of C:N:P in plant seed adopting the perspective of ecological stoichiometry was very limited. Our group has been conducting a set of tillage experiment in Northeast China (located at Changtu) since 2007. Through these experiments, we acquired a mass of N, P content and yield of maize seed. Thus, present study is to analyze the data set by the perspective of ecological stoichiometry, and intend to validate the growth rate hypothesis in maize seed. This research also could provide theoretical guidance for the agricultural production.

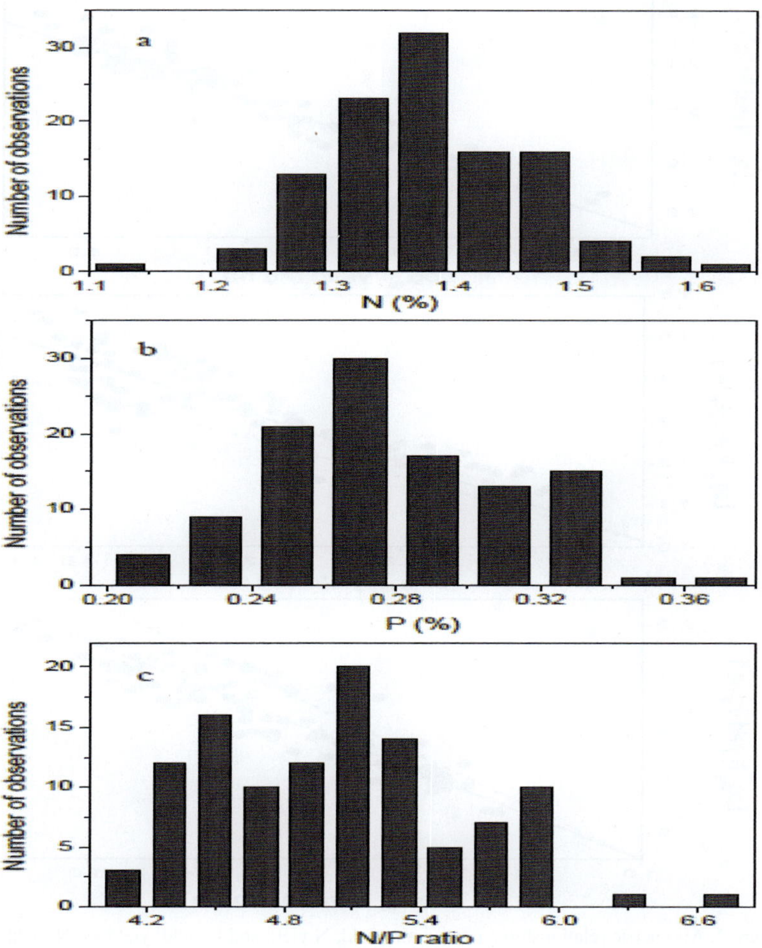

Figure 1. Frequency distributions of grain N concentration (a), P concentration (b) and N/P ratio (c).

2. Materials and Methods

The experiment design, sample and chemical analysis had been detailedly described in our previous publications (Zhao et al., 2014; Xu et al., 2015). Pearson correlation was used for quantifying relationships between two variables, and then standardized major axis regression was employed to identify the allometric relationships between two variables. Regression parameters and confidence intervals were computed using the SMATR (Standardized Major Axis Tests & Routines) library (Falsterb et al., 2006; Warton et al., 2006) in Matlab (MathWorks, Inc.).

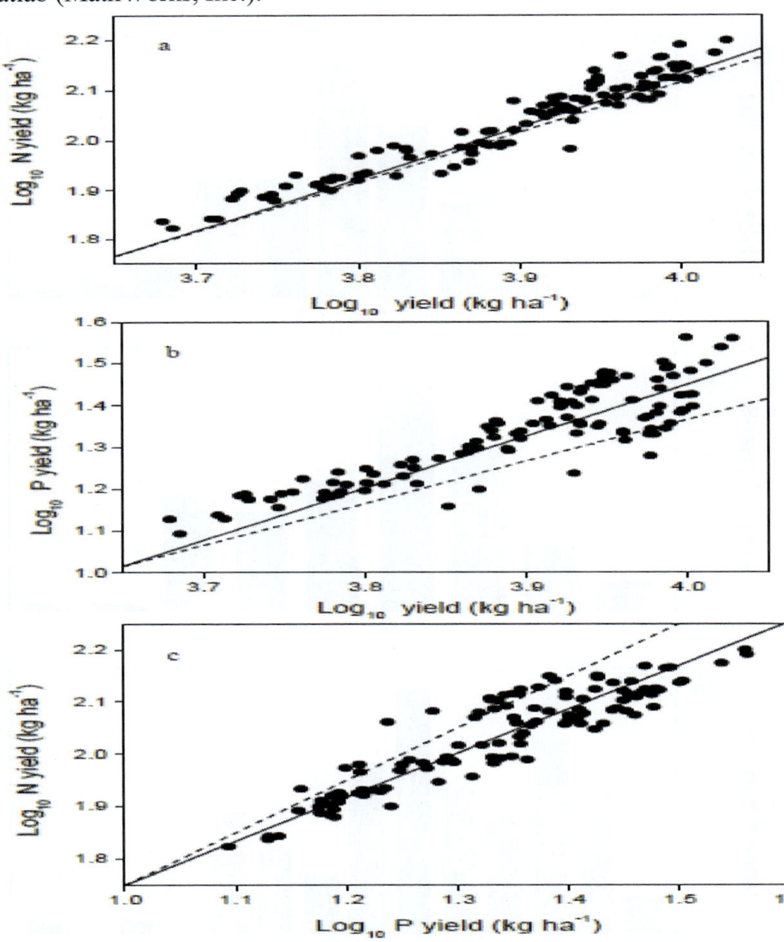

Figure 2. Allometric relationship among grain yield, N yield and P yield: yield vs. N yield (a), yield vs. P yield (b) and N yield vs. P yield (c). Solid lines were standardized major axis regression curves (for a summary of regression statistics, see Table 2). Dash line passed though the same y-axis intercept as the corresponding solid line, and the slope of the dash lines equaled 1.

3. Results

Table 1. Mean value, SD and coefficient of variation (CV; defined as SD/mean × 100) of N and P concentration and N/P ratio.

	N (%)	P (%)	N/P ratio
Mean	1.38	0.28	5.02
SD	0.08	0.03	0.53
CV	5.8	12.0	10.6

Mean values of N, P concentrations and N/P ratios across our experiment were 1.38%, 0.28% and 5.02, respectively (Table 1 and Fig. 1). The CV (CV, coefficient of variation: defined as standard deviation/mean × 100%) of P concentration (12.0%) was obviously larger than that of N (5.8%) (Table 1). There were pairwise correlations among grain yield, N yield and P yield (Table 2). However, the allometric scaling relationships among these three indices were different. There was disproportionately more P yield as grain yield increased, while N yield varied isometrically with grain yield (Table 2 and Fig. 2).

4. Discussion

Table 2. Pearson's correlation coefficients (r) and standardized major axis regression parameters relating yield, N yield and P yield.

Log_{10} Y vs. Log_{10} X	r	Intercept	Slope (mean and 95% CI)
N yield vs. P yield	0.91^{****}	0.91	0.84 (0.77, 0.91)
P yield vs. yield	0.88^{****}	-3.51	1.24 (1.13, 1.36)
N yield vs. yield	0.96^{****}	-2.03	1.04 (0.99, 1.10)

**** indicate $P<0.0001$
95% CI, 95% Confidence interval

Our analysis showed correlation coefficient between grain yield and N yield was greater than that between grain yield and P yield (Table 2), probably implying that the management of soil N appeared to be more important for achieving high grain yield in the study site. Based on the multi-year experiment, we found the CV of N concentration was much smaller than that of P, and the CV of N/P ratio was mainly caused by P (Table 1), which was consistent with the corresponding results when surveying leaf traits across large geographical scale (Wright et al., 2004; Han et al., 2005). During senescing processes of plant, the resorption of leaf P possessed relatively high CV than that of N. This might the reason of bigger CV of P concentration in seed (Aerts and Chapin, 2000). According to growth rate hypothesis, elevating P concentration is essential for increased biomass of organism because it could favor the synthesis of protein, and subsequently accelerate assimilation (Elser et al., 2003). Relatively higher CV of P concentration verified growth rate hypothesis to some extent, as suggested by Yang et al. (2010).

N and P exert different roles during the growth of organism. N is mostly distributed in protein which just represents ‘overhead’ during the growing process; P is a structural element in nucleic acids, among which rRNA is the protein-output ‘machinery’ for plant growth. Growth rate hypothesis thought that for greater biomass, more rRNA (P) was needed to synthesize protein (N) so as to support the relatively higher growth rate (Elser et al., 2000). In our data set, we detected that P increased more quickly than N, which agreed very well with the result of stoichiometric result in plant leaf (Wright et al., 2004). Our analysis, together with previous study, demonstrated that the growth of maize seed also followed the growth rate hypothesis. That to say, high grain yield was associated with investment of more rRNA (P) to produce protein (N) (Agren, 2004). Therefore, the seed collected from high yield plot usually displayed low N/P ratio.

5. Conclusions

In summary, the average of N/P ratio in maize seed was 5.02 with a CV of 10.6% according to multi-year tillage experiment conducted in Northeast China (located at Changtu). The correlation coefficient between grain yield and N yield was greater than that between grain yield and P yield. We also found disproportionately more P yield was increased as grain yield elevated, while N yield varied isometrically with grain yield. Our analysis indicated that growth of maize grain met the growth rate hypothesis. This outcome expands the scope for using ecological stoichiometry in different plant organs.

Acknowledgments

This study was financially supported by National Natural Science Foundation of China (No. 31070547, 41171242 and 41105114) and the National Key Technology R&D Program (No. 2012BAD05B01).

We are grateful to everyone who assisted with field and laboratory work. We are also grateful to anonymous reviewers for their valuable comments on earlier versions of the manuscript.

References

1. Aerts, R., Chapin, F.S., 2000. The mineral nutrition of wild plants revisited: A re-evaluation of processes and patterns. In: Fitter, A.H., Raffaelli, D.G. (Eds.), Advances in Ecological Research, Vol 30, pp. 1-67.
2. Agren, G.I., 2004. The C : N : P stoichiometry of autotrophs - theory and observations. Ecology Letters 7, 185-191.
3. Elser, J.J., Acharya, K., Kyle, M., Cotner, J., Makino, W., Markow, T., Watts, T., Hobbie, S., Fagan, W., Schade, J., Hood, J., Sterner, R.W., 2003. Growth rate-stoichiometry couplings in diverse biota. Ecology Letters 6, 936-943.
4. Elser, J.J., Fagan, W.F., Denno, R.F., Dobberfuhl, D.R., Folarin, A., Huberty, A., Interlandi, S., Kilham, S.S., McCauley, E., Schulz, K.L., Siemann, E.H., Sterner, R.W., 2000. Nutritional constraints in terrestrial and freshwater food webs. Nature 408, 578-580.
5. Falsterb, D.S., Warton, D.I., Wright, I.J., 2006. SMATR: Standardised Major Axis Tests & Routines. http://www.bio.mq.edu.au/ecology/SMATR/.
6. Han, W.X., Fang, J.Y., Guo, D.L., Zhang, Y., 2005. Leaf nitrogen and phosphorus stoichiometry across 753 terrestrial plant species in China. New Phytologist 168, 377-385.
7. He, J.S., Han, X.G., 2010. Ecological stoichiometry: Searching for unifying principles from individuals to ecosystems. Chinese Journal of Plant Ecology 34, 2-6. (in Chinese)
8. MathWorks Inc., http://www.mathworks.com/.
9. Niklas, K.J., Cobb, E.D., 2005. N, P, and C stoichiometry of Eranthis hyemalis (*Ranunculaceae*) and the allometry of plant growth. American Journal of Botany 92, 1256-1263.
10. Niklas, K.J., Owens, T., Reich, P.B., Cobb, E.D., 2005. Nitrogen/phosphorus leaf stoichiometry and the scaling of plant growth. Ecology Letters 8, 636-642.
11. Warton, D.I., Wright, I.J., Falster, D.S., Westoby, M., 2006. Bivariate line-fitting methods for allometry. Biological Reviews 81, 259-291.
12. Wright, I.J., Reich, P.B., Westoby, M., Ackerly, D.D., Baruch, Z., Bongers, F., Cavender-Bares, J., Chapin, T., Cornelissen, J.H.C., Diemer, M., Flexas, J., Garnier, E., Groom, P.K., Gulias, J., Hikosaka, K., Lamont, B.B., Lee, T., Lee, W., Lusk, C., Midgley, J.J., Navas, M.L., Niinemets, U., Oleksyn, J., Osada, N., Poorter, H., Poot, P., Prior, L., Pyankov, V.I., Roumet, C., Thomas, S.C., Tjoelker, M.G., Veneklaas, E.J., Villar, R., 2004. The worldwide leaf economics spectrum. Nature 428, 821-827.

13. Xu, Y.G., Ma, Q., Zhou, H., Jiang, C.M., Yu, W.T., 2015. Effects of straw returning and deep loosening on soil physical and chemical properties and maize yields. Chinese Journal of Soil Science 46, 428-432. (in Chinese)
14. Yang, K., Hung, J.H., Dong, D., Ma, W.H., He, J.S., 2010. Canopy leaf N and P stoichiometry in grassland communities of Qinghai-Tibetan Plateau, China. Chinese Journal of Plant Ecology 34, 17-22. (in Chinese)
15. Zeng, D.H., Chen, F.S., 2005. Ecological stoichiometry: a science to explore the complexity of living systems. Chinese Journal of Plant Ecology 29, 1007-1019. (in Chinese)
16. Zhao, Y., Zhou, H., Ma, Q., Xu, Y.G., Jiang, C.M., Yu, W.T., 2014. Soil microbial biomass C and N as affected by fertilization and cultivation systems. Chinese Journal of Soil Science 45, 1099-1103. (in Chinese)

Bioturbation effects of loach (*Misgurnus anguillicaudatus*) on the vertical distribution of sediment particles and soil animal community in paddy field

G. Sun and Y. Fang[†]

School of Life Science, Changchun Normal University, Changchun, Jilin, China
[†]*E-mail: fangyan124@aliyun.com*
www.cncnc.edu.cn

Using chemically stable glass beads as tracers, the bioturbation effects of benthic fish loach (*Misgurnus anguillicaudatus*) on the vertical distribution of sediment particles in paddy field were investigated. Under the bioturbation of *Misgurnus anguillicaudatus*, the distribution of tracers at surface and 6 cm depth of the sediment was measured. After ten days, 43.2% of the glass beads at the sediment surface is transferred downwardly, the maximum transportation depth is 7.5 cm, and the vertical transportation rate is $7.676\times10^{-4}g^{-1}\cdot cm^{-2}\cdot d^{-1}$. 39.9% and 9.9% of the glass beads at 6 cm deep are transferred upwards and downwards respectively, the maximum transportation distance is 4.5 cm and 5.1 cm respectively, and the vertical transportation rates are $7.597\times10^{-4}g^{-1}\cdot cm^{-2}\cdot d^{-1}$ (upward) and $1.894\times10^{-4}g^{-1}\cdot cm^{-2}\cdot d^{-1}$ (downward), respectively. Using profile sampling, macro-group classification and statistical method, the soil animal community in a rice-fish (*Misgurnus anguillicaudatus*) integrated ecosystem was investigated by plot experiment. The dominant groups in soil animal community are Gastropoda, Nematoda, and Enchytraeidae, the common groups are Oribatida, Collemobola, Oligochaeta, Diptera, Coleoptera, Hymenoptera, Crustacea, Lepidoptera. Compared with the control plot (without loach), there are more groups but less individuals of soil animal in the rice-loach plot. The diversity index and evenness index of soil animals in the rice-loach plot are higher than those in the control plot, while the dominance index is lower. Vertically, soil animals aggregate mainly in soil surface. The benthic fish loach increases the biodiversity and stability of paddy ecosystem. The bioturbation effects of benthic fishplay an important role in vertical distribution of sediment and soil animal community in paddy field. This work can improve our understanding of bioturbation mechanism of benthic animal, and offer evidence for rice-fish compound system and integrated agriculture.

Keywords: Bioturbation; Sediment; Soil animal; Paddy field; Benthos.

[†] This work was supported by the National Natural Science Foundation of China (31370475, 31070421), the Key Project of Science and Technology Department of Jilin Province (20140204042SF), the Project of Environmental Protection Department of Jilin Province (2010-19), the Natural Science Foundation of Jilin Province, China (201115163) and the Science and Technology Project of Educational Department of Jilin Province, China (2012-218, 2010-150).

1. Introduction

Soil animal plays an important role in matter cycle, energy flow, mineralization, regeneration in an ecosystem. Due to its sensibility to the environment, soil animal is a indispensable biological index for soil quality assessment and environmental monitoring [1]. Rice (*Oryza sativa*) is the main source of food for over half of the world population. However, the shortages of soil and water, the dependence on chemical pesticide and fertilizer are threatening rice production seriously. In the last few years, Food and Agriculture Organization (FAO) encouraged complex planting and breeding, and emphasized the production potential of aquatic animals in paddy field. In an integrated agro-system, natural resources including light, water, soil, atmosphere, heat, space are utilized more efficiently. This is a prospective approach to environmentally friendly and sustainable production of food and protein.

Bioturbation is the effect of benthic animals on chemical, physical and biological properties of aquatic ecosystem by mechanical movement and physiological activity [2]. As one of the important ecological processes in a waterbody (river, lake, wetland, etc.), bioturbation is attracting more and more interest in the fields of ecology and environmental science because of the remarkable theoretical significance and practical value [3]. Loach (*Misgurnus anguillicaudatus*) belongs to the subfamily of Cobitinae, family of Cobitidae, order of Cypriniformes. It is a common fish in freshwater, and has tenacious vitality and strong adaptability to the environment. In this work the bioturbation effects of *Misgurnus anguillicaudatus* on the vertical distribution of sediment particles and soil animal community in paddy field were investigated. The results can not only promote our understanding of bioturbation mechanism of benthic animal, butprovide scientific evidence for evaluation and manipulation of rice-fish compound system and integrated agriculture.

2. Materials and Methods

2.1. *Materials*

The paddy sediment and surface water were sampled from the Shuangyang District, Changchun City, Jilin Province (125°30′34″ E, 43°25′24″ N). The loach (*Misgurnus anguillicaudatus*) with body length of 5~7 cm was purchased from a wholesale market of aquatic product. The healthy individuals were selected for experiment. The pH value of the surface water was 7.42, the oxidation-reduction potential (Eh) was 475.74mV, the content of dissolved oxygen (DO) was 7.25 mg/L. The chemically stable glass beads with diameter of approximately 0.5 mm serve as tracers.

2.2. *Methods*

Twelve PVC culture tubes (diameter 16 cm, length 40 cm) were divided into four groups: A_1 (bioturbation group, tracers at the sediment surface), A_2 (control group without *Misgurnus anguillicaudatus*, tracers at the sediment surface), B_1 (bioturbation group, tracers at the 6 cm depth of the sediment), B_2 (control group without *Misgurnus anguillicaudatus*, tracers at the 6 cm depth of the sediment). The tracers were placed atsurface and 6 cm depth of the sediment.The paddy sediment was added into the culture tubes to 15 cm depth. Then, the in situ surface water was slowly poured into the culture tubes. All the culture tubes were placed 24h quietly under ventilation condition. After balancing, the redundant water was removed, the surface water was kept 12 cm high. The experimental system was maintained aerated at 20~24 °C. Ten days later, the sediment in the PVC tubes was divided into layers with an interval of 1 cm. The tracers in each sediment layer were counted under a stereoscopic microscope. The vertical transportation rates of sediment particles by bioturbation were calculated as: Transportation rate=transported particles (%)/ (body weight of benthos×bioturbation area×bioturbation time).

For the investigation of soil animals, two treatments were divided, including bioturbation (BIO, with loach) and control (CON, without loach). The experimental plots, with relatively homogeneous quality and fertility, were selected randomly. The area of a single plot was 80 m^2. In each BIO plot, 200 pieces of loach were released. The soil animal was collected five times in 2014 (23 June, 25 July, 24 August, 18 September, 10 October) by profile sampling method (0~5 cm, 5~10 cm, 10~15 cm). The sampling area of large soil animal, middle/small soil animal, and humidogene soil animal were 50cm×50cm, 10cm×10 cm, and 5cm×5cm, respectively. Under a stereoscopic microscope, the soil animals were classified into macro-group according to the method of Aoki [4]. Each treatment was replicated thrice. The significance of difference wastested by one-way ANOVA using the software SPSS. The quantitative characteristics of community structure of soil fauna are calculated by Shannon-Wiener diversity index (H'), Pielou evenness index (E) and Simpson dominance index (C):

$$H'=-\Sigma P_i \ln P_i \quad (1)$$

$$E=H'/\ln S \quad (2)$$

$$C=\Sigma (P_i)^2 \quad (3)$$

where, P_i is the proportion of individual number of the ith species (group) in the total number, S is the number of species (group).

3. Results and Discussion

3.1. *Distribution of tracer particles at the sediment surface*

After ten days of *Misgurnus anguillicaudatus* bioturbation, the tracer particles at the sediment surface distribute in the layers of different depths (Table 1). 43.2% of the glass beads at the surface (A_1 group) is transferred downwardly with the maximum depth of 7.5 cm. By contrast, most tracers in A_2 group (without bioturbation) are still in the original layer. For the distribution of sediment particles, A_1 group and A_2 group are of extremely significant difference (P<0.01). For A_1 group, the total body weight of *Misgurnus anguillicaudatus* is 28 g, the bioturbation area is 201 cm^2, the bioturbation time is 10 d. Thus, the vertical transportation rate of sediment particles is $7.676\times10^{-4}g^{-1}\cdot cm^{-2}\cdot d^{-1}$.

Table 1. Distribution of tracers at sediment surface under bioturbation of *Misgurnus anguillicaudatus* for ten days.

Depth (cm)	A_1 group (bioturbation)	A_2 group (without bioturbation)
0~1	56.8%	99.8%
1~2	22.8%	0.2%
2~3	9.6%	0
3~4	6.2%	0
4~5	4.4%	0
5~6	0.2%	0
6~7	0	0
7~8	0	0
8~9	0	0
9~10	0	0
Σ	100%	100%

3.2. *Distribution of tracer particles at 6 cm depth of the sediment*

After ten days of *Misgurnus anguillicaudatus* bioturbation, 39.7% and 9.9% of the glass beads at a depth of 6 cm (B_1 group) are transferred upwards and downwards respectively, the maximum transportation distance is 4.5 cm and 5.1 cm respectively (Table 2). By contrast, the tracers in B_2 group (without bioturbation) are almost still in the original layer. For the distribution of sediment particles, B_1 group and B_2 group are of extremely significant difference (P<0.01). For B_1 group, the total body weight of *Misgurnus anguillicaudatus* is 26 g, the bioturbation area is 201 cm^2, the bioturbation time is 10 d. Thus, the vertical transportation rates of sediment particles are $7.597\times10^{-4}g^{-1}\cdot cm^{-2}\cdot d^{-1}$ (upward) and $1.894\times10^{-4}g^{-1}\cdot cm^{-2}\cdot d^{-1}$ (downward), respectively.

Table 2. Distribution of tracers at 6 cm depth of the sediment under bioturbation of *Misgurnus anguillicaudatus* for ten days.

Depth (cm)	B_1 group (bioturbation)	B_2 group (without bioturbation)
0~1	0	0
1~2	0.1%	0
2~3	0.2%	0
3~4	1.4%	0
4~5	6.5%	0
5~6	31.6%	0.1%
6~7	50.3%	97.7%
7~8	6.5%	2.2%
8~9	1.9%	0
9~10	1.4%	0
10~11	0.1%	0
11~12	0	0
12~13	0	0
13~14	0	0
$\sum$	100%	100%

3.3. *Composition and number of soil fauna*

In the experimental paddy field, there are abundant sorts of soil animals. From June through October, 856 individuals of soil animal belonging to 20 groups are collected (Table 3). The individual number of soil animal in BIO (409) is less than that in CON (447). Gastropoda, Nematoda and Enchytraeidae are dominant groups (frequency>10%), accounting for 68.45% of the total. Oribatida, Collemobola, Oligochaeta, Diptera, Coleoptera, Hymenoptera, Crustacea and Lepidoptera are common groups (frequency 1%~10%), accounting for 28.98% of the total. The other 9 groups are rare groups (frequency<1%), accounting for 2.57% of the total.

3.4. *Quantitative characteristics of soil animal community*

The diversity index, evenness index, and dominance index describe the quantitative characteristics, stability and function of biological community. In the three soil profiles (0~5 cm, 5~10 cm, 10~15 cm), the diversity index and evenness index of soil animal in BIO are higher than those in CON, whereas the dominance index in BIO is lower (Table 4). In the profiles of 0~5 cm, 5~10 cm and 10~15 cm, there is insignificant difference ($p>0.05$), significant difference ($p<0.05$) and extremely significant difference ($p<0.01$), respectively, in the quantitative characteristics of soil animal between BIO and CON.

Table 3. Composition and distribution of soil fauna in the experimental site.

Group	CON (without loach) Individual number	CON (without loach) Proportion (%)	BIO (with loach) Individual number	BIO (with loach) Proportion (%)	Total individual number	Frequency (%)	Abundance
Gastropoda	137	30.65	183	44.74	320	37.38	+++
Nematoda	97	21.70	62	15.16	159	18.57	+++
Enchytraeidae	62	13.87	45	11.00	107	12.50	+++
Oribatida	35	7.83	40	9.78	75	8.76	++
Collemobola	26	5.82	22	5.38	48	5.61	++
Oligochaeta	26	5.82	10	2.44	36	4.21	++
Diptera	14	3.13	8	1.96	22	2.57	++
Coleoptera	13	2.91	7	1.72	20	2.34	++
Hymenoptera	11	2.46	7	1.72	18	2.10	++
Crustacea	11	2.46	7	1.72	18	2.10	++
Lepidoptera	5	1.12	6	1.47	11	1.29	++
Polychaeta	4	0.90	3	0.73	7	0.82	+
Diplura	2	0.45	3	0.73	5	0.58	+
Diplopoda	1	0.22	1	0.24	2	0.23	+
Chilopoda	0	0	2	0.49	2	0.23	+
Acarina	1	0.22	1	0.24	2	0.23	+
Araneae	0	0	1	0.24	1	0.12	+
Turbellaria	1	0.22	0	0	1	0.12	+
Scolopendromorpha	0	0	1	0.24	1	0.12	+
Rotifera	1	0.22	0	0	1	0.12	+
Σ	447	100	409	100	856	100	

Note: +++: Dominant group (frequency>10%); ++: Common group (1%~10%); +: Rare group (<1%).

Table 4. Quantitative characteristics of soil animal community in the experimental site.

Treatment	0~5 cm			5~10 cm			10~15 cm		
	H'	*E*	*C*	*H'*	*E*	*C*	*H'*	*E*	*C*
CON	1.95	0.87	0.18	1.14	0.68	0.21	0.96	0.44	0.25
BIO	2.04	0.96	0.16	1.52	0.82	0.19	1.31	0.69	0.23

Note: CON: without loach; BIO: with loach; *H'*: Diversity index; *E*: Evenness index; *C*: Dominance index.

Benthic fish improves the aeration, permeability and aggregation degree of paddy soil, optimizes the living conditions of soil animals [5]. Due to the feeding pressure of loach, the competitive intensity between the preys decreases, more preys including Chilopoda, Araneae and Scolopendromorpha can co-exist in the paddy field. Thus, the diversity and evenness of soil animal community in BIO increase, while the dominance decreases.

3.5. *Vertical distribution of soil fauna*

The soil animal exhibits remarkable gathering of group number and individual number in soil surface. In CON, 92.7% of group number and 75.8% of individual number gather in 0~5 cm profile. In BIO, 86.4% of group number and

68.8% of individual number gather in 0~5 cm profile (Fig. 1). The surface gathering of soil animal in BIO is lower than that in CON, which is attributed to the feeding of loach in the 0~5 cm profile.

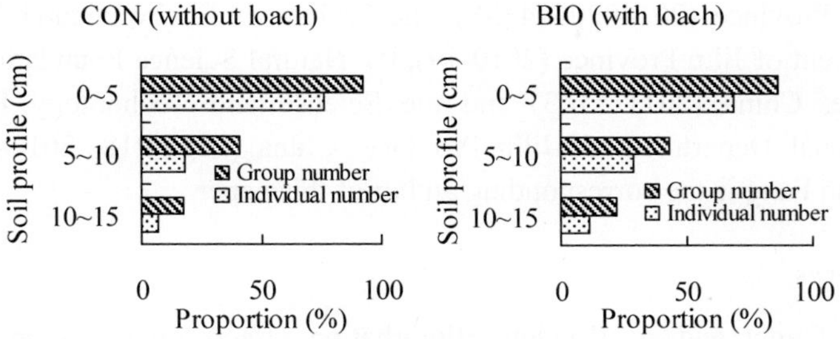

Figure 1. Vertical distribution of soil fauna in the experimental site.

4. Summary

Benthic fish can affect the vertical distribution of sediment particles in paddy field. After ten days of *Misgurnus anguillicaudatus* bioturbation, 43.2% of the glass beads on the surface are transferred downwardly with the maximum depth of 7.5 cm, and the vertical transportation rate of sediment particles is $7.676\times10^{-4}g^{-1}\cdot cm^{-2}\cdot d^{-1}$. At a depth of 6 cm, 39.7% and 9.9% of the glass beads are respectively transferred upwards and downwards, to the maximum distance of 4.5 cm and 5.1 cm, respectively; and the vertical transportation rates of sediment particles are $7.597\times10^{-4}g^{-1}\cdot cm^{-2}\cdot d^{-1}$ (upward) and $1.894\times10^{-4}g^{-1}\cdot cm^{-2}\cdot d^{-1}$ (downward), respectively. In the experimental paddy field, the dominant groups of soil fauna are Gastropoda, Nematoda and Enchytraeidae, the common groups are Oribatida, Collemobola, Oligochaeta, Diptera, Coleoptera, Hymenoptera, Crustacea, Lepidoptera. Compared with the control plot (without loach), there are more groups but less individuals of soil animal in the rice-loach plot. The diversity index and evenness index of soil animals in the rice-loach plot are higher than those in the control plot, while the dominance index is lower. Vertically, soil animals aggregate mainly in soil surface.

Bioturbation is an important ecological process in paddy field [6,7]. Benthos can affect not only the physical properties of sediment (e.g. topography, moisture, permeability, bulk density, porosity, particle size, intensity, resuspension) [8], but also the chemical properties (e.g. organic matter, phosphorus, nitrogen, pollutant, pH value, oxidation-reduction potential) and the biological properties (e.g. soil animal, soil respiration, enzyme, microorganism, fauna) [9,10]. The benthic fish promotes the stability and biodiversity of paddy ecosystem.

Acknowledgments

This work was supported by the National Natural Science Foundation of China (31370475, 31070421), the Key Project of Science and Technology Department of Jilin Province (20140204042SF), the Project of Environmental Protection Department of Jilin Province (2010-19), the Natural Science Foundation of Jilin Province, China (201115163) and the Science and Technology Project of Educational Department of Jilin Province, China (2012-218, 2010-150). Dr. Prof. Yan Fang is the corresponding author of this paper.

References

1. A. Ciutat and A. Boudou, Bioturbation effects on cadmium and zinc transfers from a contaminated sediment and on metal bioavailability to benthic bivalves, *Environmental Toxicology and Chemistry* **22**, 1574 (2003).
2. M. Gerino, M. Frignani, C. Mugnai, L. Bellucci, D. Prevedelli, A. Valentini, A. Castelli, S. Delmotte and S. Sauvage, Bioturbation in the Venice Lagoon: Rates and relationship to organisms, *Acta Oecologica* **32**, 14 (2007).
3. F.J.R. Meysman, J.J. Middelburg and C.H.R. Heip, Bioturbation: a fresh look at Darwin's last idea, *Trends in Ecology & Evolution* **21**, 688 (2006).
4. J. Aoki, *Soil Zoology* (Beilonwan, Tokyo, 1973).
5. G.H. Baker, Recognising and responding to the influences of agricultureand other land-use practices on soil fauna in Australia, *Applied Soil Ecology* **10**, 303 (1998).
6. C. Bradshaw, L. Kumblad and A. Fagrell, The use of tracers to evaluate the importance of bioturbation in remobilising contaminants in Baltic sediments, *Estuarine, Coastal and Shelf Science* **66**, 123 (2006).
7. Y. Nakamura and F. Kerciku, Effects of filter-feeding bivalves on the distribution of water quality and nutrient cycling in a eutrophic coastal lagoon, *Journal of Marine Systems* **26**, 209 (2000).
8. D.H. Shull and M. Yasuda, Size-selective downward particle transport by cirratulid polychaetes, *Journal of Marine Research* **59**, 453 (2001).
9. G. Nogaro, F. Mermillod-Blondin and M.H. Valett, Ecosystem engineering at the sediment-water interface: bioturbation and consumer-substrate interaction, *Oecologia* **161**, 125 (2009).
10. K. Timmermann, G.T. Banta and A.R. Johnsen, Effects of the polychaetes *Arenicola marina* and *Nereis diversicolor* on microbial pyrene mineralization, *Aquatic Microbial Ecology* **50**, 197 (2008).

Influence of Al-F interaction on F absorbed by rice plant*

J.H. Liu*, Y.J. Wang and L.P. Zhao
College of Resources and Environment, Jilin Agricultural University, Changchun, China.
**Email: liujinhua80@126.com*

The aim of this paper was to research the effluence of Al-F interaction on F in rice plant by biological pot experiment. The results showed that: the concentration of F in each rice plant part at different applied $Al_2 (SO_4)_3$ changed differently, and the relation of them was well fitted by quadratic equation. When the amount of $Al_2 (SO_4)_3$ applied was 0.4%-0.6%, which was equivalent to F: Al mass ratio of 6:1- 4:3, the concentration in plant was lower relatively. It indicated that its defluoridation efficiency was better. So this ratio of F: Al in soil can lower the F intake by rice plant.

Keywords: Al-F interaction, Defluoridation, F species, Rice plant parts, F concentration.

1. Introduction

F is a microelement related to human health closely. If F in environment is excessive or inadequate, it can make organic cause disease (Chai, 1995; Ammerman *et al.*, 1983; Roberta, 2001). Soil is an important medium for F circulating and redundant F in it can lead to F poisoning to human and animal through food chain. Its chemical behavior in soil was affected by many factors with a most important one of Al^{3+} because it can form compound of F and Al with different ligancy (Xie *et al.*, 1999; Anderson *et al.*, 1991; Becker *et al.*, 1950; Maclean *et al.*, 1992). Yang W J considered that the complexation of F and Al reduced the adsorption of Al and F to soil, so their biological availability increased (Yang *et a*l., 2002). Wang H H hold the opinion that acid soil could increase the degree of Al and F complex and enhance their adsorbed amount to soil, so it can lower the biological availability of F in soil (Wang *et al.*, 2006). The result of Maclean D C showed that F and Al applied together reduced the F content in aboveground part of winter wheat (Maclean *et al.*, 1992).

There are many researches about the effect of F-Al interaction on tea trees and acid soils of high level Al (Ding *et al.*, 1991; Xie *et al.*, 2008; Malea, 1995),

---

* This work is supported by the Youth research fund of science and technology of Jilin province (20150520122JH),

† Work partially supported by The science and technology development plant project of Jilin province (20130204054SF)

and its effect on the content and species changes of Al (Laura *et al.*, 2003). But the report on neutral or alkaline soil in north China is less relatively. In addition, the soda-typed saline-alkaline soil in western Jilin province where was typical water-related endemic fluorosis area has higher content of F because of its geographical and climate conditions. As a method of saline-alkali soils improvement, the ameliorant of $Al_2\ (SO_4)_3$ has better effect on improving the physical and chemical characters of the soil (Zhao *et al.*, 2001). But how the applied Al^{3+} affected the content and species of F needs further studying.

Therefore, this paper used black soil owned lower F as experimental soil, added different content of F and Al artificially to soil to simulate various F-Al interaction environments to cultivate rice plant. The objectives of this study were to characterize the influence of F-Al interaction on the concentration of F in different part of rice plant. So the paper provides a scientific basis for decreasing the biological availability of F in rice plant.

2. Materials and Methods

2.1. *Experimental soil*

This experiment was conducted at the institute of province-ministry co-constructed national emphasis laboratory of ecological restoration and management in Jilin province, the college of resources and environment, Jilin Agricultural University in Changchun. The soil derived from the suburb of Changchun city which belonged to black soil in April, 2010. It was collected from the surface layer (0 to 20 cm), air-dried, and sieved by 3mm sifter with removal of visible plant residues for biological pot experiment using. The major soil properties were as follows: pH, 7.66; organic matter (OM), 17.8 mg kg^{-1}; cation exchange capacity (CEC), 35 c mol/kg; clay grain, 25.5%, water soluble-F, 7.5 mg kg^{-1}; total F, 266.7 mg kg^{-1}.

2.2. *Pot experiment*

15kg of soil mixed with 3.9 g diammonium phosphate (American Kakichi), 6.3 g urea (Xuanhua, Hebei province), different content of $Al_2\ (SO_4)_3$ (net content was 34.0%) and KF (analytical reagent) were added to each pot in a week before transplant. The pots were watered with groundwater of no F and Al and kept 2cm water depth on the soil surface before transplant. In June 10, 2010, nine strong and unanimous rice seedlings (Ji round shaped rice 88) growing for 25 days were placed onto each pot and the roots kept 1cm below the soil surface. The experiment set three initial F content of 400, 800, 1200 mg kg^{-1} and six applied $Al_2(SO_4)_3$ levels of 0%, 0.2%, 0.4%, 0.6%, 0.8%, 1.0% of soil weight respectively. Thus, there were eighteen treatments and each treatment repeated four. The pots were placed in a greenhouse of natural light with completely random design. During the growing, the same content of water was added to each pot in each time. The plant was harvested in Oct 1 and divided into root, stem and leaf, grain. Each part was washed first by water and then by distilled water for three times, oven-dried for 15 min under 105°C to deactivate enzymes, and kept under 50°C for drying to constant weight. The grain was again divided into rice and husk by grain cleaner. All the parts were ground by grinder, then sieved through 0.63mm and put into plastic bag for chemical analysis using. In addition, the soil in each pot after harvest was mixed and collected partial soil by quartering for chemical analysis. F in plant was determined using combined extraction of acid and alkali-F selective electrode (Wu *et al.*, 1979). Every species of F was extracted by continuous fractionation and detected by F selective electrode (Wu *et al.*, 2002).

2.3. *Statistical analysis*

Analyses were conducted using the SPSS10.0 statistical package.

3. Results

3.1. *The influence of Al-F interaction on F in rice plant*

Figure 1 showed the concentrations of F in each plant part when the soil added different content of F and Al. The results indicated that husk had the highest concentration, and the order was husk>rice>root>stem and leaf which was not the same as some other researching results (Zhao *et al.*, 2005). There were two reasons probably. One was the different F source, compared with F came from air, in this study, rice growth was affected by F in soil, so the F concentration in root was higher than stem and leaf which was contrary to some researches. The other reason was that the existing Al^{3+} affected the species distribution of F in soil. Thus its transfer and distribution in rice plant was changed because compounds of F and Al with different ligancy were formed.

a

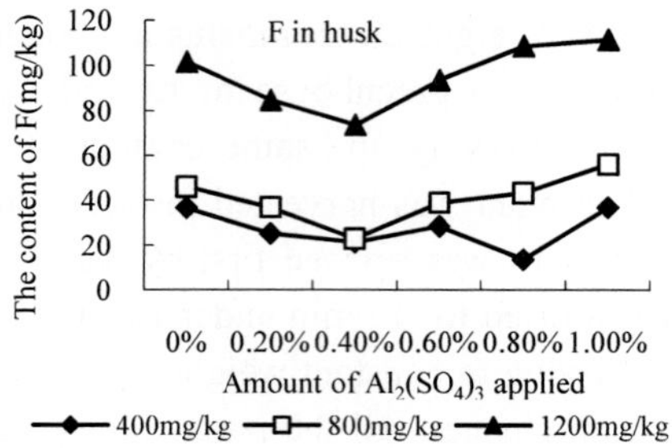
F in husk
The content of F(mg/kg)
120
100
80
60
40
20
0
0% 0.20% 0.40% 0.60% 0.80% 1.00%
Amount of Al2(SO4)3 applied
400mg/kg
800mg/kg
1200mg/kg

b

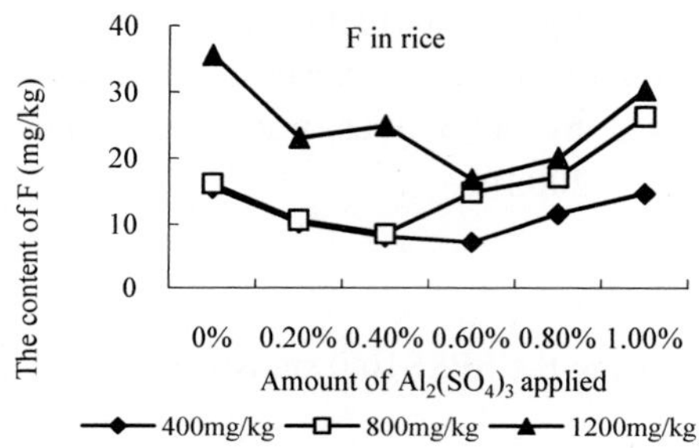
F in rice
The content of F (mg/kg)
40
30
20
10
0
0% 0.20% 0.40% 0.60% 0.80% 1.00%
Amount of Al2(SO4)3 applied
400mg/kg
800mg/kg
1200mg/kg

c

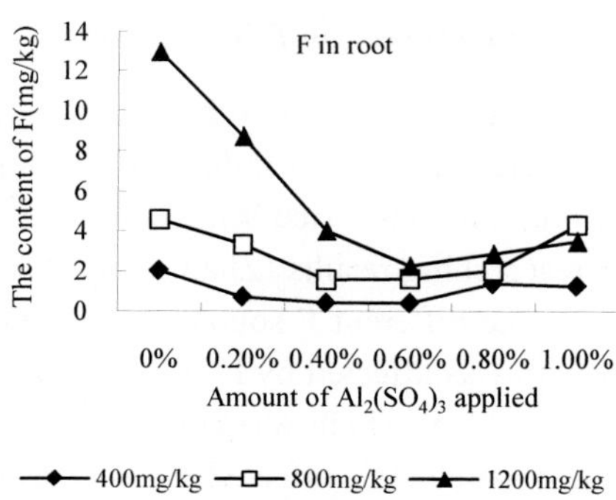
F in root
The content of F(mg/kg)
14
12
10
8
6
4
2
0
0% 0.20% 0.40% 0.60% 0.80% 1.00%
Amount of Al2(SO4)3 applied
400mg/kg
800mg/kg
1200mg/kg

d

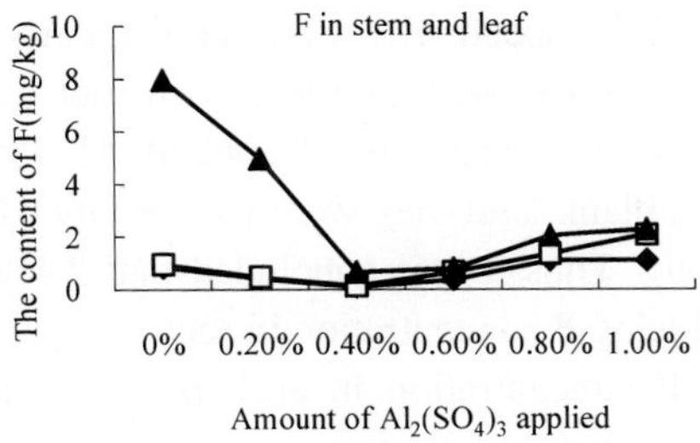

Figure 1. The content of F in each part under different $Al_2(SO_4)_3$ and F applied.

For identical initial F, the concentration of F in rice was decreasing first and then increasing with $Al_2(SO_4)_3$ increased. When the initial F was 400 and 1200 mg kg^{-1}, the minimums of F concentration in rice were 7.26, 16.86 mg kg^{-1} respectively which presented in the treatment of 0.6% $Al_2(SO_4)_3$. Compared with the treatment of blank (0%), they were decreased by 53.13% and 52.60%. But when the initial F content was 800 mg kg^{-1}, its minimum was 8.56 mg kg^{-1} with 46.96% decreasing, presented in 0.4% $Al_2(SO_4)_3$. The average reducing rate of F concentration in each $Al_2(SO_4)_3$ treatments were 34.21%, 41.56%, 37.89%, 20.65%, -14.48% respectively. Although F in rice increased with initial content of F enhancing, it could be decreased by adding $Al_2(SO_4)_3$ to soil. What more important was the quality of $Al_2(SO_4)_3$, otherwise F in rice would be higher than blank. In this paper the best treatment was 0.4-0.6% $Al_2(SO_4)_3$.

F in husk was higher than rice obviously with the similar variation trend. In the initial F content of 400 mg kg^{-1}, its minimum was 13.60 mg kg^{-1} appeared in 0.8% $Al_2(SO_4)_3$ with the decreasing rate of 63.00% compared with blank. For other two initial F, 0.4% $Al_2(SO_4)_3$ made F in husk reduce to minimums of 22.90, 73.64 mg kg^{-1}. For three initial F content, the average decreasing rate in different $Al_2(SO_4)_3$ treatments were 22.62, 39.90, 16.23, 20.66, -10.56% respectively. Thus 0.4% applied $Al_2(SO_4)_3$ could reduce more F concentration in husk.

The minimums of F concentration in stem and leaf at different initial F content were 0.10, 0.14, 0.67 mg kg^{-1} and also appeared in 0.4% $Al_2(SO_4)_3$. The average decreasing rate of F in each $Al_2(SO_4)_3$ treatment were 46.79%, 88.41%, 58.48%, 0.66%, -24.91% respectively. Compared with rice and husk, the inhibitory effect of $Al_2(SO_4)_3$ on decreasing F in stem and leaf was poor and 0.8% $Al_2(SO_4)_3$ had lost its effect.

For different initial content of F, the minimums of F concentration in root were 0.39, 1.55, 2.26 mg kg^{-1} respectively. Although 1.0% $Al_2 (SO_4)_3$ had lost its effect of reducing F in rice, husk, stem and leaf, F in root was still lower than blank in this treatment. All the treatment of applied $Al_2 (SO_4)_3$ had lower F concentration in root than blank, and they were reduced by 41.99, 71.91, 76.02, 55.58, 39.19% respectively. Thus it was concluded that 0.4%-0.6% $Al_2 (SO_4)_3$ had the best effect on reducing F concentration in root.

The relationship of F concentration in each part and the amount of $Al_2 (SO_4)_3$ applied was well fitted by quadratic equation (Table 1). Except the correlation coefficients of stem and leaf, husk, root in 400 mg kg^{-1} F reached significant level ($P \leq 0.05$), others all were extremely significant($P \leq 0.01$).

Table 1. The fitting equation of F concentration in rice and applied $Al_2 (SO_4)_3$ amount.

	Fitting equation					
	400 mg kg^{-1} F	r	800 mg kg^{-1} F	r	1200 mg kg^{-1} F	r
rice	$y=1.2x^2-8.5x+22.6$	0.9789	$y=1.6x^2-9.2x+23.1$	0.9738	$y=2.1x^2-16.1x+49.3$	0.9144
husk	$y=2.3x^2-17.2x+51.4$	0.7185	$y=3.3x^2-20.4x+62.6$	0.9102	$y=3.6x^2-21.3x+115.3$	0.8539
Stem/leaf	$y=0.1x^2-0.7x+1.3$	0.8642	$y=0.2x^2-1.0x+1.8$	0.9839	$y=0.9x^2-5.9x+13.2$	0.9615
root	$y=0.2x^2-1.4x+3.1$	0.8616	$y=0.5x^2-3.4x+7.8$	0.9719	$y=0.8x^2-7.6x+19.9$	0.9932

Therefore, it was concluded that (1) the concentration of F in each part of rice plant showed the increasing trend as the initial F content increasing, (2) when the initial F content was identical, different amount of $Al_2 (SO_4)_3$ made F concentration in each rice plant part change. The changing trend was decreasing first and then increasing in this experiment, and the minimum of F in plant appeared in the treatments of 0.4-0.6% $Al_2 (SO_4)_3$ which had better effect of defluoride.

4. Discussion

In the treatments of $Al_2 (SO_4)_3$ applied in this experiment, it was obtained that 0.4-0.6% $Al_2 (SO_4)_3$ reduced the concentration of F in each rice planting part and had the best effect of defluorine. It was because Al and F complex changed the physical and chemical properties of soil and resulted in the changing of dissolution- precipitation, adsorption- desorption, migration- transformation happened in the interface of solid and liquid which impacted the biological activity of F (Lindsay, 1979). When $Al_2 (SO_4)_3$ was added to soil, Al3+ hydrolyzed and formed $Al (OH)_x^{3-x}$ + x H+ which made the soil acidify. Simultaneously, F^- and Al^{3+}, $Al(OH)_x^{3-x}$ took place complex and formed different kind of AlF_x^{3-x}. Not only F^- was absorbed by soil, but also the complex of F and Al was absorbed. When the pH in soil decreased to a certain extent below potential of zero charge of Fe and Al oxide/hydroxide because of $Al_2(SO_4)_3$ applied, Fe and Al oxide/hydroxide was easy to accept proton and possessed positive charge which had strong adsorption capacity to F- and F-Al complex with negative charge (Farrah, 1997). Thus, the content of water soluble-F was decreased and the biological activity was reduced to a certain extent. In this experiment, the concentration of F in each rice planting part, as well as free-F and water soluble-F, were truly decreasing at lower $Al_2(SO_4)_3$ amount, but when $Al_2(SO_4)_3$ was more than 0.6%, F in plant was not in the trend of continued decreasing. In contrary, it began to increase gradually and exceeded blank in individual treatments. This can be explained by following reasons. (1) The ratio of F: Al became smaller as the amount of $Al_2(SO_4)_3$ continuous increasing which made the complex change from possessing negative charge to positive charge. Because Fe and Al oxide/hydroxide had poor absorption to F-Al complex with positive charge in acid condition as above discussed, there was more F-Al complex existed in soil solution and adsorbed by plant, which made the concentration be high. (2) With $Al_2(SO_4)_3$ increasing, the soil acidified aggravating. So part of F^-combined with H^+ and formed HF which was easier absorbed by plant than F^- and F-Al complex. This opinion was the same as the researching result of Horner J M, who considered F and acid had certain synergistic effect and inhibited the growth of root, stem and leaf when the pH in soil was rather lower (Horner *et al.*, 1995).

References

1. Ammerman, C. B. and Henry, P. R. 1983. Effects of Fluorides on Animals: Dietary and Mineral Supplement Considerations. In: Shupe, J. L. (ed.) Fluorides: Effects on Vegetation, Animals and Humans. Paragon Press, Salt Lake City. pp. 281-290.
2. Anderson, M. A., Zelazny, L. W. and Bertsch, P. M. 1991. Fluoro-aluminium complexes on model and soil exchangers. *Soil Science Society of America Journal*. 55: 71-75.
3. Becker, D. E., Griffith, J. M and Hobbs, C. S. 1950. The alleviation of fluorine toxicosis by means of certain aluminium compounds. *Journal of Animal Science*. 9: 647.
4. Chai, H. D. 1995. Sanitary Conditions in Modern (in Chinese). People education press, Beijing. pp. 616-659.
5. Ding, R. X. and Huang, X. 1991. Biogeochemical cycle of aluminium and fluorine in tea garden soil system and its relationship to soil acidification (in Chinese). *Acta Pedologica Sinica*. 28(3): 229-236.

6. Farrah H, Slavek J and Pickering W F. Fluoride interactions with hydrous aluminum oxides and alumina [J]. Australian Journal of Soil Research, 25(1):55-69.
7. Horner, J. M. and Bell, J. N. B. 1995. Effects of fluoride and acidity on early plant growth. *Agriculture, Ecosystems and Environment*. 52: 205-211.
8. Laura, F. H., Ellen, M. C. and Dharni, V. 2003. Fluoride sorption and associated aluminum release in variable charge soils. *Journal of Colloid and Interface Science.* 267:302-313.
9. Lindsay W L. Chemical equilibria in soils. [M]. John Wiley & Sons, New York, 1979.
10. Maclean, D. C., Hansen, K. S. and Schneider, R. E. 1992, Amelioration of aluminium toxicity in wheat by fluoride. *New Phytol.* 121: 81-88.
11. Robera, B. F. 2001. Phytotoxic effects of fluorides. *Plant Science.* 161: 979-985.
12. Wang, H. H., Zhu, M. X., Jiang, X. and Wang, F. 2006. Interaction of fluoride with red soil and its environmental implications (in Chinese). *Journal of Agro-Environment Science*. 25(4): 947-948.
13. Wu, F. Z., Ye, Z. J., Tang, L. Y. and Jiang, S. H. 1979. Detecting F in plant and soil by F ion selective electrode (in Chinese). *Chinese Journal of Soil Science*. 3: 22-24.
14. Wu, W. H., Xie, Z. M., Xu, J. M., Hong, Z. P. and Liu, C. 2002. Characteristics of forms of fluorine in soils and influential factors (in Chinese). *Environmental Science*. 23(2): 104-108.
15. Xie, Z. L., Sun, W. T., Chen, Z. Yin, B. and Wang, J. H. 2008. Effects of interaction of fluoride- aluminum on the adsorption characteristics and distribution of aluminum in tea garden soil (in Chinese). *Journal of Jilin Universtiy (science edition)*. 46(3): 365-370.
16. Xie, Z. M., Wu, W. H. and Xu, J. M. 1999. Translocation and transformation of fluorides in the environment and their biological effects (in Chinese). *Advances in environmental science*. 7(2): 40-53.
17. Yang, W. J., Jang, X., Xu, Y. K., Ji, G. L. and Zhao, Q. G. 2002. Sorption of aluminum and fluoride on soil as affected by complexation of aluminum and fluoride. *Acta Scientiae Circumstantiae*. 22(2): 161-165.
18. Zhao, L., Jin, B., Ma, Y. J. and Su, H. 2005. The effect of fluorine-containing emissions on rice quality and yield (in Chinese). *Journal of Agro-Enviroment Science*. 24: 52-55.
19. Zhao, L. P., Wang, Y., Ma, J., Li, C. L. and Han, X. 2001. Improvement of soda-typed saline- alkaline soil in western Jilin province (in Chinese). *Chinese Journal of Soil Science*. 32(s0): 91-96.

Chapter 2
Sustainable Energy

Optimal deployment of electric vehicle charging stations in the traffic system

D.L. Qi*, K. Wu and Y.G. Zhang

Department of Mathematics and Physics, North China Electric Power University, Baoding, 071003, P.R. China,
**Email: 923267260@qq.com*

In recent years, in order to relieve the pressure of petroleum crisis and environment deterioration, the global transportation sector began to vigorously support the development of new energy cars, especially Electric Vehicles (EVs). Charging stations as EVs infrastructures, provide energy supply services for vehicles. They become one of the key factors restricting the development of EVs. Therefore, the construction of charging stations, especially the early stage of the site selection, got high attention by researchers from all over the world. In this paper, based on the background, research the plant location problem of EVs. Through combined with the specific situation of China, we put forward a new planning and site selection model based on gas station network. And then a numerical experiment to verify the effectiveness of the model. It provides certain reference basis for the actual cases.

Keywords: Electric vehicles; Charging station; Gas station; Model; China.

1. Introduction

Nowadays, along with the rapid development of transportation industry, the consumption of non-renewable resources (petroleum) are increasing. The data show, American transport sector consumed 1,400 barrels of crude oil a day, accounting for 70% of domestic fuel consumption (EIA, 2012). And OPEC (Organization of the Petroleum Exporting Countries) predicts that by 2015 the global oil consumption will be increased by 1.21 million barrels a day (Monthly Oil Market Report, 2014-7). Among these, the transport of oil will account for more than 60% of total oil consumption (IEA). Therefore, the subject of alternative fuels for meeting the future energy demand of the transport sector has gained much attention in policy and research debate (Zhang et al. 2013).

At the same time, due to a large number of automobile exhaust emissions, caused destructive effect to the atmospheric environment. And they led to a variety of common human diseases like respiratory infections, acute poisoning, pink eye and even lead to cancer (Cheng et al. 2003). This is undoubtedly a great threat to the safety of human beings. As industrial production and vehicle emissions, the carbon dioxide (CO_2) and other strong sex of heat-trapping greenhouse gases are increasing. They has caused global warming, melting glaciers and a series of serious problem of the greenhouse effect. According to statistics, transportation emissions account for 29% of the US greenhouse gas emissions, and over 5% of global greenhouse gas emissions (Report to Congress, US Department of Transportation 2010). And a fifth of greenhouse gas emissions in Europe is due to transport (European Commission, 2011), etc. At the same time, by the International Environmental Research Center according to the latest data (Fig. 1), the sum total of Chinese CO_2 emissions beyond Europe and the United States has become the first in the world. The grim problem of atmospheric pollution, testing globe governments (especially China) governance capabilities, so how to reduce vehicle exhaust emissions and ease the air pollution situation, also became one of researchers focus on the research.

Electric vehicles (EVs), as the representative of the new energy vehicles, truly zero emissions, no pollution (Chen & Jin, 2005). And they reduce the oil energy consumption, greatly alleviate the country for oil pressure of energy depletion. Even taking into account the emissions from the power plants needed to fuel the vehicles, the use of EVs can still significantly reduce global air pollution (He, 2000). So they have received the attention from all countries in recent years. And the charging station in electrical infrastructure is one of the most important roles in the EV industry. Perfect charging infrastructure network is the basis and prerequisite for electric vehicles (Xu et al, 2009). Up to now, many countries have been committed to the construction of Electric Vehicles charging station, but they can't formed a mature system for the planning and deployment. So researchers around the world have made a lot of exploration. In China, in order to reduce construction cost and avoid the waste of land resources, some researchers suggest that adopt the way of Electric & Gas Alliance planning for the construction of the charging station. On this basis, this paper put forward a new model, based on the traditional Gas station network, to reconstruct charging stations. We choose some gas stations in the city as candidate sites, and these sites can be converted charging stations to meet the demand of charging demand area. The advantage of the model using existing network layout within the city, can simplify government early planning decisions and avoid waste of funds.

The rest paper is organized as follows. The review of the literature related will be discussed in Section 2. The model and Symbol definition will be described in Section 3. A numerical experiment is conducted to verify the model in Section 4. Then Section 5 will provide the conclusions

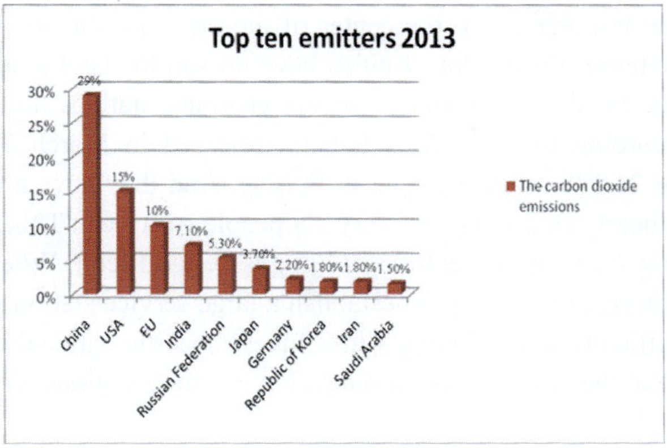

Figure 1. Global carbon emissions in 2013 ranking list.

2. Charging Station Research Status

In recent years, along with the rapid development of the Electric Vehicles industry in the world, scientific research did a lot of work surrounding the charging infrastructure construction.

Spanish scientists considered drivers daily activities to determined electric vehicles charging point location (González et al, 2014). Iranian researchers presents a Mixed-Integer Non-Linear (MINLP) optimization approach for optimal placing and sizing of the fast charging station (Payam, 2014). American researcher utilize a continuous facility location model to present methods for optimally locating electric vehicle infrastructure in highway corridors (Sathaye, 2013). Albert Y.S. and other people build a model into the target of minimizing the total cost to charging station coverage and meet the demand of charging users as constraint conditions. At the same time, they also proposed a new greedy algorithm can solve the problem of actual calculation (Albert et al, 2013). Sung Hoon Chung and Changyun Kwon formulated a multi-period optimization model based on a flow-refueling location model for strategic charging station location planning on the real traffic flow data of the Korean Expressway network (Sung & Chang, 2015). Guo Siyuan and Chen Xin charged with planning area is the target of minimizing the sum of adjacent stations of generalized distance, adopted the method of gradual optimization and found the best location (Guo & Chen, 2013).

Existing reference literature respectively from site selection principle, the constraint conditions, location model, calculation method and so on has carried on the exploration. There are a lot of advanced ideas worth our using for reference. However, the usage of gas network was less to be made use of. In

China, people live mainly in the center of the city, and the supply of land in cities is very tense. Even a lot of cities have no surplus land planning to build new facilities. So the site planning of the charging stations are difficult. For example, according to the census bulletin released in March 2014, resident population of 21.148 million people in Beijing. And they have a total of 5.371 million cars, nearly an average of every 3.9 people with a car. This makes people have a parking space in the centre of Beijing is very difficult. When considering the above factors, in the city to re-establish a large service station with charging even more difficult. So combining with Chinese specific national conditions, an model based on the gas stations for redevelop charging stations was proposed in this paper.

3. The Model

Firstly, considering the initial selection of candidate site, we give the prerequisite of this model:

(1) Infrastructure construction to conform to the city road planning, pay particular attention to the protection of city greening environment.

(2) Candidate selection of site to be easy to get power supply in the city, and meets the requirement of urban power grid planning. At the same time there should be close to the convenience of water, heat and other infrastructure, avoid too much money into the corresponding facilities.

(3) Selecting the address is in compliance with convenient transportation, such as the main trunk or facilitate access and observation of the trunk at times. Or consider the large public service facilities nearby (Shenzhen 2010, Beijing 2010).

(4) Near the residential area reconstruction in power plant, it is important to pay special attention to the fire safety problems, but also reduce battery pollution, and the effects of noise limit.

Traditional construction of gas stations, taking into account all the necessary requirements. So their network layout provide the good environment to rebuilt charging station. They greatly saved the government funding and avoided too much human waste.

Model established in this paper is based on the existing gas station network, in order to charge the cost of rebuilding the road user cost minimization goal. Road user costs include the driver from the charging demand point to the electricity cost in the process of charging stations, and other travel expenses, etc. In the model, we mainly private cars as the research object, considering the distance from the residents near the residential area to the gas station.

Here are the hypotheses of this model.

(1) All provide gas station candidate address in the literature have been through the analysis and in accordance with the various conditions of the former mentioned.

(2) Each demanding point represents a district, where demand represents the number of electric vehicles in the current area.

(3) All electric car users demand area, in a certain period of time can only accept a charging station service.

(4) The environmental cost has a simple linear relationship with the distance, which is the Euclidean distance.

Model is as follows

$$\min Z = \sum_{s=1}^{N} (Z_{1s} + Z_{2s} + Z_{3s}) \tag{1}$$

The reconstruction cost of charging stations

$$Z_{1s} = \sum_{s} \sum_{g} C_{sg} x_{sg} \tag{2}$$

In the formula

S : The set of candidate sites.

g : The grade of the charging station.

$x_{sg} \in \{0,1\}$: If we choose the candidate site S and transform it into grade g charging station is 1. Otherwise is 0.

User charging time costs

$$Z_{2s} = \sum_{s} \sum_{g} \sum_{d} x_{sg} y_{ds} T_{g} \tag{3}$$

In the formula

$y_{ds} \in \{0,1\}$: If demand point d is covered by a charging station located at S is 1. Otherwise is 0.

T_g: The time needed for the user charge the cost.

Road cost, including the driver from the charging demand point to the charging infrastructure need power consumption and other travel expenses. It is associated with the distance of demand points to the charging station,

$$Z_{3s} = \sum_{d} \sum_{s} y_{ds} h_{ds} l_{ds} \tag{4}$$

In the formula

D: The set of demand points.

l_{ds}: The distance between demanding point d to candidate site S.

h_d : The number of vehicles in demanding point d .

The constraint

$$\sum_{s} y_{ds} = 1 \qquad \forall d \in D \tag{5}$$

$$y_{ds} \le \sum_{d} x_{sg} \qquad \forall d \in D, s \in S \tag{6}$$

$$\sum_{d} h_d y_{ds} \le \sum_{g} C_g x_{sg} \qquad \forall d \in D, s \in S \tag{7}$$

$$\sum_{g} x_{sg} \le 1 \qquad \forall s \in S \tag{8}$$

The objective function (5) is one candidate site can only be transformed into one grade charging station. The objective function (6) is the service can be offered only if the charging station is located in this candidate site. The objective function (7) restricts the service capacity of the charging station. The objective function (8) one candidate site can only be transformed into one grade charging station.

4. Numerical Experiment

In this paper, based on the planning of city road network, we select the candidate site, demand point location and electric vehicles ownership, etc. Then we use a practical example to prove the validity of the model. Candidate site location are refer to the original station coordinates, and make the corresponding choice. And demand point location adopted a few large residential area of geographic coordinates.

Firstly, generate 21 demanding points and 13 candidate sites randomly from 120 x120 coordinate plane. Area demand, that is, the number of Electric Vehicles. With the flourishing degree and income situation in the region. In this article, we based on the future promotion and the development of Electric Vehicles, assuming that the demand of each area is (100,200). The related data as shown below Table 1-2.

Table 1. Candidate sites coordinates.

NO.	X	Y
1	32.80	75.12
2	71.52	53.28
3	48.48	67.28
4	27.76	44.64
5	69.44	68.56
6	37.28	60.32
7	37.28	90.16
8	88.72	62.40
9	25.60	68.32
10	29.44	50.16
11	69.68	40.40
12	91.04	84.64
13	95.92	36.40

Table 2. Demanding points coordinates and corresponding demand.

NO.	X	Y	Demand
1	49.52	100.56	120
2	28.40	91.44	140
3	59.04	93.12	130
4	33.04	81.68	200
5	65.44	84.40	180
6	68.56	80.40	160
7	24.80	70.00	190
8	50.40	74.48	160
9	63.92	75.76	140
10	59.28	66.64	110
11	34.96	58.00	150
12	23.52	54.40	170
13	54.16	54.80	130
14	69.20	53.52	115
15	85.92	50.16	200
16	90.64	78.64	190
17	95.44	67.52	185
18	105.60	86.16	120
19	117.52	68.96	175
20	90.00	45.52	190
21	25.44	46.56	200

The layout of the demanding points and candidate sites in coordinate diagram as shown below Figure 2.

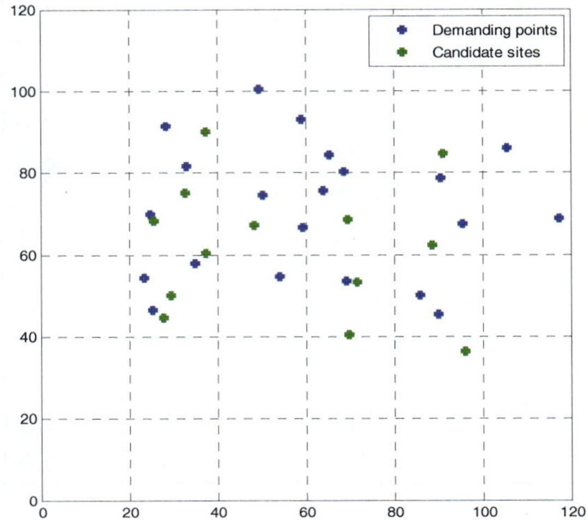

Figure 2. The layout of the demanding points and candidate.

According to Beijing pledges inspect bureau announced "Technical specifications of electricity supply and assurance for electric vehicle: Electric Vehicle charging station", there are four levels of the charging stations. And we assume the number of charging facilities within four grades of charging stations. The related data as shown below Table 3.

Table 3. Stations level.

Level	Service Capacity	Cost	Number
1	500	610	60
2	260	440	40
3	150	230	25
4	80	130	10

In the concrete algorithm, this paper assumes that the user charging time costs are fixed. Time costs and charging facilities are negatively correlated, meaning that the larger charging stations, the more number of charging infrastructures and corresponding to the user in the charging station of the less time cost. The related data as shown below Table 4.

Table 4. Time cost.

Level	Number	Time cost
1	60	10
2	40	25
3	25	40
4	10	60

According to the actual situation, the bigger the scale of the charging station, not only the number of charging infrastructure, and around the service quality, service facilities, logistics management is better than small charging station. So the charging demand users under the premise that the same distance, will be more willing to choose large charging stations for acceptance of the service site. Therefore, in order to make the model with the actual situation better, this paper set up a optimization criteria. We consider the Electric Vehicle speed and actual speed limit standard of the city, if the two charging station can be reached in ten minutes, we select the large charging station. Otherwise, we select demand point nearest the candidate points. Run the algorithm on LINGO and get the results shown in Table 5 and Figure 3. Then the total cost is 122.32336 million

Table 5. The Result.

Station	Demand point	Grade
1	4	2
2	14	3
3	8,10,13	1
4		
5	5,6,9	1
6	11	3
7	1,2,3	1
8	15,17	1
9	7	2
10	12,21	1
11		
12	16,18,19	1
13	20	3

According to the result, selecting candidate site related to demand and the users time cost. The experimental results is feasible. This model is helpful for the government when making decisions about the deployment of the charging stations.

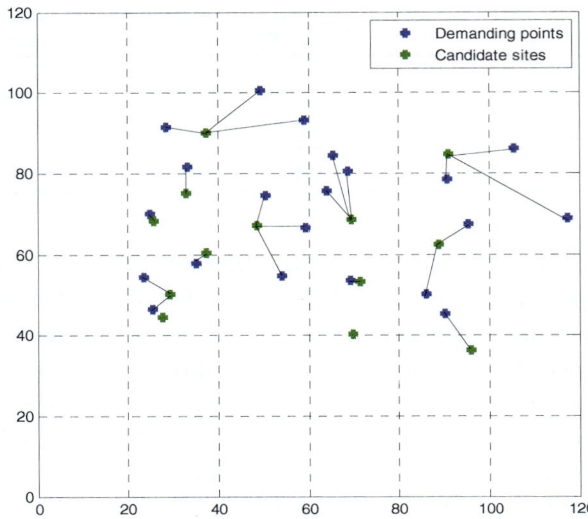

Figure 3. The allocation of the demanding points and transformed candidate sites.

5. Conclusions

This paper, on the basis of the original gas station, considering the construction of the charging infrastructure, simplify the early planning scheme, so as to greatly reduce the complexity of the problems and initial cost. Meanwhile, considering candidate positions, also can meet the demand of the Electric Vehicles owners. And we will also add the user requirements into the model, this conforms to the actual consumption. So this paper explore the construction of the charging station for Electric Vehicles, have certain research results. The establishment of the model, however, there are still some deficiencies, needs further research, such as,

(1) Because there is no exact details of gas stations in the city, so on the selection of candidate site points, has the certain assumptions.

(2) In the paper considering the distance, only is the Euclidean distance. This distance in coordinate chart is a good way to show apart. But in real life, because of the complexity of the city road network, so it can't be a good representation of the actual distance situation.

(3) Due to different cities in the building of the charging station, its size and the number of charging infrastructure will change. So in the absence of complete information, the article on the scale of the charging station has carried on the reasonable assumption.

(4) This paper focuses on the problem for the construction of the charging station, but the charging infrastructure includes charging station, charging pile construction scheme and centralized in power stations, etc. So how to perfect charging facilities in urban construction network is another topic worth studying.

Countries for promoting Electric Vehicles, cannot leave the perfect infrastructure construction. Therefore, for the study of the construction of the charging infrastructure, we need more efforts and innovation. Looking forward to the future, China can have more prominent contribution in the field of new energy vehicles.

Acknowledgements

This research was supported partly by the National Key Basic Research Project (973 Program) of China (2012CB215200), the NSFC (51277193), and the Fundamental Research Funds for the Central Universities (2014ZD43).

References

1. Albert Y.S., Lam Y.W. & Chu X.W. 2013. Electric Vehicle Charging Station Placement in Hong Kong: *IEEE Smart Grid Comm Symposium, 2013*.
2. Beijing Quality and Technical Supervision. Electric vehicle power supply and protection of technical specifications: Charging Station. In National Standard Press, *DB11/Z 728-2010, 2010.*
3. Cheng Y.B., Jin Y.L. & Liu Y.C. 2003. Progresses on adverse health effects of automobile exhaust. *Journal Of Hygiene Research, 2003.*
4. Chen Q.Q. & Sun L.Q. 2005. Present status and future trends of Electric Vehicles. *Science & Technology Review* 23(4).
5. Elsevier B.V. 2006. Appendix II: Lingo software, *Process Systems Engineering, Vol. 7, pp. 389-394.*
6. Guo S.Y. & Chen X. 2013. Site Selection Planning for Electric Vehicle Charging Stations Based on Evolutionary Optimization Method, *Urban Public Transport, Vol. 03, pp. 34-36.*
7. González J., Alvaro R. & Gamallo C. 2014. Determining Electric Vehicle Charging Point Locations Considering Drivers' Daily Activities, *Procedia Computer Science, Vol. 32, pp. 647–654.*
8. He H.W. 2000. Hybrid Electric Vehicle Power Train Reasonable Matching. *Jilin University of Technology*.
9. Payam S.B. 2014. Optimal fast charging station placing and sizing, *Applied Energy, Vol. 125, pp. 289–299.*

10. Sathaye N. 2013. An approach for the optimal planning of electric vehicle infrastructure for highway corridors, *Transportation Research, Vol. 59, pp. 15–33.*
11. Sung H.C. & Chang Y.K. 2015. Multi-period planning for electric car charging station locations: A case of Korean Expressways. *European Journal of Operational Research, Vol. 242, pp. 677–687.*
12. Shenzhen Market Supervisory Authority. Technical specification of electric vehicle charging system. In Part 2: Code for design of EV charging station and charging point, *SZDB/Z 29.9-2010.*
13. Xu F., Yu G.Q. & Gu L.F. 2009. Tentative analysis of layout of electrical vehicle charging stations, *East China Electric Power, Vol. 37, pp. 1678-1682.*
14. Zhang Q., Benjamin C.M. & Tetsuo T. 2013. A methodology for economic and environmental analysis of electric vehicles with different operational conditions, *Energy, Vol. 61, pp. 118-127.*

Applications of wind energy in concept car styling design

Z. Li
Hunan Arts and Crafts Vocational College, Yiyang, Hunan, China
E-mail: 55249746@qq.com

The paper studies automobile styling mode and designing way by the character of new energy-wind energy and solves the problems on how to grasp the characters of new-energy automobile styling and the application of new energy. It analyzes the relationship between wind energy conversion and styling in the automobile styling design through combining air flow principle and practical cases, realizes the practical application of the regenerating wind energy character in automobile headstock and body styling design of the new-energy automobile effectively, and advances the balance between aesthetics and aeromechanics in the automobile styling and its corresponding solutions.

Keywords: Green energy; Car styling; Wind energy; Sustainable.

1. Green Energy Vehicles and Sustainable Design

Green energy vehicles refer to the vehicles that use green energy or that applies green ecology concept into vehicle design and utilization. Nowadays, most of the new energy cars include: hybrid-, electric-, hydrogen- and gas-powered cars. Hybrid power refers to a car with two or more power supplies, such as battery, fuel battery, solar battery, and internal combustion engines power units. At present, the hybrid-powered car generally refers to the car with internal combustion engines generator and battery. Pure electric vehicle refers to the vehicles powered by in-car power supply, driven by electric motor and in line with various regulations and rules on road traffic and safety. Compared with traditional cars, this vehicle has smaller impact on environment, so it has a bright prospect. But the current technology is not mature yet. Hydrogen-powered car is a vehicle that can truly realize zero emission. Its emissions are pure water, which has the advantages of no pollution, zero emissions, rich reserve etc. Gas-powered car uses natural gas or compressed natural gas. Compared to traditional oil, it is more environmentally friendly and, it works better in energy saving and emission reduction. Meanwhile, it is more economical and affordable. In addition, as the reserves of traditional energy gradually decrease, clean renewable energy sources become the main force of the energy. Wind energy is inexhaustible and it is bound to be generated during the car driving process. Therefore, it has gradually attracted people's attention. Subsequently, wind energy has been applied in the design and manufacturing of vehicles as a new energy.

As a renewable new energy, wind energy follows the green-manufacturing-oriented mainstream, so that the manufacturers can place more emphasis on the sustainability of the products. Automotive wind energy (aerodynamic resistance) is an objective natural phenomenon that exists in the process of car driving and is independent of men's will. Wind-powered car mainly uses the energy arising from the operation, which is then transferred into a power source. And part of such energy can also be transferred into the braking system itself exerted by air force and air resistance. Today's resources are in shortage. Thereby, the wind-powered car is expected to be a new traffic tool for future men. The main characteristics of wind energy hybrid-powered car are listed as below: on one hand, it can effectively collect the wind energy produced in the process when the hybrid-powered car is driving at a low speed; on the other hand, it makes the consumed wind energy during the process when the car is driving at a high speed and the energy consumed by the idle engine used to some extent. Automotive wind turbine is one of the key spare parts of wind energy hybrid-powered car. In theory, there is no problem for it to be designed and produced. If it can be put into application, it can combine existing hybrid-powered cars with all the new-energy-powered cars and ordinary cars.

2. Application of Wind Energy in Automotive Design

As the "car society" gradually forms, the automotive population continues to show an upward trend. But the ultimate solution is certainly not to limit the development of the automobile industry. Rather, it is to develop alternative new energies to oil. Wind energy is a new energy and it is pollution-free and renewable. Especially, it is applied into the concept cars. With the constant development of new energies, wind energy has become an alternative energy of oil, and obtained wide attention. Meanwhile, the car styling has no more than the early-phase styling to meet only the requirements for a vehicle. Instead, it has slowly become a vehicle with commodity styling to meet both the practical requirements and aesthetic requirements of the customers.

2.1. *Working principle of wind-powered car*

The working principle of wind-powered cars in the operation is listed as below: make full use of the law of conservation of energy. The energy can only be transformed, but never disappear. Aerodynamic drag exists in the whole process of car driving. The lower the speed is, the smaller the aerodynamic drag will be and vice versa. The initial source of energy comes from fuel oil (gas). In other words, wind energy comes from fuel oil (gas). In the dynamical structure of conventional cars, the aerodynamic drag exists in the form of an energy that hinders car driving and its direct consequence is to weaken the kinetic energy of the vehicle. Namely, the wind energy is part of lost kinetic energy; and in the power structure of wind energy hybrid-powered cars, aerodynamic drag acts on the turbine of wind generator, makes it rotate (or rotate more quickly). Namely, the turbine has absorbed part of wind energy. The turbine continuously absorbs and transforms the wind energy. Therefore, in the port of the ventilation hole, eddy current effect forms. It can change airflow direction. As a result, the previously air that acts directly on the car and hinders its driving will be gathered and flow to the ventilation hole, to become the power of the generator.

2.2. *Wind-powered concept car design*

Atmospheric pressure difference causes air flow, which forms "wind". Wind power itself is a kind of clean and renewable natural energy which is rich in reserves. These advantages and characteristics can effectively help the transportation tools to achieve the lightweight design, and can better use impacts of aerodynamics on vehicles, and improve energy utilization efficiency. At present, the world first wind-powered car suitable for running on the road appeared in Germany, called "Wind Explorer". It is mainly driven by wind and kites. This racing-style roadster owns a carbon fiber body and bicycle tires, even when the battery is added, the weight in all is only about 204 kg. And without battery, the body weighs only 82 kg, far lighter than ordinary cars. And its speed per hour can reach more than 88 kilometers (Fig. 1). The whole shape is somewhat like a large kite, and the tail modelling is high lighted, which has large wind blades. Another car driven by wind energy is found in Detroit Maker Faire. It is a three-wheeled battery-driven car, where wind energy is transferred into electrical energy. The tail has a high tilting arched door, which seemingly accords with aerodynamic requirements (Fig. 2) exactly. BMW has also launched car model powered by wind energy. The car was named "Blue Dynamics" and it integrates high performance, driving pleasure and high efficiency into one. This environmental friendly concept car is a three wheeled vehicle; the rear two wheels stretch to both sides; on the top of the cockpit, there is a sail-like design. And the slender body is in the front of the cockpit. The car can drive automatically according to the wind direction, and the driver can manually adjust the "sails" at the rear of the car via the special configuration in the cockpit, and change the driving direction (Fig. 3).

Figure 1. Wind explore.

Figure 2. Aerodynamic Eco-Friendly wind-powered car.

Figure 3. BMW Three-Wheeled wind-powered car.

3. Key Points of Applying Wind Energy into Concept Car Styling

The new energy vehicles are completely different from traditional gasoline cars. Therefore, designers could not use traditional automobile design method directly into the design of new energy cars. Car styling design is a design process that can perfectly combine art, design art with technology. Also, it is an important means for automobile enterprises to participate in market competition. With the development of wind energy and the progress of technology, increase of the car speed and the attention paid by the whole society to environmental protection, the importance of automotive aerodynamics has become increasingly prominent. Car styling can directly affect the car's aerodynamic characteristics. Therefore, the styling design of wind-powered concept car needs to reflect its driving power, economy, operability and safety.

3.1. *Aerodynamics and headstock styling*

Automobile Aerodynamic mainly studies the rules of surrounding airflow and its interaction with the autos and the impacts in the driving process. It is an important performance of the vehicle, and mainly refers to the three major aerodynamic forces that include drag force, elevating force and lateral force, as well as other wind force characteristics. And because it is a bluff body running on the road, and its shapes may vary from car to car, the driving condition of the vehicle is very complex. Besides, the driving speed of ordinary car is less than the speed of sound. Thus, compared with general aerodynamics, automobile aerodynamic has its own characteristics. Accordingly, if a vehicle intends to utilize wind energy, the most direct method is to open an air intake at its head, and arrange head-on wind energy recovery device. And the windward area is not greater than the original car front windward area. Therefore, basically, this will not increase automobile driving resistance. Then, the size of the air inlet and its impacts on the auto wind drag force will become the key points of the headstock design of concept car. The frontend opening styling can help the autos to reduce the resistance force. Therefore, to take advantage of the impacts of the flow field surrounding the car and make headstock design is a critical problem to be solved.

The vehicle front-end opening can reduce its windward area, thereby reducing its resistance in the driving. It can be concluded from the vehicle driving drag coefficient that: the bigger the front-end opening is, the smaller the driving resistance coefficient will be. As shown in Fig. 4, no matter what the flow velocity is, the flow regime keeps a constant steady state. That is, it will keep a beautiful streamline. After the flow velocity increases, the flow velocity will increase sharply merely in the areas where the surface flow velocity is zero at a low speed, and no more else. Meanwhile, streamlined bodies and non-streamlined bodies will endure different resistances. As a result, while taking advantage of wind energy and distributing the proportion of hybrid powers, the designed size of the opening and the headstock should be considered. Besides, the pressure drag around a streamlined body is far more smaller than that of the other bodies of various shapes. Therefore, if the body is made into a streamlined or streamlined-similar shape, it will get smaller pressure resistance.

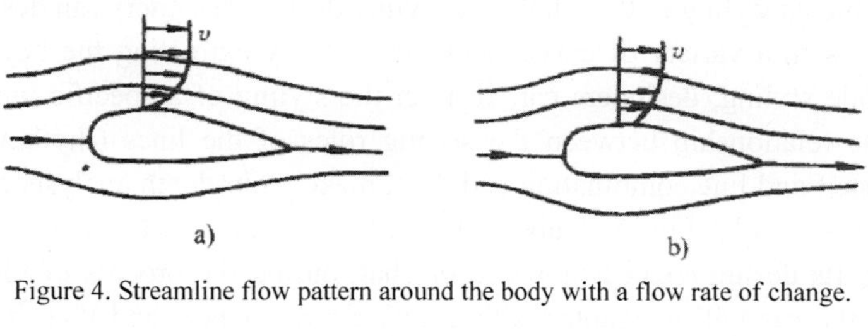

Figure 4. Streamline flow pattern around the body with a flow rate of change.

3.2. *Car body styling and air flow characteristics*

The car styling is complex; the impacts of various styling factors of the car body on the dynamic characteristics of the air are uncertain; and they will often interfere with each other. Therefore, it is hardly impossible to keep the styling style unchanged when modifying or optimizing the car. All these prevent the efficiency and level of automotive aerodynamics styling design from being improved. In the driving process, the aerodynamic force endured by the car is quite complicated. And if the characteristics of air flow are applied into car body styling, it will be better able to fit the wind power principle.

The flow fields around the car involve a kind of complex flow.In the running process of the automobile, the there are two major types of air field: one is the exterior flow of the car, including all the air flow on the surface of the vehicle body; the second is the internal flow, and it refers to the air flow that flows through the vehicle engine, vehicle exhaust system, vehicle cooling system and vehicle driver's cab. In recent years, with the rapid development of vehicle level and design level, the style of automobile styling is also showing a diversified trend. For the wind-powered car, the first problem to be solved is how to get the driving force needed by the vehicle, reduce the running resistance and improve its running speed. Therefore, the changes in styling design of the automotive front face will change the airflow volume of vehicle chassis. In the design, unsmooth and various outer convex styling of the car body should be avoided.

4. Case Study of Wind-Powered Concept Car Styling

Design a new energy concept car according to the characteristics of the wind power and energy conversion principle. The design is required to study and analyze wind power and air flow from the perspective of styling, and propose novel styling solutions.

In the initial stages of the design, with regard to the contour line in the car body elements styling in the whole car styling design, designers can describe the car body with a variety of curves and surfaces. By extracting the key lines of automobile styling, designers can abstract the styling of a specific model, and study the relationship between the styling rules of the lines (rhythm, length, tempo, etc.) and line combination and then, make an in-depth analysis of the car styling as a whole. Fig. 5 is about the early concept sketches and three view drawing. Its design concept is based on that, during the process of high-speed driving, the car will encounter with strong air resistance, and then it will use efficient wind driven generator to transfer the wind energy in the process of driving into electric energy efficiently. The opening design in the front face of the headstock fits the car body pretty well. According to the principle of air flow, the big-curve, big-opening modeling is used to fully increase the harvesting rate of wind power in the driving process. Meanwhile, the overall styling is in line with the innovative design concept of the wind-powered concept car. And its styling is novel and has a visual impact.

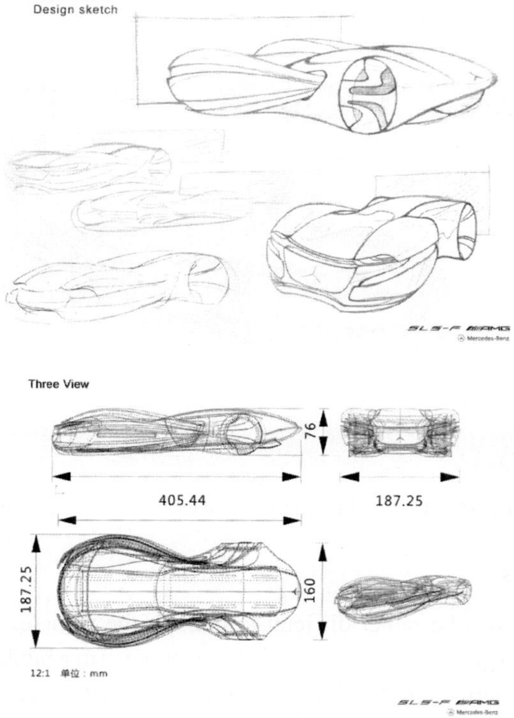

Figure 5. Mercedes Benz SLS-F wind-powered concept car.

Figure 6 shows the design sketch and energy conversion chart of the wind-powered concept car. The openings in the car body and headstock look like the streamlined blossom of the newly-born thorn apple, which can absorb wind energy effectively and use it for energy conversion. The air suite in the front of the vehicle and its door expansion is designed according to the morphology of a bloomed thorn apple. In this way, not only that the car will have a good drag coefficient, but also that the overall shape will be more free and elegant. The car body uses many carbon fiber materials, so that it will be more light-weighted as a whole. Besides, the air diversion kits in the car body and the air intake system at the body bottom keep the vehicle stable in the high-speed driving process.

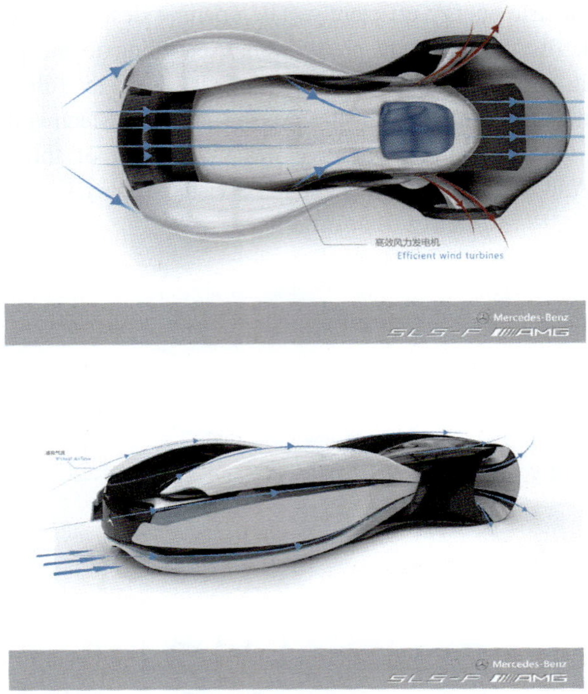

Figure 6. Mercedes Benz SLS-F wind-powered concept car energy conversion chart.

5. Conclusions

In the future, it will be rather difficult to classify the vehicles and, there will be more crossover vehicle models. Generally speaking, after bidding farewell to the constraints of internal combustion engine, the new energy vehicles will become more flexible, diversified and smart in appearance and styling. After nearly a century of development, the automobile styling has been trying to seek for a balance between aesthetics and aerodynamics. Up to now, there is great space for researches on the automobile wind energy application technology. Therefore, based on the impacts of the application of wind energy on the styling design of concept cars, it is likely to make more breakthroughs in the field of car designing.

References

1. Qing Qing Yang. *Domestic and foreign research status of new energy vehicles – with solar energy transportation as an example* [J]. Jiang Su Science and Technology Information. (19)2013.

2. Fan Liu. *Research on car styling design method based on airflow simulation* [D]. Master. Xi'an University of Technology. 2014.
3. Xian Tao Chen. *Task research and application in the field of car styling* [D]. Doctor. Hu nan University. 2009.
4. Rui Mao, Wei Zhi, Zhang Xue, Jinsong Lei. *Application prospect of wind-powered cars and methods discussion* [J]. Traffic Energy Saving and Environmental Protection. (4)2010.
5. Balouktsis, Anastasios, Chassapis, Dimitrios. Karapantsios, Thodoris. *A nomogram method for estimating the energy produced by wind furbine generator* [J]. Solar Energy, 72(3), March, 2002: 251–259.

Wind Energy as an engine for the German economy: An analysis of the contribution and development of Wind Energy to Germany's economy and environment

Bastian Heidecker

UNEP Tongji Institute of Environment for Sustainable Development, Tongji University, Shanghai, Yangpu District, 200092, China
E-mail: bheidecker@web.de
www.tongji.edu.cn

At the latest public reports of the Intergovernmental Panel on Climate Change (IPCC) or the UN Climate Change Conferences, a socio-political discussion arose over possible harmful impacts of fossil energy sources for the global climate and our environment. Also, due to the still unsolved problem of radioactive waste disposal and diverse tragic incidents in nuclear power plants within the last years, many experts criticize the use of nuclear power as a non-sustainable energy source. With regards to the finiteness and environmental harmfulness of fossil energy sources and the danger of nuclear material for present and future generations, the importance and popularity of renewable energies in Germany increases continuously. In particular, Wind Energy influences the success story of renewable energiesover the last years.

In addition to positive environmental, CO^2-saving aspects, Wind Energy can also be supportive and even fundamental in terms of growth and development for Germany`s economy. Economic growth indicators such as a rising revenue development, increasing employment rate or a high export volume are representatives for the economic importance of renewable energies and wind energy in particular.

Keywords: Energy transition; Renewable energy; Wind Energy; Economic Growth; Gemany's economy.

1. Introduction

The industrial development and a major part of the economic achievements of the 20th century has been carried by fossil energies. It can be stated that every stage of development is linked by a positive correlation to a specific amount of energy. In particular, low priced energy was hereby the key to mass production, transport, distribution and wealth in the industrial countries. However, at the latest since public reports of the Intergovernmental Panel on Climate Change (IPCC) or the UN Climate Change Conferences, a socio-political discussion arose over possible harmful impacts offossil energy sources for the global climate and our environment. The following figure shows a significant increase in the global temperature over the last 100 years and thereby an obvious relation to the increasing level of industrialization, which comes along with a higher use of fossil energies such as gas, oil and coal and therefore also a higher CO²-emission.

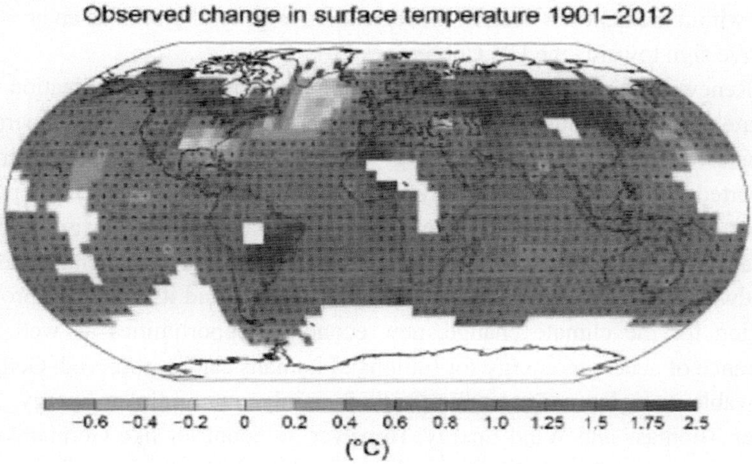

Figure 1. Climate Change illustrations by IPCC 2013[1].

The impacts of climate change such as increasing desertification of broad tracts of land on every continent, the melting of the polar caps and glaciers and heavy typhoons especially in Asia, the Caribbean and the southeastern United States are also economically too important to be ignored. Hans Joachim Schellnhuber, founding director of the Potsdam Institute for Climate Research also sees the IPCC report of 2007 as the final prove of the human caused climate change.[2] Therefore the results of the latest UN Climate Change Conferences from Copenhagen in 2009 until Lima 2014 are generally being criticized and seeing as a disappointment. The result contains only vague climate protection goals, such as limiting global warming to 2 degrees without prescribing the individual states binding targets to reduce greenhouse gases.[3] For many NGOs and environmental concerned individuals, the UN Conference 2015 in Paris is therefore the last hope to finally compile a subsequent agreement of the Kyoto-protocol. With regards to the finiteness of fossil energies, many experts see renewable energies as the only sustainable solution to reduce the CO²-emission globally without a drastic scarcity of energy. Another CO²-friendly energy source could be found in nuclear power. However with relation to the still

---

[1] Intergovernmental Panel on Climate Change (IPCC) P.6

[2] N. Podbregar, K. Schwanke, H. Frater P.29

[3] A. Mihm 2014

unsolved problem of radioactive waste disposal, this energy source shall not be considered as sustainable and future-oriented. France, in the person of François Hollande, the French President, underlines this statement by recently passing a new energy law which claims to reduce the use of nuclear energy from 75% to 50% within the next ten years. This legislation is generally seen as an ambitious, positive sign towards the UN Conference 2015.

Renewable energies could have a major impacts in the stabilization of the climate and the safeguarding of energy demand of future generations. Respective, these alternative energy sources are already been politically supported worldwide but especially in the European Union. Besides the positive impact on the climate, renewable energies can also be supportive and even fundamental in terms of economic growth and development. Therefore worldwide a rising awareness of renewable energies and its ability to provide a solution for the climate change, new economic opportunities as well as the assurance of access to energy for billions of humans can be observed. Generally, renewable energies consist among other mainly out of Solar Energy, Water Power, Biomass and Wind Energy. However, in countries like Germany due to the general weather and climate conditions and the lack of solar radiation, Wind Energy can be seen as the most efficient source. Wind Energy influences the success story of renewable energies in Germany. Today, Wind Energy has a share of about eight percent of the German power supply.[4] The use of wind as an energy source, therefore, plays a key role in the development of renewable energy towards a commercially viable and climate-friendly energy supply at reasonable prices and a high level of prosperity. The following paper shall focus on the impact of Wind Energy as representative for renewable energies on Germany`s economy. After a brief definition of the term "economic growth" the positive economic externalities of Wind Energy will be determined by the use of concrete indicators. The conclusion claims to summarize the acquired insights and give an estimation and evaluation of the use of Wind Energy for the future. Further it shall be mentioned that the term "Wind Energy" in this paper will be seen as a proper name and therefore will be capitalized.

2. Indicators of Economic Growth

The following chapter claims to give an overview of how economic growth is measurable and which aspects can indicate a positive economic effect. Usually national economic growth is measured by the increase in gross domestic product in comparison to the previous economic period. The online business dictionary defines:

---

[4] Federal Ministry of Economics and Energy 2014

"Increase in a country's productive capacity, as measured by comparing *gross national product (GNP) in a year with the GNP in the previous year."*[5]

However, a study of the effect of renewable energies on the growth cannot only be based on a possible change in GNP because by measuring the economic performance of a country, renewable energies are generally not considered in isolation. Renewable technologies (cleantech) have diverse influences on different economic areas. Therefore, in this paper concrete indicators such as revenue development, employment development and export volume shall be adduced to measure the economic importance of Wind Energy in particular. In addition to these criteria, there are also other possibilities to determine growth. An example for a rather indirect approach would be venture capital investments in wind companies and technologies.

Hereby it can be assumed that investors turn increasingly to economic areas which show or expect an economic growth. A special edition of the VC Magazine states that investments in 2010 in cleantech funds achieved EUR 66 M in Germany.[6] In terms of a clear analysis of the relationship between wind energy and growth, and with the background of the scope of this work, however, it seems sensible to limit the analysis to the above-mentioned indicators.

3. Growth Aspects of Wind Energy

In the following, the relevant aspects for individual growth through the field of Wind Energy shall be further investigated. Hereby as mentioned in Section 2, the indicators, revenue development, employment rate and export volume shall be used for a concrete measurement.

3.1. *General introduction of Wind Energy*

Wind Energy facilities can be classified into two categories. Onshore wind parks, which are located on the land and offshore wind parks which are located in the sea. The sector of Wind Energy produced in the year of 2014 an energy volume of 56 Bn. kWh, which is almost 10% of the total energy production in Germany Also the power of new installed facilities (on land) accounted 4.75 GW.[7] Globally, in 2014 Germany with 39.165 MW had the third highest amount of installed wind power after China 114.763 MW and the USA 65.876 MW. These top three installing countries claimed more than half of the total

---

[5] Business Dictionary 2013

[6] Bernd Frank 2010

[7] V.I.S.d.P. Bundesverband WindEnergie e.V. (BWE) 2015 P.38

worldwide installed power of 369.533 MW.[8] Generally it has to be stated that the technology of wind parks improved highly over the last 20 years. While an average single wind turbine in 1993 produced 260 kW, now in 2014 the amount of power multiplied by ten to 2690 kW.[9]

3.2. *Revenue development in the sector of Wind Energy*

In order to evaluate and measure the revenue performance of Wind Energy, it has to be distinguish between two origins. Firstly, the investment volume, which is revealed by the manufacturing of new Wind Energy facilities and the effects of the operation of existing facilities. Also the amount of created tax and social security contribution plays an important role in the determination of the revenue development through Wind Energy. In the year of 2012 the total volume of investment in Wind Energy in Germany amounted EUR 3.6 bn.[10] This can be separated in main investment (1.180.000 €/MW), through the costs of construction, transportation and installation, as well as additional investment costs (374.000 €/MW), such as planning, grid connection, foundations, infrastructure provision and other investment incidental expenses. After the subtraction of 13% import share, the effective domestic total investment still amounts to almost **EUR 3.3 bn.** The operation of existing wind turbines in Germany also raises significant gross value added effects. In the sense of a clear calculation of gross value added, operating costs should be deducted from the operation capita value. In our case however, operation costs and thereby caused indirect economic benefits shall be seen as an additional part of the revenue development. The total operating costs for wind turbines in Germany in 2012 resulted in the amount of **EUR 1.2 bn.**[11] In addition to the operating costs of the wind turbines that trigger economic effects in different areas such as maintenance and insurance, the profits generated by the operation of wind turbines also have economic effects. These profits are secured and subsided by the German renewable energy law (EEG) which assures a concrete amount for every produced kWh of energy. Therefore profits of energy operators in 2012 incurred were about **EUR 4.98 bn.** in revenues through the sale of electricity from wind energy on land.[12] Furthermore, the wind energy industry in 2012 also had significant effects on the public revenue of Germany. Overall, an amount of

---

[8] Volker Quaschning 2015
[9] DIW Econ GmbH 2014 P.3
[10] DIW Econ GmbH 2014 P.3
[11] DIW Econ GmbH 2014 P.5
[12] DIW Econ GmbH 2014 P.6

taxes and social security contributions in the amount of **EUR 3.39 bn.** was stimulated by the production of new wind turbines.[13]

3.3. *Employment development in the sector of Wind Energy*

Overall, the net employment effects of the energy transition in Germany in recent years are positive. Throughout the whole renewable energy sector shows a moderately positive employment development. 18,000 additional jobs per year. This corresponds to the total employment growth in the German chemical industry (+16,800) or mechanical (+16,850) in the period from 2008 to 2013.[14] A study from the German Federal Ministry of Economics and Energy claims that the gross employment rate from renewable energy sources in Germany in 2013 amounted 371400 persons.[15] The Wind Energy industry hereby, is one of the most important parts of the renewable energy sector, with almost 40 percent shows the largest employment effect. This number includes both, the people who work in businesses that provide goods and services directly for renewable energy, as well as people who are influences indirectly by working in the upstream production areas by purchasing supplies. The following table shows the composition and the development of the Wind Energy branch in the years 2013 and 2012.

Table 1. Composition of the employment rate in the sector of Wind Energy in Germany 2012/13.[16]

	Employment; Investment (incl. Export)	Employment; Maintenance & Operation	Total Employment 2013	Total Employment 2012
Wind onshore	100.800	18.200	119.000	104.000
Wind offshore	17.500	1.300	18.800	17.800

The table shows that the German Wind Energy sector in 2013 could record another successful year. This was mainly enabled through the high number of new installed wind turbines as already described in chapter 3.1. Whereas in the field of offshore wind power 595 MW were reinstalled, even 28% more power capacity than in 2012, 2997 MW were installed on land. In comparison, worldwide onshore and offshore together a total of 35.5 GW of wind capacity was newly installed. The largest global annex is carried out again in China. According to initial estimates could be reinstalled here about 16 GW.[17]

---

13 DIW Econ GmbH 2014 P.14

14 AgenturfürErneuerbareEnergien 2015

15 Federal Ministry of Economics and Energy 2014 P.7

16 Federal Ministry of Economics and Energy 2014 P.7

17 Federal Ministry of Economics and Energy 2014 P.9

Especially the sector of offshore Wind Energy is expected to provide huge opportunities for further employment development. The federal government stated that by 2020 a total of 10 GW of wind power should be built only in German waters. By 2030 even 25 GW of installed wind power capacity should be achieved in the North and Baltic Sea.[18] This can be seen as a huge potential for further employment possibilities, in the sector of academic employment (engineers, architects, urban planners, etc.) as well as for manual labor based employment (construction worker, etc.).

3.4. *Export volume in the sector of Wind Energy*

Generally, for the renewable energy sector in Germany, exports remains an extremely important pillar — despite and because of a partially economically difficult situation in the country and strong and cheap competition from China in particular in the sector of solar technology. The renewable industry and especially the German Wind Energy sector is able to draw on a wide breadth of technologies and expertise that it has continuously developed and built. The factsheet of the German Engineering Federation states that the wind industry generated an estimated turnover of about EUR 33 bn. globally in 2013.[19] This raises immense opportunities for the German export market. Further the factsheet describes:

> *"With about EUR 4.25 bn. export revenue in 2012, the wind turbine manufacturers had an export ratio of 67 % of the production volume in Germany."*[20]

This shows that the export rate stayed more or less constant over the last years despite strong competition from abroad and worldwide economic crises. Hereby the main export markets can be found in the European neighbor countries in the UK (1883 MW), Poland (894 MW), Sweden (724 MW), Romania (695 MW) and Turkey (646 MW).[21] Due to the high competition and also the big distance, which brings along high transportation costs, the markets in North America and Asia have faced some difficulties. According to the data of the international Economy Forum of Renewable Energies, more than 50 GW of wind energy capacity was reinstalled in 2014, around the world. Even though, globally the Danish wind turbine producer Vestas is leading the world market

---

[18] Bundesverband Windenergie
[19] VDMA German Engineering Federation 2014 P.1
[20] VDMA German Engineering Federation 2014 P.1
[21] VDMA German Engineering Federation 2014 P.2

with 6 GW of installed Wind Energy, also two German companies are within the Top 5.[22] Between Siemens (4,9 GW) and Enercon (3,7 GW), the Chinese manufacturer Goldwind (4,7 GW) and the US manufacturer GE (4,5 GW) occupied the 3rd and the 4th position.

4. Conclusion

Undoubtedly, as listed in Section 3 of the present paper, the Wind Energy sector in Germany shows enormous growth rates in the development of revenue, in terms of employment and also within the country's export volume. From today's perspective this level can be maintained or even more realistic be further increased in the future. The German government in the person of German Federal Minister of Economy Sigmar Gabriel agrees on this statement. In his opinion, Wind Energy does not only represent the answer to the central challenge of climate change, but also became a fascinating economic success story. Wind Energy can be seen as a "Figurehead of German engineering", as the Energy and Management Journal quoted Gabriel in 2014.[23]

Renewable energies in comprehension, as well as Wind Energy in particular can definitely be seen as one of Germany´s economic and energetic engines. However this phenomena does not only count for the German market. Diverse studies, reports and surveys during the last years come to similar results. According to the recently released Global Status Report 2015 by REN21, globally, the year 2014 has exceeded all renewable energy records with a total power capacity of 1712 GW.[24] Meanwhile a huge majority of 164 countries have policy targets to substitute and assure the immense potential of renewable energies.[25] Worldwide a rising awareness of renewable energies and its ability to provide a solution for the climate change, new economic opportunities as well as the assurance of access to energy for billions of humans still living with energy scarcities, can be observed. The REN21 report refers to two other developments: Although the global energy consumption in recent years has risen by an average of 1.5 percent and the global gross domestic product had grown by an average of 3 percent, the CO^2-emissions of 2014 remained stable compared to the previous year. That means that for the first time in four decades, the world economy grew without a parallel rising level of CO^2-emissions. The authors of the report conclude that global economic growth is increasingly independent from CO^2-emissions. This development would result mainly from two important factors.

[22] Internationales Wirtschaftsforum Regenerative Energien (IWR) 2015
[23] Angelika Nikionok-Ehrlich 2014
[24] Renewable Policy Network for the 21st Century (REN21) P.19
[25] Renewable Policy Network for the 21st Century (REN21) P.18

First of all China, as one of the leading economic forces uses more renewable energy sources and further the OECD countries committed themselves more for renewable energy, energy efficiency and in the field of sustainable economic growth. This underlines that in addition to the economic success of renewable energy, especially the impact to the global climate shall be under the focus of the public in this decade.

As mentioned in the introducing part of this paper, the coming UN Climate Conference in 2015 in Paris must therefore finally lead to a path breaking result to pass concrete laws and policies in order to reduce the CO^2-emission globally. With regards to the finiteness of fossil energy sources, their harmfulness to our environment and the missing solution for the highly radioactive waste material of nuclear plants a transition to a clean energy system with such as Solar Energy, Water Power, Biomass and of course Wind Energy can be seen as the only real future option. In a final conclusion, it has to be stated that renewable energy and Wind Energy as the energetic leader can provide both, significant ecological as well as undeniable economic benefits for Germany and also for the rest of the world.

References

1. Agentur für Erneuerbare Energien (2014): *Windkraft-Journal. Erneuerbare Energien werden zum deutschen Exportschlager und sichern Arbeitsplätze* 08.15.2014.
2. Agentur für Erneuerbare Energien (2015): *Bundesverband Windenergie. Energie wende sorgt für mehr Beschäftigung*. Berlin, 04.14.2015.
3. A.Matts (2014): DIW Econ GmbH, *Die ökonomische Bedeutung der Windenergiebranche* – Modul 1: Windenergie an Land in Deutschland, Berlin, May 2014.
4. A.Mihm (2014): Frankfurter Allgemeine, *Sorgen und Appelle Was Lima dem Klima bringt.* 12.15.2014, http://www.faz.net/aktuell/wirtschaft/wirtschaftspolitik/beschluesse-der-klimakonferenz-inhttp://www.faz.net/aktuell/wirtschaft/wirtschaftspolitik/beschluesse-der-klimakonferenz-in-lima-auf-einem-blick-13320998.htmllima-auf-einem-blick-13320998.html [Accessed 29/04/2015].
5. A.Nikionok-Ehrlich (2014): Energie& Management Online, *Windkraft als Wirtschaftsmotor und weltweiter Impulsgeber*. 23.09.2014, http://www.energie-und-

management.de/?id=84&no_cache=1&terminID=106389 [Accessed 29/04/2015].

6. *Frank* (2010): VC Magazin. Sonderbeilage, "Investing in Cleantech" *Vorsichtiger Aufschwung. Cleantech Investments sind bei Investoren gefragt, Kapital ist trotzdem knapp.* http://ecapital.de/uploads/tx_news/vc_magazin_cleantech_1.pdf [Accessed 29/04/2015].
7. Bundesverband Windenergie (BWE): *Offshore.* https://www.windhttps://www.wind-energie.de/themen/offshoreenergie.de/themen/offshore [Accessed 29/04/2015].
8. Business dictonary (2013): *Economic Growth* http://www.businessdictionary.com/definition/economichttp://www.businessdictionary.com/definition/economic-growth.htmlgrowth.html [Accessed 29/04/2015].
9. Internationales Wirtschaftsforum Regenerative Energien (2015): *Windenergie-Weltmarkt 2014: Vestas Nummer eins - Siemens verdrängt Enercon.*03.10.2014.
10. IPCC (2013): Summary for Policymakers. *In: Climate Change 2013: The Physical Science Basis*. Cambridge University Press, Cambridge, United Kingdom and New York, NY, USA.
11. J. Schiel (2014): VDMA German Engineering Federation*, Facts and Figures on Wind Energy.* September 2014, Berlin.
12. M. O'Sullivan, D. Edler, P. Bickela. (2014): Bundesministeriums für Wirtschaft und Energie, Bruttobeschäftigung durch erneuerbare Energien in Deutschland im Jahr 2013 -eine erste Abschätzung. May 2014.
13. N. Podbregar, K. Schwanke, H. Frater (2009): *Wetter, Klima, Klimawandel: Wissen für eine Welt im Umbruch,* Springer Berlin Heidelberg 2009.
14. Renewable Policy Network for the 21st Century (2015): Renewables 2015 Global Status Report, *Key Findings 2015*, Paris 2015.
15. Statista (2015): *Exportquote in der Windindustrie in Deutschland in den Jahren 2009 bis 2012.* http://de.statista.com/statistik/daten/studie/245552/umfrage/exportquote-inhttp://de.statista.com/statistik/daten/studie/245552/umfrage/exportquote-in-der-deutschen-windindustrie/der-deutschen-windindustrie/ [Accessed 29/04/2015].

16. V.i.S.d.P. Bundesverband WindEnergie e.V. (BWE) (2015): *Wind bewegt. Argumente für die Windenergie.* 6. Auflage. April 2015. https://www.wind-energie.de/sites/default/files/download/publication/wind-bewegt-argumente-fuer-die-windenergie/20150407_bwe_argumentarium_web.pdfwindenergie/20150407_bwe_argumentarium_web.pdf [Accessed 29/04/2015].
17. V. Quaschning (2015): *Weltweite installierte Windkraftleistung.* February 2015.http://www.volker-quaschning.de/datserv/windinst/index.php [Accessed 29/04/2015].

The biomass power generation electricity price calculation model and analysis in Jilin province

M.L. Zhao* and K. Li
Jilin Province Electric Research Institute Co.,LTD
Changchun, Jilin 130021, China
**E-mail: 873443433@qq.com*

Y.A. Jin
Thermal Energy Engineering Department, Jilin University
Changchun, Jilin 130025, China
Email: 378275516@qq.com

As a promising industry, the biomass power generation has a good prospect for development. This paper analyzed the current status of domestic and foreign biomass power generation, and makes an economic analysis and generation costs structure model of biomass power generation of Jilin province. The existing problems of the current biomass power generation were investigated in detail, and the development strategy of biomass power generation of Jilin province was proposed.

Keyword: Biomass; Electricity generation; Economic analysis.

1. Introduction

1.1. *General introduction*

Biomass power generation is the concrete reflection of the sustainable development path and circular economy, and biomass resources is quite rich in China, only the resource extent of agriculture and forestry waste is equal to 350 Mt coal. However, the waste straw could not be conducted effectually, and it burned in the fields would lead to atmospheric pollution and fire safety problems. The problem is long-existed and could not be settled effectually. To develop the biomass power generation projects is the efficiency measure to deal with the agriculture and forestry waste and improve environment.

1.2. *Biomass power generation development situation*

Biomass energy is concerned by national governments from the 1980s, and European countries are the first to research the biomass energy. Small-scale cogeneration used biomass has been the primary way of power generation and heating in European. America is in the leading position of biomass power generation, and its biomass power generating capacity reached 15.8 GW in 2013. China has plenty of biomass resources, and the installed capacity of biomass power generation will reach to 65 billion in 2015.

Biomass resources are extraordinarily abundant in Jilin province, China. About 94 million tons biomass energy can be collected in Jilin straw, it includes rice husk, forest wastes and household wastes. Among those, the corn straw is the most, and its amounts are up to 30 million tons. Meanwhile, 70% of those are produced in Changchun, Siping and Songyuan, and its stock density reaches to 10.87 t/h in those areas.

2. The Energy-Analysis of Biomass Power Generation

2.1. *The computing model and costs structure of biomass power generation based on Matlab/Simulink*

The costs structure of biomass power generation (COE) based on Matlab/Simulink are comprised of three parts, namely of amortization costs of total investment (COD), biomass costs (COF) and operation and maintenance costs (COM). These subsystems are built based on Matlab/Simulink in this paper.

2.1.1. *The model of COD*

In China, the COD is usually computed with depreciation by composite life method. In this paper, the formula 1 is used as the mathematical model to describe the COD.

$$\mathrm{COD} = \frac{\mathrm{TIS}}{\mathrm{P} \times \tau \times (1-\delta) \times \mathrm{a}} (\mathrm{yuan} / \mathrm{MW} \cdot \mathrm{h}). \tag{1}$$

According to the above formula, the COD mathematical model is established based on Simulink. The package model is shown in Figure 1. The input parameters are shown in Figure 1. In this figure, TIS is the total investment outlay of power station, and its unit is Yuan; The P is the net power of power station, and its unit is MW; The is the operating hours of electric generating set, and its unit is h; The is the loss of power plant grid, and its value is about 3%-7%; The a is the service life of power plant, and its value is about 20-30 years.

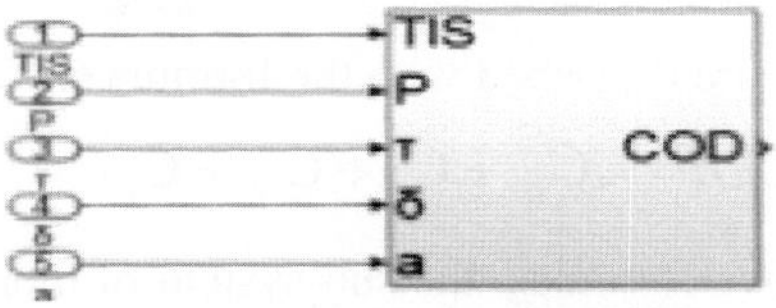

Figure 1. The system of COD.

It is shown in Figure 1 that if the TIS and decrease and the and a increase, amortization costs of total investment would decrease.

2.1.2. *The model of COF*

Because of the non-uniform straw distribution, the costs of collecting straws are hard to be computed. Therefore, it is supposed that, with the biomass power plant as the center, biomass is distributed uniformly around power station. The COF is comprised of three parts, namely of purchase costs, transportation costs and pre-processing and warehouse costs.

(1) The purchase costs of straws are that the peasant could acquire by selling straws. The straw accounts Q is defined by the formula 2.

$$Q = \pi \cdot R^2 \cdot \rho_i . \tag{2}$$

R- the radius of collecting, - the distribution density of straw

Because different straws have different heat values, it leads that the price of straws is different. Therefore, in this paper, the total costs of collecting are described by the formula 3.

$$C_g = \sum_{i=1}^{n} C_{g,i} = \sum_{i=1}^{n} Q \cdot u_i \tag{3}$$

In the formula, is the species of straws; is the unit price of straws, and its unit is Yuan/t.

(2) The transportation costs of straws.

The straw is usually transported by highway. The transportation costs are comprised of transport charges and terminal charges.

$$C_{tli} = \sum_{i=1}^{n} \iint \rho_i \gamma \beta c_n \gamma d_\gamma d_\theta . \tag{4}$$

The straws have disassembly process from collection station to power plant. The terminal charges are described by formula 5.

$$C_{t2} = \sum_{i=1}^{n} 4P_j \cdot \pi \cdot (R^2 - r^2) \cdot \rho_i. \tag{5}$$

The costs of biomass are described with the formula 6

$$COF = C_g + C_t + C_w + C_l. \tag{6}$$

According to the above formulas, the sub-system of biomass costs is found by using Simulink, and it is shown in Figure 2. From the Figure 2, it is summarized that the value of COF is influenced greatly by Pi, and R.

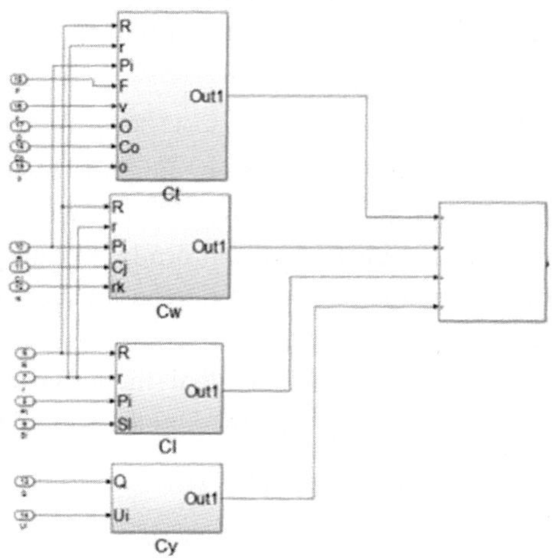

Figure 2. The model of the biomass costs.

2.1.3. *Operation and maintenance costs (COM)*

The COM is usually computed with depreciation by composite life method, and it is described with formula 7.

$$COM = (11.47\% - 15.36\%)\, COE \tag{7}$$

2.2. *The model of biomass power generation costs*

The power generation cost of power station (COE) is made up of three parts, which are total investment cost (COD), fuel cost (COF) and run and maintenance costs (COM).

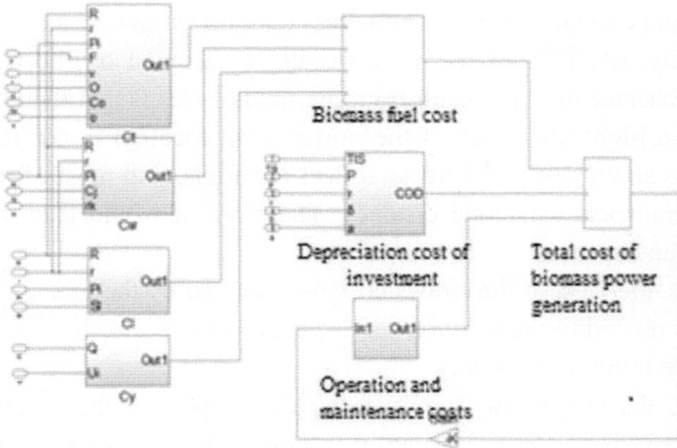

Figure 3. General layout models of biomass power generation costs.

In this paper, a biomass power plant whose generating capacity is 25 MW is used as the example to verify the model. According to the above calculation model, some boundary conditions are offered to calculate the electricity costs.

(1) the net power of this power plant is 30 MW. (2) the total investment of this power plant is 370.68 million Yuan, including of own funds 183.80 million Yuan and self-financing 200 million. (3) the depreciable life is 20 years, and generating sets operate for 7248 h at full power. The house-service consumption rate of this power plant is 11.05%. (4) the purchase price of straws is 300.49 Yuan/t, and the heat value of biomass straws is 2741.57kcal/kg. The biomass wastage because of storage is 15%. (5) the heat efficiency of biomass is 34%. (6) the operation and maintenance costs of this power plant are 43.33 million Yuan and its depreciation expense is 16 million. (7) the lending rate is 6.15%, and the repayment period is 9 years.

The purchase price of per straw unit is the biomass costs, so COF is needed to adapt to the purchase price of per straw unit × quantity of straws. According to the model, it is calculated that the generation costs of the biomass power plant is 0.79 yuan/(KW•h), and the costs is higher than the Feed-in Tariff of biomass power plant. Therefore, the biomass power plant has slid into loss, and it is in line with the real situation.

2.3. *To reduce the generation costs of biomass*

According to the mathematical measurement model, the 38% of the straws costs is occupied by the employee's profit. However, this part cannot be reduced. To this situation, the way to reduce the straws costs is to build new straws collection mode to take place of employees bundling straws mode.

Secondly, the 15% of the straws costs is occupied by the transportation expenses. Because the straws are from the fields, it leads that the transportation expenses are high. Moreover, if the power plant location is not reasonable, it would bring about that it is hard to collect materials, and most of materials are needed to transport with long distance. Therefore, it would lead to the higher transportation expenses.

Thirdly, the 12% of the straws costs is occupied by the process of bundling straws. It is related to the location service costs, so it is also hard to be reduced, and there are trends of growth.

Finally, the 9% of the straws costs is occupied by the bundling twines. Because straws have large sizes and it is easy to dilate, it needs to get a higher bundling quality. Therefore, the bundling fees will increase with the higher bundling quality. The location of storage straws is in fields, so the costs are low, and it is also hard to be reduced.

Through above analysis, the straws costs have certain compressibility. These aspects such as employees profit, the transportation expenses, and the mechanical loss fees have a higher compressibility. However, the bundling costs, bundling twines and rent costs of storage straws have low compressibility.

3. The Problems of Biomass Power Generation

Because of the zero emission of biomass power plant, it obtains the support of government, and it develops rapidly. However, there are also a lot of problems which lead to the higher generation costs.

3.1. *The lack of the mature storage system, transportation system and collection system*

Power generation in biomass power plant needs a large amount of straws, but straws distributes around the field land, large collection machinery can't run around fields to collect straws, so artificially collection is required. On the other hand, scattered straws need molding and packaging, collecting costs increase in return. In one word, it is difficult to collect straws to power plant.

Biomass straws harvest seasonally, restricting with power plant running throughout the year, so straws need to be stored. Long time store will decline heat value, and biomass straws are flammable, strict fire prevention measures are required, fuel cost increases. In the other hand, farmers can only get a little profit from straws, they have no enthusiasm to selling them and prefer to burn them, causing environmental pollution.

3.2. *The poor machine operating reliability in biomass power plant*

It is hard to collect biomass energy. For guaranteeing generation, power plant has to find alternative fuels, so it leads to more kinds of biomass materials such as rice husks, peanut shells, eucalypt wastes and corn cobs, and the species is up to 20-30. However, the biomass-burned boiler is designed for about 2 kinds of biomass, and it will decrease the boiler efficiency if too many kinds of biomass materials are used in power plant. Moreover it will decrease the boiler stability.

3.3. *Unreasonable location of biomass power plant*

Biomass power plant collects most straws nearside the plant. For normal operation, fuel collecting region should be large. If plants distribute dense, collecting regions interact with each other, straw supplying is difficult to fulfill, causing straws price increased, production cost increases in a large degree.

4. The Development Countermeasure of Biomass Power Generation

4.1. *Improve biomass collection and storage system*

Most of biomass power plants are based on the model of "farmers to agents to plants". The plants with this model negotiate with the agents directly, which is more convenient than recycling straws from farmers. But this model has great disadvantages in prices, on the one hand, the prices are controlled by agents which causes the higher costs are used in purchase material and lower benefit are gained by famers, on the other hand, the straws are added with water causes them decay in the process of poling, the straws are added with sand leads to shorten the lifetime of furnace. It is essential that looking for a new collection model, getting rid of agents in the link, the power generation enterprises connect with the village committees or large farmers nearby directly. The directional purchase contracts are signed, which is claim that the collection spots are set by biomass power plants joint committees, the operation is in charge of committees. The costs of power plants fuel would be decreased in this kind of model, at the same time the more benefits will be gained by the famers. In addition the country level biomass collection system should be established and improved quickly; the original model that collected by the agents is transformed to the model of scale collection.

In the research and analysis, the most of costs are used in the biomass transportation; long distance and complex road condition are the important influence in fuel costs. So the detailed transport schemes are made to arrange the vehicle operation reasonable, the transportation costs are reduced by scientific planning.

4.2. *Improve the scientific innovation ability*

In order to promote the healthy development of biomass power generation industry, the biomass power generation technology should be listed in the long-and-medium-term plans in technology development, the government plays a leading role in the establishment of comprehensive utilization of biomass industrial park, the biomass materials are made use of comprehensive recycle apply, the maximum degree benefit output is formed in the process. The research in biomass power generation is supported by college, which is developing well. Besides which is based on the foreign technology import, the capacity for independent innovation is improved quickly, the biomass power generation with Chinese characteristic is formed well.

4.3. *Intensify policy support*

As emerging green industries, biomass power generation is obtained proper support by the government, including improving the strength of subsides; establishing green electricity development fund in Jilin province, these policies will help plants well-developed. At the same time, the research and development of biomass power generation technology is supported by the society, the potential biomass power generation is helped to accelerate the development of biomass power generation, Jilin province will be the center of technical and the base of biomass equipment manufacturing. On the other hand, biomass resources investigation and project evaluation is done by the government, the society slack money is absorbed to the investment and construction of biomass power generation industry. The goal is to make the green industry of biomass power generation bigger and stronger.

References

1. Zhang Tiezhu. Biomass power Generation industry status quo and prospect of China [J] Rural Electrification, 20111 (8) : 56-58
2. Na Wei, Zhu Yan-Li, Liu Peng, Zhang Yong-feng. Evaluation on regional suitability of rural household biogas in the Jilin province [J]. Renewable Energy Resources, 2010 (5) : 138-142

3. Wang Zhenjiang, Zhou Changxian. Several key questions of straw direct-fired power generation [J]. Sichuan Electric Power Technology, 2006 (29) : 82-85
4. Zhuang Huiyong, Zhang Yanru. Agriculture and forestry biomass direct combustion power Industry development [J]. China Science and Technology Achievements, 2009 (13) : 17-18
5. Ke Yun. Development Strategy of China Bio Energy in 21st Century [J]. Electric Power, 2000 (9) 82-84
6. Liang Yuhong, Shen Zhongtao, Yu Ying, Zhao Biguang, Zhu Chun. Biomass Power and Benefit Analysis [C], China Electric Power Research Institute, 2007

Effect of wind shear on the loads of Wind turbine rotor*

C.X. Yang[a], X.Y. Zhang[b] and Z.T. Gao
School of Energy and Power Engineering, Lan Zhou University of Technology
Lanzhou, 730050, China
E-mail: [a]ycxwind@163.com, [b]zxy0932@163.com

The full 3-dimensional unsteady numerical computation method was used to investigate the loads of a 33 kW two-blade horizontal axis wind turbine rotor which operated in the shear and uniform inflow. The loads of the wind turbine rotor and its blades in the shear inflow were studied after confirming the validity of numerical computation method. It is found that yaw moment and tilt moment on the wind turbine rotor and its blades in periodic changes are generated in the shear inflow compared with the uniform inflow, and the amplitude of rotor torque and axial force fluctuation increase 6.87 times and 15.49 times respectively. Moreover, a phase shifting of the torque and axial force as functions of azimuth angle was found in the shear inflow.

Keywords: Wind shear; Wind turbine; Loads.

1. Introduction

In the atmospheric boundary layer, the average wind speed changes with local height, and its change law is called the wind shear [1]. With the increasing scale of wind energy development, the horizontal axis wind turbine is gradually developing towards large scale and the rotor diameter is becoming more and more large. Therefore, the non-uniformity distribution of wind speed in the wind turbine rotor swept plane caused by the wind shear is more obvious. The existence of wind shear can affect the service life and operation safety of wind turbine, so it is necessary to investigate the wind turbine aerodynamic loads in shear inflow.

Wind turbine aerodynamic characteristics and loads were studied by most previous researchers in uniform inflow [2-3]. Liu Lei etc. studied wind turbine aerodynamic loads in shear inflow, but with only a single blade as the research object [4]. Theoretical calculation method was used by Zhang Yu Liang etc. to predict the influence of wind shear of designing the wind turbine and it was suggested that the influence of wind shear can be neglected when the wind

---

* This work is supported by National Natural Science Foundation of China (No.11262011) and National key Basic Research Program of China (No.2014CB046201).

turbine is designed [5]. The investigations suggested that unfavorable aerodynamic loads and blade deflections occur as a result of the sheared inflow conditions. Meanwhile, the wake has a great influence on the aerodynamic and structural behaviors of the blade, especially at low wind speeds [6]. Lifting surface method with time marching free wake model was used to predict the periodic unsteady characteristics in shear inflow [7]. In this paper, the loads of the wind turbine rotor and its blades are studied in shear inflow by comparing in uniform inflow after confirming the validity of numerical computation method.

2. Wind Model

At present, the logarithmic law and power law are used to describe the change of the near-surface wind speed with height. The power law equation is recommended in IEC61400-1 standard which provides a formula to describe the change of wind speed with height. As shown in Eq. (1)

$$v_2 = v_1 \left(\frac{z_2}{z_1} \right)^{\alpha} \tag{1}$$

where z_2 is the height above the ground, α is wind shear coefficient, v_1 is the velocity at the hub height z_1. Most researchers also tend to use the power law equation when performing CFD calculations. First of all, the power law is relatively simple. Secondly, a survey was conducted by Zhou Liang Mao etc. proved that the change of wind speed with height is calculated by the power law is more close to the measured values [1]. Ray etc. proved that using the logarithmic law of describing the change of wind speed with height has limitations [8].

Therefore, in this paper, the distribution of the wind speed with height use the power law in the shear inflow, as shown in Eq. (1). Wind shear coefficient α is set to 0.5 and hub height z_1 is set to 15.4 m with the wind speed of 11 m/s. In uniform inflow, wind speed is set to 11 m/s.

3. Numerical Computation

3.1. *The geometry model and mesh distribution*

In the present work, the rotor of a 33 kW horizontal axis wind turbine was simulated. The NACA4415-4424 airfoils are used in the rotor, and the model is two-bladed, zero cone angle and zero tilt angle. The rotor diameter is 14.8 m with rated speed of 85 rpm. Computational domain is composed of two cylindrical subdomains and the whole calculation domains are shown in Figure 1.

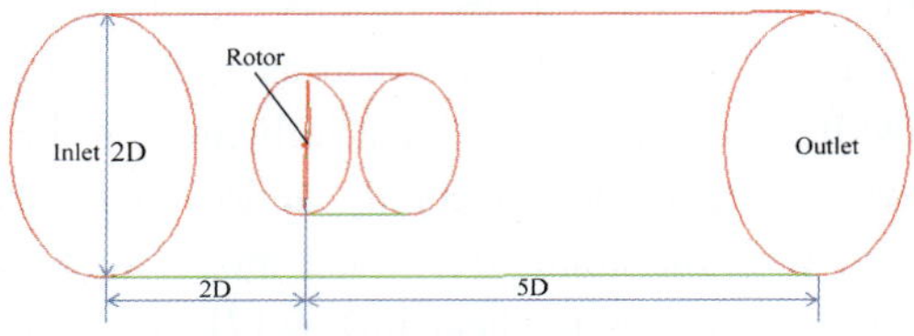

Figure 1. Computational domain size.

The hexahedral mesh was generated in this study. After verifying the mesh independence to calculation results, 110 nodes were distributed in the chordwise direction of the blade, while 210 nodes were used in the spanwise direction. The total number of nodes for all domains is about 42 million. Part of the mesh configurations of the computational domain are shown in Figure 2.

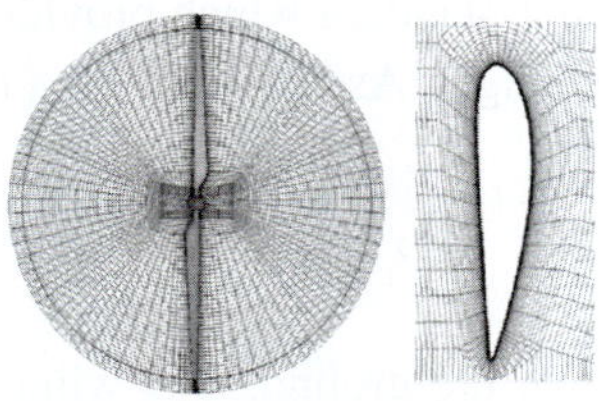

Figure 2. Mesh configurations of the computational domain.

3.2. *Numerical method and the boundary conditions*

The 3D Reynolds time-averaged Navier–Stokes equations were used to simulate the loads of the wind turbine. The K-ω SST model is chosen to turbulence model. In the shear inflow, the inlet of computational domain is set to velocity inlet boundary by a user-defined function (UDF). Both of the two inflow condition, the turbulence intensity is set to 10%. The outlet boundary condition is pressure outlet and static pressure of 0 Pa. Non-slip wall boundary condition is specified at rotor surface and rotor speed is 85 rpm. The interface between stationary domain and rotary domain is set to interface boundary condition to transfer the data. The time step of unsteady calculation is 0.00392157 s.

4. Confirmation of the Numerical Method

In order to validate the present methods, the wind turbine rotor was simulated in the uniform inflow velocity of 11.47 m/s with the rotor speed of 55.84 rpm. The compared results between the experimental pressure distributions and the numerical pressure distributions at the main load locations are shown in Figure 3. It is suggested that there is good agreement between the experimental results and the numerical results at pressure surface, but there is more deviation between the experimental results and the numerical results at suction surface, especially in the regions near trailing edge and leading edge. By analyzing the effects of mesh density on the numerical results, the results indicate that mesh density almost has no effect on the result of the calculation, so it is considered that the mesh distribution has been satisfied the requirements of the independence. The results also indicated that K-ω SST turbulence model mispredicts the position of separation point [9]. Therefore, the deviation of pressure distribution at the suction surface of the blade may be caused by unsteady flow and flow separation. In summary, the numerical method in this paper can simulate the flow characteristics of the wind turbine and the numerical results can be used as the basis for analysis.

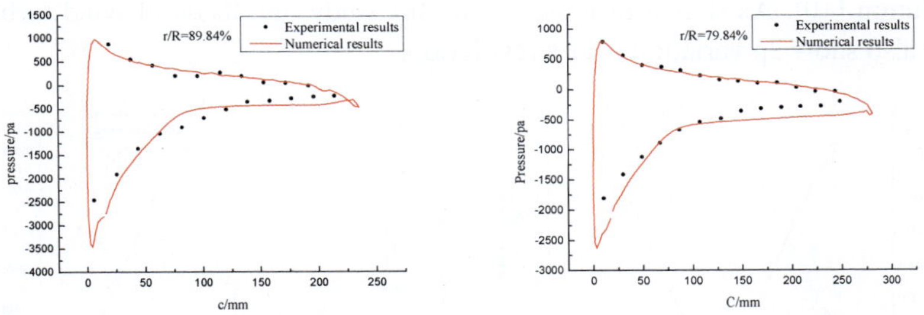

Figure 3. Static pressure distribution of different spanwise position by numerical results and experimental results.

5. Results

In this study, only wind turbine rotor was investigated, and all moment centers are defined in the center of the rotor for simply. We define 0 degree azimuth angle is at the position above the hub. Wind turbine rotor rotates in a counterclockwise direction. The schematic of rotor azimuth angle and loads are shown in Figure 4.

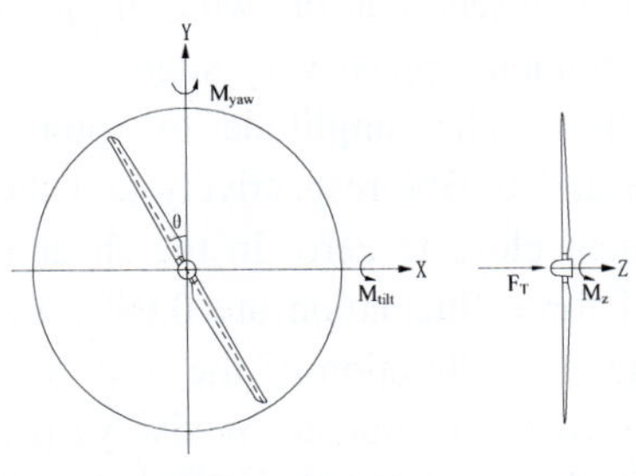

Figure 4. Schematic of rotor azimuth angle and loads.

5.1. *Blade torque, axial force, yaw moment and tilt moment*

Figure 5 shows the variation tendency of the torque, axial force, yaw moment and tilt moment of blade varies with the azimuth angle in the shear inflow. The results indicated that since the wind speed presents a non-uniformity distribution in rotor plane in shear inflow, the loads of the blade show 1p variation over a revolution. The magnitude of yaw moment and tilt moment of the blade experience positive and negative alternating changes over a revolution. That is to say, the direction of the yaw and tilt moment of the blade change with the azimuth angle. For two-blade wind turbine, each blade experience maximum wind speed and minimum wind speed during the wind turbine rotor rotates a revolution, so the loads of wind turbine rotor show 2p variation over a revolution [10]. As shown in Figure 6, in this study, the loads of wind turbine rotor also show 2p variation over a revolution.

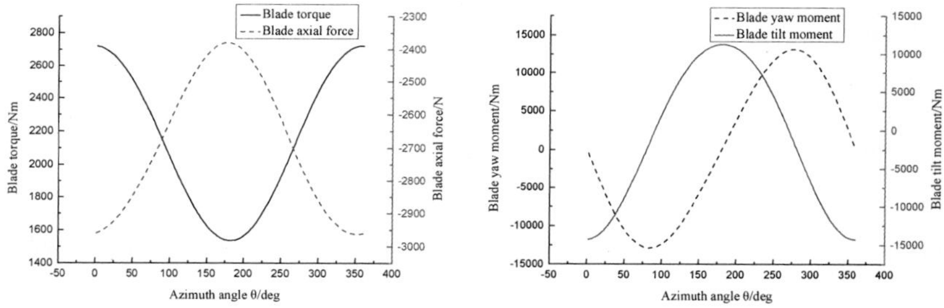

Figure 5. The azimuth variation of the blade torque, axial force, yaw moment and tilt moment in the shear inflow.

5.2. *Rotor torque, axial force, yaw moment and tilt moment*

Figure 6 shows the variation tendency of the torque, axial force, yaw moment and tilt moment of the rotor varies with the azimuth angle in the two kinds of inflow conditions. It can be seen that the wind turbine rotor torque, axial force, yaw moment and tilt moment appear very slight fluctuations with the azimuth angle in uniform inflow. The amplitude of rotor torque and axial force fluctuation are 0.08% and 0.03% respectively and the magnitude of the yaw moment and tilt moment close to zero. In the shear inflow, the amplitude of rotor torque and axial force fluctuation are 0.64% and 0.5% respectively. As well as the single blade, the direction of the rotor yaw moment changes with azimuth angle, but the rotor tilt moment is always negative because the wind speed above the hub is always higher than that below the hub. From the Figure 6, it can be seen that the shear inflow make the average of rotor torque reduced

1.6% and the average of rotor axial force reduced 1% compared with the uniform inflow.

In the shear inflow condition, it is also suggested that the rotor torque and axial force obtain the maximum value when the two blades close to the horizontal position, and obtain the minimum value when the two blades near the vertical position. While the tilt moment is just the opposite, it obtains the minimum value when the two blades is close to horizontal position, and obtains the maximum value when the two blades near the vertical position. When the two blades rotate to the horizontal or vertical position, the magnitude of the yaw moment close to zero.

By analyzing the specific calculation results, we can know that because of the induction of rotating rotor, a phase shifting between the maximum and minimum loads of rotor and the azimuth angle is discovered. The lag angle of the maximum value of the torque is 18 degrees and 24 degrees, and the lag angle of the minimum value of the torque is 20 degrees and 26 degrees. The lag angle of the maximum value of the axial force is 6 degrees, and the lag angle of the minimum value of the axial force is 8 degrees. The tilt moment and yaw moment have no obvious phase shifting.

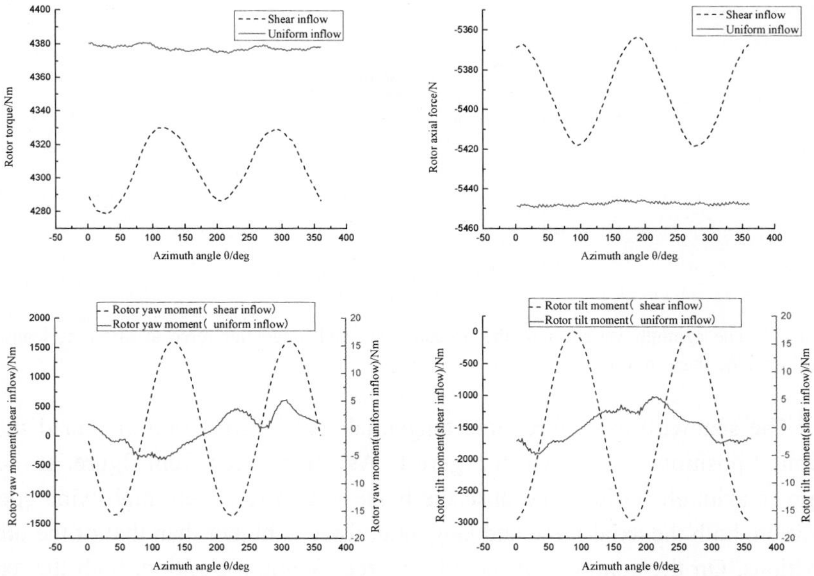

Figure 6. The azimuth variation of the rotor torque, axial force, yaw moment and tilt moment in the shear and uniform inflow.

5.3. *The axial force and tangential force*

In Figure 7, the axial force and tangential force at three span locations of blade are shown as function of azimuth angle. As well as the loads of blade and rotor, the results show that the axial force and tangential force at three span locations of blade also change in a sinusoidal way. At r/R = 19.84% span location, there is a pronounced phase lag between the axial force and tangential force fluctuation curves and azimuth angle, and the angle of phase lag is about 8 degrees and 20 degrees. The more close to the blade tip, the hysteresis effect is diminished as a result of the reducing induction effects of rotating rotor. At r/R = 96.53% span location, almost no obvious phase lag is observed.

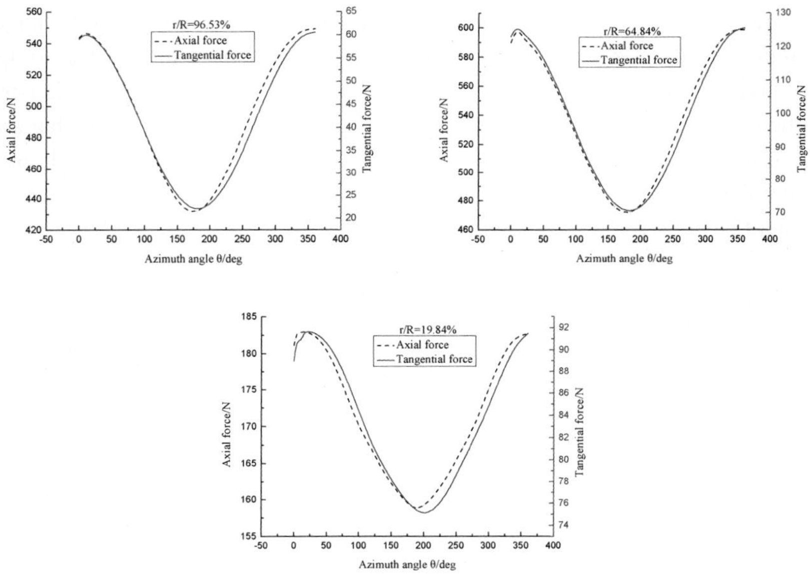

Figure 7. The azimuth variation of the normal force and tangential force at different spanwise position in the shear inflow.

The spanwise axial force and tangential force distribution at four distinct azimuth positions are shown in Figure 8. As can be seen from figure, at the 0 degrees azimuth position, because the blade is at a relatively high wind speed position, both the axial force and tangential force is higher than that of the other positions. On the contrary, at the 180 degrees azimuth position, both the axial force and tangential force is low as expected. At the 90 degrees and 270 degrees azimuth position, because the blade is almost at the same wind speed position, except the position of near the blade root (about 19.84% <r/R <49.77%), the distribution of the axial force and tangential force is basically agreement. The difference of the force at 90 degrees and 270 degrees azimuth position indicated that loads have hysteresis effect in shear inflow. However, the distribution of force along the spanwise direction at 270 degrees azimuth position is higher than that of the 90 degrees azimuth position, which is not consistent with the results of literature [11]. This may be caused by the structure and pitch angle of blade itself. The pitching axis of this blade is at the 25% chord position from the leading edge. At the 90 degrees azimuth position, the leading edge of the blade is downward and the most of the blade is at relatively high wind speed position. At the 270 degrees azimuth position, the leading edge of the blade upward and the most of the blade is at relatively low wind speed position. Near the location of the blade root, due to the blade chord becomes larger, this distribution is more obvious. Therefore, the distribution of axial force and tangential force are different at 90 degrees and 270 degrees azimuth position, and near the location of the blade root, the difference is more pronounced.

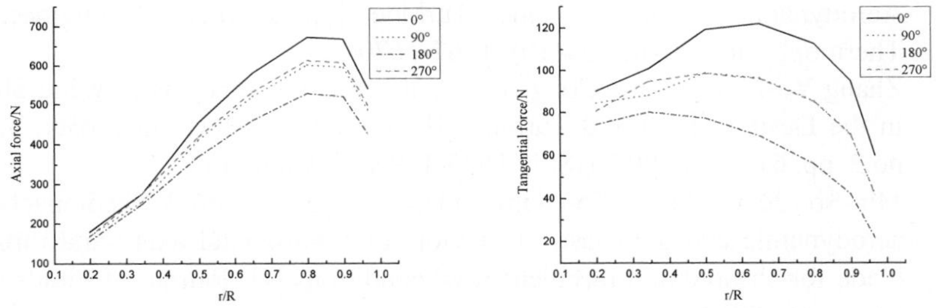

Figure 8. The azimuth variation of the normal force and tangential force along the blade span for four azimuth angles in the shear inflow.

6. Conclusions

In the shear inflow, the torque, axial force, yaw moment and tilt moment of the rotor experience periodic changes over a revolution. Compared with the uniform inflow, the amplitude of rotor torque fluctuation is increased from 0.08% to 0.64% and the amplitude of rotor axial force fluctuation is increased from 0.03% to 0.5%. A phase shifting phenomenon between the rotor torque and axial force fluctuation curves and the azimuth angle is found, and the lag angle of the maximum and the minimum load is different. The tilt moment and yaw moment have no obvious phase shifting.

When the wind speed of the hub height in the shear inflow is regard as the average speed of the uniform flow, the numerical results show that the average of rotor torque and axial force in uniform inflow is larger than that of the shear inflow.

In the shear inflow, there is a phase lag between the fluctuation curves of force at different span locations of blade and azimuth angle. The more close to the blade root, the hysteresis effect is more obvious.

References

1. He Dexin. Wind Eingineering and Industrial Aerodynamics [M]. Beijing National Defence of Industry Press, 2006.
2. Tong Gang. Computation and Analyses on the Aerodynamic Performance of a Horizontal Axis Wind Turbine Blade [D]. Lanzhou University of Technology, 2011.
3. Li De-Shun, Li Ren-Nian. Investigation of Three-Dimensional Effect on Blades of a Wind Turbine Based on Field Experiments [J]. Applied Mathematics and Mechanics, 2013, 34(10): 1073-1082.
4. Liu Lei, Shi Ke-Zhong, Yang Ke. Effect of Wind Shear on the Aerodynamic Load of Wind Turbine [J]. Journal of Engineering Thermophysics, 2010, V31(10): 1667-1670
5. Zhang Yu-Liang, Yang Cong-Xin, et al. Irrelative Analysis of Wind Shear in the Design of a Wind Turbine [J]. Journal of Xihua University. v.26, no.2. pp. 61-63, 2/2007. (ISSN 1673-159X; In Chinese).
6. Min-Soo Jeong; Sang-Woo Kim; In Lee; Seung-Jae Yoo. Wake impacts on aerodynamic and aeroelastic behaviors of a horizontal axis wind turbine blade for sheared and turbulent flow conditions [J]. Journal of Fluids and Structures. 2014.
7. Xin Shen. Wind turbine aerodynamics and loads control in wind shear flow [J]. Energy. 2011(No.3).
8. Ray, M.L., Rogers, A.L., Mcgowan, J.G., 2006. Analysis of wind shear models and trends in different terrains. In: Proceedings of American Wind Energy Association Windpower 2006, June. Pittsburgh, PA, USA.
9. Moshfeghi, M; Song, YJ; Xie, YH. Effects of near-wall grid spacing on SST-K-omega model using NREL Phase VI horizontal axis wind turbine [J]. Journal of Wind Engineering and Industrial Aerodynamics. 2012.
10. Fan Zhongyao. Numerical simulations of steady and unsteady aerodynamics in wind turbines [D]. North China Electric Power University, 2011.
11. Sørensen, N. N. and Johansen, J., “Upwind, Aerodynamics and aero-elasticity: Rotor aerodynamics in atmospheric shear flow,” Proceedings of the 2007 EWEC Conference, Milan, Italy, 2007.

Study on China's new energy development in the perspective of low carbon economy

W.Y. Feng
School of Economics and Management,
Changchun University of Science and Technology, Changchun,Jilin, China
E-mail:architecter@126.com

The main purpose of this study is to analyze the problems of China's new energy development from the perspective of low carbon economy, and to seek the countermeasures of China's new energy development. The author collected a large amount of data and literature and puts forward the formation path of China's new energy technology economic paradigm. This includes the research and development of new energy technologies, demonstration and promotion, the establishment of effective technology research and incentive mechanism, to mobilize the participation of stakeholders. To draw the conclusion that the rapid development of China's new energy needs to work together in many ways of administrative, law, taxation, finance.

Keywords: Low-carbon economy; New energy technologies; Techno-economic paradigm.

1. Introduction

The low-carbon economy and the traditional high carbon economy is the corresponding, it is a new economic model based on the low energy consumption, low pollution and low-emission. Low carbon economy by means of energy technology innovation guide the economic development mode transition to clean and efficient direction, and its essence is the efficient use of energy, clean energy development and the pursuit of green GDP, the core is innovation on energy technologies and emission reduction technology. Development of low carbon economy, Because of the responsibility to protect the environment, to fulfill the request of the national energy saving and emission reduction targets, it is simultaneously a inevitable choice about adjust economic structure, improve energy efficiency, develop new industries, construct ecological civilization, and protect resources Environmental. And To achieve the goal of a low carbon economy, the development and utilization of new energy is undoubtedly the key.

Currently various countries around the world are fierce game about global climate change and develop low carbon economy, aimed at reducing greenhouse gas emissions using, which are profoundly influence the direction of world economic restructuring and energy development. As an important support the development of low carbon economy, new energy technologies widely attention from various countries. Development and utilization of new energy technologies is an important measure to increase the energy supply, adjusting energy structure, ensure energy security and protect the ecological environment.

2. Motivation about Development and Utilization of China's New Energy

2.1. *Urgency on China's development of low-carbon economy*

Faced with global scale low carbon economy paradigm shift in the tide, China has also been active. Development of low-carbon economy and building a low-carbon society has become an important focus of our strategic direction. But for various reasons, China's low carbon economy is still in a very early stage, accelerating the transformation from high carbon economy to a low carbon economic development model is imminent, a pressing task. Specific performance in the following aspects.

2.1.1. *Respond to the global climate crisis requires China's contribution*

Since the industrial revolution, the economic development is primarily dependent on coal, oil, natural gas and other carbon-based energy, consume large amounts of fossil energy to produce carbon dioxide and methane into the atmosphere of the planet to heat up the carbon cycle imbalance, changing the Earth's biosphere energy conversion form cause catastrophic changes in the global climate. The increase in atmospheric carbon dioxide emissions are the major cause of global warming. A survey of the International Energy Agency show that carbon dioxide emissions from the United States, China, Russia and Japan, accounting for almost half of the global total in 2007, even more than half.

2.1.2. *General energy shortage*

Since reform and opening, the traditional export-oriented economic development and government investment as the main feature of the achievements of the sustained high growth of China's economy, but at the excessive consumption of resources is a heavy price. China's energy reserves are generally reflected as "more coal, oil-poor, less gas." People, February 2, 2010 reported that mainland China is only recoverable oil reserves for 11 years, the external dependence has more than 50%, natural gas situation is more optimistic, the shortage of conventional energy reality, indicating that the supply of unsustainable development new energy at the core of a low-carbon economy is an inevitable thing.

2.1.3. *From ecological destruction and environmental pollution push*

In China's energy consumption structure, coal is the main energy, accounting for 94.3% of total proven fossil energy resources, accounting for 76% of primary energy production, accounting for 69% of primary energy consumption. Has been proven recoverable reserves of fossil fuels, the structural relationship of world oil, coal and natural gas 20%, 60% and 20%, China's primary energy consumption in 2007 was 20.1% oil, coal 69.5% and natural gas 3.3%, other clean energy accounted for only 7.1%, the use of coal in the process, serious environmental pollution; a large number of mining process, to bring a series of environmental and ecological problems.

2.2. *Ensure the development of a low-carbon economy*

From the point of view of energy structure, China's coal-based energy consumption, new energy to replace fossil energy, and build a new energy economy. Development of coal resources, although a considerable extent on the support of industrialization, but the coal in human time scales are also non-renewable resources, but also in the development process, resulting in the collapse of a large area of land, destruction of arable land, pollution of rivers, groundwater systems destruction, are the main factors leading to high-carbon economy.

Therefore, new energy, represented by low carbon energy, as a low carbon economy a core content, but also more and more people's attention, abandon traditional growth model, it is a direct application of innovative technologies and innovative mechanisms in the new century, through the development and utilization of new energy sources and achieve sustainable economic and social development, is promising.

Energy development is the law of conversion from high-carbon to low-carbon, and finally to carbon-free. The future direction of energy development is clean, efficient, diversified and sustainable manner. Low-carbon development is the control of greenhouse gas emissions in the process of economic development inevitable choice, and the development of new energy is the energy efficiency of key areas. Development of new energy development policy from the perspective of administrative, legal, taxation, finance and other aspects is our unswerving choice. Therefore, new energy technology research and development point of view, demonstration, promotion, incentives, interests of participants to design new energy to form paths appear economic paradigm is particularly important.

3. Forming a Path to the Economic Paradigm of New Energy Technologies

3.1. *New energy technology research and development, demonstration, promotion*

Energy Research Center researcher Jorge Isias National University of Mexico through the study found that: new technologies mature before the need to go through a very long process. The whole process is long and uncertain, before competitive, you need to go through research and development, demonstration and promotion of three main phases. Early, mainly financed by government R&D in extremely important for the development of new energy technologies. Government activities involved in the early development of new energy technologies, can make up for the private sector to the development of new energy technologies in the early stages of the absence, for the industrialization of new energy technologies has laid a solid foundation.

R&D can be achieved through technical feasibility, but the market still faces many obstacles in terms of technology. Technology demonstration activities can further examine the technical feasibility, the purpose is to demonstrate a small amount of technical inspection, large-scale demonstration is aimed at the promotion of the technology to predict and verify the economical large-scale use of technology and operating in a commercial mode feasibility. Then select the successful demonstration of technology to promote, carry out projects to promote new technology, while indicate the cost is still high, but with the deepening of the process of technology promotion, you can achieve the opportunity to interact with the application of cost reduction, until new technology becomes economically competitive force. Government propaganda to promote the project through new technologies market scenario, it can stimulate private industry to get involved, to accelerate the technology commercialization process.

3.2. *Policy incentives*

New energy technologies is dependent on a large number of high-tech R&D and demonstration areas, in order to accelerate the demonstration of new technologies and important investment support and institutional change is necessary. Policy and technology needed to support technological development life cycle in which the characteristics of different stages of development there is a relationship. Policies should take full account of current development of technology, find the entry point to reduce policy costs and maximize revenue. It can be concluded that the development of the chain of new energy technologies, including technical supply side, the demand side as well as the government, the market, investors and other factors, including various factors about the results.

3.3. *Stakeholder participation*

Technological development of new low carbon energy economy environment using stakeholder theory analysis, in order to determine the development of new energy technologies may affect the interests of the parties concerned. For each stakeholder, it should be clear what kind of interest between themselves and the development of new energy technologies, if they have certain rights, and whether their actions will affect the development of new energy technologies.

4. Suggestions of New Energy Development and Utilization

The new energy is an important part of China's energy strategy, to meet energy demand, improve energy structure, reducing environmental pollution and promoting economic development has an important role. But generally speaking, China's new energy development is still preliminary, there are many constraints new energy development and utilization of bottlenecks need to break. In this paper, the perspective of low-carbon economy, new energy technology development chain formed path, this paper proposes the development and utilization of new energy sources suggestions from administrative, legal, taxation, finance and many other angles.

4.1. *Development and implementation of new energy development plan*

Starting in April 2009 the National Energy Board of Development and Planning in the lead organizations to develop new energy industry revitalization plan. In the new energy industry in the various sub-sectors of wind power will be the focus of future development, utilization of solar energy by technical constraints slight lag, but the new energy promotion plan was incorporated into the development and utilization of nuclear power, nuclear power is expected to accelerate. By 2020, China's total investment in the new energy field will be more than 3 trillion yuan. Will be introduced in the new energy industry development plan mainly includes the construction of new energy-related investment objectives and two parts.

4.2. *Formulate legal system on ensure that develop low carbon economy*

Experience from Western countries a low-carbon economy-related laws covering different legal sectors and different areas of society on all levels of the functional approach in terms of a low-carbon economy legislation can be classified into the following three categories: energy conservation class legislation traditional class alternative energy legislation and carbon emission control class legislation.

To a low-carbon economy is concerned! Energy conservation means that energy efficiency, reduce energy loss and reduce greenhouse gases, reducing other pollutants, meant to change the traditional model of economic development.

Refers to the replacement of traditional energy sources to clean, low-carbon energy sources to replace the use of a low-carbon energy use of traditional energy use pattern, there is the traditional low-carbon fossil energy use, development and utilization of clean new low-carbon energy in two ways. Legislative efforts to replace traditional energy class energy structure adjustment, change the traditional energy use patterns, to promote the development and utilization of green energy.

4.3. *Strengthen the government's public financial investment for new energy*

New high energy costs, investment risks, financing difficulties in the early low industrialization level of business development, government policies support is necessary. (1) the establishment of a new energy development special fund. Key support for the general investors are reluctant to invest in, market mechanisms alone can not effectively functioning solar, wind, biomass, tidal power and other new energy development projects. (2) included in the scope of public budget support, direct investment. Finance can also be put directly into the new energy-related education, training. (3) improve financial subsidies for new energy sources. First, continue to implement user subsidies, to play the expansion of the consumer market size, to further promote the leading role in the production; the second is to improve the investment subsidies, the investment subsidies and new energy business situation combine to stimulate enterprises to further improve the technology and reduce costs. (4) the implementation of a certain degree of tax incentives for new energy industry, support its development.

4.4. *Broaden the financing channels*

First, the government should direct the increase in renewable energy investment, while fiscal policy and bank credit policy cooperation. Interest payments on loans the government will increase efforts to encourage commercial banks to expand credit scale and coverage, given in term lending rates more concessions. The second is to reform the investment and financing system, new energy to diversify investment subject, allow private capital to enter the field of new energy development, while striving to create conditions for attracting foreign capital to participate in China's new energy development. Third, we should refer to foreign experience, the establishment of new energy fund system, which uses the direction of renewable energy project development, business development, product and equipment installation.

4.5. *Accelerate scientific and technological innovation*

Low-carbon technologies for the realization of human needs and sustainable economic and social development can be created and developed to achieve a lower total greenhouse gas emissions technology. Include carbon reduction techniques such as power, transportation, construction, metallurgy, chemical industry, high energy consumption , energy saving technology in the field of high emissions, the clean use of coal, oil and gas resources in exploration technology; non-carbon technologies such as nuclear, solar, wind, biomass, tidal energy, geothermal energy, hydrogen, etc. can be renewable energy technologies; go-carbon technologies such as carbon capture and storage technologies, resource utilization technology of greenhouse gases three types.

Lack of scientific research and technological innovation capability is a major obstacle of China's new energy development. Through technological upgrading of enterprises should improve the conversion rate of existing products, further reduce costs and improve competitiveness. The formation of government, business linkage mechanism to support research and development of new energy technologies. Government ensure investment in new energy technology research and development funding. The establishment of new energy technology innovation basic institutional framework. The establishment of enterprises and research institutions as the mainstay, market-oriented, industry, academia, research combining technical innovation system, the formation of independent innovation of the basic institutional framework. Selecting the right new energy technology research, development and utilization of direction and focus.

5. Conclusion

This paper analysis China's new energy development from a point of view of the technical economic paradigm in the low-carbon economic environment, the purpose is to summarize the inherent correlation of low-carbon economy and new energy technologies, the development of new energy technologies has become a strong support for a low-carbon economy. Development of new energy technologies has been one step ahead of most developed countries, to support the development of new energy technologies is very large, and mastered a certain core technologies. China is in the stage of industrialization, which demand for energy is increasing, the coal-based energy structure can not be fundamentally resolved within a short time; the development of new energy technologies has only just started, policy incentives are not perfect. Therefore, Chinese government must first have a clear direction of energy technology development plan, and constantly improve the energy technology innovation mechanism, enhance the technological innovation capability, strengthen technical cooperation with developed countries, to provide more platforms for the development of new energy technologies.

References

1. D.Y. Wang and X.Y. Zhang, *Research on the development of new energy economy based on a perspective of low-carbon*, Ecological Economy (Beijing, 2010), pp. 104-107.
2. H. Wu, *Paradigm of new energy technologies based on low-carbon economic background.* Science & Engineering, (Hefei, Anhui, 2011), pp. 101-105.
3. Islas J, *The Gas turbine: A new technological paradigm in electricity* genenration. Technological Forecasting and Social Change, (1999), pp. 129-148.
4. Johnsson F, Reiner D, Itaoka K, et al, *Stakeholder attitudes on carbon capture and storage - An international comparison*, International Journal of Greenhouse Gas Control, (2010), pp. 410-418.
5. IEA, Technology roadmap — Concentrating solar power, (Paris, France, 2010).

6. Raupach, M.R., Marland, G.P. etal. Global and regional drivers of accelerating CO2 emissions. Proceedings of National Academy of Sciences (PNAS), (2007).
7. A. S. Dagoumas, T. S. Barkera. *Pathways to a low-carbon economy for the UK* with *the macro-econometric E3MG model*, Energy Policy, (2010).
8. Sam Nader. *Paths to a low-carbon economy - The Masdar example*, Energy Procedia, (2009), pp. 3951-3958.
9. Narumitr Sawangphol, Chanathip Pharino, *Status and outlook for Thailand's low carbon electricity development*, Renewable and Sustainable Energy Reviews, (2011).
10. X.J. Chen, *Western countries measures and the effect of low-carbon economic development,* China collective economy, (Hubei, 2010), pp. 111-116.
11. D. Wu, *Experiences on International Low Carbon Economy Development and Enlightenment to China*. Economics and Management, (Hunan, 2014), pp. 270-273.
12. X. Yao, *The connotation and development trend of the new era of low-carbon Economy*, Shanxi University Education, (Shanxi, 2011), pp. 140-144.
13. S. Huang, *The new energy policy of European and American and our energy policy-oriented research in the low-carbon economy*, International economy, (Zhejiang, 2013), pp. 116-118.
14. International Energy Agency. *Clean coal technologies - accelerating commercial and policy drivers for deployment*, Sourceoecd Energy, (2009).
15. S.H. Wang, Research on the Optimization of China's Energy Structure Based on Low carbon Economy, Doctoral Dissertation, Harbin Engineering University, (2013).

Synthetic biology and sustainable energy*

Y.J. Chen
Institute of Scientific and Technical Information of China, Beijing 100038, China
E-mail: cyjmail123@163.com

Synthetic biology can be defined as a combinatorial approach where the basic elements, governing the operational functionality of a living cell, are placed together to generate life forms with novel characteristics and properties. One of the most exciting aspects of synthetic biology is the diversity of applications it may enable. New applications may be found in energy, medicine, environment and materials. In this paper, we focus on the definition of synthetic biology as well as the potential applications and key challenges, then the role of synthetic biology in producing biofuels are discussed. Finally, several key points to achieve sustainable development are summarized.

Keywords: Synthetic biology, Biofuel, Microbiology, Sustainable energy.

1. Introduction

Synthetic biology can be considered as extreme genetic engineering as it is an amalgamation of the principles of engineering and biology that provide a new paradigm for generation of novel applications in diverse areas such as therapeutics, bioremediation, waste resource management and biofuel production [1-5]. It is a multidisciplinary science in a true sense as it derives knowledge from various disciplines (such as genetics, nanotechnology, chemical engineering, robotics, systems biology and computational biology), in order to manipulate living cells to obtain desired functions.

Over the past decade, synthetic biology has emerged as an engineering discipline for biological systems. Compared with other substrates, biology possess a unique set of engineering challenges resulting from an incomplete understanding of natural biological systems and tools for manipulating them. It is speculated that with further technological advancements and understanding of the biological phenomenon, greater avenues for synthetic biology applications will emerge which would allow incorporation of the information generated through various projects based on genomics, proteomics, metabolomics and phenomics [6].

---

* This work was partly supported by the National Science and Technology Support Program of China.

Synthetic biology aims at enablement of creation of living organisms with user defined genetic circuitry and novel characteristics by modulating, regulating and redesigning the gene expression profile of the host organism, enabling its exploitation for various applications [7]. As a stepping stone towards this direction, micro-organisms (such as Escherichia coli) with minimal genome have been created [8]. Such microbes can be further transformed with genes conferring them novel functions.

On the other hand, microbiology, the multidisciplinary science of microorganisms, has a significant impact on the world. With its roots in the 19th century, it matured into a scientific discipline concerned with the structure, function, and classification of these organisms and with ways of controlling and using their activities. It continued to evolve into the major subdivisions of microbiology, which include medical, industrial, agricultural, food and dairy. Microbiology is used to treat major pandemics and frequent infections. It also has an impact on the food people consume [9], and it has been shown that the recently identified natural antitumor compounds from various natural origins including fungi and endophytic fungi [10].

The present study focuses on applications of synthetic biology in producing biofuel and achieving sustainable development.

2. Synthetic Biology: Potential Applications and Key Challenges

Synthetic biology is an emerging field combining biology with engineering but lacks a coherent and agreed definition. Synthetic biology is perhaps best understood in relation to its aims, which are described by the OECD as being "to design and build new biological parts and systems or to modify existing ones to carry out novel tasks" [11]. In other words, synthetic biology is

• the design and construction of new biological parts, devices and systems, and

• the redesign of existing, natural biological systems for useful purposes.

On the other hand, Murray did place the technologies comprising synthetic biology into four categories [12]:

• ***Advanced genetic engineering.*** Engineering bacteria to produce a precursor of artemisinin, a front-line malaria drug, epitomizes this category.

• ***DNA-based device construction.*** The BioBricks project applies the principles of electronics engineering to biology, aiming to one day build functional nano devices de novo from genes and proteins.

• ***Creating a minimal cell.*** Venter's achievement of synthesizing a complete, functional genome for Mycoplasma mycoides is the leading indicator.

• ***Creating a protocell.*** According to Murray the goal is to create a new form of life. A genuinely new living entity, not based on the biology we've known thus far. This goal remains distant, but may lead to radical advances such as silicon-based life.

For the wider community the importance of synthetic biology lies in its social and commercial potential. One estimate suggests that the global market for synthetic biology could reach US$2.4 billion by 2013, with applications ranging from medicine to agriculture [13]. Possible uses of synthetic biology include the following areas:

• ***Energy*** Custom-built microbes for generating hydrogen and other fuels, or for performing artificial photosynthesis. And the development of new pathways for producing fuel.

• ***Medicine and Health*** The manufacture of drugs, vaccines and diagnostic agents, and the creation of new tissue. Also including enhanced drug production and delivery.

• ***Environment*** The detection of pollutants, and their breakdown or removal from the environment; and including engineered dispersants and environmentally friendly materials.

• ***Chemical Industry*** The production of fine or bulk chemicals, including proteins to provide an alternative to natural fibres or existing synthetic fibres.

• ***Agriculture and Food*** Including engineered or optimized crops, and novel food additives.

Which of these applications will be first to make an impact in the marketplace is a matter of speculation, although many commentators foresee biofuel products as a likely frontrunner.

However, challenges loom at every step in the process, from the characterization of parts to the design and construction of systems. It has been pointed out that there are five key challenges [14]:

• ***Many of the parts are undefined*** A biological part can be anything from a DNA sequence. The problem is that many parts have not been characterized well. They haven't always been tested to show what they do, and even when they have, their performance can change with different cell types or under different laboratory conditions.

• ***The circuitry is unpredictable*** Even if the function of each part is known, the parts may not work as expected when put together. Synthetic biologists are often caught in a laborious process of trial-and-error, unlike the more predictable design procedures found in other modern engineering disciplines.

• ***The complexity is unwieldy*** As circuits get larger, the process of constructing and testing them becomes more daunting. For example, the researchers had to test many part variants before they found a configuration.

• ***Many parts are incompatible*** Once constructed and placed into cells, synthetic genetic circuits can have unintended effects on their host.

• ***Variability crashes the system*** Synthetic biologists must also ensure that circuits function reliably. Molecular activities inside cells are prone to random fluctuations, or noise. Variation in growth conditions can also affect behavior.

3. Synthetic Biology for Biofuels

As the fossil fuel economy grows increasingly unsustainable, it becomes more and more important that humanity develops alternative energy solutions. Climate change is occurring at an alarming pace, disrupting the biosphere, facilitating international conflicts over finite resources, and destabilizing the global economy. Synthetic biology provides scientists with a path towards rapid development of renewable fuels via biological systems and this a feasible route for realizing global energy sustainability [15]. For instance, the most widely used biofuel is ethanol produced from corn or sugar cane [16]; however, the heavy agricultural burden combined with the suboptimal fuel properties of ethanol make this approach to biofuels problematic and limited. Microorganisms engineered with optimized biosynthetic pathways to efficiently convert biomass or non-edible vegetable oils into biofuels are an alternative and promising source of renewable energy [17]. These strategies will succeed only if their production costs can be made to compete with, or even outcompete, current fuel production costs.

For biofuel production, certain microorganisms have evolved to be proficient in converting lignocellulosic material to ethanol, biobutanol and other biofuels. These native isolates possess unique catabolic activity, heightened tolerances for toxic materials and a host of enzymes designed to break down the lignocellulosic components. Unfortunately, these highly desired properties exist in pathways that are tightly regulated according to the host's evolved needs and therefore may not be suitable in their native state for production scale. A longstanding challenge in metabolic and genetic engineering is determining whether to improve the isolate host's production capacity or whether to transplant the desired genes or pathways into an industrial model host, such as E. coli or S. cerevisiae [18].

Synthetic biology might accelerate the development of second-generation biofuels that can be prepared from agricultural waste and plant residues, so avoiding competition with crops grown for food. Synthetic biology involves engineering microbes to produce specifically desired fuels, especially hydrocarbon fuels that are "drop-in" replacements for petroleum diesel and gasoline. The work that has been done thus far has been targeting the development of microbes that "eat" sugar molecules and excrete diesel-like fuel [19].

On the other hand, synthetic biology will also be key in rewiring existing regulation. The design of new regulatory pathways from synthetic genetic elements will be important in sensing the extracellular or intracellular environment and producing a programmed cellular response. For biofuel production from lignocellulosic biomass, this is of great importance as the efficient uptake of a mixture of hexose and pentose sugars simultaneously is desirable. Synthetic biology can provide tools to construct new circuits devoid of unwanted regulation from the bottom up [20] to sense the extracellular environment and produce the necessary response to digest the sugars. Ultimately, the bottom-up construction of biological circuits may extend beyond individual pathways toward all of metabolism, as the synthesis of entirely synthetic, replicable, and functional genomes has recently been accomplished [21]. In the future, this will allow synthetic biologists to build heterologous pathways into organisms devoid of unwanted pathways or properties.

4. Key Points for Sustainable Development

According to definition of sustainable energy, damage of environment is another major concern for sustainable energy development. Environmental efficiency is closely linked to working of sustainable and renewable energy development concerns. The term eco-efficiency was coined by the World Business Council for Sustainable Development (WBCSD). It is based on the concept of creating more goods and services while using fewer resources and creating lesser waste and pollution. WBCSD has pointed out seven major elements in considering eco-efficiency of developing environmental friendly products or processes for reducing environmental impacts [22].

- Reduce the material intensity of its goods and services.
- Reduce the energy intensity of its goods and services.
- Reduce the dispersion of any toxic materials.
- Enhance the recyclability of its materials.
- Maximize the sustainable use of renewable resources.
- Extend the durability of its products.
- Increase the service intensity of its goods and services.

5. Conclusions

The development and optimization of many aspects of biofuel production technology can benefit from the work already accomplished through synthetic biology. One of the current challenges facing synthetic biology is the reproducibility of the developed tools in different systems. However, current technology developed by synthetic biology has already allowed for the successful design of many heterologously expressed biofuel production and substrate utilization pathways extending beyond the most user-friendly organisms.

The advancement of synthetic biology toward new diagnostic tools and high throughput screening systems will aid in the further development of these biofuel processes for pathway optimization and enzyme discovery or improvement. The success that synthetic biology has already afforded biofuel production technology lends confidence to future synergistic developments and breakthroughs.

The desire for renewable liquid fuel replacements to petroleum has steadily increased with concerns about the current fuel economy's stability and environmental impact. The development of new biofuel production processes has sought to mitigate some of these issues. The integration of synthetic biology with the development of these processes will be significant in bringing the biofuels industry from its infancy to a commercially viable alternative to petroleum.

Acknowledgments

This work was partly supported by the National Science and Technology Support Program of China under Grant No. 2012BAC20B09.

References

1. D.F. Savage, J. Way and P.A. Silver, Defossiling fuel: how synthetic biology can transform biofuel production, *ACS Chem. Biol.* **3**, 13 (2008).
2. P.E.M. Purnick and R. Weiss, The second wave of synthetic biology: from module to systems, *Nat. Rev. Mol. Cell Biol.* **10**, 410 (2009).
3. C.W. Schmidt, Synthetic biology: environmental health implications of a new field, *Environ Health Perspect* **118**, A118 (2010).
4. D.K. Kale and P.D. Anthappan, Sustainable Treatment of Wastewater using Effective Microorganisms, *Journal of Pure and Applied Microbiology* **6 (1)**, 333 (2012).

5. R.S. Babu, S.B.C. Prasad, R. Chakrapani, M. Padmavathi, C.S.V.R. Rao and M.L. Narasu, Studies on Bioethanol Production from Spoiled Starch Rich and Cellulose Rich Vegetables by Two Stage Batch Fermentation, *Journal of Pure and Applied Microbiology* **5 (2)**, 853 (2011).
6. D. Houle, D.R. Govindaraju and S. Omholt, Phenomics: The Next Challenge, *Nature Reviews Genetics* **11**, 855 (2010).
7. P. Marguet, F. Balagadde, C.M. Tan and L.C. You, Biology by Design: Reduction and Synthesis of Cellular Components and Behaviour, *Journal of the Royal Society Interface* **4**, 607 (2007).
8. B.H. Sung, J.H. Lee and S.C. Kim, Escherichia coli Genome Engineering and Minimization for the Construction of a Bioengine, in *Systems Biology and Biotechnology of Escherichia Coli,* eds. S.Y. Lee (Springer Netherlands, 2009) pp. 19-40.
9. M.S. El-Hersh, H.A. El-Fadaly, W.I.A. Saber and A.M. El-Deeb, Human Diseases Prosecution Among Viral Infection and Food Toxins: A Review, *International Journal of Pharmacology* **9(7)**, 390 (2013).
10. S. Saeidnia and M. Abdollahi, Perspective Studies on Novel Anticancer Drugs from Natural Origin: A Comprehensive Review, *International Journal of Pharmacology* **10(2)**, 90 (2014).
11. Secretary-General of the OECD, "Synthesis Report", Symposium on Opportunities and Challenges in the Emerging Field of Synthetic Biology, *OECD & the Royal Society* 2010.
12. T.H. Murray, Society Must Weigh "Massive" Potential Benefits Against Risks of Synthetic Life. See http://www.aaas.org/news/releases/2010/1116hitachi_synth_bio.shtml
13. See http://www.easac.eu, the European Academies Science Advisory Council, January 2011.
14. R. Kwok, Five Hard Truths for Synthetic Biology, *Nature* **463(7279)**, 288 (2010).
15. Y.J. Chen, Role of Liquid Biofuels in Realizing of Global Sustainability, *Asian Journal of Chemistry* **26(11)**, 3355 (2014).
16. J.L. Fortman, S. Chhabra, A. Mukhopadhyay, H. Chou, T.S. Lee, E. Steen and J.D. Keasling, Biofuel Alternatives to Ethanol: Pumping the Microbial Well, *Trends in Biotechnology* **26**, 375 (2008).
17. G. Ramya, R. Vaithiyanathan, J.A.I. Joice, R. Sudhakar and T. Sivakumar, Application of Core-Shell Modeled Composite Catalysts for Biofuel Production from Non-edible Vegetable Oils, *Asian Journal of Chemistry* **24(12)**, 5683 (2012).

18. H. Alper and G. Stephanopoulos, Engineering for Biofuels: Exploiting Innate Microbial Capacity or Importing Biosynthetic Potential? *Nature Rev. Microbiol.* **7**, 715 (2009).
19. S.K. Lee, H. Chou, T.S. Ham, T.S. Lee and J.D. Keasling, Metabolic Engineering of Microorganisms for Biofuels Production: From Bugs to Synthetic Biology to Fuels, *Current Opinion in Biotechnology* **19**, 556 (2008).
20. N.J. Guido, X. Wang, D. Adalsteinsson, D. McMillen, J. Hasty, C.R. Cantor, T.C. Elston and J.J. Collins, A bottom-up approach to gene regulation, *Nature* **439 (7078)**, 856 (2006).
21. D.G. Gibson, J.I. Glass, C. Lartigue, V.N. Noskov, R.Y. Chuang, M.A. Algire, G.A. Benders, M.G. Montague, L. Ma, M.M. Moodie, C. Merryman, S. Vashee, R. Krishnakumar, N.A. Garcia, C.A. Pfannkoch, E.A. Denisova, L. Young, Z.Q. Qi, T.H.S. Shapiro, C.H. Calvey, P.P. Parmar, C.A. Hutchison III, H.O. Smith and J.C. Venter, Creation of a bacterial cell controlled by a chemically synthesized genome, *Science* **329(5987)**, 52 (2010).
22. L.D. DeSimone and F. Popoff, Eco-efficiency: The Business Link to Sustainable Development (MIT Press, 1997).

Study on dynamic model of amorphous silicon photocell*

Y.L. Tang†, S. Yuan, Y. Liu, X.K. Wu and J. Cai
College of Science, Guizhou University, Guiyang, Guizhou550025, China
†E-mail: tylgzu@163.com

The amorphous silicon photocell has many advantages such as simple manufacturing process, low cost and big visible light absorption coefficient. To deeper understand the working principle of amorphous crystal silicon photocell and the factors that affect solar cell performance, the equivalent circuit of photocell were simulated by using Multisim Software and AMPS-1D software in this paper. On this basis, we made some experimental simulations, and analyzed the influence of the thickness of N layer and intrinsic layer I on the open circuit voltage, short circuit current, fill factor and conversion efficiency of single crystal silicon photocell. The results show that the thickness of N layer and intrinsic layer I of amorphous crystal silicon photoelectric cell will affect on the open circuit voltage, short circuit current, filling factor and conversion efficiency. The best amorphous crystal silicon photocell is that N layer thickness is 10nm, intrinsic layer I thickness 400-600nm and the largest conversion efficiency η=9.59%.

Keywords: Amorphous silicon photocell; Dynamic model; Influence factor; Simulation analysis.

1. Introduction

Solar energy is a kind of renewable energy, which has many advantages such as green environmental protection, no fuel consumption, no regional restriction and scale flexible. In recent years, it has got the favor of all countries and will become the backbone of energy in the world. In order to develop domestic photovoltaic industry, all countries adopted various preferential policies to rapidly develop the global photovoltaic industry. Photovoltaic output reached 136.7 GW in 2013 [1].

In 1954, the first piece of valuable solar cells was made by Bell laboratory [2, 3]. Since then, the large-scale application of solar energy becomes reality.

The current solar photovoltaic cell basically has some kinds such as silicon photocell, nano-crystalline photocell, organic photocell and dye sensitized photocell [4-6]. Silicon photocell, including single crystal silicon,

* This work is supported by the National Natural Science Foundation of China program under Grant No. 11164004 and the Industrial Research Project of Guizhou province (GY[2012]3060).

polycrystalline silicon and amorphous silicon photocell, is the most mature technology. The conversion efficiency of single crystal silicon photocell is highest and occupies the main market share, polycrystalline silicon and amorphous silicon photocell taking second place among them. But the single crystal silicon photocell production cost is higher, and the cost of polycrystalline silicon, amorphous silicon photocell smaller [7, 8]. To measure the application value of a photocell, its quality factor is main parameter [5]. The conversion efficiency of amorphous silicon photocell is lower than that of crystalline silicon, but it has many advantages such as simple manufacturing process, low cost and big visible light absorption coefficient, so it has a great development potential [9,10]. As long as to solve the problem of poor stability and low conversion efficiency, it will greatly push the photovoltaic industry. According to the influence factors of photocell, this paper mainly simulate and analyze in theory the influence of the thickness of N layer and I layer for PIN structure of amorphous silicon photocell on the four performance parameters of photocell such as open circuit voltage [11], short circuit current [11], filling factor [5, 11] and conversion efficiency [5]. It could provide theoretical basis for the manufacturing of the high performance of amorphous silicon photocell and thin film solar cell.

2. The Experimental Simulation

This article only focuses on simulation of amorphous silicon photocell with PIN structure. The diagram of amorphous silicon photocell PIN structure is shown in Figure 1.

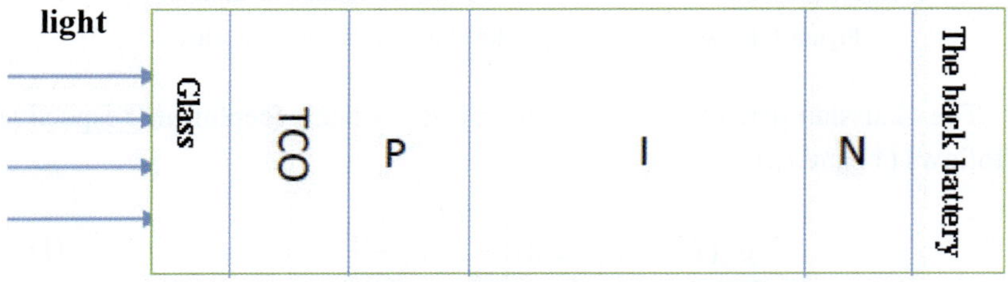

Figure 1. The structural model of amorphous silicon photocell.

Because there are lots of defects in amorphous silicon, it leads to the diffusion length of carrier in amorphous silicon photocell very short. The motive forms of carrier are mainly composed of drift motion in amorphous silicon photocell [2]. So the current in simple p-n junction amorphous silicon photocell is given priority to tunnel current and no rectification effect [12]. Therefore, you cannot use simple p-n junction structure for amorphous silicon photocell. The solution is adding intrinsic layer in the middle of the P and N layer. When light pass through the window of P layer into the intrinsic absorption layer I, the photon-generated carrier will be produced and separated by the built-in electric field through I layer. They are collected by the before electrode and after electrode to form a light voltage and current.

The before electrode of photocell is made of oxide (TCO) with transparent to visible light and good conductive [3]. The materials commonly use In, Sb, Zn and multivariate composite oxide [2] because of their wide band gap.

The DOS model for simulation was used in this article. In amorphous silicon material, electronic state is divided into extension and localized states, extended state including conduction band and valence band, localized states including tail localized state and gap localized state [12]. The tail localized state is exponential type, and gap localized state double Gaussian distribution including donor state and acceptor states respectively.

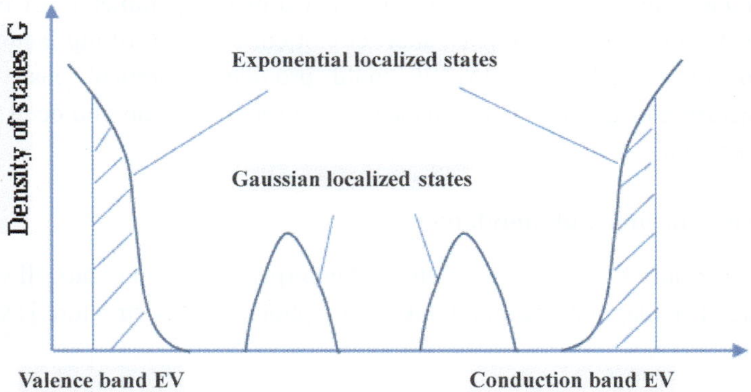

Figure 2. Localized state density distribution of amorphous silicon.

The Gaussian density of local states of donor and acceptor are respectively as follows (1) and (2):

$$g_d(E) = G_{Gd} \exp\left\{-\frac{(E - E_{pkd})^2}{2\sigma_d^2}\right\} \tag{1}$$

$$g_a(E) = G_{Ga} \exp\left\{-\frac{(E - E_{pka})^2}{2\sigma_a^2}\right\} \tag{2}$$

There E_{pkd}, E_{pka} are respectively the peak energy of donor and acceptor state, σ_d, σ_a respective standard deviation respectively.

In 1954, the first piece of valuable solar cells was made by Bell laboratory [2, 3]. Since then, the large-scale application of solar energy becomes reality. The current solar photovoltaic cell basically has some kinds such as silicon photocell, nano-crystalline photocell, organic photocell and dye sensitized photocell [4-6]. Silicon photocell, including single crystal silicon, polycrystalline silicon and amorphous silicon photocell, is the most mature technology. The conversion efficiency of single crystal silicon photocell is highest and occupies the main market share, polycrystalline silicon and amorphous silicon photocell taking second place among them. But the single crystal silicon photocell production cost is higher, and the cost of polycrystalline silicon, amorphous silicon photocell smaller [7, 8]. To measure the application value of a photocell, its quality factor is main parameter [5]. The conversion efficiency of amorphous silicon photocell is lower than that of crystalline silicon, but it has many advantages such as simple manufacturing process, low cost and big visible light absorption coefficient, so it has a great development potential [9,10]. As long as to solve the problem of poor stability and low conversion efficiency, it will greatly push the photovoltaic industry. According to the influence factors of photocell, this paper mainly simulate and analyze in theory the influence of the thickness of N layer and I layer for PIN structure of amorphous silicon photocell on the four performance parameters of photocell such as open circuit voltage [11], short circuit current [11], filling factor [5, 11] and conversion efficiency [5]. It could provide theoretical basis for the manufacturing of the high performance of amorphous silicon photocell and thin film solar cell.

Table 1. Simulation parameters of amorphous silicon photocell.

Classification	P (a-SiC)	I (a-Si)	N (a-Si)
Thickness/nm	20	0-1600	10-100
Relative dielectric constant	11.9	11.9	11.9
Electron affinity	3.7	3.8	3.8
Band gap of mobility	1.92	1.82	1.82
Optical forbidden band gap	1.82	1.72	1.78
Electron mobility	10	20	20
Hole mobility	1	2	2
Doping concentration NA	3×10^{18}	0	0
Doping concentration ND	0	0	8×10^{18}
Effective state density of conduction band NC	2.5×10^{20}	2.5×10^{20}	2.5×10^{20}
Effective state density of valence band NV	2.5×10^{20}	2.5×10^{20}	2.5×10^{20}
Characteristic energy of conduction band tail EA	0.05	0.02	0.04
Characteristic energy of valence band tail ED	0.1	0.05	0.06
GDO	1×10^{21}	1×10^{21}	1×10^{21}
GAO	1×10^{21}	1×10^{21}	1×10^{21}

Electron trapping coefficient classification of valence band tail states	1×10^{-15}	1×10^{-15}	1×10^{-15}
Hole trapping coefficient of valence band tail states	1×10^{-17}	1×10^{-17}	1×10^{-17}
Electron trapping coefficient of conduction band tail states	1×10^{-17}	1×10^{-17}	1×10^{-17}
Hole trapping coefficient of conduction band tail states	1×10^{-15}	1×10^{-15}	1×10^{-15}
Gaussian peak energy of donor state	1.27	1.12	1.2
Gaussian peak energy of acceptor state	1.17	1.02	0.8
NDG	1×10^{-16}	5×10^{-15}	1×10^{-16}
NAG	1×10^{-16}	5×10^{-15}	3×10^{-18}
Standard deviation of Gaussian donor level	0.08	0.08	0.08
Standard deviation of Gaussian acceptor level	0.08	0.08	0.15
Electron trapping coefficient of Gaussian donor	1×10^{-14}	1×10^{-14}	1×10^{-14}
Hole trapping coefficient of Gaussian donor	1×10^{-15}	1×10^{-15}	1×10^{-15}
Electron trapping coefficient of Gaussian acceptor	1×10^{-15}	1×10^{-15}	1×10^{-15}
Hole trapping coefficient of Gaussian acceptor	1×10^{-14}	1×10^{-14}	1×10^{-14}

The following result is obtained by simulation calculation. When the thickness of I layer is 380 nm and N layer thickness is 10 nm, the short circuit current I_{SC} = 13.1mA/cm^2, open circuit voltage UOC = 0.979V, filling factor FF = 0.748 and conversion efficiency η = 9.586% for amorphous silicon photocell.

3. Influencing Factors of Amorphous Silicon Photocell

3.1. *Effect of N layer thickness on the amorphous silicon photocell*

Suppose the thickness of I layer is 380 nm and thickness of P layer is 20 nm, the influence of N layer thickness on properties of amorphous silicon photocell are simulated and analyzed. An experiment simulation is carried on every 10 nm in range of 10 nm – 100 nm of N layer thickness, respectively. The four performance simulation results of photocell are shown in Figure 3.

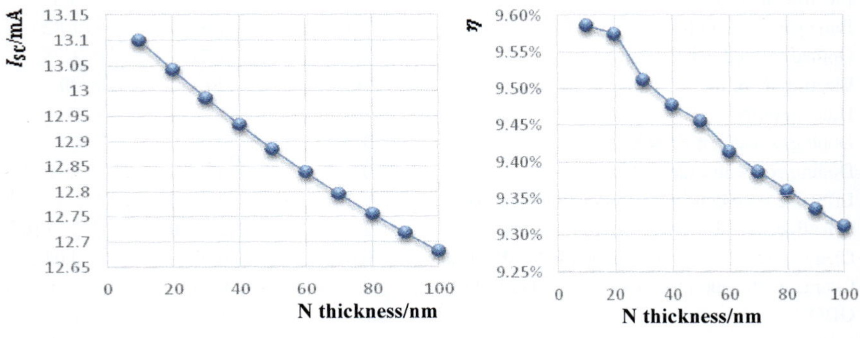

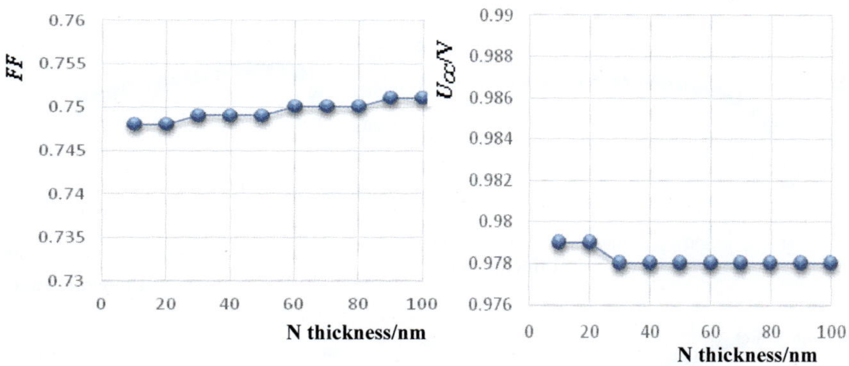

Figure 3. Effect of N layer thickness on the amorphous silicon photocell.

From Figure 3, it can be found that short-circuit current, conversion efficiency and open-circuit voltage values are varying degrees of decline with N layer thickening. Because of the increase of the thickness of the N layer to make the battery resistance increases, the short circuit current Isc decrease. Because the carrier life in amorphous silicon is short, its diffusion length is very small. Although increasing N layer thickness increases the absorption of light, the electron hole pair produced in N layer fails to spread into the built-in electric field and separate because of the small diffusion length, so it almost has no contribution to photocurrent. And N layer thickness change did not increase the size of the built-in electric field, so the filling factor and the open circuit voltage is almost no change. It was speculated that the built-in electric field produced by 10 nm thickness of N layer is enough to separate the electron hole pair, and continue to increase the thickness of N layer does not further increase the built-in electric field.

3.2. *Effect of I layer thickness on the amorphous silicon photocell*

Suppose the thickness of N layer is 10 nm and thickness of P layer is 20 nm, the influence of I layer thickness on properties of amorphous silicon photocell are simulated and analyzed. An experiment simulation is carried on every 100 nm in range of 0 nm – 1600 nm of I layer thickness, respectively. The four performance simulation results of photocell are shown in Figure 4.

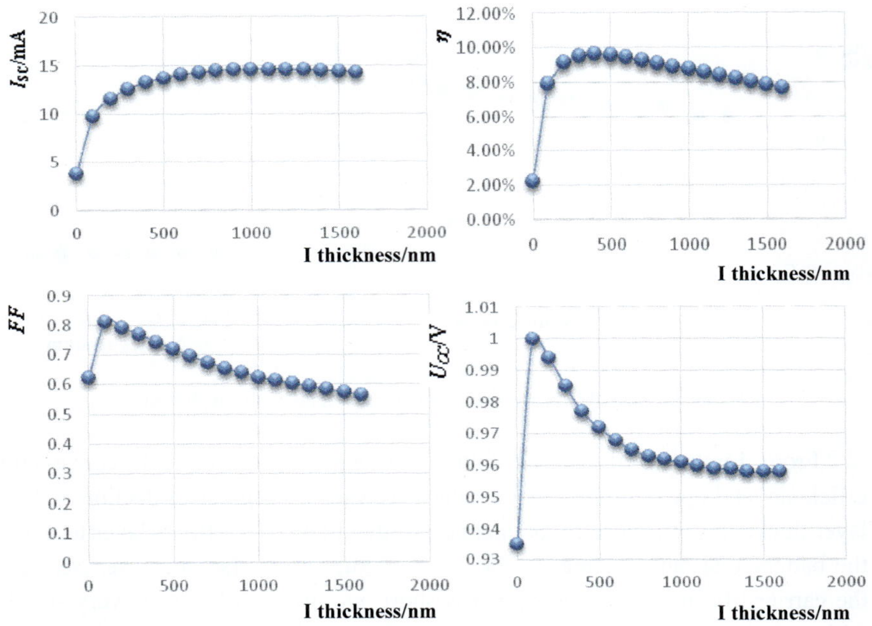

Figure 4. Effect of I layer thickness on the amorphous silicon photocell.

If the thickness of the absorbing layer I is too thin, it is not conducive to absorb incident light. If it is too thick, the defect of amorphous silicon increases, and thereby deepens the composite of carrier and increases the series resistance. Therefore, I layer thickness should be moderate.

Knowing from the above, the open circuit voltage starts to increase as I layer thickening and began to decrease at about 100 nm.

The change curve of filling factor is the same as that of open circuit voltage with first increasing then decreasing. With the continuous increase of I layer, it leads to increase the series resistance and decrease the fill factor of amorphous silicon photocell.

The short circuit current decreases with increasing the thickness of I layer. It will up to the biggest I_{SC} = 14.577 mA/cm^2 as the thickness of I layer increasing to 1.1 um. Then the short circuit current began to decrease with the increase of I layer.

According to the variation curve of conversion efficiency, the maximum conversion efficiency is up to 9.59% when the thickness of I layer was 0.4 um.

4. Conclusion

Due to the amorphous silicon, photocell has many advantages such as simple manufacturing process, low cost and big visible light absorption coefficient, amorphous silicon will be one of the preferred material of photocell along with the increase of conversion efficiency of amorphous silicon photocell in the future [13,14]. The simulation results show that the thickness of N layer and intrinsic layer of amorphous silicon photocell are affect on the open circuit voltage, short circuit current, filling factor and conversion efficiency. The best amorphous silicon photocell with PIN structure is that the thickness of N layer is 10 nm, intrinsic layer thickness 400-600nm. In this case, the maximum conversion efficiency η=9.59% and short circuit current I_{SC} = 14.577mA/cm^2.

References

1. Deep research and investment strategy consulting report of China's photovoltaic industry from 2013 to 2018. Zhiyan Consulting Group, 2013.8.
2. K. Li. Numerical study on photoelectric conversion efficiency of Amorphous silicon thin film solar battery [D]. Ningxia University, 2010.
3. G. Yang. Study on experiment and simulation of thin-film battery based on amorphous silicon [D]. University of Electronic Science and Technology of China, 2012.
4. P.M. Hu. Development and analysis of the solar cells [D]. Shanghai Jiaotong University, 2008.
5. J.F. Stephen. Solar cell device physics second edition [M]. Academic Press, 2010.
6. Z.M. Shi. Development of solar photovoltaic cells [J]. The Magazine on Equipment Machinery, 2010, 04: 69-73.
7. J.H. Zhao. Recent advances of high-efficiency single crystalline silicon solar cells in processing technologies and substrate materials, Solar Energy Materials and Solar Cells [J]. 2004, 82: 53-64.
8. S.Y. Liu, X.W. Niu, W. Shan, W. Lu, J.Y. Zheng, Y.F. Li, H.B. Duan, W.J. Quan, W.Z. Han, C.R. Wronskic, D.R. Yang. Improvement of conversion efficiency of multicrystalline silicon solar cells by incorporating reactive ion etching texturing. Solar Energy Materials and Solar Cells [J]. 2014, 127: 21-26.
9. S.D. He. Preparation of PIN-type amorphous silicon solar cells and its characteristics research [D]. University of Electronic Science and Technology of China, 2009.

10. Q.J. Cao, Z.Q. Li, Y.S. Peng, N.Y. Wu, L.X. Jia. Study on optimization of amorphous silicon solar cell using Sentaurus TCAD [J]. Power Source Technology, 2013, (1): 47-50.
11. G.Y. Zeng, Z.W. Zhang, C.L. Zhang. Photoelectric detection technology [M]. Beijing: Peking University Press, 2009.
12. E.K. Liu, B.S. Zhu, J.S. Luo. Semiconductor physics [M]. Beijing: Electronic Industry Press, 2011.
13. Z.N. Jia, X.D. Zhang, Y. Liu, Y.F. Wang, J. Fan, C.C. Liu, Y. Zhao. Improving light trapping and conversion efficiency of amorphous silicon solar cell by modified and randomly distributed ZnO nanorods [J]. Chinese Physics B, 2014, 4: 444-449.
14. L.J. Yan, C.M. Gao, B.X. Zhao, Q.M. Sun, F. Wang. Fast transmission photothermal radiometry via sampling by an internal computer sound card [J]. Journal of Electronic Science and Technology, 2015,13(1): 73-77.

Utilizing the tail energy of circulating water in coal-fired power plants: water power station

Z. Cong[a], K.M. Yang[†, b], W.F. Wang[c], Q.H. Han[a] and L. Li[a]
[a]*Zhongshi Yitong Group, Jinan, Shandong province, China*
[b]*College of Thermal Energy Engineering, Shandong Jianzhu University, Jinan, China*
[c]*Tengzhou Xinyuan Thermal Power Co. Ltd, China Huadian Corporation, China*
[†]*Email: ykm_168@163.com*

The circulating water system in coal-fired power plant is a complex system, in which certain energy can be extracted. To use this energy, the amount of energy should be computed firstly, and then suitable means to utilize it will be found out according to the local situation. A new method named water power station will be shown in this paper, which aims to increase the thermal efficiency of power plant by recycling part of hydroenergy of cooling water. Taking three different plant units, 2×1000MW, 2×600MW, 2×300MW respectively, as the research objects to be studied and analyzed. The results show that water power station can make some economical benefits and social benefits.

Keywords: Circulating water, Tail energy, Water power station.

1. Introduction

The circulating water system in coal-fired power plant is the biggest cooling system, in which there is certain energy can be extracted. With the ultimate installation increasing, the system will be bigger and bigger, and energy that can be extracted gets more and more. This system with a cooling tower called closed circle water system has been studied in great details for many years. The open loop system which draws water from river or sea and drains back directly has been mentioned barely, especially the utilizing of tail energy of the drained water. There are some examples about utilizing the tail energy of drained water with a big height of water, such as Pannan Xiangshui power plant in Guizhou and Zhangze power plant. But when the height of water is smaller than 5m, especially when the power plant is near to sea, this method has not been applied widely. There is only one example Huandian Laizhou power plant by now[1-3].

Utilizing the tail energy of circulating water to generate electricity can reduce the service-power consumption rate of the coal-fired power unit by supplying power to the whole system, increase the practical capacity of coal-fired power unit, and decrease coal consumption. It can make full use of existing infrastructure, and reduce the investment to an extreme. With the development of society and economy, the amount of energy is bigger and bigger, so energy-savingandcost-reducing have been mentioned more and more. So this method, water power station, aiming to recycle the tail energy has a necessary and bright prospect.

This article takes open loop system which draws water from river or sea and drains back directly, even has a small height of water, as the research object to study the feasibility of recycling tail energy in circulating system. Three different circulating system in different coal-fired power units were selected, 2×1000MW, 2×600MW, 2×300MW respectively, to be studied and analyzed.

2. Calculation and Discussion

Due to the limitation of natural conditions, but also for flood control, the ground of coal-fired power plant is higher than both of inlets and outlets of circulating system. The top of condenser is about 10m higher than circulating pipes, at the same time, water flow is quite large, and so the energy carried by drain water is massive[1].

Coal-fired power units with level of 1000MW generally have been equipped with three sets of circulating water pumps, and the flow of single pump about is 10 m^3/s. In winter two pumps serve one single plant unit, and three pumps for one plant unit or 5 pumps for 2 plant units in summer. When two plant units work in the same time, the flow of circulating water has a minimum 30 m^3/s and a maximum 60 m^3/s.

Coal-fired power units with level of 600MW generally have been equipped with two sets of circulating water pump, and the flow of single pump is about 11 m^3/s. In winter one pump serves one single plant unit or three pumps for two plant units and two pumps for one plant unit or 3 pumps for 2 plant units in summer. When two plant units work in the same time, the flow of circulating water has a minimum 22 m^3/s and a maximum 44 m^3/s.

Coal-fired power units with level of 300MW generally have been equipped with two sets of circulating water pump, and the flow of single pump is about 6.5 m^3/s. In winter one pump serve one single plant unit, and two pumps for one plant unit or 3 pumps for 2 plant units in summer. When two plant units work in the same time, the flow of circulating water has a minimum 13 m^3/s and a maximum 26 m^3/s.

Taking a small hydroelectric generating set (HGS) that have been put into operation as example, its design water head is 2.92m, and is about 3.0m for actual water head. Now assuming the water head of those systems are all 3.0m.

According to the formula (1), the theory output (power output) of circulating water flow can be calculated[4-5].

$$N = 9.81 \times \frac{\overline{Q_s} + \overline{Q_x}}{2} \times H \qquad (1)$$

in which, N, theory output (power output) of circulating water flow, kW

$\overline{Q_s}$, average fluid flow on upper section, m^3/s

$\overline{Q_x}$, average fluid flow on lower section, m^3/s

H, head of water, m

In the circulating system, $\overline{Q_s}$ is equal to $\overline{Q_x}$, and both of them are from normal operational conditions. According to the formula (1), calculated tail energy in different plant units are shown in Table 1.

Table 1. Tail energy in different plant units.

No.	Coal-fired power unit capacity (*MW*)	Head of water (*m*)	Number of operating pumps	Flow (m^3/s)	Power output (*kW*)
1	2×1000	3	4~6	40~60	1177.2~1765.8
2	2×600	3	2~4	22~44	647.46~1294.92
3	2×300	3	2~4	13~26	382.59~765.18

There has an energy loss during the process of transformation, so the actual power output of generator is

$$N_G = 9.81 \cdot \eta \cdot H \cdot Q \qquad (2)$$

in which, η —efficiency of hydro-generator units,

H— rated water head of plant unit, m

Q— flow of the circulating water, m^3/s

Regarding $9.81 \cdot \eta$ as a coefficient k which usually varieties from 7.5 to 9.0[6], and it equals to 8.5 constantly. So the capacities of hydroelectric generating sets installed into different plant units are shown in Table 2.

Table 2. Capacities of different HGSs.

No.	Coal-fired power unit capacity (*MW*)	Head of water (*m*)	Flow (m^3/s)	Coefficient k	Capacities of different HGSs (*kW*)
1	2×1000	3	40~60	8.5	1020-1530
2	2×600	3	22~44	8.5	561-1122
3	2×300	3	13~26	8.5	331.5-663

From Table 2, it can be drawn out that the energy of circulating water is massive and should be used.

3. Analysis of Economical and Social Benefits

Economical analysis of HGS is mainly about construction cost, operation cost and economical benefits, and it is different corresponding to each plant units. For example, the construction cost of a HGS for 2×1000MW unit is 22million RMB, and it will decrease when capacity of plant unit becomes small.

The new water power station basically realizes unattended, and a few of auxiliary equipment need to operate continuously. So auxiliary power rate is less than 2%, leading to low operating cost. Electricity generated by HGS has been used as auxiliary power directly, and that means electricity has been sold at the price same to coal-fire electricity but it is come from hydroelectric power in a sense. It is a typical conduct with low investment, but high return.

From Table 2, assuming the capacities of water power station are 1275kW, 840kW and 500kW respectively, and annual generating hours are 5500h, the amount of annual power generation and coal saving (280 grams per KWh) can be calculated, shown in Table 3.

Table 3. Annual power generation and coal saving (different capacity).

No.	Coal-fired power unit capacity (*MW*)	Capacities of different HGSs (*kW*)	Annual generating hours (h)	Annual power generation KWh	Coal saving (t)
1	2×1000	1275	5500	7012500	1963
2	2×600	840	5500	4620000	1293
3	2×300	500	5500	2750000	770

Taking 2×1000MW unit as an example, and according table 3, water power station can supply electricity about 7 million KWh, about 28 million RMB. The operating cost is about 3 million RMB, so it will take 9 years to recoup the capital outlay.

It can be seen from Table 4 that with the increase of water head, the annual power generation and coal saving increase obviously.

Table 4. Annual power generation and coal saving (different water head).

No.	Coal-fired power unit capacity (*MW*)	Head of water (*m*)	Capacities of different HGSs (*kW*)	Annual power generation KWh	Coal saving (t)
1	2×1000	3	1275	7012500	1963
2	2×1000	4	1700	9350000	2617
3	2×1000	5	2125	11687500	3272

From Tables 3 and 4, it is can be seen that water power station can make some economical benefits directly, and also can make some social benefits. Because reducing coal consumption can cut down the emissions of air pollutants, such as NO_x and SO_x, and lessen the emissions of carbon dioxide or other greenhouse gases.

4. Conclusion

The circulating water system in coal-fired power plant is a complex system, in which there is certain energy can be extracted. Utilizing the tail energy of circulating water to generate electricity can reduce the service-power consumption rate of the coal-fired power unit by supplying power to the whole system, increase the practical capacity of coal-fired power unit, and decrease coal consumption. It can make full use of existing infrastructure, and reduce the investment to an extreme.

To use this energy, the amount of energy should be computed firstly, and then suitable means to utilize it will be found out according to the local situation. A new method named water power station will be shown in this paper, which aims to increase the thermal efficiency of power plant by recycling part of hydroenergy of cooling water. Taking three different plant units, 2×1000MW, 2×600MW, 2×300MW respectively, as the research objects to be studied and analyzed.

The results shown that water power station can make some economical benefits directly, and also can make some social benefits.

Acknowledgments

This paper is supported by Research Fund for the Doctoral Program of Shandong Jianzhu University, project number 0000601337.

References

1. Meng Xilan. The hydroenergy utilization of circulating water in coal-fired power plant. Energy and Environment, 99: 32-34.
2. Liu Hua, Ding Qian. Application of circulating water tail water power generation in thermal power plant. Energy Conservation, 2015, 3 (390): 44-45.
3. Li Shun, Meng Xiongbang, Li Tailai. Design and practice about utilizing tail water of coal-fired power plant to generate electricity. Hydropower and New Energy, 2010, 2 (88): 72-74.
4. Hydroenergy design code for small hydro power projects, SL76-2009.
5. Zhong Xingang, Wu Taimeng, Feng Chuangxian. Several experiences in the design of hydropower station with tubular turbine. Guangdong Water Resources and Hydropower, 2004, 5: 29-31.

6. Li Zhongkui, Ma Jiming. Hydraulic and hydropower engineering. Beijing: Science Press, 2004.

China's shale gas policy and the effect for the industry

Q. Zhao
College of Geoscience and Surveying Engineering,
China University of Mining & Technology
Beijing, 100083, China
†E-mail: Bill.zao@163.com
www.cumtb.edu.cn

D.X. Liu* and S. Yang†
Research Institute of Petroleum Exploration and Development-Langfang, Langfang
Langfang, 065007, China
*E-mail: *Liudexun@petrochina.com.cn, †Yangshen@petrochina.com.cn*

D.M. Luo
Natural Gas Industry Mothly Issued
Chengdu, 610000, China
E-mail: ldm@trqgy.cn
www.trqgy.cn

Many shale gas policies have been issued by the Chinese government to promote shale gas resources development and utilization in 2011-2015. In order to forecast the shale gas policies in the future, the effect of the different kinds of policy has been analyzed, and the shale gas development economic benefit was analyzed in several cases with different policies assumption. Based on present cost of shale gas exploration and development, the subsidy is necessary in 2016-2020, in order to promote the rapid growth of shale gas production. Considering the environmental impact of shale gas development, safety and environmental, the protection supervision system needs to be strengthened by the government.

Keywords: Shale gas; Policy; Government; Shale gas resources.

1. Introduction

The successful development of shale gas in U.S. has changed the pattern of energy supply. Shale gas will play an important role in countries such as China, Australia, India, Canada, Indonesia and Poland. China is now actively committed to shale gas operations. Until now, 300 wells have been drilled. A lot of organizations made shale gas resources assessment of China. The result shows the geological resources are about 100-130Tcm (3500-4700Tcf) (MLR,

IHS), the recoverable resources are about 10-36Tcm(350-1270Tcf) (CAE, EIA). Marine shale gas development in Sichuan basin has gotten a big success, and the recoverable resources are 8.82Tcm (Zou *et al.*)

Table 1. Shale gas resource assessment results of different organizations.

Organizations	Geological resource ($\times10^{12}m^3$) (Tcf)	Recoverable resource ($\times10^{12}m^3$) (Tcf)	Year
IHS	100(3500)	——	2008
EIA	——	36(1270)	2011
Ministry of Land and Resources(MLR)	134(4700)	25.08(880)	2012
China AcademyEngineering	——	10-15(350-530)	2012

Shale gas recoverable resources in U.S. are about 20.7 Tcm, and 370 Bcm gas were produced in 2014. 160 Bcm gas should be produced in marine shale per annul, with about 8.82 Tcm recoverable shale gas. Considering the lower degree of prospecting and the complexity of reservoir, at least 80 Bcm shale gas should be produced per annul.

Until now, over 300 wells have been drilled in Jiaoshiba block and in Shunan area. CNPC (Petrochina) is developing the shale gas resources in Shunan area, the southern Sichuan basin, including two national pilot areas of Changning-Weiyuan and Zhaotong. SINOPEC puts emphasis on the surrounding of Sichuan Basin and is developing the shale gas in Jiaoshiba block, Fuling national pilot. CNOOC has carried out shale gas evaluation in Wuhu, Anhui Province and is now evaluating the shale gas in coal-bearing formations with CUCBM. Yanchang Petroleum Group is conducting the experiments for exploring and developing the continental shale gas, such as Yan'an national pilot.

The main shale gas policies and their effort for China's shale gas industry will be introduced in this paper. Different cases for typical shale gas well will be analysed to show the effort. The policies have made a very important effect on shale gas production. It's estimated that the shale gas production will increase to 1.5 Bcm in 2014 and 5.5 Bcm in 2015. Based on present cost of shale gas exploration and development, we think the China government will continually give a shale gas production subsidies in 2016-2020, in order to promote the rapid growth of shale gas production.

2. Shale Gas Policies and the Effort

To increase gas supply, alleviate the contradiction between supply and demand, the Chinese government actively promotes the development of shale gas resources. So the government introduced a series of industrial policies.

"Notice about shale gas development and utilization of subsidy policy" was issued by Ministry of Finance and National Energy Administration. "Shale gas industrial policy" was issued by National Energy Administration. "Notice about the work of strengthening shale gas resources prospecting & exploitation and supervision" was issued by Ministry of Land and Resources. The Ministry of Science and Technology has established technology research projects.

2.1. *Exploration and extraction rights policy*

Shale gas is appointed as a separate mineral species. In December 2011, the MLR formally approved shale gas as one of the 172 minerals. Its investment will be managed and regulated as an individual mineral. Shale gas can be prospected and produced in oil & gas blocks, but additional shale gas exploration or extraction rights should be applied for.

Since 2011, two rounds of shale gas exploration right bidding have been carried out, and in some conventional oil and gas blocks shale gas exploration right have been added.

Table 2. Shale gas exploration blocks in China.

Shale gas mining right	No.	Area (km^2)	Location
Apply for additional shale gas mining right	33	146258.38	The sichuan basin and its peripheral ordos basin
The first round of bidding	2	4236.81	Chongqing, Guizhou, Hunan
The second round of bidding	19	19504.81	Chongqing, Guizhou, Hunan, Hubei, Jiangxi, Zhejiang
Total	54	170000	Chongqing, Hunan, Hubei, Guizhou, Jiangxi, Zhejiang, Shanxi, Sichuan

2.2. *Financial subsidies policy*

"Notice about shale gas development and utilization of subsidy policy" releases 0.4 RMB/m^3 financial subsidies for shale gas produced in 2012-2015 by Ministry of Finance and NEA, in Nov. of 2012. If a shale gas company want to get the subsidies, the next five conditions must be met: (1) Gas Produced form shale reservoir with TOC>1.0%;(2) Interbedded formation ratio <20%;(3) Single interbedded formation (such as sandstone or limestone) < 1 m;(4) The shale gas should have been produced and utilized;(5) The shale gas should have been metered accurately.

A series of shale gas policy, promoted the Chinese shale gas development pace. 1.2 Bcm shale gas was produced in 2014 and the production will reach 5.8 Bcm in 2015. 77.4 million dollars subsidies will be received for shale gas produced in 2014, and the subsidies will reach 0.37 billion dollars in 2015.

In Jiaoshiba block, over 1 Bcm shale gases were produced in 2014, and the production will reach 3.2 Bcm in 2015. And about 0.28 billion dollars will be received till 2015. In Shunan area, about 0.15 billion shale gas were produced in 2014, and the production will reach 2.6 Bcm in 2015. And about 0.2 billion dollars will be received till 2015.

2.3. *Technology research projects*

Ministry of Science and Technology established the project "critical techniques for shale gas exploration and development" in 2010. Two projects in "973" project found, have been set up, which are "Southern Paleozoic shale gas resource potential accumulation mechanism and evaluation", "Basic research on Southern marine shale gas development".

Most horizontal wells got a promising initial gas producing rate. In Jiaoshiba block, the average test initial gas producing rate is about 0.33 Mcm/d, and the max test rate is 0.55 Mcm/d. In Shunan area, the average test initial gas producing rate is about 0.125 Mcm/d, and the max test rate is 0.24 Mcm/d.

Drilling cycle of horizontal wells reduced by a big margin. Changning: reduced from 140 days to 60 days.Weiyuan: reduced from 150 days to 90 days. Zhaotong: reduced from 140 days to 70 days. Jiaoshiba: reduced from 83 days to 60 days.

3. Case Study of Typical Shale Gas Well

According to the geological characteristics, three types of well with different initial gas producing rate were chozen as typical wells. The initial gas producing rates of the tree typle wells are 20 Mcm/d, 15 Mcm/d and 10 Mcm/d, and the EUR are 0.139 Bcm, 0.11 Bcm and 0.089 Bcm.

3.1. *Typical well without subsidies*

If with a 0.2 Mcm/d initial test producing rate, the IRR is 22.75%. If with a 0.15 Mcm/d initial test producing rate, the IRR is 12.13%. If with a 0.1 Mcm/d initial test producing rate, the IRR is 4.42%.

If well cost can reduce to 600 million RMB, the IRR will increase by a big margin. If with a 0.2 Mcm/d initial test producing rate, the IRR is 34.51%. If with a 0.15 Mcm/d initial test producing rate, the IRR is 19.57%. If with a 0.1 Mcm/d initial test producing rate, the IRR is 9.28%.

Table 3. Economic analysis of typical well with well cost 70 million RMB.

Parameters	Economic analysis		
Initial gas producing rate(Mcm/d)	20	15	10

EUR(Bcm)	0.139	0.11	0.089
horizontal segment(m)	1500	1500	1500
Well cost(million RMB)	70	70	70
Price(RMB/m^3)	1.6	1.6	1.6
NPV(million RMB)	24.87	1.01	-7.89
IRR(%)	22.75	12.13	4.42

Table 4. Economic analysis of typical well with well cost 60 million RMB.

Parameters	Economic analysis		
Initial gas producing rate(Mcm/d)	20	15	10
EUR(Bcm)	0.139	0.11	0.089
horizontal segment(m)	1500	1500	1500
Well cost(million RMB)	60	60	60
Price(RMB/m^3)	1.6	1.6	1.6
NPV(million RMB)	24.87	16.68	1.71
IRR(%)	34.51	19.57	9.28

3.2. *Typical well with 0.4RMB/m3 subsidies in 2014-2015*

Considering the 0.4 RMB/m3 subsidies in 2014-2015, the IRR will increase sharply. If with a 0.2 Mcm/d initial test producing rate, the IRR is 41.62%. If with a 0.15 Mcm/d initial test producing rate, the IRR is 21.61%. If with a 0.1 Mcm/d initial test producing rate, the IRR is 8.93%.

Table 5. Economic analysis of typical well with 0.4RMB/m^3 subsidies in 2014-2015.

Parameters	Economic analysis		
Initial gas producing rate(Mcm/d)	20	15	10
EUR(Bcm)	0.139	0.11	0.089
horizontal segment(m)	1500	1500	1500
Well cost(million RMB)	700	700	700
Price(RMB/m^3)	1.6	1.6	1.6
NPV(million RMB)	39.46	19.22	1.24
IRR(%)	41.62	21.61	8.93

4. Conclusions

China's government is full of confidence with shale gas development. The National Energy Administration has issued an "Action Plan for Accelerating Exploration & Development of Shale Gas", in which 30-50 Bcm shale gas will be produced in 2020 and 80-100 Bcm in 2030.

Most companies cannot enjoy the subsidies policy in 2012-2015. According to the shale gas current situation, only PetroChina and SINOPEC can obtain some subsidies of shale gas production in 2012-2015. Other companies can't realize the effect development of shale gas, and cannot enjoy the subsidy in 2012-2015. If the single well cost can achieve the expected value is still need to practice. Single well cost is 600 million RMB in the process of the economic evaluation of typical block, but the single well cost is about 700-800 million RMB. Though we think the single well cost is very probably less than 600 million RMB after 2016, it is still has some. China's shale gas resource conditions are worse than that of the US. Every shale gas reservoir is different, so different shale reservoir need different suitable technology to develop effectively. We need reduce the cost gradually and we think 2 to 3 years is necessary to have a big cut of the cost in a new shale gas block. Shale gas development has impact on the residents' living environment.

Based on present cost of shale gas exploration and development, the subsidy is necessary in order to promote the rapid growth of shale gas production in 2016-2020. Considering technological progress, single well cost has a big possibility to decrease to less than 60 million RMB. We advise government giving 0.3 RMB/m3 subsidies during 2016 to 2020, so that IRR of shale gas development project can reach 10% or higher. Considering the environmental impact of shale gas development, safety and environmental protection supervision system need to be strengthened by government. China government issued a new Environmental Protection Law on January 1, 2015. If there is a possibility of environmental pollution, precautionary measure must be done to protect against it. How to eliminate environment pollution from the source is a top priority. We advise strengthening the recycling of drilling fluid, fracturing flowback fluid, and strengthening water drilling fluid research for horizontal well, to gradually replace oil-base drilling fluidt68.

References

1. Technically Recoverable Shale Oil and Shale Gas Resources: An Assessment of 137 Shale Formations in 41 Countries Outside the United States, U.S. Energy Information Administration, 2013. http://www.eia.gov/analysis/studies/worldshalegas/.
2. U.S. Crude Oil and Natural Gas Proved Reserves, 2013. U.S. Energy Information Administration, 2014.
3. Technically Recoverable Shale Oil and Shale Gas Resources: An Assessment of 137 Shale Formations in 41 Countries Outside the United States.
4. Notice about the work of strengthening shale gas resources prospecting & exploitation and supervision, Ministry of Land and Resources.
5. Notice about shale gas development and utilization of subsidy policy, Ministry of Finance and National Energy Administration.
6. Shale gas industrial policy, National Energy Administration.

Research on influencing factors of rebound effect of energy consumption — Analysis based on provincial panel data in China*

Y. Han, J.H. Shi† and X.N. Wu
College of Business Administration, Northeastern University, Shenyang, Liaoning, China
†Email: sjhyong@163.com

This paper calculates the energy rebound effect in different provinces of China, puts forward the factors influencing the energy rebound effect, and uses the relevant indicators and panel data model to analyze the influence factors. The result indicates that the increasing proportion of industry will improve the energy rebound effect, while the increasing proportion of tertiary industry in some areas will reduce the energy rebound effect. The paper provides scientific basis for the inhibition of energy rebound effect.

Keywords: Energy intensity; Rebound effect of energy consumption; Panel data.

1. Introduction

The rebound effect was firstly described by William Stanley Jevons in 1865. According to Jevons, it is a confusion of ideas to suppose that the economical use of fuel is equivalent to diminished consumption. The very contrary is the truth.

A lot of scholars have studied on the energy rebound effect extensively. Harry D. Saunders (2007) compared and analyzed eight production and cost functions to study the energy rebound mechanism, without measurements of concrete industry or country.

Terry Barker (2007) studied the energy rebound effect of England economy from indirect effect and economy wide effects perspectives and found that the size had been increasing since 2000.

Sang-Hyeon (2007) studied the energy rebound effect of Korean domestic use of electricity from micro and macro perspectives and found two factors, products and income, influence the size. However, the domestic use of electricity doesn't represent the general energy consumption.

In conclusion, most researches mainly focus on the measurements and empirical analysis of the energy rebound effect and few documents are related to

* This work is supported by Liaoning province philosophy social sciences fund project (L13DJY089)

influencing factors of energy efficiency.

This paper calculates the energy rebound effect in China, puts forward the factors influencing the energy rebound effect, and provides scientific basis for the inhibition of energy rebound effect.

2. Build the Calculation Model of Provincial Energy Rebound Effect

The calculation of provincial energy rebound effect in China uses the Solow remainder approach, incorporating three input factors of neoclassical product functions.

$$\ln Y_{it} = \ln A_i + \alpha_{it} \ln K_{it} + \beta_{it} \ln L_{it} + \gamma_{it} E_{it} + \varepsilon_{it} \quad (1)$$

Parameter α, β, γ represents capital, labor and the output elasticity of energy respectively. According to Solow's approach,

$$\sigma = \frac{g_A}{g_Y} = 1 - \frac{\alpha g_K}{g_Y} - \frac{\beta g_L}{g_Y} - \frac{\gamma g_E}{g_Y} \quad (2)$$

Eventually, the rebound effect of technological progress is,

$$RE_t = \frac{N_t}{M_t} = \frac{\sigma_t (Y_t - Y_{t-1}) EI_t}{Y_t (EI_{t-1} - EI_t)} \quad (3)$$

EI is energy efficiency, M_t is energy saving and$N_t$is energy consumption due to technological progress.

3. Influencing Factors of Provincial Energy Rebound Effect

3.1. *Identify influencing factors of energy rebound effect*

Construction of energy: A new technology in energy production leads to a decline in energy prices. So energy construction has a certain connection to the energy rebound. It is said as energy construction investment. Industrial structure: At present the accelerating of industrialization and industry structure heavily promoting energy consumption. It is said as the proportion that industry takes in GDP. Social development advancing level: The size of rebound effect varies in different economic levels. With the development of society, the proportion of tertiary industry is increasing and it is said as the tertiary industry accounted for the proportion of GDP percentage.

3.2. *Build the model of influencing factors of provincial energy rebound effect*

Considering data availability, energy construction input, industrial structure, and social development advancing level are selected to analyze influencing factors and build the model of provincial energy rebound effect

$$R_{it} = C_{it} + m_{it} I_{it} + n_{it} T_{it} + p_{it} S_{it} + u_{it} \qquad (4)$$

R_{it} is the percentage of energy rebound of placei and timet; I_{it} is the energy construction input of placei and timet, S_{it} is the industry proportion of placei and timet, T_{it} is the tertiary industry of placei and timet,C_{it} is the constant term.

3.3. *Source and processing of data influencing factor model*

Considering consistency and availability of statistical data, Chongqing city is incorporated into Sichuan Province and Tibet is not included in this study. Data needed are energy rebound sizes, value added of provincial industries and tertiary industries and energy construction input of 29 provinces in China from 1996 to 2011, of which provincial energy construction input is based on China Energy Statistical Yearbook and the raw data of tertiary industry value added is from China Statistical Yearbook. Compared with GDP of every province, percentages of the proportion of industry and tertiary industry are drawn.

3.4. *Influencing factors of rebound effect*

The influencing factors of provincial energy rebound effect are estimated with the variable coefficient model. The result is shown in Table 1.

Table 1. Estimation result of influencing model of energy rebound effect.

Influencing factors	Input I	Industry proportion S	Tertiary industry proportion T
Beijing	-4.884	8.688	14.815
Tianjin	-0.059	11.287	14.215
Hebei	-2.256	4.778	-12.345
Shanxi	0.585	0.570	-8.610
Inner Mongolia	0.050	4.829	0.727
Liaoning	-0.373	10.203	-4.974
Jilin	-1.488	3.827	-0.214
Heilongjiang	-2.135	6.745	0.954
Shanghai	-1.433	18.509	15.959
Jiangsu	-1.439	13.319	9.608
Zhejiang	-1.009	2.075	5.206
Anhui	0.348	5.004	1.893
Fujian	-1.445	13.496	0.890
Jiangxi	-1.846	11.447	5.374
Shandong	-1.886	9.120	10.213
Henan	1.734	2.545	1.810
Hubei	-0.230	0.956	5.956
Hunan	-3.141	17.711	3.587
Guangdong	-0.196	7.116	7.236

Guangxi	-1.197	6.726	3.605
Hainan	9.735	4.446	-2.069
Sichuan	0.347	3.492	-3.736
Guizhou	5.585	2.641	-5.100
Yunnan	0.407	-0.645	-8.001
Shanxi	0.549	0.030	-10.679
Gansu	0.766	3.771	-1.360
Qinghai	0.787	-0.657	-13.054
Ningxia	6.513	-0.884	-10.164
Xinjiang	1.518	4.898	-4.452

Note: Constant terms are not listed.

4. Conclusion and Advice

The proportion of industry has positive impact on energy rebound effect, which illustrates that increasing industry proportion of most provinces promotes energy rebound effect. More influential provinces are Shanghai, Jiangsu, Fujian, Tianjin, Liaoning, Beijing and so on. Local governments are expected to strengthen management of high energy consumption, improve energy efficiency, save energy and reduce emission.

Energy construction input in some provinces has a positive influence on energy rebound, such as Shanxi, Inner Mongolia, Sichuan and so on, located mainly in the central and western regions. There are also provinces with a negative sign, such as Beijing, Tianjin, Liaoning, and Shanghai. That is because Compared with provinces in the central and western regions, the ones in eastern region make full use of energy. Inputting different energy constructions according to different regions contributes to saving energy.

Increasing the proportion of tertiary industry suppresses energy rebound effect, and it indicates social inevitable progress. As for provinces with positive influence such as Beijing, Tianjin, Jiangsu and Fujian, importance should be attached to the tertiary industry, while in provinces with negative influence, Hebei, Shanxi, Liaoning and so on, the tertiary industry is expected to be promoted, which is bound to suppress energy rebound effect.

References

1. Greening L A, David L G, Carmen D. Energy Efficiency and Consumption: The Rebound Effect-A Survey. (Energy Policy, 2000), pp. 389-401.
2. Brookes, L.G, Energy policy, the energy price fallacy and the role of nuclear energy in the UK. (Energy Policy, 1978), pp. 94-106.

3. Saunders, H.D, Does predicted rebound depend on distinguishing between energy and energy services. (Energy Policy, 2000), pp. 497-500.
4. Blair, R.D., Kaserman, D.L., Tepel, R.C. The impact of improved mileage on gasoline consumption. (Economic Inquiry, 1984), pp. 209-217.
5. Wang Weiqun, Zhou Qun. Improved model and empirical research of energy rebound effect. (Journal of Management, 2008), pp. 688-691.
6. Zhou Yong, Lin Yuanyuan. The estimation of technological progress on the energy consumption returns effects. (Economists, 2007), pp. 45-521.

Study of unbalanced load's influence on error characteristics of three-phase two-element combined transformer

G. Liu
Sichuan Electric Power Corporation Metering Center, Chengdu, China
Chongqing University, Chongqing, China
Email: lg400044400044@163.com

G.C. Lei
University of Electronic Science and Technology of China, Chengdu, China

B. Jiang, F.Z. Zhang, N. He and W. Jiang
Sichuan Electric Power Corporation Metering Center, Chengdu, China

The single-phase verification method is widely used in the verification of combined transformer at present. When combined transformer is verified, the state doesn't agree with its actual operational state. Furthermore, this method cannot reflect secondary load imbalance's influence on its error characteristics. In this paper, through the analysis of the data differences between single-phase verification method and three-phase verification method in verification of three-phase two-element combined transformers, suggesting that three-phase verification method which can simulate the operational status of combined transformer should be adopted to get more accurate results. The imbalance of secondary load has no impact on the error characteristics of three-phase two-element combined transformer.

Keywords: Error characteristics; Three-phase two-element; Combined transformer; Unbalanced load

1. Introduction

Three-phase combined transformer has been broadly used in systems of voltage class not more than 35kV as a transforming equipment from high voltage, great current signal to low voltage and small current signal. Whether three-phase combined transformer has good performance directly relates to power distribution network's security, stable operation, accuracy and reliability of measurement and vital interests of the masses. Therefore, it is really significant to ensure that three-phase combined transformer has good operating performance.

Three-phase two-element combined transformers accounts for 70%–80% of three-phase combined transformers. At present, in terms of factory test, first test, periodical inspection of three-phase two-element combined transformers, single-phase verification method has been used to inspect its current unit and voltage unit, there is no current of current unit when using single-phase verification method to inspect voltage unit, and no voltage is applied to voltage unit when current transformer is checked, that is, the high voltage and great current's influence on errors were not considered. Only the phase verified is applied load during verification, however, all phases of three-phase combined transformers are at high voltage, great current and with load during actual operation situation. Therefore, whether data checked by single-phase verification method can reflect error characteristics of three-phase combined transformer has further to be verified. Single-phase verification method is unable to obtain error characteristics when phases loads are not balance, which is exist in practice. As a result, study must be done on error characteristics of three-phase combined transformer when the loads of phases are not balanced.

According to single-phase verification method and three-phase verification method to examine three-phase two-element combined transformer, Zhang Neng et al. found that for current unit, the differences of the two methods reduced with currency increased, namely electromagnetic field's influence on error characteristics of current unit when currency was low was great (especially when currency was at 1% and 5% of rated currency) and as the currency was rising, its influence can be ignored gradually, so electromagnetic field's influence needed to be considered when transformer was verified [2]. According to single-phase verification method and three-phase verification method to test three-phase two-element combined transformer respectively, Jiang Wei, Liu Kun et al. found that influence of voltage on current unit mainly embodied at 1% and 5% of rated currency. A current unit which was qualified according to single-phase verification method might not be qualified in accordance with three-phase verification method, then the conclusion whether the combined transformer was qualified might be changed. The currency had influence on voltage unit as well and such influence was related to the design of transformer, manufacturing process, materials and other factors [3-4]. Zhou Bingshi and Song Jicheng discussed verification method and allocation of combined transformer [5-6]. However, the differences of results tested according to between single-phase verification method and three-phase verification method need further to be studied, and whether the second load unbalance of combined mutual inductor has influence on its error has not been reported yet.

In this paper, operational conditions of high voltage and high current of three-phase two-element combined transformer was simulated and three-phase verification method was used to test the combined transformer (including voltage and current). Error characteristic of three-phase combined transformer were analyzed when the second load was not balanced, which provided reference to design, manufacturing and examining of three-phase combined transformer.

2. Test Introduction

2.1. *Test object*

The version of three-phase two-element combined transformer tested herein was JLSZY-10, and its main parameters are shown in Table 1:

Table 1. Parameters of combined transformer.

Parameter type / Type of transformer	Rated primary voltage	Rated secondary voltage	Accuracy class	Rated secondary load
PT	10kV	100V	0.2	20VA
CT	20A	5A	0.2S	10VA

2.2. *Test equipment*

(1) 3 sets of 10 kV/100V, $10/\sqrt{3}$ kV/$0.1/\sqrt{3}$ kV standard voltage transformers, with an accuracy class of 0.05.
(2) 3 sets of 0A~200A standard current transformers, with an accuracy class of 0.05.
(3) Three-phase combined transformer calibrator, current loading box, current booster and voltage booster etc.

2.3. *Test principle*

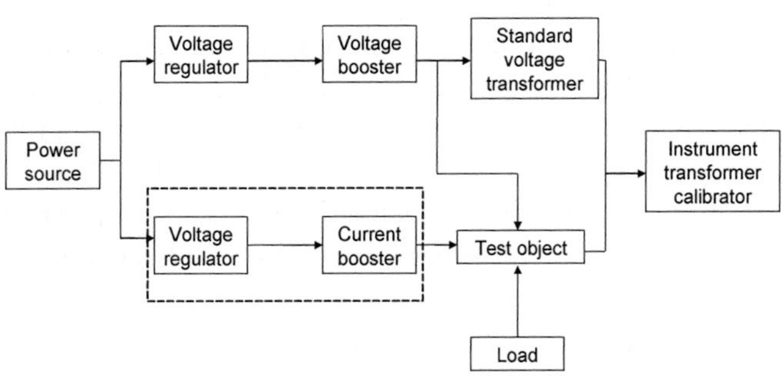

Figure 1. Principle diagram of voltage test.

Because there is similarity between principle of current error test and that of voltage error test, in this paper, only voltage error test principle is elaborated. Principle diagram of voltage test is shown in Fig. 1, when single-phase verification method was adopted to test voltage error, only voltage was applied (dotted line part is excluded), voltage was gained through voltage regulator and voltage booster, the primary voltage was put on standard voltage transformer and tested combined transformer respectively, and the secondary voltage was connected to calibration to do experiment. When in accordance with three-phase method to test voltage error, voltage regulator and current booster were used to gain rated current and put it on the tested unit, then through voltage regulator and voltage booster different voltages were obtained, the primary voltage was put on the side of standard voltage transformer and tested combined transformer respectively, and the secondary voltage was joined to calibration to do experiment. Wiring diagram of verifying combined transformer is shown in Fig. 2.

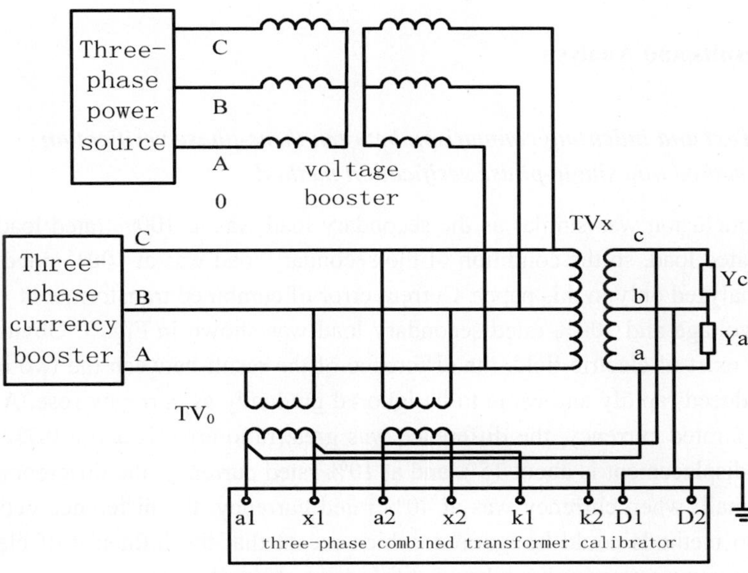

(a) Wiring diagram of voltage unit verified

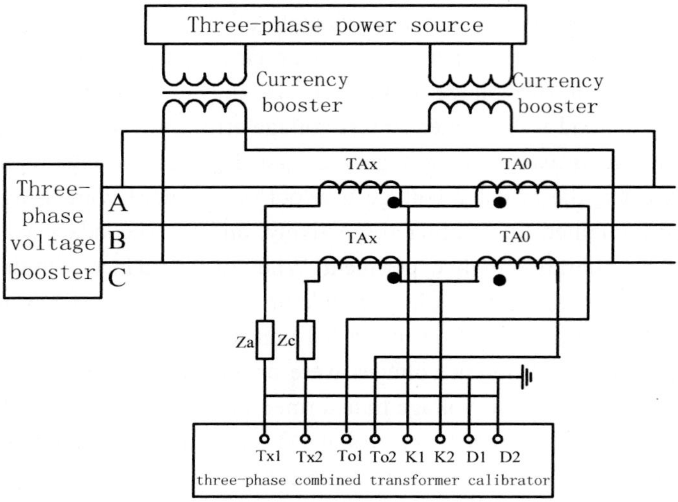

(b) Wiring diagram of current unit verified

Figure 2. Wiring diagram of examining combined transformer.

3. Results and Analysis

3.1. *Text and indenting comparison between three-phase verification method and single-phase verification method*

The conclusion was similar as the secondary load was at 100% rated load and 25% rated load, so the condition of the secondary load was at 100% rated load was analyzed only in this paper. Current error of combined transformer at 100% rated voltage and 100% rated secondary load was shown in Fig. 3. Obviously, due to exist of electric field, the difference of the result between the two methods reduced rapidly and came to be ignored gradually as currency rose. At 1% and 5% rated currency, the difference was great(ratio error is about 0.7%, and phase displacement is about 15′), and at 10% rated currency, the difference was very small, when currency was at 40% rated currency, the difference between the two methods could be ignored, which meant that the influence of electric field on current error was produced at low percentage of rated currency.

Figure 4 is the error of voltage unit of combined transformer at 100% rated currency and 100% rated secondary load, as can be seen, magnetic field did not exist using single-phase verification method, so phase displacement and ratio error data is relatively stable, only a little change took place at 100%-120% rated voltage. While the data obtained according to three-phase verification method was another change due to the influence of magnetic field, when voltage increased, the ratio error and phase displacement data changed greatly (the maximum ratio difference is about 0.1% and the phase displacement is about 5′). When combined transformer was verified according to three-phase verification method, the range of testing data was great as voltage increased.

Therefore currency (voltage) unit which was qualified when checked according to single-phase verification method may not be qualified when checked according to three-phase verification method. Hence, it is advised to adopt three-phase verification method simulating practical operation situation of combined transformer for accurately verification performance of combined transformer.

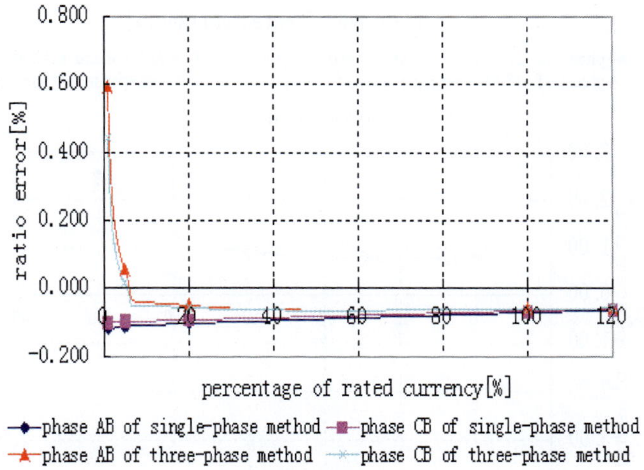

(a) Ratio error

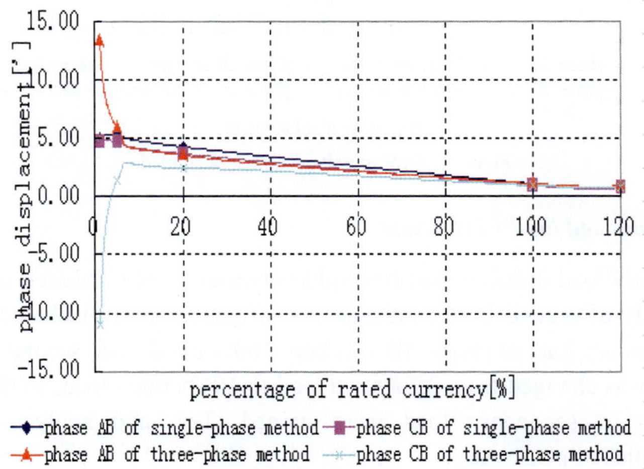

(b) Phase displacement

Figure 3. Error of current unit at rated load.

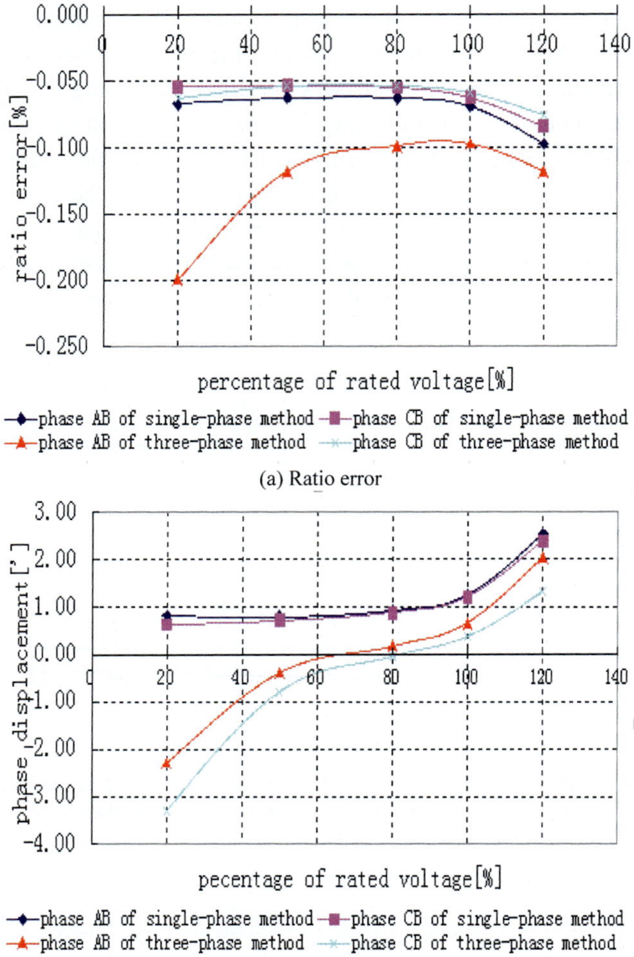

(a) Ratio error

(b) Phase displacement

Figure 4. Error of voltage unit at rated load.

3.2. *Imbalanced load's influence*

As secondary load imbalance of three-phase two-element combined transformer exists, study of second load's imbalance's influence on error characteristics is vital. Secondary load of phase AB was kept 100% rated load, secondary load of phase CB was changed from 25% rated load to 100% rated load, so that various imbalances of secondary load were gained. Test and analyze the error characteristics of phase AB.

3.2.1. *Error of voltage unit*

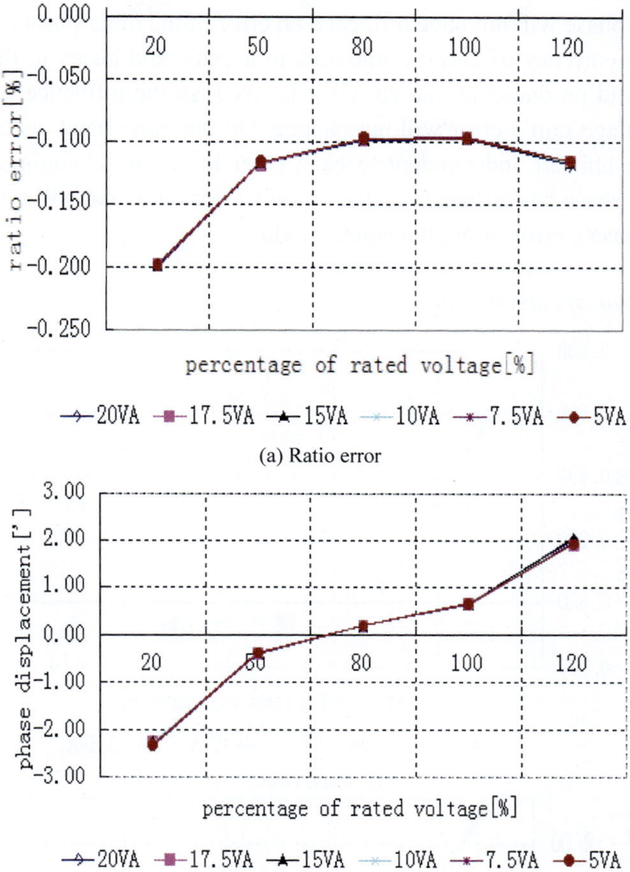

(a) Ratio error

(b) Phase displacement

Figure 5. Error of voltage unit of phase AB at different secondary load of phase CB.

Voltage unit's error of phase AB at 100% rated currency, rated load (20VA) and different secondary load of phase CB (to gain various unbalanced of secondary load) was shown in Fig. 5. It is obviously that as the voltage was rising, the ratio error of phase AB rose gradually and reached the maximum value at about 80% rated voltage to 100% rated voltage, then ratio error reduces a little as the voltage rose to 120% rated voltage. When secondary load of phase CB changed, ratio error of phase AB didn't change unless voltage changed, which meant that the ratio error of phase AB would not change as the secondary load of phase CB changed. When the secondary load of phase AB was at 25% rated load, voltage unit's error of phase AB had similar change. Therefore, as for the error of voltage unit of two-element combined transformer, change of the secondary load of one phase will not have influence on error of the other phase. The reason is when the currency of current unit rises to a value and keeps it, the value of magnetic field produced by the currency is fixed, so the influence of magnetic field on voltage unit's error will not change. On the other hand, as current unit and voltage unit are independent to each other in circuit, changing secondary load of one phase has influence on the phase's error only, and will not influence the other phase's error through magnetic field.

3.2.2. *Error of current unit*

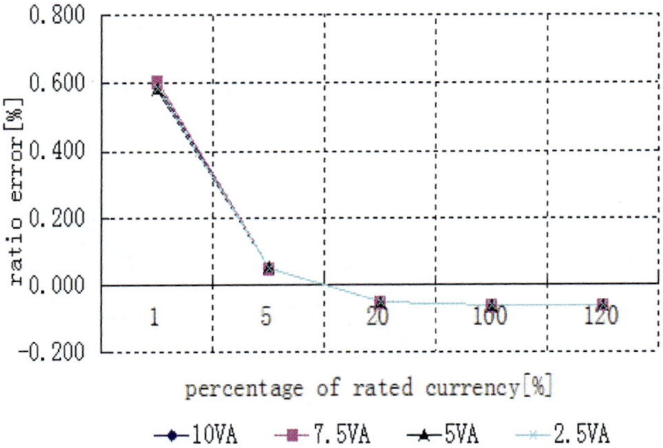

(a) Ratio error

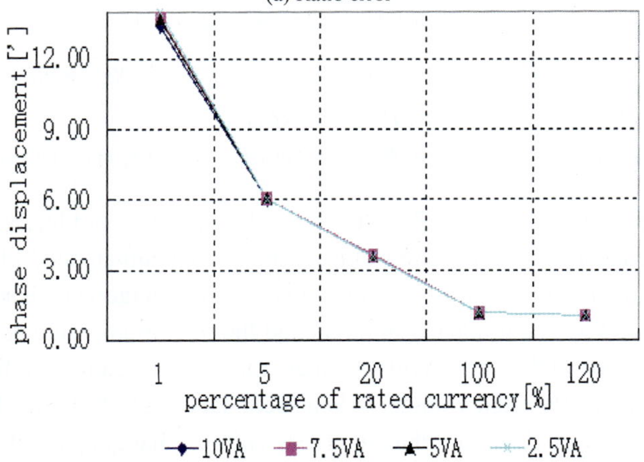

(b) Phase displacement

Figure 6. Error of current unit of phase AB at different secondary load of phase CB.

Current unit's error of phase AB at 100% rated voltage, rated load (10VA) and different secondary load of phase CB was shown in Fig. 6. It is obviously that as the current was rising, the current error of phase AB reduced. The ratio error came to be stable at 20% rated current, while phase displacement came to be stable at 100% rated current. When the secondary load of phase CB changed, the current error of phase AB wouldn't change at the same current. It meant that the current error of phase AB wouldn't change if the secondary load of phase CB changed. When the secondary load of phase AB was set at 25% rate load (2.5VA), the current error of phase AB had a similar change if the secondary load of phase CB changed. Therefore, as for current unit of three-phase two-element combined transformer, change secondary load of one phase will not have influence on the error characteristics of the other phase. The reason is similar to that of voltage unit.

4. Conclusion

a) If combined transformer was verified according to three-phase verification method which simulates the operational status of combined transformer, voltage's influence on the error of current unit happened mainly at low percentage of rated currency, when current was increasing, the influence reduced gradually to be neglected; currency of current unit had also great influence on the error of voltage unit. When three-phase two-element combined transformer was verified, the difference of the results tested between single-phase verification method and three-phase verification method were relatively great, the conclusion whether the transformer is qualified may even be changed. It is advised to use three-phase verification method to verify combined transformer.
b) As for current unit and voltage unit of three-phase two-element combined transformer, the change of secondary load of one phase will not have influence on the error characteristics of another phase, that is, unbalanced secondary load will have not influence on error characteristics of three-phase two-element combined transformer.

References

1. Jianpo Qiu, Yonghui Wang, Guoyi Yue, et al. 2010. Analysis on Problems During Error Test of Substation 10 kV Pressure Transformer and Countermeasures [J]. Hebei Electric Power. 3(29):6-8 (In Chinese).
2. Zhang Neng, Liao Hui-Min, Wu Yong-Yun, et al. 2012. Research on Three-phase Combined Transformer Verification device [J]. Automation and Instrumentation. 4:41-43 (In Chinese).

3. Liu Kun, Jiang Ying-Xia. Research on Two Calibration Methods of Three-phase Combined Transformer [J]. Sichuan Electric System Technology. 2010, 33(1):79-81 (In Chinese).
4. Jiang Wei, Yang Hua-Yun, Jiang Bo. 2011. Research on Three-phase Calibration Method of HV Three-phase Combined Transformer and its Implementation [J]. Electrical Measurement & Instrumentation. 2011, 48(547):46-49 (In Chinese).
5. Zhou Bingshi. 2011. Discussion about Verification Method for Combined Transformers [J]. Metrology & Measurement Technology. 2011, 31(4):32-34 (In Chinese).
6. Song Ji-Cheng, An Zhen, Ji Lian-Kai, An Zuo-Ping. 2009. Discussion on application and Distribution of Combined Transformer [J]. Transformer 2009, 46(7):14-15 (In Chinese).
7. Yu Xiao-Ming, Zhao Yuan, Ma Bin. 2010. The Development of Testing Device for Three-phase Voltage Transformer [J]. Shanxi Electric Power. 2010, 2(159):30-32 (In Chinese).
8. Karel Draxler & Renata Styblikova. 2009. Calibrating an Voltage Transformer to Achieve Reduced Uncertainty [C]. 2009 International Instrumentation and Measurement Technology Conference, 5-7 May 2009 Singapore.

Study on influence of voltage error by primary connection of three-phase three element combined transformer

N. He
Sichuan Electric Power Corporation Metering Center, Chengdu, China;
Email: hnsijia@163.com

L.D. Ren
Chengdu Institute of Product Quality Inspection CO.LTD/National Photovoltaic Products Quality Supervision and Inspection Center, Chengdu, China

L. Lai
State Grid Chengdu Power Supply Company, Chengdu, China

This paper focus on two types of three-phase three element combined transformer which are both used widely presently, using direct method on test comparing. Theoretical derivation is used according to test result, combined with matlab simulation analysis. At the end, this paper discusses influence on voltage error by different primary connection of three-phase three element combined transformer.

Keywords: Three-phase three element combined transformer; Primary connection mode; Voltage error.

1. Introduction

Three-phase three element combined transformer is now used widely in distribution network system, which has the advantages such as phase voltage can be possibly metered, ferroresonance is not prone to happen when earth fault occurring, which can protect the device in system, and so on. Three-phase three element combined transformer also provides signal resource for relay protection device and electric energy metering device, the good performance of which takes an important role the safe, stable operation of power system.

Both the manufacturer and measurement department in China use single phase method in testing of three-phase three element combined transformer, i.e. testing the current transformer or voltage transformer individually to make a conclusion. In 1st August, 2004, JB/T10432-2004 three-phase combined transformer is applied in China, in which, testing method of three-phase combined transformer is defined clearly, i.e. the error test, temperature rise test, mutual interference test of three-phase combined transformer must be imposed in three phase voltage combined with three phase current. However, nether manufacturer nor most measurement departments adopt the three phase method due to limitation of conditions. The shortages of single phase method are shown as follows: firstly, when testing three-phase combined transformer, current and voltage are not applied simultaneously, which is quite different from the working conditions. Secondly, the affection on voltage error and current error by both magnetic field and electric field is not considered when testing three-phase combined transformer. Thirdly, different connection mode of primary side plays an important role in voltage error in three-phase combined transformer. Subsequently, three phase testing method is important for fairness of energy trade. While, research on three phase method is rarely mentioned domestically, not to mention the three phase method applied in three-phase three element combined transformer. This paper uses three phase method to test the voltage error on two different connection modes of primary side, analyzing the influence on voltage error created by different connection modes of primary side, which provides conference for the designing, manufacture, operation and testing of three-phase three element combined transformer.

Primary connection modes of three-phase three element combined transformer have been introduced.

Three-phase three element combined transformer seals three current transformers and three voltage transformers into an enclosed space, in which the primary connection modes of those current transformer are almost the same, while the primary connection modes of voltage transformer are divided into two types: the first type is Y-yn, i.e. the neutral point N is not educed, shown as Figure 1, the other type is YN-yn, i.e. the neutral point N is educed, shown as Figure 2.

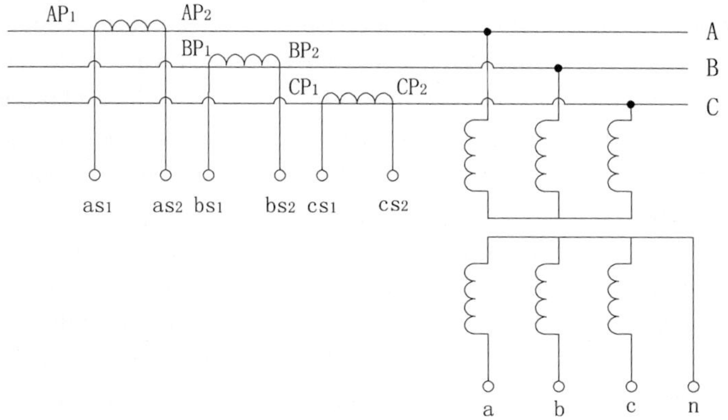

Figure 1. Y-yn connection mode of three-phase three-element combined transformer.

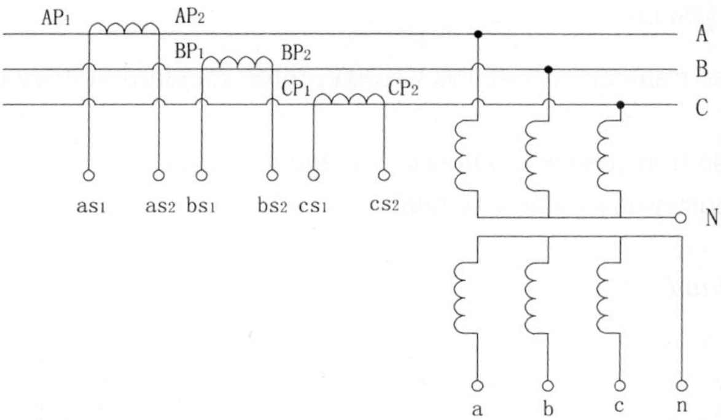

Figure 2. YN-yn connection mode of three-phase three-element combined transformer.

Most manufacturer in China adopt Y-yn connection mode for three-phase three element combined transformer, i.e. primary neutral point hanging. It is considered that primary neutral point is hardly used in installation and in use, the cost can be cut off by such design, which is an ideal design for the system that-three-phase completely balanced, i.e. $\dot{U}_A + \dot{U}_B + \dot{U}_C = 0$. However, three phase voltage are hardly balance for plenty of reasons in reality, making $\dot{U}_A + \dot{U}_B + \dot{U}_C \neq 0$, which plays an important role in the voltage error. Consequently, three-phase three element combined transformer using YN-yn primary connection mode can easily avoid such problem.

2. Test Introduction

2.1. *Introduction of test samples*

The test samples in this experiment are one three-phase three element combined transformer using Y-yn connection mode and the other using YN-yn connection mode. Both voltage error of these two transformer are qualified, shown by the manufactory as Table 1, which means both are qualified under the single phase test.

Table 1. Parameters of measured three-phase three-element combined transformer.

Primary connection mode	Type	Rated primary voltage	Rated secondary voltage	Accuracy class	Rated load
Y-yn	Voltage transformer	10000V	100V	0.2	20VA
	Current transformer	20A	5A	0.2S	10VA
YN-yn	Voltage transformer	6000V	100V	0.2S	10VA
	Current transformer	30A	5A	0.2	10VA

2.2. *Test equipment*

(1) Three-phase transformers of 10kV/100V, three standard voltage transformers of 6kV/100V.
(2) Three-phase transformer calibrator, voltage load box.
(3) Voltage booster and voltage regulator.

2.3. *Test method*

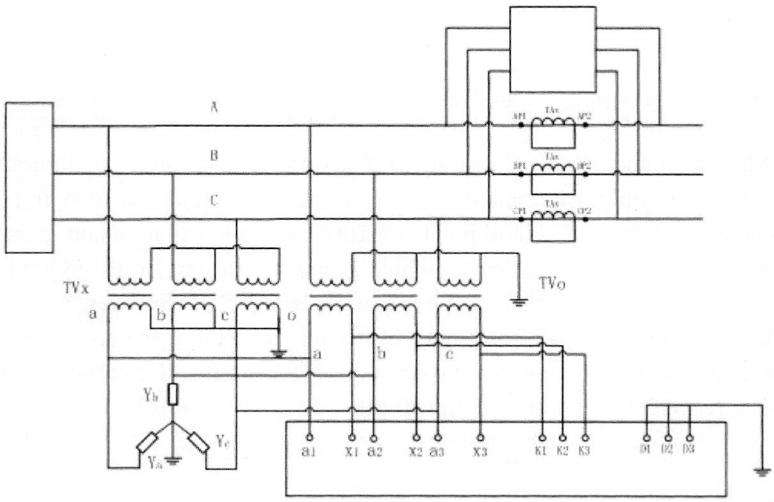

Figure 3. Circuit diagram for verification of Y-yn connection mode three-phase three-element combined transformer by three-phase method.

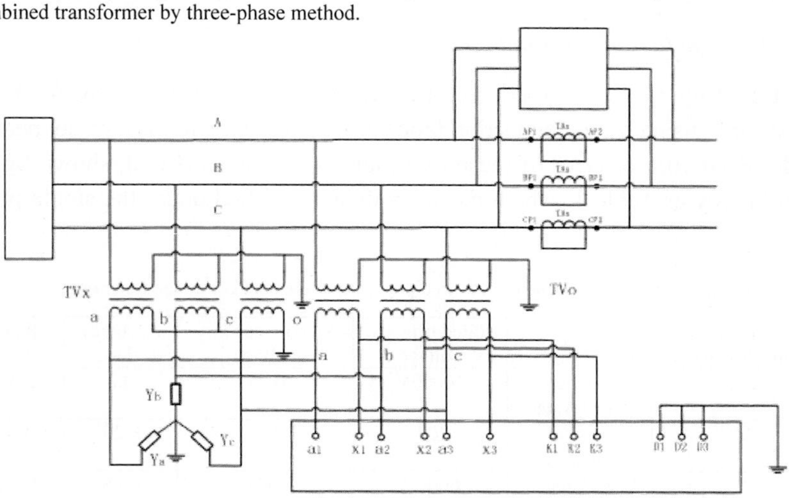

Figure 4. Circuit diagram for verification of YN-yn connection mode three-phase three-element combined transformer by three-phase method.

It can be seen from Figure 3 and Figure 4 that the only different between two types of connection model is that the neutral point of Y-yn connection mode cannot be grounded, while the neutral point of YN-yn connection mode can be grounded.

The test group is divided as follow:

The first group tests voltage error by single phase method on Y-yn connection mode transformer. The second group tests voltage error by single phase method on YN-yn connection mode transformer.

The third group tests voltage error by the connection mode shown as Figure 3 on Y-yn connection mode transformer.

The fourth group tests voltage error by the connection mode shown as Figure 4 on YN-yn connection mode transformer.

The fifth group tests voltage error by the connection mode shown as Figure 3 on YN-yn connection mode transformer.

2.4. *Analysis of test result*

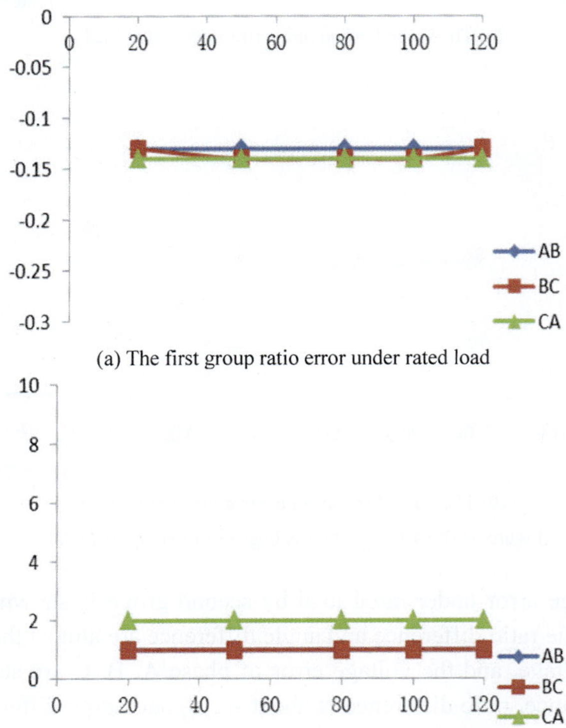

(a) The first group ratio error under rated load

(b) The first group angle error under rated load

Figure 5. The first group voltage error under rated load.

The voltage error under rated load by first group is shown as Figure 5, which means the ratio difference and angle difference are almost the same as the increase of voltage, and the voltage error of phase A, B, C are steady in same range, for instance, ratio difference is about -0.13, and angle difference is about 1.5.

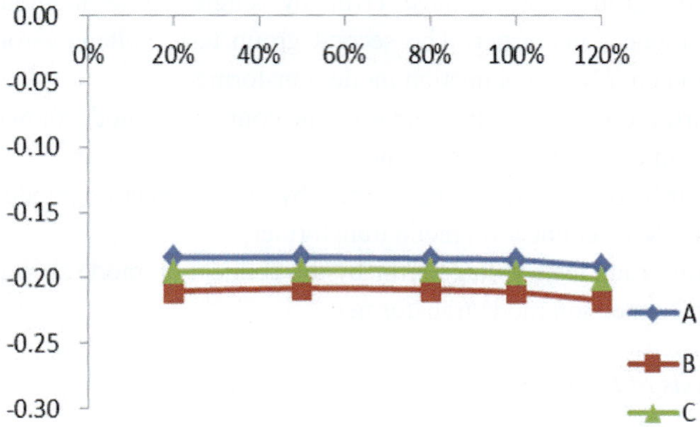

(a) The second group ratio error under rated load

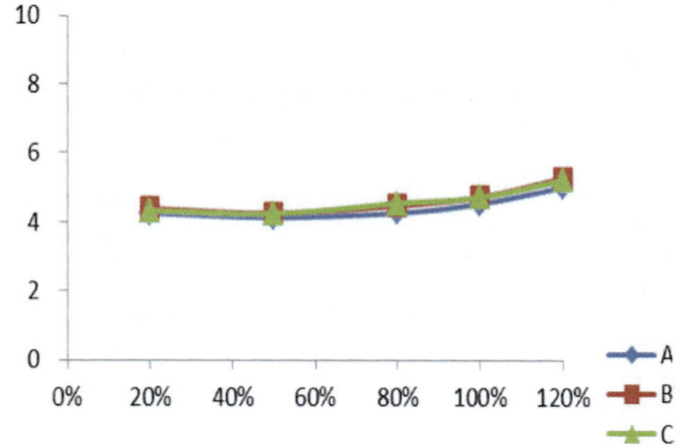

(b) The second group angle error under rated load

Figure 6. The second group voltage error under rated load.

The voltage error under rated load by second group is shown as Figure 6, which means the ratio difference and angle difference are almost the same as the increase of voltage, and the voltage error of phase A, B, C are steady in same range, for instance, ratio difference is about -0.19, and angle difference is about 4.0.

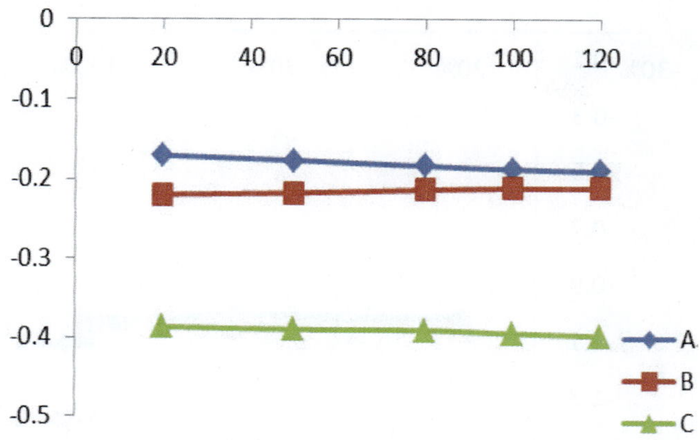

(a) The third group ratio error under rated load

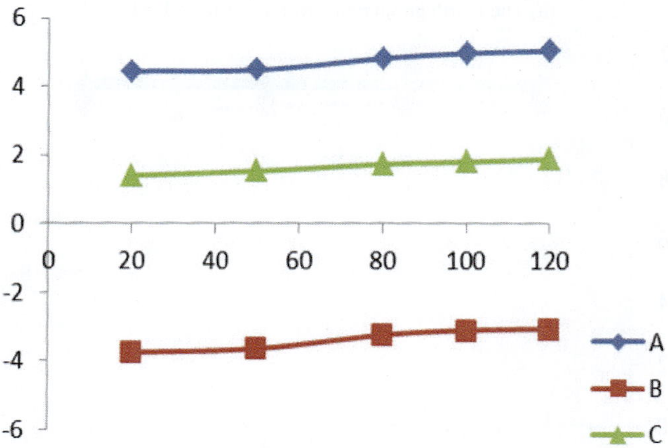

(b) The third group angle error under rated load

Figure 7. The third group voltage error under rated load.

The voltage error under rated load by third group is shown as Figure 7, which means the ratio difference and angle difference are almost the same as the increase of voltage, but theratio difference and angle difference of phase A, B, C are in totally different range, for instance, ratio difference of both phase A and B is about -0.18, ratio difference of phase C is about -0.14, angle difference of phase A, B, C is about 4, -4, 2. Compared with the first group, it is clearly that voltage error of three-phase three element combined transformer tested by three-phase method with Y-yn connection mode has migration.

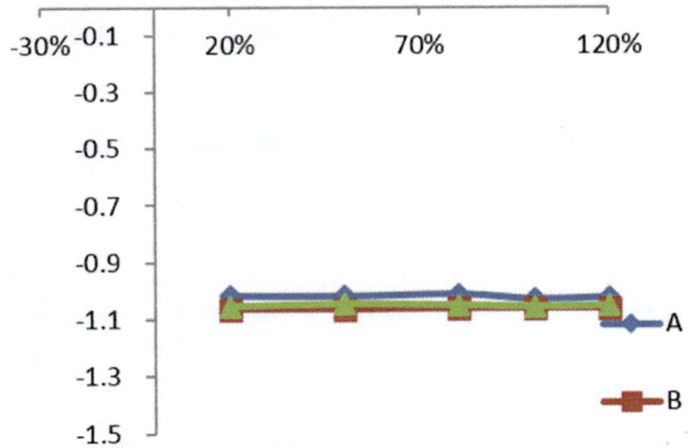

(a) The fourth group ratio error under rated load

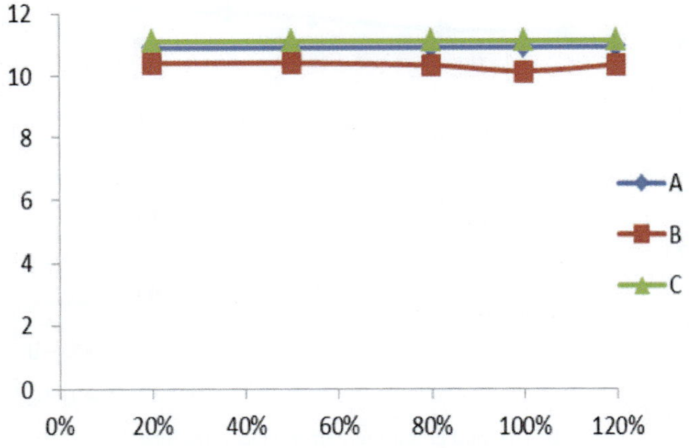

(b) The fourth group angle error under rated load

Figure 8. The fourth group voltage error under rated load.

The voltage error under rated load by fourth group is shown as Figure 8, which means the ratio difference and angle difference are almost the same as the increase of voltage, and the voltage error of phase A, B, C are steady in same range, for instance, ratio difference is about -0.9, and angle difference is about 10.0. Compared the voltage curve of Figure 8 with Figure 2, it can be seen that the voltage error of phase A, B, C in three-phase three element combined transformer by YN-yn connection mode doesn't migrate, i.e. the ratio difference and angle difference stays in a range.

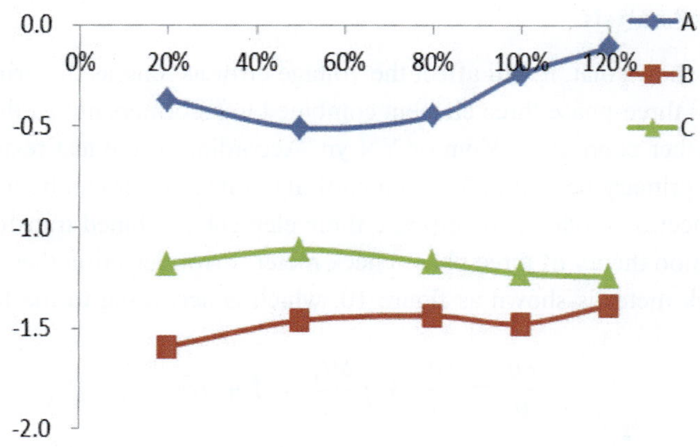

(a) The fifth group ratio error under rated load

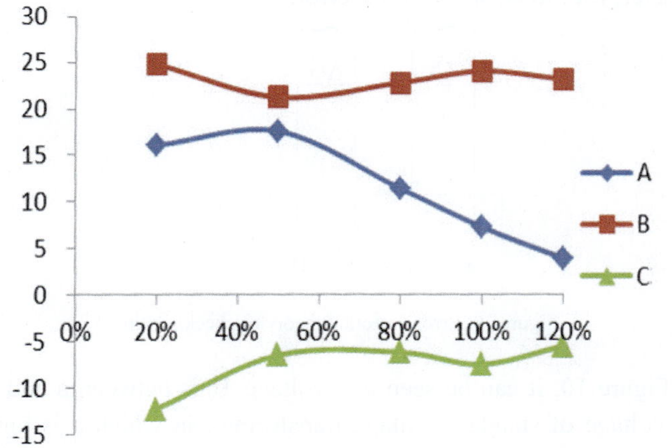

(b) The fifth group angle error under rated load

Figure 9. The fifth group voltage error under rated load.

The voltage error under rated load by fifth group is shown as Figure 9, which means the ratio difference and angle difference changes greatly as the increase of voltage, and the voltage error of phase A, B, C are not in same range. Compared the voltage curve of Figure 9 with Figure 8, it can be seen that if the transformer of YN-yn connection mode is connected as Figure 3, interruption is introduced in transformer, i.e. zero sequence voltage by primary neutral point unearthing has great influence on the voltage error of three-phase three element combined transformer, which means the influence even may result in the voltage error oversize.

3. Theory Analysis

The figure shows that, it will affect the voltage error as long as the primary neutral point of three-phase three element combined transformers not earthing effectively, whether connection Y-yn or YN-yn. According to the test result, the influence by primary neutral point N on earthing voltage error can be traced back to the connection mode of three-phase three element combined transformer and error detection theory of three phase check meter. Error detection theory of three phase check meter is shown as figure 10, which is according to the formula as follow:

$$\frac{\Delta\dot{U}}{\dot{U}} = \frac{\Delta U_x}{U} + j\frac{\Delta U_y}{U} = f + j\delta \quad (1)$$

Where, $\Delta\dot{U}$=difference value between K and D, U=voltage value showed by check meter, f=ratio error, δ=phase error.

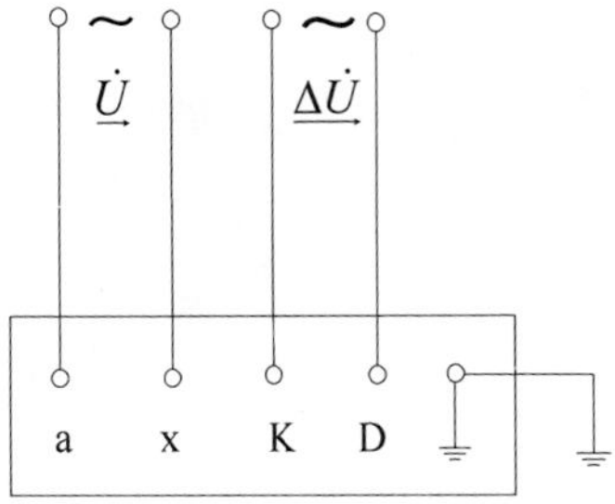

Figure 10. Error detection theory of check meter.

From Figure 10, it can be seen that voltage $\dot{U}$ is between a and x, i.e. the secondary voltage of standard voltage transformer, in which a is high voltage side and b is low voltage side. And the K, D of check meter are respectively corresponding to secondary low voltage of standard voltage transformer and test voltage transformer. Voltage $\Delta\dot{U}$ actually is as follow:

$$\Delta\dot{U} = \dot{U}_{ax} - \dot{U}_{aD} \quad (2)$$

When the three phase voltage of primary is unbalanced, the three phase voltage of three element combined transformer with Y-yn connection mode is as follow:

$$\dot{U}_A + \dot{U}_B + \dot{U}_C \neq 0 \quad (3)$$

Suppose the voltage of primary neutral point is $\dot{U}_N$, then,

$$(\dot{U}_A - \dot{U}_N) + (\dot{U}_B - \dot{U}_N) + (\dot{U}_C - \dot{U}_N) = 0 \quad (4)$$

The primary voltage of phase A, B and C is respectively $\dot{U}_{AN}$, $\dot{U}_{BN}$, $\dot{U}_{CN}$, then the corresponding secondary phase voltage is respectively $\dot{U}_{an}$, $\dot{U}_{bn}$, $\dot{U}_{cn}$, which satisfies the relationship as follow:

$$\dot{U}_{AN} = \mathrm{K}\dot{U}_{an}\ \dot{U}_{BN} = \mathrm{K}\dot{U}_{bn} \\ \dot{U}_{CN} = \mathrm{K}\dot{U}_{cn} \tag{5}$$

Since $\dot{U}_{aD} = \dot{U}_{an}$, put formula (5) into (2):

$$\Delta\dot{U} = \dot{U}_{ax} - \frac{1}{K}\dot{U}_{AN} \tag{6}$$

Put formula (6) into (1), error of phase A is as follow:

$$\frac{\Delta\dot{U}}{\dot{U}} = 1 - \frac{\dot{U}_{AN}}{\mathrm{K}\dot{U}_{ax}} = 1 - \frac{\dot{U}_{A}}{\mathrm{K}\dot{U}_{ax}} + \frac{\dot{U}_{N}}{\mathrm{K}\dot{U}_{ax}} \tag{7}$$

Similarly, voltage error of phase B and C is correspondingly as follow:

$$\frac{\Delta\dot{U}}{\dot{U}} = 1 - \frac{\dot{U}_{BN}}{\mathrm{K}\dot{U}_{bx}} \tag{8}$$

$$\frac{\Delta\dot{U}}{\dot{U}} = 1 - \frac{\dot{U}_{CN}}{\mathrm{K}\dot{U}_{cx}} \tag{9}$$

Three element combined transformer adopting YN-yn connection mode when primary three phase voltage is unbalanced, according to Figure 4, primary neutral point is earthing subsequently the sum of three phase voltage is zero, which means,

$$\dot{U}_A + \dot{U}_B + \dot{U}_C = 0 \tag{10}$$

Secondary voltage of test transformer is shown as follow:

$$\dot{U}_A = \mathrm{K}\dot{U}_{an} \\ \dot{U}_B = \mathrm{K}\dot{U}_{bn} \\ \dot{U}_C = \mathrm{K}\dot{U}_{cn} \tag{11}$$

Similarly, voltage error of phase A is as below:

$$\frac{\Delta\dot{U}}{\dot{U}} = 1 - \frac{\dot{U}_{A}}{\mathrm{K}\dot{U}_{ax}} = 1 - \frac{\dot{U}_{A}}{\mathrm{K}\dot{U}_{ax}} \tag{12}$$

Compare formula (7) with (8), it is obvious that when three phase is unbalanced, the voltage error of three phase is added correspondingly vector $\frac{\dot{U}_N}{\mathrm{K}\dot{U}_{ax}}$, $\frac{\dot{U}_N}{\mathrm{K}\dot{U}_{bx}}$, $\frac{\dot{U}_N}{\mathrm{K}\dot{U}_{cx}}$, leading to unbalance of three phase voltage error.

In another word, when the primary neutral point is unearthing, the voltage error offset of phase A is

$$\varepsilon = \frac{\dot{U}_{N}}{\mathrm{K}\dot{U}_{ax}} \tag{13}$$

Based on these three unbalanced value, this paper uses Matlab to simulate the model that, when three phase voltage angle has a small offset, for example, angle of phase A is 0.05 degree, angle of phase B is 119.95 degree, angle of phase C is -119.95 degree. The offset value of three phase voltage error can be simulated by Matlab using formula (13), shown as Figure 11.

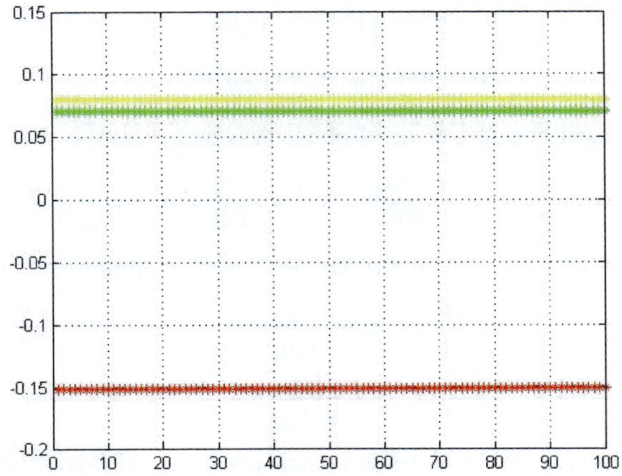

(a) The incremental ratio error

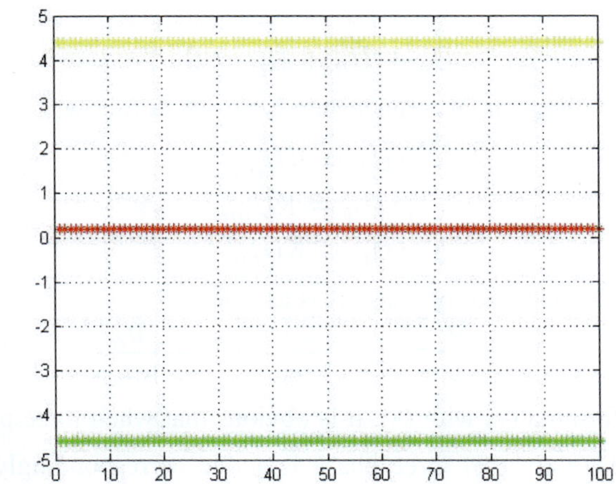

(b) The incremental angle error

Figure 11. The incremental voltage error.

According to the result of Matlab, ratio increment of A, B, C is correspondingly 0.08, 0.06, -0.15, angle increment of A, B, C is correspondingly 4.3, -4.6, 0.3. Based on the result, if the increment is added to voltage error, it will have great influence on the er-ror. Subquently, even the unbalanced voltage of pri-mary is little, it still has great influence on voltage error of three-phase three element combined trans-former by Y-yn connection mode.

4. Conclusion

In order to study the influence on the voltage error of primary connection mode of three-phase three element combined transformer, this paper takes the comparison test and theoretical analysis with method of connection Y-yn and YN-yn of transformer based on three phase. The result shows when three phase is not balanced, connection Y-yn of three-phase three element combined transformer can cause influence greatly to the voltage error without neutral point tap, and also could cause the error of the three-phase combined transformer voltage beyond the error range.

References

1. Liu Kun, Jiang Ying-Xia. Research on Two Calibration Methods of Three-phase Combined Transformer [J]. Sichuan Electric System Technology. 2010, 33(1):79-81 (In Chinese).
2. Jiang Wei, Yang Huayun, Jiang Bo. 2011. Research on Three-phase Calibration Method of HV Three-phase Combined Transformer and its Implementation [J]. Electrical Measurement & Instrumentation. 2011, 48(547):46-49 (In Chinese).
3. Zhang Neng, Liao Hui-Min, Wu Yong-Yun, et al. 2012. Research on Three-phase Combined Transformer Verification device [J]. Automation and Instrumentation. 4:41-43 (In Chinese).
4. Wu Anlan, 2008. Basic and new technology of power measurement. Press of China water conservancy and hydropower. [M]. 2008.44
5. 2004, JB/T10432. Three-phase combined transformer. [S]
6. 2004, JB/T10433. Three-phase combined voltage transformer. [S]

Simulation analysis of interphase spacers for anti-galloping in ultra-high voltage compact type power transmission lines

X.P. Zhan
Beijing University of Technology,
Beijing, Chaoyang State, China
E-mail: zhpp2008@foxmail.com
www.bjut.edu.cn

X.P. Zhan, K.J. Zhu, C.L. Liu, B. Liu and J.L. Yang
China Electric Power Research Institute
Beijing, Haidian district, China

Because phase to phase distance of the compact type power transmission line is small, phase to phase flashover caused by the icing galloping of the conductor etc danger become one of the most important disaster type for safe and stable operation of the compact type transmission line. In this paper, anti-galloping calculation model with interphase spacers has been established for the ultra-high voltage compact type continuous three span lines by finite element method, which calculates and analyzes influence of interphase spacers arrangement method on galloping characteristic under different wind speed conditions of the given span distance, the obtained calculation results have important instructive meanings for actual engineering.

Keywords: Ultra-high voltage; Compact type; Interphase spacers; Value analysis.

1. Introduction

According to operation experience in our country, icing galloping of the conductor becomes one of most key disaster type which endangers safe and stable operation of the compact type line. When it is analyzed from basic conditions on galloping occurrence, excitation conditions of the compact type line are basically same as these of the common line, there is only difference in galloping characteristics under the concrete line conditions. Following operation number increasing of the compact type line in our country, galloping issue is more and more serious. Some pertinent research works have been developed successively since 2007, some successful experience anti-galloping has been obtained. But there is no corresponding research achievement on anti-galloping design of the ultra-high voltage compact type line in domestic and foreign country.

The relevant domestic and foreign research and operation experience shows installation and configuration way of interphase spacers has obvious influence on galloping control of the transmission line. But it mostly takes actual operation experience or some simplified analysis method as principle during application at present. In China, North China, East China, Henan etc provincial companies have developed long term study in research field of interphase spacers, which have jointly compiled the first national technical condition standard for interphase spacers. China Electric Power Research Institute has also developed study on configuration way of interphase spacers in recent two years, which forwards optimum arrangement method for configuration of interphase spacers. This scheme is more detail than the foreign scheme, theoretical basis is more reasonable. The relevant operation experience shows application effect of this scheme is good in common line. For the ultra-high voltage compact type power transmission line, arrangement method of interphase spacers shall be further studied. This paper carries out simulation analysis on anti-galloping arrangement method of interphase spacers according to characteristic of the ultra-high voltage compact type line, and obtains arrangement scheme of interphase spacers under the given span distance, which has certain instructive meaning on anti-galloping of the ultra-high voltage compact type power transmission line.

2. Conductor Model of Galloping

Anti-galloping of interphase spacers is to connect the conductor of every phase mechanically by interphase spacers, and then change characteristic of the system. And movement of the conductor of every phase is involved and restrained each other so as to realize galloping control purpose. Seen from actual application effect, interphase spacers are one of anti-galloping device with good anti-galloping effect and good reliability at present, but interphase spacers can't realize galloping control completely.

Calculation models have been developed for conductor galloping without interphase spacers and with interphase spacers in this paper. Value of the conductor span distance in the middle span is 500m, both span distances at two sides are 400m and height is 0.

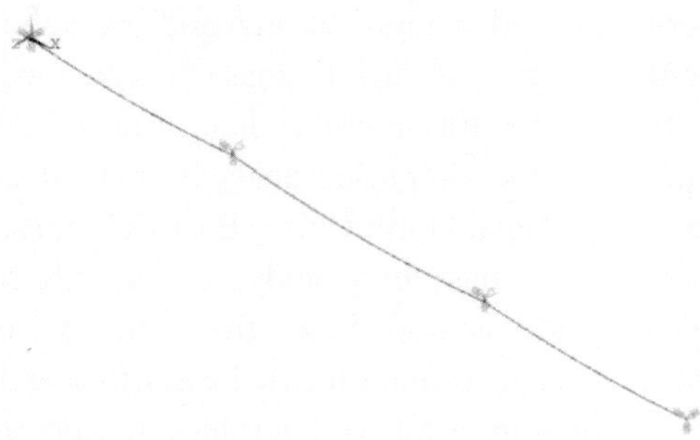

Figure 1. Single phase value calculation model.

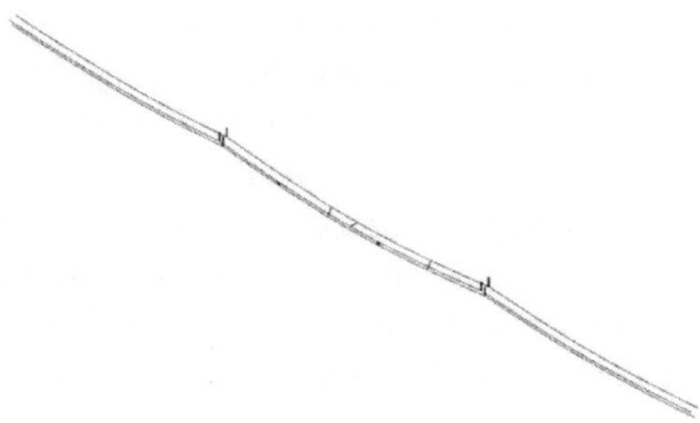

Figure 2. Value calculation model with interphase spacers.

3. Simulated Analysis with Interphase Spacers Arranged

3.1. *Line parameters*

Refer to Table 1 for line parameters of the 1000kV compact type line. D type icing is selected and ice thickness is 6mm. Table 1 provides minimum safety clearance of the actual power frequency voltage of the 1000kV compact type line to the design unit at power frequency voltage and three type of altitudes.

Table 1. Value adopted for air clearance.

Altitude H (m)		500	1000	1500
Power frequency voltage		2.7	2.9	3.1
Operation over voltage	Middle phase V type string	6.7 (7.9)	7.2 (8.0)	7.7 (8.1)

The JL/G1A-630/45 type conductor is applied, refer to Table 2 for characteristic parameters of the conductor. The three phase conductor is arranged in the tower windows, the reverse triangular arrangement way is applied. Phase to phase span is 13.856m, distances of the upper phase and the lower phase in vertical direction are 12m. Discharging occurs when phase to phase distance is less than 3.1m.

Table 2. Parameters table of conductor.

Conductor type	JL/G1A-630/45
Diameter (mm)	33.8
Unit quality (kg/km)	2079.2
Calculated breakage force (N)	182181
Elastic module (N/mm2)	63000

3.2. *Basic condition and method of dynamic calculation*

Calculation of non-linear static force and non-linear dynamic calculation are realized by the finite element software.

In order to simplify calculation and improve operation convergence speed. Because of complicity of the coupling structure, the secondary factors shall be neglected as possible when the value model is established. Compared to the actual model, main assumption conditions adopted by the value calculation model: the rod tower applies the fixed boundary; the conductor fitting is in accordance with Hooke law; large displacement and small strain geometrical non-linear mode has been considered.

Based on the above assumption, the basic dynamic movement equation is:

$$M\ddot{u} + C\dot{u} + Ku = F \tag{1}$$

where: M is quality matrix of the structure; K is rigidity matrix; C is damping matrix; $\ddot{u}$ is acceleration vector; $\dot{u}$ is speed vector; u is displacement vector; F is non-liner aerodynamic load vector including lift force, drug force and torsionalmoment.

The movement balance equationsare two order non-linear equation group. Direct integral calculation has been applied to analyze dynamic equations in this paper.

3.3. *Arrangement method of interphase spacers*

When span is 500m, conventional arrangement of interphase spacers shall be considered to control galloping of No 1-5 order. It is given the arrangement method of interphase spacers as Table 3.

For arrangement of interphase spacers under severe conditions,circulation thick arrangement methodofinterphase spacers will be applied,which will be shown as Table 3.

4. Results of galloping simulation calculation

When wind speeds are 7, 10, 15 and 18m/s respectively, calculation results of gallopingunder conventional arrangement and circulation thick arrangement way of interphase spacers are shown as following figures.

Table 3. Arrangement method of interphase spacers.

Installation number (piece)	Arrangement position (m) (distance to side with small number)			Arrangement way
	A –B (m)	A –C (m)	C –B (m)	/
5	111, 300	250	200, 389	General arrangement
7	100, 250, 410	260	110, 270, 420	Crossing and circulation thick arrangement

Notes:A-Upper left phase, C-Upper right phase, B-lower phase.

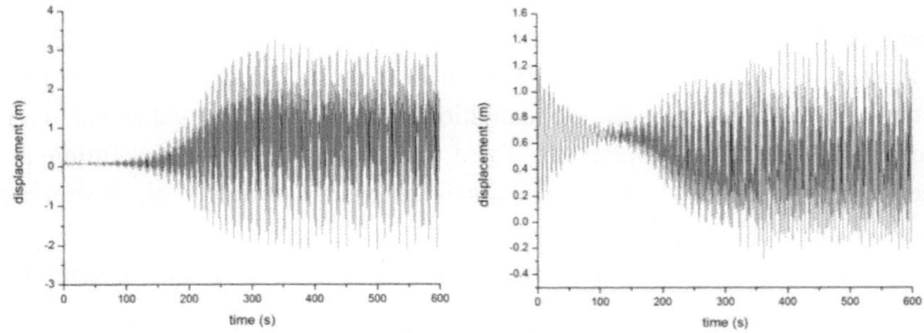

(a) Time history curves in vertical direction (b) Time history curves in horizontal direction

Figure 3. Time history curves without interphase spacers(wind speed at 7m/s).

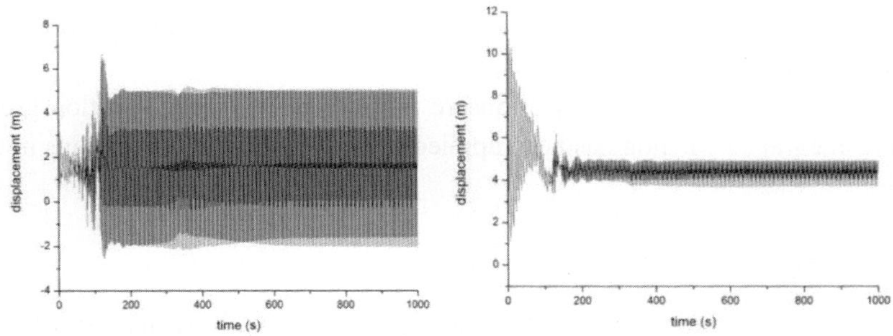

(a) Time history curves in vertical direction (b) Time history curves in horizontal direction

Figure 4. Time history curves without interphase spacers(wind speed at 10m/s).

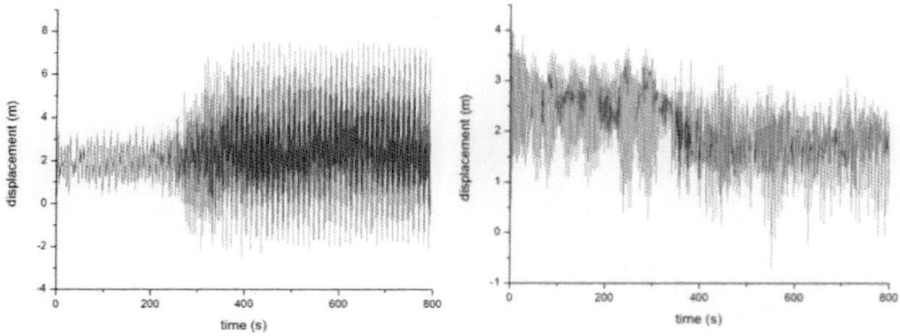

(a) Time history curves in vertical direction (b) Time history curves in horizontal direction

Figure 5. Time history curves without interphase spacers(wind speed at 15m/s).

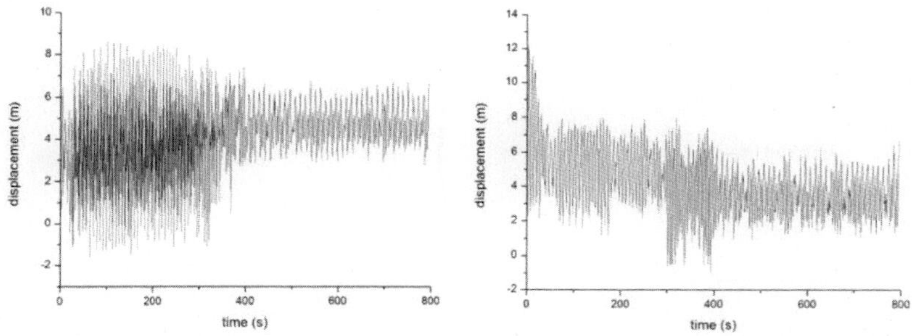

(a) Time history curves in vertical direction (b) Time history curves in horizontal direction

Figure 6. Time history curves without interphase spacers(wind speed at 18m/s).

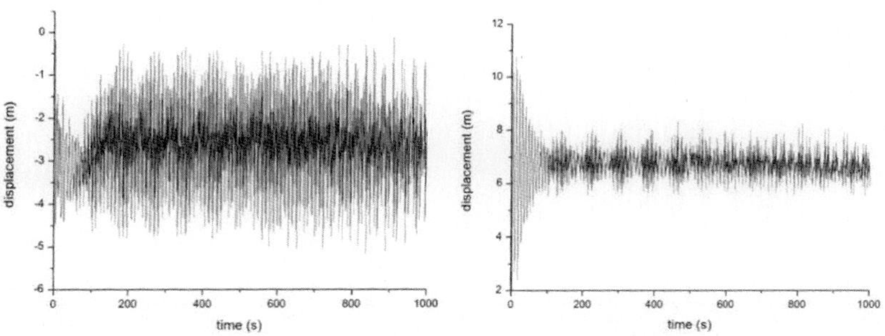

(a) Time history curves in vertical direction (b) Time history curves in horizontal direction

Figure 7. Time history curves with interphase spacers(wind speed at 7m/s).

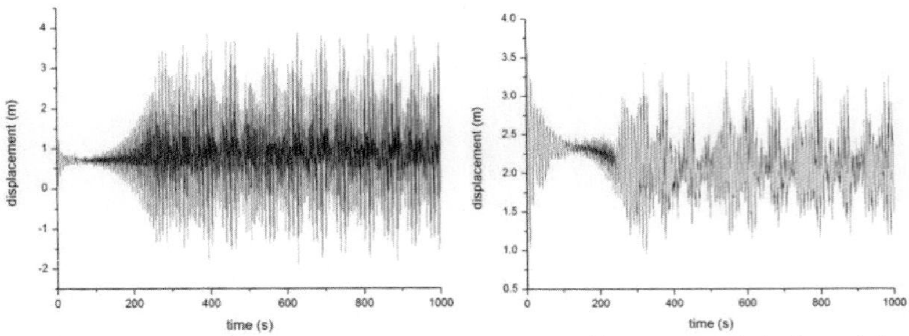

(a) Time history curves in vertical direction (b) Time history curves in horizontal direction

Figure 8. Time history curves with interphase spacers(wind speed at 10m/s).

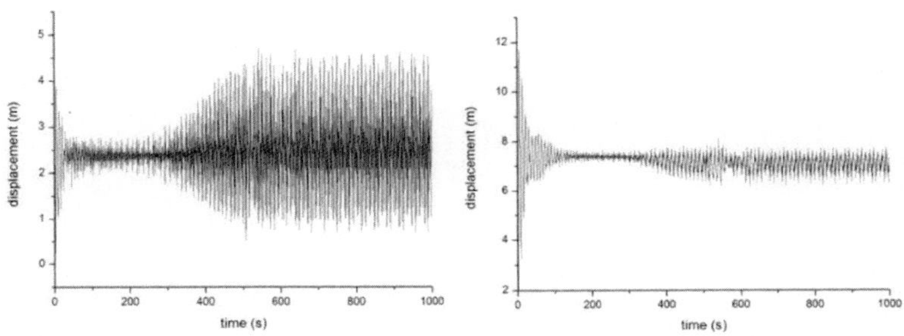

(a) Time history curves in vertical direction (b) Time history curves in horizontal direction

Figure 9. Time history curves with interphase spacers(wind speed at 15m/s).

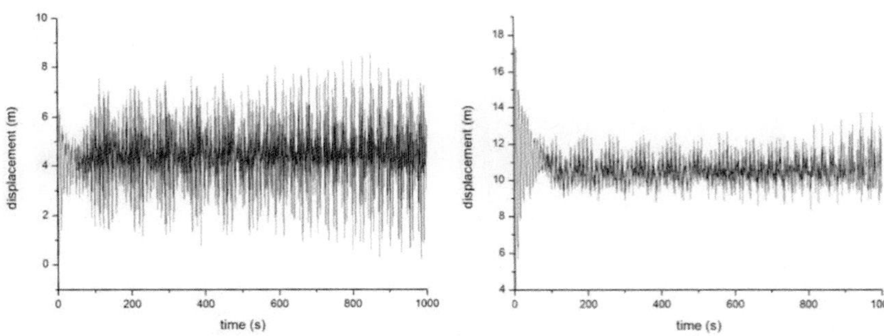

(a) Time history curves in vertical direction (b) Time history curves in horizontal direction

Figure 10. Time history curves with interphase spacers(wind speed at 18m/s).

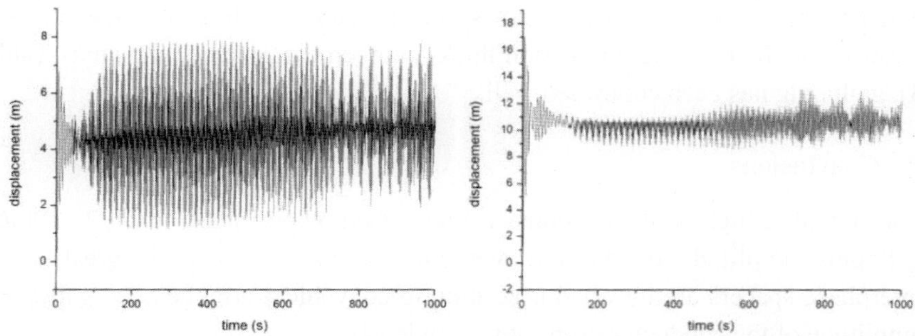

(a) Time history curves in vertical direction (b) Time history curves in horizontal direction

Figure 11. Time history curves with interphase spacers in thick scheme(wind speed at 18m/s).

The most amplitude of galloping can be shown as follow Table 4.

Table 4. The amplitude of galloping under different wind conditions.

Wind velocity (m/s)	Installation conditions of interphase spacers	Galloping amplitude (m)			Is there flashover
		Vertical	Horizontal	Minimum phase to phase distance	
7	Without interphase spacers	4.96	1.63	7.04	No
7	Conventional arrangement	4.08	1.34	7.92	No
10	Without interphase spacers	7.12	1.25	4.88	No
10	Conventional arrangement	4.92	2.26	7.08	No
15	Without interphase spacers	9.02	3.65	2.98	yes
15	Conventional arrangement	5.52	2.81	6.48	No
18	Without interphase spacers	9.82	8.03	2.18	yes
18	Conventional arrangement	8.02	4.96	3.98	No
18	Circulation thick arrangement	6.32	3.96	5.68	No

It can be found from above calculation results: when the span is 500m and wind speed is at 7, 10m/s under calculation conditions in this paper, the most amplitude of galloping in vertical direction reaches about 7m when interphase spacers aren't installed; when interphase spacers applied with the general conventional arrangement, that is no flashover for the line; when wind speed reaches 15m/s, flashover occurs in the line without interphase spacers, it meets with requirement for no flashover when interphase spacers are arranged as the conventional method; when wind speed is 18m/s, interphase spacers are arranged as the crossing circulation thick arrangement scheme (shown as Table 3), galloping has been controlled well.

5. Conclusions

For the ultra-high voltage compact line, when wind speed is at 7, 10m/s, galloping amplitude of the line without interphase spacers is great; after interphase spacers are installed according to convenient arrangement, galloping amplitude of the conductor drops down obviously.

When wind speed reaches 15m/s above etc severe weather conditions, interphase spacers are firstly arranged as convenient method. If this arrangement can't control galloping well, interphase spacers shall be applied with the circulation thinning installation. The results in this paper can provide a important guide for actual engineering.

Acknowledgements

This work was financially supported by the National Natural Science Foundation of China (51305411).

References

1. Li Yongsuang. Selection of centre line to line (phase to phase) distance of compact type line span, Power Equipment, vol 7, No. 3, 2006.
2. Fan Qinshan. Engineering Mechanical Course [M]. Beijing: High Education Press, 1998.
3. Wang Liming, Liang Xidong, Guan Zhicheng et al. Research on 500kV phase to phase composite spacer for compact lines [C]. Proceedings of the 6th International Conference on Properties and Applications of Dielectric Materials, Xi'an, China, 346-349, 2000.
4. Feng Y T, Peric D, Owen D R J. A new criterion for determination of initial loading parameter in arc2length methods [J]. Comput. Strac, 58(3):479-485.
5. DL 5217-2005 "Technical regulation for design of 220kV~500kV compact type overhead power transmission line".

Chapter 3
Sustainable Urban and Rural Development

Analysis of countermeasures for the settlement of rural land circulation disputes in Tacheng area

P.E. Wang† and L. Lin*

Research Center for Social Governance Modernization of Xinjiang
School of Political Science & Law, Shihezi University, Shihezi, China
*E-mail: †wangpeien313@163.com; *101471678@qq.com*

To analyze the countermeasures for the settlement of rural and land circulation disputes in Tacheng Area. Conducted an empirical study of Tacheng Area with peasant households involved in the local rural land circulation disputes as respondents. The basic data obtained through questionnaire survey and dispute statistics were analyzed. It is found that the rural land disputes increased on an annual basis. The analysis of the causes and settlement mechanism of the rural land disputes showed that there were many problems. Inefficient settlement of the rural land disputes can directly influence the enthusiasm of peasants in Tacheng Area about land circulation. It is imperative to put forward effective countermeasures concerning the current status of rural land disputes in Tacheng Area.

Keywords: Rural land circulation; Circulation disputes; Dispute settlement mechanism; Tacheng Area.

1. Introduction

According to the survey data of rural land circulation in 2013 in Tacheng Area, Xinjiang, it is found that disputed caused by the circulation of right to contracted management of land accounted for about 60.8%. How to effectively solve the land circulation disputes has been a top priority to safeguard rural social stability and economic development. It is of vital significance to put forward feasible countermeasures concerning the current situation. The research into rural land circulation disputes was first started by scholars in the economic field, and most of them focused on the status, causes and countermeasures of land circulation disputes.[1,2] In recent years, the legal circles have also paid more attention to land circulation disputes, focusing on the land circulation dispute settlement system and the improvement of relevant laws.[3,4] In terms of research methods, most of them concentrated on the theoretical study, and the empirical study was insufficient.[5] In particular, few scholars adopted the rural land circulation disputes in

Xinjiang as the research object. Concerning the research gap, this paper adopted rural circulation disputes in Tacheng Area, Xinjiang, as the research object, and analyzed the causes and settlement mechanism of rural land circulation disputes. Based on the analysis of problems existing in the settlement of rural land circulation disputes in the local area, feasible countermeasures were put forward.

2. Research Materials and Methods

2.1. *Research materials*

This paper selected the rural land disputes in five counties and two cities (Tacheng City, Emin County, Yumin County, Tuoli County, Wusu City, Shawan County and Mongolian Autonomous County of Hoboksar) under the jurisdiction of Tacheng Area as the research object, and conducted a questionnaire survey and data analysis of the basic situation of households involved in the land circulation disputes and the relevant problems of rural land circulation disputes, which mainly included land circulation basic procedures, causes of land circulation disputes, legal knowledge of land circulation, medical and social insurance in the local area and settlement of land circulation disputes. The data mainly came from the statistical data of Tacheng Area Agricultural Economics Bureau, including the basic situation of land circulation, the number of land circulation disputes, the number of resolved land circulation disputes, the number of personnel working in land dispute arbitrary institutions and so on.

2.2. *Research methods*

The descriptive statistical analysis method was employed to conduct an analysis of the rural land circulation and dispute status quo. Then, the case analysis method was used to classify various cases related to land circulation disputes. Based on the analysis of the questionnaire survey results, the major causes of the land circulation disputes were summarized. Household survey was adopted in the investigation area. In total, 250 questionnaires were delivered to peasant households involved in rural land circulation. The valid questionnaires collected were 229, registering an validity rate of 91.6%. At last, problems existing in the rural land circulation dispute settlement mechanism were analyzed. (See Table 1).

Table 1. Questionnaire survey samples and range.

Counties and cities	The number of valid questionnaires collected	Proportion (%)
Tacheng City	48	220.96
Emin County	33	114.41
Yumin County	29	112.66
Tuoli County	24	110.48
Wusu City	31	113.54
Shawan County	46	220.09
Hefeng County	18	7.86
Total	229	1 100

3. Results and Analysis

3.1. *Rural land circulation status of Tacheng Area: currently, Tacheng Area governs five counties and cities*

There are 64 villages and townships, and 831 village groups. The collective agricultural acreage is 36.25×10^4hm^2, of which the agricultural acreage under the household contractual management is 26.45×10^4hm^2, accounting for 72.9% of the collectively-owned land area, 6.35×10^4hm^2 of the mobile land area and 17.52% of the arable land. The households engaged in household contractual management are 11.13×10^4. By the end of 2013, 97.51% of villages had participated in rural land circulation, and 58.96% of households were involved. The total area of arable land under household contractual management accounts for 7.59×10^4hm^2 of the total. The total arable land area under household contractual management accounts for 28.70% of the total contracted arable area.

3.2. *Status quo and causes of rural land circulation disputes in Tacheng Area*

The transfer of rural land contracted management right is shortened as rural land transfer. Rural land circulation disputes refer to disputes related to rural land contracted management right. Fig. 1 shows the percentage of land circulation disputes in the total land disputes in Tacheng Area from 2009 to 2013. (Data source: Tacheng Area Agricultural Economics Bureau).

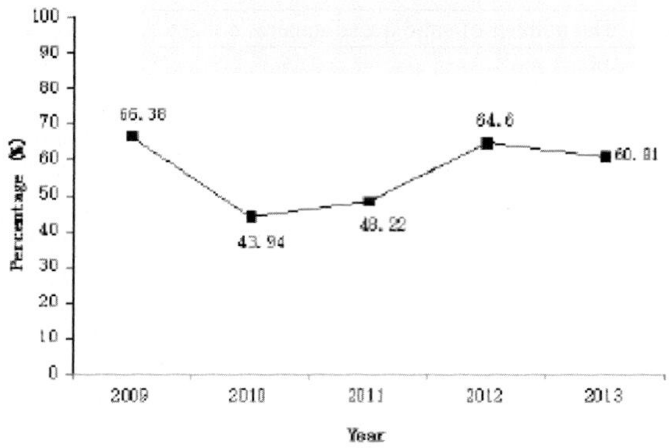

Figure 1. Statistics of rural land circulation disputes in Tacheng Area from 2009 to 2013.

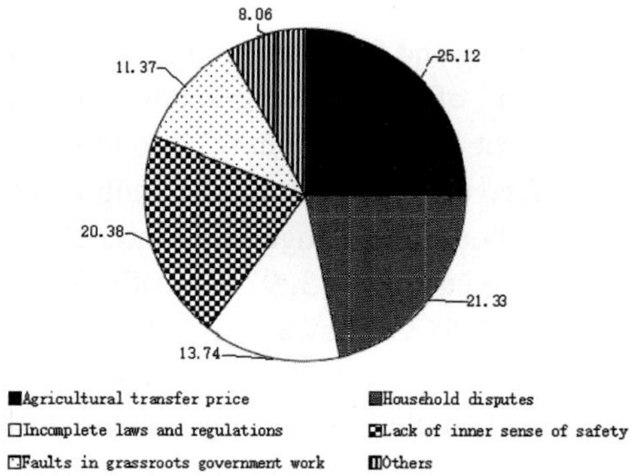

Figure 2. Statistics of main causes of the rural land Circulation disputes in 2013.

According to Fig. 2, it can be seen that major causes of rural land circulation disputes include rural land circulation price, household disputes, laws and regulations, lack of production techniques and faults in the grassroots government work. Based on the questionnaire survey statistics, this paper conducts an in-depth analysis of these causes. (1) Lack of a standard rural land circulation price evaluation mechanism. As to the question of causes of land circulation disputes, 64.19% of respondents chose “rural land circulation price”, 31.44% chose “household disputes”, 29.26% chose “incomplete laws and regulations”, 18.78% chose “incomplete social and medical insurance”, 30.13% chose “faults in grassroots government work” (Multiple choices. The total percentage is more than 100%). Thus, it can be seen that “land circulation price” is the most important cause of disputes. (2) Most relatives adopted the oral contract form. As to the question of the relationship between land circulation sub-contractors, 60.26% chosen "relatives" (including distant relatives). Those choosing the oral contract (including no written contracts but with the third party as the witness) accounted for 61.59%, and those choosing the written contract accounted for 38.41%. It can be seen that the oral contract is quite common among relatives, thus resulting in the potential danger of land circulation disputes. (3) Incomplete laws and regulations: Land Administration Law of the People's Republic of China, Law of the People's Republic of China on Land Contract in Rural Areas and Real Right Law of the People's Republic of China define the legal character of rural land contractual management differently. In particular, there have not yet been specific regulations as to the rural land circulation disputes. (4) Lack of other production techniques: As to the question of other production techniques, 36.68% chose "yes", 46.28% chose "no" and 17.04% chose "uncertain". Thus, it can be seen that peasants lack the other production techniques apart from agricultural production. After land circulation, they are unable to engage themselves into other production industries. (5) Faults in grass-roots government work: As to the question of whether the information of contractual operation right certificate is in line with the practical situation, 65.94% chose "yes", 28.38% chose "no" and 6.11% chose "uncertain".

3.3. *Problems existing in the rural land dispute settlement mechanism in Tacheng Area*

Currently, the rural dispute settlement mechanism has taken shape in various counties and cities in Tacheng Area. It is made up of four links, including negotiation, mediation, arbitration and litigation. The land circulation disputes in 2011 reached 211, of which 105 were mediated, accounting for 49.76% of the total, 79 were arbitrated, accounting for 37.44% of the total, only three were solved through litigation, accounting for 1.42% of the total (See Fig. 3).

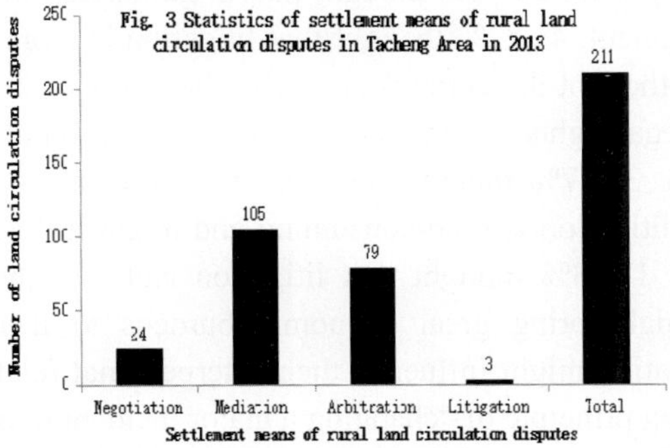

Figure 3. Statistics of settlement means of rural land circulation disputes in Tacheng Area in 2013.

Four settlement means shown in Fig. 3 coordinate with each other to solve rural land disputes, which, to some extent, safeguard peasant households' land operation rights and interests. However, due to the complexity of rural land disputes, there are some deficits existing in the rural land circulation dispute settlement mechanism. (1) Lack of legal guidance and standards inhibits the full play of the settlement mechanism. As to the question of the preferred dispute settlement, 60.26% chose "negotiation", 10.92% chose "mediation", 20.52% chose "arbitration", 8.30% chose "litigation." Thus, it can be seen that peasant households are more willing to settle disputes through negotiation. Since oral contracts outnumber written agreements, the rights and interests of the involved parties lack effective protection. Though some oral agreements are concluded, they are often not performed. (2) The mediation lacks authoritativeness, thus many agreements concluded have limited legal effect. As to the question of whether rural dispute mediation organizations play an effective role, 78.17% chose "yes", "15.72% chose "no", 6.11% chose "uncertain". It can be seen that most peasant households recognize the role of the mediation organizations. By persuading and guiding the involved parties, they can help settle land circulation disputes. However, the mediation is not compulsory and lacks authoritative. The legal effect of the agreement concluded through mediation is limited. (3) Lack of arbitration talents, office facilities and funds: In 2013, the number of arbitration committee members in various counties and cities in Tacheng Area was 58 in total, of which 23 were personnel engaged in daily chores, and only two were specialists. There were no peasant representatives. Now, Tacheng Area is faced with the problem of inadequate arbitration personnel, and serious transfer and temporary transfer. Moreover, the arbitration work has the problem of incomplete procedures to obtain evidences, inadequate funds and lack of office facilities. (4) The litigation costs high and is time-consuming, thus failing to meet the psychological demands of local peasant households. As to the question of reason to turn to litigation, 26.78% thought that litigation can offer them a objective and fair judgment, 41.82% thought that litigation is more authoritative and official, 21.50% thought the compulsory executive force of litigation can better safeguard their legal rights and interests. As to the question of the reason not to turn to litigation, 35.37% thought the litigation procedures are too complex, 27.50% thought litigation is time-consuming and might influence their agricultural production, 17.03% thought that litigation calls for the employment of lawyers, thus might bring great economic burdens to themselves, 20.09% thought that litigation might influence their interpersonal relationship and they prefer the handling principle of "changing a major incident into a minor one".

4. Conclusions

Currently, the land circulation dispute settlement mechanism in Tacheng Area is still faced with many problems. How to effectively settle the land circulation disputes has become an important factor to safeguard the social stability and economic development in the rural areas in Tacheng. Thus, it is imperative to improve the rural land circulation market, standardize the land circulation behaviors, optimize the rural circulation dispute settlement mechanism and safeguard the health and normalized development of rural land circulation.

Acknowledgment

This work was financially supported by Corps of Social Science Fund Project "period of prominent social conflicts grassroots farm disputes mediation mechanism" from Corresponding author Lin Li, China. Foundation Number was 12YB05.

References

1. Zhai Jian-Song: Collective land marketing transference problem research (2012).
2. Xiao Qiong, Cao Jian Hua, Causes and Countermeasures of the circulation of rural land disputes in China, J. Journal of Anhui Agricultural Sciences, 35 (2011) 4042-4043.
3. Cao Wu-Kun, Sheng Rong, Improve the rural land contracted management rights shares legal system, J. Management and Administration, 1 (2008) 50-51.
4. Li Chang-Jian, Liang Ju, Research on the disputes arbitration mechanism in the circulation of Agricultural land in contemporary era, J. Journal of Henan Administrative Institute of Politics and Law, 5 (2009) 144-148.
5. Qi Zong-Bao, Empirical study on the disputes for the transfer of contracted land management rights: starting from 9 litigation cases, J. Issues in Agricultural Economy, 1 (2008) 80-84.

Analysis on the principle of terrain slope impact on land suitability for urban construction in mountainous areas*

Z.S. Yang, Y.M. He and M.J. Xu†

Institute of Land & Resources and Sustainable Development, Yunnan University of Finance and Economics, Kunming 650221, China

†E-mail: xumingjun0711@163.com

Terrain slope is an important factor in evaluating land suitability for urban construction in mountainous areas. At present, the principle of the slope factor impact on land suitability for urban construction in mountainous areas is lacking a systematic discussion and refinement, and classification system and index of slope factor that impacts construction land suitability rating in mountainous area has not yet formed a unified standard or consensus. Based on the realistic need of the mountainous urban construction land suitability evaluation, the authors in this paper has analyzed and refined the basic principles of slope factor impact on mountainous land suitability for urban construction; and probed into and built the system of slope factor classification for evaluating the suitability of mountainous lands for urban construction by combining the reality of mountainous areas. The study will provide necessary foundation and technical support for promoting the strategy of "Constructing Mountainous Cities".

Keyword: Mountainous area; Urban constructive land; Suitability evaluation; Terrain slope; Impact.

1. Introduction

As one of the most important modes of urban development, constructing mountainous cities is drawing increasing attention of countries around the world. This way of urban development can save a lot of precious flat lands otherwise occupied by buildings and structures, especially plains and high-quality flat arable lands, which are a nation's very base for food security and sustainable development, and thus is becoming a prevailing trend in many mountainous areas. In recent years, Yunnan Province, along with other Chinese cities and places, for example, Lanzhou of Gansu Province, Yan'an of Shaanxi Province, Shiyan of Hubei Province, etc., where city-wide dynamic and

* Supported by National Natural Science Foundation of China (41261018).

Corresponding author: YANG Zi-sheng.

E-mail: yangzisheng@126.com.

vigorous campaigns to "Cut Mountains and Build Cities" are waged one after another (Li et al., 2014), has launched its strategy of "Constructing Mountainous Cities" (or "Cities-on-Mountains Strategy").

The first question to be answered regarding building a city (or town) on a mountain is what kind of mountain can be said qualified for urban development and what not? To answer this question, we must have a full and accurate evaluation of the mountainous land suitability for urban development, which is a precondition, and a crucial base and support, too, for the successful implementation of the "Cities-on-Mountains Strategy", and to have such a suitability evaluation, we must select and determine reasonable evaluation factors and grades for such purpose. Particularly, such evaluation factors usually involve geological structures, terrain, climate, environmental protection, vegetation and land use, disasters, covering of mineral deposits, traffic, and terrain slope, among others, is the most important factor in mountainous suitability evaluation for urban development. By far, there hasn't done yet a systematic study, exploration, even a summary or abstract, by the land science community of how the slope influences the suitability of mountainous lands for urban development, nor a standard or an agreement available as to how to grade the suitability of mountainous lands by its slope, which makes it very challenging to do a proper and accurate suitability evaluation and provide a solid theoretical base and support for the successful implementation of the "Cities-on-Mountains Strategy". In view of this, this article attempts, as necessitated by the urgent need for mountainous land suitability for urban development, to do an in-depth study and analysis, and provide an abstract, of how the slope influences the suitability of mountainous lands for urban development, and further to build and establish a slope-based suitability grading system that perfectly fits the situation of mountainous areas in China to provide a solid theoretical base and support for the successful implementation of the "Cities-on-Mountains Strategy".

2. Principles of Terrain Slope Impact on Mountainous Land Suitability for Urban Development

Slope usually refers to the gradient, or the steepness, of an inclined part of a roof, a road, or of the side of a hill or a mountain, etc., and according to Baidu Encyclopedia and 360 Encyclopedia, it tells how steep or how gentle a surface unit is.

Slope is either expressed as a percentage, or in degrees, mils, or fraction, and among them the most commonly-seen are slopes expressed as a percentage and in degrees. In most cases, slopes, denoted by *i*, are expressed as a percentage of the height difference (h) between any two given points to their horizontal distance (L), the expression is then $i = (h/L)\times 100\%$. And when expressed in degrees, a slope is usually denoted by p and computed using inverse trigonmetrical functions, as: $tanp=h/L$, then, $p=tan-1\ (h/L)$.

In mountainous areas, mountains and hills of a certain slope are viewed as natural-cultural syntheses, and in themselves are independent systems known as the mountain system or mountainous area system (Yu, 1998). Compared with their plain counterparts, mountains are very special and distinctive in some primary properties, as demonstrated by their energy gradation (potential, kinetic, and thermal, etc.), ex-situ migration of surface materials, plane heterogeneity (in horizontal direction), cut-brokenness in surface morphology, and in some secondary properties as well, as by their vertical anisotropy of natural geographic elements, environmental vulnerability and fragility, unsaturated ecological niche, etc., and in a humanistic sense, they are marginal, nearly inaccessible, isolated and sometimes conflicting (IMHE-CAS, 2000). When it comes to the resource development and ecological protection in a mountainous area, the thing or aspect requiring a serious consideration is the environmental vulnerability and fragility. The environmental vulnerability and fragility referred to herein means the fact of being sensitive to, and liable to change of a property from stable to unstable under, a disturbances or an internal or external force. By its very nature, such vulnerability and fragility is co-contributed by various forces of different nature and in different scales and directions. Take a rock as an example, when it lies on a flat ground, it is mainly under the action of its own weight and gravity, as both forces are in the same direction, the rock is normally in a stable state; However, when it lies on a slope, it is under the action of the gravity, its weight and an additional diagonal component force from its weight, they are in different directions, and the steeper a slope is, the greater the component force is and more unstable the rock is going, as shown in Figure 1. According to Professor Yu Dafu (1992), the environmental vulnerability and fragility have two basic implications: (1) Mountains have a low tolerance of biotic resources development and a weak capability of recovering from a destructive activity; (2) The geology and soil therein is poorly resistant to disturbance; These two implications show that a mountain soil-plant system in a mountainous environment is particularly vulnerable and fragile. In most cases, by mountain soil we are talking about slope soil. Slope soil, when put under the combined action of its own slide-down potential, raindrop kinetic energy, run-off scouring, and geological forces, is at risk of degenerating, for example, eroded, desertified, lithified, or going back to a young stage, etc. And talus slopes, valleys, terraces, recessed lands, etc. are particularly likely to be buried under detritus, gravel, sand and silt running down from a slope or from a upstream source, and consequently have a greatly-impaired soil fertility and degraded site conditions and agricultural production conditions. The soil erosion seen over a slope and the burial seen at the bottom of a slope (or in a valley or a low-lying land) are a best reflection of the innate vulnerability and fragility of a mountainous ecology and environment. As such, any improper development on a mountain, for example, farming, building a road, digging a channel, reclaiming, falling trees, doing urban development, etc. will make the mountain more vulnerable, fragile and unstable, and directly result in its acquired vulnerability and fragility.

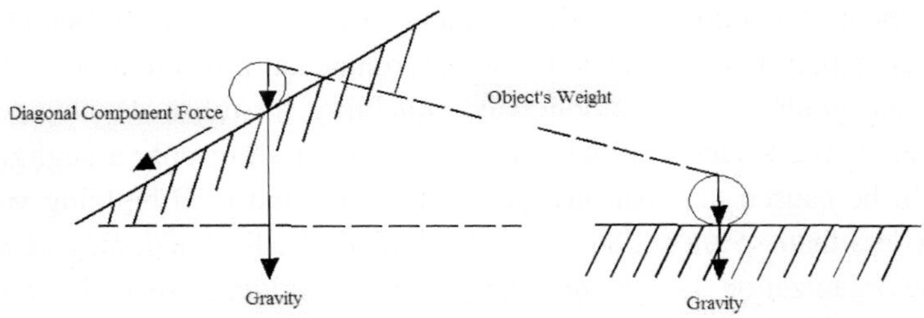

Figure 1. Diagram of stability states of an object on a slope and on a flat ground. (Source: P48, "*Overview of Mountain Science with Chinese Mountain Research*" by IMHE-CAS)

Whenever such really-intensive development activities as building cities or towns on mountains is involved, the steepness or the gradient of a slope is, and shall be taken as, one of the important influencing factors to be considered seriously. The steeper a slope is, the greater the said diagonal component force along a slope is and more unstable and risky buildings and structures on a slope will become; and on the other hand, rocks and soil on and over a slope are quite unstable and will eventually turn out into disasters, e.g. a flash flood, a landslide, or a mud-rock slide, etc. in a heavy rain or in a storm when they are subject to a combined action of raindrop kinetic energy and run-off scouring, which will definitely wreak serious destruction of ecology and environment in the development area, even heavy casualties.

To avoid such disastrous consequences, a steepness threshold (or criterion) has been suggested by some scholars and in the urban planning of some cities for slopes that are not suitable for, or shall be banned from, urban development. For example, Huang Guangyu (2006) argued in his book "Principles of Mountain Urbanology" that any slope that is steep than 55% (>28.8°) shall be banned from any development effort; "Urban Master Planning of Haikou City in 2011-2020", which was approved by the State Council in May 2011, listed slopes that are steep than 25% as slopes to be banned from any development effort; Yang

Zisheng (2013) suggested that such steepness threshold shall be 25° for Manshi City, Yunnan Province.

From the perspective of mountainous land suitability for urban development, the slope or steepness has two major implications:

(1) It, though having a different ceiling by different scholars and in different places, solely determines whether a mountainous land is suitable for urban development or not. In this sense, it is a special, mandatory and determinant factor by which the suitability of mountainous land for urban development is solely determined.

(2) For mountains and hills that have been accessed as suitable for urban development, it significantly influences and determines the quantities of surface works, construction costs, difficulty in environmental protection, as well as the degree and grade of mountainous land suitability for urban development. For example, Huang Guangyu (2006) argued that no at all or only a negligible impact will be caused to urban layout, construction and road building when the slope is less than <5% (< 2.86°), that some kind of adjustment may be made to building organization, layout and arrangement of buildings, road alignment, etc. when the slope is between 5% and 15% (2.86° and 8.53°); and that special adjustment and treatment shall be made thereto whenever the slope is between 15% and 55% (8.53° and 28.80°). Thus, to the extent of "being sustainable", the factor of slope shall be taken as a fundamental evaluation factor against which, among others forming the whole evaluation system, the degree and grade of mountainous land suitability, e.g. either perfectly, moderately or marginally suitable, for urban development is assessed and determined.

3. Slope-Based Grading of Mountainous Land Suitability for Urban Development

When doing a land suitability evaluation, the suitability of a land is graded against a range of evaluation factors. So does a mountainous area. Grading against various evaluation factors is one of the most important fundamental works to be done in suitability evaluation. When it comes to the factor of slope, there has existed a big gap between different scholars and experts regarding the slope-based grading system and grading criteria, as caused by the obvious difference in geology, topography and socio-economic status as seen in different places. The slope-based grading of a mountainous land suitability for urban development herein, as presented in this paper, has taken into thorough consideration the prevailing practice in other parts of China as well as the prevailing definition and criterion of "being gentle" for low-altitude gentle-slope mountains, and is therefore comparatively more well-targeted and more suitable for the applications under concern.

The integrated development and utilization of low-altitude gentle-slope mountains and hills is drawing increasing attention nationwide. The gentleness of a so-called gentle slope means how inclined a slope is. A gentle slope can be understood as a slope that is not steep or that is not so inclined. In other words, the basic meaning of "being gentle" or "gentleness" is not steep or is moderately inclined. But what kind of slope can be said to be “not steep” or “moderately inclined”? Here a definitely-defined criterion is required.

Most scholars agree that by “being gentle” they refer to slopes having a gradient less than 25° (Yang, 2015), that is, the steepness ceiling is agreed to be 25°, based on which, slopes that are steep than 25° shall be classified as “not suitable for, or banned from, any construction and development effort”.

Geographically, slope is an important limiting factor for resources development in a hilly or mountainous area. In such areas, the stability of materials on and over a slope usually declines as the steepness is increasing, and the slope is more vulnerable to soil erosion. Typically, a slope that is steep than 25° is extremely vulnerable to serious soil erosion, collapse, landslides, debris flows, and other kinds of geological disasters. This is why the Article 25 of Law of Soil and Water Conservation of China provides that: “Do not farm on a slope that is steep than 25°”. This paper agrees, too, that any slope that is steep than 25° shall be classified as “not suitable for any construction”.

In the meanwhile, the lower limit for “being a gentle slope” shall be a value that perfectly matches the slope-based grading system as suggested and adopted in China’s land resources survey so as to form consistent, uniform criteria for the planning, development and management of low-altitude gentle-slope mountains and hills nationwide. “The Technical Rules of Present Land Use Investigation”, developed by National Agricultural Regional Planning Commission of China in 1984, and “The Technical Rules of the Second National Land Survey”, issued by Ministry of Land and Resources of China in July 2007, both suggest to define the slope of farmlands by five grades, as: Grade I: < 2°, Grade II: between 2° and 6°; Grade III: between 6° and 15°; Grade IV: between 15° and 25°; And Grade V: > 25° (see Table 1). This grading system has been in effect for nearly 30 years and proven very effective in land-use planning, regulation, protection and management.

Table 1. Slope classification index of land survey in China.

Slope Grading	Grade I	Grade II	Grade III	Grade IV	Grade V
Criteria	<2°	2°~6°	6°~15°	15°~25°	>25°

Source: "The Technical Rules of Present Land Use Investigation" and "The Technical Rules of the Second National Land Survey".

As suggested by the slope grading system as adopted in the land survey and census of China, 6° shall be the criterion for "being a gentle slope"; Obviously, in the context of urban development suitability, a slope less than 6° is undoubtedly the most suitable and desirable. This paper thus adopts "less than 6° (<6°)" to be a general criterion for Grade-I Slope in the evaluation of a mountain's suitability for urban development in China.

In Yunnan's case, however, 8° has been conventionally taken as a dividing line separating flat areas from mountainous areas in Yunnan's flat land surveys over a long time (Yang et al., 2014). In the 2007-2009 Second Land Survey and Census of Yunnan Province, it was particularly provided that any area having a slope less than or equal to 8° over a contiguous stretch that is either longer than or equal to 1 km2 shall be classified as Flat Area (i.e. Flat Land Area). So this paper adopts "less than 8° (<8°)" to be a Yunnan-specific criterion for Grade-I Slope in the evaluation of a mountain's suitability for urban development in Yunnan. This criterion is somewhat different from its counterparts in other parts of the country, which is "<6°" in most cases.

Given the most suitable slope (<6°, namely, flat areas) and the unsuitable slope (>25°, namely, steep lands), the remaining, i.e. the slope range between 6° and 25° (or 8° and 25° in case of Yunnan), can then be further divided into two or more grades for the purpose of developing and utilizing low-altitude gentle-slope mountains and hills throughout the country. This paper, after considering the farmland slope grading criteria currently prevailing in China as well as the slope-based grading practice in the evaluation of the suitability of mountains and hills for urban development in most parts of China, further divide the slope range (8°~25°) into two more grades as: moderately suitable (any slope between 6° and 15°) and marginally suitable (any slope between 15° and 25°).

In summary, there are four slope grades, i.e. slopes less than <8°, slopes between 8° and 15°, slopes between 15° and 25°, and slopes greater than 25°, for mountainous land suitability evaluation, and by which mountains are graded as perfectly suitable (Grade I), moderately suitable (Grade II), marginally suitable (Grade III), and unsuitable for urban development (also see Table 2).

Table 2. Slope-based grading system and criteria for evaluating mountainous urban construction land suitability in China.

Grading of Suitability	Perfectly Suitable	Moderately Suitable	Marginally Suitable	Unsuitable
Slope Grading	I	II	III	IV
Slope Range	<6°	6°~15°	15°~25°	>25°

4. Conclusions

Carrying out an accurate evaluation of the mountainous land suitability for urban development will provide a strategic base as well as a crucial support for successful implementation of "Cities-on-Mountains Strategy", and terrain slope, among others, are the most important factor in such evaluation.

When carrying out a mountainous land suitability evaluation, slope is a special, mandatory and determinant factor by which a mountain's suitability for urban development is solely determined, as that slopes in a mountainous area whose gradient is greater than 25° are "not suitable for" or shall be "banned from" any construction and development effort; And a fundamental evaluation factor in the meanwhile, determining the degree and grade of a mountain's suitability for urgent development.

In China's case, slopes can be grouped by 4 slope grades, namely, slopes less than or equal to 6°, slopes between 6° and 15°, slopes between 15° and 25°, and slopes greater than 25°, corresponding to "perfectly suitable", "moderately suitable", "marginally suitable", and "unsuitable" for urban development.

Acknowledgements

This article is the result of project (No. 41261018) supported by the National Natural Sciences Foundation of China. We thank National Natural Science Foundation Committee of China and Yunnan University of Finance and Economics for their friendly helps.

References

1. Peiyue Li, Hui Qian, Jianhua Wu. Environment: Accelerate research on land creation [J]. *Nature*, 2014, 510: 29-31.
2. Yang Zi-Sheng. Analysis on the special factors for evaluating mountainous urban construction land suitability in Yunnan Province. *Research of Soil and Water Conservation*, 2015, 22(4). (in Chinese).
3. Yu Dafu. Discussion of the objects and contents of montology study —— Continuing of some opinions on the developing montology. *Journal of*

Mountain Science (Continuing *Mountain Research*), 1998, 16(1): 69-72. (in Chinese).

4. Institute of Mountain Hazards and Environment, the Chinese Academy of Sciences (IMHE-CAS). *Overview of Mountain Science with Chinese Mountain Research.* Chengdu: Sichuan Science and Technology Press, 2000. (in Chinese).
5. Yu Dafu. *Mountainous Agricultural System in Western Sichuan Province.* Chengdu: Chengdu University of Science and Technology Press, 1992: 5-12. (in Chinese).
6. Huang Guang-Yu. *Principles of Mountain Urbanology.* Beijing: China Building Industry Press, 2006: 1-159. (in Chinese).
7. Yang Zi-Sheng, Wang Hui, Zhang Bo-Sheng. Study on constructive land suitability evaluation in China's southwestern mountainous areas: a case in Mangshi, Yunnan Province In: Yang Zi-Sheng eds. *Research on Land Development & Rearrangement and Building Urban and Industrial Projects on Mountainland in China.* Beijing: Social Sciences Academic Press, 2013: 112-120 (in Chinese).
8. National Agricultural Regional Planning Commission of China. 1984. *The Technical Rules of Present Land Use Investigation.* Beijing: Surveying and Mapping Press: 5-13 (in Chinese).
9. Ministry of Land and Resources of the People's Republic of China. *The Technical Rules of the Second National Land Survey.* Beijing: China Standards Press, 2007: 1-74. (in Chinese).
10. Yang Zi-Sheng, Zhao Qiao-Gui, Xin Ling. *Land Resources of Yunnan.* Beijing: Science and Technology Press of China, 2014 (in Chinese).

Study on public service facility planning of small farm towns of tourism

Q. Yuan* and Y. Qu†
Harbin Institute of Technology, Harbin, Heilongjiang, China
*Email: *hityq@126.com; †13652992@qq.com*

This paper chooses the planning of public service facilities of small farm town of tourism in Heilongjiang province as research subject. By analyzing the reference including the classification, configuration, location and layout of the public service facilities, we summarize the rule of planning small farm town of tourism. Under the guidance of the relevant theory and practical experience, the paper select three typical farms in Heilongjiang Province; analyzes and reflects allocation and layout issues of their public service facilities. Through analyzing farm residents and tourist behavior characteristics and the satisfaction of the public service facilities, we conclude the problems of public service facilities of small farm town of tourism in Heilongjiang Province. Combining with all of the above analysis, we put forward the corresponding planning principles and planning strategy.

Keywords: Heilongjiang province; Small farm town of tourism; Public service facilities; Planning strategy.

1. Introduction

In recent years, returning to natural farm has become the best way of leisure. As a large agricultural province, Heilongjiang province provides new opportunities for the development of agricultural tourism. However, the farm public service facilities lack systematic planning. Small farm town of tourism can not provide the perfect tourism public service facilities. In the level of basic public facilities, the planning exists something missing. These lead to the sharp contradictions between tourism service facilities and public service facilities. It has become the focus of the urban planning field on how to solve this contradiction.

Some scholars mention that tourism can greatly promote the development of the towns from the angle of the relationship between tourism development and small town (Fu. 2010), some of the scholars research small town tourism development model. They consider that small town tourism should be combined with local cultural industry, the industrial towns can transform to the cultural tourist town (Murphy et al. 2006). For tourism planning of the small towns, scholars mainly put forward to establish a perfect decision-making system (Bovy. 1982), to realize combining the tourism planning and small towns planning (Costa 2001) and some other measures.

In terms of public service facilities, some scholars put forward public facility location theory and distinguish the public facility location decision-making and the individual facilities in order to take the viewpoint of arranging basic public facilities basing on the efficiency and fair (Teitz 1968). Some scholars make the cost-effectiveness analysis for the public service facility location (Austin 1974).

I select a typical small farm town in Heilongjiang province, attempt to use the research experience of the theory about tourism farm and the small towns, take the angle of public service facilities planning as the breakthrough point, analyze the existing problems on the small farm town of tourism public service facilities in Heilongjiang province, and put forward the corresponding optimization strategy of arranging and configuration of public service facilities respectively.

2. Select Research Area

Heilongjiang province has a rich resource of agriculture, has a total of 113 farms, I extract three typical farms as the research object, extraction principle as follows: 1. Belong to the Chinese top 100 farms, responding well the current public service facilities. 2. Select the type of small town tourism farm as research object. 3. Select from different travel modes. Finally, I select the Erlongshan, Hailin, Ningan small farm town of tourism as research objects (Table 1).

Table 1. Research object summary.

Name of the city	Name of the farm	Rating of the scenic spot	Tourism classification model
Five connective lakes City	Erlongshan farm	3A	Sightseeing-type
Mudanjiang City	Hailin farm	3A	Vacation-type
Mudanjiang City	Ningan farm	3A	Synthesizing-type

3. Analysis of Public Service Facility Current Situation

3.1. *The current situation of basic public service facilities*

After the investigation, it shows that these small farm towns of tourism have relatively complete basic public service facilities. It can basically satisfy the need of the small farm towns' life:

A perfect administrative facilities. The main types are the farm administrative offices, judicial law enforcement offices and community residents' committees and some other facilities. The complete configuration of basic education facilities. The schools of small tourism farm town in Heilongjiang province are generally composed of kindergarten, elementary school and junior high school, some part of the farms have elderly university. Single type of culture sports facilities. The researched culture sports facilities of small tourism farm towns are mainly composed of square, park, farm history museum, entertainment center for workers, entertainment center for aged people. Most of the health facilities provide good basic medical service, but lack of medical level. A comparatively abundant business financial facilities. Small farm business facilities mainly consists of commodity marketing and business financial facilities.

After the field survey, we easily find out that the public service facilities of each small town which present a linear spatial form arrange along the both sides of main road. The layout of all kinds of public service facilities intersperses with each other, the same kind of public service facilities distribute dispersedly (Fig. 1).

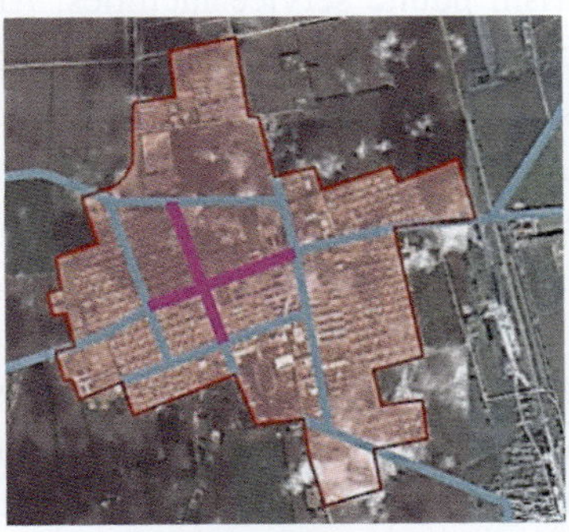

(a) Erlongshan farm

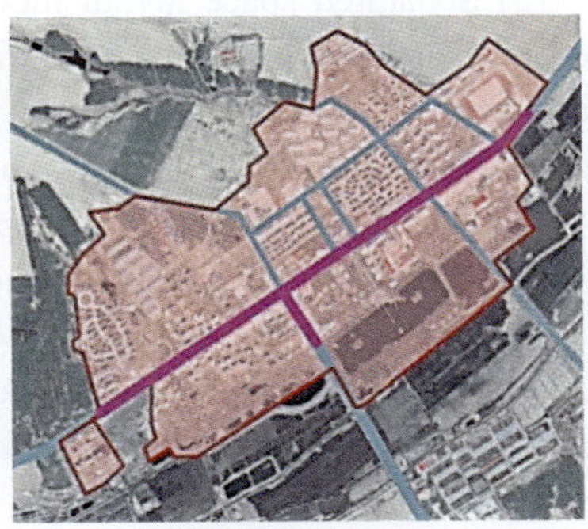

(b) Hailin farm

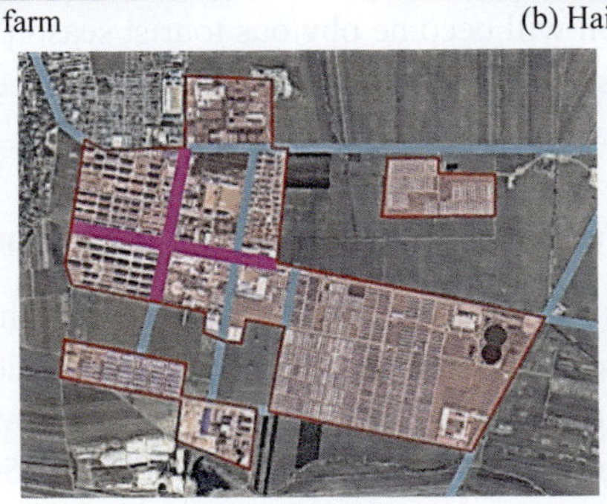

(c) Ningan farm

Figure 1. Distribution of public service facilities of small farm town of tourism in Heilongjiang Province.

3.2. *The current situation of tourism public service facilities*

(1) Single type of tourism agriculture facilities. The researched three farm are generally large scale of modern agriculture demonstration farm. But the tourist activity type and agricultural tourism facilities are relatively single. Pursuing the grand scale of agriculture demonstration becomes the construction trend of small farm towns of tourism agricultural facilities in Heilongjiang province. This trend causes the lack of tourism farm connotation, single types of tourism agriculture facilities. Single type of tourist agriculture can not attract enough tourists to small town tourism farms, even some of the farms can only attract the tourists who just visit the surrounding large scenic and pass by the area.

(2) The leave of tourism business facilities are mainly middle and low. It mainly includes two parts. One is the tourist accommodation and catering facilities. The other is tourism shopping facilities. After the field survey, we find that in the aspect of tourist accommodation, the tourism business facilities are mainly made of middle and low small hotel in Heilongjiang province, the number of business hotel or the tourist reception center in a farm is usually only one in small towns, the reception capacity is far from demanding.

(3) Separated space layout mod of tourism public service facilities. After the field survey, we find that the distribution of the tourism business facilities are generally independent of basic public service facilities in farm towns, are rarely arranged in combination. The tourism activity and layout of separated living space make it difficult to use public service facilities.

(4) The seasonal fluctuations of tourism public service facilities usage. These farms base the attractive point on picking agricultural products and agricultural sightseeing, however, farming landscape and farming activities will change when the season turns, it makes the tourism farm activities carry out only in the season which the farming landscape is rich and farming activities concentrate, the season will become obvious tourist season. Relatively, when the farming landscape is single and farming activities disperse, the season will become tourist off-season.

4. The Analysis of Farm Residents, Tourists Behavior and Satisfaction

The main objective of this research is the behavior characteristic of farms residents and tourists in Heilongjiang province, satisfaction for public service facilities, we use the questionnaire investigation and interviewing methods. The surveyed crowd which include 150 farm residents and 150 farm tourists come from three representative tourism farm towns in Heilongjiang province, each of the town will choose 50 farm residents and 50 farm tourists. Each person will be given one questionnaire. It totals 300 questionnaires. Recycles 300 questionnaires, valid questionnaires totals 300. The method of issue survey use the main way of random door-to-door questionnaire.

4.1. *Farm residents characteristics*

Based on questionnaire, the surveyed farm residents have the following two characteristics: on the one hand, the number of the young is big, and the young have a high educational level. Among the interviewee, the age ranging 20 from 30 and 30 from 40 respectively account for 42% and 22%. The age structure and high educational level make abundant farm labor in small towns and the cultivated land quantity is limited theoretically. Therefore, depending on agriculture to solve the employment problem has limited potential. At the same time, the young, middle-aged, and the old now increase the demand for leisure and entertainment facilities, and the growing demand for the medical level and leisure facilities in the future (Figs. 2a, 2b).

On the other hand, the occupational structure mainly is farm workers, the income level is generally low, the demand level of facilities is not high, the desire for increasing income is strong. 68% of the surveyed population is farm worker. Income level concentrate in the range of 20,000 to 50,000 Yuan per year, the occupation structure and low levels of income caused the low level of demand for high leave public service facilities in theory, but the desire for increasing the income is strong (Figs. 2c, 2d).

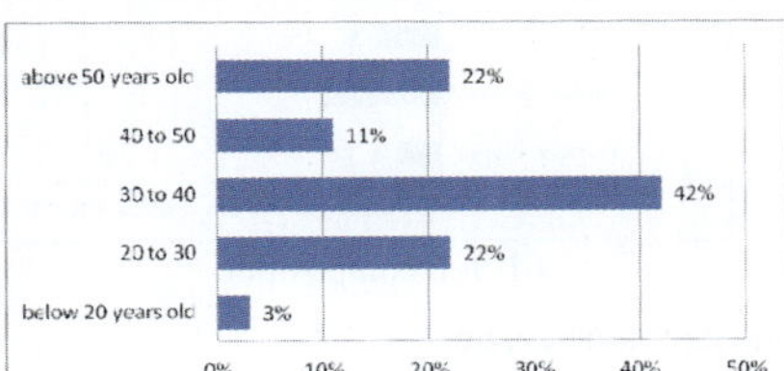

(a) Age composition

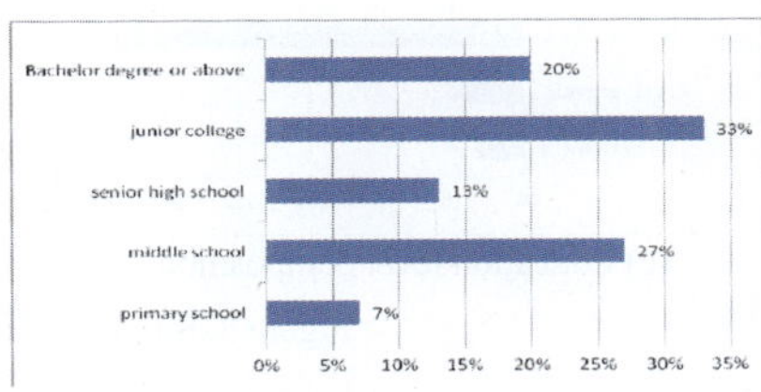

(b) Schooling composition

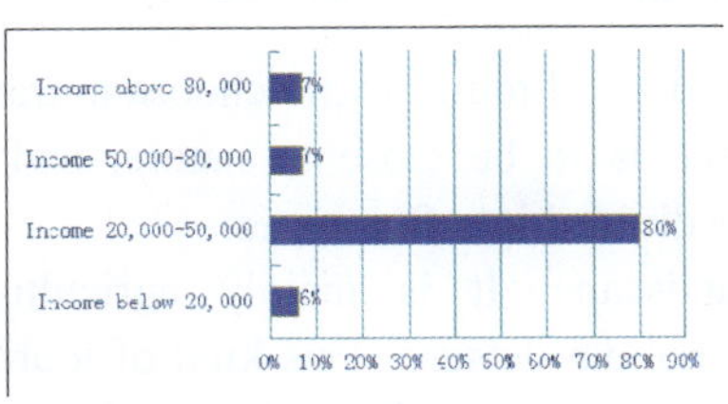

(c) Job composition

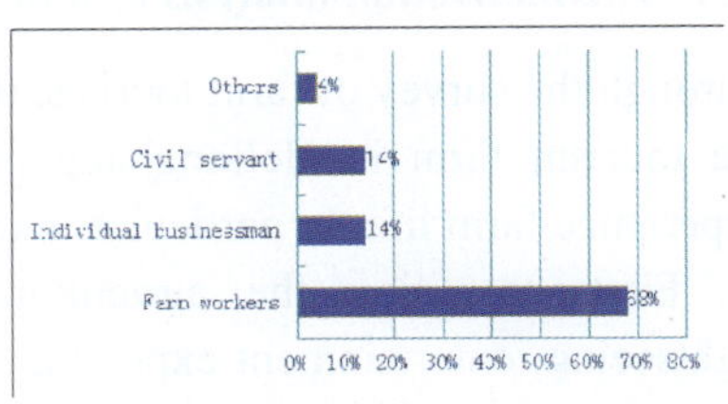

(d) level of income composition

Figure 2. Surveyed farm residents characteristics.

4.2. *Farm tourists characteristics*

Based on the questionnaire, the surveyed farm tourists have the following several characteristics (Fig. 3): According to the questionnaire data, the age of tourists who go to the small farm towns in Heilongjiang province almost covers all the age group. Ranging from 18 to 24 years old account for 42% among the surveyed tourists, ranging from 25 to 35 years old account for 34%, ranging from 36 to 45 years old account for 23%. The male female ratio of farm tourist is generally equal. According to investigation statistics, the tourists who participate in the tourism farm mainly have a high educational degree. Visitors have a junior college degree occupied 32%, bachelor degree or above account for 43%. According to the results of questionnaire survey, the tourists is mainly wage-earning and white-collar class. In addition, a few schools regularly organize the students to the farm to learn something about agricultural science.

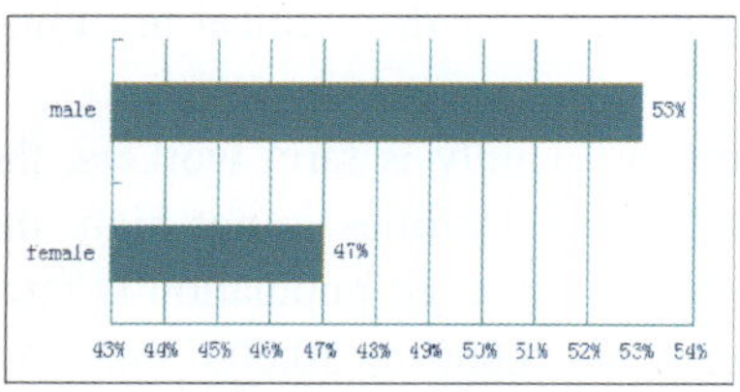

(a) Age composition

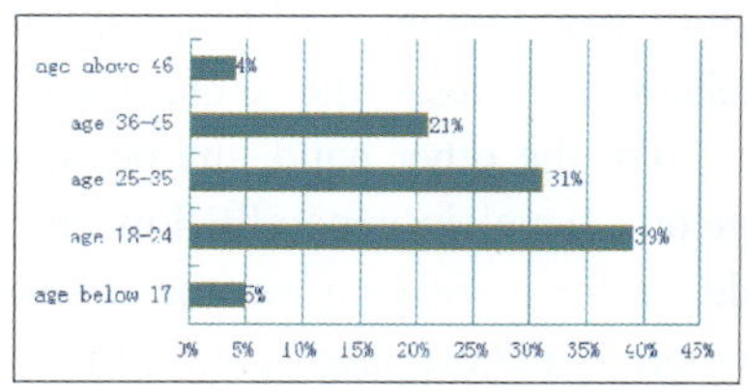

(b) Sex composition

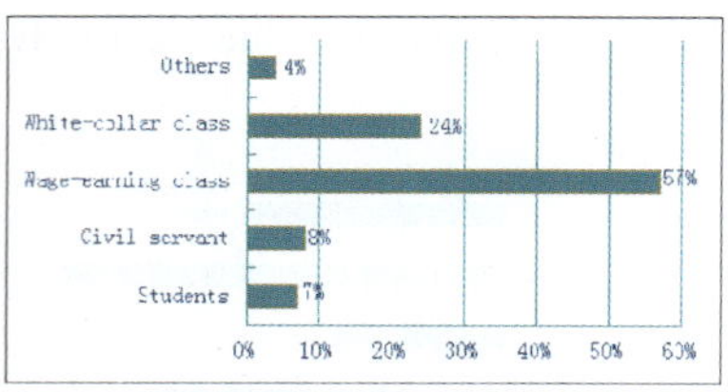

(c) Education level composition

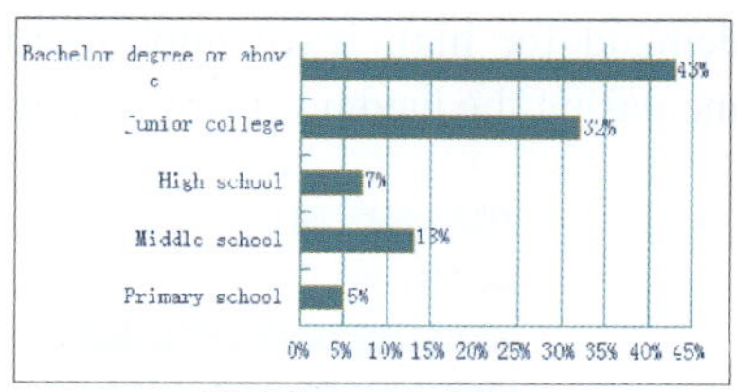

(d) Job composition

Figure 3. Surveyed farm tourists characteristics.

4.3. *The motivation analysis of tourism farm*

Through the survey of farm tourists, the motives of tourists who choose to travel the tourism farm in Heilongjiang province is to be close to nature, and to experience farm life. In particular, it can be divided into four types:

First, to enjoy the agricultural landscape. It is mainly agricultural sightseeing, entertainment experience and holiday leisure. This kind of tourists account for 69% of the surveyed tourists; Second, to study modern agricultural knowledge, broaden their knowledge, edify nature; This kind of tourists account for 23% of the surveyed tourists; Third, exhibition tourism and conference. This kind of tourists account for 3% of the surveyed tourists; Fourth, omnibus tourist motives. This kind of tourists account for 5% of the surveyed tourists.

4.4. *Behavior analysis of farm tourists*

(1) Way of traveling. Through the statistical analysis of questionnaire survey on the tourist of small farm town of tourism in Heilongjiang province, we found out that farm tourists prefer to travel together rather than follow a travel agency or the unit organization to travel. They consider that traveling with the group or unit organization is more restricted, will lose the purpose of being close to nature (Fig. 4).

(2) Travel expenses. According to the statistical analysis of the surveyed tourists to small farm towns of tourism in Heilongjiang province, they spend less than other types of travel expenses. Through the interview of farm tourists, we find that existing tourism farm activities and picking in Heilongjiang province are single, the reception level of tourism public service facilities is low. These make the tourists lack consumption and consumption desire (Fig. 5). But the surveyed tourists shows the attitude that they hope the farm will increase the typical consumer activity.

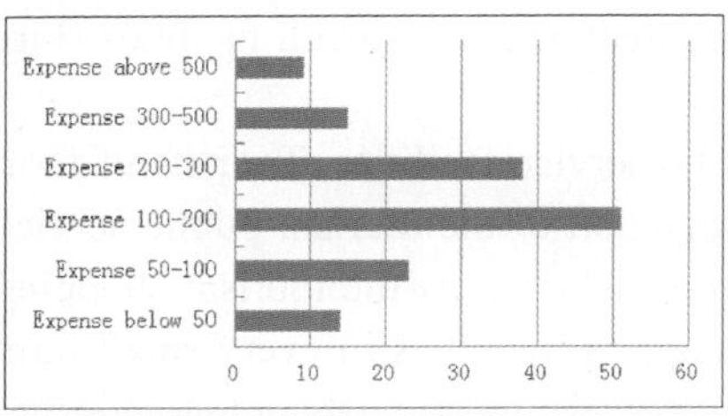

Figure 4. Trip mode.

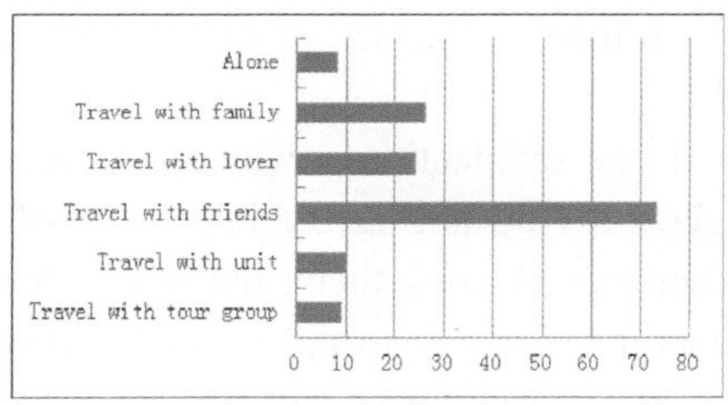

Figure 5. Travel expense.

(3) The length of traveling time. According to survey data of farm tourists, the tourists travel in the small tourism farm towns in Heilongjiang province at most for 2 days, even most of the tourists just travel for one day (Fig. 6). This phenomenon is mainly due to the lack of tourism farm public service facilities.

(4) Time quantum of traveling. Most of tourists will choose the long vocation like the International Labor Day and National Day. In addition, the weekend is a relatively popular choice for the tourism farm (Fig. 7).

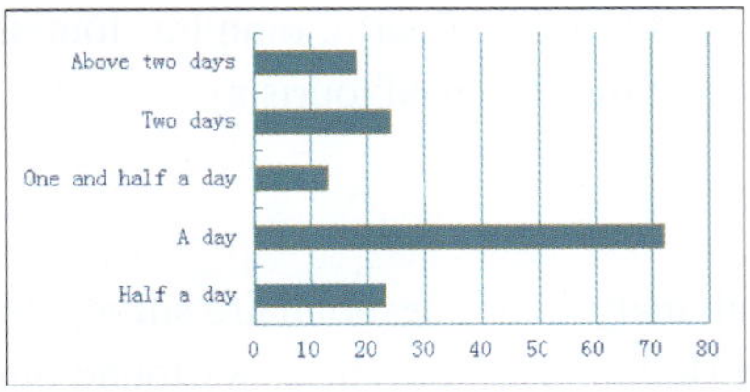

Figure 6. Time length.

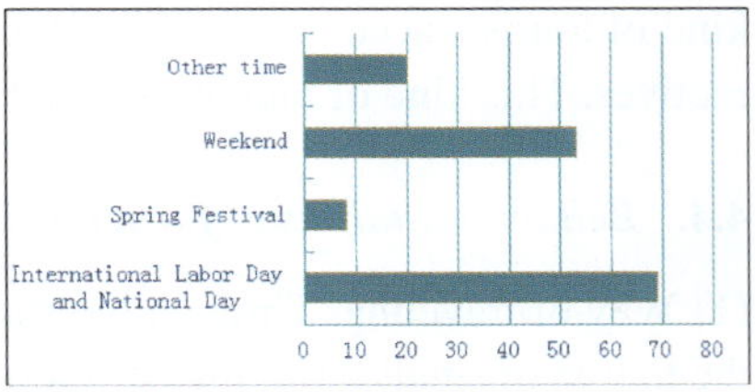

Figure 7. Traveling time.

4.5. *Satisfaction survey*

(1) The satisfaction survey of basic public service facilities. In the survey of basic public service facilities satisfaction, the highest is business financial facilities and market. Second, the farm town administrative facilities and education facilities are relatively complete, residents' satisfaction is higher. Third, as the facilities configuration of worker activity center and the elderly activity center in farm is relatively single, open time is short, it causes residents' satisfaction rate of park's entertainment facilities and square is low. Finally, due to the low level of medical treatment, it can't meet the medical needs of farm residents. As the small farm town lack the rehabilitation facilities, it reduces the farm town residents' satisfaction of medical treatment and health facilities (Fig. 8).

(2) The satisfaction survey of tourism public service facilities. Tourists of farm generally do not have a high level of satisfaction of the tourism public service facilities. Among them, the lowest level of satisfaction is the tourism shopping facility, it is due to the lack of tourism shopping facilities in every small farm town of tourism, it do not form a characteristic tourism shopping industry. Second, as the tourism public service facilities are lack of innovative and single, they also layout far from small farm towns and the traffic is inconvenience, these cause the tourists do not satisfy agriculture sightseeing and picking tourism facilities agriculture. Again, as the tourism public service facilities are lack of farming characteristic tourism catering and tourism accommodation facilities and the separation between tourism public service facilities and agricultural tourism facilities cause to use them inconveniently ,the level of satisfaction of farm tourists to the tourism catering and accommodation and other commercial facilities become low. These dates illustrate that the small farm towns of tourism in Heilongjiang province lack attention to the tourism public service facilities configuration and layout, types of tourism public service facilities lack proximity (Fig. 9).

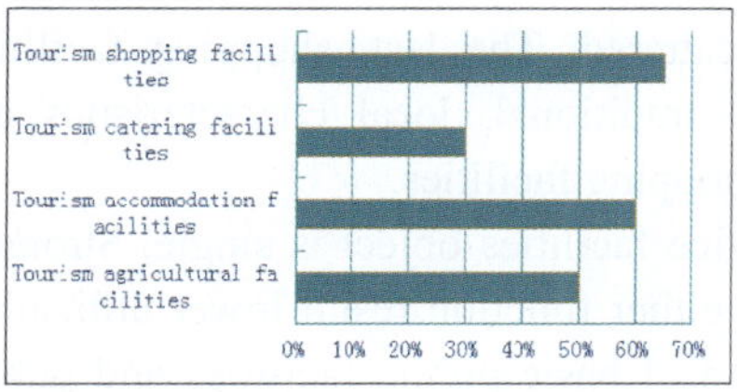

Figure 8. Basic public service facilities satisfaction.

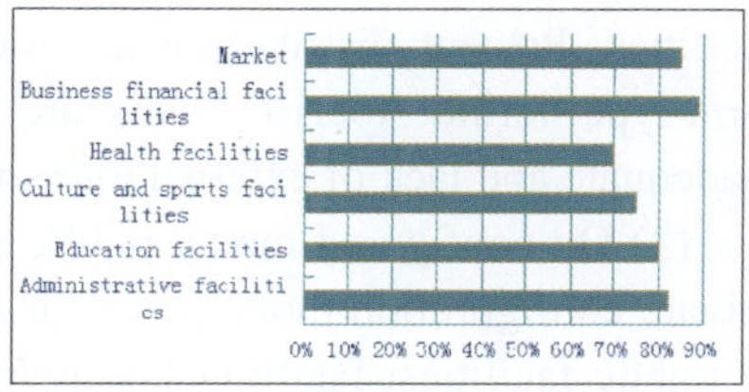

Figure 9. Tourism public service facilities satisfaction.

5. Problems of Public Service Facilities

5.1. *Facility configuration issues*

(1) Generally, basic public service facilities of small farm town of tourism in Heilongjiang province are comparatively complete, but public service facilities for tourism farm lack a lot.

Basic public service facilities are relatively complete overall, however, the low level of residents' income makes the level of demand on the farm public service facilities in small towns low. "Relative surplus" of basic public facilities resulted in lower demand of farm residents.

The planning and construction of tourist public services facilities lag behind and tourism farm development of small towns lack the support of appropriate facilities. It causes the tourists only "driving-by" the farm. Supporting construction of tourist commercial facilities also lack the farm characteristics excavation.

(2) Configuration types of public facilities do not meet the needs of residents and tourists and do not suit the demand characteristics of small farm town of tourism.

In aspects of basic public service facilities, configuration of facility does not consider the dual identity features of the farm residents as "farm workers" and "farmers". Basic public services of small farm towns stand out excessively the feature of its urbanization, lack consideration of "village" features, reducing the use of farm residents enthusiasm.

In aspect of tourism public services facilities, tourism public service facilities configuration does not meet the needs of farm tourists and do not suit the characteristics of tourism farm. First, they blindly pursue large-scale, while the configuration types is single. Second, the tourist accommodation facilities lack the farm characteristics of tourism accommodation facilities, can not urge tourists long-term stays. Again, dining facilities lack farm features. And mainly in the form of group tourism for the farm tour, lack scaled tourism and catering facilities, Present distribution are too scattered, The last, shopping facilities farm-type format tourist towns are too traditional, local characteristics are inadequate and lack of external tourism shopping facilities.

(3) On configured mode, public service facilities object is single, Stronger private configuration mode and poor use either function result lower utilization of public facilities, facilities configuration of basic public facilities and public tourism service facilities are in accordance with their respective service object strictly. Both of them lack function penetration in the configuration.

5.2. *Spatial distribution problem*

(1) Facilities site selection. Public service facilities layout space almost arrange inside the layout section, no special consideration was taken on the site selection.

(2) The layout patterns. Separation of tourism public service facilities' main part and basic public services space cause the tourism public service facilities seasonal vacancy; Separation of tourism agricultural facilities and tourism business space cause the inconvenient use of visitor.

(3) The layout morphology. Spatial layout of basic public services are basically along the field of the main road axis, segmentation of the field section is more serious, and the residents who live in the one side of the axis use the public services which on the other side of axis more inconveniently.

6. The Optimization Strategy of Public Service Facility Distribution

6.1. *The optimization strategy of different type public service facility distribution*

(1) The distribution optimization of basic public service facilities type. On the basis of inhabitant needs, ensuring the basic public service facilities overall scale, small farm town of tourism should be optimized the configuration type and scale reasonably.

(2) The distribution optimization of tourism public service facilities type. Tourism public service facilities should be prominent the characteristic of Heilongjiang Province, which would make it an irreplaceable role compared with other types public service facilities. In this way, we can keep the special advantages of farm tourism.

(3) Facilities type should be in line with local conditions.

According to their own tourism characteristic advantage, small farm town of tourism need make the travel patterns clear. Depending on the pattern, they should focus on the configuration of tourism public service facilities type, adjust measures to local conditions, and finally achieve the rational classification configuration of tourism public service facilities

(4) Public service facilities distribution types. Small farm town of tourism should enrich and innovative the public service facilities distribution types to enhance and change the tourism farm mode.

6.2. *The distribution model optimization*

The distribution model optimization of public services is mainly by breaking the boundaries between basic public service facilities and tourism public service facilities, achieving mutual penetration and object sharing between the two facilities and service. Finally, it can make efficient use of public services.

Single-function mode of tourism public service facilities has been unable to meet people's demand, tourists have a strong interest in agricultural festivals and farm living custom. And as the climate, the use of tourism public service facilities will vary with off-season and peak seasons of modern agriculture and tourism, which caused lower utilization and serious vacant phenomenon. On the other hand, tourism public service facilities distribution lack the consideration of inhabitant who is the huge 'potential' user group.

People have relatively strong demand for facilities recreations and sports facility. Since the income of resident in small farm town of tourism is generally low, they hope to use the tourist advantages of small farm town of tourism to increase revenue, leading the participating in the tourism business activities.

6.3. *Strategy for layout optimization of public service facilities*

6.3.1. *Facility location optimization*

(1) The optimization of basic public service facility location. The basic public service facilities are classified according to the service population, dividing into town level and community level. The service area of the town-level public service facilities is the entire town area, its location should be set up in the geometric center of the township and should be close to the main street. Community-level public service facilities are mainly served in the surrounding areas, and the placement of the location should reflect the fairness and should meet the basic living needs of every community in small towns.

(2) The optimization of public service facilities for tourism. According to their service object and service range, the tourism agricultural facilities and the tourism commercial facilities are divided into two types: concentrated (ones) or

disperse (ones). Those concentrate ones are mainly for tourist groups and other large-scale tourism group, which do not have to close to towns, but they must in the convenient traffic areas, so that the farm residents can take advantage of the facilities during the off-season tourism. The service objects of disperse facilities are mainly small groups such as individuals and families. These facilities should be scattered in those featured areas, which in the depths of the farm, close to the natural and have beautiful environment. In addition, these areas always have a certain distance from small farm town of tourism, but the traffic is still very convenient.

6.3.2. *Layout mode optimization*

The utilization rate of the tourism public service facilities is greatly influenced by the season, which can be improved by mixed-use. Public service facilities layout optimization is mainly achieved by mixed-use to solve the problem of low utilization rate during the off-season tourism.

From some point of view, the types of services provided by the basic public service facilities and the public service facilities in certain ways are same. Taking account of the impact of tourist peak seasons and the obvious seasonal vacancy phenomenon, organic hybrid layout is an necessary way to solve this problem.

6.3.3. *Layout shape optimization*

(1) The layout of the basic public service facilities. Change the existing traditional linear layout form into a centralized and efficient layout pattern.

(2) Optimizing the layout of public service facilities in tourism. The tourism agricultural facilities and the tourism commercial facilities should be more concentrated, at the same time, the relationship of the two kinds of facilities should be interdependent rather than separating from each other.

7. Conclusion

The public service facilities of the small farm town of tourism in Heilongjiang province has unique characteristics, and the layout pattern needs to be combined with the characteristics of the farm and the development of the tourism industry. In addition, the facilities should also to meet the needs of farm residents and tourists. In view of the existing problems in the layout of facilities, the layout form should take comprehensive consideration of the overall requirements of the towns. By optimizing the layout mode, resolving the phenomenon of the seasonal vacancy, improve the layout form, and finally realize the improvement of the public service facilities of the small farm town of tourism in Heilongjiang province. This can improve the vitality and the competitiveness of the small farm town of tourism, and put forward the feasible direction for improving the public service facilities of the towns.

Acknowledgments

This research is supported by the National Natural Science Foundation of China (Grant No. 51578176) and Natural Science Foundation of Heilongjiang Province of China (Grant No. E201310).

References

1. Fu Derong. 2010. The situation and trends of external rural tourism. *Tourism Management*: 61-65.
2. Clare Murphy, Emily Boyle. 2006. Testing a Conceptual Model of Cultural Tourism Development in the Post-industrial City: A Case Study of Glasgow. *Tourism and Hospitality Research* 6(2): 111-128.
3. Baud-Bovy. 1982. New concepts in planning for tourism and recreation. *Tourism Management* 3(4): 308-313.
4. Carlos Costa. 2001. An Emerging Tourism Planning Paradigm: A Comparative Analysis Between Town and Tourism Planning. *International Journal of Tourism Research* (3): 425-441.
5. Teitz, M. B. 1968. Toward a theory of public facility location. *Papers of the Regional Science Association* (21): 35-51.
6. Austin, C. M. 1974. The evaluation of urban public facility location: an alternative to cost benefit analysis. *Geographical Analysis* (6): 135-145.

Sustainable development evaluation on Chinese township based on principal component analysis*

B.H. Zeng, X.Y. Zhang†, F.H. Wu, Y.F. Wu, P. Chen and N.Y. Wei

Key Laboratory of Beibu Gulf Environment Change and Resources Utilization, Guangxi Teachers Education University, Nanning 530001, China

†E-mail: 191957125@qq.com

www.gxtc.edu.cn

The study tried to establish the indicator system of sustainable development with a case study of Chang Tang town. The index system included four sub-systems: economic, society, resource and environment and population, as well as 18 indices. Principal Component Analysis (PCA) was used to estimate the indicators of sustainable development during 2010~2014. The results showed that the sustainable development capacity of this town had been showing an increasing trend during 2010~2014. The comprehensive score was up to 16.85 from -17.18 which increased 1.94 times. The sustainable development level rank of the sub-system was: society>resource and environment>population> economic. Chang Tang town is going forward along the sustainable road steadily during the five years. In the future, Chang Tang should focus on ecological agriculture, efficient utilization of land resources, and the improvement of rural environmental quality.

Keywords: Sustainable development; Principal component analysis; Chang Tang Town.

1. Introduction

Sustainable development has become the mainstream development model in worldwide range. China is also committed to the implementation of sustainable development strategy. It has been 30 years since the establishment of the first sustainable development experimental zone in China [1]. With the promotion of sustainable development strategy, experimental zone will be more and more refined. With the focusing on issues of agriculture, farmer and rural area in China, the township as a unit to promote sustainable development is worth trying. So Chang Tang town as a case study had been chosen to evaluate the

* This work is supported by National Natural Science Foundation of China (41461095), Guangxi Natural Science Foundation (2011GXNSFA018094, 2014DD29046) and Guangxi high school project (2013ZD042).

† Work partially supported by Guangxi Elite Hillland of Soil Pollution and Ecological Restoration and the Collaborative Innovation Center of the Ecological Environment & Integration Development in the Xijiang River Basin.

sustainable development level by applying PCA method in this paper, for the purpose of making policy recommendations for the local government.

1.1. *Study area*

Chang Tang town is at the east gate of Nanning City, belongs to the jurisdiction of Qingxiu district of Nanning city, and has the prominent feature of urban radiation. This town is one of comprehensive reform pilot town in Nanning city, which has 265 square kilometers area with favorable natural conditions for the local agricultural production. The China-ASEAN (Nanning) Modern Agricultural Park is layout here. On the whole, the town is based on agricultural production.

1.2. *Study methods*

Sustainable development is a complex system, composed of economy, population, resources, environment, society and other subsystems. All the subsystems are mutual connection and promotion during the process of sustainable development. So the evaluation index of each subsystem is not completely independent. Based on the above considerations, the principal components analysis (PCA) method was applied to evaluate the sustainable development level of this town. Principal component analysis (PCA) is a statistical procedure that uses an orthogonal transformation to convert a set of observations of possibly correlated variables into a set of values of linearly uncorrelated variables called principal components. The method is particularly suitable for dealing with many indicators with inherent relations. This method can not only reveal the intrinsic relationship between the various indicators, and can also convert a number of indicators into several comprehensive indices. The calculation and analysis of the comprehensive indices (principal components) can realize the comprehensive level of evaluation [2, 3].

2. Evaluation of the Sustainable Development of Chang Tang Town

2.1. *Construction of index system for Chang Tang town*

Four subsystems including economy, population, resources and environment, society and 18 indices are selected to form a sustainable development evaluation system following scientific, systematic, available and feasible principles. The observation data of every index from 2010–2014 are obtained from the annual report of Chang Tang Government, as shown in Table 1.

Table 1. Evaluation index of sustainable development for Chang Tang town and the available data from 2010–2014.

Subsystem layer	Standard layer	Index layer	Unit	2010	2011	2012	2013	2014
Economy subsystem	Economic scale	J1: Industrial output value	Yuan	13500	12966	8048	2028	2124
		J2: Agricultural output value	Yuan	29107	38331	41008	43972	46171
		J3: Fixed asset investment	Yuan	5003	9464	14568	11381	20363
	Economic income	J4: GDP per capita	Yuan	5340	6328	7325	8340	9596
Population subsystem	Population status	R1: Total population	person	30253	30396	30555	31348	31867
		R2: Natural population growth rate	%	5.3	7.31	7.50	6.13	6.46
	Education level	R3: Nine-year compulsory education penetration	%	99.5	99.7	99.7	100	100
		R4: Illiteracy rate	%	4	2	2	2	1
Resource and environment subsystem	Ecological construction	Z1: Forest coverage rate	%	57.3	58.1	59.3	60.1	60.8
	Resource utilization	Z2: The area of cultivated land per capita	acre	1.24	1.21	1.21	1.17	1.14
	Environmental protection	Z3: Hazard-free treatment rate of household garbage	%	70	71.3	72	95	96
Society subsystem	Life quality	S1: The proportion of population drinking clean and safety water	%	97	98	99	100	100
		S2: Medical personnel number per thousand people	per	2.4	2.4	2.5	2.7	3
		S3: Farmer annual income per capita	yuan	5310	6328	7325	8340	9341
		S4: Residential telephone penetration rate	set	95	96	97	98	99
		S5: Radio coverage rate	%	70	80	90	100	100
		S6: TV coverage rate	%	90	95	100	100	100
	Social stability and security	S7: Employment rate of persons with disabilities	%	4	5	5	8	10

Table 2. Index data after standardized treatment.

Subsystem layer	Standard layer	Index layer	Unit	2010	2011	2012	2013	2014
Economy subsystem	Economic scale	J1: Industrial output value	Yuan	1.033	0.937	0.056	2028	-1.022
		J2: Agricultural output value	Yuan	-1.600	-0.209	0.195	0.641	0.973
		J3: Fixed asset investment	Yuan	-1.244	-0.468	0.420	-0.135	1.428
	Economic income	J4: GDP per capita	Yuan	-1.261	-0.601	-0.016	0.563	1.316
Population subsystem	Population status	R1: Total population	person	-0.909	-0.703	-0.474	0.669	1.417
		R2: Natural population growth rate	%	-1.381	0.857	1.069	-0.457	-0.089
	Education level	R3: Nine-year compulsory education penetration	%	-1.292	-0.369	-0.369	1.015	1.015
		R4: Illiteracy rate	%	1.643	-0.183	-0.183	-0.183	-1.095
Resource and environment subsystem	Ecological construction	Z1: Forest coverage rate	%	-1.274	-0.714	-0.714	0.686	1.176
	Resource utilization	Z2: The area of cultivated land per capita	acre	1.176	0.409	1.21	1.17	1.14
	Environmental protection	Z3: Hazard-free treatment rate of household garbage	%	-0.811	-0.714	-0.662	1.056	1.131
Society subsystem	Life quality	S1: The proportion of population drinking clean and safety water	%	-1.381	-0.671	0.153	0.920	0.920
		S2: Medical personnel number per thousand people	per	-0.784	-0.784	-.392	0.392	1.569
		S3: Farmer annual income per capita	yuan	-0.956	-0.558	-0.558	0.637	1.434
		S4: Residential telephone penetration rate	set	-1.267	-0.628	-0.002	0.635	1.263
		S5: Radio coverage rate	%	-1.265	-0.632	0.000	0.632	1.265
		S6: TV coverage rate	%	-1.381	-0.614	0.153	0.920	0.920
	Social stability and security	S7: Employment rate of persons with disabilities	%	-1.565	-0.447	0.671	0.671	0.671

2.2. *Standard processing of raw data*

The original data of each subsystem is processed using the standard method of Z-score in SPSS software (see Table 2).

2.3. *Correlation test*

By analyzing the correlation coefficient of the standardized indices (see Tables 3–6), the indices that have small correlation coefficients (correlation coefficient < 0.3) will be deleted [4-8].

Table 3. Correlation coefficient matrix of the economic subsystem.

	Zscore(J1)	Zscore(J2)	Zscore(J3)	Zscore(J4)
Zscore(J1)	1			
Zscore(J2)	-0.867	1		
Zscore(J3)	-0.749	0.868	1	
Zscore(J4)	-.941*	.945*	.912*	1

Table 4. Correlation coefficient matrix of the population subsystem

	Zscore(R1)	Zscore(R2)	Zscore(R3)	Zscore(R4)
Zscore(R1)	1			
Zscore(R2)	-0.071	1		
Zscore(R3)	.931*	0.13	1	
Zscore(R4)	-0.738	-0.61	-0.821	1

Table 5. Correlation coefficient matrix of the resource and environment subsystem.

	Zscore(Z1)	Zscore(Z2)	Zscore(Z3)
Zscore(Z1)	1		
Zscore(Z2)	-.946*	1	
Zscore(Z3)	0.878	-.931*	1

Tables 3 and 5 show that all the correlation coefficients between indices of economic subsystem, resources and environment subsystem (absolute value) are greater than 0.3, so it is needn't to delete any variables; Table 4 shows that the correlation coefficient between R1 and R2 of population subsystem is 0.071, and the correlation coefficient between R2 and R3 is 0.130, both are less than 0.3. So variable R2 has to be deleted; Table 6 shows that S4 and S5 have repeatability that is to mean they have the absolutely same correlation coefficients in social subsystem, one of them (S4) has to be deleted. Similarly, S6 has to be deleted in this paper.

Table 6. Correlation coefficient matrix of the society subsystem.

	S1	S2	S3	S4	S5	S6	S7
S1	1						
S2	0.827	1					
S3	0.871	.977**	1				
S4	.970**	.930*	.945*	1			
S5	.970**	.930*	.945*	1.00**	1		
S6	1.00**	0.827	0.871	.970**	.970**	1	
S7	.943*	0.658	0.69	.884*	.884*	.943*	1

2.4. *KMO value*

KMO values can be calculated by SPSS, the results of every subsystem are listed in Table 7.

If the KMO value is greater than 0.6, and the Bartlett significance level is less than 0.05, that will illustrate that there is a strong correlation between the indices of different subsystems, and the principal component analysis can be performed. Each subsystem will refuse the independent assumption of each index [9–11]. The KMO values of the four subsystems are all greater than 0.6, so the principal component analysis can be continued.

Table 7. KMO values for each subsystem.

The subsystem	KMO value	Bartlett significance level
Economic subsystem	0.682	0.022
Population subsystem	0.671	0.03
Resource and environment subsystem	0.726	0.026
Society subsystem	0.637	0.008

2.5. *Extracting the principal component*

The principal component can be extracted when the eigenvalue is greater than 1, and the cumulative variance contribution rate reached more than 80%.

Table 8. Eigenvalues, variance contribution rate, and the cumulative variance contribution rate of each subsystem.

	The principal component	The eigenvalue	The variance contribution rate(%)	The cumulative variance contribution rate(%)	The biggest variable load values
Economic	1	3.644	91.112	91.112	J4(0.995)
Population	1	2.663	88.765	88.765	R3(0.976)
Resources and environment	1	2.837	94.572	94.572	Z2(0.986)
Society	1	3.761	94.031	94.031	S5(0.991)

It can be seen from Table 8: economic subsystem has only one principal component, and the eigenvalue is 3.644 (> 1), the variance contribution rate is greater than 80%. The principal component can explain most of the original information, and can be on behalf of the overall subsystem of economic development. It can also be found that the J4 (net income per capita) has the maximum value of the load, explaining that the indicators of farmer net income per capita plays a major role in the economic subsystem. Therefore ways should be found to improve farmer's net income for Chang Tang town in the future.

Population subsystem also has only one principal component, and the eigenvalue is 2.663 (> 1), the variance contribution rate is greater than 80%. The principal component can explain most of the original information and it can represent the overall subsystem of population. The indicator R3 (nine-year compulsory education penetration %) has the maximum value of the load, explaining that the nine-year compulsory education penetration plays a major role in the subsystem of population.

Resource and environment subsystem has only one principal component and the eigenvalue is 2.837 (> 1), the variance contribution rate is greater than 80%. The principal components can explain most of the original information. It can represent the environment subsystem. The index Z2 (arable land area per capita) has the maximum value of the load, explaining that arable land per capita play a major role in the resources and environment subsystem.

Society subsystem has only one principal component, and the eigenvalue was 3.761 (> 1), the variance contribution rate is greater than 80%. The principal component can explain most of the original information, and it can represent the society subsystem. The index S5 (radio and television coverage rate) has the maximum value of the load, explaining that the index plays a major role in society subsystem.

2.6. *The comprehensive score of each subsystem*

According to the component score coefficient matrix, the linear combination of the principal components of the subsystem can be extracted, and the functions of the principal component are as following [12]:

The function of economic subsystem is:

$FJ = 0.995J4 + 0.965J2 - 0.932J1 + 0.924J3$.

The function of population subsystem is:

$FR = 0.976R3 + 0.947R1 - 0.902R4$.

The function of resource and environment subsystem is:

$FZ = -0.986Z2 + 0.968Z1 + 0.963Z3$.

The function of society subsystem is:

FS = 0.999S5 + 0.974S6 + 0.949S3 + 0.930S2 + 0.880S7.

The comprehensive score of every subsystem can be obtained after substituting the standardized data into above four formulas (see Table 9). The score trends of every subsystem are shown in Figure 1.

Table 9. Comprehensive score of sustainable development of each subsystem.

Year	F_J	F_R	F_Z	F_S
2010	-4.91	-3.60	-3.17	-5.62
2011	-2.11	-0.86	-1.78	-2.88
2012	0.51	-0.64	-0.92	-0.15
2013	2.01	1.79	2.29	3.09
2014	4.50	3.32	3.59	5.57

From Figure 1, it can be seen that the sustainable development of each subsystem in Chang Tang town during period 2010–2014 has an increasing trend. The comprehensive score of the economic subsystem increases from -4.91 in 2010 to 4.50 in 2014. After data analysis, it is found that, the total agricultural output value, farmer net income per capita, fixed assets investment have achieved a rapid growth during this period. The total growth rate of agricultural output value grows 58.6%, farmer net income per capita grows 84.4%. However, the total industrial output value of Chang Tong town reduced to 21.24 million Yuan in 2014 from 135 million Yuan in 2010. The reason is that some industries had been quitted. Despite the decline in industrial output value, the overall economy keeps on increasing. In recent years, Chang Tang town takes the China-ASEAN (Nanning) modern agricultural park as an opportunity, Chang Tang town strives to build “fruit base” and rural leisure tourism resort etc. The modern agriculture has been effectively promoted.

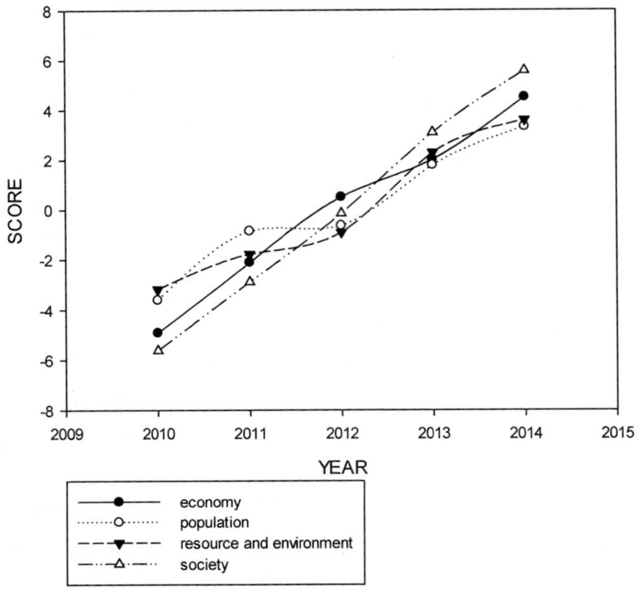

Figure 1. Score trend of each subsystem.

The comprehensive score of the population subsystem also increases from -3.60 to 3.32. The nine year compulsory education penetration rate has been to 100% since 2013. The illiteracy rate has been declining, which implies the improvement of the population quality of Chang Tang town.

The comprehensive score of the resource and environment subsystem increases from -3.17 to 3.59. Recently, under the idea of "make Chang Tang beautiful and make villages clean", Chang Tang town has strengthened the clean-up efforts to improve the rural environment quality. A number of sewage and garbage treatment projects have been completed. The people living environment has been improving under the efforts of the government's advancing the construction of new countryside. The arable land area per capita in this subsystem, which has a big load, implies that the need to protect arable land in the future.

The comprehensive score of the society subsystem increases from -5.62 to 5.37, showing the fastest growing speed among the subsystems. The good situation is related with the efforts of the livelihood improvement, including the the promotion of employment and labor security, construction of social assistance system, development of cultural broadcasting and sport, improvement of basic medical and health facilities, et al.

2.7. *Comprehensive evaluation of sustainable development of Chang Tang town*

The principal component analysis can be continued based on the comprehensive score of economic, population, resources and environment, and the society subsystem of Chang Tang town from 2010 to 2014. The comprehensive score function for the sustainable development of Chang Tang town can be obtained as following:

$$F= 0.998FS+0.991FR+0.989FJ+0.989FZ.$$

It can be seen that the social subsystem has the largest load. The comprehensive score can be calculated for the sustainable development of the town and the comprehensive score trend is shown in Figure 2.

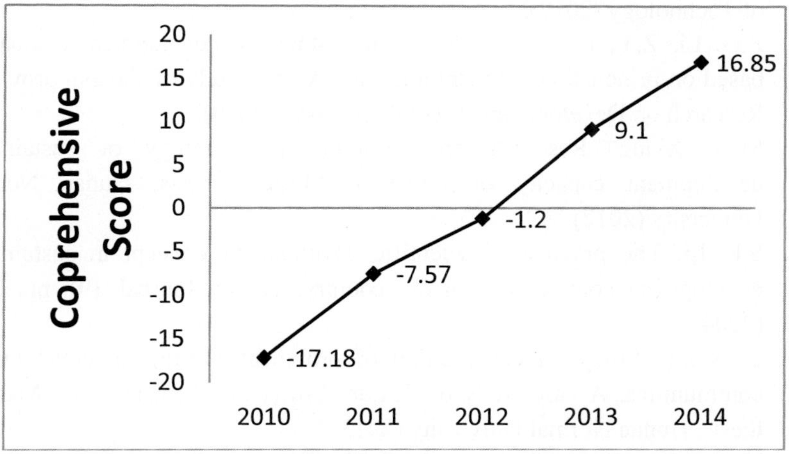

Figure 2. Comprehensive score trend of Chang Tang sustainable development.

From Figure 2 it can be seen that the comprehensive score of the sustainable development of Chang Tang town increases from -17.8 in 2010 to 16.85 in 2014, growing 1.94 times. The level of sustainable development in Chang Tang town shows a rising trend, indicating that Chang Tang town is going forward along the sustainable road steadily during the five years. Chang Tang also shows an obvious coordination development in economy, society, population, resources and environment et al. In the future, Chang Tang should focus on ecological agriculture, utilization efficiency of land resources, and the improvement of rural environmental quality.

References

1. X. Hu, Study on the Development Pattern, the Evalution Index System and Model of Low-Carbon Sustainable Development Experimental Zone, Masters thesis, Guangxi University (2014).
2. X. Li, Study on the sustainable developmental level of Tangshan city, Masters thesis, Hebei Normal University (2008).
3. J.L. Li, Study on development state assessment of sustainable communities: A case study of China national sustainable communities of yulin, Masters thesis, Northwest University (2012).
4. Y.M. Li, Comprehensive evalution of Hubei sustainable development based on principal component analysis, Masters thesis, Wuhan University of Technology (2012).
5. Z.Q. Li, Z.Y. Li, P. Liu, Regional sustainable development evaluation based on principal component analysis: A case study of Jiangsu province, Research on Development, 140(1): pp. 64-67 (2009).
6. M.D. Xiang, Research on evalution and strategy of sustainable development capacity of Zhuzhou, Masters thesis, Hunan Normal University (2012).
7. S.F. Li, The practice of scientific development concept in sustainable development communities of our country, Theory Journal, (9): pp. 8-11 (2004).
8. L. Xiang, Study on construction of provincal sustainable development communities: A case study of Suxian district of Chenzhou city, Masters thesis, Hunan Normal University (2013).
9. H.X. Zheng, Analysis and Evaluation of Regional Sustainable Development System: A Case Study of Henan Province, Masters thesis, Henan University (2001).
10. W.Q. Jiang, The construction analysis on sustainable development community of Sichuang province: A case study of Yinbin Nanxi district with the creating theme of coordinating urban and rural, Masters thesis, Sichuang Normal University (2014).
11. J. N. Li, Study on sustainable development evaluation of Loudi city, Masters thesis, Hunan University (2009).
12. N. Cao, Study on state assessment of sustainable development communities: A case study of China national sustainable communities of Yulin city, Masters thesis, Northwest University (2010).

The Chinese and foreign urban cultural landscape dynamic development analysis

Y. Li

Arts and Design School, Eastern Liaodong University, Dandong 118003, China
E-mail: ldxyliyu@163.com

Due to the geographical climate conditions in different cities, the different cultural customs and habits, designers to achieve mastery through a comprehensive study of the relevant professional knowledge on the basis of through the creative thinking, on the basis of respecting city cultural, flexible use of modern technology and new materials science and technology, skilled grasp the suitable city regional culture landscape design, this is the consensus of the cultural heritage landscape design.

Keywords: The city; Cultural landscape; Dynamic development.

1. The Regional Environment Impact on China's Cultural Landscape Dynamic

The natural environment has a great influence on the formation of the culture. China's vast territory, the uneven distribution of mountains, rivers, the domain in the west and low climate in difference is very big. Led to the garden art form the colorful style of our country. Gardening art has a different genre and theory. According to the area of the geographical position is divided into: the North garden, Jiangnan gardens, Lingnan gardens and Bashu garden, West garden and so on all kinds of different style garden. The past people used to place the Beijing landscape pattern as a representative of the Chinese classical gardens of the northern landscape, because Beijing is have been to the Forbidden City of the landscape are mostly built by the ruling class, basic characteristic is garden architectural grandeur and magnificence, north of wide area is broad, but some of the limitations of geographical environment and the natural meteorological conditions, the evergreen trees is not much, however, the river leads to blocked garden stiffness atmosphere more (as shown in Figure 1), and beautiful gentle and slightly insufficient. In addition to Beijing, Kaifeng, Luoyang, Xian and other cities also belongs to the northern landscape in the gardens.

Figure 1. The North garden.

Figure 2. Jiangnan gardens.

Transition from north to south, a gradual change of landscape scene and, Jiangnan region populated, geographical area is small, due to geographical limitations, landscape space design slightly more compact, its design features and deep (as shown in Figure 2). Because of river lake is abundant in the south of the Yangtze river, there are kinds of evergreen trees, garden tonal on simple quietly elegant, bright beautiful, although the Jiangnan landscape space feeling a little cramped, but the garden sketch delicate and exquisite. Suzhou gardens usually gardens in the south of the Yangtze river is regarded as the main representative, the other is concentrated in Hangzhou, Yangzhou, Shanghai, Nanjing and other places.

From north to south, Lingnan is a specific geographical environment, geographical environment, not only people's living habits and customs is mostly the same. Rich Lingnan garden because of subtropical plants at the age of evergreen (as shown in Figure 3), landscape conditions than the north and the south, such as clear a tranquil garden of Shunde, Dongguan's garden, Yu yu Yin mountain room and so on, have a large spacious building style and features, it is affected by geographical environment and people's living habits.

Bashu garden because of its special geography, topography ups and downs of uneven, river caused the Bashu garden has a natural and comfortable, the characteristics of the plain. Bashu garden rock pile rock mostly use natural environment landscape and the natural, of primitive simplicity generous shape. Bashu garden some memorial gardens and temples in the garden, the landscape characteristics of the artistic method is the most essential for art spirit and the unity of nature and nature worship, the famous representatives of the Du Fu thatched cottage (as shown in Figure 4), temple of marquis, Wangjiang building, etc.

Figure 3. Lingnan garden.

Figure 4. Bashu garden.

Western landscape architecture began in the Tang dynasty, Tang dynasty's great power and the attitude of opening to the outside world, makes the period of the Tang dynasty architectural structures and landscape waterscape design without exception accepted the western culture, with the western flavor material, a large number of plant species in western regions be landscape paintings on the walls of amorous feelings of the western regions statues, sculpture, building construction and decorative pattern is the western cultural factors reflected in the landscape, formed the rich western culture art landscape (as shown in Figure 5), the surviving typical representative works, longmen grottoes in Dunhuang, enough to justify grottoes temple the existence of western culture and art form.

In conclusion, we can see that garden under the natural environment of the region with the aid of natural mountain, water, stone, wood and other materials combination form in the natural climate region environment formed in the dynamic effect of the different cultural landscape. In modern times, although the urban cultural landscape convergence of urbanization construction, but different regional cities people living habits will give new and different cultural connotation, such as in Beijing had abandoned factory became the artist, in Changsha fire palace temple street as a place for people at one's leisure walk after the meal, although there is no changes in the cultural landscape in these places, city is given on the regional culture of different landscapes.

Figure 5. The west garden.

Figure 6. European garden.

2. City Cultural Landscape Dynamic Development Abroad

The development of the western landscape architecture from geometric forms to the Italian terrace garden and French classical garden to English landscape garden, and then entered into a new era of garden art development. In such a development pattern is no relationship, but this kind of evolution is based on the change of culture, regional change, a review of the western landscape dynamic evolution, to deepen our understanding of the western landscape gardening art, also can better grasp the present development of urban landscape.

European garden is also known as the west garden, it with the ancient Egyptian and Greek garden for origin development, throughout western culture, western when dealing with people and nature of thought of different Chinese culture adapt to the nature, the bud of western gardens can be dated back to ancient Egypt and Greece, it was among the first to apply geometry to the botanical garden construction, the development of the world's earliest specification style garden. After the Renaissance, humanism spirit core is people as the core, advocate the individuality liberation, the 18th and 19th century humanism theory of "victory day" to get the public recognition, autonomous pursuit the happiness in life and believe in people is the creator and master of life. Human thought of reforming the natural landscape environment can be from now on, from ancient Greece to Renaissance exquisite aesthetic rules of symmetry are constant, transformation of landscape has become a way for the modification of the natural order of (Figure 6). And because Europe focus on mathematical science, landscape form also under the influence of geometrical structure and the mathematical relationship, with the axis of the clear and geometry space relation. To express the rationality of rigorous science and landscape layout of plant shrubs cut into the pattern of spatial relations in geometrical symmetry.

European classical garden has three main periods: the Italian terrace garden and French classical garden and English landscape garden, in which the spirit grand imperial garden landscape is the most bright in the west. Leading trend in the middle of the 16th century Italy, in addition to mountain this layout is reasonable construction of the axis apparent form outside terrace garden, the city landscape on behalf of saint mark's square, ingenious use of the wrong view as well as to the irregular shape of reflecting the distribution of the grand masters, it has been known as the world artist idea of the sitting room of Europe.

At the beginning of the fifteenth century, the rise of the Italian Renaissance, the European garden style once turned to the baroque art type curve line. Landscape form appeared after the pursuit of novelty, dreamlike expression. At the end of the 15th century, can still find many Roman legacy of humanities landscape architecture in the Roman empire in Italy, the humanities landscape evoked memories of the Italian people of imperial past glory, looking forward to the Roman way of life at that time, so some noble in the scenic foothills or hillside villa built their manor. Commonly used table layout activities, in order to facilitate the plane stress in neat formation is symmetrical, in the field and the process of building relationships, buildings are often located in the central axis of the subject or the courtyard of the horizontal axis line, can also be set up on both sides of the axis. The Italian terrace garden a European mansion garden system first, axis running through it to a mountain. Several consecutive mesa, with the continuous elevation difference form exists, thus showing the European system characteristic of the magnificent, magnificent momentum, while the classical gardens also became the Renaissance artistic vitality.

The middle of the 17th century French luxurious garden style leadership trend, grave and elegant garden aristocratic momentum was under the reign of Louis XIV's primacy imperial power, the pursuit of the vast limitless space, completely artificialization space order it is the feeling of the characteristics of the largest scale and space scale. The center of the building is usually built on the highest point of terrain fails to play a role, vestibular and city avenue, backyard in size and space obey the architectural space. Statues, fountains, flower beds on both in the central axis to become the visual center of the garden landscape and dominated, so visible garden composition is of illuminative of key of central axis as the center to, other axis on both sides of the axial symmetric distribution. Waterscape is given priority to with static waterscape foil calm deep the vast, the imposing manner, the overall visual feast, give a person with symmetrical rigorous order but also embodies the authoritarian regime in the hierarchy. Since the middle of the 18th century, the British first entered the industrialization society, for the exploration of urban landscape ahead of other countries, and due to the influence of Chinese landscape art, began to pursue in the form of natural landscape, landscape no longer pursue the pattern of the typical geometry type of Europe, changed before garden style, no need pruning of the old trees, winding paths and natural pond, the appearance of the landscape is a natural landscape painting, new gardening art mature, British elegant romantic lead trend. Romantic, in the second half of the 18th century, Britain is no longer meet the natural scenery, to the original natural scenery added some new technique, the pursuit of new and profound artistic conception and more full-bodied style of poetry. Is the development trend of the overall transition from natural charming garden as pictorial gardens, so the garden has a romantic temperament, sometimes even as the cause for a strong sentimental mood and sadness things deliberately to build some ruins, drought grave, such as runners on. Finally, all of these new gardening technique greatly influence the whole of the European landscape development.

In the second half of the 19th century, under the influence of the United States all the landscape as the city society people use, landscape has evolved to support the urban population, the urban civilization and the urban landscape development created a new world. The awakening of realism, nature and science to contact to the landscape design of the city and promote the development of modern landscape design. In the 20th century, the European functionalism and the development of abstract art has remarkable influence on the landscape design, functional, abstract art and cubism brought a new vision for the design of the landscape, urban landscape from the bondage of the past in the form of classical gardens create different from classical gardens and generous dignified urban landscape form, all kinds of content to different function forms of landscape composition, a new type of modern landscape material, bold use of color, bring new vitality to the modern city.

Present landscape design, make European culture also in European integration, but culture, after all, there's a difference around the landscape architect design across the region, they have their own personal style of cultural background and habits into the works, the realized that one of the first form of urban culture, arts, customs, national spirit, etc., such as inheritance and proclaim in the future, so will gradually decline, so disappear. As a result, the European landscape architect or very respect local city culture, will work as far as possible into the local culture. Today in the wave of European integration, the big area city culture of harmony, but every culture is also exist in its own way and learn to adapt to the development of The Times and the new elements.

To sum up, affected by the development of western history and culture, the development of the western landscape in different age has different historical environment different sense, cultural differences to express embodied through the style of landscape design related elements. It is the development of a constant motion spiral way of exploration.

3. The Regional Environment of the Human Landscape Dynamic Effect in Europe

European winding coastline long and rich change, due to the inland sea of intensive inland obviously very much, create the unique natural conditions and geographical landscape of different areas, the mainland countries France and island of the British landscape can be used as a representative of the two major European garden.

Figure 7. The French garden.

Figure 8. The wind-induced natural park.

Figure 9. Contemporary and contracted style.

French geographical environment has rich natural vegetation and a large number of rivers and lakes region, due to the wide plains of habit in the ground landscape, pursue to rule type garden of artificial beauty, the composition is balanced strict symmetry, solemn momentum grand grand (as shown in Figure 7), elegant style and elegant, make free with flower beds, statues, fountains and other accessories rich landscape space.

Perhaps the glorious history and independent island form, let them according to their own culture and to the continent apart from the pursuit of style unique natural landscape, landscape type is called a natural charming garden. And well used UK region of many hills, trees, grass, forests, combine to form the natural scenery, the pursuit of natural beauty and landscape in the form of natural style, in the process of landscape preference as far as possible to erase the trace of artificial carve, blur the line between the landscape and the natural, expectations will be introduced to the natural landscape to (Figure 8), from nature to form.

Due to the geographical position different garden style also is not the same, in the Italian peninsula, villa garden is the most representative form of a garden, main villa buildings generally located in the slope of the top, below the building layers platform in accordance with the slope area of leaning, the layers of platform to push associated, with trees on either side of the central axis as the transition of landscape plants to and with the surrounding natural environment. This is Italian peninsula countries with neat type between the mainland's French garden and the island of Britain's natural scenery botanical garden and the combination of development suited to their national geographic with neat and auxiliary scenery in a form of garden.

Either way before design to study on the regional environment and space environment and dynamic of the processing of dialectical thinking, cannot be treated the same, in the form of simple sense of adjust measures to local conditions to achieve maximum sense of visual effect; Moreover is the form of regional culture element and regional landscape configuration is commonly used in simple summarize the understanding of the urban environment; Geomorphic features and fully understand the regional environment and urban skin texture, decrease as far as possible to the original man-made destruction of urban environment landscape, foreign design is to define good design like without design design it natural, no useless space, all things are in sports development, the most inferior space also may play a positive effect.

4. Diversified under the Dynamic Development of the Urban Cultural Landscape

Information globalization impact around the world now, also makes the city culture in the area of a less and less, the proportion of various forms of art and technology language showed a trend of diversified development, in the modern landscape design, how to design a space environment, create a safe, healthy, comfortable and efficient environment is the art of science and wisdom, as well as a degree at a certain stage of human society civilization development thinking of the product, is long gardening activities is the inevitable result of historical development. Human world outlook, values and ethics are also reflected in the treatment of the environment. No matter how Chinese and western landscape on the shape and style change, attention to ecological and people-oriented design concept has been throughout. Have sprung up in the 1980s, so that it is hard to go to all kinds of thoughts, unified definition of it, the design is diverse, diversified modern architecture style. Also guide the diversified development of modern landscape design, diverse forms mainly by a variety of purposes or a variety of methods to solve of, rather than a format or a method.

In the design of modern mainly has several kinds of, the first is contemporary and contracted style, by the architect Mies Fiddro far-reaching, he put forward is much less also calls the minimalism design concept, design of geometric type straight line in a simple and agile. Such kind of landscape design for the echo in the form of modern architecture, no longer follow the style of the traditional garden style, with dot, line, face as the basic element of space composition, multi-purpose trees, plant landscape layout adorned among them, the hard landscape pavement (as shown in Figure 9), transfer node local processing in abstract sculpture, art flower pot, stone, wood, bamboo, glass, stainless steel for general landscape elements in stream of people activity space, to highlight the fresh and stylish landscape color or contrast or simple single, based on efforts to use contemporary feeling extremely high quality material, high technical requirements to the construction of landscape architecture, mostly used in the commercial plaza and office space and young people to give priority to modern apartment residential projects. There is one kind is the modern natural style, its characteristics: design line is fluent, adopt modern hard landscape form and nature of the plant landscape, the combination of material use more natural stone and wood plate, etc. The use of the technique of expression through the modern organizations form to organize landscape major elements, combined with rich space use of hard landscape to create a visual focus, focus on micro topographic space form combined with soft landscape plants, with natural grass and green processes the perfect combination of modern landscape and the natural ecological harmony.

Urban culture is the fusion of history and civilization, it has a place of historical reproductivity, Chinese and western cultural differences reflected the urban regional culture and cultural differences and natural geographical features. The current economic globalization makes the world share the results of economic and technology, also prompted the urban areas of their traditional culture protection and attention. In diversity under the influence of urban cultural landscape design also appeared a kind of contradictory emotions, on the one hand, due to the urban convergence and neglect and critical reflection on the past culture produced a more intense feelings of traditional culture, on the other hand, the new development of science and technology have produced some new materials to promote fashion designers began to explore new construction craft, the exploration and development is bound to produce new artistic value and aesthetic vision. These two aspects, no matter what will eventually be back to the science and art to develop these two origins.

5. The Chinese and Foreign Urban Consensus and Blending of the Dynamic Development of Humanity

Appeared since the 1970s environmental problems, began to focus on plant photosynthesis reasonable use can not only provide rich oxygen for city environment and adjust the humidity of the air in cities is vacuuming disinfection effect such as decreased dry also can keep good water and purified water; And shelter forest in urban landscape design also can be a very good protection of urban green space system and the surrounding environment, introduces the fresh air in the suburbs to the city to improve the air quality of city environment, this kind of measures can be seen within the surrounding healthy urban ecological system can not only ensure the development of the city environment and health for people's life. As can be seen, with a nature of ecological landscape design of particular meaning to the urban environment.

Modern designers in the design, should arouse the wisdom consciously use ecological consciousness to find, nature has given us a lot of enlightenment, to observe with professional eye know site construction of the basic features of the original adjust measures to local conditions, need not deliberately innovation do STH unconventional or unorthodox. Ecology has become the essential of environmental problems of urban landscape design, modern human landscape design in fact can be interpreted as human design of urban ecosystem, namely to respect natural ecological system on the basis of the development process of design is suitable for human habitation environment system space. Landscape design not only need to identify new materials and new technology to make reasonable utilization of resources, should also be recycling energy and material, the self-sustaining development sustainable treatment technology of ground, and so on some and throughout the construction and management.

In the process of building the conscious of protecting the environment is the initial motivation of modern landscape design, respect the urban ecology and urban culture, reduce the plunder of environment and resources, reduce the destruction of urban cultural and, maintain the sustainable development of urban cultural landscape, the coordination of urban ecological and human common development is the common concern and efforts need to do, it has become the consensus of the modern landscape design.

6. Cultural Inheritance Consciousness of Landscape Design under the Consensus and Blending

Culture, is applied to environment are formed in the specific geographical environment of cultural forms, this kind of cultural forms in turn affects the environment, thus formed the environment of city characteristics. With an area and because of the change of living environment and the human migration and the movement rule of culture itself, also present in different historical stages with different morphology.

Closed early Chinese society and the superior natural resources, China's landscape is formed in such a case, and the cradle of western European civilization isolation and barren, prompted the spirit of the west, the landscape also influenced by cultural ideas and preferences change exploit natural forms, and western culture differences from philosophy and aesthetics thoughts and gardening thoughts are different. We can also see in the west garden bridges and winding path leading to a secluded spot, but very different from Chinese classical gardens artistic conception. Europeans still advocate rational, deeply loves the conquest of nature. Humble administrator's garden valve plug from France to China, you can see western gardens are handled and Oriental garden is completely different, we may temporarily from the combination of the pure geometry of the novel and rational order of visual novelty, but we are not necessarily really fully accept these strange new space form. Because long tradition of culture brand has profoundly changed our aesthetic, this difference makes us feel their own traditional culture can have a sense of belonging, thus we finally want to be able to change these exotic forms to make these new forms in accordance with our familiar appearance and scale to obtain its own identity. So we in the landscape design is more to their city culture and the traditional culture symbol of digging and now again the form of modern landscape. In the urban cultural landscape design, the construction of the natural elements adjust measures to local conditions and the cultural elements of the use of appropriate, if considering the visual, as to the cultural judgment. In the process of design, one has some kind of symbolic cultural elements and the totem culture can not only highlight the local city image, and can activate the people in the form of visual memory in the heart of the past. The familiar natural cultural elements in the urban cultural landscape characteristics.

References

1. Allen, P.M. 1999. Population growth and environment as a self-organizing system. *Discrete Dynamics in Nature and Society* 3(2-3): 81-108.
2. Gomez, F., Tamari, N., Jambalayas, J. & Green, Z. 2001, Bioclimatic Studies and human comfort in the future development of urban planning. *Landscape and Urban Planning.* 55:151-161.

3. Krause, C. 2001, Lour visual landscape. Managing the landscape under special consideration of visual aspect. *Landscape and Urban Planning.* 54: 23.
4. Macdonald, K.A. 1998, Ecology's last frontier: studying urban areas to monitor the impact of human activity. *Chron. Higher Educ.* 44: A18.
5. Nasser, M. 2001, Understanding cultural landscape transformation: Are photographic survey in Chital, Eastern Hindu Kush, Pakistan. *Landscape and Urban Planning.* 57(3-4):241-255.
6. Y.Q. Wang & X.S. Zhang 2001, A dynamic modeling approach to simulating socioeconomic effects on landscape changes. *Ecological Modeling,* 140: 141-16.

Research of choosing and developing the leading regional service industry

Q. Yang
School of Business, Central South University of Forestry and Technology, Hunan, Changsha, 410004, China
Email: 610247226@qq.com

W.W. Peng
Hunan University of Traffic Engineering, Hunan, Hengyang, 421009, China
Email: pengww888@163.com

This paper initially constructed a theoretical analysis framework of choosing the leading service industry based on results of the relevant domestic and foreign theories. The paper takes Chongqing as a case to determine its pillar service industry by the factor analysis method. From the factors of system safeguard, development orientation and platform building etc, we put forward countermeasures and proposals for cultivating the leading industry on the basis of qualitative analysis on choosing pillar industry in Chongqing.

Keywords: Factor analysis method; Evaluation index system; Pillar industry.

1. Introduction

For the accurate definition of service industry, there are various opinions in the academia. However, most definitions are derived from different organizations because of the statistics demand, and the service industry is defined as a series of specific activities[1]. In this paper, according to the availability of statistical data and the research demand, with reference to the international and domestic classification standard of industrial departments, we define the service industry as a gathering of various economic activity which include water conservancy, environment and public facilities management, social security and social welfare, technical services and geological activity, health, culture, sports and entertainment, computer service and software, lease and business service industry, scientific research, education, storage and transportation, postal service, information transmission, the financial industry and real estate industry and so on. However, no matter what definition we take, the development level and speed have become important symbols of economic and social modernization[2]. This paper takes Chongqing as a survey region and tries to establish the leading industry choosing framework based on the reality of Chongqing service industry development, choosing leading industry of Chongqing which suits for the development of modern service industry, and putting forward feasible policy suggestions for the industry development.

2. The Standard and Indicator System Construction for Choosing the Pillar Service Industry

2.1. *Choosing standard of pillar industry*

Since the reform and opening up domestic, scholars become more and more intend to research in choosing criterion about pillar industry. However, most choosing criteria are based on the standpoints from foreign economists, trying to complement it with Chinese national conditions. Among them, the relatively comprehensive one was "three benchmark theories" which put forward by Zhou Zhenhua. That is, the growth potential benchmark, flexible shortage substitution benchmark and bottleneck effect benchmark. Guan Aiping proposed the "Six benchmark theory", that is, sustainable development benchmark, market benchmark or demand benchmark, efficiency benchmark, technological progress benchmark, industrial linkage benchmark and competitive advantage benchmark. For the specific standard for choosing pillar industry, combined with actual industry development in Chongqing, we can describe roughly from the following five aspects: First, the capability of fully utilizing the natural resources, and the potential of its product market were great , market expansion ability can promote the continuous development of the regional economic. Second, relatively high economic benefits. Third, there is relatively strong industrial correlation. Fourth, there is benefit for placement of labor force. Fifth, there are more mature product technology than other industries, at the same time with strong ability to absorb new and high technology.

2.2. *Construction of the indicator system for choosing pillar service industry*

According to the strategic planning of historical task to comprehensively build a well-off society, combined with the characteristics of the pillar industry and the specific situation of socioeconomic development, this paper build a indicator system of choosing pillar industry scientifically from the five aspects.

First, the size of the market, the broad market demand are the key factors deciding whether an industry development is stable and rapid, which include industrial contribution rate of production value, flexible demand and income and increasing ratio. Second, the industrial correlation: pillar industry must has strong extension effect promoting the development upstream and downstream industries through the point to area spread of industrial chain. In this paper, we divide the industrial correlation into influence coefficient and sense degree coefficient. Third, industrial comparative advantage: tax from pillar industry is the main source of local fiscal revenue, and also is material guarantee of other industries. This paper divides the indicator of industrial comparative advantage into comparative productivity coefficient and comparative interest rate coefficient. Fourth, industrial technology development, which is important force to promote regional economic development. In this paper, its corresponding index is industrial technology development contribution. Lastly, social employment: during the development of the pillar industry, we can not only focus on the economic benefits of the one-sided pursuit, consideration should be given to social benefit. In this paper, we take comprehensive employment coefficient to indicate employment standards during choosing pillar industry.

This paper will make detailed analysis on the main service industries in Chongqing reference to the benchmarks and indicators mentioned above, in order to make a reference for choosing pillar service industry among Chongqing service industries in the future, as shown in the Table 1 below:

Table 1. Construction of indicator system for choosing pillar industry.

Benchmark level	Specific index	Index symbol	Explain
Industry market scale	Industrial output contribution value	X1	In a certain period, the proportion of one industry's output accounts for the national GDP.
	Income elastic demand	X2	The ratio of demand increase rate to national income per capita under that other conditions remain unchanged according to the comparable price.
Industrial correlativeness degree	Sensitivity coefficient	X3	When the national economic sectors adding a unit, one industry's demand sensitive due to it.
	Influence coefficient	X4	When the national department add a unit eventually use, the effect of production demand affected degree to National economic sectors due to it.
Industrial comparative advantage	Comparative productivity coefficient	X5	The ratio of total factor productivity of a certain industry to the total factor productivity of all industries.
	Comparative profit and tax rate coefficient	X6	The ratio of tax rate of a certain industry to the tax rate of all industries.
Industrial technology development	Contribution rate of industrial technology development	X7	Industrial technology development is the important power to promote the regional economic development.
Social employment	Comprehensive employment coefficient	X8	The development of the pillar industry can not only focus on the pursuit of economic efficiency, consideration should be given to social benefit.

3. The Choosing of Pillar Service Industry in Chongqing

In this paper, the relevant index and data in the evaluation indicator system during choosing the pillar industry are from "The Chongqing statistic yearbook" (2008–2010), and industrial department configuration is based on "The China input-output table" (2009). In this paper we considered 14 industries; processing data after standardization are listed in Table 2. The factor analysis steps include: use SPSS15.0 for data standardization, calculate the correlation matrix eigenvalues and eigenvectors, select principle factor according to the factor contribution, construct factor scoring model and achieve the comprehensive ranking for all samples.

Table 2. Standard value of service industry in Chongqing.

Industry name	X1	X2	X3	X4	X5	X6	X7	X8
Financial industry	0.0677	2.6945	2.6907	0.0003	0.0022	1.7485	11782.2	0.0678
Retail establishments	0.0227	2.3924	4.8869	0.0306	0.0179	1.1488	9115.7	0.0226
Road transportation	0.0436	1.8375	4.9304	0.0219	0.0179	0.819	8172.6	0.0413
Education career	0.0557	1.7723	4.8069	0.0274	0.0316	1.0814	5798.1	0.0553
Real estate industry	0.0267	1.0843	5.3827	0.1037	0.0876	0.8936	1459.3	0.0265
IT services Industry	0.0595	1.8322	5.9582	0.0905	0.0490	1.6032	1690.4	0.0591
Catering industry	0.0062	0.7060	4.5098	0.2402	0.0720	0.4804	458.3	0.0061
Health industry	0.0466	1.2937	5.6761	0.1896	0.0789	1.4328	653.22	0.0464
Research industry	0.0303	1.1352	5.4012	0.2117	0.0843	1.3508	885.8	0.0385
Tourist industry	0.0426	0.9493	4.1656	0.3034	0.0653	1.4149	546.9	0.0427
Insurance industry	0.0406	0.8992	4.1814	0.2522	0.0779	0.5572	712.8	0.0404
Railway transportation	0.0371	1.1311	4.4746	0.2156	0.0809	0.7674	970.4	0.0379
Science and technology service industry	0.0177	0.7047	4.6638	0.3936	0.1086	0.7954	463.7	0.0177
Air transportation industry	0.0677	2.6945	3.4582	0.0003	0.0022	1.7485	8788.2	0.0679

Table 3. Scores and Ranks for the 14 Service Industries in Chongqing.

Industry name	Score	Rank	Industry name	Score	Rank
Financial industry	95.26	1	Health industry	46.71	8
Retail establishments	87.51	2	Research industry	37.45	9
Road transportation	76.36	3	Tourist industry	36.12	10
Education career	64.12	4	Insurance industry	33.77	11
Real estate industry	55.29	5	Railway transportation	30.30	12
IT services Industry	50.11	6	Science and technology service industry	29.71	13
Catering industry	48.99	7	Air transportation industry	28.04	14

3.1. *Characteristic value and contribution rate of R*

At first, we collect corresponding numerical data of eight specific analysis indicators in the 14 service industries in Chongqing at 2009, and then construct the correlation coefficient matrix R and get the eigenvalue and eigenvector by using the mathematical method. In this paper the contribution rate of 3 principle factors' cumulative variance is 87.541%, which is greater than 80%. It can be concluded that the three principle factors reflect the most information of the original variables, which can be used to evaluate the comprehensive indicator, the reliability of evaluation is 87.541%.

3.2. *The calculation and sorting of comprehensive score*

In order to get a comprehensive indicator which can reflect development level of the 14 service industries in Chongqing, we use the contribution rate of weight of three principle factors to calculate the comprehensive score of the development level of the 14 service industries in 2009. As we can see from the Table 3, the higher score means the higher comprehensive development degree. The score greater than 0 means that the service level is above average, otherwise, the service level is below average.

4. The Choosing of Pillar Service Industry in Chongqing

We calculate the comprehensive score and ranking of services industries in Chongqing, then choose pillar industry according to the ranking, for example, industries of financial, retail establishments, road transportation, education career and real estate lie in the top five of the comprehensive indicator. In principle, we should choose these five industries as pillar service industry. But in view of the particularity of the service industry, we should give full consideration to the different characteristics of the choosing. Firstly, general theory divided service industry into productive service and public service.

Secondly, public service should ensure its orderly development. Therefore, when we take profit regional service industry as development guidance, public services should be taken as key industries in government plans. So, the ranking of Chongqing service industry could be adjusted as follows: financial industry, retail establishments industry, education career industry, road transportation industry, and real estate industry. The five secondary important industries are IT services industry, catering industry, health industry, research industry and tourist industry respectively. Under the background of industrial integration and innovation, we integrated all the 14 industries mentioned above. Therefore, we suggest the pillar service industries which Chongqing should be focus are: the modern financial industry, modern business trade service industry which mainly focus on wholesale and retail trade services, modern logistics industry which primarily focus on road transport and water transport, modern science and technology information service industry which mainly focus on information transmission, software and professional technology and science industry, modern tourism which is constituted by metropolitan tourism and leisure sightseeing tourism.

5. Suggestions and Countermeasures for Improving Service Industry in Chongqing

In view of basic situations of Chongqing's industry development, to support the development idea and layout in Chongqing based on the proposed train of thought for industry development, this paper puts forward the following countermeasures: First of all, release the development guidance catalogue for the regional modern service industries, and instruct the rapid and healthy development in related regional industries according to the selection of the pillar industries and the industry layout, and form the modern service industry cluster as soon as possible. Second, speed up the industrialization process as industry is an important aspect which is indispensable to support the development of modern service industry, therefore, we must strengthen regional industrial layout and rapid development. Third, speed up the urbanization process, promote the acceleration of urban construction, create a good city environment and city roads, energy, municipal administration, public information infrastructure, improve the level of people's income, all the above mentioned activities will create demand support and guarantee for the development of regional modern service industries. Fourth, speed up the opening of the service areas, at the same time, actively introduce funds and projects, draw lessons from outside the excellent experience, introduce advanced technology services, standards, and operation and management mode of well-known enterprises, to promote the formation of the service industry with comparative advantages. Fifth, enhance the construction of talent projects. Due to the lack of high-end talents for the development of modern service industry, we have to strengthen and improve the talent introduction mechanism while vigorously mining school talent in specific area, and to cultivate local talent, and pay special attention to the in-service education and again-obtain-employment education. Sixth, establish the leading enterprises in all service industries and give strong support in policy. We should cultivate a number of large scale and competitive backbone enterprises which can drive the service industry's management to step forward to be large scale, brand and network among the key pillar industries.

About the authors

Yang Qin (1989-), female, Hunan Changning, master candidate in Business School in Central South University of Forestry and Technology. Research direction: Forestry ecological and economic management. Corresponding author: Peng Wenwu, male (1970-), born in Hengyang Hunan, President of Hunan University of Traffic Engineering, associate professor, mainly engaged in education management as well as scientific research about regional economy. Zip code 421009. Email: pengww888@163.com

References

1. Zhu Y., Research of choosing and developing the leading mordern service industries in Luoyang. Masters degree thesis, Lanzhou University, 2010. (in Chinese).
2. Li M, Yang M., Review of industrial structure problems for Chinese urban industries. Industrial Technology & Economy, 2006, 2:5-8. (in Chinese).
3. Guan A., Study on benchmarks of selecting regional leading industries. Statistical Research, 2002, 12:37-40. (in Chinese).
4. Li G, Wei M., Study on selection and application of regional leading industries — A case on choosing leading industries in Shaanxi Province. Science Studies, 2004, 2:47-52. (in Chinese).
5. Wang J., On the concept, selection and mechanism of pillar industries. Jiangsu Social Sciences, 1995, 4:31-36.
6. Joachim S, Gerhard U., Regional investment incentives in [6] Germany: impacts on factor demand and growth. Annals of Regional Science, 2000, 2:173-195.
7. Solberg T G., Investment management for the long term.The Journal for Hospital Governing Boards, 1991, 01:10-19.

8. Wolff R., Efficient growth of an agglomerating regional economy. Regional Science and Urban Economic, 1985, 04:555-572.
9. Qiao J, Xu Y., Test regional economic development strategy in Chongqing municipality. Journal of Henan University, 1998.12:41-48. (in Chinese).
10. Feng J, Rong Z., Study on the standard and method for choosing regional and city leading industries. Economic Geography, 1999.06:23-26. (in Chinese).
11. Chen Y, Liu Y., Conception on Xiamen City industrial structure characteristics and the cross-century development [J], Geographical Science, 2001.01: 76-80. (in Chinese).
12. Liu Z., Modern industrial economic analysis. Nanjing University Press, 2001:26-38. (in Chinese).

Ecological risks resulting from changes in land use in Wuhan in the past 15 years[1]

J.H. Zhao[1], L.Y. Zhao[1], C. He[1,*], W.M. Du[2] and C. Huang[1]

[1]*College of resources and environment, Hubei University, Wuhan China 430062*

[2]*Zhongjing Research Center for Urban Development, Beijing China 100742*

[*]*E-mail: chao_ho@163.com*

Based on the regional quantitative calculation of five classes of ecological risks, the landscape security pattern in Wuhan has undergone significant differences in terms of ecological risks resulting from changes in land use over the past 15 years. From 1996 to 2010, the Class 1 landscape security pattern became increasingly small and transferred from the north to the west and south. The Class 2 landscape gradually shrank, and its distribution pattern transferred from a scattered form to the urban fringe. The Class 3 landscape pattern transferred from the urban center to the urban fringe and back to the urban center. The Class 4 landscape gradually became large with uneven pattern changes. The Class 5 landscape significantly transferred from the southeast to the east and north. In the urban area from 1996 to 2010, the degree of landscape ecological security in Wuhan exhibited a trend of safe to poor from downtown to the urban fringe. We can observe regular variation in landscape ecological security in Wuhan. In the main urban area, the public need for a suitable urban environment is increasing. Coupled with the controllability of human activities, landscape ecological security has gradually stabilized. However, with the acceleration of urbanization and increases in population and urban construction land, the public have gradually strengthened the development and utilization of suburban areas, particularly of cultivated land, forest, water, unused land, and other natural resources. Therefore, the degree of landscape ecological security of suburban areas has gradually declined because of human intervention.

Keywords: Wuhan; Land use; Ecological risk analysis; Spatial–temporal evolution.

Studies on ecological security have mainly focused on the concept, assessment method, and building of index systems and investigated the theoretical mechanism of ecological security and how to assess the ecological security of regional ecological systems at different spatial and temporal scales [1–4]. The early comprehensive statistical index system was built mostly by using

---

[1] Author introduction: Zhao Jin Hui, born in 1974, female, PhD, associate professor, research direction to geographical environment. E-mail: zhaojh2004@hubu.edu.cn

Funded projects: National Natural Science Fund Project (41401559), Hubei Province Science and Technology Agency project (2014CFB558), Hubei Province Education Department Project (D20141001).

[*] Corresponding author.

ecological security assessment indices in the form of the pressure–state–response paradigm. However, regional spatial heterogeneity as it exists is difficult to define accurately.

Spatial heterogeneity is significant to understanding dynamic spatial changes and the state of ecological security [8–11]. This study analyzes how to establish an appropriate index system for ecological security.

1. Calculation of Landscape Ecological Risk Index

To establish links between various landscapes in urban areas and regional ecological risks and to build an ecological risk index using the area proportions of urban landscape components to translate the spatial patterns of landscapes into spatial risk variables [5] by space sampling, if the ecological risk index (ERI) is high, the degree of landscape ecological security is also high. Conversely, if the degree of ecological security is low, the ERI is calculated as follows:

Where *ERI* is the landscape ecological risk index, *N* refers to the number of landscape types, A_i refers to the size of type *i* landscape within the area, *A* refers to the total area of the regional landscape, and R_i refers to the landscape loss index of type *i* landscape.

The formula for landscape ecological loss index is expressed as follows [6]:

Where R_i refers to the landscape loss index of the class *i* landscape component, S_i refers to the landscape disturbance index of the class *i* landscape component, and F_i refers to the class *i* landscape vulnerability index. Because of the different dimensions used, the data should first be normalized.

Based on [7] difficult it is for various types of landscapes to undergo reversed changes, the seven types of landscapes in Wuhan are ranked according to vulnerability as follows: urban construction land, “other” landscapes, unused landscapes, farmland, grassland, woodland, and water landscapes. Upon ranking, the relative weights of the vulnerability of each type of landscape, that is, the landscape vulnerability index (F_i), are obtained. For convenience in calculation and analysis, the normalized value range is [0.1, 0.9].

2. Spatial–Temporal Changes in Ecological Security Index

The landscape index is calculated through the Spatial Analyst module of ArcGIS software and landscape pattern analysis software Fragstats 3.3. According to the actual area of Wuhan, the study area is divided into a 0.5 km × 0.5 km unit sampling grid. Space division is then applied to the vector analysis diagram of Wuhan for 1996, 2006, and 2010. The image for each period is divided into 412 sampling plots, and the divided vector diagram is transferred into a grid map. Fragstats software is used to calculate the landscape pattern index of each sampling plot. The landscape security index of each sampling plot is calculated according to the abovementioned formula. The security indices obtained are used as the property values of the central point of each sampling plot. In the ArcGIS geostatistical module, the ordinary Kriging method is used for spatial interpolation to obtain the diagram for the landscape ecological security level of Wuhan.

2.1. *Changes in random coefficient of variation*

Partial_Sill/Sill is used to indicate the strength of spatial correlation. The greater the value of Partial_Sill/Sill, the stronger the spatial correlation. Accordingly, Nugget/Sill is called the base effect to express the variability between samples. Increasing Nugget/Sill increases the likelihood that the variation between samples is caused by random factors. The relevant parameters of the spherical model are shown as follows:

Table 1. Model related parameters variation function theory.

Year	Model	Nugget	Still	Nugget/ Still	Range/m
1996		0.00043	0.00081	0.53	400029
2006	Spherical model	0.046	0.131	0.35	148166
2010		0.00013	0.00194	0.71	148166

Random variation in the ecological security index of the study area was 0.53 in 1996, 0.35 in 2006, and 0.71 in 2010, showing an upward trend after a drop. This trend indicates that the small-scale random variation below 0.5 km was great in 1996 and 2010 but small in 2006. With the social and economic development from 1996 to 2006 and the increasing impact of human activities on the environment, the small-scale random variation of the ecological security index has been gradually replaced by large-scale structural variation.

2.2. *Classification of ecological security*

To evaluate ecological security, the minimum and maximum values of the ecological security index can be used as the upper and lower limits of regional ecological security.

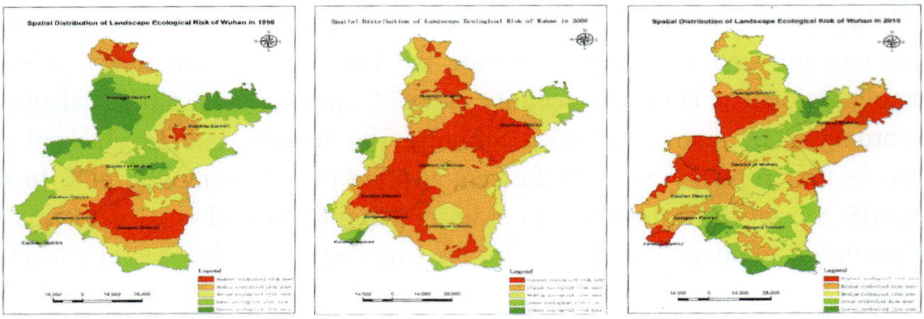

Figure 1. Diagram of Landscape Ecological Security Distribution of Wuhan in 1996 (left), 2006 (middle), and 2010 (right).

The Nature Break method in Arcgis is used to classify ecological security into five grades: the high ecological security zone (Grade 1 zone) ≤ 0.055, 0.055 < relatively high ecological security zone (Grade 2 zone) ≤ 0.071, 0.071 < medium ecological security zone (Grade 3 zone) ≤ 0.087, 0.087 < relatively low ecological security zone (Grade 4 zone) ≤ 0.103, and 0.013 < low ecological security zone (Grade 5 zone). We then calculate the ecological security index of the 412 security zones generated in this classification. Based on the model theory of variation function, a diagram of the distribution of the ecological security space grade is created by ordinary Kriging interpolation for the ecological security index of Wuhan in 1996, 2006, and 2010.

In 1996, the high ecological risk zone in Wuhan was mainly located in the middle of Jiangxia District, northwest of Huangpi District, and west of Xinzhou District. The relatively high ecological risk zone was mainly distributed in Jiangxia District. The periphery of the high ecological risk zone in Huangpi and Xinzhou District was shaped like a ring and gradually changed from high risk to relatively high risk. The relatively low ecological risk zone was mainly distributed in the southwest and southeast of Jiangxia District, west of Caidian District, east of Huangpi District, center of the Wuhan administrative area, and north-central and southeast of Xinzhou District. The low ecological risk zone was mainly concentrated in the west of Huangpi District and north of Xinzhou District and scattered in the center of the area under the administration of Wuhan. The medium ecological risk zone was widely distributed across seven districts, forming a ring surrounding the relatively high ecological risk zone. These five ecological security zones were evenly distributed in terms of area.

In 2006, the high ecological risk zone was largely distributed in the east and center of Caidian District, east of Huangpi District, and west of Xinzhou District (a substantial increase over that of 1996), with a westward trend. The relatively high ecological risk zone was mainly distributed in the center of Jiangxia

District, east of the Wuhan administrative area, and north of Huangpi District. The relatively low, low, and medium ecological risk zones were small and formed a ring in the periphery of Wuhan. Overall, the high and relatively high ecological risks were dominant in the ecological security of the study area.

By 2010, the high ecological risk zone had transferred to the west of the Wuhan administrative area, southwest of Huangpi District, and northeast and center of Xinzhou District and had greatly decreased in size compared with that in 2006. The relatively high ecological risk zone was scattered across various districts of Wuhan. The relatively low ecological risk zone was mostly distributed in Huangpi, Xinzhou, and Jiangxia Districts. The low ecological risk zone was located in the south of Jiangxia District and northwest of Xinzhou District. The medium ecological risk zone formed a ring in the periphery of the high ecological risk zone, which increased in terms of area compared with that in 2006.

A comparison of the three maps in Figure 1 shows that the ecological security of Wuhan from 1996 to 2006 deteriorated. However, the trend strengthened from 2006 to 2010. Jiangxia District underwent the most significant changes, transitioning from high-risk to medium-, low-, and relatively low-risk. Caidian District is in second place. Overall, the ecological security of the area is stable. Ecological risk only slightly changed in the entire Wuhan administrative area, indicating stable ecological security.

3. Analysis and Discussion

3.1. *Dynamic spatial changes*

From the spatial perspective, the main urban area of Wuhan, including Huangpi, Dongxihu, Caidian, Qingshan, Jianghan, and Jiangan Districts as well as northeast of Xinzhou District, exhibited stable landscape security in 1996. Ecological risk in the main urban area was saturated, and destructive changes were consistent, although certain suburban areas had ecological environments more suitable for human habitation.

In 2006, after a decade of urban development, capital expansion, population mobility, resource development, and utilization continued to improve. Landscape ecological security in Wuhan significantly declined (Figure 1).

By 2010, with the growing awareness of ecological and environmental protection in the urban development of Wuhan, ecological security in the main urban areas stabilized, but less so than in Jiangxia District, east of Xinzhou District, and north and east of Huangpi District. However, several suburban areas with poor ecological security, including Dongxihu, Caidian, and Hannan

Districts (areas with high-value resource development) as well as Jiangan and Hanyang Districts, experienced great landscape fragmentation, high ecological fragility, and weak immunity and resistance to external interference because of complex human activities.

Figure 1 shows a significant difference in the landscape security patterns of Wuhan under the same criteria in terms of landscape classification and unified parameters for landscape pattern. From 1996 to 2000, the Class 1 landscape ecological security pattern became increasingly small and shifted from the north to the west and south. The Class 2 landscape gradually shrank, and its distribution pattern transferred to the urban fringe from a scattered form. The Class 3 landscape pattern transferred from the urban center to the urban fringe and then back to the urban center. The Class 4 landscape gradually became large, with uneven pattern changes distributed primarily in 2006. The Class 5 landscape significantly transferred from the southeast to the east and north and was also widely distributed in 2006.

3.2. *Dynamic temporal changes*

By analyzing the ecological security of Wuhan in 1996, 2006, and 2010, we can observe the law of change in regional ecological security in Wuhan and find that the areas that underwent significant changes in ecological security and vulnerable security included suburban areas with high-value resource development, such as Huangpi, Dongxihu, Caidian, Hannan, Xinzhou, Jiangan, Hanyang, Jianghan, and Qingshan Districts. According to the distribution of environmentally sensitive areas and the content of ecological protection in Wuhan, five nature reserves and four small natural conservation areas (natural wetland park, Xiajiashi Reservoir, Meidian Reservoir, and other water conservation areas) were designed to protect typical lake and wetland ecosystems in Wuhan.

4. Conclusions

By 2006, after a decade of urban development, the landscape ecological security of Wuhan had significantly decreased, whereas the suburban north of Huangpi District, west of Dongxihu District, Hannan District, and south of Jiangxia District had relatively high levels of ecological security. As the development of the urban economy continues to expand the development and utilization of resources and the environment, once we focus on the extent of development rather than on improvement in quality, ecological security will certainly decrease. Thus, by 2010, with the growing awareness of ecological and environmental protection in the urban development of Wuhan, the ecological security of main urban areas had stabilized, albeit less so than in Jiangxia District, east of Xinzhou District, and north and east of Huangpi District. However, certain suburban areas with poor ecological security, such as Dongxihu, Caidian, and Hannan Districts (areas with high-value resource development) as well as Jiangan and Hanyang Districts, experienced a high degree of landscape fragmentation, high ecological fragility, and weak immunity and resistance to external interference because of complex human activities.

References

1. Yang Qingsheng, Qiao Jigang, Ai Bin. Rapid analysis of city area landscape ecological security space-time evolution [J]. Journal of Ecology, 2013, 33(4):1231-1239.
2. Wang Naiju, Zhou Taofa. Anhui Tongling city ecological security evaluation [J]. Journal of Environmental Sciences, 2012, 32(10): 2634-2640.
3. Hu Haide, Li Xiaoyu, Du Yufei. Construction of Dalian City ecological security pattern [J]. Journal of Northeastern University, 2013, 45(1):138-143.
4. Yu Rongrong, Xie Wenxia, et al. The landscape pattern of Jiaozhou Dagu Bay estuary wetland ecological security based on [J]. Chinese Journal of Ecology, 2012, 31(11):2891-2889.
5. Zang Shuying, Liang Xin, Zhang Sichong. Ecological risk analysis of land use in Daqing City based on GIS [J]. Journal of natural disasters, 2005, 14(4): 141-145.
6. Li Xiehui, Li Jingyi. Analysis of regional landscape ecological risk based on GIS — a region along the Weihe river for [J]. Arid zone research cases, 2008, 25(6): 899-903.
7. Xiao Yang, Mao Xianqiang. Landscape ecological risk analysis of [J]. China Regional Environmental Science, 2006, 26(5): 623-626.
8. Bruce K. Hope, Jeffery A. Peterson, A Procedure for Performing Population-Level Ecological Risk Assessments [J]. Environmental Management, 2000(25):281-289.

Spatial pattern of Northeast China's regional city network

M.S. Qu, X. Lu* and J.C. Yuan
Architecture and Urban Planning College, Shenyang Jianzhu University,
No. 9 Hunnan, East Road, Shenyang 110168, China,
**E-mail: luxupku@163.com*

An organizational measure of producer services' relations was used to analyze the spatial pattern of city network in Northeast China. 95 advanced producer services' head and branch offices were analyzed to investigate the inter-city linkages between 50 towns. Key result show that the morphology of the city network is featured by broad coverage, reasonable hierarchical structure, multiple centers and large inter-provincial disparity. The typical spatial pattern is strip-like central area surrounded by secondary central cities, but the three provinces show clear distinctions in spatial structure of city network. Transportation network, degree of opening-up and government management jointly play important roles in the shaping of city network.

Keywords: Producer Service; City network; Regional Spatial pattern; Northeast China.

1. Introduction

With the development of global economy, economic links between cities tend to be networked, and organizational capacity on funds, goods, labor force and other mobile factors determine the status and influence of cities in regional urban system. A famous urban sociologist, Castells proposed the "space of flows", and claimed that city nodes function as important strategic links to integrate local regions into an interconnected network. Upon a global scale, the symbolic functions of cities embody in their high-end services, e. g., production and consumption. Therefore, the level of connectivity of internationally advanced producer service (APS) corporations becomes a core element in analyzing the spatial pattern of world city networks. In recent years, world city network analysis has been introduced into China to study urban spatial pattern via APS corporation network on both national level and the urban clusters of high openness.

In the course of China's reform and opening-up, the reform in northeast China started late, and the former economic foundations of planned economy and traditional manufacture industry there were heavily shaken. In 2004, the government launched the strategy of "revitalization of northeast China", which aims at not only establishing a new industrial system but also promoting the development of service industry and urbanization in a new round of economic opening-up. On the basis of previous experiences in the research of city networks, this study tries to analyse spatial patterns of city network in northeast China using spatial distribution data of office network off producer service companies, with city connectivity as a measurement index. It provides a new perspective in studies on regional spatial form and coordinated regional development.

2. Data Sources and Research Methods

2.1. *Survey region*

The survey region of this study includes three provinces of the northeast China, Liaoning province, Jilin province and Heilongjiang province. Prefecture-level administrative units are selected as the research unit, totaling 36. Among them, 14 units are in Liaoning province, 9 in Jilin province and 13 in Heilongjiang province.

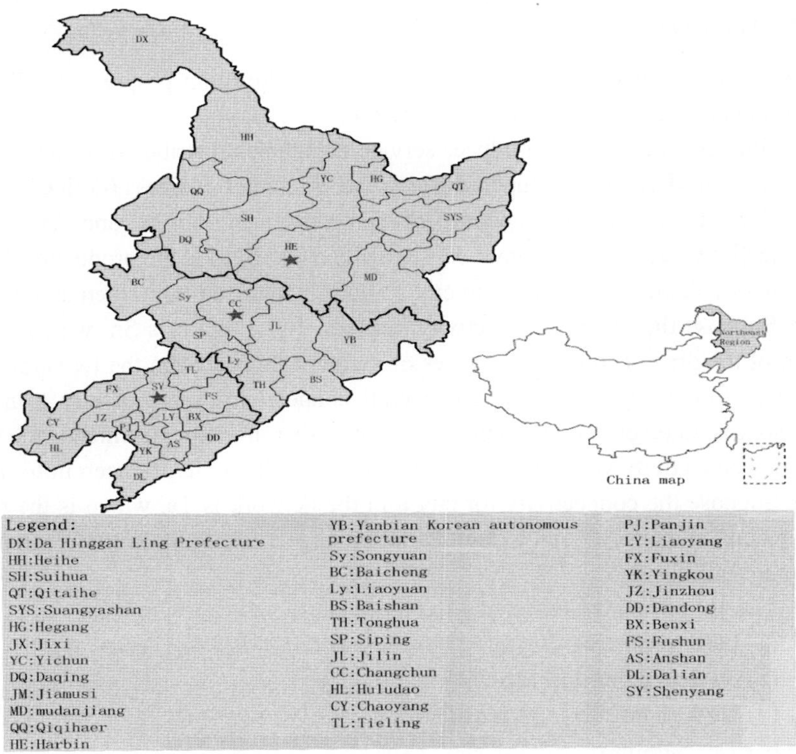

Figure 1. Survey Region and Administrative Units.

2.2. *Data sources and processing*

Producer service industry includes financial services, transportation, modern logistics, information services, high-tech services and business services. This study selected five types of enterprises in producer service industry as the object, including accounting firms, law firms, insurance companies, securities companies and logistics companies. Data were collected from corporate rankings of Chinese lawyer website, Chinese brand website, Chinese top website, Xicai website and Chinese logistics industry website. The following contents were obtained: top 100 accounting firms of 2013, China life insurance company rankings of 2014, China's top ten property insurance companies in 2013, securities company rankings of 2015(115 in total), China's top 50 logistics enterprises in 2014. In this approach, 95 enterprises of producer service industry were selected, including 15 accounting firms, 15 law firms, 19 insurance companies, 25 securities companies and 21 logistics enterprises. By browsing the official websites of these enterprises, the spatial distribution of their branches in northeast China were obtained.

2.3. *Survey methods*

The survey method is drawn from chain model algorithm proposed by Taylor (2001) in the research of world city network.

Suppose there are N producer service companies distributed in the survey region with M city units. Suppose Vai is the volume of business (defined as the distribution amount of branches) of company i in city a, and Li is the connectivity value of company i between city a and b. Li is the product of direct business volume of company i in city a and b, which can be written as formula (1). Suppose the overall connectivity between city a and b is Sn, which is the sum of the direct connectivity values of x companies between the two cities, as seen in formula (2). Suppose the overall connectivity value of city a in the network is Sa (every city is connected with another m-1 cities), which is the sum of the connectivity value of city a and that of all other cities, as seen in formula (3). Suppose the connectivity for city a in the network is Ta, which is the ratio of overall connectivity value of city a to the largest one in the survey region, as seen in formula (4).

The above-mentioned calculation formulas:

(1) $Labi = Vai \times Vbi$

(2) $Sn=L1+L2+\ldots\ldots+L(n-1)+Ln$

(3) $Sa=La1=la2+\ldots\ldots+La(m-1)+Lam$

(4) $Ta= Sa/Smax$

Based on the above formulas, a table with M lines × M columns is needed in data statistics to analyze the connectivity between M cities. In the table, the connectivity between any two cities, "a" and "b", is Lab or Lba, which are equal. Thus this is a diagonally symmetrical form. The table can clearly reflects the association between any two cities, and Sa can be obtained by calculating the sum of each line or column. According to the output results of the table, the spatial structure of city network in three provinces of northeast China is drawn with GIS based on their map.

3. Spatial Structure Characteristics of City Network

3.1. *Hierarchy of city network*

The connectivity values of 36 cities in the survey region are calculated by chain model algorithm, and the results are shown in Table 1. On this basis, cities in northeast China can be distributed into three levels. The first level includes Harbin, Shenyang and Dalian, with the connectivity value being greater than 0.7, as the central city in the urban system. The second level contains Changchun, Jilin, Qiqihar, Mudanjiang, Daqing, Yichun and Anshan, whose connectivity values fall between 0.4 and 0.7, belonging to the secondary central cities in northeast China. The other 26 cities are in the third level, with the connectivity values being lower than 0.4, and have relatively weak function and organizational ability.

The hierarchy of cities in northeast China is featured by wide coverage, reasonable hierarchical structure, multi-center phenomenon and large interprovincial differences First, the coverage of city network in northeast China is high, with the connectivity values of all cities higher than 0.15. This indicates that the branches of producer services achieve the effect of complete coverage, and even the low level cities are largely covered by these services. Second, the network level structure is reasonable. The cities are classified as high, middle and low level according to the connectivity value, with the ratio being 3:7:26, which indicates a relatively reasonable structure. The organization state of city network has become mature. Third, the city network presents a multi-center structure. The three central cities at the top of network have similar values of connectivity, which are 1, 0.84 and 0.77 in Harbin, Shenyang and Dalian, respectively. These three centers play the most important role in organization. Fourth, from the network development levels of each province, the structure of Heilongjiang province is the most reasonable, the ratio of city number in three levels being 1:4:8. The amount of cities of middle level is the least in Liaoning province, showing a dumbbell-shaped structure, with the ratio of three levels being 2:1:11. Jilin province is short of central cities with strong organization function, and the connectivity of Changchun whose connectivity is the highest is just 0.64.

Table 1. The Hierarchy distribution of city nodes based on the consumer services.

The hierarchy	Connectivity (Ta)	Number	City
First	0.70-1.00	3	Harbin, Shenyang, Dalian
Second	0.40-0.70	7	Changchun, Jilin, Mudanjiang, Qiqihar, Daqing, Yichun, Anshan
Third	0.15-0. 40	26	Yingkou, Huludao, Jinzhou, Fushun, Dandong, Chaoyang, Siping, Songyuan, Suihua, Benxi, Fuxin, Liaoyang, Panjin, Tieling, Tonghua, Baishan, Liaoyuan, Baicheng, Yanbian Korean autonomous prefecture, Jiamusi, Jixi, Hegang, Shuangyashan, Qitaihe, Heihe, Da Hinggan Ling region

3.2. *Connection features of city network*

In northeast China, the connectivity values among 36 cities are calculated through chain model algorithm. For the couples of city, the connectivity value of Harbin-Dalian (984) ranks top, followed by Harbin-Shenyang (973) and Shenyang-Dalian (877) successively. Therefore, these three cities form a triangular structure which link soundly. Besides, strong economic connection is also found between Qiqihar, Changchun, Daqing and Yichun.

By drawing the sketch of connection intensity between cities (Fig. 2), it can be noticed that the city network in northeast China manifests such a spatial pattern as strip-like central area embraced by secondary central cities. Besides, the spatial structure of city network of three provinces differs apparently. Above all, the city couples with high connectivity values are mostly among several central cities along the Harbin-Dalian Railway. A great development belt of city, based on Harbin, Changchun, Shenyang, Yingkou and Dalian, becomes the functional core of northeast China. These cities have pretty strong connections. Additionally, around the great city belt, there are many nodal cities functioned as secondary regional center, and there are obvious differences on the three provinces' network structure. In Heilongjiang province, Qiqihar-Daqing urban district, Yichun and Mudanjiang form the secondary node around Harbin in west, east and north separately. In Jilin province, the network structure is relatively loose. The connectivity value between Changchun and Jilin is just 286 and other cities also link less closely. However, in Liaoning province, it presents a typical feature that cities in coastal economic zone and the capital city Shenyang connect tightly, while the northwest zone and the center link loosely.

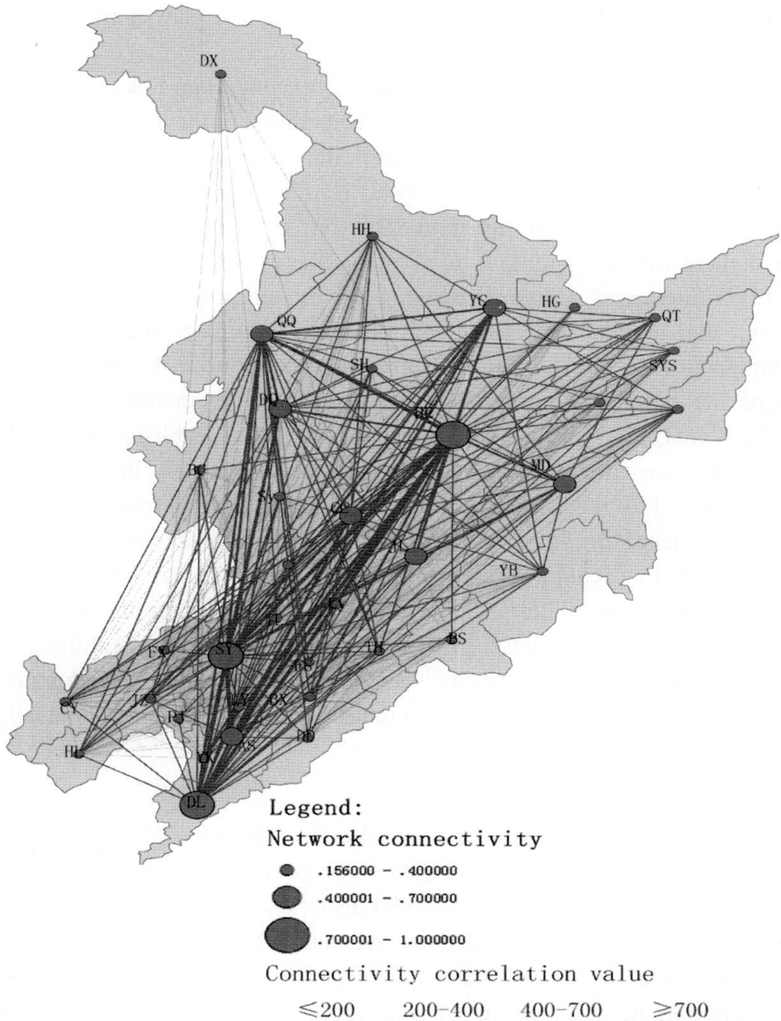

Figure 2. Connectivity Values among 36 Cities.

4. Discussion on the Genesis

From the research of producer service industry, we can see the hierarchical structure of city networks in northeast China is reasonable with high network coverage, but the development levels are different from place to place. Cities along the Harbin-Dalian Railway have high connectivity, but the development of the east and west of Jilin province and the northwest of Liaoning province is restricted. Based on the background investigation of regional development, possible causes for the formation of current city network pattern are put forward as following:

(1) Influence of traffic network: Traffic is the important link by which cities can directly connect with each other. According to the calculation by Hu in(2009) [9], 1% growth of China's transportation investment increased 0.28% of the GDP growth from 1985 to 2006. Harbin-Dalian Railway is the most important traffic corridor in northeast China, forming the city-intensive development area with higher level of integration.

(2) Influence of degree of opening-up. Centering Shenyang, Dalian, and Harbin, the middle and south of Liaoning province and the north of Heilongjiang province had higher degree of opening-up in history. They even became the frontier of northeast China in economic cooperation with Japan, DPRK, Korea, Russia and Mongolia, promoting the establishment of foreign invested service enterprises, which contributes considerably in improving the development of local service industry and interconnection of local enterprises.

(3) Influence of government control: China's government has a strong influence on the regional economic operation, including the influence on administrative rights, policies and development strategies. As the the provincial political and economic center, provincial capital has perfect facilities and the local people have high life quality, which attracts more service industries and hence enlarges the gap with cities with lower level of development gradually.

5. Conclusion

According to the research in this article, the hierarchy of city network in northeast China is featured by broad coverage, reasonable hierarchical structure, multiple centers and large interprovincial disparity. The typical spatial pattern is strip-like central area surrounded by secondary central cities, but the three provinces show clear distinctions in spatial structure of city network. Transportation network, degree of opening-up and government management jointly play important roles in the shaping of city network.

To promote reasonable development of the city network structure, the city function of core cities in the area should be enhanced by area planning, so that the capitals and talents outside the area can be attracted, and the advantageous position of the three provinces of northeast China be presented in larger area. Meanwhile, stronger support should be given to the economic development of secondary central cities, and a secondary center should be built in city network, so as to decentralize the radiation function of core cities and to alleviate excessive dependence on core cities from low-level cities. The dependence should be transferred to the adjacent secondary central city, thus realizing a good development pattern of hierarchical structure. For instance, the city group of Western Liaoning province can be built with Jinzhou as its center in the west of Liaoning province. Jinzhou should take the responsibility to drive the development of life service industry in the whole area of the west of Liaoning province, and to further increase employment opportunities and to relieve the brain drain in low-level cities.

The research of city network of northeast China in this article is an integral data analysis on various industries concerning producer services in contemporary northeast China, instead of a systemic analysis and conclusion of some specific industries, which will be the direction where further research goes.

Acknowledgment

Project supported by the National Natural Science Foundation of China (Grant No. 41301179). This article is written by graduate student Mingshu Qu, Professor Xu Lu, and Professor Jingcheng Yuan in School of Shenyang Jianzhu University.

References

1. Castells M. The Rise of the Network Society [M]. Oxford: Blackwell, 1996.
2. Peter J. Taylor, Ben Derudder. World City Network: A Global Urban Analysis [M], London, UK: Routledge, 2004.
3. Miaoxi Zhao, Zheng Liu. Research on China's City Network Based on Production Service Industry [J]. Urban Planning 2012, 36(9):23-28, 38.
4. Xu Lu. Spatial Pattern of Regional City Network Based on International Advanced Producer Services' Layout in Pearl River Delta [J]. Economic Geography. 2012, 32(4):50-54.
5. Cong Wang, Youhui Cao, Guowei Chen. Production service industry based on the Yangtze River Delta city network [J]. Geographical Research 2014, 33(2):323-335.
6. Yan Lei. Research on Producer Services Agglomeration of the Midwest Cities [D]. Hunan University Master Thesis, 2012.
7. Taylor P J. Specification of the World City Network [J]. Geographical Analysis, 2001.

Spatial analysis of retail pattern change in a fast growing city: A case study of Shenzhen, China

Y.L. Luo

Department of Land &Real Estate Management, School of Public Administration and Policy, Renmin University of China, Beijing 100872, China
E-mail: ellen.luo@ruc.edu.cn

Retail constitutes an essential part of a city's economy. In this study, we aimed to analyze the retail pattern change of Shenzhen, the most successful example of SEZ in China and explained its link with urbanization, as supermarket site had important implication for the sequence of Shenzhen's development and the influence of various factors. The panel data of area and retail sales were collected for the 6 main districts of Shenzhen from 2005 to 2014 and then analyzed using density model. Through comparison with GIS figures of retail distribution, the results confirmed the usefulness of spatial analysis with density model in a landform dependent urbanization. The retail distribution characteristic in context of Shenzhen may provide a reference indicator on evolution of retail pattern which could be used by other cities in China with similar background.

Keywords: Supermarket; retail pattern; spatial analysis; urbanization; land form.

1. Introduction

Retail constitutes an essential part of the city's economy, since it is highly correlated with the employment and economic growth. In the year 2012, retail industry contributed £151 billion to economic output in UK, 16% of the total. The retail sector employed a total of 4.3 million people, 15.9% of total in UK, which was the largest broad industrial group in the country by the number and proportion of employees.[1] In the same year, the number of retail establishments in operation in Japan totaled 1.03 million, and the number of people engaged in retail was 7.83 million.[2] Since the market-oriented economic reform in 1978, China has entered into a stage of financial deregulation and liberalization. One of the most powerful engines of the economic growth is the retailing business.

---

* Corresponding author. Email: luoyilan_ruc@163.com

[1] The retail industry: statistics and policy Standard
Note: SN/EP/06186 Last updated: 4 June 2014
Author: Chris Rhodes Section: Economic Policy and Statistics

[2] Statistical Handbook Of Japan 2014

By the end of 2012, the number of people engaged in retail business in China was 61.34 million.[3]

Closely related to employment and daily life, retail can also reflect, at least in part, the conditions of local economic development such as population density for a particular species, and the distribution change of retail facilities can be viewed as a reflective element of the city's sprawling process. To assess the dynamics of rapid urbanization process in China, the retail pattern change in fast growing cities has received extensive research attention. There have been several papers with focus on the relationship between retail patterns and urban growth of large Chinese cities in recent years, such as Beijing and Shanghai, but very little research has been done on China's young and mega cities achieving dramatic and sustainable urban development, such as Shenzhen. To investigate the characteristics of retail distribution pattern in the process of fast expansion, Shenzhen's case can be adopted as a representative one and used as one of the reference indicators in terms of sustainable urban form.

Separated only by a river, Shenzhen used to be a small farming town facing prosperous Hong Kong, and the specific location was the major reason why Shenzhen was set up as one of the Special Economic Zones (SEZ) by the central government of China, the "laboratories" for the country's transition from a centrally planned to a market economy. In the last thirty-five years, Shenzhen has accomplished many unbelievable progresses and enjoyed explosive growth, which makes it the most representative fast growing city in China. In this paper, the pattern change of Shenzhen's supermarket distribution over the last three decades was depicted using the tool of ArcGIS. Along with spatial analysis, the analysis of total retail sales data was also performed in this study to describe the features in urban districts of different types, and the reasons of those differences were discussed.

2. Literature Review

Compare with the considerable amount of studies on population growth and urban decentralization, relatively little attention has been paid to the evolving retail structure of the metropolitan communities. At present, published research material in western countries has tended to focus on two important issues. The first one concerns the timing and growth of the original fixed shop form of retailing. The related themes have been particularly important since it has focused interest on both the chronology and the nature of retail change. The second research theme has been closely associated with the locational change of

---

[3] 2013 China Retailing Industry Development Report

shops. Studies on the latter theme have forged a link between urbanization and retail patterns. The researches on the second contain mainly two topics in two periods in history: the suburbanization period post war and the downtown shopping-mall-led revival period in modern society.

Tarvers's study in 1957 [1] has shown that the relative growth in retail trade activity, both numbers of stores and value of sales, had been greater in the standard metropolitan areas (S.M.A.) than in the non-metropolitan areas of the U. S. during 1948 to 1954. Tarvers reached a conclusion that within the 168 S.M.A.'s discussed in his study, the greatest increase in retail trade had occurred in the suburban areas rather than in central cities. Furthermore, the growth in retail businesses in the S.M.A.'s with a central cities population more than 750,000 was relatively lower in the central cities than in the outlying areas, while the growth in retail businesses in the S.M.A.'s with central cities' population fewer than 750,000 was relatively greater in the central cities. Similar point has been stated by Ziskind in his study concerning retail in New York [2]. The size of central city is only one of the many elements that affect retail suburbanization, and the size of shop is considered as another element as well. According to Steinnes‘ s published work in 1982 [3], the urban development in the Midwest of America since World War II could be characterized as the suburbanization of people and jobs, and this pattern was considered similar to what had occurred in other regions of American. The suburbanization trend was also found to prevail for both large and small businesses, though in each sector the large businesses have remained more centralized.

Although suburban locations have dominated American retail activity since the late 1950s, most cities still consider retailing to be a key component of a vital downtown district. For several decades, many civic leaders and city planners have argued that a strong retail presence is crucial for the overall viability of the downtown and of the city. Since the 1960s, virtually all mid-sized and large American cities have implemented redevelopment plans which have attempted to revive downtown retailing. Robertson's paper published in 1983 [4] characterized the retailing in central business districts (CBD) of large American cities between 1954 and 1977. This study yielded some significant findings which enhanced understanding of the regression of retailing in CBD. Firstly, retailing in the CBDs of large American cities, regardless of location or population size, experienced massive absolute decline. Secondly, CBD retail sales as a proportion of standard metropolitan statistical area sales suffered heavy losses, which were greatest during the 1970s. Finally, CBD retail activity declined more rapidly in small cities than in large ones, but the former maintained comparatively higher levels of downtown retail sales. In 1997, Robertson published another article concerning the revival of downtown retail businesses [5]. As he puts it in this essay, viable downtown retail sector requires a variety of non-retail activities within close walking distance. In other words, an organic relationship to the downtown as a whole is a necessity if downtown retailing is to prosper. He also mentions that in order to compete with suburban retailing, shops in downtowns have employed four major redevelopment strategies: pedestrian malls, festival marketplaces, indoor shopping centers, and mixed-use centers.

Since the 1980s, scholars in China started to pay attention to retail location, but most of the studies back then were to explain the location decision of shops or malls. Based on central place theory, case studies on spatial pattern of commercial facilities in different cities became notable around the 1990s. In 1988, Wu and Xie analyzed the characteristics of retail commerce location of Guangzhou urban area [6]. In their article, 30 major commercial centers in Guangzhou urban area are classified by means of cluster analysis. Similar case studies of other cities have been conducted by other scholars. For example, An has published an essay on the location pattern of commercial centers of Lanzhou in 1990 [7]. By the end of the 20th century, more papers discussing the topic of suburbanization trend in major cities like Shanghai, Shenyang and Dalian. Most of the researchers at that time primarily emphasized the effects of population growth. However, few of them explored the relationship between retail distribution and urban development. When it comes to the 21st century, there appear more articles focusing on the relationship between urbanization and the development of retail system, but only limited cities of China have been analyzed from this perspective. In 2008, Qian published her work concerning the retail commercial suburbanization in Shanghai [8]. By generalizing the trends of the hypermarkets distribution of Shanghai, she visualizes the difference between developments in different types of districts. In 2012, Yu, Wang and Song managed to describe the changing retail spatial pattern in Beijing since 1984 [9], using supermarkets in Beijing as an indicator and GIS as a tool. In addition to highlighting the suburbanization trend, they emphasized the importance of planning and its influence on the development of urban area, inner suburbs and the outer suburbs. Similar studies also cover cities like Chongqing [10], but the retail change in most of the cities of China still remains unexplored.

As the dynamics of urbanization process varies from place to place, not all cities have relatively stable location categories as downtown, inner city, and suburbs like Beijing, Shanghai or Guangzhou. A young city like Shenzhen may have a brand new path totally different from other cities. Researchers interested in this topic usually pay their attention to metropolitans with longer history, and little attention has been paid to the changing retail structure of young metropolitan communities with short history. Thus, the spatial study on retail pattern of Shenzhen becomes important for optimizing researches on similar cities. As one of the most representative rapidly growing cities, Shenzhen could be set up as a model since similar situations occurs in other cities of China or may occur in other fast-developing countries when the globalization penetrates and structural changes in retail accelerate. Moreover, the results of this study may serve as a valuable reference in an indicator system for measuring urbanization process or urban sustainability. From this perspective, a detailed investigation of the Shenzhen case may provide reference indicator on evolution of retail pattern in a "time compressed" urbanization process and contribute to city design towards a more sustainable development.

3. Method and Data

3.1. *Model construction*

In order to compare and analyze the retail development of different districts[4], a density model was employed and the panel data of area and retail sales collected for the 6 main districts of Shenzhen from 2005 to 2014 was then analyzed.

Density model about population was originally presented by Clarks in 1951 [11]. Based on hypothesizes that districts in the interior have dense population (except in the central business zone) and the tendency of city to "spread itself out", a simple mathematical equation of exponential decline was generated.

$$Y = Ae\text{-}bx \tag{1}$$

x is the distance in miles from the center of the city, and y is the density of resident population in thousands per square mile. The co-efficient A measures the density and the rates of decline in density are measured by the co-efficient b. A high value of b means that density declines sharply with increasing distance from the center. Among all the forms of variants of Clark model, the semi logarithmic form is most commonly used.

$$\ln Yx = \alpha 0 + \alpha 1x + \varepsilon \tag{2}$$

In the semi logarithmic form, variable x and y have the meaning as the ones

---

[4] Since Longhua district and Guangming district were derived from Baoan district, Dapeng district and Pingshan district were derived from Longgang district after 2005, those districts of new formation were discussed as a part of its original district rather than an independent one.

in original Clark model. α_0 is the intercept, α_1 is the density gradient which in most case is a negative number and ε is the residual. While x increases, y would probably decrease, which demonstrates the hypothesis above.

Clark model is theoretically based on a city being built on perfect landform, in which the population is evenly distributed surrounding the center. Obviously, this precondition does not fit in most circumstance. Thus, scholars in later studies like Sherratt, Smeed, Newling and Cubic have adjusted the model to meet their immediate needs. Aside from the extensive empirical researches of urban population density of developed countries, Wang and Zhou construct a model which describes the urban population densities of Beijing from 1982 to 1990[12]. Using this density model as a tool, Wang and Zhou reached a conclusion that Beijing, the capital city of China, is consistent with the universal trend of suburbanization. One of the adjustments on the model is the substitution of the distance variable x. In Yu's study on retail pattern in Beijing, dummy variable is introduced in the density model to distinguish every kind of districts [9]. The introduction of dummy variables can simplify the investigation since the process of measuring distance is bypassed. Before Yu, same modification has also been done by Yiu in the study of property price gradient in 2004 [13], and Harrington's study in 1997 on the suburbanization of producer service employment [14].

Although population is used as the most important factor while analyzing the process of urbanization using Clark model, when it happens to Shenzhen, population change may not depict the whole picture of the evolving process of the city due to restricted range of isolated landform. In this case, retail could be a better indicator, since the driving forces associated with retail's locational shift include not only the trend of population growth, but also changes in many aspects such as personal mobility, disposable incomes, transportation and government land-use policies [15-17].

The model constructed in this study was similar to the one constructed in Yu's paper [9], in which density model is altered to describe the density of retail sales (retail sales/regional area) in three groups of districts, and two dummy variables D_1 and D_2 is introduced in the function to distinguish the types.

$$\mathrm{Ln}(Y+1) = \alpha_0 + \alpha_1 D_1 + \alpha_2 D_2 + \varepsilon \tag{3}$$

Y is the average retail sales density of all the main districts in Shenzhen. α_0 is the average retail density of the reference group. Because of the discrepancy lies in the area of different districts, the assumption of equal variances may have hampered a more precise evaluation. As a possible resolution, logarithmic form was used although we could hardly expect all the problems being solved completely. In this case, weighted least-squares method should be applied. Retail density data in this study was converted to logarithmic form. The logarithmic form Ln(Y+1) is used since the logarithmic form lnY would be negative when the density value is smaller than one.

The difference between Yu's study on Beijing and this study on Shenzhen is the classification of districts. Yu's method of grouping is based on the distance from the districts to the center of Beijing since the city's shape is similar to the ideal single-center city in Clark's theory. In consideration of the restricted narrow zone land form of Shenzhen located between coastline and mountain belt, as well as other factors such as socio-culture background and urban development, modification is needed when applying this method to Shenzhen. Thus, districts were grouped as central districts, industrial districts and suburban districts respectively in this study.

The Central districts of Shenzhen included Luohu and Futian. Four of all six control points connecting Shenzhen and Hong Kong were located in this region, indicating that this group is most profoundly influenced by Hong Kong among the three groups. In addition, the municipal government of Shenzhen was moved from Luohu to Futian in 2004, making Futian the administrative center of Shenzhen. The industrial districts of Shenzhen included Nanshan and Yantian where the leading industries were high-tech and tourism. Comparing with the central districts, the evolving process of these two districts is less affected by Hong Kong. Suburban districts of Shenzhen included Baoan[5] and Longgang.[6] Hasn't been turned to SEZ until 2012, Baoan and Longgang was remotely located and relatively slow in urban development comparing with other parts of Shenzhen. In fact, the location of the two districts in mountains belts was also considered the main factor of being slow down in pace of urbanization.

$$D_1=\begin{cases}1 & \text{Central District}\\ 0 & \text{non Central District}\end{cases} \qquad D_2=\begin{cases}1 & \text{Industrial District}\\ 0 & \text{non Industrial District}\end{cases}$$

In the modified density model applied in Shenzhen, the reference group would be the suburban districts. Dummy variable D1 was to determine central districts, while D2 was to determine the industrial districts. $\alpha 1$ was the average retail density of the central districts, $\alpha 2$ was the average retail density of the

[5] Baoan district in this paper refers to the old Baoan district that includes the new Baoan district, Longhua district and Guangming district.

[6] Longgang district in this paper refers to the old Longgang district that includes the new Longgang district, Pingshan district and Dapeng district.

industrial districts, and α0 was the average retail density of the suburban districts. ε was residual.

3.2. *Data description*

The data of "total retail sales of consumer goods" and district area from 2005 to 2013 was collected from "Shenzhen Statistical Yearbook" and the statistical yearbook published annually by each district from the year of 2006 to 2014. Data of 2014 was collected from "Shenzhen Statistical Bulletin" and statistical bulletin of all the districts released in 2015. The number "total retail sales of consumer goods" was modified using Consumer Price Index (CPI) each year from 2005 to 2014 to suit the study. Though the logical tool of modifying total retail sales of consumer goods" should be Retail Price Indices (RPI), the Statistics Bureau of Shenzhen Municipality stopped providing the RPI data in yearbooks years ago. To eliminate the influence of inflation on data, CPI, as the most commonly used price indices was chosen as a substitute for RPI.

Table 1. Descriptive statistics of the different groups' retail density in logarithmic form.

	Min	Max	Mean	SD
Central	1.80	2.58	2.2000	0.27707
Suburb	0.26	0.58	0.4237	0.11144
Industrial	0.59	1.13	0.8546	0.18768

Retail density in logarithmic form = ln (total retail sales/regional area+1)

Although the model itself can describe the development condition of retail system in different districts, combining it with other sorts of material can generate new method of investigation. Taking Sutton's study [18] as an example, the combination of night-time image and a relationship between area and population for urban clusters may prove to be a good starting point for the development of a more accurate method of modeling population density. In this study, while retail density model was combined with the distribution image of supermarkets in Shenzhen, GIS worked as a significant tool.

According to the classification by Ministry of Commerce, the concept of retail consists of seventeen major forms including grocery store, convenience store, discount store, supermarket, shopping mall, etc.[7] Among all the seventeen forms, supermarket is one of the most representative one because of the variety of goods and services as well as the close relationship with people's daily lives. The first supermarket in China was established in 1981 in Guangzhou Friendship Store. The presence of this supermarket was to cater for

[7] "Classification of retail formats" (GB/T18106-2004) Ministry of Commerce of the People's Republic of China

the needs of visitors from overseas. Since 1982, RMB-based supermarkets for local customers began to appear in cities like Beijing, Shanghai and Hangzhou. In the 1980s, supermarkets were not a popular choice for local Chinese. By 1987, quite a number of supermarkets were under difficult conditions. As the improvement of economic environment in the 1990s, there was a booming of supermarkets and the operators then shifted their focus to store level competition by catering to the needs of local communities. Along with these changes, supermarkets retailing had been gradually accepted by local customers in major cities and experienced a rapid growth [19, 20]. In current China, supermarkets are playing prominent roles in daily lives. However, the Chinese have shown a distinct preference for shopping behavior which is different from the western, although supermarkets are now widely accepted. As Mai and Zhao put it in their study [21], the supermarket shoppers in China are accustomed to make small purchases and frequent visits and they usually prefer not to use any form of transportation. Noting customers' preference, the supermarkets in China are likely located in residential area, commercial area or places with good transportation accessibility [22, 23]. During the process of fast urbanization, city designer could be increasingly instrumental in shaping metropolitan growth and urban land-use patterns, and the location of the shops can be viewed as an indicator that typifies the vitality and scale of the evolving townscapes.

Quite different from Shanghai or Guangzhou, there has been no traditional commerce center in Shenzhen in history. After more than three decades of rapid growth, Shenzhen is currently a metropolitan with complete retail system. The retail development occurred in Shenzhen during a thirty five year period exhibits briefly the sequence and feature of the urbanization process in different parts of the city.

The obtained data of supermarkets in Shenzhen was based on statistics presented by "Statistical Yearbook of China Chain Stores of Retail Trades and Catering Services" (2013). All of the supermarket chains (including general merchandise that involves supermarket) had sales over 10 billion yuan in Shenzhen were selected in the range of this study. They were Vanguard (including its brand Ole'), Wal-Mart, Xinyijia, Carrefour, PARKnSHOP, Shirble, Renrenle and Rainbow. Convenience stores or smaller were excluded from this study. In addition, globally and nationally influential chain stores with branches in Shenzhen like Metro, Aeon, RT-Mart, Yonghui, Tesco were also taken into account. Location information and the opening date of the stores were collected from their official websites.

4. Estimation Result and Spatial Analysis

4.1. *Estimation result*

As shown in Equation 3, the coefficients were estimated using linear regression module in SPSS 16.0, and the sample was 60 observations (6 main districts * 10 years).

The estimation results, which were obtained using ordinary least squares, were shown in Table 2, while the correlation revealed important information about the distinct nature of retail growth in different regions. Adjusted R^2 was 0.934 in this model, indicating that this model could provide good explanation. All of the coefficients' was significant under the confidence level of 0.010. The average retail density[8] of central districts was $e^{1.776}$-1=4.91 (0.1 billion yuan[9] /square kilometer), the average retail density of industrial districts was $e^{0.431}$-1=0.54, and the one of suburban districts was $e^{0.424}$-1=0.53. This estimation results were in line with realistic observations.

Table 2. The description of model 3.

	B	Std. Error	t	p
Suburban (Constant)[10]	0.424	0.064	6.579	***
Central	1.776	0.091	19.505	***
Industrial	0.431	0.091	4.731	***

***p<0.001

Attention was then paid to the conclusions which could be drawn from these coefficients implying wider interrelationships between districts. First, the results demonstrated clearly the presence of much higher retail density in central districts than in the others. The difference was depicted in the following Figure 1.

Though the model above had described the disparity between districts, the growth by time was not yet revealed. Based on the time-series data, the average growth rate of retail density (2005-2014) was then calculated. The average growth rate was 10.4% in central districts, 11.3% in industrial districts and 11.5% in suburb respectively. The average growth rate of the whole city was 10.8%. Obviously the growth rate in the less developed districts was getting faster, however the gap still existed. As it presented in Figure 2, the gap was getting even larger in recent years.

---

[8] The gap between different groups shrank due to the transformation of figures.

[9] The arithmetical unit is in accordance with the unit used in Shenzhen Statistic Yearbook.

[10] α_0 was the average retail density of the suburban districts, a constant in equation 3.

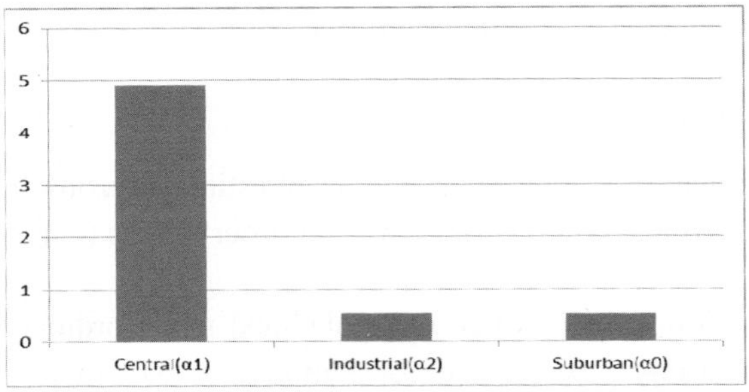

Figure 1. Average retail density of three groups (α_0, α_1, α_2).

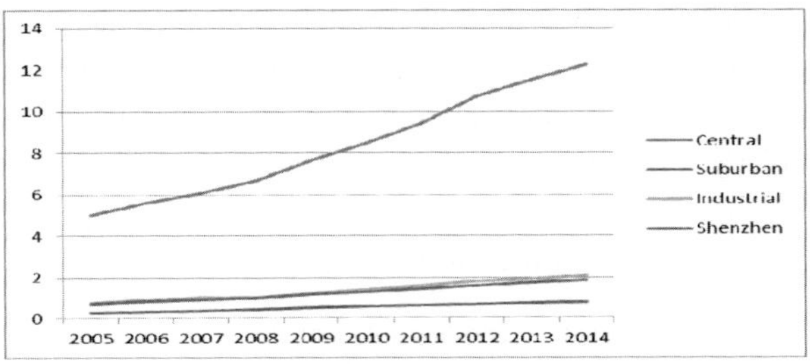

Figure 2. The growth in retail density (2005-2014).

4.2. *Supermarket development of Shenzhen*

To examine the retail distribution, the background information of 13 supermarket brands (163 stores in total) was collected before March, 2015. We noted that most of them were foreign brands, but the most rapid growth was the domestic brand "Renrenle". The added up number of opening store by the end of 1995, 2000, 2005, 2010 and 2015 (end of March) were provided in Table 3.

Table 3. The increase in the number of stores of different retail brands.

	1995	2000	2005	2010	2015
Renrenle	0	2	10	24	29
Rainbow	1	3	13	23	27
Vanguard (including Ole')	0	4	8	16	27
Xinyijia	1	2	10	19	22
Wal-Mart	0	6	11	15	21
PARKnSHOP	0	1	2	5	10
Carrefour	0	2	4	8	8
Aeon	0	0	2	5	6
Shrible	0	2	5	5	5
METRO	0	0	1	2	3
RT-Mart	0	0	0	1	2
Yonghui	0	0	0	0	2
Tesco	0	0	0	1	1
Total	**2**	**22**	**66**	**124**	**163**

The numbers of supermarkets of above selected brands in each group were shown in Figure 3, and their percentage to the total number of the selected brands shown in Figure 4.

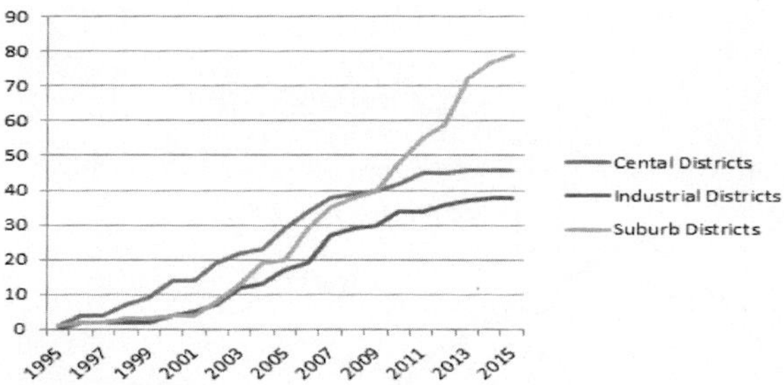

Figure 3. The numbers of supermarkets of selected brands in different groups (1995-2015).

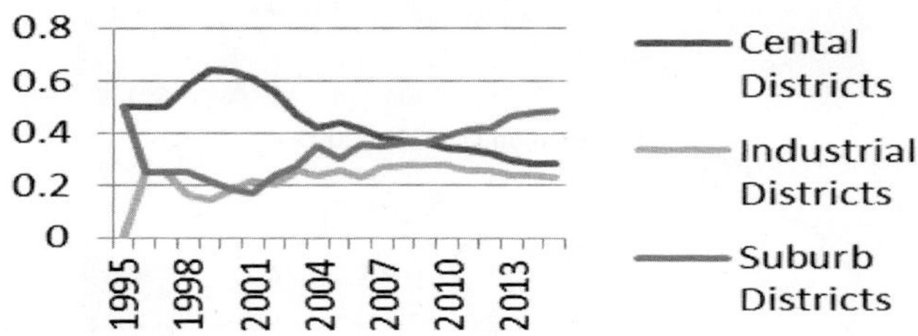

Figure 4. The percentage of supermarkets in different groups to the total number of selected brands (1995-2015).

From the figures above, it is obvious that the majority of supermarkets built before 2000 was in central districts, but since the year of 2005 growth in suburb districts had been faster, and the expansion remained at higher rate while the central districts was relatively slow in growth rate. During this period, growth rate in industrial districts was kept constant. In central districts, the number remained almost unchanged in recent years. This result exhibited a sign of fast growth in suburb districts for the coming years, and the process was just beginning. The absolute number of store alone was not enough in describing the spatial distribution of retail system of Shenzhen, the location must be taken in consideration as a factor. Thus, the distribution conditions of supermarket by the end of 2000, 2005, 2010 as well as the latest update in March of 2015 were posted in maps below for comparison.

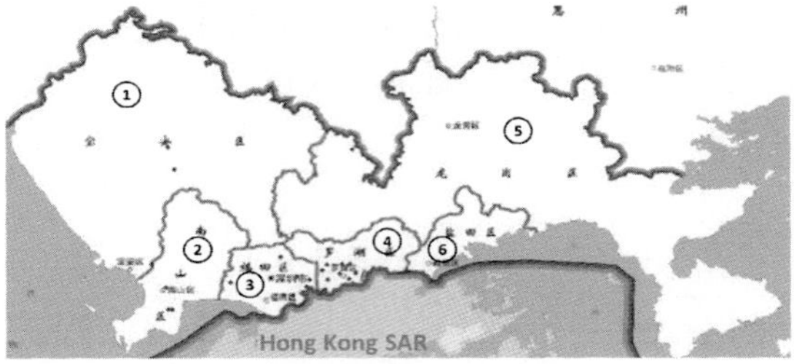

A: issued on Dec 31st, 2000

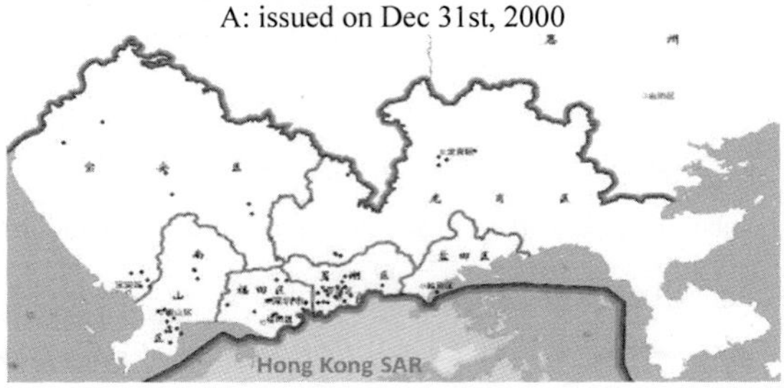

B: issued on Dec 31st, 2005

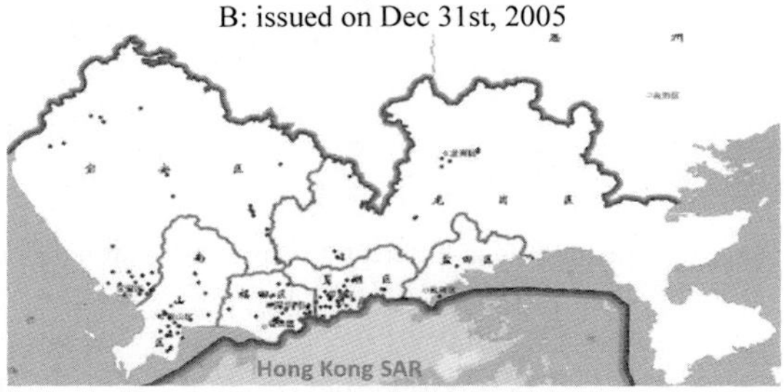

C: issued on Dec 31st, 2010

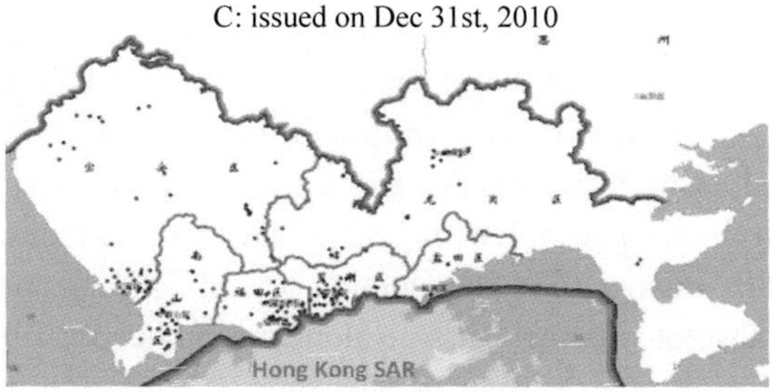

D: issued on Mar 1st, 2015

Figure 5. The distribution of supermarkets in Shenzhen in past twenty years.
1 Baoan District[11] 2 Nanshan District 3 Futian District 4 Luohu District 5 Longgang District[12] 6 Yantian District

[11] The Baoan district in the map consists of Baoan District, Longhua District and Guangming District.

[12] The Longgang district in the map consists of Pingshan District and Dapeng District.

The above figures were created using GIS. The specific addresses of all the 163 supermarkets were available in the map of Shenzhen. As shown in the four maps, most of the newly established supermarkets located in the southern part of Shenzhen near Shenzhen river and coastline. In the earlier ten years before 2005, new supermarkets were mainly built in the south-eastern part of the city. From 2010 to 2015, since the retail density was already at high level in the south (central districts and industrial districts), the faster growth rate began to move northward associated with the urbanization expanding towards mountain. Another notable logic of this decade in urban development was that even though retail system had spread out to the north, the development in flat landform still occurred prior to mountainous regions.

To summarize, retail patterns of the three types of districts in Shenzhen were as the followings.

As the earliest part of development, retail system in central districts experienced the most rapid growth before 2000. Development during this period marked the difference between this region and the other parts of the city. In the 21st century, retail expansion in central districts maintained higher growth rate. Although there was a slight falling trend from a high stand level of growth rate, the volume of retail sales and the density of retail sales remained the largest up to present.

The industrial districts were less developed than the central districts in the 20th century due to a relatively weaker connection with Hong Kong. Since the core business of industrial districts were modern manufacturing and tourism. However the result exhibited that a notable retail growth of industrial districts appeared in the years after 2000, particularly after 2005.

The suburb districts discussed in this study originally included two main districts, Baoan and Longgang, but the actual range covered six administrative districts in the later years, four of them were derived from the original two. Although the stores in suburb districts mostly located in flat areas and the gap between suburb districts and other parts of Shenzhen in retail development level was obvious, but, as seen the above figures, the suburb districts had experienced the highest growth rate in retail density among all the three groups of districts, and the number of supermarkets kept increasing at a steady pace.

4.3. *Explanation*

The results confirmed the consistency between estimation and observation. The administrative formation of three kinds of districts in Shenzhen could be attributable to the followings.

4.3.1. *Policy*

Since the year of 1978, most of the cities in China have experienced a period of rapid expansion, particularly the SEZs. As the first SEZ in China, Shenzhen is probably the fastest growing city worldwide in the last two decades of the twentieth century [24], but not all the administrative districts of Shenzhen were located within SEZ at the very beginning. Only Luohu, Yantian, Futian and Nanshan were put into the experimental practice of reform at first in 1980, and the rest districts of Shenzhen were not within SEZs until the year of 2010. Due to the thirty year variation in economic environment, huge disparity could be found between the old SEZ and new SEZ. In 2010, the GDP of Shenzhen was totaled 820.123 billion yuan. The ratio of SEZ GDP to non-SEZ GDP was 1.2:2, and the area of the SEZ part was 395km^2 while the area of the whole city was 1948 km^2.[13] These digits indicated that the non-SEZ districts' output per square kilometer only amounted to 21.2% of the SEZ districts' output per square kilometer. After the expansion of SEZ, the gap between districts has been getting smaller gradually, but it seems difficult to overcome the effect shaped in past thirty years in a short period of time.

China established several SEZs in 1979-80 which had been acting as windows to the outside world. The land-use fee was applied in Shenzhen SEZ, making it the pioneering city in China in which the land market force was firstly introduced. This decentralized land use power thus helped urban sprawl. Since then, the incoming overseas land users, who were not supposed to be entitled to free land use according to the policy designed by Chinese government in the 1950s, began to play a positive role in the process of urbanization. The land-use policies as well as urban planning eventually had profound impact on Shenzhen's development.

Overseas investment was attracted to the Shenzhen SEZ following this policy even though it was not a perfectly designed one [25]. The effect of land market force then appeared. The above mentioned gap between the SEZs and other region began to take shape, which makes it hard for suburb districts to catch up rapidly. As demonstrated in this study, the disparity in retail density exists between districts reflected, at least in part, the disparity in development level. Specifically, the authorization of the Construction Land Planning Permission Certificate mentioned in Mao's study in 2008 [26], brought about massive change in land use across districts, which led to the division between the inner city and the outer city: the land use in SEZs was considered successful, while land use of non-SEZs was comparatively ineffective. The effects of this

[13] Shenzhen Statistical Yearbook 2010

division were quantitatively reflected in the differences between retail system of central districts, industrial districts and suburb districts.

4.3.2. *Landform*

Unlike Beijing and Shanghai that located on flat land, Shenzhen has a special mountainous landform which might be an important issue that affects the feature of its urbanization. Based on study by Bu et al [27], the landform should be viewed as an essential factor that determines the characteristic of urbanization. According to their research, the influence of human activities on the natural views is weaker at higher altitudes, and there would be a belt at particular height where city view and agriculture view coexists. At the earlier phase of Shenzhen's development, this belt located in low, flat land. That explained why the degree of urbanization in districts like Futian was higher than the one of mountainous like Yantian at the beginning. As the city grows rapidly, this belt has been expanding towards higher altitude and exhibiting a featured trend of landform dependent urbanization in Shenzhen.

4.3.3. *Functional division and leading industry*

Being China's leading SEZ, commerce and finance have currently been the leading force of Shenzhen's sustainable urban development. Central districts of Shenzhen are more likely taking the administrative function, and the CBD of Shenzhen is located in this area. The mainstay industry of industrial districts are high-tech, modern manufacturing and tourism (Nanshan and Yantian). The leading industries of suburb districts are manufacturing and modern agriculture.[14]

Early in the 20th century, commerce was the original development force in Shenzhen. The primary form of retail distribution pattern laying particular stress on the south region has come into being. In 2000s, factories in the north part began to gain popularity. The immediate result was a rapid growth of labor-intensive industries and export-processing industries, mainly clothing, electronics, and toys. This trend was also explained as a north shift of export-oriented manufacturing sector from Hong Kong. Along with this migration, retail in industrial districts started its shaping. Recently, since the emphasis of industrial districts has been switched to the construction of science and technology park and service industry like tourism, manufacturing begin to move

---

[14] Reports on the situation of urban modern agriculture development, Apr 24th 2012, Economy, Trade and Information Commission of Shenzhen Municipality

to suburb districts. This process becomes the force promoting the retail development in suburb regions.

4.3.4. *Traffic*

Traffic is strongly correlated to the flows of people. In view of shop-owners, good traffic accessibility means more potential customers. Thus, traffic network is an indispensable factor when it comes to retail distribution. Traffic system of Shenzhen includes not only subway lines, bus lines or highways, but also harbors and airports which is essential to business activities. In suburb districts, newly established supermarkets scattered mostly beside the new subway lines, especially around the subway stations. Similarly, retail development in industrial districts is affected by subway lines in a large degree as well. In addition, the control points that connect to Hong Kong have profound impact on the development of the whole city. In central districts, since public transportation system is well developed in this region, density of supermarket appears to have a close relationship with the location of control points. The location of control points in other districts are also influential factors in the featured retail pattern of Shenzhen.

4.3.5. *The limitation of this study*

The limitation of this paper is that it has only focused on the retail pattern in a macroscopic view; the retail distribution feature of specific communities has not been mentioned. Themes from a microscopic view, retail pattern in urban villages as an example, could be another interesting topic for further study.

5. Conclusion

This paper described the retail pattern change of Shenzhen, the most successful example of SEZ in China. Shenzhen's path demonstrated the strong capacity in development which has been viewed as a miracle by the world. The pattern change of Shenzhen's supermarket distribution over the last three decades was analyzed and their features in urban districts of different types were described. The results confirmed the usefulness of spatial analysis with density model in a landform dependent urbanization. Being a useful indicator of the urban retail sprawl, supermarket site has important implication for the sequence of Shenzhen's development and the influence of various factors. The retail distribution characteristics in context of Shenzhen may provide a reference indicator on evolution of retail pattern which could be used by other cities in China with similar background.

References

1. Tarver J. D. Sub-urbanization of retail trade in the Standard Metropolitan Areas of the United States, 1948-54 [J]. American Sociological Review, 1957: 427-433.
2. Ziskind M. P. Labor conflict in the suburbs: Organizing retail in metropolitan New York, 1954–1958 [J]. International Labor and Working-Class History, 2003, 64: 55-73.
3. Steinnes D. N. Sub-urbanization and the "Malling of America" A Time-Series Approach [J]. Urban Affairs Review, 1982, 17(4): 401-418.
4. Robertson K. A. Downtown retail activity in large American cities 1954-1977 [J]. Geographical Review, 1983: 314-323.
5. Robertson K A. Downtown retail revitalization: a review of American development strategies [J]. Planning Perspectives, 1997, 12(4): 383-401.
6. Wu Y, Bin X, Luo C, et al. An approach to retail commerce location of Guangzhou urban area [J]. Chinese Geographical Science, 1991, 1(1): 19-29.
7. Chengmu A. Analysis of Lanzhou commercial centre's location pattern and advantage degree [J]. Geographical Research, 1990, 1: 003.
8. Qian Xuemin "A Study on the Retail Commercial Sub-urbanization in Shanghai: from the Angle of Hypermarket" E-Journal of China Urban Studies, 2008, 3(2): 79-88.
9. Yu Wei, Wang Enru, Song Jinping. Changing Retail Spatial Patterns in Metropolitan Beijing since 1984 [J]. Acta Geographica Sinica, 2012, 67(8).
10. Zheng Y, Guangxian K. Study on the retail commercial decentralization of Chongqing in 1990s [J]. Human Geography, 2002, 17(6): 53-57.
11. Clark C. Urban population densities [J]. Journal of the Royal Statistical Society. Series A (General), 1951, 114(4): 490-496.
12. Wang F, Zhou Y. Modelling urban population densities in Beijing 1982-90: Sub-urbanization and its causes [J]. Urban studies, 1999, 36(2): 271-287.
13. Yiu C. Y, Tam C. S. A review of recent empirical studies on property price gradients [J]. Journal of Real Estate Literature, 2004, 12(3): 305-322.
14. Harrington J. W, Campbell H S. "The sub-urbanization of producer service employment". Growth and Change, (1997), 28(3): 335-359.
15. Harrington J. W, Campbell H S. The sub-urbanization of producer service employment [J]. Growth and Change, 1997, 28(3): 335-359.
16. Wheeler J. O. Location, location, location-analyzing the retail environment-Jones, K, Simmons, [J]. 1988.

17. Li F, "1997 Zhongguo: lingshouye gai jizhu xie shenmo?" [What should China's retail industry remember from 1997?] Keji Zhinang, 1998, 16^8.
18. Sutton P. Modeling population density with night-time satellite imagery and GIS [J]. Computers, Environment and Urban Systems, 1997, 21(3): 227-244.
19. Wing-Chun Lo T, Lau H F, Lin G S. Problems and prospects of supermarket development in China [J]. International Journal of Retail & Distribution Management, 2001, 29(2): 66-76.
20. Sha Z Q, Li Y J. "Supermarkets in China: opportunities, problems and their solutions" [J]. Hong Kong Journal of Business Management, 1992, 10(69): 70-3.
21. Mai L W, Zhao H. The characteristics of supermarket shoppers in Beijing [J]. International journal of retail & distribution management, 2004, 32(1): 56-62.
22. Xu X, Zhou S, Lin G. The locational analysis of Guangzhou's large retail emporia [J]. City Planning Review, 2002, 7: 008.
23. Luo Xiaoguang, He Yongnan, Li Yonghe "The Location Analysis of Supermarket in City: A case Study of Harbin" Urban Studies, 2011, 18(1): I0002-I0004.
24. Wang J J, Xu J. An unplanned commercial district in a fast-growing city: a case study of Shenzhen, China [J]. Journal of Retailing and Consumer Services, 2002, 9(6): 317-326.
25. Zhu J. Changing land policy and its impact on local growth: the experience of the Shenzhen Special Economic Zone, China, in the 1980s [J]. Urban Studies, 1994, 31(10): 1611-1623.
26. Mao J, Yan X, Li Z, et al. The Effect of Urban Planning Control upon Land Use in Shenzhen [J]. Acta Geographica Sinica, 2008, 3: 011.
27. Bu X, Wang Y, Wu J. The effect of landform on landscape pattern vertical differentiation in rapidly urbanized Shenzhen City [J]. Acta Geographica Sinica-Chinese Edition-, 2008, 63(1): 75.

Perspective of Jinan creating a health city model in the low carbon economy

P.Z. Gao
School of Economics and Management of Shandong Women's University, Shandong Jinan, China
E-mail: gaopengzhaohappy@126.com

Z.W. Liu and J.H. Zhang
Research Office of Shandong Women's University, Shandong Jinan, China

The low carbon economy prompts urbanization to walk the road of sustainable development, and creating a health city is a long way to go. Based on the concept of low carbon and problems existing in the process of creating a health city, this article builds the four patterns of creating a health city, which is the collaborative model, the spiral model based on the four strategic platforms, the mode of driving the health city construction in culture, and mode of low carbon technology innovation and transformation. Lastly, the paper puts forward the countermeasures of creating a health city, firstly, unified understanding and strengthening leadership, secondly, set standards to promote standard, lastly, it is to intensify propaganda and citizen participation.

Keywords: Low carbon economy; Low carbon city; Health City Model.

1. Introduction

Since China's reform and opening up, high speed of industrialization and urbanization have completely changed the face of economy and society. It is predicted that our country's urbanization rate will be above 50% during "twelfth five-year" period, which means that China's urbanization will usher in a new round of development by leaps and bounds. However the speeding up of urbanization also brings a series of problems accordingly, such as urban and rural regime, environmental pollution, ecological destruction, traffic congestion, etc. On the other hand, the global climate change also contributed to the "heat island effect" of the city. According to statistics, the world's energy consumption of big cities account for 75% of the world, greenhouse gas emissions by 80% of the world [1], which affects the city's sustainable development, causing residents satisfaction to a certain extent. Therefore, building a resource-conserving and environmentally friendly low-carbon city is an important part of national construction a harmonious society.

Low-carbon development is an important step to achieve the development from the traditional extensive economy to intensive one. As the center of human producing and living, city plays an important role in the economic and social development. Urbanization is a way where human society pursues material and spiritual civilization relatively low-carbon. At present high-carbon urban development pattern is not desirable, while low-carbon urbanization is the inevitable choice. Low carbon city, based on the concept of scientific development, takes the path of sustainable development, and implements coordinated development, including population, resources, economic, social and environmental [2]. Low carbon city has become the common pursuit of all over the world, paying attention to cost minimization in the process of economic development, as well as the harmony between human and nature. The Chinese government pledged to achieve unit of gross domestic product (GDP) by 2020 carbon dioxide emissions 40% to 45% lower than in 2005, up to now there is a time for the last five years planning efforts. In order to promote the realization of the goal, Jinan as the provincial capital of Shandong province, through the "Creating a Health City" to realize low carbon development, and in 2013 issued "Jinan" twelfth five-year "control greenhouse gas emissions work implementation plan", to 2015 in Jinan unit GDP carbon dioxide emissions 18% lower than in 2010, the forest coverage rate of more than 35%.

2. The Present Situation of Jinan Creating A Health City Analysis

Creating a health city, an urban comprehensive function and an important symbol of civilization degree, helps to improve the investment environment of Jinan and living environment and effectively improve people's quality of life. Therefore, in the process of creating national health city, always adhere to the people-oriented scientific outlook on development; take the project of creating a health city as an important people's livelihood. What they all do has had effective achievements. Jinan officially starts the Creating a Health City work in 2004, getting people's strong support and actively cooperates with. At present, the city's life garbage treatment rate reached 99%.

However, a series of problems are worth thinking and consider at the same time. There are still absences of management problems such as environmental pollution, and urban villages, the market order, the operation is unreasonable, and so on. We must admit that the characteristics of all kinds of city construction in a certain period play a positive role. The purpose of the "Creating a Health City" is good. It helps citizen’s accomplishment of ascension and Jinan low carbon economy development. But how to make the construction of city walk the road of sustainable development is the key, therefore, to explore its low-carbon development mode is imperative.

3. Research on Model of Low-Carbon Creating A Health City in Jinan

3.1. *Collaborative creating a health city model*

Now the current urbanization process accelerated and low carbon economy becomes the trend of The Times. Each related units strengthens mutual coordination and innovation, forming a good situation of mutual support, mutual cooperation, which promotes the sustainable development of low carbon industry and low-carbon construction of Creating a Health City, mainly reflected in the following three aspects:

(a) Industrial chain synergy. It is a high efficiency and low energy consumption, low pollution multi-win situation through optimization configuration and ascension of the value chain, enterprise chain, supply chain and chain space. Its core mainly revolve around profits at both ends of the curve is through each link between the upstream and downstream. Forward end, form independent property rights at the beginning of the low carbon design preliminary. Back-end extension, use modern science and technology network for low carbon marketing in the process of product sales so as to improve the core competitiveness of products.

(b) Medium enterprise together. More and more large enterprises pay attention to their own core technology research and develop the outsourcing service industry. Therefore, on a large enterprise, they can use this opportunity for further use of consumption so as to develop the extended range of services for large enterprises energy conservation and emissions reduction of small and medium-sized enterprises [3], such as Carbon trading and financial services, other new modern intermediary service industry; etc. [4]. At present, Jinan will focus on the construction of Jinan 20 key service areas.

(c) The linkage synergies. Under the leadership of the local government, district bureau, Quality supervision, industry and commerce, health and other departments, form a good interactive mechanism, and make "innovation" which should be the effective improvement in their work. Command departments at all levels will continue to play a role of good coordination of scheduling, orderly and effectively command the creating a health city work made solid progress in the city [5].

3.2. *The strategic platform model, based on the four spiral model*

Combining its own characteristics, Jinan introduces science and technology intermediary to establish four spiral models which is the government, universities, science and technology intermediary, enterprise [6]. Jinan encourages cooperation among the key university and enterprises, science and technology intermediary with local government to increase the intensity of knowledge conversion.

(a) Government: a corresponding preferential policy is enacted to share knowledge among industry, technology and education, especially for the technology and industry highlands giving priority support. And at the same time, increase the intensity of investment in human capital, flexible the introduction of talents, build a long-acting mechanism for Creating a Health City, more effectively arouse the enthusiasm of all departments, and consolidate the expansion of Creating a Health City, to ensure that each work do not return, do not rebound.

(b) Colleges and universities: expand industry especially low carbon technology and theory knowledge of low carbon through public academic communication and meetings. with the aid of government agencies, set up their own various database, carrying on the effective management of knowledge assets and circulation. Exchange experience in the form of joint scientific research and cooperate with corresponding levels of enterprise, strengthen the industry chain each link closely linked.

(c) Science and technology intermediary: With the development of low-carbon economy, the capacity for independent innovation and low-carbon technologies are enhanced, and initially formed the technology innovation system of research institutes and the technology intermediary service system of scientific research enterprise incubators. In the process of low carbon health city construction, science and technology intermediary must pay attention to the wisdom of low carbon city to provide strategic consulting services so as to improve the conversion rate of scientific and technological achievements.

(d) Enterprise: As the center of the industrial development, enterprises play an important role in the four spiral pattern which is "administration, production, study and research, use". Relying on the government's policy support, the achievements of the scientific research colleges and universities, and technical transformation, the enterprise should make full use of creating a health city strategic platform to realize own low-carbon development model transformation and upgrade. And at the same time, the market information feedback to the relevant scientific research institutes, to form a good circulation of the low carbon innovation.

3.3. *With "culture" to promote "health" model*

First of all, Based on the Jinan low-carbon health construction of city, make the fusion of the ancient excellent cultural tradition. Relying on the national games of garden expo and other major cultural festival, do a good job in the city environment renovation, environmental protection. Second, distribute promotional materials of creating a health city understood and mobilize the masses to participate in the creating a health city further. Finally, with the aid of creating a health city, together low carbon environmental protection; improve service ability, enterprise management, technological innovation, responsibility, target and the "five habitat" to give green, energy-saving, environmental protection concept to the whole society, forming a good circulation organically.

3.4. *Model of low carbon technology innovation*

Traditional extensive economic and urbanization development pattern intensified the contradiction between energy consumption and low carbon development. However, the concept of low carbon economy provides a new way for urbanization construction, and opens up a new way of thinking. The key to promote Jinan low-carbon creating a health city is low-carbon technology innovation, transformation and upgrading of industrial structure. The only way to achieve energy conservation and emissions reduction is the combination of technology and related industries. So the focus of the construction of low-carbon health city should promote low carbon transformation of industrial structure, strive to cultivate a new economic growth point. In the process of promoting industrial structure, low carbon innovation should be encouraged and the highest goal is the continued ascent of industrial power. What the most important is on around the whole industry chain, to facilitate the construction of industry cluster, strengthen the focus on the low carbon technology research and development and promotion of the industry, and gradually to the transformation of new and high technology industries and modern service industry.

4. Summary

Urbanization construction pace is accelerating. However, the result of traffic congestion, lack of resources, environmental degradation and a series of problems ensued, which obstruct China's urban modernization. Creating health city should not only guarantee the sustainable development of urbanization, and to safeguard people's health and the improvement of life quality, therefore, grasping health city means promoting economic development and cultivating new economic growth point. However, creating a health city is a long-term task, it will not happen overnight.

First of all, unify understanding and strengthen leadership. Consider different sounds in the process of creating a health city work seriously and firm the confidence to ensure that the creating a health city work perfect organization, coordinated operation, high efficiency. Second, setting standards. On the basis of previous creating a health city, constantly to persevere, to further establish and improve the long-term supervision mechanism, clear responsibility, supervision subject, so as to achieve the national health city standards. The third is to intensify propaganda, citizen participation. The foundation of creating a health city work is the participation of citizens smoothly, give full play to the role of the creating a health city knowledge brochure, let people understand the creating a health city, concerned about the creating a health city, and take concrete actions to Jinan creating a health city work.

Acknowledgments

This work was financially supported by the key subject project of 2014 Shandong province statistical research (low carbon eco-city measure and application research of modeling and simulation based agent), and the national statistical scientific research project in 2013 (2013 ly053).

References

1. Xu Hanguo, Yang Guoan. Green turn around China's low carbon development [M]. Beijing: China power press, 2010:120.
2. Liu Zhongwen, Gao Pengzhao, Zhang Xuping. Strategic mode of low carbon city development in China [J] Science and technology progress and countermeasures, 2010, 27 (22) : 67-70.
3. Zheng Jiliang, Zeng Rong. Technology-based small micro enterprise cooperative innovation mode of circular economy and countermeasure study [J]. Science and technology progress and countermeasures, 2014, (31) : 104-107.
4. Zhou Fengqi. Shanghai resources and environment development report (2010) of low carbon city [M]. Beijing: social sciences academic press, 2010:95.

5. Jinan request accelerated the "Creating a Health City" clean up the area small advertising responsibility first in JieBan [EB/OL]. http://Creating a Health City.e23.cn/content/2014-06-01/2014060100034.html, 2014-06-01.
6. Wu Kun, Shen Zhongqin. Small micro enterprise innovation platform construction research - based on four spiral model [J]. Science and technology progress and countermeasures, 2014, (31) : 84-86.
7. Guo Wanda, Huang Jiajun, etc. Carbon Urbanization:New Coordinates of Chinese Urban Development [J]. Opening Herald, 2010, 4 (2): 21.
8. Zhang Zhongli, Zhou Fengqi. Shanghai Resources and Environment Development Report (2010) Low-carbon City [M]. Social Science Documentation Press, 2010, 2 : 218.

Planning for sustainably integrated urban rainwater management: what China should learn from developed countries

D.S. Hu[1,*], W.G. Wang[1,2] and T. Wang[1,†]
[1]School of Law, Xi'an Jiaotong University, China
[2]Shaanxi Provincial Meteorological Bureau, China
*E-mail: *deshenghu@hotmail.com, †taowangxjtu2012@hotmail.com*

Q.T. Zuo
College of Water Conservancy & Environment, Zhengzhou University, China

In an urban area, the rainwater runoff model differs greatly from the model before its urbanisation with a significant increase in surface runoff as a key phenomenon or performance due to many elements that influence the runoff model. Without a development plan taking into account of all these relevant elements and having the plan enforceable and implemented, urban rainwater mess with rainwater flood as the centre would appear in the relevant city particularly big ones. This paper analyses causes leading to the mess triggered by rainwater (including but not limited to stormwater) especially during the rainy season in China, where 62% of 351 sample cities had suffered from the mess during the period from 2008 to 2010. Lack of a long-term and stable urban development plan that considers all potential environmental impacts of rainwater in a holistic way and/or its unenforceability are the keys leading to the mess. There is a necessity to develop a plan for sustainably integrated urban rainwater management and have it legally binding for the area under urbanisation/re-urbanisation to prevent the mess and/or the urbanised area to eliminate the mess. For having it done, China has to learn lessons from developed countries.

Keywords: Planning; Urbanisation; Urban rainwater mess; Integrated urban rainwater management; Sustainable development.

1. Introduction

In the summer of 2011, one new expression came into emergence in China, that is "go to B city (any inland city) to see the sea". What does it suggest? It suggests that, when it rains in summer, many Chinese cities are covered by rainwater to a great percentage, and the cities look like the sea, in most cases of moderate rain and sometimes of above levels (for classes of rainfall in China, see Table 1). That is to say rainwater during the rainy season in China often causes urban rainwater mess, which combines rainwater flood (as the centre), little recharge of groundwater, (water) pollution, traffic block, spread of diseases, loss of property or even of life, etc., and leads to a situation full of problems in a city, in particular the big ones. In fact, according to an official survey carries out in late 2010 by the Ministry of Housing and Urban-Rural Development (Chen 2011), 62% of 351 sample cities had suffered the mess during the period from 2008 to 2010; of which, 137 cities had suffered three or more times, 74.6% had suffered with a 50 cm-depth of water, 90% had suffered with a 15 cm-depth of water, 78.9% had suffered with water covering more than a half hour, and 57 cities had suffered with water covering more than 12 hrs. However, rainwater flooding was seldom reported in 1990s in China. Therefore, it is to be of value for China to learn lessons from developed countries in dealing with urban rainwater, so as to improve planning for sustainably integrated urban rainwater management.

Table 1. Classes of rainfall in China.

Class	Criteria
Small rain	Less than both 5.0-14.9 mm/12 hrs and 10.0-24.9/24 hrs
Moderate rain	5.0-14.9 mm/12 hrs, or 10.0-24.9/24 hrs
Heavy rain	15.0-29.9 mm/12 hrs, or 25.0-49.9 mm/24 hrs
Storm	30.0-69.9 mm/12 hrs, or 50.0-99.9 mm/24 hrs
Heavy storm	70.0-139.9 mm/12 hrs, or 100.0-249.9 mm/24 hrs
Particularly heavy storm	≥140.0 mm/12 hrs, or ≥250.0 mm/24 hrs

2. Element(s) Causing Urban Rainwater Mess

Since industrialisation, the urban rainwater/stormwater issues have been the focus of scholars, professional engineers, architects, planners and policy-makers worldwide. It is widely believed that some elements, such as heavy rain, out-of-date drainage system, improper model of (rapid) urbanisation, climate change, etc., could lead to urban rainwater mess. However, considering the precipitation volume during the rainy season (mainly June, July, August and September) as well as the annual volume, data for three big Chinese inland cities (i.e., Beijing, Nanjing and Wuhan) from China Meteorological Data Sharing Service System shows that, from 1990 to 2010, the volume had not changed apparently (Figure 1, 2 and 3). See Map 1 for the locations of the three cities.

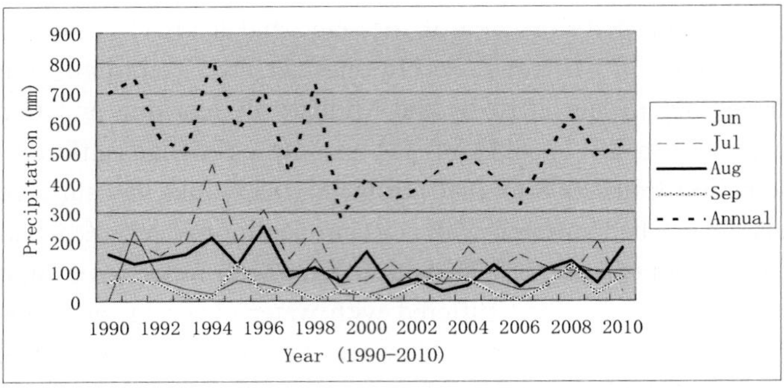

Figure 1. Precipitation in Beijing (the capital city of China).

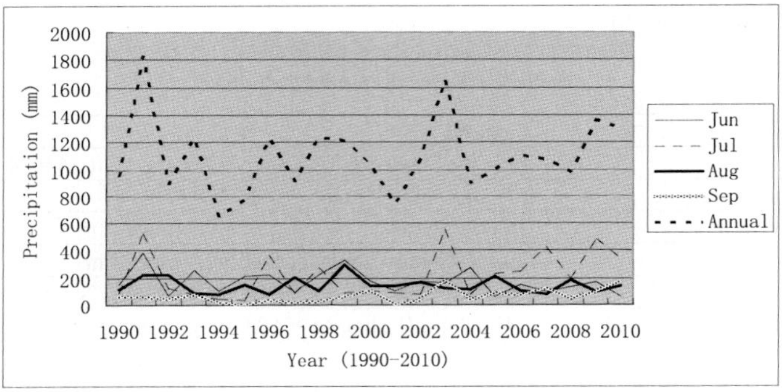

Figure 2. Precipitation in Nanjing (the capital city of Jiangsu Province).

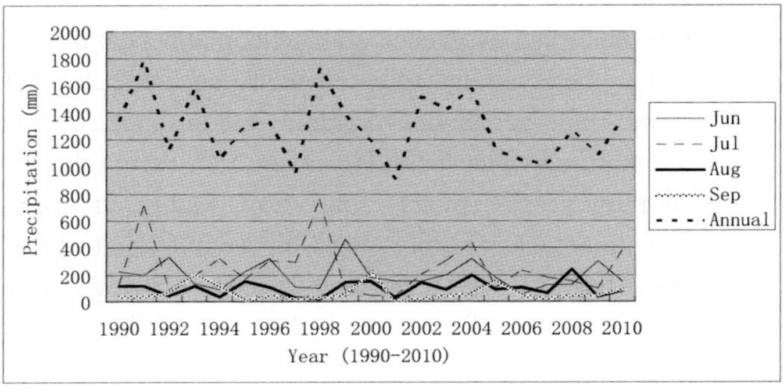

Figure 3. Precipitation in Wuhan (the capital city of Hubei Province).

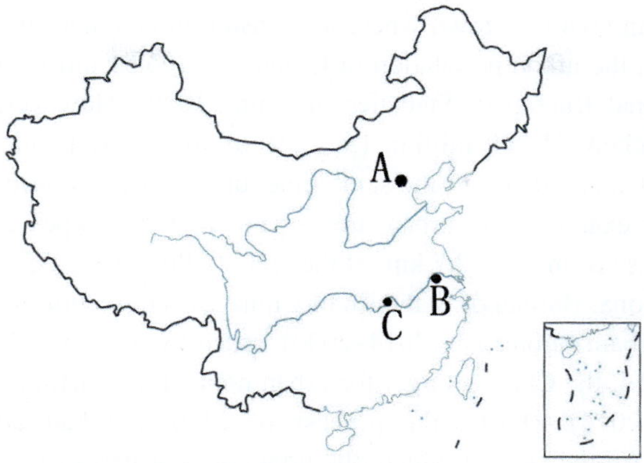

Map 1: The locations of the three cities.
A for Beijing, B for Nanjing, and C for Wuhan.

Runoff volume increases in proportion to impervious area (hard, non-absorbent surfaces). Land uses with extensive roof and paving areas create more runoff than land uses with extensive areas of absorbent soils and forest cover. According to a research on British Columbia of Canada (British Columbia Ministry of Water, Land and Air Protection 2002), in case of natural forest, the impermeable area is almost 0, and 10% of precipitation forms runoff, 50% infiltrates into the soil, and 40% evapotranspirated; in the area of single family development, the impermeable area is about 30%-50%, and 30% of precipitation forms runoff, 35% infiltrates into the soil, and 35% evapotranspirated; in the area of commercial development, the impermeable area is about 70%–100%, and 75% of precipitation forms runoff, 5% infiltrates into the soil, and 20% evapotranspirated. Other research reveals the similar findings (Public Works Department 1995). In brief, the increasing impervious area due to urbanisation decreases the rate of evapotranspiration as well as that of infiltration of rainwater in urban area, and leads to a great increase in rainwater runoff. The increased runoff could not only cause flooding in urban area and elsewhere downstream, but also impact the water cycle negatively in particular by lowering groundwater table, and bring (water) pollution, traffic block, risk of diseases spread, loss of property or even of life. It is obvious that urbanisation is the key element that leads to the great increase in surface runoff, which is at the centre of urban rainwater mess, and could further cause many other nuisances.

China is under the process of rapid urbanisation. There are four types of urban area, i.e. (1) urban area of a municipality directly under the Central Government, (2) urban area of a city where a government at prefectual level is located, (3) urban area of a city where a government at county level is located, and, (4) urban area of a town where a township government is located. On 1st July of 1990, the urban population of China was 296.51 million (26.23% of the total) (National Bureau of Statistics of China 1990). However, at the end of 2013, it reached 731.11 million (53.73% of the total) (National Bureau of Statistics of China 2014). At the same time, urban area has been expanded very quickly. For example, the urban area of the first two types enlarged with a 23.3% increase from 24,442.7 km^2 at the end of 2004 to 47,855.3 km^2 at the end of 2013(National Bureau of Statistics of China 2014). According to the National New-type Urbanization Plan (2014–2020) issued by the State Council of China in March 2014, the China's target for urban population fraction is 60% by 2020 (Bai et al. 2014). During the process of Chinese urbanisation and/or re-urbanisation (in the case of which, the urban area or part of it is re-constructed but remains as urban area), on the one hand, the leader of the city/town always believes that the city /town should pursue the bigger size forever. For example, many capital cities at provincial level want to be one of internationally biggest cities. On the other hand, the impermeable area is about 70% over of the urban area. That is to say, even without increase in the rain volume, the surface runoff will be four to six times than pre-urbanisation. Therefore, rainwater runoff from itself has been being, and will remain, one of the great challenges for urban area to face, as the source of urban rainwater mess.

3. Evolution of Management of Urban Rainwater Mess

At first glance, the part of rainwater that remains after evapotranspiration and infiltration is the key element causing the possible urban rainwater mess, and the initial and traditional approach to prevent the mess has been to remove the runoff as rapidly as possible from the developed areas, so as to prevent flooding and drainage problems. Accordingly, the traditional urban design had pursued efficiently collecting, concentrating, conveying and discharging stormwater to the nearest waterbody. When the town/city is small, the approach could work well in general.

However, firstly, flooding can occur when the sewer is overwhelmed by heavy rainfall, becomes blocked or is of inadequate capacity, and will continue until rainwater drains away. In particular, when this happens to combined sewers, there is a high risk of land and property flooding with water contaminated with raw sewage as well as pollution of rivers due to discharge from combined sewer overflows. Secondly, the rush discharging of urban rainwater runoff could cause severe downstream flooding and bank erosion in the receiving water, and measures of hardening and enlarging natural channels for preventing erosion and flooding are cost-effective and have further negative impacts, e.g., increase in downstream peak flows, provision of not un-healthy aquatic ecosystem (National Research Council 2008). Thirdly, when the percentage of impermeable area has been increasing, when the city has become even larger, not only storm could form surface runoff, but also normal rain, and further, without treatment of rainwater, pollution will occur undoubtedly. Unfortunately, the treatment ability cannot satisfy the increasing amount of rainwater sewage. Therefore, there is a cumulative negative impact of stormwater resulting from changes to the natural water balance under this approach (British Columbia Ministry of Water, Land and Air Protection 2002).

To mitigate or prevent the negative impacts, the human beings have been forced to think more about managing urban rainwater. For example, in the early 1970s, developers in the USA were required to reduce peaks of different size storms by installing flow control structures (e.g., detention basins) in the new developments (Roy et al. 2008). However, due to limitations of on-site detention, such as the risk of leading downstream flooding due to not reducing the volume, infiltration of rainwater runoff has been employed to control its volume as well as to maintain groundwater levels, stream base flows, support springs, wetlands and lakes. Then the low-impact development and conservation design appeared, with a goal of allowing development of a site while maintaining as much as of its natural hydrology as possible. Following the finding of pollution of urban rainwater, in the USA, stormwater runoff has been managed from the pollution control perspective with stormwater pollution prevention plans documenting the stormwater control measures (National Research Council 2008; Roy et al. 2008). In the UK, the approach of Sustainable Drainage Systems (SuDS) is deployed. The philosophy of SuDS aims to control urban rainwater runoff as near its sources as possible by employing a series of source control and prevention techniques (e.g., green roofs, permeable pavements, rainwater harvesting, infiltration trenches and infiltration basins), permeable conveyance systems (e.g., filter drains and swales) and passive treatment systems (e.g., filter strips, detention basins, retention ponds and wetlands) to drain urban rainwater in a more sustainable fashion than some conventional approaches (Dickie 2012). In Toronto, Canada, based on the definition of low impact development given by the United States Environmental Protection Agency in 2007, Credit Valley Conservation and Toronto and Region Conservation Authority have developed their own stormwater management strategy, with designs incorporating rainwater harvesting, green roofs, roof downspout disconnection, soakaways, infiltration, trenches and chambers, bioretention, vegetated filter strips, permeable pavement, enhanced grass swales, dry swales and perforated pipe systems, together with an effective monitoring mechanism being employed (Credit Valley Conservation 2010). In Australia, before 1990s, the solution to stormwater runoff was rapid conveyance directly to streams, although some holistic stormwater managements (for example, being incorporated into greenspace areas and/or recreational uses) adopted since 1960s. In early 1990s, some states (e.g., Western Australia) adopted the approach of water sensitive urban design, and it gained favour in late 1990s (Wong 2006). Through constantly examining the past practice, the approaches to urban rainwater management have been evolving, and the focus now is on preventing the mess at the source in a sustainable way, i.e., integrating urban rainwater management with land use planning at the centre (British Columbia Ministry of Water, Land and Air Protection 2002).

4. Approaches and Measures Dealing with Urban Rainwater (Mess) in China's Urbanisation /Reurbanization

China is under the process of rapid urbanisation/re-urbanisation. The urban area now is more than two times of that in 1990. However, it seems that the rainwater element has not been taken into (reasonable) consideration before 2010. Due to the amount of GDP is almost the only accountable index for promotion of leadership, the leader of the city/town pays little attention to the influence of urban rainwater, including the drainage system dealing with it. For example, in the majority of cities/towns, rainwater in urban areas is frequently drained into surface water sewers or sewers containing both surface and waste water. According to the data from the Office of State Flood Control and Drought Relief Headquarters, the drainage systems in the capital cities at the provincial level have been designed to cope with rainfall events that occur with only a one-in-one-year probability, or up to one-in-two-year probability, and the standard for other cities is lower. For example, the old urban area of Qingdao City, within which the drainage system was designed by the German in late 1890s, has seldom experienced urban rainwater mess in rainy season, but the new development urban areas have usually experienced. The Shenzhen City, a modern city built in 1980s, always suffers from urban rainwater mess.

Beijing, where the sewage system construction is on the top of the Chinese cities and great achievement has been done, could be taken as an example. Under Article 128 of Beijing City Master Plan (2004–2030), a high standard urban rainwater drainage system should be constructed to ensure discharging rainwater fluently, so as to prevent urban rainwater flooding. The said high drainage standard is that, one-in-one-year to one-in-three-year probability for central area and new area, and one-in-three-year to one-in-five-year probability for key urban areas, low-lying areas, rainwater discharging facilities for major road intersections and overpasses. However, the lowest standards are followed in practice. According to the data from Beijing Water Affairs Bureau (Chen 2011), at the end of 2010, it has 3709 km length of rainwater pipe and 1984 km length of combined water pipe, and the standard adopted for sewage system is one-in-one-year probability in general and one-in-three-year probability in important areas. It is reasonable for one to say that the Chinese city design almost does not consider the influence of rainwater in an integrated way, and it is inevitable that the capacities of sewers, covered urban water courses and other piped systems will be exceeded all too often. Even the cities bordering rivers, such as Wuhan and Nanjing, both bordering Yangtze River, often suffered urban rainwater mess greatly during the rainy season.

Although urbanisation/re-urbanisation has great influence on rainwater runoff, a few Chinese scholars had realised it in a thorough way before 2010. Although the changed rainwater runoff has many negative impacts, most Chinese scholars, in particular those being influential in governmental policy-making, have not understood this in a holistic way and believe that the conventional stormwater management approach is the best way to deal with urban rainwater mess. For example, regarding the key for urban rainwater mess, Baojun Yang, the associate director of China Academy of Urban Planning and Design believes that, the urban drainage system has lagged far behind, and this further causes severe "obstruction"; this is the crux of the problem (Chen 2011). It is natural that the policy-makers have paid little attention to the water-cycle and ecological aspects concerning urban rainwater. One cannot find the term of rainwater in the national-wide urban planning rules promulgated by the Ministry of Housing and Urban-Rural Development, such as Rules on Urban Planning Preparation (Dec. 2005), Rules on Preparing and Approving Provincial Urban System (Apr. 2010). In China's Urban System Planning (2005–2020) (Outline), the term of rainwater appears only once. In fact, as mentioned above, planning a site so that water is removed from it as quickly as possible can not only harm the environment, but also endanger others.

Even a few cities have taken urban rainwater issues into consideration, but the relevant planning or measures cannot be enforced. As early as on 4th March of 2003, Interim Provisions on Strengthening Rainwater Utilisation within the Site of Construction Project was issued jointly by the Beijing Planning Commission and the Beijing Water Affairs Bureau. Under the Provisions, (1) when a new project is to be constructed, or an existing project is to be extended or reconstructed, it shall be designed and constructed with rainwater utilisation being included, (2) "rainwater utilisation" means collecting, making infiltrated of, storing or using rainwater to deal with the increased runoff due to hardening the ground surface, and (3) the design and construction for rainwater utilisation shall follow the criterion of not increasing the rainwater runoff within the site and the volume of rainwater discharged out. Besides the Provisions, according to Article 118 of Beijing City Master Plan (2004–2030), approved by the State Council on 25th January of 2005, rainwater utilisation measures are required. However, almost no developer has followed these provisions, few property service suppliers have operated and maintained the relevant infrastructures/facilities, and more than 90% of residents have not known these provisions (Wu et al. 2011). In fact, a construction project named Shuangzi Residential Area, which has built, maintained and operated infiltration area and rainwater storage facilities, has not suffered rainwater mess during the storm from 16:00 of 23rd to 06:00 of 24th of June of 2011, with an average rainfall intensity of 72 mm in Beijing urban area (Wang et al. 2011). Although the construction projects not following the provisions suffered rainwater mess, the relevant unit(s) or person(s) responsible for this has/have not been punished in accordance with the law.

In the majority of cities south to the Yangtze River, there exists a severe scarcity of water in quantity, and relieving the scarcity should be one of the vital considerations for the construction of South-to-North Water Transfer Project. Unfortunately, urban rainwater has not been considered as a source of water in most cases, and made as sewage for being treated, on the one hand. On the other, the groundwater has been over extracted widely. For example, due to over extraction, ground water table had decreased 16.83 m from 1980 to 2009 in Beijing according to Beijing Water Resources Bulletin of 2009 issued by the Beijing Water Affairs Bureau.

5. Lessons for China to Learn from Developed Countries in Planning for Sustainably Integrated Urban Rainwater Management

Urbanisation/re-urbanisation has many substantial impacts, mostly negative ones, on rainwater runoff, and lead to further environmental issues, such as those concerning surface and ground water, the risk of flooding, water supply and use, wildlife and green space, pollution, waste management, land affected by contamination, etc. China's failure in dealing with urban rainwater in the process of rapid urbanisation/re-urbanisation tells us that, without a holistic approach to treat the impacts in a sustainable way, one could easily get into urban rainwater mess again and again, year by year. In the first decade of this century, some Chinese scholars (although not many) called for adoption of sustainably integrated urban rainwater management, incorporating those approaches of low impact development, water sensitive urban design and/or Sustainable Drainage Systems (Zhou et al. 2007; Hu et al. 2008; Che et al. 2009). But the Chinese government had not paid real attention, partly due to the out-of-date opinions of the influential scholars mentioned above.

However, the national wide and serious urban rainwater mess in the summer of 2011, especially in the capital of Beijing where 37 died due to rainwater flood, has forced the Chinese government to consider urban rainwater mess seriously. In 2013, the General Office of the State Council of China issued the Notice on Construction Work concerning Facilities of Urban Drainage and Anti-waterlogging (April), the State Council of China issued the Opinions on Strengthening the Construction of Urban Fundamental Infrastructures (September) and the Regulations on Urban Drainage and Sewage Disposal (October); in 2014, the State Council of China issued the National New-type Urbanization Plan (2014-2020), and Ministry of Housing and Urban-Rural Department of China issued the Technical Guidelines for Constructing a Sponge City—Low impact development stormwater system construction (Trial) (October). According to these documents, when re-developing the old urban areas and developing new urban areas, following the idea of low impact development, measures are required to be adopted and implemented to minimise adverse impacts on the original water related ecosystems. Among others, planning on construction of urban anti-waterlogging drainage facilities shall be incorporated into the urban master plan and the overall planning on urban land use, while being coordinated with urban flood control planning, under the above Notice on Construction Work concerning Facilities of Urban Drainage and Anti-waterlogging. Unfortunately, up to date, no city has revised its urban master plan to make such incorporation.

Roy, Wenger and Fletcher et al. in 2008 identified seven major impediments to sustainable watershed-scale stormwater management in the US and Australia, i.e., uncertainties in performance and cost, insufficient engineering standards and guidelines, fragmented responsibilities, lack of institutional capacity, lack of legislative mandate, lack of funding and effective market incentives, and resistance to change (Roy et al. 2008), and China has been experiencing all of them to some extent. For dealing with these impediments and others, China should learn lessons from developed countries so as to achieve the planned targets with less detours and mistakes. According to the authors' observation, lessons could be learnt in the following six aspects.

Firstly, China, including its scholars especially those being influential in policy-making, policy-makers, engineers, planners and practitioners, in the process of urbanisation/re-urbanisation, should review the practice of as well as research achievements of urbanised countries, so as to obtain good experiences or best practice from the latter, and avoid following the same old disastrous road of others. Urbanisation/re-urbanisation has direct and great impact on evapotranspiration and infiltration, and increase the runoff substantially; further, the cumulative impacts would come into appearance due to flooding in and/or out urban area, (water) pollution, damages to elsewhere downstream, infringement of ecosystems, loss of property and even life. All of these interact with each other and link together. Without understanding and considering these as a whole, one could not design a mechanism to deal with all these in a holistic way.

Second, planning is central to sustainably integrated urban rainwater management, and the planning stage is crucial to make an urban area safe, attractive and prosperous. In general, the importance of the interaction between urban water cycle and city planning and landscaping should be well aware of. Managing urban rainwater mess is a major issue for any urban planning. Although the location of an urban area can has impact on urban rainwater mess, planning for the area can have a large impact on rainwater related issues, and a multi-functional one should integrate the aesthetic, social, economic and environmental requirements. In particular, a good planning with all related elements being taken into consideration is vital to resolve the urban rainwater related issues in the process of urbanisation/re-urbanisation. Without such a planning, the policy-makers would have to take stop-gap measures, and get into the trouble of being busy to deal with constantly occurring new problems.

Third, necessary consents (including permissions, authorisations, permits and other equivalents) from the government for a particular development to go ahead are essential for the planning. These consents should be listed clearly and legally required for the developer to follow. Without getting them or fulfilling the requirements or conditions set up by them, a developer shall be sanctioned by the government's enforcement action against it, leading to be fined or imprisoned. Otherwise, no one would like to do as the law requires because of no remedies. Since 2011, China has been implementing a strict water resources management system. (Zuo et al. 2014) The requirements or conditions for this aspect could be linked with the strict water resources management system,

Fourth, reasonable conditions or requirements shall be set up for the developer by the planning. Whenever a condition or requirement is difficult to be or cannot be satisfied in practice, (most of) developers will not follow it, then the condition or the requirement will out of any practical value. For example, according to Interim Provisions on Strengthening Rainwater Utilisation within the Site of Construction Project mentioned above, within the Beijing urban area, when a new project is to be constructed, or an existing project is to be extended or reconstructed, the criterion of not increasing rainwater runoff within the site and the volume of rainwater discharged out shall be followed in designing and constructing the project. However, the criterion is too vague to be followed. This may be the root why the policies and regulations adopted in 2013 and 2014 have not been implemented and/or obeyed by local governments and most of developers.

Fifth, an effective monitoring mechanism for the whole lifetime of the planning and construction projects should be established. Without such a mechanism, the planning may be changed only or mainly for the consideration of increasing GDP, the process of applying consents from the government may be skipped by the developer intentionally or unintentionally, conditions or requirements set up by the consent may be fulfilled in part or incompletely, and the relevant infrastructures/facilities may not be operated or maintained so as to make the consents out of any practical value.

Sixth, awareness of the public is fundamental for ensuring the planning being implemented and the monitoring mechanism working well. Well understanding the benefits of sustainably integrated urban rainwater management would be helpful not only for the stakeholders to follow the planning in a conscious way, but also for the public to participate in the process of the planning as well as its implementation and monitoring. Sustainably integrated urban rainwater management involves the knowledge of different fields, it is the responsibility of the government to incorporate all of them together and spread it. It is believed that an improved understanding of the planning process and the legally-binding and scientific requirements will lead to improvements in management practices and an increasingly standardized and streamlined approach to addressing urban rainwater. In the USA, the UK and Canada, one could easily find manuals, guides, handbooks, brochures and leaflets concerning stormwater management that have been developed by the governments and put on their websites for helping the developer and the public to understand, follow or being referred. However, few of such materials could be found in China.

In conclusion, effective management of urban rainwater is critical to the continued health of streams, rivers, lakes, fisheries and terrestrial habitats (Credit Valley Conservation & Toronto and Region Conservation Authority 2010). Urbanisation/re-urbanisation doesn't have to get into urban rainwater mess. Following the development of water-related science especially environmental and biological ones and understanding of water in an era of water shortage, although the approaches and practices of managing urban rainwater will be continuing to evolve, experiences and evolutions of dealing with urban stormwater in developed countries (e.g., USA, UK and Canada) shows that, if a planning is well developed and fully followed in practice, it can actually get rid of urban rainwater mess, at least eliminate it to a great extent. In particular, this is more important now than ever before under the era of the increasing impacts of climate change. Because projections of future climate change indicate that more frequent short-duration, high-intensity rainfall and more frequent periods of long-duration rainfall could be expected (Department for Communities and Local Government 2010). For researchers and policy-makers in China, lessons in the above six aspects should be analysed thoroughly. To avoid the mess' occurrence repeatedly, a legally-binding and well-designed planning for sustainably integrated urban rainwater management together with its complete implementation is essential. These will allow urban area free of the mess in most cases, as well as avoidable of loss, money saving and more visually attractive, not a site for people to go "to see the sea".

Acknowledgment

This research was written by Desheng Hu, Weigang Wang, Tao Wang, Qiting Zuo, and was embedded in the Major Program of National Social Science Fund of China (No. 12 & ZD215), funded by the National Planning Office of Philosophy and Social Science.

References

1. Bai X, Shi P, Liu Y. 2014. Realizing China's urban dream. *Nature* 509: 158–160.
2. [Canada] British Columbia Ministry of Water Land and Air Protection. 2002. *Stormwater Planning: A Guide for British Columbia*. British Columbia Ministry of Water Land and Air Protection, Victoria.
3. Che W, Lv Fangfang, Li Junqi, Li Haiyan, Wang Jianlong. 2009. Typical stormwater and flood management systems in developed countries and their inspiration. *China Water & Wastewater* 20: 12-17 (in Chinese).
4. Chen. R. 18/07/2011. Please no longer emphasise surface but contempt underground. *People's Daily* (in Chinese).
5. [Canada] Credit Valley Conservation & Toronto and Region Conservation Authority. 2010. *Low Impact Development Stormwater Management Planning and Design Guide*. Credit Valley Conservation, Mississauga.
6. [UK] Department for Communities and Local Government. 2010. *Planning Policy Statement 25: Development and Flood Risk.* Department for Communities and Local Government, London.

7. Dickie S, Mckay G, Lons L, Shaffer P. 2012. *Planning for SuDS – making it happen*. [UK] The Construction Industry Research and Information Association, London.
8. Hu D, Ma X, Zuo Q, Hu X. 2008. Planning and Construction Mode of Green Space and Untilisation of Urban Rain-water Resources. *Water Resources and Hydropower Engineering* 9: 12-15 (in Chinese).
9. National Bureau of Statistics of China. 1990. *Bulletin of Main Data of 1990 Census* (No. 1).
   National Bureau of Statistics of China's official website.
   http://www.stats.gov.cn/tjsj/tjgb/rkpcgb/qgrkpcgb/200204/t20020404_30320.html. Accessed 1st February 2015 (in Chinese).
10. National Bureau of Statistics of China. 2014. *China Statistical Year Book 2014*.
    National Bureau of Statistics of China's official website.
    http://www.stats.gov.cn/tjsj/ndsj/2014/indexeh.htm. Accessed 2nd February 2015 (in Chinese).
11. National Research Council. 2008. *Urban Stormwater Management in the United States*. The National Academies Press, Washington, DC.
12. [City of Olympia (USA)] Public Works Department, Water Resources Program. 1995. *Impervious Surface Reduction Study* (Final Report). Public Works Department, City of Olympia (USA).
13. Roy A H, Wenger S J, Fletcher T D, Walsh C J, Ladson A R, Shuster W D, Thurston H W, Brown R R. 2008. Impediments and Solutions to Sustainable, Watershed-Scale Urban Stormwater Management: Lessons from Australia and the United States. *Environmental Management* 42:344–359. Estimating the impacts of water pricing on smallholder irrigators in North West Province.
14. Wang X, Y Liu. 04/08/2011. Rainwater is treasure, what a pity without utilisation. *People's Daily* (in Chinese).
15. Wong T H F. 2006. Water Sensitive Urban Design — the Journey Thus Far. *Australian Journal of Water Resources* 10 (3): 213-222.
16. Wu D, Geng N. 10/08/2011. Rainwater storage facilities have not used, rainwater utilisation in Beijing residential area has become a problem. *Beijing Daily* (in Chinese).
17. Zhou Y, Bing S, Zhao S, Wang M, Ding N, Zhang W, Liu J. 2007. Discussion on rainwater utilization planning in Shenzhen City. *Water & Wastewater Engineering* 2: 39-41 (in Chinese).
18. Zuo Q, Jin R, Ma J, Cui Gaotao. 2014. China pursues a strict water resources management system. *Environmental Earth Sciences* 72 (6): 2219-2222.

Brief analysis on cultural connotation of traditional residential space in Huizhou

R.Y. Liu
Architecture and Planning Institute, Anhui Jianzhu University, China

J. Hu
Baimaoling Prison of Shanghai, Shanghai Prison Administration, China
E-mail: 239635962@qq.com

This paper summarized the spatial characteristics of the ancient dwelling in Huizhou and discussed the space "Neo-Confucianism" in Confucians culture, the Taoist culture which explains the idea of "nature and humanity" and the geomantic culture of balance between Yin and Yang, hiding wind and receiving Qi in order to apply these precious cultural essence to modern architectural design practice.

Keywords: Huizhou; Ancient civilian dwellings; Patial characteristics; Cultural connotation.

1. Introduction

"Huizhou" is a historical geographical concept, in terms of its geographical space, it is located in Anhui, Zhejiang and Jiangxi Province at the junction of the three provinces, including today's Anhui Shexian, Yixian County, Jixi, Qimen, Xiuning, Huangshan City, Huizhou and Wuyuan County in Jiangxi Province. With territory mountains multiply and streams double back, Huizhou has beautiful natural landscape, like a land of idyllic beauty in the history. Huizhou experienced three large population migration in the history, in the Ching & Ming period, Huizhou social, economic and cultural achievements have reached its peak, the whole Huizhou cultural awareness was the Confucian culture as the core of the Cheng-zhu neo-Confucianism, at the same time, merged the Taoist culture, geomantic culture and other cultural forms. Huizhou ancient dwellings formed its own unique style and features, under the natural and cultural background, the rich cultural implication in its space is left for the world precious cultural heritage (see Figure 1). In this paper, the spatial characteristics of the ancient dwelling in Huizhou as the breakthrough point, the author explore the implication of Huizhou ancient dwelling in the profound connotation of culture.

Figure 1. Residential buildings of Xidi Village.

2. The "Neo-Confucianism" Space Infected by the Confucian Culture

In Huizhou honored as "state of comity" "Zou and Lu in Southeast (means culture of civilization)", the Confucian ethics of China feudal society had the lofty status in the hearts of the people in Huizhou, during the Song Dynast, the atmosphere advocating Confucianism in Huizhou is very strong, Zhu Xi became the master of the Huizhou neo-Confucianism and he Played the decisive role in the development of Confucian culture in Huizhou, at the same time Cheng-Zhu Neo-Confucianism thought was formed and rapidly developed in the Ming and Qing dynasty. Neo-Confucianism is based on confucianism, and it is a new Confucianism absorbed Taoism and Buddhism at the same time. Cheng-Zhu Neo-Confucianism regards the "Confucianism" as the origin of the universe; look the Qi as the formation of the universe material. Confucianism advocates justice and human desire are the unity of opposites, carries forward the spirit of "benevolence" in Confucianism and affirms the feudal hierarchy from the theoretical level. The most profound influence on the neo-confucianism of Huizhou architecture is to carry forward the Confucian idea of "enlightenment", Huizhou people's daily behavior are being incorporated into the specification, Huizhou neo-confucianism as the essence of patriarchal clan system has a profound influence on all aspects of Huizhou architecture and plays a crucial role in the Construction and the space layout of the Huizhou ancient dwellings.

2.1. *Interior space sequence of pecking order and distinct seniority*

It says in Huang Di House Scripture: "House is the key to the balance of yin and yang and the pattern of human relations [1]." The interior space characteristic of the ancient dwelling in Huizhou is materialization of patriarchal ethnic structure in Huizhou. The interior space sequence of the ancient dwelling in Huizhou reflects a ethical thought of pecking order and distinct seniority. The ancient dwellings in Huizhou is the space sequence longitudinally expanding to the depth along the principal axis, usually with the hall space as the central area of the dwellings; along with the courtyard connecting the halls, the hall space constitutes the central axis of the whole house. As a important ceremonial space of family, hall is usually arranged in the middle of house or the last one of house with rich space scale, open courtyard and space is capacious and bright. Wing is symmetrically arranged on both sides of the hall. Hall Storey is higher, reflects the thought that the room above was superior than the room beneath, downstairs in the hall is a place where elders hosted domestic discipline, promoted house rules. The center of the hall is decorated with a table and two chairs, this is the honour person's place in the family, the rest of the members according to rank of major and minor seat on both sides. Upstairs in the hall has the Supreme status, it is the ancestral hall for worship ancestral tablets. In the daily life, the folk house borders the door in the middle, the vestibular, for the man to parliament and met with guests and the backyard is used for womenfolk's activity area, properties of space is strictly divided and Hierarchical, Diachronic filament line and synchronic order progression exist in the ancient dwellings in Huizhou at the same time, both obviously reflecting the control role of Confucian thought in the ancient dwellings in Huizhou, actually, this is a special residential construction concept, which is fused together the level of Chinese feudal hierarchy thought and Confucian culture to form (see Figure 2).

2.2. *The courtyard space with axial symmetry and many materials*

The courtyard space of ancient houses of Huizhou also reflects the Neo-Confucianism space influenced by the Confucianism. Usually entry courtyards are located in front of main entrances of residential buildings. Plane layout of an entry courtyard is axisymmetric, tress are planted in both left and right sides, normally osmanthus trees and pomegranate trees are adopted to imply the wealthy and auspicious meaning and symbolize the idea that more children bring more happiness. Flower beds often have regular shapes, and ponds are always square and neat with thick and heavy revetments (see Figure 3). Huizhou courtyard in terms of form and material fully embody the role of guidance and control of Neo-Confucianism thought of ancient dwelling in Huizhou.

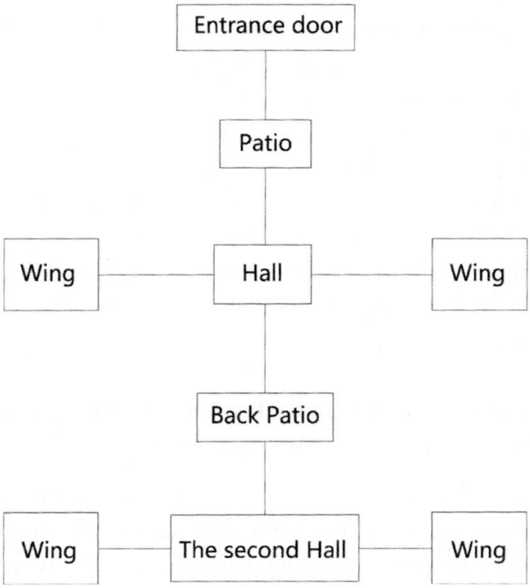

Figure 2. The space sequence of the ancient dwelling in Huizhou.

2.3. *The decorative space with the meaning of loyalty, filial piety, chastity and righteousness, along with farming, reading and being the official*

The decorative space of the ancient dwelling in Huizhou is the essence of the whole culture of Huizhou architecture. The ancient dwelling in Huizhou, under its elegant coat, actually hides so extravagant, exquisite sculpture, which is amazing. In fact, this is the proof of the Ancient People's subtle, low-key life-style. There are white walls, black tiles, seemingly plain and implicit outlines and exquisite stone carvings, brick carvings and wood carvings, all of which contain pursuit to Chinese traditional culture. Under combined influences of various factors including religious ethics, geographical environments and culture thoughts, complicated and straightforward artistic form has formed unique styles. Huizhou architectural decoration is usually composed of stone carvings, brick carvings, wood carvings, couplets, plaque and painting. It has a wide range of subjects and rich connotation and diverse forms. Among them, "loyal, filial piety, chastity, righteousness" as the theme of building decoration is the most common and widely distributed in architectural decoration. The concept is the core of Confucian ethics, "Yue Fei's mother tattoos in his back",

"24 stories of filial piety", "lying on the ice to fetch fish" and other subjects for all aspects of society set examples to the Huizhou people, and carry forward the spirit of Confucianism. Huizhou decoration still has akind of decorative theme about tolerance and "reward", which embodied benevolence and righteousness of Confucianism and the thought of respecting the imperial power. Farming-reading culture had widespread influence in Huizhou society, which was also reflected in the Huizhou decoration. For example, many Huizhou couplets promote learning and enlightenment to be an official statement. A lot of study rooms are carved with flower window in the form of ice crack, which means studying hard. Huizhou decoration shows the Huizhou people's mind to the world in an intuitive way, and reflects the core value of the people in Huizhou advocating Confucianism.

3. Syncretism between Heaven and Man and Learning from Nature — Taoist Culture in the Ancient Dwellings in Huizhou

It says in Tao-te Ching written by Lao Tzu: human learns from earth; earth learns from heaven; heaven learns from Tao; Tao learns from nature. Taoism advocates harmony between man and nature, harmonious development, and ultimately achieves the goal of "Syncretism between heaven and man". The "Harmonious" ideology advocated by the modern society and the Taoist thought come down in a continuous line. The Taoist culture of "Syncretism between heaven and man and learning from nature" in the organization of the Huizhou ancient dwellings space is mainly reflected in the Huizhou patio, the organization of Huizhou courtyard space and the use of color in the ancient dwelling of Huizhou.

3.1. *Patio space with the harmony and integration of heaven, earth and human*

As the most active element of Huizhou buildings, an open space is surrounded by four or three sides of pitched roofs in enclosed center of an ancient dwelling in Huizhou. It is so-called the patio [2]. Huizhou dwelling is surrounded by towering wall, and the tall Ma Tau wall watts is used not only in fire prevention, but also in bringing the space feeling of closure and mysterious to the interior space of Huizhou dwelling. The patio is a particular open space for communication between indoors and space in Huizhou, and it was formed by the integration of northern quadrangle dwellings and primitive dwellings in Huizhou. Huizhou residents live under skylights, look up at sky and look down to take care of the earth, which full reflects the esteem and passion of Huizhou residents to the nature, manifests in-depth explorations of Huizhou residents to relationships among heaven, earth and human, and perfectly explains the thought that man is an integral part of nature. The creation of patio makes peoples sight-line into the sky. The hall and memorial tablet of dwelling face to the patio exactly, which unifies the spirit of respecting ancients and heaven. At the same time, the patio can be seen directly from two wings through their flower windows. The life of the residents in Huizhou surrounds the patio, with the accompany of seasons, moon and stars. Earth and humans have a prefect common identity, which is leaded by the space of patio, and the residents in Huizhou put their life into nature metabolism, which forms a wonderful picture of "Syncretism between heaven and man" (see Figure 4).

3.2. *Courtyard space in Huizhou penetrating nature and borrowing scenery from heaven and earth*

The spatial layout of Huizhou courtyard is the most free and flexible space in the ancient dwelling of Huizhou. Huizhou courtyard has a flexible layout, and doesn't seek rules, generally arranged before or behind the house, sometimes arranged on both sides of the residence. In the Huizhou courtyard, people can feel the full range of the sky, the sun and the moon, rain and dew, stars and other gift of nature. Through the penetration of exterior wall window, door, scenery hole, Huizhou courtyard borrows the scenery outside the residence to the courtyard. Through the terrace, hanging fascia and partition door etc in the main building, the indoor space and outdoor space are mutual infiltration, mutual integration and interdependence. Architectural space and natural space alternately, interspersed with each other, which obviously shows the harmony thought of Taoism (see Figure 5).

Figure 5. A leaking window of a courtyard in Hong Village.

3.3. *Black, white, grey — the beauty of simple colors of the ancient dwellings in Huizhou*

The exterior wall and the roof of pink white and black color reflect the most beautiful color in the ancient dwelling of Huizhou. "The long grey tiles and eaves, the horse head and white wall" which is the true portrayal of the color in the traditional dwellings of Huizhou. Huizhou residents choosing this elegant, dry clean colors is deeply influenced by the Taoist ideas. Lao Tzu, founder of Taoism thought "five colors make people blind and five notes make people deaf", then Zhuang Zhou developed the thought of Lao Tzu. His aesthetic idea fully embodied the philosophical simplicity of the spirit of Taoism. He thinks "Simple and nothing can compete with the beauty" [3]. It was because of the Huizhou residents mind of calm, living life by accident, harmoniously with the natural that formed this "colorless" color space.

4. Balance of Yin and Yang, Hiding Wind and Receiving Qi — Geomantic Culture in the Ancient Dwelling in Huizhou

It says in San Yuan Scripture: "if the earth is good, the plants here will be lush; if the house is lucky, the people inside will be honored." Zhao Jishi at Qing Dynasty wrote in his Sending What Yuan Ji Sent: people in Huizhou pay particular attention to the theory of geomancy, and most of their usual arguments are about it. Obviously, the ancient Huizhou area especially paid attention to geomantic. The location of residence and space layout extremely paid attention to the influence of geomantic.

4.1. *Residential location model is to lean on the mountain, to surround the water and to face the barrier*

The location of the Huizhou ancient dwellings embodied the geomantic theory. The Tenth Book of Yang House says: "all the houses with running water on the left are called Green Dragon and those with long road on the right are called White Tiger; the houses with a pool in the front are called Rose finch and those with hills on the back are called Yuan Wu. These are the best locations" (see Figure 5). The environment that the majority of residents in Huizhou select to build their houses is fronting water and with mountains on the back, near the mountain and by the water, and carrying Yin and holding Yang. Huizhou residents build courtyard, patio, living room and other interior spaces above the geographical space consistent with the geomantic. Red bridge village of Huizhou in the Qing Dynasty reflected the strong sense of geomantic on site selection. The village faced a stream with mountains in the background, the left side surrounded by stream, the road on the right side was accessible. Houses in the village almost without exception rise abruptly under this excellent geomantic environment.

4.2. *The courtyard space in Huizhou with water around into the hall and without wealth exposed*

The patio space of Huizhou ancient dwellings also fully embodies the geomantic theory. Water, is the "wealth" in the geomantic. The patio space reflects the Huizhou people's desire for wealth. Water means wealth. The so-called "with water around into the hall and without wealth exposed", deeply reflects the geomantic theory of the patio space. The Qi theory of geomantic also has a profound impact on the formation of the layout in the Huizhou ancient dwellings. Buried scripture of Mr. Qing Wu says: "interior Qi produces and exterior Qi is shaped; they supplement each other and the geomancy forms automatically." People in Huizhou think that "to produce Qi is the first essence". The patio regarded as the center of the folk house, is the source of vitality of the folk house. The patio hides wind and receives Qi by connecting with the vault of heaven, and removes the filth by associating with earth. Airflow always be in circulation in Huizhou ancient dwellings. The Huizhou ancient dwellings are undertaking metabolism ceaselessly (see Figure 6).

4.3. *The entrance space of the dwellings is to exorcise evil spirits*

The handling of some taboos in the entry space also embodies the geomantic theory of Huizhou ancient dwellings. There is a record in Code of Seeing the House: good or ill luck of a house is all about the gate…The house receives Qi by gate just as people receive Qi by mouth, so the gate is called entrance of Qi. If the door of Huizhou folk dwelling face hill, wall, corner of a wall, tile etc. Generally, the treatment measure is back several paces, or turn an angle. So there are some askew doors to ward off evil. For the door straight to the T-junction, turns through the porch or courtyard space, changes the direction of the door in order to avoid the taboos (see Figure 7).

5. Conclusion

In the land of China long history of 5,000 years, the ancient Huizhou region is considered as the end of the shining star in the late feudal society, Shining in the scroll of Chinese history. Huizhou ancient residence as the material carrier of the culture of Huizhou, show the ancient Huizhou people's wisdom and incomparable skill and leave over a deep cultural connotation for us. We uphold these Huizhou People's spiritual wealth bequeathed to us, and apply them to the process of modern architecture construction. We should pay attention to the following principles: (1) Fully respect for nature, using natural, harmonious coexistence with nature in a harmonious development principle; (2) The ecological principles of saving land and suiting one's measures to local conditions; (3) The principle of focusing on the intention of the richness of space and cultural; (4) People-oriented, creating scenes of historical and cultural atmosphere of the space, highlighting the principle of spiritual places. How will our culture deeply implanted in the future building design to the modern city, as well as a residential street, a square and integrated into the community and the hearts of the people; let our culture carry forward from generation to generation, in order to solve the various defects in modern architectural design such as the lack of culture, the cultural disorder and so on, which is an important issue that we have to face in the future.

Acknowledgements

This work was financially supported by The National Science and Technology Support Program (2012BAJ08B03), Anhui Natural Science Foundation (1308085ME68), Special Fund to the Protection of Hui style Architecture (HP2013YJ101).

References

1. Zhu Yongchun. 2005. Huizhou Building. Hefei: Anhui People Press.
2. Xiao Hong. 2012. Architectural Culture of Huizhou. Hefei: Anhui Science and Technology Press.
3. Liu Renyi, Zhang Jinghua. 2010. Research on causes of Clors of flok House in Anhui Province and their Culture Connotations. *Industrial Construction*, 40(5):146–148.
4. Guo Qian. 2005. The Study of the resident Architecture and Culture of the People Department in Hunan and Jiangxi. Beijing: China Building Industry Press.
5. Zhu Xi. 1983. Notes of Chapters in The Four Books. Beijing: Zhonghua Book Company.
6. Edited by Yao Yanluan. Integration of Yang House. Edition in the seventeenth year of Emperor Qian Long.
7. Tong B. Tang. 1984. Science and Technology in China. London: Longman.
8. Robert A. M. Stern. 1990. What the Classical can do for Modern. London: Academy Group LTD.

Empirical analysis of Chinese and South Korean housing policies and their effects on housing prices change: taking Shanghai and Seoul as example

H.G. Roh* and C.F. Wu

Department of Land Management, Zhejiang University, Hangzhou, Zhejiang, China
**E-mail: siqi75@hotmail.com*

Housing policies in China and South Korea were numeralized. Meanwhile, using the loan interest rates and completed housing areas represent economic and housing market, the VAR model of housing policy, housing price, loan interest rates and completed housing areas were established respectively based on the data from 1999 to 2009 of Shanghai and Seoul, then explored the interaction mechanism among these four factors. Through quantitative analysis of the mutual influences, the conclusions were obtained. (1) There is a significant one-way causal relationship between housing policies and housing prices of Shanghai; however, at the same period, Seoul doesn't show the same trend. (2) With time passing, the contribution rate of housing policy on housing price variance is getting smaller. (3) When the contribution rate of housing policy on housing price variance showed sharply rise and fall, China and South Korea governments used different policy means and targets to cope with it.

Keywords: Housing policy; Commercial housing price; VAR model analysis; Metropolis in China and South Korea.

1. Introduction

The price of housing, as a commodity, is determined by the market. But unlike common commodities, housing prices would rocket, plunge, or be speculated if they were only subject to market mechanism without supervision. As a result, not only natural housing price fluctuations and financial market stability would be affected, even social stability would be influenced negatively. Therefore, it is an important means for state governments to regulate and control housing market by formulating corresponding housing polices. However, how much impact do such policies have on stabilizing housing prices still remains to be studied?

Considering different economic growth rates in different countries, huge regional disparities and various complexities of housing markets, it is hard to find a universal rule in a short period of time simply by studying housing policies' influences over housing price fluctuations in two or more countries.

While economic metropolis or capitals in different countries, regarded as indicators of changes in local housing market, are featured by quick response to housing policies, wide coverage of radiation and strong influence. By choosing economic metropolis of different countries and comparing the impact of the country's housing policies and other factors on housing price fluctuations, it is possible to understand the influence of housing policies and other factors over housing market from a more objective point of view. Meanwhile, by taking advantage of price change tendency and features disclosed by empirical analysis, it is more conductive to formulating policies more appropriate in the light of local circumstances.

Therefore, this study selects Shanghai and Seoul as example. With loan interest rate representing macro-economy, completed housing areas representing housing market and housing selling price data representing commercial residential housing prices, this study employs a VAR model to analyze features of mutual influences between housing policies, loan interest rates, completed housing areas and commercial residential housing prices in China and South Korea.

2. Data Sources and Variables Description

This study adopts a VAR model analysis based on quarterly data of housing policies, loan interest rates, completed housing areas and commercial residential housing prices in China and South Korea from 1999 to 2009, wherein housing policies, loan interest rates, completed housing areas and commercial residential housing prices data about China (Shanghai) comes from Soshoo.com, Shanghai Statistical Yearbook, Zhongfang Shanghai housing database, official website of People's Bank of China and China Macroeconomic Information Network. And same data about South Korea (Seoul) comes from related websites of Ministry of Land, Transport and Maritime Affairs, South Korea Real Estate Research Institute, The Bank of Korea, and associated South Korean statistical yearbooks.

2.1. *Housing policy*

Since housing policies involved in this study are not numerical variables, according to research findings of Jaeho Chung, housing policies are classified into two categories: active policies and restraining policies, which are numeralized with respective given values to generate indicators of housing policies (Chung, 2005). Specific approaches are as follows: (1) organize and classify housing policies in China and South Korea between 1999 and 2009 by standards showed in Table 1; (2) all policies issued in the two countries to reduce loan interest rates are considered policies for active housing market and are valued "+1"; (3) policies issued by Chinese government to control land supply are considered to stabilize housing prices and restrain the market, and are valued "-1"; (4) policies issued in South Korea regarding reconstruction of old urban areas are considered to stabilize housing prices and restrain the market, and are valued "-1". Housing policies issued in Q1 1999 are defined as the benchmark and valued "100". Other active and restraining policies issued every following quarter are assigned different values respectively and accumulated to generate quarterly indicators of housing policies from Q1 1999 to Q4 2009, which are used in VAR model for empirical analysis.

Table 1. Housing policies classification criteria of China and South Korea.

Categories	Contents		Character
Policy objectives	Housing market active policy		i
	Stabilize housing prices, refrain speculation		ii
Policy means	Financial policy	Housing loan	A
		Financial support	B
		Down payment ratio, Second sets of housing	C
	Administrative policy	Pay taxes	D
		Building housing	E
		Land	F
		Transformation of old area	G
		Vacant housing	H
		Trading terms	I
		Comprehensive	J
Regulate object	Real estate developers		α
	House buyers		β
	Sellers		χ

2.2. *Completed housing areas*

In order to demonstrate changes in the housing market, this study chooses completed housing areas in Shanghai of China and Seoul of South Korea for the main reason that such areas have immediate influences over housing supply and demand. In view of seasonal impact on the real estate market, it is necessary to eliminate such effects before conducting VAR model analysis. After applying natural logarithm to completed housing areas, seasonal effects are found to still remain. Comparing X12 and X11 filter processing results, it turns out X12 prevails. The completed housing areas sequence resultant from X12 filter processing proves to be free of distinct seasonal effects (Figure 1).

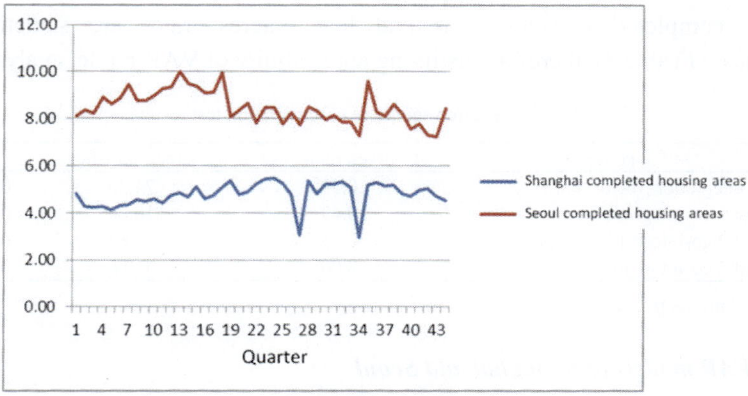

Figure 1. Shanghai and Seoul completed housing areas comparison.

3. Empirical Analyses

As stationarity of time series is extremely important to constructing measurement model, it is necessary to verify such stationarity. The standard procedure to test series stationarity is unity root. As long as the series does not have unity root, it's steady. Variables from Shanghai and Seoul are tested for stationarity with ADF (Augmented Dickey-Fuller test) approach. The findings include:

(1) Shanghai: first order difference sequence of four studied variables are all stationary series. Then a Johansen co-integration trace test is taken, which shows that: at least three of co-integration relationships of the four variables among housing policies indicators, commercial residential housing prices, completed housing areas and loan interest rates are statistically significant (Table 2), therefore justifying applicability of VAR model analysis.

(2) Seoul: The second order differences sequence of the four studied variables are all stationary series. Then a Johansen co-integration trace test is taken.

Table 2. Shanghai research sequence co-integration test results.

Null hypothesis	Characteristic root	Trace statistics	*P*
0 co - integration vector*	0.64	81.54	0.00
At least 1 co-integration vectors*	0.43	40.81	0.00
At least 2 co-integration vectors*	0.31	17.98	0.02
At least 3 co-integration vectors	0.08	3.19	0.07

Note: "*" means $p < .05$

which shows that: at least two of co-integration relationships of the four variables among housing policies indicators, commercial residential housing prices, completed housing areas and loan interest rates are statistically significant (Table 3), therefore justifying applicability of VAR model analysis.

Table 3. Seoul research sequence co-integration test results.

Null hypothesis	Characteristic root	Trace statistics	*P*
0 co - integration vector*	0.59	74.36	0.00
At least 1 co-integration vectors*	0.49	38.68	0.00
At least 2 co-integration vectors	0.25	11.64	0.17
At least 3 co-integration vectors	0.00	0.04	0.84

Note: "*" means p< .05

3.1. *VAR models of Shanghai and Seoul*

Before establishing VAR model, lag interval should be decided first. There are two things to consider when choosing lag order *p:* on the one hand, a lag interval should be big enough to fully reflect dynamic features of the constructed model; on the other hand, the bigger the lag interval is set, the more parameters need to be estimated and the less degree of freedom of the model will have (Gao, 2006). According to LR, FPE, AIC, SC and HQ standards, a fourth-order and second-order VAR model are established for Shanghai and Seoul respectively (Tables 4 and 5).

Table 4. Selection results of lag order of Shanghai's VAR model.

Lag order	Logarithmic likelihood function	LR	FPE	AIC	SC	HQ
0	-184.89	NA	0.15	9.44	9.61	9.51
1	-25.70	278.58	0.00	2.29	3.13	2.59
2	-2.78	35.53	8.43	1.94	3.46	2.49
3	15.17	24.22	8.17	1.84	4.04	2.64
4	44.38	33.60*	4.82*	1.18*	4.05	2.22*

Table 5. Selection results of lag order of Seoul's VAR model.

Lag order	Logarithmic likelihood function	LR	FPE	AIC	SC	HQ
0	-414.86	NA	14640.51	20.94	21.11	21.00
1	-253.23	282.85*	10.13*	13.66*	14.51*	13.97*
2	-243.26	15.45	14.05	13.96	15.48	14.51
3	-227.94	20.68	15.54	14.00	16.19	14.79
4	-210.33	20.25	16.37	13.92	16.79	14.95

3.2. *Granger causality test*

As disclosed by Granger causality test:

(1) Housing policies, completed housing areas, loan interest rates of Shanghai all have one-way causal relationships with commercial residential housing prices, and change in the above three factors all leaded to housing prices

fluctuations. Meanwhile, one-way causal relationship also exists between housing policies, commercial residential housing prices and loan interest rates, but the above four factors have no two-way causal relationships (Table 6).

Table 6. Granger causality test of Shanghai model.

Dependent variable	Independent variable	χ^2	p
PR	PO	10.88	0.03
	SC	14.85	0.00
	IN	15.19	0.00
	PO,SC,IN	45.51	0.00
PO	PR	2.40	0.66
	SC	2.36	0.67
	IN	6.26	0.18
	PR,SC,IN	22.82	0.03
SC	PR	6.26	0.18
	PO	4.96	0.29
	IN	2.14	0.71
	PR,PO,IN	10.22	0.60
IN	PR	12.14	0.02
	PO	23.52	0.00
	SC	4.55	0.34
	PR,PO,SC	33.58	0.00

Notes: PR means housing prices, PO means housing policy index, SC means completed housing areas, IN means loan interest rates. Same hereinafter.

(2) There is no causal relationship among housing policies, completed housing areas, loan interest rate and commercial residential housing prices of Seoul. One-way causal relationship exists between loan interest rates and housing policies. There is no two-way causal relationship between the above four factors (Table 7).

Table 7. Granger causality test of Seoul model.

Dependent variable	Independent variable	χ^2	p
PR	PO	5.62	0.23
	SC	4.72	0.32
	IN	1.91	0.75
	PO,SC,IN	13.12	0.36
PO	PR	7.20	0.13
	SC	2.42	0.66
	IN	14.95	0.01
	PR,SC,IN	23.77	0.02
SC	PR	2.67	0.61
	PO	9.02	0.06
	IN	4.43	0.35
	PR,PO,IN	24.19	0.02
IN	PR	6.86	0.14
	PO	6.24	0.18
	SC	3.08	0.54
	PR,PO,SC	11.11	0.52

3.3. *Creation of impulse response function*

By creating impulse response functions for the four variables, i.e. commercial residential housing prices, housing policies, completed housing areas and loan interest rates, and analytical results showed that: (1) Residential housing prices and housing policies in Shanghai have a short- and mid-term impact on housing prices and policies (Figures 2 and 3), while completed housing areas and loan interest rates have a short- and long-term impact on housing prices and policies (Figures 4 and 5). (2) Residential housing prices and completed housing areas in Seoul have a short- and mid-term impact on housing prices and policies (Figures 6 and 7); housing policies have a long- and short-term impact on housing prices and policies; completed housing areas has a long- and mid-term impact on housing prices and policies (Figure 8 and 9).

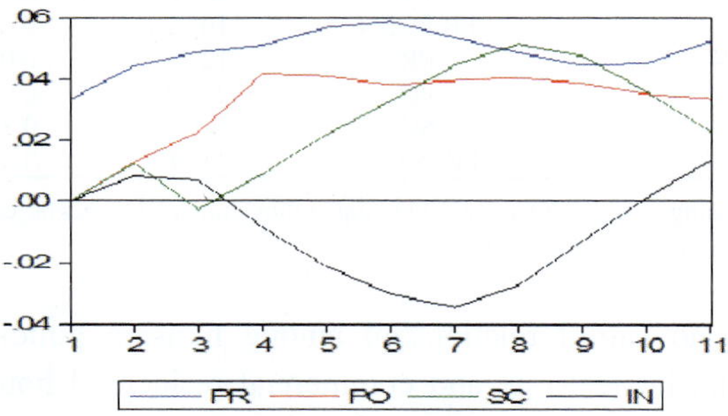

Figure 2. Impulse response of commercial housing price in Shanghai.

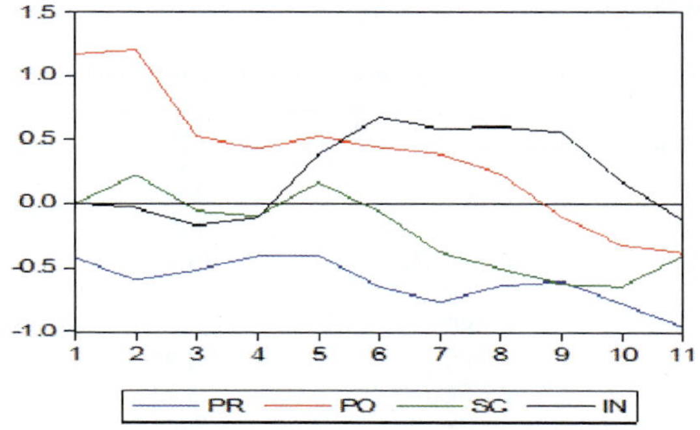

Figure 3. Impulse response of Chinese housing policy index.

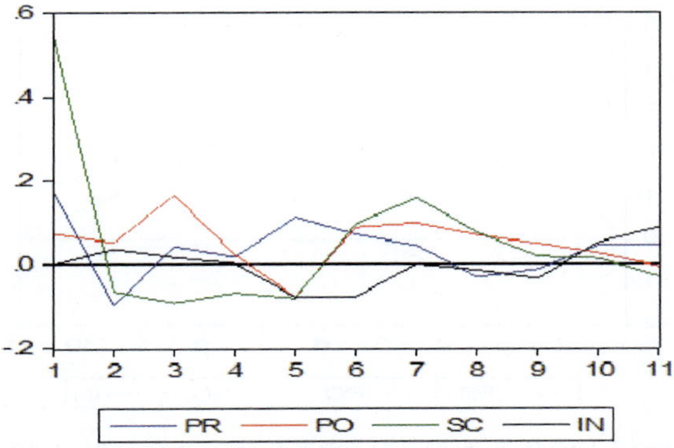

Figure 4. Impulse response of completed housing areas in Shanghai.

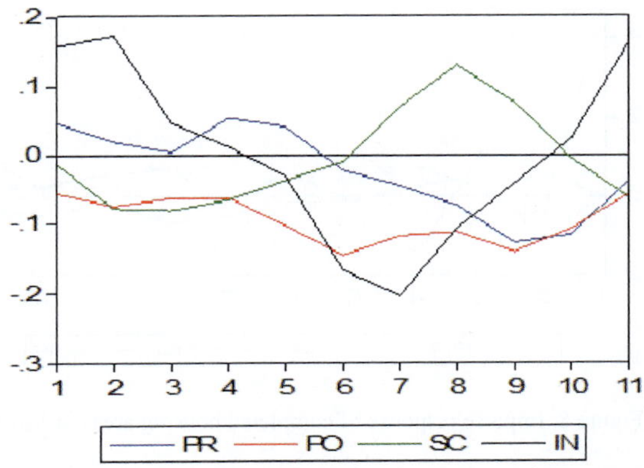

Figure 5. Impulse response of Chinese loan interest rate.

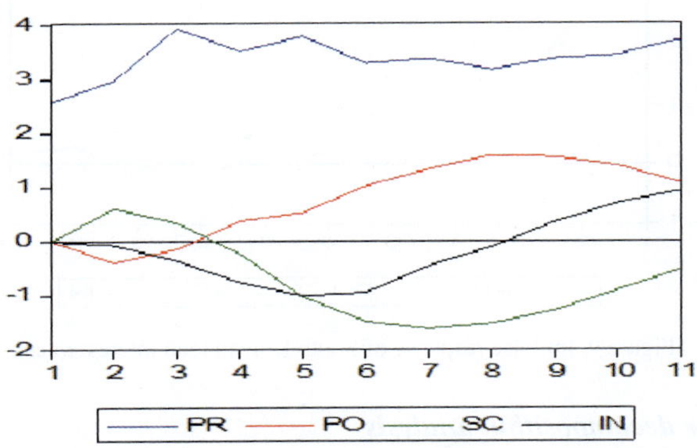

Figure 6. Impulse response of commercial housing price in Seoul.

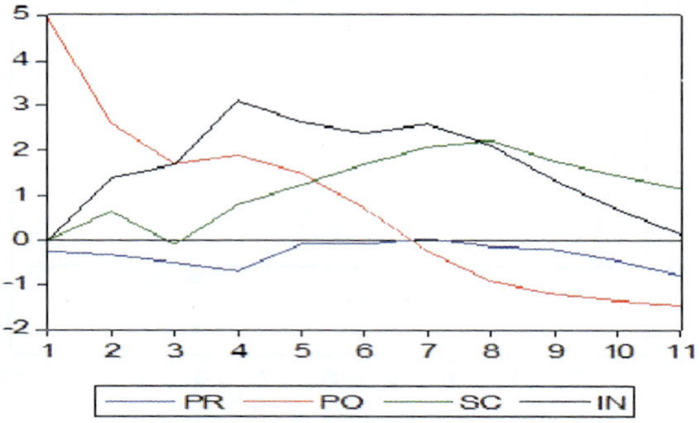

Figure 7. Impulse response of South Korean housing policy index.

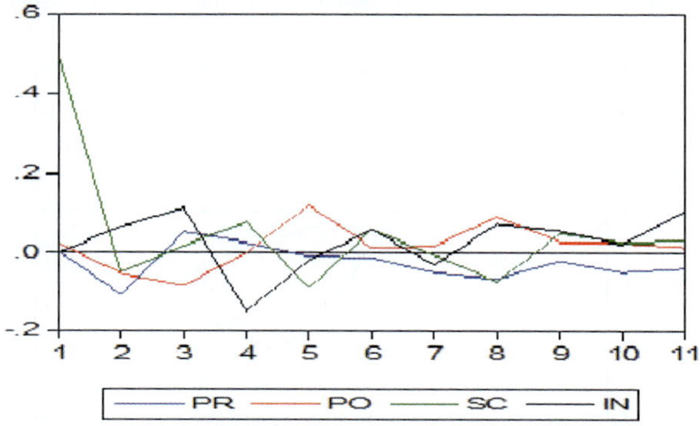

Figure 8. Impulse response of completed housing areas in Seoul.

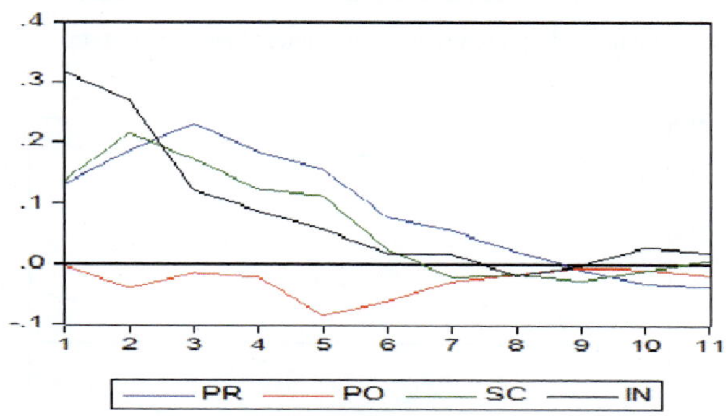

Figure 9. Impulse response of South Korean loan interest rate.

3.4. *Variance decomposition analysis*

In order to analyze each structural impact's contribution to changes in endogenous variables and further evaluate importance of different structural impacts, variance decomposition analysis is employed.

(1) Variance decomposition analysis on Shanghai variables

Variance decomposition analysis of housing policies. Housing policies and commercial residential housing prices are the most significant factors to affect variation of housing policies' variances, wherein impact of housing policies on themselves recedes with time, registering 38.41% contribution rate to total variation. Impact of commercial residential housing prices on housing policies, however, enhances with time, registering 33.28% contribution rate to total variation. Loan interest rates and completed housing areas have a relatively small impact on housing policies in the beginning, which strengthens as time goes by. The final contribution rate of loan interest rates to total variation is 16.03% (as shown in Figure 10).

Variance decomposition analysis of completed housing areas. As shown in Figure 11, completed housing areas are the most important factors to cause variation of their own variances, with a contribution rate remaining above 70% all the time. Followed by commercial residential housing prices and housing policies, loan interest rates' contribution to changes in completed housing areas' variances is insignificant all the time.

Variance decomposition analysis of loan interest rates. Loan interest rates and housing policies are the dominant factors to influence variation of loan interest rates' variances. At stage 1, the contribution rate of loan interest rates reaches 82.14%, and keeps declining afterwards. Impact of housing policies on loan interest rates grows with passage of time, with a total contribution rate of 31.45%. Commercial residential housing prices' contribution rate to variation of loan interest rates' variances is the weakest; only 13.21% (Figure 12).

Variance decomposition analysis of commercial residential housing prices. As shown in Figure 13, commercial residential housing prices and housing policies are the most principal factors to induce changes in residential building prices' variances, wherein residential housing prices have the highest contribution rate to their own variances, i.e. 54.22%. It gets as high as nearly 90% at stage 1. Contribution rate of housing policies to changes in housing prices' variances peaks at stage 5 (24.50%) and keeps declining afterwards to 10.55% at stage 44. Completed housing areas' influence over commercial residential housing prices presents a wavy form and overtakes housing policies' contribution rate at stage 18 to rank 2nd and remains so ever since. Loan interest rates' contribution rate to changes in residential housing prices' variances keeps climbing.

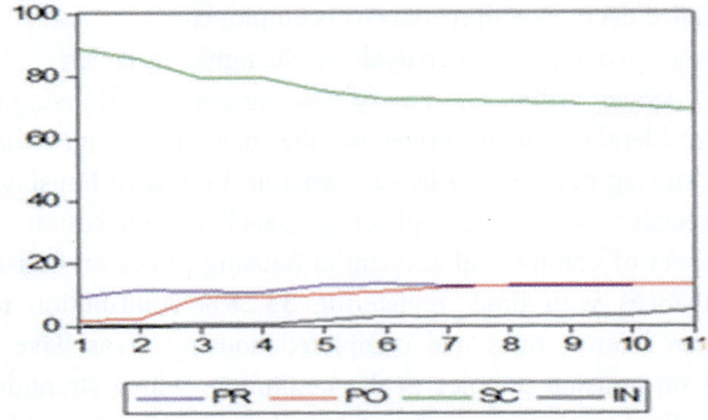

Figure 10. Variance decomposing of Chinese housing policy index.

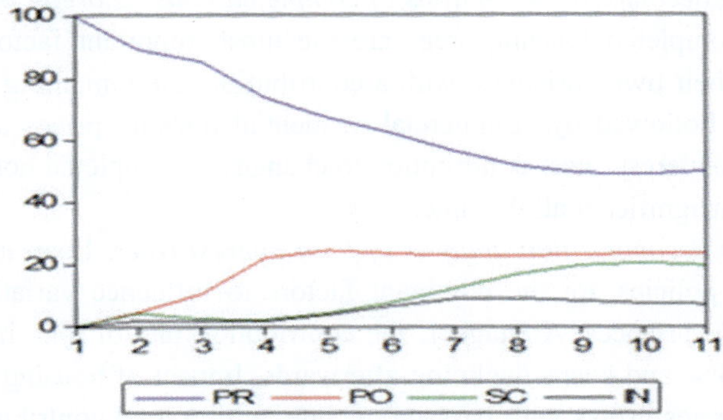

Figure 11. Variance decomposing of completed housing area in Shanghai.

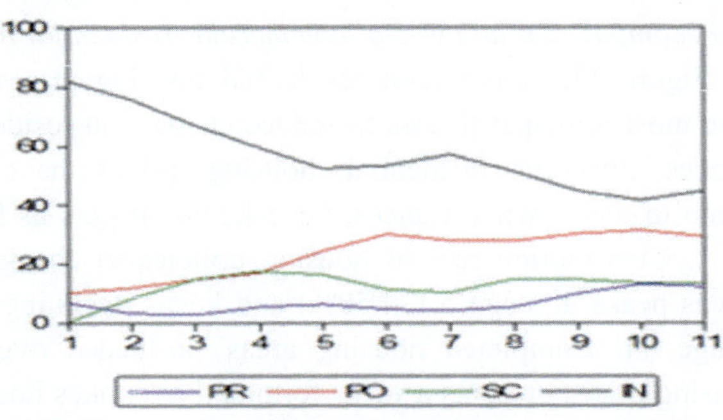

Figure 12. Variance decomposing of Chinese loan interest rate.

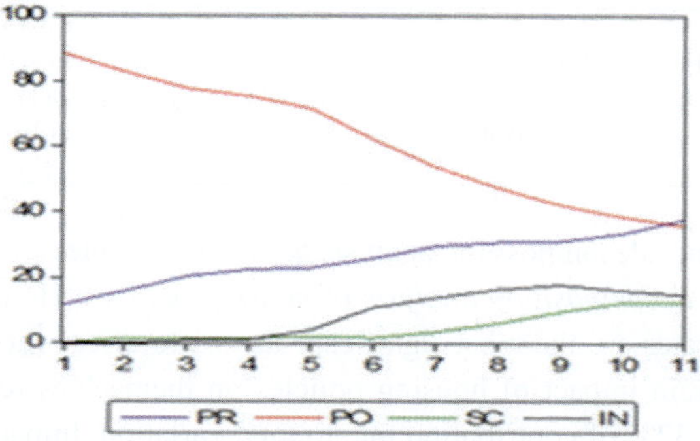

Figure 13. Variance decomposing of housing price in Shanghai.

For the purpose of further analyzing the influential relationship between Chinese housing policies and commercial residential housing prices, a second variance decomposition analysis is conducted to commercial residential housing prices between Q1 1999 to Q4 2009, which lead to the following findings: with the passage of time, housing policies' contribution rate to variation of commercial residential housing prices' variances gets smaller and smaller. During the forty-four quarters from 1999 to 2009, there are thirteen quarter-on-quarter rises of contribution rate, among which seven occurred during the active housing policies were carried out. And such rises mainly appeared during Q1 to Q12, namely, 1999 to 2001. The other six rises occurred during the active housing policies and stabilize housing prices, restrain speculation policies were carried out, concentrating during Q31 and Q35, namely Q4 2006 to 2007.

A comprehensive analysis on housing policies' contribution rate to variation of commercial residential housing prices' variances shows that when contribution rates fluctuate substantially, government normally adopts policies to adjust land policies (F), targeting at real estate developers (α). And then policies are issued to adjust the proportion of down payment, down payment for the second house (C), base of taxation (D), and targeting at house buyers (β) (Table 8).

Table 8. The impact of Chinese housing policy on the commercial housing price in Shanghai.

Quarter	Policy type	Compare to the previous quarter (variation contribution rate)	Policy object	Policy means (use number of applying times)
2-5 quarter	i	Rise	α	F(1)
			χ	D(1)
9-11 quarter	i	Rise	α	F(1)
			β	D(1)
31-35 quarter	ii	Rise	α	F(2), I(1)

			β	C(1)
44 quarter	ii	Rise	α	E(1), F(1)
			χ	D(1)
15-18 quarter	ii	Sharply decline	α	F(2)
			β	C(1)

(2) Variance decomposition analysis on Seoul variables

Variance decomposition analysis of housing policies. Housing policies and loan interest rates are the most significant factors to affect variation of housing policies, wherein impact of housing policies on themselves recedes with time, registering 42.32% of contribution rate to total variation. Impact of loan interest rates on housing policies, however, enhances with time, registering 36.74% of contribution rate to total variation. Completed housing areas' impact on variation of housing polices improves with time and peaks at stage 11 at 19.33%. Housing prices have a relatively small impact on variation of housing policies all the time at only 1.6% (as shown in Figure 14).

Variance decomposition analysis of completed housing areas. As shown in Figure 15, completed housing areas are the most important factor to cause variation of their own variance. Although the contribution rate drops with time, it remains above 69%. Loan interest rates' influence over such variation ascends to the second place after stage 3 and keeps going up, registering 15.77% contribution rate to total variation. Although influences of housing prices and housing polices over completed housing areas are on the rise, their contribution rates to total variation constantly remains below 9%.

Variance decomposition analysis of loan interest rates. Loan interest rates are the most important factors to affect variation of their own variances. The contribution rate to variation keeps falling during the first four stages and almost stabilizes till stage 5, remaining at around 39% all the time. Housing prices' influence over variation of loan interest rates moves up to the second place till stage 3 and keeps picking up afterwards, with a contribution rate to total variation staying between 30% and 35%. Contribution rate of completed housing areas' influence over loan interest rates remains constantly at around 24% since stage 3. Housing policies have the smallest contribution rate to variation of loan interest rates, only 2.73% (see Figure 16).

Variance decomposition analysis of commercial residential housing prices. As shown in Figure 17, housing prices are the most significant factor to give rise to variation of their own variances, with 82.02% contribution rate to total variation. Contribution rate of influences of housing polices, completed housing areas and loan interest rates over.

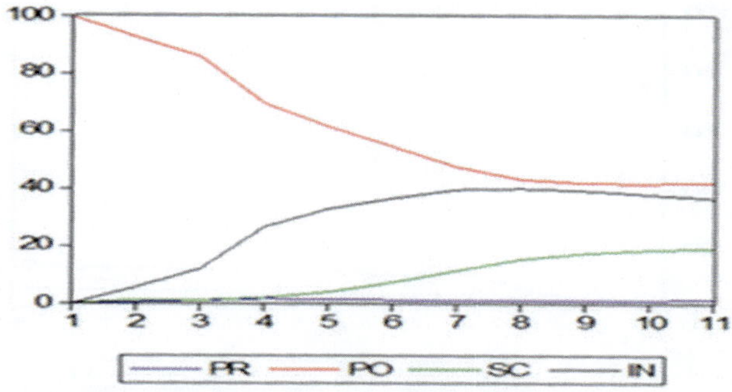

Figure 14. Variance decomposing of South Korean housing policy index.

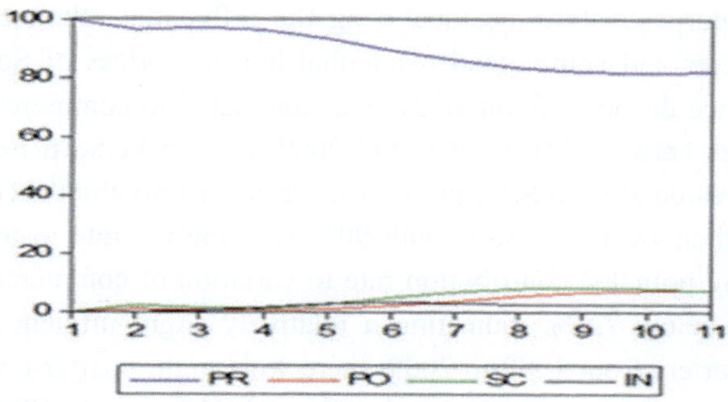

Figure 15. Variance decomposing of completed housing area in Seoul.

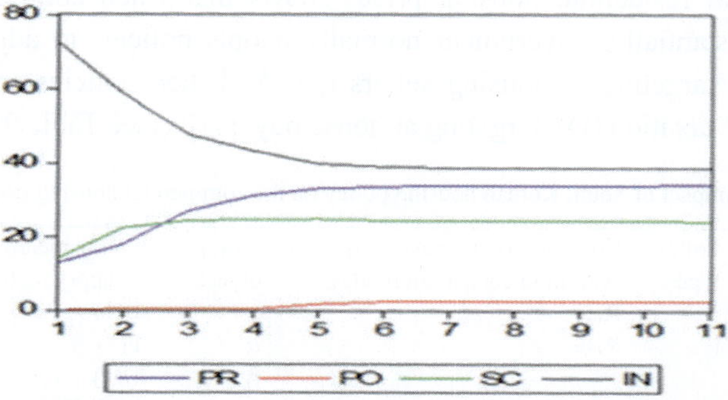

Figure 16. Variance decomposing of South Korean loan interest rate.

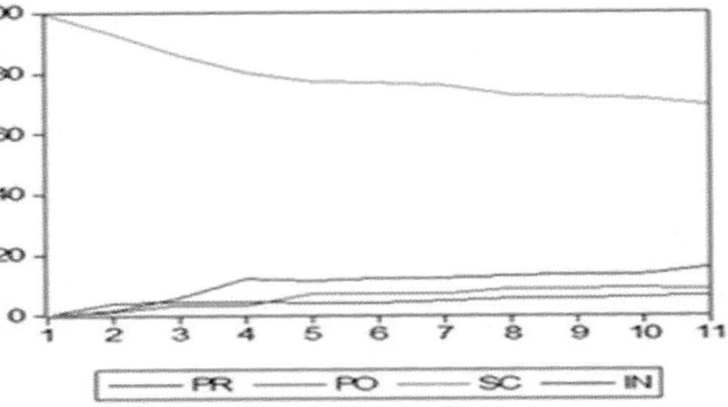

Figure 17. Variance decomposing of housing price in Seoul.

For the purpose of further analyzing the influential relationship between housing policies and commercial residential housing prices of South Korea, a second variance decomposition analysis is conducted to commercial residential housing prices between Q1 1999 to Q4 2009. As can be seen from Table 15, commercial residential housing prices are the most important factor to lead to variation of their own variances, with 90% contribution rate to total variation. While housing policies' contribution rate to variation of commercial residential prices keeps below 7.5%, indicating a relatively slight influence. During the forty-four quarters from 1999 to 2009, there were eight quarter-on-quarter rises of contribution rate, concentrating during Q1 and Q12, namely 1999 to 2001.

A comprehensive analysis on housing policies' contribution rate to variation of commercial residential housing prices shows that when contribution rates fluctuate substantially, government normally adopts policies to adjust base of taxation (D), targeting at housing sellers (χ). And then policies are issued to adjust base of taxation (D), targeting at house buyers (β) (see Table 9).

Table 9. The impact of South Korean housing policy on the commercial housing price in Seoul.

Quarter	Policy type	Compare to the previous quarter (variation contribution rate)	Policy object	Policy means (use number of applying times)
2 quarter	i	Rise	χ	B(1)
9-11 quarter	i	Rise	α	F(1)
			β	D(1)
4-10 quarter	i	Rise	α	B(1), E(1)
			β	A(1), D(1)
			χ	D(3)
13-14 quarter	ii	Sharply decline	β	D(1), I(1)
			χ	I(1)

4. Comparison of Regional Features of Housing Policies's Influcence over Commercial Residential Housing Prices in Shanghai and Seoul

4.1. *Common ground between Shanghai and Seoul*

Pulse response function analysis on Shanghai and Seoul data both show that commercial residential housing prices are the most important factors to affect their own changes, and such effects are positive.

Variance decomposition analysis on Shanghai and Seoul data both show that commercial residential housing prices are the most important factors to trigger variation of their own variances, wherein contribution rate of Shanghai data is 54.22%, and Seoul 82.02%. When housing policies' contribution rate to variation of housing prices increase, both governments issue policies to adjust base of taxation (D).

4.2. *Differences between Shanghai and Seoul*

As disclosed by Granger causality test, the housing policies, completed housing areas, loan interest rates in Shanghai China are all found to have one-way causal relationships with commercial residential housing prices, and changes in the above three factors all lead to housing prices fluctuations. While there is no causal relationship between housing policies, completed housing areas, loan interest rates and commercial residential housing prices in Seoul, South Korea.

Pulse response function analysis shows that stages when different factors affect commercial residential housing prices most powerfully are not the same. The most powerful impact on Shanghai occurred at stage 6 (2004), registering 0.059. Namely, every unit standard deviation change resultant from housing prices' impact in the previous period will lead to 5.9 units of change in housing prices of the current period. However, the most powerful impact of housing prices on themselves in Seoul occurred at stage 3 (2001), registering 3.94. There was slight fluctuation afterwards, but in general, it was on an uptrend.

Analysis on decomposition of housing prices' variances in forty-four quarters over eleven years shows that the highest contribution rate of Chinese housing policies to variation of housing prices occurred at stage 5 (i.e. Q1 2000), registering 24.50%, and kept declining afterwards till stage 44 (Q4 2009), only 10.55%. However, South Korean housing policies' contribution rate of influence over housing prices had been constantly insignificant.

When Shanghai's housing policies' contribution rate to variation of housing prices fluctuates sharply, it is highly likely that policies are issued to control land supply (F), targeting at real estate developers (α); while Seoul issues policies to regulate base of taxation (D), targeting at house sellers (χ).

Major reasons explaining the differences between empirical analysis results of Shanghai and Seoul may be that Seoul's urbanization has exceeded 98%, with no evident changes in population, therefore, commercial residential housing prices are the most important factors to arouse their own changes; while in Shanghai, urbanization ratio of suburban areas will reach 75% in 2010 (Wang, 2009), urbanization and continuous increase of population are the main reason for differences between Shanghai and Seoul during the process of empirical analysis.

5. Conclusions

Chinese housing policies have a significant one-way causal relationship with commercial residential housing prices in Shanghai. Analysis on variance decomposition of housing prices shows that housing policies' contribution rate to housing prices gets smaller and smaller, indicating there's a feedback effect between Chinese commercial residential housing prices and housing demand and supply. It's highly possible that rise in Chinese housing prices over the recent years is mainly due to rise in rigid demand. Housing prices increase will lead to speculative demand by virtue of wealth effect and investment effect. Therefore, when formulating housing policies, government may combine price control with demand regulation, e.g. taking measures to restrain speculative demand so as to stabilize housing prices.

During the same period, there is no causal relationship between housing policies and commercial residential housing prices in Seoul South Korea. Consequently, when formulating macro regulative housing polices, government should consider integrating demand and supply of indemnificatory housing like economically affordable housing. For example, to supply houses for the homeless while at the same time restrain speculative housing demand to stabilize housing prices.

When housing policies' contribution rate to variation of commercial residential housing prices rockets or plunges in Shanghai China, policies are issued to adjust land supply (F), targeting at real estate developers (α). In contrast, when contribution rate rises and falls in Seoul, policies are issued to adjust base of taxation (D), targeting at house buyers (β) and house sellers (χ).

References

1. Chung, J.H. 2005. A Study by VAR Model on the Real Estate Policy and Price. *Korean Real Estate Research*, 25: 241-251.
2. Gao, T.M. 2006. Analysis method of econometrics and modeling-E Views application and case study. *Press of Tsinghua University,* pp. 263.
3. Wang. C.L. 2009. Political sociological analysis of big cities population spatial evolution. *Shanghai People's Publishing house,* pp. 70-79.

The urbanization impacts on China's rural-urban income gap and economic growth

S.F. Qian
School of Economics and Commerce, South China University of Technology, Guangzhou 510006, China
E-mail: qian_shufang@126.com

We use the VAR model to explore two kinds of urbanization impacts on rural-urban income gap and China economic growth. Using time series data from 1978 to 2013, we come to two useful conclusions through Johansen co-integration test and impulse response function as follow: 1) Chinese urbanization has a positive effect on the urban-rural income gap; 2) urbanization improves the economic growth. We also have some suggestions for government according to the conclusions.

Keywords: Urbanization; Rural-urban Income Gap; Economic Growth; Vector Auto Regression (VAR).

1. Introduction

The ratio of the China's urban residents to total population reached 51.27% in 2011, which indicates that China's urbanization level is more than 50% for the first time. Since 2011, the ratio of urban residents to total population has been increased gradually to 52.57% in 2012 and 53.73% in 2013 respectively. The trends shows that China's urbanization level is constantly improving, and thereby the urbanization impacts on rural-urban income gap and economic growth is increasingly becoming a hot topic for research among scholars.

Lots of studies have been done in this area. For example, Wang Xiaoting, Lu Qian, and Wu Haixia (2009) employed the EG two-step co-integration and error correction model, and examined the data during 1985–2006, and came to conclusion that the rising urbanization widen the income gap between rural and urban areas. Nie Peng and Li Shiping (2011) explored the economic impact of population urbanization, and urbanization impacts on urban-rural income gap by co-integration EG two-step method, and established co-integration regression equation. Their findings also showed that the development of China's urbanization expanded the income gap between urban and rural areas. Fang Ling (2008) collected data of Shanxi Province from 1978 to 2009, and studied the relationship between urbanization and economic growth. They used Granger causality test and found that there exist the Granger causality relationship between the urbanization and economic growth, and the urbanization improved the economic growth.

By examining extant researches, we found most of the researchers agree to these two ideas: (1) increasing urbanization makes the rural-urban income gap wider, (2) urbanization has the positive effect on economic growth. We also found most of the researchers use the similar method of single equation analysis, which makes it easy to understand the reason why they would reach similar conclusions. However, we thought the single equation method was not the perfect method. This paper focus on employing more sophistication method to explore the effects of urbanization. We collected data from the period of 1978-2013, used Granger causality test, and built a vector auto-regression model (VAR), using Johansen co-integration and impulse response function analysis.

2. Selected Variables

2.1. *Urbanization (RK)*

The indicators measuring the population urbanization are the proportion of urban population, the proportion of non-agricultural population, the proportion of urban land, etc. Population urbanization refers to the dynamic process of the non-urban population migrate to urban area, so that the ratio of urban population to total population is gradually increasing. Considering the practicality data acquisition and sample size, we use the ratio of urban population to total population (a single indicator to reflect population urbanization) as indicator of urbanization, and denote the variable with RK.

2.2. *The income gap between urban and rural areas (CJ)*

According to references of the relative structure number, make a little change, the relative income of urban residents can be obtained by the following function:

$$CJ = \left| \frac{I_U}{I_U + I_R} - 0.5 \right| \tag{1}$$

I_U: The per capita disposable income of urban residents. I_R: The per capita net income of rural households. The larger of the CJ, the greater of the income gap between urban and rural.

2.3. *Economic growth (GDP)*

The income gap between urban and rural areas is got by the per capita disposable income of urban residents and the per capita net income of rural households. Considering the per capita GDP on behalf of economic growth, we denote the variable with GDP.

3. Model

3.1. *Unit root test*

Unit root test is a standard method to test the stationary time series, and it mainly includes DF test, ADF test, PP test, KPSS test, ERS testing and NP testing and other methods. The following describes the ADF test.

The ADF test adds the lag differential term of dependent variable to the right of regression equation, in order to control the high-order serial correlation, as follows:

$$\Delta y_t = c_0 + \gamma y_{t-1} + c_1 t + \sum_{i=2}^{p} \beta_i \Delta y_{t-i+1} + \varepsilon_t$$

$$\Delta y_t = c_0 + \gamma y_{t-1} + \sum_{i=2}^{p} \beta_i \Delta y_{t-i+1} + \varepsilon_t \quad (2)$$

$$\Delta y_t = \gamma y_{t-1} + \sum_{i=2}^{p} \beta_i \Delta y_{t-i+1} + \varepsilon_t$$

The null hypothesis is $H_0 : \gamma = 0$, namely the sequence exist a unit root. Before the ADF test, we must choose a reasonable lag p, and choose whether to select the tendency and constant term, from the above three model to test.

ADF test uses E-views. The data figure of three variables (RK, CJ, GDP) shows that the ratio of urban population (RK) and per capita GDP have significant upward trends, so it's advisable to logarithm, the variable names are LNRK, LNGDP.

There is a unit root at 5% significance level, so we use first-order differential AR(1), so we adopt first-order differential, after that there is no unit root at the 5% significance level, namely, DLNRK, DCJ, DLNGDP are stable.

Table 1. ADF test results.

Variables	Type(C,T,N)	ADF values	P-value test	Conclusion
LNRK	(C,T,5)	-3.112438	0.1236	Non-stationary
CJ	(C,0,4)	-0.747803	0.8182	Non-stationary
LNGDP	(C,T,1)	-3.168338	0.1092	Non-stationary
DLNRK	(C,0,0)	-4.279435	0.0021	Stable
DCJ	(C,0,4)	-3.924390	0.0059	Stable
DLNGDP	(C,0,1)	-3.296874	0.0240	Stable

Note: The C means constant term, C = 0 indicates without constant term; T means the trend term, T = 0 indicates without trend term; N represents the number of lag.

3.2. *Granger causality test*

For stationary time series x, y, if the variable x can help explain future changes of the variable y, then variable x is Granger causes of variable y. Inspection requirement is to estimate regression coefficients below:

$$y_t = \sum_{i=1}^{p} \alpha_i x_{t-i} + \sum_{i=1}^{p} \beta_i y_{t-i} + \gamma_t + \mu_t \quad (\mu_t \text{ is white noise}) \tag{3}$$

The null hypothesis is that $H_0 : \alpha_1 = \alpha_2 = \cdots = \alpha_p = 0$, and means x does not Granger cause y. If the conditions of the null hypothesis was true, the statistics $F = \frac{(RSS_R - RSS_{UR})/p}{RSS_{UR}/(n-q)}$ is the distribution of F (p, n-q). Among them RSS_R is the residual sum of squares containing x lags and after the reunification, RSS_{uR} is the residual sum of squares excluding x lags after reunification, n is the sample size, q is the lag period of y free of lag x when regression. Table 2 is the output result of the Eviews.

Table 2. Granger causality test between DLNRK and DCJ.

Null Hypothesis	Lag periods	F statistics	P-value
DLNRK does not Granger cause DCJ	1	0.22396	0.6397
DCJ does not Granger cause DLNRK	1	2.55051	0.1215
DLNRK does not Granger cause DCJ	2	0.34232	0.7134
DCJ does not Granger cause DLNRK	2	2.57247	0.0964
DLNRK does not Granger cause DCJ	3	3.81123	0.0244*
DCJ does not Granger cause DLNRK	3	2.61714	0.0765
DLNRK does not Granger cause DCJ	4	1.61162	0.2123
DCJ does not Granger cause DLNRK	4	2.05715	0.1268
DLNRK does not Granger cause DCJ	5	4.04311	0.0146*
DCJ does not Granger cause DLNRK	5	1.95697	0.1405

Note: The "*" indicates that we should reject the null hypothesis at the 5% significance level.

As can be seen from Table 2, indicating that DLNRK Granger cause DCJ, that the data of urban population ratio can help explain changes of differential data in the income gap between urban and rural.

Table 3. Granger causality test between DLNRK and DLNGDP.

Null Hypothesis	Lag periods	F statistics	P-value
DLNGDP does not Granger cause DLNRK	1	2.338110	0.1375
DLNRK does not Granger cause DLNGDP	1	0.367930	0.5490
DLNGDP does not Granger cause DLNRK	2	1.939920	0.1647
DLNRK does not Granger cause DLNGDP	2	5.828590	0.0084*
DLNGDP does not Granger cause DLNRK	3	1.005520	0.4090
DLNRK does not Granger cause DLNGDP	3	2.703780	0.0702

DLNGDP does not Granger cause DLNRK	4	0.968500	0.4476
DLNRK does not Granger cause DLNGDP	4	4.351940	0.0115*
DLNGDP does not Granger cause DLNRK	5	1.257250	0.3294
DLNRK does not Granger cause DLNGDP	5	3.417480	0.0273*

Note: The "*" indicates that we should reject the null hypothesis at the 5% significance level.

Similarly, it can be seen from Table 3 that p-value of 2,4,5-order lag less than 0.05, That means we need to reject the null hypothesis at the 5% significance level. In other words, DLNRK does not Granger cause DLNGDP, so the data of urban population ratio after logarithmic differential can help explain per capita GDP change after logarithmic differential.

3.3. *VAR model*

The mathematical expression of Vector Auto-regression Model VAR (p) is

$$y_t = A_1 y_{t-1} + \cdots + A_p y_{t-p} + Bx_t + \varepsilon_t \qquad t = 1, 2, \cdots, n \tag{4}$$

Among them: y_t is the k dimension vector of endogenous variables, x_t is d dimension vector of exogenous variables, p is the lag order, n is the number of samples, A_i is $k \times d$ dimension coefficient matrix, B is $k \times d$ dimension coefficient matrix, ε_t is the k dimension disturbance.

Using the VAR model, we need determine the lag (p) and the coefficient matrix A_i, B firstly. Test method that determine the lag order is mainly LR (likelihood ratio) test, AIC information criterion and SC guidelines and others, the selection criteria: the value of lag (p) can make AIC or SC minimum. Secondly, the root of AR can be used to determine whether the VAR model is stable, if the inverse of the all the roots are less than 1, that is located inside the unit circle, then the model is stable.

According to the AIC or SC guidelines and the size of AR roots, we can better determine the lag value (p), then establish the VAR model.

3.3.1. *VAR model of LNRK and CJ*

Considering the LR (likelihood ratio) test, AIC, SC guidelines, etc., select the lag order, as shown in Table 4.

Table 4. Select the lag value (p).

Lag	LogL	LR	FPE	AIC	SC	HQ
0	75.34506	NA	9.70e-06	-5.867605	-5.770095	-5.840559
1	166.2594	160.0092	9.29e-09	-12.82075	-12.52822	-12.73962
2	177.6321	18.19638	5.20e-09	-13.41057	-12.92302	-13.27534
3	180.0862	3.533844	6.00e-09	-13.28689	-12.60432	-13.09758
4	185.8882	7.426556	5.39e-09	-13.43105	-12.55346	-13.18765
5	190.4554	5.115261	5.47e-09	-13.47643	-12.40382	-13.17893
6	198.1548	7.391468	4.48e-09	-13.77239	-12.50475	-13.42080
7	215.4516	13.83740*	1.79e-09*	-14.83613*	-13.37347*	-14.43045*
8	216.1353	0.437580	2.92e-09	-14.57082	-12.91315	-14.11106

Note: The "*" indicates the best in the guidelines.

Table 4 shows the lag value should choose Lag(p) p=7, So our VAR model is VAR(7) and get the equation (1) and the chart of AR root (Table 5 and Figure 1). AR roots are within the unit circle, which proves that the VAR (7) is stable.

$$\begin{pmatrix} CJ_t \\ LNRK_t \end{pmatrix} = \begin{pmatrix} 1.1309 & 0.1067 \\ 0.2810 & 1.1806 \end{pmatrix} \begin{pmatrix} CJ_{t-1} \\ LNRK_{t-1} \end{pmatrix} + \begin{pmatrix} -0.0080 & 0.0578 \\ -0.4133 & -0.0822 \end{pmatrix} \begin{pmatrix} CJ_{t-2} \\ LNRK_{t-2} \end{pmatrix} + \begin{pmatrix} -0.5702 & -0.3908 \\ 0.1582 & -0.0701 \end{pmatrix} \begin{pmatrix} CJ_{t-3} \\ LNRK_{t-3} \end{pmatrix} \quad (5)$$

$$+ \begin{pmatrix} -0.1438 & 0.2494 \\ 0.4926 & 0.1771 \end{pmatrix} \begin{pmatrix} CJ_{t-4} \\ LNRK_{t-4} \end{pmatrix} + \begin{pmatrix} 0.4579 & 0.3791 \\ -0.0043 & -0.1063 \end{pmatrix} \begin{pmatrix} CJ_{t-5} \\ LNRK_{t-5} \end{pmatrix} + \begin{pmatrix} 0.0798 & -0.4938 \\ -0.3471 & -0.6244 \end{pmatrix} \begin{pmatrix} CJ_{t-6} \\ LNRK_{t-6} \end{pmatrix}$$

$$+ \begin{pmatrix} -0.3236 & 0.2821 \\ 0.1317 & 0.4741 \end{pmatrix} \begin{pmatrix} CJ_{t-7} \\ LNRK_{t-7} \end{pmatrix} + \begin{pmatrix} -0.0592 \\ 0.1007 \end{pmatrix}$$

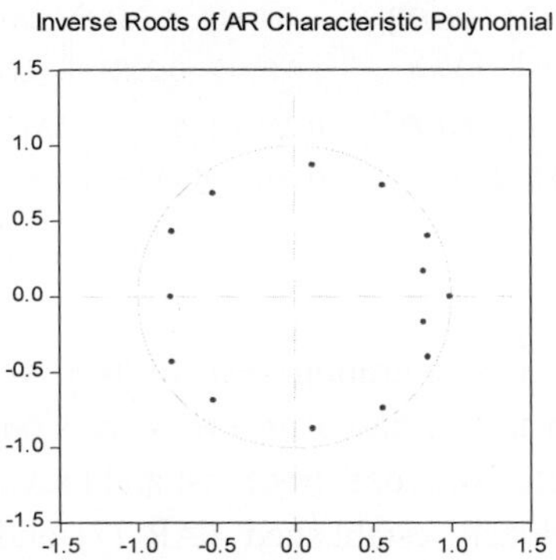

Figure 1. The AR root of VAR (7).

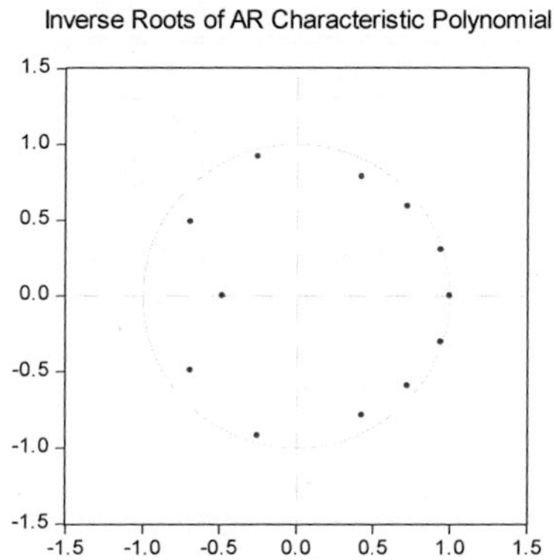

Figure 2. The AR root of VAR (6).

3.3.2. *VAR model of LNRK and GDP*

Similarly, according to the LR (Likelihood Ratio), AIC and SC criteria, etc. determine that the lag value is 6, then we can establish VAR (6), i.e., equation (2), as well as its chart of AR root (Table 6 and Figure 2), we can see that AR roots are within the unit circle, which means that the established VAR (6) is stable.

$$\begin{pmatrix} LNGDP_t \\ LNRK_t \end{pmatrix} = \begin{pmatrix} 1.6152 & -0.2672 \\ -0.0805 & 1.2151 \end{pmatrix}\begin{pmatrix} LNGDP_{t-1} \\ LNRK_{t-1} \end{pmatrix} + \begin{pmatrix} -1.5330 & -0.2788 \\ 0.1246 & 0.0671 \end{pmatrix}\begin{pmatrix} LNGDP_{t-2} \\ LNRK_{t-2} \end{pmatrix}$$
$$+\begin{pmatrix} 1.3139 & 0.9205 \\ -0.0137 & -0.4969 \end{pmatrix}\begin{pmatrix} LNGDP_{t-3} \\ LNRK_{t-3} \end{pmatrix} + \begin{pmatrix} -0.8488 & -2.0614 \\ 0.0005 & -0.2221 \end{pmatrix}\begin{pmatrix} LNGDP_{t-4} \\ LNRK_{t-4} \end{pmatrix}$$
$$+\begin{pmatrix} 0.1564 & 0.2105 \\ -0.0928 & 0.5120 \end{pmatrix}\begin{pmatrix} LNGDP_{t-5} \\ LNRK_{t-5} \end{pmatrix} + \begin{pmatrix} 0.2260 & 1.7859 \\ 0.0912 & -0.2145 \end{pmatrix}\begin{pmatrix} LNGDP_{t-6} \\ LNRK_{t-6} \end{pmatrix} + \begin{pmatrix} -0.1357 \\ -0.1357 \end{pmatrix} \quad (6)$$

3.4. *Co-integration*

We use the Johansen co-integration test to test vector regression model. Johansen co-integration test is based on the regression coefficients, which the main test contain characteristic roots trace test and maximum eigenvalue test.

According to previously established VAR (7) and VAR (6) models, that formula (1) (2), we have the below (Table 7) results of co-integration test.

Table 5. Co-integration test.

Co-integration variables	The null hypothesis	Characteristic roots	Trace statistic(P-value)	$\lambda-\max$ Statistics(P-value)
(CJ, LNRK)	no	0.583409	26.87391(0.0007)*	26.26949(0.0004)*
	not less than 1	0.019946	0.604414(0.4369)	0.604414(0.4369)
(LNGDP, LNRK)	no	0.451378	25.40410(0.0089)*	18.01037(0.0229)*
	not less than 1	0.218435	7.393724(0.1071)	7.393724(0.1071)
(CJ, LNGDP, LNRK)	no	0.750555	50.01921(0.0001)*	41.65554(0.0000)*
	not less than 1	0.243300	8.363670(0.4274)	8.363663(0.3430)
	not less than 2	0.00000024	0.00000728(0.9993)	0.00000728(0.9993)

Note: The "*" indicates that we should reject the null hypothesis at the 5% significance level.

These results indicate that there are co-integration relationship, and have a co-integration variable, namely there is a long-term stable relationship between urbanization and income gap urban–rural and economic growth.

3.5. *Impulse response analysis*

Impulse response function is used to measure the trajectories that a standard deviation of random disturbance how to affect current and future of other variables, which can portray a more intuitive and dynamic interaction effects between variables. The following Figures 3 to 6 show that the shocks impact between LNRK and CJ, LNRK and LNGDP. The abscissa represents retrospective periods, the ordinate represents the size of response that the dependent variable to the independent variable, the solid line represents the response function curve and two dashed lines represent double standard deviations confidence band.

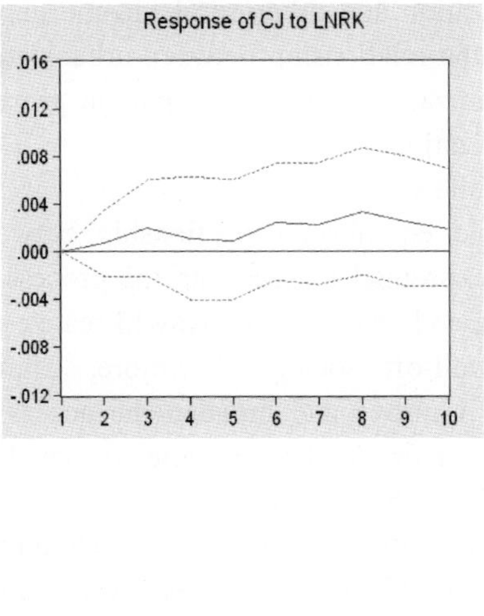

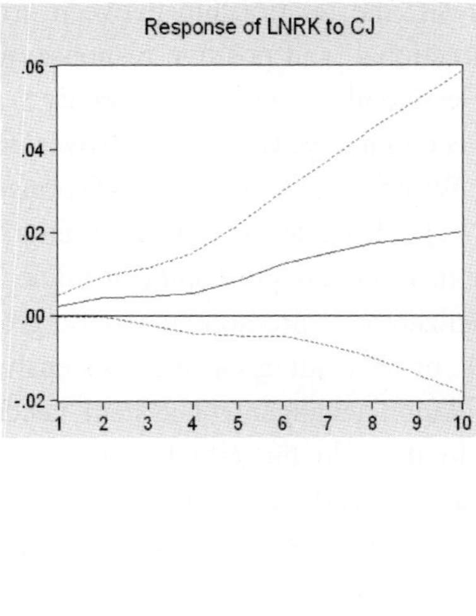

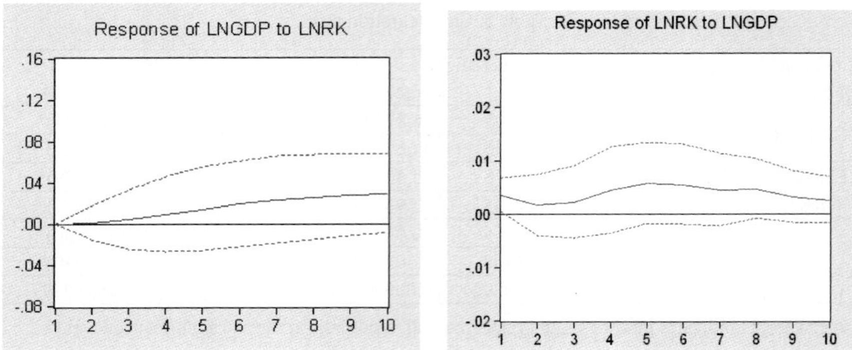

Figure 3. The Impulse response.

4. Conclusion

In this paper, using the relevant national statistical data on urbanization, and the income gap between urban and rural areas from 1978 to 2013, we constructed a VAR model of urbanization and income gap between urban and rural areas, and urbanization and economic growth. At last, we used Johansen co-integration test and impulse response function to analyze.

In the empirical research, we discovered that there existed a long-term stable relationship between urbanization and urban-rural income, urbanization and economic growth. Secondly, the Granger causality test analysis concluded that the data of urban population ratio after logarithmic differential helps explain the change data of income gap between urban and rural after differential; the data of urban population ratio after logarithmic differential helps explain per capita GDP change after differential. Finally, the conclusion of impulse response analysis: the relationship between urbanization and urban-rural income gap is bidirectional, and positive response; also the relationship between urbanization and economic growth is interactive, urbanization promotes economic growth, and economic growth also improves urbanization.

We have some useful conclusions as follows:

(1) There existed a long-term stable equilibrium relationship between urbanization and per capita of GDP. It is essential to speed up the process of urbanization to promote economic growth, and, it is the only way to realize the goal of becoming a comprehensive well-off society. Therefore, China's population management system needs to improve and broaden the access of employment in the city for surplus rural labor, so that promote the level of economic development by raising the level of urbanization.

(2) The level of urbanization has both positive and negative effects on the urban-rural income gap. On one hand, after rural labor immigrates to urban areas,

overall labor income improves, and it also reduces the surplus rural labor force, and increases the amount of resources per capita in rural areas, thereby increasing the labor remuneration and income levels; on the other hand, with the development of urbanization, urban residents' income levels upgrade significantly, which leads to the expansion of city size, and thereby more money are needed in urban infrastructure construction, thus result in corresponding reduction in funds to invest in rural areas, eventually it affects rural income growth, resulting in the expansion of the income gap between urban and rural.

Based on our analyses, we have some suggestions: (1) pay attention to promote the development of industrialization of agriculture. (2) increase the proportion of "three rural" investment, and vigorously launch agricultural technology training, develop the comparative advantage of each region; (3) enhance farmers' awareness on the concept of market, and guide farmers to take the road of market economy with Chinese characteristics. (4) encourage intellectuals and investors to develop rural areas, make contribution to their hometowns, improve the level of rural production and increase farmers' income.

References

1. Panizza, U. Income and Economic Growth: Evidence from America Data [J]. Journal of Economic Growth, 2002(7): 25-41.
2. Aghion P., Caroli E. and García P. Inequality and Economic Growth: The Perspective of the New Growth Theories [J]. Journal of Economic Literature, 1999 (08): 1615-1660.
3. Persson T. and Tabellini G. Is Inequality Harmful for Growth? [J]. American Economic Review, 1994(84): 600-621.
4. Pak Hung Mo. Income Inequality and Economic Growth [J]. KYKLOS, 2000(3): 293-316.
5. Li & Zou. Income inequality is not harmful for growth: Theory and Evidence [J]. Review of Development Economics, 1998(2): 318-334.
6. Guzhaladi. Econometrics [M]. The Press of Renmin University of China, 2000.
7. Song Yuanliang, Xiao Weidong. Dynamic econometric analysis of Chinese urbanization development and farmers' income growth [J], Quantity Technical Economics, 2005(9): 30-39.
8. Guo Song. Empirical study of urbanization impacts on economic growth [J]. Heilongjiang Foreign Trade, 2006(08): 66-67.
9. Wang Shaoping, Ouyang Zhigang. Measure of the income gap between urban and rural areas and its effect on economic growth [J]. Economic Research, 2007(10): 44-55.

10. Zhang Shiyin, Zhou Jialai. Analysis of urbanization indicators [J]. Technical and Economic, 2007(05): 84-87.
11. Zhang Lei, Fang Tiankun. Integration and causality analysis of income gap's change in urban and rural areas and economic growth [J]. Agricultural Technology Economy, 2007(3): 38-43.

Research on the construction planning of smart grid in smart city — Taking Gansu Lanzhou new district of China for example

H.L. Xu, P.F. Jin, Z.Y. Chen, Y.F. Tian and Y. Yu
State Grid Gansu Economic Research Institute, Gansu, China

M.K. Wu
North China Electric Power University, Beijing, China
E-mail: hdu_wmk@163.com

With the proposing of "Smart city" concept and its specific establishment at home and abroad, China has begun to vigorously facilitate the experimental construction of smart city. Lanzhou has become a pilot of smart city. It is of vital importance to boost the planning and construction of smart grid for the development of smart city. In this paper, a brief introduction is provided regarding the status of planning and construction of smart grid in China. Additionally, development and problems facing Lanzhou New District power grid is analyzed in detail. Based on these issues, principles regarding the planning of smart grid of Lanzhou New District are proposed. Finally, according to the objectives and characteristics of the smart grid planning, the design schemes of each construction part of smart grid are offered, which focus on the access of clean energy in the generation side, the implementation of demand response in the user side, and the construction of information communication system.

Keywords: Smart city; Smart grid; Demand side response; Construction scheme.

1. Introduction

Since entering 21st century, the problem of "Intelligentize" has attracted more and more attention. In 2008, Sam Palmisano, a CEO of American IBM, put forward the concept of "Smart earth" for the first time [1]. The construction of "Smart city" is the concrete action of "Smart earth" from idea to reality.

In foreign countries, there are some achievements in the smart grid construction of smart city, especially in the United States. For example, the government of Boulder City started the project of smart grid early. It built the city to become a city with fully integrated smart grid and mainly focused on the intelligent construction of distribution side and user side. In Europe, based on the consideration of environmental protection and sustainable development, the development of smart grid focuses on the generation side, the user side and the smart information communication, namely, the access of distributed energy and the installation of smart meters.

Because of a good reference and basis, the overall planning of smart grid construction in China has a relatively high starting point, which focuses on two aspects of "strong" and "smart". However, its construction of smart city is lagging behind, which is currently only in the initial stage. At present, China actively explores the path supporting the development of smart city, through researches on 12 smart city pilots of state-level smart new districts. Lanzhou New District is the fifth state-level new district of China, which is also the first in Northwest China. However, it also carries the arduous task of consuming and delivering the large-scale new energy of Gansu Province. According to the analysis regarding current installed capacity and consumption and transmission ability of Gansu Hexi power grid, the existing transmission channel is unable to meet the need of transferring the surplus power of the power generation enterprises. Renewable energy electricity accounts for a large proportion in the power structure of Gansu Province, so it is not a small challenge to address the problem of renewable energy use. In addition, how to realize the scale of external supply through the development of smart city is closely related to the construction of smart grid. Therefore, designing the construction scheme of smart grid well is one of the current research emphases and key issues.

The research on detailed construction plan of smart grid is still relatively fragmented, which is mainly aimed at each part of grid [2–3]. But there are many studies on the overall development plan of smart grid. An overall plan on the development of smart grid in China was proposed in [4]. In [5] and [6], we can see that in order to meet the strategic needs of China's national energy development, the construction of smart grid in China must be combined with UHV power grids and power system reform. The development driving force and construction foundation of smart grid in China was introduced in [7]. Combining with the plans of urban power networks and the development practice of smart grid in China, Tong Jun [8] took Yuhang power grid for example to show the plan of substation, distribution, electricity consumption and communication part of the urban power network.

In this paper, a brief introduction about the construction situation of smart grid in China or Lanzhou New District is given. Additionally, development and problems facing Lanzhou New District power grid is analyzed in detail. Finally, the design principles and content of construction plan for smart grid of Lanzhou New District are proposed.

2. Status Quo of China's Smart Grid Planning and Construction

Smart grid is an important platform for the implementation of new energy strategy and the optimization allocation of energy resources. It is of great significance to implement the major technology industrialization project of smart grid for adjustment of energy structure, energy saving and emission reduction, and response to climate change. State Grid Corporation of China has implemented a number of representative pilot projects in the pilot construction stage of smart grid, which has a wealth of experience on the construction of smart grid.

At present, there are 12 state-level new districts in the plan of China, such as Shanghai Pudong New District, Tianjin Binhai New District, Chongqing Liangjiang New District, Gansu Lanzhou New District, and so on. Each new district has taken some measures for boosting the construction of smart grid. For example, in 2011, Shanghai Pudong New District vigorously developed smart grid, accelerating the investment of new energy, guiding demands reasonably, and promoting the development of low-carbon city. In 2012, Chongqing Power Company took "quality", "green", and "happiness" as core goals and closely combined the construction idea of smart grid and strategic positioning of Liangjiang New District to promote the comprehensive construction project of smart grid. At present, Chongqing Power Company has completed nine projects, such as smart community, smart home, distributed photovoltaic generation, 95598 interactive platform, emergency command center, electric quality monitoring, deepening application of smart building, charging facilities of electric cars, and smart substation.

3. Situation Analysis of Lanzhou New District Power Grid

Full support for the purchase of renewable energy power and giving priority to use it are pointed out in the supporting documents of the 9th document of China's new electric reform. The installed capacity ratio of renewable energy is more than 50% in Gansu. So as a state-level demonstration pilot project, the smart grid project of Lanzhou New District undertakes a large social responsibility of new energy consumption.

At present, the production and life electricity consumption within the scope of Lanzhou New District is mainly in rural area. From the aspect of electricity structure, there are several kinds of users. The concrete structure of electricity users is shown in Figure 1. Most of the enterprises entering the new district are in the construction period and the new load is mainly the load of temporary construction, so the large-scale use of electric power will be reflected in the next 2–3 years.

With the further acceleration of development and construction in Lanzhou New District power grid, the problem of power supply shortage and bad power supply reliability will become increasingly prominent. For example:

The construction standard of power grid is relatively low. With the start of constructing Lanzhou New District, the existing power grid has been unable to meet the electricity needs of urban construction. So the planning and layout of power grid should be carried out with the combination of overall planning of the new district.

At present, the 35kV Qinchuang substation and Xicao substation undertaking the main task of power supply for the region are fully loaded, which cannot meet the demand of load development.

(1) The power supply capacity of the 31.5MVA main transformer of 110kV Konggang substation has been unable to meet the surging electricity demand of the new load in core area of Lanzhou New District.

(2) The 110kV Xicha substation and 110kV Yuantai substation are far from the core area of the new district, which are difficult to provide support for the recent electricity consumption of the new district.

Combined with the current situation of power system and policy requirements, promoting the large-scale consumption and transmission of renewable energy in smart city is the most important problem in the construction planning of smart grid. In all parts of the construction planning, making plans should be combined with the planning objectives and closely related to this key problem.

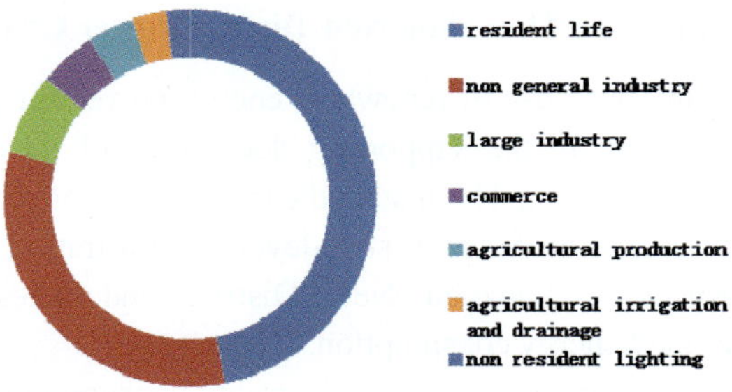

Figure 1. The diagram of electricity users structure.

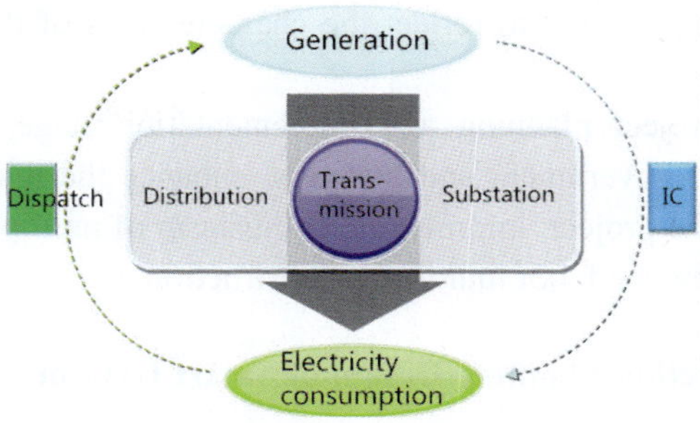

Figure 2. The diagram of construction parts.

4. The Planning Principles of Constructing Smart Grid in Lanzhou New District

The overall goal of Lanzhou New District has been specific in The Aggregate Planning of Lanzhou New District, which is boosting integrated development of advanced manufacturing industry and modern service industry, and building a city with strong industry, ecological environment, beautiful lakes and modern style.

The construction planning of smart grid in Lanzhou New District should be taken combination of the overall goal. Based on this, the overall planning principles for constructing smart grid in the new area are explicated as following:

(1) Design the planning in strict accordance with the principle of "unified planning, unified standard, unified construction" determined by State Grid. And build the smart grid with close focus on the developing strategic objectives and overall idea of State Grid's strong and smart grid construction.

(2) Base on the stage orientation of meeting needs; cater to the goal of increasing values and profits; play the role of appropriate technology demonstration; take "promoting the consumption and delivery of renewable energy" as the key; take "serve the economic development, serve energy saving and emission reduction, serve people's livelihood construction" as the fundamental goal; start from the actual development needs of Lanzhou New District and closely integrate with the existing projects of smart grid; combine with the latest technology; strive to obtain the maximum comprehensive social benefits of the smart grid construction.

(3) As for the construction content, to fully utilize the latest intelligent technology and properly process the coordination relations among key technology research, pilot projects construction, application and promotion. For the construction process, to pursue overall coordination as well as different

focus of different part, and to reflect the characteristics of the regional power grid.

(4) In the project planning and implementation stage, to focus on the communication of government and business, enhance the understanding of all walks of life for the project, improve the enthusiasm of multiparty participation, and explore the new model of multiparty construction.

5. The Construction Planning Design of Smart Grid in Lanzhou New District

The overall goal of smart grid construction in Lanzhou New District is:

(1) Build Gansu's first construction project of smart grid, leading the construction of Gansu's smart grid.

(2) Carry out the construction and research of smart grid in many areas, meet the needs of users, play a role in technology demonstration, drive the development of smart grid of Gansu Province, and achieve the increase of values and profits.

(3) Construct a regional smart grid matched the development positioning and status of Lanzhou New District, serving the development of regional economic, energy saving and emission reduction and the construction of people's livelihood. In addition, to establish a grid management mode with a clear business scope and smooth business processes, supporting the harmonious development of grid and city.

Design the comprehensive construction planning of smart grid, through following the idea of unified planning, taking the promotion of renewable energy consumption and transmission as the key, taking the supply system and service system of smart grid as the functional positioning, taking "quality", "green", and "livable" as the core, and from the perspective of the unified fusion of power flow, information flow and business flow.

The construction idea of Lanzhou New District smart grid is around three core features, which can be expressed as follows:

(1) Take "quality" as the core to upgrade the regional power grid.

Carry out the construction of smart grid with the core of "quality", including smart substation, distribution automation, management system of distribution network fault repair, information communication platform, visualization platform, electric quality monitoring and other sub projects, in order to fully improve its intelligent level and the ability to control the regional power grid. What's more, construct a high-quality electricity supply system in urban local areas, with a strong grid structure, safe and reliable power supply, fast response, good power quality and strong resilience.

(2) Take "green" as the appeal to improve energy efficiency.

Taking "green" as the ultimate appeal, taking smart grid as the visible reflection, and taking service of energy saving and emission reduction as the goal, the implementation of this project tightly combines the targets of "green". A series of projects, such as smart substation and distribution automation, will be carried out in the grid side to improve the efficiency of power transmission and reduce network losses. Through the access of clean energy, the operation control of micro grid, smart park, energy management, electric vehicle charging facilities and other sub projects constructed in user side, to optimize energy structure, improve efficiency of terminal energy utilization, realize environmental-friendly use of electric energy, promote the new ideas of green energy, and help the realization of green Lanzhou.

(3) Take "livable" as the goal to provide electricity services.

Taking the needs of people's livelihood as the guide, taking providing colorful services for users and building livable new Lanzhou as the goal, to strengthen the construction of distribution side and user side, and highlight the service function on livelihood construction. The sub projects of smart grid are supposed to be constructed with targets, including DA, distribution network fault repair management system, electric information acquisition, smart business hall, energy management, and smart grid visualization platform. Then, to strengthen the two-way interaction between grid and user, and change the service concept and service form. In short, to create a high-quality, active and two-way power service system.

In addition, according to the above thinking, the concrete construction plan for Lanzhou New District smart grid can be designed from several parts, such as generation, transmission, substation, distribution, electricity consumption, information communication and dispatch, which is shown in Figure 2.

(1) Generation part.

Lanzhou is located in the west of China, which is abundant in renewable energy resources, such as wind-force and PV. So it is necessary to focus on improving the proportion of renewable energy electricity with grid connection and reduce the waste of resources. Considering the implementation of new electricity reform policy and based on the environmental protection and land status of the construction of Lanzhou New District, the PV generation system can be considered to build, which can improve the proportion of electric sources using renewable energy in the new area and promote clean energy use and low-carbon development of the smart city.

(2) Transmission part.

For the construction of transmission part, the use of smart transmission technology is the key consideration, such as the application of wide area measurement technology.

(3) Substation part.

According to the planning load needs, the smart transformation of existing substations should be completed and the new smart substations also should be built. By using the related technology of smart substation, to realize the intelligentization of primary equipment, the networking of secondary equipment as well as the condition monitoring and self-diagnosis of all kinds of equipment. What's more, by building the comprehensive data platform and advanced application mechanism, to realize sharing the running status information of all kinds of equipment, so as to improve the level of smart analysis and decision for the substation part.

(4) Distribution part.

In the distribution part, the planners should pay attention to the construction of distribution automation, power quality monitoring and so on. The completed DA master station of Lanzhou DA pilot project can be used as the DA master station of the new area. At the same time, the supporting construction of DA terminal and communication network also should be carried out. As for power quality monitoring, online and smart monitoring system should be adopted to automatically monitor the power quality of distribution network, which can help the grid realize power quality accident analysis, interference source location, and power quality management decision-making. In short, to improve the safe operation level of the power grid.

(5) Electricity consumption part.

In the user side, the smart construction needs the features of information, automation and interaction. The planners should pay attention to the construction of electric information collection, energy efficiency management, and so on.

The electric users' information acquisition system includes implementation of remote communication channels, acquisition terminals, smart meters and others.

As to the user's energy efficiency management, the technology of user demand response should be implemented. The user demand response includes two functions: using electricity orderly and demand response (DR). Using electricity orderly can be divided into designing the plan of using electricity orderly and analyzing the effect of its implementation. For DR, the enterprises' history of electricity consumption should be comprehensively considered. And to predict the future electric load trend via the load forecasting model. Then, based on this data, to analyze and arrange combining the controllable load list of enterprises' interiors to formulate the plan of DR, which can guide users to actively participate DR. Lastly, to analyze the effect.

(6) Dispatch and communication part.

Building smart information communication network with the features of automation, information and large capacity, is the goal of the information communication system construction, which contains external communication system construction, information network system construction, administrative telephone exchange system construction and other content. In the first floor of Lanzhou New District business hall, a smart grid visualization platform should be built according to IEC and standards established by State Grid, realizing the data exchange, data sharing and application integration with provincial company's production management system, GIS system, marketing management system, 95598 system, electric information acquisition system, dispatching automation system, electric vehicle charging plants, smart park management system and other application systems. In short, to display the new area smart grid fully, dynamically, vividly, multi-dimensionally and visually.

6. Conclusion

Based on the problems, the related objectives and the tasks of expected planning, the planning principles and planning methods of the comprehensive construction design scheme of smart grid in Lanzhou New District are proposed. Furthermore, in accordance with the construction parts, the overall design scheme of smart grid planning in Lanzhou New district is described. As a model for cities, if Lanzhou New District completes the construction conforming to the planning, the grid development of Lanzhou City and even Gansu Province will be driven. What's more, the efficient and optimized energy utilization will be promoted to achieve the demonstration and leading role of Lanzhou New District.

References

1. Palmisano S J. CEOs Deliver Remarks on the Economy and Stimulus Package [EB/OL]. (2009-01-28)[2012-08-08]. http://www.ibm.com/ibm/ideasfromibm/us/news_story/20090130/index.shtml.
2. The world energy power network. French Power Company tested smart grid to improve the use of wind power.[EB/OL]. 2009-02-03. http://www.wefweb.com/news/200923/1017348758_0.shtml.

3. M. Jiang. Compilation of "smart grid" and "smart energy" [EB/OL]. 2009-03-17. http://power.nengyuan.net/200903/17/25057.html.
4. Yang De-Chang, Li Yong, C. Rehtanz, Liu Ze-Hong, Luo Long-Fu. Study on the Structure and the Development Planning of Smart Grid in China. Power System Technology 2009, 20:13-20.
5. Q. Shuai. Innovative development and construction of smart grid—the practice of the East China's advanced dispatching center projects construction. China Power Enterprise Management, 2009(4): 19-21.
6. The Zhongdian network. State grid releases the smart grid planning [EB/OL]. 2012-04-24. http://www.eccn.com/news_20120424095948l8.html.
7. J. Zheng. Review of the current situation of domestic and international smart grid construction. Shanghai Electric Power, 2011, 03:256-259.
8. J. Tong. Research on modern urban power network planning under the background of smart grid. Zhejiang University, 2010.

Dynamic relation analysis on employment, wages and income gap of urban residents in China for panel data model as the tool*

†W. Wang and W.J. Cao
School of Economics, Harbin University of Commerce,
Harbin, Heilongjiang, P.R.China
†E-mail: wangwei8786@126.com

J.M. Chen
School of Finance and Public Management, Harbin University of Commerce,
Harbin, Heilongjiang, P.R.China
E-mail: zdw2042@126.com

The dynamic relationship of employment, wages and income gap of urban residents in China was analyzed by using the panel data model, regarding the eastern, central and western of China as research objects. We selected three indicators, the number of employees at the end of the year, the total wages and the Gene coefficient as explanatory variables and explained variables. The results show that the effect of wages for improving or worsening the income gap is greater than the effect of employment, and the effect of employment for income gap is gradually decreasing, the effect of wages for income gap is increasing from eastern to western of China from 2002 to 2012.

Keywords: Urban residents; Employment; Wages; Income gap; Panel data model.

1. Introduction

In recent years, the labor force has been showing a trend which is flowing from the mid-west to the east on China labor market, the income between different regions is not balanced due to different status of employment, meanwhile differences of development space has been formed in all regions. China urban employment in industry, trade, employment stability and non-regular employment are important factors which affect wage differences, wage level has a direct impact on the income gap between the regions, employees also flow in

* This work is partially supported by Grant 12YJC790190 of Humanities and Social Sciences Fund of Ministry of Education, Grant 13AZD071 of SSFC, Grant 2013GXS2D020 of National Soft Science Research Project in China, Grant 1253G029 of Young Academic Backbone Support Program in Heilongjiang Province, Grant LBH-Q14094 of Postdoctoral Scientific Research Initiation, Grant GC13D101, QC07C117 of Soft Science Technology Foundation in Heilongjiang Province in China.

different regions and industry because of the level of wages, which will affect the total amount of employment officers in the region eventually. It is clear that employment, wages and income gap of urban residents will contact and influence each other. How much influence will bring due to income gap in urban employment exactly? It needs to be analyzed and verified by appropriate practical research.

There is a long history of research on employment and wages in the field of economics, and has formed their own different academic perspectives. In many economic indicators which affect employment, wage levels can be described as the most significant and the most direct indicator. Classical economics and neoclassical economics agreed that wages and employment negatively correlated. John Barton believes that the monetary value decline will lead to the growth rate in commodity prices higher than the rate of wage growth, falling in real wages increases profits and provides more job opportunities (John Barton, 1991). Keynesians took the opposite attitude that wages and employment is positively correlated.

Keynes argued that effective demand is a key factor to determine social employment, so employment will not increase because of the decline in wages, on the contrary, effective demand will increase because of higher wages and promote employment (Rong Mo, Jun Liao, 2010). Phillips select British unemployment and money wage rate from 1861 to 1957 as the research indicators, empirical analysis proves that unemployment and money wage rate of change has a reverse relationship (Phillips A W., 1958).

By contrast insufficient consumption theory, Bhaduri and Marglin summed up an analysis framework that higher real wages will have a positive impact on employment due to the effective demand (Bhaduri A, Marglin S, 1990). The new wage theory since the mid-20th century, such as the efficiency wage model (Shapiro C, Stiglitz J E, 1984), insider and outsider model (Lindbeck A, Snower D J, 1987) and the implicit contract model (Baily M N, 1974) believes that changes in real wages and employment is not relevant.

Compared with foreign countries, the research of the relationship between wages and employment in China started relatively late, and how to research is a single narrow,most of empirical studies based on the perspective of micro individuals. Only some of them research the relations of total employment and wages from the macro perspective. Jun-Qing Yang has proposed a model, rising wages and employment, in this way raising wages can not only makes workers feel wages rise, it is also stimulate the enthusiasm of workers, so as to improve the marginal productivity of labor, and increase the manufacturers for the production and operation of capital amount eventually, which is advantageous to solve the employment problems under the labor demand (Junqing Yang, 2005). Guang-Xin Wang and Xian-Guo Yao think that minimum wage will increase the total revenue of workers, which can improve the level of total social consumption, as the demand of companies for labor is caused by increasing the demand for enterprise products,expanding the scale of production, and it will improve the demand for social labor eventually (Guangxin Wang, Xianguo Yao, 2014). But Liu Junze thinks that the elasticity of China's employment has showing a steady rise trend gradually since 1996, which also objectively reflects the employment absorption capacity of the economic growth in China. The falling of the rate of employment growth caused by national economic growth is because that the statistical index selected is not accurate, so that there are no underlines the hidden unemployment in our country's economic system that has Chinese characteristic factors (Jun Liu, 2014).

At present, the most relevant research is also relied on theoretical analysis, quantitative analysis is not a general method, basically give priority to with time series or cross section data,combined with vector auto regressive model can not reflect the change rules absolutely, in order to most fully discuss the relationships of the wages of our country in different areas, obtain employment and the income gap, we put to use the panel data model in this article, combining with the relevant statistical data from 2002 to 2012, modeling to get the static and dynamic analyses of the east, west and central regions. In order to make the results we get more complete, efficient and meaningful practically, the training, education degree, age, gender and some other factors of individual workers in microcosmic variable which can be on the impact of wage differences have been ignored. And we get the dynamic relationships of national employment, wage and income gap from the macro perspective.

2. Panel Data Model of Employment, Wages and Income Gap of Urban Residents in China

2.1. *Variable selection and data description*

In this paper, the gini coefficient (Lngini), the number of the employees in the end of the year (Lncyr), the total wages (Lngzze) were selects as explanatory variables and explained variables. From the macroscopic level, wages, namely the total labor compensation, is the total labor income paid to the employees for their work as well as the most important part of the income of Chinese urban residents, moreover, it is the most important source of income gap. As a measure of income inequality, the gini coefficient demonstrates the changing of urban residents income gap in our country, so it was selected as an empirical analysis of explained variables.

This study collected the panel data of Chinese urban residents from 2002 to 2012 of 29 regions from *China Statistical Yearbook*, because the data of Gansu and Hunan provinces are not comprehensive, the empirical analysis was processed ignoring the two provinces. In order to ensure the authenticity of the panel data and reduce the hetero scedasticity, all the data are operated logarithm cally.

It needs to check whether every variable is stable or not before the estimation of the panel data. The unit root test can ensure the stability of the data, and prevents the spurious regression phenomenon, otherwise it will produce negative influence for the stability of the regression analysis. In the unit root test of the selected panel data, if the results show that the original sequence is unstable, the data should be operated difference processing. In this paper, the Eviews6.0 unit root test was used. It can be obtained from the results of unit root test, the stability of the three groups of time sequence, namely the Gini coefficient, the number of the employees in the end of the year and the total wages was significant, which illustrated that the three sets of time series can be regressed respectively.

2.2. *Determining the type of panel data model by F statistic test*

In order to verify the above three groups of time series is suitable for which type of panel data, no individual impact of constant coefficient model, Variable intercept model, Variable coefficient model. Based on the analysis of co-variance test, that is F statistic test, in order to establish the panel data model more better (Haiyan Cheng, 2012; Weiya Zhao, Shoukun Chen, Jin Sun, 2013).

Hypothesis 1,

$$H_1 : b_1 = b_2 = \cdots = b_N ,$$

Hypothesis 2,

$$H_2 : a_1 = a_2 = \cdots = a_N ; b_1 = b_2 = \cdots = b_N .$$

First of all, Under the hypothesis H_2, the test statistic F_2 obey F distribution that under the appropriate degrees of freedom, that is

$$F_2 = \frac{(S_3 - S_1)/[(N-1)(k+1)]}{S_1/[NT - N(k+1)]} \sim F[(N-1)(k+1), N(T-k-1)].$$

If the F_2 statistic value is less than the critical value at the specified level, that is $F_2 < F_\alpha$, then accept the assumption that H_2, we regard that sample data conform to the same intercept-constant coefficient model. On the contrary, if $F_2 > F_\alpha$, then continue to test hypotheses H_1.

Secondly, under the hypothesis H_1, the test statistic F_1 obey F distribution that under the appropriate degrees of freedom, that is

$$F_1 = \frac{(S_2 - S_1)/[(N-1)k]}{S_1/[NT - N(k+1)]} \sim F[(N-1)k, N(T-k-1)].$$

If the F_1 statistic value is less than the critical value at the specified level, that is $F_1 < F_\alpha$, then accept the assumption that H_1, we regard that sample data conform to variable intercept-constant coefficient model. On the contrary, if $F_1 > F_\alpha$, we regard that sample data conform to variable intercept-variable coefficient model (Yun Wang, Zhihe Long, Qingqing Chen, 2012).

In the paper, Eviews 6.0 can be used to analyze the three types of panel data, that is variable coefficient, variable intercept, hybrid model parameter estimation. Calculating variable intercept variable coefficient model of the sum of squared residuals, S_1, variable intercept constant coefficient model of the sum of squared residuals, S_2, the same intercept constant coefficient model of the sum of squared residuals, S_3.

To find out the S_1, S_2, S_3 in F statistic, we can get

$$F_1 = 3.421 > F_{0.01}(84261) = 1.652$$

$$F_2 = 9.915 > F_{0.01}(56319) = 1.434.$$

We can safely come to the conclusion that reject the hypothesis H_2, and reject the hypothesis H_1. The result showed that it's more reasonable that establish a variable coefficient panel data under the 1% significant level.

2.3. *Determining the type of panel data model by Hausman test*

In order to determine the panel data model is applied to the fixed effect model or random effect model, this paper chooses Hausman test to inspection method (Yun Wang, Zhihe Long, Qingqing Chen, 2012).

Hypothesis 1 presents individual effect uncorrelated with the explanatory variables, Individual random effect regression model.

Hypothesis 2 presents individual impact associated with the explanatory variables, Individual fixed effects regression model.

If we accept the hypothesis 1, the result showed that individual effect uncorrelated with the explanatory variables, we should chooses the individual random effect regression model. If we accept the hypothesis 2, the result showed that individual impact associated with the explanatory variables, we should chooses the individual random effect regression model.

In the paper, we use Eviews 6.0 to test Hausman. The resulting the value of the Houseman statistic is 5.9604, the corresponding probability is 0.0537, The test results shows that rejected the null hypothesis of random-effect model. Therefore, annual panel data of 29 regions from 2002 to 2012 in China for individual fixed effects regression model.

Based on the analysis, the result showed that it's more reasonable that establish a fixed effect panel data model under the 1% significant level.

Using Eviews 6.0 to estimate the fixed effects regression model with variable coefficients (Jingshui Sun, 2012), the results are shown in Table 1.

Table 1. Estimation results of fixed effect panel data model with variable coefficients.

Variable	Coefficient	Std. Error	t-Statistic	Prob.
C	-5.892473	1.127910	-5.224240	0.0000
LNCYR	0.649198	0.147954	4.387850	0.0014
LNGZZE	1.052876	0.117212	8.982639	0.0000

It can be obtained from Table 1, the elastic coefficient of the number of the employees in the end of the year and the total wages for the Gini coefficient were 0.649 and 0.649, respectively, which showed that if the number changes of the employees in the end of the year of Chinese urban residents was 1%, the Gini coefficient would change 0.649% at the same time, if the changes in total wages was 1%, the Gini coefficient would change 1.053% at the same time. In further discussion, under the condition of the constant of other factors, if the Gini coefficient adjust 1% expected, it will cause an adjustment of 1.54% (1% / 0.649%) of Chinese urban employment. The change of the Gini coefficient will cause an adjustment of total wages by 0.95% (1% / 1.053%) correspondingly. It shows that the effect of the increase or decrease of the Gini coefficient for the total wages is greater than that of the number of workers. If considering the combination of the two effects, it would be more complex.

The elastic coefficient of total wages is 1.053, which is a positive number greater than 1, which illustrated that the wages level plaid a positive role in promoting income gap. But it is difficult to consider the effects of wages level nationwide, it also involves many factors which cause wage imbalance, such as areas, engaged in the industry, jobs and so on, even in the same positions, the wages level may be different that caused by other objective factors.

2.4. *The panel data model of eastern, central and western regions*

This paper selects the eastern regions, Heilongjiang, Jilin, Liaoning, Hebei, Beijing, Tianjin, Shandong, Jiangsu, Zhejiang, Fujian, Guangdong, Hainan and other twelve provinces. The central regions include Henan, Shanxi, Hubei, Anhui, Jiangxi, Guangxi, Inner Mongolia and other seven provinces. The western regions include Shaanxi, Chongqing, Guizhou, Yunnan, Sichuan, Gansu, Ningxia, Qinghai, Xinjiang, Tibet and other ten provinces (Naiquan Liu, Xiujun Zheng, Yanli Jia, 2005). Using Eviewes 6.0 to estimate the panel data model of the eastern, central and western regions in China (Badi H. Baltagi, 2010), the results are shown in Table 2 and Table 3.

Table 2. Estimations of fixed effect panel data model with variable coefficients in different regions.

Regions	Variable	Coefficient	Std. Error	t-Statistic	Prob
	C	-6.913473	2.819711	-2.451672	0.0398
Eastern	LNCYR	1.013198	0.217255	4.662729	0.0002
	LNGZZE	1.232876	0.224451	5.488940	0.0400
	C	-3.806473	1.362418	-2.793561	0.0234
Central	LNCYR	0.390198	0.191621	2.035265	0.0444
	LNGZZE	0.566876	0.061925	9.140063	0.0000
	C	-45.06943	1.789093	-3.293288	0.0014
Western	LNCYR	-1.014198	0.010437	10.53662	0.0000
	LNGZZE	15.46576	0.248419	4.295151	0.0000

Table 3. Statistics of fixed effect panel data model with variable coefficients in different regions.

Regions	R-squared	Adjusted R-squared	Durbin-Watson stat	F-statistic
Eastern	0.973012	0.971134	2.054458	421.83502
Central	0.981034	0.980189	1.868422	567.47720
Western	0.903065	0.895146	1.997469	108.04420

Seen from Table 2 and Table 3, the parameter T statistics declined to the original hypothesis of coefficient to be 0 on the significant level of 1% , the value of R2 of Chinese eastern, central and western regions were 0.973012, 0.981034, 0.903065, which were greater than 0.9, it showed that the reason for the income gap being more than 90% of the three regions were the total employment and total wages.

To compare and analysis the estimated results of eastern, central and western regions of China, the coefficient of elasticity of the number of employees and total wages all are positive at the end of the year, it declares that the influence of these two indexes on the Gene coefficient is in the same direction. That is number of employees and total wages can reduce or enlarge with the Gene coefficient at the same time, combined with the actual situation of economic and social development, it shows that this conclusion has good accuracy and stability.

3. Conclusions

The paper has provided the dynamic analysis of the employment, wages and income gap of urban residents in our country which mainly explore the dynamic relationships between the number of employees, the total wages of workers and the Gini coefficient by constructing panel data model. Through empirical analysis of the influence mechanism and the effect of the dynamic relationships between the employment, wages and income gap, conclusions are derived as follows.

Firstly, the employment and wages of urban residents both have a positive effect on the income gap. What's more, the overall effect of wages to enlarge the income gap is greater than that of the employment. However, it is difficult to consider the effects of all levels of wages across the country which involves unbalanced factors that affect wage.

Secondly, the influence degree of urban employment and wages upon the income gap varies areas to areas. Our results suggest that the influence degree of the employment upon the income gap reduces whereas that of wages upon the income gap increases gradually from the east to the west section.

Thirdly, the influence degree upon the income gap and the Gini coefficient by urban employment and wages is relatively balanced in the east and the middle of our country. However, the intensity and the difference are obviously increased and the effect is not consistent in the western region.

Therefore, feasible and personalized solutions appropriated to different regional economic development levels which aimed to improve the income gap can be laid down on the basis of our conclusions mentioned above. In view of the east and the middle of our country where economic development level is higher, promoting employment can be adopted to narrow the income gap; According to the low level of economic development in the western region, measures can be done by adjusting the level of wages in order to balance the income of different levels and achieve the aim of improving the income gap. Due to the influence factors such as economic development, policies and regulations, labor supply and demand, social security and so on differs between regions, the research about the relationships of the employment, wages and income gap of urban residents in China is much more complex in practice than that is mentioned in this paper.

References

1. John Barton, Observations on the circumstances which influence the condition of the labor, China Commercial Press (Beijing, 1991), pp. 52.
2. Rong Mo, Jun Liao, How to see the impact of labor prices on Employment, China Development Observation, Vol. 12 (2010), pp. 19-21.
3. Phillips A W, The Relation Between Unemployment and the Rate of Change of Money Wage Rates in the United Kingdom, 1861-1957, Blackwell Publishing Ltd, Vol. 25 (1958), pp. 283-299.
4. Bhaduri A, Marglin S, Unemployment and the Real Wage: The Economic Basis for Contesting Political Ideologies, Cambridge Journal of Economics, Vol. 14 (1990), pp. 375-393.
5. Shapiro C, Stiglitz J E, Equilibrium Unempioyment as a Worker Discipline Device, American Economic Review, Vol. 74 (1984), pp. 433-444.
6. Lindbeck A, Snower D J, Cooperation, harassment and involuntary unemployment: an insider-outsider approach, C.E.P.R. (1987).
7. Baily M N, Wages and employment under uncertain demand, Review of Economic Studies, Vol. 41 (1974), pp. 37-50.
8. Junqing Yang, On the rise in wages of workers' motivation and Employment, a model of the employment of Keynes's Theory, Chinese Journal of Population Science, Vol. 5 (2005), pp. 54-57.
9. Guangxin Wang, Xianguo Yao, Minimum wage and employment based on dynamic equilibrium, Journal of Zhejiang University of Science and Technology, Vol. 4 (2014), pp. 258-264.
10. Jun Liu, Price level, foreign capital utilization and employment situation: the interaction mechanism and the performance of China, Shanghai Economic Review, Vol. 9 (2014), pp. 57-68.
11. Haiyan Cheng, Research on testing methods in panel data models, Economics Press (Beijing, 2012).
12. Weiya Zhao, Shoukun Chen, Jin Sun, Type identification test of panel data model.Statistics and Decision, Vol. 18 (2013), pp. 4-7.
13. Yun Wang, Zhihe Long, Qingqing Chen, Fixed effect tests of spatial panel model, Journal of Business Economics, Vol. 12 (2012), pp. 85-92.
14. Jingshui Sun, Guide to the application of EViews in Econometrics, Tsinghua university press (Beijing, 2012).
15. Naiquan Liu, Xiujun Zheng, Yanli Jia, An analysis of the regional policy focus shift and its comprehensive effects in China, Journal of Finance and Economics, Vol. 1 (2005), pp. 25-37.
16. Badi H. Baltagi, Econometric analysis of panel data, Machinery Industry Press (Beijing, 2010).

Sustainable development approach of Chinese home inspection industry based on SWOT analysis

Y. Xiao
Chongqing Technology and Business University, Chongqing, China
E-mail: 237197830@qq.com

On the base of learning from foreign experience, this paper aims to find out a sustainable development path for Chinese home inspection industry. In order to gain professional understanding of home inspector and home inspection industry, the author uses a questionnaire survey method and SWOT analysis tool. This paper presents the results of original empirical research utilizing data drawn from interviews and survey questionnaires. The results show that there are significant differences between Chinese and foreign home inspection industry, despite the huge demand of home inspection, Chinese home inspectors still have a long way to go. The sustainable development approaches presented in the paper contribute to the application of home inspection enterprises in China.

Keywords: Real estate market; Home inspection; Professional association.

1. Introduction

With the growing up of China's middle class, there is rigid demand in real estate investment. However, since the majority of ordinary investors are lack of professional knowledge, and due to information asymmetry between investors and developers, it is reported that there are a lot of disputes and cases about houses, especially the second-hand houses. These disputes focus primarily on the quality problem of the houses or the products don't conform to the contracts. Taking Zhejiang province as an example, from 2003 to 2012, the courts had received 118,761 cases of real estate disputes, over 54.09% of the disputes were about unqualified houses. In 2006, according to Beijing Consumer Association's consumers' satisfaction survey, 23.1% of consumers encounter housing quality problem.

Thus, domestic buyers need professionals' professional suggestions and based on this realistic background, home inspectors and home inspection business began to emerge in China. However, due to the uneven qualifications of practitioners in home inspection business, it is urgent to call for experienced and well-trained inspectors, so that the rights and interests of both developers and property buyers can be maintained.

2. The Origin and Development of North America and Britain's Home Inspection Industry

2.1. *The origin of home inspection business*

Home inspection is limited, non-invasive examination of the condition of a home, often in connection with the sale of that home. Home inspections are usually conducted by home inspectors who have the training and certifications. Home inspectors are professionals who use their knowledge to inspect house and deliver a written report to the clients, which describe the condition of the home at the time of inspection. The first known home inspection company, Home Equity Loss Protection Services dba/H.E.L.P.S.(Glen Ellyn, IL), was originally founded in America by Christopher P. Nolan and Loyola Professor.

With the growing of home inspection industry, home inspection industry associations and special license for inspectors were introduced in the United States. It did not take a long time when home inspection business spread in Canada. In 1992, a professional home inspection association was founded in Canada. In Britain, it was originally anticipated in 2004 that the Home Condition Reports would be included in the Home Information Pack during house transactions. Up to now, home inspection industry in a few developed countries has developed steadily; therefore this article's research is mainly based on the experiences of North America and Britain.

2.2. *The main operations of Home inspection business in North America and Britain*

2.2.1. *Home inspection business in North America*

Home inspection profession originated in the 50s of last century in the United States, and then spread to Canada. In Canada and the United States, a contract of buying a house may include a contingency that the contract is not valid until the buyer, through a home inspector or other agents, has had an opportunity to verify the condition of the property. Home inspectors also provide services such as inspections for wood destroying insects, radon testing, septic tank inspections, water quality, mold (or excessive moisture which may lead to mold), and private well inspections are sometimes a part of home inspector's services if duly qualified.

In many states and provinces, home inspectors are required to be licensed, but in some states the profession is not regulated. In America, there are 40 states that require inspectors to be licensed. Other states like Maine and Michigan have no current licensing requirement on home inspectors. Typical requirements for obtaining a license are the completion of an approved training course and/or a successful examination by the state's licensing board. Several states and provinces also require inspectors to periodically obtain continuing education credits in order to renew their licenses. Regulations and certification requirements about inspectors in some states in America are shown in Table 1 and Table 2.

Table 1. Regulations of inspectors in some states in America.

State	Texas	South Carolina	Oregon	Washington
Related regulations	Texas Licensure	South Carolina Licensure	Oregon Certification	Washington State Licensure
Year of enacting	1991	1996	1997	2009
Regulation agency	The Texas Real Estate Commission	The Residential Builder Commission	The Construction Contractors Board	The Department of Licensing

Table 2. Certification/ license requirements in some states in America.

Texas: The following requirements apply to professional inspector: 1) 18 years of age or older 2) U.S. citizen or legal alien 3) Have served as a Real Estate Inspector for one year and 175 home inspections under direct supervision of a Professional Inspector 4) Successful completion of 30 hours of approved education 5) Successful passage of an approved examination *No insurance requirements apply
South Carolina: 1) Provide evidence that the home inspector is a certified member of a home inspector association recognized by the commission, which requires successful completion of a written examination to become certified 2) Has home inspection experience of no less than one year and 50 inspections under direct supervision * No insurance requirements apply
Oregon: 1) Successful completion of 20 eligibility points 2) Successful passage of an approved examination 3) General liability insurance – $300,000 or $500,000
Washington: 1) Successful completion of high school or equivalent 2) Successful completion of 120 hours of approved education; 3) Successfully completion of up to 40 hours of field training supervised by a licensed home inspector 4) Successful passage of the written exam approved by the state

The practical standards for home inspectors are those enacted by professional associations, such as the International Association of Certified Home Inspectors (INACHI) in the United States, the American Society of Home Inspectors (ASHI), and the National Association of Home Inspectors (NAHI); and in Canada, the Canadian Association of Home and Property Inspectors (CAHPI) and the Professional Home & Property Inspectors of Canada (PHPIC).

2.2.2. *Home inspection business in Britain*

Home inspectors in the United Kingdom (mainly in England and Wales) were certified to carry out the Home Condition Reports that became part of the Home Information Packs on 1 June 2007. Home inspectors were required to complete the ABBE (Awarding Body for the Built Environment) Diploma in Home Inspection to show they meet the standards. The government had suggested that between 7,500 and 8,000 qualified and licensed home inspectors would be needed to meet the annual demand of nearly 2,000,000 home information packs. In the event, many more than this entered training resulting in a massive over supply of potential inspectors. However, with the cancellation of Home Information Packs by the coalition Government in 2010, the role of Home Inspector became permanently redundant.

3. The Key Points of International Home Inspection Industry

Overall, the main features of international home inspection industry can be summarized as follows:

3.1. *Highly institutionalized*

The home inspection industry in most western countries is regulated in some form based on laws and regulations. Professional associations also play a critical role in maintaining high standards of professional practice by setting up entry qualification requirements and organizing continuing education.

3.2. *Highly flexibility*

Although home inspection industry is regulated by different professional associations, each professional association is relatively independent and it can make its own regulation depending on its own situation. On the other hand, professional associations can also be highly collaborative when they face with common problem that will influence the whole industry.

3.3. *Professional, precise, independent and impartial*

There are specific laws and regulations about home inspection. For example, housing inspection tools and methods are well defined in the regulations. It highly values the impartiality of home inspection reports because those reports may cause potential conflict of interest among various parties in trade.

4. SWOT Analysis of Chinese Home Inspection Industry

Home inspection industry appeared in China as a brand-new profession in 2006, and then it spread quickly from top cities like Shanghai and Beijing to other big cities. As a branch of real estate service business, home inspection industry is highly related with the development of real estate market in China. The increasing demand in housing investment as well as economic boom and political stability foster the demand of home inspection.

This research is based on a questionnaire survey conducted from March 2015 to July 2015. Three hundred and fifty questionnaires were sent out and the survey received 300 responses, including 200 Internet questionnaires (67%) and 100 paper questionnaires (33%). The web questionnaires were published on a famous survey website (http://www.sojump.com) and the respondents were randomly chosen. The paper questionnaires were handed out to citizens in big cities such as Beijing, Shanghai and Guangzhou. After the questionnaires were collected, statistical analysis was made by using MS EXCEL. The feedback captured some significant views, reflecting the current status quo about home inspection industry as well as providing a meaningful reference for inspectors. The results of this survey are discussed below.

4.1. *The strengths of home inspection industry*

4.4.1. *Market advantages*

Huge market demand is the foundation for a profession to survive. In China's current real estate transactions, there is professional house inspection during the delivery of new houses or second handed houses. The quality of the houses is mainly depended on good faith ethics of real estate developers, and it will be house buyers themselves who conduct home inspection and determine the condition of a structure.

According to the survey, about 71% of 300 respondents do not think they can do home inspection themselves and would like to hire a home inspector in time of needing professional inspection. For the respondents who believe they can conduct home inspection themselves, there is still a percentage of 28.77%

who have the will to hire home inspectors (see Table 3). This result shows that since house buyers are not professional inspectors, there is huge demand for reliable, impartial and professional home inspectors to do inspection for them. The inspection report may provide a kind of proof in case house buyers claim losses caused by inferior quality of the houses or future disputes related with house quality. However, home inspectors are in short supply in China.

Table 3. Survey on the demand of home inspectors.

	Capable of home inspection	Not capable of home inspection
Willing to hire a home inspector when needed	28.77%	71.23%
Not willing to hire a home inspector when needed	33.33%	66.67%

The survey also investigated the types of houses people want to inspect (see Table 4). According to the survey data, it can be concluded that most people (87%) need home inspection services when they buy residential properties, followed by commercial properties (59%). People have the least willing to inspect houses they rent (only 29%). It is predicted that in the future, with fast-paced urbanization, there will be increasing demand for commercial properties, especially in regional center areas. Also, due to the diversity of employment patterns and population shift, there will be increasing demand for renting houses. It is positive that home inspectors have a lot of opportunities to explore residential and commercial property market and broaden their horizons.

Table 4. Survey on the demand of home inspectors.

Types of houses need inspection	% of respondents (N=300)
Residential property (buying)	87%
Commercial property (buying)	9%
Residential property (renting)	29%
Commercial property (renting)	37%

4.4.2. *Talents advantages*

Concerning whether should home inspectors are required to be licensed, the result of the survey shows that about 52% of the respondents agree that inspectors should be certificated and licensed. About 44% of the respondents consider that other professionals such as real estate agents, real estate appraisers or building engineers are qualified to conduct home inspections. The last 4% of the respondents believe that inspectors do not need certificates, as long as they have the capability to conduct home inspection (see Table 5). The present situation is that there is huge supply of human resources of real estate agents, appraisers and engineers. For example, there are over 47,000 registered real estate appraisers in China. All of these talents are potential talent pool for home inspection industry and they are solid base for the future development of this industry.

Table 5. Survey on the qualification of inspectors.

Whether should home inspectors are required to be licensed or qualified?	% of respondents (N=300)
They should get national inspector certification.	52%
Other professionals such as real estate agents, real estate appraisers or building engineers are qualified to conduct home inspections too.	44%
They do not need certificates, as long as they have the capability to conduct home inspection.	4%

4.2. *The weaknesses of home inspection industry*

4.2.1. *Lack of appropriate law regulations*

Since home inspection is an emerging industry, laws and regulations about this industry is almost a blank. First of all, there is not a clear definition of home in national or local laws and regulations. Secondly, there are some other questions unsolved. Which legal provisions should be used to regulate the behavior of inspectors? Is the home inspection report legally effective and how can people use the report? Can house sellers say no to home inspection reports? All of these questions need legal answers.

4.2.2. *Unorganized management of inspectors and the industry*

"Home inspection division management association", "China home inspection division association", "home inspection association of China" and other home inspection associations can be searched on the Internet, but none of them are qualified to provide inspector training and continuing education, or issue qualification certificates. Some so-called associations are even considered as fraud. Some people who engaged in home inspection business do not have necessary knowledge about inspection, but their misbehavior will not be punished because there is no legal binding.

4.2.3. *Non-uniform standard and fees for home inspection*

After investigating several home inspection companies in Wuhan, Hangzhou, Guangzhou and Shanghai, it is found out that these companies do not have a uniform standard and methods of home inspection. Some companies even do not know how to write a complete housing inspection report. The fees charged on home inspection vary from company to company and city to city. Fees are charged by square meter or charged by house prices, which makes consumers confused.

4.2.4. *Low business reputation and social recognition*

The result of the survey shows that ordinary people are not familiar with home inspection profession or inspectors. About 35% of the respondents never heard of inspectors. 53% of the respondents know a little of inspectors. Only about 3% of the respondents claimed that they know about inspectors very well but never contact with them. The last 9% have some kind of contact with inspectors (see Table 6).

Table 6. Survey on the recognition of inspectors.

The degree of recognition about inspectors	% of respondents (N=300)
Never heard	35%
Know a little	53%
Know very well but never contact	3%
Have some kind of contact	9%

4.3. *The opportunities of home inspection industry*

4.3.1. *Opportunity brought by economic growth*

According to a report issued by Credit Suisse in October 2014, the population of Chinese middle class has accounted for 1/3 of the world. The middle class in china has become the main force in real estate investment market. With the quick development of urbanization and industrialization, there will be a lot of opportunities for home inspection industry.

4.3.2. *Opportunity brought by innovatory ideas and concepts of new generation*

Nowadays, generation Y and Z have grown up and become the base of this society. With the trend of small-scale family, young people would like to leave their parents and look out for job opportunities in some other places, which makes young people the main force of leasing, buying and selling houses. It is no doubt this generation will become the main clients of home inspection industry in the next 10 years.

4.4. *The threats of home inspection industry*

4.4.1. *Malignant competition inside home inspection industry*

Due to unorganized regulations of home inspection industry, it is common that there is malignant competition among different home inspection companies. Low-price competition and poor service are often reported publicly.

4.4.2. *Threats from new policy*

To ensure the quality of houses, in some provinces such as Jiangsu, the government demands that every single new house should be inspected either by builders or the certified third party. The inspection report is legally effective to both the sellers and buyers. If the builders accredit the third-party inspectors to do this job, the third party should be qualified and be able to conduct the inspection work. However, people tend to trust the "official institutions"—builders instead of the third party, which makes it very difficult for inspectors to compete with builders.

5. Sustainable Development Approach for Chinese Home Inspection Industry

5.1. *The positioning and expansion of the home inspection market*

Home inspection business is a branch of real estate services. It seems that home inspection has similar business part with real estate appraisal and real estate consulting, but home inspection has irreplaceable uniqueness. Home inspection companies can provide services including homebuyers inspection, home sellers inspection, foreclosure inspection, pre-delivery inspection, structural inspection, plumbing inspection and so on. The purpose and positioning of home inspection business could different. Either the clients who need inspection service or the inspectors who provide the service should bear it in mind. Home inspection companies should position their business areas based on their strengths.

5.2. *Regulate home inspection industry in all areas*

5.2.1. *Laws and regulations*

With the increasing demand of home inspection, it is urgent that the government should formulate a series of laws and regulations to make this industry well organized and recognized by the public. Firstly, the threshold requirement for approving inspectors should include a successful completion of approved education and a successful passage of an approved national uniform examination. Secondly, approved inspectors should past an annual test and attend continuing education organized by professional associations, so as to maintain high standards of professional practice. Thirdly, the legal validity of home inspection reports should be made clear and defined in laws. Lastly but not least, regulations of home inspection process and criteria should be published.

5.2.2. *Unified standard and fees for home inspection*

According to the survey, most respondents tend to buy home inspection services with lower prices. In addition, respondents who come from the more developed regions tend to expect high fees for home inspection (see Table 7).

Table 7. Expected fees in different regions.

Regions	Charge as 3-5 Yuan/m^2	Charge as 5-7 Yuan/m^2	Charge as 7-9 Yuan/m^2	Charge as more than 9 Yuan/m^2
Municipalities and richer coastal areas	37.50%	29.17%	29.17%	4.16%
Prefecture-level cities	30.30%	27.27%	25.21%	17.22%
County-level cities	13.67%	33.33%	53.00%	0.00%

5.3. *Clean up professional associations*

Different kinds of home inspection associations arouse much confusion among consumers and it is not helpful for the development of this industry, therefore, the authorities should take measures to clean up the industry by banning unqualified associations. The commissions of professional associations should include inspectors' appraisal, supervision and continuing education.

5.4. *Collaborative development between home inspection industry and other industries*

Collaboration between home inspection industry and other industry such as real estate appraisal and consulting is helpful for home inspection industry to broad its business area. For example, building inspection reports can be used as a reference for real estate appraisal or real estate consulting, or can be used by real estate agencies in the process of house leasing or transactions. In this way, it will create a harmony environment for home inspection industry and encourage cooperation between different companies.

6. Conclusions and Suggestions

The birth of home inspection industry is an inevitable outcome during the developing process of real estate market in China. There are more and more real estate disputes accompanied with diversified property investment demand. Home inspection provides a way to solve the dispute, maintain the rights and interests of all parties, and improve the transparency of the market.

However, home inspection industry has just started, it faces with a lot of challenges such as lack of appropriate law regulations, unorganized management of inspectors and the industry, non-uniform standard and fees for home inspection and low business reputation and social recognition. Based on the experience of overseas home inspection industry, Chinese home inspection industry will find a sustainable development way after a series of reforms.

References

1. Chang H C, Ho C H. 2011. A note on solving the EOQ model with imperfect quality subject to in-house inspection [J]. IMA Journal of Management Mathematics, 22(3): 301-306.
2. Konstantaras I, Goyal S K. 2007. Papachristos S. Economic ordering policy for an item with imperfect quality subject to the in-house inspection [J]. International Journal of Systems Science, 38(6): 473-482.
3. Badali R, Dennis G, Meyer B. 2002. Appraisal and home inspection process: U.S. Patent Application 10/140,008[P].
4. Skidmore B, Sumner D. 2002. Integrated home inspection and home warranty system: U.S. Patent Application 10/202,596[P].
5. O'Malley K. 1993. House inspection in the USA and Canada [J]. Structural Survey, 11(3): 252-257.
6. Excerpt from Maryland home inspection statute". Michie.lexisnexis.com. Retrieved 2012.
7. “8 Common Plumbing Questions". API Plumbing. Retrieved 5 August 2014.
8. "Excerpt from Maryland home inspection statute". Michie.lexisnexis.com. Retrieved 2012-05-29.
9. "Mold Removal and Real Estate and Property Managers Responsibility", Environmental Protective Solutions, Retrieved 6 July 2014.

Expanding research of moment distribution method in solving statically indeterminate structure*

Y.Q. Jiang
College of Civil and Environmental Engineering, Anhui Xinhua University, Hefei, Anhui, China, 230088
E-mail: 59198533@qq.com

Z.J. Geng
Anhui Transport Consulting & Design Institute Co., Ltd., Hefei, Anhui, China, 230088
E-mail: 441854242@qq.com

In order to simplify the calculation of statically indeterminate structure which has middle hinge, spring support or the rod has EI=∞ part, this paper summed up the shape constant table and the load constant table which will be used in the three situation when using the moment distribution method for solving. It is also provide materials for cultivating students' innovation ability and the implementation of innovative education. College students and teachers can draw lessons from the ideas in this paper, and also can be used for further development.

Keywords: Structural mechanics, Moment distribution method, Joint line displacement.

1. Introduction

At the beginning of the twentieth century, the second industrial revolution in Europe promoted the extensive use of frame structures in civil engineering structures, in order to solve the internal forces of the frame structures, displacement method [1] [2] is proposed by mechanical workers for solving statically indeterminate structure to overcome the shortcomings, such as too many unknown quantities and the heavy workload of force method for solving the frame structures. But the displacement method results in a set of simultaneous equations of unknown displacements. The number of simultaneous equations will be equal to the number of unknowns to be evaluated. Thus one needs to solve these simultaneous equations to obtain displacements and beam end moments. The calculation is too complex without computer.

* This work is supported by grant 20101149 of Quality engineering in Anhui Province. Work partially supported by grant KJ2015A336 of Anhui University Natural Science Foundation.

The moment-distribution method is proposed by Hardy Cross in 1932, it is used for solving continuous beams and No-sway frames, actually solves these equations by the method of successive approximations [3]. In this method, the results may be obtained to any desired degree of accuracy. Until recently, the moment-distribution method is very popular among engineers. It is very simple and being used even today for preliminary analysis of small structures. In 1950s, No-shear Distribution method is proposed by qian lingxi which is used to analyse Specific Frames. In this paper, three kinds of other special frame has been analysed by the moment distribution method.

2. Derive Rotational Stiffness S, Carry-Over Factor C, Fixed-End Moments M^F

According to the force method [3] [4], the rotational stiffness S, carry-over factor C and fixed-end moments M^F of three kinds of structures can be calculated. For the convenience of readers, the rotational stiffness S and carry-over factor C are given in Table 1 called shape constant, the fixed-end moments M^F for various load cases are given in Table 2 called load constant.

Table 1. Shape constant.

Serial No	Calculation diagram	Rotational stiffness S_{AB}	Carry-over factor C_{AB}
1	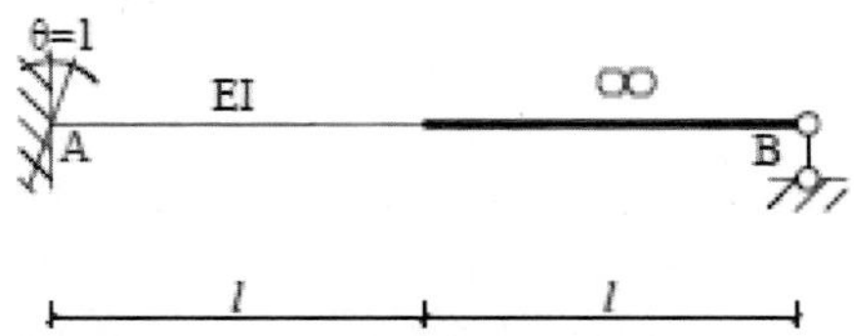	$\frac{12EI}{7l}$	0
2	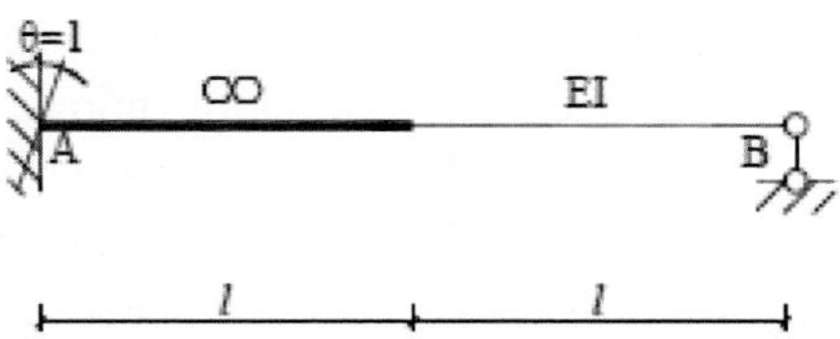	$\frac{12EI}{l}$	0
3	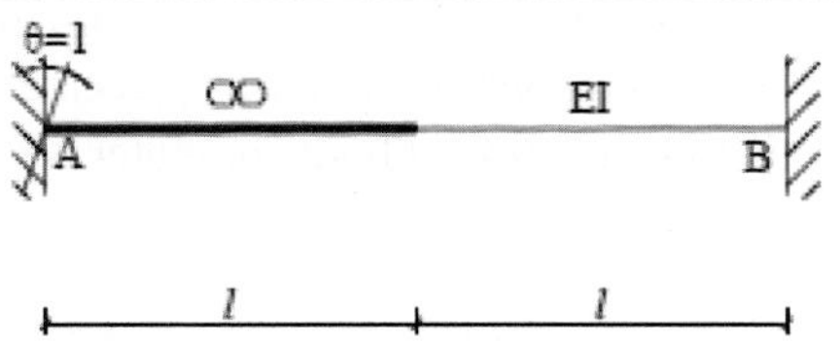	$\frac{28EI}{l}$	$\frac{2}{7}$

4	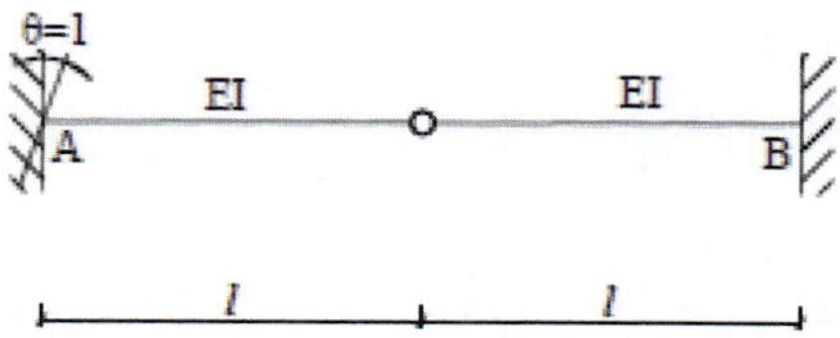	$\frac{3EI}{2l}$	1
5	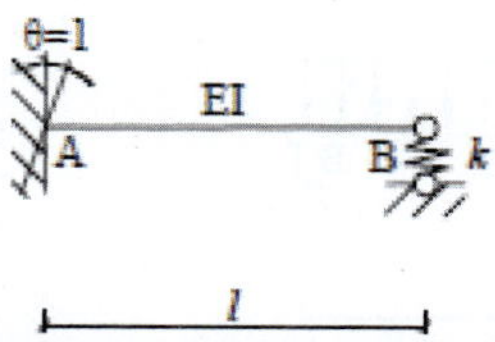	$\frac{3EIkl^2}{kl^3+3EI}$	0

Table 2. Load constant.

Serial No	Calculation diagram	Fixed-end moment (clockwise moments are positive) M^F_{AB}	M^F_{BA}
1	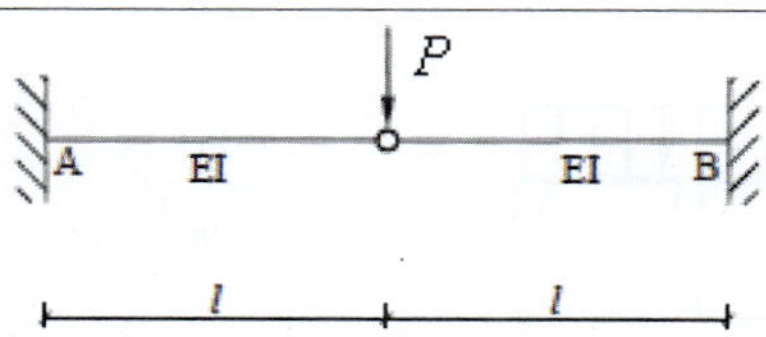	$-\frac{1}{2}Pl$	$\frac{1}{2}Pl$
2	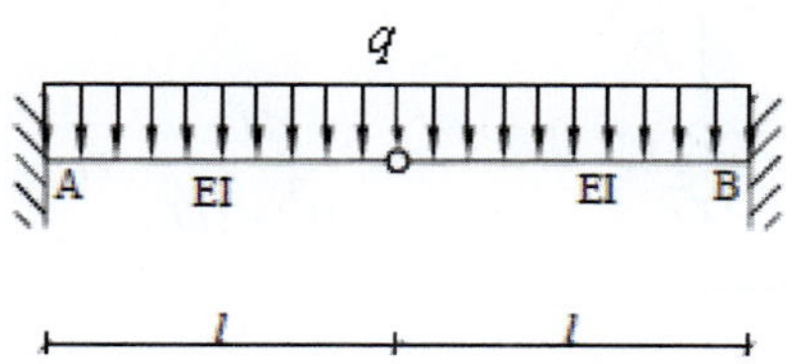	$-\frac{1}{2}ql^2$	$\frac{1}{2}ql^2$
3	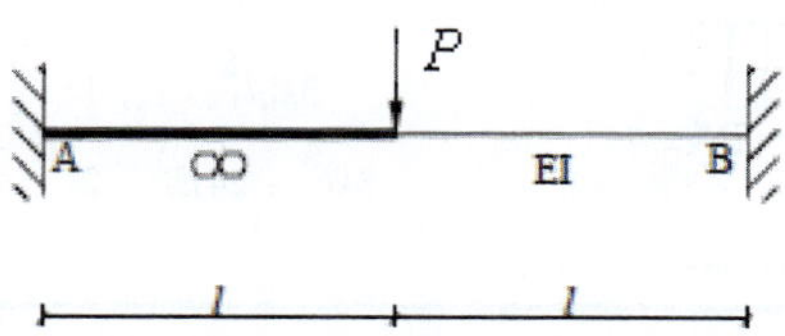	$-Pl$	0
4	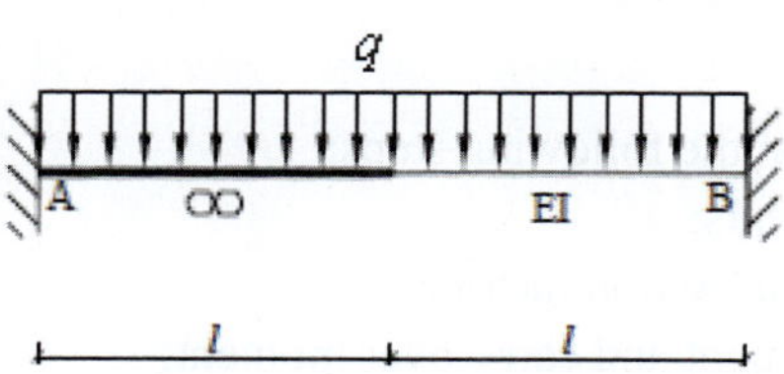	$-\frac{13}{12}ql^2$	$\frac{1}{12}ql^2$

5	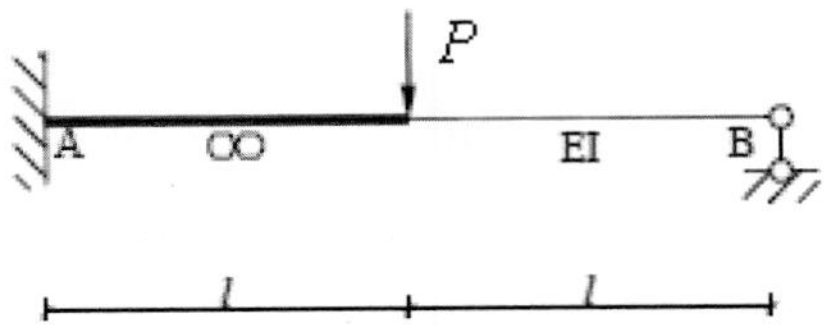	$-Pl$	0
6	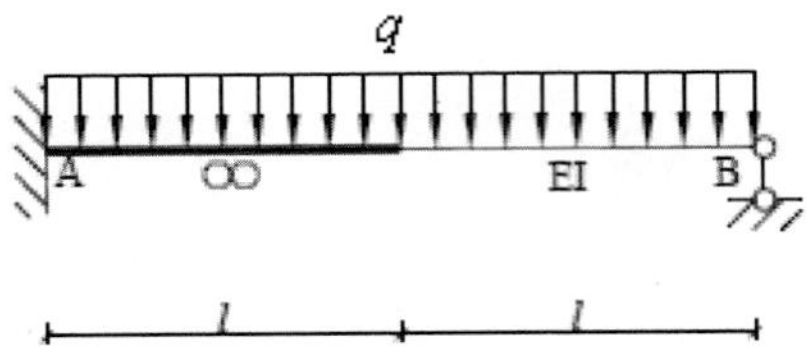	$-\frac{5}{4}ql^2$	0
7	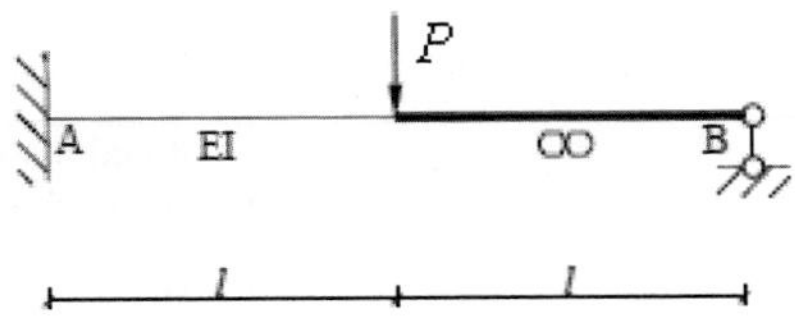	$-\frac{2}{7}Pl$	0
8	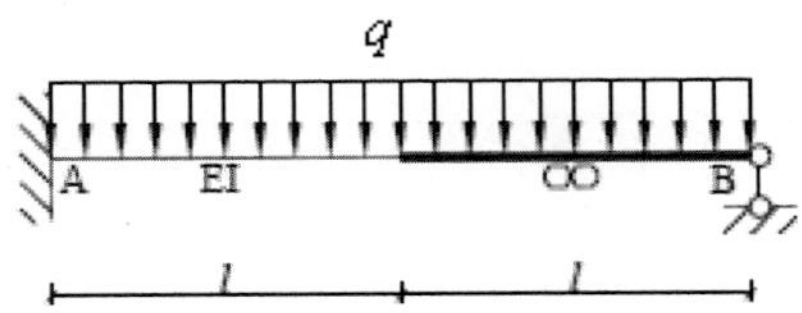	$-\frac{11}{28}ql^2$	0
9	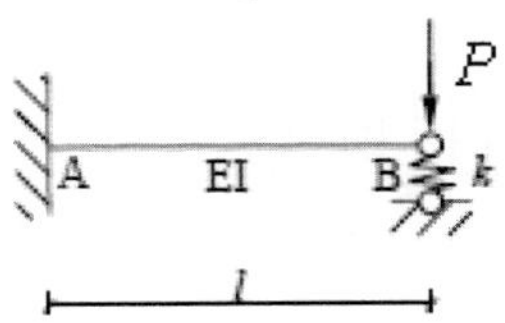	$\frac{kPl^4}{kl^3+3EI}-Pl$	0
10	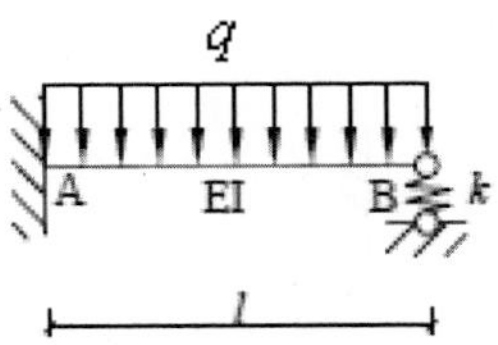	$\frac{3kql^5}{8kl^3+24EI}-\frac{1}{2}ql^2$	0

3. Examples

All the examples can be solved in the following steps:
Step 1: calculate fixed end moments;
Step 2: calculate stiffness and distribution factors;
Step 3: compute distribution moment and carry-over moment;
Step 4: calculate Final Moments.
Example 1: Analyse the rigid frame has middle hinge shown in Fig. 1(a).

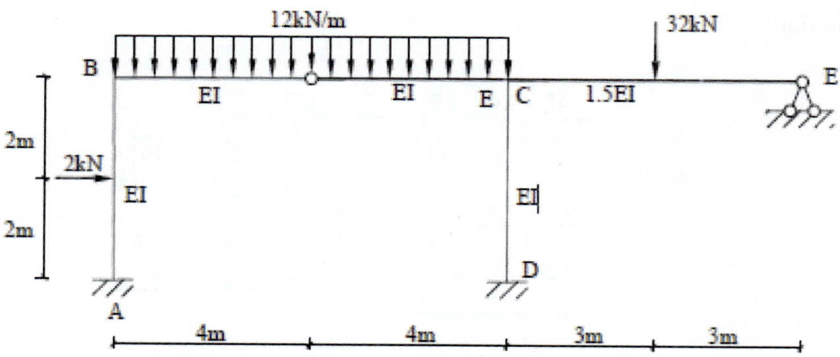

Figure 1(a).

Solution:

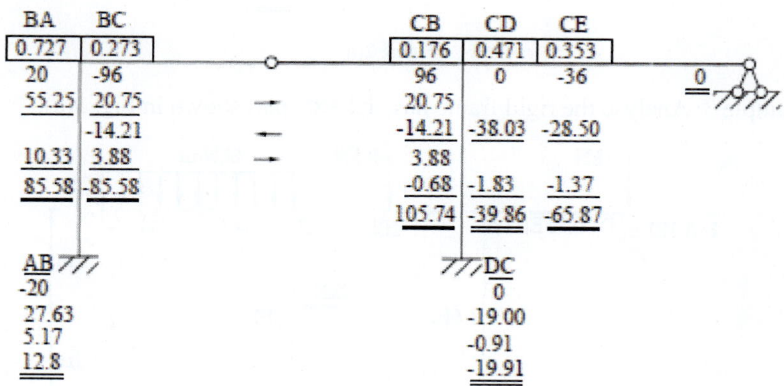

Figure 1(b).

Example 2: Analyse the rigid frame has spring support and middle hinge shown in Fig. 2(a).

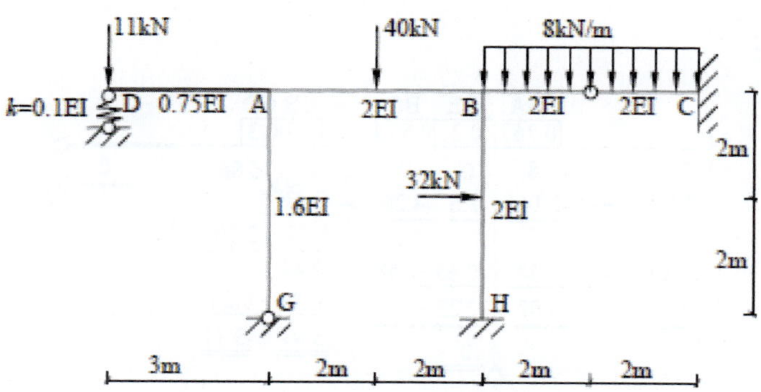

Figure 2(a).

Solution:

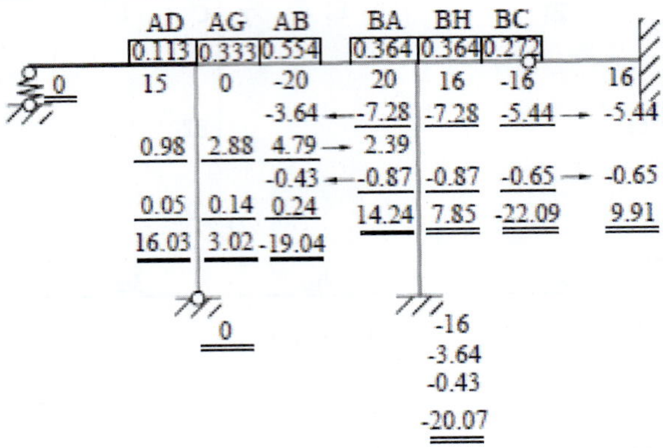

Figure 2(b).

Example 3: Analyse the rigid frame has EI=∞ part shown in Fig. 3(a).

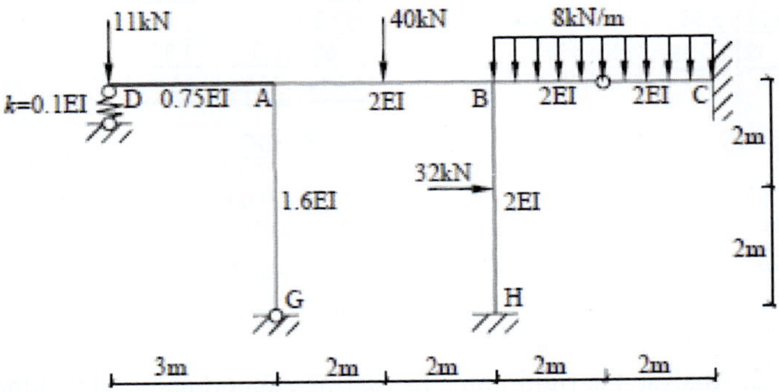

Figure 3(a).

Solution:

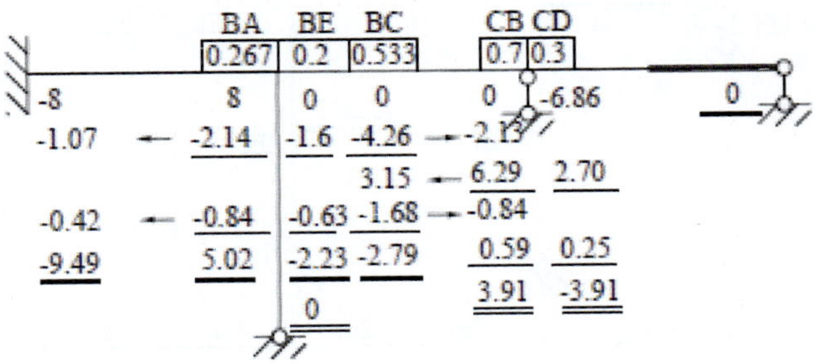

Figure 3(b).

4. Conclusion

Compared with force method and displacement method, the statically indeterminate structure which has middle hinge, spring support or the rod has $EI=\infty$ part could be solved very easily using the above method. There are many structure calculation methods, and which one will be used to solve the problem always depends on the structural characteristic. Along with the advent of high-level, ultra-high-level and a variety of complex structures, the computer has already been widely used in structural engineering, which reduced the importance of the simplified structure algorithm, but it should be noted that, in the process of personnel training, educators can not only just focus on the use of procedures, but also to carry out the research of calculation method, and to strengthen the basic theory of education and learning, and to play a good role in developing students' ability to innovate and in implementing innovative education.

References

1. Bao Shi Hua, Gong Yao Qing. Structural Mechanics [M]. Wuhan: Wuhan University of Technology Press, 2006:370-374.
2. Junarkar S B, Shah H J. Mechanics of Structures: II [M]. Anand: Charotar Publishing House, 1999:307-312.
3. Samuelsson A, Zienkiewicz O C. History of the Stiffness Method [J]. International journal for Numerical Methods in Engineering, 2006, 67:149-157.
4. Long Yu Qiu, Bao Shi Hua. Mechanics of Structures: I [M]. Beijing: Higher Education Press, 2006: 209-324.

Study on the popular science tourism comprehensive evaluation index system of the Geopark*

W. Luo
School of Economics and Management, Wuhan Polytechnic University, Wuhan, 430023, China
E-mail: luowei0723@163.com
www.whpu.edu.cn

L. Huang and Z.W. Yan
School of Economics and Management, China University of Geosciences, Wuhan, 430074, China
E-mail: 3757328@qq.com

One of the most important goal of establishing Geoparks is popularizing the knowledge of Geo-sciences. China's Geoparks construction started early and developed quickly. The number of Geoparks in China, including international and national ranked highly in the world. However, there still existed some problems such as "thinking highly of declaration while neglecting construction", which may restrict the development of popular science tourism. Based on AHP and experts' interview, this study established the comprehensive evaluation system of science tourism in Geoparks, including 8 evaluation index and 20 evaluation factors, and all factors had specific weight and non dimensional treatment. By using this system, not only the government can check and guide the development of Geoparks, the Geopark itself also can carry out the construction of science tourism to promote the sustainable development of Geoparks construction.

Keywords: Geoparks; Popular science tourism; Comprehensive evaluation index system.

1. Introduction

Popular science tourism is a new form of tourism, which is a combination of knowledge, education, amusement and entertainment. It is the product of the development of science and technology at a certain stage, and the performance of a high level tourism industry development. Geopark which possesses special geological significance, rare natural attribute and high aesthetic value is a unique natural areas with certain scale and distribution of geological vestiges landscape as the main body, including other natural landscape and humanities landscape [1].

* This work is supported by the Scientific Research Plan Key Project Foundation (No. 2013d08) and the Talent Introduction Research Start-up Project Foundation (No. 2014rs05) of Wuhan Polytechnic University, China.

Therefore, the Geopark is an important position for the development of popular science tourism and an important carrier to improve the scientific and technological content of tourism resources. Until now, the World Geopark Network (GGN) supported by UNESCO has a total number of 111 members, distributed in 32 countries and regions around the world, including our country 31 which has the most Geoparks in the world. China attached great importance to the construction and development of Geopark. Since 2001, the Ministry of Land and Resources has approved the seven batches, a total 240 National Geopark construction which played an very important positive role in popularizing science and promoting local economic development.

In the development of China's Geopark, there are also many problems, such as "weight declaration and income while light protection and popularization" and so on. In January 2013, UNESCO gave yellow card warning to three International Geopark of China: Zhangjiajie in Hunan, Lushan in Jiangxi and Wudalianchi in Heilongjiang, the reason is the lack of popularizing geosciences to public [3]. This event alarmed the development and construction of Geopark. During the 13 years from the 16th National People's Congress of the Party, China's tourism industry has been developed very fast, the national per travel rate raised from less than one time to more than twice and the era of mass tourism has arrived. The 18th National People's Congress of the Party has established the scientific concept of development as the guiding ideology, tourism industry must firmly grasp this theme and transform develop form to maintain a sustained, stable and rapid development. As a complex of geological tourism resources with high level of technology and science, "popular science tourism" is a breakthrough in geopark tourism development. The geoparks in China has entered a new period, which transfer into scientific direction from quantity growth.

2. Domestic Researches of Popular Science Tourism in Geopark

Geopark of China developed rapidly since its construction. Experts and scholars studied it from resources evaluation and tourism development to tourist line design and customer satisfaction [2–4]. Along with the development of technology and the improvement of people's cultural quality, science tourism of Geopark has gained attention gradually. For example, "China National Geological Park Construction Guide" issued by the Ministry of Land and Resources made it clear that one of the main objectives of the establishment of a Geopark is popularizing geo-sciences, and Land Resources Department (2008) 126 document proposed that Geopark can also be declared as land and resources science base. It can be seen that in order to better protect the geological relics, to accelerate the development of the geopark tourism, scientific tourism has become the top priority in the construction of geopark. Through the literature research, Chinese scholars' research of science tourism in geopark is mainly manifested in three aspects.

First, it is about the experience of foreign countries, especially the development of popular science tourism in the United States of America. Such as the United States attached great importance to the various types of park management and construction work, strengthen government's planning and constraint mechanism to construct and manage the park through the establishment of a unified management model to get sustainable utilization of resources. The interpretive system of American national park lasted for centuries, from it we can get that the important value of interpretation should be fully aware of, and the most important goal of it is protection. Lastly, interpretive programs and education should adjust measures to local conditions [5–6].

Second, it is about the case and countermeasure researches on the popular science tourism in China. For example, Hua Peng et al. (2005) studied the popular science tourism of Danxia landform by taking Danxia Mountain World Geopark as an example, and they thought the popular science tourism of Danxia landform should combine with sightseeing tourism in order to seek further development of the popular science tourism while promoting the quality of sightseeing at the same time. Mingtai Zhao (2008) analyzed the composition of Geopark popular science education system, and pointed out problems existing in present Geopark popular science education through Taimu Mountain National Geopark, then analyzed the causes of these problems, put forward the specific countermeasures. Tingfeng Yang (2009) analyzed the scientific value of karst geological tourism resources and studied the current situation and existing problems, then put forward development countermeasures for karst geological popular science tourism [7]. Xiaoying Dong (2010) had quantitative analysis to Cuihua Mountain Park of popular science tourists' perception through establishing popular science tourists perception assessment index system, and gained knowledge about tourists' perception to guide the Geopark popular science tourism development.

Third, special researches on popular science tourism in Geopark. On the development model side, Kairui Chen, Xuebin Zhong and Zhiguo Sun (2010) put forward primary, secondary and senior development methods and content of popular science according to different ages and education based on the analysis of Xianning karst geological resources; Xuejian Yu, Xiaoxia Yang and Yongling Cheng (2012) put forward and analyzed 3 popular science development models: local science education, teaching practice and popular science education of the general tourism science education. On the popular science carrier and program side. Guangwei Shi analyzed popular science tourism route design by using Dijkstra Algorithm and fuzzy mathematics; Luoyang Qian and Yan Wang studied construction of Geopark science interpretation system, Mei Wu, introduced foreign service quality evaluation model to evaluate the quality of Geopark science interpretation system and put forward countermeasures; Tianming Qu (2011) and Yaomei Guang (2011) studied the design of the Geopark museum and the explanation of the tour guide respectively.

3. Construction of Comprehensive Evaluation Index System of Popular Science Tourism in Geopark

3.1. *Selection of evaluation factors*

Evaluation factor is the key to the accuracy of evaluation. This article used AHP to establish comprehensive evaluation index system of popular science tourism in Geopark. In the evaluation index factor, it follows the standard and principle of system, hierarchy, representation and importance. In the selection of factors, the article took a reference to the Ministry of Land and Resources' "China National Geopark Construction Guide" (2006), "National Geopark Construction Standards" and "National Geopark Qualified Acceptance Criteria". Then, it used Delphi method and questionnaire survey to experts and scholars in the files of geology, tourism science, Geopark and geological tourism. Finally, it determined the Geopark popular science tourism comprehensive evaluation of 8 evaluation index, 20 evaluation factors (Table 1), the 8 evaluation index are geological museum, interpretation identification board, science movie hall, site of Geopark and popular science books, guide interpretation system, geosciences popular route, popular science activities, etc.

Table 1. Comprehensive evaluation of popular science tourism in Geopark.

Index layer	Factor layer
Geological museum	Exhibition area, Collection content, Commentary, Opening time
Interpretation identification board	Numbers of signs, Position, Commentary content, Sign style
Science movie hall	Numbers of seats, Play content
Geopark website	Site content, Display design
Popular science books	Types of books, Books content
Guide interpretation system	Professional training, Commentary content
Geosciences popular route	Line number, Route content
Popular science activities	Number of times, Number of participants

3.2. *Determine the weight of evaluation factors*

In the AHP process, the weights of the evaluation factors can be determined by establishing the discriminant matrix. In the discriminant matrix, the importance of the factor is determined by the comparison of different layers. This study used AHP process to analyze data of opinions from investigating of experts to get the absolute weight of factors, and then used advanced process to get relative weight of each factors. Its specific practice: according to Discriminant matrix "A", which composed of comparison of N factors of the same layer (x_1 $x_2..x_n$) between the upper layer's Y factor, and was recorded as $w_1,w_2...w_n$, named as hierarchical single order weight vector of "A". In "A", w_i means the weight that NO.i factor to upper Y factor, from this we can get hierarchical single order. Finally, it should pass through consistency test, if the test result is less than or equal to 0.1, show that a judgment matrix, the result could be accepted. On the contrary, the results should be modified until the test results accord with the requirements.

According to the comprehensive evaluation index and the factor weights, combined with the basic principle of AHP, general arrangement of each factor also need consistency test, and the result is less than 0.1, discriminant results can be accepted. Therefore, we can get the weight of each evaluation factor (Table 2).

Table 2. Comprehensive evaluation factors weight of Geopark popular science.

Evaluation Index	Weight	Evaluation Factors	Weight
Geological museum	0.15	Exhibition area	0.0375
		Collection content	0.0375
		Commentary	0.0375
		Opening time	0.0375
Interpretation identification board	0.15	Number of signs	0.0375
		Position	0.0375
		Signs content	0.0375
		Sign form	0.0375
Science movie hall	0.1	Number of seats	0.05
		Play content	0.05
Popular science books	0.1	Types of books	0.05
		Books content	0.05
Site of Geopark	0.1	Website content	0.05
		Display design	0.05
Guideinterpretation system	0.15	Professional training	0.075
		Commentary content	0.075
Geosciences popular route	0.1	Line number	0.05
		Line content	0.05
Popular science activities	0.15	Number of times	0.075
		Number of participants	0.075

3.3. *Non- dimensional evaluation of the index factors*

In the comprehensive evaluation process of Geopark popular science, each factor has its own dimension, which makes no comparability. Hence, in order to clear evaluation, we have to give non-dimensional treatment to evaluation factors after established evaluation factor system. This kind of evaluation can have a unified basis.

4. Conclusion

In China, popular science practice started early, however, relative research is lagging behind. As a new thing, the research of the National Geopark began after 2000, the theoretical research lagged behind the practice. The construction and development of Geopark popular science tourism is still in the exploratory stage, especially on the Geopark popular science tourism basic principle of the method, measures. What's more is that implementation planning and evaluation has not formed the unified understanding. Qualitative research, such as case study and question countermeasure has become the main body of the science tourism research of the Chinese Geopark. The construction and development of Geopark popular science tourism is a systematic work, the evaluation is restricted by many factors, that is why quantitative research of special study and introduction of mathematical model method can become theoretical research and development trend in the future.

This article established comprehensive evaluation index system of popular science tourism in Geopark composed of 8 evaluation index and 20 evaluation factors by expert interview survey and AHP. Among them, geological museum, interpretation board and science movie hall are hard power index; site of Geopark, popular science books, guide interpretation system, geo-science popular route and geo-science popular activities are soft power index. Geological museum, interpretation board, guide interpretation system and geo-science popular activities accounted for a large proportion up to 0.6, which means they are the most important parts of Geopark construction. At the same time, science movie hall, popular science books, geoscience popular route and site of Geopark are important influencing factors of Geopark construction. By using this system, not only the government can check and guide the development of Geopark, the Geopark itself also can carry out the construction of science tourism to promote the sustainable development of Geoparks construction.

Needs to be pointed out is, Geopark popular science tourism is a systematic project, it has a very high demand for innovation. With the construction and development of Geopark, a series of new measures and methods will continue to emerge and they will greatly promote the rapid development of Geopark popular science tourism. We have reason to believe that Geopark will plays an very important role in popularizing public science in the future.

Acknowledgments

This project is supported by the Scientific Research Plan Key Project Foundation (No. 2013d08) and the Talent Introduction Research Start-up Project Foundation (No. 2014rs05) of Wuhan Polytechnic University, China.

References

1. D.Y. Wei, Development Trend and Global Network of Geoparks and Discussion on Several Relative Questions, *Progress in geography.* 30, 515 (2011).
2. A.Z. Chen, Some Problems About the Construction of National Geological Park of China, *Resources and industries.* 5, 58 (2003).
3. T. Xu, On the Study Progress of Tourism System of China's National Geoparks and Several Problems Required to be Urgently Studied, *Tourism Tribune.* 25, 84 (2010).
4. S.M. Fang, Hong Kong's Typical Geological Heritage Resources and Geopark Building, *China Population Resources and Environment.* 21, 147 (2011).
5. H.Z. Xie, The Properties of National Park of America and Its Significance used for Reference. *Carsologica sinica.* 22, 73 (2003).
6. Y. Sun, The Rise of Interpretation in U.S. National Parks and Its Inspirations, *Chinese garden.* 6, 110 (2012).
7. T.F. Yang, Development of Popular Science Tour on Karst Geology, *Journal of Geological Hazards and Environment Preservation.* 20, 140 (2009).

Chapter 4
Ecologically Sustainable Development

Research on land degradation and management countermeasures in Qinghai-Tibet plateau

J.P. Yan
School of Geographic and Environmental Sciences, Northwest Normal University, Lanzhou, Gansu, 730070
E-mail: zhl.grase@163.com

Qinghai-Tibet Plateau is viewed not only as the cradle of a few great rivers in East Asia and Southeast Asia, but also as the promoter region of climate in the Northern Hemisphere. It plays a vital role in ecosystem service. However, it has been suffering a lot from serious land deterioration in recent years and is in desperate need to seek for countermeasures against soil loss and degradation. After making an analysis on the current situation of land degradation in Qinghai-Tibet Plateau and the influences caused by the issue, the thesis put forward some suggestion and countermeasures, such as, the establishment of integrated ecosystem management system, formulation of different management strategies against land loss in different regions and the strategy combined with the improvement of livelihood of farmers and herdsmen.

Keyword: Qinghai-Tibet Plateau; Land deterioration; Management countermeasures.

1. Introduction

Since "International Geosphere Biosphere Programme (IGBP)" and "International Human Dimensions Programme on Global Environmental Change (IHDP)" put forward "Land-Use and Land-Cover Change (LUCC)" hand in hand in 1995, the research work which unifies land-use and land-cover change with land quality changes has become one of frontier fields in LUCC study [1]. A certain set system of indexes is used to measure the changes of land quality and then ecological and environmental effects caused by land-use changes are reflected by the help of this method.

Qinghai-Tibet Plateau is not only the cradle of a few great rivers in East Asia and Southeast Asia—the Yellow River, the Yangtze River, Langcang River, Nujiang River, the Brahmaputra and the Indus, but at the same time, it is the promoter region of climate in the Northern Hemisphere as well. Climate changes in the eastern and southwestern parts of China are directly driven by it; and what is more, it also influences climate changes upon the Northern

Hemisphere and even on the whole world with obvious adjustability and foresight [2].

However, under the effect of natural factors and human factors [3], land in Qinghai-Tibet Plateau has suffered such severe degradation that the area of exploitative land resources has greatly reduced accompanying with the downfall of land productivity, the decline in agriculture and animal husbandry production and the destruction of producing and living facilities, which has led to great damage to the sustainable development of society and economy. What is worse, its fucntion as "ecological source" and "river source" has been impaired with the decay of biological diversity. Hence, the ecological security in the area has been challenged seriously; as a result, not only has the protection and construction of primitive ecological environment in Qinghai-Tibet Plateau and the development of society and economy been restricted, but the sustainable development of Qinghai-Tibet Plateau and even that of China have been in peril as well. Therefore, comprehensive assessment on the degree of land degradation in the area should be carried through, which not only belongs to one of the fundamental jobs in formulating ecological environmental planning and land-use planning, but also meets the urgent need agaisnt land loss. Based on international land degradation assessment standards, this study, on one hand, established the comprehensive assessment index system aimed at the land degradation on Qinghai-Tibet Plateau, and then made evaluation of degradation degrees in various kinds of land and comprehensive land degradation degrees, but more important put forward corresponding management measures and strategies to fight agaginst land degradation in Qinghai-Tibet Plateau, and such work aimed at providing reference for preventing land degradation in Qinghai-Tibet Plateau as well.

2. Current Situation of Land Degradation in Qinghai-Tibet Plateau and Its Main Influences

2.1. *Current situation of land degradation in Qinghai-Tibet Plateau*

According to the latest monitoring data on desertification and certain research data on grassland resource, after the deduction of overlapped areas which can be calculated both as grassland and deserted land, the total area of degraded land in Qinghai-Tibet Plateau reaches to 1270767.3 km^2 accounting for 56.98% of the total land area in Qinghai-Tibet Plateau. In these 1270767.3 km^2, area of desertification is 363563.35km^2, area of water erosion 181395.55km^2, salinization land 27726.89 km^2, area of freeze-thaw desertification 61365.49 km^2 and grassland degradation 636700km^2. These five land degradation types respectively account for 28.61%, 14.27%, 2.19%, 4.83% and 50.10% of the total degradation land area. It is quite obvious that deserted land and degraded grassland are the two main degradation types in Qinghai-Tibet Plateau, and after them follow water erosion land and salinization land and at last is freeze-thaw desertification land whose area and per cent are quite small and low (Table 1).

Successively areas of degraded land are 654863.02km^2, 323026.35km^2, 131420.23km^2, 146542.37km^2 and 14915.31km^2 in Tibet Autonomous Region, Qinghai Province, Gannan Plateau in Gansu Province, the southern foot of Qilian Mountain and Western Sichuan Plateau accounting respectively for 51.53%, 25.42%, 10.35%, 11.53% and 1.17% of the total degradation land area. From the data, it's quite clear that land degradation in Tibet is the most serious one and then follows Qinghai Province whose degradation area is not small at all, either (Table 1).

Table 1. Degraded land area in Qinghai-Tibet plateau.

Province (Region)	Desertification land (km^2)	Water erosion land (km^2)	Salinization land (km^2)	Freeze-thaw desertification land (km^2)	Degraded grassland (km^2)	Total (km^2)
Tibet	202327.85	69706.69	9632.89	47895.59	325300	654863.02
Qinghai	105009.73	53136.72	18110	13469.9	133300	323026.35
Gansu	48767.52	7552.71	–	–	75100	131420.23
Sichuan	7406.2	39136.17	–	–	100000	146542.37
Yunnan	52.05	11863.26	–	–	3000	14915.31
In total	363563.35	181395.55	27742.89	61365.49	636700	1270767.3

2.2. *The main influences caused by land degradation in Qinghai-Tibet Plateau*

2.2.1. *Sharp decline in ecosystem service*

Qinghai-Tibet Plateau is not only the cradle of a few great rivers in East Asia and Southeast Asia—the Yellow River, the Yangtze River, Langcang River, Nujiang River, the Brahmaputra and the Indus, but at the same time, it is the promoter region [2] of climate in the Northern Hemisphere as well. Climate changes in the eastern and southwestern parts of China are directly driven by it; and what is more, it also influences climate changes upon the Northern Hemisphere and even on the whole world with obvious adjustability and foresight [4]. Moreover, Qinghai-Tibet Pateau with Tibet Plateau as its main body is also an important center of global biological diversity. However, land degradation occurred in this area has already led to a series of changes among regional ecological factors and further brought about the alteration in climate and hydrology [5]. Such changes and alterations exert great impact upon biotic community and the habitat species to live and then make eco-environement in "source region" worse further. Not only does land degradation cause damage to surface vegatation and impact on the supply, recycle and redistribution of underground water in water conservation area and convergence zone, which then futher influences the ecological balance downstream, but it makes surface bare area larger and larger resulting in higher surface reflectivity whose rise further leads to climate changes in surrounding areas, such as, decrease of precipitation, temperature build-up and severe drought. Then later forms a vicious circle between land desertification and regional climate changes. Degradation of frozen land, exsiccation of land and vegatation degradation have made environment in Qinghai-Tibet Plateau worse and degraded; as one of sand storm frequency-occurring districts, Tibet Plateau is the second main source of sand storm and land degradation, in turn, will no doubt widen the range of sand storm and its influence intensity [6].

In addition, land degradation also makes survival environment of surface vagatation worse and vegatational community evolves toward the direction of simplication and high insect resistance resulting in the decline of biological diversity and even the loss of the diversity [7]. According to monitoring data, with grassland degradation becoming worse and worse, not only the ratio of fine species of forage grasses in the community is declining successively, but species and biological community also exhibit marked changes, for example, under the condition of non-degradation, dominant species of Kobresia humilis meadow community including Kobresia humilis, Kobresia capillifolia and Kobresia pygmaea meadow and their degree of dominance can reach to above 90%; when light degradation happens, although species of Kobresia still gets the upper hand, its dominance keeps going down with the increasing growth of forbs; at the time of medium degradation, the degree of dominance of Kobresia species of course continues falling off while at the same time the importance of Mesophyte vegetation gradually grows and co-edificato between Mesophyte vegetation and Kobresia species begins to form; at the phase of severe degradation, great changes are made in the composition of grassland communities, for its dominant species are replaced by Mesophyte vegetation and xerophilous plants and the number of community species drops to 10%–25% of the primary phytocoenosium. This point can also be proved by another example: in wide valley region in Shigatse, the psammophytic shrub community species can reach to fifteen or twenty more in fixed sand land and partial bare sand ground; and less than ten in half fixed sand land and only have five or seven in half flowing sand ground.

2.2.2. *Severe threat posed upon agriculture and animal husbandry production*

Total area of degraded land in Qinghai-Tibet Plateau reaches to 1274696.32 km^2 accounting for 49.3% of the whole land area of Qinghai-Tibet Plateau.Because extreme severe land degradation and severe land degradation make biological productivity lost, it's generally difficult to be recovered as farming and animal breeding land; as for the land which suffers medium or light land degradation, it's also required that a long-term natural recovery or man-made vegetation planting be done to help biological productivity of the land under recovery. Then one result people have to face is the decline of available land area and the shrinking of development space in agriculture and animal hunbandry industry, which can be seen clearly through the example happened in Gannan District [8]. Since 1980, almost one hundred hectares of farmland in Zhanang County have been deserted as the result of sand erosion and a large amount of fine grassland in Nagqu Prefecture and Gannan Plateau also has been converted into useless land under the action of desertification.

Land degradation not only makes available area of land shrink, but more important it makes the quality of land gradually lower and worse and then further leads to the slipping in developing quality of agriculture and animal husbandry. Naturally, the livelihood of farmers and herdsmen is under its seriousinfluence. To be more specific, land degradation causes organic materials, many nutrious elements like nitrogen, phosphorus, kalium, mealy sand and clay components existing in the surface soil incessantly to be blown away under the action of wind. The heavy loss of nutrition and organic matters, in turn, makes fertility and quality of soil down and the land barren with the dropping of land potential productivity [9].

It's estimated that 109.63kg of organic materials, 5.36 kg of nitrogen, 22.56 kg of total phosphorus and 628.04 kg of total kalium in per mu Tibetan land are lost annually because of wind erosion. As for other types of land, the loss of above-mentioned materials and elements every year in per mu land respectively reaches to 88.7kg, 5.23kg, 9.8kg and 244kg in sandy soil and 69.3kg, 4.3kg, 3.9kg and 108.3kg in grassland. According to the findings of our investigation, leaving alone other areas and districts, only in Tibet has total area of farmland under the threat of wind and sand already reached to $3.07\times10^4 km^2$ accounting for 13.8% of the total land area of Tibet Prefecture and caused reduction of crop output as much as 3.08×104t every year. In addition, in Tibetan desertification region, about $3.33\times104 hm^2$ of farmland which accounts for 14% of the total Tibetan land area is also threatened by sand and the loss of crop output can be as much as 3×104t every year. Degraded alpine meadow grassland in Qinghai-Tibet Plateau is about 162000km^2 accounting for 32.40% of the total grassland degradation in the district with an average vegetation coverage of 46%. The ratio of fine grass simply is 25% while ratio of weeds and noxious weeds can be as high as 75%. Output of the herbage only reaches to one seventh of that of in fine non-degraded grassland. All these make the incomings of farmers and herbsmen shrink and worsen their poverty situation.

3. Management Strategies and Countermeasures against Land Degradation in Qinghai-Tibet Plateau

3.1. *To establish an integrated eco-system management system*

The integrated ecosystem management method refers to a kind of ecological resources management method whose emphasis is put on the link between ecological functions of ecosystem and its service and the link between ecosystem and human society, economy and production systems [10]. The core concept is to admit the interdependency between human beings and natural resources in a direct or indirect way rather than regarding natural resources as an independent one. Therefore, after choosing ecosystem factors in an integrated management way, a cross-industry, trans-department or interdisciplinary comprehensive management framework [11,12] can be established.

That is to say, when drawing up prevention and control planning against land degradation, we should take the correlation among various factors into consideration and view eco-environment as wholeness and then introduce multi-department management into the planning of natural resources and put it into effect. Seeking for new ways of resources optimization and fund distrubution should be enouraged as well as the innovation of management system and operation mechanism. By doing these, new methods then will be found out to prevent and control land degradation from the root. Not only ought cooperated coordination mechanism among different departments to be built up with great passion in every district of Qinghai-Tibet Plateau, but project exemplary sites should be set up in every county as well. Under the guide of certain policy and regulations, integrated prevention and control strategies and action planning against land degradation from the level of province then can be formulated after participative training and by making an integrated assessment on organization ability, natural environment, social economy and land degradation. Land degradation information service center also should be set up to carry out demonstration activities fighting against land degradation.

3.2. *To formulate management countermeasures on the base of different situations in different districts*

From the data of integrated land degradation in Qinghai-Tibet Plateau, we can see that as the extreme severe degradation regions, areas of Qaidam Basin and the southern foot of Qilian Mountain suffer a lot from desertification, salinization and freeze thawing, especially from desertification and salinization, for both of them play a major role in land degradation; Subregion in Southeast Tibet, Plateau Lake Basin in Southern Tibet, subregion of Sanjing river valley in Estern Tibet and subregion of middle reaches of Yalutsangpo River belong to light degradation areas and therefore various types of degradation are low or even free from degradation. On the base of its characteristics in land degradation in Qinghai-Tibet Plateau, such as, various types, great intensity, large covering areas and apparеant regional differences and because prevention and control of land degradation is a kind of system engineering work, it means that we must adjust meaures according to different local conditions and then more scientific, reasonable and workable strategies and countermeasures can be put forward. Integated management planning with obvious regional features will be made, for example, formulating province-levelled prevention and control planning against land degradation, as well as the community's planning and participation framework.

3.3. *To combine with the improvement of farmers and herdsmen's livelihood*

A number of researches and studies show that the poor are not only the main victims of environmental degradation, but also the catalyst of environmental degradation. Many issues contributes to the environmental problems, for example, high income putting more pressure on environment, incomplete property rights strengthening the vicious cycle between environment and poverty and rising population worsening poverty and environment. Because climate conditions and human survival environment in Qinghai-Tibet Plateau are quite harsh, it's hard for prevention and control work to be in effect in a short period of time. However, land degradation and poverty are just like a pair of twin brothers who are hard to be parted from each other, for poverty leads to disarranged land clearing and planting, overgrazing and other predatory behaviors upon land resource definitely cause the land degradation at the same time. Land degradation, in turn, brings about lower productivity and lower income.In this way, farmers and herdsmen are cornered into the vicious cycle [13] of "land degradation and poverty". Naturally, it comes to a conclusion that prevention and control work in Qinghai-Tibet Plateau must be connected with the improvement of livelihood of farmers and herdsmen tightly and to help them get rid of the poverty. The study has already shows that diverty of livelihood with non-agricultural activities is not only good at lessening the fragility of livelihood and the threat of famine, but more important it can strengthen response capability to the changes of eco-environment and relieve the pressure upon the ecology [14,15,16] as well. So we must actively push farm households to improve their living ability [17] by participating into secondary and tertiary industries and their livelihood diversity and non-agriculturization will also be strengthened by such a transition.

References

1. Y.S. Liu, B.M. Chen. A Study on Chinese Sustainable Development and Land-Use/Land-Cover Change. Geographical Research, 2002, 21(3): 324-330.
2. M.C. Tang, C.Q. Li. An Analysis on "Qinghai-Tibet is the Climate Promoter Region", Paper Compilation on the First Academic Colloquium of Chinese Qinghai-Tibet Plateau Seminar. Science Press, Beijing, 1992.
3. X.Y. Zhao. The Influence of Humanity Factors upon Environment in Gannan Pasturing Areas. Journal of Geographical Science, 2010, 65(11):1411-1420.
4. S.J. Lausanne. Environment and Development in Qinghai-Tibet Plateau. China Tibetology Press, 1996.
5. W.S. Shen, X.Z. Cao, F.Y Shen. Classification and Grading of Land Degradation in China. Journal of Ecology and Rural Environment, 2006, 22(4):88-93.
6. N. Xu, X.D. Guo, Y.T. Hong, et al. A Study on Land Degradation Assessment Index System on the Base of Document Analysis. Journal of Geographical Science, 2008, 28 (3):425-430.
7. X.Z. Wang, S. Li, Q.R. He, et al. Dynamic Analysis on Desertification and Land Degradation in Tibet and its Assessment. Soil and Water Conservation of China, 2009, (7):25-28.
8. Z.X. Wen, J.J. Qu, W.M. Zhang. Types of Land Degradtion in Contiguous Area of Jin-Shan-Mong and its Assessment. China Desert, 1994, 14(2):90-94.
9. X.Z. Wang, Q.R. He, S. Li, et al. Types of Land Degradation in Qinghai-Tibet Plateau and Assessment on its Degradation Degrees. Research on Soil and Water Conservation, 2009, (4):14-18.

10. S. Li, Y.X. Dong, G.R. Dong, et al. Division of Land Desertification in Qinghai-Tibet Plateau. China Desert, 2001, 21(4):418-427.
11. Y.D. Li, C.C. Wang. A Research on Integrated Ecological Management System of Land Degradation in Qinghai-Tibet Plateau. Journal of Qinghai Univeristy (Natural Science Edition), 2009, 27(4):82-84.
12. W. Guo, H.J. Dang. From Idea to Legislation: Integrated Ecological System Management and Comprehensive Legislation Mode. China Population, Resources and Environment, 2009, 19(3):41-45.
13. H.Y. Feng, Y.G. Liu. From the Perspective of Documentary analysis: A Study on the Relation between Environmental and Land Degradation and Poverty. Qinghai Social Science, 2012, (3):37-41.
14. B.C. Glavovic, S. Boonzaier. Confronting coastal poverty: building sustainable coastal livelihoods in South Africa [J]. Ocean and Coastal Management, 2007, 50(1-2): 1-23.
15. J.Z. Yan, Y.L. Zhang, H.Y. Zhu, et al. Responding Reaction of Inhabitants upon Environmental Degradation in Different Districts of Upper Reaches of Dadu River. Journal of Geographical Science, 2006, 61(2):146-156.
16. C. M. Shackleton, S. E. Shackleton, E. Buiten, et al. The importance of dry woodlands and forests in rural livelihoods and poverty alleviation in South Africa [J]. Forest Policy and Economics, 2007, 9(5): 558-577.
17. X.Y. Zhao. The Influence of Livelihood Capital upon Life Satisfaction of Farmers and Herdsmen—A Case Study of Gannan Plateau. Geographic Research, 2011, 30(4):687-698.

Effects of forest degradation on soil microbial communities in Daxing'an mountains

H.H. Zhang[1,2], H.S. Pang[1], H. Ao[1], X. Li[1], Z.Y. Teng[1], J.W. Hu[1], G.Y. Sun[1] and X.L. Zhang[1*]

[1]College of Life Science, Northeast Forest University, Hexing Road 26, Xiangfang District, Harbin 150040, China

[2]College of Resources and Environment, Northeast Agricultural University, Mucai Road 59, Xiangfang District, Harbin 150030, China

[*]E-mail: xtwfwf@126.com

It is a common phenomenon of forest degradation that forest is shifting to grassland in Daxing'an Mountains. Analyzing the microbial functional diversity in soils is thinked great significance for restoration of forest. Therefore, by using Biolog technique, this paper studied soil microbes' ability to use a variety of carbon sources on different levels of vegetation degenerations (forest, shrubland and grassland). The results show that: there are significant differences between three types of degenerated soil microbial in the ability of using carbon sources, which, the average well color development (AWCD) fluctuated between 0.25 to 0.39, the order is: shrubland> forest >grassland, the values between shrubland and grassland reached significant differences level, Shannon index (H) and richness index (S) presented similar tendency. Differences in carbon utilization patterns exist between three types of vegetations, amino acids utilization is forest>shrubland>grassland, in contrast with the use of amines/amides. Showing that vegetation degeneration in Daxing'an Mountains lower the functional diversity of rhizosphere microbial communities, suppress the quantity and activity of soil microorganisms, decrease the microbial capability of using single carbon substrates.

Keywords: Daxing'an Mountains; Forest degradation; Soil microbial community; Biolog; Carbon utilization.

1. Introduction

As an important composition of terrestrial ecosystem, forests have played important roles with different functional properties in conserving soil and water, maintaining the productivity of land and sequestrating biochars. Daxing'an Mountains have been one of the largest national forest regions, which had provided a numerous forest products including timers in its early developed process [1]. However in the past two decades, forests had been degradated for

* This work was financially supported by The "Twelfth Five-Year-Plan" National Science and Technology Support Program of China (2011BAD08B02-3).
Corresponding author.

biodiversity decreased, soil erosion and fertility in Daxing'an Mountains. At present, forests degradation have turned to be a serious ecological problem, which could induce profound influences on the ecosystemand economic development in local, thus affecting the quality of life for millions of people living in this region [2].

Nowadays, soil biodiversity and ecosystem function have been concerned by researchers in ecological science. While, most of the studies have focused on the changing of soil microbial community and physical and chemical parameters with anthropic activities such as prescribed burning, logging and forestry managements [3-5].

Soil microbes play an important role in terrestrial ecosystem, which related to lots of fundamental processes, such as decomposing organic matters, cycling the nutrient, transferring energy, forming the structure of soil and humus [6]. However, soil microbes were sensitive to ecological factors and were affected immediately, and then which actives were often regarded as being mainly limited by carbon availability. Which were considered that could, these characteristic make microbes become a well indicator of soil quality [7,8].

The metabolic diversity of soil microbes were effects by environmental and interactions between microbial species [9]. Thus, these physiological properties were regarded as a potential indicator for changes of soil quality. Due to the fact that the physiological profile mainly reflect activity of copiotrophic bacteria, The Biolog Ecoplate was considered a useful method that could detect dynamics of microbial communities in utilizing the carbon substrates [10]. Microorganism are the dynamic component of soil, and the soil microbial quantity was the source or libraries of soil nutrients. The flow of material and energy cycle were performed synergy by different species of microbial community. Several reported that community diversity were more representative in the aspect of soil quality than quanlities of microbial [8].

In this paper, we examined changes in functional diversity of soil microbial community from three degradation stages by analyzing Biolog data. The results of experiment could is to provide theoretical support for restoring forest vegetations in Daxing'an Mountains regions.

2. Materials and Methods

2.1. *Site description and soil sampling*

The experiment was conducted at the Forest District of Jiagedaqi located in the southeast slope of Daxing'an Mountains (50°09′-50°35′N,123°45′-124°26′E) with a mean annual air temperature of −1.2°C and mean annual precipitation of

494.8 mm. Frost-free period of 85–130 days per year. The main types of soil are brown coniferous soil and grey forest soil.

The soils of three regions (forest, shrubland and grassland) were evaluated the impact of forest degradation in a randomized design with three replicates in July 2013. Three composite soil samples were gathered from each site. A composite sample was a mix of ten cores (0–10cm depth). The rocks, plant debris and soil fauna had been removed and homogenized the soil samples, which were stored at 4°C until use.

2.2. *Biolog method*

The BIOLOG Eco technology is based on the utilization of sole carbon source by the culturable fraction of soil microorganisms and aims to provide information on the functional diversity of the microbial community, which analyses were conducted following the method originally used by Garland and Mills [11] with slight modifications There were three replicates for each soil sample. Soil samples were extracted by shaking 10 g of soil with 90 mL 0.85% NaCl. After shaking, samples were centrifuged for 10 min at 2600 g. The supernatant was diluted by tenfold, then 150 μL dilution liquid was used to inoculate the BIOLOG Eco plates (BIOLOG, Hayward, CA, USA). The plates were incubated at 25 °C in the dark and color development was measured as absorbance in 590 nm at 24 h intervals by microplate reader would record the process. The Biolog assay could provide information-rich data rapidly, inexpensively and reproducibly. The rate of carbon sources being utilized is measured by reduction of tetrazolium, a dye which will change from colorless to purple.

The average well color development (AWCD) was calculated as follows: AWCD $=\sum(Ci- R)/31$, where Ci represent OD value of well number i, R represent OD value of control well. Soil microbial diversity indices were calculated according to Staddon [12].

2.3. *Data analysis*

Data statistical analysis, principal component analysis and relative analysis were all conducted with SPSS 16.0 software.

3. Results

3.1. *Community-level physiological profiles*

Average well color development (AWCD) developing in the Biolog-Eco microplates were used to determine metabolic activity of soil microbial community, the profile of AWCD values generally is a sigmoidal pattern with the incubation time. In this study, the AWCD values of each sample were shown in Fig. 1. The first 24 h, AWCD value curves increased slowly, no significant difference was observed among the rhizosphere soils of the three vegetations, indicating that the metabolic activity of soil microorganisms is depressed and few substrates in microplates could be utilized; thereafter, the AWCD soared in the next 96 h, reached the plateau at 120 h, and kept a high level thorough the remaining time, indicating that large amount of substrates were used by soil microorganisms effectively. After 120 h, values of AWCD ranked in the following order: shrubland>forest>grassland. There was significant difference in AWCD values between shrubland and grassland soil at the 96 h time (*P* value=0.05).

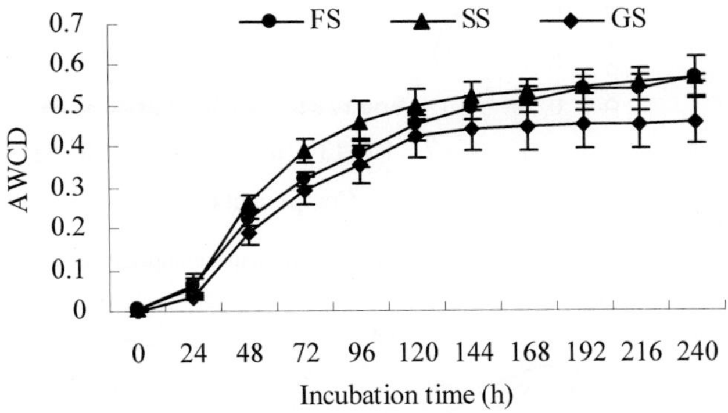

Figure 1. Average well color development (AWCD) changes with incubation time of different soil samples. FS, forest soil; SS, shrubland soil; GS, grassland soil. The same below.

3.2. *Functional diversity indices of soil microbial community*

Table 1. Analysis of AWCD for soil microbial communities and diversity indices from in different soil samples.

Treatment	AWCD	Shannon index (*H*)	Richness index (*S*)	Evenness index(*E*)
FS	0.32±0.02ab	2.96±0.08cd	14.67±1.53bc	1.10±0.04ab
SS	0.39±0.03a	3.16±0.05a	21.00±2.65a	1.04±0.03b
GS	0.29±0.03b	2.86±0.07d	15.33±2.89bc	1.06±0.06b

Note: Data in the table are mean±S.D, values followed by different small letters mean significant difference ($p<0.05$) between treatment and control in the same.

The changes of soil microbial diversity indices of soils at different soil sample were summarized in Table 1. The Shannon index in soil samples from shrubland exhibited a higher level, which increased by 10.50% and 6.76%, respectively, compared with the grassland and forest soil. Meanwhile, there was significant difference between shrubland and grassland soil. Great variation had existed in substrate richness index, the highest value was in the soil of shrubland, which increased by 36.99% than that of grassland soil. No significant difference was observed in Shannon index between forest and grassland soil. The evenness of three soil samples presented similarly, there was no difference among them.

3.3. ***Principal component analysis of carbon utilization patterns***

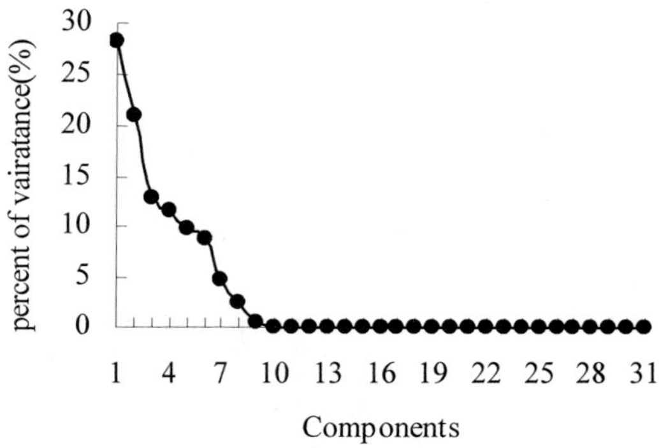

Figure 2. Variance percent of 31 principal components.

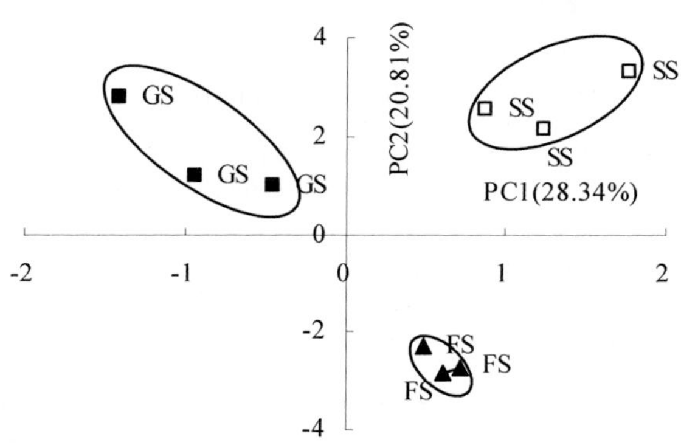

Figure 3. Principal component analysis of carbon sources utilization of soil microbes under different treatments.

Principal component analysis was conducted to reduce the dimensionality in the data and demonstrate carbon sources utilization patterns of microbes in different soil samples [13]. In this study, two principal components were extracted: PC1 and PC2, explained variance 28.34% and 20.81% of all data in ECO microplates, respectively. The catabolic diversity of soil microbial communities in three samples (FS, SS, GS) could be distinguished easily by the first two principal components. Scores of the three most positive and most negative carbon substrates on PC1 and PC2 are listed in Table 2. γ-hydroxybutyrate and 4-hydroxybenzoic acid were detected most positive by PC1 and PC2 scores, respectively. While D-mannitol and glycyl-L-glutamate were discriminated most negatively related to PC1 and PC2, respectively. Combing all the results determined by PCA, we found that functional diversity and structure of soil microbial communities from different soil samples be considerably separated from each other, and carboxylic acids, ie. γ-hydroxybutyrate, itaconic acid and D-malic acid, and polymers, ie. α-cyclodextrin and α-methyl-glucoside, are responsible for the variation in soil microbial communities among three soil samples.

Table 2. Most discriminant 3 carbon substrates as determined by PCA about data of carbon source utilization on community level.

Substrate no.	PC1	Score	PC2	Score
1	γ-hydroxybutyrate	0.905	4-hydroxybenzoic acid	0.947
2	itaconate	0.783	L-arginine	0.882
3	D-malic acid	0.775	α-D-lactose	0.763
1	D-mannitol	-0.771	glycyl-L-glutamate	-0.91
2	α-cyclodextrin	-0.752	putrescine	-0.779
3	tween 80	-0.491	D-xylose/aldopentose	-0.662

4. Discussions

The benefited influence of vegetation on soil quality has been admited in last decades [14], especially, different plant species could prevented or ameliorated forest degradation. Community-level physiological profile reflect metabolic activity of soil microbial communities, which make it available to indicate the whole dynamic of microbes. In this study, the AWCD values ranked as followed: shrubland>forest>grassland, which demonstrated that in the aspect of efficiency to utilize carbon sources, soil microbial communities of grassland were the lower than that of forest and shrubland. The functional diversity of soil microbial community is mainly related to soil organic C content [15]. In comparison with grassland and forest, shrubland has the largest amount of plant species, resulting in an accumulation of total and labile organic C. Accompanied by root exudates, flora and fauna litter, soil microbes have sufficient carbon source to provide energy for proceeding their vital movement. Forest degradation lead to lost of organic matter and decomposable organic C fractions [16]. As a consequence, evenness, richness and diversity index values become decreasing. But, there are potential drawbacks of the Biolog technique include the fact that the carbon sources in plates are not necessarily those found in soil. And it had been shown that primarily selected microbes are mainly made up of fast-growing copiotrophic bacteria, only a small proportion of the whole community [17,18].

5. Conclusions

In this study, the results showed that forest degradation could led to a variety of effects on functional diversity of soil microbial community, we observed substantial differences in ability to metabolize a series of carbon substrates in soil from different levels of degradation. Shrubland tend to have the highest microbial metabolic activity, evenness, richness and diversity index values. Therefore, in synthesis arrangement of regenerating forest vegetation in Daxing'an Mountains, we could suggest that shrubland constructing is key step.

References

1. Zhou, Y.L., Vegetation of Daxing'an Mountains, China [M]. Beijing: Science Press, 1991.
2. Wei, Y.C., Ouyang, Z.Y., Miao, H., Zheng, H., 2009. Exotic Pinus carbaea causes soil quality to deteriorate on former abandoned land compared to an indigenous Podocarpus plantation in the tropical forest area of southern China. J. For. Res. 14, 221–228.
3. Epelde, L., Becerril, J.M., Barrutia, O., Gonzalez-Oreja, J., Garbisu, C., 2010. Interactions between plant and rhizosphere microbial communities in a metalliferous soil. Environ. Poll. 158, 1576–1583.
4. Giai, C., Boerner, R.E.J., 2007. Effects of ecological restoration on microbial activity, microbial functional diversity, and soil organic matter in mixed-oak forests of southern OH, USA. Appl. Soil Ecol. 35, 281–290.
5. Leckie, S.E., 2005. Methods of microbial community profiling and their application to forest soils. Forest Ecol. Manage. 220, 88–106.
6. Kennedy AC. Bacterial diversity in agroecosystems. Agr. Ecosyst. Environ. 1999;74:65–76.
7. Pennanen T, Frostegård Å, Fritze H, Bååth E. Phospholipid fatty acid composition and heavy metal tolerance of soil microbial communities along two heavy metal-polluted gradients in coniferous forests. Appl. Environ. Microbiol. 1996;62:420–8.
8. Liao M, Xie XM. Effect of heavy metals on substrate utilization pattern, biomass, and activity of microbial communities in a reclaimed mining wasteland of red soil area. Ecotox Environ Saf 2007;66:217–23.

9. Zak, J.C., Willig, M.R., Moorhead, D.L., Wildman, H.G., 1994. Functional diversity of microbial communities–a quantitative approach. Soil Biol. Biochem. 26, 1101–1108.
10. Sullivan, T.S., Stromberger, M.E., Paschke, M.W., Ipollito, J.A., 2006. Long-term impacts of infrequent biosolids applications on chemical and microbial properties of a semi-arid range land soil. Biol. Fertil. Soils 42, 258–266.
11. Garland, J.L., Mills, A.L., 1991. Classification and characterization of heterotrophic microbial communities on the basis of patterns of community-level sole-carbon-source utilization. Applied and Environmental Microbiology 57, 2351–2359.
12. M. P. Maila, and P. Randima, "Soil microbial communities: Influence of geographic location and hydrocarbon pollutants," Soil Biology & Biochemistry., 2006, vol. 38, pp. 303–310.
13. Campbell, C.D., Grayston, S.J., Hirst, D.J., 1997. Use of rhizosphere carbon sources in sole carbon utilisation tests to discriminate soil microbial communities. Journal of Microbiological Methods 30, 33–41.
14. Binkley, D., 1994. The influence of tree species on forest soils: Processes and patterns, in: Mead, D.J., Cornforth, I.S. (Eds.), Proceedings of the Trees and Soil Workshop, Canterbury, New Zealand, 28 Feb.–2 Mar, 1994. Lincoln University Press, Canterbury, New Zealand, pp. 1–33.
15. Degens, B.P., Schippers, L.A., Sparling, G.P., Vojvodic-Vukovic, M., 2000. Decreasesin organic C reservesin soils canreducethe catabolic diversity of soil microbial communities. Soil Biol. Biochem. 32, 189–196.
16. Haynes, R.J., Beare, M.H., 1996. Aggregation and organic matter storage in meso-thermal, humid soils. In: Carter, M.R., Stewart, B.A. (Eds.), Structure and Organic Matter Storage in Agricultural Soils. CRC Press, Boca Raton, pp. 213–262.
17. Smalla, K., Wschtendorf, U., Heuer, H., Liu, W., Forney, L., 1998. Analysis of Biolog GN substrate utilization patterns by microbial communities. Appl. Environ. Microbiol. 64, 1220–1225.
18. Soil Classification Working Group, 1991. Classification. A taxonomic system for South Africa. Soil and Irrigation Research Institute, Department of Agricultural Development, Pretoria.

Impacts of absorption and desorption of biochar to ammonium in soil clay*

Q.F. Zheng†

Center of Analysis & Measurement, Inner Mongolia University for Nationality, Tongliao 028000, China
†E-mail: Zhengqf2006@126.com

J. Zhao

College of Life Science, Inner Mongolia University Hohhot, 010021, China
Inner Mongolia Key Laboratory of Environmental Pollution Control & Waste Resource Reuse, Hohhot, 010021, China
E-mail: nmgzhaoji@yahoo.com

There is growing interest in using biochar to improve soil productivity and contribute to food security and carbon fixation. In this experiment, we used corn stalk biochar to study different adsorption mechanisms using Fourier Transform Infrared Spectroscopy (FTIR) analysis. The study investigated the nitrogen absorption/desorption properties of various types of biochar, and explored these properties in degraded soil. Studying soil ammonium (NH_4^+) adsorption and desorption and the kinetic characteristics of clay and clay minerals helps reveal NH_4^+ fixation and release mechanisms. We studied nitrogen loss, retention, and bioavailability in biochar-amended soils fertilized with NH_4^+ using leaching and pot experiments. Study results may inform nitrogen fertilization and utilization, soil productivity, and crop production techniques.

Keywords: Biochar; Soil Clay; Ammonium.

1. Introduction

Biochar is a carbon-rich product generated using low-temperature oxygen -starved combustion (also known as pyrolysis) of carbonaceous biomass, such as crop residues, stall bedding, cull timber, and sawmill wastes (Joseph et al., 2009). When used as a soil amendment, biochar may boost soil fertility and improve soil quality by raising soil pH, increasing moisture holding capacity, attracting more beneficial fungi and microbes, improving cation exchange capacity (CEC), and retaining soil nutrients (Lehmann et al., 2006; Lehmann, 2007). Biochar is also attractive as a soil amendment because it sequesters carbon from the atmosphere-biosphere pool and transfers it to soil (Winsley, 2007; Guant and Lehmann, 2008; Laird, 2008). Biochar may persist in soil for millennia, because it resists microbial decomposition and mineralization. Biochar persistance depends strongly on its properties, which are shaped both by pyrolysis conditions and the feedstock used in its production. Previous studies suggest that a bioenergy strategy that includes biochar in soil leads to a net sequestration of CO_2 (Woolf et al., 2010), and may decrease emissions of other greenhouse gases such as N_2O and CH4 (Spokas et al., 2009).

The effect of biochar on soil function is shaped by biochar characteristics (e.g. chemical composition, surface chemistry, particle and pore size distribution), and the physical and chemical stabilization mechanisms of biochar in the specific soil it is placed within. Understanding biochar interactions and changes over time within the soil system is critical to understand its role in improving (and avoiding degredation of) soil properties, processes, and functioning.

Biochar has been promoted as a potential soil amendment (Joseph et al., 2009) to enhance soil fertility and water management. Its chemical characteristics are highly variable in formative properties, based on: feedstock, particle size, temperature, method of pyrolysis, time of retention, and furnace oxygen level (Benaddi et al., 2000; Strelko et al., 2002; Antal and Gronli, 2003; Guo and Rockstraw, 2007). A particular characteristic of interest is biochar's adsorption/desorption reactions with externally added nitrogen.

In tropical weathered soils, nitrogen bioavailability is an important control for crop production (Brady & Weil, 2002), because nitrogen bioavailability for crops is restricted by nitrogen-fixing soil constituents, such as iron and aluminum. With some predicting that economically mineable nitrogen reserves will be exhausted within 50–100 years (Gunther, 2005; Cordell et al., 2009), it is vital to understand and manage soil nitrogen fixation and bioavailability (Van Rotterdam et al., 2009). While much work has focused on nitrogen sorption and fixation due to high iron and aluminum oxides, less attention has been paid to nirogen desorption (Barros et al., 2005). Better understanding both sorption and desorption would help better understand and manage nitrogen mineralization, soil solution nitrogen concentration, and nitrogen bioavailability.

Organic matter additives have long been used to enhance nitrogen cycling and bioavailability, by reducing nitrogen sorption sites (Ohno et al., 2007; Ohno and Amirbahman, 2010). Biochar is expected to act in the same way, while also adding a limited supply of essential nutrients to the soil (Glaser et al., 2002; Lehmann et al., 2003). Nutrient fate, particularly soil nitrogen, is poorly understood when biochar is involved.

In biochar formation, as pyrolysis temperatures increase, the following effects occur: volatile compounds in the biochar matrix are lost; biochar yield decreases; surface area increases; and surface functional groups that provide exchange capacity decrease (Rutherford et al., 2004; Guo and Rockstraw, 2007; Lehmann, 2007; Joseph et al., 2009). These effects, in turn, influence nitrogen adsorption and desorption. Mixing biochar with mineral soil should result in competition between the mineral soil adsorption complex and the biochar's nitrogen sorption. If biochar acts like naturally occurring soil organic matter (Bhatti et al., 1998), soil nitrogen bioavailability should increase, by covering or binding nitrogen adsorption sites.

This study's specific objective was to determine the effect of biochar on nitrogen adsorption and desorption in soil. This study supports a broader goal of better understanding the mechanisms concerning biochar in soils and the environment.

2. Materials and Methods

2.1. *Soil clay and clay minerals*

The soil clay used in this study was typical black soil, taken from the City of Nen River, Heilongjiang Province. Clay minerals were illite and montmorillonite, provided by Hebei Lingshou Tianjiang Mines Ltd and Zhejiang Sanding Technology, Ltd.

2.2. *The preparation of biochar*

This paper used the raw material corn (Zea mays l.) straw to create biochar. The corn straw was taken from Tongliao City, around a village. Biochar was prepared using oxygen pyrolysis. After being placed in an airtight crucible, the clean dry straw was crushed. A muffle furnace was used for preheating; the material was heated anaerobically for 2 hours at 300 °C, to obtain the pyrolysis temperature of biochar samples (marked as BC300). The samples were uniformly cooled and grinded, filtered through 100 mesh sieving, and stored in a desiccator.

FTIR is a tool that reflects the structure of matter and measured infared biochar. This allows for the effectively reflection of active functional groups and adsorption sites. In this study, biochar was studied using a Nicolet5700 FTIR at Inner Mongolia University for Nationallity, China. Amendment powders were blended with spectroscopic grade KBr at the ratio of 1:10 in an agate mortar. Spectra of the samples were scanned between 4,000 and 500 cm^{-1} at a resolution of 4.0 cm^{-1}.

2.3. *Preparation of samples*

2.3.1. *Adsorption process*

Simulated biochar samples were prepared using biological carbon from corn stalks (generated at 300°C), divided into six different proportions of biochar to soil clay or clay minerals (0%, 5%, 10%, 20%, 50%, and 80%) into six separate 50 ml tubes containing 1 g clay and 1 g clay minerals. Next, 0.1 mol·L^{-1} NH_4Cl and 0.1 mol·L^{-1}KCl (each at 5 mL) were added to each sample; followed by 10 mL NaCl. After stirring and mixing, samples were shocked in the cyclotron water bath oscillator at 25±1 °C. In the adsorption stage, samples were treated in a centrifuge for 10 min, 30 min, 60 min, 120 min, 240 min, 360 min, 480 min, 720 min, and 1,440min at 5,000 r/min^{-1}. Supernatant samples of 0.50 ml were collected after 5 minutes and at each subsequent time interval listed above. The NH_4^+ level was determined using ion chromatography for each sample.

2.3.2. *Desorption process*

After the centrifuge step, we discarded the adsorbant supernatant, and then added 0.1mol·L^{-1}NaCl20 ml. After stirring and mixing, the samples were shocked in the cyclotron water bath oscillator at 25±1 °C. During the desorption stage, samples were treated in a centrifuge for 10 min, 30 min, 60 min, 120 min, 240 min, 360 min, 480 min, 720 min, and 1,440 min at 5,000 r/min^{-1}. Supernatant samples of 0.50 ml were collected at 5 minutes and at each subsequent time interval listed above. The NH_4^+ level was determined using ion chromatography for each sample.

2.4. *Statistical analysis*

Adsorption and desorption data were analyzed using Microsoft Excel 2003 and Origin Pro8.0 software. The Levenberg-Marquardt (LM) algorithm was used to support nonlinear fitting; several iterations were performed to assess fit. R^2 values indicate model fit.

3. Results and Discussion

3.1. *Biochar changes before and after the adsorption*

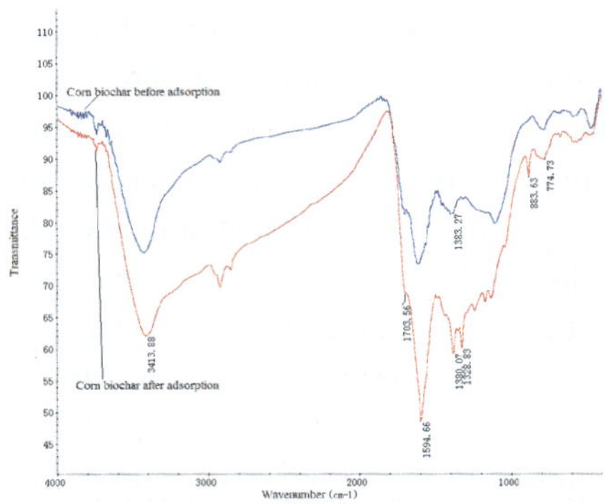

Figure 1. Biochar changes before and after adsorption.

Figure 1 presents FTIR spectra of the corn straw biochar before and after adsorption. The broad and strong absorption peaks in the 3700 ~ 2900 cm-1 were assigned the free or associated -OH stretching vibration. This band shows that the biochar has a large number of carboxyl, hydroxyl, and carboxyl-containing oxygen functional groups, providing the basis for biochar surface ion exchange. The bands between 1703 and 1594 cm-1, respectively, were assigned to -C=O- stretching of carboxy, esterfunction oraldehyde group, and -C=C- stretching of aromatic ring. The band at 1120cm-1 was assigned to -C-O stretching of ethers; and the bands bettwen 800cm-1 and 700cm-1were assigned the rings that are strongly stretching in pyridine, furan and so on. The results suggest that the large proportion of aromatization and heterocyclization groups areimportant in absorping cations in corn straw biochar. The sources of surface alkaline groups are relatively wide. Most studies have suggested that biochar nitrogen content is not high; the main reason for its alkaline nature is the surface of its conjugated aromatic structures (Radovic, 2001; Boehm H, 1994).

Biochar surfaces have the same structures as other γ-pyrone ring or heterocyclic ring structures, and together form a continuous layer of graphite, with a high concentration of π electron cloud electronic structure. As such, biochar surfaces can act as a Lewis base with water molecules to form an electron donor-acceptor (Electron Donor-Acceptor, EDA), making the biochar appear to be alkaline. This process formula can be expressed as (Radovic et al, 2001; Rivera-Utrilla et al, 2003; Sanchez-Polo, 2002):

$$C\pi + 2H_2O \longrightarrow C\pi\text{-}H_3O^+ + OH^-$$

Contrasting the same species of carbon adsorption FTIR spectra before and after the experiment, we can see that oxygen-containing functional groups with a peak characteristic of BC300 significantly change. Wave numbers become larger, peak intensity is lower, amplitude decreases, and peaks broaden. The 3423, 2360cm-1 absorption is significantly weakened, and migrate to 3641cm-1, generating a small peak. Given that the ammonia adsorption-OH is occupied, -OH of the intramolecular hydrogen bonding decreases. Ion exchange mode plays an important role in biochar ammonia adsorption.

3.2. *Study of kinetic on the adsorption dynamics of ammonium by clay*

3.2.1. *Kinetic curves of adsorption and desorption*

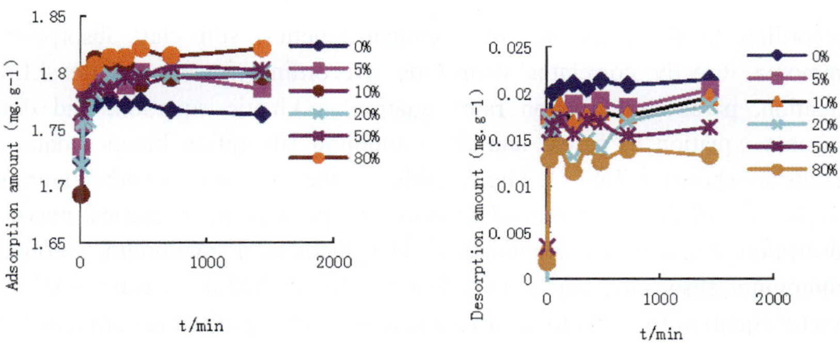

Figure 2. Comparison of ammonium adsorption and desorption in soil clay.

Figure 2 shows that for tests of ammonium mineral adsorption on clay, the content of ammonium is 80%>50%>20%>10%>5%>0% at the different experimental levels. At first, the adsorption tends to increase, then it reduces and ultimately achieves an even level. Adding biochar after this results in the significant increase in the adsorption of ammonium content. This shows that biochar has a significant impact on and promotes soil adsorption capacity.

For tests of ammonium ions mineral desorption on clay, the content of ammonium is 0%>5%>10%>20%>50% at the different experimental levels. Adding biochar into clay leads to a decrease in ammonia desorption. For some types of biochar, nitrogen fixation is relatively strong, retaining ammonia during desorption. However, for the control samples without biochar, the absorption of ammonia is perfectly desorbed. The sample of 80% was not tested, because it was partially contaminated during the experiment. Adsorped reaction of the clay on ammonium ion occurs in the initial hour, and then weakens after 1 hour to reach a steady state. This shows that the soil clay quickly adsorbed exogenous NH_4^+ with nitrogen application.

3.2.2. *Fitting relationship of the model with kinetic curves*

Table 1. R2 of different kinetic models on ammonium adsorption and desorption in soil clay.

Clay	Adsorption ($R_{0.05}^2$=0.3993, $R_{0.01}^2$=0.5846)				Desorption ($R_{0.05}^2$=0.4667, $R_{0.01}^2$=0.6342)			
	Elovich equation	Parabolic diffusivity equation	A kinetic equation	Double constant equation	Elovich equation	Parabolic diffusivity equation	A kinetic equation	Double constant equation
0%	0.2725	0.0726	0.6859	0.2948	0.5649	0.0966	0.8451	0.3284
5%	0.7804	0.5228	0.9796	0.747	0.4256	0.2455	0.9512	0.4991
10%	0.7081	0.4525	0.9642	0.6747	0.2735	0.1909	0.9139	0.4391
20%	0.5364	0.3079	0.9937	0.5236	0.239	0.1337	0.8709	0.4115
50%	0.9586	0.806	0.9671	0.8733	0.7081	0.4525	0.9642	0.6747
80%	0.8657	0.811	0.8972	0.6743	0.1287	0.0554	0.7104	0.2948

According to the principles of chemical kinetics, soil clay absorption of ammonia directly correlates with time. Accordingly, we used the Elovich equation, parabolic diffusion rate equation, a kinetic equation, and double constant equation to assess and fit ammonium adsorption kinetics data. The results are shown in Table 1. The variable y is the amount of adsorbed ammonia, t is the time of the adsorption of ammonium, and A is the theoretical maximum adsorption capacity of ammonium. The theoretical maximum amount of ammonium absorption was obtained using Origin Pro8.0 software, applying a kinetic equation function to achieve a nonlinear fitting of the experimental data. We then statistically analyzed and compared the output to the results. The kinetic equation R^2 reached 0.01 significance level, and R^2 values were arranged in descending order as follows: a kinetic equation>Elovich equation> double constant equation> parabolic diffusivity equation. The results show that the kinetic equation is a good fit with the test soil clay ammonium adsorption kinetics data, and results in a high correlation coefficient R^2 value. The kinetic equation's R^2 value is significantly greater than the other equations. These results can help us deepen our knowledge about the ammonium ion adsorption kinetics of soil clay.

3.3. *Study of kinetic on the adsorption dynamics of ammonium by clay minerals*

3.3.1. *Kinetic curves of adsorption and desorption*

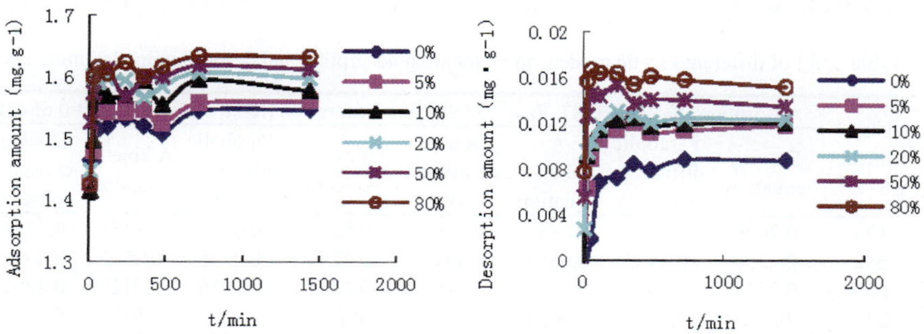

Figure 3. Comparison of ammonium adsorption and desorption by clay minerals.

Figure 3 shows that the test of illite and montmorillonite clay minerals on ammonium ions mineral adsorption yields an ammonium content of 80%>50%>20%>10%>5%>0% across experimental treatment levels. The control sample adsorption tends to increase, and then decrease, and ultimately level out. Adding biochar has significant role in promoting ammonia adsorption. Adding different proportions of biochar results in response curves trends that are basically the same: the curve showed a sharp increase at 1h and then levels off after 1h. The results show that the cumulative clay mineral absorption ammonium adsorption increased with increased time periods, but the unit time of amount of ammonia absorption decreases with the same time extension.

The test of clay minerals on ammonium ions mineral desorption, shows the content of ammonium is 80%>50%>20%>10%>5%>0% over different treatment levels. After adding biochar, the desorption of ammonium content increased significantly. From this, it appears that the mineral itself contains ammonia fixing ability. For the ammonium desorption rate, the clay achieves a higher rate of release in the first hour, then declines, and then levels off.

3.3.2. *Fitting relationship of the model with kinetic curves*

Given the results to this point, we used the Elovich equation, parabolic diffusion rate equation, a kinetic equation, and double constant equation to fit ammonium adsorption kinetics data based on the clay soil added. The results are shown in Table 2. The statistical analysis and fitting results demonstrate that the kinetic equation R^2 reached a 0.01 significance level. R^2 values are arranged in descending order as follows: a kinetic equation>Elovich equation> double constant equation> parabolic diffusivity equation. The results show that the kinetic equation can be well fitted for the ammonium adsorption kinetics data, and result in a high correlation coefficient R^2 value. These results can help us deepen our knowledge about the ammonium ion adsorption kinetics of soil clay.

Table 2. R2 of different kinetic models on ammonium adsorption and desorption in clay minerals.

Clay minerals I:S=2:1	Adsorption ($R_{0.05}^2$=0.3987, $R_{0.01}^2$=0.5934)				Desorption ($R_{0.05}^2$=0.3796, $R_{0.01}^2$=0.6542)			
	Elovich equation	Parabolic diffusivity equation	A kinetic equation	Double constant equation	Elovich equation	Parabolic diffusivity equation	A kinetic equation	Double constant equation
0%	0.3649	0.1966	0.8451	0.3784	0.5649	0.0966	0.8451	0.3284
5%	0.4256	0.2455	0.9512	0.4991	0.4256	0.2455	0.9512	0.4990
10%	0.2735	0.1909	0.9139	0.4491	0.2745	0.1919	0.9129	0.4291
20%	0.239	0.1337	0.8659	0.4216	0.239	0.1337	0.8709	0.4115
50%	0.228	0.0527	0.8709	0.4115	0.1357	0.0571	0.7135	0.2946
80%	0.1297	0.0564	0.7204	0.2848	0.1297	0.0564	0.7204	0.2848

4. Conclusion

This study investigated biochar's usefulness as a soil amendment, and concludes that its usefulness depends largely on biochar characteristics. Biochars vary significantly in their sorption and desorption properties. Experimental results showed that biochar formation makes feedstock nitrogen more soluble. When added to the soil, biochar releases nitrogen, which is adsorbed on the soil surface and fills sorption sites. Some of this sorption is not reversible. Adding biochar to soil clay or the clay minerals, in which adsorption and desorption of subtances such as ammonia are advantageous, results in higher nutrient availability. As the proportion of biochar to soil clay increased from 5% to 20%, the sorption content of ammonia was higher when compared with CK (0%), so biochar may be better for maintaining soil fertility. These results establish a strong theoretical basis for future study.

Other questions remain about the utility of biochar as a soil amendment, including its leachate toxicity and salinity, pH effect on nutrient bioavailability, adddition rates, the influence of biochar on native soil organic matter mineralization (Wardle et al., 2008), and mineralization interactions with soil type and fertility. While these questions remain, this study demonstrates that biochar has the potential to aid soil nitrogen management in degraded or low fertility soils.

Acknowledgments

This study was supported by the Natural Science Foundation of Inner Mongolia (2013MS0606), Universities Science Research Project of Education Department of Inner Mongolia (NJZY11015), initiate of scientific research project of Inner Mongolia University for Nationalities (BS262), Education Research project of Inner Mongolia University for Nationalities (MDYB2013079), National Natural Science Foundation of China (31260108, 21261015).

References

1. Antal, M.J. & Gronli, M. 2003. The Art, Science, and Technology of Charcoal Production. Industrial & Engineering Chemistry Research, 42, 1619–1640.
2. Brady, N.C. & Weil, R.R. 2002. The nature and properties of soils. Prentice Hall, Upper Saddle River, NJ.:394-397.
3. Benaddi, H., Bandosz, T.J., Jagiello, J., Schwarz, J.A., Rouzaud, J.N., Legras, D. & B_eguin, F. 2000. Surface functionality and porosity of activated carbons obtained from chemical activation of wood. Carbon, 38, 669–674.
4. Boehm H. Some aspects of the surface chemistry of carbon blacks and other carbons. Carbon, 1994, 32 (5) : 759-769.
5. Cao X, Harris W. Properties of dairy-manure-derived biochar pertinent to its potential use in remediation. Bioresource Technology, 2010, 101(14): 5222-5228.
6. Downie A, Crosky A, Munroe P. Physical properties of biochar Lehmann J, Joseph S. Biochar for environmental management: Scienceand technology, London: Earthscan, 2009: 13-32.
7. Gaunt, J., Lehmann, J., 2008. Energy balance and emissions associated with biochar sequestration and pyrolysis bioenergy production. Environ. Sci. Technol. 42, 4152-4158.
8. Guo, Y. & Rockstraw, D.A. 2007. Physicochemical properties of carbons prepared from pecan shell by phosphoric acid activation. Bioresource Technology, 98, 1513–1521.
9. Joseph, S., Peacock, C., Lehmann, J., Munroe, P., 2009. Developing a Biochar Classification and Test Methods. In: Biochar for Environmental Management: Science and Technology (Eds. Lehmann, J. & Joseph, S.), Earthscan.

10. Laird, D. A., 2008. The charcoal vision: A win-win-win scenario for simultaneously producing bioenergy, permanently sequestering carbon, while improving soil and water quality. Agron. J. 100, 178-181.
11. Lehmann, J., Joseph, S., Ed. 2008 Biochar for Environmental Management Science and Technology, Earthscan: Sterling, VA.
12. Lehmann, J., 2007. Bio-energy in the black. Front. Ecol. Environ. 5, 381-387.
13. Lehmann, J., Gaunt, J., Rondon, M., 2006.Bio-char sequestration in terrestrial ecosystems-A review.Mit. Adapt. Strat. Global Chang 11, 403-427.
14. Radovic L R, Moreno-Castilla C, Rivera-Utrilla J. 2001. Carbon materials as adsorbents in aqueous solutions Radovic L R. Chemistry and physics of carbon, New York: Marcel Dekker.: 227-405.
15. Rivera-Utrilla J, Sanchez-Polo M. 2003, Adsorption of Cr (III) on ozonised activated carbon: Importance of Cπ-cationinteractions. Water Research, 37 (14) : 3335-3340
16. Sanchez-Polo M, Rivera-Utrilla J. 2002, Adsorbent-adsorbate interactions in the adsorption of Cd (II) and Hg (II) on ozonized activated carbons. Environmental Science and Technology, 36 (17) : 3850-3854
17. Spokas, K. A., Koskinen, W. C., Baker, J. M., Reicosky, D. C., 2009. Impacts of woodchip biochar additions on greenhouse gas production and sorption/degradation of two herbicides in a Minnesota soil, Chemosphere 77, 574-58
18. Winsley, P., 2007. Biochar and bioenergy production for climate change mitigation. New Zealand Sci. Review 64, 5-10.
19. Woofl, D., Amonette, J.E., Street-Perrott, F.A., Lehmann, J., Joseph, S. 2010. Sustainable biochar to mitigate global climate change. Nature Comm. 1, 56-64.

Study of unsaturated undisturbed loess constitutive model based on damage theory*

Z.Y. Wang
College of Geology & Environment, Xi'an University of Science and Technology, Xi'an 710054, China
E-mail: wangzhaoyang1999@126.com

Based on damage theory, paper proposed the unsaturated undisturbed loess nonlinear damage constitutive model, which can reflect undisturbed unsaturated unique mechanical properties. The model contains a total of 13 parameters, each parameter can be measured by the test. The model revealed the inherent law of some stress-strain behavior of unsaturated soil, and then pushed the constitutive model of unsaturated to a new level.

Keywords: Unsaturated soil; Constitutive model; Loess; Damage theory.

1. Introduction

Unsaturated soil widely distributed, many different types, and different types of unsaturated soils different characteristics, thus establishing unsaturated soil constitutive model is a complex and difficult task. Coleman in 1962 first proposed the unsaturated soil constitutive relationship, but in fact still are studies related to volume change and shear constitutive equation (Coleman, 1962). Blight previously has been studied constitutive equations body change, which is to follow Bishop effective stress equation research (Blight, 1965). However Fredlund and Morgenstorn stress is based on two independent variables to establish the constitutive equations (Fredlund, 1976).

As a new discipline damage theory has been widely used in materials science, but in soil mechanics application only a few years ago. Academician of Shen Zhujiang has proposed elastoplastic damage model and nonlinear mechanics model damage, which may reflect structural saturated soil dilation and softening properties at low confining pressures (Shen, 1993). Desai also put forward a disturbed state concept model similar structural soil. (Desai et al., 1997). On the basis of experimental studies, Wang et al., (2008) obtained unsaturated loess considering density and temperature of the soil – water

* This work is supported by cultivation fund of Xi'an University of Science and Technology (2014011).

characteristic curve of expression. Liu et al., (2007) proposed a complex computational model for deformation volume of undisturbed loess of unsaturated and saturated.

Undisturbed unsaturated is a typical structural soil, which has a large number of pore development and structural strength. Triaxial testing process CT scan results showed that both before and after test shearing undisturbed loess, soil structure damage evolution of relatively clear. (Chen et al., 2001; Wang et al., 2006). Therefore, based on the theory of nonlinear damage proposed by academician of Shen Zhujiang, combining unsaturated soil strength and deformation characteristics of research and testing of advanced CT technology, a nonlinear damage constitutive model of unsaturated loess, which can reflect the undisturbed unsaturated complex deformation characteristics.

2. Establish Constitutive Model

Shen Zhujiang (1993) proposed soil deformation process is regarded as a mixture of undisturbed soil and damage soil of the two materials, and with double – spring model to describe both the share of stress. In accordance with the concept, the soil stress σ has been shared by the adding part of undisturbed soil σ_i and damage soil share part σ_d, that is,

$$\{\sigma\} = (1-\omega)\{\sigma_i\} + \omega\{\sigma_d\} \tag{1}$$

Where ω called damage ratio, that is the proportion of damage soil; $\{\sigma_i\}$ and $\{\sigma_d\}$ respectively, undisturbed soil and damage soil stress.

The type of incremental form

$$\{\Delta\sigma\} = (1-\omega)[D_i]_t\{\Delta\varepsilon\} + \omega[D_d]_t\{\Delta\varepsilon\} - (\{\sigma\}_i - \{\sigma\}_d)\Delta\omega \tag{2}$$

Where $[D_i]_t$ is tangent stiffness matrix of the undisturbed soil, and $[D_d]_t$ is tangent stiffness matrix of the damage soil.

First, assume undisturbed loess ideal elastomer, the stress tensor and tensor ball into partial tensor, then

$$\sigma_{oi} = K_i\varepsilon_i \tag{3}$$

$$\sigma_{si} = 2G_i\varepsilon_s \tag{4}$$

Where K is the bulk modulus, and G is the shear modulus. $[D_i]$ is an elastic matrix, then

$$[D_i]_t = [D_i]_s = [D_i] \tag{5}$$

After loess structural damage completely destroyed, its properties are similar to silt. We can apply the concepts of unsaturated soil, using the following power and hyperbolic functions to describe the volume compression curve and shear curve.

$$\begin{cases} \varepsilon_v = \dfrac{\sigma_m - u_a}{kP_{atm}}\left[1 + \left(\dfrac{\sigma_m - u_a}{P_{atm}}\right)^m\right] & (6) \\ \varepsilon_s = \dfrac{\sigma_1 - \sigma_3}{G_0\left(1 - \dfrac{\sigma_3 - u_a}{\sigma_{sf}}\right)} & (7) \end{cases}$$

$$\begin{cases} K_s = \dfrac{kP_{atm}}{1 + \left(\dfrac{\sigma_m - u_a}{P_{atm}}\right)^m} & (8) \\ G_s = G_0\left(1 - \dfrac{\sigma_3 - u_a}{\sigma_{sf}}\right) & (9) \end{cases}$$

Where:

$$\sigma_m = \frac{1}{3}(\sigma_1 + \sigma_2 + \sigma_3) \tag{10}$$

$$G_0 = kP_{atm}\left(\frac{\sigma_3 - u_a}{P_{atm}}\right)^n \tag{11}$$

$$\sigma_{sf} = \frac{2(\sigma_3 - u_a)\sin\varphi'}{1 - \sin\varphi'} \tag{12}$$

Where G_0 and σ_{sf} are the initial tangent modulus and soil damage when the principal stress difference, respectively; P_{atm} is atmospheric pressure, k and n is a constant test. Injury soil tangent modulus literature formula (Wang, 2004), namely

$$E_t = \left[1 - \frac{R_f(1 - \sin\varphi')(\sigma_1 - \sigma_3)}{2[c' + (u_a - u_w)\tan\varphi^b]\cos\varphi' + 2(\sigma_3 - u_a)\sin\varphi'}\right] \cdot kp_{atm}\left(\frac{\sigma_3 - u_a}{p_{atm}}\right)^n \tag{13}$$

Correspondingly:

$$\begin{cases} G_t = \dfrac{E_t}{2(1 + \mu)} & (14) \\ K_t = \dfrac{E_t}{3(1 - 2\mu)} & (15) \end{cases}$$

Where G_t and K_t are tangent shear modulus and tangent bulk modulus respectively.

The literature (Wang, 2004) Turkey's cohesion $(u_a - u_w)\tan\varphi^b$ and structural strength of the relationship $\tau_s = \lambda q_s$ into (13), there is

$$E_t = \left[1 - \frac{R_f(1-\sin\varphi')(\sigma_1 - \sigma_3)}{2[c' + \lambda q_s]\cos\varphi' + 2(\sigma_3 - u_a)\sin\varphi'}\right] \cdot kp_{atm}\left(\frac{\sigma_3 - u_a}{p_{atm}}\right)^n \quad (16)$$

Tangent modulus E_t formula was introduced in the structural strength of the soil to determine the parameters easier.

The literature (Wang, 2004) study results, soil damage variable axial strain relationship can be expressed as follows:

$$D_{ME} = a\varepsilon_1^2 + b\varepsilon_1 + c \quad (17)$$

Since the damage variable soil reflects the rate of change in the density of the process of soil damage, and damage to the soil than the soil deformation process reflects the proportion of soil damage, injury changes also reflect the proportion of soil bulk density rate cut in the process of change, so that we can use to damage variable soil damage than the equivalent in place, thus:

$$\omega = D_{ME} = a\varepsilon^2 + b\varepsilon + c \quad (18)$$

Where a, b and c may damage variable from CT and axial strain curve obtained triaxial test.

From Eq. (18), we have

$$\Delta\omega = (2a\varepsilon + b)\Delta\varepsilon \quad (19)$$

Substituting Eq. (3), (4), (8), (9), (14), (15) into Eq. (2), it will have the soil stress-strain increment expression:

$$\{\Delta\sigma\} = (1-\omega)[D_i]_t\{\Delta\varepsilon\} + \omega[D_d]_t\{\Delta\varepsilon\} - ([D_i]_s - [D_d]_s)\{\varepsilon\}\Delta\omega \quad (20)$$

Substituting Eq. (19) into Eq. (20),we have

$$\{\Delta\sigma\} = (1-\omega)[D_i]_t\{\Delta\varepsilon\} + \omega[D_d]_t\{\Delta\varepsilon\} - ([D_i]_s - [D_d]_s)\{\varepsilon\}(2a\varepsilon + b)\{\Delta\varepsilon\}$$
$$(21)$$

The formula can be expressed as follows in a simplified form

$$\{\Delta\sigma\} = [D]_{nd}\{\Delta\varepsilon\} \quad (22)$$

Where Eq. (22) is the expression of unsaturated soil nonlinear damage model.

Where $[D]_{nd}$ is nonlinear damage matrix, expression of

$$[D_{nd}] = (1-\omega)[D_i]_t + \omega[D_d]_t - ([D_i]_s - [D_d]_s)\{\varepsilon\}(2a\varepsilon + b) \quad (23)$$

Where $[D_i]_t$ is the stiffness matrix of undisturbed soil, expressed as

$$[D_i]_t = \begin{bmatrix} K_t + \frac{4}{3}G_t & & & & & \\ K_t - \frac{2}{3}G_t & K_t + \frac{4}{3}G_t & & & & \\ K_t - \frac{2}{3}G_t & K_t - \frac{2}{3}G_t & K_t + \frac{4}{3}G_t & & & \\ 0 & 0 & 0 & G_t & & \\ 0 & 0 & 0 & 0 & G_t & \\ 0 & 0 & 0 & 0 & 0 & G_t \end{bmatrix} \tag{24}$$

$[D_i]_s$ undisturbed soil secant matrix, the expression with $[D_i]_t$. Soil composition damage stiffness matrix with $[D_d]_t$ identical $[D_i]_t$. $[D_d]_s$ is secant matrix damage soil, expressed as

$$[D_d]_s = \begin{bmatrix} K_s + \frac{4}{3}G_s & & & & & \\ K_s - \frac{2}{3}G_s & K_s + \frac{4}{3}G_s & & \text{symmetry} & & \\ K_s - \frac{2}{3}G_s & K_s - \frac{2}{3}G_s & K_s + \frac{4}{3}G_s & & & \\ 0 & 0 & 0 & G_s & & \\ 0 & 0 & 0 & 0 & G_s & \\ 0 & 0 & 0 & 0 & 0 & G_s \end{bmatrix} \tag{25}$$

3. How to Determine the Parameters of the Model

Model parameters as follows:

Undisturbed soil: K_i and G_i

Damage soil: c' , φ' , φ^b , G_0 , σ_{sf} , G_t , K_t , n and R_f

Damage evolution parameters: a and b

K_i and G_i may be inferred from the slope of the test curve and no side compression confined compression test curve of the initial segment of the undisturbed soil. Since the sampling perturbations, so the determined K_i and G_i are generally small. The best conditions in situ measurement, for example, Penetration test measures K_i by static, by vane shear test measures G_i .

The basic parameters of soil damage c' , φ' , φ^b , n and R_f and the determination described in the literature (Wang, 2004), the compression curve and shear curve is determined from the volume of soil G_0 and σ_{sf} . Determined by unsaturated soil triaxial test from the formula (13), (14) and (15).

Damage evolution parameters a and b may damage variable by CT tests to determine the strain curve.

Among tangent modulus structural strength represented need not be determined φ^b , λ and q_s only need to identify the specific determination method described in the literature (Wang, 2004).

4. Preliminary Verification Model

In this paper, we used undisturbed loess preliminary document (Wang, 2006) verification using the proposed constitutive expression. Comparative literature (Wang, 2006; Wang, 2004) in the test results to determine the model parameters are listed in Table 1.

Table 1. The parameters of model.

c' /kPa	φ' /(°)	φ^b (°)	G_0	σ_{sf}
11.4	26.7	16.6	1.62	380
G_t /kPa	K_t /kPa	n	R_f	K_i
4 856	15 891	0.65	0.87	200
G_i	a	b		
100	-0.000 06	0.005 3		

Substituting the above parameters into (22), the calculation method use the initial stress iteration by literature (Shen, 1999). Calculate the stress-strain curve and comparison of test results obtained in Figure 1.

Figure 1, sample 1, sample 2 and sample 3 confining pressure was 50 kPa, 100 kPa and 200 kPa when shearing, suction experiments were 100 kPa, which is not marked in the curve model results. Calculation results of the model reflects the same suction conditions under shear stress value with the confining pressure increases, while also reflects the damage model of unsaturated shearing hardening. As can be seen from Figure 1, the proposed model can basically reflect the characteristics of the shear curve.

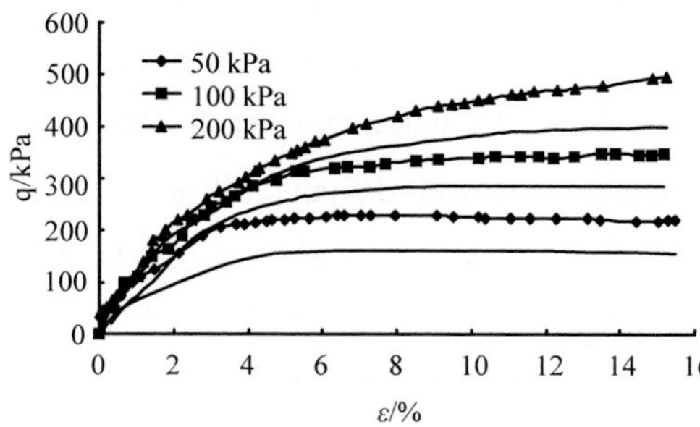

Figure 1. The comparison between the test result and the model prediction of triaxial shear test.

5. Conclusions

Nonlinear damage constitutive mode of unsaturated undisturbed loess is simply and each parameter can be a direct contact with the soil physical, soil deformation mechanism. Although damage model has extensive application prospect, but any kind of model can not take into account all factors that affect the soil, which can not apply to all soil types and loading conditions. Soil damage model established in this paper has the following characteristics.

(1) In this paper, nonlinear models, CT of saturated soil testing techniques applied to structural strength of unsaturated soil nonlinear damage model is a great progress on the model.

(2) While the parameters of the model needs to be isotropic soil triaxial compression test unidirectional compression test, triaxial shear test suction control soil and soil CT triaxial shear test, test workload, many parameters, parameter the need for further optimization, but the model reveals the inherent variation of unsaturated soil stress-strain characteristics.

(3) The proposed model for the same suction conditions and three different confining pressure triaxial tests performed to confirm the model calculation results with the experimental values are close, but the model presented in this paper is only a preliminary verification, we need to practice later in continuous improvement.

References

1. Coleman J D. Stress-strain relations for partly saturated soil. Geotechnique, 1962, 12(4): 348-350.
2. Blight G E. A study of effective stresses for volume change; Proceedings of Symposium on Moisture Equilibrium and Moisture Changes in Soils Beneath Covered Areas. Butterworths: Sydney, 1965: 259–269.
3. Fredlund D G, Morgenstorn N R. Constitutive relation for volume change in unsaturated soils. Canadian Geotechnical Journal, 1976, 13(3): 261-276.
4. Z.J. Shen. An Eiasto-piastic Damage Model for Cemented Clays. Chinese Journal of Geotechnical Engineering, 1993, 15(3): 21-28.
5. Z.J. Shen. A nonlinear damage model for structured clay. Hydro-Science and Engineering, 1993, (4): 247-255.
6. Desai C S, Basaran C, Zhang W. Numerical algorithms and mesh dependence in the disturbed state concept. Int J Num Meth Engrg, 1997, 40: 3059-3083.

7. T.H. Wang, J. Lu, C.K. Yu. Soil-water characteristic curve for unsaturated loess considering temperature and density effect. Rock and Soil Mechanics, 2008, 29(1): 1-5.
8. X. Liu, L.M. Wang, L.Z. Chen, et al. A volumetric model of loess based on microstructure instability and generalized suction theories. Rock and Soil Mechanics, 2007, 28(12): 2517-2523.
9. Z.H. Chen, Z.H. Lu, Y.B. Pu. The matching of computerized tomography with triaxial test apparatus for unsaturated soils. Chinese Journal of Geotechnical Engineering, 2001, 23(4): 387-392.
10. Z.Y. Wang, W.K. Ni, Y.B. Pu. Microscopic experiment under triaxial test conditions of loess structure characteristic. Journal of Xi'an University of Science and Technology, 2006, 26(1): 51-5.

Assessment and study on the ecological sustainability of general land use planning based on modified ecological footprint model*

Y.Z. Liu†, L.L. Zhou and Y. Chen
School of Resources and Environment Engineering, Wuhan University of Science and Technology, Wuhan, China
†E-mail: Liuyanzhong@wust.edu.cn
www.wust.edu.cn

Q. Li[1] and Z. Zhang[2]
[1]Three Gorges Geotechnical Consultants Co., Ltd., Wuhan, China
[2]School of Business, Hubei University, Wuhan, China

This paper has used and improved the ecological footprint model from the perspectives of the classification of ecologically productive area and the calculation of ecological footprint.An assessment of ecological sustainability is implemented based on the improved model from two aspects: implementation and predictability of the general land use planning (2006-2020) of Linxiang City. The results show that the per capita ecological consumption profit/loss will gradually turn from a profit in 2005 to a deficit in 2013 and it is predicted that the per capita ecological consumption deficit will still go up by 2020. Meanwhile, the per capita sustainable ecological loss will fall due to fast population growth. Although the implementation of the general land use planning is beneficial to improve the ecological capacity, the city will still remain at an ecological unsustainable state. The research concludes that the major way to reduce ecological deficit and promote ecological capacity of Linxiang City is to control excessively fast population growth, change the current consumption patterns and economic growth mode and optimize the land utilization structure and layout.

Keywords: Modified Ecological Footprint Model; General Land Use Planning; Ecological Sustainability Assessment.

1. Introduction

General Land Use Planning refers to the temporal and spatial overall layout and arrangement for land development, utilization, treatment and conservation in response to requirements for the national economy and sustainable social development in a specific area or region. Only a slight direct impact will be produced by the formulation and operation of General Land Use Planning to local ecological environment, whereas all changes triggered by implementation

* This work is supported by National Natural Science Foundation of China (41201600, 41201164).

of the planning in land utilization will bring a profound, extensive and complicated impact to the ecological environment. Moreover, such impact is irreversible [1]. The General Land Use Planning is generalized as a significant approach to rational land allocation and land resource utilization and featured by a close tie with ecological environment protection. Assessment and Study on the Ecological Sustainability of General Land Use Planning in the view of sustainable ecological development have not only followed requirements for national land resource conservation but also played a decisive role in keeping ecological security and realizing regional sustainable development.

2. Study Approaches

2.1. *Overview of ecological footprint model*

Ecological footprint model (*EFM*) was proposed by the Canadian ecological economist William Reese in 1992 and specific algorithms [2–3] were contributed by Wackernagel in 1996. Now *EFM* has been referred to as a major approach [4] to measure impacts brought by human activities to ecosystem. In this model, the difference between demand of human activities for ecosystem service and supply of services from ecosystem is cited to assess sustainable development [5] of an area. It has been spread fast and applied widely [6–9] in the assessment and study on sustainable development. Some scholars have tried to assess and study General Land Use Planning [5,10–14] based on *EFM*.

2.2 *EFM modification and calculation*

2.2.1. *Modified EFM*

In *EFM*, a consumption matrix of land utilization is formulated to interpret the relation [12] between human consumption and land resource. In the traditional *EFM* calculation, all kinds of ecological resource and energy consumption are converted into six types of ecologically productive area, including farmland, grassland, forestland, building land, fossil energy land and water area. According to actual land utilizations of the city, ecologically productive areas shall be divided into seven types, including farmland, grassland, forestland (including garden land), water area, fossil energy land (including mining land), construction land (excluding mining land) and natural reserved land.

(1) There may be higher sustainability in more undeveloped areas with lower living standards [9] in the traditional ecological footprint assessment and study. In order to avoid such a fact, ecological footprints (*EF*) are divided into consumption ecological footprint and output ecological footprint. In detail, the former type refers to the acquisition of ecologically productive land for resource

consumption in an area, whereas the latter type represents acquisition of ecologically productive land for resource output in this area. The ecological footprint calculation model is shown below.

$$EF = N * ef = N * r_j \sum_{i=1}^{n} a_i = N * r_j \sum_{i=1}^{n} (\frac{c_i}{p_i}) \tag{1}$$

In the formula, N is the number of population, ef indicates per capitaecological footprint, r_j refers to equivalence factor corresponding to type j of ecologically productive land (by dividing average ecological productivity of type j of ecologically productive land with average ecological productivity of all types of ecologically productive land in the world), a_i means area of ecologically productive area converted from type i of per capita consumption. During the consumption footprint calculation, c_i is the consumption of item i per capita. As for output footprint calculation, c_i is per capita output corresponding to item i, p_i is average productivity of item i and n is the number of consumption item.

(2) Ecological capacity (EC) refers to actual area of ecologically productive area owned by population of an area. Its formula is shown below.

$$EC = (1-12\%)N * ec = (1-12\%)N \sum_{i=1}^{n} (a_j r_j y_j) \tag{2}$$

In the formula, N is number of population and ec is the per capita ecological capacity, a_j is area of type j of per capita ecological production land and r_j is equivalence factor, y_j indicates output factor corresponding to type j of ecologically productive land (the ratio of average productivity of a specific type of land in an area to average productivity of similar type of land in the world). In actual calculation, biodiversity reserve area shall be deducted at the rate of 12%.

2.2.2. *Consumption profit and deficit*

The difference between ecological capacity and consumption ecological footprint is defined as consumption profit and deficit. Consumption profit ($ER_{consumption}$) or consumption deficit ($ED_{consumption}$) reflects natural land utilizations by regional population. Profit will emerge when the ecological capacity exceeds consumption ecological footprint, in other words, demands of human activity for ecological resources in an area are beyond ecological supply capacity of the ecosystem. Otherwise, consumption deficit will be incurred.

2.2.3. *Sustainable ecological loss (EL)*

The difference between ecological capacity and output ecological footprint is defined as output deficit (ED_{output}) and the gap between consumption profit/ deficit and output deficit is cited as the sustainable ecological loss.

$$EL=|\ ED_{output}\ |\ \mp ER_{consumption} \tag{3}$$

Sustainable ecological loss represents the tolerable sustainable loss of ecologically productive land since the time of study. When consumption profit prevails or consumption deficit is lower than output deficit, ecologically productive land of the area cannot only guarantee local economic development but also relieve external ecological pressure based on resource output and indicates that the area development is in a relative sustainable state, and the general land utilization planning is rational and feasible. As consumption deficit is ascending and even goes beyond output deficit, ecological capacity of the area cannot guarantee the local development any longer. External resources are needed to relieve internal ecological pressure. Under such a circumstance, the planning shall be properly adjusted to promote ecological capacity of local land.

3. Assessment on the implementation of general land use planning of Linxiang City based on EFM

3.1. *General descriptions of the data*

Data for ecological footprint calculation are excerpted from statistical yearbook (2005 to 2013), General Land Use Planning (2006 to 2020) and land usealteration data (2005 to 2013) of Linxiang City and predictions of related data for the year 2020.

Ecological footprint accounts of the city are divided into biological resource account and energy account. The biological resource account contains 4categories, including agriculture, forestry, animal husbandry, fishery products, a total of 17 types of products. The energy account contains electric power, coal, liquefied petroleum gas and gasoline.

Most of main parameters involved in ecological footprint calculation in this paper are determined by referring to equivalence factors and output factors of Wackernagel traditional calculation model. In detail, the equivalence factors of farmland, grassland, forestland, building land, fossil energy land and water area are 2.8, 0.5, 1.1, 2.8, 1.1 and 0.2, the output factors of grassland, forestland, building land, fossil energy land and water area are 0.19, 0.91, 2.19, 0.91 and 1. The ratio of average productivity of farmland in the city to average productivity of farmland in the world is selected as farmland's output factor, the output factor is 1.73. Lai Li's equivalence factor algorithm for unused land is utilized as equivalence factor of natural reserved land, the equivalence factor is 0.12 and the output factor is 1 [1].

3.2. *Ecological footprint calculation and result analysis for the implementation of General Land Use Planning of Linxiang City*

3.2.1. *Analysis of dynamic change of ecological capacity*

As shown in Table 1, total available ecological capacity of the city has been increasing year on year in the planning period (2005-2013) and the increment is 3821 hm^2. But the growth rate of total available ecological capacity is 1.23%, lower than the population growth rate of 8.14%. As a result, the per capita available ecological capacity has declined by 0.0409 hm^2 per capita. In this sense, population is a major factor influencing per capita available ecological capacity.

According to Table 2, ecological supply of the city largely depends on farmland, forestland and construction land. From 2005 to 2013, although the total farmland area has risen, its proportion in the supply of ecological capacity has dropped by 0.43%. In addition, the area and proportion of forestland contributing to ecological capacity has declined slightly. Total area of construction land has been increasing year on year and its proportion in the supply of total ecological capacity has risen by 0.87%. Total area of fossil energy land was also enlarged and its proportion has gone up by 0.05% at the growth rate of 35.71%.

Table 1. Comparisons of Per Capita "Consumption-Output" Ecological Footprint, Per Capita Ecological Capacity and Per Capita Ecological Profit & Deficit. (Unit: hm^2/per capita, per capita)

Year	2005	2006	2007	2008	2009	2010	2011	2012	2013
Per capita available capacity	0.6393	0.6378	0.6291	0.6205	0.6117	0.6022	0.593	0.6215	0.5984
Per capita consumption footprint	0.4769	0.503	0.6361	0.5007	0.5375	0.5992	0.4877	0.6085	0.6309
Per capita output footprint	1.2602	1.4178	1.4224	1.3598	1.2981	1.3057	1.3517	1.4586	1.3913
Per capita consumption profit & deficit	0.1624	0.1348	-0.0069	0.1198	0.0742	0.003	0.1052	0.0129	-0.0325
Per capita output profit & deficit	-0.621	-0.7801	-0.7933	-0.7393	-0.6864	-0.7035	-0.7587	-0.8371	-0.7928
Per capita sustainable ecological loss	0.7833	0.9148	0.7863	0.8591	0.7606	0.7065	0.864	0.85	0.7603
Population	486500	487755	495011	502112	510170	518929	528354	504621	526079

Table 2. "Consumption-Output" Total Ecological Footprint and Ecological Capacity of Each Type of Ecologically Productive Land in Linxiang City. (Unit: hm^2, %)

Type of ecologically productive land	Total ecological capacity				Ecological footprint of total consumption			
	2005	Proportion	2013	Proportion	2005	Proportion	2013	Proportion
Farmland	184700	46.66	185224	46.23	101855	43.9	158916	47.88
Grassland	6	0.001	6	0.001	60328	26	78608	23.68
Forestland	96611	24.41	95926	23.94	270	0.12	992	0.3
Water area	2490	0.63	2482	0.62	32478	14	38023	11.46
Construction land	68867	17.4	73192	18.27	1471	0.63	4523	1.36
Fossil energy land	558	0.14	753	0.19	35614	15.35	50856	15.32
Natural reserved land	179	0.05	170	0.04	-	-	-	-
With 12% off as reserved area for biodiversity	42409	10.71	42930	10.71	-	-	-	-

3.2.2. *Analysis of dynamic change of per capita ecological consumption footprint*

As reflected by related data in Table 1, per capita ecological footprint of consumptions in the city was fluctuating but rising as a whole. The per capita consumption footprint has risen by 0.154 hm^2 per capita in nine years at the growth rate of 32.29%.

As reflected by related data in Table 2, the demand of human consumptions in the city for six types of ecologically productive land has been increased to some extent from 2005 to 2013. Most of the consumption demands are brought from farmland, grassland, water area and fossil energy land. To be specific, the proportion of farmland in ecological footprint has risen by 3.98%, but the proportion of grassland, water area and fossil energy land has decreased by 2.32%, 2.54% and 0.03% respectively, whereas the proportion of forestland and construction land has ascended drastically by 0.18% and 0.73% at the growth rate of 150% and 116% respectively.

3.2.3. *Analysis of dynamic change of per capita ecological consumption profit & deficit*

According to related data in Table 1, the per capita ecological profit of consumption has declined year on year from ecological profit 0.1624 hm^2 per capita in 2005 to ecological deficit 0.0325 hm^2 per capita in 2013. Total decrement in nine years was 0.1949 hm^2 per capita and the variation rate was 120.01%.

3.2.4. *Analysis of sustainable ecological loss*

As indicated by related data in Table 1, sustainable ecological loss of the city was fluctuating all the time and the highest rate was 456478 hm^2 in 2011. It went down in next two years. Total sustainable ecological loss was 399993 hm^2 in 2013, increasing by 4.96% in comparison with 381094 hm^2 in 2005. Such growth rate is lower than the population growth of 8.14%. As a result, the per capita sustainable ecological loss has decreased from 0.7833 hm^2 in 2005 to 0.7603 hm^2 in 2013 at the rate of 2.94%.

4. Ecological footprint prediction and analysis on general land use planning objectives (Year 2020) of Linxiang City

4.1. *Prediction of ecological footprint for planning objectives (year 2020)*

Annual average growth rate is utilized to predict consumption ecological footprint of the city in 2020 according to ecological footprint and ecological capacity data from 2005 to 2013. The ecological footprint is 454049.48 hm^2 for consumption and 854600.86 hm^2 for output and the population is 563343. Meanwhile, area of all types of land in the year 2020 is predicted according to the supply of all types of ecologically productive land in all recorded years to estimate the ecological capacity. Related prediction data are shown in Table 3.

Table 3. Comparisons between Predicted Value and Planned Value of All Types of Ecologically Productive Land in the Year 2020 for Linxiang City. (Unit: hm^2, %)

Type of ecologically productive land	Predicted value	Planned value	Difference	Growth rate
Farmland	38332.87	37410.14	922.73	2.47
Grassland	59.53	59.60	-0.07	-0.11
Forestland	95235.64	96103.40	-867.76	-0.90
Water area	12375.00	12459.31	-84.31	-0.68
Construction land	12589.55	13116.37	-526.82	-4.02
Fossil energy land	977.90	322.33	655.57	203.38
Natural reserved land	1346.14	1213.20	132.94	10.96

Table 4. Comparisons between Predicted Value and Planned Value of Ecological Footprints in 2020 for Linxiang City. (Unit: hm^2/per capita, %)

Type	Predicted value	Planned value	Difference	Growth rate
Per capita consumption footprint	0.8060			
Per capita output footprint	1.5170			
Per capita ecological capacity	0.5652	0.5508	0.0145	2.63
Per capita consumption deficit	-0.2408	-0.2552	0.0145	-5.67
Per capita output deficit	-0.9518	-0.9663	0.0145	-1.50%
Per capita sustainable ecological loss	0.7110	0.7110	0	0%

4.2. *Analysis of ecological footprint for planning objectives (year 2020)*

As predicted according to related data in Table 4, the per capita consumption footprint is 0.8060 hm^2, the per capita output footprint is 1.5170 hm^2 and the per capita ecological capacity is 0.5652 hm^2 in target year of the planning. In light of results of calculation for all kinds of land in target year determined by the General Land Use Planning, the target value of per capita ecological capacity is 0.5508 hm^2 in 2020. As reflected by the comparison between predicted value and target value, the per capita ecological capacity will rise by 0.0145 hm^2 in 2020 compared with the planned value and the growth rate will reach 2.73%. Meanwhile, the per capita consumption deficit will decrease by 5.67%.

By following the current development trend, ecological capacity of the city will rise and the consumption deficit will fall to some extent in the target year 2020. In view of the supply of all related types of ecologically productive land, the predicted value of ecological capacity is higher than the target value because of the increase of farmland and fossil energy land and the proper decrease of forestland, water area and construction land. Although ecological capacity is mostly contributed by farmland, forestland and construction land, infinite enlargement of these three types of land will not promote ecological capacity of land efficiently. Therefore, the city is proposed to reasonably allocate land resources in the period from 2014 to 2020.

5. Conclusions and Proposals

(1) The control over fast population growth plays an important role in reducing per capitaecological consumption footprint, increasing area of per capita ecological capacity and finally promoting sustainable ecological development for all types of land.

(2) Continuous expansion of consumption footprint will inevitably increase ecological pressure in the area. Consumption demands of people can be relieved only by changing the current consumption mode and economic development pattern. Here the city is proposed to call for sustainable consumption, place much emphasis on the production and utilization of energy-saving products, save energy, develop circular economy, protect ecological environment, eliminate extensive economic development, create intensive, energy-efficient and ecological production and consumption modes and ensure the harmony of economic development with population, resource and environment.

(3) Promoting ecological capacity is an important task for the General Land Use Planning and also one of approaches to regional sustainable ecological development. Proposals are presented for the city to efficiently develop

ecological capacity. These proposals include (1) strictly keeping the dynamic equilibrium of total farmland, promoting farmland quality, improving soil condition and crop variety to enhance biological productivity of farmland, (2) turning wild grassland, bare land and other types of unused land among natural reserved areas into grassland or forestland through land development and reclamation to promote their contribution to ecological capacity and meet demands of people for the consumption of meat and (3) protecting water resources to prevent water resources being polluted and destroyed to guarantee the supply of water products.

(4) During compiling the next round of General Land Use Planning, the city is proposed to take the sustainable ecological utilization of land resources as criteria and give priority to coordinating the restructuring and layout optimization of land utilization and transforming from quantitative increase in the past to the optimization of land utilization structure and layout. At the same time, efforts shall be made to ensure farmland promotion in both quantity and quality, gradually enhance the saving and intensive use of land, reasonably enlarge construction land and promote utilization rate of construction land. Also, protection is proposed as primary objective and development as secondary objective to develop unused land in a proper sequence.

References

1. Lai, L. and X. J. Huang (2005). Assessment of ecological footprint of national general land use planning in China. China J. Transactions of the CSAE. 21(2):66-71.
2. Rees, W. E (1992). Ecological footprint and appropriated carrying capacity: What urban economics leaves out. UK. Environment and Urbanization. 4(2):121-130.
3. Rees, W. E. and M. Wackernagel (1996). Our Ecological Footprint: Reducing Human Impact on the Earth. Canadian Journal of Urban Research. 5(2):303.
4. Liu, Y. Z., J. F. Li, Z. Zhang and P. Hu (2008). Application and enlightenments on sustainable land utilization assessment based on ecological footprint approach in China. China J. Geography and Geo-Information Science. 24(1):81.
5. Cai, Y. Q., Y. Yu, Y. B. Yuan and J. H. He (2010). Environmental impact assessment of land use planning based on ecological footprint. China J. Scientific and Technological Management of Land and Resources. 27(4): 62-66.

6. Xi, J. C., Q. S. Ge, S. K. Cheng and C. H. Lu (2004). Ecological footprint of tourism consumption–A case study of Beijing foreign tourist arrivals. China J. Journal of Natural Resources. 19(2): 224-229.
7. Fan, X. Q. (2005). Study on the principle of water resources ecological footprint and application. M.Sc. thesis. Dept. of Water Resources and Environment, Hohai University.
8. Wang, Y. X., Y. Wang and X. Y. Yao (2015). Study on ecological footprint and ecological capacity in Guangdong province. China J. Chinese Agricultural Science Bulletin. 31(4):245-249.
9. Zhang, Z. Q., Z. M. Xu,G. D. Cheng and D. J. Cheng (2001). The ecological footprints of the 12 Provincesof West China. China J. Acta Geographical Sinica. 56(5):600-610
10. Shao, L., W. X. Du and J. X. Yang (2011). Environmental impact assessment of Xinjiang land use planning based on ecological footprint. China J. Rural Economy and Science-Technology. 22(5): 62-66.
11. Bai, Y. (2012). Assessment of eco-efficiency of Tianjin land use master plan based on ecological footprint model. China J. Economic Geography. 32(10): 127-132.
12. Zheng, Y. R., Y. D. Zheng, J. F. Ge and C. X. Xu (2014). Implementation evaluation of the general land use planning of Hebei province based on the ecological footprint model. China J. Research of Soil and Water Conservation. 21(5) :230-235.
13. Yin, S. H., Y. Z. Liu and X. H. Wang et al (2013). The EIA study on the ecological footprint of structure adjustment of the overall land use planning in Linxiang city. China J. Journal of Anhui Agri. Sci. 41(36):14110-14114.
14. Zhang, Y. (2011). A case study on Shanxi province: evaluation of regional ecological security based on ecological footprint. China J. Resources and Industries. 13(S1):57-64.

Study on the construction of eco-control technique system for sustainable landscape design*

L.C. Huang†

Art & Design School, Dalian Polytechnic University,
Dalian, 116034 / Liaoning, China
†E-mail: baomason@163.com
www.dlpu.edu.cn

J.F. Zuo

Bioengineering Institute, Hainan College of Vocation and Technique,
Haikou, 570216 /Hainan, China
E-mail: 421756031@qq.com

Sustainable landscape design is the effective way resolving the contradiction between human and land resources for the subject of landscape architecture, and ecol-control technology is the guarantee of sustainable landscape design while the ecological evaluation system is the verification of sustainable landscape design. There is no uniform definition of the sustainable landscape design now, and there is only the partial, fragmented research on ecol-control techniques, it is necessary and urgent to make a complete eco-control technological system for sustainable landscape design. The situation, problems and development trends of the domestic and international modern landscape design were analyzed. The connotation, extension and characteristics of the sustainable landscape design were defined. The eco-control technology system for sustainable landscape design was constructed, based on the whole design process of the landscape "full-life-cycle", the two dimensions of theory and practice, the three different scales of macro-medium-micro scales, a verification of post design, whose construction was “1+2+3+3” which means one process, two dimensions, three scales and three sides. And the macroscopic-scale control refers to planning control and the total capacity control, the medium-scale is to control by management such as human resources, managing model and public participation etc, micro scale control, the main part, is the design & engineering techniques, which is related to the sustainability directly. And the post-landscape design evaluation includes economy cost control, local culture heritage, and the sustainable developing environment.

Keywords: Eco-control; Sustainable Landscape; Landscape Design; Technology System.

---

* This work is supported by grant 31270747 of the National Natural Science Foundation of China.
† This work is also supported by grant 2015lsktwx-07 of the Liaoning Provincial Federation Social Science Circle 2015 Economic and Society Development Projects Foundation.

1. Introduction

1.1. *Overseas review of sustainable landscape design*

In the last century 60–70's, Rachel Carson's "silent spring" waked up the people from the wealth dream of industrial ages, Lynn White revealed that the origin of environmental crisis is western culture, Garrett Hardin's "tragedy of public resources" told the nature of the capitalist economy and the human, Donella Meadows, on the other hand, warned the crisis of human survival by calculating the earth resource limitation [1], which changes the focus attention of all the designers from beauty, form and superior culture to nature, to the relationship between human and nature of other cultures. In 1969, McHarg published a book of "Design with Nature" [2], which suggested that the establishment of the ecological Design idea, Sim Vander Ryn and Stuart put forward twelve principles of ecological design, Lyle also gave the design and regeneration principles of human ecology [3], Thayer studied the principles of sustainable landscape, visual ecology and ecological city [4], the restoration technology and methods about river, biological, vegetation were also studied [5–9], Melby and Cathcart's offered a technology way of sustainable landscape design in their "sustainability: landscape design technology" [10].

1.2. *Domestic research review sustainable landscape design*

In China, as the national strategy, "people-oriented, ecological civilization, the industrial structure, growth mode and consumption pattern of saving energy & resources and protecting the environment, constructing resource-saving and environment-friendly society, were put forward in the 17th National Congress of CPC " [11]. On August 17 in 2006, the economical urban landscape green space was defined by the Ministry of Construction: With the least land, the least amount of water, the least amount of funding, the least disturbance on the environment, according to the reasonable and recycling rules of resources, improving the utilization rate of resources, to plan and design the most efficient urban ecological security.

As to Academics, the scientific definition of landscape design, the introduction absorption and innovation of foreign landscape design theory, landscape planning and design, components and engineering design & practice were researched in three scales. On a macro scale, Kongjian Yu, designed firstly the ecological infrastructure of sites, named "opposite-planning", and took Taizhou Park as an example. Yu-Yao Ye [12], studied on the urban space structure ecology-oriented. The concept of low-carbon city was discussed by

Shu-Hua Wang [13] and other scholars. Medium scale, the ecological network system of city green space was studied. Plant community and many other landscape elements such as water system, industrial wasteland, and landfill were also studied [14–16]. On a Micro scale, bio-restoration technology was discussed. Yuanxiang Wu discussed on approaches of minimized cost based on the landscape ecology theory [17], Xu Xi, Binyi Liu studied on the landscape cost control between China and America. Leichang Huang explored the effective ways of sustainable landscape design by ecological model, the three-dimensional ecological planning model [18–20]. Na Wu [21] designed "low-carbon water purification landscape" of Shanghai Post-beach Park of world expo, a good case combining the theory with practice.

In general, there is no clear definition of sustainable landscape design and evaluation standard in China, no systematic ecological control technology system. There is still some phenomenon of imitation, blind adoration to foreigner, concept speculation of "people-oriented", "green and ecological design", ignoring the cost control, subjective decision-making, and lack of public participation and so on.

2. Sustainable Landscape & Eco-Control

2.1. *Sustainable landscape*

Landscape, in its Dutch origin, landscape meant to adapt cultural and natural processes to create new territory. As the word moved into English and other languages, it took on visual meaning as well [22].

Sustainable development, as defined by the United Nations World Commission report, is development that meets the needs of the present without compromising the ability of future generations to meet their own needs. Sustainability is a broad and optimistic concept that addresses the central issues of our society. It related to many subjects such as ecology, society and economic etc, applying in a variety of spatial and temporal scales and many fields. Now that more people are taking the idea of sustainability seriously, there is a need to define specific practices and approaches that will move us in the direction of more sustainable landscapes.

What does the term "sustainable landscape" mean? There are varying definitions but sustainable landscape should include an attractive environment that is in balance with the local climate and requires minimal resource inputs, such as fertilizer, pesticides, gasoline, time, and water. A sustainable landscape saves water, money, and labor. It's easier to live with and it's better for the environment. A sustainable landscape is affordable, beautiful, and good for the planet. It works right, demands little, looks great, and rewards its users every day.

A sustainable landscape's long-term is always accomplished through series of short-term goals. The sustainable development agenda should be more geared toward learning processes than projected future outcomes [23].

2.2. *Eco-control*

Eco-control is to adapt the sustainable development in the world including all industries, especially the landscape architecture, which refers to wide fields of human, aesthetics, engineering, vegetation, ecology, etc [24]. Landscape architecture, as a maturing academic discipline urgently needs to develop its methodological repertoire to generate new knowledge, has become quite persistent over the last decades [25–27].

3. Eco-Control System of Sustainable Landscape Design

3.1. *Construction of the eco-control system*

The eco-control system has a construction of "1+2+3+3". "One" refers to a process control, to control in the whole life-recycle process of landscape design; "Two" refers to the two dimensions both theory and practice; "Three" means that the system includes macro-medium-micro scales control. The 2nd three means three sides of economy, culture and nature, each of them responses to cost, public participation and environment (Fig. 1). And all the above makes a complete cycle, with the development of new technique, the new cycle will move toward.

3.2. *Basic research*

It is necessary and basic to analyze the current status and influencing factors of modern sustainable landscape design, to explore the existing problems in the modern landscape design, developing trend and impaction on human. Define the concept of sustainable landscape design, sustainable landscape design characteristics, attributes, and relationships with the landscape design.

3.3. *Eco-control theory system*

It includes macrocosm of planning & layout control, medium control of design management, and microcosm of engineering control three scales. The system runs through the whole life cycle of the design process, they are interacted and mutual support, the macro-medium control needs micro engineering technology, the macro control makes it possible for medium and micro control, guarantying the order of the whole design, the micro control is the direct stage to realize the sustainable landscape design.

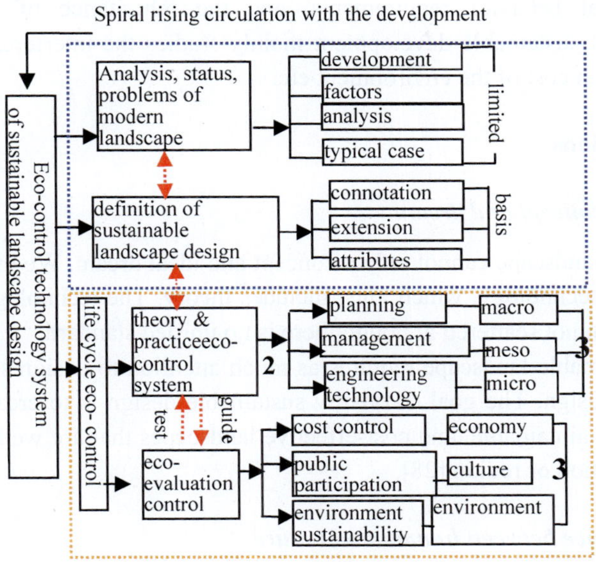

Figure 1. The eco-control system of sustainable landscape design.

3.4. *Eco-control technology system*

The macro eco-control mainly includes "3S" technology, regional eco-security evaluation, urban ecological planning, ecological network system, etc; Micro mainly studies the cost control and human participation technology; Microscopic eco-control mainly studied the integration of all kinds of engineering and technology, such as multidisciplinary technology: phys-technology, biotechnology, wetland restoration technology, closing to the nature technology and so on. The four sustainable landscape element subsystems of terrain, water, plant and hard landscape were researched.

3.5. *Eco-evaluation control*

It is the way to verify the effect of the eco-control system, which includes three sides of the comprehensive cost control, regional cultural inheritance and environment sustainable development. And this stage is the post-life-cycle process of sustainable landscape design the eco-control after the completion of the works, the main target is to regulate and amend the before –control process, which will be the basis for the next cycle of sustainable.

Comprehensive cost control includes economic, environmental and human cost; culture control includes public participation, the local culture, the public's psychological behavior requirements, and the inheritance of site history; Environment sustainable development mainly studies the interference strength, the restoration cost of the environment etc.

4. Conclusions

4.1. *Not a concept and dream*

Sustainable landscape cannot only a concept and ideal dream. We must realize it by concrete technology, which must includes theory. The technology must be a whole system not scattered although there is no unified standard.

A sustainable landscape requires as much attention to design specifics as a traditional design. The goal of quality sustainable design is to create aesthetic, functional, maintainable, and cost-effective landscapes that are well suited for a specific location or region [28].

4.2. *A balance between human and nature*

Whether your interest is mainly in landscape or mainly in sustainability, whether it is mainly as a landscape theoretician or practitioner, or even in another disciplinary area [29], it must be a balance between human and nature, and human must control their action. It is the duty of all the world people; we must obey it without exception.

4.3. *The eco-control is open and developing*

Sustainability should be seen as a direction, rather than a concrete goal [30]. The control must be an open and developing system. It will be develop with the progress of society and technology.

Acknowledgment

The authors want to thank the anonymous reviewers for their helpful comments and suggestions. The authors are grateful to the support of the National Natural Science Foundation of China (Grant no. 31270747), and this is also supported by grant 2015lsktwx-07 of the Liaoning Provincial Federation Social Science

Circle 2015 Economic and Society Development Projects Foundation, the same thanks to them.

References

1. K.J. Yu, D.H. Li, J.P. Li, Landscape and ecological design of the city: concepts and principles, *J. Chinese Lands. Architecture*, 6, pp. 3–9(2001).
2. McHarg I. *Design with Nature.* John Wiley & Sons, Inc. (Garden City, NY, Natural History Museum Press, 1969).
3. Lyle J.T., *Regenerative design for sustainable development.* (John Wiley & Sons, Inc. 1994).
4. R. L. Thayer, J. Gray World, *Green Heart, Technology. Nature and the Sustainable Landscape*, (John Wiley & Sons Inc.1993).
5. E.S. Bernhardt, M.A. Palmer, J.D. Allan, et al. Synthesizing US river restoration effort. *J. Science*, 308, pp. 636–637(2010).
6. J.C. Marks, G.A. Haden, M. O'Neill, et al. Effects of flow restoration and exotic species removal on recovery of native fish: Lessons from a dam decommissioning. *J. Restoration Ecology*, 18, pp. 934–943(2010).
7. M.A. Palmer, E.S. Bernhardt, J.D. Allan, P.S. Lake. Standards for ecologically successful river restoration. *J. Journal of Applied Ecology*, Vol. 42, 2, pp. 208–217(2005).
8. S. Postel, B. Richter, *Rivers for Life: Managing Water for People and Nature*. (Washington DC: Island Press, 2003).
9. P.B. Shafroth, M.K. Briggs. Restoration ecology and invasive riparian plants: An introduction to the special section on Tamarix spp. in western North America. *J. Restore Ecology*, Vol. 16, 1, pp. 94–96(2008).
10. Y. Zhang, Y.Y. Li, (America) P. Melby, T. Cathcart. *Sustainability: landscape design technology*. (Beijing, Mechanical Industry Publishing Housem, 2005).
11. B.X. Chou. *To carry out the economical landscape, promoting the sustainable development of city*, in the national economical landscape scene's speech at the meeting, on August 17(2006).
12. Y.Y. Ye, H.O. Zhang, "ecological orientation" of urban spatial structure research review, *J. City Planning*, Vol. 32, 5, pp. 69–74+82(2008).
13. S.H. Wang, Based on the concept of low carbon city landscape ecological design study. *J. Ecological Economy*, 12, 192–195(2010).
14. J.R. Gao, Close to nature management — Based on the landscape ecology of the waste stream control engineering, *J. Journal of Beijing Forestry University*, Vol. 21, 1, pp. 80–85(1999).

15. P.F. Wang, C. Wang, F. Qian, J. Qian, J.R. Zhou, Urban water ecological system construction model research progress, *J. Journal Of Hehai University (Natural Science Edition)*, Vol. 3, 5, pp. 485–489(2003).
16. H.L. Liu, The ecological restoration of mining wasteland and sustainable landscape design. *J. Acta Ecological Sinica*. Vol. 24, 2, pp. 323–329(2004).
17. Y.X. Wu, S. Yu, Based on the theory of landscape ecology, landscape cost minimization approach, *J. Journal Of Harbin Institute Of Technology (Social Science Edition)*, Vol. 13(3), 82–88(2011).
18. L.C. Huang, S.H. Ye, et al, A sustainable landscape ecosystem design — A case study, *J. Annals of the N. Y. Academy of Sci.* 1195 (S1), E154–E163 (2010).
19. L.C. Huang, Z.Q. Fan, et al. Three-dimensional ecological planning — An effective approach to control the urban sprawl, *IEEE*, 4302–4305(2008).
20. L.C. Huang, X. Gu, et al. The Ecological Control method of the Highway Construction — A Case Study, *ESIAT*, 1, 171–174(2009).
21. Na Wu, Low carbon ecological technology in the application of landscape engineering, *J. Green Technology*, 3, 32–343(2011).
22. F. Steiner, Landscape ecological urbanism: Origins and trajectories, *J. Landscape and Urban Planning*, 100, pp. 333–337(2011).
23. Tony, M.R.G, Planning for sustainability as a learning concept. *J. Ecol. Econ*. 26, pp. 121–137(1998).
24. L.C. Huang, F.C. Cao, et al. Study on the Eco-Control System of Sustainable Expressway Landscape, *J. NEPT*, Vol. 12, 4, pp. 611–614(2013).
25. J.F. Benson, On research, scholarship and design in landscape architecture. *J. Landscape Research*, Vol. 23, 2, pp. 198–2042(1998).
26. R.D. Brown, & R.C. Corry, Evidence-based landscape architecture: The maturing of a profession. *J. Landscape and Urban Planning*, Vol. 100, 4, pp. 327–329(2011).
27. E.M. Deming & S. Swaffield, *Landscape architecture research, inquiry, strategy design*. (New Jersey, John Wiley & Sons, Inc.2011).
28. T. W. Cook, A. M. Van Der Zanden, *Sustainable Landscape Management*, (Hoboken, New Jersey, John Wiley & Sons, Inc. 2011).
29. J.F. Benson & M. Roe, *Landscape And Sustainability (Second Edition)* (London and New York, Taylor & Francis group press, 2007).
30. R.T.T. Forman, *Land Mosaics: The Ecology of Landscapes and Regions* (New York, Cambridge University Press, 1995).

Analysis of the spatial difference amounts of acid deposition in Urumqi, NW, China*

J. Wei*
School of Resources & Environment Science, Xinjiang University,
Urumqi 830002/XinJiang, China
E-mail: cd95hj@sina.com
www.xju.edu.cn

X.H. Yang
School of Geography Science and Tourism, Xinjiang Normal University,
Urumqi, 830054, China
E-mail: 294858242@qq.com
www.xjnu.edu.cn

In order to reveal the differences objectively and accurately of amount of acid deposition in different areas of Urumqi, the monitoring sites of acid deposition have been laid in three different areas of the city, monitoring of acid deposition in accordance with the technical specifications of collecting samples for analysis. The results showed that pollution of acid deposition in Urumqi still remains serious, and the value of the acid deposition in different regions are significantly different, which of southern area is decreasing, while the northern area shows the increasing trend of acid deposition value. From the layout of monitoring sites, the northern area, as the reference point of the urban monitoring sites background value of acid deposition at first that the value was obviously higher than other regions, so, it almost lost the meaning, which suggests that the local environment department should take steps to keep the valid monitoring data of background site.

Keywords: Acid Deposition; Spatial Difference; Ion Concentration; Air Pollution.

1. Introduction

Natural precipitation plays an important role in removing the pollutants in the atmosphere [1]. Recently, the researchers have evaluated and judged the atmospheric environmental quality and the precipitation mechanism through analyzing the chemical composition of precipitation samples. [2–4] From the view of the current domestic research, the ion precipitation mainly includes $SO4^{2-}$, NO_3^-, NH_4 + and Ca^{2+}, presenting that the north is higher than that in the

This work is supported by National natural Science Foundation of China (41161074).

south which still exists limitation. The evaluation focused on the discussion and analysis for precipitation pH and the frequency of acid rain, not on the depth discussion for acid deposition. However, the foreign studies on Acid Deposition are based on settlement which are looked as the important basis of evaluating pollution [7–9]. Acid Deposition includes not only acid rain frequency and precipitation pH, it also covers the settlement, an important indicator. As for precipitation pH, it is influenced by the atmospheric environment of combined action of acid and alkaline substances. Because of the different geographical position, there is a big difference between the atmospheric concentrations of acid and alkali material. Therefore it is the ion settlement that has an effect on the environmental ecosystem, not the ion concentration basing on a certain precipitation, which reflects the absolute amount of chemical constituents in each precipitation. It not only considered the effect of rainfall, but also solved the properties of the ion concentration which can not be accumulated in past research [10]. Therefore, acid deposition seems more scientific in the research of ecological and environment effects.

Urumqi is a typical valley city with a large amount of pollutant emissions during economic and social development, which formed a severe atmospheric environment pollution in partial areas [11]. But as the urban extension and industrial structure adjustment, the original pattern of atmospheric pollution is broken and pollutant concentration in space along with the run of the city presents the trend to north [12]. To explore the spatial difference of acid deposition, this experiment was set in three monitoring points including the North, the South and the middle area and through the calculation of the acid precipitation quantity to discuss its difference in the process of urban development.

2. Research Method

2.1. *Monitoring sites and methods*

According to the site principles of technical specifications for acid precipitation monitoring, we set three monitoring sites in the southeast area (SA) of Urumqi, middle area (MA) and the north area (NA) respectively (Figure 1). The sampling sites were set according to the national standard, which were away from the local pollution sources and without obstacles in 2 meters. Sampling method was made according to the specification of the acid precipitation monitoring technology (HJ/T165-2004) regulation, using the wet deposition sampler to collect information about the atmospheric precipitation samples.

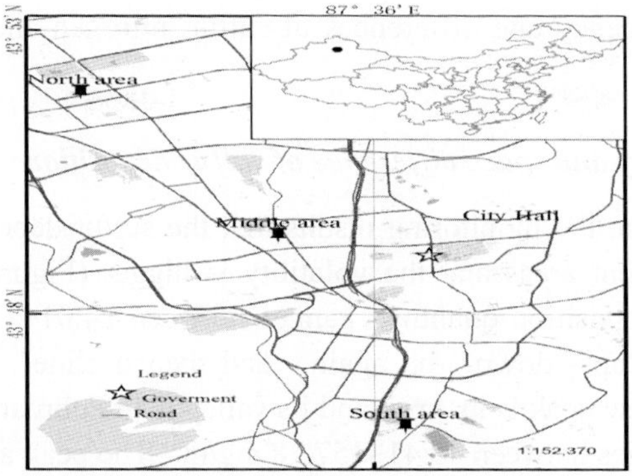

Figure 1. Location of monitoring sites.

2.2. *Monitoring projects*

According to the national standard method of precipitation samples for laboratory analysis, the project includes precipitation, SO_4^{2-}, NO_3^-, Cl^-, NH_4^+, Ca^{2+}, Mg^{2+}, K^+ and Na^+. Data testing adopts three-step examining and verifying to make sure the correctness of the data.

2.3. *Index calculation method for wet deposition*

The average of each ion concentration was calculated by using the rainfall weighted average while wet deposition quantity was calculated by the general method. That is to say, the annual wet deposition quantity for any ion was calculated as a certain component settlement to 1m^2 soil in the sampling time [10].

$$D_{wi} = \sum_{j=1}^{n} C_{ij} Q_j$$

In the formula: Cij means the concentration of ion i in jth precipitation (mg/l), Qj (mm) is jth rainfall, n is for the rainfall total number; Dij is the wet deposition quantity for the ith ion.

The contribution to Sulfur flux is wet deposition quantity of SO42- multiplies a third and that to Nitrogen flux is wet deposition quantity of NO3- and NH4+ multiplies by 0.23 and 0.78 respectively.

3. Research Results

Through the data, it shows that ions of the atmospheric precipitation in Urumqi are mainly SO4^{2-}, NO_3^-, NH_4^+ and Ca^{2+} [13–15] which have an obvious contribution on acid deposition. Therefore, this research was concentrated on analyzing the tendency and differences of sulfur, nitrogen and calcium in acid precipitation.

3.1. *The tendency and space difference of sulfur deposition*

It can be seen from the monitoring results that the sulfur deposition quantity is different in different areas and the volatility is bigger (Figure 2). In southern area, the sulfur deposition quantity changes between 24.01~130.68 mg/m^2 and the trend shows "up - down - up again - and then decline". The whole trend presents “M” whose peak is in April and its valley is in February. In middle area, its volatility changes between 45.41~157.28 mg/m^2. The peak appears in January, then shows “decreasing - increasing - turning down” trend while its valley is in May with a small uptrend afterwards. The biggest change takes place in northern area fluctuating between 38.95~236.01 mg/m^2 as “M”. Its peak is in April at 236.01 mg/m^2 and its secondary summit appears in June at 181.31 mg/m^2 with a down trend till December reaching the valley. From the space change of sulfur deposition quantity, it is lower in Southern area as a whole than that in other two districts while it is much higher in Northern area from March to July than that in southern and Middle area and then after August it goes down obviously lower than that in other two areas. It can be seen from calculating formula that the settlement relates to rainfall and ion concentration. It shows through the correlation test that the sulfur deposition in southern, middle and Northern area has little relationship with rainfall and it still exists differences with significant correlation of ion concentration, the correlation index is ($R = 0.698$, $p < 0.05$ and $R = 0.712$, $p < 0.01$ and $R = 0.903$, $p < 0.01$). That is to say, as the difference of geographical position and meteorological conditions in different regions, the settlement is different. The correlation between acid settlement and the ion concentration present obvious increasing trend from south to north.

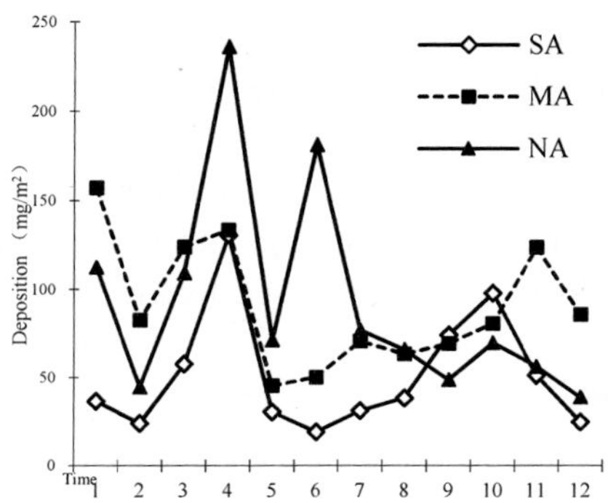

Figure 2. Trends of sulfur deposition in different areas.

3.2. *The tendency and space difference of nitrogen deposition*

From the trend of nitrogen deposition, it is almost the same with a certain difference in volatility in different areas. Variations in southern, middle and northern area is 16.93 ~ 85.26 mg/m^2, 21.46 ~ 51.65 mg/m^2, and 15.51 ~ 44.72 mg/m^2 peaking in September, January and April respectively (Figure 3). Due to the settlement of TN is made up of NO_3^-- N and NH_4^+ -N which has a close relation with rainfall. It is concluded from the correlation test (Table 1) that TN settlement in Southern area has significant correlation with NH_4^+ -N, NO_3^-- N and rainfall while that in Middle area is only related with NH_4^+ -N. The correlation between TN and NH_4^+ -N, NO_3^-- N is more significant while that is not obvious with rainfall.

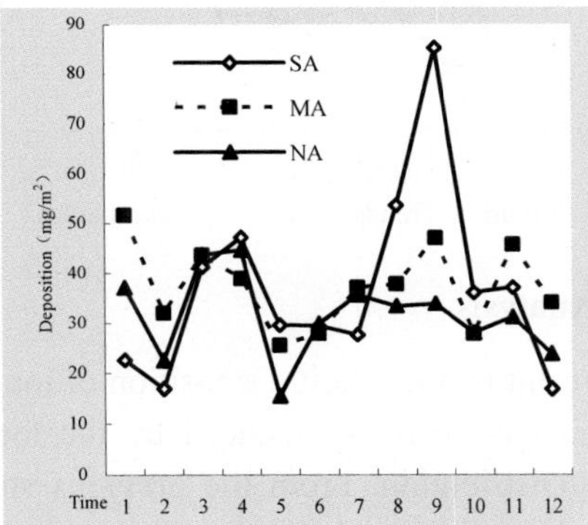

Figure 3. Trends of nitrogen deposition in different areas.

Table 1. The correlation coefficient of TN in different areas.

TN	NO_3^--N	NH_4^+-N	Rainfall
Southern area	.963**	.974**	.972**
Middle area	0.248	.798**	0.317
Northern area	.624*	.825**	0.361

*. At the 0.05 level (bilateral) significant correlation.
**. At the 0.05 level (bilateral) significant correlation.

3.3. *The tendency and space difference of calcium deposition*

As the presentation in the settlement curve of Ca_2^+, you can see that the relatively minimum 46.56 ~ 46.56 mg/m^2 is in southern area with the highest and lowest point in August and February respectively; Middle area is larger between 19.11 ~ 216.57 mg/m^2 with its peak and valley appear in August and December; Northern area held the maximum 89.78 ~ 248.3 mg/m^2 whose peak and valley appear in April and December (Figure 4). Through the curve trends, southern and Middle areas are very similar and the apparent difference is just on the order of magnitude. The variations of Ca_2^+ settlement in Northern area is larger than that in other two districts during March and July while the rest of the time shows alternation. Through the correlation test, in addition to the significant correlation between Ca_2^+ settlement and Ca_2^+ ion concentration ($R = 0.618$, $P < 0.05$), factors in other areas have no significant correlation with each other.

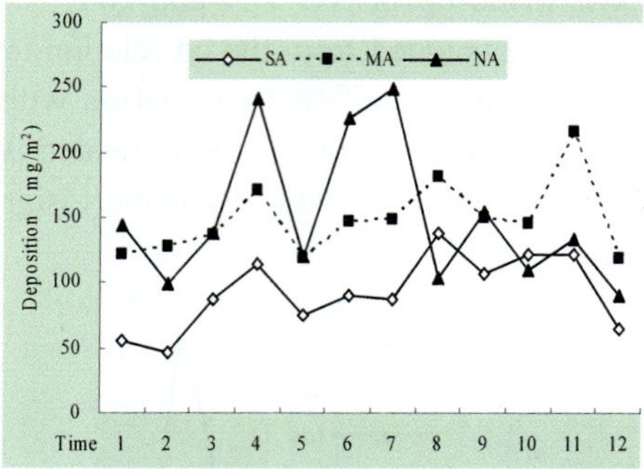

Figure 4. Trends of Ca_2^+ in different areas.

4. Discussion and Analysis

We can see from the result that the Acid Deposition of ion differs from different areas of the city, which is deeply influenced by the locating urban function region and Industrial Distribution. From the previous studies we can draw a conclusion that the air pollutants vary greatly as the changes of seasons. However,similar law isn't found in the studies of Acid Deposition. In the past, it's usually consider that the south area of the city suffers the most serious pollution of SO_2 [12]. However, we can see from the Monitoring that sulfur deposition here is much low than that of the middle area and north area of the city. This also indicates that this maybe have necessary relations with urban industry restructuring and city distributing. Primary heavy-polluted corporate had removed from the old town, south area of the city, besides, companies didn't move also reduce the emission of sulfur pollution by taking measures like improving technology and energy conservation. North area of the city, on the other side, as the development of the city goes further, more and more companies gather here, together with the existed urban-rural fringe small companies, emission of sulfur pollution increases severely, which also cause the increasing of the sulfur deposition. We can see from the related inspection that different sulfur deposition has less about natural precipitation. Sulfur deposition mainly concerns about space ion concentration. What's more, the difference between sulfur deposition volume and ion concentration shows that from south to north, sulfur deposition has an intimate connection concentration with space ion concentration.

TN settlement also presents different characteristics in different regions. It shows from the correlation test that TN settlement is greatly correlated with NH_4^+-N, NO_3^--N and rainfall explaining that Nitrogen settlement is influenced or interfered more in this area while TN settlement is related to NH_4^+-N significantly and associated with NO_3^--N and rainfall, but not significantly, showing that the regional nitrogen deposition mainly originates from the contribution of NH_4^+-N. TN settlement in northern area is mainly from NH_4^+-N and NO_3^--N. From the point of correlation coefficient, the contribution of NH_4^+-N is bigger than that of NO_3^--N, which is associated with the large number of farmland. It is known that the farmland contributes lots of ammonia to the atmospheric environment, deriving from using a large amount of ammonia in the process of growing crops and vegetation. It changes with seasons, and travels to the surface through natural precipitation, which is consistent with previous research conclusion [13,17].

The settlement difference of Ca_2^+ shows the same characteristics with the former, that is to say, different regional settlement curve has obvious difference. From the previous studies, the concentration of Ca_2^+ is not only related to fossil fuel combustion emission, but also has a certain relationship with the physical and chemical properties of soil in different regions [15,18]. The monitoring results show that Ca_2^+ settlement were higher in northern area which has a close relationship with the geographical position. The surrounding large farmland and the bare surface are likely to be the contribution of Ca_2^+ sources. But the low correlation between Ca_2^+ settlement and the space ion concentration show that the Ca_2^+ in atmosphere may spread out with the atmospheric movement or travel to the surface in the form of dry deposition.

5. Conclusion

We found that air environmental pollution in Urumqi is still serious by analyzing the data of Acid Deposition in different areas which have different deposition features. The results showed that the original serious polluted area Southern area becomes better and it has lower Acid Deposition and Ca_2^+ deposition than other districts while the known Northern area with better environmental quality is turning to have higher deposition. The results indicate that the atmospheric environment of Northern area has been polluted severely in the process of city development and industrial structure adjustment, therefore it is important to pay more attention to its environmental protection in the future work.

The data of Acid Deposition monitoring sites can objectively reflect different districts' atmospheric pollution. The monitoring site in northern area was the control points (or as background values) for Acid Deposition without any change afterwards. But based on the expanding of Northern area and the changing of industrial distribution and locating of the urban function, it is meaningless as the morning data has exceeded others. It is very important to choose the background values under effective treatment. Therefore, the environmental protection department should adjust the control point site in time in the procedure of planning urban and industry development in order to judge the pollution status of atmospheric environment scientifically.

References

1. W.X. Wang, G.A. Ding (1997). The Geographical Distribution of Ion Concentration in Precipitation over China [J], Research of Environmental Sciences 10(2):1–6.
2. W.X. Wang, Z.W. Feng, X.Y. Tang, et al. Acid Deposition and Its Ecological Impact, "Eight Five" National Scientific and Technological Project 852912201 [R] Beijing: Chinese Research Academy of Environmental Sciences, 1996, 65–100.
3. L. Jin, Q. Xu, A.G. Lin, et al. 2006. Trend analysis of wet deposit ion between 1987–2004 in Beijing City [J]. Acta Scientiae Circumstantiae, 26(7): 1195–1202.
4. Driejana, D.W. Raper, D.S. Lee, et al. 2001. Wet deposition to an up load area of England in 1988 and 1999: Measurements and modeling [J]. Water, Air, and Soil Pollution, 130: 1625–1630.
5. W.X. Wang, P.J. Xu (2009) Research Progress in Precipitation Chemistry in China [J], Process in Chemistry, 21(23): 266–281.
6. F.M. Yang, K.B. He, F. Lei, Ma YL etc (2004), Chemical character of atmospheric precipitation in Beijing years of 2001~2003. [J]. China environment science, 24(5): 538–541.
7. Driejana, D.W. Raper, D.S. Lee, et al. Wet deposition to an up load area of England in 1988 and 1999: Measurements and modeling [J]. Water, Air, and Soil Pollution, 2001, 130: 1625–1630.
8. E.P. Farrell, J. Aherne, G.M. Boyle, et al. Long term monitoring of atmospheric deposition and the implications of ionic inputs for the sustainability of a coniferous forest ecosystem [J]. Water, Air, and Soil Pollution, 2001, 130: 1055–1060.

9. A.G. Pillai, S.N. Medha, G.A. Momin, et al. Studies of wet deposition and dustfall at pune, India [J]. Water, Air, and Soil Pollution, 2001, 130: 475–480.
10. X.Y. Tang. Acid Deposition Chemistry, Atmospheric Chemistry [M] Beijing: Higher Education Press, 1990, 226–286.
11. J. Wei, G.H. Wang, Q. Ren (2011) Study on Quantitative Models of Concentrations of Air Pollutants over Urumqi (J) Arid Zone Research, 28(5): 896–899.
12. X.G. Chen, J. Wei, Q. Ren, K.L. Zhang (2013) Effects of the Urban Expansion on the Spatially Varying Trends of Distribution in Air Pollutants Concentration: A Case Study of Urumqi. Ecology and Environmental Science. (J) 22(6): 1015–1019.
13. J. Wei, Z.H. Liu, C.X. Zhai (2012) Characteristics Analysis of Ion Wet Deposition Fluxes of Atmospheric in Urumqi, Journal of Ningxia University (Natural Science Edition), 01:116–119.
14. X.F. Lv (2013) Study on Temporal and Spatial Changes of Chemical Composition of Rainfall in Urumqi, Xinjiang. [J] Arid Environmental Monitoring. 27(2):55–59.
15. J. Wei (2012) Analysis on Wet Deposition in Urumqi during the Period from 2000 to 2009, Arid Zone Research, 01:116–119.
16. J. Wei (2012) Air pollution and spatial variation of pollutants in Urumqi, Journal of Arid Land Resources and Environment. 26(1): 67–70.
17. H.Y. Zuo, H. Gao, J. Chen (2006) Measurement of NO2 Concentration in Atmosphere in Chengdu by Solar Spectra (J) Spectroscopy and Spectral Analysis 26(7)1356–1359.
18. Y.F. Han, G.N. Sun, Q. Li, S.J. Chen (2006) Analysis on temporal changes of acid rain and its chemical composition in Xi'an City, [J] Journal of Shanxi Normal University (Natural Science Edition) 34(4): 109–113.

The effects of light intensity on feather pecking and other behavioral activities of laying hens*

C.P. Wang

College of Life Science, Jiamusi University,
Jiamusi, 154007, China
E-mail: wcp100200@163.com
www. jmsu.org

Y. Wang

College of Food and Biological Engineeriny, Qiqihar University
Qiqihar, China
E-mail: wangyan0468@126.com

C.B. Wei

College of Animal Science and Veterinary Medicine,
Heilongjiang Bayi Agricultural University,
Da qing, China
E-mail: weicb298@163.com

From 10 week to 12 week, 160 Hailan layers were assigned at random to groups of 5 individuals and distributed among 32 pens. They were kept in pens with four different light intensity (5lux, 10lux, 15lux, 20lux). We observe the feather pecking, feeding groveling, preening, foraging, drinking, standing and walking. We find that light intensity had significant effects on rate of feather pecking. With the light intensity increasing, the occurrence of feather pecking, preening, foraging, drinking, standing and walking increased, the occurrence of feeding and groveling decreased. The score of damaged feather of laying hens was increased with the increasing of light density, but the damage of feather was little. When feather pecking occurred, the main parts of body pecked by laying hens was followed by tail (38.7%), back (23.6%) and head (19.8%). The behavior of the pecked laying hens was groveling (33.5%), standing (30%), feeding (30%).

Keywords: Light intensity; Feather pecking; Behavior activities.

1. Introduction

Feather pecking (FP) is a type of abnormal behaviour in poultry that consists of pecking at feathers of other birds, sometimes pulling the feathers out and often

---

* This work was funded by Natural Science Foundation of China (31302008) and Natural Science Foundation of Heilongjiang Province (QC2012C104; QC2013C031).
Corresponding author: C.B. Wei

eating them (Blokhuis and Wiepkema 1998). It poses major welfare problems for the hens, and constitutes an economical burden for the farmer. Hens peck and pull at the feathers of conspecifics, causing damage to the plumage and loss of feathers. This adversely affects the costs of egg production, since loss of feathers results in considerable higher energetic needs and, consequently, higher feed requirements of hens as a result of increased body heat loss (Leeson and Morrisson, 1978; Tullet et al., 1980). Furthermore, feather pecking, which is painful in itself (Gentle and Hunter, 1990), may result in severe damage to the integument of the birds, including wounds of the skin. Wounded birds may become the victim of cannibalism (Allen and Perry, 1975; Blokhuis et al., 2000; Hughes, 1982). This may result in a loss up to 15% of the birds per production cycle (in alternative housing systems) (Keeling et al., 1988).

It has been shown that many factors such as stocking density, group size, dietary composition, light intensity and genetic background can influence the occurrence of feather pecking (Hughes & Duncan, 1972; Blokhuis, 1989; Vestergarrd, 1994; Savory, 1995).

2. Material and Methods

2.1. *Subjects and husbandry*

The experiment was carried out with a total of 160 Hailan layers assigned at random to groups of 5 individuals and distributed among 32 pens. The birds were not beak-trimmed and had been reared in an cage system, the cage floor was covered with nothing. Feed and water were given ad libitum. Incandescent light gave different light intensity. Providing different light intensity (5lux, 10lux, 15lux, 20lux) through adjust the number of incandescent light. No animal died during the experiment.

2.2. *Procedures*

Adopting airtight manual light in farm house. Incandescent light gave different light intensity. Providing different light intensity (5lux, 10lux, 15lux, 20lux) through adjust the number of incandescent light. Also using luminometer to control light intensity. The farm house was separated to four districts by using black plastic. The row of 32 pens was subdivided into 4 blocks.

2.3. *Behavioral observations*

Focal animal observations were carried out to investigate behaviour of Feather pecking, feeding, groveling, preening, foraging, drinking, standing and walking in more detail. Chosing a focal animal in each group, and the focal animals were marked individually with black paint. Data collection by one persons, ran from week 10 to week 12. The observer sit on the chair, which distance to the ground about 2 m. Thus the observer can scan the focal animal in each group. Hens were observed 160 min every day, in the morning between 09:30 and 12:10. Each group was observed for 10 min. Recording one behaviour data in 5 second. Thus we can obtain 120 behaviour datas on the focal animal in each group. The percentage of chicks engaged in each behaviour was calculated from the total number of chicks observed during all scan samples in each pen.

2.4. *Behaviour parameter*

We observed the behaviour of focal animal in each pen, we adopted scan methods to obtain behaviour data. The all occurrences sampling was briefly interrupted every 5 second for a scan sample of the activity of chicks. For each scan we recorded the number of chicks engaged in eight mutually exclusive activities: feather pecking, feeding, groveling, preening, foraging, drinking, standing and walking. They were defined as follows: (1) feather pecking: feather pecking interaction was attributed to 1 of the following 4 types of behaviour (increasing intensity): a) pecking a feather without pinching, b) pinching a feather and pulling weakly, c) pulling at a feather with a vigorous backward movement of the head and d) plucking a feather. (2) feeding: the chick was by the feeder with its head above the food; (3) groveling: the chick showed no other defined behaviour while its chest feathers touched the ground. (4) preening: the chick nibbled, stroked or combed its plumage with its beak (Kruijt 1964) or stretched its wings or legs; (5) foraging: the hen is pecking at the floor or at other parts of the pen (but not at the food) or it is standing/moving with its head in a lower position than the rump. (6) drinking: the chick's bill was oriented to, and within 5 cm of the cup drinkers; (7) standing: standing on both feet without showing another defined behaviour; (8) walking: locomotion without showing another defined behaviour.

2.5. *Partitions of the body pecked by laying hens*

Focal birds from each pen were individually scored at 10 weeks and 11 weeks. The body was divided (Gunnarsson et al., 1995) into 11 regions: head, upper neck (back side of the neck), back (part between wings), rump, tail, belly (abdomen), breast, under neck (front side of the neck), wings–primary feathers, wings–coverts and legs.

2.6. *The score of feather damage*

The plumage was scored for damage at 4 and 5 weeks old. Each of these areas was scored separately as follows: 0, no damage; 1, slight feather damage/loss with no bare skin; 2, $<1cm^2$ skin denuded; 3, up to $5\times5cm^2$ bare skin, or $<1cm^2$ bare skin with minor haemorrhage; 4, $> 5\times5cm^2$ bare skin, or up to $5\times5cm^2$ bare skin with $<1cm^2$ haemorrhage; 5, 1 to $2cm^2$ haemorrhage, or $> 5\times5cm^2$ bare skin, with $<1cm^2$ haemorrhage.

2.7. *Statistical analyses*

The pens were treated as independent units (n=32) in all analyses. Mean values for the behavioral activities, score of feather damage were calculated for each pen. The analyses were performed using SPSS and Microsoft Excel. All statistical tests are 2-tailed with an alpha level of 0.05.

3. Result

3.1. *Behavioral observations*

Light intensity had significant effects on rate of feather pecking, which was highest in hens provided with high level of light intensity ($P<0.01$).

As to other 7 behaviour, With the light intensity increasing, the occurrence of preening, foraging, drinking, standing and walking increases, the occurrence of feeding and groveling decreases. (Table 1)

3.2. *Partitions of the body pecked by laying hens*

From the total of 313 feather pecks (gentle and severe) observed throughout the 32 pens, 121pecks were targeted towards tail (38.7%), 74 on back (23.6%), 62 on head (19.8%), 16 on wings-primary feathers (5.1%), 15 on neck (4.7%), 14 on belly (abdomen) (4.5%), 4 on rump (1.3%), 2 on legs (0.64%), 2 on breast (0.64%), 2 on under neck (front side of the neck) (0.64%), 1 on wings-coverts (0.32%).

3.3. *The score of feather damage*

The score of injured feather of the laying hens was increased with the increasing of light intensity (Figure 1) but the damage of feather was little.

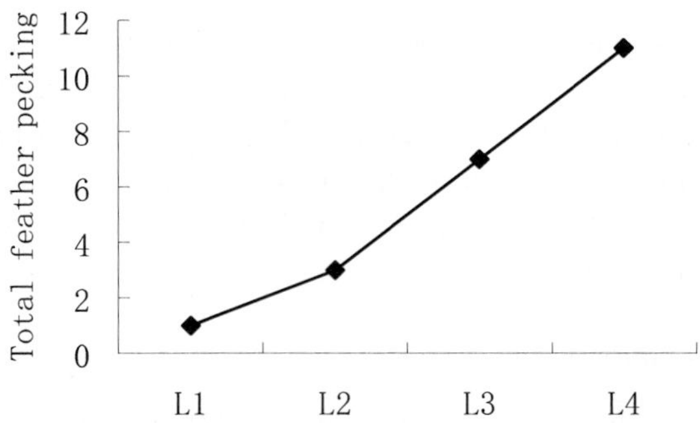

Figure 1. The influence of light density to the record of the total score of feather damage.

Table 1. Effects of light intensity on feather pecking, feeding groveling, preening, foraging, drinking, standing and walking, (per 4 pens per 40 min). Means as well as minimum and maximum values of 4 pens each per housing condition are given. P values derived from ANOVA.

	Light intensity				
Behaviour	5lux	10lux	15lux	20lux	P values
Feather pecking	0.83±0.17	1.25±0.18	1.67±0.41	2.42±0.36	P<0.01
Feeding	51.67±2.71	49.83±1.12	38.08±4.35	35.50±2.53	P<0.05
Groveling	35.67±2.66	32.17±6.70	29.58±3.72	26.75±3.21	P<0.01
Preening	16.75±1.41	18.42±3.37	23.17±3.42	24.92±1.23	P<0.05
Foraging	4.17±0.37	7.08±1.56	8.67±0.72	9.58±1.03	P<0.05
Drinking	4.08±0.48	4.42±0.56	5.08±0.94	5.25±1.33	P<0.05
Standing	29.08±1.84	31.67±1.49	34.33±2.09	36.83±4.02	P<0.05
Walking	2.33±0.50	2.50±0.47	3.08±0.68	4.58±0.43	P<0.01

3.4. *Behaviour of the pecked laying hens*

When the behaviour of feather pecking occur, the pecked laying hens were groveling (33.5%), feeding (30%), standing (30.0%), drinking (2.6%), preening (2%), foraging (1%), walking (0.64%).

4. Discussion

Our experiment provide four different light intensity levels: 5lux, 10lux, 15lux, 20lux to verify how the intensity of light influences the incidence of feather pecking in detail. The intensity of light influences the incidence of feather pecking. It is generally agreed that increasing the brightness of the light increases the level of feather pecking (Allen and Perry, 1975; Hughes and Black, 1974; Kjaer and Vestergaard, 1999).

It is important to understand how light can affect birds. Factors that are important to laying hens production include intensity (brightness), photoperiod (duration), and wavelength (color). But our experiment merely consider the influence of the light intensity. So our next experiment is to consider the influence of photoperiod (duration), and wavelength (color).

Keeling (1994) divided feather pecks according to the force of the peck and the reaction of the pecked bird into 'gentle' (light pecks, often in bouts) or 'severe' (forceful pecks, pecked bird moves away). Leonard et al. (1995) used the term allopeck for all types on the other bird and subdivided allopecking into allopecking (light touch, nibbling motion of the bill), light pecking (as previously but without the nibbling motion), aggressive pecks (forceful pecks), pulling (grasping and pulling of feathers) and toe pecks. But our experiment did not divided feather pecks. So we did not distinguish which form of feather pecking is relation to plumage damage.

The different peck time of different partitions of the body pecked by laying hens is mainly due to the structure of different body part and the degree of easy to be pecked. Wood-Gush & Rowland (1973), in an experiment with pen-housed Brown Leghorns, reported that most feather pecks were delivered to the rump, followed by the tail and back. Savory & Mann (1997) observed that most feather pecks were deliver on the back in a group of Hisex hens and on the thigh in White Leghorns. Norgaard-Nielsen et al. (1993) found feather pecking on the breast and back of White Leghorns was the most pronounced and developed 1 St. B. Bilcik & L. J. Keeling (1999) find most feather pecks were on the tail, followed by the rump and back. Hierden et al. (2004a) find most feather pecks were on the tail and belly. In our experiment, our experiment find that when the feather pecking occurrences, the pecking chick mainly pecks the tail plumage of pecked chick (38.7%), next is back (23.6%) and head (19.8%). Tail plumage is the eastest to be pecked, which is due to feedstuff is provided adequately, feeding time is short; the space of movement is small, laying hens mainly grovel, stand and feed, thus tail plumage is very easy to expose, it make pecking chick to peck tail plumage easily. Back and head plumage is the easier to be pecked, which is due to laying hens mainly grovel in cage system, thus back and head plumage is very easy to expose, it make pecking chick to peck back and head plumage easily.

Assessment of plumage condition using feather scoring has been used by several authora as an alternative to behavioural observations. Hughes & Duncan (1972) estimated feather pecking damage by giving a single score (0 to 4) for the whole body. The same 5 point scale, but given separately for 4 body parts, was used by Allen & Perry (1975). Tauson et al. (1984) used a 4 point scale on 5 body parts. An even more detailed scoring method together with a photographic guide was published by Gunnarsson et al. (1995). They scored feather condition on 11 body parts, using a 3 point scale. Our experiment adopt the method of Gunnarsson et al., using a 5 point scale.

As to the behaviour of the pecked laying hens, when feather pecking occurrences, the most behaviour of the pecked chicks are groveling (33.5%), standing (30.0%) and feeding (30.0%). It is because of when the laying hens are groveling, standing, and feeding, they are easy to be pecked, also their body is exposed to other chicks.

References

1. L. Lamport, *LaTeX, A Document Preparation System*, 2nd edn. (Addison-Wesley, Reading, MA, 1994).
2. Allen, J., and G. C. Perry, 1975. Feather pecking and cannibalism in caged layer flocks. Br. Poult. Sci. 16:441–451.
3. Appleby, M.C. 1998. Modification of laying hen cages to improve behaviour. Poultry Science 77: 1828–1832.
4. Bilcik, B, Keeling, LJ 1999: Changes in feather condition in relation to feather pecking and aggressive behaviour in laying hens. Br Poult Sci 40: 444–451.
5. Blokhuis, HJ 1989: The development and causation of feather pecking in the domestic fowl. PhD. Thesis, Agricultural University, Wageningen 109 p.
6. Gentle, MJ, Hunter, LN 1990: Physiological and behavioural responses associated with feather removal in Gallus gallus var domesticus. Res Vet Sci 27: 149–157.
7. Gunnarsson, S., Odenk., Algers. B., Svedberg, J. & Keeling. L (1995) Poultry health and behaviour in a tiered system for loose housed layers. Report 35, Swedish University of Agricultural Sciences, Department of Animal Hygiene, Skara, Sweden.
8. Hughes, Bo 1982: Feather pecking and cannibalism in domestic fowls. In: BESSEI, W (Ed).: Disturbed behaviour in farm animals. Verlag Eugen Ulmer, Stuttgart, pp. 138–146.
9. Hansen, I. & Braastad, B O 1994. Effect of rearing density on pecking behaviour and plumage condition of laying hens in two types of aviary. Applied Animal Behaviour Science, 40, 263–272.
10. Hughes, Bo, Black, A.J., 1974. The effect of environmental factors on activity, selected behaviour patterns and 'fear' of fowls in cages and pens. British Poultry Science, 15, 375–380.

11. Hughes, Bo, Duncan, IJH 1972: The influence of strain and environmental factors upon feather pecking and cannibalism in fowls. Br Poult Sci 13: 525–547.
12. Keeling, L.J., Hughes, B.O., Dun, P., 1988. Performance of free-range laying hens in a polythene house and their behaviour on range. Farm Building Progress, 94, 21–28.
13. Keeling, LJ 1994: Feather pecking — who in group does it, how often and under what circumstances? Proceedings 9th European Poultry Conference, Glasgow, pp. 288–289.
14. Kjaer, JB, Vestergaard, KS 1999: Development of feather pecking in relation to light intensity. Appl Anim Behav Sci 62: 243–254.
15. Koelkebeck, KW, Amoss, MS Jr, Cain, JR 1987: Production, physiological, and behavioral responses of laying hens in different management environments. Poult Sci 66: 397–407.
16. Kruijt, J. P. 1964. Ontogeny of social behaviour inBurmese red junglefowl (Gallus gallus spadiceus). Behaviour, 12 (Suppl.), 1–201.
17. Leeson, S. & Morrison, W. D. 1978. Effect of feather cover on feed efficiency in laying birds. Poultry Science, 57, 1094–1096.
18. Leonard, ML, Horn, AG, Fairfull, RW 1995: Correlates and consequences of allopecking in White Leghorn chickens. Appl Anim Behav Sci 43: 17–26.
19. Nørgaard-Nielsen, G., J. Kjær, and H. B. Simonsen, 1993. Field test of two alternative egg production systems the Hans Kier System and the BOLEG II Aviary. Research report no. 9, National Institute of Animal Science, Research Centre Foulum, pp. 1–89 (Danish with English summary).
20. Savory, CJ 1995: Feather pecking and cannibalism. World Poultry Sci J 51: 215–219.
21. Savory, CJ, Mann, JS 1997: Behavioural development in groups of pen housed pullets in relation to genetic strain, age and food form. Br Poult Sci 38: 38–47.
22. Simonsen, HB, Vestergaard, K, Willeberg 1980: Effect of floor type and density on the integument of egg-layers. Poult Sci 59: 2202–2206.
23. Sokal, R.lo R. & Rohlf, F. J. 1981. Biometry. 2nd edn. New York: W. H. Freeman.
24. Tauson, R., T. Ambrosen, and K. Elwinger, 1984. Evaluation of procedures for scoring the integument of laying hens—Independent scoring of plumage condition. Acta Agric. Scand. 34:400–408.
25. Tullett, S.G., Macleod, M.G. & Jewitt, T.R. (1980) The effects of partial defeathering on energy metabolism in the laying fowl. British Poultry Science, 21: 241–245.

Advances on regional water environment carrying capacity

R.M. He*, J.Y. Zhang, X.J. Wang and G.Q. Wang
State Key Laboratory of Hydrology-Water Resources and Hydraulic Engineering, Nanjing Hydraulic Research Institute, Nanjing, China
**E-mail: hrm0928@126.com*

Water is essential for all creatures on the earth and the Water Environment Carrying Capacity (WECC) is a critical factor that extensively influences the pace and the scale of economic and society development. The preconditions and the foundation to protect water environment and utilize water resource include: Developing a WECC info system, analyzing the features of each factor and the interrelations among them, revealing the mechanism and condition of water environment sustainable carrying capacity, and implementing an accurate WECC assessment are. In this paper, the basic theory and the research background of WECC are firstly introduced, followed by the review and investigation on the state of the art assessment methods of WECC. Based on the systematic analysis on the features of various methods, the existing problems are proposed, and some suggestions for the further study in this area are given by the authors.

Keywords: Water environment carrying capacity; Assessment methods; Development trend; Problems.

1. Introduction

Due to the growing problems of water resource shortage, environmental pollution and ecology system depravation, Water Environment Carrying Capacity (WECC) began to draw world-wide attentions. The key issues for this problem are to maintain a stable and corporative relationship between the human society and the nature, and to build a reliable system for flood prevention, stable water supply and healthy environment. With the rapid growth of economy, the need for water use is significantly increased. Correspondingly, it caused a huge increase on the amount of wastewater. However due to the limited capacity of wastewater treatment and the imperfect management system of water resource, the pollution problem in river basins became extremely severe, which is regarded as one of the bottlenecks for regional economic and society developments.

2. Water Environment Carrying Capacity

2.1. *The origin of carrying capacity and definition of WECC*

Basically, carrying capacity is a physical term which means the maximum affordable load before the object is destroyed (David, 1999). In the study on regional systems, the term is used to describe the maximum bearing capacity that a regional system reacts to external environment changes. Ecology is one of the subjects firstly to adopt the concept of carrying capacity (Seidl & Tisdel, 1999). Carrying capacity refers to the maximum number of organisms in one species which can survive under certain environmental conditions (Carey & Daniel, 1993; Dushoff, 2000; Gao, 2001). Nowadays, carrying capacity is a common technical term to describe the degree of environmental development.

Carrying capacity refers to the number of individuals who can be supported in a given area within natural resource limits, and without degrading the natural social, cultural and economic environment for present and future generations. The carrying capacity for any given area is not fixed. It can be altered by improved technology, but mostly it is changed for the worse by pressures which accompany a population increase. As the environment is degraded, carrying capacity actually shrinks, leaving the environment no longer able to support even the number of people who could formerly have lived in the area on a sustainable basis. No population can live beyond the environment's carrying capacity for very long.

After the concept of environment carrying capacity was firstly proposed in China in the 1990s, the study on water environment carrying capacity started in China. So far, there has no formal definition for WECC which has gained wide acceptance. Currently, several ways can be identified for its definition: (1) Pollutant carrying capacity: the maximum wastewater or pollutant carrying capacity of water area when it can maintain a sustainable water system and ecosystem (Wang, 2001); The maximum pollutant carrying capacity of a water body that its system functions can sustain (Liao, 2002). (2) Supporting capacity: the capacity threshold value of a given area water environment to support economic development and living requirement at a certain time (Cui, 1998; Ma, Li & Kang 2002). (3) External load: the maximum external load that basin water environment can afford whilst without causing a damage to the key structure and functions of the water environment system (Zeng, 1997).

The definitions consider WECC as the inherent property of water environment, which are relative to technology level, time and location. At various conditions, the implied meaning and constraint conditions of WECC are different.

The deficiencies of these definitions can be generalized as follows: (1) failure of giving precise evaluation standards, especially from the economic and social aspects; (2) failure of reflecting the relation among population, economy, society, resource and environment, and the environment-human interaction; (3) failure of presenting conditions of WECC, leading to an indistinct extension of WECC.

According to the writer's opinion, WECC should be defined in theoretic level and operational level respectively. Theoretic definition (i.e. wide-sense WECC) refers to the threshold value of society-economic development and life requirement that can be supported in a given water environment, with keeping normal water system functionality and a sustainable ecosystem. It includes carrying capacity of water resource and pollutant carrying capacity. The purpose is to explain the restriction that human bears from water environment in philosophy, economics, sociology and resource subject, and to arouse people to protect water environment. The operational meaning (i.e. arrow-sense WECC) is precise definition of region extent, eco-social development level, sci-tech standard and region water environment conditions, special emphasis on sustainable region eco-environment. According to the operational meaning a certain region WECC can be evaluated. In this paper, operational meaning is adopted, which means pollutant carrying capacity of water environment.

With the development of society and economy, it is not enough that only WECC is considered in the view of the natural resource property in the earth ecosystem and without respect to its social features. It decides the sustainable and dynamic character of society, economy and ecology. The following factors should be included:

(1) The carried main body of WECC is water environment system, and the carried objects include human beings, society-economic system and its associated ecosystem.

(2) It should follow the principle of sustainable development, and all involved factors have to be developed synchronously.

(3) We should place emphasis on the importance of ecology environment. The ecological water consumption and environment water demand will be key issues of the further study in this area.

2.2. *Characteristics of WECC*

Essentially, WECC embodies existential natural law that human being follows and reflects changes of water environment resource ideas and values with the development of society-economy. WECC has the following characteristics: object inherence, relative limit, dynamic adjustment and multiple objectives

(Feng, 2000; Liao, 2002; Long, Jiang & He, 2004; Sang, 2001; Tang & Ye, 1998; Zhang & Yang, 2002).

(1) Object inherence. It is the natural and socioeconomic limit that a given water body supports, showing that in a given region, the WECC of a certain water body has a bearing limit (carrying threshold value) during socioeconomic development period.

(2) Relative limit. With the restriction of natural conditions and social factors, WECC have a maximum carrying upper limit (i.e. possibly maximal index) under a certain historic stage. WECC is not absolute limit under any technological and administrative levels anytime, but relative and conditional limit restricted water environment, society economy and eco-environment conditions.

(3) Dynamic adjustment. WECC is related to historical development phases. The main body and object of WECC, i.e. the water environment and society economy system, is dynamic, the quality and quantity of water environment will change for man's subjective initiative, leading to mutative carrying capacity.

(4) Multiple objectives. The precondition of WECC is preventing deterioration of water environment structure against human's survival and realizing sustainable water environment system functionality. Under the condition of special water environment, through multi-objective investigation of human activity, a threshold limit of scale and intensity is carried out.

3. Assessment Methods of WECC and Existing Problems

3.1. *Assessment methods of WECC*

The assessment methods of WECC are related to populations of region, economic development state, environment, ecology and intergeneration prospect, which is broad and complicated, but currently there is no common method gaining wide acceptance. The existing methods are generalized into four classes: background analysis, multiple objectives optimization, system dynamics, and index system assessment methods (Cui, 1998; Feng, 2000; Fu & Ji, 1998; Li, Lu & Yang 2004; Liao, 2002; Tang & Ye, 1998; Zhang & Yang, 2002).

(1) Background analysis method is to estimate regional potential WECC through analogizing two similar regions in natural and social background. This method normally only accounts one or several limited factors which are independent each other and easy for use. However restricted to its static historic background and separating the interactions between economy, society, resource and environment, it is not suitable to apply WECC for complicated systems.

(2) Multi-objectives optimization optimizes a vector function regarding to various optimization targets whilst satisfying the problem constraints. It divides

regional water resources and society-economic systems into several subsystems, simulates it in a mathematical model and connects each subsystem with multiple-objectives key-model conjunctive variables.

(3) System dynamics is an approach to modeling the dynamics of population, ecological and economic systems, which usually interact strongly with each other. Systems dynamics was founded in the early 1960s by Jay W. The developed program tools, like DYNAMO, enable us to simulate systems easily. System dynamics is fitted to the study on WECC that is high order, nonlinear, multivariable, multi-feedback, mechanism complicated and time varying.

(4) Index system assessment is another quantizing method widely used, including vector module, fuzzy comprehensive evaluation, principle component analysis methods, etc.

Vector module method considers WECC as vector composed of n indexes that every index has m development plan states, normalizing these vectors, and evaluating WECC with normalizing vector.

The multivariate statistical method of principle component analysis is a very effective and convenient tool for reducing the number of variables in a data set and for obtaining useful two-dimensional views of a multi-dimensional data set. The purpose of principle component analysis is data reduction, and interpretation through a few linear combinations of the original variables.

3.2. *Existing problems of WECC evaluation methods*

The above mentioned evaluation methods have many advantages, and the deficiencies are the follows:

(1) Lack of a common understanding on WECC. As introduced before, there are various existing definitions of WECC, but the corresponding understandings are from different aspects. Furthermore, the understanding is superficial. It is difficult to precisely define the concept, connotation and influencing factors of WECC.

(2) The evaluation theory of WECC is under construction. The present study on WECC is still at its early stage. Its theory and methodology needs to be developed in subject crossing or technology method innovation.

(3) Weak operability and generality. General WECC models are built based on certain conditions and catchments. Because the characteristics of each basin are different, the model representing complexity, randomness and fuzziness of WECC is scarce. The study on the qualitative index quantizing WECC model is insufficient, so there is no systematic and efficient qualitative index quantization method, leading to Weak operability and generality model.

(4) Insufficient study on self-cleaning capacity of water bodies. The current WECC quantization models adopt economy and population as targeting objectives. However, only are there a few study on self-renew and self-cleaning, including how to quantify the water quantity. This aspect has to be emphasized in the future study.

4. Further Studies of WECC

WECC is the important and fundamental issue, which deal with regional economy-society-environment sustainable development. However, influenced by various factors, multi-subjects and their intersections, the theory system and the quantification assessment method of WECC are under development. It is difficult to make a systematic and fair evaluation for regional WECC status and sustainable utilization mode, leading to potential randomness and blindness in using water resource and protecting water environment. Based on the present situation in China, it has scientific significance and a practical value to engaging in the study of theory system and evaluation method of WECC for interrelation collaboration in region economic-social development and water environment sustainable carrying capacity. Nowadays, further studies of WECC are the following:

(1) Completing basic theory. The basic theory study on WECC must be completed systematically, including the determination of study goal, distinguishing standard, service object and research scope, and completion of concept system, connotation and characters.

(2) Establishing a suitable evaluation system. The principles include: definite concept, vivid exposition with theory foundation and statistics meaning; authority and universality, comparing WECC with the other similar region; stability, sensitivity and objectivity; facile data, handy computation and great information content; clear logic and hierarchy of inter-index; reflecting sustainable idea and Chinese national conditions. According to the above principles, it is essential to build scientific, fair, impartial and authorized WECC index embodying sustainable idea.

(3) Strengthening study on ecology and environment water demand. To realize sustainable development, it should be the study target of WECC not to cause deterioration or degradation of eco-environment and eco-system of the region. The scientific research, appropriate estimation, accurate calculation and satisfaction ecology and environment water demand are the base to maintain ecological balance, assort economy, society with eco-environment and realize a sustainable development through the harmonization of the human and the nature.

(4) Building universal evaluation models. The existing models are mostly restricted to a certain regional condition. The universal evaluation models meeting complexity, randomness and fuzziness of WECC should be studied, and they can be applied to different types of regions.

(5) Strengthening study on subject crossing. WECC is related to many features of nature, economy and society, so we should not only explore inherent rule of water environment but also research the interrelation between water environment system and external factors. With the basic principles of hydraulics, water environment subject and ecology, economy, systematic and sustainable development, theory system of WECC is established to analyze the connotation and interrelation of carrying capacity factors, so as to reveal the mechanism and condition of water environment sustainable carrying capacity.

5. Conclusions

Water environment is a key element of social and economic systems, and the carrying capacity plays an important role in the regional development. The key to solve regional water environment problems in China lies on the problem how to handle the interrelation between water environment and economic development. When ensuring a steady development of economy, it needs to set an appropriate water environment utilization goal, balance between short-term and long-term interests, and deal with the relation between regional and national development. Furthermore, it is also essential to guarantee sufficient finical, technical and material supports for water environment sustainable utilization and protection.

Acknowledgment

Authors are grateful to the Foundation of State Key Laboratory of Hydrology-Water Resources and Hydraulic Engineering (No. Y514010) and the National Natural Science Foundation of China (No. 41330854) for their financial support of the research. Special thanks to the anonymous reviewers and editors for their constructive review.

References

1. Carey, & Daniel, I. 1993. Development based on carrying capacity: c strategy for environmental protection. Global Environ. Change. 3(2), 140–148.
2. Cui, F. J. 1998. The carrying capacity of municipal water environment and its case study. J. Natural Resour. 13(1), 58–62. (in Chinese)

3. David, P. 1999. Carrying capacity reconsidered. Population and Environ. 21 (1), 5–27.
4. Dushoff, J. 2000. Carrying capacity and demographic stochasticity: scaling behavior of the stochastic Logistic model. Theoretical Population Biology. 57, 59–65.
5. Feng, S. Y. 2000. Water resource sustainable utilization and management. Science Press, Beijing, China. (in Chinese).
6. Fu, X. & Ji, C. M. 1998. A comprehensive evaluation of the regional water resource carrying capacity. Resour. and Environ. in Yangtze Basin. 8(2), 230–237. (in Chinese).
7. Gao, J. X. 2001. Sustainable development- ecologic capacity theory and its application. China Environmental Sciences Press, Beijing, China. (in Chinese).
8. Li, J. Q., Lu, G.H. & Yang, X.H. 2004. DPPIM for comprehensive assessment to regional Water resources bearing capacity. J. Hohai University (Natural Sciences). 32(1), 1–4. (in Chinese).
9. Liao, W. G. 2002. Water environment carrying capacity and the assessment system. J. China Institute of Water Resource and Hydropower Research, 6(1), 1–8. (in Chinese).
10. Long, T. R, Jiang, W. C. & He Q. 2004. The new perspective of water resources carrying capacity. J Water Resour. 1, 38–45. (in Chinese).
11. Ma, W. M., Li, S. X. & Kang, J. H. 2002. Research advance of analysis method on city water environmental carrying capacity on arid region of northwest in China. J. Ningxia Agricultural College, 23(4), 68–86 (in Chinese).
12. Sang, Y. H. 2001. Primary analysis of environment carrying capacity. J Graduation of Sun Yat-Sen University. 22(3), 63–68. (in Chinese).
13. Seidl, I., & Tisdel, C. A. 1999. Carrying capacity reconsidered: from Malthus' population theory to cultural carrying capacity. Ecolog. Econ. 31, 395–408.
14. Tang, J. W. & Ye, W. H. 1998. Study of environmental bearing capacity and its quantification. China Environ. Sci. 18(3), 227–230. (in Chinese).
15. Wang, S. C. 2001. Water environment carrying capacity analysis and adjustment. Academic Report in 70 Anniversaries of China Academy of Water Resources (in Chinese).
16. Zeng, W. H. 1997. Preliminary suggestion on integrated planning of aquatic environment. J Hydraul. Engng. 10, 77–81. (in Chinese).
17. Zhang, W.G. & Yang, Z. F. 2002. The assessment of ground water environment carrying capacity based on index system. J. Environ. Sci. 22 (4), 541–544. (in Chinese).

Chapter 5
Economic and Environmental Policy and Sustainable Development

Overview: the impact of China's high-speed rail on economic development

X.Y. Wang and J.Y. Chai
Economic Management College of Dalian University, Dalian, Liaoning, China
E-mail: Wangxieyong@dlu.edu.cn

Y. Sun
Policy Research Office of Dalian Jinzhou New Area, Dalian, Liaoning, China

As a modern means of transport, high-speed rail has a significant influence on the transformation and upgrading of economic and sustainable development. This paper mainly elaborate the impact from four aspects, which including the macroeconomic management, the regional economic development, the spatial economic landscape and the optimization development of the industry, so as to sum up and analyze the impact on economic about the existing literature. Finally, proposing the prospects for the future research in connection with the problems of the existing research.

Keywords: High-speed rail; Economic development; Overview.

1. Introduction

High-speed rail refers to the new railway line that the speed can reach 250km/h or more, or the original line that the speed can reach 200km/h after alteration. It has become the world important means of transport to relieve the traffic pressure and it is identified as one of the important strategic emerging industry in China. The opening of the Beijing-Tianjin Intercity Railway in 2008 represent that China has entered the era of high-speed formally. By the end of 2014, China's high-speed rail has opened 34 lines, operating mileage has reached more than 16,000 kilometers, and covering about 28 provinces even over 160 prefecture-level cities. The rapid development of the domestic high-speed rail make the impact on economic development become a hot spot for research. So in this paper, the literature retrieval time is limited to "2008-2015", the database is limited to "China Academic Journals Full-text Database", the subject is limited to the field of "economic and scientific management", the theme is "high-speed rail", and the journal level is defined as "core journals", and 533 articles were presented. Then we selected 122 articles that is related to the economic development, and analyze the representative paper to provide theoretical reference for the future study.

2. The Impact on the Macroeconomic Management

The operation of high-speed rail can produce huge social and economic benefits, and stimulate the national economic development. Most scholars analyzed the macroeconomic effects from urbanization, social benefits and so on. On the one hand, the speed railway affect the pattern of urbanization across the agglomeration of urban population to the cities along the railway or the scale structure of the city was olive, which has a significant effect on the population growth [1]. On the other hand, it will cause city effect to stimulate the emergence of the new growth point of politics, economic and culture, which will greatly promote the development of employment and industry [2]. Finally, discussing the market benefits, cost effectiveness as well as the macroeconomic factors of the high-speed rail in the perspective of economics and clearing its market positioning and utility target. For example, Sun empirical analyzed the effcet of high-speed rail on more than 85 Prefecture-level cities by using Hedonic price model, and the results showed that high-speed rail can significantly enhance the value of the city and greatly enhance the market potential of second and third-tier cities. And putting forward proposals for high-speed rail site selection over the view of improving the social efficiency [3].

In addition, due to its improvement on technology, management, and institutional arrangements, etc, the high-speed rail is more energy saving compared to the normal-speed rail, which greatly facilitating the protection of environmental along the region and creating a sustainable economic development atmosphere [4]. We should draw lessons from the foreign governments that by using the macroeconomic methods, such as urbanization, sustainable development and integrated transport system to draft new transport strategy, so as to provide reference for the green ecological construction of high-speed rail and the sustainable development of the economic [5]. But we should also know that as a transport with high fixed structure, only when the level of the passengers' demand is above the equilibrium level, the high-speed rail can achieve the goal of economic balance, and its construction and development can realize economic and sustainability. For example, Ding started with the ecological environment, studied the possibility of influence of the Beijing-Shanghai high-speed rail on regional ecological environment, and proposed the effective improvements, so as to provide reference for the green ecological construction of high-speed railway and the sustainable economic development [6].

3. The Impact on the Development of Regional Economy

The development of high-speed rail can stimulate regional economic growth from GDP growth, increasing employment and expanding domestic demand, and it is necessary to develop high-speed rail in the region. First, the high-speed rail provide an efficient transportation support for regional economic development and enhance the closeness and unity of the regional markets. It can also reduce the regional disparities and promote the appearance of regional economic zone and corridors, even to promote the regional economic integration more significantly. Second, Yang studied the Beijing-Shanghai case and pointed out that the high-speed rail mainly from the importance of regional transportation, accelerating the adjustment of industrial space layout range, speeding up reshape regional spatial structure thus three aspects affect the region economy development [7]. But empirical studies have shown that the impact is both positive and negative. For example, Wang used the method of DID empirically analyzed the data of 287 cities above prefecture level in 2006-2010, and the result showed that under the background of the current slowdown economic situation, high-speed rail will not spur economic growth in the short term [8].

The grand development of the high-speed rail network will also promote the extension of urban functions to speed up the pace of urbanization, and to help the city grow to the sub-center of economic or a new economic growth pole in the future. Mainly as follows: First, the high-speed rail will strengthen the economic ties between the cities along the rail, and improve the city's ability to absorb the elements. It can narrow the relative differences but not affect the absolute difference between cities [9]. Second, the high-speed rail has accelerated the city's urbanization process, expanded the economic scale and made contribution to the urban development and pattern reconstruction of the city along the rail [10]. Chen just considered that high-speed rail is a necessary but not sufficient factor to affect the city's economic growth,and developing the high-speed rail will help adjust the system of urban agglomeration , which will bring the market structure effect and the labor pool effect [11]. Finally ,the construction of high-speed rail has narrowed the scope of urban circle and promoted the integration of the city along the economic circle, greatly inspired the development potential of the city.

4. The Impact on the Space Economy

The basic impact of the construction of high-speed railway is changed the traffic location of the region along the rail, and promoted the region's traffic advantages transform into economic advantages to improve the overall regional accessibility, such as Zhao pointed out that high-speed rail improve the accessibility level of the Yangtze River Delta region comprehensively. And by promoting the development of metropolitan circle and economic shaft, finally promoted the equalization of the regional spatial structure in Yangtze River Delta [12]. High-speed rail will reduce the time and space distance between the city, and the developed and underdeveloped regions will be connected together. It will also change the spatial pattern and realize the integration development between the city. For example, Fang's study showed that high-speed rail strengthen the spatial correlation of the transport network in Yangtze River Delta and the distribution of the spatial concentration is more concentric. The superposition of different levels of transport network greatly optimize the spatial structure of transport network of the city group in Yangtze River Delta [13]. Yang's study suggest that the high-speed rail will further enhance the center of Beijing and Shanghai in the railway transportation, promote the regional integration and the scope of Yangtze River Delta and Beijing-Tianjin-Hebei area, and stimulate the formation of more secondary economic growth zones [14].

In empirical research, Wang build a time cost matrix to study the relation intensity with foreign economic and the city's features on space distribution of the four municipalities and 333 prefecture-level administrative city under three conditions. The result significantly the effect of high-speed rail on the regional space reconstruction [15]. Chen using ArcGIS software analyzed the impact of high-speed rail on the change of the space pattern in the Yangtze River Delta region, the result showed that the whole area is changing from single-function center to multi-purpose center, and Shanghai's space node function have weakened [16].

5. The Impact on the Industrial Structure Development

The construction and operation of the high-speed rail is helpful to the industrial development, especially to promote the upgrading of the related industrial structure. And that the accessibility and convenience of the high-speed rail changing the environment of industrial development, the three effects also strengthen the power of industry restructuring. For example, Wang Gang measured how the high-speed rail boost the development of industry economy, studied the economic impact on the relevant key industries so that to provide transportation technology support for the industrial development [17]. Xu pointed that the improvement of the high-speed rail network promote the flow of the production factors in Yangtze river delta region, and the reducing transportation cost compensates for the rising resource cost. It also enhance the industrial strategy and regional advantages to speed up the industrial transfer and structural adjustment of the cities in the region [18].

For example, the Beijing-Shanghai railway affect the innovation and creative industries in Shanghai, Beijing and the cities along after the opening as well as the impact on the utilize of the creative industries to promote industrial restructuring , and it promote the virtuous circle of the industrial economy along [19]. Zhang focused on the five cities along the Zhengzhou-Xi'an high-speed rail, using the shift-share space structure model to analyze the impact on the economic development as well as the second and third industrial structure of the city. The result showed that Zhengzhou, Xi'an become the core city along the line and industrial structure of most city's in the region can be optimized, and that the secondary and tertiary industries showed a trend of rapid development [20]. Feng using the method such as grey prediction and data envelopment to analyze the change of the value of various industries before or after the high-speed rail in Hubei Province, and the influence on the industry after the operations across the province. The result indicated that the high-speed railway pull the development of the whole industry in Hubei province significantly [21].

6. Conclusion and Prospect

6.1. *Conclusion*

This paper mainly focused on the impact of high-speed rail on the economic development and got the following main conclusions on the basis of predecessors' research results:

First, the theoretical research mainly gathered in all kinds of impact on the economic development, and most scholars believe that the high-speed rail affect the economic development through two major paths. On the one hand, affecting the economic growth and industrial structure by the characteristics it possessed, thereby affecting the overall economic policy and macroeconomic management. On the other hand, strengthen the construction of the transportation network by improving the relations with other modes of transportation, thereby promote the regional transport integration and strengthen the regional accessibility so as to provide infrastructure security for the coordinated economic development of the cities in the area.

Second, most empirical studies analyzed the economic effects of a specific high-speed railway, and most scholars primarily using regression analysis and gray correlation models to empirical analyze the economic effects of high-speed railway. The results showed that high-speed rail and economic development is mutually reinforcing, its effects is significantly different at the same time. It widens the economic gap between different regions as well as the cities and sites along the rail.

Third, on the whole, most of the studies have shown that the development of high-speed rail generate a positive and multifaceted impact on the economic development and it plays an important role in increasing economic vitality and meeting the needs of sustainable development in China. The high-speed rail will become the backbone way in the modernization comprehensive transportation system in current or future. We should further planning for the construction of the high-speed rail network, to enrich the transportation infrastructure for the "One Belt and One Road" and to speed up the pace of China's high-speed rail to go out, and finally to build the traffic source of power for the sustainable development of economy.

6.2. *Prospect*

First, due to the development of high-speed rail is later in China. The research literature about the high-speed rail is less before 2008, and the number of the relevant literature review is only two after 2008, lacking of theoretical support. Meanwhile, the study focused on the economic impact of the high-speed rail in the eastern region of China. Should pay attention to guide the scholars develop academic history consciousness, standardized the academic research. And it should not limited to the eastern part but also to extend the study to the Midwest so as to support the implementation of the "One Belt and One Road" strategy, vigorously promoting western China open for Eurasian.

Second, the studies emphasized more on the macro economic impact on the overall economy, but less on the microscopic impact. Thus making the policy carried is relative macroeconomic but no real guidance. The subsequent research should pay more attention to the impact on the city along, the core city as well as the edge of the city to narrow regional difference and to maximize the effectiveness of the economic zone along the high-speed rail. At the same time, pay attention to the financial industry, the logistics industry, business services and other services to meet the needs of modern development and to maximize the effectiveness of the economic zone along the high-speed rail.

Third, facing the rapid development of high-speed rail, it is necessary to draw the foreign relatively mature methods to compensate for the deficiencies of the breadth and accuracy of the existing research methods. Building a scientific evaluation system so that the empirical research can reference the model and choose the indicators more scientific. Meanwhile, strengthen the prediction study about the impact ability of the high-speed rail which is planing to build in the future rather than analyze the impact of the existing high-speed rail.

Acknowledgements

Thanks are due to Y. Sun for assistance with the subject and valuable discussion.

[1]Xie Yong Wang, Professor of Economics, MA student adviser, engaged in economic management research.

[2]Ji Yang Chai, Master Candidate, study in regional economics.

[3]Yi Sun, division chief of the policy research office, study in regional economics.

References

1. Tan, C.L. & Zhu, Y.L. & Zhong, Z.H. 2014. High-speed rail network's impact on China's urbanization pattern. Urban Problems (9).
2. Wang, S.H. 2010. The development of China's high-speed railway and the analysis of its economic impact. Southwest Jiaotong University (Social Science edition) (5).
3. Sun, C. & Zheng, S.Q. & Zhang, Y.J. 2014. External influences on the Chinese urban economy of the high-speed rail. Guangdong Social Science (5).
4. Zhang, H.B. 2011. The low-carbon comparative advantages of China's high-speed rail. Macro Economic Research (7).
5. Lv, Z.Y. & Li, W.X. 2013. The Foreign high-speed rail development enlightenment for the sustainable development of high-speed rail in China. Logistics Technology (3).
6. Ding, X.L. & Yang, L.Z. & Wang, Z.H. 2010. The impact of the construction of Beijing-Shanghai high-speed railway (Xu Shanghai Section) on the ecological environment and the protection measures. Anhui Agricultural Science (8).
7. Yang, W.F. 2011. Impact of Shanghai-Beijing high-speed railway on regional economic development of China. Ecological Economy (7).
8. Wang, Y. & Nian, M. 2014. High speed railway driving the development of the regional economy. Shanghai Economic Research (2).
9. Tan, C.L. & Huang, X.Y. 2014. The change of economic ties within the cities along the railway. Economic Fabric (4).

10. Luo, L. 2013. Effect of High Speed Rail on urban development along. Southwest University for Nationalities (Humanities and Social Science edition) (5).
11. Chen, J.J. & Zheng. G.J. 2014. Agglomeration perspective of high-speed railway and the urban development. Jianghuai Forum (2).
12. Zhao, D. 2012. The evolution of accessibility spatial pattern under the influence of the high-speed rail in Yangtze River Delta city group (4)
13. Fang, D.C. & Yang, Y.W. 2013. The characteristics of the spatial structure of the transportation network in Yangtze River Delta in the high-speed rail era. Areal Research and Development (2).
14. Yang, W.F. 2010. The impact of high-speed railway on the regional spatial structure. Beijing Social Sciences (6).
15. Wang, J.E. & Jiao, J.J. & Jin. F.J. 2014. Influence of the high-speed railway on China's urban space interaction intensity. Geographical Journal (12).
16. Chen, J.J. & Zheng, G.J. & Liu, Y. 2014. High-speed rail's effect on the space structure evolution in Yangtze River Delta. Economic Geography (8).
17. Wang, G. & Gong, L.T. 2013. The industry economic effects of the construction and investment of the high-speed railway. Macro economic Research (6).
18. Xu, Y.P. 2012. The construction of the high-speed railway promote the industrial transfer and restructure in Yangtze River Delta. Railway Transport and Economy (12).
19. Shi, W.D. & Sun, X.L. 2008. Impact of the construction of Shanghai-Beijing high-speed railway on the development of creative industries on both sides or cities along. Economic and Management Research (10).
20. Zhang, Y.P. & Yan, Y. & Wang, J.G. 2014. The industrial space structure adjustment in the cities along the Zhengzhou-Xi'an rail. Railway Transport and Economy (7).
21. Feng, B. & Zheng, L.L. & Zhou, Y.R. 2014. The high-speed rail's influence on the development of the regional industry. Hubei Social Science (5).

Research on China's financial risk indicators and precaution system

Y. Kuang
Hunan Institute of Technology, Hengyang, Hunan, China, 421002

Z. Zhang
Business school of Central South University of Forestry and Technology, Changsha, Hunan 410004
E-mail: zhangzhe7222100@126.com

Financial stability is one of the most important conditions ensuring national economy of a continuous, fast, and healthy development. Based on the international research efforts, this paper is intended to build precaution index system from macroscopic and microcosmic aspects, which focus on the handling of financial risks before and afterwards. This serves as another contribution to the study on financial stability.

Keywords: Financial crisis; Early warning system; Index system.

1. Introduction

With the development of the globalization of the world economy, the development of the Chinese economy and the world economy are increasingly linked, in such a background, the China share in economic globalization and financial deepening bring great interests, but also will inevitably face from foreign financial tsunami hit. To guard against financial crises, ensure that China's financial security, establish a adapt to China's economic and financial development early warning mechanism, in order to form or expand in crisis before it to take effective control, so as to avoid crisis or mitigate the crisis broke out on the national economy of China of the adverse impact of economic development, financial security and social stability has a very important realistic and practical significance.

2. Review on the Early Warning Theory and the Index System of Financial Risk

The research on financial risk began in Mexico, Argentina and other Latin American countries. In the age of 20th century 70-80s, these countries have erupted in monetary crisis. and the financial crisis occurrence frequency climbed since the 1990s, the crisis not only to local economic and social development caused serious harm, but also to the global financial market and the development of the world economy, this bad situation has caused the academic attention on the causes of the financial crisis, infections and warning index system, and other aspects of the wide attention and in-depth research. On the financial crisis warning index system for research, foreign research mainly from the following several aspects of: First, through to the cause of the crisis and its change for qualitative factors discussed to determine early-warning index system; Second, through the "group" and "control group" comparison method to determine which measures help to predict the probability of a crisis; Third, applied econometrics model, the model to estimate in the future or a few period of the probability of the crisis. In the above three kinds of study, based on the objectivity of the factors such as consideration, the third type of academia and research of the mainstream consensus, which more representative of several research results were Kaminsky, and Goldstein Abiad, the results for representative. All in all, after a long period of time the development, foreign academia in the financial crisis warning system development had done a lot of theory and model. Relative to character, because the domestic financial risk early warning the later start, as the current research status is concerned, although have a some empirical research achievements, but most of the work as stay on foreign financial crisis warning on the introduction of the theories and models, from the whole financial system to consider China's overall financial risk system the research results of the does not appeared.

3. The Specific Construction of Financial Risk Early Warning Index System

Predict the risk of financial crises, must through the selection of some representative index, by using the quantitative analysis method, accurately as possible, scientific use of financial risk early warning model measure include each aspect, and comparison and analysis of the different areas of the changes of the financial risk.

3.1. *The financial stability of the warning index selection principle*

Absorbs the domestic and foreign research results in the foundation, on the financial risk early warning index system design should follow the following principles:

Firstly, systematic, financial risk early warning is a complex system of contact each other by numerous subsystems namely constitution components.

But the system is not simple addition each subsystem, the optimal each subsystem that may not be able to achieve integral optimal, but to make each subsystem to integrate science to ensure system "integral of the best".

Second, orientation, choice of each specific statistical evaluation index to comply with the requirements of the warning of the risks of improve the financial, belongs to the category of the financial risk, can be in a certain degree of financial risk reflected on connotation and the characteristic, to improve the financial risk early warning goal is a clear orientation, to the improvement of the modern financial system has positive guidance and function.

Third, maneuverability, the choice of each index data have reliable source, the calculation method of index to science, facilitate the computer to carry on the data processing. But often some important factors is difficult to be directly quantitative statistical data integrity or lack. So in the study design early-warning index system should be in order to reflect the key factors of the index, as far as possible the flexible methods for obtaining index data.

3.2. *The index system selection process of early warning financial risk*

Establish a financial safety warning index system for one purpose, it is to make supervision department comprehensive as possible, completely, specific about the safety of the financial district of the influence factors, so as to have targeted and take effective measures. Influence of financial stability numerous factors, and various factors of the relative importance of and the interaction for the development level of a nation also, openness, economic size, economic structure and the influencing factors of different vary widely. Financial risk from countries all over the world experience, with the risk of China's financial industry the truth, China regional financial safety warning index system can be roughly from the macro and micro prudent index of leading indicators will be built two levels. So, China's financial crisis warning evaluation index system is established process can summarize for: decomposition to describe China financial crisis warning evaluation index system always goals–establish corresponding standards-creating pass class time structure—build primary index collection–screening index–setting up the index system (the following Figure 1).

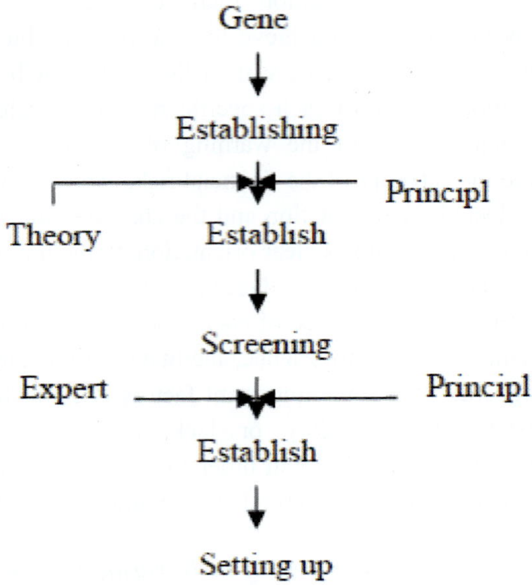

Figure 1. The basic construction process of financial crisis warning evaluation index system.

3.3. *The establishment of the financial risk warning index system*

Through the above analysis, we clear the financial risk pre-warning system concrete structure and elements, according to the domestic and foreign experts and scholars on financial risk characteristics description, use frequency statistics method and theory analysis, the choice in the financial risk early-warning system used in high frequency, and financial risk the connotation and features of the related index, constitute a primary index collection.

Considering the recognition as financial risk root and financial risk early warning of the foundation of the theory of paper, and combining the current financial risk data and statistics for the particularity of the possibility, we from the macroeconomic, financial system, bubble risk and the global economy from four aspects such as selection index established 18 Chinese financial risk monitoring index system, as the warning system alternative index. Through the above analysis, we in the primary index set based on making the simple adjustment, final financial risk pre-warning system index system, the index system of four subsystems, total 18 a specific measure indicators. Among them, the macro economic indicators mainly reflects the macroeconomic environment stability; The financial system index mainly includes monitoring bank crisis and currency crisis indicators reflect the financial market stability; Bubble risk index, asset prices changes in risk; While the global economic index is the main consideration major foreign economic entity changes on the domestic influence. (The following Table 1 shows Specific warning index system).

Table 1. China's financial risk early warning index system.

Type	Index name	Type	Index name
Index of Macro economic	GDP growth index	Index of financial system	Domestic credit/GDP
	The inflation rate foreign net assets/GDP		Foreign net assets/GDP
	Value added of industry growth speed		M2 multiplier (M2 / M1)
	Export rate		loans/deposit
Bubble risk index	The share price index rate (the Shanghai index)		savings deposit/M2
	Real estate investment growth rate		interest rates
	Construction loans/bank loans		M2 / foreign exchange reserves
The index of global economic	international crude oil prices state		foreign exchange reserves growth rate
	The United States economic growth rate		foreign direct investment/GDP

Note: GDP refers to GDP, CPI refers to the consumer price index, M1 refers to the broad money, and M2 refers to narrow currency.

3.4. *The understanding of Macro economic indicators*

Firstly, Track definition: Macro economic indicators mainly reflects the macroeconomic environment stability, in the index system mainly by the following "GDP growth", "Value added of industry growth speed", "inflation" and "export rate" and so on several specific indicators. First, "GDP growth" and "value added of industry growth speed", "GDP growth" and "value added of industry growth speed" is mainly used in the measure of a national economy the rate of economic growth, in a sense, the economic development too quickly possible overheating situation, and development is slow could mean the economy into recession stage; Second, "inflation", "the inflation rate" is usually measured in macro economic stability of the important sign, can measure the macro economic control for the financial system produce negative influence; Third, "export rate", "export rate" is the measure of a country foreign trade stability index, a lot of financial crisis are accompanied with the sharp decline in exports. Financial crisis spread theory suggests that one country foreign trade is often the main financial crisis spread channel.

Secondly, financial system index meaning: The financial system index mainly includes monitoring bank crisis and currency crisis indicators reflect the financial market stability, mainly by "domestic credit/GDP", "Foreign net assets/GDP", "The actual interest rate", "Foreign direct investment/GDP", "M2 /

GDP", "M2 multiplier", "Loan/deposit" and "savings deposit/M2" and so on several specific indicators,

The "domestic credit/GDP" measure a country's credit expansion, excessive credit expansion on one hand the circulation of currency excessive, cause inflation pressure, on the other hand the fast growth in lending caused the banks' bad loans probability increase; The "M2 / GDP" called the money supply and inflation. The large increase in the money supply, not only can cause inflation pressure, but also will lead to low efficiency of monetary policy. The index of the high proportion of May is the sign of the financial deepening, also may be the financial risk warning signs of growth; the "foreign net assets/GDP" to reflect on its foreign net assets of the influence of the financial market; the "the actual interest rate" is eliminating price increases factors of after interest rates. The index can reflect the capital supply and demand situation of a country and capital inflows appeal. High real interest rates show that one country will be or is under attack speculation. It also often used as a measure of financial institutions robustness; the "foreign direct investment/GDP", index to measuring a country's economic dependence on foreign capital. High ratio of foreign capital means that the country depends on too high if foreign capital into the slowdown will be the country's economic growth to pose a serious blow; the "M2 multiplier", money multiplier improve means the base currency is currency derivatives amplifier, economic bubble more; the "loan/deposit", high or the loan growth than may indicate the savings-bank system more fragile, not enough liquid assets to cope with the impact of foreign appear; Eighth, "savings deposit/M2" mainly reflect the bank risk probability change, that family and enterprise for the trust bank system, if the ratio is low, show the degree of trust is low; Ninth, "China's foreign exchange reserves growth" and "M2 / foreign exchange reserve" central bank reserves enough to measure the degree of index.

Thirdly, Bubble risk and global economic part index meaning: The accumulation of financial risks and release will through the stock market and housing markets reflected, outstanding performance for abnormal fluctuations in asset prices, if the asset prices tumbled, will lead to financial assets shrink, financial asset bubble burst, form a lot of financial institution non-performing assets, worsening economy, cause financial crisis. Bubble risk index, consider the risk of changes in asset prices, in the main index system by "stock index rate (the Shanghai index)", "The real estate investment growth", "Construction loans/bank loans" and several other indicators, and the global economic index is the main consideration major foreign economic entity changes on the domestic influence, the main "international crude oil prices rate" and "the economy growth" in the two indicators.

4. Conclusion

In order to design better index to reflect China's financial security situation, this paper, from the Angle of different influence factors of financial security, the paper builds open financial security warning index system. The study conclusions promoted the Chinese financial security warning development scientific and standard process and the financial security of early warning that will provide premise condition and necessary preparations, but also as a China in different periods financial security situation from review basis, to make China's financial industry executives can more reasonable using Chinese domestic limited resources to carry on the financial crisis prevention, maintain financial stability. However, this study is limited to the warning system alternative index of choice, the choice for index further without making further exploration, and in the next research, we should through such as such as the construction of mathematical model Logit, Granger test was used to determine if the various subsystems in warning index, the index of the crisis of the early warning for that role, and use the final selection index by factor analysis method can reflect the synthesis of China's financial crisis levels of specific risk index, and planning the corresponding risk level, according to China's financial risk overall depict fluctuations, and according to the historical experience of China's economic development, and the characteristics of different development stages reference, determine the index value and constructs the alarming China's financial risk early warning signal system, early warning signal output system shows the financial risk state and describe the operation condition of financial hot and cold.

About the authors

Yu-Kuang, male (1967), born in Hengyang Hunan, Associate professor. Mainly engaged in the research and teaching of risk management, cost management. E-mail: kuangyu@tom.com; Corresponding author: Zhe-Zhang, male (1988-), born in Hengyang Hunan, graduate student of Central South University of Forestry and Technology, mainly engaged in rural area development and tourism management research, Code: 410004. E-mail: zhangzhe7222100@126.com, phone code: 15273443191

References

1. Kaminsky, G.L., S. Lizondo and C.M. Reinhart. Leading Indicators of Currency Crises [R]. IMF Staff Paper, 1998, Vol. 45, No. 1.
2. Goldstein, Morris, Graciella L. Kaminsky, and Carmen M. Reinhart. Assessing Financial Vulnerability: An Early Waring System for Emerging Markets [R]. Washington DC: Institute for International Economics, 2000.
3. Abiad, Abdul. Early Warning Systems for Currency Crises: A Survey and a Regime-Switching Approach [R]. IMF Working Paper, 2003, 03/32.
4. Wen Yuechun, Huang Funing. Overview of domestic financial risk early warning system and EWS framework of integrated management [J], China money.

Chinese outward direct investment in other BRICS members from the traditional economic and institutional perspective

Y.L. Wang
School of Economics and Management, Beihang University, Beijing 100191, China
E-mail: wyljn08125@163.com

Y.M. Shao
Academy of Mathematics and Systems Science,Chinese Academy of Sciences, Beijing 100190, China
Center for Forecasting Science, Chinese Academy of Sciences, Beijing 100190, China
E-mail: yanminshao@amss.ac.cn

This paper aims to develop a basic conceptual framework of Chinese outward direct investment (ODI) in other BRICS members, integrating the traditional economic factors with the institutional perspective. As demonstrated in this paper, the traditional economic factors are indispensable. However, the institutional ones count more. Hence, both types are expected to be taken into account when ODI firms tend to conduct a cross-boundary investment.

Keywords: Chinese ODI; BRICS; Traditional economic factors; Institutional perspective.

1. Introduction

Chinese government has speeded up the implementation of "going global" strategy in the 12th five-year plan, actively encouraging qualified enterprises to conduct foreign investment, thus leading to enterprises increasing internal motivation to integrate into economic globalization. According to 2012 Statistical Bulletin of China's Outward Direct Investment, ODI net flows reached $87.8 billion, increased by 17.6% compared to 2011; ODI flows and stocks in 2012 accounted for a share of 6.3% and 2.3% globally (UNCTAD, 2013). There are 178 destination host countries, and 85.7% concentrating in Asian and Latin American. This paper focuses on the investment in other BRICS members.

Chinese multinationals, with a unique mixture of government involvement and private initiatives, are a distinguishing characteristic of 21st century globalization (Zhang & Alon, 2010). Chinese multinationals, like any others, have to meet the institutional requirements of the host country (Rui & Yip, 2008).

However, in some countries they face greater institutional barriers than companies from other countries have to overcome, due to reluctance stemming from the state ownership of many Chinese firms, and the possible search for political objectives rather than commercial ones (Globerman & Shapiro, 2009). Traditional FDI theories, on the basis of perfect market mechanism of the western developed countries, ignore the effect of institutional factors on the investment in emerging economies. Thus, institutional theories provide a new visual angle and analysis framework for the research on Chinese FDI in emerging markets.

This paper aims to integrate the traditional economic factors with the institutional perspective, making an attempt to develop a basic conceptual framework of Chinese ODI in BRICS members. As demonstrated in this article, the traditional economic factors are indispensable; However, the institutional ones count more. Hence, both types are expected to be taken into account when ODI firms tend to conduct a cross-boundary investment.

The structure of remaining paper is organized as follows. The insights from the previous empirical literature and the research hypotheses are presented in section 2. Section 3 provides a detail description of the data, variables and the method. Section 4 discusses the results and Section 5 concludes by suggesting the limitations.

2. Theory and Hypotheses

2.1. *Traditional economic perspective*

Four types of investment motivations are produced from traditional FDI location choice theories, including market-seeking, natural resource-seeking, efficiency-seeking and strategic asset-seeking (Dunning, 1993). While, more and more researches also believe that four motivations above are also adapt in the investment of Chinese multinationals.

A large number of empirical studies have clarified the positive correlation between the host country market scale and the FDI flow (Bevan and Estrin, 2004). Meanwhile, comparing to the economies of slow or stagnant growth, BRICS members of rapid development can provide multinational companies more chances to gain an advantage. Rapid economic growth is accompanied by high levels of aggregate demand in product and service, which cause more FDI inflows. Economic openness is also an important influence factor of Chinese ODI. As the BRICS with higher degree openness, the economies tend to be adapt to global development pattern of produce and commerce. Thus, openness is also used to evaluate the potential of market-seeking investment. Secondly, because of the scarcity of Chinese natural resources, especially comparing to mineral reserves of South Africa and Brazil, crude oil of Russia and tropical forest of Brazil, more and more Chinese firms invest for seeking natural resource. Lastly, according to the latest research, one of the main reasons of Chinese firms' ODI behavior is to upgrade their own competitive disadvantage and compete with developed countries. So, we put forward the following hypotheses.

H1a: The possibility of Chinese firms' investment will increase in host country with larger market scale, faster growth and higher openness.

H1b: The possibility of Chinese firms' investment will increase in host country with rich natural resource endowment.

H1c: The possibility of Chinese firms' investment will increase in host country with more strategic assets.

2.2. *Institutional perspective*

In the past few years, the institutional perspective has become one of the most suitable theoretical frameworks for analyzing strategic decisions made by companies from emerging or transition economies (Peng, Wang & Jiang, 2008). Institutions are more than just background conditions. Institutional theory makes it possible to establish solid grounds to explain the internationalization of companies from emerging economies entering other emerging economies and the markets of more developed countries alike (Wright, Filatotchev, Hoskisson & Peng, 2005).

Doing business is particularly important for multinationals operating in more than one institutional context (Meyer, Estrin, Bhaumik & Peng, 2009). The formal and informal rules affect not only how a company chooses to enter an economy, but the very decision on whether or not to set up in a particular country. Institutional factors play an increasingly important role in location decisions (Zheng, 2009). Institutional factors alter the cost of doing business in one nation rather than another, which affects every aspect of the multinational's behavior, including location choice (Henisz & Swaminathan, 2008).

Institutions establish the rules of the game that structure interactions, and organizations are the players limited by these rules, which can be both formal (laws and regulations) and informal (customs, traditions, or codes of conduct) (North, 1990). According to institutional theory, companies make their strategic choices based on the interaction between institutions and the organization itself, and attempt to obtain institutional legitimacy in terms of the host country's rules and regulations (Cui & Jiang, 2010).

H2a: Chinese firms tend to invest in host country of favorable doing business.

H2b: Chinese firms tend to invest in host country with higher politic and law risk.

H2c: Chinese firms tend to invest in host country with similar cultural background.

3. Variables and Method

3.1. *Variables*

A detailed description of variables and variable sources are given in Table 1. The variables of GDP and GDP growth (GDPG) are often used to measure the market influence to assess whether Chinese firms tend to be attracted to large market size and market growth potential. Even though they are usually deemed to raise the amount of ODI, the extant literature has not yet reached definitive results about the market-seeking motivation of Chinese MNEs per sector.

Openness: The variable of openness (FDI net flow to GDP) is added to capture its application for location decision, which is expected to be positive with Chinese firms' ODI motivation.

Energy: To proxy for the natural resources endowment, the percentage of energy net import to energy usage is employed, as the fact that most Chinese firms in study concentrated on the oil subsector. Positive index means the host country regarding as energy-import country; and negative index means energy-export country. For natural resource-seeking, Chinese firms are more likely to invest in energy-export countries with high resource endowment.

Patent: Patents possessing are seemed to be the strategic assets in the host country. World Development Indications offer the registered patents of residents and nonresidents, and the latter are usually applied to proxy for the strategic assets. Chinese firms investing abroad for strategic assets should be more attracted to countries with high property rights, while it may be significant in case of strategic asset-seeking ODI to enhance competitive advantages.

Business: Based on Doing Business Index ranking, the higher doing business ranking is, the more suitable for the engagement in business activities in the host country; conversely, it is less suitable for the business activities. And the variable is expected to be positive relation with Chinese ODI for the reason of seeking institutional protection.

Prisk: Political risk of the host country is proxy for the political risk rating of International Country Risk Guide (ICRG). This rating assigns risk points to a pre-set group of factors, termed political risk components. In each case the lower the risk point total is, the higher is the risk; and the higher the risk point total is, the lower is the risk.

Cultural: this paper adopts the equation by Kogut and Singh to evaluate the cultural distance between host country and China.

Biltrade: From institutional approach, bilateral trade indicates the intensity of transactional dealings and choice in terms of exporting destinations, and importing sources by firms from the home country. Thus, the variable of bilateral trade between China and host economy is used as a proxy for mimic pattern of FDI location choice. Unlike the variable which was Buckley used, we used the percent of trade with China to the total trade volume of the host country. Data for this variable were sourced from the State Statistical Bureau of China (SSBC) and UNCTAD.

Table 1. Lists of variables and sources of data.

Variables	Proxy for	Data sources
ODI	Chinese stock ODI	SSBC
GDP	Host country GDP (US dollar, constant)	World Bank Development Indicator
GDPG	Host country GDP growth	World Bank Development Indicator
Openness	FDI net flow (% of GDP)	UNCTAD
Energy	Energy imports, net (% of energy use)	World Bank Development Indicator
Patent	Patent applications, nonresidents	World Bank Development Indicator
Business	Ease of doing business index (1=most business-friendly regulations)	World Bank Development Indicator
Prisk	Political risk rating of the host country	ICRG
Cultural	Cultural difference between host country and China	www.geert-hofstede.com
Biltrade	The percent of trade volume between China and host country	SSBC and UNCTAD

3.2. *Method*

The sample period is from the Statistical Bulletin of Chinas Outward Foreign Direct Investment, which provides the Chinese ODI net stocks to each host country covering from 2003 to 2012. With panel setting, we propose the following estimated equation:

$$\begin{aligned}\ln ODI_{jt} &= \alpha_{jt} + \beta_1 \ln GDP_{jt} + \beta_2 GDPG_{jt} + \beta_3 openness_{jt} \\ &+ \beta_4 energy_{jt} + \beta_5 patent_{jt} + \beta_6 business_{jt} \\ &+ \beta_7 prisk_{jt} + \beta_8 cultural_{jt} + \beta_9 biltrade_{jt} + \mu_j + \varepsilon_t\end{aligned} \tag{1}$$

with j=1,2,3 to denote the host country j; t=2003,…,2012 denote the sample period.

With panel setting, firstly, random-effects models are estimated; while the Hausman tests, which test the random-effects assumption that time invariant factors are uncorrelated with the explanatory variables, by and large reject the assumption. We therefore focus on the fixed-effects estimation of panel data.

4. Modeling Results and Discussions

Table 2 displays the panel results of our sample. From the estimated results, almost every variable passed the significance test and the roles on Chinese ODI are in line with the expectations according the theories in sector 2. Especially, all institutional factors passed through the test at 1% significant level, which denotes that the institutional regulations of the host country do has a decisive influence on our enterprises to invest in BRICS members.

Table 2. Modeling results.

	Variables	Coefficient	
Traditional Economic Factors	GDP	0.342	(0.723)
	GDPG	0.002	(1.235)
	Openness	3.467*	(1.876)
	Energy	-0.481**	(-2.014)
	Patent	0.033	(1.267)
Institutional Factors	Business	-0.673***	(-3.178)
	Prisk	0.579***	(7.273)
	Cultural	-3.612***	(-3.857)
	Biltrade	1.286***	(2.333)
R^2		0.785	
D-W		3.463	
F		53.567	

Note: t-statistic are in parenthesis. ***,** and* denote the 1%, 5% and 10% significant level.

Firstly, we try to discuss the role of traditional economic factors on Chinese ODI of BRICS members. For the variable of openness, the evident positive relation and significance at 10% level indicate that all the BRICS members have higher openness economics. And the higher openness level promotes Chinese firms to invest in the host countries with intensive export-orienting. Comparing with traditional market capacity variables of host country (market size and growth, GDP and GDPG), Chinese outward enterprises on BRICS should pay attention on the domestic market, but more on the market expansion capacity and aiming at broader external market with higher openness.

The results also confirm the natural resource-seeking motivation of Chinese ODI as the coefficient of Energy in Table 2 is significantly negative and significant at 5% level. The results reveal that Chinese firms are more willing to invest in energy-export countries with high resource endowment, regardless of any development level of the host country and firm's ownership structures. The speed of economic growth that China has experienced over the last three decades and the projected growth in the future means continued demand for natural source. For instance, since 2009 China has become the largest market of passenger cars, which means an increased demand for oil in addition to the energy required to power its industries.

An interesting and original finding is related to strategic asset-seeking motives with the variable of patent. The estimation reveals that the availability of patent is not positively associated with the probability of Chinese ODI in those countries. The insignificant positive coefficient of patent in the model perhaps is caused by the fact that China is the largest developing economy with the advantage of strategic asset, which conversely highlights the importance of market-seeking in those less developed economies.

Then, we try to discuss the role of institutional factors on Chinese ODI of BRICS members. It is worth noting that the coefficients of Business are significantly negative in the model, conveying the favorable business environment to be an attractive of Chinese ODI, while conversely in the developing host country. And it also suggests that Chinese firms' attempt to take the first-mover advantage in the new market investing in developing countries with unfavorable business environment as the coefficients of the variable in host country model positively be link with the location choice.

With respect of the politic risk, the significant sign of the variable in the model suggests that the risks of the host country do affect Chinese multinational in unconventional way. A reasonable explanation might be from the concentrated political power of the home country and increase the investment risk. Comparing with multinational companies from the developed countries, Chinese companies find themselves in the political environment within the country risk, and more familiar with the government for frequent macroeconomic regulation and control of environment, better access to the legitimacy of the local system.

For Hypothesis 2c, which stated that cultural distance would negatively affect Chinese outward FDI location choice, we obtained mixed results. One possible reason for our result may be that the influence of cultural distance may depend on the Chinese company's objectives. While market-seeking investments might well have been initially aimed at countries in which this distance was smaller, strategic asset-seeking investments have been mainly aimed at developed countries in North America and Europe, which are culturally more distant. Also, many Chinese companies do not seem to shy away from cultural distance, perhaps aided by the alliances they have made in China with multinationals from host countries.

The result of biltrade reveals that a particular destination choice for trade is repeated in high frequency over time so that it is institutionalized in Chinese investor's mindsets, expanding trade destination choice to FDI location choice will become a way to gain legitimacy.

5. Conclusions

This paper mainly aimed to study the influence of host country traditional economic and institutional factors on the location decisions made by large Chinese multinationals investing in other BRICS members. Our findings show certain characteristics that differ from the conventional wisdom of multinationals. In spite of the traditional economic factors which influence our country enterprise to carry out ODI behavior in other BRICS members, the empirical results show that the effect of institutional factors indeed play the important role and not be allowed to ignore. Before the foreign investment, the system environment of the host country should be considered carefully.

The main empirical contribution of our research lies in having provided new evidence regarding a phenomenon that is acquiring ever-increasing institutional importance when invest in BRICS members. In any case, when interpreting our results, the limitations inherent to the nature of our empirical research should be taken into account. The main limitation may be that certain variables that may also have an influence on location decisions could not be included, such as the specific reason for each outward FDI—looking for economic efficiency. Depending on what the objective is for Chinese companies in each country, the institutional factors linked to each location may play a very different role. For example, institutional restrictions that may arise when a Chinese company makes an investment to access a resource considered strategic for the host country may not be applied when investments are made in that same country for the purposes of accessing its market.

References

1. Cui, L., & Jiang, F. 2009a. FDI entry mode choice of Chinese firms: A strategic behavior perspective. *Journal of World Business* 44(4): 434–444.
2. Cui, L., & Jiang, F. 2009b. Ownership decisions in Chinese outward FDI: An integrated research framework and research agenda. *Asian Business & Management* 8(3): 301–324.
3. Globerman, S., & Shapiro, D. 2009. Economic and strategic considerations surrounding Chinese FDI in the United States. *Asia Pacific Journal of Management* 26(1): 163–183.
4. Henisz, W., & Swaminathan, A. 2008. Institutions and international business. *Journal of International Business Studies* 39(4): 537–539.

5. Meyer, K. E., Estrin, S., Bhaumik, S. K., & Peng, M. W. 2009. Institutions, resources, and entry strategies in emerging economies. *Strategic Management Journal* 30(1): 61–80.
6. Peng, M. W., Wang, D. Y. L., & Jiang, Y. 2008. An institution-based view of international business strategy: A focus on emerging economies. *Journal of International Business Studies* 39(5): 920–936.
7. Rui, H., & Yip, G. S. 2008. Foreign acquisitions by Chinese firms: A strategic intent perspective. *Journal of World Business* 43(2): 213–226
8. Wright, M., Filatotchev, I., Hoskisson, R. E., & Peng, M. W. 2005. Strategy research in emerging economies: Challenging the conventional wisdom. *Journal of* Management Studies 42(1): 1– 33.
9. Zheng, P. 2009. A comparison of FDI determinants in China and India. *Thunderbird International Business Review* 51(3): 263–279.
10. Quer, Claver, E., & Rienda. 2012. Political risk, cultural distance, and outward foreign direct investment: empirical evidence from large Chinese firms. *Asia Pacific Journal of Management* 29: 1089–1104.

Research on Xinjiang credit support from the perspective of low carbon development

C.Y. Liu* and Z.G. Gao†

Xinjiang University of Finance and Economics, Urumqi 830012, China
*E-mail: *163.liuchenyue.com@163.com, †gaozhg1206@163.com*

We firstly based on the IPCC international national greenhouse gas inventory calculation 1995–2012 Xinjiang annual carbon emissions, and then selects the Xinjiang 1995–2012 years of carbon emission intensity as a low carbon economy development index, total loans of financial institutions accounted for the proportion of GDP as the index of financial and credit support, empirical analysis of the test with the relationship between the two. Found that there are long-term stable cointegration relationship between carbon emission intensity in Xinjiang and financial credit support among the variables, the loan balance of financial institutions accounted for the proportion of GDP growth by 1 percentage point each, the intensity of carbon emissions will rise by 0.137 percentage points, Xinjiang financial and credit support and not to curb carbon emissions intensity increases, and based on this puts forward some enlightenment and suggestions of the development a low carbon economy in Xinjiang.

Keywords: Low carbon; Credit support; Carbon emission intensity.

1. Introduction

Global warming and low carbon economy is one of the most concerned problems, low carbon and sustainable development is the eternal theme of human development, China began more than American became the world's largest country of carbon emissions from 2007, as a responsible big country, we are faced with enormous pressure on energy saving. The Chinese government attaches great importance to the greenhouse gas emission reduction and low carbon economy development, in the Copenhagen conference, China government promised that by 2020 carbon emissions per unit of GDP than in 2005 decreased 40%–45%, developing green low carbon cycle economy has become the best choice for the sustainable development of economy in our country, but large scale development of low carbon economy is a long-term and difficult task, It not only needs the participation of citizens, enterprises and government, more in need of the strong support of the credit and financial industry.

The current research of low-carbon economy, carbon finance and credit document is not much, among them Tang Changbao [1] (2011) believe that our country commercial bank practice low carbon credit is a gradual process, to steadily push forward, realize the healthy, steady, sustainable development in the gradual adjustment; Zhou Yan [2] (2010) think the process of low carbon credit system bank for enterprises to develop a set of standardized, service in the carbon finance can effectively promote the enterprise into the carbon finance, and puts forward to the bank as the "network symbiosis of carbon finance key"; Wu Yan etc. [3] think that China should draw lessons from international experience, to encourage commercial Banks to participate in carbon trading market; Wang Zhouwei [4] think that China city should vigorously develop low carbon credit, accelerate carbon financial innovation. The above documents are mostly conducted a preliminary exploration and development from the perspective of qualitative analysis of low carbon credit, the theory and method of many beneficial to the development of low carbon credit mechanism and implementation, this paper based on the previous achievements, the path and mechanism of the method of quantitative analysis of Xinjiang regional development of low carbon credit, in order to provide the necessary reference of developing low carbon economy in Xinjiang.

2. Evaluation and Analysis of Xinjiang Carbon Emission

2.1. *Estimation method of carbon emissions*

Carbon emissions mainly comes from the mass burning fossil fuels, Xinjiang species fossil energy consumption mainly including coal, coke, crude oil, gasoline, kerosene, diesel oil, fuel oil and natural gas, such as eight kinds of energy, according to the number over the years all kinds of energy consumption in Xinjiang, multiplied by the corresponding energy of CO_2 emission coefficient, add up get the annual carbon emissions from 1995 to 2012 in Xinjiang. All kinds of energy of CO_2 emission coefficient are shown in Table 1, carbon emissions estimation formula is as follows:

$$CO_2 = \sum_{i=1}^{8} A_i \times \begin{bmatrix} \text{low calorific value} \times CO_2 \text{ Conversion} \\ \text{coefficient} \times \text{Carbon oxidation rate} \times \dfrac{44}{12} \end{bmatrix} \tag{1}$$

Among them, CO_2 for carbon emissions, A_i for the first class I energy consumption.

In this paper, based on the above estimation method to calculate the carbon emissions from 1995 to 2012 in Xinjiang, the following data are derived from over the years, Xinjiang statistical yearbook, and after finishing the proceeds.

Table 1. Kinds of energy consumption of the CO2 conversion coefficient.

Energy	Emission factor (tc/TJ)	Carbon oxidation rate (%)	Low calorific value (MJ/t, km^3)	CO_2 Conversion coefficient
Coal	25.8	98.0	20908	1.94
Coke	29.5	98.0	28435	3.01
Crude oil	20.0	99.0	41816	3.04
Gasoline	18.9	99.0	43070	2.95
Kerosene	19.6	100.0	43800	3.15
Diesel oil	20.2	99.0	42652	3.13
Fuel oil	21.1	99.0	41816	3.20
Natural gas	15.3	99.5	38931	2.17

Note: tc for tons of carbon, TJ for calorific value, unit for gigajoule, MJ for mega joule. Table data from the IPCC international greenhouse gas (GHG) listing guidelines.

2.2. *The calculation results and characteristics analysis*

According to the results of our calculations we plotted in figure 1995–2012 Xinjiang carbon emissions per capita carbon emissions and carbon intensity trend shown in Figure 1. Figure 1 shows that since 1995, carbon emissions in Xinjiang shows a tendency of rising sharply, the average annual growth rate as high as 9.53%, and is bounded by 2003, prior to 2003 the annual growth rate of 4.83 percent, rose rapidly after 2003, the growth rate is as high as 13.71%; Per capita carbon emissions are also showed similar trend, the overall average annual growth rate of 7.65%, and also in 2003, for industry, before the average growth rate of 2.85%, after 2003 was 11.91%, and the overall total carbon emissions and per capita carbon emissions into the rapid rise of the "turning point" at the same time; can also be seen from Figure 1, Xinjiang carbon intensity trend showing a wider at the bottom of the "V" type, from 1995 to 2003, carbon intensity, on the whole, begin to continue to fall, started to rise again after 2003, carbon emissions began to rise faster than the speed of economic growth rate, Xinjiang is facing very serious situation to reduce emissions.

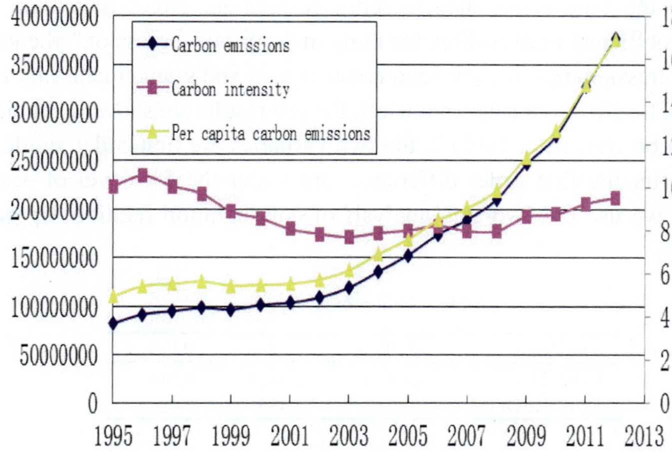

Figure 1. Xinjiang 1995–2012 carbon emissions, per capita carbon emissions and carbon emission intensity trends.

3. Low Carbon Economy Development in Xinjiang and the Empirical Analysis of Credit Support

Carbon finance and credit is the future low carbon economy reliable guarantee for the development of source of funds, financial institutions through the capital loans to enterprises take the lead in development of low carbon economy, at the same time, for those who are in energy intensive and highly polluting industries limited loan and credit scale, in order to achieve the purpose of reducing carbon emissions. Therefore, research on financial credit support to the mechanism of the low carbon economy development in Xinjiang has become the focus lies.

3.1. *The selection of indicators and the regression equation set*

Based on the needs of the research, we choose carbon intensity in Xinjiang from 1995 to 2012 as explanatory variables, to reflect the condition of low carbon economic development in Xinjiang, for CEI, selection in Xinjiang from 1995 to 2012 the balance of loans of financial institutions accounted for the proportion of GDP to reflect the region CREDIT support variables of Xinjiang, for CREDIT.

To eliminate the influence of different variance, we investigated two variables take logarithm, remember to lnCEI and lnCREDIT, so we build bivariate regression equation is as follows:

$$\ln CEI = \beta_0 + \beta_1 \ln CREDIT + u_t \quad (2)$$

Due to the time series data is serial correlation, direct regression makes regression coefficient of distortion, resulting in "spurious regression" phenomenon, so before regression must be unit root, cointegration and granger causality test, we use eviews8. 0 software for the above test, the test results are shown in Table 2.

We can be seen from Table 2, the two variables are first-order single whole sequence, after the first order difference, are under the 5% level of stationary series, then we use EG two-step analysis of cointegration relationship between two variables.

Table 2. Unit root test results.

Variable	(C,T,K)	T value	P value
D(lnCEI,1)	(0,0,0)	-2.5418	0.0147
D(lnCREDIT,1)	(0,0,0)	-3.4157	0.0021

3.2. *The cointegration test*

EG two-step method is the first step on the original equations directly regression residual sequence;

The second step is to unit root test of residual error sequence, if the residual is smooth, it shows that there is a cointegration relationship between the two variables, the opposite does not exist.

First we use eviews8.0 software on OLS regression testing, the original sequence for estimating equations as follows:

$$\ln CEI = 2.17 + 0.282 \ln CREDIT + u_t \qquad (3)$$

By the further available: $u_t = \ln CEI - 0.282 \ln CREDIT - 2.17$. For unit root test of this residual again, result in the following Table 3.

As can be seen from Table 3, the residual sequence of ADF statistic for 2.393, under the level of 5% and 10% are stationary series, residual smoothly, thus between lnCEI and lnCREDIT have long-term stable cointegration relationship, satisfy the prerequisite to establish error correction model.

Table 3. Residual sequence unit root test.

Sequence	Type(C,T,K)	T valure	P value	Conclusion
U_t	(0,0,0)	-2.393	0.0207	smooth

3.3. *The error correction model*

According to the analysis above, we establish error correction model to modify the original model, the estimation equation is as follows:

$$\ln CEI = \beta_0 + \beta_1 \ln CREDIT + u_t + \beta_2 ecm \tag{4}$$

Among them, $ecm = u_{t-1}$. Then we get to OLS estimates estimation equation is as follows:

$$\ln CEI = 2.164 + 0.137 \ln CREDIT + 0.901 ecm \tag{5}$$

(238.12) (2.382) (8.368)

This equation $R_2 = 0.87$, DW value = 0.87, F statistic = 46.62, the fitting effect is good, all the variables through the test of significance of 5%, the result is credible.

We can be seen from the error correction model, Xinjiang current financial credit support with respect to the reduction of carbon intensity didn't have inhibition effect, but the exceptional performance of carbon intensity, the balance of loans of financial institutions in Xinjiang every 1% rise in the proportion of the GDP in Xinjiang region, the carbon intensity will rise by 0.137%. This means loans to financial institutions in Xinjiang is not effective to promote energy conservation and carbon reduction, but the most of the funds to support is easy to cause carbon industry in the economic development of Xinjiang at the same time, it must cause our enough attention, at the same time, we can also see the previous financial support credit level of current carbon intensity also has certain positive influence.

4. Discussion

We can be seen from the above do empirical analysis, financial credit support in Xinjiang did not reduce the carbon intensity of Xinjiang, instead of helping the carbon intensity of bucks, so what is hidden in the mechanism behind this?

First of all, the enterprises to carry out low carbon R&D facing financial constraints and risks alone bear problems, and the financial credit support can well solve the above problems, low carbon development for the enterprise to provide convenient financing channels, however, in turn, can also be interpreted as if the financial credit is not well supported by the development of a low carbon economy, instead of promoting the high-energy, high-polluting industries to further expand the scale, in order to seek the rise of the regional economy in the short term, then it causes carbon intensity rise, not fall is perfectly normal. The rise in carbon intensity in Xinjiang and the financial credit flowing to strong relationship, Xinjiang financial credit resources to support too many energy-based industries, such as heavy-duty, lack of support for low-carbon types of trades,financial credit to promote the internal mechanism of carbon intensity increases.

Secondly, from the consumer level, financial institutions can be optimized allocation of funds to guide consumers pay more attention to low carbon environmental protection products, to voluntarily to buy energy saving low carbon consumer durable goods give more preferential loan rates, to achieve the purpose of the shift towards low carbon economic development model.

5. Conclusion and Enlightenment

In this paper, by constructing error correction model, the empirical test of Xinjiang financial credit support and the relationship between the carbon intensity, found no effective Xinjiang financial credit limit the rise in carbon intensity on the development of low carbon economy in Xinjiang did not play a proper role.

Xinjiang is located in the border areas of west China, a vast, rich in resources, is the key to the future in China building energy strategic reserve base, to ensure the supply of energy in our country is of great significance, as an important energy, low carbon economy development has difficulty and more urgency. Xinjiang ecological environment fragile, once destroyed, recovery cost will be very high, so we must pay more attention to Xinjiang economy development and ecological environment coordinated, instead of passing, obtain immediate short-term interests, finally will do more harm than good.

However the phases of development and the people improve the urgency of life determines the Xinjiang to go low-carbon path is difficult, How to reduce low carbon economy brought about by the rise in production costs for the Xinjiang such underdeveloped areas is particularly important. Through low carbon credit adjustment, optimization and support the development of low-carbon economy is a market-oriented path, can effectively reduce the costs and risks of low-carbon development. This is because the financial credit for companies to focus on low-carbon development and to raise funds, spread risk, but can also guide financial credit consumer spending low-carbon products, the formation of a low-carbon consumption concept, and thus from both production and consumption to reduce carbon emissions and can create low-carbon development margins, reduce carbon emissions and achieve a low-carbon economy.

In short, when combined with a low-carbon and credit, carbon credits and carbon finance have come into being, the current international and interregional trade gradually direction transition to a low-carbon and environmental protection, low carbon future competitiveness will reshape the global trade competition, facing the severe situation, Xinjiang and China must go ahead, vigorously develop low carbon financial personnel, reserve more low carbon people intelligence library, for the future of our country and the xinjiang region of low carbon economic development can not standing in place for adequate credit support.

Acknowledgments

This work was financially supported by the National Social Science Fund Project (13AZD083), Graduate Research and Innovation Key Project in Xinjiang University of Finance and Economics (cdyjk2014035).

References

1. Shang Changbao, Ying-Yu Wu. Low carbon economy under the background of our country commercial bank credit behavior selection [J]. Journal of Zhejiang Finance, 2011 (11) : 45–48.
2. Zhou Yan. China's commercial Banks based on the symbiosis network carbon finance business development research [J]. Journal of Financial BBS, 2010 (S1) : 36–41.
3. Wu Yan, Chang-De Zheng. Low carbon financial restraint effect of the development of the research in China [J]. Journal of Southwest University for nationalities (Humanities and Social science edition), 2011 (8) : 151–155.
4. Wang Zhouwei, The construction of modern carbon financial services system of China study [J]. Journal of regional financial research, 2014 (7) : 42–46.
5. Guang-Pu Wei, Chun-Ling Zhao. Low carbon economy under the perspective of credit support research in jiangsu, for example [J]. Journal of Nanjing University of finance and economics, 2014 (3) : 55–63.

An analysis on the determinants of cash dividend payout policies of Chinese listed companies – based on financing cost and governmental interference

L.L. Shi[1,†] and C.H. Kang[2]
School of Economics, Wuhan University of Technology, Wuhan, Hubei, China
[†]E-mail: shilelai@hotmail.com

Based on the discussions about the relation between the cash dividend payout ratio and the solvency, firm size, investment opportunity, net cash flow of companies, this paper analyzes the effects of financing cost and governmental interference on the cash dividend payout policies of Chinese listed companies. The results are as follows: the financing cost significantly affects the cash dividend payout ratio, and the higher the financing cost is, the smaller the cash dividend payout ratio is. Likewise, the governmental interference also exerts significant effects on the cash dividend payout ratio. Furthermore, according to the empirical results, the net cash flow per share is positively correlated to the cash dividend payout ratio but the solvency, investment opportunity and size of companies are in negative correlation with the cash dividend payout ratio.

Keywords: Cash dividend payout ratio; Financing cost; Governmental interference.

1. Introduction

As an important field of corporate finance, the theories of dividend policies have been hot topics for the scholars from home and abroad all the time. Reasonable dividend payout policies cannot only set up a good corporate image but also arouse the passion of investors to further invest in the corporate, thus the corporate can make great progress. However, due to special national conditions, the property right structure and governance structure of Chinese listed companies as well as the regulation mechanism for security market have many deficiencies, and accordingly the dividend policies of Chinese listed companies are nonstandard or disordered.

Some data show that China's A-share companies seldom paid cash dividends and more than two thirds of listed companies didn't distribute cash dividends before 1999. To urge them to form reasonable profit distribution mechanisms for investors and "force" them to distribute cash dividends, the

---

[1] Lelai Shi is a Ph.d candidate of School of Economics, Wuhan University of Technology.
[2] Canhua Kang is a professor of School of Economics, Wuhan University of Technology.

government announced *the Decision about Amending Several Rules on Cash Dividends of Listed Companies in 2008 and the Notice about Further Enforcing the Related Matters of Cash Dividends of Listed Companies* in 2012. The cash dividend behaviors of listed companies can be known through the studies and analysis about influential factors on the cash dividend payout of listed companies to some extent. Meanwhile, there exist some non-efficiency factors in the capital market, like asymmetric and insufficiently transparent information, the cash dividend payout helps to reduce the degree of information asymmetry faced by companies. Therefore, it is of extraordinary theoretic and practical significance to include the cash dividend policies and the influential factors into a common system for empirical studies.

In terms of related empirical studies, Allen & Michaely (1995) found that the indicators like the company size, the shareholding ratio of major shareholders, the financial leverage risk and the investment opportunity significantly affected the cash dividend payout ratio. Mitton (2004) showed that the companies of more reasonable governance structures tended to pay higher cash dividends and only when it was against the background of investor protection, could the dividend payout reduce the agent cost. Amidu & Abor (2006) pointed out that the dividend payout ratio was positively correlated to the solvency and cash flow making ability of companies and national tax rate but in negative correlation with the risk, shareholding volume and book value of companies. Hellström & Inagambaev (2012) pointed out that the factors to be considered were different when the dividend payout policies were decided by companies of different sizes and the future investment opportunity was also a key factor in determining the dividend payout ratio.

The main innovations of this paper are as follows: Firstly, existing analysis framework for influential factors on cash dividend distribution behaviors of companies is expanded and the financing constraints are included into influential factors on the decisions of cash dividends distribution of companies in form of financing cost, and Tobin's Q is employed to represent the investment opportunity and analyze the decisions of cash dividend distribution of companies. Secondly, the event study is applied to analyze the effects of Chinese government's policy interference on the decisions of cash dividend distribution of companies.

2. Research Methods

2.1. *Hypotheses*

In this paper, the investment opportunity, solvency, net cash flows and size of companies are chosen as explanatory variables to analyze the decisions of cash dividend distribution of listed companies.

H1: The larger the investment opportunity faced by the companies is, the lower the cash dividend payout ratio is. That is to say, the investment opportunity and the cash dividend payout ratio are in negative correlation.

H2: The weaker the solvency of companies is, the lower the cash dividend payout ratio is. To put it another way, the solvency and the cash dividend payout ratio are positively correlated.

H3: The larger the net cash flows are, the higher the cash dividend payout ratio is. That is to say, the net cash flows and the cash dividend payout ratio are positively correlated.

H4: The larger the company size is, the higher the cash dividend payout ratio is. That is to say, the company size and the cash dividend payout ratio are positively correlated.

H5: The higher the financing cost is, the lower the cash dividend payout ratio is. That is to say, the financing cost and the cash dividend payout ratio are negatively correlated.

H6: Due to the government's policy interference, the cash dividend payout ratio is significantly raised in the same year when a statute is issued.

2.2. *Sample data*

In this paper, China A-share listed companies from 2006 to 2013 are chosen as the research samples. The sample data are processed as follows: (1) the financial companies are excluded; (2) the companies without complete data and PT and ST companies are eliminated; (3) to reduce the effects of outliers, all variables are winsorized. After the sample data are filtered as above, a total of 5090 data in 1056 groups which meet the requirements are derived, and the data type is the unbalanced panel data. The data mainly come from China Stock Market Accounting Research (CSMAR) Economic and Financial Database.

2.3. *Modeling*

To verify whether the financing cost, investment opportunity, debt-to-asset ratio, net cash flow per share and size of companies affect the cash dividend distribution behaviors, a basic model is constructed to validate the hypotheses above.

$$drp_{it} = \alpha + \beta_1 cofcs_{it} + \beta_2 tbq_{it} + \beta_3 da_{it} + \beta_4 ncfps_{it} + \beta_5 \text{In } cs_{it} + \varepsilon_{it} \quad (1)$$

In this paper, eight yearly virtual variables are added to measure the incidence of the government's policy interference on the cash dividend payout ratio. The fixed effect regression of unbalanced panel data is conducted for the models.

$$drp_{it} = \alpha + \beta_1 cofcs_{it} + \beta_2 tbq_{it} + \beta_3 da_{it} + \beta_4 ncfps_{it} + \beta_5 \text{In } cs_{it} + \sum_{i=0}^{7} \beta_{6+i} y_i + \varepsilon_{it} \quad (2)$$

The names and definitions of the variables involved in this paper are presented in Table 1.

3. Empirical Results

3.1. *Empirical results and analysis*

The individual fixed effect regression of unbalanced panel data is conducted for the models (1) and (2), and the regression results are as shown in Table 2.

The difference between the models (1) and (2) lies in whether the time virtual variables are taken into consideration. According to the regression results of the model (2), the regression coefficient of the unit financing cost (cofcs) is -1.965757 and the significance t value is -5.32, which suggests that the cash dividend payout ratio significantly decreases by 1.965757% as the financing cost increases by 1%. Hence, Hypothesis 5 is tenable, that is, the financing cost is negatively correlated to the cash dividend payout ratio of companies.

The regression coefficient of investment opportunity (tbq) is -0.0155794 and the significance t value is -4.18, which suggests that the cash dividend payout ratio significantly decreases by -0.0155794% as Tobin's Q value increases by 1%. Therefore, Hypothesis 1 is tenable.

The regression coefficient of the debt-to-asset ratio (da) is -0.0775546 and the significance t value is -2.97, which suggests that the cash dividend payout ratio significantly decreases by -0.0775546% as the debt-to-asset ratio increases by 1%. Hence, Hypothesis 2 is tenable.

The regression coefficient of the net cash flow per share (ncfps) is 0.0168063 and the significance t value is 5.25, which indicates that the cash dividend payout ratio significantly increases by 0.0168063% as the net cash flow per share increases by 1%. Hence, Hypothesis 3 is tenable.

The regression coefficient of the company size (lncs) is -0.032435 and the significance t value is -4.50, which indicates that the cash dividend payout ratio significantly decreases by 0.032435% as the logarithm of total assets at the end of period increases by 1%. Hence, Hypothesis 4 is untenable.

It can be found from the regression results of the model (2) that the t values of most yearly virtual variables are very significant, in which y_2 and y_6 are the most significant and equal to 6.59 and 4.15 respectively. The broken line charts of regression coefficients and t values of y_2 and y_6 are shown in Figure 1 and Figure 2 respectively.

y_2 and y_6 denote 2008 and 2012 respectively, and their regression coefficients are maximal and equal to 0.0547506 and 0.0444126 respectively, and their significance levels are the highest. This suggests that the cash dividend payout ratios of listed companies in 2008 and 2012 were increased by 5.47506% and 4.44126% respectively in comparison with those in 2006. Hence, it can be inferred that Chinese policy interference significantly affects the cash dividend distribution of listed companies. However, the regression coefficients of yearly virtual variables also show that these policies have very significant short-term effects and the effects were the most significant in the same years when the statutes were published. Consequently, Hypothesis 6 is tenable.

3.2. *Robustness test*

The average interest-rate spread between the one-year and five-year debts is chosen as Premium during the sample period. In Eq. (3) and Eq. (4), r1 refers to the interest rate of the one-year debt, and the total financing cost is obtained, followed by the average financing cost which serves as a variable to measure the incurred from financing for companies.

$$\text{Total Financial Cost} = \text{Short-term Debt} * r1 + \text{Long-term Debt} * (r1 + \text{Premium}) \quad (3)$$

$$\text{Average Financing Cost} = \text{Total Financing Cost} / \text{Total Asset} \quad (4)$$

After the unit financing cost in Eq. (1) and Eq. (2) is substituted with the average financing cost in Eq. (4), the fixed effect regression for panel data is performed and results are shown in Table 3.

The debt-to-asset ratio cannot completely reflect the solvency of companies, while the times interest earned is an indicator to measure the capability of companies to pay interests on liabilities (the capability to repay the debt interest). The larger times the profits before tax and interest earned by production and operation of companies are of the interest charges, the stronger the capability of companies to pay interest charges is. Hence, the debt-to-asset ratio is substituted with the interest cover ratio of companies and then the regression analysis is conducted and results are shown in Table 4.

It can be seen from the regression results that the results of robustness test are consistent with the research findings of our main models and it can be inferred that the research findings are of good robustness.

4. Conclusion

Unlike other scholars, this paper tries to explain the decisions of cash dividend distribution of companies from the micro level, i.e. the financing cost. The empirical results are as follows. Firstly, the higher the financing cost is, the stronger internal financing motivations the companies have, which leads to the lower cash dividend payout ratio of companies. Secondly, the larger Tobin's Q is, the better investment opportunities the companies are faced with, which also results in the lower cash dividend payout ratio of companies. Thirdly, the higher the debt-to-asset ratio is, the weaker the solvency is, which leads to the lower cash dividend payout ratio of companies as well, so the solvency and the cash dividend payout ratio are positively correlated. Finally, the higher the net cash flows are, the larger the possibility of cash dividend distribution is.

References

1. Alli, K. Khan, Q. & Ramirez, G.G. 1993. Determinants of dividend policy: A factorial analysis. The Financial Review. 28(4): 532–547.
2. Miller, M. H. & Modigliani, F. 1961. Dividend policy, growth, and the valuation of shares. The Journal of Business. 34(4): 411–433.
3. Lintner, J. 1956. Distribution of incomes of corporations among dividends, retained earnings, and taxes. American Economic Review. 46(2): 97–113.
4. Fama, E. F. & Babiak, H. 1968. Dividend policy: An empirical analysis. Journal of the American Statistical Association. 63(324): 1132–1161.
5. Richardson, G. Sefcik, S. E. & Thompson, R. 1986. A test of dividend irrelevance using volume reactions to a change in dividend policy. Journal of Financial Economics.17: 313–333.
6. Smith, C. Jr. & Warner, J. B. 1979. On financial contracting: An Analysis of Bond Covenants. Journal of Financial Economics. 7(2): 117–161.
7. Myers, S. C. & Majluf, N. S. 1984. Corporate financing and investment decisions when firms have information that investors do not have. Journal of Financial Economics. 13(2): 187–221.
8. Allen, F. & Michaely, R. 1995. Dividend policy. Handbooks in Operations Research and Management Science. 9: 793–837.
9. Mitton, T. 2004. Corporate governance and dividend policy in emerging markets. Emerging Markets Review. 5(4): 409–426.
10. Amidu, M & Abor, J. 2006. Determinants of dividend payout ratios in Ghana. The Journal of Risk Finance. 7(2): 136–145.
11. Hellström, G. & Inagambaev. G. 2012. Determinants of Dividend Payout Ratios: A Study of Swedish Large and Medium Caps. Umeå School of Business and Economics.

Table 1. The names, symbols and definitions of variables.

Name of variables	Symbol	Definition
The cash dividend payout ratio	dpr	The annual cash dividends/the earnings per share
The unit financing cost	cofcs	The financial expenses/the total assets at the end of period
The investment opportunity	tbq	Tobin's Q
The debt-to-asset ratio	da	The total debts at the end of period/the total assets at the end of period
The net cash flow per share	ncfps	The net cash flows of operation/the number of common shares
The company size	lncs	The nature logarithm of total assets of company at the end of period
The yearly virtual variable	yi	The eight yearly virtual variables from 2006 to 2013
The long-term debt	sl	The long-term loan of company at the end of year
The short-term debt	lll	The short-term loan of company at the end of year
The number of times interest earned	ntie	The earnings before tax and interest/the interest expenses

Note: sl, lll and ntie are the variables involved in the robustness test in this paper.

Table 2. Regression results of empirical models.

Variable	Model 1	Model 2
cons	0.567097*** (6.03)	0.8201444*** (5.30)
cofcs	-1.761346*** (-4.80)	-1.965757*** (-5.32)
tbq	-0.0174186*** (-5.33)	-0.0155794*** (-4.18)
da	-0.061936*** (-2.39)	-0.0775546*** (-2.97)
ncfps	0.0169886*** (5.35)	0.0168063*** (5.25)
lncs	-0.0196266*** (-4.56)	-0.032435*** (-4.50)
y0		0
y1		0.0134246* (1.64)
y2		0.0547506*** (6.59)
y3		0.0267308*** (3.04)
y4		0.0141319 (1.47)
y5		0.0289598*** (2.84)
y6		0.0444126*** (4.15)
y7		0.027694*** (2.46)
ADJ-R2	0.2731	0.2143
F Value	21.19	14.09

Note: *** stands for the significance at 1% level, ** for the significance at 5% level and * for the significance at 10% level.

Table 3. The robustness test of regression results of models.

Variable	Model 1	Model 2
cons	0.5999075*** (5.91)	0.8972205*** (5.33)
cofcs2	-1.851734*** (-4.76)	-2.110998*** (-5.40)
tbq1	-0.018318*** (-4.88)	-0.015868*** (-3.71)
da	-0.0589695** (-2.05)	0.0799982*** (-2.76)
ncfps	0.0163234*** (4.72)	0.0159745*** (4.58)
lncs	-0.02087*** (-4.51)	-0.035925*** (-4.60)
y0		0
y1		0.0142427 (1.61)
y2		0.0618158*** (6.89)
y3		0.0300564*** (3.16)
y4		0.0173211* (1.65)
y5		0.0353557*** (3.18)
y6		0.0510571*** (4.36)
y7		0.0314088*** (2.56)
ADJ-R2	0.2475	0.1927
F Value	18.59	13.50

Note: *** stands for the significance at 1% level, ** for the significance at 5% level and * for the significance at 10% level.

Table 4. The robustness test of regression results of models.

Variable	Model 1	Model 2
cons	0.5362496*** (4.91)	0.8684041*** (4.81)
cofcs	-2.77757*** (-6.71)	-3.042331*** (-7.30)
tbq1	-0.0186621*** (-4.54)	-0.0172432*** (-3.67)
ntie	0.000286*** (4.47)	0.0002923*** (4.60)
ncfps	0.0165696*** (4.33)	0.0163871*** (4.26)
lncs	-0.0158181*** (-3.29)	-0.0319054*** (-3.90)
y0		0
y1		0.0162557* (1.75)
y2		0.0612136*** (6.51)
y3		0.0323843*** (3.19)
y4		0.0193226* (1.73)
y5		0.0386821*** (3.27)
y6		0.0547417*** (4.39)
y7		0.0328813** (2.53)
ADJ-R2	0.2937	0.2292
F Value	23.22	14.85

Note: *** stands for the significance at 1% level, ** for the significance at 5% level and * for the significance at 10% level.

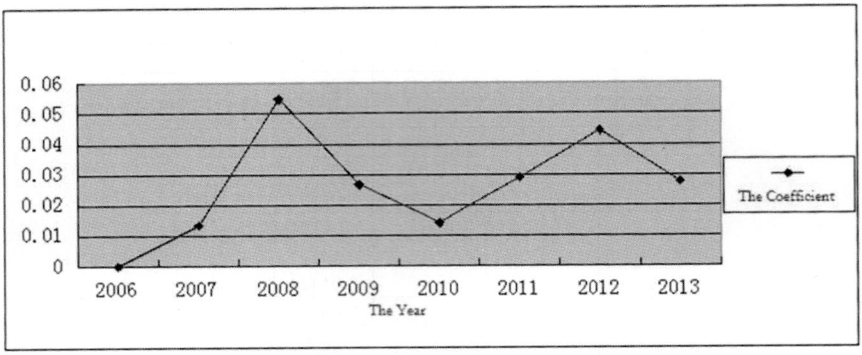

Figure 1. The regression coefficients of yearly virtual variables.

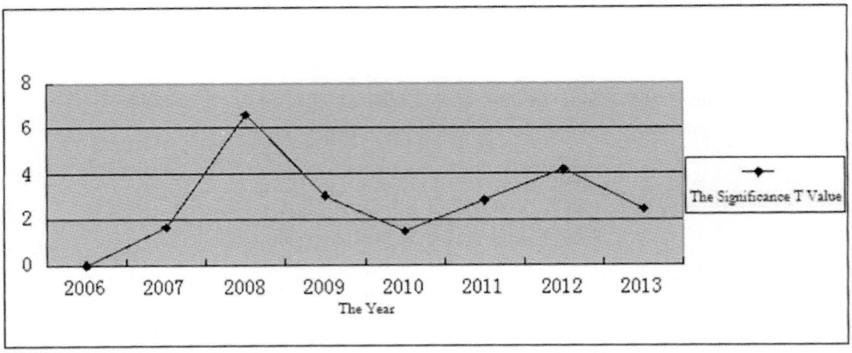

Figure 2. The T values of yearly virtual variables.

Measure of the development level of science and technology talents of Guangxi based on entropy and TOPSIS model[*]

F. Wei[‡] and P. Li[†,§]

School of Management, Guangxi University of Science and Technology, Liuzhou, Guangxi, China

E-mail: [‡]fwei916@163.com, [§]1017064040@qq.com

Based on evaluating the development capability of the Guangxi scientific and technological talent, this paper constructs the evaluation index system of the scientific and technological talent development level, which uses entropy method to empower evaluation indexes and then uses Topsis method to study the development level of the scientific and technological talent of our country's 31 provinces empirically. The results show that: the development level of the Guangxi scientific and technological talent that ranks 23 nationwide is relatively low. By contrasting the conditions of Guangxi and other provinces, we can put forward measures and suggestions to improve the development level of the Guangxi scientific and technological talent.

Keywords: Scientific and Technological talents; Development Level of Talents; Entropy method; Topsis method.

1. Introduction

Research on the development of science and technology talent, the literature at home and abroad are mainly concentrated in two aspects: first, the research on scientific and technological personnel investment gains. Use the method of data

---

* **[Project]** 1. "The empirical analysis of Guangxi scientific and technological personnel policy benefits and comparative study with other provinces" (Guangxi association [2014] B - 02);

2. "Transformation of the West River Economic Belt automotive innovation-driven deepen" (Liuzhou association 20140104);

3. <Construction of the university philosophy social science research base of the whole district>—"Urban cultural competitiveness research in the path of ascension based on the construction of social science research base";

4. "Research of teaching research universities' graduate student education serving local economic" (JGY2014119);

5. "Research of sustainability governance framework of traditional manufacturing supply chain of the western region based on the view of ecological innovation Angle" (15BJY077).

† **[Author's brief introduction]** F. Wei, professor of Guangxi University of Science and Technology, master tutor, the research direction of management science and engineering. P. Li, graduate student major of enterprise management, research directions of the human resource management.

envelopment analysis (DEA), and focused on the efficiency study of scientific and technological personnel input and output[1]. So Young, Sohn, etc[2] researched on the field of R&D by using DEA method and the decision tree, evaluated the efficiency of the commercialization of research results, thus provided a reference for the decision-making of the government or a company; B.M. Liu[3] when evaluating the development level of science and technology talents of Henan province, build evaluation model of input-output efficiency of the scientific and technological personnel firstly, then used DEA method to evaluate it comprehensively; Liu[4] analyzed the regional environmental factors that affect the efficiency of the science and technology talent development, and used DEA method for 31 provinces (municipalities) of science and technology talent development efficiency, and revealed that the scale of science and technology enterprise and technology industry agglomeration in the important role of talent development; X. Fan, S.C. Ma, etc[5] analyzed the efficiency of the activities of science and technology input and output of Liaoning province by using DEA and the rank and analogy method, and the results showed that: the efficiency of Liaoning province is relatively low, and there are problems like redundant of science and technology human resources and low use of science and technology funds. Second, the evaluation about the talent development level of science and technology. Based on the qualitative analysis of science and technology talent, and adopted the analytic hierarchy process (AHP) when empower the evaluation indexes. Barbara L Ran and Gary A Adam[6] studied the influence factors of talent flow comprehensively, and pointed out that age, gender, job satisfaction, welfare treatment, work environment, training, promotion, business incentives, and the complexity of the work are closely related to the mobility; F. Chen[7] when evaluating development level of science and technology talents of Shandong province, used the AHP method to determine index weight, and analyzed the development level of science and technology talents of 17 cities in Shandong province empirically by using Topsis method, and divided results into three categories: high, medium and low; C.H. Zhang, J. Sun, etc.[8] constructed the evaluation index system of science and technology talent development level, and analyzed the development level of science and technology talent of 31 provinces by using the AHP and Topsis model, thus put forward suggestions to promote China's science and technology talent development level.

Research results from the current point of view, relevant scientific literatures mainly focused on the output efficiency and technology talents qualitative analysis, and most of AHP for Index Weight of a comprehensive and objective evaluation of a particular area of science and technology talents, thus exists some limitations. This study attempts to use entropy method for objective evaluation of each empowerment, thus avoid the subjectivity of AHP's empowerment to some extent, and use TOPSIS model to measure the development level of Guangxi Technicians.

2. Entropy Weight Method and TOPSIS Model

2.1. *Determining the weights of factors impacting on the development level of science and technology talents based on entropy method*

To avoid the arbitrariness of human factors, the study attempts to take advantage of entropy method to empowerment factors that influence the development level of scientific and technological personnel objectively.

According to the interpretation of the basic principles of information theory, the information is a measure of the degree of ordering of the system, entropy is a measure of the degree of disorder of the system. The larger the entropy, the higher the degree of disorder metrics, the lower the utility of the information. Based on this, this paper introduce the information entropy into the weights determining of factors that influence the development level of science and technology talents, firstly, collect the raw data of each factor to form the data sequence, and then calculate the entropy of the data sequence, if the difference between the data sequence level is more larger, give it greater weight, on the contrary, then give it less weight.

2.2. *Construction of TOPSIS model*

2.2.1. *First, formate the data matrix*

Assume that the number of evaluation indexes and objects are m and n, the multi-index evaluation matrix is like this:

$$R = \begin{pmatrix} r_{11} & r_{12} & \cdots & r_{1n} \\ r_{21} & r_{22} & \cdots & r_{2n} \\ \vdots & \vdots & & \vdots \\ r_{m1} & r_{m2} & \cdots & r_{mn} \end{pmatrix}_{\mathrm{m}\times n} \tag{1}$$

2.2.2. *Data normalization*

Since the dimension data are different, there exist large differences between values, so the data must be normalized before evaluating so that the processed data can focused on [0,1]. Command

$$z_{\mathrm{ij}} = y_{ij} \Big/ \sqrt{\sum_{i=1}^{m} y_{ij}^2} \quad \mathrm{i=1,2,\cdots,m;j=1,2,\cdots n} \tag{2}$$

Then we can obtain standardized evaluation matrix like this: $z_{ij} = (z_{ij})_{m\times n}$,
$z_{ij} \in [0,1]$.

2.2.3. *Clearly the best and worst Vector*

Find the optimal and worst value of each indicator, then the optimal value together constitute the optimal vector Z^+, and the worst value together constitute the worst vector Z^-:

$$Z^+ = (z_1^+, z_2^+, \cdots z_m^+) \quad ; \quad Z^- = (z_1^-, z_2^-, \cdots z_m^-) \tag{3}$$

$$z_i^+ = \max(z_{i1}, z_{i2}, \cdots z_{in}) \text{ i=1,2,}\cdots\text{m;} \quad z_i^- = \min(z_{i1}, z_{i2}, \cdots z_{in})\text{i=1,2,}\cdots\text{m.}$$

2.2.4. *Calculate distance*

Calculate the weighted euclidean distance between the optimal solution and each evaluated project, and calculate the weighted euclidean distance between the worst solution and each evaluated project.

$$D_i^+ = \sqrt{\sum_{j=1}^{m} [W_j(Z_{ij} - Z_{ij}^+)]^2} \quad ; \quad D_i^- = \sqrt{\sum_{j=1}^{m} [W_j(Z_{ij} - Z_{ij}^-)]^2} \tag{4}$$

W_j is the weight of each index, $i = 1, 2, \cdots n$

2.2.5. *Calculate how close each evaluated project with the best solution*

$$C_i = \frac{D_i^-}{D_i^+ + D_i^-} \quad \text{i} = 1, 2, \cdots n, \; C_i \in [0,1]. \tag{5}$$

If C_i is more close to zero, this indicates that the i-th evaluation object is closer to the worst level; On the contrary, the closer the value to 1, indicating the i-th evaluation object is closer to the optimal level. That is, the value of C_i in [0,1] greater, the evaluation results more excellent.

3. Empirical Research of Guangxi Scientific and Technological Personnel Development Level Measurement

3.1. *Select indicators and collect data*

When constructing evaluation index system of science and technology talent development level, this paper respectively from two aspects: the government investment, science and technology talent investment income. Specifically, a review of government investment mainly consider the emphasis on science and technology talents in the area, incentives for scientific and technological personnel, including R & D Funds investment, the number of colleges and universities, and scientific and technological personnel fees and so on; a review of technology professionals investment income mainly consider the richness of human resources, personnel developments and market-oriented scientific and technological personnel, including the construction of the existing scientific and technical personnel, the amount of patent applications and acceptions, the number of international and domestic papers, per capita GDP, and technical market transactions and so on. (Table 1).

Various secondary indexes using the data in 2012, the original data from the 2013 China statistical yearbook of science and technology and national statistics yearbook in 2013. Considering the "Number of contracts traded technology market" and "technical market transactions" are missing in the statistical yearbook, this research adopts the average of the western region instead of it. It is worth mentioning that, the western region here includes nine provinces: Inner Mongolia, Guangxi, Shanxi, Guizhou, Yunnan, Qinghai, Ningxia, Gansu, Xinjiang.

3.2. *Index weight determination*

Because of the importance of each indicator varies, we need to give different weights for each index, this paper use entropy method to empower each indicator objectively, and use Matlab software to calculate the weight of each index. (Table 1).

Table 1. Evaluation index system of science and technology talent development level and the weight of each secondary index.

Level indicators	The secondary indicators	Weight
Emphasis on scientific and technological personnel	R&D expenses within budget	0.058293
	External R&D funds expenditure	0.055823
	R&D expenditure intensity	0.024096
	the number of colleges and universities	0.016159
Incentives for scientific and technological personnel	scientific and technological personnel fees	0.071221
	Construction costs of fixed assets	0.056484
Richness of talent resources	Number of R & D personnel	0.049454
	R&D personnel of full-time equivalent	0.054479
	Number of research and development institutions	0.018693
	Population density of talent resources	0.038325
	Area density of talent resources	0.151234
Talent development efforts	The domestic and international publications	0.040486
	amount of patent applications and acceptions	0.087765
	amount of invention patent applications and acceptions	0.072246

Market-oriented scientific and technological personnel	per capita GDP	0.010279
	Number of contracts traded technology market	0.071768
	technical market transactions	0.123197

3.3. *Data processing*

Make preprocess of data from evaluation index system of science and technology talent development level by using the method of vector norm, the transform formula is the formula (2).

3.4. *The empirical results by using TOPSIS method*

To determine positive ideal solution WZ^+ (with right) and negative ideal solution WZ^- (with right) based on TOPSIS method, namely 17 secondary indexes' respective optimal solution and worst solution after empowering.

Calculate D_i^+, $D_i^-$and C_i by using Matlab software, and sort by the value of C_i, if the value of C_i is great, it is suggested that the development level of science and technology talents of region i is optimal. The calculation results are shown in Table 2.

Table 2. Calculation results by using topsis method.

District	D_i^+	D_i^-	C_i	Rank
Beijing	0.079708	0.163467	0.67222	1
Shanghai	0.121536	0.135249	0.5267	2
Jiangsu	0.146736	0.104306	0.415494	3
Guangdong	0.157038	0.093492	0.373178	4
Zhejiang	0.171243	0.060868	0.262238	5
Shandong	0.176678	0.054931	0.237171	6
Tianjin	0.16852	0.051395	0.233703	7
Shanxi	0.188515	0.028945	0.133104	8
Hubei	0.189661	0.027874	0.128137	9
Liaoning	0.190112	0.026667	0.123015	10
Sichuan	0.193803	0.025094	0.114637	11
Anhui	0.19403	0.022615	0.104389	12
Henan	0.196977	0.020996	0.096325	13
Fujian	0.196054	0.020733	0.095637	14
Hunan	0.19844	0.018758	0.086365	15
Hebei	0.200694	0.015411	0.071311	16
Heilongjiang	0.201221	0.013476	0.062766	17
Chongqing	0.200586	0.013051	0.06109	18
Jilin	0.204343	0.010457	0.048685	19
Shanxi	0.204523	0.009781	0.045643	20
Jiangxi	0.204896	0.008833	0.041326	21
Inner Mongolia	0.205438	0.007603	0.035686	22
Guangxi	0.206638	0.007641	0.035661	23
Gansu	0.205655	0.007226	0.033945	24

Yunnan	0.206247	0.006673	0.03134	25
Xinjiang	0.208579	0.004531	0.021262	26
Guizhou	0.2083	0.004039	0.019022	27
Xizang	0.208484	0.004021	0.018924	28
Ningxia	0.209314	0.002935	0.013826	29
Qinghai	0.209648	0.001996	0.009433	30
Hainan	0.209458	0.001983	0.009381	31

3.5. *Analysis of the empirical result*

The development level of scientific and technological personnel of our country's 31 provinces can generally be divided into three categories.The first category: regions of C_i> 0.2, including seven provinces and we can find these seven provinces and regions are highly developed, and feathered by higher content of science and technology talents, more government investment, and remarkable scientific and technological achievements. The second category: regions of 0.1 <C_i <0.2, including five provinces, which are characterized by middle development level of the science and technology personnel, less content of science and technology talents, and less scientific and technological achievements. The third category: areas of C_i <0.1, including 19 provinces, most of which are underdeveloped and are characterized by backward development level of the science and technology personnel with the scientific and technological talents and technological achievements are significantly less than other provinces. Guangxi ranked 23rd nationwide, far behind of the developed eastern region and central regions. Investigate its reason, we can find that:

(1) Guangxi's emphasis on science and technology talent is not enough. In 2012, Guangxi' R&D expenses within budget is only 9.715 billion, less than a third of the average of the national 31 provinces, while at the same time Beijing, Jiangsu, and Guangdong's R&D expenses within budget are respectively Guangxi 11 times, 13 times, and 12.7 times; The R&D funds external spending nationwide on average is 1.885 billion, 3.6 times of Guangxi 518 million; About R&D funds investment intensity, Guangxi is only 0.75%, is one of the 12 provinces that less than 1%. By the end of 2012, Guangxi only have 70 colleges and universities, not only does not meet the national average, and colleges and universities are relatively concentrated, mainly distributed in Nanning and Guilin.

(2) Incentive on scientific and technological personnel of Guangxi is not strong enough. In 2012, "scientific and technological personnel fees" and "Construction costs of fixed assets" of Guangxi are 2.3 billion and 1.4 billion, compared with the average of national 31 provinces (8.5 billion, 4 billion), 3.7 times and 2.9 times of Guangxi.

(3) Resource of science and technology talent of Guangxi is far from enough. In 2012, Guangxi have R&D staff 65000, and the R&D personnel of full-time equivalent 40000, yet, average of national 31 provinces are 150000 and 100000. In addition, the number of research and development institutions of Guangxi is 123, which is slightly higher than the national average but still far lower than the number of eastern developed areas. In late 2012, every ten thousand people of Guangxi with R&D personnel 14, and per square kilometers of Guangxi has R&D staff 2700, compared with the average (35,29000), we can find that Guangxi is serious shortage of R&D personnel.

(4) Guangxi' development effort of scientific and technological personnel is not significant. In 2012, the average number of patent applications of national is 4.5 times of Guangxi. Besides, the average number of invention patent acceptions of national is 4424, which is 4.9 times of Guangxi. The number of papers published in the international and domestic of Guangxi is 12726, lags behind the average level, and lags far behind the eastern developed areas.

(5) Guangxi' marketization of scientific and technological personnel is inadequate. In 2012, Guangxi's per capita GDP is only 27952, not just behind of the average, but slightly higher than that of Xizang, Yunnan, Gansu and Guizhou. In addition, "Number of contracts traded technology market" and "technical market transactions" of Guangxi are 423 and 252 million, not only far lower than the same period of the national average, but also lower than other parts of the west except Hainan.

4. Suggestions of Improving the Development Level of Guangxi Science and Technology Talents

4.1. *To raise the attention of the scientific and technological personnel*

(1) Fully aware of the development of talent is directly related to the progress of science and technology, and related to the construction of the new Guangxi which is rich, powerful, democratic, civilization and harmonious, it is also the key to construct innovative Guangxi.

(2) To implement the idea that "human resources is the first resource". In 2014, Nanjing issued regulations of "Nanjing Zijin science and technology talents entrepreneurship special community," which is the first local laws and regulations to promote the development of scientific and technological personnel pioneering and science and technology park. Guangxi can draw lessons from it, formulate and perfect the relevant regional scientific and technological personnel policies, rules, including talent introduction policy, talent training, talent development policies and specific industries and regional

scientific and technological personnel policy. Also should fully implement the strategy of "talents strong Guangxi", and increase funds for scientific research and experimental development of external and internal spending, so as to improve the R&D spending accounts for GDP of Guangxi, striding into the national average (1.47%), and gradually to Jiangsu (2.38%), Shanghai (3.37%), Beijing (5.95%) in the shortest possible time.

4.2. *Perfect the incentive mechanism of scientific and technological personnel*

(1) The Government of Guangxi should increase human capital investment of science and technology talents. Because of science and technology talent is a talent, is the result of human capital investment[9]. Currently, wages of Guangxi technology personnel are generally lower[10], and this has seriously hindered the enthusiasm and creativity of them. For example, in 2012, “scientific and technological personnel fees” and “Construction costs of fixed assets” of Guangxi are less than 2.3 billion and 1.4 billion, while the average of nationwide are 8.5 billion and 4 billion. Therefore, the government should take important measures to increase investment and make efforts to improve working conditions of science and technology personnel.

(2) Encourage and support scientific and technological personnel’ innovation and entrepreneurship. In 2014, Fujian promoted a plan of scientific and technological innovation and mechanism reforming, which mentioned that “traditional evaluation standard of qualified personnel in science and technology is relatively single, we should establish and perfect the scientific classification and innovate performance oriented evaluation mechanism according to the characteristics of the different types of scientific research activities”. The government can draw lessons from it and strengthen policy guidance of science and technology reward, and optimize the structure of award, and increase the indicators’ weight of application, economic benefit and contribution to economic and social development.

4.3. *Increase cultivation and introduction for scientific and technological personnel, especially high-level talent*

By the end of 2013, “small talent highland in Guangxi” were introduced more than 6500 high-level personnel, but total high-level personnel are still less, and resources of science and technology talent are not rich.

In 2013, MianYang city issued implementing rules, aims to accelerate the introduction of high-level innovative entrepreneurial talent. Guangxi can draw lessons from it, and introduce qualified high-level innovative entrepreneurial talent and team surrounding it's key industries, and give them financial aid and relocation allowance. The following two aspects must be completed after introduction: First, carry out a dynamic management system. Require them to submit annual work report, the results obtained, and the annual work plan at the end of each year, and cancel the current year's financial support, installation assistance or post allowance if they does not meet the conditions. Second, optimize the environment of talent development. Human Resources and Social Security Department shall establish units compiled without limit and super job hiring measures for high-level personnel who entered the business unit. According to introduced talents' household and children schooling, public security and education departments shall respectively formulate concrete operating method, etc.

4.4. *Encourage talents of science and technology to innovate, and promote the transformation of scientific and technological achievements*

(1) The government should reform of disposition and usufruct of scientific and technological achievements. Within the time limit prescribed by the state, institution can decide to transfer, licensing, and investment cooperation to the scientific and technological achievements they have, and all revenues to unified management, autonomous regulations or agreed technology staff income proportion, transformation of scientific and technological achievements to promote scientific and technological achievements into use.

(2) Make full use of existing science and technology talent resources, to explore the potential of science and technology human resources, and strive to make all the efficiency of human resources in Guangxi area reached maximum play. Currently, most of the scientific and technological personnel of Guangxi are in institutions and universities[10], so the government should encourage teachers and researchers to go to the market and start a business, thus improving the conversion rate of the application technological achievements. At the same time, encourage enterprises to cooperate with universities and research institutes, and promote the construction of technology alliance, thus enhance the science and technology innovation ability of Guangxi.

References

1. C.H. Liu. Based on TOPSIS method of talents of science and technology development level evaluation research in China [D]. Qingdao: Ocean University of China, 2011.
2. So Young Sohn, Tae Hee Moon. Decision tree based on data envelopment analysis for effective technology commercialization [J]. Expert Systems with Applications, 2004, 26(2): 279–284.
3. B.M. Liu. Science and technology talent development and evaluation research of Henan province [D]. Tianjin: Tianjin University, 2008.
4. Z. Liu, J. Sun. The influence factors of regional science and technology talent development environment in China [J]. Social Sciences in Guangdong, 2013 (3): 29–35.
5. X. Fan, S.C. Ma, J.Z. Zhu. Regional activities of science and technology input and output efficiency of the comparative analysis [J]. Science and Technology Management Research, 2013, (24): 79–83.
6. Barbara L Ran, Gary A Adams. Attracting retirees to apply: desired organizational characteristics of bridge employment [J]. Journal of Organizational Behavior, 2005, 26(6):649–695.
7. F. Chen. Science and technology talent development evaluation research of Shandong province [D]. Qingdao: Ocean University of China, 2013.
8. C.H. Zhang, J. Sun, C.H. Liu. Measure of talent development level of science and technology research in China — Based on TOPSIS model of 31 provinces (municipalities and autonomous regions) to measure [J]. Science & Technology Progress and Policy, 2012, (12): 137–140.
9. Z. Yang. Science and technology talent agglomeration mechanism and empirical research in China [D], Wuhan: Wuhan University of Technology, 2012.
10. M. Wu. Guangxi high-level innovative talents of science and technology development system and mechanism innovation research [J]. Journal of Guangxi College of Education, 2013 (02): 44–49.

Research on the evaluation index system of China's trade facilitation*

J. Lv[‡] and S. Jiang[†]
School of Management and Economics, Beijing Institute of Technology, Beijing, 100081, China
E-mail: [‡]luj@most.cn, [†]jiangshanyou@163.com

In this paper, a more efficient index system of trade facilitation has been set up and has been utilized in measuring China's major trading partners. The index system selected certain indicators from five themes, namely, port efficiency, tariff environment, infrastructure and service, information and communication technology and business environment. Analytic Hierarchy Process (AHP) has been used to construct a much more comprehensive index system measuring trade facilitation under open environment and Delphi Method has been used to identify the weight of each index. The empirical results of trade facilitation on China's 25 major trading partners has showed China's level of trade facilitation remains at a low and middle level worldwide currently.

Keywords: Trade Facilitation; Efficient Index System; China.

1. Introduction

As Premier Li Ke-Qiang has pointed out, China is facing a much more complex and challenging environment, with exports and imports slowing down in the past few years. Hence, China's government needs to set up some innovative mechanisms to enhance trade facilitation [1] and promote enterprises' competitiveness. The trade facilitation is all about how procedures and controls governing the movement of goods across national borders can be improved to reduce associated cost [2] burdens and maximize efficiency while safeguarding legitimate regulatory objectives.

Maskus and others (2001) [3] address some more important empirical methods and challenges in quantifying the gains of trade facilitation in the area of harmonized regulations. John S. Wilson and Catherine L. Mann (2002) [4] put forward a measurement of Trade Facilitation Index system, including seven indicators to measure the benefits of trade facilitation, which are, the difficulty of the port logistics, customs procedures, regulatory environment, standard consistency, labor mobility, electronic commerce adoption, Specialization and

* This paper is supported by Jing Lv and Shan Jiang. Shan Jiang is Corresponding author.

transparency of the Asia Pacific Economic Organization Administration. John S. Wilson and Catherine L. Mann (2003) [5] built up 4 indexes, which are, port efficiency, customs environment, regulations environment, and electronic commerce.

2. The Evaluation Index System of China's Trade Facilitation

Upon all previous evaluation index, we have tried to construct a new trade facilitation of evaluation index system under principles of scientific, operational, digital and integrated. Ours data sources include World Economic Forum (WEF), International Institute for Management Development (IMD), World Bank (WB), United Nations Conference on Trade and Development (UNCTAD), International Trade Centre (ITC), International Air Transport Association (IATA), International Telecommunication Union (ITU), Global Express Association (GEA) and Transparency International (TI).

At the same time, in order to make the research results more objective and accurate, we specially invited experts and professors from General Administration of customs of China, Ministry of Commerce of China, Ministry of transportation of China and municipal customs to join in the researches of selection and evaluation of the indexes by the Delphi Method.

After repeated screening, the expert group finally selected 5 first-level indicators and 33 second-level indicators from a large number of selectable indicators. Different from previous studies, the index system covers all of the contents of trade facilitation.

For example, as the infrastructures in the Southeast Asian countries are not perfect and power supply often been interrupted, thereby the power supply index (C_{48}) has been introduced in the system to reflects this easily overlooked factor, which will also substantially impact facilitate trade. Another example is the corruption index (C_{57}), which normally hidden under the table. The introduction of this index more accurately describes the influence of different countrie's social atmosphere and officials clean degree to trade facilitation.

The whole test system is shown in Table 1:

Table 1. The trade facilitation index system.

First level indicator	Second level indicator	Code	Weight
Port efficiency (B_1)	Customs services index	C_{11}	0.0255
	Efficiency of the clearance process	C_{12}	0.1377
	Irregular payments in exports and imports	C_{13}	0.0481
	Time predictability of import procedures	C_{14}	0.1043

	Customs transparency index	C_{15}	0.0585
Tariff environment (B_2)	Tariff rate	C_{21}	0.0462
	Complexity of tariffs index	C_{22}	0.0462
	Share of duty-free imports	C_{23}	0.0462
Infrastructure and services (B_3)	Available int'l airline seat km/week, millions	C_{31}	0.0276
	Quality of air transport infrastructure	C_{32}	0.0237
	Quality of railroad infrastructure	C_{33}	0.0298
	Liner Shipping Connectivity Index	C_{34}	0.0326
	Quality of port infrastructure	C_{35}	0.0188
	Paved roads	C_{36}	0.0118
	Quality of roads	C_{37}	0.0143
	Ease and affordability of shipment	C_{38}	0.0212
	Logistics competence	C_{39}	0.0136
Information and communication technology (B_4)	Mobile phone subscriptions/100 pop.	C_{41}	0.0479
	Individuals using Internet	C_{42}	0.023
	Fixed broadband Internet subscriptions/100 pop.	C_{43}	0.0159
	Active mobile broadband Internet subscriptions/100 pop.	C_{44}	0.0061
	B2B	C_{45}	0.0105
	B2C	C_{46}	0.0116
	Government Online Service Index	C_{47}	0.0161
	Getting electricity	C_{48}	0.0516
Business environment (B_5)	Protection of property index	C_{51}	0.0177
	Efficiency and accountability of public institutions index	C_{52}	0.0036
	Access to finance index	C_{53}	0.0085
	Openness to foreign participation index	C_{54}	0.0046
	Index of physical security [6]	C_{55}	0.0376
	Resolving insolvency	C_{56}	0.0076
	Corruption index	C_{57}	0.0237
	Getting credit	C_{58}	0.0079

3. Weight Determination and Model Construction by AHP

As shown in Figure 1, the Hierarchical Model for the Comprehensive Evaluation of Trade Facilitation is established.

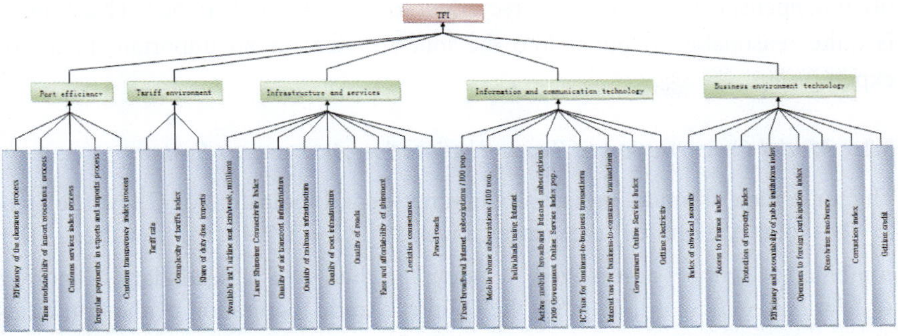

Figure 1. A hierarchical model for the comprehensive evaluation of trade facilitation.

3.1. *The Determination of Judgment Matrix and Weight*

By referring to the literature related to trade facilitation [7], according to the different influence degree of trade facilitation of each index to promote the bilateral trade volume, combining with the Delphi method results with hierarchical analysis method by Yaahp software, we establish each element of the B layer relative to the overall goal of TFI pairwise judgment matrix and weight which are shown in Table 2.

Table 2 TFI Judgment Matrix

TFI	Port efficiency	Tariff environment	Infrastructure and services	Information and communication technology	Business environment	Wi
Port efficiency	1.0000	3.0000	3.0000	2.0000	2.0000	0.3741
Tariff environment	0.3333	1.0000	0.5000	0.5000	3.0000	0.1385
Infrastructure and services	0.3333	2.0000	1.0000	1.0000	2.0000	0.1935
Information and communication technology	0.5000	2.0000	1.0000	1.0000	1.0000	0.1827
Business environment	0.5000	0.3333	0.5000	1.0000	1.0000	0.1112

In addition, we obtain the weight of trade facilitation evaluation index system which is shown in last column of Table 1. It can be seen that in all the Trade Facilitation Index, customs clearance efficiency is regarded as the most important, the weight is 0.1377. Several other very important index in turn is Time predictability of import procedures, Customs transparency index and Getting electricity. In western developed countries, the power supply won't be a problem, but for Asian countries such as India and ASEAN, power supplies often happens, causing very adverse effect to the normal clearance. Therefore, it is quite reasonable to put it into the index system as an important factor by expert group.

3.2. *Model Construction of Trade Facilitation Assessment System*

The main idea of modeling is: In obtaining second indicators of C level relative to the total weight of the target layer W_{ij} ($i = 1, 2, \cdots, 5$; $j = 1, 2, \cdots, 9$), combined with normalized index value of secondary indexes C_{ij} ($i = 1, 2, \cdots, 5$; $j = 1, 2, \cdots, 9$), the country's trade facilitation evaluation index TFI is established by the weighted sum method.

$$TFI = \sum_{i=1}^{5} \sum_{j=1}^{9} W_{ij} C_{ij}. \tag{1}$$

4. Empirical Analysis

Given data availability, according to the customs statistics, 25 countries or regions have been selected, to which all are China's most important trading partners. The data source includes "The Global Competitiveness Report 2014-2015" from World Economic Forum, "Doing Business 2015" from World Bank, "World Competitiveness Yearbook 2013" from International Institute for Management Development and "Corruption Perceptions Index 2013" from Transparency International.

4.1. *Raw Data Standardization Processing*

Taking into account the measurement standards and the scope of the original data is not consistent, in order to facilitate the comparison and analysis, it is required to make every data index for Z-score standardization.

New data = (original data - mean) / standard deviation

There are three steps in Z-score standardization by Excel:

(i) Calculate the arithmetic mean of each variable (index) X_i and standard deviation S_i ;

(ii) Standardization.

$$Z_{ij} = \frac{X_{ij} - X_i}{S_i} \tag{2}$$

Z_{ij} is the value of the variable after the standardization; X_{ij} is the actual value of the variable.

(iii) Exchange the plus or minus sign before the inverse index.

After standardization, the value of the variable Z_{ij} fluctuated around 0.

4.2. *Evaluation Results and Analysis*

After Z-score standardization of the original data by SPSS, substitute it into trade facilitation comprehensive evaluation model:

$$TFI = \sum_{i=1}^{5}\sum_{j=1}^{9} W_{ij}C_{ij} \tag{3}$$

Then we obtain Table 3.

Table 3. The Trade Facilitation Evaluation Index and Ranking of 26 Countries and Regions in 2013.

Rank	Country or Region	Trade Facilitation Evaluation Index
1	Singapore	0.859095492
2	Hong Kong	0.84573401
3	Netherlands	0.610840556
4	Japan	0.562042872
5	New Zealand	0.561286788
6	Switzerland	0.552124142
7	United Kingdom	0.551115643
8	Germany	0.4918723
9	United States	0.419728676
10	Taiwan	0.395668158
11	Australia	0.339151954
12	South Korea	0.272883599
13	France	0.226696401
14	Canada	0.209973611
15	Malaysia	0.134534866
16	China (Mainland)	-0.121096908
17	Italy	-0.31510352
18	South Africa	-0.329257864
19	Thailand	-0.562372408
20	Vietnam	-0.585281707
21	Indonesia	-0.704585433
22	Brazil	-0.772992906
23	Philippines	-0.807755172
24	India	-0.828559853
25	Pakistan	-0.955538944
26	Russia	-1.050204352

5. Conclusion

The comprehensive evaluation index of the TFI scores could be divided into four grades: 0.4 above is very convenient, -0.40 to 0 is convenient. -0.8 to 0 is inconvenient, 0.8 or less is very inconvenient. Therefore, come to the conclusion:

Singapore, Hong Kong, Netherlands, Japan, New Zealand, Switzerland, United Kingdom, Germany and United States are very convenient for trade; Taiwan, Australia, South Korea, France, Canada and Malaysia are convenient for trade; China (Mainland), Italy, South Africa, Thailand, Vietnam, Indonesia and Brazil are inconvenient for trade; Philippines, India, Pakistan and Russia are very inconvenient for trade. Evaluation results show that the developed countries such as Europe and the United States and "four little dragons in Asia" are convenient or very convenient in trade. India and almost all ASEAN nations are developing countries, and they are inconvenient or very convenient in trade. At the same time, due to Russia's big obstacles in trade liberalization, its trade facilitation index ranked near the bottom.

Table 3 shows that Singapore is the world's most open and convenient trade country with TFI score 0.859. Russia has become the most inconvenient country with TFI score -1.05. We should pay attention to urge Russia to improve the level of trade facilitation, in order to adapt to the continuous development of bilateral trade relations.

Hong Kong's TFI is 0.845, only slightly behind Singapore, ranked second in the world. China should actively strengthen customs cooperation with Hong Kong, draw lessons from Hong Kong's free trade policies, and implement CEPA "Closer Economic Partnership Arrangement" (including the follow-up 8 supplementary agreement), which will benefit both two economies.

The tables are designed to have a uniform style throughout the paper? It does not matter how you choose to place the inner lines of the table, but we would prefer the border lines to be of the style shown in Table 1. For the inner lines of the table, it looks better if they are kept to a minimum. The caption heading for a table should be placed at the top of the table.

Acknowledgments

This paper is supported by Jing Lv and Shan Jiang.

References

1. L. Liu and C. Y. Yue, Investigating the impacts of time delays on trade, *Food Policy*. 39, pp. 108–114 (2013).
2. S. Patricia and P. Richard, Measuring international trade costs, *World Economy*. 35, pp. 740–756 (2012).
3. K. Maskus and E. Keith (eds.), *An empirical ramework for analyzing technical regulations and trade*, in *Quantifying the Impact of Technical Barriers to Trade: Can It Be Done?* eds. K. Maskus and J. S. Wilson (University of Michigan Press, Ann Arbor, 2001) pp. 127–133.

4. C. Mann and Y. P. Woo (eds.), *Trade facilitation: a development perspective in the Asia-Pacific region Asia Pacific Economic Cooperation* (Asia Pacific Economic Cooperation, 2002).
5. Wilson and C. Mann (eds.), Trade facilitation and economic development: a new approach to measuring the impact, *World Bank Economic Review*. 17, pp. 367–389 (2003).
6. D. A. Baldwin, The concept of security, *Review of International Studies*. 23, pp. 13 (1997).
7. T. Nitipon and S. Suthathip (eds.), An Analytic Hierarchy Process (AHP) Approach to evaluate factors that influence cross border trade facilitation: a case study of East-West economic corridor route, in *10th International Conference on Service Systems and Service Management (ICSSSM'13)*, pp. 857–862 (2013).

Does environmental information disclosure increase the firm value and investors' confidence?

T.M. Yuan
Department of Accounting, Linyi University, Shandong, 276005, China
E-mail: yyytm@126.com

Use all A-share listed companies of main board market in 2012 as samples, this paper investigate the influence of environmental information disclosure behavior on the firm future market value and investors' confidence in capital market. The research conclusion indicate that: (1) The listed companies' environmental information disclosure behavior index and environmental information disclosure quality index are both significantly positive correlation with the company's next seasonal added value; (2) The disclosure quality significantly positive correlation with investor confidence. In short, the high quality of disclosure of environmental information can significantly improve the company value and investors' confidence in Chinese capital market.

Keywords: Environmental Information Disclosure; Firm Value; Investors' Confidence.

1. Introduction

Enterprise is the direct creator of social wealth, also is main producers of the environment pollution, therefore, environmental pollution governance requires enterprises actively participate in. Enterprise full disclosure of environmental information, it is the precondition of environmental governance, is also the important way that investors understand company and important reference for investment decisions. Accounting information usually have certain information content and value for investors, and different types of information disclosure has a different influence on stock prices and stock trading volume (Wang, Y.T. et al, 2001). At the same time, the quality and level of the environmental information disclosure can reflect the corporate implementation of social responsibility. So, strengthen the environmental information disclosure research has important theoretical and practical significance. This article with all of the 2012 a-share listed companies as research samples, in-depth analysis of the influence of environmental information disclosure behavior on the firm future market value and investors' confidence in capital market. Aiming to promote the continuous improvement of environmental information disclosure and disclosure quality level, to meet the information needs of stakeholders, and promote the sustainable development of economy in China.

2. Literature Review

Foreign research on environmental information disclosure originated the social responsibility information disclosure in the 1940s, become a research hotspot after 70s. In the cause of the environmental information disclosure, previous scholars divided it into two aspects to study: companies choose active disclosure of environmental information disclosure and the interference factors. The causes of companies choose to take initiative and proactive to disclose environmental information are to avoid political cost, prevent reputational damage and fulfill social responsibility, etc. Maltby (2004) research shows that: the enterprise can show their special and important value through the environmental information disclosure. So enterprises actively improve its value to society not only response to the demand of the society. Interference factors of corporate disclosure of environmental information mainly includes the industry sensitivity (Comier et al, 2005; Frost, 2007), external control (Porteiro, 2008), company size, financial condition, debt levels, and corporate governance mechanism.

Domestic scholars have made the preliminary discussion on the market reaction and economic consequences of environmental information disclosure, but have not a unified direction. Shen H.T. (2010) theoretically discussed the market response and the economic consequences caused by the environmental information disclosure, and impact on the equity capital cost. Tang Y.J. & X.S. Cheng (2005) empirical study found that the environmental information disclosure has a certain improvement, but not particularly significant impact on corporate performance and enterprise value. Tang, G.P. & L.H. Li (2011) study found that the company environmental information disclosure is negatively related to the investors' confidence, environmental information disclosure index is negatively related to the value of the company.

Because Chinese scholar have not easy channel to achieve environmental information disclosure information, the environmental disclosure information data used by early study are collected manually. It is not sufficient that research of capital markets reaction for the environmental information disclosure, and there is still no exact significantly conclusions in China. Therefore, it is necessary to conduct a comprehensive in-depth research of all listed companies environmental information disclosure market reaction and economic consequences in China, in order to provide empirical evidence for improvement our country's environmental information disclosure.

3. The Research Hypothesis and Research Model

3.1. *Theory analysis*

The theory, used to interpret environmental information disclosure market reaction, are mainly voluntary disclosure and signaling hypothesis, and modern behavior finance theory, etc.

(1) The voluntary disclosure and the signaling hypothesis.

Voluntary disclosure theory regard that, although external pressure theory has certain reasonable and persuasive, but it does not explain enterprise voluntary disclosure of environmental information. This view think that inadequate disclosure of corporate information can discount enterprise investment value for the information users, even list it in the worst, it would do more harm to the enterprise financing and the future development. Signal transmission is used to solve a series of problems caused by information asymmetry problem, such as adverse selection problem. When an information is not timely disclosed, investors tend to think of it as a bad news. In order to avoid the occurrence of adverse selection problem, companies with "good news" more willing to take the initiative to disclose information, separate himself apart from the "bad news" enterprise (Dye, 1985; Verrecchia, 1983). Maltby (2004) research shows that the enterprise can show its special and important value through the social environment information disclosure. So enterprise is not only in social pressure to disclosure information, in order to show its not the worst, at the same time, in order to avoid the information user make a bad guess in the case of lack information, willing to take the initiative to open more information. In addition, the holding company's managers have incentive to signal transfer undervalue through voluntary disclosure, make their stock is worth more (Healy & Palequ, 2001).

(2) The modern behavioral finance theory.

The modern behavioral finance theory, basis on the effective financial market theory, using cognitive psychology analysis investor behavior, is important theory to effective explain the investor behavior. The theory regard that human economic behavior under the influence of psychological factors, such as confidence, remorse, experiences, feelings, attitudes, opinions, attention and expectation, etc. Changing people's expectations and market decision-making behavior. Can be used to explain the investors market reaction for information disclosure. Chun Wang, C. & Xu, L.B. (2009), Long Yuee (2010) argues that investor confidence, as investors' specific psychological phenomenon, will affect the company stock trading price and trading volume through the investor beliefs and behavior change, and thus affect the value of the company.

3.2. *Research hypothesis put forward*

In today's increasingly prominent environmental problem in China, investors and other stakeholders have started to pay close attention to the environmental protection, environmental information disclosure quality and disclosure level can reflect the corporate implementation of social environment responsibility. According to the modern behavioral finance theory, the company's disclosure of environmental information must have certain information content for investors, changing people's expectations and market decision-making behavior, affect stock market price and trading volume, and thus affect the company value. Therefore put forward:

H1: Environmental information disclosure behavior are positive related to the added market value of listed companies.

H2: Environmental information disclosure quality are positive related to the added market value of listed companies.

H3: Environmental information disclosure behavior can enhance investors confidence.

H4: High quality of environmental information disclosure can enhance investors confidence.

3.3. *Research model*

Reference to previous research model, and according to the increase of variables this study, designed the following regression model to verify the above hypothesis:

$$\Delta MVS_{it+1}=\beta_0+\beta_1 EDI_{it}+\beta_2 Size_{it}+\beta_3 Lev_{it}+\Sigma\gamma_j IND_j+\varepsilon \qquad \text{(Model 1)}$$

$$TurnM_{it+1}= a_0+a_1 Disc_{it}+a_2 Size_{it}+a_3 Lev_{it}+a_4 INST_{it} +a_5 Loss_{it}+\Sigma\lambda_j IND_j+\varepsilon \qquad \text{(Model 2)}$$

$$TurnM_{it+1}=a_0+a_1 EDI_{it}+a_2 Size_{it}+a_3 Lev_{it}+a_4 INST_{it} +a_5 BTM_{it}+a_6 Loss_{it}+\Sigma\lambda_j IND_j+\varepsilon \qquad \text{(Model 3)}$$

3.4. *Variable selection and measurement*

All the index and the index connotation explanation used in this paper listed in Table 1.

Table 1. Variables and variable definition.

Variables	Variables definition
Disc	= Disclosure environmental information assignment 1, otherwise the value 0. indicate companies disclose environmental information or not.
EDI_1	= Sum up environmental information disclosure entry score, reaction company disclosure of environmental information content quantity.
EDI_2	= Weighting value of each entries content (adopt 0, 0.01, 0.02 standard for evaluation), reaction company disclosure of environmental information content quality.
MV	= The market value of listed companies, standardized by total assets.
⊿MVS	= The stock market value of listed company value compared with the previous quarter in the third quarter of 2013, standardized by total assets.
INST	= Proportion of institutional investor share holding.
BTM	= Book value to market value.
Turn Mj	= Stock monthly turnover rate of jth month (Investor Confidence Index, j=1~10). The j month total trading volumes/ the total shares outstanding at the beginning of the year.
Size	= Natural logarithm of total assets.
Lev	= Debt level.
ID	= Industry dummy variable.
t	= 2012.
ε	= Error term.

4. The Sample Data

4.1. *Sample selection criteria and data sources*

This paper selects 2012 a-share listed companies in the Shanghai and ShenZhen stock exchange as the overall sample, then, deletes the not complete data companies and extreme value companies, the final samples is 2098 companies. This study of financial data and market value index derived from the CSMAR database, The environmental information disclosure data is obtained by Manual sorting according to the "corporate social responsibility report" in CSMAR database; Institutional investors holding data taken from Wind database.

4.2. *Descriptive statistics*

Table 2 shows the descriptive statistics result of all samples. The mean of environmental information disclosure index Disc, EDI1, EDI2 are 0.193, 0.0296, 0.0159 respectively in 2012. Proportion of disclosure environmental information companies is only 19.3%, and there are larger differences in environmental information disclosure level between differences enterprises. The mean of ⊿MV3 is 0.2937 in the third quarter of 2013, The mean of stock monthly turnover is 0.4775, 0.2946, 0.4331, 0.4837, 0.4284, 0.4042 respectively in may, June, July August September, October, Shows that market turnover trend growing from June to August in 2013, Volatile 4-5 months, September to

October showed a downward trend. That is to say, after the environmental information disclosure in April, the recent stock market active period is in the third quarter. It also suggests that the market influence of environmental information disclosure should be embodied in the third quarter in China market in 2013.

Table 2. All samples descriptive statistics.

Variables	Mean	Standard deviation	Minimum	Maximum
Disc	0.1930	0.3948	0.0000	1.0000
EDI_1	0.0296	0.1003	0.0000	1.0000
EDI_2	0.0159	0.0564	0.0000	1.0000
⊿MV3	0.2937	0.6928	-3.9149	16.2768
TurnM 10	0.4042	0.3960	0.0002	4.3172
TurnM 9	0.4284	0.4274	0.0001	6.0947
TurnM 8	0.4837	0.4921	0.0002	6.4669
TurnM 7	0.4331	0.4898	0.0030	5.1348
TurnM 6	0.2946	0.3311	0.0022	4.8049
TurnM 5	0.4775	0.4451	0.0001	4.5756
TurnM 4	0.2806	0.3060	0.0015	3.3288
Size	22.0562	1.4798	15.7294	30.4956
LEV	0.4761	0.2334	0.0164	1.8973

5. Regression Results and Analysis

5.1. *Regression results of model1-H1,H2 be verified*

From Table 3 can be seen: (1) The listed companies environmental information disclosure behavior index is significantly positive with the company's next seasonal added value at 5% level. This suggests that the listed company disclosure of environmental information compared with the do not to disclose environmental information listed company, could improve the market value of listed companies. (2) The listed companies environmental information quantitative index and disclosure quality index are all significantly positive with the company's next seasonal added value at 5% level. In short, Capital market response strongly to the environmental information disclosure. This from another Angle shows that environmental information disclosure quality of listed companies can more greatly improve the listed companies market value in the period of stock market active. Consistent with our expectations, hypothesis 1, and 2 is verified.

Table 3. Regression results of model 1.

	⊿MV3 t(p)		⊿MV3 t(p)		⊿MV3 t(p)
inter	3.923*** (14.179)	inter	3.865*** (14.491)	inter	3.895*** (14.496)
Disc	0.085** (2.092)	EDI_1	0.566** (2.095)	EDI_2	0.352** (2.284)
Size	-0.15*** (-11.693)	Size	-0.147*** (-11.959)	Size	-0.148*** (-11.982)
LEV	0.046 (0.663)	LEV	0.04 (0.576)	LEV	0.039 (0.569)
IND	Control	IND	Control	IND	Control
Adj R^2	0.131		0.131		0.131
F(P)	19.585(.000)		19.586(.000)		19.643(.000)
MaxVIF	1.655		1.655		1.655
N	2098		2098		2098

Note: (1) Model:$\triangle$MVS it+1= β0+β1EDI it+ β2 Sizeit+β3 Levit +ΣγjINDj+ε; (2) The definition of all variables are shown in Table 1; (3) ***, ** and * is 1%, 5% and 10% significant level.

5.2. *Regression results of model2 - H3 cannot be verified*

From regression results of model 2 Table 4 can be seen: The listed companies environmental information disclosure behavior index is not significantly positive with the company's stock monthly turnover rate. This suggests that there was no significant difference between the listed company disclosure of environmental information and do not disclose. Hypothesis 3 cannot be verified.

Table 4. Regression results of model 2.

	Turn M 5 t(p)	TurnM6 t(p)	TurnM7 t(p)	Turn M8 t(p)	Turn M9 t(p)	Turn M 10 t(p)
constant	2.084*** (11.676)	1.344*** (9.808)	2.038*** (10.406)	2.266*** (11.564)	1.714*** (9.795)	1.510*** (9.366)
Disc	-.031 (-1.190)	-.018 (-.893)	-.020 (-.691)	-.042 (-1.478)	-.022 (-.878)	-.036 (-1.570)
Size	-.064*** (-7.603)	-.037*** (-5.747)	-.049*** (-5.400)	-.065*** (-7.056)	-.042*** (-5.172)	-.036*** (-4.79)
LEV	-.204*** (-4.529)	-.158*** (-4.595)	-.306*** (-6.206)	-.231*** (-4.665)	-.136*** (-3.079)	-.196*** (-4.844)
INST	-.305*** (-8.462)	-.208*** (-7.511)	-.307*** (-7.778)	-.342*** (-8.602)	-.321*** (-9.020)	-.298*** (-9.132)
loss	-.103*** (-3.311)	-.036 (-1.501)	-.087** (-2.555)	-.108*** (-3.153)	-.070** (-2.283)	-.066** (-2.368)
R2	.177	.132	.185	.181	.135	152
F(p)	24.397	17.446	21.568	24.930	17.880	20.343

	(0.000)	(0.000)	(0.000)	(0.000)	(0.000)	(0.000)
Max VIF	1.938	1.932	1.927	1.918	1.921	1.929
N	2098	2098	2098	2098	2098	2098

Note: (1) Model: TurnMit+1= a0+a1Discit++ a2 Sizeit+a3Levit +a4INSTit+a5Lossit+ Σλj INDj +ε; (2) The definition of all variables are shown in table 1; (3) ***, ** and * is 1%, 5% and 10% significant level.

5.3. *Regression results of model3 - H4 be verified*

From regression results of model 3 Table 5 and Table 6 can be seen: (1) The listed companies environmental information disclosure quantitative index and disclosure quality index are all significantly positive with the company's may, June, July, August stock monthly turnover rate. Especially during the stock market "active" period, July and August, the significance level reached 5% and 1%. This suggests that environmental information disclosure quantitative and quality can significantly increase the stock turnover rate, and improve investor confidence. This confidence with our anticipation, Hypothesis 4 is verified.

Table 5. Regression results of model 3(1).

	Turn M4 t(p)	Turn M 5 t(p)	TurnM6 t(p)	TurnM7 t(p)	Turn M8 t(p)	TurnM9 t(p)	TurnM10 t(p)
constant	1.63^{***} (7.118)	2.766^{***} (8.535)	1.458^{***} (7.71)	2.375^{***} (7.483)	2.376^{***} 9.305	2.057^{***} (7.945)	1.619^{***} (6.03)
EDI_1	0.134 (1.19)	0.281^{*} 1.771	0.159^{*} (1.722)	0.318^{**} (2.072)	0.339^{***} 2.731	0.136 (1.08)	0.128 (0.982)
Size	-0.09^{***} (-6.257)	-0.116^{***} -7.525	-0.056^{***} (-6.231)	-0.09^{***} -5.781	-0.09^{***} -7.446	-0.071^{***} (-5.798)	-0.06^{***} (-4.659)
LEV	0.056 (0.715)	0.08 (0.72)	-0.018 (-0.281)	-0.05 (-0.456)	0.036 (0.42)	-0.121 (-1.372)	-0.038 (-0.417)
INST	-0.063 (-1.424)	-0.147^{**} (-2.343)	-0.076^{**} (-2.076)	-0.158^{**} (-2.566)	-0.167^{***} (-3.394)	-0.129^{***} (-2.578)	-0.133^{**} (-2.563)
BTM	0.119^{***} (3.616)	0.222^{***} (4.782)	0.089^{***} (3.279)	0.09^{**} (1.989)	0.103^{***} (2.822)	0.124^{***} (3.373)	0.091^{**} (2.388)
loss	-0.076^{*} (-1.905)	-0.078 (-1.38)	-0.025 (-0.756)	-0.071 (-1.289)	-0.051 (-1.155)	-0.059 (-1.308)	-0.086^{*} (-1.829)
IND	control	control	control	control	control	control	control
Adj R^2	0.191	0.274	0.212	0.232	0.311	0.412	0.375
F(sig.)	5.72^{***}	8.563^{***}	6.342^{***}	6.977^{***}	10.093^{***}	14.93^{***}	12.95^{***}
Max VIF	3.36	3.36	3.36	3.36	3.36	3.36	3.36
N	405	405	405	405	405	405	405

Note: (1) Model: TurnMit+1 = a0+a1EDIit++ a2 Sizeit+a3Levit +a4INSTit +a5BTMit +a6Lossit + ΣλjINDj +ε; (2) ***, ** and * is 1%, 5% and 10% significant level.

Table 6. Regression results of model 3(2).

	TurnM4 t(p)	Turn M5 t(p)	TurnM6 t(p)	TurnM7 t(p)	TurnM8 t(p)	TurnM9 t(p)	TurnM10 t(p)
constant	1.645^{***} 7.173	2.787^{***} 8.592	1.464^{***} 7.726	2.388^{***} 7.509	2.379^{***} 9.286	2.06^{***} 7.941	1.611^{***} 5.984
EDI_2	0.097 1.466	0.188^{**} 2.017	0.095^{*} 1.766	0.192^{**} 2.14	0.187^{***} 2.576	0.079 1.074	0.056 0.732
Size	-0.069^{***} -6.315	-0.117^{***} -7.584	-0.056^{***} -6.249	-0.088^{***} -5.809	-0.091^{***} -7.429	-0.072^{***} -5.797	-0.059^{***} -4.612
LEV	0.053 0.676	0.074 0.671	-0.021 -0.318	-0.055 -0.503	0.032 0.374	-0.122 -1.392	-0.039 -0.424
INST	-0.064 -1.44	-0.148^{**} -2.362	-0.077^{**} -2.086	-0.158^{**} -2.579	-0.168^{***} -3.401	-0.129^{***} -2.584	-0.133^{**} -2.559
BTM	0.12^{***} 3.643	0.223^{***} 4.812	0.089^{**} 3.296	0.091^{**} 2.012	0.103^{***} 2.833	0.125^{***} 3.381	0.091^{**} 2.38
loss	-0.077^{*} -1.92	-0.079 -1.398	-0.025 -0.769	-0.072 -1.302	-0.052 -1.173	-0.059 -1.316	-0.086 -1.834^{*}
IND	Control	Control	Control	Control	Control	Control	Control
Adj R^2	0.192	0.276	0.212	0.232	0.31	0.412	0.374
F(sig.)	5.767^{***}	8.63^{***}	6.352^{***}	6.996^{***}	10.031^{***}	14.93^{***}	12.922^{***}
Max VIF	3.371	3.371	3.371	3.371	3.371	3.371	3.371
N	405	405	405	405	405	405	405

Note: (1) Model: TurnMit+1 = a0+a1EDIit++ a2 Sizeit+a3Levit +a4INSTit +a5BTMit +a6Lossit + ΣλjINDj +ε; (2) ***, ** and * is 1%, 5% and 10% significant level.

6. Research Conclusions and Contributions

This paper inspects market reactions of environmental information disclosure. The research gain the following conclusions: (1) The listed companies` environmental information disclosure behavior index and environmental information disclosure quality index both are significantly positive with the company's next seasonal added value. (2) The disclosure quality significantly positive correlation with investor confidence.

In short, the higher quality of disclosure of environmental information can higher significantly improve the company value and investors' confidence in Chinese capital market. Indicate that environmental information disclosure has significantly increased the value of the company. Capital market response strongly to the environmental information disclosure. Environmental information disclosure behavior and disclosure quality played the role of market mechanism, and exerted positive influence on investors decision-making and company's market value in a certain extent, which is beneficial to promote enterprise voluntary disclosing environmental information.

References

1. Gary. 1992. A new ideological of western accounting theory in the 90s – Green accounting theory, Accounting Research (5):23–28.
2. Geng, J.X. & Jiao, R.J. 2002. Primary exploration of environmental accounting information disclosure of listed companies. Accounting Research (1):43–47.
3. Li, J.F. 2002. Our country enterprise environment report: the present situation, demand and future, Journal of accounting research (1):42–50.
4. Shen, H.T. & Li Yu, X.L. 2010. The analysis of environmental information disclosure of listed companies of heavy pollution industry in China, Journal of securities market leader(6):51–57.
5. Shi, J.Y., Sun, X.T. & Liu, F.F. 2011. Investors overconfidence and corporate investment behavior. Journal of northeastern university (1): 33–37.
6. Tang, Y.J. & Cheng, X.S. 2005. Information disclosure mechanism evaluation, information disclosure index and enterprise performance – Based on the survey of 931 listed companies. Journal of Management Review (10):8–15.
7. Wang, Y. T. Sun, Z. & Chen, S.M. 2001. Accounting reform and the quality of accounting information-empirical evidence from China's securities market. Journal of accounting research (7):16.

Assessing the economic impact of China's carbon tax policy through a static computable general equilibrium analysis

C.B. Qin, J.N. Wang*, C.Z. Ge, Y.T. Duan and C.Y. Tian
Chinese Academy for Environmental Planning (CAEP), Beijing 100012, China
**Email: wangjn@caep.org.cn*

In this study, the General Equilibrium Analysis system for Environment (GREAT-E) is applied to analyze the effects on macro-economy, carbon emission, income distribution, sectoral production, trade structure and factor demand through five different tax rates of carbon tax charge scenarios. The simulation results show that the imposition of carbon tax is of limited impact on China's macro-economy, in which the reduction in GDP can be made within the affordable range. Relatively, the emission reduction effect of the imposition of carbon tax on CO2 is much greater than its negative effect on the economic development. It suggests that imposing carbon taxes can lead to important shifts in production, consumption, value added, and trade patterns. However, Introducing carbon tax can significantly affect income distribution, which will result in negative effects in residential welfare improvement. In order to reduce negative impacts of carbon tax, it is suggested to raise the tax rate step by step and follow the principle of tax neutrality. The government can reduce the adverse impact in the imposition of carbon tax by such means as to relieve the value added tax or to provide subsidies to the vulnerable groups.

Keywords: China; Carbon tax; CGE; Economic impact; Climate change.

1. Background

Based on the agreement of Bali Roadmap, developed country should assume measurable emission reduction obligations while developing country should also take some actions on greenhouse gas emission reduction. This is the first time that a policy document of global climate change requires developing countries to take action on measurable emission reduction. From 2000 to 2010 China's energy consumption attained year-on-year growth of 120%, the proportion of the world increased from 9.1% to 20% and the proportion of carbon dioxide emissions rose from 12.9% to 23%. The per capita carbon dioxide emissions have already exceeded the world average. China is the signatory country of the United Nations Framework Convention on Climate Change (UNFCCC) and it has the largest greenhouse gas emissions in the world. For China, the pressure of emissions reduction is getting bigger and bigger. To face the international pressure and to maintain the image of a responsible big country, China has to do more research on greenhouse gas emissions and make corresponding policies.

What's more, China's society and economy development are in the period which faces severe environmental constrains. Chinese government attached great importance to climate change and put forward the target that carbon dioxide emissions per unit of GDP in 2020 decline 40% to 45% compared to 2005. In the 12th Five-Year Plan, dealing with climate change has been officially brought into mid-term and long-term programs for national economic and social development. It is set that per unit of GDP energy intensity and carbon dioxide emissions intensity to decline 16% and 17% relatively during 12th Five-Year Plan period. When making policy for emissions reduction and thinking about dealing with climate change in China, carbon tax should be considered as an important choice. China could set up proper carbon tax policy, make an effort to eliminate the side effect on economy, and then effectively use carbon tax to promote China's emissions reduction and greenhouse gas emissions control.

Imposing carbon tax could promote enterprise energy saving, reduce carbon dioxide emissions and encourage the development of sustainable energy. However, it will increase the burden on enterprise as well as influence China's economy competitiveness. Therefore, these impacts of the carbon taxation policies need to be carefully instigated: What is the reasonable level of carbon tax rate? How would the carbon tax impact China's CO2 emission? To what extent would it impact China's macro-economy? How does it influence China's economic structure and trade structure?

As an effective policy analysis tool in the field of economics, the Computable General Equilibrium Model (referred to as the CGE model) can provide support in answering the above-mentioned questions and provide quantitative analysis for the economic influence and environmental impact of imposing the environmental tax. Xie (1996) used a static CGE model (including 7 production sectors and 3 pollution control sectors) to analyze the effect on environment and economy of China's carbon tax policy, also evaluated the economic impact of achieving 80% treatment rate of industrial effluents. Wu and Xuan (2002) used a static CGE model to analyze the impacts of China's sulfur tax policies. The results showed that the imposition of sulfur tax can lead to the negative effect on China's GDP, but it would be beneficial to the adjustment of energy structure and economic structure to greatly reduce the emission of sulfur dioxide. Wang et al. (2005) discovered that the carbon tax could make the production of coal and natural gas decline through simulation analysis using CGE mode. Based on "Energy-Economy-Environment" CGE model, Pang et al. (2008) simulated the economic impact on China's imposition of fuel tax. Based on the regional green social accounting matrix from China, Guangdong and Guizhou in 2002, Jin et al. (2008) built a "Region-Economy-Environment" CGE model (REECGE) to find out that carbon tax would take better effect than energy tax. Qin et al. (2012) developed the GeneRal Equilibrium Analysis sysTem for Water (GREAT-W), an economy-wide static Walrasian CGE model with water as a production factor, to assess the likely effects of water tax charges on the Chinese economy. The GREAT-W model is further extended into the GeneRal Equilibrium Analysis sysTem for Environment (GREAT-E) to assess the economic impacts of China's environmental taxation reforming (Qin et al., 2015).

Many scholars have focused on proving the existing problems of double-dividend with different implementation way on environmental policy when analyzing environmental policy with CGE model. The revenue of carbon tax could be used to reduce distortionary tax, and then produce other benefits besides the environmental benefit. And this idea has already been widely discussed in some national environmental policies. Bovenberg and Mooij (1994) used static CGE model to show that only if the labor supply function bends backwards, strong double-dividend would be possible. Jorgenson and Wilcoxen (1993) used dynamic model to find out that only if revenue of environmental tax is used to reduce capital tax, strong double-dividend could exist. With dynamic model of Norway economy, Bye (2000) found out that the neutral exchange between increases of environmental tax and reductions of labor income tax could increase welfare.

In order to analyze the impact of carbon tax on economy competitiveness and greenhouse gas emissions reduction, this article built a China's carbon tax impact analysis CGE model based on the GREAT-E model. It also simulated the impact of carbon tax on macro-economy, resident welfare, carbon dioxide emissions, production, consumption and trade structure. This research provided scientific basis for China's carbon tax rate set and levy approach.

2. Analytical Framework

2.1. *GREAT-E model for carbon taxation analysis*

In this study, the General Equilibrium Analysis System for Environment (GREAT-E) is used to analyze the likely effects of carbon tax charges on the Chinese economy with carbon emission as a production factor. The model is developed by using the mathematical program system for general equilibrium (MPSGE), which is a general algebraic modeling system (GAMS) extension developed by Rutherford (1998), with the MCP GAMS solver. Its theoretical structure is typical of most static CGE models, and consists of equations describing producer demands for produced inputs and primary factors; producer supplies of commodities; demand for capital investment; household demand; export demand; government demand; relationships of basic values to production costs and to purchaser prices; market-clearing conditions for commodities and primary factors; and numerous other macro-economic variables and price indices (Robinson 1999). The more detailed description of the model refers to Qin et al. (2011, 2012, 2013). Here, only a general description of the production structure is given in the following analysis.

Figure 1 presents a diagrammatic overview of the structure of the model. In this paper's standard neo-classical CGE model, each activity is assumed to maximize profits, defined as the difference between revenue earned and the cost of factors and intermediate inputs. As full competition and constant returns to scale are assumed, no excess profits can be reaped and the maximum-profit condition diminishes to a least-cost condition. Profits maximized are subjected to a production technology. The model uses multi-level nested production functions to determine the level of production. At the top level, the technology is specified by a constant elasticity of substitution (CES) function of two quantities: value-added and aggregated intermediate inputs. The aggregate intermediate input is determined by a Leontief function of disaggregated intermediate inputs, whereas value-added is itself a nested CES function of primary factors. Capital and labor are combined by a CES function at the bottom level, and this capital-labor composite is subsequently linked with carbon emission by a CES function. This combination of composite primary inputs is the same across production sectors. However, this does not imply the same composite factor endowment combination for every product because shares of inputs and the elasticity parameters between inputs are not the same across the production sectors.

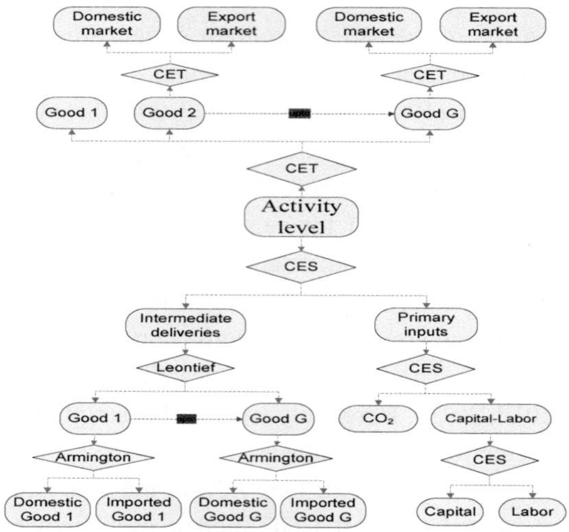

Figure 1. The structure of carbon tax CGE model.

Each activity is assumed to only produce one commodity, which is used to satisfy intermediate demand and final domestic (private consumption, capital accumulation and public consumption) and foreign demands (exports). The revenue of the activity is determined by the level of the activity, yields, and commodity prices at the producer level (Phuwanich and Tokrisna, 2007). Factors are assumed to be freely mobile across sectors. The capital and labor markets are closed by assuming that the demand for each of these factors is equal to their supply. In contrast, the total carbon emission cannot be greater than the total carbon emission; moreover, emission pricing may differ across sectors. Therefore, these assumptions imply that labor and capital is fully employed whereas carbon emission demand may not fully exhaust supply.

The model also assumes imperfect substitution between goods differing in origin or destination. Aggregated domestic output is allocated between domestic sales and exports by maximizing revenue for any given output level. The constant elasticity of transformation (CET) function is used to formalize this concept of imperfect substitution between domestic consumption of sectoral output and exports. Domestic market demand is made up of the sum of demands stemming from household consumption, government expenditure, investment, and intermediate inputs. All domestic market demands are for a composite commodity made up of imports and domestic outputs; the quantification of the demands for imported commodities is derived on the assumption that domestic demanders minimize cost subject to imperfect substitutability. In this study, the Armington (1969) CES-function form is used to determine demand composition between domestic outputs and imported goods. The demands for imported commodities are derived by the assumption that domestic demanders minimize cost subject to imperfect substitutability.

Households receive direct and indirect income from factor endowments, enterprises, and transfer payments from other institutions (i.e. government and rest of the world). Household incomes are expended on direct taxes, savings, consumption, and transfers to other institutions. Each household's consumption choices are characterized by the linear expenditure system (LES) based on net income after subtracting remittances and savings.

Enterprises do not consume any commodities. Apart from this, the payments to and from enterprises are modeled in the same way as the same payments to and from households. The government collects taxes, receives emission permit revenues, and lump-sum transfer payments from the rest of the world (ROW). Final demands of government for different goods are determined by present constant ratios with respect to the aggregated institutional income (revenues, savings and total consumption aggregate, respectively).

2.2. *China's social accounting matrix 2010*

Due to the lack of an official social accounting matrix (SAM) published by the government, we need to build a SAM by combining data from various sources into a consistent SAM framework. In this study, SAM for the Chinese economy 2010 includes 48 production sectors – 1 agricultural sector, 31 industrial sectors and 16 service sectors.

The data on activities, commodities and import and export accounts are based on the national input-output table of China's economy for the year 2010. The revenue of expenditure accounts for the government came from the Finance Yearbook of China 2011 (MOF 2011) and tax data came from the Tax Yearbook of China 2011 (SAT 2011). Household and government revenue and expenditure are adjusted based on the flow-of-funds accounts of the China Statistical Yearbook 2011 (NBS 2011). Emission accounts of carbon dioxide are estimated through emission coefficients for different energies. Energy consumption data comes from the China Energy Statistical Yearbook 2011 (NBS 2011). In order to add the carbon emission into the CGE framework, we assume a very small tax rate charged on carbon emission (0.1 CNY/ton) in the benchmark scenario. The simple version of China's SAM for the year 2010 developed by the authors is shown in Table 1.

Table 1. The simple version of Chinese SAM for the year 2010 (unit: 108 CNY).

Total	Saving	Institution						Factor		Commodity			Activity				
		ROC	Government	CO_2 tax*	Enterprise	Urban	Rural	Capital	Labor	Service	Industry	Agriculture	Service	Industry	Agriculture		
67053			74	0.08				1787	36406	4479	15087	9220				Agr.	Activity
879635			40602	8.09				79722	76686	91732	550427	40460				Ind.	
303268			19623	0.63				70892	75011	61125	72384	4233				Ser.	
71156		4103													67053	Agr.	Commodity
969619		89983												879635		Ind.	
310572		7304											303268			Ser.	
188103						141078	47026									Labor	Factor
152400					152400											Capital	
62244	29378		290							12779	14054	5742				Rural	Institution
189890	73203		4547							56772	48940	6428				Urban	
157512	88247		17697			38676	12892									Ent.	
8.8			8.8													CO_2	
82842	19065				3880	6733	1192			51474	0	498				Gov.	
101390	-16290				1232	3403	1134			17126	93940	845				ROC	
193604										15086	174787	3731				Invest	
	193604	101390	82842	8.8	157512	189890	62244	152400	188103	310572	969619	71156	303268	879635	67053	Total	

* In order to add the carbon emission into the CGE framework, we assume a very small tax rate charged on carbon emission (0.1 CNY/ton) in the benchmark scenario.

3. Experimental Simulation Scenarios

In the world, most of the countries which have carbon tax are using fixed rate for carbon tax. Generally, countries will start with a lower rate period, and then promote the rate step by step. For example, when Sweden introduced carbon tax in 1991, the tax for CO2 was 27 EUR per ton, but the tax for CO2 had been promoted to 114 EUR per ton in 2009. Now, the price of carbon permits in the EU-ETS is about 7 EUR per ton. The price of the international CDM carbon permits has fallen from 20 EUR per ton in 2008 to 0.35 EUR per ton. According to the public information in June 2013, the relevant departments in China have delivered 'The people's Republic of China Environmental Protection Tax Law (Draft)' to related high-energy-consumption industry associations, including steel, electricity, nonferrous metals and coal industry associations. Initially, the draft sets 10 CNY per ton as lower limit for the carbon tax rate. As there is a purpose of encouraging energy saving and emission mitigation while the task of dealing with climate change will become harder, the draft sets 100 RMB per ton as upper limit for the carbon tax rate. The carbon tax rate adjustment is possible with the lower limit and the upper limit.

In China, the current consumption tax is taxing product oil while the resource tax is taxing oil, natural gas and coal. The combustion of these energies is the important source of CO2 and SO2. As a result, the energy consumers will spend more if we introduce carbon tax in China. Therefore, a lower-rate carbon tax will be used in the early stages after us in-troducing carbon tax. Later on, the carbon tax rate will be promoted gradually, which will depend on the economic development and the request of energy saving and emission mitigation. To simulate the effects on the economic competence based on different carbon tax rate, this article sets low tax rate, medium – low tax rate, medium tax rate, medium – high tax rate and high tax rate as five classes of tax rate. These five classes of tax rate are set by 10, 30, 50, 70 and 100 CNY per ton in each scenario, respectively. Therefore, the CGE model will be used to simulate the influences on social economy if introducing carbon tax.

4. Results and Discussions

4.1. *The macro-level results*

Introducing carbon tax will affect the real GDP slightly. When the carbon tax is 10, 30, 50, 70 and 100 CNY per ton, the simulation result indicated that the real GDP would just reduce by 0.022%, 0.066%, 0.110%, 0.153% and 0.216%, respectively (see Table 2). Introducing carbon tax will promote the production costs of firms to some extent. Especially when the tax rate is increasing fast, the GDP will be increasingly affected. But overall, even when introducing high rate carbon tax, the scale of the economy will be affected slightly, which suggests that the effects on macro-economy by introducing carbon tax are limited.

Consumer Price Index (CPI) will be impacted by charging carbon tax. When the carbon tax is 10, 30, 50, 70 and 100 RMB per ton, the simulation result indicated that the CPI will increase by 0.046 %, 0.14%, 0.23%, 0.32% and 0.45%, respectively (see Table 2). It shows that introducing higher rate carbon tax will increase the production cost on firms. In turn, the increased production cost will increase the price of goods and services. At last, the consumers will pay the cost, which will have negative effect on residential welfare. Therefore, the government needs focus on the effects on production cost by carbon tax when introducing carbon tax. The principle of tax neutrality can be taken into consideration. For example, value added tax can be reduced to reduce the burden of firms. These will result in the double dividend of the environment and economy.

Though introducing carbon tax will significantly affect the income distribution. Through the carbon tax, the government income increases considerably while the residential disposable income decreases considerably. From Table 2, we can find that the total residential income will reduce by 0.48%, 1.43%, 2.35%, 3.25% and 4.57% respectively, when the carbon tax is 10, 30, 50, 70 and 100 RMB per ton. The rural residential income will reduce by 0.56%, 1.66%, 2.73%, 3.78% and 5.30% while the urban residents total income will reduce by 0.46%, 1.36%, 2.25%, 3.11% and 4.36%, respectively. If the tax rate of carbon tax is too high, the residential income will be affected greatly, especially the reduction on rural residential income is larger than the urban residential income. It suggests that the vulnerable groups will be affected more. According to the estimation from Table 2 when the carbon tax is 10, 30, 50, 70 and 100 RMB per ton, the government revenue will increase by 0.96%, 2.84%, 4.67%, 6.46% and 9.07%, respectively. The result shows that introducing carbon tax will has significant negative effects on income distribution issue. The government gets a large number of tax revenue, but the transferred carbon tax costs significantly reduce the residential welfare. At the same time, the carbon tax will promote the price of consumptions goods, but the vulnerable groups are weaker to bear the price rising. Therefore, measures are needed to hedge the negative effects on income distribution caused by carbon tax. The increased government income enables the government to spend financial resources to reduce the negative effects on residential welfare caused by carbon tax, including reducing value added tax, reducing income tax or subsidies the vulnerable groups.

Table 2. Results from simulations at the macro-level (%).

	Low	L-M	Middle	M-H	High
Real GDP	-0.022	-0.066	-0.11	-0.153	-0.216
Total export	0.046	0.138	0.23	0.321	0.456
Total import	-0.07	-0.20	-0.32	-0.44	-0.61
Wage rates	-0.07	-0.22	-0.36	-0.49	-0.68
Capital rents	-0.07	-0.21	-0.35	-0.48	-0.68
All Household income	-0.38	-1.11	-1.82	-2.51	-3.52
Rural household	-0.48	-1.43	-2.35	-3.25	-4.57
Urban houshold	-0.56	-1.66	-2.73	-3.78	-5.30
Gov. income	-0.46	-1.36	-2.25	-3.11	-4.36

4.2. *The influences on CO2 emission*

Compared with the slightly reduction on GDP, introducing carbon tax will obvious reduce the CO2 emission. Figure 2 points out the simulation result that when the carbon tax is 10, 30, 50, 70 and 100 RMB per ton, the total CO2 emission will reduce by 0.21%, 0.62%, 1.01%, 1.39%, 1.94%. The coefficient of elasticity of CO2 mitigation is 9.5, 9.4, 9.2, 9.1, 9.0, which suggest that introducing carbon tax will reduce CO2 emission more than nine times as the negative effects on GDP reduction. Relevant tax neutral policies are needed to be designed and introduced, including reduction on value added tax and income tax, especially differential tax reduction on low-energy-consumption industries. Therefore, it is possible for us to keep economic growth, while restraining the expansion on high-energy-consumption, adjusting industrial structure to ensure large scale of CO2 emission mitigation. At last the double dividend of energy saving, emission reduction and economic development will be achieved.

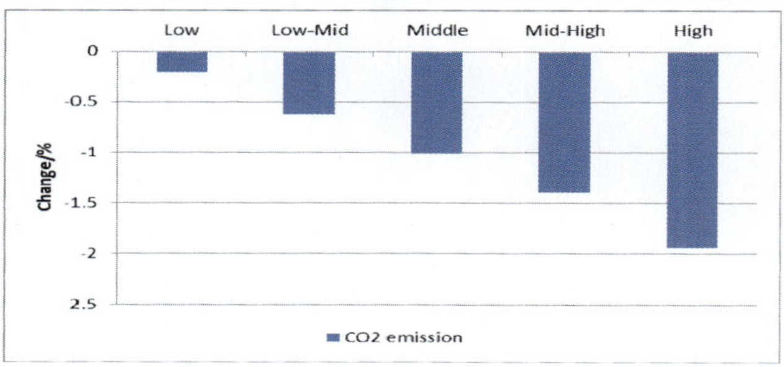

Figure 2. Influences on CO2 emission by charging carbon tax.

4.3. *The sectoral results*

4.3.1. *The impacts on production structure*

Carbon tax is favorable for industrial restructuring, which means the expansion of high energy-consuming industries will be inhibited while low energy-consuming industries will account for more proportion in national economy. Figure 3 shows the production scale changes of different sectors when carbon tax is collected. In terms of the level of outputs, imposing carbon tax restrains the consumption intensity as well as the industries with large carbon dioxide emission. And the higher the tax rate the greater the influence. Contrarily, levying carbon tax will promote the development of low energy industries. A big drop on output appeared in some sectors including coal mining, oil and natural gas extraction, ferrous metal mining, non-ferrous metal mining, food and beverage industry, tobacco, textiles, leather, oil processing and coking, ferrous metal smelting, electricity and heat, water supply and other high energy-consuming industries. Carbon tax increases industry production costs and inhibits the expansion of production capacity. Relatively, output increased in some sectors, they are telecom and electronic equipment, postal services, research, comprehensive technical services, public facilities management, education, health and social security, culture and entertainment, public administration. Although the level of output declines in some low energy industries, the proportion in national economy is getting higher. Since the expansion of high energy-consuming industries are restrained, production factors including capital and labor are transferred to low energy industries, especially the high technology and modern service sector. If the tax relief policy could be implemented in low energy-consuming industries, allocation of resources would be better optimized to promote transformation and upgrading of industrial structure, as well as to change the pattern of economy development effectively.

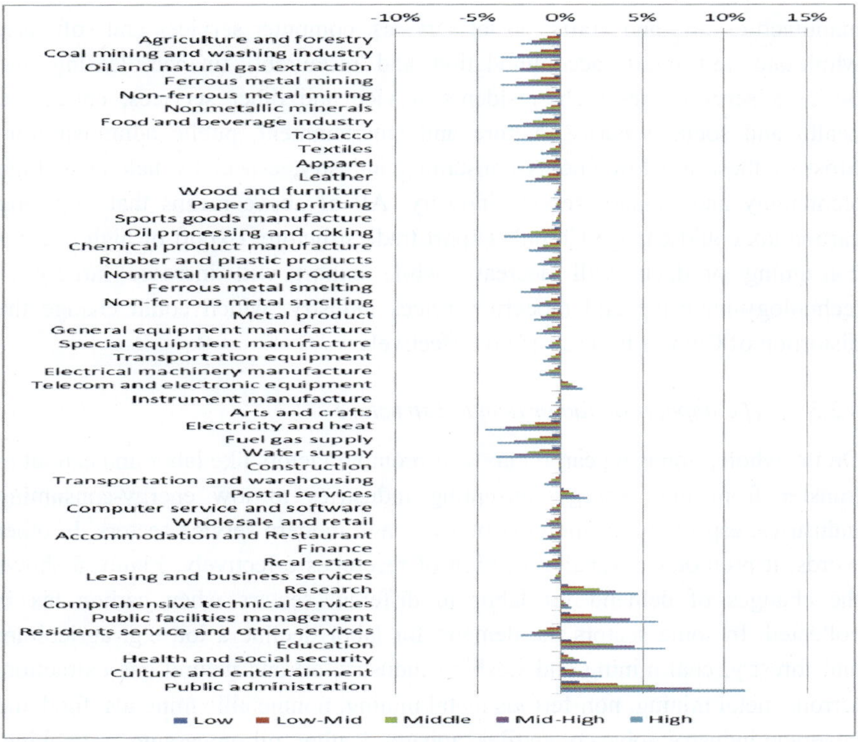

Figure 3. Influences on sectoral production by charging carbon tax.

4.3.2. *The impacts on export structure*

Imposing carbon tax will restrict export in high energy-consuming industries while the export competitiveness in low energy-consuming industries will be improved. Also the influence of trade surplus on China's energy consuming and carbon dioxide will decline. Figure 4 shows the export changes of different sectors when carbon tax is collected. Export volume declines a lot in some sectors including coal mining, oil and natural gas extraction, ferrous metal mining, non-ferrous metal mining, nonmetal mineral products, oil processing and coking, chemical product manufacture, rubber and plastic products, nonmetal mineral products, ferrous metal smelting, non-ferrous metal smelting, metal product, general equipment manufacture, special equipment manufacture, electricity and heat. All of these sectors belong to high energy-consuming industry. Levying carbon tax rise production costs and the increase of price weaken the export competitiveness. Also in some sectors the export increases, they are tobacco, textiles, leather, telecom and electronic equipment, instrument manufacture, arts and crafts, postal services, computer services and software, wholesale and retail, accommodation and restaurant, finance, leasing and business services, research, residents service and other services, education, health and social security, culture and entertainment, public administration. Most of these are low energy-consuming industry, especially belong to high technology and modern service industry. All of above means that imposing carbon tax could change China's export trade structure. Export of high energy-consuming products will decrease while export will increase largely in technology-intensive and modern services industry, which could change the distortion of China's trade structure effectively.

4.3.3. *The impacts on factor demand structure*

On the whole, imposing carbon tax will promote factors like labor and capital to transfer from high energy-consuming industries to low energy-consuming industries, especially technology-intensive and modern service sectors. In other words, it promotes optimal allocation of resources effectively. Figure 5 shows the changes of demand for labor in different sectors when carbon tax is collected. In some sectors the demand for labor decline a lot, e.g. agriculture and forestry, coal mining and washing industry, oil and natural gas extraction, ferrous metal mining, non-ferrous metal mining, nonmetallic minerals, food and beverage industry, tobacco, textiles, apparel, leather, oil processing and coking, chemical product manufacture, rubber and plastic products, nonmetal mineral products, ferrous metal smelting, non-ferrous metal smelting, metal product, general equipment manufacture, special equipment manufacture, transportation equipment, electrical machinery manufacture, electricity and heat, fuel gas supply, water supply, real estate. Most of these sectors belong to high energy-consuming industries. Imposing carbon tax restrains the expansion of these sectors, which leads to decrease of demand for labor. In other sectors the demand for labor shows a trend of increase, e.g. telecom and electronic equipment, instrument manufacture, postal services, research, comprehensive technical services, public facilities management, education, health and social security, culture and entertainment, public administration. As these sectors consume low energy, the freed labor from high energy-consuming industries could be received for their own development. Imposing carbon tax promotes development of these low energy-consuming industries.

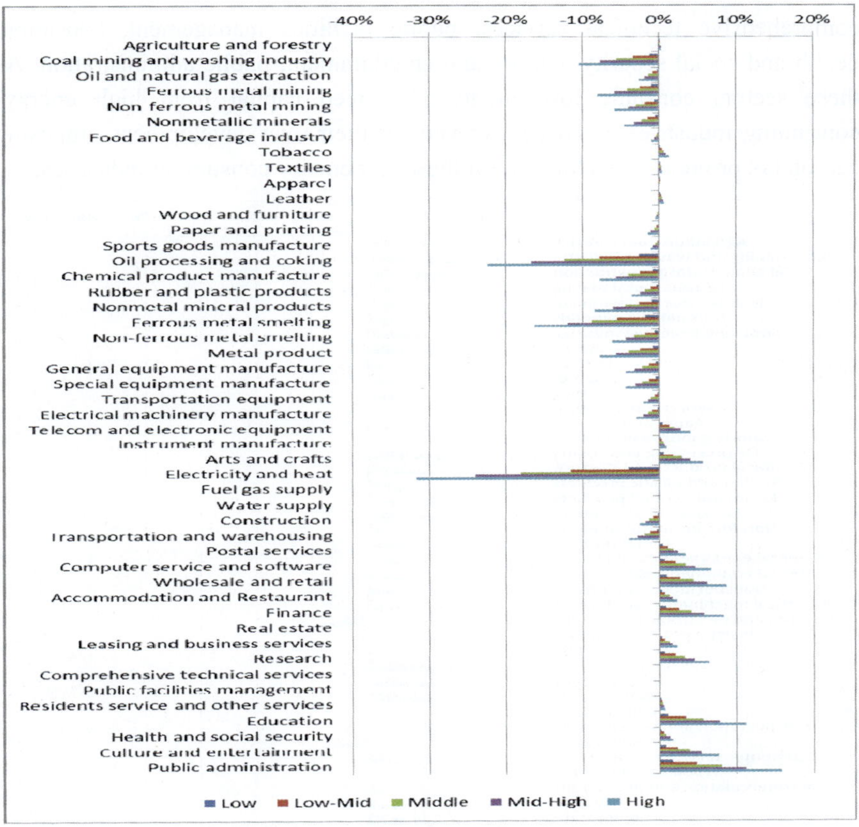

Figure 4. Influences on sectoral exports by charging carbon tax.

The changes of capital are similar with that of labor structure. After levying carbon tax, capital element also exhibits the trend that it transfers from high energy-consuming industries to low energy-consuming industries, especially technology-intensive and modern service sectors. Figure 6 shows the changes of demand for capital in different sectors when carbon tax is collected. In some sectors the demand for capital decline a lot, e.g. coal mining and washing industry, oil and natural gas extraction, ferrous metal mining, non-ferrous metal mining, nonmetallic minerals, food and beverage industry, tobacco, oil processing and coking, electricity and heat, fuel gas supply, water supply, real estate. Most of these sectors belong to high energy-consuming industries. Imposing carbon tax restrains the expansion of these sectors, which leads to decrease of demand for capital. In other sectors the demand for capital shows a trend of increase, e.g. telecom and electronic equipment, instrument manufacture, postal services, leasing and business services, research, comprehensive technical services, public facilities management, education, health and social security, culture and entertainment, public administration. As these sectors consume low energy, the freed capital from high energy-consuming industries could be received for their own development. Imposing carbon tax promotes development of these low energy-consuming industries.

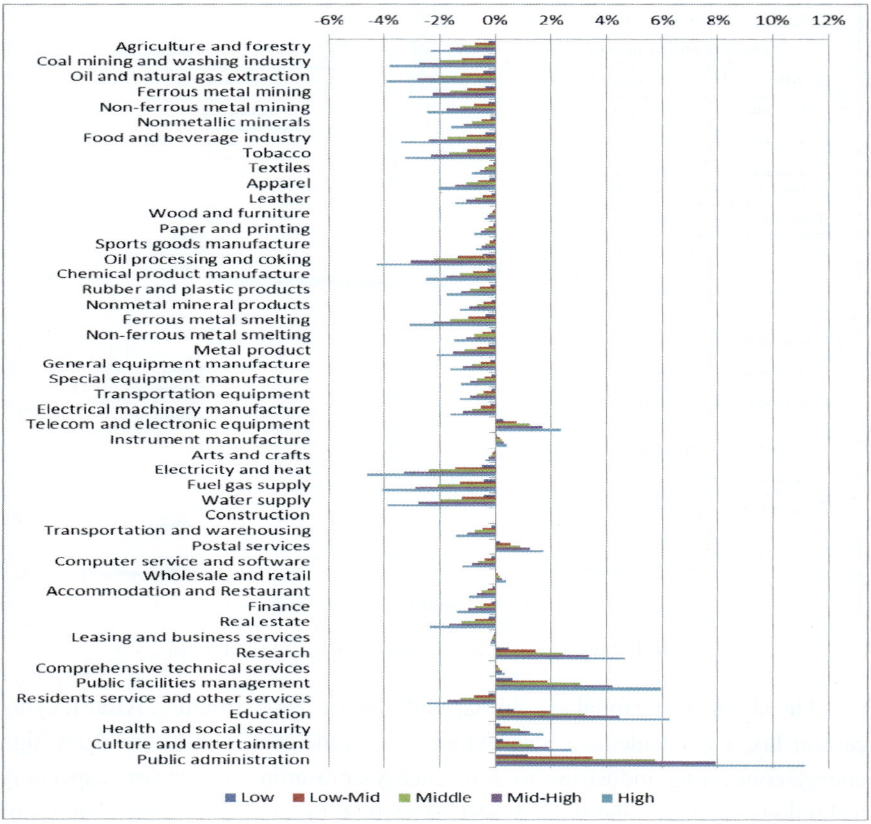

Figure 5. Influences on sectoral labor demand by charging carbon tax.

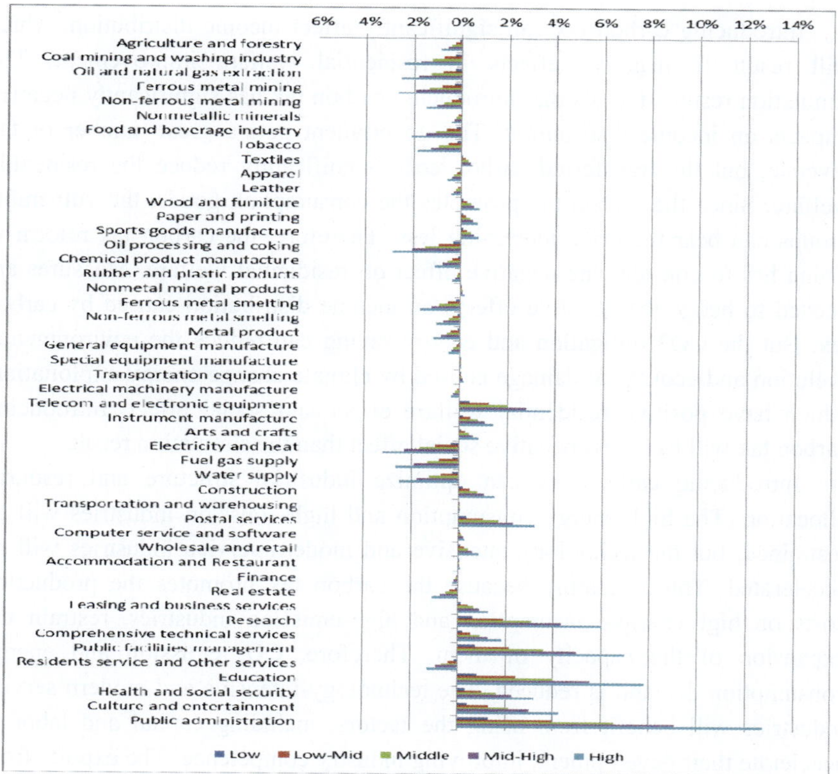

Figure 6. Influences on sectoral capital demand by charging carbon tax.

5. Conclusions and Policy Implications

5.1. *Main conclusions*

In this study, GREAT-E model is applied to analyze the effects on macro-economy, carbon emission reduction, income, industry structure, trade structure and factor demand through five different tax rates of carbon tax charge scenarios.

Introducing low rate carbon tax will affect the economy in China slightly, but will restrain CO2 emission significantly. The simulation result points out that introducing carbon tax will have limited effects on the Chinese macro-economy, the GDP reduction is sustainable. Relatively, the CO2 emission reduction effect and energy saving effect are far more than the restraining effect on economic development by introducing carbon tax. A higher rate of carbon tax can drastically reduce CO2 emission.

Introducing carbon tax can significantly affect income distribution, which will result in negative effects in residential welfare improvement. The simulation result suggests that introducing carbon tax has significantly negative impacts on income distribution. The government gets a great number of tax revenue, but the transferred carbon costs significantly reduce the residential welfare. Since the carbon tax promotes the consumption goods, the vulnerable groups can bear the price increasing less. Therefore, the carbon tax reform in China has to consider the negative effect on residential welfare. Measures are needed to hedge the negative effects on income distribution caused by carbon tax. But the CO2 mitigation and energy saving can reduce the environmental pollution and ecological damage caused by climate change energy exploitation, which have positive residential welfare effect and social effect. Introducing carbon tax will have less negative social effect than the simulation result.

Introducing carbon tax can optimize industrial structure and resource allocation. The high-energy-consumption and high-emission industries will be restrained, but the technology-intensive and modern service industries will be accelerated. This is mainly because the carbon tax promotes the production costs on high-energy-consumption and high-emission industries, restrain the expansion of the capacity of them. Therefore, CO_2 emission and energy consumption demand is reduced. The technology-intensive and modern service industries will benefit from using the factors, including capital and labor to accelerate their development, improving industry competence. The exports from high-energy-consumption and high-emission industries will be restrained, and the export competence of the technology-intensive and modern service industries will be improved. It can efficiently change the distorted trade structure of 'trade surplus, energy deficit, CO_2 emission deficit'.

5.2. ***Main policy suggestions***

The determination of carbon tax rate needs to fully consider the bear capacity of firms and the effects on social economy. The carbon price need to scientific and reasonable, fully consider the national condition and economic development condition. The carbon price cannot be too high, or it will damage the international competence of firms, affect the social stability and economic development continuous and healthy.

The formulation of tax rate needs to be step by step. In the background of economic globalization, the domestic and international experiences point out that introducing carbon tax will affect the international competence of firms. Therefore, many countries have introduced policies to reduce tax rate for high-energy-consumption industry, and policies to encourage energy saving. China is a developing country, the most important target is keeping economy continuous and stable developing, and to improve individuals' living standard. Therefore, the formulation of carbon tax plan has to be step by step.

The formulation of carbon tax policy needs to follow the principle of tax neutrality. Since introducing carbon tax will significantly affect the income distribution. To reduce the effects on country competence and residents' welfare caused by introducing carbon tax, relevant tax reduction policies are needed to hedge these negative effects. In the background of national finance and tax reform, following the principle of tax neutrality, we suggest the government need promote carbon tax revenue and reduce tax rate on value added tax and income tax.

To fully benefit from the carbon tax policy in terms of social effects, the carbon tax needs to be taxed in consumption relief, which is better to stimulate consumers reducing their energy consumption. But it is easier to operate to tax in sale relief, in terms of actual management and operation. However, it will just be a promotion on energy purchasing price to consumers, which will fail to fully benefit from the carbon tax policy in terms of social effects. Besides, the tax rate for carbon tax in recent years will be low. So the tax in sale relief can lead to total loss on social effects. Therefore, carbon tax should be taxed in consumption relief, can be with environmental tax or pollution fee by the same departments, in order to reduce social management costs.

Acknowledgements

The research received financial support from the National Social Science Fund of China Key Project (12AZD040), the emergency project of National Natural Science Fund (71441029), and the Public Utility Research Program of Ministry of Environmental Protection, China (201309043). The comments and revision suggestions by the reviewers and editors on the manuscript are also very appreciated.

References

1. Armington, P. A. (1969). A theory of demand for products distinguished by place of production. IMF Staff Papers, 16(1), 159–178.
2. Bovenberg, L.A., de Mooij, R.A., (1994). Environmental Levies and Distortionary Taxation, American Economic Review 94, 1085-1089.

3. Bye, B., (2000). Environmental Tax Reform and Producer Foresight: An Intertemporal Computable General Equilibrium Analysis, Journal of Policy Modeling, Vol. 22, No. 6, 719-752.
4. Jin, Y., Lei, M., Huang, T., 2008. The Impact of Environmental Tax on Different Regional Economic Environment. Economic Science, 2007(3): 104-112.
5. Jorgenson, D.W., Wilcoxen, P.J., (1993). Reducing U.S. Carbon Emissions: An Econometric General Equilibrium Assessment", Resource and Energy Economics, 14, 243-268.
6. MOF (Ministry of Finance), 2011. *China Financial Yearbook 2011*, Beijing: China Financial Publishing House.
7. NBS (National Bureau of Statistics), 2011. *China Statistical Yearbook 2011*, Beijing: China Statistics Press.
8. NBS (National Bureau of Statistics), 2011. *China Energy Statistical Yearbook 2011*, Beijing: China Statistics Press.
9. Pang J., Zou J., Fu S., 2008. The impact analysis of imposing fuel tax in China through computable general equilibrium model. Inquiry into Economic Issues, 2008 (11).
10. Phuwanich, L., and Tokrisna, R., 2007. Economic policies for efficient water use in Thailand. Soc. Sci. 28: 367-376.
11. Qin, C., Bressers, H.J.A., Su, Z., Jia, Y., Wang, H., 2011. Economic impacts of water pollution control policy in China: A dynamic computable general equilibrium analysis. Environmental Research Letters 6 044026 doi: 10.1088/1748-9326/6/4/044026.
12. Qin, C., Jia, Y., Su, Z., Bressers, H.J.A., Wang, H., 2011. The Economic Impact of Water Tax Charges in China: A Static Computable General Equilibrium Analysis. Water International, 37:3, 279-292.
13. Qin, C., Su, Z., Bressers, H.J.A., Jia, Y., Wang, H., 2013. Assessing the Economic Impact of Mitigating North China's Water Scarcity Strategy: a Multi-Regional Water Computable General Equilibrium Analysis. Water International, 38:6, 701-723.
14. Qin, C., Wang, J., Ge, C., Gao, S., Liu, Q., 2014. Economic and Emission Impact Analysis of Reforming China's Environmental Taxes System. China Population Resources and Environment, 2015 (03). (In press)
15. Robinson, S., et al., 1999. From stylized to applied models: building multi-sector CGE models for policy analysis. The North American Journal of Economics and Finance 1, 5–38.

16. Rutherford, R.T., 1998. Economic Equilibrium Modeling with GAMS: An Introduction to MS/MCP and GAMS/MPSGE. Draft monograph. www.gams.com/doc/solver/mpsge.pdf.
17. SAT (State Administration of Taxation), 2011. *China Tax Yearbook 2011*, Beijing: China Taxation Publishing House.
18. Wang C., Chen J., Zou J., 2005. Impact assessment of CO2 mitigation on China economy based on a CGE model. Journal of Tsinghua University (Science and Technology), 45(12).
19. Wu Y., Xuan X, 2002. Environmental Taxation Theory and Application Analysis in China. Beijing: Economic Science Press.
20. Xie, J., (1996). Environmental Policy Analysis. Aldershot: Ashgate Publishing Company, 1996.

A bottom-up method for providing sustainability practice facilities in university: A case study in China

Y. Tang
College of Architecture and Environment, Sichuan University,
Chengdu, Sichuan Province, China
E-mail: tangyan618@126.com

Public and stakeholder participation has been considered as a new principle of sustainability. The students are key stakeholders in the course of the sustainable development. In this paper, we discuss how to provide sustainability practice facilities in university with a bottom-up method, and offer example cases from Sichuan University. It is our hope that this approach will be useful for readers in order to better identify the students' opportunities in participation of sustainable issues.

Keywords: Sustainable development; Bottom-up method; Participation; Practice facilities.

1. Introduction

The literature devoted to researching on the university sustainable development is immense. There are many studies on investigating the student's awareness and perceptions of sustainable development. The previous studies found that students showed willingness to support and participate in sustainable initiatives. The students play a key role in sustainable development in universities [1]. However, there has been lack of research on the students participation in the sustainable university decision-making issues. At present, about the university sustainability decision-making issues, the Chinese universities usually take top-down approach while ignoring the students' perceptions. As the previous study shows that the sustainability goals will not be achieved without the cooperation and participation from all stakeholders [2]. Both theoretical literature and practical experiences reveal that higher education institutions can contribute to sustainable development in many ways. Public and stakeholder participation has been considered as a new principle of sustainability. Especially, the students are key stakeholders in the course of the sustainable development [3].

The study showed that student opportunities is important to achieve the sustainable development goals in university. Especially the factor of providing sustainability practice facilities is vital to student to participate the sustainable issues [4]. Therefore, this research focused on providing sustainability practice facilities using a bottom-up method. First, the study investigated the student's awareness and behaviors on sustainable development in university. Second, the study discussed how to providing a practice facilities using brain-storming method. At last, we analyzed how to push the authority leaders take action using the bottom-up method. This study would give the readers an interesting method to push the stakeholders to participate the sustainable issues.

The bottom-up approach plays a critical role to raise awareness across the higher education institution and encourage interactions between various stakeholders [5]. As one of the key stakeholders in universities, students play a key role in the bottom-up approach of sustainable university projects [6]. Most Chinese universities take top-down approach to address sustainability issues in universities where students' awareness and perceptions are not reflected in decision making [7].

2. A case study in Sichuan University(SCU) in China

2.1. *A bottom-up method model*

In this model, as shown in Figure 1, we starting the sustainable management issue from the investigation of the students' behaviors and awareness. Then we discuss the inclusions of the investigation in a students seminar. We use the brain-storming method in the expert workshop, and analyze the students' behaviors and awareness about sustainable issues. Meanwhile, put forward the suggestions of providing practice facilities.All this above, provide the basis for authorities.

2.2. *Student questionnaire survey*

Based on the review of the literature, we designed the questionnaire survey to understand what SCU students do and how they think about sustainability. In order to investigate the students' behaviors of sustainability issues, a large-scale questionnaire survey was conducted between March and July 2014 in SCU.

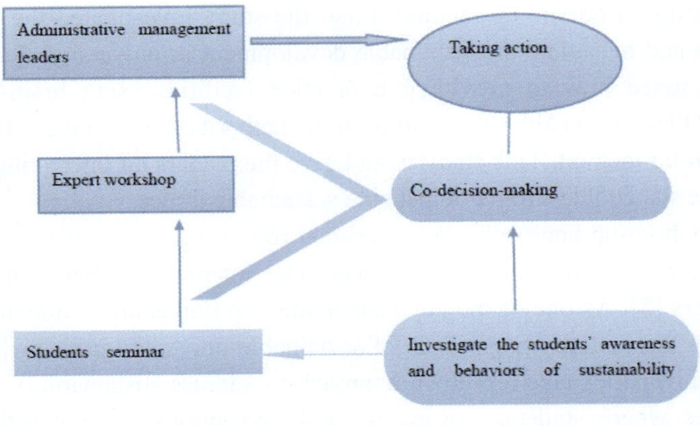

Figure 1. A bottom-up models for students participate in the sustainable issues.

The goal of the questionnaire survey is find out the obstacles of the sustainable development and how to provide the students environmental sustainability practice facilities. The questionnaire include four section, and focus on energy conservation, waste prevention, environmental protection, and the SCU sustainability practice facilities.

The first section is about transportation, the second is about energy conservation and recycling, the third is about climate change, and the fifth is about providing a sustainable practice platform in the university.

There were 21 questions in the survey, and these questions were measured using a 5-point Likert scale which ranged from 1 that meant strongly disagree to 5 that meant strongly agree. Students were asked to assign a number from 1 to 5 to indicate the level of their agreement on this statement. Higher the number, the greater level of supportive opinion of students towards the statement.Four example questions are provided as follows:

"Do you agree with that people should conserve water and electricity even if it is inconvenient"?

"We should recycle even if it is inconvenient".

"People should reuse things".

"We should buy environmentally-friendly products".

2.3. *Results*

The students were invited through the classroom and the internet, a total of 605 valid questionnaires were returned. The students covered different campuses and grades.

In the students survey, 605 students from divisions of the Sichuan University were invited to participate in this study. As the Figure 2 showed, we ranks the factor categories of student's behaviors and awareness about sustainability in SCU. Survey showed that students generally had a perceive to have sustainable curricula. Students largely showed interests on recycling and waste reduction.

In the survey and workshop, we found that the students generally think that the lack of understanding of the concept of sustainable development. The students hope universities start sustainable curricula. Besides that, there lack of sustainable activities in the university, especially there lack of the information of the related activities even some activities had been taken. Therefore, in the brain-storming workshop, we give a suggestion that making a platform for transmission and communication of the sustainable issues.Taking account this suggestion, the authorities made a decision to put forward the construction of an ecological cafe bar. The cafe bar was built as a platform of participation the sustainable issues, such as recycling what interests the students most.

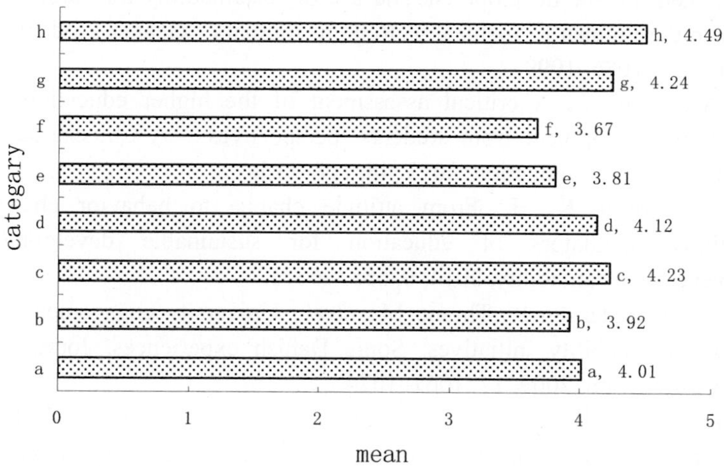

Figure 2. Ranking of factor categories of student's behaviors and awareness about sustainability in SCU.

a-Energy Conservation; b-Water Conservation; c-Sustainable Food; d-Sustainable Transportation; e-Green Purchasing; f-Recycle Waste; g-Sustainable Curricula.

3. Conclusions

At present, most Chinese universities take top-down method to make the decisions of sustainable issues. This method is unfavorable to the students involved in the sustainable issues. This study adopted a bottom-up method to discuss how to provide sustainable practice facilities for the students in the university. The research investigated the behaviors and awareness of the students, then analyzed the students' perceptions on university sustainability in SCU. The survey show that the students generally had a perceive to have sustainable curricula. Meanwhile, the students largely showed interests on recycling and waste reduction. The bottom-up method can provide students opportunities to participate in the sustainable issues.

References

1. Beringer, A., The Luneburg sustainable university project in international comparison: And assessment against North American peers. International Journal of Sustainability in Higher Education, 2007, 8(4), 447–461.
2. Wright, T., University presidents' conceptualizations of sustainability in higher education. International Journal of Sustainability in Higher Education, 2010, 11(1), 61–73.
3. Michwitz, P., Melanen, M. The role of co-operation between academia and policymakers for the development and use of sustainability indicators – A case from the Finnish Kymenlaakso Region. Journal of cleaner Production, 2009, 17(12), 1086–1099.
4. Yuan, X.L., Zuo J.. A critical assessment of the higher education for sustainable development from students' perspectives – A Chinese study, Journal of Cleaner Production, 2013, 48, 108–115.
5. Ismael V., Marie K. H., From attitude change to behavior change: Institutional mediators of education for sustainable development effectiveness. Sustainability, 2014, 6, 6553–6575.
6. Martin L., Christensen P., Thrane M., et al., University engagement and regional sustainability initiatives: Some Danish experiences, Journal of Cleaner Production, 2009, 17, 1067–1074.
7. Wass T., Huge J., Block T., et al. Sustainability assessment and indicators: Tools in a decision-making strategy for sustainable development. Sustainability, 2014, 6, 5512–5534.

Chapter 6
Pollution and Environmental Protection

The cause and countermeasure of haze

C.Y. Chen and B.F. Meng*
Changchun University of Science and Technology, Jilin, China
**E-mail: 1360423902@qq.com*

Controlling haze is a challenge for us. After collecting the latest relevant data and conducting a depth analysis, we found some man-made causes. Accordingly, the countermeasures are put forward, such as cutting down on the vehicle exhaust, transforming the pattern of economic development, optimizing the structure of energy consumption, etc. These measures are targeted and effective. Also, they will be helpful to promote the healthy development of economy and society, making contributions to the construction of beautiful China.

Keywords: Haze; Man-made causes; Countermeasures; Targeted and Effective.

Haze is the mixture of fog and haze, usually appears in the relative humidity between 80%–90%. The fog is milky white, which consists of a large number of tiny droplets floating in the air. The horizontal visibility is usually less than 1 km when it appears. Nevertheless, haze is very dry dust particles floating in the air. The horizontal visibility is less than 10 km. Haze is also called the ash haze, the main ingredients of which is smoke, dust, sulfuric acid, nitric acid and organic hydrocarbons. They can make the atmosphere cloudy so that we are in obscured vision.

1. The Analysis of Haze Formation

"Green Book of climate change: climate change report (2013)" pointed out that the human social and economic activities of human factors combined with the impact of climate factors caused by climate change in recent years, resulting in China's hazy weather increased.

1.1. *Artificial effect*

1.1.1. *Vehicle exhaust*

Emission of auto refers to vehicle exhaust which consists of CO (carbon monoxide), NOx +HC (hydrocarbons and nitrogen oxides), PM (particles, soot), and other harmful gas. When it is not conducive to air diffusion, such as temperature inversion or static wind, it will form hazy weather. That the sharp increase in cars and low fuel quality is the main reason of vehicle exhaust pollution. In 2014, the number of automobiles in china reached 154 million, 6.4 times that of 2004, took up 60% of the number of motor vehicles. Vehicle exhaust causes increasing of pollutants discharge and urban suspended solids, lead to the formation of urban hazy weather directly. That the quality of fuel is not high also contributes to the occurrence of haze, the quality of oil is very important for the discharge of pollutants. Although stage IV emission regulation has been implemented in most parts of China, sulfur content is still 5 times the current standard of the European Union. If the quality of fuel is not good, it must let out lots of pollutants that will harm the environment.

1.1.2. *Fugitive dust pollution*

The component of dust is PM10. Dust pollution is the result of large-scale urban expansion. In recent years, many parts of our country are in the process of urbanization, just pursuing business interests blindly and ignoring the environmental capacity of the city. On the one hand, the real estate project launched a large number of old city demolition and renovation project constructions that leads to blossom of the construction site everywhere. The backfilling of earthwork in the construction site, construction of excavation and sediment transport, construction waste, sand and dust pollution of bulk cargo vehicles, which will produce a lot of dust. On the other hand, with the expansion of cities, a large amount of farmlands have been used, many trees have been destroyed. But it is known to all that the green plants help to purify air. The disappearance of arable land makes the trees and the green plants that attached to land disappear during the process of urban expansion, and the ecological function of the city is greatly weakened. New green urban district or hedgerow cultivation is far from enough to meet the needs of urban ecology because the growth speed is low and planning is not in place. The strength of haze and dust increased obviously because of lacking of green plants' dust removal and purification.

1.1.3. *Coal pollution*

The energy characteristics of China — "rich coal, lean oil, less gas" decides that the energy consumption structure is coal dominated that will remain in the short term. PM2.5 is the direct internal cause of haze formation, and sulfur dioxide, nitrogen oxides and soot which of PM2.5, which become the important pollutants. Half of the world's coal burns in China, the vast majority of sulfur dioxide, nitrogen oxide and soot emissions are from coal consumption. In the coal consumption of China, about half is used for power generation, although the current technology of coal-fired generating power and reducing emission is already quite mature. Annual emissions are still very alarming due to the huge base of coal consumption. Sulfur dioxide and nitrogen oxide emissions accounted for almost half of the total emissions. Except for power generation, the rest of China's coal consumption is basically used for industrial manufacture and heating. Statistics show that hazy weather is obviously seasonal variations, when the northern cities have entered the heating season, the coal-fired heating in traditional way increases pollutants. For example, Changchun started heating on 25th October in 2014 however, after a few days severe air pollution and hazy pollution occurred. Not only in the northeast, but in Beijing-Tianjin-Hebei region, heating haze is also a frequent visitor. It's reported that monitoring data showed the Air Quality Index in most parts of Beijing reached 500 last year, the maximum pollution level.

1.1.4. *Industrial emissions*

China now has the world's largest category, the most complete industrial system. Industry occupies the dominant position in the economic development of our country, industry for economy and society to provide raw materials and products at the same time, also consumes a lot of resources and energy, a large amount of pollutant discharge. It is reported that industrial production is the important source of atmospheric pollution. As mentioned above, China's coal used to generate electricity and heating in winter, basically used in industrial manufacturing. Low level smoke-controlling of industrial sector, electric heating, coal consumption industries, iron and steel, and mineral products of carbon's emissions of the country's is nearly 70% one year. Clearly, huge industrial emissions exhaust smoke dust (powder) is the important cause of hazy weather.

1.1.5. *Environmental pollution monitoring is not in place*

"Three" does not reach the designated position in pollution control, which are the most important reasons for the fog haze continues to appear. First, enterprise one-sidely pursuit of economic efficiency, regardless of the national energy conservation and emissions reduction policy constraints, privately discharges, lacking of social responsibility. Second, some local governments pursuit GDP one-sidely, relax the requirement of energy conservation and emissions reduction, boost their investment blindly in energy-intensive industries. Third, the environmental pollution treatment is not in place, lacking of pollutants discharge monitored, atmospheric pollution control and schedule can't keep up with the growth of the pollutants discharge rate.

1.2. *The climate factors*

Environment issues that are caused by a variety of pollutants, adverse weather conditions can also hinder the dilution, diffusion and the clear of pollutants, leading the fog haze pollution broke out in a row.

1.2.1. *The weak cold air activities*

In recent winter years, cold air activity on the China is weak, the wind speed is lower. The arrival of the cold air will lead to the corresponding wind, rainfall, snowfall, air purification effect. Because of the weak cold air, the elements of rain formed at the time of cold weak in high altitude can also burn off before coming to the ground so that the pollutants can't be settled.

1.2.2. *Atmospheric temperature inversion*

In general, in the lower atmosphere, the temperature is decreased with an increase of height. But in autumn and winter season, sometimes the temperature rises with the increase of the height, this phenomenon is called inversion. In the inversion layer, the air is warm and light in the air cooler and heavy, like a "cover" over the city, seriously hinders the air convection movement. The water vapor, dust, and all kinds of harmful gas of the air is in desperate straits due to "Inversion". Only floating on the surface of the atmosphere, thus forming the fog hazy weather.

1.2.3. *Static unstable weather*

According to the different causes and pollution characteristics of heavily-polluted weather can be divided into two categories, static stability, sand and dust. Static stability type means due to the heavy pollution weather lasting for diffusion meteorological conditions resulting in widespread accumulation of pollutants, eventually particulate matter (PM10) heavy pollution level. It reduces radiation to reach the ground after haze's formation. The stability of the atmosphere increases when the static weather is stable, and it is conducive to the constantly accumulation and aggregation of aerosols, causing more droplets, so forming a "vicious circle" leading the haze pollution last few days. In some areas, in addition to the meteorological factors, fog may also be affected by high low terrain. Such as Beijing, Tianjin and Hebei surrounded on three sides by mountains face the pollutants coming after the mountains block in piedmont, diffusion is not easy to form fog.

2. The Measures of Haze Pollution Reduction

2.1. *Reduce automobile pollution, to vigorously develop public transport and improve the quality of fuel*

The development of public transportation can reduce exhaust emission and control air pollution, also solve the problem of traffic congestion. The government should improve urban programming, roads, bus service and other aspects of the full range, and provide truly convenient public transportation, which is convenient for people to reduce the chance of using private cars. At the same time, referring to the thorough public transport system in major cities of the world, drawing on the experience of urban planning and construction. Improving the quality of fuel, using high-quality oils in the country uniformly and accordingly accelerating the development and promoting the clean fuel and energy.

2.2. *Prevent and control dust pollution, strengthen the management of urban construction and promote the construction of forest city*

Too fast or excessive urbanization would cause irreversible adverse consequences and is not conducive to the sustainable development of urbanization. In the new urbanization process, we must have a long-term vision of the rational use of urban land. The first is to promote the energy-saving emission reduction of urban construction, to plan reasonably in advance, to vigorously promote environmental friendly buildings, to audit control in building materials selection, control the selection of the building materials, construction design, construction standards, building environmental protection, etc., and to increase the intensity of the building energy-saving and environmental protection. The second is to institutionalize the implementation of dust management, and to establish the urban construction environmental protection law enforcement inspection and normalized operating system of environmental protection, making the department of environmental protection have a way to keep the construction sector. The third is that "Forest City, making the city to embrace forest" which is an important measure to improve the city ecological environment. In situations of severe air pollution in our country, it is necessary to rethink the mode of urban development, and that urban construction "To reflect respect for nature, comply with the nature, the concept of harmony, relying on the existing landscape context and other unique scenery build the city into the nature, and so that the residents could see mountain, see water, remember nostalgia".

2.3. *Control of coal consumption, reform energy structure, enhance the utilization of coal*

The coal-based energy consumption situation will not change for a long time. If we want to reduce the emission of air pollutants once and for all, the structure of energy consumption should be optimized, meanwhile, the consumption of coal emissions beasts should be strictly controlled.

Firstly, conscientiously implement "the 12th Five-year Plan for National Energy Technology Development", "Development of Sustainable Energy Sources in the 12th Five Year Plan". "Long-term Planning of Renewable Energy" and other related planning to ensure control of the total coal consumption and accelerate the development of new energy. Secondly, change energy development, transfer development mode and adjust the structure. The government encourages to use energy as the main energy, clean energy as auxiliary energy into clean energy as the main energy, fossil energy strategy as the auxiliary energy. The key is to reduce the proportion of coal consumption, and vigorously develop clean energy, optimize the power structure and layout. We should increase the proportion of fossil energy generation, strengthen the exploration and exploitation of clean energy. Thirdly, promote the system to enhance the efficiency of coal use, promote clean coal utilization and reduce the coal direct combustion use, increase the proportion of coal washing, expand the scope of the ban on the high polluting fuel in the city. The implementation of natural gas or electricity substitute for coal gradually, encourage rural areas of northern to build clean coal distribution center and promote the use of clean coal. To change the energy structure of coal, and to improve the efficient use of coal, we must also formulate concrete and feasible system or regulations according to law.

2.4. *Revise the industrial emission standards, adjust the economic structure and transform the economic development mode*

Due to various reasons, our country's air quality standards is implemented in phases currently, and it will not be implemented nationally until January 1st, 2016. During this period, how to prevent and control PM2.5 has become a problem. It has been a long time that the industrial emission standards promulgated in 1996 and 2006 can only reflect the environmental and economic conditions of that time. Therefore, the government should amend the industrial emissions standards as soon as possible, and it is necessary to be incorporated into the environmental legislation. In addition, China should accelerate the transformation of the pace of economic development, adjusting the economic structure from excessive reliance on the second industry to the third industry of modern service industry, tourism and other low emission transformation. On the one hand, it can promote industry to turn into the green, low carbon by eliminating backward production capacity, strengthening the energy-saving management of the key enterprises, controlling of high energy consumption and high emission projects strictly, promoting energy efficiency benchmarking standards, and strengthening energy-saving water-saving technological transformation; on the other hand, government should speed up the pace of industrial upgrading, lead the resource to agglomerate reasonably, and construct three new development pattern of industry as well as improve the proportion of the third industry in the national economy.

2.5. *Increase the intensity of the hazy weather environment supervision*

In response to the continuous hazy weather, the government should strengthen supervision. The high pollution and high emissions enterprises will be punished seriously so that the quality of the environment will be effectively improved. What's more, cracking down on the violation in all kinds of poison. For example, environmental administrative law enforcement is not in place, the government responsibility is not implemented and it costs too little for enterprises. It is suggested that our country adjust evaluation index of provincial leaders, take PM2.5 as the provincial and municipal leaders' assessment of the binding targets. The governments should test steel, cement, chemical products frequently and sudden raids regularly. Especially the night and non-workdays, some companies should be on-line dynamic monitoring that discharge pollutant secretly. And emissions of various violations are to be prohibited. Constituting emergency predict scheme. Developing emergency plan in the hazy weather is necessary. Stopping or limiting the enterprises which discharge of pollutants. Increasing the intensity of the emergency disposal and cutting off the pollution emissions sources.

Everything has a source, if we want to solve the problem of haze, the key is from the source. Not only does environmental protection bureau and the weather bureau should continue to introduce and improve relevant policies and regulations, but they also accelerate the transformation of the mode of economic development and continue to change the way of life. At present, how to treat the haze in a scientific way is a difficult task, also, it is a serious challenge to mankind.

References

1. Hong-Wei Xiao. The analysis and cures of Hazy causes [J]. Journal of Macroeconomic management, 2014 (7).
2. Ya-Ping Li. The depth analysis and cures of hazy seasons for the future. The 2014 National Seminar of Environmental and Resources Law.
3. Fog Agent for China: Industrial emissions and vehicle emission pollutants coexist, the Qinghai journal, 2014, 1–17 (7).
4. Jia-Xing Can, Xue-Liang Guo. Artificial atmospheric pollutants impact on development of a winter dense fog forming study [J]. Journal of Atmospheric Sciences, 2012 (5).

An engineering design and application of treatment emulsion liquid wastewater

Q. Huang
China University of Geosciences, Beijing, China
East China Jiao Tong University, Nanchang, China

H.H. Li and F.P. Hu*
East China Jiao Tong University, Nanchang, China
**E-mail: Hufengping1968@126.com*

This project wastewater is emulsion wastewater of high COD and oil, mainly from semi-synthetic metalworking fluid production process of automobile gear factory machining parts. The wastewater treatment using equalization tank + vibration membrane + anaerobic fluidized bed + moving bed biofilm reactor(MBBR) + substrate-enhanced ozone degradation of organics (SODO) + 863 filter + activated carbon treatment process. Production and running results show that, the treated effluent COD<60 mg/L, oil<20 mg/L, Effluent indicators have reached "Integrated wastewater discharge standard" (GB 8978-1996). This has realized the waste water "resource", reducing environmental pollution.

Keywords: Emulsion; Vibration membrane; MBBR; SODO.

1. Introduction

An automobile gear factory machining parts, in the semi-synthetic metalworking fluid production process liquid during the production can produce high COD, oil emulsion wastewater and cleaning wastewater. The chemical composition of wastewater mainly includes: water, the base oil (mineral oil, vegetable oil, synthetic ester or their mixture), surface active agent, antirust additive (naphthenic acid zinc, petroleum sulfonate (and the emulsifier), petroleum sulfuric acid barium, benzotriazole, sorbitol single oleate, aluminum stearate), extreme pressure additives (containing sulfur, phosphorus, chlorine and other elements of polar compounds), friction improver (anti-friction or oiliness additive), antioxidant and so on. The typical feature of the wastewater is high oil content, high organic matter content and high chroma, and because of the presence of emulsifier in wastewater, which makes the oil in the wastewater can stable dispersed in the water, forming a stable emulsion. If the wastewater without treatment but direct emissions, will affect aquatic organisms normal growth, make the aquatic animals and plants has oil or toxicity, and smell of water, or soil can't normal metabolism and microbial metabolism, crop losses due to bad or death, damage the ecological environment. Therefore, must be handled only after discharge, on the other hand, the oil in the water can be recycled.

The wastewater from the factory is difficult to remove the pollutants. Therefore, conducted a series of experimental study on the wastewater, and the results of study are applied to the practical engineering, the treated wastewater COD is less than 60 mg/L, the oil is less than 20 mg/L, solved this kind of emulsion wastewater treatment, realize the waste water "resource", guarantee the qualified enterprises drainage, reduce environmental pollution.

2. Design Scale, Inlet and Outlet Water Quality

2.1. *Design scale*

Daily treatment capacity of the Project emulsion wastewater generated from the production of $30m^3$, so the engineering design process and water to $30m^3/d$.

2.2. *Inlet and outlet water quality*

Water emulsion wastewater are shown in Table 1, the waste water treated water should achieve "Integrated Wastewater Discharge Standard" (GB8978-1996) level standards, emission standards in Table 2.

Table 1. Emulsion wastewater quality.

Water Quality Index	COD (mg/L)	Oil (mg/L)	pH	Visual
Emulsion	200000	58000	6–9	Cream thick mixture of oil and water, viscous

Table 2. Wastewater discharge standards.

Water Quality Index	COD (mg/L)	Oil (mg/L)	pH	Chromaticity
Emulsion	60	20	6–9	60

3. Treatment Process

3.1. *Water quality analysis*

Compared Table 1 influent and effluent quality with Table 2, the key point of waste water treatment is removal of oils and organic. High oil content, oil dispersed in water is very stable and poor organic biodegradable were difficulty of the process.

3.1.1. *Emulsified wastewater with high oil content*

The waste water belongs to emulsified wastewater with high oil content, floating oil, dispersed oil, emulsified oil and dissolved oil co-exist, and at the same time has a lot of impurity like dust or sand in wastewater. Floating oil and dispersed oil relatively easy to remove. Owing to "water in oil" emulsion, more organic additives and high stability, emulsified oil must be demulsification than removed by using the settlement method. The dissolved oil volume is less than 0.1μm hard to remove. It is difficult to remove by sedimentation separation method; the more common method is to use adsorption or filtration to remove. Before biochemical treatment should reduce oil at a fit value, otherwise it will affect the activity of bacteria in the sludge.

3.1.2. *Refractory organics*

The organic content of the garbage is high per liter of hundreds of thousands of milligrams. It results from a part of oil in waste, and part of refractory biochemical degradation macro molecule organic of arious types of emulsifiers, surfactants at wastewater product. In order to meet the organic discharge standard, must make these refractory organic molecules scission into small biodegradable organic molecules biodegradable. Therefore we should use the water treatment technology which combining anaerobic and aerobic.

3.2. *Technical flow process*

According to the difficulties facing production wastewater quality characteristics, treatment and discharge requirements, the processing process as shown in Figure 1. The pretreatment and biochemical treatment is the core of the whole system. For removal of COD and the oil is difficult.

With a mature and stable pre-treatment + biochemical treatment + post-processing can reach the stable standard. Each processing unit key techniques described below.

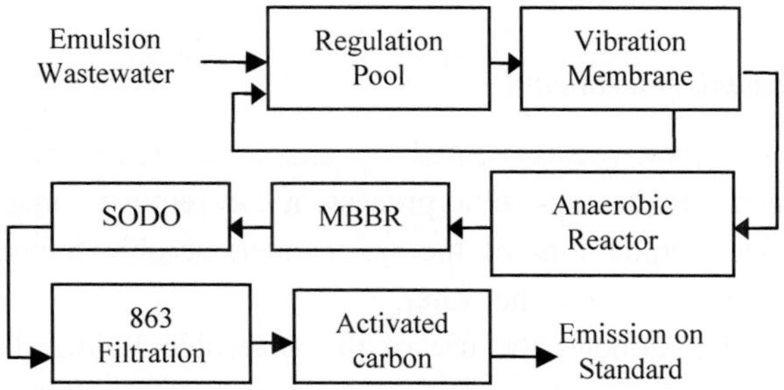

Figure 1. Wastewater Treatment Process.

3.2.1. *Pre-treatment*

(1) Regulation Pool
The amount of emulsion wastewater automobile plant emissions is cyclical. Wastewater first enters adjusting tank to regulate water quantity and quality. COD in waste water emulsion content of up to 200 000 mg/L. Follow-up by the high concentration of waste water after vibrating membrane technology concentrated into the regulation pool, water mixed with raw water COD can reach 600 000 mg/L, and later entered the diaphragm process. For the underground wastewater regulation pool type steel reinforced concrete structure, through the high and low level control, PLC program control mode.

(2) Vibration Membrane

By high frequency vibration mechanical vibration membrane, an enormous sine wave shear stress on the membrane surface of membrane separation technology "dynamic". The operation uses the cross flow pattern. With the help of a membrane module produced by ultrasonic vibration, effectively prevent the accumulation of pollutants in the membrane surface, weakened the concentration polarization effect, greatly reduce the possibility of congestion and pollution, reduce the pollution "static" membrane separation technology, the problem such as blocking, can greatly increase the water production efficiency at the same time, and reduce membrane cleaning cycle, prolong the service life of the membrane. Vibration Membrane technology has a strong anti-clogging ability, high energy utilization, covers an area of less, wide application range and low operating cost advantages. Abroad have this kind of "dynamic" membrane separation technology is applied to many fields, and achieved good effect.

Single set of diaphragm device covers an area of about 1.5 m^2, membrane area of 150 m^2, operation and automatic cleaning operations. Membrane module is independent and diaphragm using PVDF material, folding type, flow channel width of 2 mm. Membrane module can change according to the production requirements, by changing the upgrade of the whole treatment process can be realized.

3.2.2. *Biochemical treatment*

The treated wastewater was treated by anaerobic treatment and treated by aerobic treatment. In the anaerobic process, it can remove some pollutants and improve the biodegradability of the wastewater, aerobic treatment to further remove the organic matter in the water.

Biochemical treatment: the use of the anaerobic fluidized bed + MBBR combination to remove most organic pollutants and other pollutants.

(1) Anaerobic fluidized bed

Anaerobic biological treatment is a complex process of Microbial Chemistry, under anaerobic conditions, by facultative bacteria and designed of anaerobic bacteria after three successive phases (both hydrolysis acidification phase, hydrogen producing acetogenic and methanogenic phase) degradation of organic matter, the final product is carbon dioxide and methane, anaerobic biological processing has characteristics of high efficiency and low energy consumption. Anaerobic biological treatment was mainly influenced by pH value, temperature, alkalinity, carbon, nitrogen, phosphorus ratio and trace element etc.

(2) MBBR Technology

Aerobic treatment has been developed many new technology, in addition to the traditional activated sludge system, there are contact oxidation process, SBR process, oxidation ditch process, etc. In this scheme, the selection of aerobic treatment system was compared with the combination of MBBR with activated sludge and contact oxidation. MBBR Process core part is the proportion close to water the suspended packing direct investment added to aeration tank as microbial activity carrier, relying on enhances the role of the aeration tank aeration and water in a flow state. Different from previous fillers, suspended filler can frequently contact with sewage frequently, so it is called "mobile biological film". MBBR is the carrier of the growth biofilm, in the wastewater of the biological contact oxidation process; its purpose is to remove the COD/BOD of sewage. It absorbs the advantages of the traditional fluidized bed and biological contact oxidation method both, and the formation of a highly effective wastewater treatment method is suspended growth activated sludge and attached growth biofilm process combining a process, refractory wastewater has good treatment effect. MBBR Process with small space, expansion, simple operation and management, no clogging, biological easy to restore vitality, type and structure of the pool with the common contact oxidation pool like and suitable for renovation project and other advantages.

The filling rate (volume ratio) of MBBR in the aeration tank is good for 30–70%. Aerobic treatment need nutrients, the proportion of the general relationship between BOD: N: P=100:5:1, if inadequate nutrients in the process of operation, through is arranged in front of the dosing device. In addition to the foam generated in the biochemical process, through the good oxygen pool and the foam agent, to achieve a good oxygen tank foam cuts.

3.2.3. *Subsequent treatment*

Subsequent treatment unit mainly technology includes SODO advanced oxidation +863 filter + activated carbon to further remove residual organic matter, suspended solids and others in the water ,which ensures to attach water quality discharge standards.

(1) SODO ozone contact oxidation pond

SODO (Substrate-enhanced Ozone Degradation of Organics Process) technology is a water advanced treatment technology, by introducing ozone scavenger SODO particles in the based of traditional method of ozone oxidation to increase the effective concentration of ozone in the water. SODO particles has the ability of adsorption and capturing ozone molecules, which preferentially capture ozone to oxidize organic matter by ozone-enriched water to solve the water contains residual COD emission standards than permit in the aqueous solution.

Oxidation using ozone generator discharge air or oxygen into ozone, ozone discharge preparation principle is gap discharge, the power supply unit provides high voltage electric field leaving flowing oxygen generator in this field through. Ozone generator inside the tank itself and the discharge chamber is a ground electrode; the high voltage is applied to the metal electrode insulator, which coated with a special external insulation material, so the high voltage electric field was formed between the insulating layer and the ozone generator tank grounding, oxygen by the high-voltage corona discharge into ozone.

Ozone generating chamber structure using the tank, discharge body is in a honeycomb distribution in the tank body, which using high titanium ceramic and has two concentric tubes, each set of discharge body is equipped with a separate high-voltage fuse, the role of high-voltage fuse is fusing itself to protect the continued normal operation of the ozone system to ensure that a single set of discharge failure without affecting other body discharge body work when partial current is too large in the ozone generation chamber. Ozone contacting tank design size is $1.0m^3/h$.

(2) 863 filter

863 filter removals the suspended solids in raw water by adsorption and mechanical entrapment by using fiber filter, in particular adsorption properties of the oil. It belongs to ordinary quick filter equipment, which is called pressure filter of making the water to be filtered through the boost (about 0.6MPa) through the filter layer to achieve filtration. 863 high efficiency fiber filter is an advanced structure, excellent performance pressure-type fiber filter device, which uses a new type of bundle wadding-fiber filter as a filter element, which filter's diameter can be to several tens of microns even a few micrometers, a large surface area,

filtration resistance and other advantages, which solved the granular media filtration precision filter size restrictions by other issues. Its replacement product of quartz sand and other granular media filtration equipment.

The structural characteristic of the device is in the upper end of the filter layer of the filter can be changed with the fiber density adjusting means. Water run from top to bottom through the filter layer. In this case, the fiber density adjustment means to promote fiber downward movement, after being pressurized filter layer, the density gradually increased, so that the filter layer along the direction of water flow cross-section is gradually reduced, and the corresponding filter pore diameter and pore decreases, which achieved a depth filtration. When the contaminated filter layer needs clean and regeneration, clean water from under the supremacy through the filter layer. In this case, the fiber density adjustment means automatically opened fiber filter layer in a relaxed state, which achieved the desired cleaning effect.

The project chooses 2 devices of 863 high efficiency filter of 300mm diameter, each single treatment capacity of 1 m^3/h. The device only costs 1–3% of the water cycle, which can be backwashed by the raw water.

3.3. *Treatment effect*

Wastewater treatment process by the above, after debugging, the process section of the water quality of the Table 3.

Table 3. Each process section of effluent quality.

	COD (mg/L)	Oil (mg/L)
Influent	200000	600000
Regulation Pool Effluent	58000	10000
Vibration Membrane Effluent	10000	20~40
Anaerobic Effluent	2000	<20
MBBR Effluent	≤200	-
SODO Effluent	60	-
863 filter Effluent	<60	-
Activated Carbon Effluent	<60	-
Total removal rate %	99.97	99.96

4. Conclusions

The project based on the produced emulsion wastewater characteristics, using equalization tank + vibration membrane + anaerobic fluidized bed + moving bed biofilm reactor (MBBR) + substrate-enhanced ozone degradation of organics (SODO) + 863 filter + activated carbon treatment process. The wastewater influent COD concentration of 200 000 mg/L, oil for 58 000 mg/L case, treated effluent COD < 60 mg/L, the oil content < 20 mg/L, the removal rate were as high as 99.97% and 99.96%. This technology has the advantages of good treatment effect, high impact load and safety, etc. It is suitable for the treatment of oil emulsion wastewater in certain metal processing or manufacturing industries.

Acknowledgement

This work was financially supported by the National Natural Science Foundation (61272197), Landing Plan of Scientific and Technological Project of Jiangxi Provincial Colleges and Universities (KJLD2013037) and Cultivation Plan of Leadership for Excellence Jiangxi Province and Poyang Lake 555 Engineering (S2013-57). Natural Science Foundation of Jiangxi Province (20151BAB203032)

References

1. Feng Liu. 2008. Application study on extracting and separating the herbs of the honeysuckle and gardenia in Qingkailing injection by Vibratory Shear enhanced Processing membrane filtration technique. *Guangdong College of Pharmacy*.
2. Chao Chen, Qingyang Wu, Lin Wan, et al. 2012. High Frequency Vibration Membrane Treating Oily Wastewater. *Guangzhou Chemical Industry* 40 (2): 84–88.
3. Xiaohua Zhen. 2013. Study on high Recovery Rate Process for Desalination of Industrial Brine. *Lanzhou Jiaotong University*.
4. Jingxian Li, Lin Luo, Huixia Yang. 2007. Status quo of Application and Research Progress of MBBR Process. *Sichuan Environment* 26(5):97–101.

The landfill leachate treatment research of basing on the two levels of A/O biological technology

F. Chen[1,2,*] and Z.W. Cao[3]

[1]*China University of Geosciences (Beijing), Beijing 100083, China*
[2]*North China Institute of Aerospace Engineering, Heibei 065000, China*
[3]*North China Institute of Aerospace Engineering, Department of Mechanical Engineering, Heibei 065000, China*
[*]*E-mail: chf_chenfeng@126.com*

The concentration of garbage leachate is high, the water quality and amount change fastly, and the treatment is a great problem at home and abroad. This paper introduces the characteristics and the commonly used treatment technology of landfill leachate, analyzes the characteristics of various processing methods, and proposed two-stage biological treatment technology, introduce the feasibility of this technology, the technological process and the main equipment technical parameters for the engineering simulation studies, and finally, the paper points out the trend of waste leachate treatment.

Keywords: Garbage leachate; Two levels of A/O; Biological technology.

As the continuous development of economy and the living standard, the output of municipal waste is increasing. The disposal of urban garbage has became a big problem in the modern cities. Garbage landfill are the mainly disposal methods at present,the biggest problems of garbage landfill is the garbage Leachate. The composition of landfill leachate is complex,and contains a variety of pollutants. Landfill leachate is a kind of organic wastewater of high concentration. If it is not handled properly, it will cause serious pollution to the environment.

1. The Characteristics of Landfill Leachate

The collection mode of our living garbage is the mixed collection in our country, which makes the leachate mainly has the following characteristics: (1) The concentration of organic pollutants is high. (2) The composition is complex. (3) The ammonia nitrogen content is high. (4) The ratio of N/P disorders and phosphorus is deficiency. (5) Water quality change largely with landfill time and stability degree[1-3].

2. The Common Technology of Landfill Leachate Treatment

At present, The common technology of landfill leachate treatment included biological method, physico-chemical method, the land treatment method. The biological method is the most economic method and MBR, UASB, SBR technology is widely used in recent years. The cost of physico-chemical method is higher than biological method, but for some special pollutants in the leachate it is necessary. The more commonly used methods are: removing method, MPA precipitation, membrane technology.

Biological method which has the advantages of good treatment effect and low operating cost, is the best methods in landfill leachate treatment compared with biological treatment. Because of higher cost of physical-chemical method,it is not suitable for treating a large number of leachate. Physical-chemical method is not affected by the changes of water quality, and it has good treatment effect for poor biochemical leachate. The physical-chemical method is usually as a preprocessing or deep processing technology of leachate. Land treatment method, that is, under the condition of artificial control, through the land-plant system and through the synthesis reaction of physical-biological-chemical, make the leachate purified. This method which is suitable for vast land area is simple and economicand the buffer capacity is big enough.

3. Two Levels of A/O Biological Technology

A1/O1/A2/O2 (two level of A/O) technology is the combination process of aerobic biochemical, external type ultrafiltration and nf/ro membrane deep processing. The A1 is on behalf of the level of denitrification; O1 is on behalf of the level of nitration; A2 is on behalf of the secondary denitrification, O2 is on behalf of the secondary nitration. Nitrification and denitrification can be carried out on the secondary reaction through the control of the process conditions, and the nitrogen can be treated in maximum.

3.1. *The feasibility analysis of technology*

Landfill leachate belongs to high concentration organic wastewater pollutant, and most of them are difficult to fully degraded. This pollutant is biodegradable humus, further mildew acid substances. To ensure the removal rate of COD, BOD, ammonia nitrogen and total nitrogen at any time is the main content of leachate treatment. In all kinds of processing technology, biological method is the lowest cost of operation, because of its lowest consumption on chemicals, while it can remove most of the ammonia nitrogen. The biochemical method assist complementarily with appropriate post-processing of comprehensive treatment is the well and stable effect in recent years.

3.2. *The technical process*

Leachate are pumped into the regulating pool and then pumped into the filter to remove large particles. Leachate is token into the nitrification and denitrification in the secondary treatment system after the impurities, including level of denitrification, level of nitration, secondary nitrification and denitrification. The secondary external ultrafiltration system, through the role of aerobic microbial removal of organic matter in water and through the secondary nitrification and denitrification removal the ammonia nitrogen in leachate. The separation of spate, sludge intercept, raising microorganism concentration and removal efficiency of the reactor improving of biochemical processing system is achieved by the outer membrane system ultrafiltration membrane mud. Ultrafiltration membrane system through the booster pump, water into nanofiltration apparatus for further removal of organic matter and ammonia nitrogen. When nanofiltration effluent is standard, qualified effluent discharge into the labor pool; when the water is not up to standard, water from ultrafiltration membrane system directly got into the reverse osmosis treatment devices removing of organic matter and ammonia nitrogen, and qualified water discharge into the pool.

3.3. *The characteristics of process*

The process adopts many new technologies, such as improvement of nitrification and denitrification in the secondary system (meet the needs of the different microbial growth environment), the external type ultrafiltration membrane system, anti fouling (wrong streaming) nanofiltration, reverse osmosis membrane system, disc dehydration system and so on.

The complete set process has advantages of high efficiency, low energy consumption, impact resistance, stable operation and easy management. The specific performance as follows: (1) The treatment effect is stable and reliable. (2) Shock load capability is strong. (3) Strengthen the ability of biological denitrification. (4) Deodorant is effective. (5) Degree of automation is high. (6) Process adaptability is strong.

3.4. *The main equipment technical parameters*

3.4.1. *Pretreatment system*

Pretreatment system is mainly to intercept large pieces of floating suspended solids in water, adjust the water PH value, calculated the water inflow of the leachate treatment system for measurement at the same time

3.4.2. *First level of nitrification and denitrification*

The leachate after pretreatment come into the first level of denitrification, and mixed liquor from first level of nitration return into the pool level of denitrifying pool. NO_3-N in backflow of mixture is used as carbon source material of denitrifying bacteria under the action of carbonaceous organic matter in raw sewage on denitrifying pool for denitrification reaction. The level of denitrifying pool water come into the level of nitrification. There is carbon oxidation of organic matter in denitrifying pool. The nitrogen from first level of denitrification is token off from first level of nitrification.

3.4.3. *Second level of nitrification & denitrification*

Nitrification and denitrification in the secondary level, the first level denitrifying pool water go into the secondary denitrifying pool and denitrifying bacteria using the mixture of endogenous metabolites occur the further nitration. Secondary effluent go into the secondary denitrifying pool, which occur further oxidation of organic compounds containing carbon. In the secondary denitrifying pool the residual nitrogen ammonification and nitrification of ammonia nitrogen happens and organic matter produced by the secondary denitrifying pool nitrogen nitrification in pool can be take off.

3.4.4. *The cooling system of high temperature*

Due to the high temperature in summer and leachate high concentration, it produces a significant amount of heat in the carbide and ammoniation process. Water temperature rise to higher than the limit temperature of the microbial growth because of combination of both nitrification and denitrification. Setting up the cooling system ensure the water temperature to meet the requirements of microbial metabolism, by the means of the heat exchanger.

3.4.5. *The ultrafiltration system*

In ultrafiltration system activated sludge is pumped to the outer membrane module in filtering as the liquid is discharging. Activated sludge is intercepted by ultrafiltration membranes and returned in order to improve the sludge concentration in the pool to continue their biological treatment. Microbial bacteria can be separated through efficient ultrafiltration system, leaving microbes and sterile in the water, and there is not microbes and suspension in effluent.

3.4.6. *The nanofiltration system*

The interception ability of nanofiltration system is between 19–98%, intercept molecular weight is 150–300 Dalton. The interception membranes to ion is selective, allowing monovalent salt through the membrane, with medium through salt rate. Compared with the reverse osmosis, nanofiltration has lower operation pressure and higher recovery rate.

3.4.7. *The reverse osmosis system*

The separation size of reverse osmosis system is generally less than 0.1 nm. Its separation particle level can reach level of ion. It is the most sophisticated liquid membrane separation technology. It can block all soluble salt and organic molecular weight greater than 100. It can remove the soluble metal salts, organic pollutants, bacteria, fever, colloid particles and the desalination rate is greater than 99%. The removal rate of COD, ammonia nitrogen and total nitrogen can reach more than 95%, with effluent water quality stability.

3.4.8. *The sludge dewatering system*

The sludge removing from Leachate treatment system discharge to the sludge pit, then drying after dehydration and landfill dumps.

3.4.9. *The deodorization system*

The stink from the nitrification system, denitrification system and pool of sludge pit transports to biological deodorization system for soil deodorization.

4. Summary

In order to solve the problem of leachate pollution, the key is to take out necessary treatment for leachate reaching emissions standards. Two levels of

A/O biological treatment technology is the improvement and optimization of A/O process. Although our country has a distance in leachate treatment compared with western developed countries, as the development of biotechnology, our country will be solve the waste leachate treatment problem very well.

References

1. Su K Z, Yu H Q. Formation and characterization of aerobic granules in a sequencing batch reactor treating soybean-processing wastewater [J]. Environmental Science and Technology, 2005, 39 (8): 2818–2828.
2. Wang S G, Liu X W, Gong W X, et al. Aerobic granulation with brewery wastewater in a sequencing batch reactor [J]. Bioresource Technology, 2007, 98 (11): 2142–2147.
3. Wang H L, Yu G L, Liu G S, et al. A new way to cultivate aerobic granules in the process of papermaking wastewater treatment [J]. Biochemical Engineering Journal. 2006, 28 (1): 99–103.

The numerical simulation research on the combustion characteristics of waste incinerator under variable load*

B. Yang†, M.D. Li, J.F. Ni, K. Ma, X.C. Wang and F. Xue
Guangzhou Special Pressure Equipment Inspection and Research Institute, Guangzhou, 510100, China
†E-mail: boycs@126.com
http://www.gzbpvi.org

Using the computational fluid dynamics simulation software FLUENT, the numerical simulation research on combustion characteristic in a waste incinerator under variable load condition was studied. The results showed that the model is reasonable and credible after the verification with the actual operation of waste incineration. When the load changed, the oxygen supply of waste incinerator was adequate. With decreasing the load, the high temperature zone in the grate furnace shifted to an earlier grate. Meanwhile the velocity decreased which would not obviously affect the normal flow of flue gas in the incinerator. In order to ensure a stable combustion, it is necessary to start the auxiliary burner of the boiler when the load was reduced to 50% or start the small power auxiliary burner of the boiler when the load was reduced to 65%. Under the condition of low load (50% and 65% load), reducing the primary air of the fourth grate and increasing correspondingly the primary air of the second and third grate will make the high temperature flue gas to form a significant cyclone in furnace, which will promote the heat transfer and increase the residence time of flue gas. In this condition, reducing the grate speed will increase the thickness of waste, which will postpone the waste incineration region, so that the high temperature area will be mainly in the second and third grate.

Keywords: Variable load; Waste incineration; Numerical simulation; Combustion characteristics.

1. Introduction

The production of municipal solid waste increases continually with the improvement of economic development and living standards (Hu Zhifeng et al., 2012). So that the problem of municipal solid waste is becoming day by day an important issue for the city development and people life. Compared with other methods of waste disposal, waste incineration has many advantages, such as: large subtracting capacity, high processing speed, destroying bacteria by high

* This work is supported by General Administration of Quality Supervision, Inspection of Public Projects (20140159) and Guangdong Province Key Laboratory of Efficient and Clean Energy Utilization (2013A061401005).

temperature and generating electricity, etc (Liu Junwei et al., 2012). Consequently more and more developed countries and regions adopt waste incineration as the main method of waste disposal. In general, the waste incinerators are operated under variable load condition, which will affect the combustion efficiency of waste incinerator. At present, many researchers studied the combustion characteristics of waste incinerator under different air distribution scheme (Li Longjun et al., 2013; Lin Hai & Ma Xiaoqian, 2011; Wang Lingling et al., 2013; Zhou Chunyan et al., 2003). However, the study on the variable load is less.

Therefore, this paper was validated with the actual operation conditions. And then adopting the reasonable model, the paper studied the combustion characteristic of temperature, velocity, component emission and particle path in a waste incinerator under variable load (100%, 80%, 65% and 50%), which would provide reasonable measures for the incinerator under the actual operation conditions of variable load.

2. Computational Conditions

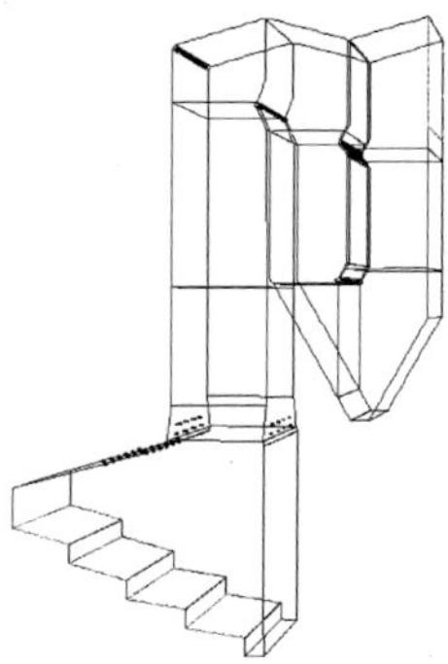

Figure 1. The whole structure of waste incinerator.

The research object of this paper is a 750 t/d grate furnace incinerator in Guangzhou, China. On the basis of the actual waste incinerator, the whole structure of "furnace + three flue" was built, which was shown in Figure 1. According to the actual operation of waste incinerator plant, the meshing of the incinerator was divided into 10 regions, which were furnace and three zones of each flue. The mesh system was refined at the important locations, such as grate, secondary air nozzles and over-fire air nozzles. The waste considered in this study was municipal solid waste in Guangzhou. The elemental analysis and proximate analysis of waste in Guangzhou is shown in Table 1.

Table 1. The elemental analysis and proximate analysis of waste in Guangzhou.

Elemental analysis(daf, %)		Proximate analysis(ar, %)	
C	57.625	Volatile	26.18
H	8.453	Ash	18
O	31.5	Fixed carbon	5.82
N	0.234	Moisture	50
S	0.469		

3. Numerical Simulation Methods

The commercial computational fluid dynamics (CFD) software Fluent was utilized to study the combustion characteristics of waste in this paper. The software FLIC was used to calculate the combustion on the grate, which could connect with Fluent (Stopford, 2002). The mass and energy models of FLIC was used to calculate the combustion on the grate. The results exported from the FLIC were imported into the Fluent as the velocity inlet boundary. The species transport model was used to simulate the fuel combustion in the waste incinerator. The standard modeling equations were solved by simple arithmetic (Xu et al., 2000). The standard κ-ε turbulence model was utilized to simulate the turbulence flow in the incinerator (Lin Hai et al., 2010). The P-1 radiation model was used to calculate the radiation heat transfer, which affected the combustion and particulate (J., 2002). This model considered the influence of gravity. The secondary air boundary was velocity inlet, while the flue outlet boundary was pressure outlet. The boundary conditions of the furnace and flue were set up according to the actual operation conditions.

4. Model Validation

This paper was validated with the actual MCR operation conditions. The comparison of the simulation results and actual results based on the furnace outlet parameters were shown in Table 2.

Table 2. The comparison of the simulation results and actual results based on the furnace outlet parameters.

Parameter	Actual result	Simulative result	Error
Temperature of furnace outlet	1306.25 K	1234.64 K	5.48%
Temperature of the first flue outlet	1186.65 K	1086.97 K	8.40%
O_2 content	7%	7.18 %	-2.57%
H_2O content	18.8 - 26.3%	19.96 %	With the actual range

As shown in Table 2, compared with the actual results in terms of the furnace outlet temperature, the first flue outlet temperature and the oxygen content of the furnace outlet, the errors in the simulation results were 5.48%,

8.40% and -2.57%, respectively. Moreover, the H_2O content of the furnace outlet was within the actual operation. The above parameters were far less than the engineering error range (20%), which could satisfy the demands for simulating. Therefore, this model was accurate and credible.

5. Results and Discussion

During the process of waste combustion, the incinerator load often changes. Therefore the paper studied the combustion characteristic of temperature, velocity, component emission and particle path in a waste incinerator under variable load (100%, 80%, 65% and 50%).

5.1. *Temperature profiles*

The temperature profiles of waste incinerator furnace under different load was shown in Figure 2. And the detailed parameters for the simulation results under different load was shown in Table 3.

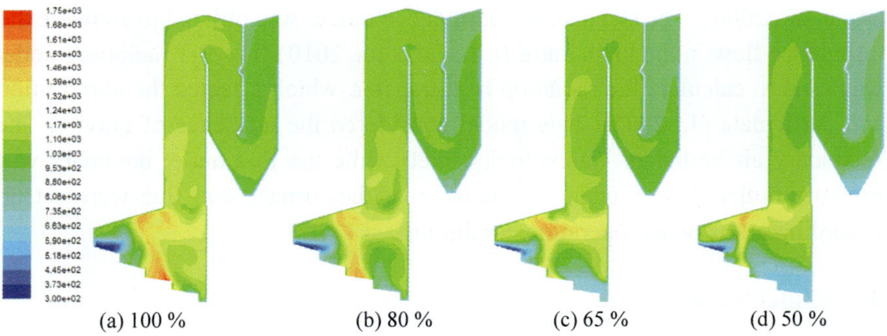

Figure 2. The temperature profiles of waste incinerator furnace under different load (Unit: K).

Table 3. The detailed parameters for the simulation results under different load.

Different load	Temperature of furnace outlet(K)	Temperature of the first flue outlet(K)	H_2O content (%)	O_2 content (%)
100 %	1234.64	1086.97	19.96	7.18
80 %	1171.47	1056.10	20.45	6.49
65 %	1171.16	1064.37	21.35	6.28
50 %	1143.40	991.51	20.78	7.53

As shown in Figure 2, With decreasing the load, the high temperature zone in the grate furnace shifted to an earlier grate. It is mainly because the grate speed was constant and the load was reduced, which would decrease the waste thickness on the grate. So that waste was easier to decompose and release moisture, volatile matter and then burn under the high temperature. Thus the combustion of fixed carbon shifted to an earlier grate. After combustion the fourth grate was the cooling area. In this condition, the cryogenic cooling effect was caused in the fourth grate, which was not conducive to maintain the high temperature of incinerator.

As shown in Table 3, the O_2 contents of four kinds load were similar, which indicated the oxygen supply of waste incinerator was adequate under different loads. Moreover, the H_2O contents of four kinds load were also similar. It is mainly because of the high water content (up to 50%) in waste. However, the furnace temperature was different when the load changed. When the load was 50%, the temperature of the first flue outlet was quite low, which could not guarantee the demands of combustion and electricity generation in the incinerator. In addition, it could guarantee the requirement of flue gas stayed at above 850°C for 2 s when the load reduced to 65%. However, it could not meet the requirement of combustion when the load reduced to 50%. Therefore, when the load reduced to 50% in the actual operation, the auxiliary burner must be started. In order to ensure a stable combustion, it is suggested to start the small power auxiliary burner of the boiler when the load was reduced to 65%.

5.2. *Velocity profiles*

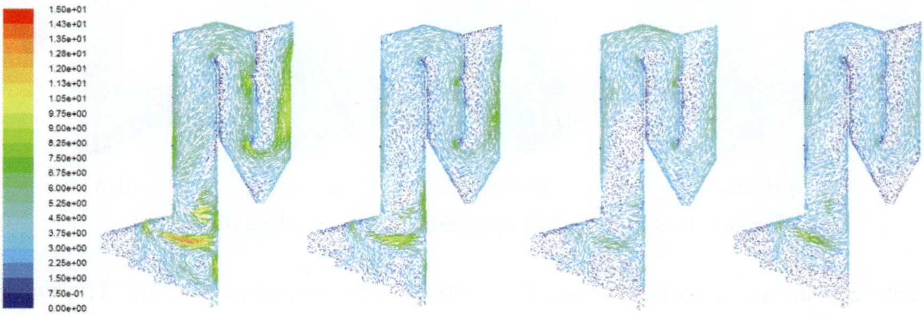

(a) 100 % (b) 80 % (c) 65 % (d) 50 %

Figure 3. The velocity profiles of waste incinerator furnace under different load (Unit: m/s).

As shown in Figure 3, the velocity profiles of waste incinerator was similar with decreasing the load. The high speed area concentrated in the burner area, which could form a strong tangential cyclone in the central region. In addition, with decreasing the load, the amount of waste incineration reduced accordingly, so that the amount of the actual required air reduced. Therefore, Figure 3 could be clearly shown that the speed of the secondary air and over-fire air reduced obviously and the cyclone formed in the furnace became weakened.

Furthermore, the velocity decreased, which would not obviously affect the normal flow of flue gas in the incinerator.

5.3. *Component emission profiles*

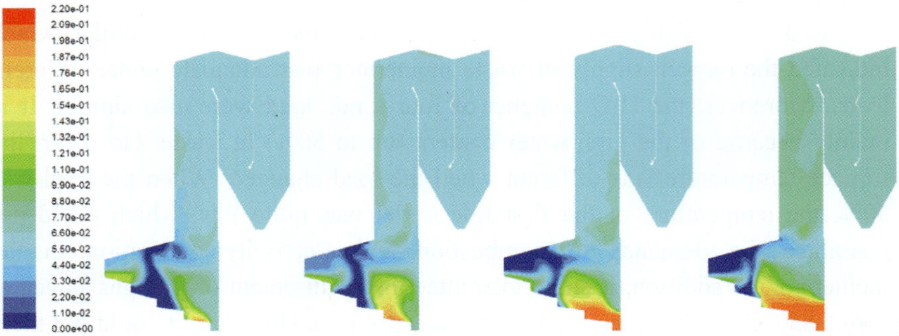

(a) 100 % (b) 80 % (c) 65 % (d) 50 %

Figure 4. O_2 profiles of waste incinerator furnace under different load.

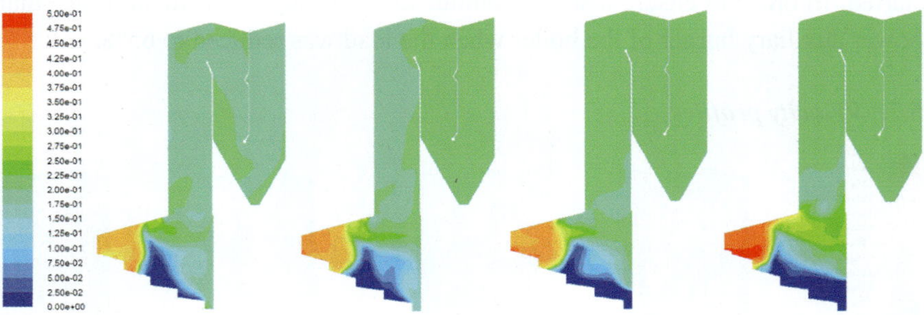

(a) 100 % (b) 80 % (c) 65 % (d) 50 %

Figure 5. H_2O profiles of waste incinerator furnace under different load.

There is an association between O_2 content and temperature profile. The waste combustion would consume a large amount of O_2 in the high temperature region, so that the low O_2 concentration region appeared. However, the combustion process of fixed carbon had been completed in the low temperature region of the second half of the grate. Then this process hardly consume O_2. Therefore, the high concentration of O_2 appeared in this region. As shown in Figure 4, under the condition of low load (50% and 65% load), it should reduce the primary air of the fourth grate and increasing correspondingly the primary air of the second and third grate. In this condition, reducing the grate speed will increase the thickness of waste, which will postpone the waste incineration region, so that the high temperature area will be mainly in the second and third grate.

The H_2O contents of different load were similar. The high concentration region was in the first grate, which was because water in the waste evaporated into H_2O under high temperature. The H_2O content gradually reduced in the second grate. Meanwhile, with the assistance of the primary and secondary air, H_2O evenly distributed in the furnace and the flue.

5.4. *Particle path*

The particle path and residence time of flue gas in waste incinerator under different load was shown in Figure 6. The figure showed that the particle path of flue gas was similar. After injected into the furnace, the primary air was sprayed to the back wall with the help of the secondary air. A part of flue gas sprayed to the lower right corner of furnace and then formed the cyclone. The other main part of flue gas sprayed upward into the flue. Meanwhile the flue gas formed a strong tangential cyclone under the influence of over-fire air, which would promote the heat transfer and increase the residence time of flue gas.

With decreasing the load, the cyclone in the lower right corner of furnace disappeared gradually. It was mainly because the waste in the second half of the grate did not consume oxygen, so that the strength of primary air in this region was strong, which would directly spray the secondary air upward into the flue. Therefore, the cyclone strength in the lower right corner of furnace became weakness. As shown in Figure 6, under the condition of low load (50% and 65% load), reducing the primary air of the fourth grate and increasing correspondingly the primary air of the second and third grate will make the high temperature flue gas form a significant cyclone in furnace, which will promote the heat transfer and increase the residence time of flue gas.

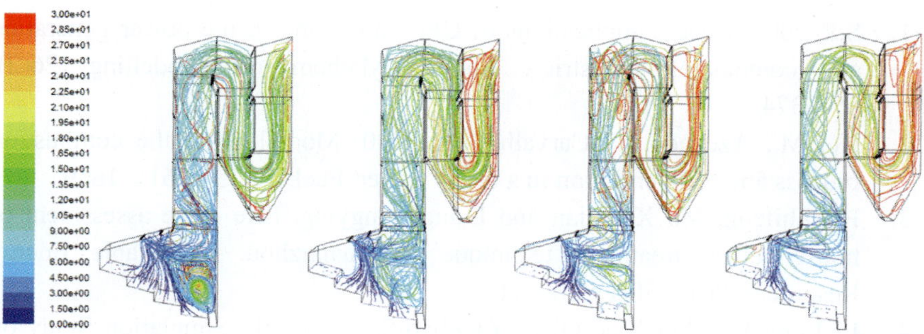

(a) 100 % (b) 80 % (c) 65 % (d) 50 %

Figure 6. The particle path and residence time of flue gas in waste incinerator under different load (Unit: s).

6. Conclusions

(1) With decreasing the load, the high temperature zone in the grate furnace shifted to an earlier grate.

(2) When the load changed, the oxygen supply of waste incinerator was adequate.

(3) In order to ensure a stable combustion, it is necessary to start the auxiliary burner of the boiler when the load was reduced to 50% or start the small power auxiliary burner of the boiler when the load was reduced to 65%.

(4) With decreasing the load, the velocity decreased which would not obviously affect the normal flow of flue gas in the incinerator.

(5) Under the condition of low load (50% and 65% load), reducing the primary air of the fourth grate and increasing correspondingly the primary air of the second and third grate will make the high temperature flue gas form a significant cyclone in furnace, which will promote the heat transfer and increase the residence time of flue gas.

(6) In this condition, reducing the grate speed will increase the thickness of waste, which will postpone the waste incineration region, so that the high temperature area will be mainly in the second and third grate.

Acknowledgements

This study is supported by the General Administration of Quality Supervision, Inspection of Public Projects (20140159) and Guangdong Province Key Laboratory of Efficient and Clean Energy Utilization (2013A061401005).

References

1. S.P. 2002. Recent applications of CFD modelling in the power generation and combustion industries. Applied Mathematical Modelling, 26(2), 351–374.
2. Xu, M., Azevedo, J., Carvalho, M. 2000. Modelling of the combustion process and NOx emission in a utility boiler. Fuel, 79(13), 1611–1619.
3. Hu Zhifeng, Ma Xiaoqian and Liang Zengying. Life cycle assessment of house refuse treatment technique in Guangzhou. Renewable Energy Resources, 2012. 30(1): 106–112.
4. Li Long-Jun, Ma Xiao-Qian, Yu Zhao-Sheng, et al., Simulation Study on the Impact of Temperature Field in MSW Incinerator under Air Distribution. Environmental Science & Technology, 2013.10: 39–42.

5. Lin Hai, Ma Xiao-Qian. Optimal Layout of OFA for Large-scale Municipal Solid Waste Incinerator. East China Electric Power, 2011. 39(11): 1911–1915.
6. Lin Hai, Ma Xiao-Qian, Yu Zhao-Sheng, Numerical Simulat ion of Large-scale Municipal Sol id Waste Incinerator. Journal of Chinese Society of Power Engineering, 2010, 2: 128–132.
7. Liu Junwei, Lei Tingzhou, Yang Shuhua, et al., Garbage-fired Power Generation in China Today and Development Trends. SINO-GLOBAL ENERGY, 2012. 6: 29–34.
8. Wang Ling-Ling, Yu Xiao-Feng, Zhu Hong-Fang, R&D on the Furnace and Air Distribution System of Incineration of Municipal Solid Waste. Industrial Boilers, 2013.6: 18–20.
9. Zhou Chun-Yan, Ma Xiao-Qian, Mao Kai, Analysis of Influence of Air Distribution on Enhancing Combustion Efficiency of Waste Incinerator. Gas & Heat, 2003. 23(7): 395–399.

Treatment of Zinc-bearing waste water by flocculating*

C.Y. Chao*, Y.L. Yang, G.F. Li and D.J. Ma
Department of Environment North China Institute of Science and Technology, Beijing, China
**E-mail: chaochunyan@ncist.edu.cn, kiddey@126.com*

In this paper we studied the effect of PAC, PAM, FeSO4 as flocculant in zinc-bearing water. The zinc-bearing water was disposed with flocculant, and control the test conditions, such as pH, coagulation mixing intensity and time, flocculant dosing quantity, then determinate the content of zinc in water by Flame Atomic Absorption Spectrometry and removal rate, by reaction conditions and the relationship between the removal rate curve, draw all kinds of flocculant for waste water removal effect is best when required by the optimum condition, and through the contrast test to select the best remove effect of flocculating agent.

Keywords: Flocculant; Zinc-bearing water; Removal.

1. Introduction

Flocculants can be used to settlements the pollutant from water. The flocculation technology has been widely used in water supply, waste water treatment and sludge dewatering. Flocculation is one of the important methods of water treatment. Traditional flocculants for water treatment which are low molecular weight inorganic salts, has gradually been replaced by polymer flocculant due to the large cast t and poor effect. Polymer flocculant have become a modern major flocculant in water treatment for its strong turbidity bleaching ability, and ease of operation. Polymer flocculant can be divided into three broad categories: inorganic polymer flocculants, organic polymer flocculant and bioflocculant.

Inorganic flocculants are aluminum sulphate, aluminum chloride, ferric sulfate, ferric chloride and their polymers. Organic polymer flocculants are polyacrylamide (PAM), sodium polyacrylate, polyethylene sulfonate. We studied the effect flocculant such as ferric chloride, PAC, polyacrylamide, starch, aluminum sulfate, ferrous sulfate in zinc-bearing water treatment, research their

* This work is supported by:
Foundamental Research Funds for the Central Universities (3142014017)
Fund Project of North China Institute of Science (2011B010)

flocculation, application, flocculation mechanism and preparation of excellent performance composite flocculant.

2. Experiment

2.1. *Materials*

PAC, Aluminum chloride $[Al_2 (OH) \cdot Cl_{6-m}]_n$ solution, 10g / L;
Aluminum sulfate $Al_2 (SO_4)_3 \cdot 18H_2O$ solution, 10g / L;
PAM,Polyacrylamide solution 10g / L;
Ferrous sulfate ($FeSO_4$) solution 10g / L;
Starch solution 10g / L;
HCl solution 10%;
NaOH solution 10%.

2.2. *Methods*

Take zinc water (concentration is 100mg / L) 700ml, divided into 7 parts, add flocculant and adjust the pH from 5.0 to 11.0 at room temperature. The concentration of zinc ion were measured so that we can find the best pH which the concentration of zinc was lowest and the removal rate of zinc is the highest. Then study the best quality of flocculant and the best stirring speed with the same method.

3. Results and Conclusions

3.1. *The effect of PAC in zinc-bearing water*

We fix PAC flocculant 0.7g respectively, and adjust the pH from 5.0 to 11.0 at room temperature. We measured the concentration of zinc ion. With the increase of pH, the removal rate of Zn in the water is increased, the (Figure 1). The removal rate of Zn changed little when the pH more than 8.0. Then we measured the concentration of zinc ion at different quantity of PAC in the pH 8.0, the removal rate of Zn increased as the increasing of the quantity of PAC(Figure 2).

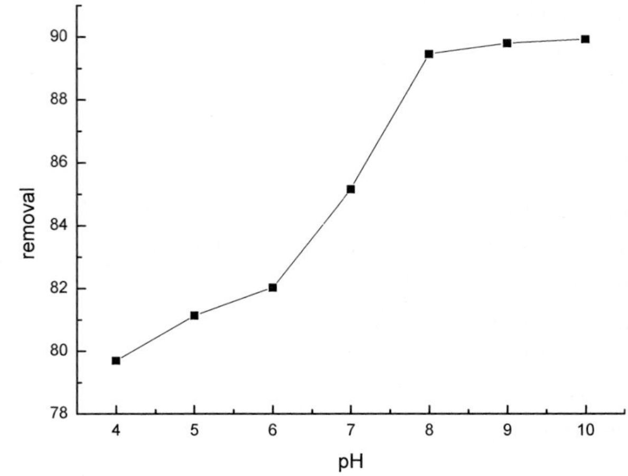

Figure 1. Effect of pH on the removal with PAC.

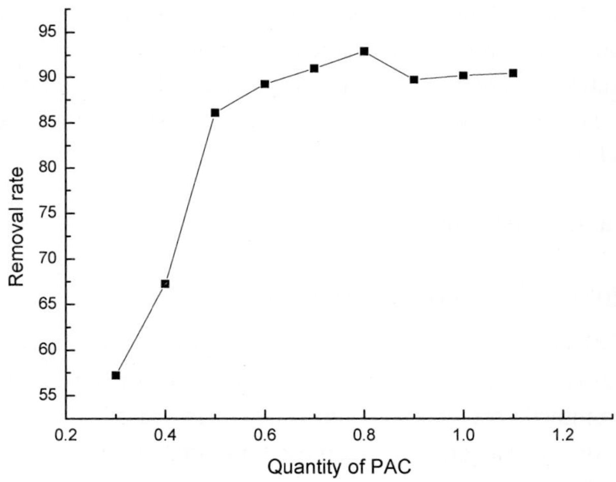

Figure 2. Effect of quantity on the removal with PAC.

3.2. *The effect of PAM in zinc-bearing water*

We study the best pH and quantity with the same method. From Figure 3 and Figure 4 , we can see the best pH of PAM is 6.0, the removal rate of Zn increased as the increasing of the quantity of PAM.

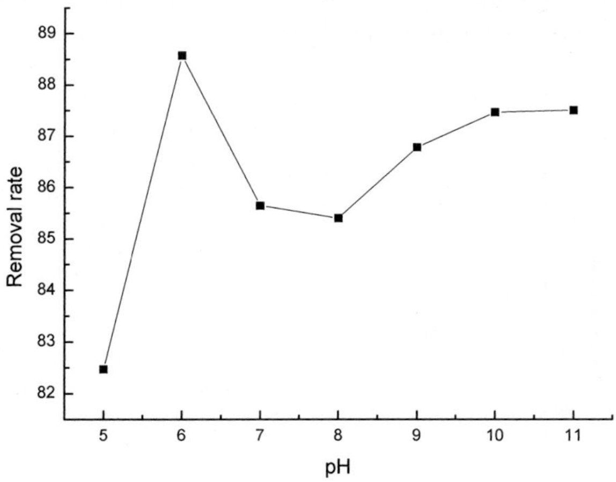

Figure 3. Effect of pH on the removal with PAM.

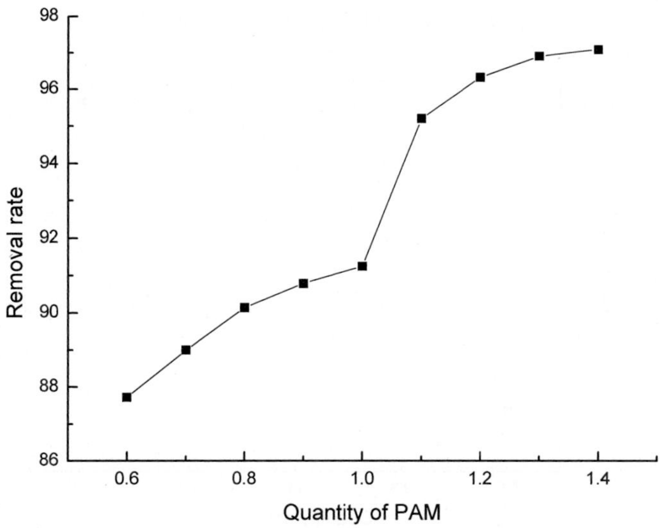

Figure 4. Effect of quantity on the removal with PAM.

3.3. *The effect of $FeSO_4$ in zinc-bearing water*

As to FeSO4, the pH has a great influence on the removal rate, the removal rate of Zn is worst when pH is 6.0, then the removal rate of Zn increased as the increasing of pH and quantity of FeSO4. The stirring speed can also affect the removal rate of Zn which showed in Figure 7.

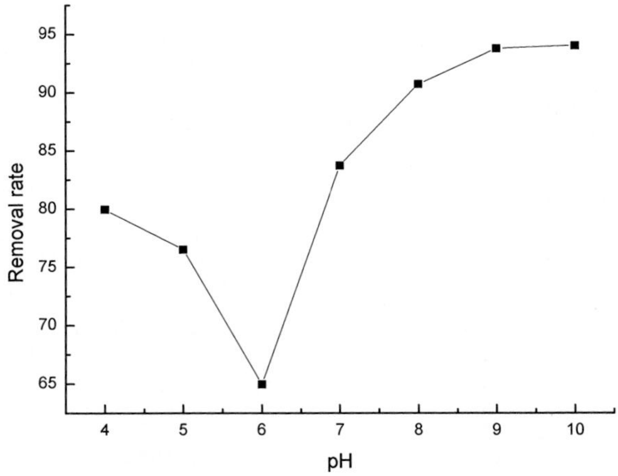

Figure 5. Effect of pH on the removal with FeSO4.

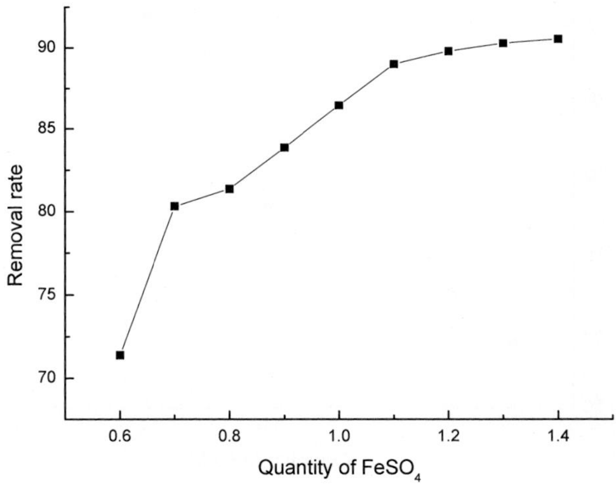

Figure 6. Effect of quantity on the removal with FeSO4.

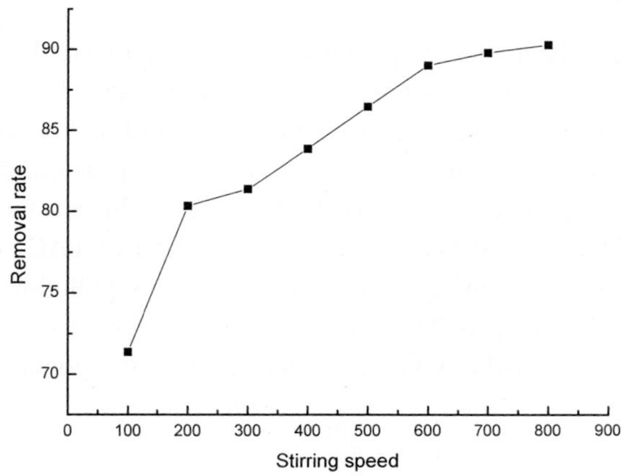

Figure 7. Effect of stirring speed on the removal with FeSO4.

3.4. *Orthogonal test*

From single-factor test we can see that, pH, quantity, and stirring speed was the main factor which influenced the removal rate of Zn, so we carried out Orthogonal test with three factors with flocculant FeSO4. Each of them was tested at three levels. The result are shown in the Table 1. The results showed that the Zinc in water was removed effectively when pH=7.0, quantity=1.0g, stirring speed=500r, the order of influencing factors is pH, quantity, stirring speed.

Table 1. Orthogonal test of flocculant $FeSO_4$.

	Stirring speed	Quantity of $FeSO_4$	pH	removal rate of zinc
	300r (1)	0.9g (1)	7.0 (1)	85.18
	300r (1)	1.0g (2)	8.0 (2)	86.85
	300r (1)	1.1g (3)	9.0 (3)	96.02
	400r (2)	0.9g (1)	9.0 (3)	94.24
	400r (2)	1.0g (2)	7.0 (1)	87.42
	400r (2)	1.1g (3)	8.0 (2)	92.17
	500r (3)	0.9g (1)	8.0 (2)	88.92
	500r (3)	1.0g (2)	9.0 (3)	91.66
	500r (3)	1.1g (3)	7.0 (1)	84.96
K_1	0.1047	0.1055	0.1415	
K_2	0.0872	0.1136	0.1069	
K_3	0.1149	0.0895	0.0603	
R	0.0277	0.0241	0.0812	
The best conditions	500r	1.0g	5.0	
Order of factors	pH, quantity, stirring speed			

3.5. *Conclusions*

This research mainly studies the treatment effect of inorganic flocculant, organic flocculant, on the treatment of the zinc in waste water. Determine the best flocculant and the best conditions. By single factor test and orthogonal test, the optimum conditions for treating zinc in waste water was determined. The result shows that: The quantity, and stirring speed can influence the effect of the removal rate. The removal rate of Zinc in waste water by PAC and PAM is the best, but the residual aluminum ion would cause new pollution. The effect of FeSO4 was not as good as PAC, but when we use the best conditions which determined by orthogonal test, the removal effect was also good.

References

1. Cristina Veronica Gherasim, Gelu Bourceanu. Removal of chromium(VI) from aqueous solutions using a polyvinyl-chloride inclusion membrane: Experimental study and modelling [J]. Chemical Engineering Journal. 2013.
2. Yue Wen, Zhiru Tang, Yi Chen, Yuexia Gu. Absorption of Cr(VI) from aqueous solutions using chitosan-coated fly ash composite as biosorbent [J]. Chemical Engineering Journal. 2011.
3. Evans, Johanna, Davids. Kinetics of cadimium up-take by chitsan based crab shells. Water Research. 2002.
4. Rozada F, Otero M, Mor n A, et al. Activated carbons from sewage sludge and discarded tyres:Production and optimization. Journal of Hazardous Materials. 2005.
5. George A, Ciorba C, Ilie V, Laura P. Correlation between organic component and electrode material: Consequences on removal of surfatants from wastewater. Electrochimica Acta. 2000.
6. Li Xiong Gu, Zhi Fang Liu, Zhong Yong Xu. Phytoremediation of chromium and lead using water lettuce pistia stratiotes L. [J]. Applied Mechanics and Materials. 2013.
7. Hassan M. Al-bishri, Tarek M. Abdel-Fattah, Mohamed E. Mahmoud. Immobilization of [Bmim + Tf 2 N −] hydrophobic ionic liquid on nano-silica-amine sorbent for implementation in solid phase extraction and removal of lead [J]. Journal of Industrial and Engineering Chemistry. 2012.

Chemical composition of $PM_{2.5}$ emitted from the burning of agricultural residue in Sichuan Province, China*

Z.M. Xiao[1,‡], J.H. Wu[2] H. Du[2], Y.F. Zhang[2], X.H.Bi[2] and K. Chen[1]

[1]*Tianjin Environmental Monitoring Center, No.19 Fukang Road, Tianjin 300071, China*

[2]*State Environmental Protection Key Laboratory of Urban Ambient Air Particulate Matter Pollution Prevention and Control, College of Environmental Science and Engineering, Nankai University, No. 94 Weijin Road, Tianjin 300071*

[‡]*E-mail: xiaozhimei01@163.com*

Wheat straws and rape stalks were collected from rural agricultural fields in Sichuan Province, and $PM_{2.5}$ were sampled in a closed combustion chamber designed to simulate the burning of wheat straws and rape stalks under the flaming and smoldering conditions. Inductively coupled plasma (ICP) analysis was used to determine the compositions of 18 elements, and 4 water-soluble inorganic ions were analyzed using ion chromatography, organic carbon (OC), and elemental carbon (EC) were measured using a thermal optical carbon analyzer. Organic compounds were measured by gas chromatography-mass spectrum analysis (GC-MS). The results indicated that the type of burning significantly influenced the composition of $PM_{2.5}$ emitted from wheat straws and rape stalks. OC contributed the largest fraction in fine particulate matters, followed by EC. Among water-soluble species, Cl^- and K^+ were found to be the most common ion and cation. Na, K, and Ca were the main elemental components of $PM_{2.5}$, and the relative amount of each element was higher under flaming conditions than under smoldering conditions. β-sitosterol made the highest contribution to $PM_{2.5}$ of any organic compound and can be considered as organic tracer of particles produced by the burning of agricultural residues. Levoglucosan also made up a relatively large fraction of $PM_{2.5}$.

Keywords: Biomass burning; Wheat straws; Rape stalks; $PM_{2.5}$.

1. Introduction

The burning of biomass has drawn considerable attention in recent years due to its significant roles in regional air pollution and its adverse effects on climate and human health [1–3]. The burning of agricultural field residues, as an important type of open biomass burning, commonly takes place during harvest seasons in the crop-producing regions of world. On a global scale, the estimated annual total dry matter combusted due to agricultural burning is around 540Tg [4]. In China, the estimated amount of agricultural residue burned every year is about 110-157.5 Tg [5–6], accounting for about 23% of all agricultural residue

* This work was supported in part by National Natural Science Foundation (21207069, 31370700).

burning in the world.The importance of emissions produced by the burning of agricultural residues to the deterioration of local and regional air quality have been recognized in China [7–10]. Several studies have been conducted in order to assess the impact of agricultural residue burning on air quality [7,11–15]. However, most of these articles have focused on emission factors of pollutants from the burning of agricultural residues and established emission inventories of biomass burning [7,11–14]. Only a few studies have investigated the chemical composition of fine particulate matter ($PM_{2.5}$) produced by agricultural burning [15], even though this is essential to better understanding of the influence of these practices on local atmosphere in the view of receptor-oriented model, such as chemical mass balance model.

In this study, wheat straw and rape stalk, which are widely used in agricultural regions in southwestern China, such as those in Sichuan Province, were chosen as representatives of agricultural residues. The chemical compositions of $PM_{2.5}$ emitted from agricultural residues burning were analyzed to provide new data on agricultural burning source profiles.

2. Materials and Methods

2.1. *Site description*

Wheat straws and rape stalks were collected from rural agricultural fields in Sichuan Province. At each location, 10 random sampling points were situated far away from areas with large amounts of anthropogenic pollution. Approximately 1–2kg of agricultural residue was collected per sample. Samples were sealed in clean paper bags. A closed combustion chamber was designed to simulate the combustion of agricultural residues under controlled conditions. An adjustable electric heater and two $PM_{2.5}$ samplers were placed in the combustion chamber, and clean air was supplied from a vent. A schematic diagram of the open burning experiment is presented in Figure 1. The type of burning (flaming vs. smoldering) was controlled by an adjustable electric heater. External voltage of 220V was used to burn the agricultural residues under flaming conditions, and 110V was used for smoldering conditions [15]. The moisture content and combustion types of cereal straw are listed in Table 1. Agricultural residues were cut into small pieces (about 3–5cm) and added to the heater gradually. Smoke sampling commenced after burning status had remained stable for 10 minutes.

Table 1. Collection of cereal straw samples.

Biomass type	Moisture content (%)	Combustion type
Wheat straws	7.4	flaming vs. smoldering
Rape stalks	10.2	flaming vs. smoldering

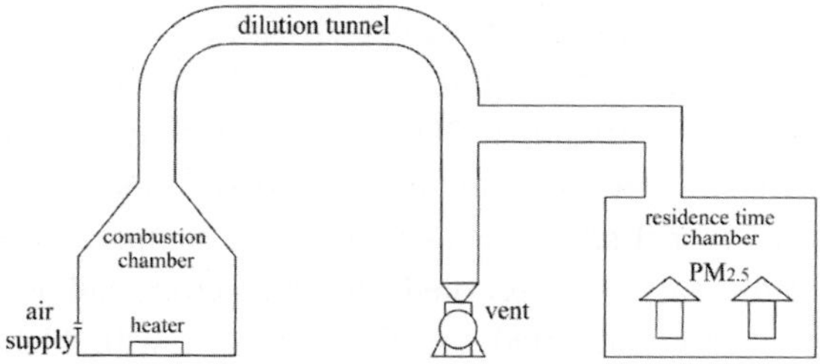

Figure 1. Burning system.

Two parallel samplers, both medium-volume $PM_{2.5}$ samplers (Wuhan Tianhong Intelligence Instrumentation Facility, TH-150 Medium Volume Sampler), were used to collect $PM_{2.5}$. A quartz-fiber filter (90mm diameter, type 2500QAT-UP, Pall Life Sciences, U.S.) was used for subsequent water-soluble inorganic ions and carbon component analysis, a polypropylene-fiber filter (90mm diameter, Beijing Synthetic Fiber Research Institute, China) was used for elemental analysis. Loaded and unloaded filters were conditioned for 24h at about 40% RH and 25°C in an air-conditioned room and weighed on a microbalance with a resolution of 10μg.

2.2. *Chemical analysis*

Inductively coupled plasma (ICP) analysis was used to determine concentrations of Na, Mg, Al, Si, S, K, Ca, Ti, V, Cr, Mn, Fe, Co, Ni, Cu, Zn, As, and Pb in samples collected using polypropylene-fiber filters. Water-soluble inorganic ions were analyzed using ion chromatography (DX-120, Dionex Ltd., U.S.) after sample extraction with deionized water. Organic carbon (OC) and elemental carbon (EC) were measured using the thermal optical carbon analyzer (Desert Research Institute (DRI), Model 2001, Atmoslytic Inc., Calabasas, CA, U.S.). A $0.5cm^2$ punch of each sample was analyzed for eight carbon fractions in accordance with the IMPROVE protocol [16].

Particulate organic matter (POM) collected on the quartz fiber filters was analyzed using GC-MS. This method consists of extraction and GC-MS analysis. Samples were extracted three times with 30 ml dichloromethane/methanol (3:1 v/v) in thick-wall, sealed bottles (pre-baked at 550°C for 4.5h) using a mild ultrasonic bath maintained at room temperature. The extracts were pooled and then reduced to about 5ml using a rotary vacuum evaporator. Then the extract was concentrated to 1ml using mild nitrogen (ultra-purified>99.9) blowing. It was then derivatized with BSTFA (BSTFA/TMCS, 99:1, Supelco) to replace the active hydrogen with trimethylsilanized derivant. The derivatized fractions were analyzed using a

Thermo Finnigan PolarisQ GC-MS equipped with a DB-5MS column (60m in length, 0.25mm diameter, 0.25μm film thickness, Agilent) and oven temperature programming (60°C for 10min to 300°C for 38min at 6°C/min). Multipoint calibrations were conducted using authentic standards spiked with the same final concentrations of internal standards as used in the samples. The quantification standards included sugars, methoxylated phenols, sterols, and other materials.

Laboratory blanks (subjected to POM analysis with blank filters) were added to the analytical batches to evaluate possible contamination. In the present study, the measurement of laboratory blanks in organic analyses showed that the concentrations were very low and could be ignored in organic analysis. Spiked filters were analyzed to determine the overall efficiency and accuracy of the measurements. Average recoveries and deviations of the spike samples were 61.8–105.1% and 5.6–15.4%, respectively, which met the accuracy requirements of this study. The coefficients of correlation (R^2) of standard compounds calibration curves were 0.990±0.016. The target compounds were not quantified if the single to noise ratio (peak area) was less than 10. Method detection limit of the target compounds in this study was 0.1-4ng/μl (1μl splitless injection of total 1ml extractions).

3. Results and Discussion

Levels of OC, EC, soluble species, elements, and organic compounds in $PM_{2.5}$ were measured, and the chemical compositions of $PM_{2.5}$ emitted from agricultural residues subjected to different types of burning are listed in Table 2.

Table 2. Chemical compositions of PM2.5 emitted from agricultural residues from different types of burning (units %).

Chemical Species	Wheat straws		Rape stalks	
	Flaming (n=6)	Smoldering (n=8)	Flaming (n=8)	Smoldering (n=8)
Na^+	0.01±0.00	0.01±0.00	0.45±0.11	0.06±0.01
NH_4^+	0.86±0.15	0.17±0.05	3.99±0.64	0.13±0.03
K^+	1.33 ± 1.10	0.33 ± 0.44	6.04 ± 5.28	0.51±0.08
F^-	0.14±0.03	0.33±0.06	0.03±0.01	0.11±0.02
Cl^-	3.06±0.37	1.05±0.13	14.11±2.57	0.95±0.22
NO_3^-	0.04±0.01	0.04±0.01	0.16±0.03	0.05±0.01
SO_4^{2-}	0.15±0.03	0.10±0.02	0.23±0.07	0.00±0.00
OC	47.62± 4.38	47.71±4.30	36.76±10.04	47.22± 4.61
EC	4.41 ± 1.76	1.53 ± 0.75	10.19 ± 2.93	2.50 ± 2.33
Na	2.23 ± 1.49	0.80 ± 0.67	4.90 ± 4.63	0.86 ± 0.63
Mg	0.20 ± 0.15	0.08 ± 0.08	0.49 ± 0.55	0.08 ± 0.07
Al	0.58 ± 0.39	0.21 ± 0.18	1.21 ± 1.23	0.23 ± 0.18
Si	0.67 ± 0.48	0.29 ± 0.19	0.77 ± 0.52	0.18 ± 0.13
S	0.00 ± 0.00	0.00 ± 0.00	0.00 ± 0.00	0.00 ± 0.00
K	1.60±0.59	0.33±0.06	6.74±1.79	0.71 ± 0.65

Chemical Species	Wheat straws		Rape stalks	
	Flaming (n=6)	Smoldering (n=8)	Flaming (n=8)	Smoldering (n=8)
Ca	1.19 ± 0.83	0.49 ± 0.44	2.36 ± 2.28	0.43 ± 0.32
Ti	0.00 ± 0.00	0.00 ± 0.00	0.00 ± 0.00	0.00 ± 0.00
V	0.00 ± 0.00	0.00 ± 0.00	0.00 ± 0.00	0.00 ± 0.00
Cr	0.00 ± 0.00	0.00 ± 0.00	0.00 ± 0.00	0.00 ± 0.00
Mn	0.00 ± 0.00	0.00 ± 0.00	0.01 ± 0.01	0.00 ± 0.00
Fe	0.25 ± 0.16	0.05 ± 0.03	0.21 ± 0.18	0.03 ± 0.02
Co	0.01 ± 0.00	0.01 ± 0.01	0.05 ± 0.14	0.00 ± 0.00
Ni	0.00 ± 0.00	0.00 ± 0.00	0.01 ± 0.03	0.00 ± 0.00
Cu	0.04 ± 0.03	0.04 ± 0.03	0.10 ± 0.08	0.03 ± 0.02
Zn	0.01 ± 0.01	0.01 ± 0.01	0.05 ± 0.07	0.01 ± 0.00
As	0.00 ± 0.00	0.00 ± 0.00	0.00 ± 0.00	0.00 ± 0.00
Pb	0.00 ± 0.00	0.00 ± 0.00	0.00 ± 0.00	0.00 ± 0.00
Guaiacol	0.01 ± 0.01	0.01 ± 0.00	0.00 ± 0.00	0.01 ± 0.00
Syringaldehyde	0.02 ± 0.01	0.02 ± 0.01	0.03 ± 0.01	0.03 ± 0.01
Methyl vanillate	0.01 ± 0.01	0.02 ± 0.01	0.02 ± 0.00	0.01 ± 0.00
Isoeugenol	0.03 ± 0.02	0.04 ± 0.01	0.08 ± 0.02	0.04 ± 0.01
Mannosan	0.01 ± 0.00	0.03 ± 0.01	0.17 ± 0.07	0.04 ± 0.02
Levoglucosan	0.05 ± 0.01	0.13 ± 0.04	0.22 ± 0.12	0.07 ± 0.01
Vanillin	0.01 ± 0.00	0.01 ± 0.00	0.05 ± 0.03	0.01 ± 0.00
Stigmasterol	0.01 ± 0.00	0.02 ± 0.01	0.02 ± 0.01	0.03 ± 0.01
β-sitosterol	0.12 ± 0.03	0.19 ± 0.04	0.27 ± 0.07	0.14 ± 0.04

3.1. *Distribution of carbon fractions in different agricultural residues*

3.1.1. *OC and EC*

Carbonaceous components were found to be dominant in the $PM_{2.5}$ emitted from biomass burnings. In the present study, the relative amounts of OC in $PM_{2.5}$ emitted from wheat straws subjected to flaming and smoldering were similar, 47.62% and 47.71%, respectively, which indicated that temperature has a moderate impact on the conversion of organic substances into OC in wheat straw. For rape stalks, these values were 36.76% and 47.22%. This indicates that lower temperatures favor the conversion of organic substances into OC for the rape stalks. OC/$PM_{2.5}$ ratios of the wheat straws and rape stalks subjected to smoldering were similar, but wheat straws subjected to flaming had a higher OC/$PM_{2.5}$ than rape stalks. The OC/$PM_{2.5}$ of the wheat straws in this study was higher than that reported by Li et al. [17], but lower than that reported by Zhang et al. [15]. OC/$PM_{2.5}$ in the flaming status of the rape stalks in this study was lower than that reported by Li et al. [17] and Zhang et al. [15], but for smoldering OC/$PM_{2.5}$ was higher than reported by Li et al. [17], lower than reported by Zhang et al. [15].

EC/$PM_{2.5}$ ratios of wheat straws were found to be 4.41% and 1.53% in flaming and smoldering straws. In rape stalks, EC/$PM_{2.5}$ ratios were 10.19% and 2.50% in the flaming and smoldering stalks, respectively, both of which are higher than for wheat straws. This may due to the different compositions of wheat straws and rape stalks. A larger $EC/PM_{2.5}$ ratio was observed upon flaming than upon smoldering. This may be due to the fact that the higher temperature of flaming causes more conversion of organic substances into EC. Similar results were reported by Zhang et al. [15].

Ratio of OC/EC was an important indicator of pollution characterization and source of fine particulate matter in the atmospheric. According to study, the ratio of OC/EC was about 3 in the fossil fuel-dominated combustion emissions cities [18], while the ratio was 1.1 in the atmospheric which affected by the diesel vehicle emissions [19]. Average OC/EC of $PM_{2.5}$ emitted from agricultural residues in open burning was 8.1reported by Li et al. [17], and OC/EC values of $PM_{2.5}$ were about 6.5–8.3 in different time during agricultural residues burning season in Gwangju, Korea [20–21], ratio of OC/EC was 5.7±0.6 near the source of biomass combustion [22]. In this study, OC/EC was 10.8 and 3.6 in the flaming for straws and rape stalks, respectively. There was great variation for the ratio of OC/EC in all studies, which may be influenced by straw types and combustion conditions and other factors, this phenomenon also illuminated that the localization research was very necessary for providing new data on regional agricultural burning source profiles.

3.1.2. *Distribution of eight carbon fractions in different agricultural residues*

According to the Interagency Monitoring of Protected Visual Environments (IMPROVE) TOR protocol, carbon content can be separated into seven thermally defined carbon fractions OC1, OC2, OC3, OC4, EC1, EC2, and EC3. These correspond to temperature plateaus, and there is one fraction of optically detected pyrolyzed carbon (OPC). The IMPROVE protocol defines OC as OC1+OC2+OC3+OC4+OPC and EC as EC1+EC2+EC3-OPC. Carbon abundances in each of these fractions differ by carbon source [23].

The distribution of eight carbon fractions emitted from the burning of different agricultural residues can provide some information. The carbon fractions of wheat straws and rape stalks after different types of combustion are shown in Figure 2. For wheat straw, average abundances of OC1, OC2, OC3, OC4, OPC, EC1, EC2, and EC3 in TC were 22.92%, 12.25%, 5.09%, 0.58%, 7.41%, 11.69%, 0.06%, and 0.06%, respectively, for flaming and 22.65%, 11.10%, 5.17%, 0.50%, 8.29%, 9.70%, 0.08%, and 0.04%, respectively, for smoldering. For rape stalks, these values were 16.25%, 7.43%, 6.37%, 2.57%, 4.15%, 14.08%, 0.13%, and 0.13%, respectively, for flaming and 23.35%, 12.06%, 4.23%, 0.61%, 6.69%, 9.34%, 0.08%, and 0.04%, respectively, for smoldering. In this study, OC1, OC2, OPC, and EC1 were generally more abundant than other species and there was very little OC4, EC2, or EC3 in TC of the wheat straws and rape stalks. Previous studies have shown OC1 to be the most abundant species in biomass burning [24–25], OC2 was the most abundant carbon fraction in coal combustion samples [25], OC3 and OC4 were enriched in the road dust profile [24], and EC2 was the most abundant species in diesel exhaust [23,26]. OPC was found to increase with water-soluble organic carbon content [27]. In this study, contributions of OC to TC were lower than those reported by Cao et al. [25], but the contribution of EC to TC was higher than that reported by Cao et al. [25]. The different distributions of eight carbon fractions in different regions was probably due to the different components in different regions.

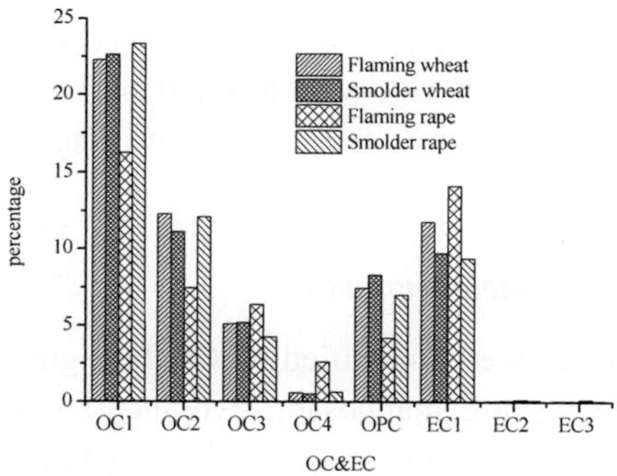

Figure 2. Distribution of eight carbon fractions in the wheat and rape.

3.2. *Distribution of water-soluble species and elements*

Among water-soluble species, Cl^- and K^+ were the most common anions and cations, respectively in $PM_{2.5}$ from burning of both wheat and rape. Cl^- accounted for 3.06% of the material released during flaming and 1.05% of that released during smoldering of wheat straws. For rape stalks, these values were Cl^- 14.11% and 0.95%. K^+, which is often used as a marker of total biomass burned, made up 1.33% of the total material released during flaming and 0.33% of that released during smoldering of wheat straws. For rape stalks, these values were 6.04% and 0.51%. Cl^- and K^+ made up more of the total $PM_{2.5}$ produced by flaming rape stalks than flaming wheat straws, but the contributions of Cl^- and K^+ to $PM_{2.5}$ were comparable for the smoldering wheat straws and rape stalks. Cl^-/K^+ ratios for flaming and smoldering wheat straws were 2.3 and 3.2, respectively. For rape stalks, these values were 2.3 and 1.9. The ratios of Cl^-/K^+ reported by Li et al. [17] were 1.4 and 2.7 for wheat straw and maize stover, respectively. Zarate [28] reported about 0.8 for cereal waste. Hays reported [29] ratios of 1.1 for wheat straw and 2.8 for rice straw. It can be concluded that Cl^-/K^+ ratios vary considerably, possibly because of varying levels of K and Cl in the individual fuels.

Each element made up more of the $PM_{2.5}$ released during flaming status than during smoldering. For wheat straws, the most common element was Na, accounting for 2.23% of total $PM_{2.5}$, followed by K and Ca, accounting for 1.60% and 1.19% of total $PM_{2.5}$, respectively. For rape stalks, K was the most common element, followed by Na and Ca, these three elements accounted for 6.74%, 4.90%, and 2.36% of total $PM_{2.5}$, respectively. All three made up less than 1% of the $PM_{2.5}$ released by smoldering of wheat straws, Na made up 0.80%, Ca made up 0.49%, and K made up 0.33% of total $PM_{2.5}$. For rape stalks, Na made up 0.86%, K made up 0.71%, and Ca made up 0.43%.

The ratios of $Na/PM_{2.5}$ and $Ca/PM_{2.5}$ were higher than those reported by Li et al. [17] but the ratios of $K/PM_{2.5}$ were lower than those reported by Li et al. [17]. The difference may due to different content in agricultural residues or to burning.

3.3. *Distribution of organic compounds*

Nine organic compounds were quantified by GC-MS, guaiacol, syringaldehyde, methyl vanillate, isoeugenol, mannosan, levoglucosan, vanillin, stigmasterol, and β-sitosterol. The distribution of organic compounds emitted from wheat straws and rape stalks subjected to flaming and smoldering was different. For wheat straws, each organic compound made up less of the total amount of material released during flaming than during smoldering. However, for the rape stalks, each organic compound made up more of the total amount of material released during flaming than during smoldering. The most common organic compound in the wheat straws and rape stalks was β-sitosterol, accounting for 0.12% of the material released during flaming and 0.19% of the material released during smoldering for the wheat straws. For rape stalks, these values were 0.27% and 0.14%. Levoglucosan was the second most common organic compounds, making up 0.05% and 0.13% of the material released by flaming and smoldering wheat and 0.22% and 0.07% of material released by flaming and smoldering rape stalks. The relative amounts of β-sitosterol were higher than those reported by Zhang et al. [15], but the relative amounts of levoglucosan were lower than those reported by Zhang et al. [15], Fine et al. [30], or Schauer et al. [31]. This difference may be attributable to bias from the ratio of cellulous to hemicellulose or from the burning. β-sitosterol and levoglucosan were present in higher concentrations than other organic compounds, For this reason, they are suitable for use as organic tracers of the agricultural residues left by burning.

4. Conclusions

In this study, wheat straws and rape stalks were collected from rural agricultural fields in Sichuan Province and subjected in to flaming and smoldering in the laboratory. $PM_{2.5}$ emitted from combustion was sampled, and 18 elements, 4 water-soluble species, OC, EC, and organic compounds were measured.

Results showed that among the fine particles emitted, OC/$PM_{2.5}$ made up 47.62% of those produced by flaming and 47.71% of those produced by smoldering for wheat straws. For rape stalks, these values were 36.76% and 47.22%, respectively. EC/$PM_{2.5}$ made up 4.41% of the particles produced by flaming and 1.53% of those produced by smoldering of wheat stalks. For rape stalks these values EC/$PM_{2.5}$ were 10.19% and 2.50%, respectively. OC1, OC2, OPC, and EC1 were generally more abundant in the distribution of TC, and there were very few OC4, EC2, and EC3 particles in TC of the wheat straws and rape stalks. Cl^- and K^+ were important anion and cations in $PM_{2.5}$ from burning of agricultural residues. Cl^- accounted for 3.06% of the particles produced by flaming and 1.05% of those produced by smoldering of wheat straws. For rape stalks, these values were 14.11% and 0.95%. K^+ made up 1.33% of the particles produced by flaming 0.33% of those produced by smoldering of wheat straws. For rape stalks, these values were 6.04% and 0.51%. Each element made up more of the total amount of material released during flaming than during smoldering, and Na, K and Ca were the three most important elements. The distributions of organic compounds in $PM_{2.5}$ produced by wheat straws and rape stalks were different. For wheat straws, each organic compound made up less of the total amount of material released during flaming than during smoldering status. However, for rape stalks, each organic compound made up more of the total amount of material released during flaming status than during smoldering. The most prevalent organic compound in the wheat straws and rape stalks was β-sitosterol, and the second-most prevalent was levoglucosan.

Acknowledgments

We would like to thank the Environmental Protection Agency of Sichuan province for their support of this study.

References

1. P.J. Crutzen and M.O. Andreae, *Biomass burning in the tropics: Impact on atmospheric chemistry and biogeochemical cycles*. Science, Vol. 250 (1990), pp. 1669–1678.

2. J.S. Levine, W.R. Cofer and D.R. Cahoon, *Biomass Burning: A Driver for Global Change*. Environ Sci Technol, Vol. 29 (1995), pp. 120–125.
3. K. Torigoe, S. Hasegawa, O. Numata, S. Yazaki, M. Matsunaga, N. Boku, M. Hiura and H. Ino, *Influence of emission from rice straw burning on bronchial asthma in children*. Pediatrics. Vol. 42 (2000), pp. 143–150.
4. M.O. Andreae and P. Merlet, *Emission of trace gases and aerosols from biomass burning*. Glob. Biogeochem. Cycles, Vol. 15 (2001), pp. 955–966.
5. G. Cao, X. Zhang and F. Zheng, *Inventory of black carbon and organic carbon emissions from China.* Atmos. Environ, Vol. 40 (2006), pp. 6516–6527.
6. X. Yan, T. Ohara and H. Akomoto, *Bottom-up estimate of biomass burning in mainland China*. Atmos. Environ, Vol. 40 (2006), pp. 5262–5273.
7. B. Zhu, Emission Characteristic of $PM_{2.5}$ from Crop Straw Burning. Ph.D. thesis, Peking University, Beijijng, China (2004).
8. S. Yang, H. He, S. Lu, C. Chen and J. Zhu, *Quantification of crop residue burning in the field and its influence on ambient air quality in Suqian. China*. Atmos. Environ, Vol. 42 (2008), pp. 1961–1969.
9. Y.H. Zhang, M. Hu, L.J. Zhong, A. Wiedensohler, S.C. Liu, M.O. Andreae, W. Wang and S.J. Fan, *Regional integrated experiments on air quality over Pearl River Delta 2004 (PRIDE-PRD2004): Overview.* Atmos. Environ, Vol. 42 (2008), pp. 6157–6173.
10. H. Guo, H.R. Cheng, Z.H. Ling, P.K.K. Louie and G.A. Ayoko, *Which emission sources are responsible for the volatile organic compounds in the atmosphere of Pearl River Delta?* J. Hazard. Materi, Vol. 188 (2011), pp. 116–124.
11. X.H. Li, Characterization of Air Pollutants Emitted from Biomass Burning in China. Ph.D. thesis, Tsinghua University, Beijing, China (2007).
12. H.F. Zhang, X.N. Ye, T.T. Cheng, J.M. Chen, X. Yang, L. Wang and R.Y. Zhang, *A laboratory study of agricultural crop residue combustion in China: emission factors and emission inventory*. Atmos. Environ, Vol. 42 (2008), pp. 8432–8441.
13. G.F. Shen, Y.F. Yang, W. Wang, S. Tao, C. Zhu, Y.J. Min, M. Xue, J.N. Ding, B. Wang, R. Wang, Z.H. Shen, W. Li, X.L. Wang and A.G. Russell, *Emission factors of particulate matter and elemental carbon for crop residues and coals burned in typical household stoves in China.* Environ Sci Technol, Vol. 44 (2010), pp. 7157–7162.
14. Y.S. Zhang, M. Shao, Y. Lin, S.J. Luan, N. Mao, W.T. Chen and M. Wang, *Emission inventory of carbonaceous pollutants from biomass burning in the*

Pearl River Delta Region, China. Atmos. Environ, Vol. 76 (2013), pp. 189–199.

15. Y.X. Zhang, M. Shao, Y.H. Zhang, L.M. Zeng, L.Y. He, B. Zhu, Y.J. Wei and X.L. Zhu, *Source profiles of particulate organic matters emitted from cereal straw burnings*. J. Environ. Sci. Vol. 19 (2007), pp. 167–175.
16. B. Hou, G.S. Zhuang, R. Zhang, T.N. Liu, Z.G. Guo and Y. Chen, *The Implication of Carbonaceous Aerosol to the Formation of Haze: Revealed from the Characteristics and Sources of OC/EC over a Mega-city in China.* J. Hazard. Materi, Vol. 190 (2011), pp. 529–536.
17. X.H. Li, S.X. Wang, L. Duan, J.M. Hao, C. Li, Y.S. Chen and Y. Liu, *Particulate and Trace Gas Emissions from Open Burning of Wheat Straw and Corn Stover in China.* Environ Sci Technol, Vol. 41 (2007), pp. 6052–6058.
18. X.Y. Zhang, Y.Q. Wang, X.C. Zhang, W. Guo and S.L Gong, *Carbonaceous aerosol composition over various regions of China during 2006.* J. Geophys. Res. Vol. 113 (2008), DOI: 10.1029/2007JD009525
19. J.G. Watson, J.C. Chow and J.E. Houck, *$PM_{2.5}$ chemical source profiles for vehicle exhaust, vegetative burning, geological material, and coal burning in Northwestern Colorado during 1995*. Chemosphere, Vol. 43 (2001), pp. 1141–1151.
20. S.Y. Ryu, J.E. Kim, H. Zhuanshi, Y.J. Kim and G.U. Kang, *Chemical composition of post-harvest biomass burning aerosols in Gwangju, Korea.* J. Air Waste Manage. Assoc. Vol. 54 (2004), pp. 1124–1137.
21. S.Y. Ryu, B.G. Kwon, Y.J. Kim, H.H. Kim, and K.J. Chun, *Characteristics of biomass burning aerosol and its impact on regional air quality in the summer of 2003 at Gwangju, Korea.* Atmos. Res., Vol. 84(2007), pp. 362–373.
22. M.T. Chuang, C.C.K. Chou, K. Sopajaree, N.H. Lin, J.L. Wang, G.R. Sheu, Y.J. Chang and C.T. Lee, *Characterization of aerosol chemical properties from near-source biomass burning in the northern Indochina during 7-SEAS /Dongsha experiment.* Atmos. Environ, Vol. 78 (2012), pp. 72–81.
23. J.G. Watson, J.C. Chow, D.H. Lowenthal, L.C. Pritchett, C.A. Frazier, G.R. Neuroth and R. Robbins, *Differences in the carbon composition of source profiles for dieseland gasoline-powered vehicles*. Atmos. Environ, Vol. 28 (1994), pp. 2493–2505.
24. J.C. Chow, J.G. Watson, H.D. Kuhns, V. Etyemezian, D.H. Lowenthal, D. Crow, S.D. Kohl, J.P. Engelbrecht and M.C. Green, *Source profiles for industrial, mobile, and area sources in the big bend regional aerosol*

visibility and observational (BRAVO) study. Chemosphere, Vol. 54 (2004), pp. 185–208.
25. J.J. Cao, F. Wu, J.C. CHow, S.C. Lee, Y. Li, S.W. Chen, Z.S. An, K.K. Fung, J.G. Watson, C.S. Zhu and S.X. Liu, *Characterization and source apportionment of atmospheric organic and elemental carbon during fall and winter of 2003 in Xi'an, China*. Atmos. Chem. Phys, Vol. 5 (2005), pp. 3127–3137.
26. J.C. Chow, *How does the sample affect the measurement of different carbon fractions?* International Workshop for the Development of Research Strategies for the Sampling and Analysis of Organic and Elemental Carbon Fractions in Atmospheric Aerosols. Durango, Colorado. (2003).
27. H. Yang, and J.Z. Yu, *Uncertainties in charring correction in the analysis of elemental and organic carbon in atmospheric particles by thermal/optical methods*. Environ Sci Technol Vol. 36 (2002), pp. 5199–5204.
28. I.O. Zarate, A. Ezcurra, J.P. Lacaux and P.V. Dinh, *Emission factor estimates of cereal waste burning in Spain*. Atmos. Environ, Vol. 34 (2000), pp. 3183–3193.
29. M.D. Hays, P.M. Fine, C.D. Geron, M.J. Kleeman and B.K. Gullett, *Open burning of agricultural biomass: physical and chemical properties of particle-phase emissions*. Atmos. Environ, Vol. 39 (2005), pp. 6747–6764.
30. P.M. Fine, G.R. Cass and B.R.T. Simoneit, *Chemical characteriation of fine particle emissions from fireplace combustion of woods grown in the northeastern United States*. Environ Sci Technol, Vol. 35 (2001), pp. 2665–2675.
31. J.J. Schauer, M.J. Kleeman, G.R. Cash and B.R.T. Simoneit, *Measurement of* emissions *from air pollution sources. 3.* C_1-C_{29} *organic compounds from fireplace combustion of wood*. Environ Sci Technol, Vol. 35 (2001), pp. 1716–1728.

Study on sodium hypochlorite disinfection of reclaimed water in China

W.B. Chen*, X.L. Song, F.Z. Qin and Y.F. Sun
School of Municipal and Environmental Engineering, Shandong Jianzhu University, China
**E-mali: cwb929@sdjzu.edu.cn*

The research object has been performed in a secondary effluent of urban sewage plant of Jinan in China. By investigating the dosage of sodium hypochlorite and inactivated effect on the total number of bacteria and E.coli during disinfection contact time, the main objective of this article is to analyze the attenuation of residual chlorine and the correlation of the CRT values and inactivated rate of coliform bacteria. The results shown that when the chlorine dosage and the contact time reached to 50mg/L and 30min or 40mg/L and 120min respectively, microbial quantity and amount of total residual chlorine could meet the regulations of "standards of reclaimed water quality" (SL368-2006) in China. According to the correlation analysis of the CRT values and inactivation rate of coliform bacteria, when the sodium hypochlorite dosage and contact time reached to 50mg/L and 30min respectively, the result could also meet "standards of reclaimed water quality" (SL368-2006). The results were applied successfully to engineering practice.

Keywords: Sodium hypochlorite disinfection; Reclaimed water; Total number of E.coli.; Residual chlorine; The CRT values.

1. Introduction

In recent years, the serious shortage of available water resources has seriously restricted the sustainable economic development in China, so sewage recycling utilization has become one of the main ways to solve the worldwide problem. Municipal wastewater water quality has greatly improved by secondary biological treatment, but the absolute number of bacteria is still considerable, and there is the possibility of pathogens, potentially harmful to our health. Based on "the reuse of urban recycling water-water quality standard for urban miscellaneous water consumption" (GB/T18920-2002), when chlorine disinfection contact time is 30 min, the total number of E.coli is less than 3/L, and the residual chlorine is not less than 1.0 mg/L. For this, the reclaimed water disinfection must be strengthen for ensuring the security and healthy of the reclaimed water. Sodium hypochlorite is a highly effective broad-spectrum, safe and stable strong sterilization disinfectant. It has the similar disinfection effect to chlorine, but it produces less disinfection by-products than liquid chlorine. Sodium hypochlorite used as disinfectant is relatively safe. The application of sodium hypochlorite as a disinfectant in the sewage treatment plant gradually increased in recent years [1]. A secondary urban sewage treatment plant in Jinan of China is used as the research object by author, the correlation of the C_RT values and inactivated rate of coliform bacteria is explored, with carrying on the experimental study of reclaimed water disinfection by investigating the sodium hypochlorite disinfection dosage and contact time on the inactivation effect of total number of bacteria and E.coli, analyzing the residual chlorine attenuation change, and applying to the engineering practice.

2. Materials and Methods

2.1. *Materials*

Trial is performed on a sewage treatment plant in Jinan, the core biological treatment using "A^2O" process, the main processing is: inflow water→mechanical coarse grid cyclone grit chamber→biological anaerobic tank pool→efficient secondary settling tank sedimentation tank→high efficiency fiber filter→ disinfection→the plant effluent.

Experimental raw water is the high efficiency fiber filter effluent of sewage treatment plant, raw water quality indicators are shown in Table 1. Sodium hypochlorite was commercially chemical disinfection agent used in the experiments. The density of sodium hypochlorite solution is 1.28g/cm^3 and available chlorine content is greater than 10%.

Table 1. Water quality of test water sample.

E.coli (CFU/L)	TPC (CFU/mL)	NH4+-N (mg/L)	TOC (mg/L)
10^6~10^7	10^4~10^5	0.5~0.7	26~30

2.2. *Test Methods*

Factors that influence efficiency of sodium hypochlorite disinfection include sodium hypochlorite dosage, disinfection contact time, temperature, pH, etc. For the engineering application, adjusting the pH and temperature to strengthen the disinfection effect has a significant limitations. The experiment mainly studied the effect of sodium hypochlorite dosage and contact time on disinfection. A 250ml conical flask is used for taking 100 ml water samples five copies, adding respectively 1g/L of sodium hypochlorite solution (0.5, 1, 1.5, 2, 2.5, 3, 3.5, 4, 4.5, 5mL) to water samples, contacting reaction, respectively, 30, 60, 90, 120min. Residual chlorine is measured immediately after each contact time reaction, simultaneously terminating reaction by 0.3% sodium thiosulfate 2ml chloride, respectively, measuring the total number of bacteria, total coliform count and residual chlorine decay before and after disinfection.

The experimental ways of water quality indicators (include total bacteria, total coliform count, residual chlorine) is based primarily on "the reuse of urban recycling water-water quality standard for urban miscellaneous water consumption" (GB/T18920-2002) and "standards of reclaimed water quality" (SL368-2006).

3. Results and Discussion

3.1. *Disinfection effect of sodium hypochlorite*

Disinfectant dosage is the important factor influencing disinfection effect [2,3]. Under the different disinfection condition, the inactivated effect of sodium hypochlorite on the total number of bacteria in secondary effluent is shown in the Figure 3.1.

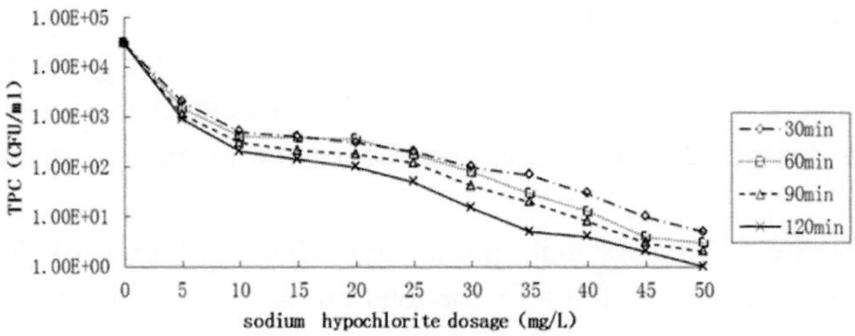

Figure 3.1 Inactivated effect of sodium hypochlorite on the total number of bacteria.

Sodium hypochlorite generated hypochlorous acid (HClO) after it is added to secondary effluent [4]. HClO is a kind of small neutral molecule and it can spread to the negatively charged surface of the bacteria, HClO can penetrate the body and damage enzyme system by oxidation, leading finally to the bacteria die. At the same time, the part of hypochlorous acid is further broken down to new ecological oxygen [O], the strong oxidizability of new ecological oxygen make the protein of bacteria and virus denature and kill pathogenic microorganisms, achieving the effect of disinfection [5]. As can be shown from the Figure 3.1, after the sodium hypochlorite was added to the sample, the total number of bacteria decreased with the increase of sodium hypochlorite dosage under the same contact time; When the sodium hypochlorite has the same dosage, the total number of bacteria has a tendency to reduce with the extension of contact time.

Under the different disinfection condition, the inactivated effect of sodium hypochlorite on the total number of E.coli in secondary effluent is shown in Figure 3.2.

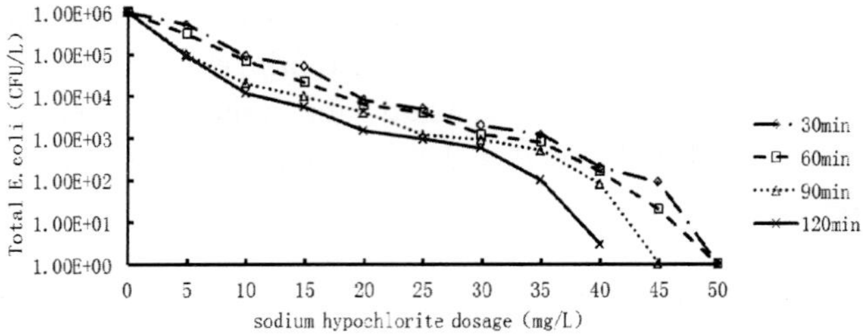

Figure 3.2 Inactivated effect of sodium hypochlorite on the total on the total number of E.coli.

The results have shown that the impact of sodium hypochlorite dosage for disinfection effect is very obvious. Increasing the dosage of sodium hypochlorite disinfectant can effectively shorten the residence time and improved the disinfection efficiency. At the same time, the contact time has a significant effect on the disinfection sterilization. The number of E.coli has a downward trend with the extension of disinfection contact time, and under reaching the same disinfection effect, prolonging disinfection contact time can effectively reduce the dosage of disinfectant, and reduce the amount of disinfection by-products

The Figure 3.2 shows that sodium hypochlorite dosage of 20 mg/L, the disinfection contact time of 30min, the total number of E.coli can meet the level 1B standards of "discharge standard of pollutants for municipal waste water treatment plant" and the class water requirements of "the reuse of urban recycling water-water quality standard for scenic environment use" (GB/T18921-2002). Sodium hypochlorite dosage of 10mg/L and disinfection contact of 120min can also meet the above requirements. When sodium hypochlorite dosage reached to 50mg/L and the disinfection contacted 30 min, the number of E.coli can reach "standards of reclaimed water quality" (SL368-2006) and "the reuse of urban recycling water–water quality standard for urban miscellaneous water consumption" (GB/T 18920-2002).

3.2. *Total chlorine attenuation changes in disinfection process*

Residual chlorine refers to the rest content of the available chlorine after a certain contacting time in water, dividing into compound residual chlorine and residual chlorine free [6]. Dosing active chlorine compounds and the ammonia compound generated for residual chlorine compounds, including NH_2Cl, $NHCl_2$

and $NHCl_3$; Free residual chlorine refers to ClO^-, HOCl, Cl_2, etc. [7], its speed and sterilization effect is good, but not stable; Compounds of chlorine and free residual chlorine is the total residual chlorine. Total residual chlorine is an important indicator to monitor if the water quality meets requirements and the dosing amount of disinfectant, and is also the guarantee the sustained disinfection.

Under the different initial concentration of sodium hypochlorite, the change of total residual chlorine with contact time is shown in Figure 3.3.

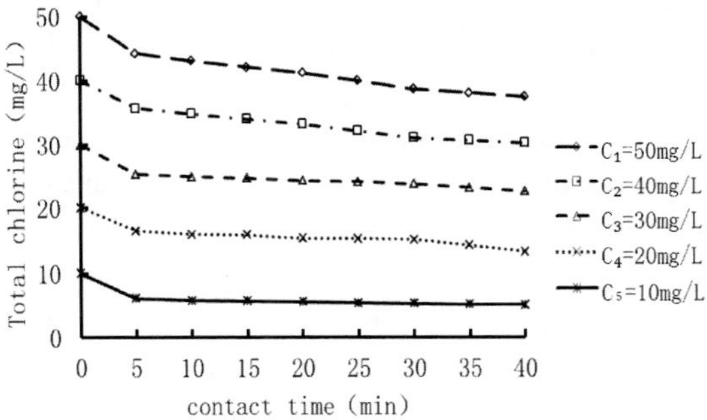

Figure 3.3 Change of total residual chlorine under different initial concentration of sodium hypochlorite.

After sodium hypochlorite was added to water samples, residual chlorine decreased fastest within the first 5min, the attenuation tendency steadily decreased after 5min, the main reason is that after chlorine dosage was added to water samples, within the initial 5min, it reacted quickly with microorganisms and organic in water samples, chlorine rapidly declining, most of the organic matter in water samples was consumed in this process, most free microorganisms in water are inactivated; Residual chlorine is mainly used to inactivate microorganisms attached to the inside particulate matter after 5min, since the chlorine penetrating into particles is a slow process, so the tendency of residual chlorine become gradually gently [8].

Figure 3.3 has shown that the higher the initial amount of sodium hypochlorite dosage, residual chlorine consumption rate increases accordingly, within 40min contact reaction, the attenuation of residual chlorine (C_5=10mg/L) is 4.9 mg/L, but the residual chlorine attenuation increases to 12.5mg/L (C_1= 50mg/L. Literature [9,10] indicate that chlorine oxidation capacity enhanced with the increase of initial dosing quantity, the more organic reaction with chlorine in water, resulting in the higher consumption.

Under the condition of different initial concentration of sodium hypochlorite, after contact reaction 30min, total residual chlorine meet "the water quality standard of reclaimed water in China" (SL368-2006) and "cities sewage recycling of mixed water quality" (GB/T 18920-18920) regulation that total residual chlorine is not less than 1.0 mg/L.

The C_RT value is defined as the product of residual disinfectant (chlorine) concentration and the contact time T, unit: mg.min/L. it is used to represent the amount of disinfectant and can be used as a evaluation index of disinfection efficiency. E.coli inactivation rate is defined as the log value that is the ratio of the number of E.coli before and after disinfection, showing with Log (N_0/N_t) (N_0 and N_t is respectively the number of coliform bacteria before and after disinfection), inactivated rates can objectively describe the changes of the number of E.coli [11].

In this experiment, the relation of C_RT values and the inactivated rate of coliform bacteria was studied by changing the disinfectant dosage and contact time, the results are shown in Figure 3.4.

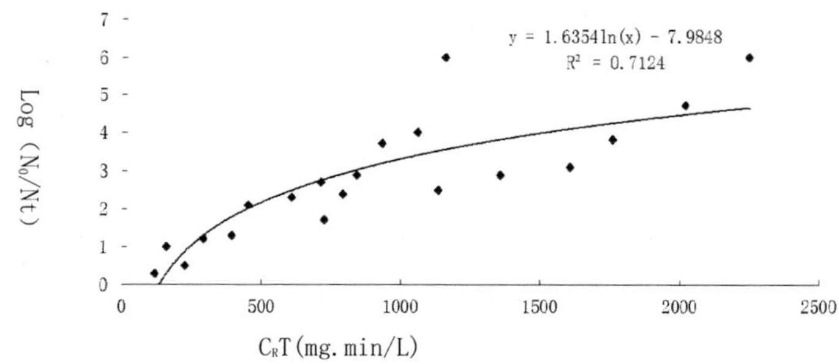

Figure 3.4 Inactivated rate of E.coli Log (N_0/Nt) of recycling water under different disinfection dosage C_RT.

As the Figure 3.4 shown, along with the increase of the C_RT, E.coli inactivation rate is on the rise. The correlation coefficient of experiment data regression analysis is 0.7124, so the C_RT values and the rate of E.coli inactivated present certain relevance. According to the regression relationship between the C_RT values and the inactivated rate of coliform bacteria, when the number of E.coli is less than 3/L, C_RT has a value of 1500 mg.min/L, namely disinfectant dosing 50 mg/L, disinfection contact when 30 min. It is basically identical with the inactivated effect test results of sodium hypochlorite dosage and contact time on E.coli.

The regression relationship of C_RT values and the inactivated rate of coliform bacteria is applied to reclaimed water disinfection processing sewage treatment plant, basing on "standards of reclaimed water quality" (SL368-2006) and "the reuse of urban recycling water-water quality standard for urban miscellaneous water consumption" (GB/T18920-2002) standard value of E.coli, which calculated the inactivation rate of E.coli. According to the relation curve of inactivation rate of E.coli and the C_RT value 2.4, the minimum C_RT values can be found. Then, according to the sewage treatment plant disinfection facilities case, the contact time T can be determined, and residual chlorine concentration can be calculated, thus determined whether residual chlorine meet the requirement of reclaimed water standard. At the same time, according to the relationship of residual chlorine and dosage graph 2.3, the concentration of disinfectant dosage was determined. Or according to the standard limit value requirements of residual chlorine on reclaimed water and the relationship of residual chlorine and dosage graph 2.3, the concentration of disinfectant dosage was determined; At the same time, calculating the required disinfection contact time, the time was used to guide the sewage treatment plant disinfection. Therefore, the economy optimization and control of sewage treatment plant of reclaimed water disinfection process can be realized.

4. Conclusion

Sodium hypochlorite has a good disinfection effect on the secondary effluent of urban sewage plant, and it can effectively and rapidly inactivate pathogenic microorganism in reclaimed water. Disinfection effect was better when sodium hypochlorite dosage became larger and the contact time was longer. Results have shown that the chlorine dosage and the contact time reached to 50mg/L and 30min or 40mg/L and 120min respectively, microbial quantity and amount of total residual chlorine could meet the regulations of "standards of reclaimed water quality" (SL368-2006) and "the reuse of urban recycling water–water quality standard for urban miscellaneous water consumption" (GB/T 18920-2002). According to the correlation analysis of the C_RT values and the inactivation rate of coliform bacteria, when the sodium hypochlorite dosage and the contact time reached to 50mg/L and 30min respectively, the results could also meet required water standards. Therefore, C_RT values could be used to guide reclaimed water disinfection process. On the premise of that water quality met the national standard, the appropriate C_RT values could be determined and in basis of this C_RT values, the best disinfectant dosage and contact time could be determined in the engineering practice. The C_RT values is of great significance in economic and security of the reclaimed water.

References

1. Hsu, C. S., & Huang, D. J. 2015. Disinfection of herbal spa pool using combined chlorine dioxide and sodium hypochlorite treatment. Environmental Monitoring & Assessment, 187(2), 1–11.
2. Cheng Lihua, Li Luyao, Bi Xuejun, etc. 2013. Sodium hypochlorite disinfection effect of the secondary effluent and the impact on water quality [J]. China water supply and drainage, 29 (21): 104–106.
3. Zhen Hua Li. 2011. Urban sewage plant secondary treatment effluent sodium hypochlorite disinfection test study [J]. Water treatment technology, 37 (5) : 110–112.
4. Yan Xu, Fan Jinchu. 2002. Water supply engineering [M]. Beijing: China building industry press.
5. Zhang Huihua. 2012. Sewage treatment plant by using sodium hypochlorite disinfection experiment research [J]. Journal of Chemical Industry in Guangdong. 39(12).
6. Li Jia, Zhu Baiquan, Zhu Ming, etc. 2012. Fold point chlorination reaction in the application of sodium hypochlorite disinfection water [J]. Journal of Environmental Engineering, 6 (11): 4069–4073.
7. Qin Ye. 2014. Sodium hypochlorite disinfection process in the application of water [J]. Water Supply Technology, 8(3): 21–23.
8. Zeng Zhengren. 2008. Water supply pipe network residual chlorine decay law study [D]. Hunan University.
9. Liu Cheng. 2012. Urban reclaimed water pipeline residual chlorine decay rule and influence factors of experiment research [D]. Tianjin University.
10. Gao Peng, Liu Yuan. 2014. Water supply network of changes of residual chlorine decay in the main influence factors analysis [J]. Journal of Building Materials Development Direction, (3): 150–151.
11. Quan Wei. 2006. Disinfection technology in wastewater recycling process of experimental study [D]. Harbin Industrial University.

Analysis of heavy metal pollution prevention, control and management in Nandan County – A case study of Nanfang Non-ferrous Metals Group

Q.C. Deng[1] and X.F. Wang[2,*]

[1]*Crawford School of Public Policy, Australian National University, Canberra 541004, Australia*

[2]*College of Light Industry and Food Engineering, Guangxi University, Nanning 530004, China, Environmental Monitoring Station of Guangxi Zhuang Autonomous Region, Nanning 530028, China*

[*]*Email: wangxiaofei26@163.com*

Nandan is the largest non-ferrous metal mining and smelting area in Guangxi, which suffers from serious heavy metal pollution caused by the mining and smelting activities. This paper makes an in-depth analysis of the current heavy metal pollution of Nandan, with Nanfang Non-ferrous Metals Group as the typical individual case. It exposes the heavy metal pollution, puts forward the preventive, control and management measures for the same, to provide scientific basis for effective pollution control and prevention in the county.

Keywords: Smelting; Heavy Metal Pollution; Control and Prevention; Management.

1. Introduction

With a domestically high proved reserve of Sn, Sb, W, Al, Pb, Zn and other minerals, Guangxi is a national key non-ferrous metal production region and reputed as "Home of Non-ferrous Metal". In Nandan, the largest non-ferrous metal mining and smelting area in Guangxi, due to poor technology, management and other factors, mining and smelting activities have caused serious heavy metal pollution. Improvement of the polluted environment, for the sake of public health, is vital to the sustainable socioeconomic development of Nandan County. Analysis of the prevention, control and management measures against heavy metal pollution is of practical significance. For this purpose, we select Nanfang Non-ferrous Metals Group as the study subject, to provide scientific basis for the heavy metal pollution prevention and control in Nandan County.

* This work is supported by Guangxi Science and Technology Department development project (14125001-2-1) and Guangxi sudden pollution emergency distinguished specialist positions project.

2. Current Heavy Metal Pollution in Nandan

Situated in northwest of Guangxi, Nandan County is a multi-nationality county with rich mineral resources — over 20 non-ferrous metals, e.g., Sn, Sb, Zn, Au, Ag, Cu, Fe, Im, and W etc. are minerable, among which the Sn reserve of over 1.44 million ton ranks the 1st in China, bringing Nandan the honors of "Home of Non-ferrous Metal", "Tin Capital in China" and "Paradise of Mineralogists". Nandan, with an independent system of non-ferrous metal mining, dressing and smelting, turns out the highest yield of tin and zinc and is considered the largest tin and zinc production base in China. In the county, major non-ferrous metal production bases are established in Dachang Town and Chehe Town. In Chehe Town, a town with population of 17,500 and intensive industry of non-ferrous smelting, there are nine large smelting factories and sixteen large mineral dressing factories. While products of smelting factories include zinc calcine, zinc oxide, electrolytic zinc, sulfuric acid, zinc ingot and indium ingot etc., the dressing plants mainly turn out ores of lead, zinc and tin etc. Dachang Town, with a population of 28,300, is situated in a mountainous basin, with the resident houses built in the bottom of basin. Attracted by the rich vein under the town, multiple dressing and smelting enterprises have been established on or near the mountain, and the underground has been connected by crossing tunnels. In the long history of mining and smelting, the slag and waste ores accumulated by years have worsened the environmental pollution in ore mining areas. Especially from the beginning of 1980s to the end of 1990s, large scale of disorderly mining was carried out, tailings and tailing sand were stocked everywhere, which caused serious soil erosion, vegetation deterioration and cultivated land pollution, just name a few. In ensuing years, local government adopted a series of treatment measures, but it was hard to resolve the old problem of environmental pollution — content of heavy metal in surface water, soil, atmosphere and crops was still high or excessively high. In the Guangxi heavy metal pollution prevention and control program, the treatment of heavy metal pollution in Nandan is the priority among priorities.

3. Smelting Plant's Impact on Environmental Pollution — An Individual Case Analysis

Nandan Nanfang Non-ferrous Metals Co., Ltd., situated in Hechi • Nandan Non-ferrous Metal and New Material Industrial Park in Chehe Town, is a comprehensive enterprise integrating smelting and recycling of non-ferrous metal. In 2003 when the company was founded, phase I project of 60,000 t/a electrolytic zinc technical reform was built, and in 2006, required by the national industrial policy, a 125,000 t/a zinc-indium technical reform project was built. In the past 10 years, volatilizing kiln slag treatment project, indium and cadmium recycling procedure, diluted acid purification from sulfuric acid system and system for recycling of heavy metals of arsenic, cadmium, lead and zinc have been developed. It turns out every year 185,000 tons of zinc ingot, 320,000 tons of byproduct sulfuric acid, 60 tons of indium ingot, 1,000 tons of cadmium ingot and 8,000 tons of zinc oxide. In 2012 the total industrial output stood at RMB 2.5 billion Yuan. As a key enterprise for Guangxi non-ferrous industry, cleaner production enterprise and safe production standardization enterprise in the autonomous region, a national pilot unit for recycling economy and the largest of the 9 smelting plants in Nandan that boasts the best performance in environmental protection, Nanfang Non-ferrous Metals Co., Ltd. is ideal for the case study of this paper.

3.1. *Increasing the input to meet the access conditions*

The *Access Conditions for Lead-Zinc Industry* promulgated in 2007 provides specific requirements on the corporate layout, scale, external conditions, process and equipment, energy consumption, integrated use of resources and environmental protection etc. With annual zinc and lead smelting capacities of 185,000 tons and 80,000 tons respectively, Nanfang Non-ferrous Metals Co., Ltd. meet the access conditions of 185,000 tons/a and 80,000 tons/a of zinc and lead smelting scales. Though in the current double-conversion and double-absorption acid production process, technology and equipment applied by most Nandan zinc smelting companies substantially meet the access conditions, and some manufacturers provide integrated resource use and waste heat recovery systems, the existing sintering — blast furnace process for lead smelting is obsolete and the production equipment seriously lags behind those applied in leading companies in China. The fume generated by such process can not be used in acid production due to low content of SO_2, and causes serious pollution to local atmosphere when discharged into air after being desulphurized and dedusted. Oxygen bottom blowing lead melting process or oxygen top blowing lead melting process, and double-conversion and double-absorption or other double-absorption acid production system, which are advanced lead smelting technology currently used in lead smelting companies in China, are not yet applied in Nandan. While Nanfang Non-ferrous Metals Smelting Co., Ltd. provides "single-conversion and single-absorption" acid production + tail gas absorption system for the lead smelting and "double-conversion and double-absorption" acid production + tail gas absorption + waste gas recovery system for the zinc smelting, other lead and zinc smelting companies provide no acid production and waste heat recovery systems.

3.2. *Source control, three wastes treatment*

Wastewater treatment. Attaching importance to controlling the wastewater source, Nanfang Non-ferrous Metals separates the clean water from sewage, and reuses in production the workshop sewage collected in recycle pool and water recycling equipment. Over 94% of process water is reused. The company realizes separation of rainwater and sewage as well. An initial rainwater collection pool with a total storage capacity of 7,500 m^3 has been built, which can hold the rainwater of first 30 minutes of local maximum rainfall with a return period of 20 years and deliver the same to sewage treatment station. Inside the terminal water treatment station with a capacity of 5,000 t/d, "lime-ferric salt two-step and three-stage process" is applied to treat the sewage from acid production, other industrial sewage and initial rainwater. In 2013, a wastewater advanced treatment system was developed, to further treat and purify the production wastewater with processes of membrane filtration and reverse osmosis, so as to facilitate the recycling use of production system.

Waste gas treatment. In Nanfang Non-ferrous Metals, baked fume from zinc smelting fluidized bed boiler is fed into the waste heat boiler to recover the waste heat and collect the dust, passes through cyclone dust collector and electrostatic precipitator for further dust collection, and enters into the two-level dynawave packed tower for scrubbing purification. The demisted and dried fume will enter the double-conversion and double-absorption acid production system, and the exhaust gas, after acid production and undergoing desulfurization and alkaline spraying absorption, will be emitted on standard through the 60m chimney. Fume from volatilizing kiln and multiple hearth furnace will be further treated by the desulfurization system, besides the existing dust collection system, to reduce the sulfur emission. After treated by the "lime double spraying — compressed air oxidization" desulfurization process, the exhaust gas can be emitted on standard.

Waste residue treatment. Standardized storage bin and slag dump are built, to avoid the air storage of raw materials and slag, and renders the environmental management of waste residue to meet the national requirements. Intermediate slag including leaching residue, Co-Ni residue and sewage residue etc., is sent into the volatilizing kiln production line, to allow for recovery of valuable metal and its own transformation into the harmless kiln slag through high temperature solidification of the harmful elements contained. To realize full use of kiln slag and avoid waste residue discharge, a kiln slag recycling and treating system is built, for further recovery of valuable elements from the waste residue and production of iron slag, coal granule and carbonaceous mud. In 2013, the technology of transforming the copper slag into the environment-friendly copper sulfate was put into commercial use, to fully recover the copper slag generated.

3.3. *Intensification of environment management*

(1) Cleaner production inspection, for "energy saving, consumption and emission reduction, and efficiency increase"

Carrying out cleaner production inspection in 2009, Nanfang Non-ferrous Metals managed to substantially reduce the annual emission of SO2, the annual discharge of lead, cadmium and mercury and increased the yearly output of sulfuric acid by 8,400t, drastically raising the recycle and use rate of resources and bringing obvious environmental and economic benefits. Zinc smelteries in Nandan, by applying the hydrometallurgical process, reach the cleaner production standards with fine performance in resource & energy consumption, process application, pollutant emission and integrated use. While the lead smelteries, by producing electrolytic lead through the sintering — blast furnace melting — electrolytic refining process, lag behind the other smelteries in China applying oxygen bottom blowing melting process in performance of resource & energy consumption, technical application, pollutant emission and integrated use, and their cleaner production is generally rated as Level III.

(2) Having a circulatory economy, means realizing "consumption *reduction and recycling use as resources"*

Nanfang Non-ferrous Metals innovates the intensive development mode of such non-ferrous metals as Pb, Zn, Sb and Cu, vigorously promotes the circulatory economy, intensifies the recovery and reuse of resources, enhances the technical content and added value of the products and reduces resource consumption and pollutants emission and discharge, so as to realize the corporate benefit, ecological benefit and social benefit.

Circulatory economy projects such as tailgas desulfurization system, sewage removal and separation project and sewage treatment station with daily capacity of 5,000 m^3 have been built, to reuse the limited resources, reduce the production cost and contain the consumption and waste of social resources.

A circulatory lead, zinc smelting industrial chain coexisting with cleaner production and energy management has been formed inside the system, and is expanded from seven aspects, namely raw material supply, residual heat use, integrated recycling and use of symbiotic, associated metal resources, acid production with smelting fume, sewage treatment and circulatory use, production extension and circulatory use of lead, zinc waste slag, to realize the closed circulatory economy system of object, energy integration.

4. Causes of Heavy Metal Pollution in Smelting Industry of Nandan County

The continuity and development trend of smelting industry of Nandan County is grim, and it's urgent to upgrade and transform the industry.

4.1. *Low production scale and poor facilities*

In Nandan there are now 9 lead and zinc smelteries with a certain production scale, including Nanfang Non-ferrous Metals Smelting Co., Ltd. with zinc productivity of over 100,000 t/a and lead productivity of over 60,000 t/a.

The sintering – blast furnace process that they apply for lead smelting fails to meet the requirement of equipment in the access conditions and falls into the outdated category in Industrial Structure Adjustment Directly 2011, therefore must be obsoleted.

4.2. *Stagnant pollution control process and technology*

The emission of pollutants from lead & zinc smelting in Nandan generally meets the current provisions of *Emission Standard of Pollutants for Lead and Zinc Industry* (GB25466-2010). From 2012, a more stringent revision of this standard was enforced, and those lead & zinc smelteries refusing to improve their pollution control ability would fail to satisfy the new standard in pollutant emission.

(1) Waste gas treatment and emission

Pollutants generated from lead and zinc smelting mainly include SO_2, soot, nitric oxide and heavy metal. Nandan industry generally applies the process of roasting by fluidized furnace – fume acid production, and conventional hydrometallurgy for electrolysis. Typical technical process is as follows: fume of fluidized furnace passes through the electrostatic precipitator for dust collection – dilute acid scrubbing – demisting – drying – transformation etc., and over 99.5% of the SO_2 in the fume can be transformed into SO_3 for acid production, with an absorption rate of over 99.7% and a total use rate of sulfur over 99%, and an SO_2 content in exhaust gas of <400mg/m3 after the tailgas from acid production undergoes the alkaline spraying. Leaching slag from the electrolytic process is delivered back to the rotary kiln for treatment, and after the fume of such rotary kiln undergoes desulfurization and dedusting, the emission content of its SO_2 and that of the acid production tailgas can satisfy the provisions of *Emission Standard of Pollutants for Lead and Zinc Industry* (GB25466-2010).

Limited by the sintering – blast furnace reduction process applied in lead smelting, the SO_2 content in the fume is low and will hamper the acid production. While Nanfang Non-ferrous Metals Smelting produces acid with the "single conversion and single absorption" system, other 8 lead smelteries all emit the sintering and blast fume after desulfurization and dedusting, and when such desulfurization and dedusting system breaks down, excessive SO_2 will be emit and cause serious pollution to the atmosphere. Besides, traditional process is applied for lead smelting, leading to high random emission of dust and seriously polluting the local environment.

(2) Wastewater treatment and discharge

Wastewater generated from lead and zinc smelting in Nandan is mainly heavy metal – bearing acid industrial wastewater. After treated by the chemical neutralization and sedimentation process, some of the wastewater will be recycled and used as supplementary water for cooling and slag flushing, other will be discharged.

Established before 2010, the wastewater treatment stations for lead and zinc smelting all met the provisions on pollutant discharge of Integrated Wastewater Discharge Standard (GB8978-1996). In October 2010, with the promulgation of *Emission Standard of Pollutants for Lead and Zinc Industry* (GB25466-2010), the lead and zinc industry will face more stringent provisions on discharge limit of water pollutants.

In recent years, the environment protection authority has intensified the management of non-point water pollution source, and basically all smelteries have built the initial rainwater collection pools. Nevertheless, some of them still have to improve their facilities and management, for their collections pools are small in volume, collection system is incomplete and early rainwater is not separated from late rainwater etc.

(3) Pending solid waste treatment problems

Most of the waste slag from lead and zinc sintering, namely the rotary kiln slag and furnace slag, can be recycled and sued as cement, raw materials of brickworks etc., and valuable metal can be extracted from waste slag generated from other production processes, such as leaching slag, Cu-Cd residue, anode mud etc.

With defective anti-seepage measures, the heavy metal contained in leachate of stacked smelting waste slag can contaminate the ground water. Such leachat discharged due to poor collection measures, will pollute the surrounding soil and surface water.

(4) Nanfang Non-ferrous Metals Co., Ltd. is situated in the Non-ferrous Metal and New Material Industrial Park in Chehe Town. The narrow layout, structure of the industrial park, while failing to form aggregation effect, plus the poor infrastructure for sewage treatment and solid waste concentrated disposal, will despoil the sustainable development of the whole industrial park.

5. Prevention, Control and Management Measures Against Heavy Metal Pollution in Nandan County

5.1. *Strengthening policy research, improving prevention, control and management measures against heavy metal pollution and promoting the business transformation and upgrading*

Prevention and control measures against heavy metal pollution shall be taken, and local regulations, systems on closing down outdated production facilities and prevention, control of heavy metal pollution be established and improved.

Relevant loan and taxation policies on encouraging the upgrading and reconstruction of smelteries, and removing outdated production facilities shall be formulated, while special fund support for technical modification and pollution control, prevention shall be enhanced.

5.2. *Promoting strategic restructuring, optimizing productivity layout and pushing the industrial transformation and upgrading*

The industry access policy shall be stringently enforced, while the industrial structure and production technology of non-ferrous metal mining, dressing and smelting industry in key areas shall be optimized and upgraded, to promote the coordinated development of such industry. Meanwhile, to reverse the chaotic competition among small and scattered smelteries in Nandan, business reform shall be deepened, big group strategy implemented, integrated operation realized or strategic alliance formed through combination among mining, dressing, smelting and processing companies, internationally competitive large enterprises or enterprise groups established and professional development of SME shall be encouraged.

5.3. *Intensifying cleaner production inspection, developing circulatory economy and improving the clean production and pollution prevention, control ability of the smelting industry*

National industrial policy and environmental protection regulations shall be stringently enforced, new cleaner production process and equipment shall be adopted and natural resources reasonably used, to realize non increase or decrease of pollution along with production increase. All pollutants of industrial source shall be emitted or discharged on national or local standards. Outdated production facilities and business causing heavy pollution shall be resolutely closed down. Companies are encouraged to adopt new technology and equipment, upgrade the technology and carry out cleaner production, and intensify the recycling use of process waste gas, wastewater and waste slag. Industrial structure and layout adjustment in key areas of Nandan shall be accelerated, with circulatory economy concepts guiding the regional development and industrial transformation.

5.4. *Carrying out region-based environment improvement, intensifying park concentration management*

For heavy metal pollution control, the comprehensive prevention, control guideline of "source reduction – process control – terminal treatment" must be upheld, meanwhile the environment management responsibilities shall be assigned to regions and social awareness must be strengthened. The treatment model of closed, isolated "tree wastes treatment" shall be transformed into the modal of comprehensive environment management in serious pollution areas (mining areas or large combined enterprises) under overall social arrangement. Measures recommended include integration of same categories, park concentration management, intensified and concentrated pollutants treatment, construction of regional heavy metal pollution prevention and control facilities and improving the waste residue, wastewater and waste gas treatment technology of the companies concerned.

5.5. *Accelerating the modification of environmental protection facilities, to make sure emission on standard*

Emission Standard of Pollutants for Lead and Zinc Industry (GB25466-2010) was in effect from 2012, requiring that the advanced heavy metal pollution control process and technology be introduced, renovation of sewage, waste gas treatment facilities be fastened, so as to make sure the pollutants are emitted or discharged on standard.

5.6. *Intensifying the treatment of pollution site and soil contamination*

Management of contaminated site and soil environment shall be intensified. The heavy pollution sites shall be more seriously supervised, managed and blocked, so as to avoid the transport and shift of pollutants and the occurrence of pollution accident.

5.7. *Improving the environment supervision and management ability and emergency response ability*

Supervision on business's pollutant discharge and emission must be intensified. Besides the conventional monitoring elements such as SO2 and nitric oxide, the mercury, chromium, cadmium, lead and mettaloid arsenic must be monitored as well. Heavy metal pollution accident warning and emergency system shall be improved, to improve the response to the warning of environment accident. Sound environment warning system must be provided to key enterprises, areas in Nandan and Diaojiang River for the sake of risk prevention.

5.8. *Intensifying the publicity and encouraging the public and media for participation and supervision*

Harms, prevention, control, treatment and protection after heal related to heavy metal can be publicized through such news media as broadcast, TV, newspaper and internet. A public interaction platform can be built, to inform the news media and public more about heavy metal pollution prevention and control and involve them in the supervision.

Acknowledgments

This work was financially supported by Guangxi Science and Technology Department development project (14125001-2-1) and Guangxi sudden pollution emergency distinguished specialist positions project.

References

1. D.J. Cui, Y.L. Zhang. Current Situation of Soil Contamination by Heavy Metals and Research Advances on the Remediation Techniques [J]. Chinese Journal of Soil Science, 2004, 35(3):366–370.
2. K. Deng, Z.G. Hu, C.Q. Guo. The study on mineral resources and sustainable development of Nandan county [J]. Mineral Resources and Geology, 2010, 24(6):552–556.
3. Guangxi Research Academy of Environmental Sciences. Soil Background Value Study Method and Soil Background Value of Guangxi [M]. Nanning: Guangxi Science and Technology Press, 1992.

4. Department of Environmental Protection of Guangxi Zhuang Autonomous Region. Environmental Quality Report of Guangxi Zhuang Autonomous Region 2006–2010 [R].
5. L.L. Han, H.C. Cao, S.J. Dai, S.L. Gao. Heavy metal Pollution and Development of Treatment Technology Research [J]. Non-ferrous mining and metallurgy, 2011, 27(3):94–97.
6. L. Li. Analysis on Pollution of Heavy Metal to Soil and Its Prevention and Treatment of Kaifeng City [J]. Journal of Yellow River Conservancy Technical Institute, 2011, 23(4):27–29.
7. X.H. Liu, X.J. Yu. Study on Characteristics and Countermeasures of Heavy Metal Pollution in Yangtze River Delta [J]. Ecological Environment, 2010, 10(230):164–166.
8. X.H. Miao, L.P. He, H.X. Tao, H.B. Quan. Study on Heavy metal pollution prevention situation and control countermeasure in Gansu province [J]. Gansu Science and Technology, 2010, 26(23):36–38.
9. F.S. Wei, J.S. Chen, Y.Y. Wu, C.J. Zheng. Study on the Background Contents of Soils in China [J]. Chinese Journal of Environmental Science, 1991, 12(4):12–19.
10. Y.H. Zhang, Z.Y. Wen. Study of Policy for Environmental Protection in Mining Areas [J]. China Mining Magazine 2002, 11(6):57–60.
11. Y.M. Zhao, Z.K. Li. Study on the control countermeasures in soil heavy metal pollution of Industrial zone [J]. Inner Mongolia Science Technology and Economy, 2001, 3:33–34.

Impacts of pollution control technologies on employment: Evidence from China*

W.J. Yan†
School of Economics and Management, Xi'an Technological University, Xi'an, 710032, China
†E-mail: mayywj0528@163.com

Y. Xiong
School of Economics and International Trade, Zhejiang University of Finance & Economics, Hangzhou, 310018, China

Pollution control technologies include the clean production technologies and end-of-pipe technologies. There has been much debate on the economic and ecological implications of the adoption of clean production technologies and end-of-pipe technologies. A wide range of research efforts have also been devoted to address the connection between pollution control technology and employment in developed countries. Nevertheless, the results are complex, and the employment issues of clean technologies and end-of-pipe technologies have not been on the research agenda for a long time in China. In this paper, an empirical test about the effect of clean production technologies and end-of-pipe technologies on employment is made by using SYS-GMM method based on the sample of 26 provinces and three centrally-controlled municipalities over a ten-year period from 2003 to 2012. The regression results indicate that end-of-pipe technologies have positive impact on the employment, while cleaner production technologies have negative impact on the employment, which is consistent with theoretical analysis. Since cleaner production technologies are major pollution control technologies at present stage and in the future, our government should give full consideration to the employment effects of cleaner production technologies in the promotion of such technologies.

Keywords: Clean production technologies; End-of-pipe technologies; Employment effects; SYS-GMM method.

1. Introduction

Pollution control technologies generally include clean production technologies and end-of-pipe technologies. End-of-pipe treatment is, by definition, not pollution prevention which scarcely affects production processes or the organization of the corporations operating those processes, as end-of-pipe

---

* This work is supported by the Major Project of National Social Science Fund of China (grant no. 13 & Zd158).

technologies are always appended to existing processes. With the end-of-pipe technologies environmental problems are perceived as being separate from the production process. However, such technologies do not address the root of the environmental problem. Therefore, where possible it seems preferable to change the production process in order to avoid generating unwanted byproducts. A change in the production process that reduces emissions is often referred to as pollution prevention or clean production. In such cases, addressing the environmental problem is integrated within the production process itself. Clean production technologies take environmental considerations in the entire production process into account. By 2014, the total discharge of major pollutants show a decline trend in China. Compared with the pollutants of “eleventh five-year plan”, the chemical oxygen demand (cod) emissions fell by 8.21% and sulfur dioxide emissions by 10.98% in 2014. In recent years, there has been much debate on the economic and ecological implications of the adoption of clean production technologies as alternatives to end-of-pipe technologies. While the ecological impacts of clean production technologies and end-of-pipe technologies are generally well known, the impact of clean production technologies and end-of-pipe technologies on employment is indeterminate from a theoretical point of view and also a question which needs to be answered empirically. The purpose of this paper is to investigate the quantitative impact of clean production technologies and end-of-pipe technologies on employment from the perspective of China sample.

Over the past few years, the employment implications of the adoption of clean production technologies as alternatives to end-of-pipe has been analyzed empirically in several studies over the past years by foreign scholars. Early research on the electric power industry by Gollup and Roberts (1983) [1] found significant job loss associated with increased environmental regulations. While more recent work find no evidence that environmental regulation decreased labor demand (See e.g. Tietenberg, 1992, Goodstein, 1995, 1996, Morgenstern and Pizer, 2002, Axel Marx, 2000, Berman and Bui, 2001, Pfeiffer and Rennings, 2001, Getzner, 2002, Bezdek, Wendling and Diperna 2008, ILO, 2009) [2–11]. Within the empirical literature on employment impacts of environmental innovations, most studies have specifically analyzed environmental innovations which are employment beneficial. Goodstein (1995) [3] analyzed how many people were laid-off and how many companies had to close down owing to environmental regulation between1987 and 1990 based on a database in the USA which was set up to register the main reasons for firing people. Morgenstern and Pizer (1999) [5] researched the impact of environmental spending on industry-level employment based on four heavily polluting industries: pulp and paper mills, plastic manufacturers, petroleum refiners, and iron and steel mills. Axel Marx (2000) [6] used a survey in EU to measure how many companies had to fire people and/or close down a unit because of environmental policy. Berman and Bui (2001) [7] analyzed the data from the South Coast Air Basin comparing petroleum refineries in the Los Angeles area to all other U.S refineries. Pfeiffer and Rennings [8] (2001) conducted an industry survey of environmentally oriented firms in Germany. Bezdek [10] analyzed data from national and regional. All these studies found that employment impacts of environmental innovations were positive. These results have been confirmed by an international study Getzner [9] and ILO [11]. Getzner made a survey of companies in five European countries and concluded that the quantitative impact on employment in companies adopting clean technologies is slightly positive, and the main effects of clean technologies lie in the qualitative sphere. On the one hand, there are significant improvements in qualifications and physical elements of the working environment. On the other hand, changes in the working organization are also significant. Moreover, a novel analytical tool was developed by International Labor Organization (ILO) which found that the moving to a low-carbon economy also presents major employment opportunities, by opening new markets, and by stimulating eco-innovation and investment in more efficient production techniques.

In China, however, most of the research in this field has been limited to the economic implications and ecological implications of cleaner production and end-of-pipe technologies (See e.g. Gao, Qin, Wang, Sun, and Zhang) [12–16]. Gao and Tong analyzed the environmental performance of cleaner production and end-of-pipe treatment based on the sample of Jilin Province [12]. Qin, Zhao and Wan investigated the relationship between economic implications and ecological implications of cleaner production technologies using the data of China metal products industry survey in 2009 as the empirical basis [13]. Moreover, Wang and Yu, Sun, and Zhang have studied the current situation, countermeasure and trend of clean production technologies and end-of -pipe technologies [14–16]. To the best of our knowledge, the employment effect of cleaner production and end-of-pipe technologies have been neglected for a long time in China.

Generally speaking, adoption of clean production technologies as alternatives to end-of-pipe are labor saving or labor enhancing, the answer to such problem is not uniform. On the one hand, the adoption of clean technologies might rationalize the production process as resource and energy inputs are reduced and waste is minimized. However, as the production function of many companies suggests, the reduction in material inputs also reduces labor input. On the other hand, more cost-efficient production processes enhance the competitive position of the company in the market. That includes the chance to increase the company's output which in turn drives up labor demand [17–18].

This paper is organized as follows: in Section 2, we give a brief introduction of the data and empirical methods; in Section 3, we present the primary results and robustness tests; and in Section 4, we reach the conclusion.

2. Methods and Data

The purpose of this paper is to evaluate the effectiveness of clean production technologies and end-of-pipe technologies two core variables on employment in China. In addition, we introduce five control variables. Considering the dependent variable is influenced by some independent variables as well as dependent variables are lagged one year, we rely on a ten-year (2003 to 2012) sample and a SYS-GMM regression model is employed as the modeling strategy. The mathematical formula is:

$$\begin{aligned} Empl_{it} &= \alpha Empl_{it-1} + \beta_1 Mdzl_{it-1} + \beta_2 \ln K + \beta_3 ldscl \\ &\quad + \beta_4 gdp + \beta_5 wrqd + \beta_6 gybz + v_i + \mu_{it}, \end{aligned} \tag{1}$$

$$\begin{aligned} Empl_{it} &= \alpha Empl_{it-1} + \beta_1 Qjsc_{it-1} + \beta_2 \ln K + \beta_3 ldscl \\ &\quad + \beta_4 gdp + \beta_5 wrqd + \beta_6 gybz + v_i + \mu_{it}. \end{aligned} \tag{2}$$

Where i denotes three direct-controlled municipalities and 26 provinces, t denotes time (2003–2012), μ_{it} is the error term, v_i is the intercept for each entity, β and α are parameters to be fitted. $Empl_{it}$ is the dependent variable which represents the number of employees, namely average number of employees in ten thousand, and $Empl_{it\text{-}1}$ represents the dependent variable lagged one period. Unit is million people. These two core variables are clean production technologies and end-of-pipe technologies, respectively, which are important technologies of pollution control. Based on the method of Chen and Li [19], the end-of-pipe technologies are represented by the pollution discharge rate, namely the ratio of pollution emissions and pollution generated, and clean production technologies are represented by pollution generated rate, namely pollution produced by per unit of output. The sulfur dioxide is used as the pollution representative when calculating clean production technologies and end-of-pipe technologies these two indicators, while the chemical oxygen demand is used as the pollution representative when make robustness test. In order to make the estimation more accurate and consider the other factors that affect the employment comprehensively, the following control variables are added. Capital stock is represented by net value of fixed assets and unit is billion RMB. Labor productivity is represented by the ratio of industrial output value and employees of up-scale enterprises. Many foreign scholars have also studied the relationship between labor productivity and employment.

The scale of economies, namely GDP, and unit is billion RMB. Nationalization features, which is represented by the ratio of total output value of state owned industry and output value of up-scale industrial. Industrial structure is represented by the tertiary industry output value divided by the GDP. The effect of international trade is represented respectively by the "foreign trade dependency" (the total import plus export divided by GDP) and "foreign investment dependency" (the ratio of provincial foreign direct investment and national foreign direct investment).

The data of foreign investment dependency come from Chinese economy database, and the data of foreign trade dependency come from China Statistical Yearbook. We collected the basic data from websites of environmental database in China Statistical Year book to calculate clean production technologies and end-of-pipe technologies. The data of fixed assets, the average employment, labor productivity, and state-owned share come from China Industrial Economic Statistical Yearbook, except data in 2004, the data in 2004 come from China Economic Census in 2004 Yearbook. The basic data of the scale of economies and industrial structure come from China Statistical Year book.

Based on a panel data regression analysis on a sample of 26 provinces and three centrally-controlled municipalities over a ten-year period from 2003 to 2012 (Excluding Tibet and Hainan), we find that empirical support for the positive employment impacts of clean production technologies and negative employment impacts of end-of-pipe technologies.

3. Estimation Results

3.1. *Primary regression*

According to model setting, we use stata10.0 to run the OLS regression. Estimation results are shown in Table 1 which shows that most estimated parameters are statistically significant at convention levels. Model 1 and model 3 are primary regression results, in which independent variables only include dependent variable lagged one period, clean production technologies, and end-of-pipe technologies. Model 2 and model 4 introduce control variables, and the regression results of the four models show that number of employees lagged one period is positively related to the number of employees, which is a process of accumulation and adjustment.

Table 1. Primary regression result.

Variables	Model 1	Model 2	Model 3	Model 4
L.ljyry	1.002***	0.611***	1.006***	0.614***
	(0.00138)	(0.0480)	(0.00602)	(0.0426)
lmdzllag	0.0235***	0.0697***		
	(0.00702)	(0.0126)		
lqjsclag			-0.136***	-0.0426***
			(0.00431)	(0.0148)
fdi		0.505***		0.367***
		(0.118)		(0.127)
lgdzc		0.109***		0.183***
		(0.0368)		(0.0321)
jck		0.100***		0.115***
		(0.0180)		(0.0229)
ldscl		-0.00481***		-0.00444***
		(0.000603)		(0.000656)
gybzl		-0.213**		-0.140
		(0.0860)		(0.0932)
lgdp		0.274***		0.221***
		(0.0325)		(0.0367)
scgdp		-0.756***		-0.871***
		(0.123)		(0.176)
Constant	0.0381***	-0.696***	0.478***	-0.680***
	(0.00924)	(0.150)	(0.0199)	(0.241)
Arellano-Bond test for AR(1)	-2.7879	-2.1341	-2.6527	-2.1133
	(0.0053)	(0.0328)	(0.008)	(0.0346)
Arellano-Bond test for AR(2)	1.0265	0.40758	1.0562	0.21928
	(0.3047)	(0.6836)	(0.2909)	(0.8264)
Sargan test	0.3730	0.4799	0.3384	0.4916

Note: Table notes: *, ** and *** indicate statistical significance at 10%, 5% and 1% levels, respectively.

In Model 1 and model 2, end-of-pipe technologies lagged one year to allow enough time for such technologies to take effect. Regression coefficient of end-of-pipe technologies lagged one period is positive and significant under 1% significance level, which indicates that end-of-pipe technologies have a positive influence on the number of employees. End-of-pipe technologies are not discouraged due to their environmental effect, but any technologies would also remain a bit of pollution even if the production technologies which were considered relatively advanced were adopted, so the end-of-pipe technologies should be used anyway.

In Model 3 and model 4, clean production technologies lagged one year to allow enough time for such technologies to take effect. Regression coefficient of clean production technologies lagged one period is negative and significant under 1% significance level, which indicates that clean production technologies have negative influence on the number of employees. The results are consistent with the reality and the existing theoretical analysis. Environmental effect of clean production technology was widely recognized and appreciated, but the use of clean production technology could squeeze out of the labor force demand because the clean production technologies have a substitution effect. Our empirical test results are that the clean production technologies are not conducive to social employment according to sample in China.

A main cause of high number of unemployment and layoffs is that some important industry or key areas, such as banking, telecommunications, petrochemical and other large state-owned enterprises in the field of monopoly, have begun to lay off massive workers. Foreign investment dependency and foreign trade dependency have significantly positive impact on employment which is consistent with the reality of the number of practitioners in foreign enterprises has increased year by year from 1998.

In the journal of world investment report in 1994, the United Nations Conference on Trade and Development believed that foreign direct investment in the host country's production and business activities can not only create the employment opportunities directly, but also create employment opportunities indirectly. The ratio of the tertiary industry and GDP has a negative impact on employment, which is contrary to our theoretical expectations. One possible explanation is that China's tertiary industry development is weak and the creation of more employment opportunities are limited, so the main power of absorbing employment are related with manufacturing industry and other sectors in the secondary industry.

3.2. *Robustness tests*

Instead of sulfur dioxide, chemical oxygen demand (cod) is used to compute end-of-pipe technology and clean production technologies these two indicators. The regression coefficient of end-of-pipe technology is positive and significant under 1% significance level when introducing control variables. The regression coefficient of clean production technologies is negative and significant under 1% significance level. The description of regression coefficient and the significance of the control variables are the same as which in Table 1. Robustness tests results is omitted.

4. Conclusions

Creating jobs and pollution control are two urgent problems to be solved in China. Can pollution control and employment be reconciled? This paper concluded that the end-of-pipe technologies are conducive to employment creation and clean production technologies have a weakening effect on employment based on the sample of 26 provinces and three centrally-controlled municipalities over a ten- year period from 2003 to 2012. The clean production technologies as main pollution control technology will be employed in the future in China. However, if a technology can only produce good environmental effect, which caused massive unemployment, the applicability of the policy is worth considering. We believe that the status quo that crowding out of jobs by cleaner production technology at present stage will reverse with the transformation of China's industrial structure and the way of economic growth. It is worth noting that, although the test of our results indicated that the cleaner production technology is not conducive to employment, the cleaner production technology is the first choice for prevention and control of environmental pollution. The influence of the technology on employment is also affected by the national macroeconomic policy, and there is no one-to-one relationship between pollution control technology and employment. There was a big difference between China's industrial structure and which in developed countries. China's long-term economic development depends on high energy consumption and high pollution industry, while the tertiary industry accounted for a large proportion of the national economy in developed countries. With the development of China's pollution control technology and the environmental protection industry, we believe that the cleaner production technology will produce a positive effect on employment.

References

1. F.M. Gollup and J.R. Mark, Environmental regulations and productivity growth: The case of fossil-fueled electric power generation, *J. Polit. Econ.* 91, 654(1983).
2. T. Tietenberg, *Environmental and Natural Resource Economics*, (Harper Collins, New York, 1992).
3. E. Goodstein, Jobs or the environment? No trade-off. *Challenge* 01, 41 (1995).
4. E.B. Goodstein, Jobs and the environment: An overview. *Environ. Manage.* 20, 313(1996).

5. R.D. Morgenstern, A.M. Pizer, and J.S. Shih, Jobs versus the environment: An industry-level perspective, *J. Environ. Econ. Manage.* 43, 412 (2002).
6. A. Marx, Ecological modernization, environmental policy and employment: Can environmental protection and employment be reconciled?, *Eur. J. Soc. Sci.* 13, 311–325(2000).
7. E. Berman and L.T.M. Bui, Environmental regulation and labor demand: Evidence from the south coast air basin, *J. Public Econ.* 79, 265 (2001).
8. F. Pfeiffer and K. Rennings, Employment impacts of cleaner production-evidence from a German study using case studies and surveys, *Bus. Strateg. Environ.* 10, 161(2001).
9. M. Getzner, The quantitative and qualitative impacts of clean technologies on employment, *J. Clean. Prod.* 10, 305(2002).
10. R.H. Bezdek, R.M. Wendling, and P. Diperna, Environmental protection, the economy, and jobs: National and regional analyses, *J. Environ. Econ. Manage.* 86, 63(2008).
11. ILO. Green Policies and Jobs: A Double Dividend? in ILO (ed.), the Global Jobs Crisis and Beyond, World of Work Report, 2009, Geneva 22, Switzerland.
12. Y.C. Gao, L.J. Tong, Y.J. Ma, *et al.*, Environmental performance analysis of clean production and end-pipe treatment, *Geogr. Res.* 30, 505(2011).
13. P.H. Qin, L.X. Zhao, and J.B. Wan, The impact of cleaner production techniques on manufacturing firm economic and environmental performance: Evidence from Chinese metal manufacturing survey in 2009, *Ecol. Econ.* 30, 49(2014) (in Chinese).
14. L.P. Wang and X.L. Yu, The development of clean production and end treatment. *China Popul., Resour. Environ.* 3, 428(2010) (in Chinese).
15. X.F. Sun, J. Li, and X.P. Li, The study of status and development trend of cleaner production in china, *Environ. Sci. Manage.* 11, 185(2010) (in Chinese).
16. Y.H. Zhang, Present situation countermeasures of cleaner production implementation in China, *Pollut. Control Technol.* 3, 75(2006) (in Chinese).
17. Y. Lu, Green policies and jobs in china: A double dividend? *Econ. Res. J.* 7, 42 (2011) (in Chinese).
18. B. Matteo, Employment impacts of cleaner production: Theory, methodology and results, *ZEW Econ. Stu.* 21, 3(2003).
19. Y.Y. Chen, K.W. Li, Factor decomposition and influencing factors of industrial sulfur dioxide pollution density in China, *Manage. World*, 3, 14(2010) (in Chinese).

Environmental protection in clothing industry

D.M. Wang
Fashion School of Dalian Polytechnic University,
#1 Qinggongyuan, Ganjingzi, District, Dalian Liaoning China
Email: wangdongmeidl@sina.com

Twenty-first century is a time with rapid development of science and technology, and with progress and prosperity of the society. But it is also a time with all kinds of problems, such as the rapid increase in population, a large number of energy consumption, forest reduction, greenhouse effect, pollutants emissions, and the reduction of biological diversity. The protection of the environment has become one of the most important issues of human beings in twenty-first century. China is the largest textile and clothing producing and exporting country in the world. The production of clothing is a complicated process, which involves many steps and adds a lot of pollution to the environment.

Keywords: Environmental protection; Pollution; Textile industry; Clothing production and processing.

1. Introduction

Twenty-first century is a time with rapid development of science and technology, and with progress and prosperity of the society. But it is also a time with all kinds of problems, such as the rapid increase in population, a large number of energy consumption, forest reduction, greenhouse effect, pollutants emissions, and the reduction of biological diversity. The protection of the environment has become one of the most important issues of human beings in twenty-first century.

Textile is an industry with high labor intensity and large external dependence in China. China is the largest textile and clothing producing and exporting country in the world. The downstream industries of textile include fashion industry, home textiles and industrial textiles, etc. The production of clothing is a complicated process, which involves many steps, such as the plantation and reserve of materials, the processing and dyeing, the production of clothing and the transport of finished products. When clothing is presented in the consumption process, people can only see the beautiful and comfortable side of it. But actually the production of clothing, even including some famous clothing brands, causes environmental pollution in its productive chains.

Meanwhile, parts of the pollution remains in the clothes, which will cause harm to human body.

Figure 1. Cotton (Natural Fiber).

Figure 2. Hemp Fiber (Natural Fiber).

Figure 3. Acrylic Fiber (Chemical Fiber).

2. Analysis of Clothing Production

Clothing production requires a variety of fabrics and accessories. (Figure 1, Figure 2)The raw materials of clothing can be divided into natural fiber and chemical fiber. Natural fiber refers to the textile fiber which comes from the plants or animals either natural or artificially cultivated. It is an important material source of textile industry. Since ancient times, human life has been closely related with the fiber. The history of human beings' weaving and spinning with natural fiber can be traced back to 3000 years ago. In China, Egypt, South Africa's early cultures, there are records of the use of natural fiber in spinning and weaving. Chemical fiber (Figure 3) is a kind of fiber that human artificially make by using a variety of different raw materials, such as coal, oil and so on through chemical treatment..

Fiber is the most basic one of clothing materials. Its performance and appearance will directly affect the wearing, storage and processing performance of clothing.

2.1. *Fiber*

In human survival and developing process, natural fiber is the earliest clothing fiber that human being discovered and recognized, which has a history of thousands of years.The natural fiber existed in nature mainly includes cotton, hemp, silk and animal fur. Of which, the main molecular composition of cotton and hemp fiber is cellulose and the main molecular composition of silk and wool is protein (existed in the form of polyamide polymer). The cotton fiber has the largest production and can be used in a very wide scope. Hemp fiber of good quality can be used for clothes. The production of wool and silk is much lesser than cotton and hemp, but they are excellent textile raw materials. Wool fiber can be made into wool fabric and silk fiber can be made into silk goods. When they are made into clothes, they are magnificent and solemn, which are beloved by the people. With the development of human society, the family of natural fiber is also expanding. The natural fiber is widely loved by the people because its good absorption, comfort and environmental friendliness. Although the natural fiber is more environment friendly than chemical fiber, it also causes some pollution in the planting process by using the pesticides.

Chemical fiber enriches the clothing fabrics because of its good shape-maintenance and flexibility, and its easy-washing design and diversified species. According to its different materials, chemical fiber can be divided into regenerated fiber, semi-synthetic fiber, synthetic fiber, inorganic chemical fiber. The raw materials used in the regenerated fibers are cellulose and protein in nature. Some fibers are made of the derivatives of natural polymer through chemical treatment, such as acetate fiber. The synthetic fiber is made of coal, petroleum, natural gas through polymerization and processing. So the synthetic fiber has the meaning of "poly", such as acrylic fiber, polyester fiber. The use of solvents and medium in the producing process of chemical fiber will cause pollution to the environment.

2.2. *Analysis of pollution in productive process*

Of all the waste water discharged by the textile industries, including cotton, wool, silk, hemp, knitting, dyeing and other productive processes, the maximum one is the waste water discharged by dyeing and printing industry. China is the world's most important textile exporting country, and also the most important fiber consumption country. 70%–80% of China's artificial fiber are consumed by domestic uses. China Environmental Status Bulletin of 2001 shows that 70% of the rivers, lakes and reservoirs in China have been polluted in different degrees. 20% pollutants are industrial waste water. As an important industry in

China's economic development, the textile industry consumes a large number of chemicals, which are durable and hazardous. The textile industry has become a major source of water pollution (Figure 4).

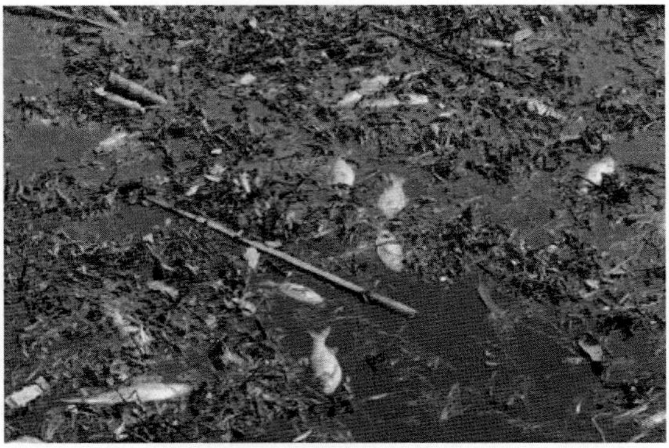

Figure 4. Waste water discharged by printing enterprises.

Textile waste water mainly includes printing and dyeing waste water, chemical fiber production waste water, wool scouring waste water, among which printing and dyeing waste water is the main pollution source. According to incomplete statistics, every day domestic printing and dyeing enterprises discharge about 3000–4000 thousand ton waste water. Printing and dyeing factory process every 100m fabric, the production of waste water is 3–5t. The traditional printing and dyeing process not only produces a large number of poisonous waste water, but some poisonous substances will also be attached to the fabric, such as formaldehyde, fluorescent whitening agent, soft agent, which will affect body health.

An environmental protection organization recently reported that, clothing suppliers including a number of well-known brand discharged environment hormone to the Chinese rivers. This survey has conducted more than 1 year, and all the data have been through sampling investigation. This environmental protection organization has surveyed the waste water discharged by the textile industry city of Ningbo, Zhejiang Province and a dyeing enterprise which is located in Zhongshan city, Guangdong Province several times. And the samples were sent to the Green Peace Research Laboratory of the Exeter University and the Omegam independent environmental analysis laboratory in Holland for testing. The results are that the main toxic and harmful substances in the waste water are for nonylphenol (NP), perfluorooctane sulfonate (PFOS) and perfluorooctanoic acid (PFOA). These substances interfere with the internal secretion of the organism and are poisonous to the reproductive system, and have an effect on the immune system and liver. Studies have indicated that these substances, even though they are rare, are highly hazardous.

Since the establishment of the textile industry, there have been a wide range of pollution problems. Recently, the use of the toxic and harmful substances in textile industry has become a serious threat to the ecological system and human health. Environmental protection department pointed that in 2010 the total amount of textile waste water is 11.6% of the all industrial waste water. Dyeing and printing process is the key part of the textile industry chain, which also has the highest technology in the textile industry chain. But it is also the part of the largest pollution. The emissions of waste water account for 80% of the total emissions of textile industry. It is the focus of the environmental protection in textile industry.

3. The Current Situation of Environmental Protection in the Clothing Industry

3.1. *Textile industry's emphasis on environmental protection*

First of all, Chinese Environmental Protection Department has paid attention to the problem of pollution in the textile industry. The ecological fashion forum of 2014 was held in Beijing. The experts and scholars called attention to China's textile industry pollution problems and hoped we can develop a more environment friendly textile industry. Currently, China's textile industry pollution problem has reached a point of urgency. "Energy-saving and emission-reduction" is still the major direction for the textile industry to change its development in the next period of time. It is also an important part of the ecological development of textile industry.

The implementation of the Water Pollutant Discharge Standards of Textile Dyeing and Printing Industry in January 1, 2013 triggered a shake. China Textile Industry Association found that a large number of enterprises can not meet the standard's requirements. But the developing trend of the textile industry is producing with concept of environmental protection. The textile industry not only needs to develop, but also needs to protect the environment. Constructing the printing and dyeing waste water treatment facilities, researching and developing the technology to reduce the resource consumption and pollution, are the focus of our future work.

3.2. *The design concept of green clothing*

Nowadays people are attaching more and more importance on the concept of environmental protection, so they are also demanding more on clothing's environmental protection. Green clothing refers to the clothing used natural fibers which don't pollute the land while planting as its raw materials. Its printing and dyeing process used the chemicals and coloring materials which are healthy and strictly control the harmful substances such as formaldehyde and halogenated dyeing carrier. Green clothing completely eradicates the use of the 22 kinds of carcinogen, the corresponding 100 kinds of fuel additives and more than 10 kinds of harmful heavy metals.

The concept of green design began in the 1980's. In February 1997 the Germany Dusseldorf exhibition for the first time displayed the green clothing, thus made green, leisure and healthy a world language. Ecological clothing use natural plants or animals as its raw materials. It not only reflects the environmental protection concept in its dyeing and printing process, but the accessories like buttons, zippers also adopt pollution-free way to produce. From the production of the raw materials to its processing, the green clothing totally starts from the concept of environmental protection and avoids using the environment-harmful substances such as chemicals and resin. Colored cotton refers to a new kind of cotton which is cultivated through modern high-tech. In its opening stage, it has the natural color. Its color is simple and elegant, which doesn't need chemical bleaching. Colored cotton is no pollution to the environment, thus it is recognized as the green product in the world. Colored cotton is widely used in children's clothing and underwear. (Figure 5, Figure 6)

Figure 5. Colored cotton.

Figure 6. Clothing made by colored cotton.

Environmental protection is the voice of the time. The development and progress of the society should not cost the environmental damage as the expense. The rapid development of the society has made people realized the importance of environmental protection. If the environment is polluted, it is human being who get hurt. In such a situation, green clothing is becoming the developing trend of clothing industry.

Green textiles generally has the functions of deodorization, antibacterial, anti-inflammatory, anti-UV, anti-radiation, anti-itching, humidification, etc. Nowadays, China has begun to develop such products, mainly in the production of underwear. This kind of textile is good for human health, thus it is popular with consumers.

Environmental protection is the issue that we should pay attention in this era. We can develop our society only when we can protect our mutual home. The development of textile industry should also focus on the sustainable development, thus creating a good development trend.

References

1. http://baike.haosou.com/doc/4933393-5153602.html
2. Chen Dongsheng, Gan Yingjin, *Fabric for Fashion design.* (China Light Industry Press)
3. http://photo.39.net/Release/A/CT100045/CT1500000459/2011-08-24/TK1500017144.html#p=PT1500017164
4. http://baike.haosou.com/doc/4594724-4806361.html
5. http://www.tnc.com.cn/info/c-001004-d-3528635.html
6. http://www.chinairn.com/news/20140410/160411950.shtml
7. http://www.d1ep.com/news/show.php?itemid=7574

Ammonium removal from drinking water by *Acinetobacter* sp. SFA 10 at low temperature*

D.Y. Zhang†, W.G. Li, X.J. Gong and P. Wang

School of Municipal and Environmental Engineering, Harbin Institute of Technology, Harbin, China

†E-mail: duo0314@163.com

www.hit.edu.cn

Acinetobacter sp. SFA 10 was isolated from river water and found to remove ammonium at 5 °C. The ammonium removal rate was 2.05 mg NH_4^+-N/l/h when the initial ammonium concentration was about 5 mg/l. When the initial ammonium concentration was higher than 45 mg/L, the ammonium removal rate of Acinetobacter sp. SFA 10 decreased. With the ammonium removing by Acinetobacter sp. SFA 10, NH_2OH was produced without nitrite or nitrate accumulation. Ammonia monooxygenase (AMO) and hydroxylamine oxidase (HAO) were the key enzyme for nitrification. In *Acinetobacter* sp. SFA 10 cells, *amo* and*hao* genes were amplified and about 600bp. Nitrate reductase gene (*nap*A), nitrite and nitrate reductase enzymes were not detected in Acinetobacter sp. SFA 10. The results in this study showed that Acinetobacter sp. SFA 10 oxidized ammonium to NH_2OH, then convert NH_2OH to nitrate. Acinetobacter sp. SFA 10 didn't have the ability of denitrification.

Keywords: Ammonium; Drinking water; Low temperature; Heterotrophic nitrification; Enzyme.

1. Introduction

Ammonium in raw surface water used for drinking water production is undesirable, because it might cause taste and odor problem, as well as decrease disinfection efficiency. While ammonium pollution in surface water in China was common. 28.3% of surface water was seriously polluted which could not be used as drinking water resource (The Report of Ministry of Environmental Protection of the P. R. China, http://datacenter.mep.gov.cn/). The ammonium was one of major pollutants. Songhua River was the main drinking water source for citizens of Harbin, China. In 2014, the highest ammonium concentration was above 2 mg/l. For guaranteeing drinking water safety, the Ministry of Health of

* This work is by Heilongjiang Provincial Science Foundation for Distinguished Youth Scholar (Grant No. JC200708).

† Work partially supported by grant 2014M551250 of China Postdoctoral Science Foundation.

the P. R. China has regulated ammonium below 0.5 mg/l in Standards for Drinking Water Quality (GB5749-2006).

Break-point chlorination is a classical process used for ammonium removal. However, the increase of chlorine content in water could stimulate the formation of undesirable chlorinated by-product. Heterotrophic nitrification is the process that ammonium is oxidized to NO_2^--N or NO_3^- -N by heterotrophic microbes [1–3]. Till now, *Alcaligenesfaecalis*, *Acinetobacter* sp., *Bacillus* sp., *Diaphorobacter*sp., *Microbacterium* sp., *Providenciarettgeri*, *Pseudomonas stutzeri* and *Rhodococcus* sp. were found to remove ammonium via heterotrophic nitrification pathway [4–13]. Some of these microbes can simultaneously covert NO_2^--N or NO_3^- -N to N_2O and/or N_2. *Microbacterium* sp. was found to perform autotrophic nitrification with the ratio of 0.11 mg NH_4^+-N/l/h at 5 °C [10].

For heterotrophic nitrification, ammonia monooxygenase (AMO) and hydroxylamine oxidase (HAO) are two important enzymes. AMO catalyzes the oxidation of NH_4^+-N to NH_2OH and HAO catalyzes the oxidation of NH_2OH to NO_2^--N. The membrane-associated AMO was purified from *Paracoccuspantotropha*. The enzyme consists of two subunits with molecular masses of 38 and 46 kDa. The purified enzyme is a quinol oxidase, is inhibited by light and a variety of chelating agents and is activated by cupric ions [14]. The periplasmic HAO in *Paracoccuspantotropha* was a soluble 20 kDa monomer. This enzyme can use the periplasmic redox proteins, cytochrome c551 and pseudoazurin as electron acceptors, both of which can also donate electrons to denitrification enzymes. *amo* and *hao* gene encoding AMO and HAO enzymes were also widely used to study the heterotrophic nitrifying bacteria. Most of heterotrophic nitrifying bacteria could simultaneously reduce the nitrate to N_2 or N_2O via denitrification under aerobic condition. The nitrite reductase and nitrate reductase play important role in this process.

Acinetobacter sp. SFA 10 was isolated from river water. The nitrification ability of the strain SFA 10 was studied. The effect of the temperature on the ammonium removal efficiency was investigated. The ammonium removal pathway of the strain SFA 10 was also discussed in the study. The results of the study could supply a novel solution for ammonium removal from drinking water.

2. Materials and Methods

2.1. *Isolation and identification of heterotrophic nitrifying bacteria*

The basic medium used in this study contained 0.5g/l NH_4Cl, 1.0g/l CH_3COONa, 0.05g/l $MgSO_4{\cdot}7H_2O$, 0.2g/l K_2HPO_4, 0.12g/l NaCl, 0.01g/l $MnSO_4{\cdot}4H_2O$, 0.01g/l $FeSO_4$, pH7.0. The isolation process was following the method described by Zhang et al. [10]. The sequences of the 16S rDNA fragments were extracted and sequenced by TAKARA biotechnology (Dalian) Co. Ltd. (Dalian, China). The 16S rDNA sequences were compared with that of other microorganisms by way of BLAST (http://blast.ncbi.nlm.nih.gov/Blast.cgi).

2.2. *Assessment of heterotrophic nitrification characteristics*

The isolated bacterium was inoculated into the basic medium and cultured at 10 °C. After cultivating for 0.5, 2, 4, 6, 8, 10, 15, 20, 25 and 30 hours, the liquid samples were collected respectively to detected total nitrogen (TN), NH_4^+-N, NH_2OH, NO_2^--N, NO_3^- -N and TOC concentration. The bacterial contents were determined by measuring OD600.

2.3. *Assessment of ammonium removal abilities*

The isolated bacterium was inoculated into the basic medium and cultured at 10 °C for 24–48 hours when the population amount reached to more than 10^7 cells/ml. One liter liquid culture was collected and centrifuged at 6000 rpm for 5 min. The sedimentation were collected and washed with sterile pure water for 3 times. Then the collected bacteria were inoculated to triplicate 50 ml sterile basic medium. The ammonium concentration in the basic medium was adjusted to 0.5, 1. 2. 5, 10, 20, 45, 70, 100 mg/l. The basic mediums without inoculation were used as control. After incubated for 30 min, the ammonium removal rate were detected. The calculation equation was Rammonium=$(S_1-S_0)/t$, where $S_1$was the ammonium concentration detected after 30 min incubation, S_0 was the initial ammonium concentration at the start of experiment, t was the incubation time (0.5 hours).

2.4. *Effect of temperature on the ammonium removal efficiency*

The isolated bacterium was inoculated into the basic medium and cultured at 10 °C for 24–48 hours when the population amount reached to more than 10^7 cells/ml. One liter liquid culture was collected and centrifuged at 6000 rpm for 5 min. The sedimentation were collected and washed with sterile pure water for 3 times. Then the collected bacteria were inoculated to triplicate 50 ml sterile basic medium. The ammonium concentration in the basic medium was adjusted to 5 mg/l. The basic mediums without inoculation were used as control. The isolated bacterium was cultivated at 5, 8, 10, 12 and 15 °C for 30 min, then the ammonium removal rate were detected.

2.5. *Amplification of amo and hao genes*

The genome DNA was extracted with the TAKARA kit (Dalian, China). The primers for *amo* gene was 5'-GGGGTTTCTACTGGTGGT-3' and 5'-AGCTTGTGAGTATATTTAGAACG-3'. The primers for *hao* gene was 5'-CATGGTCTGGCCGAACGGACAAT-3' and 5'-TTCTCTGGAATGCCCGGAAC-3'. PCR amplification of *amo* gene was: initial denaturation of double-stranded DNA for 5 min at 94°C; 40 s at 94°C, 40 s at 40°C, and 40 s at 72°C, and extension for 7 min at 72°C. PCR amplification of hao gene was: initial denaturation of double-stranded DNA for 5 min at 94°C; 30 s at 94°C, 30 s at 40°C, and 30 s at 72°C, and extension for 7 min at 72°C. All amplification products were analyzed by electrophoresis in 1.0% (wt/vol) agarose gels, followed by ethidium bromide staining (1.2 mg/lethidium bromide in 1×Tris-acetate-EDTA).

2.6. *Analytical methods*

The growth of isolates was tested by spectrophotometry at a wavelength of 600 nm (OD_{600}). The ammonium concentration was determined colorimetrically according to Water quality-Determination of ammonium-Nessler's reagent colorimetric method (GB7479-87). NH_2OH was analyzed by indirect spectrophotometry [15]. The nitrate and nitrite were analyzed using N-(1-naphthalene)-diaminoethane photometry and phenol disulphonic acid methods according to the State Environmental Protection Administration of China (Ministry of Environmental Protection of the People's Republic of China, 2002). TOC and TN were analyzed by using an Aurora Combustion Total Organic Carbon Analyzer 1030C (OI, America).

3. Results and Discussion

3.1. *Indentification of the isolated bacterium*

Strain SFA 10 was selected based on ammonium removal at 5 °C. The cells are short rods, Gram-negative, catalase-positive, oxidase-negative, non-motile and non-spore-forming. The 16S rDNA sequence of strain SFA 10 was 1422 bp and submitted to NCBI with accession No. HM486418. After blast, the 16SrRNA sequence of the strain SFA 10 showed 98% similarity with Acinetobacterbouvetii, and 91%-97% with other species of the genus of Acinetobacter. The results proved that the strain SFA 10 was the species of the genus of *Acinetobacter*. Bootstrapped neighbor-joining relationship was estimated with MEGA version 4.1 (Molecular Evolutionary Genetics Analysis [http://www.megasoftware.net]). The phylogenetic tree is shown in Figure 1. The strain SFA 10 was putatively identified as *Acinetobacter* sp. Their role in heterotrophic nitrification was fully investigated in this study.

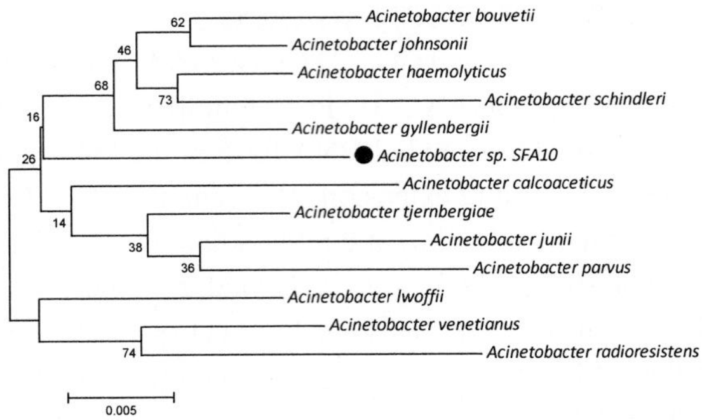

Figure 1. Phylogenetic tree based on 16S rRNA gene sequences. The position of *Acinetobacter* sp. SFA 10 is shown with respect to other closely related species. The tree was generated using the neighbor-joining method.

3.2. *Nitrification characteristics of Acinetobacter sp. SFA 10*

For evaluating the nitrification characteristics of *Acinetobacter*sp. SFA 10, sodium acetate was added as carbon source (C/N=4) and the temperature was enhanced to 10 °C. The production of intermediates was tested, such as NH_2OH, NO_2^--N andNO_3^- -N. The results are shown in Figure 2. With the ammonium decreasing, the microbial growth and the variation of TOC were detected. The results are shown in the Figure 3. After 20 hours, 10 mg/l ammonium were utilized by the strain SFA 10. The ammonium removal rate was 0.47 mg NH_4^+-N/l/h. With the utilization of ammonium, NH_2OH was produced. At 5 hours, the NH_2OH was the highest, then decreased slightly (Figure 2). The results indicated that the ammonium was oxidized by the strain SFA 10. Unlike many heterotrophic nitrifying bacteria, there were no nitrite or nitrate accumulation during the ammonium removal by *Acinetobacter*sp. SFA 10. Many heterotrophic nitrifying bacteria showed simultaneously aerobic denitrificaction ability. It might also exist in the strain SFA 10.

With the decrease of TOC and ammonium, the strain SFA 10 showed growth (Figure 3). *Acinetobacter*sp. SFA 10 was a heterotrphic bacterium that need organic carbons to provide the energy for growth.

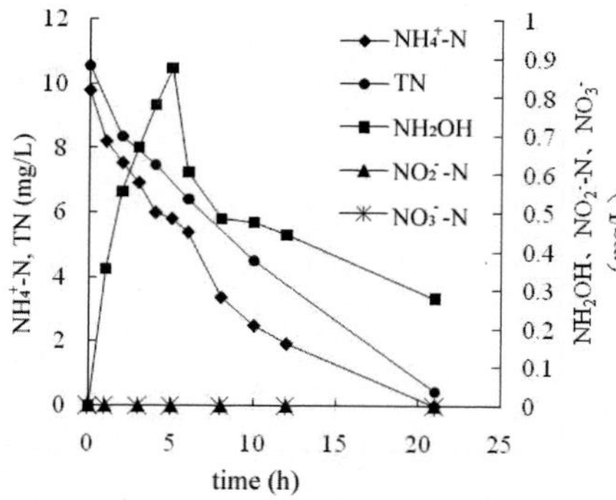

Figure 2. The changes of ammonium, total nitrogen, hydroxylamine, nitrite and nitrate.

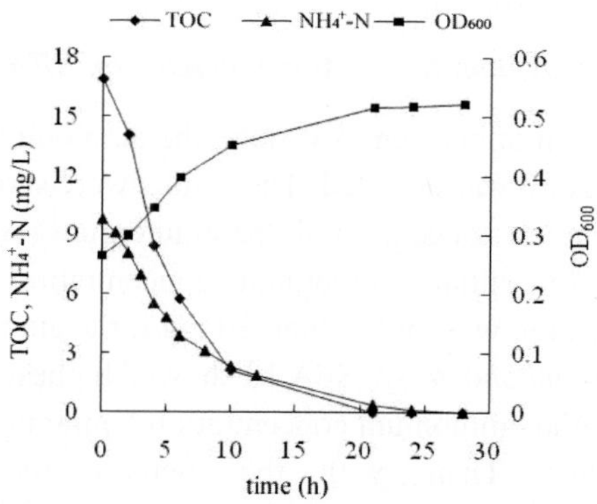

Figure 3. The changes of TOC, ammonium and the bacterial growth.

3.3. *The effect of temperature on ammonium removal efficiency of Acinetobacter sp. SFA 10*

The ammonium removal by *Acinetobacter*sp. SFA 10 at different temperatures was investigated in ammonium solution (about 5 mg/l) at C/N 4. With the increase of temperature, ammonium removal rate of Acinetobactersp. SFA 10 was enhanced. At 5 °C, the ammonium removal rate was 2.05 mg NH_4^+-N/l/h, which was higher than the reported heterotrophic nitrification rate [10]. The ammonium removal rate was the highest at 15 °C (3.48 mg NH_4^+-N/l/h) in this study. The results indicated that*Acinetobacter*sp. SFA 10 showed higher ammonium removal ability for drinking water treatment at low temperature.

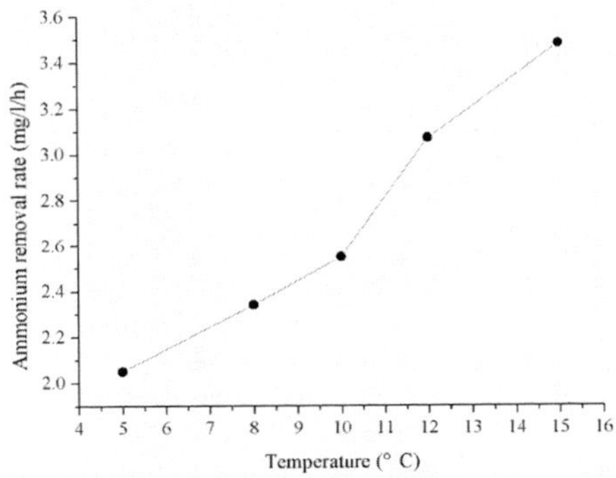

Figure 4. The effects of temperature on the ammonium removal rate of Acinetobacter sp. SFA 10.

3.4. *Ammonium removal ability of Acinetobacter sp. SFA 10*

For confirming the ammonium removal span, the ammonium removal ability of *Acinetobacter*sp. SFA 10 was detected. The results were showed in the Figure 5. At low initial ammonium concentration, the ammonium removal rate increased with the increase of initial ammonium concentration. When the initial ammonium concentration was higher than 20 mg/l, the ammonium removal rate increased slightly. *Acinetobacter*sp. SFA 10 showed highest ammonium removal rate at 45.94 mg/l initial ammonium concentration (ammonium removal rate was 10.63 mgNH_4^+-N/l/h). Then, with the increase of initial ammonium concentration, the ammonium removal rate decreased. In this study,*Acinetobacter*sp. SFA 10 was isolated from river water and fit for oligotrophic environment, so high ammonium inhibited the activity of it. The results indicated that it might be fit for drinking water treatment at low temperature as *Microbacterium*sp. described before [10].

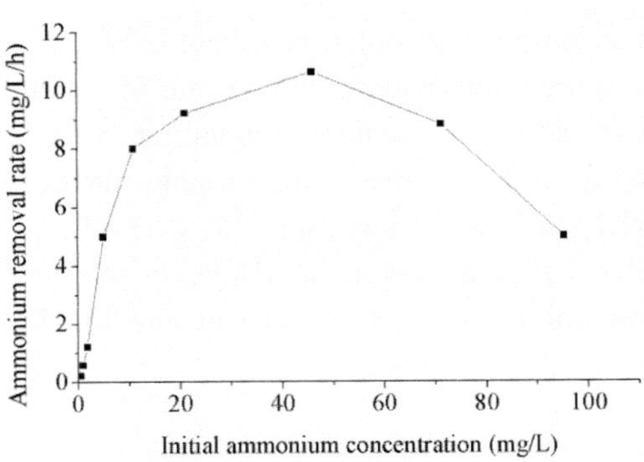

Figure 5. The ammonium removal ability of Acinetobacter sp. SFA 10.

3.5. *Amplification of amo and hao genes*

For understanding the pathway involved in the ammonium removal process by strain SFA 10, a preliminary studies on *amo* and *hao* genes amplificationwere conducted. The results are shown in Figure 6. The results proved that *amo* and *hao* genes were existed in *Acinetobacter* sp. SFA 10. The molecular weight of *amo* and *hao* gene in *Acinetobacter* sp. SFA 10 were about 600 bp. The results proved that *Acinetobacter* sp. SFA 10 oxidized ammonium to NH_2OH, then converted NH_2OH to nitrate. That also explain the phenomenon that NH_2OH accumulated during ammonium removal process (showed in Figure 2).

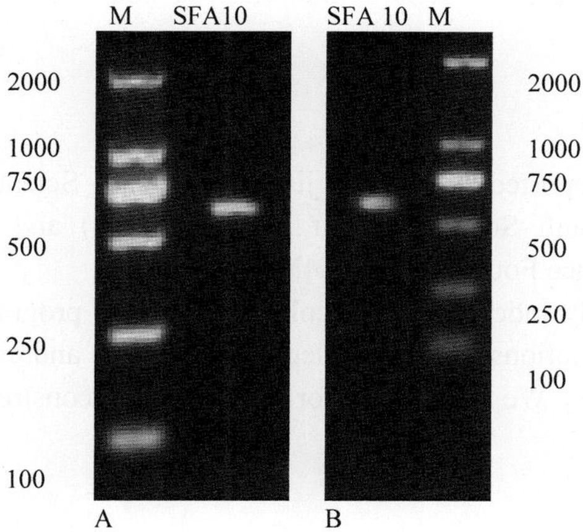

Figure 6. PCR amplification results for *hao* (A) and *amo* (B) genes. M was the molecular marker used in this study, which molecular weight was from 100-2000 bp.

3.6. *The denitrifying ability of Acinetobacter sp. SFA 10*

For detecting the denitrifying ability of*Acinetobacter* sp. SFA 10, the nitrite and nitrate was used as sole nitrogen source in the basic medium. The strain SFA 10 was inoculated in the basic medium for 30 min. no nitrite or nitrate elimination occurred (the data were not shown here). The nitrate reducatase gene (*nap*A) was not amplified in this study. Neither nitrite nor nitrate reductase activity detected in *Acinetobacter* sp. SFA 10 (the data were not shown here). The results indicated that*Acinetobacter* sp. SFA 10 didn't possess the ability of denitrification.

4. Conclusion

For removing ammonium in drinking water at low temperature, Strain SFA 10 was isolated form Songhua River and identified as *Acinetobacter* sp.After 20 hours, 10 mg/L ammonium were utilized by the strain SFA 10 without NO_2^--N andNO_3^- -N accumulation. At 5 °C, the ammonium removal rate was 2.05 mg NH_4^+-N/l/h, which was higher than other reported strains. When the initial ammonium concentration was higher than 45 mg/L, the ammonium removal rate of *Acinetobacter*sp. SFA 10 decreased. NH_2OH was detected during NH_4^+-N degradation. amo and haogenes were 600 bp and amplified from *Acinetobacter*sp. SFA 10. NO_2^--N andNO_3^- -N could not be utilized as sole nitrogen source.

Acknowledgment

This study is supported by Heilongjiang Provincial Science Foundation for Distinguished Youth Scholar (Grant No. JC200708) and funded by China Postdoctoral Science Foundation (2014M551250).

We are greatly indebted to our colleagues in this program that have made significant contributions to the collection, analysis and/or interpretation of samples and results. We are grateful for the useful and constructive comments of reviewers.

References

1. L. A. Robertson, E. D. W. J. Van Niel, R. A. M. Torremans and J. G. Kuenen. Simultaneous nitrification and denitrification in aerobic chemostat cultures of *Thiosphaerapantotropha*. *Applied and Environmental Microbiology* 54, 2812–2818 (1988).
2. W. Ludwig, G. Mittenhuber, and C. G. Friedrich. Transfer of *Thiosphaeapantotropha*to *Paracoccusdenitrificans*. *Int. J. Syst. Bacteriol* 43, 363–367 (1993).
3. Rainey, F. A. D. P. Kelly, E. Stackebrandt, J. Burkhardt, A. Hiraishi, Y. Katayama, and A. P. Wood. A reevaluation of the taxonomy of Paracoccusdenitrificans and a proposal for the creation of *Paracoccuspantotrophus* comb. nov. *Int. J. Syst. Bacteriol* 49, 645–651 (1999).
4. H. Joo, M. Hirai, and M. Shoda. Characteristics of ammonium removal by heterotrophic nitrification-aerobic denitrification by *Alcaligenesfaecalis* No. 4. *Journal of Bioscience and Bioengineering* 100, 184–191 (2005).

5. B. Zhao, Y. L. He, J. Hughes, X. F. Zhang. Heterotrophic nitrogen removal by a newly isolated *Acinetobactercalcoaceticus* HNR. *Bioresource Technology* 101, 5194–5200 (2010).
6. X. F. Huang, W. G. Li, D. Y. Zhang, W. Qin. Ammonium removal by a novel oligotrophic *Acinetobacter* sp. Y16 capable of heterotrophic nitrification-aerobic denitrification at low temperature. *Bioresource Technology* 146, 44–50 (2013).
7. X. P. Yang, S. M. Wang, D. W. Zhang, L. X. Zhou. Isolation and nitrogen removal characteristics of an aerobic heterotrophic nitrifying-denitrifying bacterium, *Bacillus subtilis* A1. *Bioresource Technology*. 102, 854–862 (2011).
8. B. Zhao, Y. L. He, X. F. Zhang. Nitrogen removal capability through simultaneous heterotrophic nitrification and aerobic denitrification by *Bacillus* sp. LY. *Environmental Technology* 31, 409–416 (2012).
9. A. A. Khardenavis, A. Kapley, H. J. Purohit. Simultaneous nitrification and denitrification by diverse *Diaphorobacter* sp. *Appl Microbiol Biotechnol* 77, 403–409 (2007).
10. D. Zhang, W. Li, X. Huang, et al. Removal of ammonium in surface water at low temperature by a newly isolated *Microbacterium* sp. strain SFA13. *Bioresource Technology* 137, 147–152 (2013).
11. S. M. Taylor, Y. He, B. Zhao, J. Huang. Heterotrophic ammonium removal characteristics of an aerobic heterotrophic nitrifying-denitrifying bacterium, *Providenciarettgeri* YL. *Journal of Environmental Sciences* 21, 1336–1341 (2009).
12. J. Zhang, P. Wu, B.Hao, Z. Yu. Heterotrophic nitrification and aerobic denitrifcation by the bacterium *Pseudomonas stutzeri* YZN-001. *Biouresource Technology* 102, 9866–9869 (2011).
13. P. Chen, J. Li, Q. X. Li, Y. Wang, S. Li, T.Ren, L. Wang. Simultaneous heterotrophic nitrification and aerobic denitrification by bacterium *Rhodococcus* sp. CPZ24. *BioresourceTechology* 116, 266–270 (2012).
14. J. W. B. Moir, L. C. Crossman, S. Spiro, D. J. Richardson. The purification of ammonia monooxygenase from *Paracoccusdenitrificans*. *FEBS Letters* 387, 71–74 (1996).
15. Y. Ming, Y. Fengke, Indirect Spectrophotometry of Hydroxylamine Hydrochloride. *Journal of Instrumental Analysis* 18, 64–66 (1999).

The study on the black carbon emission control from international shipping*

Q. Wang[1,a], P. Tao[2,b] and L.J. Wang[1,c]
[1]Maritime College, Shandong Jiaotong University, NO.115, Xinwei Road, Weihai, Shandong, 264200, China
[2]Weihai Yuanhang Technology Development Co. Ltd., Weihai, Shandong, 264200, China
E-mail: [a]qian_nancy@163.com, [b]1390631@163.com, [c]jkdy11@163.com

This paper reviews the negotiation process of the international community on the reduction of Black Carbon emission from international shipping and provides a general introduction on the status of the emission reduction. It also analyses the tendency of regulatory work of IMO concerning emission reduction by the international shipping and its possible impact. This paper provides some relevant suggestions.

Keywords: Black carbon; Emission control; IMO; International shipping.

1. The Concept of the Black Carbon and the Impact to International Shipping

Black Carbon (BC) is a kind of strong light absorption carbonized aerosol due to incomplete combustion of petroleum fuel, and is supposed to be component of particulate matters. It is the second contaminant of Sulfur Oxide (So_x) emission in atmosphere. Black Carbon represents a continuum from part coking matters to graphite and soot granules.

Expect being harmful to human health, Black Carbon granules also effects the climatic change. When the matters attaches on the surface of snow or ice in the North Pole or low latitude area, they make shiny side darken and absorb heat energy to result in the melt of snow or ice.

From the environmental influence of the Black Carbon, the sea ice of the Arctic region is a focused issue. The ship's Black Carbon emission is on the increase, and then accelerates the melt of ice of the Arctic regions. This problem is focused by International Maritime Organization (IMO).

2. Work Progress of the International Organization

International community paid attention to emission pollution problems from

* This paper is finished by Qian Wang, Peng Tao and Lijuan Wang.

international shipping for a long time. IMO focuses on the Black Carbon at the 58th meeting of Maritime Environment Protection Committee (MEPC) in October 2008. The meeting discussed the article submitted by International Friend of Earth (FOEI), which summarized and analyzed many kinds of emission reduction methods which can strongly effect emissions for the climatic change, especially included the information of Black Carbon emission.

2.1. *The 15th meeting of BLG*

The problem of Black Carbon has explicit progress on the 15th meeting of Sub-committee of Bulk Liquid and Gas (BLG) in February 2011. Clean Shipping Coalition (CSC) submitted documents for the emission lists of Black Carbon (BC), Organic Carbon (OC) and Sulfur Oxide (SO_x) in the Arctic region and the assessment report for information about Black Carbon which made by the committee of experts; The documents submitted by Secretariat introduced the background information including the usable measurements for solving the Black Carbon and the summary of all kinds of measurements. After exchanging of views, BLG request the MEPC to give clear guiding for the topic of Black Carbon. Therefore BLG invited member countries and international organizations to submit specific proposals to MEPC, including the definition of Black Carbon emission.

2.2. *The 16th session of MEPC*

It is held in the headquarter of IMO in Paris from 11 to 15 July 2011. The session adopted the amendment of Maritime Agreement Regarding Oil Pollution (MARPOL), in which set the America Caribbean Sea as a new emission control area, set Baltic Sea as special waters to prevent the sanitary waste pollution. IMO also adopted the compulsory rules about Greenhouse Gases emission reduction measures from international shipping.

MEPC adopted a decision on solving the Black Carbon aerosol emission from international shipping in the Arctic Region and instructed the BLG finish the following work: giving clear definition for Black Carbon emission from international shipping; examining Black Carbon measuring method and determining the most proper test method to the Black Carbon emission from international shipping; finding proper control measurements for Black Carbon emission from international shipping to reduce the impact on Black Carbon emission from international shipping in the Arctic Region. Then BLG should submit a final report to the 65th meeting of MEPC which will be held in 2014.

2.3. *The 65th meeting of MEPC*

The second topic in discussion in the 65th meeting of MEPC was problem of air pollution, in which discussed the Black Carbon emission in the Arctic Region. The meeting deliberated the progress of Black Carbon emission in the Arctic Region of the 17th meeting of BLG and the 57th meeting of sub-committee on ship Design and Equipment (DE) and agreed that the related work of DE should start after BLG reached a result. The meeting emphasized on suggestions of Black Carbon emission to Arctic impact of "ships in Arctic", not "international shipping". Because the impact of Black Carbon is only regional and the ships outside the Arctic Region had little influence to Arctic.

2.4. *The 67th meeting of MEPC*

The meeting is from 13 to 17 October 2014 decided to bring the issue of the impact on the Black Carbon emission to Arctic to IMO Sub-Committee on Pollution Prevention and Response 2nd session (PPR2) to continue discuss and finally develop a single Black Carbon definition.

3. Black Carbon Definition of PPR2

IMO Sub-Committee on Pollution Prevention and Response 2nd session (PPR2) is held in headquarter of IMO in Paris from 19 to 23 January 2015. Since MEPC67 failed to develop a definition of Black Carbon in the Arctic region, PPR2 continues to discuss this topic.

The sub-committee concurred with a working group's view that the "Bond et al." definition for "Black Carbon" should be recommended to the MEPC for approval as the definition of Black Carbon for international shipping, as this definition is measurement-method neutral, and widely supported by the scientific community.

The "Bond et al." definition describes Black Carbon as "a distinct type of carbonaceous material, formed only in flames during combustion of carbon-based fuels.

It is distinguishable from other forms of carbon and carbon compounds contained in atmospheric aerosol because it has a unique combination of the following physical properties:

It strongly absorbs visible light with a mass absorption cross section of at least 5 m2g-1 at a wavelength of 550nm (nanometer).

It is refractory; that is, it retains its basic form at very high temperatures, with vaporization temperature near 4000K.

It is insoluble in water, in organic solvents including methanol and acetone, and in other components of atmospheric aerosol; and it exists as an aggregate of small carbon spherules."

The Sub-Committee also noted that there was a need for measurement studies to be carried out and at this stage the focus should be on using the agreed definition to support data collection, to identify the most appropriate measurement methods for Black Carbon emissions from international shipping.

4. Measuring Methods of Black Carbon

At present, the measuring Black Carbon particle concentration methods mainly are:

4.1. *Optical measurement*

Optical measurement is the method that utilizes the optical theory to measure the carbonic matter in aerosol. The most common is commercial black carbon meter. It tests the Black Carbon particles concentration by testing the Black Carbon particles absorptivity. This measurement has great uncertainty because other carbonaceous aerosol also has certain absorption.

4.2. *Thermal decomposition measurement*

This measurement is the method that first heats under the inert gases to remove organic carbon molecule, then tests the black carbon particles concentration by oxidation combustion under oxygen gases. But, because a part of organic carbon will be carboned, the testing result actually includes Black Carbon and Charcoal that car boned by Organic Carbon.

By experimental contrast, scholars drew a conclusion that they can solve the problem of organic carbon carbonization disturbance in process of measuring the black carbon particle concentration using thermal decomposition measurement by controlling temperature, time and gas. Using thermal decomposition measurement can almost measure the black carbon particle concentration.

4.3. *Optothermal measurement*

Optothermal measurement is the method that combines the optical and thermal decomposition measurement to measure the carbonic material of particles more exactly. This is the most exact measurement now, but its disadvantage is that equipment is too expensive and difficult to popularize.

Other domestic studies about Organic Carbon, Elemental Carbon (Black Carbon), measurement mainly made by equipment aboard. For example, Hong Sun, Qingchen Guo used 240C elemental analyzer made in Perkin-Elmer of America, to carry out relevant studies on aerosol material respectively. Xugang Chi studied on the concentration level and seasonal change of Organic Carbon, Elemental Carbon, using MT-5 C, H, N elemental automatic analyzer made in Japan. Xinmin Liu studied on carbonic material used carbon analysis meter made by Sunset Co. of America.

5. Possible Control Measures

IMO reached an agreement on Black Carbon definition and measuring methods for a long time which had many times negotiations and discussions. The achievement is very rare.

Although international shipping is considered little effect to global Black Carbon emission in recent meetings, the process for the Black Carbon topic is propelled by cautious manner. With the increasing of international maritime transportation, especially because of the ice dissolution speeding up the open of new sea route to increase the transportation in the Arctic Region, some people think it is extremely urgent and imperative to take measurements to reduce Black Carbon emission. The simplest way is to forbid ships to use heavy oil.

It is reported that IMO plans to forbid using heavy oil completely in worldwide in 2020. Meanwhile it is said that although the agreement about Black Carbon effects now reached is only regional, focuses on the Arctic Region emphasis, Europe and America may put further pressure on IMO, so controlling Black Carbon emission may promote around the world based on the Arctic.

References

1. Schmidt M. W. I., Noack A. G. *Black carborboninsoils and sediments: analysis, distribution, implications, and current challenges*. Global Biogeochemical Cycles, 2000, 14(3), 777–793.
2. Mingyu Jiang, ZhuLing Chen, *Thermal decomposition and thermal analysis measurement of Black Carbon aerosol,* The 9th Aerosol Meeting Technical Seminar.
3. Marissa S. Liang, Timothy C. Keener, M. Eileen Birch, Richard Baldauf, Jill Neal, Y. Jeffrey Yang, *Low-wind and other microclimatic factors in near-road black carbon variability:* A case study and assessment implications. Atmospheric Environment, 2013, Vol. 80.
4. http://www.imo.org/
5. http://www.imla.co/

Study of the VUV photocatalytic treatment on *Escherichia coli* bacteria in ballast water

L. Zhang*, L.H. Dong, T. Liu, D.S. Wang, C.H. Fan, Y. Zhou and Y.S. Yin
College of Ocean Science and Engineering, Shanghai Maritime University, Shanghai, China
**E-mail: zhangli@shmtu.edu.cn*

A novel system (named VUV Photolysis) was presented in this paper for the treatment of ballast water. The microstructure and composition of photocatalytic film were investigated by scanning electron microscopy, atomic force microscopy. The antibacterial effect of VUV (Vacuum UltraViolet) Photolysis was proved by plate count method and fluorescence microscopy. The results showed that 850cfu/ml Escherichia coli in ballast water were completely eliminated by the system in 60 seconds. The effectiveness exceeds the regulation of International Maritime Organization (Escherichia coli less than 250cfu per 100 millilitres). Moreover, the antibacterial mechanism of VUV Photolysis was analyzed by comparing with traditional UV disinfection.

Keywords: Ballast water management system; Photocatalyst; Photocatalysis; Bacteria.

1. Introduction

Because of the large volumes and frequency of use, ballast water is currently the most frequently cited method for the worldwide transference of non-indigenous organisms. Combating invasive species in ballast water is a matter of global urgency. International agencies such as the International Maritime Organization (IMO), United States Coast Guard (USCG), and U.S. federal and state governments are combating the problem by mandating ballast water exchange in the open ocean or utilizing ballast water treatment systems. In 2004, the IMO adopted a new convention called the International Convention for the Control and Management of Ships' Ballast Water and Sediments. This convention will phase in ballast water treatment standards over the coming years, beginning in 2009 with requirements for smaller new ships. And now worldwide regulations are set to take effect.

The National Research Council evaluated a variety of approaches or systems for removing indigenous organisms from ballast water. Among these systems, Ultraviolet (UV) irradiation system was recommended as one of the most promising technology [1]. However, the biggest concern raised with regard to UV systems is the threat of organisms can sometimes repair and reverse the destructive effects of UV after retention in the ballast system through a "repair mechanism", known as photoreactivation, or in the absence of light known as "dark repair"or "re-growth" [2–4].

Heterogeneous photocatalysis, as a promising advanced oxidation process for water and wastewater treatment, has attracted interests of many researchers. Its superiority was demonstrated by numerous investigations for purification of widely organic contaminants [5,6]. Such as it can work well under room temperature, it involves nearly none chemicals except the photocatalyst which is stable and reuseful, it has no selectivity to organic pollutants (including bio organisms), it is low cost and significantly low energy consumption. Especially it can completely kill the microorganisms by cell degradation of the whole, which overcomes the shortcomings of "dark repair" in traditional UV disinfection.

In this work, a novel ballast water management system (named VUV Photolysis BWMS) by using the photocatalytic technology is developed for ships for the control and treatment of ballast water to minimize the risk of organisms and pathogens invasion. The design of our VUV Photolysis system and bacterial tested results using a prototype VUV Photolysis system under the onshore testing achieved so far are also presented.

2. Experimental Section

2.1. *Principles of VUV Photolysis system in treatment of ballast water*

The principles of the VUV Photolysis system is based on the traditional well established UV photocatalysis. But it is an enhanced photocatalysis by improvement to the core light source.

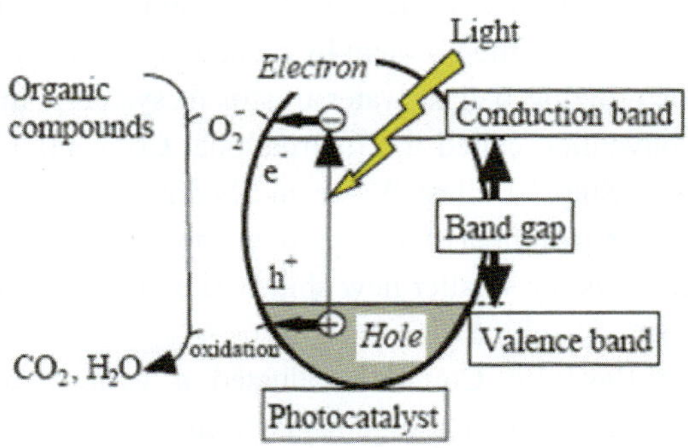

Figure 1. Mechanism of traditional photocatalysis.

According to the requirement of IMO G8 and G9 regulations, the VUV Photolysis BWM system was designed for fulfillment the lab test and onshore test of the novel VUV photocatalysis process. Its treatment capability can be adjusted from 20 m^3/hr to 60m^3/hr.

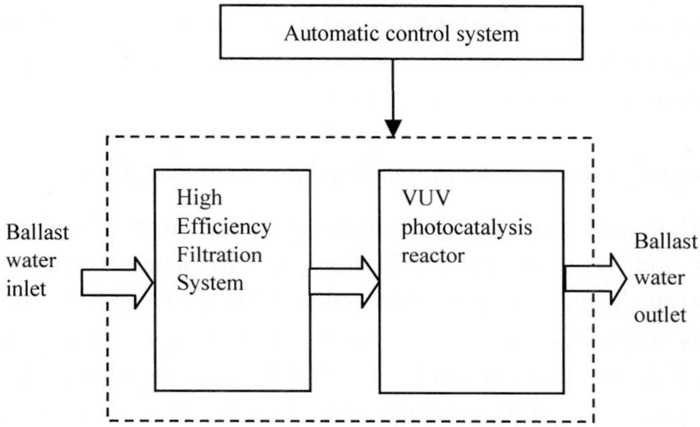

One module of VUVPhotolysis BWM system

2.2. *Characterization of TiO_2 coating*

The surface morphologis of new coated TiO_2 film were observed by a JSM—6301F scanning electron microscopic (SEM) with 15kV as the accelerating voltage and a SPA-400 model AFM (SPI3800N Serial, Seiko) in tapping mode.

2.3. *Sterilization experiment using VUV Photolysis system*

The *E. coli* concentration was measured using the APHA standard method 9222 (Clesceri et al., 1998) which involves membrane filtration. Serial dilutions of irradiated and control samples were performed. The samples were filtered, cultured on mFC-BCIG medium (Ciebin et al., 1995), and the number of colonies was counted after a 24-hour incubation period. The dose response curves were generated as a plot of the number of colony forming units (CFU) per 100mL of the samples versus the applied dose of UV or gamma irradiation. Fecal coliforms were cultured on mFC medium and were analyzed using the same procedure (Clesceri et al., 1998) as the one used for *E. coli.*

E. coli growth rates with and without treatment by VUV Photolysis system were determined by measuring the optical density at 600nm using a spectrophotometer (Eppendorf BioPhotometer) for 24 hours.

3. Results and Discussion

It is well known that the short lifetime of the immobilized photocatalyst increases the cost of its application for frequent replacement of the photocatalyst. Thus, a highly photoactive TiO_2 film catalyst with long-term stability is urgently needed in our system. After trying the coating of TiO_2 film photocatalysts on different kinds of substrate, such as organic fiber, pumice stone, polymer film, glass, ceramics and so on, we finally succeed in prepared long life-time TiO_2 film on stainless steel.

3.1. *Characterization of the immobilized photocatalyst films*

It is well known that the short lifetime of the immobilized photocatalyst increases the cost of its application for frequent replacement of the photocatalyst. Thus, a highly photoactive TiO_2 film catalyst with long-term stability is urgently needed in our system. After trying the coating of TiO_2 film photocatalysts on different kinds of substrate, such as organic fiber, pumice stone, polymer film, glass, ceramics and so on, we finally succeed in prepared long life-time TiO_2 film on stainless steel.

Figure 2a showed the new coated film under the SEM chapped like a droughty field. But the AFM images of the new coated TiO_2 film indicated that the TiO_2 crystal grains were uniform spheres and compactly bonded together all over the surfaces, as shown in Figure 2b. The average crystal size of the TiO_2 was 21nm.

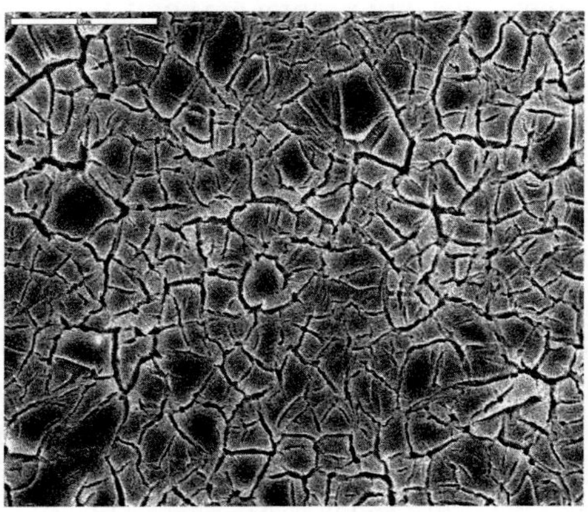

(a)

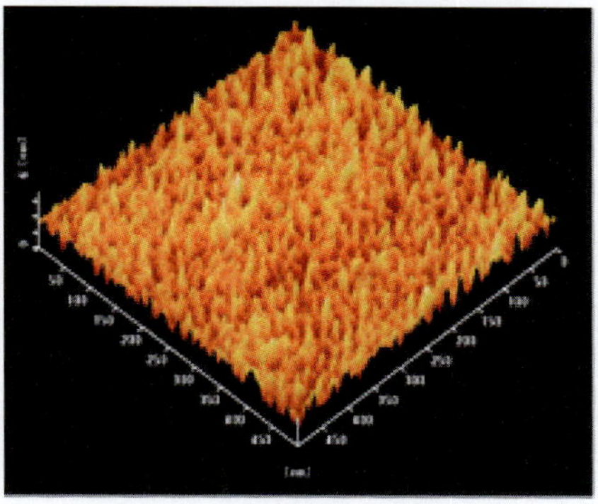

(b)

Figure 2. SEM micrograph (a) and AFM micrograph (b) of new coated TiO_2 film.

3.2. *Photocatalytic activity of TiO_2 film*

The photocatalytic activities of new coated TiO_2 film and after long term aging in sea water were characterized by electrochemical methods. The Photovoltaic of new coated TiO2 film and after 8000 hours of aging in sea water were detected with a CHI660B (CH Instruments Inc.) electrochemical workstation in a classical three-electrode electrochemical cell. A saturated calomel electrode (SCE) was as the reference electrode. The photopotential of open-circuit was observed under the irradiation of a 3 W low-pressure mercury lamp. Figure 3 showed the photopotential of open-circuits were observed under the irradiation of a 3 W low-pressure mercury lamp.

It was obviously shown in Figure 3 that the absolute value of photo-potential of 8000h aged TiO_2 film changes a little comparing with the new coated TiO_2 film. This indicated that the TiO_2 film had excellent photocatalytic activation and long-term stability. It could be interpreted as follows. For the electrode with a semiconductor film under dark conditions the carriers on the electrode only originated from thermodynamic excitation. However, when the TiO_2 film was illuminated with UV light, in addition to the thermodynamically excited carriers, a large number of photogenerated carriers accumulated and moved toward one direction under the action of space charge electric field, resulting in the increase of the concentration of electrons in the electrode. As a result, the potential of the semiconductor TiO_2 film decreased significantly and a Photo-potential was generated. Furthermore, the higher of Photo-potential reflected the larger concentration of photogenerated electron-hole pairs separated on the electrode interface and the higher of photocatalytic activity of TiO_2 film. If the Absolute value of Photo-potential changed a little, the photocatalytic activity of TiO_2 film changed a little.

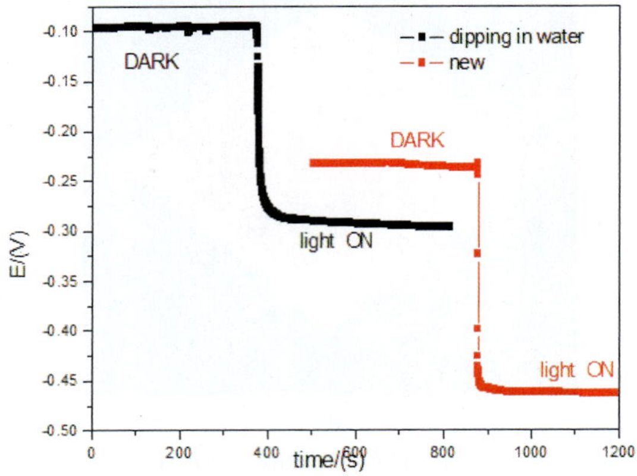

Figure 3. Plots of UV light generated potential of coated TiO_2 films. (dark) New prepared film; (red) caged TiO_2 films after dipping in sea water for 8000h.

3.3. *Antimicrobial efficacy of the VUV Photolysis*

We investigated the antimicrobial efficacy of the VUV Photolysis system by supplementing Luria–Bertani (LB)-agar media. The ballast water with the *E. coli* concentration of 850cfu/mL was tested by MPN method. The 0.1mL untreated and treated with VUV Photolysis system ballast water were added to the molten LB-agar solution and the mixture was allowed to solidify at 34°C for 24h. As shown from the LB-agar plates, treatment by VUV Photolysis for just 60s can inhibit the formation of colonies for *E. coli* absolutely (Figure 4b).

The effect of VUV Photolysis system on the *E. coli* growth kinetics in liquid media was studied (Figure 4c). The E. coli growth was monitored by measuring the optical density at 600nm (OD600) based on the turbidity of the cell suspension. Without treatment, the V. natriegens growth curve increased obviously as the bacteria reproduced normally in 24 hours in culture. However, the growth curve of treated ballast water kept constant as the VUV Photolysis could totally inhibite E. coli growth permanently for less than 10s.

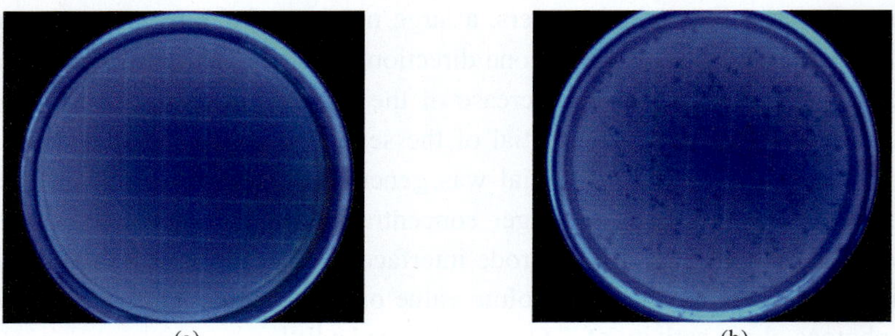

(a) (b)

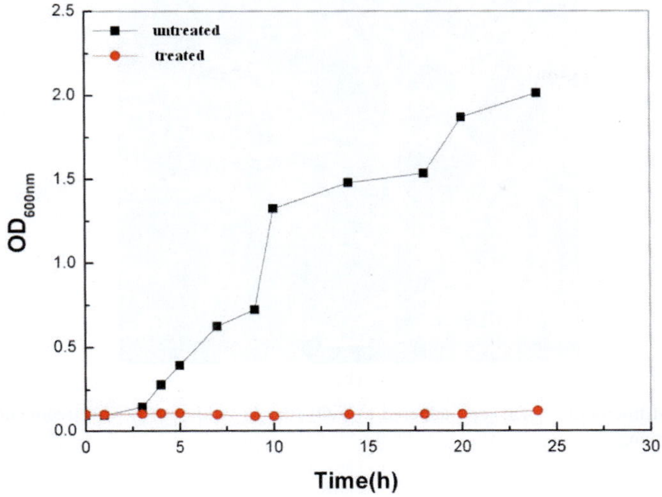

(c)

Figure 4. LB-agar inoculated with E. coli showing variable numbers of colonies when supplemented (a) without (b) with treatment by VUV Photolysis system and (c) bacterial growth curve in LB Lennox liquid media.

The TiO_2 films were taken out of the untreated and treated ballast water after 1 day of exposure to the bacteria, lightly rinsed in sterile distilled water, and then dried for SEM analysis. In Figure 5, it is clear that the size of *E. coli* was about 5 μm × 600 nm × 500 nm, which is consistent with previous reports [7]. After immersion for 1 day, many *E. coli* cells were attached on the TiO_2 film (Figure 5a). After VUV Photolysis for 60s, however, only one adhesive bacterium was found on the film, and the *E. coli* cell was obviously bread into pieces. It is proved that the VUV Photolysis method can destroyed the bacterial structure.

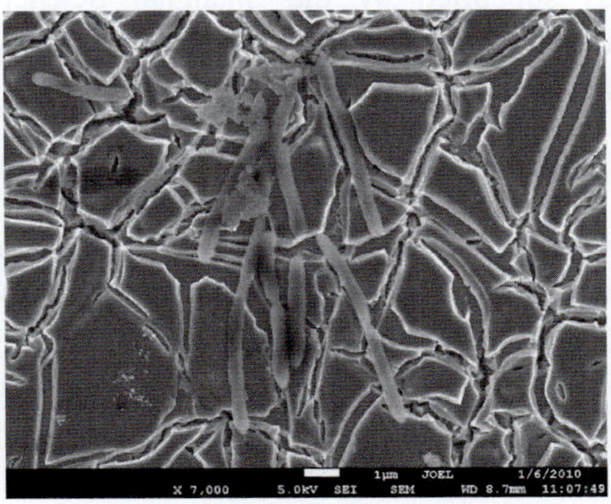

(a)

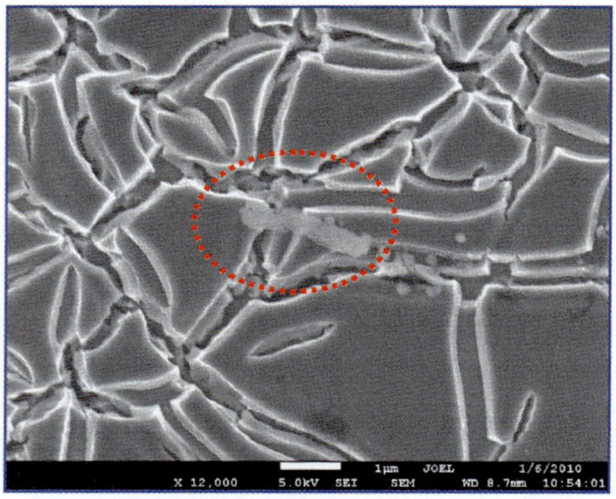

(b)

Figure 5. SEM micrograph of *E. coli* adhered to TiO_2 film (a) without (b) with treatment by VUV Photolysis system.

4. Conclusions

In conclusion, the microstructure and composition of photocatalytic film were investigated by scanning electron microscopy, atomic force microscopy. The antibacterial effect of VUV (Vacuum UltraViolet) Photolysis was proved by plate count method and fluorescence microscopy. The results showed that 850cfu/ml Escherichia coli in ballast water were completely eliminated by the system in 60 seconds. The effectiveness exceeds the regulation of International Maritime Organization (Escherichia coli less than 250 cfu per 100 millilitres). Moreover, the antibacterial mechanism of VUV Photolysis is destructive and the *E. coli* cannot revive.

Acknowledgment

This research was supported by the Shanghai Science and Technology Innovation Action Plan (No.14DZ1205302) and the National Natural Science Foundation (No.51101098).

References

1. Dunstan, P. K. & Bax, N. J. 2008. Management of an invasive marine species: defining and testing the effectiveness of ballast-water management options using management strategy evaluation. *ICES journal of marine science,* 65(6): 841–850.
2. Nebot-Sanz, E. Dávila, I.S. Balao, J. A.A. & Quiroga, J. M. 2007. Modelling of reactivation after UV disinfection: Effect of UV-C dose on

subsequent photoreactivation and dark repair. *Water Research,* 41 (14):3141–3151.

3. Harm, W. 1968. Dark repair of photorepairable uv lesions in Escherichia coli. *Mutation Research/Fundamental and Molecular Mechanisms of Mutagenesis,* 1(1):25–35.
4. Goosen, N. *2008*. Moolenaar, G. F. Repair of UV damage in bacteria. *DNA Repair.* 7(3):353–379.
5. Kabra K, Chaudhary,R. Sawhney, R.L. 2004. Treatment of hazardous organic and inorganic compounds through aqueous-phase photocatalysis: A review", *Ind Eng Chem, Res,* 43(24): 7683–7696.
6. Chin, P. Yang, L.P. Ollis, D.F. 2007. Kinetic Modeling of Photocatalyzed Soot Oxidation on Titanium Dioxide Thin Films. *Ind Eng Chem, Res*, 46 (23):7598–7604.
7. Taghipour, F. 2004. Ultraviolet and ionizing radiation for microorganism inactivation. *Water Research,* 38(14):3940–3948.

Chapter 7
Environmental Materials

Photooxidation aging behavior of PLA/PBAT films with different PBAT contents*

P. Si, J. Zou†, J.F. Ying, Y.H. Zhang, F.F. Wang and Y.Z. Qi
School of Materials Science and Engineering, Jiangsu University of Science and Technology, Zhenjiang, 212000, China
†E-mail:zj_881996@163.com, 278785037@qq.com
www.just.edu.cn

In order to investigate the photooxidation aging behavior of polylactide/poly (butyleneadipate-co-terephthalate) (PLA/PBAT) films, the films were prepared by melt extrusion and blow molding with benzoyl peroxide (BPO) as the interfacial compatibilizer. Their mechanical properties, molecular weight and dynamic thermo mechanical properties during UV irradiation progress have been investigated by universal tensile testing machine, gel permeation chromatography (GPC) and dynamic mechanical analysis (DMA). The results showed that the degradation rate of the composites had accelerated noticeably with increasing the content of PBAT. During irradiation, the elongation of the films decreased continuously, while the tensile strength increased originally and followed by a strong decrease; the molecular weight reduced meanwhile the polydispersity index enlarged; the lower storage modulus and higher T_g appeared at the same time.

Keywords: Polylactide; PBAT; Photooxidation aging; Mechanical property.

1. Introduction

In the market of degradable materials, polylactide (PLA) is no doubt one of the most promising candidates for further developments due to its high mechanical strength, good processing performance and excellent biological compatibility. Nonetheless, one of the major drawbacks, brittleness, severely limit its range of applications. Accordingly the toughening modification is extremely essential to the practical application of PLA[1-2]. Until now, various polymers have been used as toughening modifiers for PLA. Among them, Poly (butyleneadipate-co-terephthalate)(PBAT) is a very popular one[3-4], not only for its outstanding flexibility but also owning to its biodegradation property.

Now days, PLA/PBAT composites are receiving considerable attention for conventional applications such as packaging and mulching films in agriculture. Because most of these products are outdoor used, the controlled photo-oxidation of them represents a laboratory approach of assessing their applicability and the deterioration of their functional properties under real conditions[5].

On this basis, a UV accelerated degradation test was conducted in this paper to evaluate the mechanical properties, dynamic thermo mechanical properties and molecular weight change of the samples during irradiation, the content of PBAT on the influence of degradation rate is also investigated.

2. Experimental

2.1. *Materials and preparation of films*

Polylactic acid (PLA) (Mn=74,000 g mol^{-1}, Mw/Mn =1.4, D-isomer content=4.1%, density=1.25g/cm^{-3}) was supplied from Nature Works (4032D) in pellet form.

Poly (butyleneadipate-co-terephthalate) (PBAT) (Mn=81,000 g mol^{-1}, Mw/Mn =1.6, specific gravity =1.21g/ cm^{-3}) was purchased from BASF (Ecoflex_F_BX7011).

Benzoyl peroxide (BPO) was purchased from Akzo Nobel N.V (Perkadox L-W75).

Prior to extrusion, PLA pellets were dried at 80 °C for 1 day in a convection oven and the PBAT pellets were oven-dried at 75 °C overnight. Melt blending was performed using a corotating twin- screw extruder (Nanjing GIANT SHJ-30) at a screw speed of 100 rpm. The temperature profile of the extruder barrels was 155/165/175/180/185/185/180/170°C from the first heating zone (next to feeding throat) to die, respectively. The formulations of PLA/PBAT/BPO composites are listed in Table 2.1. After being oven-dried, the extruded pellets were blow-molded (Ruian XINQIAO GQ-45) into films of 10μm thickness.

2.2. *UV-light irradiations*

UV-light irradiation was carried out for up to 192 h at wavelengths higher than 313 nm in a climatic chamber (Shanghai JIEYING LUV-2), in the presence of oxygen (under air), at 60 °C. The irradiation intensity was about 125 W/m^2, and the films were placed on a rotating carousel positioned in the centre.

Table 2.1. The formulations of PLA/PBAT/BPO composites.

Sample	PLA (phr)	PBAT (phr)	BPO (phr)
1	55	45	1
2	45	55	1
3	35	65	1

2.3. *Methods of investigation*

The tensile properties were evaluated using an tensile testing machine (CMT-4304). The experiments were conducted at room temperature with the deformation rate of 10 mm/min according to GB/T 1040.3—2006. Experimental data were averaged on five samples.

Gel permeation chromatography (GPC) (Wyatt Technology WH2-04) were carried out on a system equipped with a size exclusion chromatography, an eighteen angle laser light scattering instrument and a refractive index detector. The measurements were achieved at 25 °C, using THF as eluent (0.5 mL/min) and polystyrene standards as references. All samples before testing were previously dissolved in THF for 12 h. After dissolving samples in THF some gels were observed. In order to separate gels from PLA/PBAT, solutions were choulved with two layersof filter papers and then filtered three times using nanosieves Whatman before injection.

Dynamic mechanical analysis (DMA) of the different samples was carried out by means of a dynamic thermal mechanical analyzer (NETZSCH DMA 242). Film samples with dimensions 5×15 mm were tested by applying a variable tensile deformation. The displacement amplitude was set to 100 μm, whereas the measurements were performed at the frequency of 1 Hz. The range of temperature was from 30 °C to 140 °C at the scanning rate of 5 °C/min.

3. Results and Discussion

3.1. *Mechanical characterization*

Figure 3.1 reports the mechanical properties evaluated from the stresse-strain curves for all the samples. It is evident that the tensile strength of the composites decreased with the increase in PBAT content from 45 to 65 phr. The minimum value of about 19.35 MPa was observed at a blend ratio of 35/65 (PLA/PBAT). This behaviour is probably due to the presence of soft elastomeric phase that reduced the crystallinity of PLA which is confirmed by Mukesh Jiang et al.[6]. However, addition of flexible PBAT to PLA was successful in increasing the flexibility in the blend. Moreover, the presence of PBAT within PLA reduces its brittleness characteristics which is evident from the increase of elongation from 75.15% when the content of PBAT is 45 phr to 158.98% in case of PLA/PBAT blend at 35/65 ratio.

UV irradiation produces a strong deterioration of both tensile strength and elongation. The elongation of the samples show a continuous decrease with the UV irradiation. In contrast, the composite samples show an initial increase of the tensile strength at short exposure time, followed by a strong reduction for long exposure time. As reported previously[7], the photo-degradation produces both chain scission and crystallization. At short irradiation time, the physical crosslinking produced by the crystals prevails with a consequential increase in tensile strength. For long exposure time the molecular weight reduction became dominated and the tensile strength starts to decrease.

Obviously, the addition of PBAT allows an acceleration of the degradation rate. It can be explained with the truth that PBAT could directly absorb the light in the range of 300-330 nm, which is likely to provoke Norrish cleavage[8].

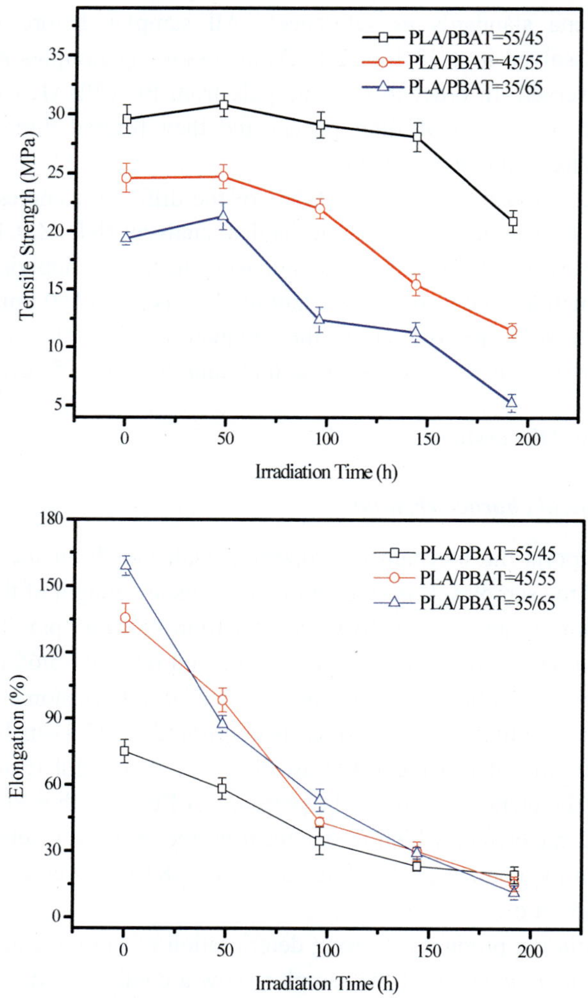

Figure 3.1. Mechanical properties of the various samples as a function of the irradiation time.

3.2. *Molecular weight characterization*

Table 3.1 shows the variations in the molar weight and polydispersity index of PLA during photooxidation.

Table 3.1. Molecular weights and polydispersity of PLA/PBAT composites samples measured by GPC under various UV irradiation times.

Sample	Irradiation Time(h)	M_w	M_n	PDI	$M_W(t)/M_w(0)$
PLA/PBAT=55/45	0	105400	65060	1.92	-----
PLA/PBAT=55/45	96	66430	28270	2.65	63%
PLA/PBAT=55/45	192	32700	9160	3.87	31%
PLA/PBAT=45/55	0	120640	72240	1.77	-----
PLA/PBAT=45/55	96	71400	27670	2.58	59%
PLA/PBAT=45/55	192	28010	7150	4.02	23%
PLA/PBAT=35/65	0	127700	73820	1.73	-----
PLA/PBAT=35/65	96	72130	26520	2.72	56%
PLA/PBAT=35/65	192	21320	5060	4.21	17%

One can observe a decrease in the molar weight clearly, which indicated that photooxidation of the composites provokes chain scissions that are responsible for the formation of low molecular weight oligomers.The reduction in molecular weight seems to increase with increasing the PBAT content. For example, the reduction in Mw for PLA/PBAT blend at 35/65 ratio after 192 h of exposure was higher than 80% whereas after 192 h of exposure the sample with 45phr content of PBAT shows a reduction of only 69%. The polydispersity index (PDI) of all the samples showed a continued increase of values during irradiation, and there is a significant increase with the aging time especially for the samples with higher concentration of PBAT, this result is attributable to a larger Mn decrease registered for these samples. All these changes above are responsible for the loss of mechanical and other physical properties of PLA/PBAT composites.

3.3. *DMA analysis*

DMA analysis helped to better characterize the mechanical properties and the glass transition temperatures (Tg) of PLA/PBAT blends before and after UV irradiation treatment. For example, in Figure 3.2 (I) is compared the storage modulus (E') of PLA/ PBAT blends with different contents of PBAT before and after UV irradiation treatment. It can be seen that with PBAT content increased, the value of E' of the blends decreased, which indicated that the blends with less PBAT content were more elastic than those with more PBAT content. Simultaneously, the content of PBAT played a significant role on the reduction extent of E' value during UV irradiation progress. For the sample whose PBAT conten is 65 phr, E' at 30 °C was reduced dramatically from about 798 MPa to 454 MPa by 192 h of UV treatment. In contrast, the corresponding reduction in E' due to UV irradiation treatment of PLA/PBAT blend at 55/45 ratio is less significant (from about 1405 MPa to 1178 MPa after 192 h of UV treatment).

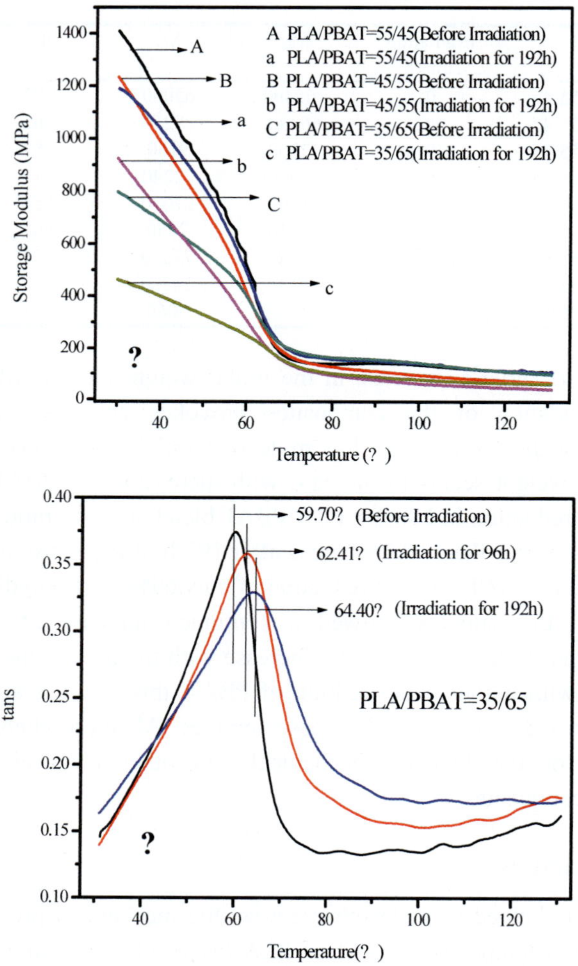

Figure 3.2. DMA thermograms of PLA/PBAT blends before and after irradiation.

As we can clearly see from Figure 3.2 (II), with the extension of irradiation time, Tg of the samples (Take the composite with the content of PBAT is 65 phr as an example) showed a slight increase, meanwhile, the peek became shorter and wider. This phenomenon can be explained as follows: When UV irradiation treatment was performed, the degradation makes PLAmolecular chains much shorter and the degree of chain entanglement drops dramatically. In this case, the function of crystals as physical crosslinking points becomes pronounced, as a result the T_g became higher. At the same time, chain scission forms more small chains and increases the molecular mobility, leads to more diverse chain movement and easier relaxation transition.

4. Conclusions

The photodegradation behavior of PLA/PBAT composites, at different PBAT loading, was studied. All the experimental results showed that the increase of PBAT content improves the degradation rate of PLA/PBAT composites. In particular, the tensile testing results indicated that with UV treatment the elongation of composites decreased continuously, whereas the tensile strength of the composites increased initially, then a strong decrease was observed. All the samples, their molecular weight reduced meanwhile the polydispersity index enlarged. From DMA characterization, a lower storage modulus was observed for all the composites after UV exposure. In addition, the higher T_g, more abundant chain movement and easier relaxation transition appeared as a result of UV irradiation.

References

1. Zhou L, Zhao G, Feng Y, et al. Toughening polylactide with polyether-block-amide and thermoplastic starch acetate: Influence of starch esterification degree [J]. Carbohydrate polymers, 2015, 127: 79–85.
2. Pluta M, Piorkowska E. Tough and transparent blends of polylactide with block copolymers of ethylene glycol and propylene glycol [J]. Polymer Testing, 2015, 41: 209–218.
3. Arruda L C, Magaton M, Bretas R E S, et al. Influence of chain extender on mechanical, thermal and morphological properties of blown films of PLA/PBAT blends [J]. Polymer Testing, 2015, 43: 27–37.
4. Al-Itry R, Lamnawar K, Maazouz A, et al. Effect of the simultaneous biaxial stretching on the structural and mechanical properties of PLA, PBAT and their blends at rubbery state [J]. European Polymer Journal, 2015, 68: 288–301.
5. Pablos J L, Abrusci C, Marín I, et al. Photodegradation of polyethylenes: Comparative effect of Fe and Ca-stearates as pro-oxidant additives [J]. Polymer degradation and stability, 2010, 95(10): 2057–2064.

6. Jiang L, Wolcott M P, Zhang J. Study of biodegradable polylactide/poly (butylene adipate-co- terephthalate) blends [J]. Biomacromolecules, 2006, 7(1): 199–207.
7. Torikai A, Takeuchi A, Nagaya S, et al. Photodegradation of polyethylene: effect of crosslinking on the oxygenated products and mechanical properties [J]. Polymer photochemistry, 1986, 7(3): 199–211.
8. Grossetete T, Rivaton A, Gardette J L, et al. Photochemical degradation of poly (ethylene terephthalate)-modified copolymer [J]. Polymer, 2000, 41(10): 3541–3554.

Large scale recycling technique of old asphalt mixture

S.Q. Li[1,2] and H. Li[2,3,*]

[1]*Highway School, Chang'an University, Xi'an 710064, China*
[2]*Highway Traffic Safety and Emergency Protection Technology and Equipment Transportation Industry R & D Center, Guangzhou 510420, China*
[3]*Guangdong Hualu Communications Technology Co, Ltd. Guangzhou 510420, China*
[*]*E-mail: hao_li_20007@163.com*

This paper attempted the large scale recycling technique which increasing the content of old asphalt mixture by 30%, 45%, 60%. Experimenting with a large number of high temperature stability and water stability tests of recycling mixture in different contents indoors. The results of rutting test, GTM test and the single penetration test show that recycling mixtures has a good high temperature performance, and there is a linear relationship between dynamic stability and penetration of asphalt; The results of immersion marshall test, freeze-thaw split test and cantabro test show that water stability of recycling mixtures first rises and then drops, and the content of RAP should be in a reasonable range to keep a good water stability performance; GTM test shows that the compaction of recycling mixtures is easier than the new asphalt mixture. The higher the content, the easier the compaction. Large scale recycling technique of old asphalt mixture was used in entity engineering in China for the first time, the application effect was good.

Keywords: Road engineering; Large scale hot recycling; High temperature stability; Water stability; Compaction properties.

1. Introduction

China has entered into the large-scale road maintenance time. Annual maintenance mileage will amount to 800,000 kilometers. The rational use of old asphalt mixture can save resources and protect environment, old asphalt mixture recycling technology was born under the background.

It is required that the content of reclaimed asphalt pavement (referred to as RAP) should be in the range of 15%–30% in Technical Specifications for Highway Asphalt Pavement Recycling [1]. But the NCAT loop attempted 20% RAP and 45% RAP tests, the results showed that 45% RAP can also provide excellent anti rutting performance and durability [2]. The A-140 highway maintenance of Spanish Huesca province even used 60% RAP [3]. Therefore, carrying out the application and evaluation of large scale recycling technique is very necessary in China, which has a strong practical and theoretical guidance.

The Guangzhou-Huizhou Expressway asphalt pavement structure layer is upper layer of 4cm AK-16A+ middle layer 5cm AC-20I+ lower layer 6cm thick AC-25I. Upper layer, middle layer used modified asphalt, lower layer used the ordinary asphalt.

According to the survey, Guangzhou-Huizhou (referred to as GH) Expressway was opened to traffic for more than ten years, asphalt pavement was still keep in a high intensity, weak spot occurred only locally, but most of asphalt pavement appeared various distress in early days such as crack, deformation and water damage. In this case, large scale recycling technique described can work with the maintenance of GH Expressway.

2. The Content of RAP

The recycling technique of GH Expressway used central plant hot recycling technique. Milling and planing upper, middle and lower layer of GH Expressway. The AC-13 and AC-20 in upper layer and middle layer which using framework-dense structure are used to improve the structure stability and water stability. The lower layer AC-25 used suspended dense structure to improve anti cracking ability and durability. Based on the function of each layer, the paper attempted to increase the content of old asphalt mixture by 30%, 45%, 60%. Using Marshall design method, AC-13 asphalt mixed material added 0%, 45%, 60% RAP, AC-20 asphalt mixed material added 0%, 30%, 45% RAP, AC-25 asphalt mixed material added 0%, 30%, 45%, 60% RAP respectively. Ordinary asphalt used Dong guan East AH-70 heavy traffic asphalt.

Table 2.1. The content of RAP.

AC-13	AC-20	AC-25
Modified asphalt AC-13+0% RAP	Ordinary asphalt AC-20+0% RAP	Ordinary asphalt AC-25+0% RAP
-	Ordinary asphalt AC-20+30% RAP	Ordinary asphalt AC-25+30% RAP
Ordinary asphalt AC-13+45% RAP	Ordinary asphalt AC-20+45% RAP	Ordinary asphalt AC-25+45% RAP
Ordinary asphalt AC-13+60% RAP	-	Ordinary asphalt AC-25+60% RAP

3. High Temperature Stability

Rutting test [4], GTM design method [5–7], the single penetration test [8,9] are used to study high temperature stability performance. As rutting test cannot evaluate high temperature stability comprehensively [10], this paper adopted GTM design method and the single penetration test to study high temperature performance of central plant hot recycling mixtures in different ways.

3.1. *Rutting test*

The results of rutting test are shown in Table 3.1.

Table 3.1. Rutting test results of central plant hot recycling mixtures unit: time/mm.

AC-13	Dynamic stability	AC-20	Dynamic stability	AC-25	Dynamic stability
Modified asphalt AC-13+0% RAP	5138	AC-20+0% RAP	1709	AC-25+0% RAP	1862
-	-	AC-20+30% RAP	3324	AC-25+30% RAP	3972
AC-13+45% RAP	4217	AC-20+45% RAP	3807	AC-25+45% RAP	4290
AC-13+60% RAP	4769	-	-	AC-25+60% RAP	4812

Dynamic stability of AC-13+45% RAP was 82% of modified asphalt mixture, AC-13+60% RAP was 93% of modified asphalt mixture. Dynamic stability of AC-20+30% RAP was 194% of the ordinary asphalt mixture, AC-20+45% RAP was 223% of the ordinary asphalt mixture. Dynamic stability of AC-25+30% RAP was 213% of the ordinary asphalt mixture, AC-25+45% RAP was 230% of the ordinary asphalt mixture, AC-25+60% RAP was 258% of the ordinary asphalt mixture.

For AC-13 mixture, high temperature stability performance of AC-13+45% RAP and AC-13+60% RAP was closed to modified asphalt mixture. For AC-25 and AC-20 mixture, the content of rap increased, so did dynamic stability of central plant hot recycling mixtures, which was 2-2.5 times of the new common asphalt mixture, the growth rate was obvious.

The content of asphalt in RAP, the content of RAP and mixed the optimum asphalt content can determine the proportion of new asphalt and old asphalt. The penetration of asphalt mixture can be calculated by using the harmonic formula:

$$\lg\lg\eta_{mix} = (1-\alpha)\lg\lg\eta_{old} + \alpha\lg\lg\eta_{new} \tag{1}$$

α is the proportion of the new asphalt, η_{mix} is viscosity of asphalt under 60°C after mixing, η_{old} is viscosity of old asphalt under 60°C before mixing, η_{new} is viscosity of new asphalt under 60°C before mixing.

Research found that there was a linear relationship between dynamic stability and penetration of asphalt.

For AC-20 and AC-25 hot recycling mixtures, functional relation was expressed as follows respectively.

y = -68.213x + 6471, the correlation coefficient R=0.996(AC-20)

y = -72.658x + 7003.6, the correlation coefficient R=0.992(AC-25)

It was known to all that the more the content of rap, the bigger the proportion of old asphalt. Figure 3.1 showed that smaller the penetration, the higher dynamic stability. Maximum was almost twice of minimum. Rutting test showed that hot recycled mixture can meet the standard requirements, and had a higher resistance to rutting than that of ordinary asphalt mixture.

The research of Yang Ping [11] also showed that dynamic stability and penetration of hot recycled mixture had a good linear relationship. The same rules showed that this is no coincidence, and its innate law and characteristic.

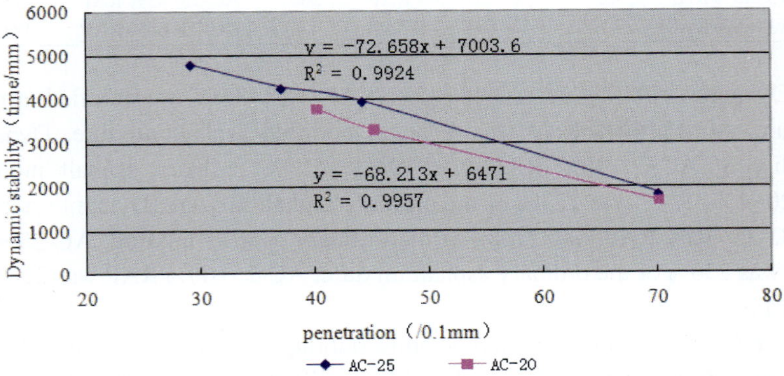

Figure 3.1. The relationship between dynamic stability and penetration.

3.2. *GTM test*

The results of GTM test are shown in Table 3.2.

Table 3.2. GTM test results of AC-13, AC-20, AC-25.

Type	Compaction laps (turn)	GSI	GSF	GSI value	GSF value	Static compression modulus (kPa)	Static shear modulus (kPa)
Modified asphalt AC-13+0% RAP	266	0.97	1.56	≤1.05	≥1.3	21030.5	64711.5
AC-13+45% RAP	299	1.05	1.62			20257.0	60771.0
AC-13+60% RAP	238	1.04	1.64			18031.0	54093.5
AC-20+0% RAP	318	0.95	1.64	≤1.05	≥1.3	23758.0	71275.0
AC-20+30% RAP	278	1.03	1.64			19759.5	59279.0
AC-20+45% RAP	268	1.03	1.69			24153.5	72460.0
AC-25+0% RAP	261	0.99	1.61	≤1.05	≥1.3	22567.3	67702.0
AC-25+30% RAP	260	0.96	1.77			27826.0	83478.5
AC-25+45% RAP	257	1.02	1.74			25065.5	75196.5
AC-25+60% RAP	231	0.92	1.71			24135.0	72405.0

GSI and GSF met the requirements of GTM. Molding temperature is 150°C and 165°C for ordinary asphalt and modified asphalt respectively. For AC-13, compaction effect was similar between 150°C and 165°C; For AC-20 and AC-25, compaction effect was also similar in the same temperature. The results showed that construction temperature of hot recycled mixture can agree with the ordinary asphalt.

For AC-13, AC-20, AC-25, GSF which mixed RAP is bigger than that which did not mix RAP. It could draw a conclusion that anti shear strength and high temperature stability performance of hot recycling mixture was better.

3.3. *The single penetration test*

The results of the single penetration test are shown in Table 3.3.

Table 3.3. The single penetration test results of AC-13, AC-20, AC-25.

Type	Shear strength (MPa)	Increase ratio of shear strength (%)
AC-13+0% RAP	0.590	-
AC-13+45% RAP	0.937	59%
AC-13+60% RAP	0.933	58%
AC-20+0% RAP	0.555	-
AC-20+30% RAP	1.110	100%
AC-20+45% RAP	0.910	64%
AC-25+0% RAP	0.801	-
AC-25+30% RAP	0.983	23%
AC-25+45% RAP	0.894	12%
AC-25+60% RAP	1.177	47%

For AC-13, anti shear strength which has been added RAP increased by 60% compared with AC-13 modified asphalt mixture. For AC-20, anti shear strength which has been added RAP increased by more than 60% compared with AC-20 ordinary asphalt mixture. Both high temperature stability and anti rut ability are improved.

For AC-25, compared with AC-25 ordinary asphalt mixture, anti shear strength which has been added RAP increases by 12%–47%.

4. Water Stability

Using the immersion marshall test, freeze-thaw splitting test, Kentabao water dispersion test to study water stability performance in the JTG E 20-2011 "Standard Test Methods of Bitumen and Bituminous Mixtures for Highway engineering".

4.1. *Immersion Marshall test*

The soaking remnants stability is used to evaluate the water stability performance in Immersion Marshall Test. The results of soaking remnants stability are shown in Table 4.1.

Table 4.1. The immersion marshall test results of central plant hot recycling mixtures unit: %.

AC-13	Soaking remnants stability	AC-20	Soaking remnants stability	AC-25	Soaking remnants stability
AC-13+0% RAP	98.9	AC-20+0% RAP	97.1	AC-25+0% RAP	94.7
-	-	AC-20+30% RAP	96.6	AC-25+30% RAP	96.4
AC-13+45% RAP	94.3	AC-20+45% RAP	83	AC-25+45% RAP	93.5
AC-13+60% RAP	95	-	-	AC-25+60% RAP	86.3

For AC-13, compared with AC-13 modified asphalt mixture, the soaking remnants stability of hot recycled asphalt mixture which has been added 45%, 60% of RAP reduces by 4.6% and 3.9% respectively.

For AC-20, compared with AC-20 ordinary asphalt mixture, the soaking remnants stability of hot recycled asphalt mixture which has been added 30%, 45% of RAP reduces by 0.5% and 14.1% respectively.

For AC-25, compared with AC-25 ordinary asphalt mixture, the soaking remnants stability of hot recycled asphalt mixture which has been added 30%, 45%, 60% RAP increase by 1.7%, reduce by 1.2% and reduce by 8.4% respectively.

For AC-25 and AC-20 , the water stability performance is best when the content of RAP is 30%. For AC-13, the water stability performance of hot recycled asphalt mixture is lower than modified asphalt mixture.

4.2. *Freeze thaw splitting test*

The tensile strength ratio is used to evaluate the water stability performance in Freeze-thaw Split Test. The results of tensile strength ratio are shown in Table 4.2.

Table 4.2. The freeze-thaw Split test results of central plant hot recycling mixtures unit: %.

AC-13	Tensile strength ratio	AC-20	Tensile strength ratio	AC-25	Tensile strength ratio
AC-13+0% RAP	97.3	AC-20+0% RAP	84.1	AC-25+0% RAP	79.1
-	-	AC-20+30% RAP	92	AC-25+30% RAP	91.7
AC-13+45% RAP	90.1	AC-20+45% RAP	89	AC-25+45% RAP	80.4
AC-13+60% RAP	83.5	-	-	AC-25+60% RAP	89.1

For AC-13, compared with AC-13 modified asphalt mixture, the tensile strength ratio of hot recycled asphalt mixture which has been added 45%, 60% of RAP reduces by 7.2% and 13.8% respectively.

For AC-20, compared with AC-20 ordinary asphalt mixture, the tensile strength ratio of hot recycled asphalt mixture which has been added 30%, 45% of RAP increase by 7.9% and 4.9% respectively.

For AC-25, compared with AC-25 ordinary asphalt mixture, the tensile strength ratio of hot recycled asphalt mixture which has been added 30%, 45%, 60% RAP increase by 12.6%, 1.3% and 10% respectively.

For AC-25 and AC-20, the water stability performance is best when the content of RAP is 30%. For AC-13, the water stability performance of hot recycled asphalt mixture is lower than modified asphalt mixture.

4.3. *Cantabro test*

According to a large number of field sampling test results, the ability to resist water damage was divided into five levels, as shown in Table 4.3.

Table 4.3. Evaluation criteria of the ability to resist water damage.

Level	Excellent	Good	Medium	Inferior	Bad
Flying loss (%)	<10	≥10,<20	≥20,<40	≥40,<60	≥60

The flying loss is used to evaluate the water stability performance in cantabro test. The results of flying loss are shown in Table 4.4.

Table 4.4. The results of flying loss.

Type	RAP content (%)	Ratio	Flying loss (%)	Level
AC-25	0	100:0	29.1	Medium
	30	69:31	11.6	Good
	45	60:40	19.9	Good
	60	35:65	21.2	Medium
AC-20	0	100:0	32.2	Medium
	30	70:30	10.8	Good
	45	60:40	17.4	Good
AC-13	0	-	10.9 (modified)	Good
	45	60:40	15.9	Good
	60	47:53	17.9	Good

As shown in Table 4.4, the flying loss decreased at first and increased later, the flying loss is lowest when the content of RAP is 30%. And the results are in line with the result before.

5. Compaction Properties

Firstly, the Marshall method and GTM method are used to design central plant hot recycling mixtures and confirm rational distribution range and the optimum proportion of asphalt respectively. Then compare mixture void fraction of the two methods. The change of the void fraction can be used to evaluate the degree of compaction of the mixed materials. The greater the change, the greater the degree. As shown in Table 5.1.

Table 5.1. The results of void fraction [12].

Type	Void fraction /%		
	Marshall	GTM	Difference
Modified asphaltAC-13	3.7	2.8	0.9
Ordinary asphaltAC-13+45% RAP	3.9	1.1	2.8
Ordinary asphaltAC-13+60% RAP	3.6	0.4	3.2
Ordinary asphaltAC-20	4.5	4.1	0.4
Ordinary asphaltAC-20+30% RAP	4.0	1.1	2.9
Ordinary asphaltAC-20+45% RAP	4.3	1.8	2.5
Ordinary asphaltAC-25	4.6	3.8	0.8
Ordinary asphaltAC-25+30% RAP	4.1	1.8	2.3
Ordinary asphaltAC-25+45% RAP	4.8	1.8	3.0
Ordinary asphaltAC-25+60% RAP	4.4	0.9	3.5

In order to study the compaction rate of mixed materials, the calculation method was used to evaluate the compaction rate.

Calculation method: GTM test system collects rich data automatic in the process of test. The calculation results of these data can reflect the mixed material compaction rate to a certain extent. Firstly, the bulk density is used as standard density in the Marshall design method, and the density collected in the process of GTM test is transformed as compaction degree of GTM. Then there is a linear relationship between both during natural logarithm processing of the rotation numbers. The slope between starting point and 97% compaction degree is K, which shows the rate of reaching 97% compaction degree. The results are summarized in Table 5.2.

Table 5.2. The results of compaction rate [11].

Type	Compaction times under 97% degree of compaction (/time)	Compaction times under balance (/time)	K value
Modified asphaltAC-13	—	266	3.952
Ordinary asphaltAC-13+45% RAP	58	277	4.732
Ordinary asphaltAC-13+60% RAP	39	238	5.388

Ordinary asphaltAC-20	—	318	3.203
Ordinary asphaltAC-20+30% RAP	68	278	3.774
Ordinary asphaltAC-20+45% RAP	53	268	4.854
Ordinary asphaltAC-25	261	261	3.257
Ordinary asphaltAC-25+30% RAP	130	260	3.592
Ordinary asphaltAC-25+45% RAP	72	257	4.454
Ordinary asphaltAC-25+60% RAP	40	231	4.945

For AC-25, compared with AC-25 ordinary asphalt mixture, both the compaction times and K values of hot recycled asphalt mixture which has been added RAP reduce. The rule applies to AC-13 and AC-20, which shows the greater the value of K, the easier the compaction.

6. Engineering Application

December 2010 to January 2011, large scale recycling technique of old asphalt mixture is used in Guangzhou-Huizhou Expressway. The test road is 4080 metres long. This is the first time that large scale recycling technique of old asphalt mixture is used in entity engineering in China.

Three kinds of structure scheme are used in test road, that is middle layer C-20+30% RAP+ lower layer AC-25+30% RAP, middle layer C-20+30% RAP+ lower layer AC-25+45% RAP, middle layer C-20+45% RAP+ lower layer AC-25+45% RAP. The test road keeps flat and dense through two years observation.

7. Conclusion

It is the first attempt to increase the content of old asphalt mixture by 30% and 45% and apply to real engineering. Through high temperature stability, water stability test and engineering application, test road is in good condition. The main conclusions are as follows:

(1) The high temperature stability of central plant hot recycling mixtures was evaluated by rutting test, GTM test and the single penetration test. The results showed that central plant hot recycling mixtures has good high temperature performance, and there was a linear relationship between dynamic stability and penetration of asphalt.

(2) The results of immersion marshall test, freeze-thaw split test and cantabro test showed that water stability of central plant hot recycling mixtures first rise and then drop. The content of RAP should be in a reasonable range to keep a good water stability performance.

(3) GTM test showed that compaction of central plant hot recycling mixtures was easier than the new asphalt mixture. The higher the content, the easier the compaction.

References

1. JTG F41-2008, Technical specifications for Highway Asphalt Pavement Recycling [S]. Beijing: People Traffic Press, 2008.
2. Richard Willis, etc. Phase III NCAT Test Track Findings [R]. NCAT Report:09-08. Auburn, Alabama, Dec 2009.
3. David H. Timm. Design, Construction And Instrumentation Of The 2006 Test Track Structural Study [R]. NCAT Report:09-01. Auburn, Alabama, Feb 2009.
4. JTG E 20-2011 Standard Test Methods of Bitumen and Bituminous Mixtures for Highway Engineering [S]. Beijing: China Communication Press, 2011.
5. DB13/T 978-2008, Rotating compaction shear experiment method (GTM) asphalt mixture design and construction technical specification [S]. Hebei province, Hebei province bureau of quality and technical supervision, 2008.
6. GTM method of asphalt mixture design and construction technical specification [S]. Tianjin: Tianjin municipal highway administration, 2011.
7. Wu Chuanhai, Liu Yangshao. Proportion Design of Anti-rutting Asphalt Mixture Based on GTM [J]. Journal of Wuhan University of Technology (Transportation Science & Engineering), 2012, 36(1):51–55.
8. Tan Wei, Zhou Gang. The comparison on the stability of asphalt mixture at high temperature [J]. Petroleum Asphalt, 2009, 23(4):4–8.
9. Sun Lijun. Asphalt pavement structural behavior theory [M]. Beijing: People Traffic Press, 2005.
10. Cao Lintao, Li Lihan, Sun Daquan, Chen Jianjun. Research of Evaluation Indexes for Asphalt Mixtures Wheel Tracking Test [J]. Journal of Wuhan University of Technology, 2009, 29(11):14–17.
11. Cha Xudong, Yang Ping. Recycled asphalt mixture design penetration range [J]. Journal of China & Foreign Highway, 2007, 27(3):176–179.
12. Fang Yang, Wu Chuanhai. Compaction Characteristics of Central Plant Hot Recycling Mixture With High Content RAP Based on GTM [J]. Journal of Wuhan University of Technology (Transportation Science & Engineering), 2014, 38(1):181–183.
13. Zhang Zhengqi, Bian Xiuqi, Du Qunle, Shi Xin. Study on Factors Effecting on Compaction Property of Asphalt Mixture [J]. Journal of Wuhan University of Technology, 2012, 34(6):36–41.

Utilization of spent brewer's yeast for selenium-enriched yeast

N.X. Sun†, Y.H. Liu and Y.X. Wang
Biological Science and Technology College, University of Jinan, Jinan, Shandong, China
†E-mail: chm_sunnx@ujn.edu.cn
www.ujn.edu.cn

The aim of the study was to utilize the spent brewer's yeast for the preparation of the selenium-enriched yeast by using Plackett-Burman design and Box-Behnken design. The Plackett-Burman multifactorial design was first employed to screen the significant factors in the fermentation for the selenium yield, and subsequent use of the response surface methodology was further optimized for the selenium yield by Box-Behnken design. The important factors in the culture medium, identified by the initial screening method of Placket-Burman, were sodium selenite, H_3PO_4 and the biomass of the spent brewer's yeast. The optimal amounts for maximum selenium yield were: sodium selenite 18.0 mg/L; H_3PO_4 1.8 g/L in 250 mL flask and the biomass of the spent brewer's yeast 20.2 g/L. Using this statistical experimental design, the selenium yield by the spent brewer's yeast under optimal condition reached about 1679.32±31.87 μg selenium /g dry cell.

Keywords: Spent brewer's yeast; Selenium yield; Plackett-Burman design; Box-Behnken design.

1. Introduction

Spent brewer's yeast is a by-product of beer production and the amount produced by the brewery industry can reach even 2.5% of beer produced. Most commonly it is used as protein-rich animal feed supplement, although its share in the feed is limited due to its bitter taste, the rest is discarded as industrial waste.[1] So the spent brewer's yeast revaluation is necessary and very important as an environmental protection measure. Some potential applications for spent brewer's yeast utilization have been made in the pharmaceuticals production, extracts production and β-glucan extraction.[2–5] Besides the production of industrial products, spent brewer's yeast has a great potential for enriching with minerals such as selenium.[6] In that case, spent yeast could earn an extra income for preparing selenium-enriched bakers' yeast as a high-value product.

† Work financially supported by the Shandong Science and Technology Planning Project in Colleges and Universities (J12LF10).

Selenium (Se) is an essential trace element with beneficial properties as nutrient for humans and animals health.[7] It is a component of a number of important selenoproteins and enzymes required for such functions as antioxidant defense, reduction of inflammation, thyroid hormone production, DNA synthesis, fertility and reproduction, inhibition HIV progression to AIDS, and immune function.[8,9] In the last two decades, there has been increasingly progress in our knowledge and understanding of the biological roles of Se and its importance in human or animal nutrition.

It is generally believed that the ingestion of organic Se compounds is better and safer than that contained inorganic Se.[8] Studies on the effects of Se have suggested that inorganic Se is more toxic and less bioavailable than its organic forms, and one potential reason for differences in genotoxicity observed among selenocompounds is their distinct metabolism. In the present the best organic selenium sources are selenomethionine and selenium yeast, and the selenium yeast can be used as human dietary supplements and therapeutic product as well.

Under appropriate conditions yeasts are capable of accumulating large amounts of selenium and incorporating them into organic compounds, mostly in the form of selenium-containing proteins and selenium-containing polysaccharides.[8] In the preparation of selenium-enriched yeast, fresh yeast, *Saccharomyces sp.*, is often used and fermented to transform the inorganic selenium, selenite, to organic selenium compounds. While in this study the spent brewer's yeast discarded after four fermentation processes was reused to incorporate the inorganic Se to organic selenium-containing yeast. The objectives of this paper were to select the fit conditions by Plackett-Burman design and determine their optimal concentrations in the culture medium using a factorial Box-Behnken design methodology.[10] With these statistical designs, a quadratic polynomial regression model could be substantially established to increase the Se yield, so that to provide scientific basis for its industrialization in the future.

2. Material and Methods

2.1. *Microorganism*

The yeast used was a spent brewer's yeast slurry (a strain of *Saccharomyces cerevisiae*), a by-product from a brewery, with a solid content of ~20% provided by a Brewery from Qingdao. The spent yeast was washed three times with three water volumes. After washing by centrifugation the yeast biomass moisture was 17.5% for yeast Generation 4 (after four fermentation cycles).

2.2. *Selenium yeast by spent brewer's yeast preparation*

Spent brewer's yeast was cultivated in a glass vessel with 100 mL culture medium for yeast growth and selenium solution in different conditions. The medium used for selenium-enriched yeast was composed of (g/L): $(NH_4)_2SO_4$ 8; Glucose 60; the amount of H_3PO_4, the sodium selenite and the biomass of the spent brewer's yeast were added according to the statistical experimental design. The microorganism was cultured in 250 mL Erlenmenyer flasks and incubated on a rotary shaker, where the culture temperature and the rotation rate were discussed in the experiment. After 24 h cultivation, the mycelium was separated from the broth by centrifugation at 10,000× g for 15 min to obtain the selenium-enriched yeast.

2.3. *Determination of the Se yield*

The yeast cells were obtained and then washed with deionized water by centrifugations (1,900× g, 10 min), and the precipitate were dried under vacuum to a constant weight. As only the inorganic form of Se compounds can be detected, the Se yield in yeast has to be digested thoroughly and transformed to the inorganic compounds. The inorganic Se determination was carried out using a fluorescence spectrometric method with some modifications. About 300 mg portion of the sample were digested (100°C, 6 h) with 20 mL of a mixture of concentrated HNO_3, $HClO_4$ and H_2SO_4 (v:v:v = 10:4:5) in a digestive flask, until the organic Se form in yeast was completely digested to inorganic Se. To avoid the volatilization loss of Se, a reflux equipment was employed during digestion process.

After digesting, the solution was cooled and made a constant volume with water. The blank was digested in the same way. The mixture was centrifuged at 8,300× g for 15 min, and then the supernatant was filtrated and the filtrate could be analyzed directly. The results were presented as µg/g, the quantity of Se per 1 g yeast cell dry biomass.

2.4. *Optimization procedure*

The optimization of the cultivation for preparation of the selenium-enriched yeast by the spent brewer's yeast was carried out in two stages:

Plackett-Burman design based on the first order model: This was used to screen the important variables that influence the Se yield during cultivation. Each variable is represented at two levels, high and low which are denoted by (+1) and (-1), respectively. The coded level of each variable is shown in Table 1 and the experimental design with twelve trials for screening six variables is shown in Table 2. All experiments were performed in duplicate and the mean values are given. The variables, whose *p*-values were lower than 0.05 in Table 1, were considered to be the significant factors.

Response surface methodology (RSM) was used to optimize the screened components for enhancing the Se yield. A Box-Behnken factorial design with three factors and three levels, including five replicates at the centre point, was used for fitting a second order response surface. The factors, their values, and the experimental design were given in Table 3, respectively. This methodology allows the modeling of a second order equation that describes the process. The Se yield of the Se-enriched yeast was analyzed by multiple regression through the least squares method to fit the following equation 1:

$$Y=\beta_0+\sum \beta_i x_i+\sum \beta_{ij} x_i x_j+\sum \beta_{ii} x_i^2. \qquad (1)$$

Where Y is the measured response variable; β_0, β_i, β_{ij}, β_{ii} are constant and regression coefficients of the model, and x_i, x_j represent the independent variables in coded values.

Data from the Box-Behnken factorial design for the optimization of the Se yield in Se-enriched yeast was subjected to a second-order multiple regression analysis using the least squares regression methodology to obtain the parameter estimators of the mathematical model. The regression analysis and analysis of variance (ANOVA) were carried out using the response surface regression analysis (RSREG) procedure of the Design Expert v7.1.6 to fit second order polynomial equations for all response variables. Canonical analysis, which is used to predict the shape of the curve generated by the model, was carried out as well.

2.5. *Statistical analysis*

All trials were carried out in triplicate and all the data were reported as means standard deviation (SD). The statistics significance was evaluated using Student's *t*-test and $p < 0.05$ or 0.01 was taken as significant.

3. Results and Discussions

3.1. *Screening of important factors for Se yield by the spent brewer's yeast*

The maximum and minimum levels of variables chosen for trials in Plackett-Burman experimental design represent a big variation in nutrient sources. The design for 12 trials with two levels of concentrations for each variable with the Se yields is shown in Table 1. The variables represent six different affecting factors. The results of statistical analysis shown in Table 1 suggested that the *p*-value of X_7 (H_3PO_4), X_9 (sodium selenite) and X_{11} (the biomass of the spent brewer's yeast) were less than 0.05, which are considered to significantly influence Se yield. Other independent variables with *p*-value above 0.05 were generally considered insignificant. After the first experiment, the conditions were reduced to three major variables by the Plackett-Burman experimental design, suggesting that Plackett-Burman design is a powerful tool for screening fermentation factors. The exact optimal values of the individual factors were still unknown but could be determined by the subsequent Box-Behnken design.

Table 1. Variables and their levels employed in Plackett-Burman design for screening of culture conditions affecting Se yield by spent brewer's yeast.

Variable code	Variable	Value		*p*-Value
		-1	+1	
X_2	pH	4.4	6.4	0.0510
X_4	Glucose (g/L)	3.0	6.0	0.0530
X_6	$(NH_4)_2SO_4$ (g/L)	0.4	0.8	0.0701
X_7	H_3PO_4 (g/L)	0	10.0	0.0244
X_9	Sodium selenite (mg/L)	10.0	20.0	< 0.0001
X_{11}	Spent brewer's yeast (g)	20.0	30.0	< 0.0001

Table 2. Plackett-Burman experimental design matrix for screening of factors affecting Se yield by spent brewer's yeast.

Run	X_1	X_2	X_3	X_4	X_5	X_6	X_7	X_8	X_9	X_{10}	X_{11}	Y
1	+1	-1	-1	-1	+1	-1	+1	+1	-1	+1	+1	888
2	-1	-1	-1	+1	-1	+1	+1	-1	+1	+1	+1	836
3	+1	+1	-1	-1	-1	+1	-1	+1	+1	-1	+1	1552
4	-1	+1	+1	+1	-1	-1	-1	+1	-1	+1	+1	528.4
5	+1	-1	+1	+1	+1	-1	-1	-1	+1	-1	+1	1142.8
6	-1	+1	-1	+1	+1	-1	+1	+1	+1	-1	-1	817.6
7	-1	-1	+1	-1	+1	+1	-1	+1	+1	+1	-1	1536
8	+1	+1	-1	+1	+1	+1	-1	-1	-1	+1	-1	1163.6
9	+1	+1	+1	-1	-1	-1	+1	-1	+1	+1	-1	1609.6
10	-1	-1	-1	-1	-1	-1	-1	-1	-1	-1	-1	919.6
11	-1	+1	+1	-1	+1	+1	+1	-1	-1	-1	+1	445.6
12	+1	-1	+1	+1	-1	+1	+1	-1	-1	-1	-1	327.6

3.2. *Optimization of screened sources for Se yield by spent brewer's yeast using response surface methodology*

Three variables of X_1 (Sodium selenite), X_2 (the biomass of the spent brewer's yeast) and X_3 (H_3PO_4) were chosen for further study based on Box-Behnken design. Table 3 shows the 17 experimental runs with different combinations of the three variables along with experimental responses. As shown in Table 3, a considerable variation in the Se yield was found, depending on the levels of the three variables in the medium. The maximum Se yield was found to be 2062.253 μg(Se)/g(dry yeast) in run number 7 and the minimum 1008.803 μg(Se)/g(dry yeast) in run number 1. For estimation of the RSM model errors, the centre point in the design was repeated four times.

Table 3. Box-Behnken experimental design matrix with experimental values of Se yield by spent brewer's yeast.

Runs	Sodium selenite (X_1)(mg/L)	Biomass (X_2)(g/L)	H_3PO_4 (X_3) (g/L)	Se yield (μg/g)
1	12.00	23.00	1.70	807.0
2	18.00	20.00	1.70	1549.8
3	24.00	17.00	1.70	1595.2
4	18.00	23.00	2.00	1232.2
5	18.00	23.00	1.40	1254.7
6	12.00	17.00	1.70	1053.7
7	18.00	17.00	1.40	1649.8
8	24.00	20.00	2.00	1569.0
9	18.00	17.00	2.00	1596.9
10	18.00	20.00	1.70	1507.1
11	18.00	20.00	1.70	1495.3
12	18.00	20.00	1.70	1437.7
13	18.00	20.00	1.70	1471.7
14	12.00	20.00	2.00	894.2
15	12.00	20.00	1.40	869.8
16	24.00	20.00	1.40	1567.6
17	24.00	20.00	1.70	1523.4

By applying multiple regression analysis on the experimental data, the following second order polynomial equation was found to explain the Se yield by only considering the significant terms and is shown below:

$$Y = 1865.5 + 411X_1 - 168.5X_2 - 7.5X_3 + 54.5X_1X_2 - 7X_1X_3 + 9.5X_2X_3 - 285X_1^2 - 24.5X_2^2 - 49X_3^2 \quad (2)$$

Where Y is the predicted response; X_1, X_2, X_3 are coded values of Sodium selenite the biomass and phosphoric acid, respectively.

Analysis of variance (F-test) showed that the model is well adjusted to the experimental data. The coefficient of variation (CV) indicates the degree of precision with which the treatments are compared. The lower value of CV (2.20) indicates a greater reliability in the experiments performed. The determination coefficient (R_2), which is commonly used to check the goodness of the model, implies that the sample variation of 99.4% for Se yield is attributed to the independent variables in the current study. The value of R (0.9871) for Equation

(2) being near to 1.0000 indicates a close agreement between the experimental results and the theoretical values predicted by the model equation. Linear, cross product were significant at the 1% level. Therefore, the quadratic model was selected in this optimization study.

Equation (2) clearly demonstrated that X_2 and X_3 had positive effects on total Se yield, while X_1 had a no significant effect on it. The absolute value for the coefficient of X_1 is larger than those of X_2 and X_3, which indicated that the linear term influence of X_1 was more significant than those of X_2 and X_3. The function was not linear and indicated the importance of interactions among X_1, X_2 and X_3. The coefficients of X_1X_2 are obviously larger than that of X_1X_3 and X_2X_3.

The statistical optimal values of variables are obtained when moving along the major and minor axis of the contour and the response at the centre point yields the maximum Se yield. These observations were also verified from canonical analysis of the response surface. Canonical analysis revealed a minimum region for the model. The stationary point presenting maximum Se yield had the following critical values: sodium selenite, 15.8 mg/L; glucose, 40.2 g/L; the liquid volume 120 mL; the other conditions included $(NH_4)_2SO_4$, 8.0 g/L; $MgSO_4$ 5.00 g/L, K_2HPO_4 1.00 g/L, KH_2PO_4 1.00 g/L; the natural pH about 5.0; the inoculum size 106; and the temperature 28°C. The predicted Se yield for these conditions was 1661.82 μg/g. A repeat fermentation by the spent brewer's yeast under optimal condition was carried out to verify the optimization. The maximum Se yield obtained was 1679.32±31.87 μg/g, which was close to the predicted value.

4. Conclusion

Statistical optimization of the nutrient sources in a medium using Plackett-Burman and Box-Behnken design appears to be a valuable tool for the yield of the enriched Se by *Saccharomyces sp.* III. Plackett-Burman design made it possible to consider a number of variables. As Se was a biomass related product, the increase of biomass enhance the activity of the most important selenic enzyme — Glutathione Peroxidase (GSH-Px), which is critical for survival and function of aerobic organisms because it can effectively inhibit oxidation and prolong life of the organism. Sodium selenite, H_3PO_4 and the biomass of the spent brewer's yeast were chosen as the conditions for Se yield by the spent brewer's yeast in the Plackett-Burman design. The main factors and their concentrations in the medium were further optimized by Box-Behnken design. The predicted and verifiable Se yield under optimal conditions was

1679.32±31.87 μg/g. The next experiment in Se-enriched yeast by the spent brewer's yeast would be adjusting the medium compositions or conditions for fermentation carried out in a 10-L bioreactor.

References

1 J. Dunarea and D. Danube, Selenium yeast from spent brewer's yeast, *J. Agr. Pro. Technol.* 16(3): 341–345(2010).

2 V. Barbosa and A. Nadja, Followed extraction of β-glucan and mannoprotein from spent brewer's yeast and application of the obtained mannoprotein as a stabilizer in mayonnaise. *Inno. Food Sci. Emer. Technol.* 23:164–170(2014).

3 W. Pietrazk and J. Kawa–Rygielska, Utilization of spent brewer's yeast for supplementation of distillery corn mashes, *Pol. J. Chem. Tech.* 15(4):102–106(2013).

4 I. Ferreira and O. Pinho, Brewer's *Saccharomyces yeast* biomass: characteristics and potential applications. *Trends Food Sci. Technol.* 21:77–84(2009).

5 A. Halasz and A. Lasztity, Use of yeast biomass in food production, CRC Press (1991).

6 G. Marinescu and A. Gabriela, Industrial nutrient medium use for yeast selenium preparation, *AUDJG-Food Technol.* 35(1):45–53(2011).

7 T. Nagodawithana and F. Gutmanis, Method for the production of selenium yeast, *Patent* US 4530846(1985).

8 M. Rayman, The importance of selenium to human health, *The Lancet.* 356(9225):233–241(2000).

9 T. Kaur and M. Bansal, Selenium enrichment and anti-oxidant status in bakers yeast *Saccharomyces cerevisiae* at different sodium selenite concentrations, *Nutri. Hospital.* 21(6):707–708(2006).

10 H. Yin and Z. Chen, Optimization of natural fermentative medium for selenium-enriched yeast by D-optimal mixture design, LWT-Food Sci. Technol. 42:327–331(2009).

One-bath dyeing of wool/acrylic blend using a model quaternised S-triazinyl cationic reactive dye

Y. Zhang and W. Zhang†
Light Chemical Engineering Department, Yancheng Institute of Industry Technology, Yancheng, Jiangsu 224005, China
†E-mail: zhangwlsh@163.com

A model quaternised S-triazinyl cationic reactive dye was dyed on wool/acrylic blend in one-bath and its dyeing and fastness properties were investigated. The results indicate that the reactive cationic dye having this structure give a high level of exhaustion and fixation on wool, acrylic and wool/acrylic blend. Levelling, and fastness properties on the wool/acrylic blend were also in general very good.

Keywords: Wool/acrylic blend; One-bath dying; Reactive dye; Quaternised S-triazinyl.

1. Introduction

Acrylic fiber is a less expensive alternative to wool and has, in many respects, replaced wool. The outstanding chemical and physical properties of acrylic fibers, such as high strength, good abrasion and insect resistance, have led to many applications in apparel as well as in the production of wool/acrylic fibers, which are widely used in knitwear.

Conventionally, acrylic fibers, which are spun from a copolymer of acrylonitrile containing 1–15 percentage mass of several vinyl comonomers containing carboxylate or sulphonate groups, are dyed with cationic dyes [1, 2]. In contrast, wool fibers are conventionally dyed with acid and reactive dyes because of the presence of functional groups, such as amino groups in the fibers [3], which would interact mainly with ionic and covalent bonds with these dyes, respectively. A one-bath process is cheaper and less time consuming, but is possible only if reaction between the two types of dye in the dyebath can be avoided. This normally entails the use of large amounts of surfactant which is unwelcome in the effluent. Therefore, the dyeing of wool/acrylic fibers is traditionally conducted in a two-bath process, one for acid or reactive dyeing and the other for cationic dyeing of wool and acrylic components, respectively.

Alternatively, a one-bath dyeing method using one dye would be of interest, as it offers the desired union shade, together with energy saving. Reactive cationic dyeing is a possible alternative [4–5], and a recent work has been reported using monofluorotriazine reactive cationic dyes on wool/acrylic blends [2]. The monofluorotriazine reactive cationic dyes have integrated structure of cationic dyes and reactive dyes, which can ionic band with acrylic fiber and covalent bond with wool fiber. Moreover, the special reactive group, namely, monofluoro-striazine (MFT) group, will fix efficiently to wool fiber, which is readily damaged by alkali, at the boil under acidic or neutral condition.

In the present paper, wool/acrylic blends were dyed in one-bath by using a model quaternised S-triazinyl cationic reactive dye. Like monofluorotriazine reactive cationic dyes, the model quaternised S-triazinyl cationic reactive dye also has integrated structure of cationic dye and reactive dye. It is noteworthy that quaternised S-triazinyl cationic reactive dyes possess an alkali-free fixation group, namely, quaternised S-triazinyl reactive group [6–8]. The model quaternised S-triazinyl cationic reactive dye, Dye1, was designed and synthesized by the authors' research group [9]. The molecular structure of Dye1 is shown in Figure 1. Its behavior on a wool/acrylic blend has been studied using a one-bath dyeing technique. The dyeing performance of the dye was assessed by determining its exhaustion and fixation on wool and acrylic fiber, both separately and in a blend. Its levelling, build-up and fastness properties have also been determined.

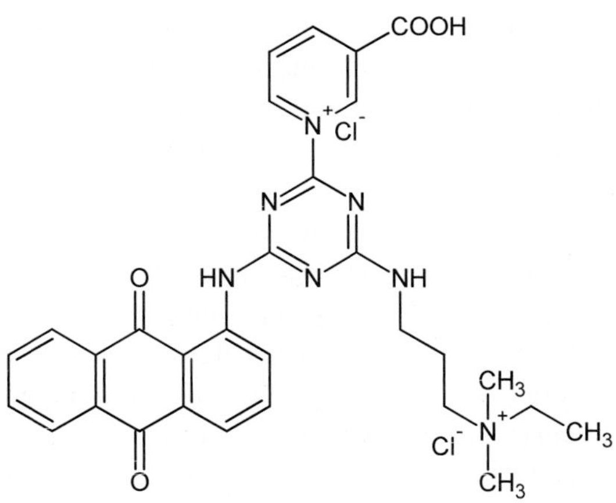

Figure 1. Molecular structure of Dye1.

2. Experimental

2.1. *Materials*

Cotton fabrics for color fastness tests (GB/T 7568.2-2008) and acrylic fabrics for color fastness tests (GB/T 7568.5-2002) were obtained from Shanghai textile industry institute of technical supervision and used for the dyeing experiments. Knitted wool/acrylic fabric blend (50:50) was obtained from Changshu Changhe Knitting & Textile Co., Ltd. The quaternised S-triazinyl cationic reactive dye, Dye1, was synthesized in previous work [9]. Other chemicals were from Shanghai Chemical Reagent Plant, China.

2.2. *Dyeing of fabrics*

Wool, acrylic and wool/acrylic fabrics were dyed in an IR dyeing machine (Pyrotec-2000) at liquor ratio 40:1.

For wool and acrylic fabrics dyeing, the dyebath containing 1% owf of Dye1 was maintained at pH 4, 5, 6 and 7 using an acetic acid–sodium acetate buffer solution, respectively. The substrates were immersed in the dyebath at 40°Cand the temperature increased to 70, 80, 90 and 100°C at 1°C /min. Dyeing was continued at this temperature for 60 min. After dyeing, each dyed sample was squeezed to recover residual liquor and rinsed with water until the rinsing water was clear. The rinsing water was added to the residual liquor and the absorbance measured.

For wool/acrylic blends dyeing, the dyebath containing 1.0, 2.0, 4.0, 6.0, 8.0 10.0% (on weight of fabrics, owf) of Dye1 was maintained at pH 7. The substrates were immersed in the dyebath at 40°C and the temperature increased to 100°C at 1°C /min. Dyeing was continued at this temperature for 60 min. After dyeing, each dyed sample was squeezed to recover residual liquor and rinsed with water until the rinsing water was clear. The rinsing water was added to the residual liquor and the absorbance measured.

2.3. *Exhaustion and fixation rate measurement*

Hitachi U-2000 UV–vis spectrophotometer (Hitachi, Japan) was used to measure the UV–vis absorbance of the dye solutions before and after exhaustion. The concentration of dye was calculated based on a calibrated absorbance-concentration relationship at the λmax of the dyes. After dyeing, unfixed dye from the samples was extracted by boiling water and also measured by the UV–vis spectrophotometer. The exhaustion rate E of the dye on the fibers was calculated by Eq. (1).

$$E = \left(\frac{A_0 - A_1}{A_0} \right) \times 100\% \tag{1}$$

where A_0 is the initial the absorbance of the dye solutions before exhaustion, A_1 is the absorbance of the dye solutions after exhaustion.

The fixation rate F on the fabric was calculated by Eq. (2).

$$F = \left(\frac{A_0 - A_1 - A_2}{A_0 - A_1} \right) \times 100\% \qquad (2)$$

where A_2 is the absorbance of the dye in the extracted solution.

2.4. *Color measurement*

The relative color strength of dyed fabrics, expressed as K/S, where K and S are the absorption and scattering coefficients, respectively, was measured by the light reflectance technique using the Kubelka–Munk Eq. (3).

$$K/S = \frac{(1-R)^2}{R} \qquad (3)$$

The reflectance (R) and the levelling properties of the dyed samples were measured using a HunterLab (USA) Ultra Scan PRO spectrophotometer interfaced with a personal computer. In order to achieve correct readings, the spectrophotometer was set to exclude any specular components. Five readings per sample of one layer (moving the sample in between) were averaged.

The levelling properties of dyed fabrics were assessed by measuring the color differences with in each sample at five separate points and the average color difference (ΔE) between these points was determined.

2.5. *Fastness testing*

Color fastness was determined according to the respective international standards: fastness to washing, ISO 105-C04; fastness to perspiration, ISO 105-E04; fastness to rubbing, ISO 105-X12; and fastness to light, ISO 105-B02.

3. Results and Discussion

3.1. *Dyeing of wool and acrylic*

To investigate the dyeing of quaternised S-triazinyl cationic reactive dye on wool/acrylic blends, dyeing properties were firstly investigated on wool and acrylic separately.

3.1.1. *Effect of temperature*

A series of dyeings using 1% owf Dye 1 were carried out under different temperatures at pH 7. Table 1 shows the effect of temperature on fixation and exhaustion on wool and exhaustion on acrylic with Dye 1. It can be seen that the exhaustion of Dye1 on acrylic is very low below 90°C, and the exhaustion at 100°C is satisfactory. These phenomena is related to the glass transition temperature (Tg, 80-85°C) of acrylic fabric and the difficult diffusion of dye molecules inside the acrylic fiber at lower temperatures. Table 1 also shows the effect of temperature on exhaustion and fixation on wool fabric. Excellent exhaustion of Dye1 on wool emerged at temperature 90°C and 100°C. The phenomena is attributed to the open temperature (above 80°C) of scale layer on wool fiber surface and the difficult diffusion of dye molecules inside the wool fiber at lower temperatures. Moreover, fixation was also greatly affected by temperature. The results suggest that quaternised S-triazinyl was able to combine covalently with nucleophiles on the wool, under high temperature (100°C), in agreement with the literature reported [7]. Based on the above analysis, it is recommended that wool/acrylic blends should be dyed under 100°C.

Table 1. Effect of temperature on fixation and exhaustion.

Temperature (°C)	Fixation (%)	Exhaustion (%)	
	Wool	Wool	Acrylic
70	32.1	49.5	30.9
80	34.5	64.1	32.4
90	45.3	91.2	67.4
100	77.8	90.5	87.6

3.1.2. *Effect of pH*

A series of dyeings using 1% owf Dye 1 were carried out under 100°C at pH 4, 5, 6 and 7. Table 2 shows the effect of pH on fixation and exhaustion on wool and exhaustion on acrylic with Dye 1. Exhaustion of the dye on wool was very low at pH 4. After dyed at pH 5, the exhaustion markedly increased to 88.7% from 32.1%. Then the exhaustion rates increased slowly with the pH increasing. That is because, if the pH value is higher than the isoelectric point of wool (4.2–4.8), $-OOC-W-NH_2$ ions are formed and the wool fiber becomes negatively charged. This assists the exhaustion of cationic dyes.In contrast, the exhaustion of the dye on acrylic decreased slowly with the pH increasing. But for all that, excellent exhaustion rates were observed at each pH value. It can be seen that the fixation of the dye on wool was greatly influenced by pH value. The fixation rates were very poor below pH 7, indicating that quaternised S-triazinyl reactive groups did not well combine covalently with nucleophiles on the wool. After dyed at pH 7, the ideal fixation rate of 77.4% was obtained. This was since this coincided with the neutral-fixing characteristic of quaternised S-triazinyl [7].

Based on the above analysis, it is recommended that wool/acrylic blends should be dyed at pH 7.

Table 2. Effect of pH on fixation and exhaustion.

pH	Fixation (%)	Exhaustion(%)	
	Wool	Wool	Acrylic
4	25.3	32.1	98.2
5	24.9	88.7	97.8
6	35.7	90.1	89.1
7	77.4	90.8	87.8

3.1.3. *Effect of time*

A series of dyeings using 1% owf Dye 1 were carried out under 100°C at pH 7. The exhaustion rates of Dye1 on wool and acrylic at different time were tested at interval of 10 min. The exhaustion curves for Dyes 1 on wool and acrylic were shown as Figure 3. As shown in Figure 3, there was little dye exhaustion when the time was below 40min (80°C). The exhaustion rates of the dye increased remarkably when the temperature exceeded 80°C. Within 20 min of holding time at 100°C, the dye exhaustion became constant, indicating that the sorption equilibrium had been reached. For wool fabric, the exhaustion rates of the dye were low when the time below 50min (90°C). Then the exhaustion rates increased slowly with the dyeing time increasing. Within 60 min of holding time at 100°C, the dye exhaustion became constant, indicating that the exhaustion equilibrium had been reached. Based on the information in Figure 2, a temperature profile for dyeing can be recommended. The initial dye bath temperature may be raised quickly to 80°C, after that, the temperature should be raised slowly. And the recommended holding time at 100°C was about 60 min to allow the complete diffusion of dye molecules into the fiber interior.

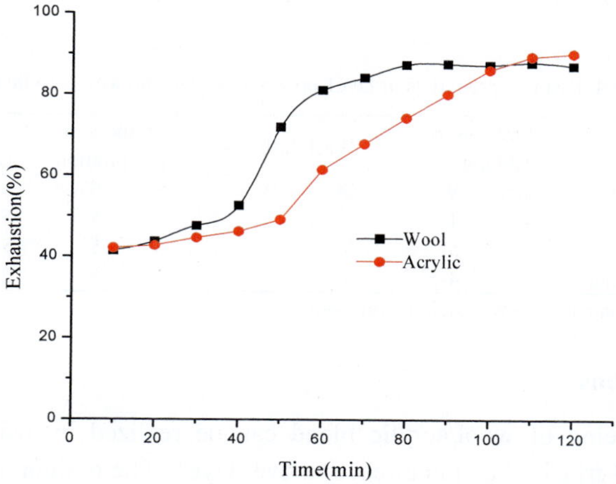

Figure 2. The exhaustion curvesfor Dyes 1 on wool and acrylic.

3.2. *One-bath Dyeing of wool/acrylic blend*

Table 3 shows the exhaustion of Dye1 on the wool/acrylic blend. The results clearly confirm the good exhaustion of these dyes on the wool/acrylic blend. Under the conditions used for dyeing, the dyes reacted not only with thiol and amide groups on the wool but also underwent hydrolysis by the water. The hydrolysed dyes, however, remained cationic in character and were thus able to dye the acrylic. The levelling properties of wool/acrylic fabrics dyed with Dye1 are also given in Table 3. It is clear that the average color differences (ΔE) of the dyed fabrics show good levelling properties in all concentrations(ΔE < 1).

Table 3. Exhaustion and leveling properties of Dye1 on wool/acrylic blend.

Concentration (%, owf)	Exhaustion (%)	Levellingproperties (ΔE)
1	90.7	0.31
2	90.1	0.44
4	88.7	0.63
6	85.5	0.54
8	81.1	0.49
10	78.6	0.61

3.3. *Fastness properties*

The fastness results for Dye1 on wool, acrylic and wool/acrylic fabric blend are summarised in Table 4. The results show that the fastness to rubbing, washing, perspiration and light on wool, acrylic and wool/acrylic fabric blend were good throughout.

Table 4. Fastness properties of Dye1 on wool, acrylic and wool/acrylic blend.

Fabric	Fastness to rubbing		Wash fastness		Fastness to perspiration		Light fastness
	Dry	Wet	SC	SW	SC	SW	
Wool	4	4	5	5	5	5	5
Acrylic	5	5	5	5	5	5	5-6
Wool/acrylic blend	4-5	4-5	5	5	5	5	5

[a] SC = staining on cotton; SW = staining on wool

4. Conclusions

One bath dyeing of wool/acrylic blend can be realized by using a a model quaternised S-triazinyl cationic reactive dye, Dye1. The optimal dyeing process of this dye for wool/acrylic blend was holding temperature at 100°C for 60 min. The results of exhaustion, fixation, levelling properties, and fastness properties on the wool/acrylic blend were in general very good. One bath dyeing process of wool/acrylic blend has the advantages of simple process, simple dyeing recipes, union shade, together with energy saving.

References

1. L. K. El Gabry, Effect of mineral acids on the properties of acrylic fabrics. Color Technol., 236, 120(2004).
2. K. L. Xie and A. Q. Hou, One-bath dyeing of wool/acrylic blends with reactive cationic dyes based on monofluorotriazine. Color Technol., 307, 120(2004).
3. X. Zhu, X. Q. Zhou, W. Zhang. One-pot reductive amination of araldehydes by aniline using borohydride with $CeCl_3.7H_2O$ as catalyst. J. Chem. Res., 390, 39(2015).
4. C. L. Zheng, A. L. Yuan, H. B. Wang and J. Sun, Dyeing properties of novel electrolyte-free reactive dyes on cotton fibre. Color Technol., 204, 128(2012).
5. K. Srikulkit and P. Santifuengkul, Salt-free dyeing of cotton cellulose with a model cationic reactive dye. Color Technol., 398, 116(2000).
6. N. Morimura and M. Ojima, New reactive dyes fixable at neutral pH. Dystuff Rep., 28, 74(1985).
7. S. N. Croft, D. Lewis, R. Orita, and T. Sugimoto, Neutral-fixing reactive dyes for cotton. Part 1 – synthesis and application of quaternised S-triazinyl reactive dyes. Color Technol., 195, 108(1992).

8. J. A. Bone, T. T. Le, D. A. S. Phillips and J. A. Taylor, One-bath dyeing of polyester/cotton blends with disperse and bis-3-carboxypyridinium-*s*-triazine reactive dyes. Part 2: Application of substantive reactive dyes to cotton at 130 °C and neutral pH. Color Technol., 117, 124(2008).
9. Y. Zhang and W. Zhang, Clean dyeing of cotton fiber using a novel nicotinic acid quaternary triazine cationic reactive dye: salt-free, alkali-free, and non-toxic by-product. Clean Technol. Envir., 563, 13(2015).

The aerodynamic performance and acoustic property of wings with tubercles along the leading edge

S.S. Li, B. Wang and X.X. Zhang
Department of Automation, University of Science and Technology of China, Hefei, Anhui, China
Institute of Intelligent Machines, Chinese Academy of Science, Hefei 230031, Anhui, People's Republic of China

L.K. Zhu and D.Y. Kong*
State Key Laboratory of Transducer Technology, Institute of Intelligent Machines, Chinese Academy of Science, Hefei 230031, Anhui, People's Republic of China
**E-mail: kongdy@iim.ac.cn*

A.P. Liang
Institute of Zoology, Chinese Academy of Sciences, Beijing, China

Inspired by the silent flight of owls, airfoils with tubercles along the leading edge were found to show acoustic superiorities compared with the oneswithout tubercles. Based on this study, an airfoil with tubercles along the leading edge was modeled and fabricated. The differences of aerodynamic performance and acoustic property between the baseline airfoil and modified one were shown by the simulations based on the DES solver and ACTRAN. By analyzing the flow field around the airfoils, a mechanism of noise reduction could be concluded.Besides, tubercles along the leading edge could delay stall and enhance the lift coefficient after stalling with little drag penalty. Such a structure could be widely used in many engineering fields, such as wind power generators and turbines, whichcost less energy and benefited noise reduction.

Keywords: Engineering bionics; Tubercles; Aerodynamic performance; Acoustic property.

1. Introduction

In the past few decades, small unmanned aerial vehicles (UAVs) were widely used in a lot of fields [1-3], such as surveillance, reconnaissance and detection of biological, chemical, or nuclear materials. The noise of UAVs was high when they were used to execute reconnaissance low-altitudemissions, this could result in a failure. Therefore it is a hot issue to search for the potential methods to reduce the aerodynamic noise radiating from the UAVs for a long time. Creatures in the nature have evolved for a long time, they optimized the adaptability and unique function features to fit well with living environment. For instance, owls have the distinguished ability of a silent flight [4]. It was found that the morphology of non-smooth leading edges of Asiootus was of vital importance to lower the air flow noise during their fights [5–8]. L.Q. Ren [5] et al. found that the noise increase can be up to 7dB with the leading edges covered by plastic film. As is well known in the fluid dynamic literature, relatively small modifications near a leading edge can substantially alter, delay or reduce separationeffects. This is most often achieved by tubercles such as vortex generators that enhance boundary layer attachment on wings [9–10]. To some extent, the rise of lift means the noise reduction for equal lift value. The non-smooth leading edge as a flow control of passive mechanisms was easy to be fabricated [11], so it was a reasonable method to improve the performance of wings of UAVs.CAA (Computational Aeroacoustic) was widely used to simulate the aeroacoustic field along with CFD (Computational Fluid Dynamics), the results had a good agreement with the experimental tests [7]. In this study, tubercles were employed for NACA2412 airfoil to obtain the differences between the baseline airfoil and the modified one. Numerical computations and experimental tests were implemented to explore the essential mechanisms of noise reduction.

2. Models Creations

It is well known that long-eared owls have such a silent flight that their prey is unaware of the danger, the ability of flying silently is dependent on its special structure, such as the non-smooth morphology along the leading edge. This structure alters the flow pattern around the airfoils, this may be the reason why they could fly quietly. Figure 1 is the close-up of an owl, obvious tubercles were found along the leading edge. Because of the asymmetry of owl's wings, the models obtained by the scanning system could not be used directly to conduct experiments. NACA2412 was regarded as a baseline airfoil, the non-smooth leading edge could be modified by tubercles along the leading edge. They were shown in detail in Figure 2. The tubercles were defined by the amplitude A and the wavelength λ.The number N of tubercles could be calculated by the formula:

$$N = S/\lambda$$

Pro-Engineer, a computer-aided design (CAD) program was used to design the models, the wings were fabricated with PLA material by a 3D printer. It was necessary to ensure that all the wings had the same planform area. The airfoils in this study had a chord length of c=40mm and a span of s =120mm. Here 2mm and 12mm were selected as the amplitude and the wavelength of the protuberances respectively.

Figure 1. Non-smooth leading edge of owls

Figure 2. Baseline airfoil and the airfoil with tubercles.

3. Wind Tunnel Experiments

3.1. *Experimental equipment and description*

The experiments of aerodynamic performance of airfoils were conducted in the FO-series low-speed wind tunnel in Institute of Electronics, Chinese Academy of Science. The measurements of noise reduction of wings were implemented in the semi-anechoic room of Institute of Intelligent Machines, Chinese Academy of Science. The semi-anechoic room has a length of 8.3m, a width of 4.2m and a height of 3.6m. The radius of free field is 3.5m, the limiting value of the low frequency is 20Hz, the total SPL of the background noise is 28dB. The measurement system consists of AWA6290M Spectrum Analyzer, precise impulse sound level meter of AWA5661, which has a range of 27 to 140 dB and an accuracy of ±0.7dB, its resolution ratio is 0.1dB. The measurement method adopted here abides by GB/ T2888-1991 standard. The experimental platform of testing the aerodynamic performance and acoustic property of both airfoils were set up according to GB1236-1985 standard. The monitoring points of noise generated by the airfoils had a distance shown as Figure 4. We regarded noise spectrum of one-third octave and total sound pressure level as the final indexes to evaluate the properties. The Reynolds number based on the chord length is 1.0×10^5, which corresponds to a Mach number of 0.10.

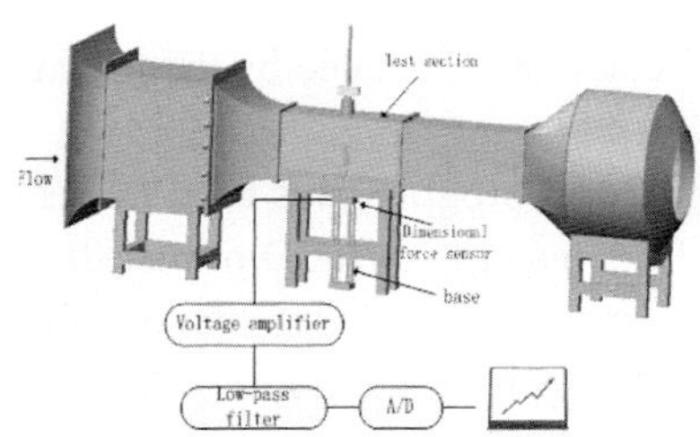

Figure 3. Schematic of experimental facility.

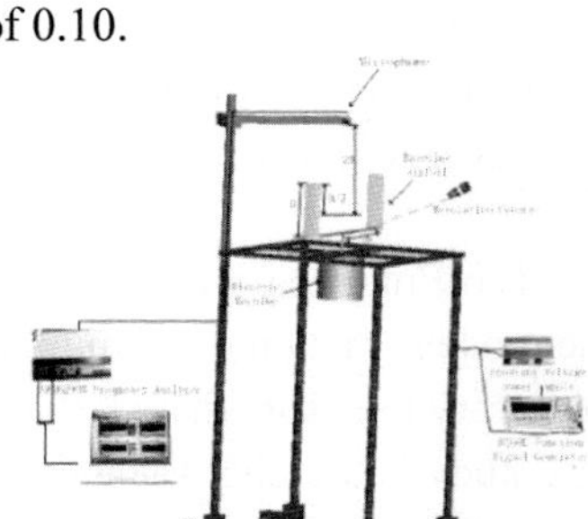

Figure 4. Platform of testing rotational noise.

3.2. *Experimental scheme and analysis*

In this study, lift and drag coefficients as a function of angle of attack (AOA) were compared between the baseline airfoil and the modified one. The sound pressure level of the airfoils was also considered as an important index to evaluate the acoustic property. The results of the wind tunnel testswere shown in Figure 5 and Figure 6.

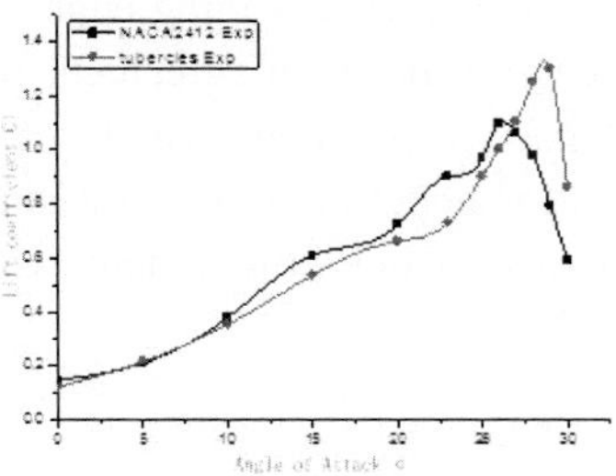

Figure 5. Lift coefficient versus AOA.

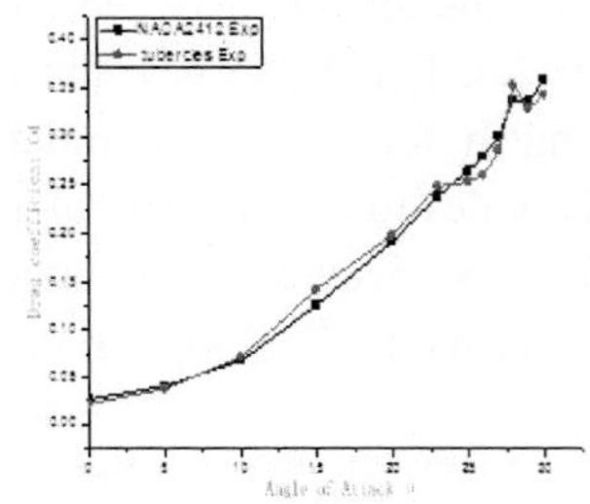

Figure 6. Drag coefficient versus AOA.

From the figures above, a superior performance could be obtained. The stall was delayed for the airfoil with tubercles compared with the baseline one. The maximum lift was shown to be somewhat greater for the model with waved leading edge. When AOA ranged from 0° to 10°, the modified airfoil almost had the same lift coefficient with the baseline one. When the modified airfoil wasat stall, it would get a higher lift coefficient and have a smoother decline. The modified airfoil nearly had the same drag coefficient with the baseline one. It was concluded that the structure of tubercles could enhance the aerodynamic performance without or with a little drag penalty. For detecting the sound pressure level around the airfoils, the sound collected by the microphone was shown at AOA of 29° asa form of 1/3 octave spectra in Figure 7. Among the most range of the frequency spectrum, the SPL was declined. The detailed reasons why this phenomenon occurred were explained later.

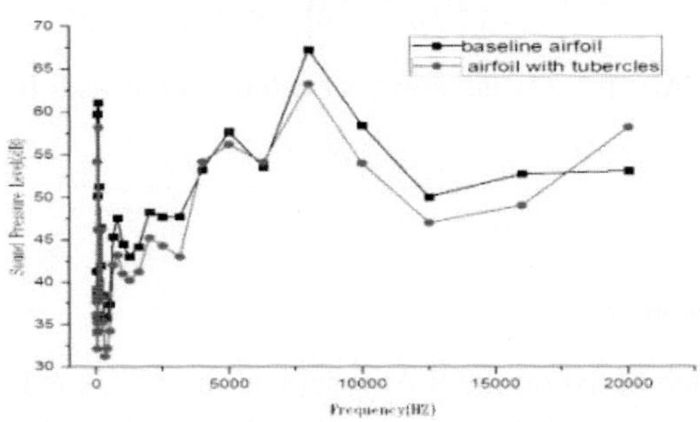

Figure 7. The comparison of the frequency spectrum of both airfoils at 29°.

4. Numerical Simulation

There were some sources of noise stemming from the wing models, such as leading edge noise generated from the interaction between the incoming flow and the leading edges of wings, the turbulent layers and trailing edges even the shed vortices by the trailing edges. In this research, the mechanisms of noise reduction could be concluded by analyzing the differences of the near flow around the airfoils between the smooth leading edge and the leading edge with tubercles. For ACTRAN, the mesh for CFD need not be in an agreement with the mesh for acoustic simulation. There is conservative integration from the CFD nodes to the acoustic mesh, all the information about the fluid flow can be saved, the accuracy can be ensured. The process about how Fluent works together with ACTRANwas displayed in Figure 8.

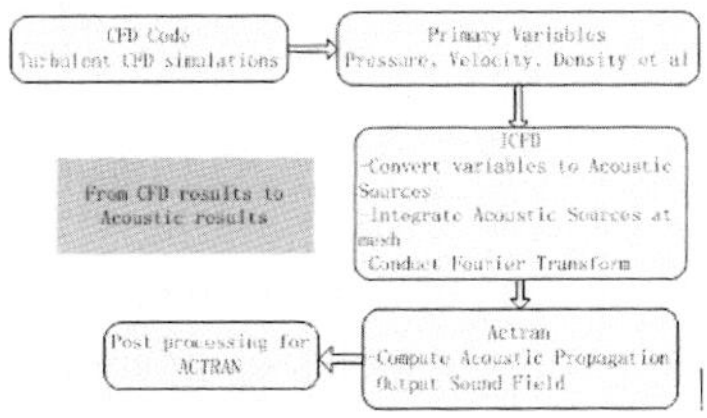

Figure 8. The process of all the simulations about the fluid and acoustic field.

4.1. *Governing equation*

Here, numerical analysis was performed using structural grids and solving the N-S equations, which comprised the conservation of mass, momentum and energy as shown in the following equations respectively:

$$\frac{\partial\rho}{\partial t} + \nabla.\,(\rho U) = 0 \tag{1}$$

$$\frac{\partial(\rho U)}{\partial t} + \nabla.\,(\rho U \times U + p\bar{\bar{I}}) = \nabla.\bar{\bar{\tau}} \tag{2}$$

$$\frac{\partial(\rho U)}{\partial t} + \nabla.\left[U(\rho e_t + p)\right] = \nabla.\,(U.\bar{\bar{\tau}}) - \nabla.Q \tag{3}$$

As to ACTRAN, the key equation for calculation is as following:

$$\underbrace{\omega^2\int_V N_a\rho dV - \int_V \frac{\partial N_a}{\partial x_i}c_0^2\frac{\partial\rho}{\partial x_i}dV}_{\text{Acoustic propagation}} = \underbrace{i\omega\oint_S N_a\rho v_i n_i dS}_{\text{Surface Source}} + \underbrace{\int_V \frac{\partial N_a}{\partial x_i}\frac{\partial T_{ij}}{\partial x_j}dV}_{\text{Volume Source}} \tag{4}$$

4.2. *Turbulence model*

The DES approach based on the S-Amodel and the SST k-ω model was used in the current research to solve the flow characteristics over the wings. The DES models were often referred to as a hybrid of RANS and LES, where the RANS was employed in the near-wall region but the LES was employed in the core turbulent region. The computational cost for the DES approach was lower than the LES approach and greater than the RANS approach. In the simulations, the computational domain had a velocity inlet of 36.5m/s with variable AOAs, there was an outflow boundary of the outlet, the surface of the wing was non-slipping wall. Separated implicit solver based on pressure was adopted here, second order upwind scheme had better accuracy and stability, so it was a very suitable scheme to be used for better results. The highest frequency of this study was 5000 Hz, sampling frequency was at least twice the highest frequency of the signal according to Nyquist Sampling Theorem. So during the computation of fluent, 0.0001s was chosen as the value of time step. The sampling time would be 0.0364 second, this time calculated would ensure that the incoming flow could go through the total tunnel. Hence, while the unsteady computations were conducted, the iteration number was set as 400, this not only provided a guarantee for precise results but also cost less time and computational sources.

4.3. *Computational models and mesh generation*

In the following simulations, all conditions were the same, including the same parameters, the same boundary conditions and so on. The computational domain of the airfoil was shown in Figure 9. The inlet boundary away from the center of the baseline was 16c (c represents the chord length) and the outflow boundary from the trailing edge was also 16c in order to reduce the turbulence reflected from it. The front and back away from the both ends of the airfoil was 3c. The top and bottom of the tunnel away from the surfaces of the airfoil was 4c. ICEM CFD was used to generate the structural and unstructured mesh shown as in Figure 10 and Figure 11. By adjusting the ratios, different numbers of mesh cells could be obtained. So a grid independence study was significant in numerical analysis because the grid resolution should be sufficiently fine to resolve all important flow parameters, such as lift and drag coefficients. Three cases were considered in this study: first grid(G1) with 1,684,575 cells, second grid(G2) with 3,621,736 cells, and third grid(G3) with 6,759,435 cells in the domain. The lift coefficient values of different grids for the modified one were depicted in Figue 12. Here AOA was set as 0° to find the variation of the lift coefficient. It was observed the lift coefficient remained almost constant all the way. Therefore 6,759,435 was chosen as the final number of grid cells.

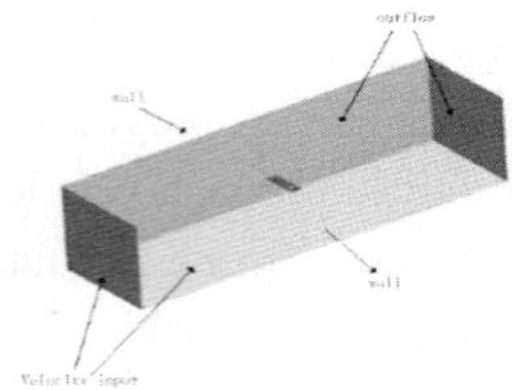

Figure 9. Computational region of the model.

Figure 10. Mesh for baseline airfoil.

Figure 11. Mesh for modified airfoil.

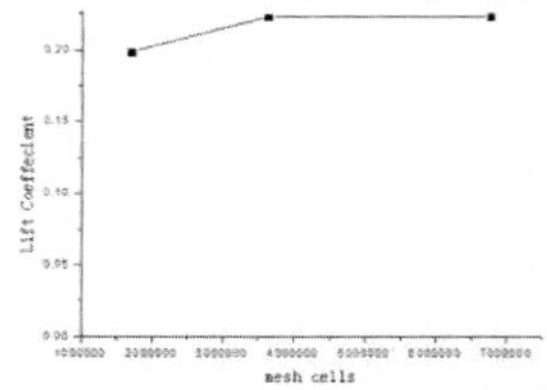

Figure 12. Lift coefficient versus different numbers of cells.

4.4. *Computational results and comparisons*

Lift coefficient,drag coefficient and lift-to-drag ratio as a function of AOA were compared and analyzed. From Figure 13, Figure 14 and Figure 15, we could find a better performance of the airfoils with tubercles. Simultaneously the acoustic comparison between the baseline airfoil and the modified one by ACTRAN was also shown in Figure 16. By the post-processing of the steady iterations, the contours of pressure coefficient over the surface of both airfoils and contours of the velocity around both wings were presented in Figure 17 and Figure 18.

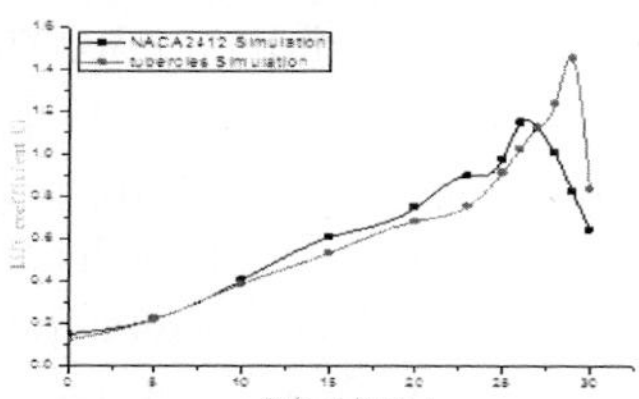

Figure 13. Lift coefficient versus AOA.

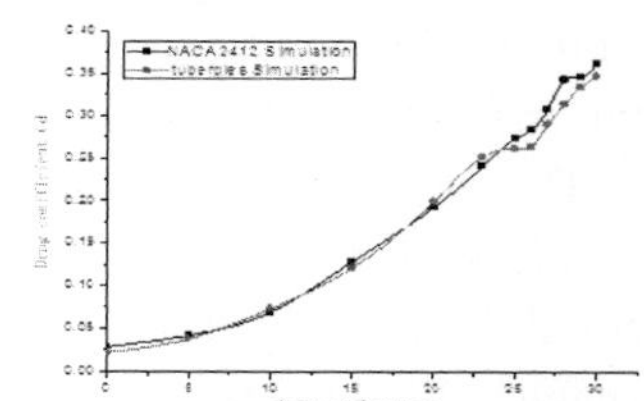

Figure 14. Drag coefficient versus AOA.

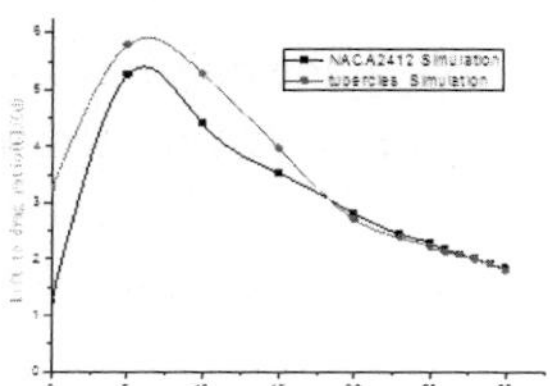

Figure 15. Lift-to-drag ratio versus AOA.

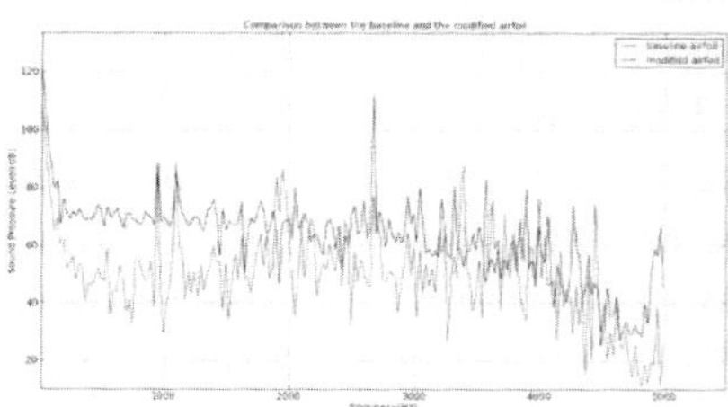

Figure 16. Frequency Spectrum comparison of both airfoils.

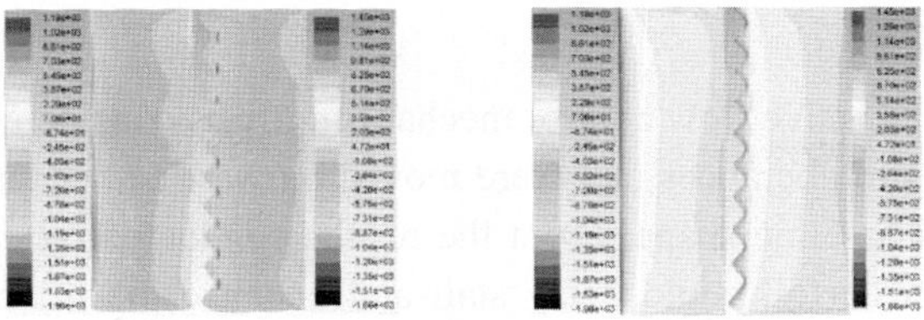

Figure 17. The contours of pressure coefficient over the surfaces of both airfoils.

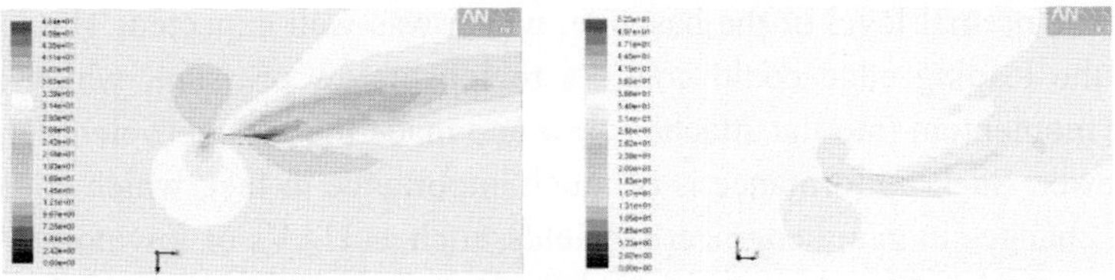

Figure 18. The contours of velocity around the baseline and modified airfoils.

4.5. *Analysis of mechanisms of noise reduction*

From Figure 5 and Figure 13, Figure 6 and Figure 14, the experimental results agreed with the simulations well. The airfoil with tubercles was advantageous over the one with a smooth leading edge because the flow over it was well organized due to troughs formed by the tubercles, whereas flow over a smooth flipper was disorganized into eddies at steeper angles of attack. The streamwise vortices originated by the tubercles enhanced the flow attachment over the suction surface of the modified one, which resulted in a higher lift coefficient after stall. Besides, the streamwise vortices made the flow much smoother, which could reduce the noise from the broken vortices to an extent. The contours of the simulations on wavy airfoils showed periodic variation in the wake width across the wing span, which were the dark blue regions at troughs along the leading edge in Figure 17. A wide wake occurred where the wing protruded downstream and a narrow wake occurred wherethe body protruded upstream. The vortices stemming from the tubercles also caused low pressure in the valleys of tubercles. Due to the wavy geometry of the modified airfoil, the vortices stemming from the tubercles rotated in a way that caused them to move towards the valleys causing a coalescence of fluid between tubercle peaks. Because of low pressure inside these vortices, they combined to create low pressure in the troughs of the tubercles. A leakage of fluid from high pressure to low pressure resulted in spanwise flow and the formation of tip vortices. The differences between the two surfaces of the wing would impart kinetic energy to the fluid, which resulted in induced drag as the lift was generated.

5. Conclusions

In the case of an active flow control mechanism it was considered an easier task to employ the protuberances that were moved outward rather than outward and inward from a baseline shape. From the results conducted by experiments and simulations, the airfoil could delay stall and enhance the lift coefficient after stalling. Besides, when reaching the maximum lift coefficient, the SPL was below the normal level of the baseline, which was well expected. The tubercles along the leading edge could create a region of low pressure, which brought higher momentum into the attached flow and made the lift coefficient larger and SPL lower. Such performance is of much importance in fact, which could bring lots of changes in many engineering fields, such as UAVs of low noise, "silent" fan, more efficient wind turbines and so forth, providing much motivation for sustainable development.

Acknowledgments

This work was supported by the National Basic Research Program of China (973 Program) (No. 2011CB302102, No. 2011CB302104) and National Natural Science Youth Foundation of China (No. 51405001).

References

1. Frank E. Fish, Hydrodynamic Design of the Humpback Whale Flipper, *Journal of Morphology*, Vol. 225 (1995), pp. 51–60.
2. S.M. Sun, C.Y. Xu, Experimental research on noise reduction of bionic axial fan blade and mechanism analysis, *Journal of Jilin University*, Vol. 2, pp. 382–387.
3. B.M.G, Numerical Analysis of Wind Turbine Airfoil Aerodynamic Performance with Leading Edge Bump, *Mathematical Problems in engineering*, (2014), pp. 1–8.
4. C. Xin, Research on Hydrodynamic Performance of Three-dimensional Airfoil with Tubercles on Leading-edge, *Applied Mechanics and Materials,* Vol. 575 (2014), pp. 405–413.
5. L.Q. Ren, Noise reduction mechanism of non-smooth leading edge of owl wing, *Journal of Jilin University*, Vol. 38 (2008), pp. 126–131.
6. M.M. Zhang, Experimental study of flow separation control on a low-Re airfoil using leading-edge protuberance method, *Experimental Fluids*, Vol. 55 (2014), pp. 1710–1723.

7. Ahmed Farouk Abdel Gawad, Utilization of Whale-Inspired Tubercles as a Control Technique to Improve Airfoil Performance, *Transaction Series on Engineering Sciences and Technologies*, Vol. 2 (2013), pp. 212–218.
8. Alessandro Corsini, On the role of Leading-Edge Bumps in the Control of Stall Onset in Axial Fan Blades, *Journal of Fluid Engineering*, Vol. 135 (2013), pp. 1–9.
9. C.A. Ozen, Control of vertical structures on a flapping wing via a sinusoidal leading-edge, *Physics of Fluids*, Vol. 22 (2010), pp. 1701–1703.
10. Ernst A. vanNierop, How Bumps on Whale Flippers Delay Stall: An Aerodynamic Medol, *Physics Review Letters*, Vol. 100 (2008), pp. 4502–4506.
11. H.S. Yoon, Effect of the wavy leading edge on hydrodynamic characteristics for flow around low aspect ratio wing, *Computer & Fluids*, Vol. 49 (2010), pp. 276–289.

Ionic liquid–based dispersive liquid–liquid microextraction combined with high performance liquid chromatography for the simultaneous determination of 11 sulfonamide in honey[†]

C.Q. Wu[*], Y.H. Chen, D.Y. Chen and J. Gong

College of Environmental Science and Engineering, Guangzhou University, Guangzhou 510006, China

Key Laboratory for Water Quality Security and Protection in Pearl River Delta of Ministry of Education and Guangdong Province, Guangzhou 510006, China

Guangdong Provincial Key Laboratory of Radio Nuclides Pollution Control and Resources, Guangzhou, 510006, China

[*]*E-mail: wucuiqin@gzhu.edu.cn*

L.Z. Xu

Department of Obstetrics and Gynecology, the Second West China Hospital, Sichuan University, Chengdu, China

A simple, rapid, and sensitive method for the quantitative monitoring of eleven sulfonamides in honey was developed by coupling ionic liquid-based dispersive liquid–liquid microextraction (DLLME) to high performance liquid chromatography with diode-array detection. Optimum DLLME conditions (5 mL of honey solution at pH=3, 50 μL $[C_6MIM][PF_6]$ as extraction solvent and 0% (w/v) NaCl) allowed the repeatable, accurate and selective determination of 11 sulfonamides (sulfacetamide, sulfapyridine, sulfamerazine, sulfamethazine, sulfameter, sulfachloropyridazine, sulfadoxine, sulfamethoxazole, sulfisoxazole, sulfadimethoxine and sulfachinoxalin). The method was validated by means of the obtention of calibration curves of the whole method as well as a recovery study at two levels of concentration. The LODs in honey were in the range 0.006–0.015mg/kg, and the recoveries in spiked honey sample were between 85.3% and 97.4%.

Keywords: Ionic liquid; Dispersive liquid–liquid microextraction; Sulfonamides; High-performance liquid chromatography; Honey.

1. Introduction

Honey is used worldwide in the agro-industry and consumed throughout the world for its sweet taste and nutritive qualities. Honeybee products are considered

[†] This work is supported by the National Natural Science Foundation of China (No. 21207022, 41473097) and Guangdong Provincial Natural Science Foundation of China (No. S2013010013632).

to be natural and healthy, able to prevent infections and used to treat surface wounds in medicine (honey therapy) [1]. Among the contaminants encountered in beekeeping, the most important are substances used to control bee diseases. Sulfonamides are employed in apiculture for the treatment of bacterial diseases, notably American and European foulbrood, due to their low cost [2]. The sulfonamide residue determination in honey is of concern because of their potential carcinogenic character and the possibility of development of antibiotic resistance in humans [3]. In the European Union, maximum residue limits (MRLs, 0.1mg/kg) have been established for sulfonamide compounds in various food stuffs of animal origin [2]. In some countries (Switzerland, UK and Belgium), MRLs have been set for sulfonamides (0.01–0.05 mg/kg) [4].

In the last decade, many research groups have focused on method, to analyze sulfonamide residues in honey by high performance liquid chromatography using UV/visible detectors [5] and fluorescence detectors [6], mass detectors [1,7]. MS detection is the preferred choice but it is responsive to matrix effect because of high concentration of sugar in honey. Fluorescence detection needs derivatization with derivatization reagent. Thus, HPLC separation coupled with UV-Vis is fast, simple, and easily accessible.

Due to the very low concentrations of SA residues and the complexity of the matrices, the analyte enrichment and sample cleanup are of great importance for the determination of SAs. Many sample preparation methods, such as liquid–liquid extraction (LLE) and solid-phase extraction (SPE) were applied for SAs. However, in LLE, large volume of organic solvents was required, and in SPE, SPE cartridges were usually expensive. These methods were very time-consuming. In recent years, microextraction has been applied in the analytical chemistry. Solid-phase based microextractions, such as solid-phase microextraction (SPME) and stir bar sorptive extraction (SBSE), provide a simple and solventless sample preparation and have been applied for extracting SAs from food samples [8, 9]. However, SPME and SBSE are time-consuming and the coated fibers or bars are generally expensive and easily destroyed. Compared with the solid-phase based microextractions, liquid phase microextraction (LPME) has the benefits of both economic and convenience, and has been applied for the extraction of SAs in environment water [10]. However, LPME still has some drawbacks, such as instability of the microdrop, operational difficulty and bubble formation during the extraction.

Recently, a novel microextraction method termed as dispersive liquid–liquid microextraction (DLLME) based on the use of an appropriate extraction solvent and disperser has been developed [11]. This method has many advantages, including simplicity of operation, rapidity, low cost, high recovery and enrichment factor, and obtains success in many applications. However, the commonly used high-density extraction solvents, such as chlorobenzene, chloroform, and carbon tetrachloride were typically highly toxic and environment-unfriendly. Ionic liquids (ILs) can be used as green extraction solvents instead of organic solvents due to their unique physicochemical properties, such as negligible vapor pressure, good solubility for organic and inorganic compounds, and high thermal stability. The IL has been used in DLLME as extraction solvents with good results [6]. However, no reports have been published on simultaneously extracting the eleven of sulfacetamide, sulfapyridine, sulfamerazine, sulfamethazine, sulfameter, sulfachloropyridazine, sulfadoxine, sulfamethoxazole, sulfisoxazole, sulfadimethoxine, sulfachinoxalin from honey by ionic liquid based dispersive liquid-liquid microextraction followed by determination with HPLC-DAD.

In the present study, ionic liquid-based dispersive liquid–liquid microextraction (IL-based DLLME) combined with HPLC was first applied for the determination of eleven sulfonamides in honey. The aim of this work is to simplify the analytical step, reduce the consumption of toxic solvents and improve the sensitivity. The effects of various experimental parameters were studied and optimized. The developed method was successfully applied to the analysis of real samples.

2. Experimental

2.1. *Instrumentation*

The HPLC was performed using an Agilent 1200 liquid chromatographic system (Agilent Technologies, USA) including a photodiode array detector (DAD), and a Rheodyne7725i injector (Cotati, CA, USA) equipped with a 20 μL loop. Separation was accomplished using an Agilent Extend-C18 column (250 mm × 4.6 mm, 5 μm). A binary mobile phase composed of acetonitrile (phase A) and 0.2% acetic acid in water (phase B). The mobile-phase gradient profile was as follows: 0-12 min, 15% A; 12-20 min, 15-20% A; 20-28 min, 20% A; 28-35 min, 20-30% A. The flow rate was held at 1.0 mL/min. The injection volume was 5 μL by a Hamilton syringe (702NR, 25μL). The wavelength of the photodiode array detector was adjusted to 270 nm. The dispersive liquid liquid microextraction solutions were centrifuged by an Anke TDL-4 centrifuge (Shanghai, China).

2.2. *Chemicals and samples*

Sulfacetamide (SAAM, 99%), sulfapyridine (SPD, 99%), sulfamerazine (SMR, 99.2%), sulfamethazine (SMZ, 99%), sulfameter (SM, 99%), sulfachloropyridazine

(SCP, 100%), sulfadoxine (SDX, 99%), sulfamethoxazole (SMX, 99.5%), sulfisoxazole (SIX, 98.7%), sulfadimethoxine (SDM, 99.5%), sulfachinoxalin (SQX, 97%) were purchased from Dr. Ehrenstorfer Co. (Germany). Standard stock solutions were prepared at a concentration of 1000 mg/L in methanol. Working solutions were prepared weekly by an appropriate dilution of the stock solutions with methanol. Standard stock solutions were stored at -18°C and working solutions were stored at 4°C. Chromatographic grade acetonitrile were purchased from ANPEL Scientific Instrument (Shanghai) Co. Ltd. (Shanghai, China). 1-butyl-3-methylimidazolium hexafluorophosphate ([C_4MIM][PF_6]), 1-hexyl-3-methylimidazolium hexafluorophosphate ([C_6MIM][PF_6]), and 1-octyl-3-methylimidazolium hexafluorophosphate ([C_8MIM][PF_6]) were purchased from Chengjie Chemical Co., Ltd. (Shanghai, China). All other chemicals used were of analytical-reagent grade.

Honey samples were purchased from a local supermarket. 2.0 g of honey sample was diluted to 5 mL with water (pH 3) and then spiked with the mixed working solution of SAs. The resulting solution was referred to as sample solution, filtered through 0.45μm filters, and then stored at 4°C.

2.3. *IL-based DLLME*

The solution previously obtained was added into a 10 mL glass centrifugal tube. 50 μL [C_6MIM][PF_6] (extraction solvent) was rapidly injected into the sample solution to induce a cloudy solution composed of fine drops of [C_6MIM][PF_6] dispersed entirely into the sample solution. The mixture was then centrifuged at 4000 rpm for 10 min and the extraction solvent precipitated to the bottom of the centrifuge tube. An aliquot of 5 μL of extraction solvent was withdrawn and injected into the HPLC system for analysis.

3. Results and Discussion

3.1. *Optimization of IL-based DLLME*

To obtain high sensitivity and precision, several experimental parameters affecting the extraction efficiency were studied, including type and volume of extraction solvent, and pH of sample solution as well as ionic strength. A pre-optimization for the experimental parameters was performed and based on the preselected optimum parameters, the effects of experimental parameters were investigated again. The spiked samples were prepared by spiking the standard solution in the pure water and used in the experiments. All the experiments were performed in triplicate and the concentration of SAs in the spiked samples was 0.2 μg/mL.

3.1.1. *Type of extraction solvent*

The type of the extraction solvents is important parameters in the DLLME procedure. The choice of the extraction solvent is based on some criteria, such as high extraction efficiency, good chromatographic behavior, low solubility in water and higher density than water. According to these criteria, three kinds of hydrophobic imidazolium-ILs, including $[C_4MIM][PF_6]$, $[C_6MIM][PF_6]$, and $[C_8MIM][PF_6]$, were investigated. As shown in Figure 1 (A), the peak areas obtained with $[C_6MIM][PF_6]$ and $[C_4MIM][PF_6]$ are larger than those obtained with $[C_8MIM][PF_6]$. But when $[C_4MIM][PF_6]$ was used, the repeatability was worse. This may be due to the higher solubility of $[C_4MIM][PF_6]$ in water (1.88 g/100 mL) than that of $[C_6MIM][PF_6]$ (0.75 g/100 mL) at 25°C [6], which led to more difficultly to get the IL phase at the bottom of the tube. Accordingly, $[C_6MIM][PF_6]$ was selected as extraction solvent in the subsequent experiments.

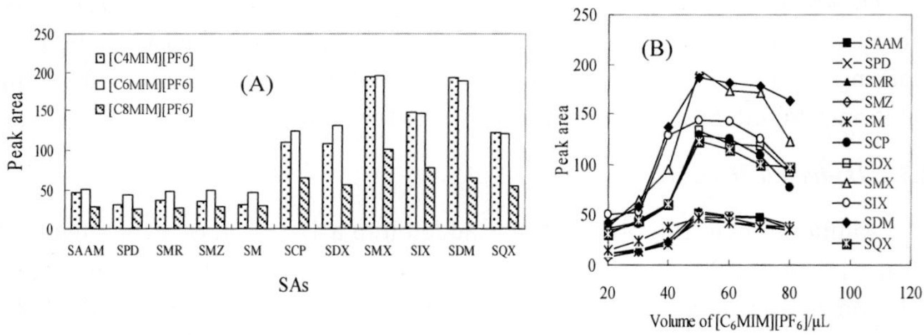

Figure 1. Effect of type (A) and volume (B) of extraction solvent on extraction efficiency.

3.1.2. *Effect of the volume of extraction solvent*

Different volumes of $[C_6MIM][PF_6]$ (20, 30, 40, 50, 60, 70 and 80 μL) were used as extraction solvents to estimate the influence of extraction solvent volume on extraction efficiency of SAs. As shown in Figure 1 (B), the peak areas of the SAs increase with the increase of the volume of $[C_6MIM][PF_6]$ from 20 to 50 μL, and then slightly decrease with the further increase of the volume of $[C_6MIM][PF_6]$, because the increase of the IL phase volume can result in the dilution of the analytical solution. Thus, 50 μL of $[C_6MIM][PF_6]$ was selected in the further studies.

3.1.3. *Effect of pH of sample solution*

The pH value of the sample solution determines the present state of analytes and thus affects the extraction efficiency. In this experiment, different pH value (1, 2, 3, 4, 5, 6) of sample solution was investigated, and the results are shown in Figure

2 (A). It can be seen that the peak areas of SAs are highest when pH is 3. Therefore, pH 3 was chosen in the subsequent experiments.

3.1.4. *Effect of salt concentration*

Sodium chloride was added into the sample solution to increase the ionic strength of the sample solution. The experimental results shown in Figure 2 (B) indicate that the peak areas of SAs decrease with the increase of NaCl concentration. The reasons were that (1) the high ionic strength decreases solubility of extraction solvent in aqueous sample which leads to increase in volume of the extraction solution and causes the concentration of analytes to be decreased, and (2) the high ionic strength increases viscosity of aqueous phase which leads to decrease in diffusion coefficients of analytes [12]. Based on the above experimental results, no salt was added in the following experiments.

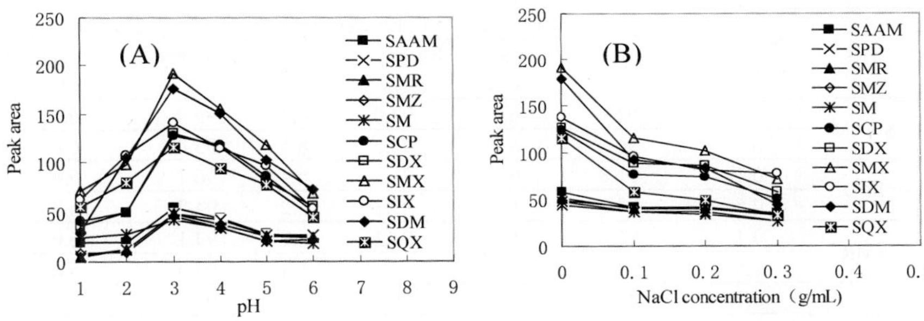

Figure 2. Effects of pH value (A) and NaCl concentration (B) of sample solution on extraction efficiency.

3.2. *Characteristics of the analytical method*

Under the optimized conditions, a series of experiments were performed for obtaining linearity, precision, and the limits of detection (LODs). All the experiments were performed in triplicate. The linear regression equations and correlation coefficients are listed in Table 1. The correlation coefficients (r) ranged from 0.9988 to 0.9992 for all the analytes. The limits of detection (LODs, S/N = 3) for the eleven SAs were in the range from 0.006 to 0.015 mg/kg. The relative standard deviation (RSD) varied between 4.9% and 6.2%, which was calculated from five replicate determinations of standard solutions at the concentration of 0.04 μg/mL for each SA.

Table 1. Linear ranges, regression equations, LODs and RSDs of the eleven SAs.

SA	Linear range / (μg/mL)	Regression equation	r	LOD/ (mg/kg)	RSD/% (n=5)
SAAM	0.01-0.4	Y=283.79X-0.0181	0.9991	0.008	5.6
SPD	0.01-0.4	Y=248.23X+0.1908	0.9992	0.012	5.9
SMR	0.01-0.4	Y=265.64X-0.8317	0.9990	0.009	6.1
SMZ	0.01-0.4	Y=218.81X+3.6065	0.9991	0.009	5.3
SM	0.01-0.4	Y=209.00X+2.3015	0.9990	0.012	6.2
SCP	0.01-0.4	Y=585.73X+4.7328	0.9989	0.008	5.9
SDX	0.01-0.4	Y=625.01X+0.3043	0.9991	0.006	6.1
SMX	0.01-0.4	Y=897.67X+1.4506	0.9989	0.006	5.0
SIX	0.01-0.4	Y=667.04X+0.4643	0.9988	0.006	5.3
SDM	0.01-0.4	Y=830.13X+8.4254	0.9990	0.006	4.9
SQX	0.01-0.4	Y=561.28X+0.6473	0.9992	0.015	5.7

Y: peak area; X: mass concentration, μg/mL

Table 2. Recoveries and their precisions (RSDs) of the eleven SAs spiked in real honey samples (n=5).

SAs	Spiked / (mg/kg)	Winter honey		Wildflower honey		Longan honey	
		Recovery / %	RSD / %	Recovery / %	RSD /%	Recovery / %	RSD / %
SAAM	0.1	92.9	6.2	92.0	6.3	95.4	6.5
	0.5	94.8	5.9	96.2	6.5	93.0	6.6
SPD	0.1	91.8	6.6	91.4	6.4	96.7	6.4
	0.5	94.2	6.5	93.3	6.2	94.2	6.3
SMR	0.1	92.8	6.8	93.2	6.8	94.1	6.6
	0.5	93.6	6.0	96.7	6.4	94.9	5.8
SMZ	0.1	91.7	6.6	88.9	6.5	91.7	5.4
	0.5	94.5	6.2	92.0	5.8	92.5	5.2
SM	0.1	92.1	7.3	92.1	6.1	92.6	6.3
	0.5	94.3	6.5	97.4	6.1	94.4	6.4
SCP	0.1	88.1	6.8	86.5	6.7	93.6	6.2
	0.5	92.3	6.4	88.7	6.4	91.9	5.3
SDX	0.1	90.3	6.3	85.3	6.6	92.1	6.3
	0.5	92.0	5.8	88.8	6.2	93.0	5.9
SMX	0.1	90.6	6.7	87.9	6.2	92.0	5.4
	0.5	93.8	6.1	90.4	5.9	92.3	5.6
SIX	0.1	92.1	6.6	91.2	6.4	92.5	6.5
	0.5	95.7	6.2	92.7	5.6	93.1	5.7
SDM	0.1	89.0	6.8	92.8	5.9	92.0	6.1
	0.5	92.6	5.8	93.6	5.4	94.9	5.8
SQX	0.1	91.3	6.9	88.0	6.1	88.5	6.1
	0.5	93.9	6.3	90.7	5.8	91.3	5.9

To evaluate the applicability of the proposed method, some real samples, including wildflower honey, longan honey and winter honey were analyzed. The SAs in the samples are not detectable. To evaluate precision and accuracy of the proposed method, the spiked samples were analyzed. The analytical results are shown in Table 2. The RSDs nd recoveries are in the range of 5.2–7.3% and 85.3–97.4%, respectively. It can be considered that the present method provides acceptable recoveries and precision for the determination of SAs in real samples. The typical chromatograms of the blank and spiked sample are shown in Figure 3.

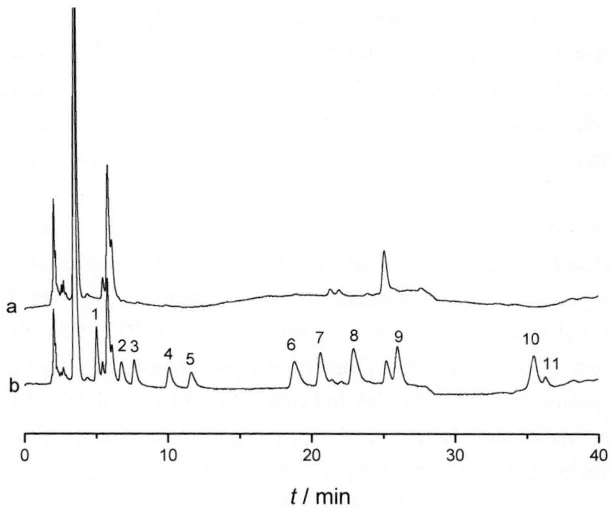

Figure 3. Chromatograms of (a) blank winter honey, (b) winter honey spiked with the eleven SAs (0.25 mg/kg) 1.SAAM 2.SPD 3.SMR 4.SMZ 5.SM 6.SCP 7.SDX 8.SMX 9.SIX 10.SDM 11.SQX.

4. Conclusions

The present method based on IL-DLLME was an effective method in extracting SAs in honey. Compared with conventional liquid-liquid extraction, the present method avoided excessive consumption of organic solvents, extraction and clean-up can be performed in one-step. The recoveries, precision, and LODs obtained were satisfactory.

References

1. E. Dubreil-Chéneau, Y. Pirotais, E. Verdon, D. Hurtaud-Pessel, Confirmation of 13 sulfonamides in honey by liquid chromatography–tandem mass spectrometry for monitoring plans: Validation according to European Union Decision 2002/657/EC, J. Chromatogr. A 1339 (2014) 128–136.
2. T. A. Catelani, I. V. Tóth, J. L.F.C. Lima, L. Pezza, H. R. Pezza, A simple and rapid screening method for sulfonamides in honey using a flow injection

system coupled to a liquid waveguide capillary cell, Talanta 121 (2014) 281–287.

3. G. C. Bedendo, I. C. S. F. Jardim, E. Carasek, A simple hollow fiber renewal liquid membrane extraction method for analysis of sulfonamides in honey samples with determination by liquid chromatography–tandem mass spectrometry, J. Chromatogr. A 1217 (2010) 6449–6454.
4. Z. Bargańska, M. Ślebioda, J. Namieśnik, Determination of antibiotic residues in honey, Trends Anal. Chem. 30 (2011) 1035–1041.
5. R. H. M. M. Granja, A. M. M. Niño, F. Rabone, A. G. Salerno, A reliable high-performance liquid chromatography with ultraviolet detection for the determination of sulfonamides in honey, Anal. Chim. Acta 613 (2008) 116–119.
6. X. Xu, R. Su, X. Zhao, Z. Liu, Y. P. Zhang, D. Li, X. Y. Li, H. Q. Zhang, Z. M. Wang, Ionic liquid-based microwave-assisted dispersive liquid–liquid microextraction and derivatization of sulfonamides in river water, honey, milk, and animal plasma, Anal. Chim. Acta 707 (2011) 92– 99.
7. R. Galarini, G. Saluti, D. Giusepponi, R. Rossi, S. Moretti, Multiclass determination of 27 antibiotics in honey, Food Control 48 (2015) 12–24.
8. Y. Wen, M. Zhang, Q. Zhao, Y. Q. Feng, Monitoring of five sulfonamide antibacterial residues in milk by in-tube solid-phase microextraction coupled to high-performance liquid chromatography, J. Agric. Food Chem. 53 (2005) 8468–8473.
9. X. J. Huang, N. N. Qiu, D. X. Yuan, Simple and sensitive monitoring of sulfonamide veterinary residues in milk by stir bar sorptive extraction based on monolithic material and high performance liquid chromatography analysis, J. Chromatogr. A 1216 (2009) 8240–8245.
10. Y. Tao, J. F. Liu, X. L. Hu, H. C. Li, T. Wang, G. B. Jiang, Hollow fiber supported ionic liquid membrane microextraction for determination of sulfonamides in environmental water samples by high-performance liquid chromatography, J. Chromatogr. A 1216 (2009) 6259–6266.
11. M. Rezaee, Y. Assadi, M. R. Milani Hosseini, E. Aghaee, F. Ahmadi, S. Berijani, Determination of organic compounds in water using dispersive liquid–liquid microextraction, J. Chromatogr. A 1116 (2006) 1–9.
12. M. A. Farajzadeh, D. Djozan, P. Khorram, Development of a new dispersive liquid–liquid microextraction method in a narrow-bore tube for preconcentration of triazole pesticides from aqueous samples, Anal. Chim. Acta 713 (2012) 70–78.

Technology and practice of full size utilization of steel slag in asphalt concrete road construction*

X.D. Xiang†, C.H. Lee, Y.F. Chang, and H.M. Fang
School of Resource and Environmental Engineering,
Wuhan University of Science and Technology
Wuhan, Hubei, P. R. China, 430081
†E-mail: drxxd@163.com
www.WUST.edu.cn

A great amount of steel slag has been produced every year in China. In order to solve the utilization problem of the steel slag, the preparation technology and its application of steel slag asphalt concrete for road construction has been investigated. The steel slag aggregate crushing and washing technologies are developed. The stone mastic asphalt (SMA), open graded asphalt friction course (OGFC), semi rigid base, as well as flexible base of steel slag asphalt concrete road have been prepared. It is found that if the counterattack crusher in traditional aggregate preparation technology is replaced by the cone crusher, the size of the steel slag aggregate becomes more uniform. Moreover, when the wet rolling screen is introduced into the steel slag aggregate preparation system, the irregularity of the steel slag aggregate can be enhanced. According to the test results, performances of steel slag SMA and OGFC developed in this work are better than that of the common basalt SMA and OGFC. The dynamic stability and the pendulum friction coefficient of the steel slag SMA are 7910 time/mm and 65 respectively. The permeability coefficient, Marshall residual stability, dynamic stability, and freeze-thaw splitting strength ratio of the steel slag OGFC are 4680 mL/min, 92 %, 4670 time/mm, and 86% respectively. It is also proved that the semi rigid base and flexible base can be made from steel slag entirely. The more steel slag in semi rigid base or in flexible base, the better the mechanical property is.

Keywords: Steel Slag; Asphalt Concrete Road; Stone Mastic Asphalt (SMA); Open Graded Asphalt Friction Course (OGFC); Semi Rigid Base; Flexible Base.

1. Introduction

Steel slag is the by-product of the steelmaking. There are three main steel-milting furnaces, Martin furnace, electric furnace, and converter. Since the converters are commonly used in steel mill in China, this project is aimed at the investigation and application of steel slag from converter in asphalt concrete road construction.

---

* Work partially supported by grant 2013010803010378 of the Scientific Research Achievement Transformation Project of Wuhan City.

The amount of steel slag is about 15% of the crude steel. More than one hundred million tons of the steel slag had been produced in 2013 in China [1]. These huge quantities have exceeded the steel slag treatment ability. Therefore, some serous environmental problems appear. For instance, a huge land space is occupied; the underground water is polluted by the steel slag. Especially in rainy seasons, the alkali material and heavy metal are etched out from the steel slag.

Nowadays, on the other hand, the natural stone is shrinking very fast due to the rapid development of road transportation. It is predicted that the limestone will be exhausted next 15 years at recent utilization speed [2]. Therefore, if the steel slag is used for road construction, at lest, 50 million tons of natural stone can be saved.

In fact, there are many advantages of the steel slag for road material. (1) Good resistance to abrasion because the steel slag contains some metal elements, such as the iron. (2) Fine edged shape since the steel slag is irregular, the embedded structure can be formed when the steel slag is used for asphalt concrete aggregate. (3) Strong adhesion because the asphalt can stick on the steel slag tightly to make the road be more durable.

At present, although, 40%–60% of steel slag was used for road construction in Europe [3], America [4], and Japan [5], a series of basic research and integrated technology are still required to meet need of the high performance of the asphalt concrete aggregate road, such as, to develop the integrated technology of the slag crushing, screening and cleaning, to optimize the performance of steel slag asphalt concrete mixture.

This research began in 2003. Now, the integrated technology of steel slag aggregate production and steel slag asphalt concrete preparation for road construction has been formed. The objective of full size utilization of steel slag has been achieved. More than 400 million annual economic benefits from the steel slag application had been obtained in WISCO by this technology.

2. Preparation Technology of Steel Slag Aggregate

2.1. *Combination crushing technology of steel slag aggregate*

In the traditional aggregate preparation technology the jaw crusher and the counterattack crusher are used, as shown in Fig. 1a. This crushing technology is not suitable for steel slag crushing process because the counterattack crusher leads to the steel slag to be over break easily. Then, the counterattack crusher is replaced by the cone crusher in aggregate preparation, as shown in Fig. 1b.

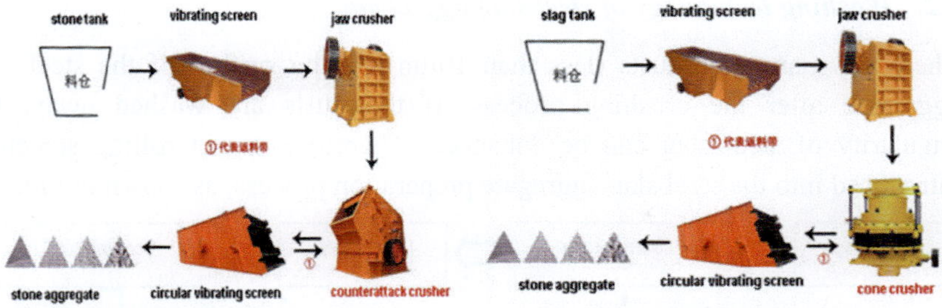

(a) Traditional aggregate preparation system (b) New aggregate preparation system

Figure 1. Substitution of counterattack crusher to cone crusher in aggregate preparation.

The measurement results of the steel slag aggregate size distribution in traditional aggregate preparation (Fig. 2a) and in new aggregate preparation technology (Fig. 2b) have shown that the aggregate produced by combination of jaw crusher and cone crusher is more uniform. It is important to ensure good performance of the steel slag asphalt concrete road.

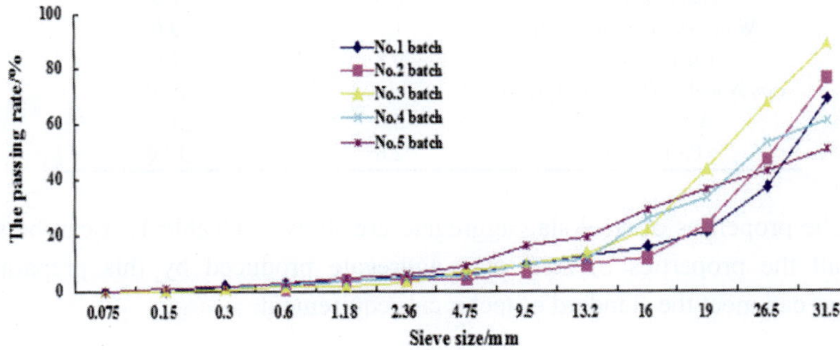

(a) Jaw +counterattack crusher

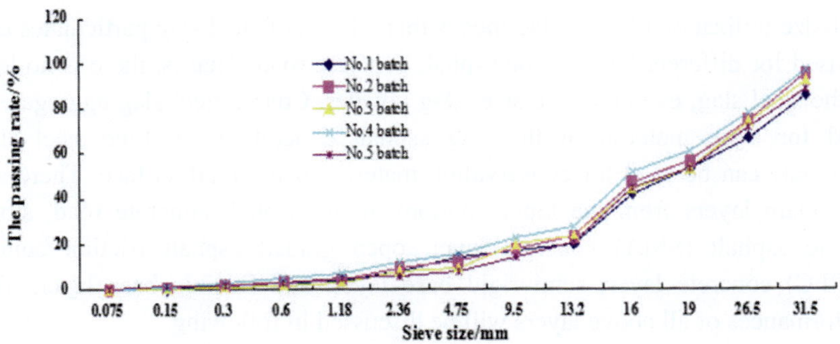

(b) Jaw +cone crusher

Figure 2. Comparisons of steel slag aggregate distribution produced by different crushing technology.

2.2. *Washing technology of steel slag aggregate*

There are many fine dusts (less than 10μm) on the surface of the steel slag aggregate after the crushing process. If the dusts are washed away, the angularity of steel slag can be enhanced. Therefore, a wet rolling screen is introduced into the steel slag aggregate preparation process, as shown in Fig. 3.

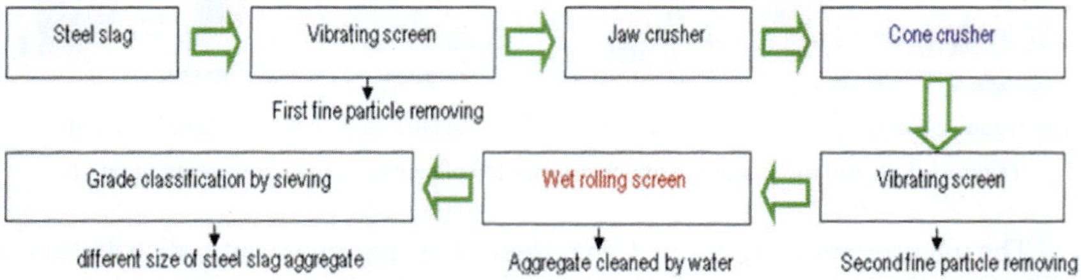

Figure 3. Multi-stage process of slag aggregate preparation.

Table 1. Properties of steel slag aggregate.

Property	Experiment results	Technical requirement
Relative density	3.3	≥2.6
Water absorption rate/%	1.3	≤3.0
Dust content/%	0.8	≤1.0
Los Angeles abrasion value/%	13.2	≤28.0
Adhesion level	5.0	≥4.0
Fastening/%	2.6	≤12.0

The properties of steel slag aggregate are shown in Table 1. It can be seen that all the properties of steel slag aggregate produced by this preparation process can meet the standard of technical requirement.

3. Full-size Utilization of Steel Slag in Asphalt Concrete

Full-size utilization of steel slag means that all size of steel slag particulates can be used for different layer of the asphalt concrete road. That is, there is no loss of the steel slag, even the fine steel slag powder. Coarse steel slag aggregate is used for main material of the slag asphalt concrete road. Fine steel slag aggregate can be used for conservation material of the road surface. There are four main layers from the top to bottom of the asphalt concrete road, stone mastic asphalt (SMA) concrete layer, open graded asphalt friction course (OGFC) concrete layer, semi rigid base layer, and flexible base layer. The performances of all above layers will be discussed in following.

3.1. *Steel slag SMA*

SMA (Stone Mastic Asphalt) is the mixture of fine powder, small aggregate, polyester, and asphalt. SMA can get into the coarse aggregate to increase the temperature stability, water stability, durability, anti-skid, and smoothness of the road. The mineral mixture composition scheme is selected in Table 2. The composition of new SMA and traditional SMA is given in Table 3. The test results are given in Table 4.

Table 2. Mineral mixture composition scheme.

No.	1#	2#	3#	4#	5#
Size	16-10mm	10-5mm	5-3mm	<3mm	<5mm
Material	Slag	Slag	Basalt	Basalt	Fine powder

Table 3. Composition of new and traditional asphalt concrete SMA.

Basic parameter	New SMA in this research	Traditional SMA
Grad	1#:2#:3#:4#:5#=38:36:8:8:10	1#:2#:3#:4#:5#=34:38:10:8:10
Polyester fiber	0.3%	0.3%
Steel slag proportion	74%	72%
Asphalt-stone ratio	6.4%	6.4%

Table 4. Performance test results of new and traditional asphalt concrete SMA.

Performance	New SMA in this research	Traditional SMA
Dynamic stability (time/mm)	7910	5020
Pendulum friction coefficient	65	59

According to the test results in Table 4, new steel slag SMA has higher dynamic stability. One reason is possibly that the steel slag aggregate cleaned by wet rolling screen has more edges. New steel slag SMA also has greater pendulum friction coefficient. This means that the steel slag SMA has good anti-skid property.

3.2. *Steel slag OGFC*

OGFC (Open Graded Asphalt Friction Course) asphalt concrete is mainly used to high way construction because it has high porosity to prevent the water from keeping on the road (good drainage properties). Here, OGFC-16 is taken as an example in Table 5.

Table 5. Material composition.

Aggregate size	10-16m	5-10mm	<5mm	Packing
Composition	45%slag	40%limestone	14% limestone	1% Ore

Note: Asphalt- stone ratio is 4.4

The performance test results of steel slag OGFC is given in Table 6. It can be seen that the performance of steel slag OGFC is better than that of the common basalt OGFC and meets the technical requirements.

Table 6. Performance test results of steel slag OGFC.

Index	Permeability coefficient (mL/min)	Marshall residual stability (%)	dynamic Stability (time/mm)	freeze-thaw splitting strength ratio (%)
Value	4680	92	4670	86
Note	Common basalt OGFC is 3000-4000	Technical requirement ⩾80%	Standard requirement ⩾3000	Standard requirement ⩾70%

3.3. *Semi rigid steel slag base*

The steel slag aggregate can also be used in semi rigid base. It is necessary to discuss the performance of the semi rigid steel slag base when the material is totally substituted by steel slag. The gradation of mineral composition is shown in Table 7.

Table 7. Gradation of mineral composition.

Sample 1	Sample 2	Mixture ratio
Steel slag (25-16mm)	Steel slag (25-16mm)	25
Steel slag (16-10mm)	Steel slag (16-10mm)	20
Steel slag (10-5mm)	Steel slag (10-5mm)	15
Steel slag (5-3mm)	Basalt (5-3mm)	10
Steel slag (3-0mm)	Basalt (3-0mm)	30

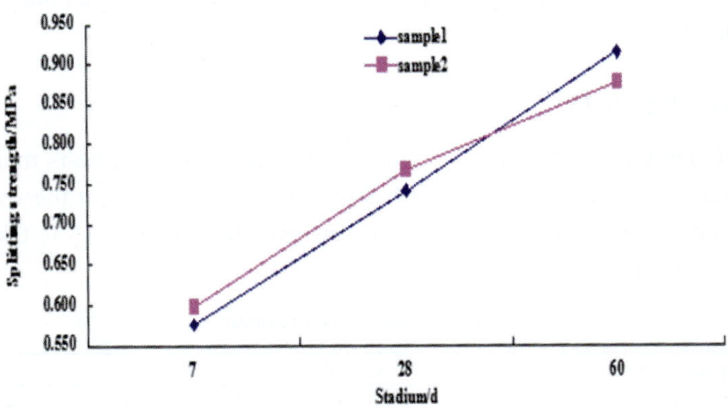

Figure 4. Experimental comparisons of splitting strength.

The comparisons of the splitting strength of sample 1 and sample 2 are shown in Fig. 4. The results indicate that the mechanical property is getting better when 100% of steel slag is used in semi rigid base.

3.4. *Flexible steel slag base*

The performances of flexible steel slag base are also investigated. The material composition with different amount of steel slag and basalt is given in Table 8.

Table 8. Mixture material composition.

Size (mm)	Sample 1#	Sample 2#	Sample 3#	Sample 4#	Ratio (%)
25-16	Steel slag	Steel slag	Steel slag	Basalt	42.0
16-10	Steel slag	Steel slag	Steel slag	Basalt	16.0
10-5	Steel slag	Steel slag	Basalt	Basalt	12.0
5-3	Steel slag	Basalt	Basalt	Basalt	20.5
3-0	Steel slag	Basalt	Basalt	Basalt	9.5

The test results of residual stability ratio (Fig. 5) and freezing-splitting strength ratio (Fig. 6) indicate that the water stable performance of the flexible steel slag base is stronger than that of the flexible basalt base.

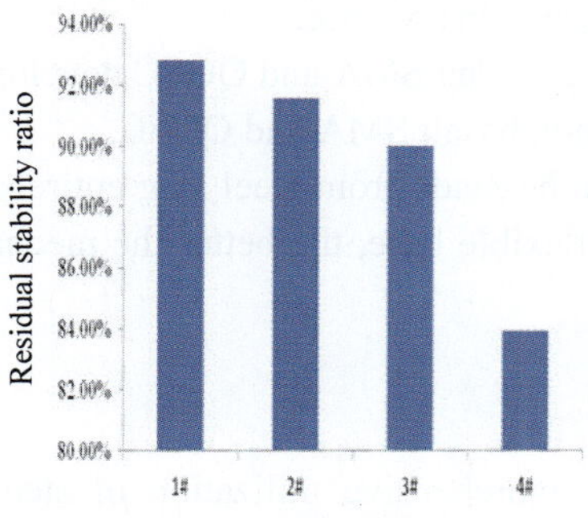

Figure 5. Residual stability.

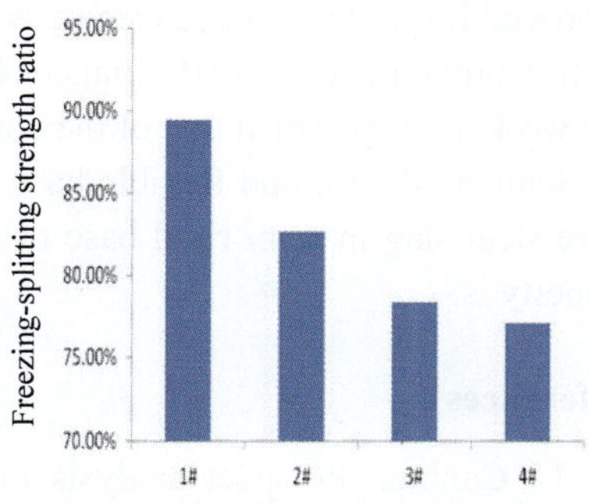

Figure 6. Freezing-splitting strength.

4. Applications

More than 10 pilot steel slag asphalt concrete roads had been paved since 2011, such as Guanggu fourth road in Wuhan (Fig. 7a), WNBM service road (Fig. 7b), WISCO ring road (Fig. 7c), Wu-Huang highway (Fig. 7d), etc. The applications of steel slag in road construction are successful. For example, in Guanggu fourth road (the design length is 2.5km, width is 16m, and thickness is 130mm) construction in 2013, three grades of steel slag aggregate (31.5–19.0mm, 19.0–9.5mm, 16.0–2.36mm) are used. All the performances meet the technical requirement of the highway construction. The paved road is smooth and dense. There are almost no defects in the road till now.

(a) Guanggu Fourth Road (b) WNBM Service Road (c) WISCO Ring Road (d) Wu-Huang Highway

Figure 7. Examples of steel slag asphalt concrete road.

5. Conclusions

(1) When the counterattack crusher is replaced by the cone crusher in the steel slag aggregate preparation, the uniformity of the steel slag aggregate distribution is increased.

(2) It is important to introduce the wet rolling screen into the steel slag aggregate preparation process. It is clear that the irregularity of the steel slag aggregate is increased if the dusts on the steel slag aggregate surface are removed by the wet rolling screen. Furthermore, the water stability, porosity, durability, dynamic stability, and anti-skid property of the steel slag asphalt concrete road are improved if the steel slag aggregate is cleaned in advance.

(3) It is proved that the performances of steel slag SMA and OGFC developed in this work are better than that of the common basalt SMA and OGFC.

(4) Semi rigid base and flexible base can be made from steel slag entirely. The more steel slag in semi rigid base or in flexible base, the better the mechanical property is.

References

1. Li Canhua. Prospect analysis on comprehensive utilization of steel slag under the background of Iron and steel enterprise restructuring [C]. The 10th annual meeting of Chinese society for metals metallurgy technology economic society, 2009, Beijing.
2. Zheng Huirong. It is inaccurate to say that China's cement limestone is enough for 30 years [N]. China Building Materials, (2005), http://www.chinaim.com.cn/_private/zxrn/wgslshygy.htm
3. Wu Xuequan, Zhu Hong, Hou Xinkai, et al. Study on steel slag and fly ash composite Portland cement [J]. Cem Coner Res, 1999, 29(7) :1103.
4. Altuna,Yilmai. Study on steel furnace slags with high MgO as additive in Portland cement [J]. Cem Concr Res, 2002, 32(8):1247.
5. Motz,Geiseler J. Products of steel slags all opportunity to save natural resources [J]. Waste Manage, 2001, 21(3):285.

Research on local green materials and Tibetan buildings on the basis of green

L.P. Li[†] and S. Fan[‡]
Department of Architecture, College of Architecture and Urban Planning, Kunming University of Science and Technology,
Kunming City, Yunnan Province, China
E-mail: [†]llping402@sina.com, [‡]fs1204@sina.com
www.university_name.edu

Based on the field testing and analysis of the Tibetan dwelling in Shangri-La, this article analyzes the local buildings of Tibetan people on the basis of green building and gives some advices to the local people on improving the living environment. The research results reveal that Tibetan dwellings is to build the house within the bearing capacity of environment, and during construction, use low consumption and low pollution environment-friendly building materials as much as possible to build low energy consumption and minimize the resource utilization.

Keywords: Local green materials; Tibetan dwellings; Temperature; Measured data and analysis.

1. Introduction

Under the circumstances of backward production tools and technologies, varied and graceful local buildings are built by using the local materials with the wisdom of the working people. Under the local climate and environment, unique building types that are harmonious with the natural environment are built. In these buildings, there are many "green" construction ideas and technologies. In the previous researches on local buildings, it is analyzed mainly from the formation of ancient settlement, the types of buildings, landscape and the manifestation of buildings as the carrier of culture, etc. As for the survey on the local buildings of the Tibetan nationality in Shangri-La this time, these dwellings are analyzed and researched mainly from building technology and the indoor thermal environment testing of building, etc. Meanwhile, green building system is taken into account for overall consideration. It researches some issues in local buildings on the basis of physical quantitative analysis in the hope of finding out the excellent parts — tradition worth being inherited, which are ignored in previous researches. Hoping to provide new ideas and enlightenment for present green building design — how to make the buildings meet the requirements of green buildings better with low technologies and low consumption, meanwhile hoping it can offer guidance to the local people who will build the house by themselves to build more comfortable new dwellings with local characteristics instead of the houses with brick wall and tile roof newly built by the villager in comparatively developed traffic areas, which do not have the styles and features of the local traditional buildings, as well as the "green" of traditional buildings.

2. Analysis on the Measured Data of Indoor Thermal Environment

In order to understand the current situations of indoor thermal environment of Tibetan-style dwellings of different materials in Shangri-La, we tested on such parameters as indoor temperature of Tibetan-style dwellings with rammed earth walls, and the surface temperature of building envelope on the winter solstice in 2008, and found out the changes rules and features of the test parameters through comparative analysis, which provides a necessary actual measurement basis for improving the comfort of indoor thermal environment of Tibetan-style dwellings in winter.

We carried out tests on several typical dwellings in Shangri-La to research and compare the thermal insulation properties of traditional rammed earth walls. The buildings tested is a Tibetan traditional rammed earth dwelling in the east-west direction, which is of square shape in plane and has three rooms. The traditional rammed earth dwelling is as shown in Fig. 1, Fig. 2. The three walls of this dwelling are rammed earth walls, and the other side is a corridor installed with glass. The first floor of this dwelling are living room and animal house, this is because changes in living style of the county town have taken place and the Tibetans no longer take the stock raising such as yak as their main economic source. The corridor on the 1st floor is installed with a steel frame glass door to form an adjacent greenhouse, which is beneficial for the utilization of solar energy and heat insulation in winter. The 2nd floor is main space where the Tibetans live, which is of square shape in plane with a living room and two bedrooms. The corridor on the 2nd floor is also installed with glass windows. The roof is formed in such a manner that a double-pitch wooden roof is arranged on the flat soil roof through supports.

As shown in Fig. 2, master-bedroom is in the west orientation and bedroom is in the west orientation on the 2nd floor. As shown in Fig. 3, the air temperature of the master-bedroom reached the peak of 5.2°C at 2:49 p.m., while the air temperature of the bedroom arrived at the maximum of 11.66°C at 6:19 p.m. The temperature of the bedroom is 6.46°C higher than the one of the master-bedroom. The minimum air temperature was 0.7°C, 0.32 lower than the minimum air temperature of the corridor which was 1.02°C. The humidity of the master-bedroom reached the peak of 48.2% at 9:49 a.m., while the one of the bedroom arrived at the maximum of 42.5% at 11:49 a.m. The maximum of the humidity is 5.7% higher than the one of the bedroom. The minimum humidity was 25.7%, 5.2 higher than the minimum humidity of the bedroom which was 20.5%. According to the analysis of measurement data, we think that the indoor heat environment quality of Tibetan dwellings is relatively low. It will be to research some new methods to improve the indoor heat environment quality of Tibetan dwellings.

Figure 1. Tibetan-style dwelling with rammed earth walls.

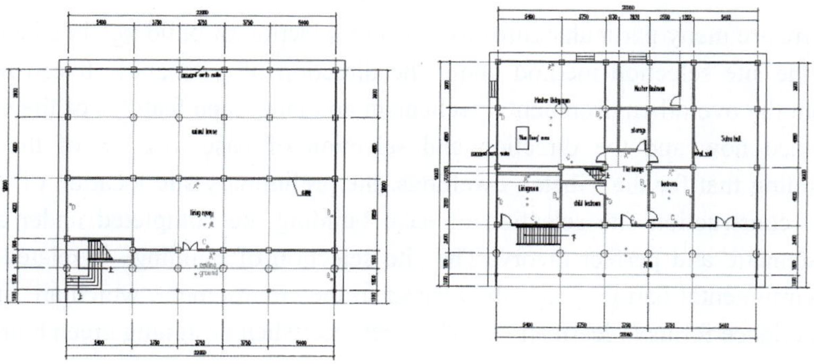

Figure 2. Plan of the 1st Floor and the 2nd Floor of Tibetan-style dwelling.

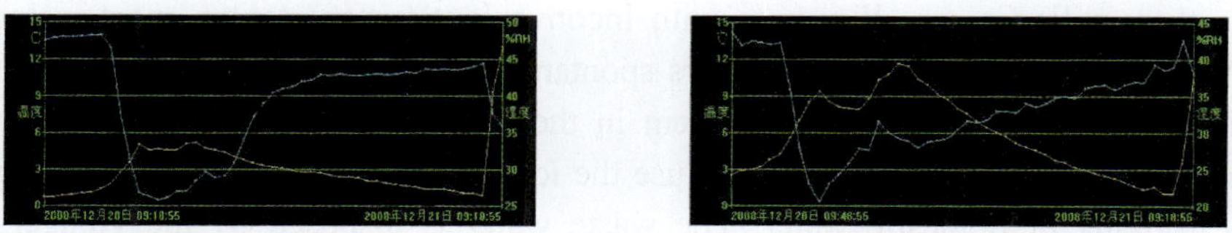

(a) Master Bedroom (b) Bedroom

Figure 3. Comparison of the air temperature of the master-bedroom with the bedroom.

3. Equations Analysis on Local Green Materials and Tibetan Dwellings

It was research and analyze each aspects concerning Tibetan dwelling on the basis of green buildings in this paper in the hope of finding the instructive and significant parts for green building design and construction tools from the building shape of Tibetan people with a long history and the research on building technology. Meanwhile, knowing the insufficiency of Tibetan dwellings well can be instructive for villagers to rebuild and build the dwellings, can improve the housing condition of Tibetan people and improve their living environment. Besides, it can be more ecological and environment-friendly.

Our understanding of green building is to build our house within the bearing capacity of environment. During construction, use low consumption and pollution environment-friendly building materials as much as possible to build low energy consumption and minimize the resource utilization. The followings are our rough analysis on Tibetan dwelling in accordance with several existing assessment documents concerning green building.

3.1. *Selection of construction site*

There are many particular cares for the site selection of buildings in China, such as the site selection method under the direction of geomancy. It can be seen from the overall environment, mechanism and landscape features of the villages formed now and the direction and selection of base location of the single building that for the Tibetan dwellings, the preliminary site location of villages and construction site selection of each building are completed under a quite systematic and perfect theory. The site selection of buildings is related to the environmental harmony and the respect to the environment, which are of value experience for us to learn and use for reference when designing green buildings.

3.2. *Water-saving*

As these villages are all remote with inconvenient transportation, most water sources are searched by the villagers spontaneously and the water is led to the villages to solving the water problem in the whole villages. The villages we investigated and surveyed this time use the local or the nearby mountain spring as drinking or domestic water. The waste water is disposed by the villagers casually in a habitual way and there is no concept of waste water treatment and recycling. In the upriver villages, if the waste water without treatment is discharged into the river directly, a pollution source will be formed.

It is suggested to construct waste water treatment and recycling project and rainwater collection and use facilities after the completion of drinking water project in the future construction of the villages to save water and reduce pollution.

3.3. *Energy-saving and using solar energy with local materials*

In the villages, firewood is the main energy. According to the villagers , they will burn dozens of carriages of firewood. They build methane tanks and hot water supply equipment with solar energy by providing funds.

In the past when the technology level was low, for the selection of construction materials, out ancestors simply obtained the materials from local places and use them to build their dwellings after simple processing and such cases can been seen everywhere, such as the stone slab house in Buyi nationality of Guizhou, the bamboo buildings in Xishuangbanna and the cave dwelling in the Loess Plateau. These principles are also the principles for the green buildings today, i.e. obtaining materials from local places and using the recyclable and the fast-growing materials.

In the Tibetan dwellings, the main construction materials are earth and wood and sometimes stone. The wall of earth houses is rammed earth wall, which has less influence on the environment compared with other materials (bricks and concrete blocks) and besides, after the abandoning of the buildings, the earth rammed wall will become earth after being scattered, which will not generate building garbage. Meanwhile, the heat-insulation performance and thermal stability of the earth rammed wall are relatively good (as is seen from the measurement analysis above). Surveyed with the method of modern “full life cycle analysis”, rammed earth wall is also a kind of relatively excellent construction material.

4. Conclusion

Based on the system standard of green buildings, we combed the Tibetan dwellings in Shangri-La roughly and mainly carried out measurement and analysis to the indoor heat environment of the dwellings. We think that the important parts to be improved in Tibetan dwellings include the following:

(1) It is required to construct a perfect waste water treatment and recycling system, rainwater recycling application system and garbage collection system in Tibetan villages to reduce the influence and pollution to the environment.

(2) To increase the utilization of clean and renewable energy such as solar energy, improve the firewood burning stoves, reduce the firewood use amount and design and improve the smoke exhaust system and air intake and heat exchange equipment uniformly to improve the indoor air quality while reduce heat loss.

(3) The heat-insulation performance and heat stability of rammed earth wall is good; from the aspect of “full life cycle analysis”, the rammed earth wall almost has no influence on the environment, so it should be reserved but not replaced by brick and concrete structure. It is also an unacceptable method to change the walls of main faces into wood plate walls. If it passive solar sunlight rooms are added externally, the changing is feasible in theory.

(4) Doors and windows are the largest heat loss parts of buildings with bad air tightness and large heat loss amount. It is suggested improve the structures and improve air tightness and adopt the measures of double-layer glass with night heat insulation plate if the condition permits to reduce the heat loss of doors and windows and improve the heat insulation performance of the whole building.

(5) The stairways of the several dwellings we investigate and survey almost directly lead to outdoors; although the balcony is generally in the lounge, the heat loss of the whole building is increased. It is suggested add a wall and door in the place where the stairway leads to outdoors to improve the heat insulation performance.

References

1. Standard of Evaluation standard for green building (GB/T 50378—2006), in Chinese.
2. Green Building Evaluation System [M] Beijing: China Architecture and Building Press, 2002.
3. [British] Tom Woolley et al. Green Building Manual 1 [M] translated by Tang Po and Xu Shen et al. Beijing: Machinery Industry Press, 2006.
4. [American] Brian Edwards Green Buildings [M] translated by Zhu Ling and Zheng Zhiyu, Shenyang: Liaoning Science and Technology Press, 2005.
5. Green Buildings compiled by Xi’an University of Architecture and Technology [M] Beijing: China Planning Press, 1999.

The clinical analysis use of amlodarone in the treatment of congestive heart failure

W. Lin* and C.L. Mei
Affiliated Hospital of Jilin Medical College, Jilin 132013, China
Beihua University, Jilin, 132013, China
**E-mail: lin2784879@sina.com*

To study of amiodarone in treatment of congestive heart failure arrhythmia of clinical efficacy and safety. Methods 80 patients with CHF complicated with ventricular arrhythmia patients,were randomly divided into control group and treatment group,the control group was given routine anti heart failure treatment, the treatment group was treated with amiodarone therapy, comparative analysis of two groups of patients in improving cardiac function, arrhythmia, left ventricular ejection fraction (LVEF), rhythm of the heart failure readmission rates and adverse drug reaction and other negative situation. Results Antiarrhythmic amiodarone group, improvement of heart function, left ventricular ejection fraction, rhythm of the heart failure readmission rates and other aspects were significantly better than the control group, the difference was statistically significant ($P<0.05$). Conclusion Amiodarone is the treatment of congestive heart failure and ventricular arrhythmia of effective drugs, safe use. Adverse reaction to light

Keywords: Congestive heart failure; Ventricular arrhythmia; Amiodarone.

1. Introduction

Severe ventricular arrhythmias can affect blood flow dynamics, ventricular arrhythmia for independent risk factor for sudden death caused by heart failure, sudden death rate as high as 50%–60% [1]. Refractory ventricular tachycardia (vt) is a fatal arrhythmia, especially in cardiac insufficiency, high mortality rate in a critical condition [2]. The control of congestive heart failure with ventricular arrhythmia, can decrease the mortality. Amiodarone has good inhibition of sympathetic activation, fight ventricular arrhythmia and dilate blood vessels, improve myocardial ischemia, can safely used in severe heart failure, reduce sudden death and reduce mortality [3]. Amiodarone is the preferred drug for the treatment of refractory ventricular tachycardia [4]. Our clinical application of amiodarone treatment for heart failure and severe ventricular arrhythmias of 80 cases obtained satisfactory curative effect, the report is as follows.

2. Materials and Methods

2.1. *The general information*

This group of 80 patients, all in accordance with the "2007 China to guide the diagnosis and treatment of chronic heart failure" criteria, the average age (54 +/- 7). 28 cases of female patients, 52 cases of male patients, 41 cases of coronary heart disease, hypertensive heart disease in 27 cases, 12 cases rheumatic heart disease, cardiac function class II–IV. 68 cases of frequent ventricular premature beat (35 cases pairs of ventricular premature beat), 12 cases paroxysmal ventricular tachycardia, continuous cardiac in more than 15 times of ventricular tachycardia and cardiac when motion of five or more ventricular tachycardia in 5 cases. Eliminate II and II degrees above atrioventricular block and sinus bradycardia, sick sinus syndrome, severe lung disease, severe liver disease, abnormal thyroid function, severe electrolyte disorder. All patients were randomly divided into control group (40 cases) and treatment group (40 cases), two groups in age, sex, degree of illness there was no significant difference ($P > 0.05$), had comparability.

2.2. *Therapeutic method*

All 80 patients before treatment discontinuation original anti-arrhythmic drugs at least five half-life, conventional heart failure treatment after admission (beta blockers, and aldosterone receptor antagonist, diuretics, ACEI class, etc.), Holter, echocardiography before and after the treatment, routine physical check, etc. Treatment group in the control group on the basis of giving oral amiodarone in moderation. The first week: 0.2 g each time, 3 times a day; In the second week: 0.2 g each time, 2 times a day; The third week: 0.2 g per day, 1 time a day. Continuous treatment of 12 weeks to watch. All two group patients were given routine checkups and adjust dosage according to heart rate, QT interphase. Such as significantly prolong period between QT (QT 0.56 or higher) and tardy arrhythmia (HR < 50 times/min) to stop the use of amiodarone.

2.3. *Therapeutic effect evaluating*

According to the dynamic electrocardiogram (ECG) for 24 h before and after the treatment results and NYHA heart function classification standard: (1) Excellent the number of frequent ventricular premature beat were markedly improved by an average 70% or higher; The decrease in the number of pairs of ventricular premature beat 80% or higher; Short a ventricular tachycardia disappear to 90% or higher, continuous cardiac in more than 15 times of ventricular tachycardia and cardiac when motion of five or more ventricular tachycardia disappear completely; Cardiac function improves cardiac function in more than class II or to reach to class I, symptoms and signs disappeared basically. (2) Efficacious the effective number of frequent ventricular premature beat an average loss of 70% or higher or verge of reducing 90% or higher; Cardiac function improves cardiac function in class I or did not meet class I, signs and symptoms improve. (Invalid) can not meet the above criteria.

2.4. *Statistics*

SPSS software was used for statistical analyses and the values are presented as the means +/- SD. Measurement data using t test, counting data by chi-square test. P values of less than 0.05 were taken to indicate statistically significant differences.

3. Results

3.1. *Therapeutic outcome*

To observe the end during hospitalization in cardiac function improvement, control group: improve the level I 27 cases, achieve grade I, signs and symptoms disappeared basically in 11 cases, improve heart function class II above 18 cases; Treatment group: to improve the grade I 36 patients, improve heart function class II above 29 cases; Reach the level I, signs and symptoms disappeared basically 18 cases, comparing the two groups was statistically significant difference ($P < 0.05$).

3.2. *Adverse reactions*

Treatment group, 40 patients in the hospital during the period of taking amiodarone, gastrointestinal reaction 3 cases, 2 cases in intractable chronic cough, disappeared after the drug was stopped, there were 3 cases of tardy arrhythmia, discontinuation after 1 week or reduce dose ease, face and aroud the eye skin pale blue change in 1 case.

4. Discussion

Chronic heart failure with the development of the disease and is aggravating, ventricular wall tension increases, the sympathetic nervous system, renin angiotensin aldosterone system activation, the combination of a variety of factors such as water and electrolyte balance disorder, spontaneity and complex significantly increased the occurrence of ventricular arrhythmia, is one of the important reasons of death in patients with heart failure [5]. Congestive heart failure in ventricular arrhythmia, the hemodynamic change often aggravate heart failure, and even life threatening. So choose good answer law effect and influence on cardiac function in small drug, for correcting the hemodynamics, improve prognosis has positive significance [6]. Foreign literature, in the treatment of senile heart failure with the arrhythmia is easy to change for the worse due to the rapid arrhythmia, exist a greater risk of treatment [7]. I, II, IV anti-arrhythmic drugs all have certain negative inotropic effect, application in congestive heart failure, can make the worsening heart failure, the patient case fatality rate increases, thus anti-arrhythmic treatment in patients with heart failure is difficult.

Amiodarone was originally used in clinical as expansion of coronary artery drug. Reason is that it can be a non-competitive blocking receptor alpha and beta receptors, expansion of coronary artery, increase the blood supply, expanding peripheral artery, reduce the external resistance, lower blood pressure, reduce myocardial oxygen consumption, and no obvious influence cardiovascular and discharge. Therefore, amiodarone has the effect of heart failure. This determines the amiodarone in the treatment of heart failure anti-arrhythmic unique role, as well as improve the arrhythmia matrix in heart failure, and has direct anti-arrhythmic effects. The treatment of patients with heart failure, after the ideal, rather than when implanted ICD, choice of amiodarone replace ICD inhibit symptomatic ventricular arrhythmia. Recommend II class b, C grade evidence [8].

Amiodarone for anti-arrhythmic drugs has been increasingly used in the treatment of heart failure, it can not only improve left ventricular function, increase exercise tolerance, and can make with cardiac death and hospitalization for the event at the end of joint rate dropped significantly [9]. GESICA experiments further confirmed such as amiodarone is suitable for the patients with heart failure, and reduce the incidence rate of ventricular arrhythmia, improve survival quality [10]. Amiodarone treatment index, the characteristics of long half-life, fewer drugs, anti-arrhythmic variety. Especially can control the general anti-arrhythmic drugs ineffective malignant ventricular arrhythmia [11]. In recent years, studies have shown that oral amiodarone can improve heart failure in patients with ventricular fibrillation threshold, reduce ventricular fibrillation occurred [12]. To complete some of the more recent prospectie, randomized, double-blind, placebo-controlled large-scale clinical trials, For amiodarone fight ventricular arrhythmia efficacy and safety of provides a reliable basis [13]. CASCADE [14], the study is a randomized clinical trial will result in 228 cases of the myocardial infarction patients with ventricular fibrillation were randomly divided into amiodarone group 113 cases (183 mg/d) and conventional drugs group 115 examples, the results of amiodarone group 2, 4, 6 years of mortality and the incidence of sustained VT group were significantly lower than that of conventional drugs (all $P < 0.\ 01$). Domestic some reports, amiodarone in chronic heart failure with sustained ventricular arrhythmias or fast ventricular rate of persistent atrial fibrillation, ventricular tachycardia control was 75%, and the fast ventricular rate of atrial fibrillation total successful rate was 71% [15]. Foreign literature reports, amiodarone can be safely applied to the severe heart failure [16]. It can reduce the damage of the phospholipids in ischemic myocardium, adjust the ischemic myocardial energy metabolism in cells, free radicals, protecting cell membrane and organelles structural and functional integrity, against lipid peroxidation and has a good effect against myocardial ischemia.

Amiodarone is a benzene furan derivatives with mild non-competitive adrenaline receptor blockers, belong to class III anti-arrhythmic drugs. And mild I and IV class anti-arrhythmic drug properties. Inhibit atrial and cardiac conduction fiber fast sodium ion flow, amiodarone extend all kinds of action potential of myocardial tissue and effective refractory period, and can slow conduction velocity. Reduce the sinoatrial node self-discipline. The inhibition of atrioventricular bypass forward conduction is greater than the reverse. The resting membrane potential and action potential height had no effect. Due to the complex extremely excessive extension, electrocardiogram (ecg) have a Q - duration between T and T wave change. Amiodarone intravenous injection with mild negative inotropic effect, but usually not inhibit left ventricular function. Amiodarone is directly for coronary and peripheral vascular expansion effect, can affect the thyroid hormone metabolism. Foreign reports, amiodarone long-term treatment of heart failure, through its unique selectivity of cardiac sympathetic nerve function, and show the unique benefits of heart failure [17]. Amiodarone is the treatment of congestive heart failure in ventricular arrhythmia and effective drugs, the use of safe. Adverse reaction, it is necessary to liver and kidney function, electrolyte, periodic review of thyroid function in order to early detection and treatment related side effects.

Acknowledgments

The project supported by Educational Department of Jilin province in China. (No.2015153, 201339033, 2013ZC029). Chunli Mei is the corresponding author.

References

1. China heart pacemaker and ECG physiological branch of Biomedical Engineering Institute. The Chinese Medical Association Branch of Cardiovascular disease. The Cardiovascular Disease Magazine Editorial Board. etc. Amiodarone anti-arrhythmic treatment application guide, *Chinese Journal of Cardiology*, 2004, 12(32): 1065–1071.
2. Szabo BM, Crijns HJ, Wiesfeld AC, et al. Predictors of mortality in patients with sustained ventricular tachycardias or ventricular fibrillation and depressed left ventricular function: Importance of beta-blockade, *Am Heart J*, 1995, 130:281–286.
3. Roy D, Talajic M, Dorian P, et al. Amiodalone to prevent reeulienee of atrial fibrillation. *N Engl J Med*, 2002, 342(13):913–920.
4. Doval HC. Class III antiarrhythmic agents in cardiac failure: Lessons from clinical trials with a focus on the Grupo de Estudio de la Sobrevidaen la Insuficienda Cardiaca en Angentina (GESICA), *Am J Cardiol*, 1999, 84 (9A)109R–114R.
5. Jiang-Hua Zhong, Cun-Tai Zhang, Li Ying, et al. Oral amiodarone of rabbit dilated cardiomyopathy is the influence of the homogeneity across the ventricular wall., *Chinese Journal of Cardiology*, 2003, 31(1):57–60.
6. Gonzalezer, Kannewurfbs, Ornatojp. Intravenous amiodarone for ventricular arrhythmias overview and clinical use, *Resuscitation*, 1998, 39:33–42.
7. Khnn IA. Pharmacological cardioversion of recent onset atrial fibrillation, *Eur Heart J*, 2004; 25(15) : 1274–6.
8. Jin-Tao Xiang. In 2008 the ESC diagnosis and treatment of acute and chronic heart failure guidelines, *Chinese journal of cardiac pacemaker and ecg physiological*, 2008, 22(6):526.
9. Massie BM, Fisher SG, Radford M, et al. Effect of amiodare on clicinal status and left ventricular function in patients with congestive heart failure. *CHF-STAT investigators, Circulation*, 1996, 93(12):2128–2134.
10. Clelandj GF, Dargie HJ, Findlay IN, et al. Clinical heamodynamic and antiarrhythmic effects of long term treatment with amiodarone of patients in heart failure, *Br Heart J*, 1987, 57:436.
11. Xin Zheng. A review of clinical application of amiodarone, *Chinese journal of cardiac pacemaker and ecg physiological*, 2001, 15(5) (Supplement):1–4.

12. Ross R. The pathogensis of atherosclerosis M. In: Eugene Braunwald ed. Heart disease: A textbook of cardivascular medicine. *5 ed, Philad ephiasaunders,* 1999, 1105–1126.
13. Bansal BC, Gupt RR, Buadal MR, et al. Senum lipids and uric acid relationship in ischemic thrombotic, cerebro vascular disease, *Stroke*, 1975, 6:304–307.
14. The CASCADE investigators. Randomized antiarrhythmic drug therapy in survivors of cardiac arrest (the CASCADE study), *Am J Coll Cardiol*, 1993, 72:280–285.
15. Jian-Ming Shao, Min-Min Xu, Qian-Min Tao. Amiodarone in chronic heart failure with the application of cardiac arrhythmias, *Chinese Journal of Cardiology*, 2003, 31(6):460.
16. Nicklas M, Mckenna WJ, Stewart RA, et al. Prospective double-blind placebo-controlled trial of low-dose amiodarone in patients with severe heart failure and asymptomatic frequent ventricular ectopy, *Am Heart J*, 1991, 22:1016–1021.
17. Kaye DM, Dart AM, Jennings GL, et al. *J Am Coll Cardiol*, 1999, 33(6):1553–1559.

Activin receptor type IIA nerve protection mechanisms research on Oxygen-Glucose deprivation in PC12 cells

C.L. Mei, L.J. Zhang* and H.M. Liu
Beihua University,
Jilin, 132013, China
**E-mail: gaoda2002@126.com*

ActRIIA protects neurons against toxicity and plays a vital role in regulating the brain's response to injury. In this study, The mechanisms involved in the neuroprotective effects of ActRIIA were investigated in a model of ischemic stroke. The results show that ActRIIA plays an important role in transferring neuronal protective signals inside. We found that ActRIIA effectively increases the survival rate of PC12 cells and relieves OGD damage. To clarify the neuroprotective mechanisms of ActRIIA, the effects of ActRIIA on the ActA/Smad pathway and on the up-regulation of inducible nitric oxide synthase (NOS) and superoxide dismutase (SOD) were investigated using OGD in PC12 cells. ActRIIA increases the expression of Smad3, Smad4 and P300 and that 50 ng/mL and 100 ng/m of ActRIIA reduces NO levels and increases SOD activity, respectively.These results suggest that the neuroprotective effects of ActRIIA in ischemia are related to the activation of the ActA/Smad signaling pathway.

Keywords: ActRIIA; Neuroprotective effects; OGD.

1. Introduction

Ischemic stroke is one of the leading causes of death, serious and long-term disability[1]. It results in oxygen and glucose deprivation of neurons and leads to necrotic loss. This kind of injury can trigger endogenous activation of some neuroprotective factors and related signal transduction. Recent attention has focused on the function and mechanism of cellular signal transductions involved in the ischemic brain injury[1–2].

ActRIIA is a member of the transforming growth factor-beta (TGF-β) superfamily, and its signaling is an important regulator of many cellular process. ActA can bind with ActRIIA and initiate the intracellular transmission of signals[3–4]. After ligand binding, ActRIIA phosphorylates and thereby activates and phosphorylates ActRIA, as well as intracellular R-Smads. Then the latter forms a heterodimer, which will regulate target genes by translocating into the nucleus and recruiting transcriptional co-activators or co-repressors. To further define the role and mechanism of endogenous ActRIIA and its signaling in response to acute ischemic damage, we used an OGD model in PC12 cells to simulate ischemic injury on neurons in vitro[5].

In this study, the oxygen-glucose deprivation of PC12 cells was used to establish a cerebral hypoxia-ischemia model to investigate the mechanism of ActRIIA neuroprotection. In addition, we measured the amount of NO and the induction of SOD in an attempt to elucidate possible mechanisms that underlie ActRIIA-mediated protection against OGD in PC12 cells.

2. Materials and methods

2.1. *Cell culture and Oxygen-Glucose deprivation*

Rat adrenal pheochromocytoma (PC12) cell line was obtained from Gold Amethyst Pharm & Bio-Tech Co. Ltd (Beijing, China). Cells were grown in complete Dulbecco's modified Eagle's medium (DMEM, Gibco, Carlsbad, CA, USA) supplemented with 10% fetal bovine serum (FBS, Gibco), in a humidified atmosphere of 5% CO2 and 95% air at 37°C, subcultured with NGF for 6 days routinely. Glucose-free DMEM (Gibco) supplemented with deoxygenated 1.0 mmol/L sodium dithionite (Na2S2O4) was used for the induction of OGD as previously described[7–8]). PC12 cells seeded at a density of 1.0 × 108/L were washed three times and then cultured in the OGD medium at hypoxic conditions (37°C, 5% CO2 and 95% N2).

2.2. *Cell survival rate assay*

The MTT method was used to assess the cytotoxic effects of ActRIIA. The cells were grown to a density of 5 × 104 cells/well and were then treated with 10 ng/mL, 20 ng/mL, 30 ng/mL, 50 ng/mL and 100 ng/mL ActRIIA in a 96-well plate for 24 hours. At the end of the treatment, the ActRIIA-containing medium was carefully removed and the cells were treated with OGD for 16 h. The culture medium was removed and 200 μL medium containing 20 μL MTT (3-[4,5-dimethylthiazol-2-yl]-2,5-diphenyl tetrazoliumbromide) (5 mg/mL in PBS) (St. Louis, MO, USA) was added to each well. After 4 hours of incubation at 37oC, the medium was removed and 100 μL DMSO was added to each well. The optical absorbance (A) of each well was read at 490 nm. The percentage of viable cells was calculated as follows: (A of experimental group / A of control group) × 100%).[5]

2.3. *Western blotting analysis*

After treatment with ActRIIA and OGD for 16 hours , the cells were washed twice using cold PBS and 1×106 cells were lysed using RIPA buffer (50 mmol/L Tris (pH 8.0), 150 mmol/L NaCl, 0.1% SDS, 1% NP40 and 0.5% sodium deoxycholate) containing protease inhibitors (1% cocktail and 1 mmol/L PMSF). Total proteins were separated using 15% SDS-PAGE and were transferred to a PVDF membrane. The membrane was blocked using Tris-buffered saline with 0.1% Tween 20 (pH 7.6, TBST) for 1 hour at room temperature and was incubated with the primary antibody solution (1:1000) at 4 oC overnight. After two washes in TBST, the membrane was incubated with the HRP-labeled secondary antibody (Santa SC-2073) for 1 hour at room temperature and was washed three times with TBST. The final detection was performed using enhanced chemiluminescence (ECL) western blotting reagents (Amersham Biosciences, Piscataway, NJ) and the membrane was exposed to Lumi-Film Chemiluminescent Detection Film (Roche). Loading differences were normalized using a monoclonal ß-actin antibody. The antibodies used in the study included P300 (Santa SC-585), Smad3 (Santa SC-101154), Smad4 (Santa SC-73040), SOD (Santa SC-18503), NOS (Santa SC-49055) and ß-actin (Santa SC-2021)[5].

2.4. *Measurement of NO*

Nitrite production, measured by the Griess reaction, was used as a measure of NO production. Briefly, 100 μL of culture supernatant was incubated with an equal volume of Griess reagent (1 part 0.1% naphthylethylene-diamine, 1 part 1% sulfanilamide in 5% H3PO4) in 96-well tissue culture plates for 10 min at room temperature in the dark. The absorbance at 540 nm was determined using a microplate reader (SpectraMAX 340, Molecular Devices, Sunnyvale, CA) [6].
Measurement of SOD

Cells (1×105) were harvested and after treatment with OGD for 16 h and ActRIIA plus OGD for 16 hours, the cells were washed twice with cold PBS and were lysed in ice-cold 0.1M Tris/HCl (pH 7.4 containing 0.5% Triton X-100, 5 mM β -ME, 0.1 mg/mL PMSF). The cell lysates were centrifuged at 14,000 rpm for 5 minutes at 4°C and the cell debris was discarded. The total SOD activity in the supernatants (contain the cytosolic and mitochondrial fractions) were measured using the Superoxide Dismutase (SOD) Activity Assay Kit (Biovision k335-100, San Franciso, USA) in accordance with the manufacturer's instructions[6].

2.5. *Statistics*

SPSS software was used for statistical analyses and the values are presented as the means +/- SD. ANOVA was used to compare the mean values. P values of less than 0.05 were taken to indicate statistically significant differences.

3. Results

3.1. *Effects of ActRIIA on cell proliferation*

The cytotoxicity of OGD for 16 hours and ActRIIA plus OGD for 16 hours treatments were determined by examining their effects on the proliferation of PC12 cells. PC12 cells were treated with 10 ng/mL, 20 ng/mL, 30 ng/mL, 50 ng/mL and 100 ng/mL ActRIIA for 24 hours before OGD for 16 hours (Figure 1). MTT assays showed that treatment with OGD for 16 h effectively inhibited the growth of PC12 cells by 21.32%. Compared to the OGD for 16 h-treated group, the ActRIIAplus OGD for 16 h-treated group inhibited the growth of PC12 cells by 17.72%, 13.31%, 9.46%, 6.43% and 5.81%, in dose-dependent manner, and the survival rates were higher than in the OGD for 16 h-treated group. Because there was no significant difference (P>0.05) between the survival rates of the 50 ng/mL and 100 ng/mL ActRIIA-treated cells, 50 ng/mL ActRIIA was used in all subsequent experiments.

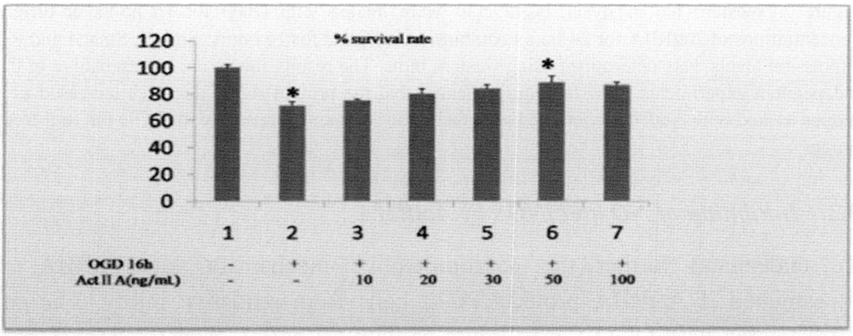

Figure 1. MTT assay showing the growth inhibition of PC12 cells treated with ActRIIA for 24 hours plus OGD for 16 hours. The cells were grown to a density of 5 × 104 cells per well in a 96-well plate for 24 hours. The results show the growth of PC12 cells following incubation in 96-well plates for 24 hours with 10 ng/mL, 20 ng/mL, 30 ng/mL, 50 ng/mL and 100 ng/mL ActRIIA plus OGD for 16 hours, compared with OGD for 16 hours. * denotes a significant difference from 50 ng/mL ActRIIA plus OGD for 16 hours, P<0.05. The data represent the means +/- SD obtained from three separate experiments that were performed in triplicate.

3.2. *Effects of ActRIIA on ActA/Smad pathway*

The ActA/Smad pathway plays a protective role in ischemia following its activation. To investigate the neuroprotective mechanisms of ActRIIA, the expression of Smad3, Smad4 and P300 were examined using western blots (Figure 2). Compared with the controls, the Smad3, Smad4 and P300 expression levels in PC12 cells increased in the OGD for the 16 h-treated group. Compared with this group, the expression levels of Smad3, Smad4 and P300 increased in the groups treated with ActRIIA for 24 h combined with OGD for 16 hours. In this group, as the concentration of ActRIIA increased, the expression levels of Smad3, Smad4 and P300 increased in a dose-dependent manner. Therefore, ActRIIA may mediate the neuroprotective effect of the ActA/Smad signaling pathway.

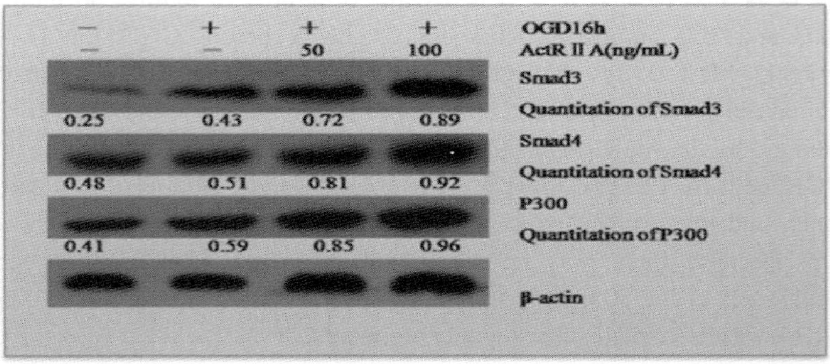

Figure 2. Western blot analysis. PC12 cells were treated with OGD for 16 hours or different concentrations of ActRIIA for 24 hours combined with OGD for 16 hours. Smad3, Smad4 and P300 expression levels were determined using western blots. The results shown are representative of three independent experiments. NIH imaging indicated that the protein signal densities increased in the groups treated with ActRIIA combined with OGD for 16 hours compared with OGD for 16 h-treated groups.

3.3. *Inhibition of NO and iNOS by ActRIIA*

To understand further the neuroprotective mechanisms of ActRIIA, we investigated if ActRIIA protects PC12 cells from oxidative injury. The cells were treated with OGD for 16 hours or with 50 ng/mL ActRIIA for 24 hours plus OGD for 16 hours, and NO production was assayed by measuring the levels of a stable NO metabolite, nitrites, in the conditioned medium. Incubation with ActRIIA (50 ng/mL) for 24 hours effectively inhibited NO production in OGD-stimulated PC12 cells (Figure 3). To determine if the inhibitory effect of ActRIIA on NO production was due to a decrease in NOS protein levels in the cytosol, the levels of NOS protein were examined by western blot analysis. As shown in Figure 4, the treatment with 50 ng/mL ActRIIA resulted in a significant decrease in NOS protein levels.

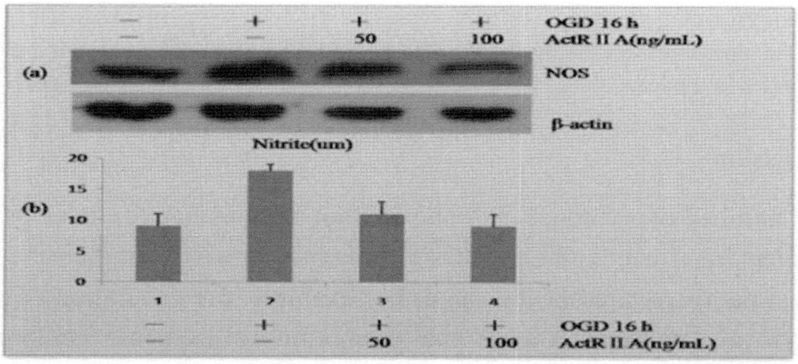

Figure 3. Effect of ActRIIA on oxidative stress in OGD for 16 h-induced PC12 cells. Cells were OGD-induced by incubation with 50 or 100 ng/mL of ActRIIA for 24 hours combined with OGD for 16 hours. The nitrite content was measured using the Griess reaction. The values indicate the nitrite production in cells that were exposed to the culture supernatants collected from cells that were treated with OGD alone or cells that were exposed to OGD plus ActRIIA. The data represent the means +/-SD from three independent experiments that were performed in triplicate. Western blot analysis of OGD-induced cells or cells treated with increasing concentrations (50 and 100 ng/mL) of ActRIIA for 24 hours combined with OGD for 6 hours. The western blot shows the levels of NOS protein expression. β-actin was used as an internal control.

3.4. *Increasing the activity of SOD by ActRIIA*

The effect of ActRIIA on SOD activity and protein expression were examined (Figure 4). As shown in Figure 5a, compared with the controls, the SOD activity in PC12 cells decreased in the OGD for 16 h-treated group. Compared with the OGD for 16 h-treated group, the SOD activity increased following treatment withdifferent concentrations of ActRIIA for 24 hours combined with OGD for 16 hours. Furthermore, the treatment with 50 ng/mL or 100 ng/mL ActRIIA led to a significant increase in SOD protein levels (Figure 4b).

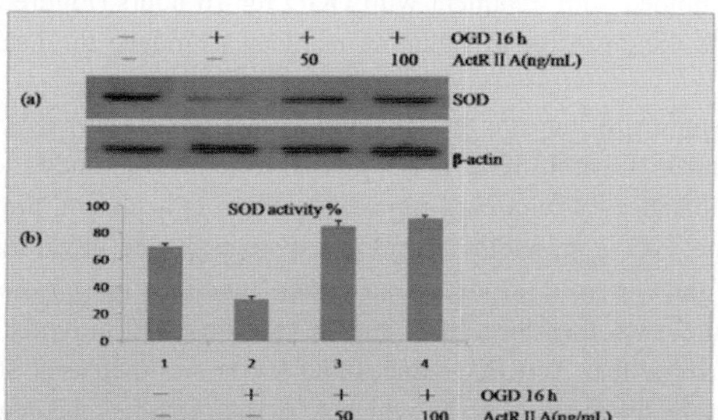

Figure 4. The effect of ActRIIA on SOD activity. (a) Western blot analysis of cells treated with OGD for 16 h or cells treated with OGD for 16 hours in combination with increasing concentrations (50 and 100 ng/mL) of ActRIIA for 24 hours. The western blot shows the SOD protein expression. β-actin was used as an internal control. (b)Cells were OGD-induced or were treated with 50 or 100

ng/mL of ActRIIA for 24 hours plus OGD for 16 hours. The effect of ActRIIA on SOD activity was assessed. The data represent the means +/- SD obtained from three independent experiments that were performed in triplicate.

4. Discussion

A large number of evidence has shown that various factors take part in the neuroprotective effects in the process of ischemic damage, but their mechanisms and signal pathways are not the same. For example, the generation of SOD and NO plays an important role in the promotion of ischemic cerebrovascular disease. ActinA can activate the ActinA/Smads signal transduction pathway, promote astrocyte gumnosis and regulate the anti-apoptosis mechanism in neurocytes. Existing animal models are affected by many factors. In order to control the experimental conditions, a stable cell model of cerebral ischemia was used in this study[7]. OGD was applied to PC12 cells and the results show that the OGD for 16 h-treated group effectively inhibited the growth of PC12 cells, indicating that the cell model causes cell damage. When compared with the OGD for 16 h-treated group, the ActRIIA plus OGD for 16 h-treated groups effectively increased the survival rate of PC12 cells (Figure 1). Therefore, ActRIIA has a protective effect on OGD-induced cell injury.

Smad3, Smad4 and P300 participate in the ActA/Smad pathway. The later can enter the nucleus to influence gene transcription and the activation of a series of biological functions. Our results demonstrate that treatment with OGD for 16 hours increases the expression of Smad3, Smad4 and P300. When compared with OGD 16 h-treated group, Smad3, Smad4 and P300 expression levels increased following treatment with different concentrations of ActRIIA for 24 h combined with treatment with OGD for 16 hours (Figure 2). ActRIIA may mediate the ActA/Smad signaling pathway to induce the neuroprotective effect.

Many studies[8] indicate that oxidative stress plays a key role in ischemic cerebrovascular disease. SOD plays a vital role in the body's oxidant and antioxidant balance by removing superoxide anion radicals and protecting cells from damage. NO is an another important messenger and effector molecule in vivo, and plays a role in neurotransmitter functions in physiological and pathological events, therefore, SOD and NO are involved in oxidative damage. The results show that ActRIIA reduces NO levels and increases SOD activity, and this may be the mechanism by which ActRIIA promotes the neuroprotective effect (Figure 3 and Figure 4).

5. Conclusion

In conclusion, our results demonstrate that ActRIIA protects PC12 cells in an OGD model by the activation of the AcA/Smads pathway, the inhibition of NO production and the increased activation of SOD. Further study is required to understand fully the beneficial role of ActRIIA in post-ischemic injury and will eventually lead to clinical interventions that will salvage brain cells that are at risk in ischemic stroke.

Acknowledgments

The project supported by Educational Department of Jilin province in China. (No.2015153, 201339033, 2013ZC029). Lijuan Zhang and Huimin Liu are the corresponding authors.

References

1. S.E. Chiuve, K.M. Rexrode, Spiegelman D, Logroscino G, Manson JE, Rimm EB. Primary prevention of stroke by healthy lifestyle. Circulation, 6(2008) 947–54.
2. U. Dirnagl, C. Iadecola, M.A. Moskowitz, Pathobiology of ischemic stroke: An integrated view,Trends Neurosci, 9(1999)391–7.
3. F.H. Ebner, S. Mariotto, E. Darra, Use of STAT1 inhibitors in the treatment of brain I/R injury and neurodegenerative diseases, Agents Med Chem, 11(2011)12–7.
4. H. Taniguchi, Function of prostaglandin E2EP receptors in the acute outcome of rodent hypoxic ischemic encephalopathy, Neurosci Lett 31(2011)185–90.
5. C.L. Mei, Nerve growth factor (NGF) combined with oxygen glucose deprivation (OGD) induces neural ischemia tolerance in PC12 cells, Afr J Biochem Res, 5(2011)315–20.
6. J.T. He, Neuroprotective effects of exogenous Activin A on oxygen-glucose deprivation in PC12 cells, Molecules, 17(2012)315–27.
7. J. Mang, Endogenous Protection Derived from Activin A/Smads Transduction Loop Stimulated via Ischemic Injury in PC12 Cells, Molecules, 18(2013)12977–86.
8. C.L. Mei, The Effects of Smad6 via Small RNA Pathways on the Ischemic Injury in PC12 Cells, Clinical lab, 12(2014)2063–2071.

Analysis on diversity of intestinal aerobic bacteria from 3rd instar larvae of *Dendrolimu kikuchii* Matsumura

R. Shi, Z. Xiong and N. Zhao
Key Laboratory of Biodiversity Conservation in Southwest China, State Foresty Adiministration, Southwest Forestry University, Kunming 650224, PR China

M. Ye
Key Laboratory of the Ministry of Education for Agro-Biodiversity and Pest Management,
Yunnan Agricultural University, Kunming 650201, PR China

B. Li*
Key Laboratory of the Ministry of Education for Agro-Biodiversity and Pest Management,
Yunnan Agricultural University, Kunming 650201, PR China
Yunnan Reascend Tobacco (Group) Technology Co., Ltd., Kunming 650106, PR China
E-mail: biaobiao201@163.com

In order to study the diversity of intestinal aerobic bacteria in the 3rd instar larvae of *Dendrolimus kikuchii* Matsumura, 16S rDNA was PCR amplified by the generic primers. The 16S rDNA amplicons (about 1.5 kb) were digested by four restriction enzymes, and the characteristic ARDRA fingerprint was obtained. The results of digestion showed that the diversity of intestinal bacteria could be reflected by the results of all four restriction enzyme digestions. Phylogenetic analysis based on sequencing showed that *Klebsiella pneumoniae* and *Bacillus cereus* were the dominant microflora, and *Enterobacter* sp., *Bacillus* sp., *Citrobacter* sp. and *Staphylococcus* sp. were not abundant microflora in the intestinal tract of 3rd instar larvae of *Dendrolimus kikuchii.*

Keywords: *Dendrolimus kikuchii* Matsumura; Gut microbiology; Diversity.

1. Introduction

The larvae of *Dendrolimus*, *Lasiocampidae*, *Lepidoptera* are distributed mainly in Yunnan, Jiangxi, Fujian, Anhui and Zhejiang provinces of China. The *D. kikuchii* Matsumura larvae are one or two generation per annum, and seven instars are included in all *D. kikuchii* life in Yunnan province, China. The host range of *D. kikuchii* includes *Pinus kesiya* var. *langbianensis* (A.Chev) Gaussen, *Pinus yunnanensis* Faranch, *Keteleeria evelyniana* Mast and *P. massoniana* Lambert (Institute of Zoology of The Chinese Academy of

Sciences, 1979). *D. kikuchii* is a primary pest of *Pinus kesiya* var. *langbianensis* (A.Chev) Gaussen, and is often referred as the "smokeless fire" (Gninenko et al, 2002; Mellec et al, 2008). The poison fur of *D. kikuchii* could threaten human health, causing physical disabilities and blindness when *D. kikuchii* bloomed (Wang et al, 1999). Regarding the researches on *D. kikuchii*, several previous studies focused on the control of *D. kikuchii* (Boguś et al, 2007), while few papers reported the trophic value of *D. kikuchii* (Tong & He, 2010). So far there is no study focused on the intestinal aerobic bacteria from the larvae of *D. kikuchii*.

The intestinal tracts of *D. kikuchii* comprises complex microbial ecosystems, and the complexity is attributed to the differences in anatomical features, diet, and the health of the *D. kikuchii* (Moselio, 2006). With the development of methodology, particularly 16S rDNA based molecular techniques, the composition and function of insect gut microbiota were gradually delineated in recent years. So far the gut microbiota of *Ceroplastes japonicus* Green (Bei et al, 2005), *Tenebrio obscurus* and *T. molitor* (Zhang et al, 2006), *Eupolyphaga sinensis* Walke (Liu et al, 2007a), *Locusta migratoria manilensis* Meyen (Liu et al, 2007b) and *Clanis bilineata tsingtauica* (Lv et al, 2009) have been characterized, but the gut microbiota of *D. kikuchii* has never been studied and reported. In this paper, the diversity of intestinal aerobic bacteria from 3rd instar larvae of *D. kikuchii* Matsumura was characterized with amplified ribosomal DNA restriction analysis (ARDRA), and the intestinal aerobic bacteria were indentified through sequencing of 16S rDNA. The results of this study could provide the theory foundations for studying the interactions between the gut microbiota and *D. kikuchii*, and may contribute to the control of *D. kikuchii* as well as providing microbial resources.

2. Materials and Methods

2.1. *Materials*

The 3rd instar larvae of *D. kikuchii* were collected from *Pinus yunnanensis* in Wenshan, Yunnan, P.R.China.

NA solid medium: CMC-Na 20g, Zhengzhou Chengwang Chemical Co., LTD, Chemically Pure (CP). Na_2HPO_4 2.5g, Zhengzhou Chengwang Chemical Co., LTD, Analytical Reagent (AR). KH_2PO_4 1.5g, Zhengzhou Chengwang Chemical Co., LTD, AR. Peptone 2.5g, Shandong Sunkeen Biology Co., LTD, CP. Agar 20g, Shandong Sunkeen Biology Co., LTD, CP. Distilled water 1000ml, pH 7.0–7.5.

2.2. *Isolation of bacteria from D. kikuchii*

The collected health 3rd instar larvae of *D. kikuchii* were starved two or three days for the purpose of clearing excrement in their intestinal tract. They were cleansed with 75% ethanol for surface sterilization for ten seconds, followed with 10% sodium hypochlorite cleansing for ten seconds, and then sterile water washing for five times. The intestinal tracts of 3rd instar larvae of *D. kikuchii* were took under sterile conditions, and the pH value of the intestinal tract was tested. The intestinal tracts were triturated with 1 mL sterile water in sterile centrifuge tube, and 5 serial ten-fold dilutions were made. 0.5 ml of each dilution was spread on NA solid agar plate and incubated at 28°C overnight. Colonies were purified and the strains were preserved as slants at 4°C.

2.3. *DNA extraction*

Chromosomal DNA was extracted using TIANamp Bacterial DNA Kit (Tiangen Bioech, Beijing, China) according to the manufacturer's instructions. The concentration of purified DNA was determined using a spectrophotometer Smart-Spec 3000 (Bio-Rad, Marnes la Coquette, France).

2.4. *16S rDNA sequencing and theoretical restriction profile determination*

Universal primers 27F (5'-AGA GTT TGA TCC TGG CTC AGA ACG AAC GCT-3') and 1492R (5'-TAC GGC TAC CTT GTT ACG ACT TCA CCC C-3') (William et al, 1991) (Sangon, Shanghai, China) were used for amplification of the 16S rDNA from about 50 ng genomic DNA with the following amplification conditions: 94°C for 5 min, 30 cycles of 94°C for 1 min, 56°C for 1 min, 72°C for 3 min with a final extension 72°C for 10 min. PCR aliquots (5μl) were run on a 0.8% (w/v) agarose gel (Invitrogen) in 1X TAE with a 100 bp ladder (Tiangen Bioech, Beijing, China) and visualized by ethidium bromide staining using a GelDoc 2000 (Bio-Rad).

2.5. *Amplified 16S rDNA restriction analysis (ARDRA) of bacterial isolates*

Restriction enzyme digestion of the PCR amplified 16S rDNA products were carried out at 37°Cfor 12 hours in a final volume of 25μl with *Hind* III and *Hae* III (Sangon, Shanghai, China), and after 65°C for 4 hours in a final volume of 25μl with *Hinf* I and *Taq* I (Sangon, Shanghai, China). Restriction fragment patterns were analysed using a 1.0% (w/v) agarose gel (Invitrogen) in 1X TAE with a 100 bp ladder (Tiangen Bioech, Beijing, China) and visualized by ethidium bromide staining using a GelDoc 2000 (Bio-Rad).

2.6. *DNA Sequencing and alignment*

PCR products were sequenced in both forward and reverse directions by Sangon (Shanghai, China). Forward and reverse sequences were checked, edited and aligned using ContigExpress Software (USA). Complete edited sequences from all samples were aligned manually.

2.7. *Clustering and phylogeny analysis*

DNA gel images were converted into binary matrixes (1, presence of band and 0, absence of band) and clustered by UPGMA method using Sorensen's coefficient in the MVSP 3.0 Package (Kovach Computing Services, Pentraeth, Isle of Anglesey, UK). According to the clusters of the elected typical strains, the phylogenetic relationships of typical strains were reconstructed in MAGE4.0 through neighbor-joining methods, and the robustness of trees was determined by 1,000 (neighbor-joining).

3. Results

3.1. *The result of ARDRA community profiles*

Twelve intestinal aerobic bacteria were isolated from the 3rd instar larvae of *D. kikuchii*, and pH value was 7.2 in the larvae intestinal tract. Figure 1 shows that the 16S rDNA of twelve isolated intestinal aerobic bacteria were amplified with the length of ~ 1500bp. PCR products of 16S rDNA were co-digested with *Hind* III, *Hae* III and *Hinf* I, *Taq* I. The agarose gel electrophoresis fingerprint of ARDRA is shown in Figure 2. The left fingerprint in Figure 2 shows the full-length sequences of 16S rDNA digested with *Hind* III and *Hae* III, which includes 45 restriction fragments and four band types marked as A, B, C and D. These fingerprints indicated that the combination of two average cutting frequency restriction enzymes (*Hind* III, *Hae* III) could not fully digest and distinguish the 16S rDNA PCR product of twelve intestinal aerobic bacteria tested. The right fingerprint in Figure 2 shows the full-length sequences of 16S rDNA digested with *Hinf* I and *Taq* I, which includes 56 restriction fragments and six band types marked as a, b, c, d, e and f. Similarly, the right fingerprint in Figure 2 indicated that the combination of two higher cutting frequency restriction enzymes (*Hinf* I, *Taq* I) could not fully digest and distinguish the 16S rDNA PCR products of twelve intestinal aerobic bacteria tested. These results showed that by using combined fingerprints (*Hind* III + *Hae* III and *Hinf* I + *Taq* I), more fragments with larger size were obtained, and the polymorphism of twelve intestinal aerobic bacteria from the 3rd instar larvae of *D. kikuchii* were successfully revealed.

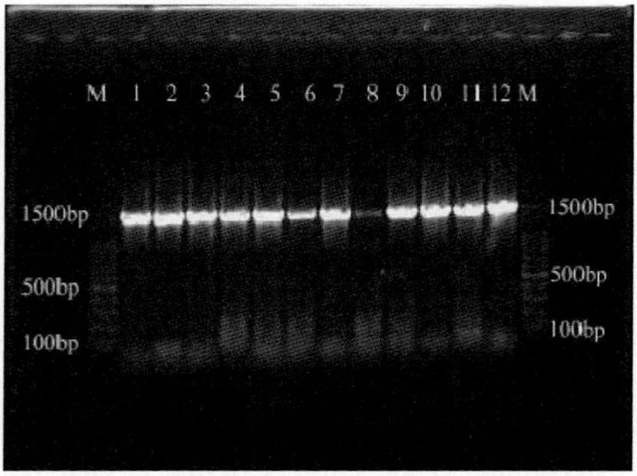

Figure 1. Agarose gel electrophoresis of 16S rDNA PCR.

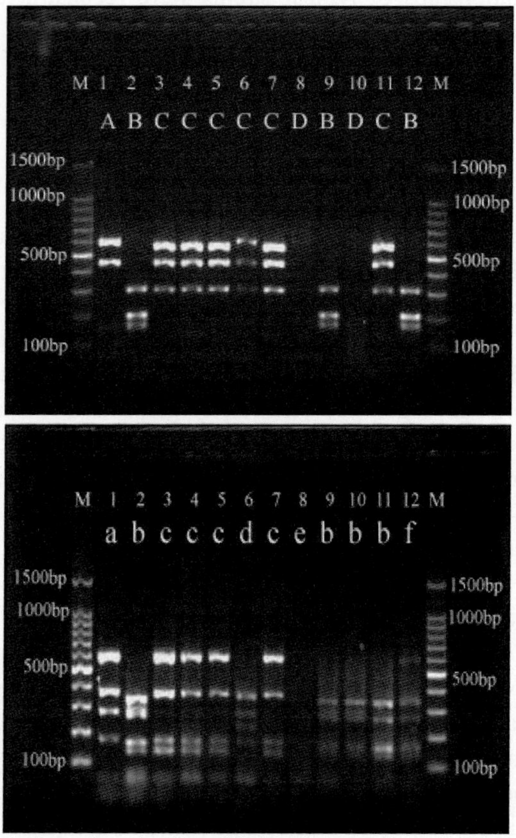

Figure 2. Agarose gel electrophoresis of amplified 16S rDNA digested with Hind III, Hae III and Hinf I, Taq I from gut flora.

Note: "M" in Figure 2 is 100 bp DNA ladder, other letters are different electrophoresis band type, numbers in Figure 2 are different strains.

3.2. *The result of clustering analysis and phylogenetic analysis*

The 16S rDNA restriction enzyme digestion fingerprints of the twelve intestinal aerobic bacteria isolated from the 3rd instar larvae of *D. kikuchii* were clustered by the UPGMA method, as shown in Figure 3. Clustering analysis revealed that twelve intestinal aerobic bacteria were clustered into seven taxonomic operational units (OTUs) under ninety percent similarity. These OTUs were OTU1, OTU2, OTU3, OTU4, OTU5, OTU6 and OTU7 in Figure 3. Only one bacterial strain was present in OUT1, OTU2, OTU4, OTU5 or OTU7, suggesting that these strains were not the dominant bacteria. One explanation could be that these bacteria didn't dominate nutrition possession in intestinal tracts of 3rd instar larvae of *D. kikuchii*. It's also possible that these bacterial strains only settled in intestinal tracts for a short time, or these bacterial strains were not adapted to the intestinal environment. There were three bacterial strains in OUT4 and four bacterial strains in OUT6. These bacteria accounted for 25.00% (OUT4) and 33.33% (OUT6) of all isolated bacteria; in other words, OUT4 and OUT6 were the dominant microflora. Dominant microflora plays an important role in providing colonization resistance in order to inhibit harmful microbes. Moreover, dominant microflora can produce metabolites that can either inhibit the invasion of exogenous microbes, or trigger host immune responses that could control the number of symbiotic intestinal bacteria to inhibit the proliferation of exogenous bacteria. On the other hand, the dominant microflora competes with exogenous microbes on nutrients and occupies the ecological niche to inhibit the harmful microbes.

The result of phylogenetic analysis based on of the representative strains in OTUs were shown in Figure 4. *Enterobacter* sp., *Bacillus* sp., *Citrobacter* sp. and *Staphylococcus* sp. were isolated and identified inform the intestinal tracts of 3rd instar larvae of *D. kikuchii*, and these intestinal aerobic bacteria were not the dominant microflora. The 16S rDNA sequence of 3I02 in OUT3 displayed 100% sequence similarity with the *Klebsiella pneumonia* type strain (AY043391.1), hence was identified as *K. pneumoniae*. The 16S rDNA sequence of 3I03 in OUT6 displayed 100% sequence similarity with the *B. cereus* type strain (GU056811.1), therefore 3I03 was identified as *B. cereus*. Since *K. pneumoniae* and *B. cereus* were the dominant microflora, they may play an important rolein supporting the health of the 3rd instar larvae of *D. kikuchii*, and may produce enzymes to digest cellulose or protein host ate. The 3rd instar is a very important instar of *D. kikuchii* larvae. Many pine needles were ate by the 3rd instar of *D. kikuchii*, therefore large quantity of cellulose was digested in its intestinal tract. In this paper we didn't look at whether the dominant microflora can digest cellulose from pine needles.

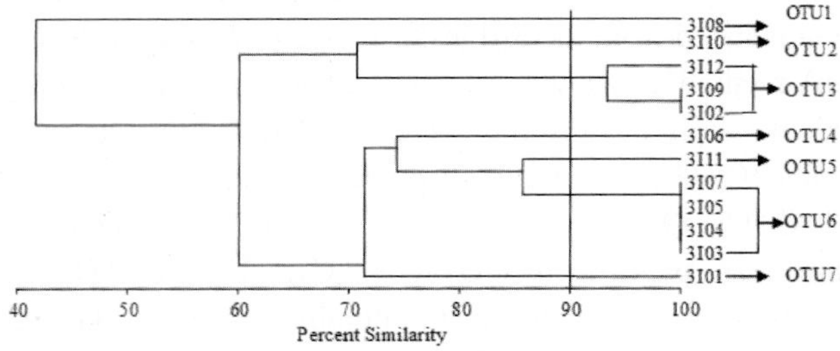

Figure 3. The results of ARDRA in dendrogram.

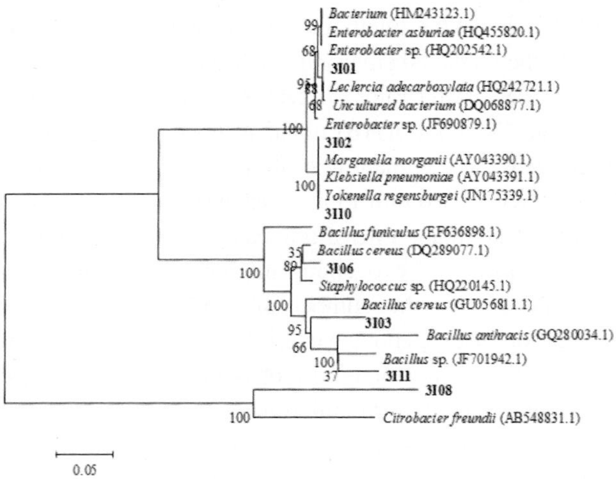

Figure 4. Phylogenetic tree of representative strains and related reference strains based on the full-length of 16S rDNA sequence.

4. Discussion

Amplified ribosomal DNA restriction analysis (ARDRA) is a molecular technique that combined restriction fragment length polymorphism (RFLP) and Polymerase chain reaction (PCR) together. ARDRA was widely used in the analysis bacterial diversity (Wu et al, 2001; Kemp & Aller, 2004; Coton et al, 2005; Chen & Fang, 2006; Li et al, 2008; Ning et al, 2008; Paolo et al, 2009), sequencing to representative strains after ARDRA so that reduce sequencing for atypical strains (Sun et al, 2008). The full-length sequences of bacteria 16S rDNA are highly conservative in evolution and are convenient to sequence. Therefore the sequence the bacteria 16S rDNA are widely used in taxonomy studies (Zhang, 2005). In this paper we identified bacteria strains through the homology alignment of sequenced 16S rDNA with the database of GenBank (Morio et al, 2002).

This study reported the development of a rapid and reliable method based on ARDRA analysis for characterizing the diversity of intestinal aerobic bacteria in intestinal tracts of the 3rd instar larvae of *D. kikuchii*. The result of phylogenetic analysis suggested that *K. pneumoniae* and *B. cereus* were the dominant microflora, while *Enterobacter* sp., *Bacillus* sp., *Citrobacter* sp. and *Staphylococcus* sp. were not dominant microflora in intestinal tracts of the 3rd instar larvae of *D. kikuchii.*

The interactions between the gut microbial flora and the host are complex that both competitions and collaborations are present. Gut microbes play important roles in both host nutrition and protection against the invasion of exogenous microbes. Furthermore, they are involved in multi trophic interactions between the host and other biological factors, and are also able to enhance the host's immune response (Moselio, 2006). A cooperative relationship predominates in *D. kikuchii* whose food is primarily pine leaves that contain relatively small amount of protein, fat, and digestible carbohydrates (starch and sugars) but large amount of plant fiber, with cellulose and hemicelluloses as the main components. *D. kikuchii* has a neutral intestinal tract, in which aerobic bacteria are inhibited, allowing the *D. kikuchii* enzymes to digest the foods first and absorb the nutrition.

The intestinal tract of *D. kikuchii* larvae is the ecological niche for intestinal microflora. The intestinal microflora plays an important role for the growth and development of *D. kikuchii*, and is also involved in the host interactions with food and natural enemies. Intestinal microflora of *D. kikuchii* could be extensively diverse due to different instar, food and living environment. The diversity and function of intestinal bacteria in *D. kikuchii* will continue to be studied as new methods are developed. Dillon's (2005) outcome is consistent with the predictions of community ecology theory that species-rich communities are more resistant to invasion than species-poor communities. However, there are few papers about bioroenosis evolutionary laws of insect intestinal microflora. So we should study the intestinal microflora of *D. kikuchii* on following aspects: ecological evolutionary laws of gut bacteria in intestine, gut microecology microscopic ecology theory. Furthermore, we should research the characteristics of enzyme production, enzyme activity, interactions with host and the application of gut bacteria in *D. kikuchii*. By thoroughly study and analyze the gut bacteria in *D. kikuchii*, we may be able to control *D. kikuchii* through the control of intestinal bacteria in *D. kikuchii*.

Acknowledgments

This research was supported by Key Laboratory of Forest Disaster Warming and Control in Yunnan Province Open Fund (No. ZK14A106 and ZK14SB01), and Forestry industry research special funds for public welfare projects (No. 201304105), and Academic Award for up-and-coming Doctoral Candidate of Yunnan Province (No. kx141156), and Key Laboratory of Conservation of Rare, State Forestry Administration Open Fund 2014, and The Science and Technology Planning Project of Yunnan Province (Grant 2011FA013).

References

1. Bei, S.G., Liu, Y.S., Cui, G.X., et al. 2005. Isolation and identification of bacteria from gastrointestinal tract of *Ceroplastes japonicus* Green. Journal of Shandong Agricultural University (Natural Science), 2, 209–212.
2. Boguś, M.I., Kędra, E., Bania, J., et al. 2007. Different defense strategies of *Dendrolimus pini*, *Galleria mellonella*, and *Calliphora vicina* against fungal infection, Journal of Insect Physiology, 9, 909–922.
3. Chen, M. & Fang X. 2006. Isolation and ARDRA analysis of cucumber entophytic antagonists against *Ralstonia solanacearum*. Acta Microbiologica Sinica, 6, 984–987.
4. Coton, M., Laplace, J.M., Coton, E. 2005. Zymomonas mobilis subspecies identification by amplified ribosomal DNA restriction analysis. Letters in Applied Microbiology, 40, 152–157.
5. Dillon, R.J., Vennard, C.T., Buckling, A., et al. 2005, Diversity of locust gut bacteria protects against pathogen invasion. Ecology Letters, 8, 1291–1298.
6. Gninenko, Y.I. & Orlinskii A.D. 2002. Dendrolimus sibiricus in the coniferous forests of European Russia at the beginning of the twenty-first century. EPPO Bulletin, 3, 481–483.
7. Kemp, P.F. & Aller J.Y. 2004. Bacterial diversity in aquatic and other environments: What 16S rDNA libraries can tell us. FEMS Microbiology Ecology, 47, 167–177.
8. Li, L.B., Liu, Z.J., Yang K. 2008. Soil microbial ARDRA analysis of Tanggula Mountain Pass along the Qinghai–Tibet railway. Acta Ecology Science, 11, 5482–5487.
9. Liu, Y.S., Li, M.L., Liu J.Z., et al. 2007b. Study on the intestinal bacteria in *L. migratoria manilensis* Meyer. Chinese Journal of Microecology, 1, 34–36, 39.
10. Liu, Y.S., Ye, B.H. Gao, H.Q., et al. 2007a. Study on the intestinal bacteria of *Eupolyphaga sinensis* Walker. Chinese Journal of Microecology, 2, 158–162.

11. Lv, F., Liu, Y.S., Wang Z.P. 2009. Pilot study on intestinal bacteria of *Clanisb ilineata tsingtauica*. Chinese Journal of Microecology, 1, 36–39.
12. Mellec, A.D., Anne, M.H., Michalzik, B., et al. 2008. Impact of a pine lappet (Dendrolimus pini) mass outbreak on C and N fluxes to the forest floor and soil microbial properties in a Scots pine forest in Germany. Canadian Journal of Forest Research, 7, 1829–1841.
13. Morio, I., Shihomi, I., Yasushi Y. 2002. Paralio bacillus ryukyuemsis gen, nov, a new Gram-positive, slightly halophilic, extremely halotolerant, faculantative anaerobe isolated from a decomposing marine alga. Journal of General Applied Microbiology, 48, 269.
14. Moselio, S. 2006. The desk encyclopedia of microbiology. Beijing Science Publishing House, 514–525.
15. Ning, H., Zhang, R.X., Chen, H. 2008. ARDRA analysis of Bacillus spp. isolated from Lake Dianchi and their algae-lytic effect. Journal of Lake Sciences, 20, 675–680.
16. Paolo, C., Elena, V., Diego, P. 2009. Soil humic compounds and microbial communities in six spruce forests as function of parent material, slope aspect and stand age. Plant Soil, 315, 47–65.
17. Sun, L., Qiu, F.B., Zhang, X.X. 2008. Endophytic bacterial diversity in rice (Oryza sativa L.) roots estimated by 16S rDNA sequence analysis. Microbial Ecology, 5, 415–423.
18. Tong, Q., He, J.Z. 2010. Evaluation on nutritional value of four species of pine caterpillars. Journal of Northeast Forestry University, 4, 98–101.
19. Wang, Y.W. & Jiang W. 1999. Research on Ocular Injury Caused by Dendrolimus. Chinese journal of Ocular Trauma Professional and Eye Disease, 3, 187–188.
20. William, G., Weisburg, Susan M. 1991. Lane 16S Ribosomal DNA Amplification for Phylogenetic Study. Journal of bacteriology, 1991, 2, 697–703.
21. Wu, S.H., Zhang, H.W., Xiong Z. 2001. Biodiversity of Frankia strains in nodules from Alnus and Hippophae by ARDRA. Chinese Journal of Applied Ecology, 12, 883–886.
22. Zhang, L., Liu, Y.S., Liu, D.W., et al. 2006. Comparison of bacteria isolated from gastrointestinal tact of Tenebrio obscurus and T. molitor. Entomological Journal of East China, 15, 17–21.
23. Zhang S.Q. 2005. Microbial Molecular Ecology. Beijing Science Publishing House, 236.

Chapter 8
Technology For Resource Development and Protection

Simulation research of out of the coal silo model based on witness

L.Z. Liu*, D. Wang and X.Y. Fan
Economic and Management Institute, Yanshan University, Qinhuangdao, Hebei, P.R.China
**E-mail: llz7111@sina.com*

With the development of society, environmental problems were paid attention to by people increasingly. Coal silo is the main development direction of large port in the future, with the respect to environmental protection, reducing the loss of coal, increasing the number of coal storage. This article takes silo storage as the research object in Qinhuangdao port, simulates coal silo transport model based on Witness simulation platform, determines a variety of simulation program, finds the weakness and the causes of the weakness, through the simulation, and then takes corresponding measures to optimize the coal operation process.

Keywords: Silo storage; Qinhuangdao port; Witness software.

1. Introduction

With the rapid development of the national economy, the demand of energy has been more and more urgent, which led to the rapid development of the coal terminal, but also brought serious environmental pollution problems. At present, the main way to prevent dust both at home and abroad are: spray/sprinkle water, windproof net, airtight, green belts, silo storage, etc. (Liu et al. 2011), which have different advantages and disadvantages. But in the point of long-term development of port, silo storage is the main development direction of the future. Especially in terms of Qinhuangdao Port, which is located in the north of China, bearing an acute shortage of water, low temperature in winter, months of frozen and high wind, the efficiency of spraying to prevent dust is low (Qinhuangdao Port). Under this condition, silo storage is the best choice to prevent dust.

As to build silos, it must be equipped with the complete loading and unloading process system which is matched with the production requirements, but there are lots of difficulties to solve these problems by traditional mathematical modeling method. With the help of computer simulation technology, we can cross these obstacles and validate the reliability and accuracy of the model, analyze the output and get some meaningful conclusion which has practical significance to guide the production and management.

2. Witness Software

Witness is a simulation software which aiming at production, transportation, planning, it can be used in the simulation of discrete time system and continuous fluid. The simulation of witness builds the corresponding model of the practical system which contains the same things, then sets a certain operation time on the computer and simulates the production process, finally we can get every metrics report which needed in solving practical problems, the simulation data can provide quantitative basis for planning and running of the production line. Witness can display the whole simulation process of production and transportation system, every problem is visible which can help the user to adjust the model parameters timely and achieve the optimization of logistics system configurations (Ma 2010, Liu 2013, Chen 2009).

3. The Principle of Silo Storage

Silo storage is one kind of modern storage warehouse with high degree of mechanization and automation, which not only make the management of coal and yard clearly, but also greatly improve the working efficiency of the coal blending and the accuracy of coal blending, the workload associated with coal blending is less. At the bottom of each row silo layout 2 warehousing out belt. In the loading operation without coal blending demand, a silo discharge the coal and two belt machine below the silo can supply a shipping line. In the loading operation with coal blending demand and two coal storage in a row, the control room could blend coal in the 2 belts machine for warehousing-out or in the shipping line downstream according to the customer demand proportion, while the two coal storage silo are in different rows. If the blending ratio is 1 : 1 that one of warehousing-out belt for each silo need to be started, and blending operation to been done in the shipping line. If the blending ratio is not 1 : 1 that all warehousing out belt of the silo with higher proportion and one warehousing-out belt of the silo with lower proportion need to been started and blending operation to been done in the shipping line (Liu 2013).

The model is so complicated that the closed silo storage model is divided into warehousing-in and warehousing-out. because of the limited space, we only discuss the model simulation of warehousing out for the limited space.

4. Elements Design for Simulation Model of Warehousing-Out

Coal:

Type: passive

Input to model: FLUID = beans

CONTENTS = IUNIFORM (4 000, 5 000)

Container:

Type: active

Inter arrival: POISSON (91.8,4)

Lot size: 6 000

To: push to Empty container yard

Silo extraction:

Input from: IF Storage tank < 550

PULL from coal out of WORLD

ELSE

Wait

ENDIF

Output to: PUSH to SCRAP

Fluid rules: contents

To: connect to pipe

Storage tank:

Initial fluid: beans

Fill from: connect pipe

5. Operation Analysis of Warehousing-Out Model Text and Indenting

Port operations for a period of 350 days, the basic unit of simulation clock is minute, The simulation time is set to 350*24*60=504 000 simulation clock (Wang 2011).

5.1. *The initial operation analysis of warehousing-out model*

The operation results of warehousing-out model in the initial conditions as shown in Table 1, Table 2:

Table 1. The use of the equipment.

Item	Idle	Busy	Filling (%)	Empty	Blocking	Output
Silo extraction	94.00	0.02	0.00	5.98	0.00	100 times
Yard crane	0.00	76.92	0.00	0.00	23.08	387 685 containers
Quay crane	23.08	76.92	0.00	0.00	0.00	387 684 containers
Filling(1)	0.00	88.83	4.44	0.00	6.73	223 854 containers
Filling(2)	0.00	88.83	4.44	0.00	6.73	223 853 containers

Table 2. The statistical indicators of horizontal transportation system.

Item	Idle	Busy	Blocking	Output
		(%)		
Empty container(1)	0.00	1.51	98.49	379 905 containers
Empty container(2)	0.00	0.27	99.73	67 823 containers
Full container(1)	3.10	64.60	32.3	223 855 containers
Full container(2)	3.10	64.60	32.3	223 851 containers
Electric drum	76.92	23.08	0.00	387 685 containers
Belt	6.06	93.94	0.00	452 163.62 tons

As shown in the table, the equipment efficiency and the horizontal transportation system can completely meet the coal loading capacity in the current situation of ship arrived at the Qinhuangdao port. The role of silo extraction is to serve as transportation functions between belt conveyor and soil. In order to achieve the optimal principles (Li 2007, Ying 2009), it should start the belts at least as far as possible in the actual operation, and take into account the equipped with storage tank for filling, silo extraction idle rate should be maintained at a high level and 94% is normal in the model. As show in Table 1, all equipment are in normal busy rate level except yard crane and filling machine are blocking occasionally. As show in Table 2, empty containers extraction line and full containers conveying line has blocking phenomenon, the empty containers extraction line blocking is serious, which is the main problem to be solved in this model.

5.2. *Proposal of warehousing-out model*

(1) Change the container arrival time interval

Containers arrival time intervals obey Poisson (300, 1), the simulation results are shown in Table 3 and Table 4:

Table 3. Simulation results of equipment with Poisson (300, 1).

Name	Idle	Busy	Blocked
		(%)	
Silo extraction	93.47	0.02	0.00
Filling(1)	0.00	95.23	0.00
Filling(2)	0.00	95.23	0.00
Quay crane	23.08	76.92	0.00
Yard crane	0.01	76.92	23.08

Table 4. Simulation results of horizontal transportation system with Poisson (300, 1).

Name	Empty	Move	Blocked
		(%)	
Empty container(1)	0.00	1.90	98.09
Empty container(2)	0.00	0.00	100.00
Full container(1)	4.77	95.23	0.00
Full container(2)	4.77	95.23	0.00

As can be seen from the table, blocking of filling disappear, but other congestion problem has not been alleviated. Therefore, changing the arrival time interval can not solve the blocking of the model.

(2) Change the numbers of equipment

Experiment 1: Review the simulation results of model, Yard crane slightly blocked with the rate of 23.08%. In order to solve the blocking problem, increase the quantity of yard crane to 2, the time interval to Poisson (300, 1), other mechanical configurations are the same above, and the simulation results are shown in Table 5 and Table 6.

Table 5. Simulation results of increase the numbers of yard crane.

Name	Idle	Busy	Blocked
		(%)	
Silo extraction	93.47	0.02	0.00
Filling(1)	28.51	71.48	0.00
Filling(2)	9.50	90.49	0.00
Quay crane	23.08	76.92	0.00
Yard crane(1)	9.50	90.50	0.00
Yard crane(2)	9.50	88.12	2.38

Table 6. Simulation results of increase the numbers of horizontal transportation system.

Name	Empty	Move	Blocked
		(%)	
Empty container(1)	0.00	14.26	85.73
Empty container(2)	0.00	4.75	95.24
Full container(1)	32.31	67.69	0.00
Full container(2)	32.31	67.69	0.00

According to the data, in the machinery, the blocking problem of yard crane have already been solved, the slight blocking of filling have been solved completely. In the horizontal transportation, empty containers extraction line blocking problem is still serious, but the empty container transportation line blocking problem are solved.

Experiment 2: The blocking problem in empty containers extraction line should been solved after experiment 1.

In order to configure filling line and empty containers extraction line, we set containers arrival time intervals to Poisson (300, 1), the number of silo extraction is 1, yard crane increase to 2, number of full containers conveying line is 2. The simulation scheme are shown in Table 7.

Simulation and analysis for Scheme 1 and Scheme 2, the results are in Table 8, Table 9, Table 10, Table 11.

Table 7. Simulation scheme of solve the empty container extraction line blocking.

Scheme	Filling	Empty containers extraction line	Full containers extraction line	Yard crane	Silo extraction
1	4	4	2	2	1
2	2	4	2	2	1

Table 8. Simulation results of the horizontal transportation in Scheme 1.

Name	Empty	Move	Blocked
		(%)	
Empty container(1)	0.01	84.73	25.02
Empty container(2)	0.01	84.71	25.04
Empty container(3)	0.01	84.71	25.04
Full container(1)	66.07	33.93	0.00
Full container(2)	32.31	67.69	0.00

Table 9. Simulation results of equipment in Scheme 1.

Name	Empty	Move	Blocked
		(%)	
Silo extraction	6.53	93.47	0.00
Filling(1)	29.22	70.78	0.00
Filling(2)	10.08	89.91	0.00
Filling(3)	1.78	98.22	0.00
Filling(4)	0.11	99.89	0.00
Quay crane	80.04	19.96	0.00
Yard crane(1)	9.98	89.36	0.67
Yard crane(2)	9.98	87.94	2.08

Table 10. Simulation results of the horizontal transportation in Scheme 2.

Name	Empty	Move	Blocked
		(%)	
Empty container(1)	1.02	70.40	28.58
Empty container(2)	1.02	70.37	28.61
Empty container(3)	1.02	70.37	28.61
Empty container(4)	1.02	70.36	28.62
Full container(1)	67.69	32.31	0.00
Full container(2)	64.97	35.03	0.00

Table 11. Simulation results of equipment in Scheme 2.

Name	Empty	Move	Blocked
		(%)	
Silo extraction	3.53	96.47	0.00
Filling(1)	28.66	71.37	0.00
Filling(2)	9.61	90.38	0.00
Quay crane	80.99	19.01	0.00
Yard crane(1)	9.50	90.50	0.00
Yard crane(2)	9.50	88.12	2.38

By contrast, there is little difference between the two schemes, and they all ease the blocking significantly. Given the principle of "as little as possible to use the resources and reduce the energy consumption" (Cai 2009), Scheme 2 is the optimal solution which use less filling than Scheme 1.

6. Conclusion

In this paper, we have built the warehousing-out model of coal silo with Witness simulation software based on the operation parameters of Qinhuangdao port. We can find the problem of the model and the actual operation, optimize nodes of the model and the actual operation, so that the model has better application and persuasion, and also provide some countermeasures and suggestions for the reconstruction project: the Qin Port relocation of the west to the east.

Acknowledgement

The work was partly supported by Research and Development Program on Science and Technology of Hebei Province (Program ID: 14457667D).

References

1. Cuilian Liu, Toulan Yu. 2011. The construction of green port in China, *Wuhan University of Technology (Social Science Edition)*, 24 (3): 328-331.
2. Qinhuangdao port [EB/OL]. *http://baike.so.com/doc/6051597.html*
3. Zhenxin Ma. 2010. The study of Qinhuangdao port coal yard operation based on the Witness. Qinhuangdao: Yanshan University.
4. Jinpeng Liu. 2012. The simulation and optimization of coal port logistics based on the Witness. Shandong: Shandong University of Science and Technology.
5. Chen Juan. 2009. The method study of coal wharf logistics capability by simulation and evaluation based on Witness. Wuhan: Wuhan University of Technology.

6. Lizhuo Liu. 2013. Research on operation system of port-logistics. Qinhuangdao: Yanshan University press.
7. Dan Wang. 2011. Study and innovation of green logistics system of Qinhuangdao port coal. Qinhuangdao: Yanshan University.
8. Yunjun Li. 2007. Modeling and analysis of closed yard coal wharf logistics system. Wuhan: Wuhan University of Technology.
9. Ying Tang. 2009. Study on the key production strategy of environment-friendly coal wharf based on professional coal blending function. Wuhan: Wuhan University of Technology.
10. Zhide Cai. 2009. Study on equilibrium optimization of container yard operation under the uncertain environment. Dalian: Dalian Maritime University.

Interactive effect of siderite and hematite and its mechanism in flotation

D.H. Wang[a], X.M. Luo[b], W.Z. Yin[a,*], Y.Q. Ma[a] and X.M. Li[a]
[a]*College of Zijin Mining, Fuzhou University, Fuzhou, Fujian Province, China*
[b]*Faculty of Land and Resource Engineering, Kunming University of Science and Technology, Kunming Yunnan Province, China*
[*]*E-mail: yinwanzhong@163.com*

The interactive effect of siderite and hematite and its mechanism were investigated through flotation tests, dissolved components logarithm diagram, calculations of EDLVO theory and SEM. The results of flotation test show that the hematite recovery can be declined in some degree by adding various particle fractions of siderite, which the reason is that some collector will be consumed by various particle fractions of siderite, what's worse, the HCO_3^- and CO_3^{2-} dissolved from siderite can compete with oleate ions on the surface of hematite. In addition, the recovery of fine siderite can be improved by adding hematite. Calculations of EDLVO theory and SEM indicate that some part of the fine grade siderite cover on the surface of hematite after the addition of coarse hematite, as a result, the apparent grain size of fine siderite as well as the possibility of adhesion between fine siderite and bubble are incresed, what's more, the environment of flotation is optimized as well

Keywords: Siderite; Hematite; Sodium oleate; Interactive effect.

1. Introduction

In complex ore flotation system, there exists interactive effect between minerals such as activating and depressing, which will have an impact on flotation separation (W.Z. Yin et al., 2014). Dissolution of minerals play a major role in determining the nature of the interactions taking place on the mineral surfaces (Q.W. Qin & Y.H. Hu, 1999), in which case adsorption competing or activation by dissolved irons may occur (N. Luo et al., 2012; D.Z. Wang & Y.H. Hu, 1987). Mineral dissolution produce a lot of unavoidable irons, so it's important to investigate the presence of irons and their behavior on mineral surface, in order to improve concentrate quality and control flotation processes better (G.S. Liu et al., 2005). Adhesion and coagulation between minerals, which is also called "carrier flotation", is another main reason for mineral interaction (J.W. Lu et al., 2015).

In recent years, mineral processors dealing with hematite have faced a challenge that the flotation results of hematite are poor in presence of carbonate, such as siderite, dolomite and so on (Z.Y. Zhang et al., 2008; B. Yang, 2010; X.M. Luo et al., 2012). The Fe grade of iron concentrate decreases with the increase of carbonate content in iron ores. Carbonates are salt-type minerals. Their dissolution characteristics play a major role in determining the nature of the interactions taking place in the bulk solution or on the mineral surfaces (P. Van Cappellen et al., 1993; G. Chen & D. Tao, 2004). The dissolved mineral species can undergo the reactions such as hydrolysis, complexation, adsorption, and surface and bulk precipitations, which could inhibit the selective interaction between reagents and minerals (A.P.L. Nunes et al., 2011). In recent years, the content of carbonate minerals in Donganshan iron ore has been increasing, leading to deteriorated reverse flotation performances or even "unclear separation between concentrate and tailing" (R.L. Lu, 2004). It has been reported that the dissolution behavior had a crucial effect on collector and modifier, thus affecting the mineral separation (Y.H. Hu, 1989). The investigation of solution chemistry was particularly important for salt-type mineral flotation.

Z.Y. Zhang et al. (2008) in their laboratory study demonstrated that the adsorptive cover of siderite that is weakly depressible by starch on the hematite and quartz is the main cause for difficult reverse flotation of Donganshan iron ore containing carbonate minerals. Therefore, the key in realizing an efficient separation between the iron minerals and the quartz is to separate the siderite prior to the reverse flotation or to find a flotation agent capable of effectively activating or depressing siderite. Y.A. Tian & B.Q. Sun (2010) investigated the use of a new-developed anionic collector TS and the modified sodium silicate to separate mono-mineral siderite and hematite. A good concentration result was obtained by the use of these two agents in flotation tests on artificial ores of siderite and hematite. Y.G. Zhu et al. (2009) investigated the autogenous-carrier effect in fine ilmenite (0-20 μm) flotation. Compared with the fine ilmenite separating flotation, the recovery rate of fine ilmenite with size of 0-20 μm increases from 52.56% to 61.96% by using autogenous-carrier flotation for Panzhihua ilmenite ore. Ucbas Y et al. (2014) used magnetic carrier to separate chromite from serpentine in fine sizes as a function of pH, dodecylamine (DDA) dosage, and amount of magnetite. Experiments revealed that chromite could be separated from serpentine with a 96.8% recovery under the experimental conditions of $4x10^{-4}$MDDA, 25 mg magnetite, and at pH 12.

Researchers have dealt with the problem of carbonate minerals from flotation processes, flotation reagents, and flotation equipments with some progress, moreover, there has been very little published literature investigating the reason and generation mechanism of interaction effect in siderite and hematite system. In this study, the pure mineral flotation tests were conducted using sodium oleate as collector to discuss the influence of siderite on hematite and impact of hematite on siderite. The aim of the present study was to investigate the interactive effect mechanism of siderite and hematite by means of solution chemistry, EDLVO theory, ultraviolet spectrophotometry (UV), SEM analysis and EDS energy spectrum.

2. Materials and Methods

2.1. *Materials*

The siderite and hematite minerals used in this study were obtained from Anshan, Liaoning province, China. Each lump ore sample was handpicked to select pure minerals. The two minerals were sieved to -106+45μm, -45+18μm and -18μm size fractions separately as samples after a sequence of processes, crushing, sorting, ball grinding, shaking table and high-intensity separation. It can be seen from the analytical results that both siderite and hematite has a purity over 95%.

2.2. *Reagents*

In this work, sodium oleate with chemical purity was used as the anionic collector, and sodium hydroxide and hydrochloric acid with analyzed grade were as pH modifiers. Distilled water was used for preparation of solutions throughout this study.

2.3. *Method*

Flotation experiments were carried out in a 30mL XFG batch flotation cell with the impeller at a speed of 1500 rpm. The mineral suspension was prepared by adding 2.0g mineral sample according to the test design to 25mL distilled water in flotation cell agitating for 1min. After that, suspension pH was adjusted using the solutions of HCl or NaOH and recorded, followed by 2 min conditioning. Then the collector sodium oleate was added and left to condition for further 2 min. The flotation time was fixed for 3 min after which the floated product and non-floated product were collected, filtered, dried, weighed and analyzed respectively.

The sodium oleate adsorption capacity on minerals was measured by ultraviolet spectrophotometer (Spectro Flex 6600). And the surface of floating products were observed by S-3500n scanning electrical Microscopy (Shimadzu company, Japan) and Inca Energy Dispersive Spectrometer in order to check out the adhesion between siderite and hematite.

3. Results and Discussions

3.1. *Influence of siderite with different particle size on hematite flotation*

Flotation tests were carried out on hematite of -106+45μm and siderite of different particle size, -18μm, -45+18μm, and -106+45μm. The pH of the suspension was adjusted to 9, and sodium oleate dosage was 120mg/L. The recovery curves of hematite under different particle size and amount of siderite added are presented in Figure 1.

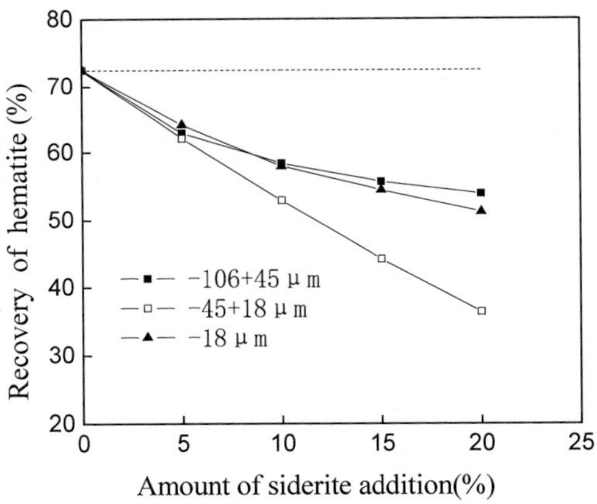

Figure 1. Influence of different particle size siderite on flotation of hematite.

From Figure 1, it's obvious that the hematite has a lower recovery with siderite of different particle size, especially that of -45+18μm. The presence of siderite adversely affected hematite flotation when sodium oleate was used as collector, thus making the separation between the two minerals difficult.

3.2. *Mechanisms of siderite negatively affecting on hematite flotation*

3.2.1. *Influence of sodium oleate dosage on hematite flotation in hematite and siderite system*

Based on the above-mentioned results, at pH 9 and sodium oleate 120mg/L, the recovery of hematite decreased in the presence of siderite of different particle size. Therefore further flotation tests were carried out on hematite in a size range of -106+45μm under different sodium oleat dosage. Different particle size of siderite were chosen, including -106+45μm, -45+18μm, -18μm and the content of siderite was kept to 20%. The pH was also adjusted to 9. The hematite recovery obtained in the presence of siderite under different dosage of sodium oleate is presented in Table 1.

The hematite recovery increased significantly as the amount of sodium oleate addition increased in the presence of siderite of different particle size. It meant a higher sodium of collector was needed to promote the recovery of hematite in hematite and siderite system. For example, to achieve a recovery about 93%, the amount of sodium oleate added was 120mg/L in the absence of siderite, but when siderite of 20% was added, it increased to 200mg/L or even more, demonstrating the main cause of low hematite recovery in the presence of siderite might be the great consumption of sodium oleate by siderite. Therefore ultraviolet spectrophotometer was adopted in order to verify the huge adsorption of sodium oleate on siderite. Every sample of 2.0g was conditioned at pH 9 for 5min after adding 120mg/L sodium oleate, followed by UVS investigation. Sodium oleate adsorption on different minerals is presented in Table 2.

Table 1. Influence of sodium oleate dosage on flotation of hematite in binary minerals system.

Minerals	Siderite addition		Sodium oleate	Hematite
	Particle size(μm)	Content(%)	dosage(mg/L)	recovery(%)
Hematite (-106+45μm) and siderite (pH=9)	——	0	120	93.05
	-106+45	20	120	73.61
			200	91.88
	-45+18	20	120	50.44
			200	94.64
	-18	20	120	59.38
			200	74.27
			400	94.30

Table 2. Sodium oleate adsorption capacity on different minerals.

Mineral	Particle size fractions(μm)	Adsorption capacity of sodium oleate(mg/g)
hematite	-106+45	0.56
siderite	-106+45	0.46
	-45+18	0.58
	-18	1.17

As it is shown in Table 2, at pH 9, the sodium adsorption capacity on siderite with -18μm was much higher than that of hematite, while other two particle size showing no obvious difference unexpectedly. So it can be concluded that the decrease of hematite recovery in the presence of siderite was not only due to sodium oleate consumption by siderite, but also might be other reasons. Siderite, a salt-type mineral, would dissolve some ions in the pulp. The mineral dissolution would be talked about later, although Table 1 indicated the increase of hematite recovery was large with increasing sodium oleate dosage.

3.2.2. *Influence of siderite dissolution on hematite flotation*

Siderite is easily dissolved in water. Different composition concentration dissolved from siderite as a function of pH according to all the equation related is shown in Figure 2. Figure 2 showed that HCO_3^- and CO_3^{2-} in the solution were the main ions in alkaline condition and Fe^{2+}, and $FeOH^+$ were the principal ions in neutral and acidic condition.

It can be seen from Figure 2 that at pH 9 under which flotation tests was carried out, CO_3^{2-}, HCO_3^- were dominant as a consequence of siderite dissolution. These irons adsorbed onto hematite competing with sodium oleate, thus making the surface of hematite hydrophilic and decreasing hematite recovery. Therefore the decrease of hematite recovery in the presence of siderite using sodium oleate as collector at pH about 9 was mainly due to two reasons. On the one hand, part of sodium oleate was consumed by siderite for the adsorption capacity difference of sodium oleate between two minerals, which resulted in decreasing the adsorption of sodium oleate onto hematite. On the other hand, the competitive adsorption onto hematite between ion (HCO_3^- and CO_3^{2-}) dissolved from siderite and sodium oleate occurred, which was probably the most important reason.

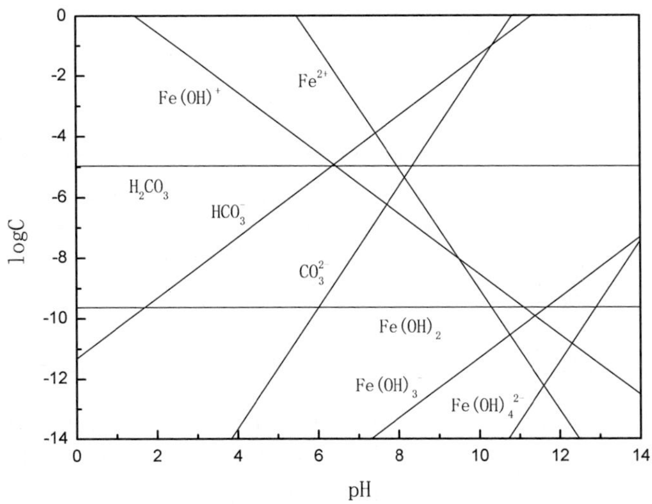

Figure 2. Dissolved components logarithm diagram of siderite.

3.3. *Influence of hematite of different particle size on siderite flotation*

The recovery curves of siderite of -18μm under different paritcle size and amount of hematite are presented in Figure 3. The sodium oleate dosage was 400mg/L and pH was adjusted to 9.2. Unlike the negative effects of siderite on hematite flotation, hematite improve siderite flotation by increasing the recovery only using sodium oleate as collector.

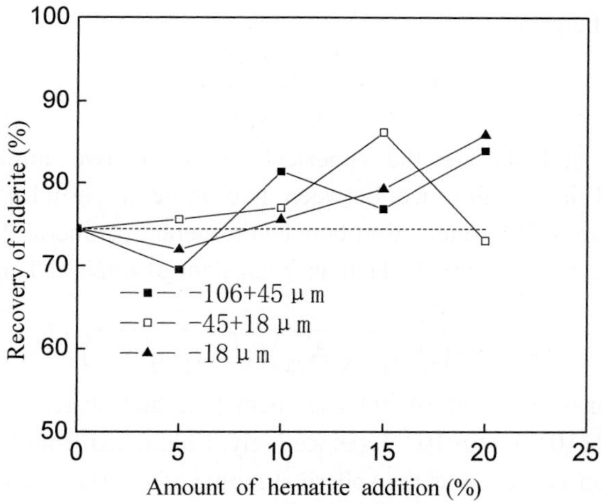

Figure 3. Influence of different particle size hematite on flotation of -18μm siderite.

3.4. *Mechanism of hematite improving siderite flotation*

The above results indicated there was an increase in siderite recovery using sodium oleate as collector when hematite of different particle size was added at pH 9. So mechanism of hematite with different particle size on siderite flotation was further studied with extended DLVO theory (EDLVO theory) to calculate interaction energy between the two mineral particles. The repulsive and agglomeration between the two mineral particles are mainly determined by Van der Waals forces, electrostatic forces and hydrophobic interaction forces as follow.

$$V_T^{ED} = V_W + V_E + V_H \quad (1)$$

Where V_T, V_W, V_E, V_H are the total interaction energy, Van der Waals energy, electrostatic interaction and hydrophobic interaction energy, respectively.

3.4.1. *Van der Waals energy*

When there is a big difference in size between two mineral particles, one is coarse and the other is fine, The model is based on the interaction between the sphere and the plate:

$$V_W = -\frac{A_{132}R_1}{6H} \tag{2}$$

When particles have a relative size, model is based on the interaction between two spheres:

$$V_W = -\frac{A_{132}}{6H}\frac{R_1R_2}{R_1+R_2} \tag{3}$$

Where R_1 and R_2 are the spherical radius of two mineral particles respectively; H is the distance between two mineral particles; A_{132} is the Hamaker constant of the interaction between mineral 1, mineral 2 and media 3 in vacuum; A_{11}, A_{22}, A_{33} are the Hamaker constant of mineral 1, mineral 2 and media 3 in vacuum respectively.

$$A_{132} \approx (\sqrt{A_{11}}-\sqrt{A_{33}})(\sqrt{A_{22}}-\sqrt{A_{33}}) \tag{4}$$

The Hamaker constant of siderite, hematite, and water in vacuum are 6.3×10^{-20}J, 23.2×10^{-20}J, 4.0×10^{-20}J, respectively. Calculated from Eq. (4), A_{132} is 1.44×10^{-20}J. 5μm is chosen as the spherical radius of siderite R_1, while hematite particle size R_2 has five values, 5μm, 10μm, 30μm, 50μm and 80μm. The interaction model based on spheres is adopted when R_2 is 5μm or 10μm, conversely another model is adopted when 30μm, 50μm, 80μm are taken.

3.4.2. *Electrostatic interaction energy*

When there is a big difference in size between two mineral particles, one is coarse and the other is fine, the model is based on the interaction between the sphere and the plate:

$$V_{ER} = 4\pi\varepsilon_a R_1 \varphi_{01}^2 \ln[1+\exp(-\kappa H)] \tag{5}$$

When particles have a relative size, model is based on the interaction between two spheres:

$$V_{ER} = \frac{\pi\varepsilon_a R_1 R_2}{R_1+R_2}(\varphi_{01}^2+\varphi_{02}^2)\left[\frac{2\varphi_{01}\varphi_{02}}{\varphi_{01}^2+\varphi_{02}^2}p+q\right] \tag{6}$$

Where

$$p=\ln\left[\frac{1+\exp(-kH)}{1-\exp(-kH)}\right] \tag{7}$$

$$q=\ln\left[1-\exp(-2kH)\right] \tag{8}$$

Where $\varepsilon_a=\varepsilon_0+\varepsilon_r$, ε_0 and ε_r are the permittivity of vacuum (8.854×10^{-12} $C^{-2}J^{-1}m^{-1}$) and solution (water is 78.5 $C^{-2}J^{-1}m^{-1}$) respectively, so ε_a is 6.95×10^{-10} $C^{-2}J^{-1}m^{-1}$; φ_{01} and φ_{02} are the surface potential of two mineral particles respectively, which is substituted by zeta potential in calculation; H is the distance between two mineral particles; k^{-1} is the Debye length of $0.104nm^{-1}$.

5μm is chosen as the spherical radius of siderite R_1, while hematite particle size R_2 have five values, 5μm, 10μm, 30μm, 50μm and 80μm. At pH 9, zeta potential of siderite and hematite are -15.09mV and -38mV respectively.

3.4.3. *Hydrophobic interaction energy*

When there is a big difference in size between two mineral particles, one is coarse and the other is fine, the model is based on the interaction between the sphere and the plate:

$$V_H=2\pi R_1 h_0 V_H^0 \exp(-\frac{H}{h_0}) \tag{9}$$

When particles have a relative size, model is based on the interaction between two spheres:

$$V_H=2\pi\frac{R_1R_2}{R_1+R_2}h_0 V_H^0 \exp(-\frac{H}{h_0}) \tag{10}$$

Where h_0 is attenuation length, which is usually 1~10nm, the value of which is 2.5nm here; V_H^0 is the constant of hydrophobic interaction energy of $-44.5mJ/m^2$. If $V_H>0$, there exist repulsion between mineral particles and it is expressed as V_{HR}; if $V_H<0$, there exist attraction between mineral particles and it is expressed as V_{HA}.

3.4.4. *The total interaction energy*

EDLVO curves between fine siderite and different particle size hematite at pH 9 in sodium oleate solution are presented in Figure 4. It can be seen that the interaction energy between fine siderite and hematite is shown as a strong attraction after leaping over a weak potential energy barrier. On condition that an amount of mechanical energy is put into the system, fine siderite could adhere to coarse hematite or agglomerate with fine hematite, thus increasing the superficial particle size of fine siderite, which favor the adhesion with bubbles and increase siderite flotation recovery.

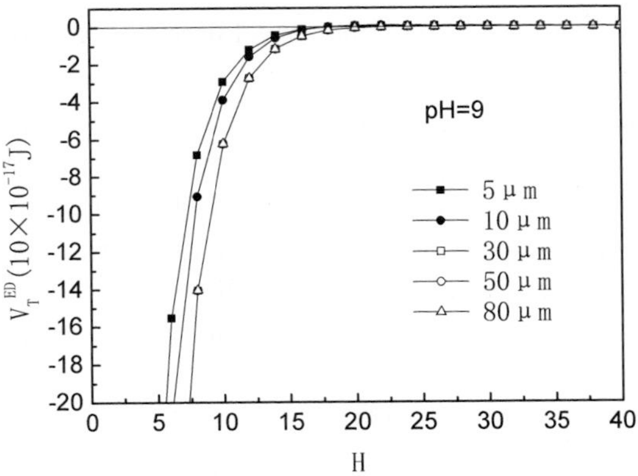

Figure 4. EDLVO curves between fine particle size siderite and different particle size hematite.

To verify part of fine siderite coating on the hematite surface, scanning electrical microscopy and energy dispersive spectrometer were adopted to analyze the floated product. In the flotation test, siderite of -18μm was mixed with hematite of -106+45μm accounting for 15% in weight, pH was adjusted to 9 and the sodium oleate dosage was 400mg/L. SEM analysis and EDS energy spectrum results were shown in Figure 5, from which it is clear that fine siderite coated on the hematite surface, hematite performed as a carrier in siderite flotation using sodium oleate as collector.

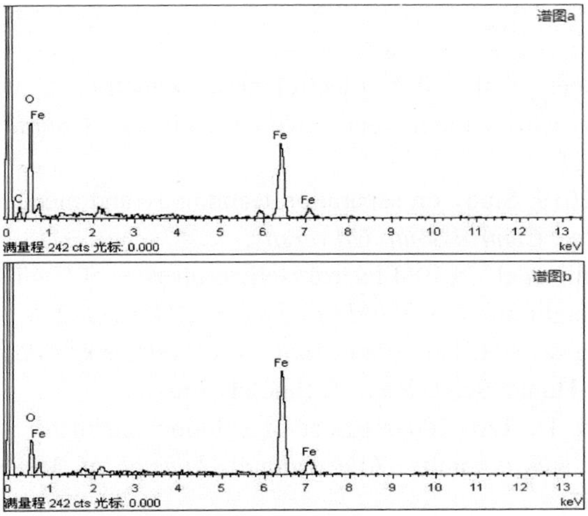

Figure 5. SEM analysis and EDS energy spectrum results of flotation tailings of mixed minerals with -106+45μm quartz and -18μm siderite.

4. Conclusion

In the present study, the interactive effect and its mechanism were determined for the siderite–hematite system by solution chemistry, EDLVO theory. The obtained results indicated that hematite recovery decrease in the presence of siderite, however the addition of hematite could increase the siderite recovery.

The hematite recovery decrease in the presence of siderite to some extents, partly due to the consumption of sodium oleate by siderite for the adsorption capacity difference of sodium oleate between two minerals, however the main cause is the competitive adsorption onto hematite between ion (HCO_3^- and CO_3^{2-}) dissolved from siderite and sodium oleate, also decreasing the adsorption of sodium oleate onto hematite.

The siderite recovery in concentrate increase as the amounts of hematite addition increase only using sodium oleate as collector. The results indicate that part of fine siderite coat on the coarse hematite surface. Hematite perform as a carrier in the flotation, increase the superficial particle size of siderite and possibility of adhesion with bubbles, hence optimizing the flotation environment and increasing siderite recovery.

Acknowledgment

This work was financially supported by the National Natural Science Foundation of China (Grant No.51374079).

References

1. A.P.L. Nunes et al., 2011. Electrokinetic properties of wavellite and its floatability with cationic and anionic collector. *Colloid Interface Sci*, 361:632–638.
2. B. Yang, 2010. Study on separation technology and mechanism of siderite and hematite. *Central South University*.
3. D.R. Vučinić et al., 2010. Electrokinetic properties of Hydroxyapatite under flotation conditions, *J. Colloid Interface Sci*, 343:239–245.
4. D.Z. Wang & Y-H. Hu, 1987. *Solution chemistry of flotation.* Changsha: Changsha: Hunan Science and Technoloty Press.
5. G. Chen & D. Tao, 2004. Effect of solution chemistry on flotability of magnesite and dolomite, *International Journal of Mineral Processing*, 74:343–357.
6. G.S. Liu et al., 2005. Influence and mechanism of copper ions and nickel ions on flotation of talc. *Journal of The Chinese Ceramic Society*, 33(08):1018–1022.
7. J.W. Lu et al., 2015. Effects of magnetite on magnetic coating behavior in pentlandite and serpentine system. *Mineral Engineering*. 72:115–120.
8. N. Luo et al, 2012. Study on the Flotation Behavior and Mechanism of Rhodochrosite and Calcite. *Nonferrous metals (Mineral Processing Section)*, 4:41–45.
9. P. Van Cappellen et al., 1993. A surface conplexation model of the carbonate mineral-aqueous solution interface, *Geochim. Cosmochim. Acta*, 57:3505–3518.
10. Q.W. Qin & Y-H. Hu, 1999. Dissolution of sparingly soluble minerals and its effect on floatability of mineral. *Mining and Metallurgical Engineering*. 19(2):30.
11. R.L. Lu, 2004. Study on Characteristics of Technological Mineralogy of Donganshan Iron Mine's Ore. *Metal Mine* 9: 40–41.
12. Ucbas Y et al., 2014. Separation of Chromite from Serpentine in Fine Sizes using Magnetic Carrier. *Separation Science and Technology*. 49(6): 946–956.
13. W.Z. Yin et al., 2014. Effect of flotation reagents on direct reverse flotation of carbonate-containing iron ores. *Journal of University of Science and Technology Beijing*. (36)2:154.
14. X.M. Luo et al., 2012. New flotation technology research on carbonate-containing hematite. *Advanced Materials Research*, 454:210–215.
15. Y.A. Tian & B-Q. Sun, 2010. Experimental Research on Flotation Separation of Siderite and Hematite. *METAL MINE*, 4: 58–63.

16. Y.G. Zhu et al., 2009. Autogenous-carrier flotation of fine ilmenite. *The Chinese Journal of Nonferrous Metals*, 19(3): 554–560.
17. Y.H. Hu, 1989. Research on solution chemistry and floatability of salt-type minerals. *Central South University*.
18. Z.Y. Zhang et al., 2008. Influence of the Siderite in Donganshan Iron Ore on Reverse Flotation. *METAL MINE*, 10:52–54.

Application and study of the power oil technology in the oil production two plant of North China Oilfield

Z.W. Tong*, W.X. Wang, X.B. Meng, Y. Tao, C.X. Tian, J.L. Zhang, J. Wang, H. Ming and Y.H. Li

No.2 Oil Production Plant
Huabei Oilfield Company, PetroChina, Bazhou, Hebei, China
**E-mail: cy2_sqthw@ petrochina.com.cn*

Combined with the diagrams of the amount of oil in Chaheji oil field of application, a brief description of the diagrams of the amount of oil technology principle and system components. Through the analysis of the application effect of different types of oil wells, the advantages and problems of the power oil technology are described, and the management of the oil application is discussed and discussed. Through Chaheji oilfield 111 wells field application shows that, the diagrams of the amount of oil technology can well reflect the well condition fluctuation and real Nissan levels.

Keywords: Liquid production quantity; Separator; Power chart quantity oil; Error analysis; Management.

Most of the domestic oil well metering method mainly for glass tube oil quantity, turn the bushel oil, mobile calibration car, mass flowmeter and ultrasonic flow instrument method and widespread application device and process is too cumbersome, input as well as many other problems, difficult to adapt to the ground system to simplify the optimization requirements. With the wide application of new gas and liquid flow metering technology, the metering method of oil well production is more and more easy to use and convenient in time. In recent years, with the development of the continuous progress of diagnosis technology and automation technology, merit chart measurement technology of the validation study is becoming more and more mature and the continuity of data acquisition and storage also make management personnel at the scene of production development and the change rule is easier to grasp.

According to the oilfield company launched Chaheji oil field is simplified and optimized deployment of the overall work, traditional tertiary cloth station mode will be to a cloth station transition mode, metering station will gradually be canceled, to simplify the process and reduce the investment for the construction of the ground, to solve the problem of oil well measurement, we and Mining

Research Institute introduced diagrams of the amount of oil, in oil production plant is Chahe in field and test and application.

1. Power Chart Quantity and Oil Technology Remote Measurement Principle and Function

1.1. *Technical principle*

Diagram of the oil well metering principle is according to the measured light bar shown in diagram, borehole data, pumping oil machine, rod combination and dynamic liquid level and many related data from wells, rod pump pumping oil system of three-dimensional mechanical and mathematical model. By using the calculated pump dynamometer card, through recognition of the pump dynamometer card access to pump the effective stroke, and then draw the oil well liquid producing capacity. The variation of the load and the displacement and the time of the suspended load at the wellhead is used to establish the pumping rod, the finite element model and the liquid column difference model.Then solve the deep well pump at the entrance of load displacement relationship (pump dynamometer card), through the recognition of computer calculated effective stroke of the pump, the wellhead fluid production in amounts similar to pump stroke and pump piston annular space volume product.

The technology is the rod pumping system is regarded as a complex vibration system, its model is a vibration wave equation, and the boundary conditions of the surface dynamometer card is the vibration equation and by the boundary condition is surface dynamometer card for pump dynamometer card in directional wells, so as to obtain the output (see Figure 1).

This technology is an Technology of realizing liquid production measurement based on automatic admission and monitoring of oil wells, and can automatically calculate the amount of liquid produced by single well.

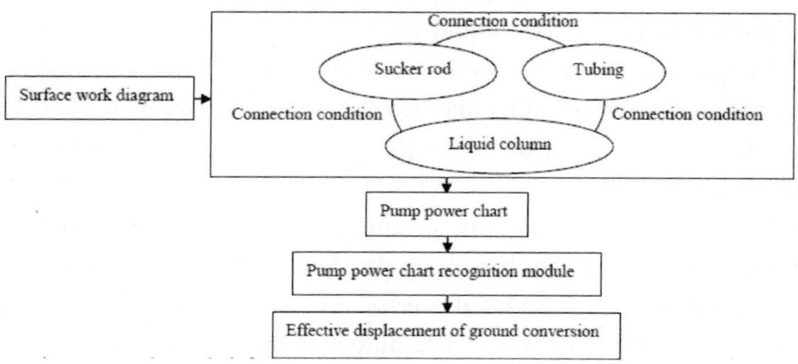

Figure 1. Power chart method of measurement technology diagram.

1.2. *System components*

The oil system of power diagram is mainly composed of hardware and software.

Hardware is mainly composed of RTU, load sensor and pressure sensor, etc. All the sensors to collect the data through a small wireless transmission to RTU, RTU is transmitted through GPRS network to the server, server to receive all the data from wells, automatic calculation, calculation results through the browser in real-time release, 1 hours to upload a diagram, realize the real-time monitoring of oil well production. Data maintenance, query can be remote operation, the user can track the production of a single well and work conditions through the network timely.

Software mainly by remote automatic measurement and analysis system client, oil well remote monitoring and measurement system, remote measurement and analysis optimization system, warning software four parts of the software.

Remote automatic measurement and analysis system of the client is mainly used to build up the pumping well based accounting; oil well remote control and measurement system is mainly used for pumping set communication parameters of oil wells and daily oil well management, browse; remote measurement and analysis optimization system is mainly used to pumping oil well data and calculate the to web forms issued to query; alarm software mainly on rod break, stop well diagnosis of abnormal oil wells to provide real-time alarm to remind management personnel timely detection and tracking, which greatly facilitated the well daily management.

2. The Application Diagram in Oil Technology in Oilfield Cha He

2.1. *Error analysis of power chart quantity oil*

Currently oil extraction factory branch north oil production operation area oil three stations, oil station six, eight production station total 111 wells and the production method of oil well liquid producing capacity of diagram, from the view of the operation, diagrams of the amount of oil system stable operation and diagrams of the amount of oil technology can reflect the oil well yield volatility and condition change, most of the wells in the accuracy of basically meet the production requirements.

North fork oil production zone produced fluid volume ≥5t/d oil port (84), will be the diagram of the liquid production wells and metering separator calculation results relative ratio, the error <15% wells has 79, accounted for 94% of the total number of wells. The output liquid volume of the oil well has 27 wells, the average error of the oil is 39.1%. It is not hard to see that the error of the high yield liquid oil well is obviously lower than that of the low output liquid (see Figure 2).

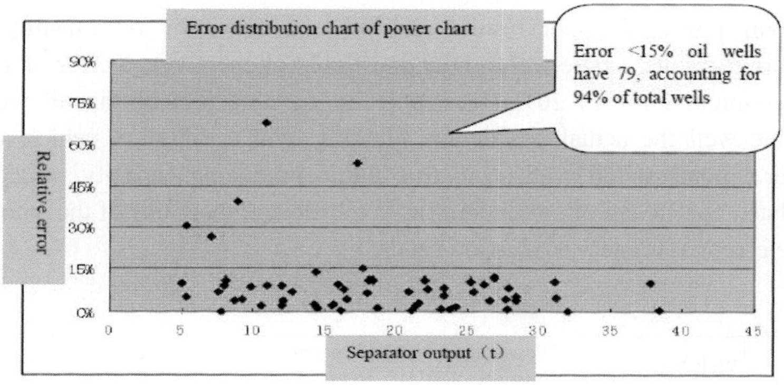

Figure 2. Error distribution of oil production in the oil production area of the branch north of the branch.

2.1.1. *Error analysis of high yield liquid oil produced by liquid yield > 10t/d (the calculation of the following data are based on the separator)*

Liquid yield > 10 tons / day of high-yield wells a total of 49, power diagram results and oil and separator oil quantity after comparing results, single well production fluid volume average error of 4.8%, the maximum error of 10%, this kind of well dynamometer diagram is always energetic, small deviation, within the allowable error range, basically meet the production demand (see Table 1).

Table 1. Comparison of liquid production volume of liquid yield and 10t/d.

Serial number	Well No.	Power chart quantity oil (t/d)	Separator quantity oil (t/d)	Poor liquid (t/d)	Error (%)
1	cha19-24	84.4	86	-1.6	-1.9
2	cha 74-121	52.13	49	3.1	6.4
3	cha 48-125	46.88	46	0.88	1.9
4	cha 30-58	42.6	40.0	2.6	6.6
6	cha 19-22	38.0	39	-1.0	-2.6
7	cha 19-102	33.0	33	0.0	0
8	cha 31-63	33.1	32.5	0.6	1.8
9	cha 30-153	32.5	32.0	0.5	1.6
10	cha 19-108	32.4	32	0.4	1.1
11	cha 30-156	22.41	22	0.41	1.9
12	cha 19-213	29.0	30	-1.0	-3.3
13	cha 74-128	23.03	23	0.03	0.1
14	cha 74-125	22.69	24	-1.31	-5.5
15	cha 105-2	31.9	29	2.9	10
……					

2.1.2. *Error analysis of the middle liquid oil well produced by 10 - 5t/d*

Middle liquid oil wells in liquid producing capacity of 10 5t/d total of 35 wells that will power diagram results and oil and separator oil quantity after comparing results, single well production fluid volume average error of 14.5%, the maximum error of 20% (see Table 2), for this part of the oil well we combine well the actual production situation, with calibration vehicle of the nuclear volume of oil wells, and true liquid producing capacity, single well calculated coefficient of correction, fixed coefficient, diagrams of the amount of oil the essential to meet production needs.

Table 2. Liquid output of 10 - 5t/d of oil production volume of oil well.

Serial number	Well No.	Power chart quantity oil (t/d)	Separator quantity oil (t/d)	Poor liquid (t/d)	Error (%)
1	cha19-5	11.8	10	1.8	18
2	cha 19-9	11.2	10	1.2	12
3	cha 30-136	8.85	9	-0.2	-1.7
4	cha 48-127	9.09	9	0.09	1
5	cha 48-74	10.43	9	1.4	15.9
6	cha 74-213	10.87	12	-1.13	-9.4
7	cha 30-150	8.7	9.0	-0.3	-2.9
8	cha 31-65	10.2	9.8	0.4	4
9	cha 30-151	6.4	8.0	-1.6	-20
10	cha 30-152	5.9	7.2	-1.3	-17.6
11	cha 115	8.1	7	1.1	16
12	cha 8	8.2	7	1.2	17.1
13	cha 30-51	5.1	6.0	-0.9	-15
14	cha 30-61	6.9	6.0	0.9	14.3
15	cha 30-162	8.88	9	-0.12	-1.3
……					

2.1.3. *Error analysis of low yield liquid oil well produced by liquid output 5t/d*

There are 27 wells in the low yield liquid of 5t/d. After comparing the results of the power oil results with the results of the separator, the average error of the single well is 39.1% and the error is 544% (see Table 3). This kind of well due to the low yield and most of the wells for intermittent oil, formation fluid supply instability, separator due to measurement in a period of a day, there chance and can not fully reflect the real oil well output level, yield volatility is large, so it can not by simple adjustment coefficient to revise it, for this part of the oil well, due to the condition of diagrams of the amount of oil monitoring wells of 24 consecutive hours, better reflect well the real situation

Table 3. Comparison of liquid produced liquid volume produced in liquid volume 5t/d.

Serial number	Well No.	Power chart quantity oil (t/d)	Separator quantity oil (t/d)	Poor liquid (t/d)	Error (%)
1	cha79-29X1	2.26	3	-0.7	-24.7
2	cha 31-73	3.0	3.0	0.0	0
3	cha 30-135	3.0	3.0	0.0	0
4	cha 48-136	4.75	2	2.75	137.5
5	cha 74-206	0.19	2	-1.8	-90.5
6	cha 79-25	2.51	2	0.5	25.5
7	cha 31-38	0.0	2.0	-2.0	-100
8	cha 19-106	0.4	2	-1.6	-79
9	cha 19-109	2.8	3	-0.2	-6.7
10	cha 48-132	1.54	3	-1.46	-48.7
11	cha 79-17	1.74	2	-0.26	-13
12	cha 19-100	2.0	2	0.0	0
13	cha 48-76	1.66	2	-0.34	-17
14	cha 79-112	6.44	1	5.4	544
15	cha 31-206	2.21	2	0.2	10.5
……					

The results of the cha19-110 well diagram are fluctuating between 1 and 10t/d, and the oil of the separator is 3t/d. The results of the separator amount of oil can be measured in the fluctuation range of the oil results, and the measurement can more correctly reflect the changes of oil production volume due to the 24 hours on-line measurement.

2.2. *Error analysis of oil well error in power chart fluctuation*

Oil production fluctuation to through simple adjustment coefficient to revise it, for this part of the oil well, after a large number of field tests, we believe that due to the diagrams of the amount of oil condition monitoring wells of 24 consecutive hours, so diagrams of the amount of oil technology to more precisely reflect well the real situation.

2009 April 25 8 03, Cha 48-128 separator results relatively greater volatility in the previous day, liquid production by the fluctuations in the 29t/d to 40t/d. At the same time diagrams of the amount of oil analysis system also accurately capture to the period, the power fluctuation map. Because the power chart of oil is the 24 hour transmission power chart, it can be seen that, for the power chart fluctuations big oil well, the results of the power oil results more close to the true level of oil wells.

3. Power Chart Quantity and Oil Process Management

Diagrams of the amount of oil process will in oil production plant is fully implemented, as a new technology, diagrams of the amount of oil involved reservoir geology, production technology and equipment more professional departments, how to achieve timely and effectively solve the problems exposed in the process of application and management of late stage is very important.

Seriously organize training. Diagrams of the amount of oil technology training is an important link in the application of the process after the success of, engineering, geology and other related departments to carefully prepare training content, do detailed, easy to understand, and seriously organize the related staff learning, so that everyone can operate.

By the calibration of the car to the single well production liquid quantity calibration, and constantly corrected single well power diagram calculation and production coefficient, improve the power chart calculation of production accuracy.

To develop the oil management method and the operation guide, and to the factory system file to go. The function diagram of oil management approach to work diagram of oil based data acquisition and update, data comparative analysis, equipment installation, maintenance and checking were detailed provisions, and strive to achieve fast and effectively solve the problems exposed in the diagrams of the amount of oil technology application process, to meet the needs of the work of oil and gas field development.

4. Conclusions

The diagram the amount of oil process integrated oil well automatic monitoring, condition of automatic diagnosis and automatic yield measurement technology, realize the timing acquisition, automatic calculation, without human intervention, more objectively reflects the oil well liquid producing capacity, excluding the separator of human error, equipment error, error conditions such as error.

For the high yield liquid oil well, after many times of comparison, the results of power chart and the separator amount of oil results are in good agreement, the basic energy can correctly reflect the oil well conditions and real daily output level.

The production characteristics of oil well determine it can not be continuous uniform out liquid. Indicator diagram is a 24-hour tracking well liquid producing amount of change, so the intermittent oil, liquid production fluctuations larger wells, diagrams of the amount of oil process than to the separator more accurately reflect well the real production.

Diagrams of the amount of oil technology should not be confined to the comparison of the accuracy of data and diagrams of the amount of oil itself is a management concept, enhance. It not only simplifies the surface metering process, reduce the engineering investment, and good real-time tracking of oil well production status in a timely manner to the managers to provide accurate information of production wells, and reduce the labor intensity of workers, improves the management level of the oil well.

References

1. Shi Mingyi et al. Application of power chart method in the field of oil field. Oil planning design, 67, 2001 (3): 9–12.
2. Cui Zhenhua. Rod pumping system [M]. Beijing: Petroleum Industry Press. 1994.

Coupling the EFDC and CE-QUAL-ICM models to simulate water quality of shallow lake in Inner Mongolia, China

X. Li[1,*], S.L. Zhang[2], L.X. Wang[3,4], M.M. Gou[1] and X.Q. Xu[1]

[1]*Inner Mongolia Engineering Research Center for Water-saving Agriculture, Inner Mongolia Normal University, Hohhot 010022, China*

[2]*Online Monitoring Center on Pollutants in Inner Mongolia Autonomous, Hohhot 010011, China*

[3]*College of Environment and Resources, Inner Mongolia University, Hohhot 010021, China*

[4]*School of Earth and Environmental Sciences, University of Adelaide, SA 5005, Australia*

**E-mail: 549645024@qq.com*

In this study, a novel aquatic eutrophication model was presented by coupling the Environmental Fluid Dynamics Code (EFDC) and three-dimensional eutrophication models (CE-QUAL-ICM). The meteorological factors (wind speed, evaporation and radiation) and emergent plant factors (spatial distribution, density, height and diameter) were substantially accounted for in the model to reflect the effects of adsorption and degradation of pollutants by emergent plant and the changes of bottom stress and flow field. The proposed model was used to simulate the inter-annual and seasonal changes of water quality in Wuliangsuhai Lake which is a shallow eutrophication lake surrounded by the emergent plant in Inner Mongolia, China during 2004–2009. The simulated concentrations of total nitrogen, total phosphorus, chlorophyll-a and chemical oxygen demand with consideration of the presence of emergent plants were satisfactorily agreement with the measured ones. The simulation results indicate that it is very important to reduce the pollutant load into the lake, which can directly change the concentration of the water quality in the lake area. The information of water quality in shallow and dense emergent plants covered aquatic area where in-site observation is unavailable could be obtained from the model. Therefore, it appears that the developed model is an effective approach to simulate water quality of shallow grass-algae lakes, and can be used as a useful water environmental planning and management tool in these regions.

*Keywords: W*ater quality; Shallow lake; EFDC; CE-QUAL-ICM; Emerged plant.

1. Introduction

The water quality of lake plays an important role in the whole aquatic environmental system [1–4]. The aquatic environment is a complex system [5–7], which is affected by many factors, such as hydrology, meteorology, topography, geomorphology, physical, chemical, biological, ecological processes and many uncertain nonlinear factors in the water column. As a result of the above factors, there is a little theoretical solution for simulating the aquatic environment. Most of simulations should rely on numerical solution. Therefore, the mathematical model becomes a major approach to study the variations of water quality and quantity under different hydrodynamic conditions, and to estimate the water quality of area in which in-site observation is unavailable. The mathematical model can be also useful for predictions of future water environment situation. Some mathematical models, such as CE-QUAL-W2, Mike, Delft3D and Wasp, have been used in water quality, flow field and temperature in rivers, estuaries, lakes, bays, reservoirs and other water bodies [8–10].

The distribution of emergent plants in the water is an important factors affecting water quality. Density and morphological indexes of emergent plants can directly affect the changes process of hydrodynamic, turbulent coefficient and bottom shear stress [11]. At the same time, emergent plants continuously absorb nutrients from the surrounding environment for their own growth requirement [12]. Therefore, the spatial distribution and growing processes of emergent plants should be considered adequately in simulating water quality. Also, it is hard to sample water column to monitor and assess the water quality when covered by the emergent plants. However, the effects of the aquatic emergent plant on water quality are ignored in the most of mathematical eutrophication models.

2. Materials and Methods

2.1. *Study area*

Wuliangsuhai Lake (40°36′-41°03′N, 108°43′-108°57′E) is located in Urat Front Banner of Bayannaoer city, Inner Mongolia autonomous region, China (Fig. 1). It is an important part of Hetao Basin, which has a gentle south-eastern slope, with elevation ranging between 1,007 and 1,060 m above sea level. The largest lake in the Yellow River watershed, was formed after the Hetao basin was cut off from the main channel of the Yellow River due to a natural river diversion ca. 1850. The lake has an average water depth of 1.5 m, is 35–40 km long, with a total area that has varied between 305 and 344 km^2 since 1986. The semi-arid to arid climate has an annual average air temperature of 8.1 °C, an average annual precipitation of 216 mm (mostly occurring between May and October), and an estimated annual evapotranspiration of 1,511 mm. The lake water is usually frozen from mid-November to mid-April.

Wuliangsuhai Lake reed cover has accounted for 46% of the total area of the lake according to the interpretation results of satellite images in 2010. With the heaviest agricultural activity in the watershed of Yellow River named Hetao

Irrigation District and local industrial development, various kinds of sewage water were discharged into Wuliangsuhai Lake before into the Yellow River [13,14]. As a result, algae showed abnormal growth and algal bloom broke out frequently, which accelerated the growth of emergent plants, aggravated the degree of eutrophication and deteriorated water quality. The average total nitrogen content (TN) and total phosphorus content (TP) of water in lake inlet is 9.54 mg/L and 0.28 mg/L, respectively, which are 48 times and 14 times as large as that of respective international eutrophication standards. The annual amounts of TN and TP discharging into the Wuliangsuhai Lake were about 1090 t and 66 t, respectively.

The objectives of this study are to develop a coupled aquatic eutrophication model with the emergent plants considered, and then to simulate the water quality of Wuliangsuhai Lake.

2.2. *Data collection*

Based on the hydrodynamic conditions and distribution characteristics of pollutants in Wuliangsuhai Lake and survey standard of wetland ecosystem, the whole lake is divided into 2 km × 2 km square size grids. The sample monitoring points were arranged in plum-shape and positioned in each grid intersection using GPS. The water samples were obtained from May to October with an interval of half a month during 2006–2009. Three replicates were taken for each monitoring point. The water samples were sent to laboratory in time for tests. Water quality indicators such as total nitrogen, total phosphorus, chlorophyll-*a* and chemical oxygen demand were performed with specific detection methods according to supplemented edition of "water and wastewater monitoring and analysis methods". During the study, in some regions where there are dense emergent plants or shallow water, it is difficult to go there and obtain water samples.

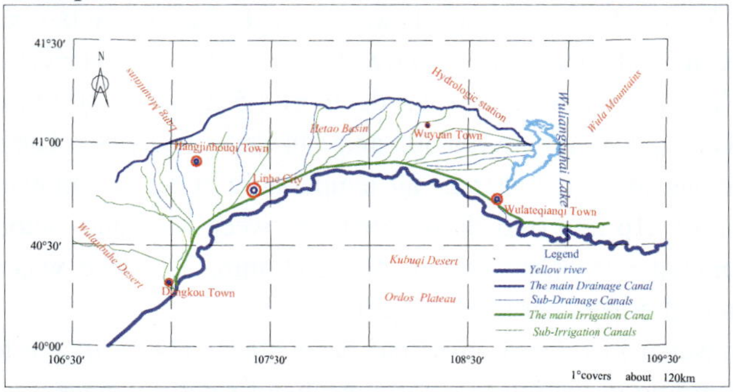

Figure 1. The irrigation and drainage system of Hetao Irrigation District and location of Wuliangshuai Lake.

3. Results

3.1. *The dynamics of pollutants and algae in the Wuliangsuhai Lake*

In the study, using measured water quality data of multiple sample points in lakes, total nitrogen, total phosphorus, chlorophyll, and chemical oxygen demand from 16 April 2006 to 14 Oct. 2009 to verify adaptability of the model. The results show that relative error between the simulated and measured values is less than 30%. Select O_{10} located in sampled center point in Wuliangsuhai Lake, the total nitrogen, total phosphorus, chlorophyll and chemical oxygen demand within four years are simulated (Figs. 2–5). Water quality indicators including total nitrogen, total phosphorus, chlorophyll-*a* and chemical oxygen demand in 4 years are simulated to find whether to consider the existence of emergent plants in the model. Other input files are fixed in the established model, and based on class V Surface Water Environmental Quality Standard (GB3838-2002) the value of dissolved oxygen, chemical oxygen demand, BOD, ammonia nitrogen, total phosphorus and total nitrogen are taken as input file of pollution load, to reflect various water quality changes in lakes.

It can be seen from the simulation results (Figs. 2–3) that the dynamics of total nitrogen and total phosphorus are similar. The maximum value of total nitrogen and total phosphorus concentration appear in spring. There are two major reasons. On one hand, with the changes of climatic factors, such as sunlight enhanced, temperature rise and wind increased, remains of plants and animals began to be hydrolyzed and mineralized, exchange in water body is enhanced, and sediment release rate is accelerated, resulting in the increase in the concentrations of TN and TP. On the other hand, in winter there is no agricultural irrigation in Hetao Irrigation District, so there is no agricultural drainage discharged into Wuliangsuhai Lake. Then the industrial wastewater and domestic sewage play a main role on the quality and quantity of the lake, causing increases in concentration of TN and TP. However, a downward trend is showed on concentration of TN and TP in October and November each year when the huge quantity of water was irrigated into the agricultural land to decrease its content of salt in Hetao Irrigation District. That huge agricultural drainage into Wuliangsuhai Lake significantly reduced the concentration of pollutants.The content of algae in the model is represented by using concentration of chlorophyll-*α*. The peak of chlorophyll-*α* in Wuliangsuhai Lake appears in July and August when algal absorb various nutrients in water and quickly breeding which lead to TN and TP appears lower value (Fig. 4).

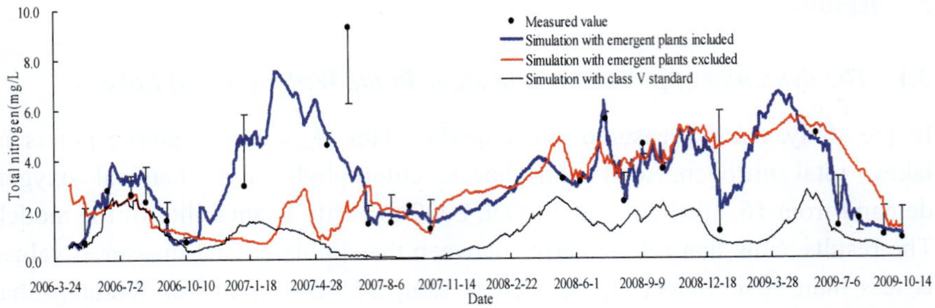

Figure 2. Comparison between the measured and simulated total nitrogen contents under different conditions.

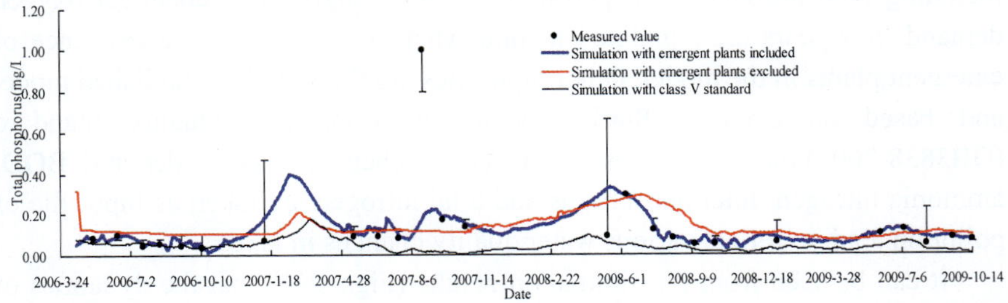

Figure 3. Comparison between the measured and simulated total phosphorus contents under different conditions.

The concentration of chlorophyll-α is lower on the autumn and winter because of lower temperature, weakened light intensity, and reduced algal photosynthesis. It can be shown that from the entire four-year changes that the peak of chlorophyll-α decreased year by year and annual change process also showed a certain periodicity.

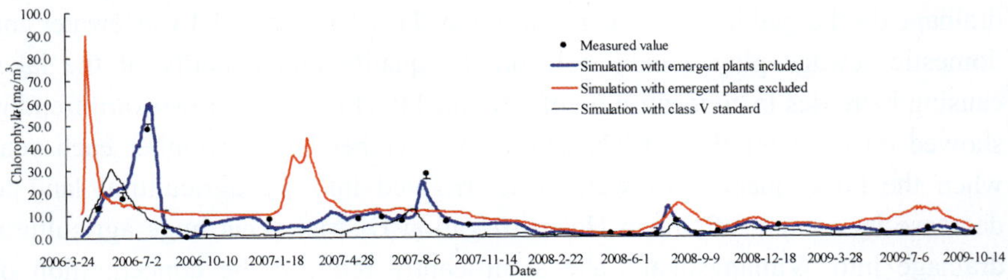

Figure 4. Comparison between the measured and simulated chlorophyll-a contents under different conditions.

The dynamics of organic pollution can be shown from Fig. 5. The peak value of COD appear in July and August when aquatic plants and zooplankton frequently active because of a large amount of oxygen in the water column being consumed by their metabolic. In addition, the concentration of COD increased in winter because of the deficit of dissolved oxygen when ice covered.

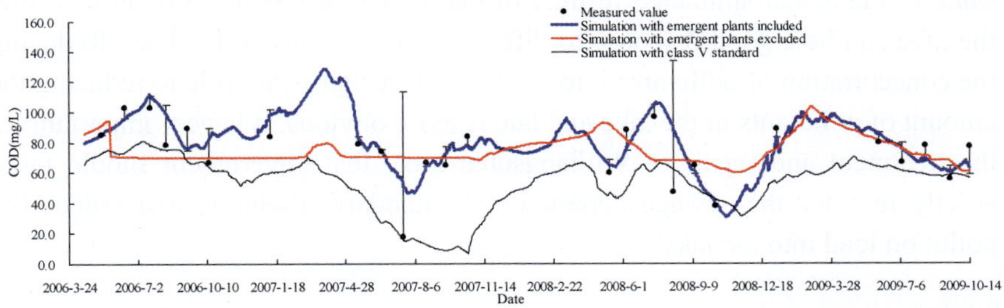

Figure 5. Comparison between the measured and simulated COD contents under different conditions.

Figures 2 to 5 show that the growth process of emergent aquatic plant seriously affects the water quality of the lake. In eutrophication model without considering the emergent plants, the simulated relative error between the simulated and measured values is substantially larger than 30%. The overall level of error is about 50%. Furthermore, the simulated results do not reflect the actual dynamics of water quality.

It can be seen from Figs. 2–5 that simulated values can basically reflect the trend of the measured values. The error bars of various graph show that the relative error more than 30% accounted for 11% of the total. The relative error from 20% to 30% accounted for 19% of the total. The relative error from 10% to 20% accounted for 24% of the total. The relative error from 5% to 10% accounted for 11% of the total. The relative error less than 5% accounted for 11% of the total. Overall, the coupled model takes into account the spatial distribution of the emergent plants, which the error of major simulated water quality variables is less than 30%, and the better simulation results than without considering the emergent plants. In winter, relative errors between simulated and measured values are large, which are more than 30%. The main reason is that in the frozen period hydrodynamic condition and boundary condition of the lake water have taken place in fundamental changes which are not taken into account in the coupled model. The simulation results are dissatisfied in the period.

3.2. *The accuracy and predictability of the coupled model*

Figures 2 to 5 show that using class V standards to control pollution load into the lake simulated value in the model is less fluctuate than measured value, for taking the daily certain same water quality indicators as the input file. Because the measured value of those water quality parameters are higher than the class V standard value of Surface Water Environmental Quality, simulated value is basically less than the measured value and the simulated dynamic process is the same to the actual simulated results. In the model, different pollution load into the lake can be a good response to different water quality in the lake. Reducing the concentration of pollutants into the lake has a significant role to reducing the amount of pollutants in the lake and late is more obvious. It is very important for the treatment and repair of Wuliangsuhai Lake that government should more strictly monitor the sewage department to standard discharge and reduce the pollution load into the lake.

4. Discussion

4.1. *The spatial distribution of emergent plants impacting on water quality*

To simulate the water quality of shallow water in which distributed a number of emergent plants, some researchers think small number of emergent plants may be ignored, or did generalizability to directly give parameters on reaction absorb nutrients quality. But in fact, the spatial distribution of emergent plants in water column can hinder the process of hydrodynamic change, change critical shear stress for sediment resuspension and affect dynamic changes of diffusion coefficient and turbulent coefficient. In addition, aquatic plants in different periods affect the absorption intensities of different nutrient substance, and change the distribution of nutrient substance in water column [15]. Thus, the distribution of emergent plants great impact on the water quality processes, especially for large area distribution of emergent plants. From this study showed that coupled model simulation results considering the distribution of emergent plants was better agreement with the actual situation.

4.2. *Dynamic characteristics of emergent plant growth affects on water quality*

In this paper, the characteristics of emergent plants can be shown from these parameters, such as density, growth height and plant width in different positions. It was too simple that using these parameters to describe the growth of emergent plants. Factors on emergent plants are not dynamic in the model, and water quality change is complex in grass-algae-type lakes. Therefore, the coupling model did not fully simulate the actual situation. The future development of the model must be added the dynamic growth of emergent plants, and to reflect actual situation.

5. Conclusions

In this paper, the EFDC model and the CE-QUAL-ICM model are coupled by accounting for the spatial distribution of emergent plants, changes of meteorological factors, dynamic conditions for phytoplankton growth. The established eutrophication model was applied to simulate the inter-annual and seasonal changes of water quality variables in Wuliangsuhai Lake. The calibrated eutrophication model can satisfactorily describe the measured water quality variables with the simulated relative error less than 30%, and most of which was less than 20%. Eutrophication coupled model better simulate and predict some information areas of not entering in, for example shallow or dense emergent plants.

The comparison between the simulation result with and without consideration of the existence of emergent plants, considering the spatial distribution of emergent plants in the coupled model can make the simulation results consistent with the measured ones. It can be seen that emergent plants have a great impact on water quality. When water surface is covered by aquatic plants, the model is of great significance to water quality monitoring, simulation and water body restoration.

It can be seen from simulation scenarios of reducing pollution load into the lake that changes of into the lake pollution load directly affect changes of water quality in the lake. Therefore, reducing the concentration of polluting substances into the lake can repair the eutrophication problem in Wuliangsuhai Lake.

Acknowledgments

This article was supported by the National Natural Science Foundation, No. 51469026; Inner Mongolia Science and Technology Support Project, No: 20121601 and No: 20130422.

References

1. Yongsik Sin, Bongkil Hyun, Byungkwan Jeong, et al. Impacts of eutrophic freshwater inputs on water quality and phytoplankton size structure in a temperate estuary altered by a sea dike.Marine Environmental Research, 85: 54–63 (2013).
2. Natalia Kuczyńska-Kippen. Environmental variables of small mid-field water bodies and the presence of rotifera groups of different ecological requirements. Pol. J. Environ. Stud. 23(2): 373–378 (2014).
3. Nguyen Thi Thuy, Justin Chun-Te Lin, Yaju Juang, et al. Temporal variation and interaction of full size spectrum Alcian blue stainable

materials and water quality parameters in a reservoir. Chemosphere, 131: 139–148 (2015).

4. Aina G. Gómez, Bárbara Ondiviela, Araceli Puente, et al. Environmental risk assessment of water quality in harbor areas: A new methodology applied to European ports. Journal of Environmental Management, 155: 77–88 (2015).
5. Boney A D, Corner E D S. The effect of light on the growth of sporelings of the red algae Antithamnion plumula and Brongniartella byssoides. Journal of the Marine Biological Association of the United Kingdom, 43: 319–325 (1963).
6. Yamilett Carrillo-Guerrero, Edward P. Glenn, Osvel Hinojosa-Huerta. Water budget for agricultural and aquatic ecosystems in the delta of the Colorado River, Mexico: Implications for obtaining water for the environment. Ecological Engineering, 59: 41–51 (2013).
7. Vasiliki S. Thomaidi, Athanasios S. Stasinakis, Viola L. Borova, et al. Is there a risk for the aquatic environment due to the existence of emerging organic contaminants in treated domestic wastewater? Greece as a case-study. Journal of Hazardous Materials, 283: 740–747 (2015).
8. Vassilios Pisinaras, Christos Petalas, Georgios D. Gikas, et al. Hydrological and water quality modeling in a medium-sized basin using the Soil and Water Assessment Tool (SWAT). Desalination, 250(1): 274–286 (2010).
9. Charalampos Doulgeris, Pantazis Georgiou, Dimitris Papadimos, et al. Ecosystem approach to water resources management using the MIKE 11 modeling system in the Strymonas River and Lake Kerkini. Journal of Environmental Management, 94(1): 132–143 (2012).
10. Gilioli G, Pasquali S, Tramontini S, et al. Modeling local and long-distance dispersal of invasive chestnut gall wasp in Europe. Ecological Modeling, 263(10): 281–290 (2013).
11. Ji, Z G. Hydrodynamics and water quality Modeling Rivers, Lakes, and Estuaries. In: A John Wiley & Sons, editors. INC., p. 1–15 (2007).
12. Chen B, Chen G Q, Hao F H, et al. The water resources assessment based on resource exergy for the mainstream Yellow River. Communications in Nonlinear Science and Numerical Simulation, 14: 331–344 (2009).
13. Li X, Gou M M, Wang Y. Analysis on the Exceeding the Risk Ratio of Mouth Water Quality for Wuliangsuhai Lake in Inner Mongolia. Journal of Agro-Environment Science, 30: 1638–1644 (2011).
14. Daniela B, Giulia M, Giovanni B, et al. Analyses of three native aquatic plant species to assess spatial gradients of lake trace element contamination. Aquatic Botany, 83: 48–60 (2005).

15. Felipe Segala Ferreira, Alyne Bezerra Tabosa, Raimundo Bemvindo Gomes, et al. Spatiotemporal ecological drivers of an aquatic plant community in a temporary tropical pool. Journal of Arid Environments, 115: 66–72 (2015).

Ecological efficiency for non-ferrous metals industry in China based on bulk materials flow analysis

Q. Yue*, J.K. He, H.M. Wang, C. Zhang, F. Liu and X.Z. Hu
SEP Key Laboratory of Eco-Industry, Northeastern University, China
**E-mail: yueqiang2005@aliyun.com*

Bulk materials flow analysis for non-ferrous metals system of China was carried out based on bulk materials flow analysis method. Ecological efficiency for non-ferrous metals industry in China during the period 1995-2009 was analyzed. Resource efficiency reached its peak value in 2005 and resource economic efficiency reached its peak value in 2008. Energy efficiency reached its peak value in 2009 and energy economic efficiency reached its peak value in 2008. Based on the results of bulk materials flow analysis for the non-ferrous metal system of China, decoupling analysis was further carried out according to the decoupling indicator. Value of decoupling indicator for domestic extraction used was 0.720 during the period 1999-2008, which is bigger than that of domestic processed output, direct materials input and total materials requirement. It seems the decoupling indicator of domestic extraction used got the higher value during this period due to the shortage of domestic non-ferrous ore output and the dependence on imported resources.

Keywords: Non-ferrous metals industry; Ecological efficiency; Bulk materials flow analysis; Decoupling analysis; Materials input.

1. Introduction

Since the beginning of 1990s, China has encountered an excessive growth of non-ferrous metals outputs. Especially in the period 2000-2007, the average annual growth rate of non-ferrous metals outputs reached 17.12 percent, and it was higher than the average annual growth rate of GDP about 7 percent (National Bureau of Statistics of China 1996-2010; Editorial board of China non-ferrous metals industry yearbook 1996-2010). As a result, while the GDP doubled, the non-ferrous metals outputs increased from 783.81 million tons to 2370.05 million tons and it increased more than three times. The growth rate of non-ferrous metals outputs is a little higher than the growth rate of GDP even in the period of rapid industrialization under normal circumstances. Due to the influence of the global financial crisis, the non-ferrous metals outputs didn't increase too much in the period 2008-2009. However, it had entered a rapid growth since the year 2010 (Editorial board of China non-ferrous metals industry yearbook 1996-2010). The high non-ferrous metals outputs and the rapid growth rate of non-ferrous metals outputs have been an indisputable fact in China, which will bring a lot of problems, such as the issues of excessive resources (including energy) consumption and pollutants emission. Average percentage of energy consumption in the non-ferrous metals industry accounted to the total energy consumption was 3.26 percent in the period 1995-2009 (National Bureau of Statics Division of Energy 1996-2010), and the materials consumption, pollutants emission also accounted to a large proportion. But average percentage of the industrial added value of non-ferrous metals industry accounted to China's GDP was only 0.87 percent (National Bureau of Statistics of China 1996-2010; Editorial board of China non-ferrous metals industry yearbook 1996-2010). Therefore, non-ferrous metals industry belongs to "high energy consumption, high materials consumption, and high pollution" industry. So it is necessary to study on ecological efficiency for non-ferrous metals industry.

Eco-efficiency refers to the efficiency of ecological resources meeting human needs and it is the ratio of output to input. The "output" refers to the value of products and services provided by the enterprise, the industry or the economy; the "input" refers to the use of resources and energy in the enterprise, industry or economy and environmental pressure caused by them (Vogtlander et al. 2002). Therefore, eco-efficiency indicators include resource efficiency, energy efficiency and environment efficiency, etc. Eco-efficiency measurement methods for the analysis of different research objects have been proposed. In particular, there are more evaluation and research on a large scale, e.g. the circular economy construction in a city, a region or a nation (Yin et al. 2012). Eco-efficiency analysis for metals industry, especially for the non-ferrous metals industry is still very limited. Among them, literature (Mao et al. 2010) is based on the analysis of industrial composition and the relationship between industrial system and external environment, chooses energy consumption and the main environmental emissions, such as wastewater, solid waste, sulfur dioxide, smog and dust as the environmental impact categories and defines eco-efficiency corresponding to those environmental impacts. Literature (International Council on Mining & Metals 2014) put forward the evaluation indicator of eco-efficiency for the metal mining industry. Eco-efficiency is characterized by using energy consumption or waste emission per unit output of mining products. And it discusses the ways to improve eco-efficiency from the perspective of management. Literature (Liu et al. 2007) converts ecological occupies of non-ferrous metals industry into ecological footprint and it analyzes the variations of ecological footprint's size, structure and intensity for the non-ferrous metals industry of China. Literature (Hilson 1999) proposes an evaluation framework of eco-efficiency for the non-ferrous metals mining industry, and it discusses the ways of how to reduce resource and energy consumption, waste emission and simultaneously reduce the production cost of the enterprises. Literature (Wu 2007) describes the index system of sustainable development for British non-ferrous metals industry, and it screens a series of indicators based on the economic, social and environmental sustainability, which used to monitor and evaluate the sustainable development process for the non-ferrous metals industry. The above studies don't systematically investigate the relationships between resource consumption of non-ferrous metals industry, the losses of non-ferrous metal containing materials and the economic indicators of non-ferrous metals industry. Based on this circumstance, this paper uses the bulk materials flow analysis method to study on the ecological efficiency for the non-ferrous metals industry of China, and the similar studies have not been carried out.

2. Methodology

Bulk materials flow analysis model used in this paper is shown in Figure 1 (Eurostat 2009; Wang et al. 2012; Dai & Chen 2010; Moriguchi 2007; Schandl & Turner 2009). As study object is China's non-ferrous metals industry, the investigative object is non-ferrous metals containing materials and don't contain other materials when applying this model.

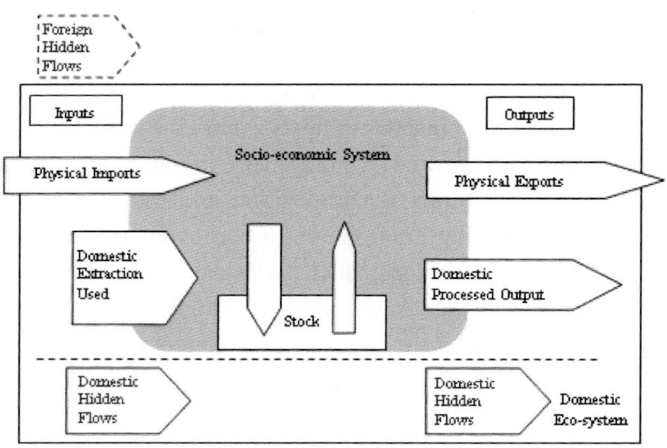

Figure 1. Bulk materials flow analysis model.

In this model, the flows across the socio-economic system boundary belong to direct flows. For the inputs, it includes physical imports and domestic extraction used; for the output, it includes physical exports and domestic processed output; in the socio-economic system, there are inflows and outflows of metal in the stock, the net increment of them is the net increment of the stock. Taking the analysis of bulk materials flow analysis for the non-ferrous metals industry as an example, the meanings of each flow are as follows:

a. Domestic Extraction Used (DEU): the non-ferrous metals mineral resources obtained from the nature annually, which is put into the production process of the socio-economic system.

b. Physical Imports and Physical Exports (PI and PE): it contains the non-ferrous metal resources and products imported to domestic or exported to foreign countries through the import and export trade.

c. Net Additions to Stock (NAS): it is a measure of physical growth of the economy system, which is equal to the difference of the increment and elimination among the buildings, infrastructure, machinery, vehicles and other goods, etc.

d. Domestic Processed Output (DPO): it refers to the material quantities discharged into environment, which includes loss of non-ferrous metal mineral resources of domestic mining and imported resources of non-ferrous metals containing materials during the process of production, fabrication & manufacture and consumption, etc. It mainly includes the waste discharged into the air, the water and the land and the dissipative emission from the using of products.

e. Hidden Flows (HF): it refers to the excavations that must be excavated in mining process, but not enter market and production process. For example, in order to mine the non-ferrous metal ore, a lot of rock must be stripped, the latter does not enter non-ferrous metal production process and it also does not enter the consumption process as a commodity.

According to above indicators, two important indicators are defined as follows:

a. Direct Materials Input (DMI): DMI=DEU+PI.

b. Total Materials Requirement (TMR): TMR=DMI+DHF+FHF.

Based on bulk materials flow analysis model in Figure 1, we analyzed the bulk materials flow for non-ferrous metals system of China in the period 1995-2009. Based on the data of the bulk materials flow analysis for the non-ferrous metal system, we analyzed the resource (economic) efficiency, the energy (economic) efficiency and the decoupling indicators.

Resource (economic) efficiency indicators, energy (economic) efficiency indicators and decoupling indicators reflect ecological efficiency for non-ferrous metals industry respectively from different aspects.

Resource (economic) efficiency studies the relationships between non-ferrous metals outputs (non-ferrous metals industrial added value) and consumption of non-ferrous metal ore. Energy (economic) efficiency studies the relationships between non-ferrous metals outputs (non-ferrous metals industrial added value) and non-ferrous metals industrial energy consumption. The increment of these indicators reflects the production efficiency's improvement of the ore, energy and other natural resources for the non-ferrous metals industry. Decoupling analysis studies the relationships between the resource inputs of non-ferrous metals containing materials and the waste emission for the non-ferrous metals industry and the non-ferrous metals industrial added value, their improvement reflects the promotion on the resources conservation and the environmental protection for non-ferrous metals industry.

3. Analysis of Resource Efficiency and Energy Efficiency

3.1. *Collection of relevant data*

By consulting the "Chinese Non-ferrous Metals Industry Yearbook" and the "China Energy Statistics Yearbook", etc (Editorial board of China non-ferrous metals industry yearbook 1996-2010; National Bureau of Statics Division of Energy 1996-2010; Wang 2011), we got the relevant data of Chinese non-ferrous metals industry in the period 1995-2009 and it was shown in Table 1.

Table 1. Relevant data of China non-ferrous metals industry in 1995-2009 (unit: 104t).

	DEU	DHF	PI	FHF	Metals Output	Energy consumption (10^4tce)	Industrial added value (10^9yuan)	Apparent consumption[1]	PE
1995	8545	10574	768	14130	497	3399	266	490	246
1996	8753	10139	768	20604	523	3655	239	519	140
1997	9424	11922	747	21882	581	3755	243	512	221
1998	9300	10817	826	25799	616	3761	213	556	203
1999	9599	10761	993	29087	695	3899	305	659	221
2000	10506	10908	1365	37328	784	4505	410	804	235
2001	10233	10679	1680	41468	884	4470	528	877	241
2002	11610	14732	1919	43123	1012	4928	680	1025	286
2003	12690	15757	2421	56212	1228	6159	966	1252	375
2004	14358	18009	2940	59055	1431	7027	1400	1440	425
2005	14963	19992	3559	73663	1639	8085	2175	1673	410
2006	17744	24002	4906	89707	1915	9600	3641	1861	568
2007	20904	29609	7933	175105	2370	11697	5117	2454	446
2008	23232	31885	8030	152633	2551	12151	5874	2713	432
2009	23287	33527	8672	121336	2604	12234	5730	3155	287

1. Non-ferrous metals apparent consumption = Non-ferrous metals outputs + Imported quantities of non-ferrous metals - Exported quantities of non-ferrous metals.

As shown in Table 1, domestic non-ferrous metal ore outputs increased rapidly during 2002-2008. Although growth rate was only 0.2 percent as a result of the impact of the financial crisis, the ore outputs have reached 232.87 million tons in 2009. The total amount of China's non-ferrous mineral resources was large, and percentage of the discovered amount of resources in China to the total world resources was about 12 percent. But most of them couldn't be available for utilization. Small metal varieties, such as tungsten, tin, antimony, molybdenum and rare earths, etc, have great advantages in mineral reserves. While the main mineral reserves, such as copper, aluminum, lead and zinc, etc, that the percentage of their outputs to the total non-ferrous metal outputs is about 94 percent, are inadequate (Gou & Zuo 2004). In addition, China's non-ferrous mineral resources have the characteristics of poor quality, more lean ore, less bonanza, few large deposits, high utilization costs, difficult treatment and weak market competitiveness. Among the identified reserves, percentage of the valuable reserves to the total amount of resources is less than 30 percent (Gou & Zuo 2004). Therefore, the supply of the non-ferrous metal mines in China is in severe shortage.

3.2. *Bulk materials flow analysis for non-ferrous metals industry*

Materials flow analysis results for non-ferrous metals system of China in 1995 were shown in Figure 2.

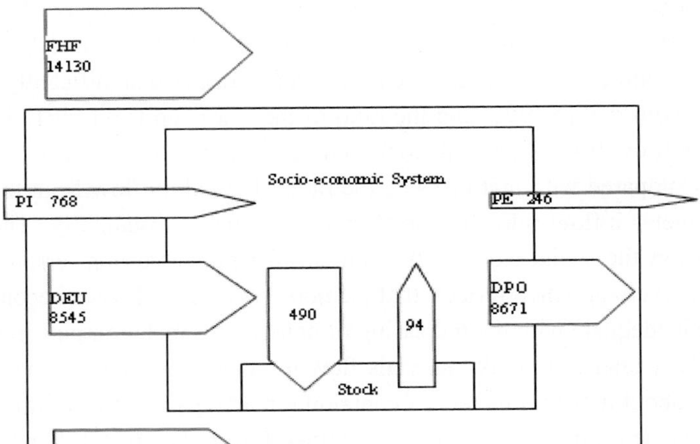

Figure 2. Materials flow analysis diagram for the non-ferrous metals system of China in 1995 (unit: 10^4t).

Figure 2 showed the ratio of PI and DEU was 1:11.13, and the ratio of metal outflow from social stock and inflow into social stock was 1:5.21.

Materials flow analysis results for non-ferrous metals system of China in 2009 were shown in Figure 3.

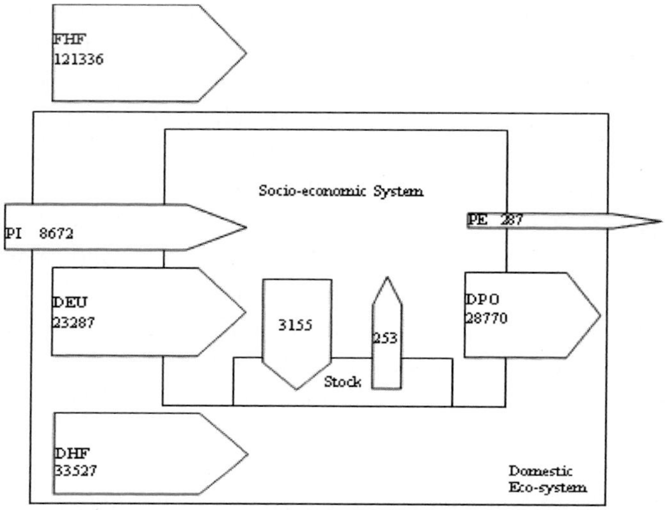

Figure 3. Materials flow analysis diagram for the non-ferrous metals system of China in 2009. (unit: 10^4t).

As shown in Figure 3, although non-ferrous metal ore outputs of domestic mining reached 232.87 million tons, the ratio of PI to DEU reached 1:2.70. The imported resources of various non-ferrous metals increased significantly in 2009 compared with that in 1995, and the ratio of them was up to 11.29:1. The ratio of metal outflow from the social stock to metal inflow into the social stock was 1:12.47. Compared with that in 1995, the ratio of metal outflow from the social stock to metal inflow into the social stock declined substantially. The metal outflow from the social stock is the source of depreciable non-ferrous metals scrap for recycling, which reflects that the non-ferrous metal scrap become more and more inadequate with the increasing of non-ferrous metal outputs in China.

Similarly analyzing bulk materials flow in other years, we get the relevant indicators about non-ferrous metals containing materials input and emission for non-ferrous metals industry of China in 1995-2009, as shown in Table 2.

Table 2. Relevant indicators of non-ferrous containing materials input and emission for the non-ferrous metals industry of China in 1995-2009.

	DEU/10^4t	DMI/10^4t	TMR/10^4t	DPO/10^4t		DEU/10^4t	DMI/10^4t	TMR/10^4t	DPO/10^4t
1995	8545	9313	34017	8671	2003	12690	15111	87080	13600
1996	8753	9521	40264	8940	2004	14358	17298	94362	15542
1997	9424	10171	43975	9510	2005	14963	18522	112177	15950
1998	9300	10126	46742	9422	2006	17744	22650	136359	20382
1999	9599	10592	50440	9779	2007	20904	28837	233551	26120
2000	10506	11871	60107	10904	2008	23232	31262	215780	28329
2001	10233	11913	64060	10875	2009	23287	31959	186822	28770
2002	11610	13529	71384	12302	Growth rate	7.42%	9.20%	12.93%	8.94%

3.3. *Analysis of resources (economic) efficiency and energy (economic) efficiency*

3.3.1. *Resources (economic) efficiency*

Resources efficiency is defined as the ratio of the non-ferrous metal outputs to the consumption of non-ferrous metal ore.

The higher resource efficiency is, the larger is non-ferrous metal output per unit non-ferrous metal ore input, and vice versa. That is to say, the higher resource efficiency means that producing same non-ferrous metal outputs will generate less tailing and have smaller impact on environment correspondingly. As shown from the curve in Figure 4, the resource efficiency rose gradually before 2005 and reached its peak value in 2005 (0.0885t non-ferrous metal/t ore); and then the resource efficiency for the non-ferrous metals industry declined gradually with the rapidly increasing of non-ferrous metals outputs.

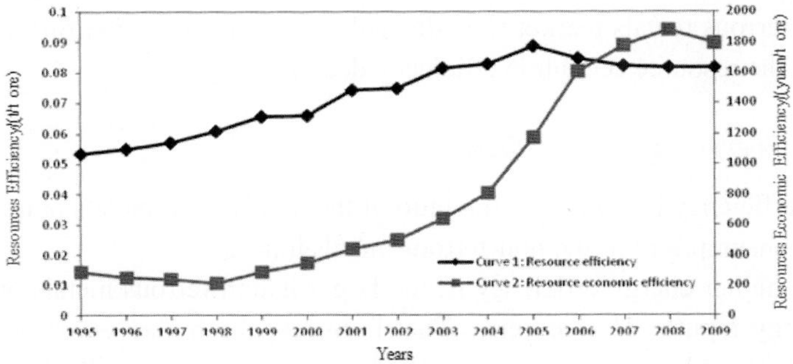

Figure 4. Resource efficiency and resource economic efficiency for the non-ferrous metals industry of China in 1995-2009.

Resource efficiency declined gradually in 2005-2009. And the main reason was that a large part of the non-ferrous metal scrap resources came from the obsolete non-ferrous metal products. Among the raw materials (the non-ferrous metal ore and the non-ferrous metals scrap) in non-ferrous metals production, however, the quantity of old scrap for recycling in China is inadequate at present. Because the non-ferrous metal products have an "average life span", the non-ferrous metals outputs and the non-ferrous metal products consumption before an average life span which accounted for only a small share comparing to now. Non-ferrous metals outputs of China have increased rapidly in recent years, and the growth rate of non-ferrous metals scrap resources is slower than the growth rate of non-ferrous metals outputs, so non-ferrous metals scrap resources is very inadequate. Production of non-ferrous metals is largely dependent on the non-ferrous metal mineral resources and resource efficiency for the non-ferrous metals industry declined slightly. Under the circumstances of the non-ferrous metals outputs remaining basically stable or declining slowly in the future, percentage of the non-ferrous metals scrap resources to the non-ferrous metals production will rise gradually and resource efficiency for the non-ferrous metals industry will also be improved gradually.

Resources economic efficiency is defined as the ratio of the non-ferrous metal industrial added value to the consumption of non-ferrous metals ore.

The higher resource economic efficiency is, the larger is non-ferrous metals industrial added value created per unit non-ferrous metal ore input, and vice versa. That is to say, the higher resource economic efficiency means that obtaining the same non-ferrous metals industrial added value will consume less non-ferrous metal ore and have the smaller impact on environment correspondingly. As shown from the curve in Figure 4, resource economic efficiency reached its peak value in 2008(1878.96 yuan RMB/t ore). Because the non-ferrous metals market was affected by the impact of the financial crisis in 2009, the resource economic efficiency declined slightly.

3.3.2. *Energy (economic) efficiency*

Energy efficiency is defined as the ratio of the non-ferrous metals outputs to the energy consumption for the non-ferrous metals industry.

The higher energy efficiency is, the larger is non-ferrous metals output per unit energy input, and vice versa. That is to say, the higher energy efficiency means that producing the same non-ferrous metal outputs will consume less energy and have less emission to the environment correspondingly. As shown from the curve in Figure 5, energy efficiency increased in general and reached its peak value in 2009 (0.2129t non-ferrous metal/tce).

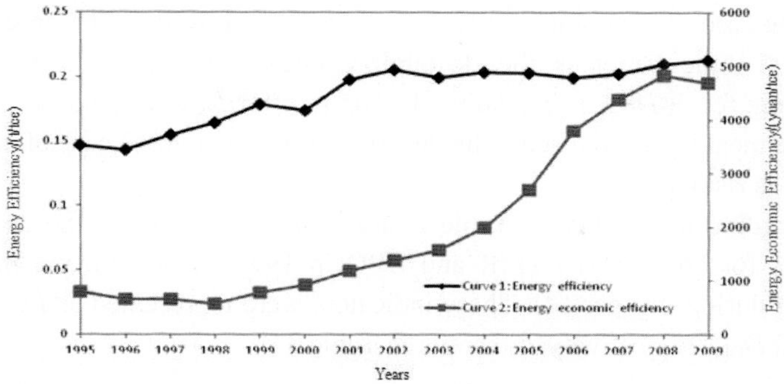

Figure 5. Energy efficiency and energy economic efficiencyfor the non-ferrous metals industry of China in 1995-2009.

Energy economic efficiency is defined as the ratio of the non-ferrous metal added value to the energy consumption for the non-ferrous metals industry.

The higher energy economic efficiency is, the larger is non-ferrous metals added value created per unit energy input, and vice versa. That is to say, the higher energy economic efficiency means that obtaining the same non-ferrous metals industrial added value will consume less energy and have less emission to the environment correspondingly. As shown from the curve in Figure 5, energy economic efficiency basically escalated and reached its peak value in 2008 (4834.17 yuan RMB/tce). As the non-ferrous metals market was affected by the impact of the financial crisis in 2009, the energy economic efficiency declined slightly.

As shown in Figure 4 and Figure 5, resource economic efficiency and energy economic efficiency for the non-ferrous metals industry of China declined gradually during the period of 1995-1998, which was due to the non-ferrous metals industrial added value basically decreased during this period.

4. Decoupling Analysis

Decoupling indicator can be used to measure the relationships between resource use or waste emission and economic indicator, and its formula is as follows (Lu & Yue 2010; Wang et al 2013):

$$D = \frac{t}{g} \times (1+g) \tag{1}$$

Where D is decoupling indicator, t is the annual decreasing rate of resource use (or waste emission) per unit economic indicator, g is the annual variation rate of economic indicator.

In the case of economic indicator is rising (g is positive), the bigger of the D value is, the better is the decoupling degree. For $D \geq 1$, it is absolute decoupling; for $0 < D < 1$, it is relative decoupling; for $D \leq 0$, it is non-decoupling. It is more meaningful to discuss the decoupling problem in the case of economic indicator is rising.

According to the data in Table 2 and formula (1), we get the decoupling indicators for DEU, DMI, TMR and DPO in 1999-2008(economic indicators increased during this period). These indicators were represented as D_{DEU}, D_{DMI}, D_{TMR} and D_{DPO} respectively, as shown in Table 3.

Table 3. Decoupling indicators of input and emission for the non-ferrous metals industry of China in 1999-2008.

	D_{DEU}	D_{DMI}	D_{TMR}	D_{DPO}
1999	0.926	0.894	0.818	0.913
2000	0.726	0.650	0.444	0.666
2001	1.091	0.988	0.771	1.009
2002	0.532	0.528	0.602	0.544
2003	0.779	0.722	0.478	0.750
2004	0.708	0.678	0.814	0.682
2005	0.901	0.872	0.659	0.953
2006	0.724	0.669	0.680	0.588
2007	0.561	0.326	-0.759	0.305
2008	0.248	0.432	1.514	0.429
Average value	0.720	0.676	0.602	0.684

It can be seen from Table 3 that the value of D_{DEU} was bigger than that of D_{DPO}, D_{DMI} and D_{TMR}, and it indicated that DEU's decoupling degree was relatively the best.

The D_{DEU} was relatively the highest, at the same time D_{DPO} was also relatively higher. This was closely related that the production for the domestic non-ferrous metals depended heavily on imported non-ferrous metal ore and the non-ferrous metal scrap (self-produced and imported). Although the quantity of domestic mining of non-ferrous metal ore increased during the period, the increment was not large. Correspondingly the amount of generated tailing in the process of the domestic non-ferrous metal ore production was not large, so the value of D_{DPO} was relatively high.

5. Discussion and Suggestion

Resource efficiency of non-ferrous metals industry had a gradually descending trend during the period of 2005-2009. The main reason lied in non-ferrous metals output increased rapidly, while growth rate of scrap non-ferrous metals couldn't keep pace with growth rate of non-ferrous metals output. The non-ferrous metals scrap resources were very inadequate and the production of non-ferrous metals was largely dependent on non-ferrous metal ore resources, so resource efficiency for non-ferrous metals industry declined slightly. But we can expect that percentage of the non-ferrous metals scrap resources to the production of the non-ferrous metals will rise gradually under the circumstances of the non-ferrous metals output remaining basically stable or declining slowly in the future. And resource efficiency for the non-ferrous metals industry will be improved gradually. However, resource economic efficiency for the non-ferrous metals industry was always rising during the period. The main reason was that although the production for the non-ferrous metals was largely dependent on ore resources, the increasing range of non-ferrous metals market price was also great and this made the non-ferrous metals industrial added value increase much more, therefore the resource economic efficiency was improved greatly.

Energy efficiency was on the rise in the whole during the period 1995-2009 although it had some twists and turns. At the same time, energy economic efficiency was basically escalating. This showed that energy utilization efficiency for the non-ferrous metals industry was improved gradually, which was closely related to the non-ferrous metals enterprises have taken the measures of implementing cleaner production, developing circular economy, etc. in recent years. Non-ferrous metals enterprises realized the cascade utilization and recycling of energy and reduced the loss and waste of energy. We can foresee that the energy (economic) efficiency will be further improved in the future.

Based on the results of bulk materials flow analysis for the non-ferrous metal system of China, it seems D_{DEU} got the higher value during this period due to the shortage of domestic non-ferrous ore output and the dependence on imported resources.

We should further improve the recovery rate of the non-ferrous metals ore mining and milling, the comprehensive utilization rate of mineral resources and reduce the loss of non-ferrous metals ore resources. And we should also enhance the utilization of the non-ferrous metals ore resources abroad. At the same time, we should enhance the recovery and utilization of metal scrap for the non-ferrous metals industry and improve the recovery rate of non-ferrous metals scrap as far as possible. It will also reduce the consumption of mineral resources and the energy consumption in the production of non-ferrous metals. Energy efficiency should be further improve by taking the measures of energy cascade utilization and recycling and reduce carbon dioxide emission, sulfur dioxide, smoke and dust, etc.

6. Conclusions

(1) Bulk materials flow analysis method can be applied to the analysis of the ecological efficiency for the non-ferrous metals industry. By using this method, we can understand the resource efficiency, the energy efficiency and the decoupling for the non-ferrous metals industry. Effective measures and suggestions can be put forward to improve the ecological efficiency for the non-ferrous metals industry.

(2) In 1995-2009, resource efficiency for the non-ferrous metals industry of China reached its peak value in 2005 (0.0885t non-ferrous metal/t ore) and resource economic efficiency reached its peak value in 2008 (1878.96 yuan RMB/t ore). Energy efficiency was generally in an ascendant trend and reached its peak value in 2009 (0.2129t non-ferrous metal/tce). Energy economic efficiency basically escalated during the period and reached its peak value in 2008 (4834.17 yuan RMB/tce). Due to the impact of the financial crisis, the resource (energy) economic efficiency declined in 2009.

(3) The value of D_{DEU} was 0.720 during the period 1999-2008, which was bigger than that of $\mathrm{D}_{\mathrm{DPO}}$, D_{DMI} and D_{TMR}. It seemed D_{DEU} got the higher value during this period due to the shortage of domestic non-ferrous ore output and the dependence on imported resources.

Acknowledgements

This research was supported by the National Natural Science Foundation of China (71373003, 41401636, 71003018) and the Consulting Project of Chinese Academy of Engineering (2015-XY-11).

References

1. National Bureau of Statistics of China. 1996-2010. China Statistical Yearbook. Beijing: China Statistics Press.
2. Editorial board of China non-ferrous metals industry yearbook. 1996-2010. China Non-ferrous Metals Industry Yearbook. Beijing: China Non-ferrous Metals Industry Yearbook Press.
3. National Bureau of Statics Division of Energy. 1996-2010. National Bureau of Energy Statistics of China. Beijing: China Statistics Press.
4. Vogtlander, J. G., Bijma, A., Brezet, H. C. 2002. Communicating the eco-efficiency of products and services by means of the eco-costs/value model. *Journal of Cleaner Production* 10(1): 57-67.
5. Yin, K., Wang, R.S., Zhou, C.B., Liang, J. 2012. Review of eco-efficiency accounting method and its applications. *Acta Ecologica Sinica* 32(11): 3595-3605.

6. Mao, J.S., Zeng, R., Du, Y.C., Jiang, P. 2010. Eco-efficiency of industry sectors for China. *Environmental Science* 31(11): 2788-2794.
7. International Council on Mining & Metals. 2014. Materials stewardship: Eco-efficiency and product policy. https://www.icmm.com/document/19.
8. Liu, J.X., Wang, Q., Sun, P., Gu, X.W., Li, G.J. 2007. Ecological footprint of China's non-ferrous metal industries. *Resources Science* 29(1): 155-159.
9. Hilson, G. 1999. A proposed framework of eco-Efficiency for the nonferrous metal mining sector. *Journal of Environmental Systems* 27(3): 191-208.
10. Wu, C.G. 2007. Leading indicators for sustainable development of nonferrous metal industry in Britain. *Land and re*sources information 8: 19-22.
11. Eurostat. 2009. Economy-wide material flow accounts: Compilation guidelines for reporting to the 2009 Eurostat questionnaire (Version 01, June 2009). European Statistical Office, Luxembourg, 11-20.
12. Wang, H. M., Hashimoto, S., Moriguchi, Y., Yue, Q., Lu, Z. W. 2012. Resource use in growing China: Past trends, influence factors, and future demand. *Journal of Industrial Ecology* 16(4):481-492.
13. Dai, J. & Chen, B. 2010. Materials flows analysis of fossil fuels in China during 2000-2007. *Procedia Environmental Sciences* 2:1818-1826.
14. Moriguchi, Y. 2007. Material flow indicators to measure progress toward a sound material-cycle society. *Journal of Material Cycles and Waste Management* 9(2):112-120.
15. Schandl, H. & Turner, G. M. 2009. The dematerialization potential of the Australian economy. *Journal of Industrial Ecology* 13(6):863-880.
16. Wang, H.M. 2011. Bulk- Materials flow analysis for China. Shenyang: Doctoral dissertation of Northeastern University.
17. Gou, H.S. & Zuo, L. 2004. Discuss on Status and Developmental Countermeasure of China's Nonferrous Industry. *Non-ferrous metals (extractive metallurgy)* (6): 2-4.
18. Lu, Z.W. & Yue, Q. 2010. I_eGTX Equation and I_eGT_e Equation: Analysis of SO_2 and COD emissions during economic growth in China. *Research of Environmental Sciences* 23(9): 1101-1107.
19. Wang, H. M., Hashimoto, S., Yue, Q., Moriguchi, Y., Lu, Z. W. 2013. Decoupling Analysis of Four Selected Countries: China, Russia, Japan, and the United States during 2000–2007. *Journal of Industrial Ecology* 17(4): 618-629.

The sedimentary evolution characteristics of black shale in Chongqing, SW China

Y. Zhu[1,2,3,4,*], D.H. Li[1,2,3], S.R. Li[1,2,3,4], Z.G. Zhou[4] and C.G. Chen[1,2,3,4]

[1]*Key Laboratory for Shale Gas Resource & Exploration, Ministry of Land and Resources, Chongqing Institute of Geology and Mineral Resources, 400042 Chongqing, China*

[2]*Chongqing Engineering Research Center for Shale Gas Resource & Exploration, Chongqing Institute of Geology and Mineral Resources, 400042 Chongqing, China.*

[3]*Chongqing Shale Gas Research Center of State Key Laboratory of Petroleum Resource and Prospecting, 400042 Chongqing, China*

[4]*Schlumberger Zhongyu Shale Gas Technical Services (Chongqing) Co., Ltd., 400043 Chongqing, China*

**E-mail: fishzy68@163.com*

In this paper, using strata combination, lithology, fossil, marker bed, formation thickness and geochemical indicators to divide black shale into two third-order sequences (SQ1 and SQ2), and each sequence is further divided into two system tracts, TST and HST. Through analyzing sedimentary facies and the distribution, combining well tie profiles, black shale distribution and geochemical elements, the results show that Lower Cambrian black shale are neritic deep shelf deposit, and shallow shelf deposit as well as arenaceous shelf deposit.

Keywords: Shale; Sedimentary evolution; Chongqing.

1. Introduction

The area of Chongqing is in the joint part of thrust-fold belt of Sichuan Mesozoic foreland basin and southern margin of Yangtze landmass, structurally forming asymmetric anticline and syncline which are dense in EW and parallel in general.

The black shale in this study area developed in continental margin, intracratonic depression and the sedimentary environment of foreland flexural basin, which is the shale gas accumulation horizon with deep burial and big changes of lithofacies combination.

2. Regional Geological Background

The exploration area located in intracontinental miogeosyncline and northern marginal miogeosyncline environment where the lacunae forming after the Yangtze cratonic uplift and earthcrust rupture.

In late proterozoic and early neoproterozoic, because of Jinning movement and the regional dynamometamorphism in the same period, all the rock stratums became transmutative and metamorphous, forming holomorphic folds, single-phase metamorphic rock represented by killas and graywacke, constituted the regional fold basement.

3. Sedimentary Environment

Marine organic rich shale formed in continental shelf setting, of which deposition rate is fast, geological condition is blocked and organic matter is prolific, usually associated with large-scale transgression process. In the whole Yangtze region, there happened a biggest transgression in early Paleozoic, basically formed a unity sea area in Cambrian, thus entered a relatively stable platform development stage.

Profile L (Figure 1) is located in the eastern part of the study area, basically represents the overall development characteristics of the strata in the area. The sedimentary environment of the early Cambrian in this study area is turbulent, regression and transgression were frequent, so the change of lithofacies combination is larger.

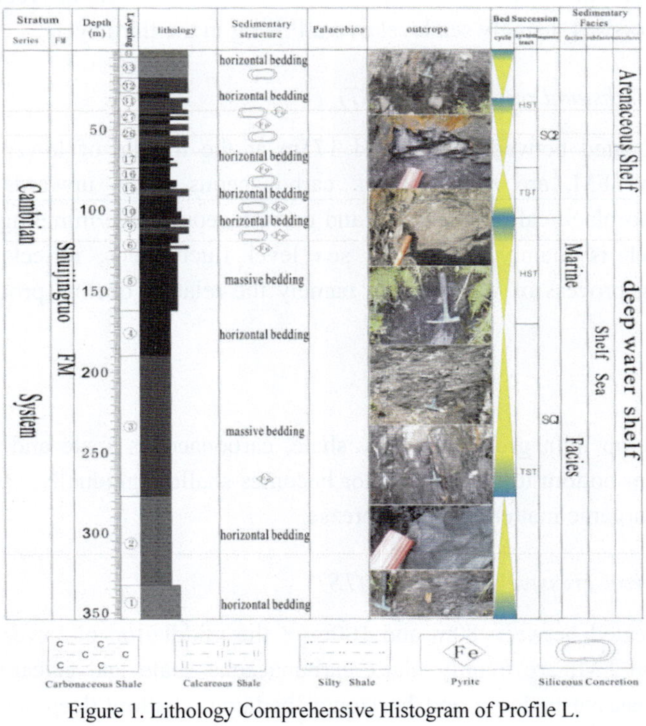

Figure 1. Lithology Comprehensive Histogram of Profile L.

According to strata combination, lithology, fossil, marker bed, formation thickness and geochemical indicators, the Shuijingtuo FM is divided into two third-order sequences (SQ1 and SQ2), and each sequence is further divided into two system tracts, TST and HST.

3.1. *SQ1*

The lithology at bottom are mainly calcareous shale and carbonaceous shale which belong to deep shelf facies, the top are mainly carbonaceous shale, calcareous shale mingling with thin silty shale. From bottom to top, lithofacies present complete cycle variation characteristics.

3.1.1. *Transgressive system tract (TST)*

TST is located between 175m and 355m at the bottom of lower Cambrian Shuijingtuo FM, are mainly black thin-thick calcareous shale and carbonaceous shale, develop massive bedding and horizontal bedding, a set of calcareous rock deposits at the bottom. TST is a set of black mud shale deposit under deep-water environment, of which the increase speed of accommodation space is slightly bigger than the sediment supply rate, forming a set of retrogradation parasequence set, namely sea level overall rising in the third cycle.

3.1.2. *Highstand system tract (HST)*

HST is located between 100m and 175m at the middle of lower Cambrian Shuijingtuo FM, are mainly black carbonaceous shale, upwards gradually change into black calcareous shale and carbonaceous shale mingling with silty shale. HST is mainly caused by sea level fluctuations, reflects the slow shallowing process of water depth, namely the relative decline process of sea level.

3.2. *SQ2*

Most develop light grey-black silty shale, carbonaceous shale and calcareous shale. From bottom to top, rock color becomes shallow gradually, sand content is higher, organic matter content decrease.

3.2.1. *Transgressive system tract (TST)*

TST is located between 80m and 100m at the middle-top of lower Cambrian Shuijingtuo FM, are mainly black carbonaceous shale and calcareous shale, develop massive bedding and horizontal bedding, shallow deep shelf deposit.

TST is a set of black organic rich shale deposit under deep-water environment, of which the increase speed of accommodation space is slightly bigger than the sediment supply rate, forming a set of retrogradation parasequence set, namely a brief dropping under the background of sea level overall rising in the third-order cycle, foreseeing that sea level rising is "step" or cyclical.

3.2.2. *Highstand system tract (HST)*

HST is located between 0m and 80m at the top of lower Cambrian Shuijingtuo FM, are mainly black thin carbonaceous shale and silty shale, upwards gradually change into beige-dark grey silty shale, siltstone content gradually increase, the color becomes shallow, organic matter content decrease at the same time. It can be seen obvious progradation sequence from stratigraphic profiles, reflected the slow shallowing process of water depth, namely the relative decline process of sea level.

4. Sedimentary Facies

Through the analysis of the profile sedimentary facies, Lower Cambrian Shuijingtuo Formation in Northeast area of Chongqing are mainly a set of deep shelf deposit, shallow shelf deposit and sandy shelf deposit and these three facies has distinct characteristics in color, lithology, rock structure and total organic carbon content.

4.1. *Deep shelf facies*

Deep-water shelf deposit in Northeast of Chongqing area mainly developed in middle-early Shuijingtuo, lithology are mainly black carbonaceous shale and calcareous shale (Figure 2), a set of siliceous rocks deposition at the bottom (Figure 3), mingling with thin siltstone and silty mudstone. Meanwhile the strata accidentally developed silty and siliceous rocks concretion (Figure 4) which can be used as the markers of deep-water shelf facies in this area. Rock composition of this black shale mainly include quartz, clay mineral, a small amount of feldspar and carbonatite, in addition, developing pyrite generally, distributing sporadically (Figure 5), occasionally developing the schistose concretion (Figure 6). Shuijingtuo black shale mainly developed block massive bedding and horizontal bedding (Figure 7).

Figure 2. Black carbonaceous shale at bottom of profile D.

Figure 3. Black siliceous shale at bottom of profile D.

Figure 4. Small sandy rock concretion from profile D.

Figure 5. Sandy rock concretion at bottom of profile L.

Figure 6. Pyrite concretion of profile L.

Figure 7. Dispergated pyrite at bottom of profile L.

4.2. *Shallow shelf facies*

Shallow-water shelf deposit in this area mainly developed in late Shuijingtuo, lithology are mainly light gray-gray silty shale and siliceous shale (Figure 8), develop a set of light gray dolomitic limestone on the top. Rock composition mainly include quartz, clay mineral, a small amount of feldspar and carbonatite, silty content is relatively high, rock color becomes shallow, organic matter content decreased significantly, mainly develop current bedding and cross bedding (Figure 9).

Figure 8. Grey brown silty shale of profile L.

Figure 9. Current bedding at middle of profile D.

4.3. *Sandy shelf facies*

Sandy shelf facies deposit mainly developed in middle-late Shuijingtuo, lithology are mainly gray-dark gray medium-thick silty shale or argillaceous siltstone (Figure 10), develop a set of light gray dolomitic limestone on the top. Sedimentary structure development is relatively limited, mainly developing current bedding, small cross bedding and corrugated bedding (Figure 11).

Figure 10. Grey silty shale of profile D.

Figure 11. Shallow gray corrugated bedding of profile D.

Based on completing sedimentary division of detailed profiles, from west to east choosing profile D, profile L and profile P (Figure 13) to analyze, it can be seen that this shale in Northeast of Chongqing overall relatively stably developed and large thickness. The area around of profile D, of which the thickness of organic rich shale is large and is the deposition center of the period, the thickness of the black shale is also the most developed. Eastern profile P relatively developed organic rich shale, and is the center of sub-deposition.

5. Distribution Characteristics of Sedimentary Facies

5.1. *Geochemical elements*

Modern sediments and ancient geochemical characteristics of rock stratum can be used to distinguish the sedimentary environment. Measuring organic carbon content and sulfur content of pyrite, and using correlation of organic carbon-sulfide (mainly pyrite) to distinguish the forming environment [1,2]. In the anoxic marine environment (euxinic environment), organic matter, no experiencing oxidation and decomposition, soon sank down to the bottom of waters which are anoxic and H2S containing, thus it is beneficial to the preservation of organic matter and participate in reduction.

In this kind of sedimentary environment, pyrite can be directly formed in reaction process of H2S between the interface of terrigenous iron and water-sediment [3]. Therefore, pyrite which developed in the anoxic marine sediment are the product of contemporaneous effect, the quantity of pyrite are controlled by iron mineral content [4]. Therefore, the organic carbon content and pyrite content are no obvious correlation. TOC and pyrite content of Lower palaeozoic organic rich shale in this study area are no obvious correlation (Figure 12), therefore it can be considered to an anoxic marine sedimentary environment at the time.

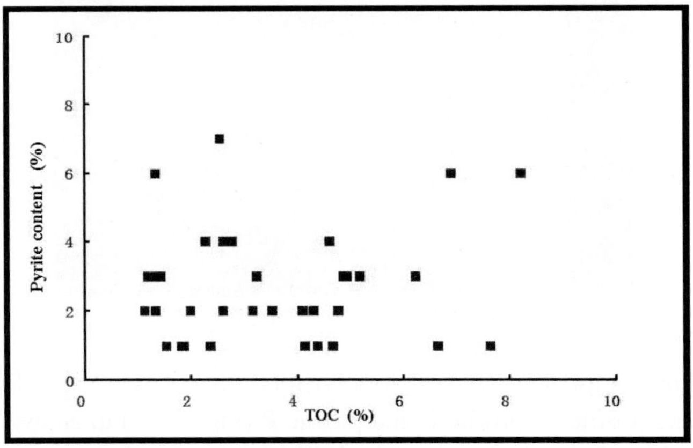

Figure 12. Correlation of Pyrite Content and Organic Carbon Content.

5.2. *Distribution of sedimentary facies*

Through various systems tracts of three detailed profiles including thickness of silty shale, formation thickness, thickness of black mudstone and grey mudstone, combined with all kinds of facies marks and the statistical analysis results to make analysis and classification for distribution of sedimentary facies in the plane.

From contour map of strata thickness, combined with distribution of sedimentary system in the plane, it can be seen that the facies types in that period are mainly shallow deep-water shelf facies, shallow shelf facies and sandy shelf facies. Around north of profile D region are deposition and subsidence center of black shale, the thickness of the organic rich shale is relatively the largest, and in the east, organic rich shale of profile P is also relatively development and is the center of sub-deposition (Figure 13).

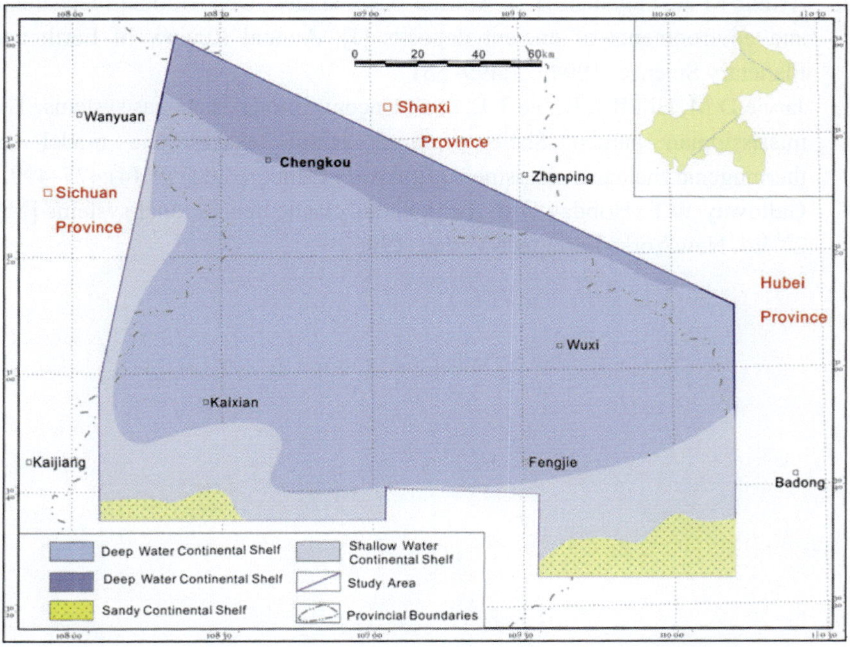

Figure 13. Diagram of Organic Rich Shale Sedimentary Facies in Lower Cambrian Shuijingtuo Formation.

6. Conclusions

Based on comprehensive research and analysis, it concludes that north and east are two black shale depocenters which are in EN-WS distribution, stable and continuous. The thickness is between 500m and 700m, most of it is more than 600m. Continental shelf facies of Lower Cambrian Shuijingtuo is good for development of black shale and has the characteristics of high TOC, fine grain, deep colour, pure lithology, and large thickness. Therefore, there is a wide prospect for shale gas exploration in this area.

Acknowledgements

We would like to thank Chongqing Institute of Geology & Mineral Resources and Schlumberger Zhongyu Shale Gas Technical Services (Chongqing) Co., Ltd. for financial and technical support.

References

1. Morford J L, Emerson S. The geochemistry of redox sensitive trace metals in sediments [J]. Geochim. Cosmochim. Acta, 1999, 63:1735–1750.
2. Arthur M A, Sageman B B. Marine black shales: Depositional mechanism and environments of ancient deposits [J]. Annual Review of Earth and Planetary Science, 1994, 22:499–551
3. Jarvie D M, Hill R J, Ruble T E, et al. Unconventional shale gas systems: The mississippian barnett Shale of north-central texasas one model for thermogenic shale gas assessment [J]. AAPG Bulletin, 2007, 91(4):475–499.
4. Galloway W E, Hobday D K. Terrigenous clastic depositional systems [M]. 2nd Ed. New York: Springger-Verlag, 1996.

Application of comparable transaction method in mineral asset valuation

Z. Yang
China University of Geosciences, Beijing, China
E-mail: charlie.z.yang@live.com

This paper reviewed market approach in mineral asset valuation and examined the most widely used and best understood methodology under the market approach – Comparable Transaction method. The importance and uniqueness of Comparable Transaction method were identified in this article. This paper further discussed important procedures and characteristics of applying comparable transaction method. To a certain extent, subjectivity presences in all mineral asset valuation using Comparable Transaction method. Accuracy should not outweigh valuers' judgments and Comparable Transaction method remains a popular method in valuing mineral assets across the world.

Keywords: Mineral valuation; Market approach; Comparable Transactions; VALMIN; CIMVAL.

1. Valuation Approaches

1.1. *Mineral asset valuation*

Mineral valuation is the process of determining the monetary value of a mineral asset. As with any other business valuation, mineral valuation is a difficult task. While general business sense and industry knowledge is required in conducting any valuation, mineral valuation demands much more than that. An in-depth understanding of exploration and mining operations is just as important as sophistication in valuation techniques when it comes to mineral valuation.

1.2. *Three main valuation approaches*

For any valuation subject, there are three generally accepted valuation approaches, namely Cost Approach, Market Approach, and Income Approach.

The *Income Approach* is based on the principle of anticipation of benefits and includes all methods that are based on the income or cash flow generation potential of the mineral asset.

The *Market Approach* is based primarily on the principle of substitution. The subject mineral asset valued is compared with the transaction value of similar mineral assets or projects, transacted in an open market.

The *Cost Approach* is based on the principle of contribution to value and usually involve analysis of exploration expenditures for their contribution to the exploration potential of the mineral assets.

Basic valuation techniques and all three approaches apply to the valuation of a mineral asset, yet more specialised and advanced "tricks" are required in order to complete a mineral valuation assignment. General valuation approaches and methods also apply to mineral valuation with some modifications to key parameters to make it specific for the nature of mineral assets. Each approach has two widely used and accepted methods while other methods are rejected or considered secondary.

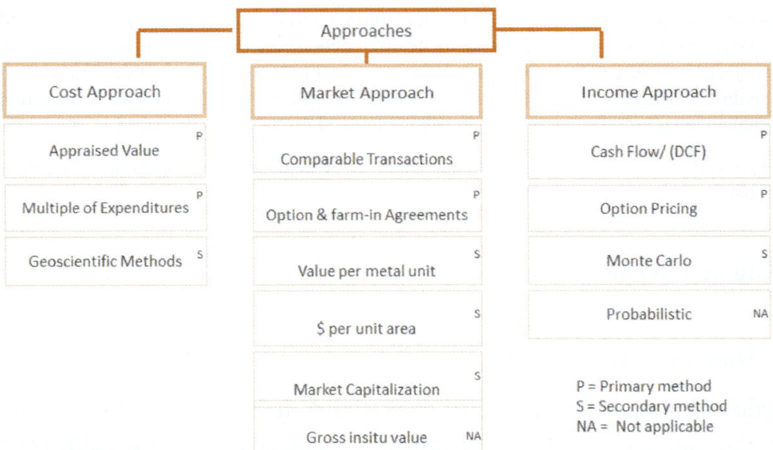

Figure 1. Valuation methodologies under each approach recommended in CIMVAL standard.

1.3. *Methodologies under Market Approach*

As identified in Figure 1, there are six common methodologies under the market approach. Comparable transactions is the most commonly used method in the industry and is conceptually easy to understand and mathematically uncomplicated to execute. Lacking comparable transactions and information of each transaction can make the adoption of this method difficult and in some cases inapplicable. Option and farm-in agreements on the other hand are not as intuitive and can be difficult to compute. These two methods are considered the primary methods under the Market Approach. Value per metal unit, value per unit area, and market capitalisation methods are based on similar concepts as price multiple and are often considered as secondary methods. Although simple in nature and computation, gross in-situ value method is considered too simple to reflect important differences of mineral assets and lacking fundamental theoretical basis. Such method is generally not allowed in valuation.

2. Market Approach

2.1. *Market opinion*

Having said the above, similarity in valuation methods does not mean mineral valuations are identical to other valuations. In fact, several differences can be identified. At the present, most mineral valuations conducted by industry practitioners are based on Net Present Value of cash flows or Discounted Cash Flow method of the Income Approach and most mineral valuations only provide an economic NPV or Investment Value based on specific investment parameters. This creates a problem. Lack of valuation conclusion based on inputs derived from markets and the absence of an estimated Market Value and solely relying on estimation from Income approach may result in misleading of the true value of the mineral asset and hence making the mineral asset illiquid at unrealistic price. Hence, the application of market approach on mineral valuation is reviewed by comparing and contrasting several unique advantages of the approach in the valuation of mineral assets.

2.2. *Selection Decision*

One of the reasons that market approach should be deployed in mineral valuation whenever applicable is that the selection of applicable approaches and methods in mineral valuation often depends on the development of the subject mineral asset. Most of the mineral assets can be categorised into one of the four broader development stages described below.

– **Exploration Stage**: mineral asset that has been acquired, or is being explored, for mineral deposits but for which economic viability has not been demonstrated and mineralization may or may not have been identified.

– **Mineral Resource Stage**: advanced stage exploration properties with identified mineral resources but not yet demonstrated to be economically viable by a feasibility study.

– **Development Stage**: mineral asset that mineral resources have been identified and is being prepared for mineral production and for which economic viability has been demonstrated by a feasibility study or prefeasibility study, including those that are not yet financed or under construction.

– **Production Stage**: a mineral asset is said to be in production stage when it is with an operating mine, with or without processing plant, which has been fully commissioned and is in production.

Mineral Resources Stage is generally considered equivalent to Advanced Exploration Stages sets out in Australian VALMIN in definition and description. Australian VALMIN also subdivides Development Stage into Pre-Development Stage and Development Stage depending on whether a decision to proceed with development and construction has been made or not. The general concept remains the same.

From the below chart, market approach is applicable throughout the entire life span of a mineral asset. The importance of market approach is already addressed above.

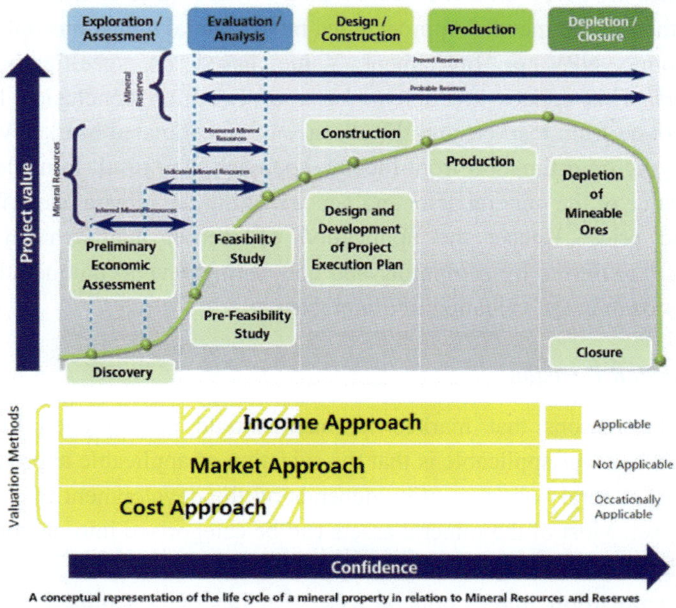

Figure 2. Valuation approaches against development stages of a mineral asset.

2.3. *Market data is everywhere*

To a certain extent, market information is used or employed in all three approaches and discount rate in the DCF method in the Income Approach comes from market information, although not transaction based, information from publicly listed company forms part of market information. Multiples adopted in the Cost Approach involve market data too. Without relying on market information, no valuation can be reasonably justified. However, screening of many comparable transactions to find several ideal comparables is challenging, in-depth knowledge in the mining industry and the good understanding of mineral valuation of a valuer will always be demanded.

2.4. *Valuation parameters*

Unlike valuation on other business or assets where various price multiples derived from accounting data (earning, book value, or sales revenue) are considered to be the comparing parameter in comparing transactions, dollar per unit of metal or reserve or resource is often the most widely adopted. Under certain circumstances or junior mineral asset without resource data may be estimated using dollar per unit of area method with big assumptions. Conversion between different levels of resources and/or reserves is commonly applied to ensure apples are compared to apple. Some local authorities may restrict the use of certain resource into the valuation. For example, SEHK forbids the use of inferred resources to be included in mineral valuation.

2.5. *Large adjustment*

Factor adjustment is a key process in carrying out comparable transaction valuation of mineral assets. From time to time, there are valuations where small adjustments (10% to 30%) are adopted which are only appropriate for certain other valuation such as the valuation of real estate properties. However, large value adjustments (sometimes > 100%) are necessary for mineral assets comparisons. Typical adjustment factors include tonnage, grade, and risk. Adjustments may be offsetting or cumulative depending on the direction of the adjustment (positive upwards or negative downwards). Total adjustments may be moderate or significant (sometimes greater than 10-fold).

3. Mineral Valuation, Science or Arts

3.1. *Quantitative and qualitative adjustments*

Common adjustment factors to reflect the difference in two comparable transactions include:

- Time and Price adjustment
- Development Status adjustment
- Reserves vs. Resources balance adjustment
- Deposit/project size adjustment
- Deposit Grade adjustment

– Open Pit vs. Underground Mining adjustment
– Operating Cost (including energy price factors) adjustment
– Country Risk adjustment
– Other Risk adjustment, including Project Specific Risk, Taxes, Royalties and Levies Difference, Infrastructure and other significant CAPEX Requirement.

Some of them are common to valuations of other business or assets while others are specific to mineral valuation.

It is not difficult to notice that some of those adjustments are based on quantitative measurements while the others are qualitative.

Time and price adjustment is a typical example of quantitative adjustment. It adjusts for change in the price of underlying metal/mineral. Such adjustment is usually arrived by finding the ratio of the price of the underlying mineral at the time of the transaction to that at the effective date of valuation. These quantitative adjustments can be relatively objective and less likely to be manipulated.

On the other hand, project specific risk or deposit grade adjustment are examples of qualitative adjustment and are more likely to be subjective to each valuer and more likely to face the risk of manipulation. Ambiguities occur both in determining whether an adjustment is needed (e.g. what is considered to be a specific risk event) and what is the correct size of adjustment (e.g. how much adjustment for such specific risk).

Certain adjustments such as country risk adjustment can be treated as either quantitative or qualitative depending on how it is defined and measured. Some valuations use country risk premium as a proxy for country risk and determine the size of the adjustment quantitatively while others use a more qualitative measure by comparing it to that of a reference country.

3.2. *Key size measurement*

When applying comparable transactions, a value per unit of resource or reserve or metal must be calculated as the key measurement of the size of each mineral asset and compare it to that of the valuation subject to draw valuation result. The selection depends on the circumstance of both the valuation subject and the underlying mineral assets of the comparable transactions and availability of information.

Ideally, the most appropriate measurement should be used consistently but valuers tend to find it too difficult due to the lack of available information, especially for those private transactions. The amount of resources and reserve in each category may not be disclosed publicly or are estimated and reported according to different standard (think of comparing an Australian project with JORC-compliant reserves and inferred resources and a Chinese project with 332 and 333 resources). As a result, resources and reserves are sometimes converted to ensure apple-to-apple comparison. Such conversions are highly subjective depending on a valuer's experience and knowledge of both the reporting standards and the mineral assets.

3.3. *Value range and preferred value*

There are two distinct voices in forming valuation conclusion. One seeks a definitive value to be determined to minimise ambiguity and subjectivity while the other requires a value range to be formed (and may ask to identify a preferred value within this value range) to address an important nature of valuation – a valuation is a process of best estimation and uncertainties will always present. Regardless of the approach, there will always be a certain degree of subjectivity present when drawing value conclusions and it is not always as simple as a rigorous scientific calculation.

4. Conclusion and Recommendation

In general, there is no generic valuation approach that can be applied to all mineral assets. CIMVAL recommends that development stage of a mineral asset must be considered while choosing valuation approach and adopt primary methods under each approach whenever applicable. Practically, Comparable Transaction method remains a popular method in valuing mineral assets across the world. In addition, all valuation approaches must be considered and used as many approaches and methodologies as possible to ensure an unbiased estimation of the value of the mineral asset is determined. At a minimum, more than one valuation methods (may be under different approaches) must be used and reconciliated to arrive at the value conclusion. Unique circumstance of the subject mineral asset must be reflected in both the valuation approaches or methodologies and the valuation process. Subjectivity and professional judgment is required along the process. The valuers have the best knowledge of the mineral asset and are in the best position to express their opinion on it.

References

1. Special Committee of CIMVAL. 2003. *The CIMVAL Code, Standards and Guidelines For Valuation of Mineral Properties.* Canada.

2. The VALMIN Committee. 2005. *Code for the Technical Assessment and Valuation of Mineral and Petroleum Assets and Securities for Independent Expert Reports - The VALMIN Code*. Australia.
3. Rudenno, V. (3rd Edition) 2009. *The Mining Valuation Handbook – Mining and Energy Valuation for Investors and Management.* Sydney: Wrightbooks. 69–72.
4. Runge, I. 1998. *Mining Economics and Strategy*. Littleton: Society for Mining, Metallurgy and Exploration, Inc. 37–40.
5. Torries, F. (1st Edition) 1998. *Evaluating Mineral Projects: Applications and Misconceptions*. Littleton: Society for Mining, Metallurgy and Exploration, Inc. 14–15.
6. Smith, G. V. & Parr, R. L. (1st Edition) 2007. *Intellectual Property: Valuation, Exploitation, and Infringement Damages*. London: Wiley. 50–65.
7. Wellmer, F., Dalheimer, M. & Wagner, M. 2008. *Economic Evaluations in Exploration*. New York: Springer. 89–100.
8. Pratt, S. P., Reilly, R. F. & Schweihs, R. P. 2000. *Valuing A Business – The Analysis And Appraisal of Closely Held Companies*. Chicago: McGraw-Hill. 26–33, 77–90, 210–256.
9. Lawrence, M. 2001. An Outline of Market-Based Approaches for Mineral Asset Valuation Best Practice. *VALMIN '01 – Sydney – Australia*. 1–6.
10. Lawrence, R. 2002. Valuation of mineral properties without mineral resources: A review of market-based approaches. *CIM Bulletin (vol. 95)*. 2–5.

Research on the influence of coal mining on aquifer and its prevention technology in Shanxi Province

Z.X. Zhang*, Y.B. Zhang, X.T. Fu, K. Wang, Y.L. Shi and X. Shen
Taiyuan University of Technology, Taiyuan, Shanxi 030024, China
**E-mail: zhangzx666666@126.com*

To ensure safety production of coal resources and effective protection of aquifer, it is necessary to research the influence of coal mining on aquifer. Based on field investigation and collection of geology, hydrogeology etc. of coal mining area in Shanxi Province, the influence of coal mining on aquifer was analyzed from three aspects such as aquifer structure, groundwater resources and groundwater quality. It showed that (1) coal mining caused severe damages to aquifer structure; (2) coal mining caused continuous decrease of groundwater resources in aquifer; (3) coal mining caused serious pollution of groundwater quality. Combing with coal mining status, geological and hydrogeological conditions of coal mining area in Shanxi, this paper put forward prevention technologies for aquifer damage control under the condition of coal mining. The results provide important reference values for sustainable exploitation of coal resources and aquifer protection of coal mining area in Shanxi Province.

Keywords: Coal Mining; Aquifer Damage; Groundwater Pollution; Goaf; Prevention Technology.

1. Introduction

China is a country with severe water shortage, years of economic development not only caused the decrease of water resources in most of the region, but also water resources suffered all kinds of pollution. Especially in coal mining area, this kind of phenomenon is more outstanding under the condition of coal mining[1]. Coal resources is the basic energy of China, it has an important strategic position in the development of economy and society. The large-scale mining of coal resources, on the one hand pushed the continuous and stable development of the national economy, on the other hand caused serious damage of the natural occurrence condition of groundwater in China, and it intensified the contradiction between supply and demand of regional water resources[2]. According to statistics, there are 70% water shortages in the national 86 key state-owned coal mining area, and 40% belong to the serious shortage of water. With the further increase of the intensity and the depth of coal mining, the influence on groundwater in coal mining area will become increasingly worse.

Nearly 30 years, the influence of coal mining on groundwater received great attention by scholars both at home and abroad. Foreign scholars such as Stoner J D[3], Lines G C[4], Booth C J[5], et al researched the hydrogeological effect caused by coal mining. Domestic scholars such as Han Baoping[6], Zhang Fawang[7], Wu Qiang[8], Li Tao[9], et al studied the influence of coal mining on groundwater from different point of view. Due to the differences of geological, hydrogeological conditions and coal seam occurrence status in various regions, the study on the influence of coal mining on groundwater is far from comprehensive, especially in the influence of coal mining on aquifer. In most coal mining area of China, coal mining caused aquifer damage, which not only increases the cost of coal production, but also causes groundwater waste. In recent years, the technology of coal mining safety and prevention of water disasters attracts high attention by Chinese scholars. Nan Shiqing[10] et al studied safety mining technology on ore body adjacent to water-bearing crushed zone and fault, and formulate mining technology scheme guarantying safety in production; Liu Li[11] et al pointed out that green mines should have the characteristics of operation in accordance with the law and so on; Shi Xianxin[12] et al researched prevention measures against water and sand inrush and their application in Shendong mining area. The results of these studies are of high practical value, and have references for safe production of coal enterprise in China.

Shanxi Province is China's important production base of coal energy, the influence of coal mining on aquifer in coal mines become more and more serious for many years, it not only threatens the safety of coal mine production, but also contributes to the trend of water shortage. To ensure safety production of coal resources and effective protection of aquifer, it is necessary to research the influence of coal mining on aquifer in Shanxi Province. Based on field investigation and collection of geology, hydrogeology etc. of coal mining area in Shanxi, the influence of coal mining on aquifer was analyzed. Combing with coal mining status, geological and hydrogeological conditions of coal mining area in Shanxi, prevention technologies were put forward for aquifer damage control under the condition of coal mining.

2. Background Materials and Methods

2.1. *General situation of the study region*

Shanxi Province is an inland Province of China, it is located between 110°14′–114°33′ E and 34°34′–40°43′ N. The length from north to south is 680 km, the width from east to west is 380 km with a total area of 156271 km^2. Shanxi is northeast high-lying, southwest of the low, complex terrain. It is a temperate continental monsoon climate, and the mean annual precipitation is 500 mm. The rivers belong to Yellow River basin and Haihe River basin. Coal resources is extremely rich in Shanxi, there are six coalfields such as Datong, Ningwu, Xishan, Qinshui, Huoxi, Hedong from north to south.

2.2. *Geological conditions*

The outcropped strata in Shanxi coalfield are mainly Precambrian, Cambrian, Ordovician, Carboniferous, Permian and Triassic, Datong coalfield also has Jurassic and cretaceous, the overlying strata are Eogene, Neogene and Quaternary. The main coal bearing strata are mainly Taiyuan group and Shanxi group of Carboniferous, followed by Datong group of Jurassic. Coal measure strata of Taiyuan group and Shanxi group are widely distributed in the provincial coalfields, and the workable coal seams are two to seven layers and one to three layers, respectively. Coal measure strata of Datong group is mainly distributed in several regions of Datong coalfield, the workable coal seams are one to eight layers. All coal seams are flat-gently inclined coal seams, and the dip angle is commonly 3° to 6°. Most of the buried depth of coal seam is below 300 to 500 metres. The geological structure in six coalfields of Shanxi Province is mainly syncline structure, only Hedong coalfield as monocline structure. In general, coalfield only has a small amount of rolling fold and some less dense fault or collapse column. On the whole, the geological structure in Shanxi coalfield is simple.

2.3. *Hydrogeological conditions*

The coal measure aquifer in northern coalfield of Shanxi is mainly sandstone fracture aquifer of Datong group, Shanxi group and Taiyuan group, the overlying aquifer is sandstone fracture aquifer of Yungang group and Shihezi group, the underlying aquifer is karst fissure aquifer of Ordovician. The coal measure aquifer in southern coalfield of Shanxi is mainly sandstone fracture aquifer of Shanxi group and Taiyuan group, the overlying aquifer is sandstone fracture aquifer of Shihezi group, the underlying aquifer is karst fissure aquifer of Ordovician. Groundwater recharge in Shanxi coalfield is mainly the precipitation recharge, river leakage recharge and lateral groundwater recharge. Groundwater runoff is primarily the laminar flow movement. Groundwater discharge is mainly spring discharge, lateral discharge and artificial discharge.

2.4. *Research methods*

By using comprehensive analysis methods of geology, hydrogeology, mining engineering, water chemical, and the influence of coal mining on aquifer in Shanxi Province was researched. Combing with previous theoretical studies and the practical conditions of coal mining area, this paper put forward prevention technologies for aquifer damage control under the condition of coal mining.

3. Influence of Coal Mining on Aquifer

3.1. *Severe damages to aquifer structure*

During construction period, the excavation of mine shaft exposed the upper aquifer and aquitard, formed crisscross transportation alleys and return air alleys in mine, and the aquifer structure was damaged to a certain degree. For the coal mining area, the influence on aquifer structure was local, and the damage was one-off.

During production period, with the coal mining working face goes forward constantly, the goaf area increased gradually, which led to the movement and damage of the overlying strata, the damage area can be divided into caving zone, fractured zone and bend zone from bottom to top, the caving zone and the fractured zone is known as water flowing fractured zone. When the water flowing fractured zone spread to the upper aquifer, it caused the damage of aquifer structure, increased the permeability of aquifer, and the groundwater infiltrated into the goaf, then the groundwater was discharged gradually as mine water for safety production. For underlying aquifer of coal seam, coal mining only had a certain disturbance to the floor aquitard, and had no damage to underlying aquifer structure. In this period, the influence on aquifer structure was widespread, the damage was continuity.

After service period, a large number of goaf was formed in the coal mining area, then the mine was abandoned, and the wellbore was blocked. In the initial stage, the groundwater from the overlying aquifer still continued to the convergence of goaf, the rise of groundwater level was at a rapid speed. As time continued, the growth of the groundwater level became more and more small. Under the influence of groundwater and the residual subsidence of overlying rock of coal seam and so on, the subsidence of overlying rock of goaf was continuously, it further damaged the aquifer structure, and the damage didn't stop until the residual subsidence of overlying rock was over.

3.2. *Decrease of groundwater resources in aquifer*

Affected by the damage of overlying aquifer structure, groundwater movement in aquifer from horizontal to vertical, and the groundwater seeped into goaf and became mine water. In order to ensure safety production, many of the mine water were expelled. The mine water discharge increased the groundwater discharge of aquifer. Within the influence scope of coal mining on aquifer, groundwater depression cone was formed. Along with mine water discharge, the groundwater level continued to decline, groundwater depression cone was continually expanding, and it caused continuous decrease of groundwater resources in aquifer. In some coal mining area, the groundwater in overlying aquifer seeped into goaf through water transmitting fault, it reduced the groundwater resources of aquifer. According to the survey, the damage of overlying aquifer by coal mining caused many water supply wells dry in mountains, which led to drinking water difficulties in many areas.

In coal mining area of Shanxi, water yield property of Ordovician karst aquifer is better. When the coal seam floor is lower than the groundwater level of karst aquifer, the coal mining area is under pressure, and coal mining may be threatened by karst water. As the effective thickness of water-resisting layer is less than the damage thickness, water-inrush from coal floor may occur. In order to avoid water-inrush from coal floor and reduce the water pressure, a large amount of karst water were discharged, which caused the decline of karst water level. As the karst water was expelled, there caused continuous decrease of groundwater resources of underlying aquifer. According to the survey, due to large number of discharge of karst water in coal mines of Shanxi Province, there caused attenuation of karst spring flow that contact with karst aquifer, the most serious is Jinci spring of Taiyuan City, it was dried up in April 1994, so far no pouring.

3.3. *Pollution of groundwater quality*

Due to high sulfur content in coal seam and its surrounding rock in Shanxi coal mining area, when the coal seam was extracted, along with the increase of the air, the originally reductive environment turned into oxidation environment. In the process of infiltration, the groundwater of overlying aquifer contact with sulfide and air, after a series of oxidation, hydrolysis reaction, sulfide was oxidized into sulfuric acid, and formed the acid mine water. When acid mine water had neutralization reaction with carbonate minerals, alkaline mineral in the surrounding rock, there increased the dissolve of soluble minerals, and formed a highly mineralized mine water. In some coal mining area, during the process of mine water flow, a large amount of suspended solids such as coal and rock powder etc. were brought into mine water, and formed mine water of containing suspended solids, the mine water is mostly black-grey, the turbidity is quite high, and the PH value is neutral. In individual coal mining area, due to the enrichment of the element such as Cu, Pb etc. was higher, when heavy metals were leached by acid mine water, they were released from the rock and entered into the mine water, and formed the mine water containing heavy metal.

In addition, many untreated mine water were discharged directly to the surface or rivers, when the polluted water seeped into groundwater, it caused groundwater pollution of aquifer. When mine water in goaf infiltrated into the underlying aquifer, it caused water quality pollution of aquifer. In some coal mining area, due to the improper disposal of solid wastes, under the precipitation eluviation, the pollutants transferred into aquifer, and caused groundwater quality pollution. After the mine was abandoned, as groundwater level of goaf was rising, the adsorption, dissolve and exchange effect between groundwater and the surrounding rock were strengthened, and it caused the pollution of mine water quality continuously. When mine water level was higher than groundwater level of aquifer, the recharge of mine water caused water pollution of aquifer.

4. Prevention Technology

4.1. *Strengthen hydrogeological exploration*

Before coal mining, it should be comprehensively to use traditional methods such as hydrogeological exploration, hydrogeological drilling, hydrogeological prospecting, hydrogeological test, hydrogeological observation etc. and the new technology such as remote sensing, global satellite positioning, geographic information system etc., to find out the coal seam and its roof aquifer and floor water-resisting layer, then determine the water abundance of aquifer and water resisting property of water-resisting layer, and detect water bearing property and permeability of the main fault and the collapse column, to provide accurate information for the sustainable development and overall plan of coal mining area.

4.2. *Design safety waterproof coal pillar*

When coal mining is under the aquifer or near fault, it must grasp the hydrogeological and engineering geological conditions, mining method, mining height and roof control method of coal mine, to design safety waterproof coal pillar with certain width or height according to the national rules and regulations. This technology can result in the loss of partial coal resources, but through design safety waterproof coal pillar, it can effectively protect the integrity of aquifer structure, and prevent the destruction of groundwater and water bursting in mine.

4.3. *Mining with filling in goaf*

Mining with Filling in Goaf is the technology by using relevant materials to fill goaf, to prevent or mitigate goaf overburden deformation and failure, and achieve the purpose to protect the overlying aquifer structure. According to the actual situation of Shanxi coal mining area, to select sandstone, coal gangue, fly ash, slag, etc. as the filling material, and fill the goaf by use the way such as manpower, gravity, mechanical. In the process of coal mining, the filling quality of goaf must be ensured, and it can avoid the damage of aquifer structure and increase the degree of coal mine safety.

4.4. *Coal seam strip mining*

Coal seam strip mining is the technology to divide a certain size strip along the seam strike or direction within the scope of the mining, extract one coal seam strip, keep one coal seam strip, and the two is in alternate arrangement. Through the strip pillar to support overburden rock, it can reduce the damage of aquifer structure. For the coal mining area of Shanxi, the stripe width must be reasonable designed, it can ensure the stability of coal pillar. Although the coal production of strip mining is low, the overburden movement caused by coal mining is not obvious, and the influence of coal mining on aquifer structure is small.

4.5. *Overburden seperation grounting*

Overburden Separation Grouting is the technology by use the separation hole formed in the process of the overburden movement after coal seam mining, through the ground drilling, pour some kind of packing materials into the separation zone, control the degree of overburden subsidence, and reduce the damage degree of aquifer structure. Combining with the practice of coal mine in Shanxi Province, the fly ash, coal gangue powder from thermal power plant can be used as grouting materials. Before grouting, it must determine reasonable grouting horizon and arrange reasonable grouting hole. When grouting, it must choose suitable grouting process to insure the grouting quality.

4.6. *Floor water-resisting layer grounting*

Floor water-resisting layer grouting is the technology to fill water flowing fracture of coal floor water-resisting layer, reinforce the coal seam floor water-resisting layer, and control the influence of mine pressure and water pressure of water-resisting layer after mining, to increase water resistance ability of floor water-resisting layer, and cut off hydraulic connection between the aquifer and the coal seam. In order to effectively protect karst water resources, the cement can be used as grouting material. When grouting, the grouting hole must be layout according to the grout diffusion radius, this can ensures the grouting effect.

5. Conclusions

The influence of coal mining on aquifer in Shanxi was huge, it not only caused severe damages to aquifer structure, but also caused continuous decrease of groundwater resources and serious groundwater quality pollution of aquifer. Combining with the practical conditions of coal mine in Shanxi Province, this paper put forward some prevention technologies such as strengthen hydrogeological exploration degree, design safety waterproof coal pillar, mining with filling in goaf, coal seam strip mining, overburden separation grouting and floor water-resisting layer grouting for control the aquifer damage under the condition of coal mining. The results provide important reference values for the sustainable exploitation of coal resources and the aquifer protection of coal mining area in Shanxi Province.

Acknowledgements

This work is financially supported by the National Natural Science Foundation of China (51309173) and the Natural Science Foundation of Shanxi Province (2014011001-2). The authors would like to thank Doctor Li Jianwen of Taiyuan university of technology for providing helpful suggestions.

References

1. Zhao Zhihuai, Lu Yuanzhao, Lu Jiahe, et al. Regularity of Water Drainage in Mining in Shanxi Coal Mine, Journal of China Coal Society, Issue 5, 1995, pp. 556–560. (in Chinese)
2. Zeng Qingming, Shi Longqing. Analysis of the Influence of Coal Mining on Water Resources in Shandong Province and Its Countermeasures,

Journal of Shandong University of Science and Technology, Vol. 28, Issue 2, 2009, pp. 42–46. (in Chinese)
3. Stoner J D. Probable Hydrologic Effects of Subsurface Mining, Groundwater Monitoring Review, Vol. 5, Issue 1, 1983, pp. 51–57.
4. Lines G C. The Ground-water System and Possible Effects of Underground Coal Mining in the Trail Mountain Area, Central Utah, U. S. Geological Survey Water-Supply Paper, Vol. 22, Issue 59, 1985, pp. 5–30.
5. Booth C J. Strata-movement Concepts and the Hydrogeological Impact of Underground Coal Mining, Groundwater, Vol. 24, Issue 4, 1986, pp. 507–515.
6. Han Baoping, Zheng Shishu, Xie Kejun, et al. An Approach to the Hydrogeologic Effect Induced by Underground Coal Mining, Journal of China University of Mining & Technology, Vol. 23, Issue 3, 1994, pp. 70–77. (in Chinese)
7. Zhang Fawang, Li Duo, Zhao Hua. The Destruction of Groundwater Resources and Its Control under the Circumstance of Coal Mining, Journal of Hebei College of Geology, Vol. 19, Issue 2, 1996, pp. 115–119. (in Chinese)
8. Wu Qiang, Dong Donglin, Fu Yaojun, et al. Research on Water Pollution Induced by Coal Mining, Journal of China University of Mining & Technology, Vol. 31, Issue 1, 2002, pp. 19–22. (in Chinese)
9. Li Tao, Li Wenping, Chang Jinyuan, et al. Similar Simulation of Dynamic Dive Table with Approximate Shallow Buried Coal Mining in Northern Shaanxi Province, Journal of Coal Society, Vol. 36, Issue 5, 2011, pp.722–726. (in Chinese)
10. Nan Shiqing, Zhang Juanxia, Guo Xianzhang. Study of Safety Mining Technology on Ore Body adjacent to Water-bearing Crushed Zone and Fault, Disaster Advances, Vol. 3, Issue 4, 2010, pp. 428–431. (in Chinese)
11. Liu Li, Chen Liqun, Tang Jianxin. Present Situation and Future Prospects of Geologic Environment Issues in Mines in China, Disaster Advances, Vol. 3, Issue 4, 2010, pp. 563–566. (in Chinese)
12. Shi Xianxin, Li Lin. A Study on Prevention Measures against Water and Sand Inrush and Their Application in Shendong Mining Area, Disaster Advances, Vol. 5, Issue 4, 2012, pp. 1129–1135. (in Chinese)

Improved method research on free flowability for large particle size rockfill

K.C. Sun, L.F. Liu and Q. Li
College of Hydraulic and Environmental Engineering, Three Gorges University, Hubei, China
E-mail: liulinfeng0408@126.com

Y.Q. Tan
Yichang Design Branch of Mid-south Design & Research Institute for Hydroeletric, Yichang 443002,China

The free flowability of granular meterial is usually studied by measuring the repose angle. In this paper, an improved measuring method of repose angle for large particle size rockfill is presented. An experimrntal measurement device model, which can simulate the natural accumulation condition of large particle size rockfill well, is first established. To obtain the precise repose angle, this paper clearly and accurately define when and whether the accumulation is in the critical state for large particle size rockfill based on the experiments and experimental simulation effects. At last, according to the experiments of two measuring methods including the improved method and the traditional method, the experimental measurement angles and the actual natural repose angles are compared. It turns out that the measured results of improved method are closer to the natural repose angle and the improved method has great applicability. This research results are extremely important to accurately judge the free flowability of granular mixtures slope. In addition, it is of great significance for the stability of loose body and prevention of similar slope failure.

Keywords: Large particle size rockfil; Free flowability; Critical accumulation; Repose angle.

1. Introduction

The granular mixture materials are widely exist in geotechnical engineering, like crushed rock mass, natural gravel, rockfill material in the projects of slope, tunnel, underground cavities, railroad bed and hydraulic structure. Slope collapse owing to its instability causes great harm to the slope, dams, roads, railways and other projects. It can caused inestimable losses to people's lives and property. Therefore, as the important parameter to reflect the stability of granular mixtures slope, the study of the repose angle is of great importance. According to the definition, the repose angle is the biggest angle of free surface relative to the horizontal plane when the rockfill falls freely and continuously and the accumulation is in the state of limit equilibrium.

2. An Overview of Measuring Methods of The Repose Angle for Loose Body

At present, there are two main methods for the measurement of repose angle: Fixed-funnel method and the Slope staking-capillary method. Xiaohong Tian measured the repose angle of general grain crops by the fixed-funnel method, but this method has an inevitable defect. The impact of falling rockfill caused by gravity can lead to peak clipping phenomenon on top of the accumulation, so the result exists some system error. By improving the traditional measuring method, Jianbao Liu put forward a new measuring method for the repose angle of desert sand. To a certain extent, the improved method can reflect the actual repose angle of desert sand and reduce system error. In medicine, Yuxiu Dong improved the method of measuring the repose angle of powder drug by analyzing the reasons of system deviation caused by fixed-funnel method. Chenhua Wang experimented with self-made sand-sliding slope device. Sands slide down along the slope and form the accumulation. But they didn't explain when the accumulation can reach a critical state, so this method can’t guarantee that the measured angle is the repose angle. All of these methods have provided rich experience in the measurement of the repose angle for loose body. On this basis, our research further improves the traditional method to make it adapt the characteristics of coarse particles.Copying old text onto new file

2.1. *Traditional slope staking-capillary method*

The commonly measuring method of repose angle is slope staking-capillary method. Its experimental model device is as shown in Figure 1.

The experiment introduction is as follows. The inclination of the upper half part of the device (Section AB) is 70 degrees, a little bit larger than the natural repose angle of general granular mixtures. The lower half part (Section BC) is used to simulate the slope toe section and its inclination is about 10 degrees. Specific operation steps are as follows. Firstly, the fixed–funnel should be fixed in point A. Then, the rockfills slide down through the funnel to form a free accumulation near point B. Last, the angle of the free accumulation can be measured through the transparent glass panel. This angle is the repose angle of the rockfill accumulation.

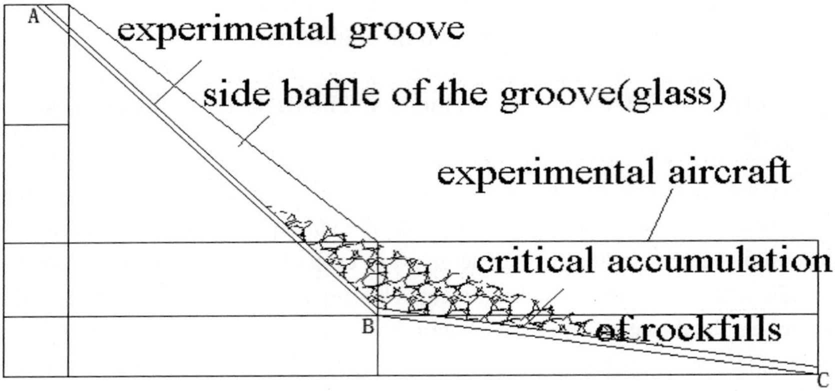

Figure 1. Device of slope staking-capillary method.

The shortcomings of traditional slope rolling measure method are as follows; The literature based on this measurement method didn't make it clear when and whether the rockfill accumulation is on the critical state. Besides, some literature only explain that when a certain amount of rockfill fall through the fixed funnel model and form a accumulation on the bottom board of model tank, the measured tilt angle of the slope is the repose angle. But it can be confirmed that the accumulation is on the critical state. In other words, the angle measured may not be the maximum angle of the free surface relative to horizontal plane.

2.2. *Improved slope staking-capillary method*

Considering the imperfection of free accumulation formed by the traditional method, the improved slope staking-capillary method method is put forward in this paper, and the experimental moder is shown in Figure 1.

It is proved by experiments that when a 13cm-diameter funnel is used to measure the repose angle of rockfill, the rockfill of 25-50mm particle size can't fall freely and continuously. Because good interlocking action exists between the rockfill particles and can form arch effect. So the falling conditions for this particle-size rockfill can't be met and rockfills are unable to form a critical accumulation. The fixed-funnel method can't apply to measure the repose angle for the larger particle size rockfills. So the fixed-funnel of simulated device is removed. We scoop up the rockfill and make sure that it roll down along the slope surface freely without initial motive force. The initial drop height of the rockfill in the whole test must be controlled to avoid that the distance between the rockfill and the top of the rockfill accumulation is too large. Otherwise, the falling rockfill have obvious peak clipping effects on accumulation, affecting the accuracy of repose angle. At the same time, in the falling process of rockfill, when the rockfill material is observed to stagnate on the top of the accumulation and doesn't roll down to the center and the slope toe of the accumulation, the pouring of rockfill should stop immediately. This moment, the angle is the maximum angle of the free surface relative to horizontal plane. This method can ensure that it's just at or close to the critical state and the measurement value is close to the nature value.

Finally, for the formed accumulation, rockfills in the foot of the slope should be removed, which contact loosely and obviously pose no threat to the stability of the accumulation. The horizontal length can be obtained by measuring the horizontal distance from the slope crest to the slope toe where the rockfill contact closely. For the vertical height measurement, we take a relatively stable point (not necessarily the highest point) on the top of the accumulation to be the calculation point of the vertical height. And the distance between this point and the horizontal plane of slope toe is the vertical height. Through the above methods, we can measure the angle of the slope line relative to horizontal plane. And this measuring angle is the angle of the critical accumulation called the repose angle.

3. Example

By sampling the homogeneous granular mixtures slope of three kinds of particle sizes in Duodigou, Tibet autonomous region, physical experiments are carried out. The experiments show that the moisture content and viscous coefficient of specimen material are 0.3% and zero respectively. Therefore, the impact of viscosity on repose angle can be ignored. Then, through using the slope rolling stacking-capillary methods including the improved and the traditional, we measure the repose angle of accumulation body respectively for the rockfill of with different particle size range. There are three kinds of particle size ranges: 20mm–25mm, 25mm–40mm and 40mm–50mm. According the experimental measure results and field measure test results of natural repose angle, we can make a comparison as shown in Table 1, Table 2 and Table 3.

Table 1. The comparison table of the repose angle for rockfills of homogeneous particle size between 40mm and 50mm.

Repose angle/°	Group 1	Group 2	Group 3	Group 4	Group 5	Average value
Former method	43.5	43.1	43.3	43.2	43.4	43.3
Improved method	42.9	42.6	42.8	42.7	42.7	42.7
Natural repose angle	42.1	42.1	42.1	42.1	42.1	42.1

Table 2. The comparison table of the repose angle for rockfill of homogeneous particle size between 25mm and 40mm.

Repose angle/°	Group 1	Group 2	Group 3	Group 4	Group 5	Average value
Former method	41.9	41.6	41.4	41.5	41.7	41.6
Improved method	41.3	41.2	40.9	41.0	41.3	41.1
Natural repose angle	40.6	40.6	40.6	40.6	40.6	40.6

Table 3. The comparison table of the repose angle for rockfill of homogeneous particle size between 20 mm and 25mm.

Repose angle/°	Group 1	Group 2	Group 3	Group 4	Group 5	Average value
Former method	40.3	40.1	39.9	40.1	40.7	40.2
Improved method	39.3	39.4	39.6	39.5	39.7	39.5
Natural repose angle	38.5	38.5	38.5	38.5	38.5	38.5

The variation tendency of repose angle and the difference of tested data are shown in Figure 2, Figure 3 and Figure 4.

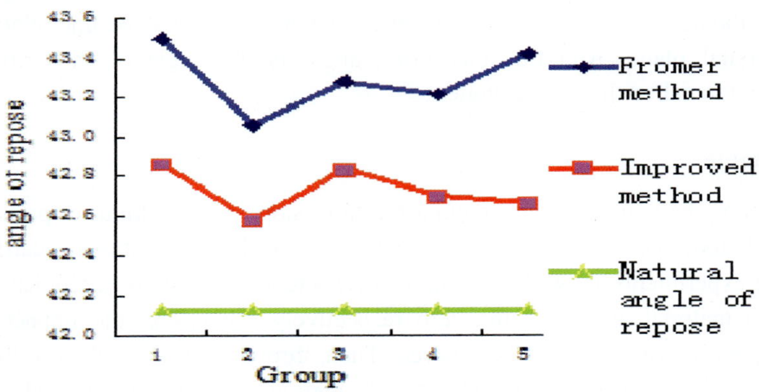

Figure 2. The comparison figure of the repose angle for rockfill of homogeneous particle size between 40mm and 50mm.

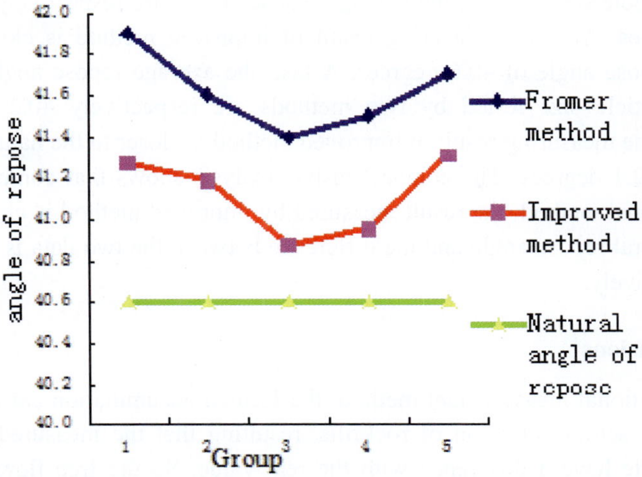

Figure 3. The comparison figure of the repose angle for rockfill of homogeneous particle size between 25mm and 40mm.

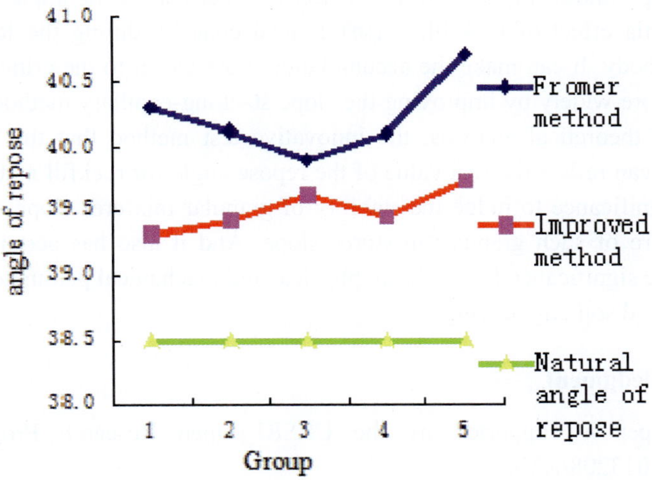

Figure 4. The comparison figure of the repose angle for rockfill of homogeneous particle size between 20mm and 25mm.

The results of the above analysis can be known. By successively using the traditional method and improved method to test rockfill repose angle of 40–50mm particle size, the average repose angles of the five experiments are respectively 43.3 and 42.7 degrees. The measuring result of improved method is closer to the natural repose angle of 42.1 degrees. In the same way, for 25–

40mm particle size rockfill, the average repose angles are respectively 41.6 and 41.1 degrees. And the measuring result of improved method is closer to the natural repose angle of 40.6 degrees. A last, the average repose angles of 20–25mm particle size tested by two methods are respectively 40.2 and 39.5 degrees. The measuring result of improved method is closer to the natural repose angle of 42.1 degrees. The comprehensive analysis shows that compared with the traditional method, the result measured by improved method is closer to the actual natural repose angle and the difference between the two data is small and stable relatively.

4. Conclusions

In the traditional measurement method, the formed accumulation can't simulate the natural actural situation of rockfills, resulting that the measured value of repose angle have a difference with the real value. So the free flowability of large particle size rockfill isn't reflected well. The deviation is due to the backward measurement method of the repose angle and the undesirable sampling without engineering reality. Besides, because of the impact of rockfall and inertia effect of rockfill, it isn't a ideal cone by during the formation of rockfill body. It can make the accumulation more closer to the critical state and apply more widely by improving the slope stacking-capillary method. From the point of theoretical analysis, the innovative test method that this paper puts forward can reflect the real value of the repose angle for rockfill material. It has great significance to judge the stability of granular mixtures slope and prevent the failure of such granular mixtures slope. And it also has certain scientific reference significance for the basic physical and mechanical parameters research of rock and soil engineering.

Acknowledgment

This paper is supported by the CRSRI Open Research Program (No. CKWV2013208/KY).

References

1. Ma G, et al. 2010. Study of anchorage mechanism of granular mixture. *Chinese Journal of Rock Mechanics and Engineering*, 29(08):1577–1584.
2. Hu Chao, et al. 2014. Study of anchoring effect of anchor spacing in granular mixture. *Rock and Soil Mechanics*, 35(07):2088–2094.

3. Tian X.H., et al. 2011. Study on influence factors of grain angle of repose by funnel method. *Journal of the Chinese Cereals and Oils Association*, 26(10):108–113.
4. Liu J.B., et al. 2010. A new experiment measurement for the angle of repose of desert sand. *Journal of Lanzhou university (natural science)*, 46(04):37–41.
5. Dong Y.X., et al. 2008. Perspectives on the measurement of angle of repose. *Journal of China Pharmaceutical University*, 39(4):317–320.
6. Wang C.H., et al. 2007. Formation and basic characteristics of sand-sliding slope composed of granular clasts: Part I of sand-sliding slope series. *Rock and Soil Mechanics*, 28(01):29–35.
7. Meng Z. & Yang W.J. 2012. Preliminary exploration of difference between submarine angle of repose and internal friction angle of sediment particles. *Journal of Sediment Research*, (04):76–80.
8. Zhou Y. & Zhang G.Q. 2009. Effect of granular size on the angle of repose in the chute. *Science in China Series G: Physics, Mechanics and Astronomy*, 52(4):563–565.
9. Shi Y.L., et al. 2007. Calculation of angle of repose and dry density of sediment. Engineering Journal of Wuhan University, 40(03):14–17.
10. Zhou Z. Y., et al. 2014. Angle of repose and stress distribution of sandpiles formed with ellipsoidal particles. *Granular Matter*, 16:695–709.

Drilling rig monitoring system design and application within the oil company in Republic of the Congo

T. Ngoma-Foutou* and R.J. Ma

School of Electrical Engineering, University of Jinan, Jinan 250022, China

**E-mail: ngomated@yahoo.fr*

Based on the drilling process, we designed a set of rig time monitoring systems according to the required monitoring parameters and designed the hardware and software for the system. The design of hardware monitoring is necessary for drilling operation. In order to achieve the design of hardware features, the design of the software need real-time operation and a good real-time scheduling and optimization. A human-computer interface is designed, as well as a real-time dynamic display. The system already meets the engineering staff's convenient operation, visualization, telecommunications and real-time demands. The drilling monitoring system has been used in offshore oil field of Minsala in Congo. It has achieved good results.

Keywords: Oil rig; Drilling process; Monitoring and control system; Simulation; Design.

1. Introduction

In recent years, the level of the security monitoring rig has been improved greatly. The monitoring system greatly improves the reliability of drilling equipment and drilling projects, while ensuring the quality of the well itself. The drilling process uses real-time monitoring of the process parameters in order to achieve security monitoring and fault diagnosis of rig equipment. This helps the drilling process parameters to be optimized. The principle of drilling is the real-time feedback to the control, which establishes a connection between real-time monitoring and the control system. This also allows a remote terminal to establish contacts for remote monitoring. Thus, the traditional technology of drilling is the control by a computer to combine the technology of measurement and control system parameters during drilling, which is the key of the drilling rig equipment to achieve leapfrog development.

In this paper, a monitoring system is designed; the process parameters were recorded, stored and analyzed during drilling, to achieve a successful oil well drill. The design of system monitoring includes hardware, software and HMI.

Through real-time monitoring and analysis, it can also detect any abnormalities during the drilling process to ensure safety and efficiency.

2. Drilling Monitoring System Architecture

2.1. *Different drilling monitoring parameters*

With some sensors installed in different places, real-time observation of drilling parameters is ensured. The main parameters include: ROP (drilling rate), forward speed, WOB, the tool load, RPM (revolution per minute rate), the rotation speed, FLOW in: input flow rate, SPP: inner standpipe pressure, line pressure, TORQE. (Operation of the distillation column), SPM (storage per minute). It explains the three icons in the drilling penetration rate and the relationship between different drilling parameters (WOB / RPM / TORQUE).

The Rig Monitoring Software aims at helping the operator to control the production process (dynamic HMI); and decide providing data for the completion of the production. The monitoring system provides the immediate production unit for all devices, dynamic and comprehensive information, so that the operator can interact and affect the production process.

The oil platform monitoring system, which main functions are monitoring and control, is installed in a centralized control room. There are lots of analog variables and digital variables needed to complete the tasks, such as the device control and monitoring, fault detection and diagnosis, compensation and correction.

Power Actuators is in the middle of the rig and pumps mud: 100 meters away from the rig and 50 meters away from the pump mud. In the control center, in order to realize the remote control of the signal, we can control the draw works (M1), M3 and M4 operating panel can control two motors of pumps mud. And M2 for drilling, the DCC it is the cabin of drilling control. One basic schematic of the power actuators for the rig is presented in Figure 1. The Control operates high power motors of drilling process.

The issued software oriented monitoring system improves the total performances of the traditional drilling rig control system using the PLC based monitoring system with a Lab VIEW designed software for computer level human interaction. Figure 2 describes the principle of a drilling installation monitoring starting with the heavy equipment level (on the bottom), going up to the programmable logic (PLC) and supervising software.

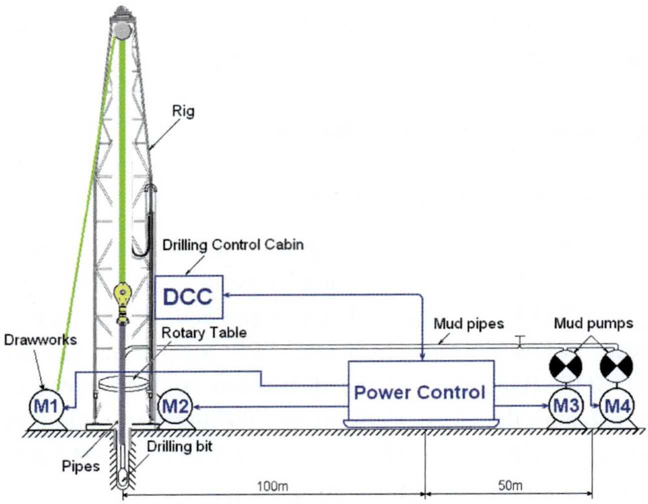

Figure 1. Position of chief driller desk.

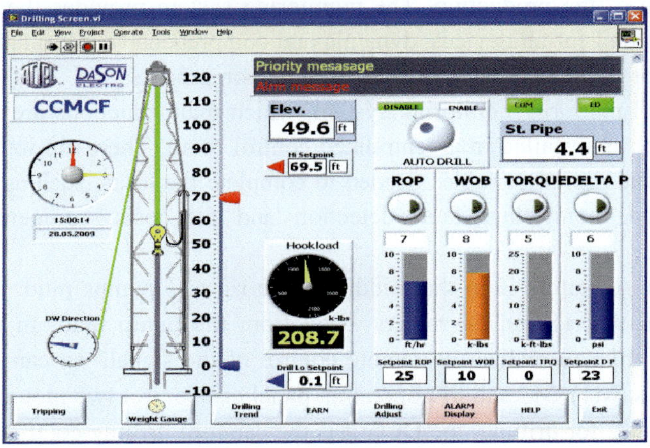

Figure 2. Computer based monitoring of a drilling rig.

2.2. *Parameters to be monitored*

The equipment which are monitored at this stage are (Figure 3):

Diesel generators having 1500kVA at 600V;

- Connection switches to main power supply distribution used by generators;
- Motor drive converter;
- Connection switches to main power supply distribution used by converter;
- Excitation converter for motors;

- Connection switches to main power supply distribution used by excitation converters;
- Connection switches to auxiliary power supply distribution used by transformers;
- Electrical drive motors status (visual symbols).

Figure 3 is a drilling process monitoring: the host communicates with the LAN device via a proprietary protocol, and sends the corresponding command to the network on the receiving device. Equipment according to the acceptance of data returns the corresponding value in the PC, which analyzing the return of content. Data processing can realize the function of data acquisition for the drilling process of real-time monitoring. Generators Motors, Auxiliary supply and drilling rig data and other data are collected.

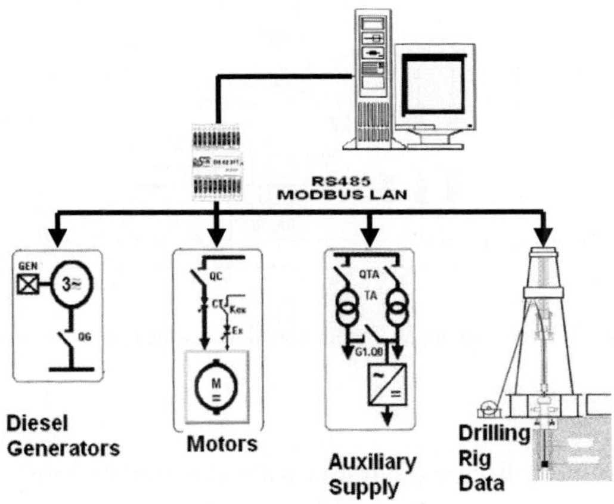

Figure 3. Drilling process monitoring.

3. Hardware and Software Design

The monitoring system is split in two directions: Hardware and Software.

3.1. *Hardware*

The monitoring system is dictated by the general design of the drilling rig dictated. Some parts are important: The Diesel generators forming a local power plant and the numerical drives for rigs power motors. These are connected to the monitoring system using a PLC as shown in Figure 4.

In Figure 4 the GEN 1500kVA of Caterpillar type 600V diesel generator, can provide continuous power. QG is used for the connection switch of generator power distribution. QC for the connection switch, can change the distribution of the AC power supply. Ex for the KEX and of the converter is used to drive the power distribution of the connection switch. The CT is regard as a rotary regulator.

The hardware structure of the rig monitoring system DAS helps the operator of generators data and drive motors acquire, and to quantify it: then they can take action. In the application process of the monitoring system, the hardware is essential. This product applies to the seabed and ground operations.

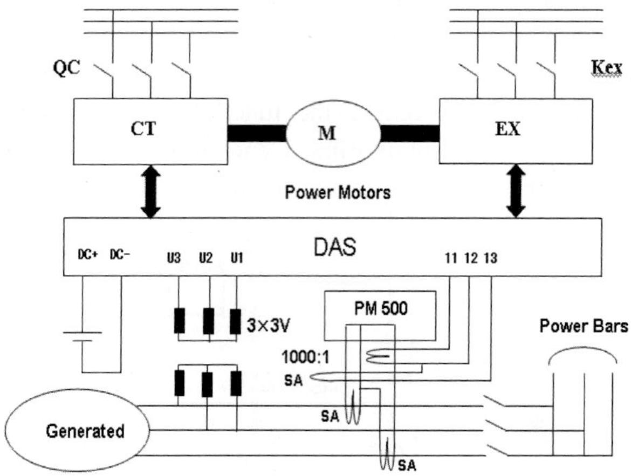

Figure 4. Drive motors and generators data acquisition system (DAS).

3.2. *Software*

Six main parts in the software diagram power the application:

Initialization- responsible for hardware and software setup;

Data acquisition- bringing data to the central processing unit;

Measurements and DSP- performing basic measurements of the system parameters and advanced digital analyses of data;

Database- on line storage of data, trends and events;

Trends, events and alarms- the SCADA like environment for the rig;

Report- report generator and state publisher on the net using secured protocols.

The application panel backup in Lab VIEW is the software graphical diagram in Figure 5.

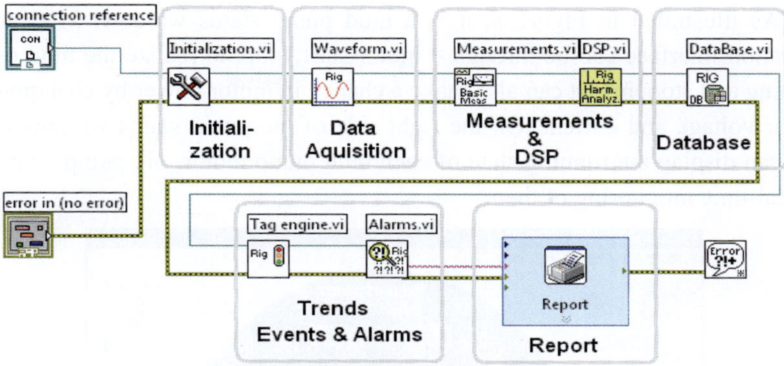

Figure 5. Lab VIEW application software diagram.

3.3. *HMI*

The human-machine interface (HMI), named DBS drilling rigs operating system, usually installed in the driller's control room, provides the scene of most of the equipment for remote control and complete drilling operations. Design the touch screen interface is the main operation, from Draw-works Status Window, hook charging status window, mud pump status window, generator parameters and mud system parameter have a display interface.

The operator can start, stop, control the speed of the winch and mud pump motors in the HMI. The touch screen can also display hanging heavy, WOB, ROP (rate of penetration), mud return flow, mud pumping, standpipe pressure, wheel speed, wheel torque, hook speed, traveling height, tension and other parameters.

As illustrated in Figure 6, it is a Draw-works Status Window. This interface can complete the tripping operation, single motor and double motor tripping in two ways. The four windows on the left side of the diagram show the voltage and current values of the two motors. MTI button and MT2 button control the two motors to open and close. On the right side of the window shows the winch motor, brake and advance of the bit automatic running state.

As illustrated in Figure 7, it is a hook charging status window. Two sets of readings are displayed in the middle of the window. The inside side of the display is hook-load, the unit is k-lbs, the outer side of the display is on bit weight, the unit is k-lbs, can be achieved by the WOB UP button, the DOWN Zero button and the Zero WOB button, to achieve the lifting of the pressure, and to reduce the pressure and the function of the remove WOB. The circular dial in the lower left corner shows the time. The exit button at the bottom right is the exit function.

As illustrated in Figure 8, it is a mud pump status window. Mud pump operation interface can be through 4 motor start, stop, to realize the mud pump running and stopping. It can also make a change in motor power by changing the motor voltage and current. On the right side of the pump status window mud, you can display the running state of each mud pump motor, oil, pump vent, can be real-time monitoring of them.

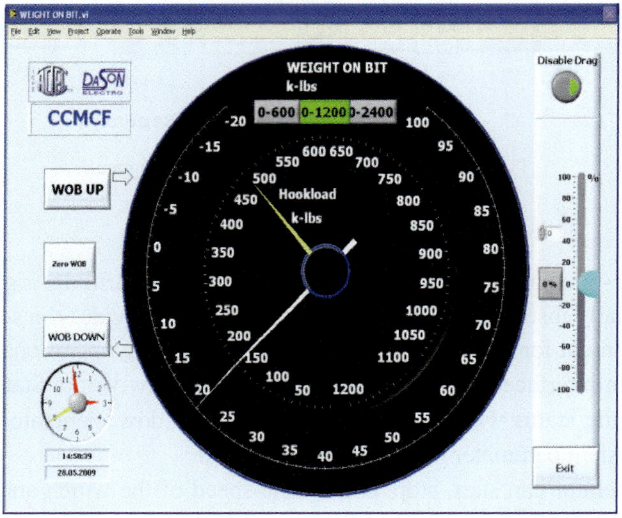

Figure 6. Hook charging status window.

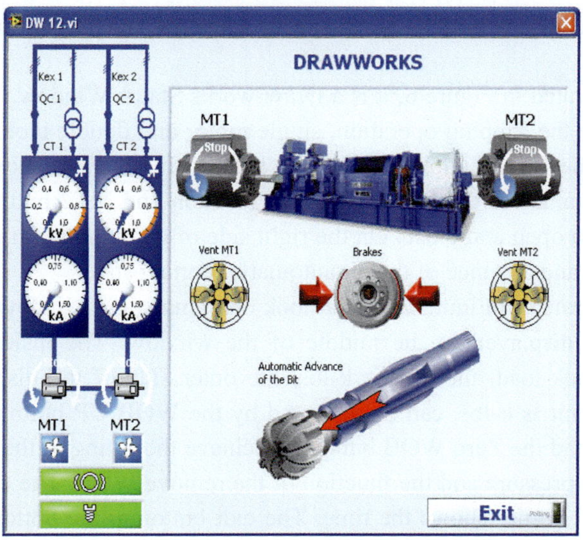

Figure 7. Draw-works Status Window.

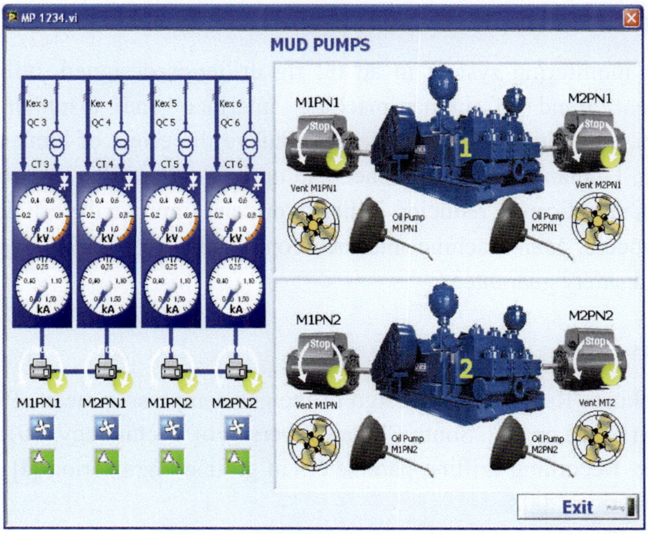

Figure 8. Mud pump status window.

There are drawing elements from the control windows are active. The left side instrument allows global inspection of parameters at power supply level and each equipment status is shown in the right side of display. This greatly improves the efficiency of the operation.

4. Drilling Monitoring System In the Application of Congo-Brazzaville

In Congo, drilling monitoring system applications are as follows:

The drilling monitoring system has been used in offshore oil field of Minsala and Congo Minsala deepwater oil exploration wells more than 2000m in depth, which belongs to the deepwater oil field. The temperature at the bottom of the sea bed is only about 10°C. Congo Minsala deepwater reservoir permeability 9850mD~1170mD, porosity 23.8%~31.1%, belong to high porosity and high permeability reservoir; reservoir temperature to 56 DEG C, the pressure is 17.9mpa; rock composition is dominated by single crystal quartz sandstone, the formation sand grain size median average value for 128.33tmi.

It has achieved good results that it has accumulated half billion dollars of investment and increased the output of three percent. The current oil field output of up to 5000 barrels of oil, fourteen million cubic feet of natural gas.

5. Summary

A drilling monitoring system in an oil rig driller is designed, utilizing rapid development trend of human-machine interface and displaying drilling parameters. The touch screen is used to achieve the effect of human-computer interaction. Human-machine interface can optimize the driller control structure and layout, significantly reducing drillers work and improve the efficiency of the drilling process. Man-machine interface control also realizes intelligent driller control and safety guarantee.

References

1. Fu Jiakuan. Research and design of the gas compressor means PLC control system based on [D]. South China University of Technology, 2012.
2. Pfister. Recording drilling parameters in ground engineering [J]. J. Ground Engineering, 2008.
3. Society Eni Congo [J]. http://www.eni-congo.com.
4. Mechbal, M. Verge. Hydraulic process of an Oilfield Drilling System [M]. September 2008.
5. Society Boscongo [M]. http:// www.Boscongo.com.
6. Liu Qiang. Automation Research and Design Oil Rig [D]. Northeast Petroleum University, 2013.

Research on total dissolved solids variation of deep confined fresh water under exploitation conditions in Dezhou

J. Feng*, Q.S. Zhao, Y. Teng and M. Tian
Department of Environmental Science, Qingdao University, Qingdao 266071, China
**E-mail: qdufengjuan@126.com*

The distinct characteristics of water environment is comprised of salt water and fresh water in Yellow River downstream alluvial plain so that the natural aquatic environment condition of underground fresh water is rather sensitive and fragile. The continuous exploitation of the groundwater brings great changes in the hydrodynamic force field and hydrochemical field and intensified the trend of freshwater salinization in this area. Freshwater salinization of deep groundwater is one of the major water environment security being faced by the study area. Based on hydrogeological condition and the dynamic monitoring data of groundwater quality in Dezhou from 2006 to 2009, the paper establishes prediction models for TDS variation of deep confined fresh water. Confined water depression cone is selected as simulated zone as well as Total Dissolved Solids taken as the simulated factor which can reflect freshwater salinization in the study area well. Across the whole region from 2009 to 2015, areas with Total Dissolved Solids of less than 1000 mg/L will decreased by 14.10 km^2 as well as areas with Total Dissolved Solids of more than 1200 mg/L will increased by 4.58 km^2. Total Dissolved Solids of observation wells show an increasing trend. With the groundwater levels falling and the pressure drop by over-exploitation of the groundwater resource, sharp hydraulic gradient causes damage to natural balance between salt and fresh water, and then Total Dissolved Solids of the deep confined fresh water in aquifers tends to increase gradually. It indicates that saline water aquifer on the top of deep fresh water aquifer can invade fresh water by leakage under continuous exploitation condition. According to variation of Total Dissolved Solids, deep confined fresh water should be exploited by the stages and layers with the aim to give it enough breathing space for restoration.

Keywords: Total Dissolved Solids; Freshwater Salinization; Numerical Simulation; Exploitation Condition.

1. Introduction

The distinct characteristics of water environment is comprised of salt water and fresh water in Yellow River downstream alluvial plain so that the natural aquatic environment condition of underground fresh water is rather sensitive and fragile [1]. Deep fresh groundwater and middle salt groundwater keep steady pressure balance under natural state. But since 1970s, the continuous exploitation of the groundwater and excess exploitation have caused rapid decline of groundwater levels with that the area of pore confined aquifer system groundwater levels depression cone has been thousands of square kilometers in this area [2]. With center of groundwater exploitation form new potential sink by over-exploitation of the groundwater, new groundwater flow systems that streamline points to the center of extraction which damage to natural balance between salt and fresh water have been formed, and then freshwater salinization of deep groundwater become one of the major water environment security. The continuous exploitation of deep groundwater in area of salt water further change natural balance of groundwater system, that water head of deep water sharply falling and overlying salt water from jacking by fresh water to integrally shift down. The over-exploitation of the groundwater brings great changes in the hydrodynamic force field as well as hydrochemical field and intensified the trend of freshwater salinization in this area [3]. Formation and development of deep underground fresh water levels sharp fall and large area depression cone is process of deep confined water mining and supplement imbalance as well as storage resource increasingly exhausted, which is destroying severely limited and precious deep fresh groundwater resources that form through a long geological periods [4].

The center of deep confined water depression cone in Dezhou that is typical deep confined water depression cone in north china plain is located in Chengzhuang Township Zhangzhuang, which cone center water level is 147.91 m in buried depth and -126.80 m in water elevations in 2009. Dezhou exploit deep confined water form 1965 and deep confined water level is 2.00 m in buried depth. Duing to over-exploitation of the deep confined water, deep confined water levels depression cone was preliminarily formed in 1970 and deep confined water level is 10.00 m in buried depth. Cone center water level is 30.00 m in buried depth in 1973, and it becomes to 41.50 m in 1978. In the mid of 1980s, the amount of deep wells showed an increasing trend by the process of the industrialization and urbanization which brought the deep confined water levels rapid falling. In 1990, the deep confined water levels depression cone center water level was -55.23 m in water elevations and was 76.23 m in buried depth. Deep confined water levels depression cone center water level was -84.30m in water elevations and was 105.30 m in buried depth in 2000. Deep confined water levels depression cone in Dezhou is formed regional groundwater depression cone centres on Decheng, which extend to Wucheng, Pingyuan, Ling County, Ningjin and Wuqiao, Jing County, Guchengzhu County in Hebei Province at present. In 2009, the entire Decheng district which is 539.00 km^2 within the scope of dezhou deep confined water levels depression cone centres on Chengzhuang Township Zhangzhuang, which center water level was 147.91 m in buried depth and fell 145.91 m compared with 1965.

Now, the study concentrate on groundwater resources evaluation model, the exploitation and utilization of groundwater resources, the ecological effects of groundwater exploitation, the study on depression cone caused by groundwater exploitation, the study on land subsidence caused by groundwater exploitation and so on. But the study on freshwater salinization of deep confined water under groundwater exploitation is insufficient [5–10].

The down-move of salt and fresh water interface refers to the upper portion of the shallow depression of high TDS saltwater moves downward, showing the deep confined freshwater elevated TDS [11]. Confined water depression cone is selected as simulated zone as well as Total Dissolved Solids taken as the simulated factor which can reflect freshwater salinization in the study area well. Based on hydrogeological condition and the dynamic monitoring data of groundwater, the paper establishes prediction models for TDS variation of deep fresh groundwater [12–13].

2. Regional Environment Profile

2.1. *Physical geography*

The study area is located in Central North Shandong Plain, south across the Yellow River and Jinan. It is adjacent to Binzhou, Jinan as well as Liaocheng in the east and northwest and bounded with Zhangweixin River and Wei Cancal in the north. The area is also adjacent to Cangzhou and Hengshui of Hebei Province.

The area belongs to semi arid monsoon climate of warm temperate zone and has four distinct seasons. The city's average precipitation is 525.57 mm for many years. Characteristics of rainfall are few and concentrated and annual rainfall is uneven. Its average evaporation volume is 1706.72 mm for many years. Surface water system in the region more developed and major rivers which from south to north are Yellow River, Tuhai River, Dehuixin River, Majia River and Zhangweixin River and so on.

The study area belongs to typical Yellow River downstream alluvial plain which terrain flat, and the general trend slowly tilt which from southwest to northeast. The ground elevation from 34 m in Qihe County southwest corner to 7 m in Qingyun County northeast corner and the slope is from 1/1000 to 1/10,000. Under the effect of natural and anthropogenic factors, the area forms landscape that has interphase highlands, slopes and depressions in the south west and north east to distribution, which is based on Pan-Siltation that had resulted in the frequently changing Yellow River's course and alluvial–diluvial which form Zhu River in southern mountains.

2.2. *Regional geology*

The study area belongs to North China Platform in tectonic. Fracture of Qihe–Guangrao divides this area into two secondary structural units. South of fracture is Luxitai anticline and north is Liaojitai syncline. Under the effect of otherness vertical movement, deep strata distribution in the area has a distinct partition. In bulge area, general lack of Paleogene formations and Neogene formations directly cover in Archean, Paleozoic and Mesozoic Formations. In depression area, Neogene is more complete and the thickness is greater than 3000.00 m.

In study area, formation from old to new since the Neogene is the Neogene Guantao Formation, the Neogene Minghua Town Formation and the Quaternary.

3. Division and Characteristics of Salt, Fresh Aquifers in Study Area

The study area belongs to the pore water hydrogeological region in Yellow River downstream alluvial plain and groundwater stored in pore of unconsolidated sediments in Quaternary and Tertiary. Under the effect of stage and otherness vertical movement since Cenozoic, aquifer is complex structure, overlapping staggered on spatial distribution and groundwater has obvious zonation. According to the difference in the vertical direction of the groundwater burial conditions and the chemical composition of water, groundwater is less than 800 m for the three distribution areas in shallow freshwater, middle salt water and deep confined water.

In shallow freshwater distribution area, bottom interface of shallow freshwater is usually from 20.00 m to 60.00 m in buried depth. Lithology of aquifer mainly is fine sand of middle-upper part in Holocene Series and Upper Pleistocene sub system and thickness is from 10.00 m to 25.00 m. Water yield of single well is from 300.00 m^3/d to 1000.00 m^3/d and hydrochemical type $HCO_3 \cdot Cl$-$Na \cdot Mg$ water.

In middle salt water distribution area, top interface of middle salt water is 60.00 m average and floor is from 190.00 m to 250.00 m in buried depth. Lithology of aquifer mainly is fine sand of pleistocene series and thickness is from 20.00 m to 30.00 m. Hydrochemical type mainly comprised of $Cl \cdot SO_4$-Na water. The groundwater in this formation has not been exploited at present.

In deep confined water distribution area, the deep confined water is located in the under of middle salt water distribution area. Lithology of aquifer mainly is fine sand, medium sand and medium coarse sand of Neogene fluvio-lacustrine deposits. Aquifers' thickness and burial characteristics are controlled by the base structure, top interface of middle salt water and ancient geographical environment. Roof of aquifer is 200.00 m average and floor is from 600.00 m to

800.00 m in buried depth. This fresh water formation is main water supply of this area.

4. Prediction Models for TDS Variation of Deep Confined Fresh Water

4.1. *Conceptual hydrogeological model*

According to the hydrogeological conditions of the simulation area, aquifer III is the main water resource to be explored which depth to water table is more than 300 m. The lower part is Aquifer IV which depth to water table is from 500 m to 800 m. The mesosphere is aquiclude. Groundwater drop funnel is defined as calculated and simulated region with a 539.00 km^2 area, which boundary is defined as the second flow boundary. Conceptual hydrogeological model is shown in Figure 1.

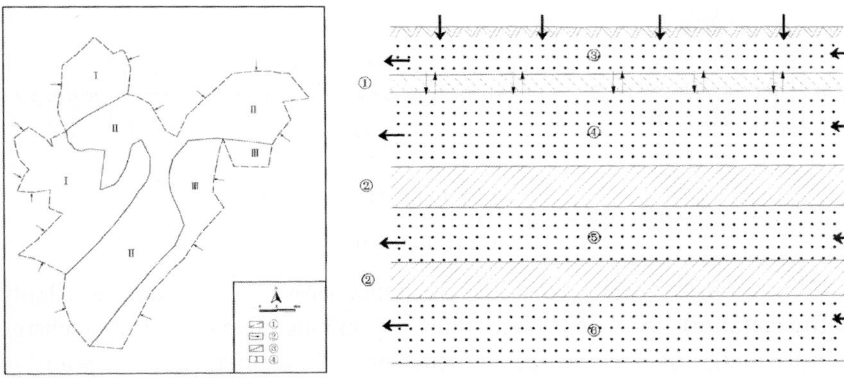

① Second flow boundary
② Flow direction
③ Boundary of divisional parameters
④ Number of divisional parameters

a. (planar graph)

① Leaky aquifer ② Aquiclude ③ Sallow phreatic water- semi-confined water ④ Mesosphere confined water ⑤ Aquifer III
⑥ Aquifer IV

b. (profile graph)

Figure 1. Schematic diagram of conceptual hydrogeological model.

4.2. *Prediction model for TDS variation*

According to Conceptual Hydrogeological model above, the hydrodynamic dispersion model of two-dimension in the region was constructed as follows. The mathematical model can be expressed as following:

$$\mu \frac{\partial h}{\partial t} = div\left(\text{Tgrad}h\right) + \varepsilon \qquad \left(x, y\right) \in \Omega, \quad t > 0$$

$$h\left(\mathrm{x,y,t}\right)\Big|_{t=0} = h_0\left(x,y\right) \qquad \left(x,y\right)\in\Omega$$

$$\left(\mathrm{Tgrad}h\right)\cdot\vec{n}\Big|_{\Gamma_2} = q_1\left(x,y,t\right) \qquad \left(x,y\right)\in\Gamma_2,\quad t>0$$

$$V = -\frac{k}{n}\,\mathrm{grad}h$$

$$\frac{\partial}{\partial t}\left(M\cdot n\cdot C\right) = div\left(M\cdot n\cdot \mathrm{Dgrad}\,C\right) - div\left(M\cdot n\cdot V - \mathrm{C}\right) + \varepsilon \mathrm{C}_\varepsilon$$

$$(x,y)\in\Omega, t>0$$

$$C\left(x,y,t\right)\Big|_{t=0} = C_0\left(x,y\right) \qquad \left(x,y\right)\in\Omega$$

$$M\cdot n\left\{\left[V_xC-\left(D_{xx}\frac{\partial C}{\partial x}+D_{xy}\frac{\partial C}{\partial y}\right)\right]\cos(n,\hat{x})+\left[V_yC-\left(D_{yx}\frac{\partial C}{\partial x}+D_{yy}\frac{\partial C}{\partial y}\right)\right]\cos(n,\hat{y})\right\}\Bigg|_{\Gamma_2}$$

$$= q_2(x,y,t)$$

$$(x,y)\in\Gamma_2, t>0$$

Where h = water distribution function(m); T = transmissibility coefficient(m^2/d); ε = algebraic sum of mining intensity and recharge intensity(m/d); μ = storage coefficient of aquifer; n = porosity; C = concentration of TDS(mg/L); D = dispersion coefficient(m^2/d); M = aquifer thickness(m); C_ε = TDS concentration in leakage recharge(mg/L); q_1 = unit width flux on the second flow boundary (m^2/d); q_2 = concentration flux of TDS on the second flow boundary (mg/($m^2 \cdot$d)); Γ_2 = second flow boundary; and n = unit outward normal direction of boundary.

4.3. *Model identification*

The simulation zone was divided into 185 within nodes and 59 boundary nodes and the cells were 309 same triangular elements with 548.85 km^2. The borders were defined as the second flow boundary. Systematic test data of water levels and water quality were used for fitting and testing the mathematical model. Recharge rate of class II boundary was determined according to equal water level line by Darcy formula. Exploitation quantity of deep confined water is

2047×10^4 m^3/a in 2006, 2062×10^4 m^3/a in 2007, 2089×10^4 m^3/a in 2008 and 2100×10^4 m^3/a in 2009. It was based on survey results.

Hydrogeological parameters that requires of this time numerical model is established include storage coefficient of aquifer, transmissibility coefficient, longitudinal dispersion, horizontal dispersion and porosity. The initial values of parameter partition, storage coefficient of aquifer, transmissibility coefficient and porosity are determined according to parameters which calculated from geological exploration and pumping test data, which based on hydrogeological partition. It is shown in Figure 1 (a). With upstream Shijiazhuang, Hebei district adjacent to the simulation zone and both belong to Pore water hydrogeological region in Yellow River downstream alluvial plain, hydrogeological conditions are very similar. Initial values of longitudinal dispersion and horizontal dispersion are determined according to dispersion test data of Shijiazhuang.

According to dynamic data of groundwater, the simulation period was selected from 1 June 2006 to 1 June 2008. TDS in 1 June 2006 was defined as initial value.

Fitted values of hydrogeological parameters are got though Trial-and-error method for parameter adjustment and root mean square of fitting error as evaluation standards is shown in Table 1. Plane distribution rule of hydrogeological parameters is gradually increased from west to east, which is suitable for actual hydrogeological conditions.

Table 1. Partition and values of hydrogeological parameter.

partition	μ	T (m^2/d)	α_L (m)	α_T (m)	n
I	0.0044	360.00	7.30	1.42	0.23
II	0.0025	210.00	6.40	1.18	0.21
III	0.0020	150.00	5.30	1.05	0.20

4.4. *Model verification*

The simulation was dated from 1 June 2002 to 1 June 2008. The average fitting error of TDS on the 244 compute nodes is 6.82 mg/L, which is quite qualified comparing to require of water quality error accuracy. The outcome of the fitting number is satisfied with the standard of *Groundwater resources management model work requirements (GB/T 14497-93).* It is shown that our fitting accuracy and efficiency is quite ideal. During the three years' modeling process, no conspicuous error accumulate and enlarge has happened, which shows that established mathematical model is suitable for the actual situation of the study area in generalization, boundary conditions and parameter determination of hydrogeological conditions.

5. Prediction of TDS Variation on Deep Confined Fresh Water

5.1. *Analysis on balance of water resource*

Analysis on balance of water resource get data from the deep confined fresh water level and qualified test from 1 Jun. 2006 to 1 Jun. 2008, as well as data of hydrogeological parameters that already tested. The outcome of calculate shows that total recharge rate is 1892×10^4 m^3/a and total excretion amount is 2100×10^4 m^3/a of aquifer in simulation zone. Recharge rate is 208×10^4 m^3/a less than excretion amount, which shows negative equilibrium and groundwater is over-exploitation.

5.2. *Prediction of TDS variation*

According to the verification and identification of mathematical model, we input initial conditions, boundary condition and source sink term of prediction model to predict variation of TDS. At period of prediction, the current exploitation quantity is 2100×10^4 m^3/a. Period of prediction is from 1 Sept. 2009 to 1 Sept. 2015.

Table 2. Variation enclosed area by contour of TDS.

TDS (mg/L)	2006	2009	2015
<1 000	100.49 km^2	107.58 km^2	93.48 km^2
>1 200	18.91 km^2	16.64 km^2	21.22 km^2

According to the prediction, variation trend of TDS isoline closed-circle area in simulation zone is shown in Table 2. It shows that areas with Total Dissolved Solids of less than 1000.00 mg/L would increase by 7.09 km^2, as well as areas with Total Dissolved Solids of more than 1200.00 mg/L would decrease by 2.27 km^2 from 2006 to 2009. Areas with Total Dissolved Solids of less than 1000.00 mg/L would decrease by 14.10 km^2, as well as areas with Total Dissolved Solids of more than 1200.00 mg/L would increase by 4.58 km^2 from 2009 to 2015.

The forecasting result of TDS indicated that TDS of peephole in simulation zone would decrease from 2006 to 2009, but TDS would increase from 2009 to 2015. For instance, TDS of ZK9 peephole would decrease from 821.00 mg/L in 2006 to 819.00 mg/L in 2009, and then it would increase to 873.00 mg/L in 2015. In conclusion, TDS of deep confined fresh water in simulation zone would have different degree of increase with the passage of time and seasonal variation would be smaller.

6. Conclusions

It is the first time to establish prediction models for TDS variation of deep fresh groundwater which quantificationally reflect the trend of freshwater salinization in the study area, and utilizing the variation and trend in the space and time scales of deep confined fresh water TDS under current exploitation quantity which is simulated and predicted by established model. Comparing to 2006, the TDS of observation wells in simulation zone would decrease in 2009, but would increase from 2009 to 2015. With the groundwater levels falling and the pressure drop by over-exploitation of the groundwater resource, sharp hydraulic gradient causes damage to natural balance between salt and fresh water, and then Total Dissolved Solids of the deep confined fresh water in aquifers tends to increase gradually. Saline water aquifer on the top of deep fresh water aquifer, which hydrochemical type is $Cl \cdot SO_4$-Na and TDS is between 5.00 g/l to 20.00 g/l. Saline water aquifer on the top of deep fresh water aquifer can invade fresh water by leakage under continuous exploitation condition.

In study area, the shortage recharge and the slowly run of level of deep confined fresh water which speed of hydrologic loop and alternate renewal is lower comparing to the shallow confined water, so its exploitation type should different from the shallow confined water which joined nowadays water loop actively. At the same time, the study area belongs to multi aquifer system which comprised of salt water and fresh water. So this type of groundwater should be exploited by the stages and layers with the aim to give it enough breathing space for restoration.

Due to the determinate environmental condition, we acquired inadequate monitoring data of groundwater quality in the transitional area between salt and fresh water. We cannot accurately determine the hydraulic characteristics and the problem of salt water leakage between fresh water aquifer and salt water aquifer, which is also the key of continue work and study in coming future.

7. Acknowledgments

This work was supported by 'The research on freshwater salinization evolution under exploitation conditions in alluvial plain of the Low Yellow River' of National Natural Science Foundation in China (Grant No. 40772151). The authors are extremely grateful to reviewers and editors for constructive comments on the original manuscript.

References

1. Q.S. Zhao et al. Shallow Groundwater Quality Evolution in Dezhou City. *Journal of Jilin Unviersity: Earth Science Edition*, 2010, 40(5):1075–1082.
2. Z.H. Zhang et al. Evolution and development of groundwater environment in North China Plain. Beijing: Geology Publishing House, 2000.
3. D.C. Wang et al. *General Hydrogeology*. Beijing: Geology Publishing House, 1995.
4. Z.H. Zhang et al. To carry out research on evolution and development of groundwater environment under the influence of human activities. *Hydrogeology and Engineering Geology*, 1992, 19(5):1–2.
5. Q.Y. Liu et al. Summarize of water environment status in Dezhou City. *Haihe Water Resources*, 2009(2):25–26.
6. X.F. Qi & F. Liu. Exploitation, Utilization and Protection of Groundwater Resource in Dezhou City. *Resources Environment & Engineering*, 2007, 21(6):752–754.
7. Z.J. Sun & H.F Wang. Influences and Countermeasure to Ecological Environment Dezhou Groundwater Mining Creates Change and Countermeasure. *Ground water*, 2009, 31(4):30–32, 79.
8. J. Feng. Study on the Dynamic Change and Simulation of Deep Groundwater in Dezhou City. Qingdao: Qingdao University, 2008.
9. J. Feng et al. Application of Visual MODFLOW to the dynamic change and simulation of deep groundwater in Dezhou City, China. *Advanced Materials Research*, 2010, 113–114:1025–1030.
10. L.Z. Yang et al. Relationship between deep groundwater resources composition and land subsidence under the condition of exploitation in Dezhou, Shandong, China. *Geological Bulletin of China*, 2010, 29(4):589–597.
11. Z.L. Shen et al. *General hydrogeochemistry*. Beijing: Geology Publishing House, 1999.
12. O.C. Zinekiewicz & R.L. Taylor. *The Finite Element Method*. Beijing: Tsinghua University Press, 2008.
13. D.Q. Wang et al. Mathematical Methods for Surface and Subsurface Hydrosystems. Beijing: High Education Press, 2006.

Study on cracks of vertical water wall outlet in supercritical pressure boiler

K. Li* and R. Zhang

Jilin Province Electric Research Institute Co., LTD
Changchun, Jilin 130021, China
**E-mail: 873443433@qq.com*

Aiming at the problem of boiler shutdown caused by several crack appeared on vertical supercritical operation of a certain power plant boiler water wall outlet. In this paper, the characteristics of the crack and boiler operation parameters curve are analyzed. The study found that the crack in boiler vertical water wall outlet is caused by larger temperature deviation between tubes and thermal stress produced by the pipe temperature change quickly. The thermal stress usually appeared in low load condition after a cadre of boiler state. It also have a corresponding relations between the change of the boiler feed water flow and the vertical water wall tube temperature fluctuations. Pointing at the cause and characteristic of the crack, the paper suggests that as far as possible to reduce the long-term low load operation type boiler after a cadre state. Under a cadre state in boiler. Under low load condition, operators should strengthen the monitoring of the vertical water wall outlet wall temperature and adjust immediately when a big deviation emerged. Further improve the automatic feed water and automatic feeding coal to water ratio adjustment ability in the boiler operation and prevent water coal ratio imbalance in the process of load change.

Keyword: Thermal stress; Vertical water wall; Temperature difference; Supercritical boiler.

1. Introduction

1.1. *General introduction*

China power station is currently with high efficiency, low coal consumption as the goal towards a new energy utilization and environmental protection, equal emphasis on the sustainable development of the times. The important sign is that the supercritical and ultra supercritical units are developing rapidly [1]. With the application of supercritical and ultra supercritical technology in boiler, the design and operation of the boiler has changed correspondingly. Using the structure of the coil water wall in the radiation zone of the furnace and the vertical tube screen structure in the low heat intensity region of the radiant area of the furnace has been the common use of supercritical and ultra supercritical boiler [2]. The new design and operation mode of supercritical and ultra supercritical boiler show new features. It is also arise the phenomenon of vertical water wall stress change with the load [3]. There are many cracks in the outlet of the vertical water wall in a power plant supercritical boiler. It lead to the vertical water wall failure and boiler outage. In this paper, we make a analysis of operating parameters curve for the characteristics of the power plant boiler vertical water wall at the exit of the crack and vertical water wall at the outlet of the tube wall temperature, boiler water, coal quantity and separator outlet pressure in order to find out the cause and feature of the crack. And then, the countermeasures of controlling the crack are proposed. It provide a technical support and assistance to the normal operation and reconstruction of the boiler.

1.2. *Equipment introduction*

The boiler of plant is 350MW supercritical once though boiler and it's model is HG-1110 / 25.4-HM2 type, the boiler use Π type arrangement, balanced ventilation, singe reheat, solid slag, steel frame, full suspension structure.

Boiler is single furnace which wide is 14.6273m, deep is 14.6273. In BMCR condition, boiler superheater outlet steam pressure is 25.4MPa and temperature is 571 °C; reheat steam outlet pressure is 4.287MPa and temperature is 569 °C; water temperature is 284.1 °C.

Boiler milling system for the medium speed mill is direct blow pressurized system configured six HP863 type pulverizes, coal fineness R90 = 37%. SOFA burner is arranged in the main combustion chamber that is above the four corners area, in order to achieve staged combustion to reduce NOX emissions.

Boiler soda process by a built-in steam separator into double process, boiler start-up system is composed of built-in steam separator, water storage tank, water control valves, etc. Water wall membrane imported from the cold until the middle of the mix hopper between the collector box for the Spiral Wall, the intermediate header tube screen transition is converted to vertical, and is formed on the furnace front wall, side walls, hanging back wall and the water pipe,a temperature measuring point is installed in the vertical water wall outlet every eight root canal, water wall outlet header by small connection pipe to drop the entrance, the decline in the folding Angle of flame tube into the decoration of header, divided into two road, respectively by the folding flame Angle into the side water wall convection tube bundle and the horizontal flue inlet header into the horizontal flue side walls, from the horizontal flue side walls and the export of convection bank header into the steam separator.

2. The Characteristics of the Crack in the Vertical Water Wall Outlet

The leakage of the vertical water wall outlet of a supercritical boiler in a power plant is found in the operation, it cause the failure of vertical water wall and boiler outage. Before examination revealed shutdown, rear, left and right side wall at the exit of the vertical water wall multiple transverse cracks appear,as shown in Figure 1 circle.

Figure 1 (a). Crack Vision Photo.

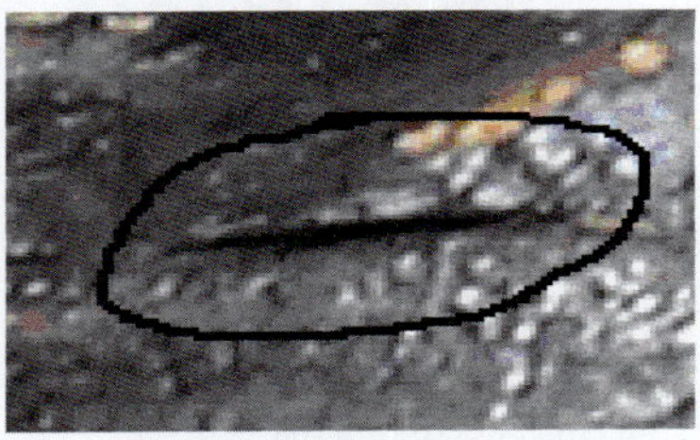

Figure 1 (b). Cracks close-range photograph.

Figure 1 (a) is a vertical water wall at the exit cracks vision pictures, Figure 1 (b) to crack the close-range pictures. Cracks appeared in the vertical water wall outlet, set near the weld box and located in the non-heated boilers. All cracks position located between the two tubes, Transverse state, left and right sides all present and steam blown out from the cracks, blowing thin tube neared the Wall.

3. The Cause Analysis of the Crack and the Countermeasures to Control

3.1. *The cause analysis of the crack*

To find the cause of boiler water wall at the exit of the vertical cracks reasons, we obtain a temperature curve of temperature measurement points from 50% BMCR unit adjacent to 35% BMCR conditions slippery pressure shutdown process running vertical water wall at the exit and corresponding to the amount of water, the amount of coal and the separator outlet pressure curve, Curve shown in Figure 2.

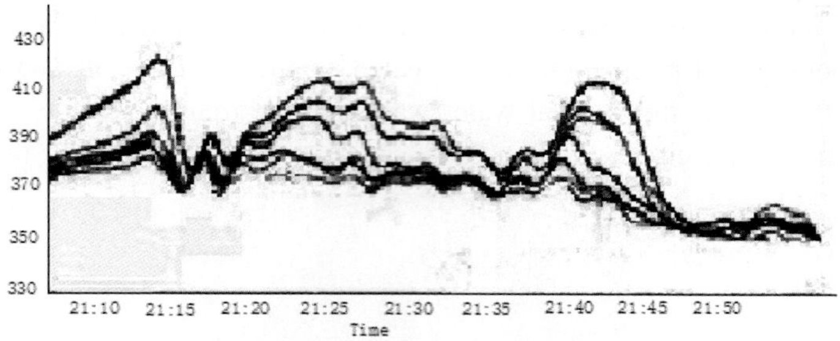

(a) Vertical Water wall at the outlet temperature graph

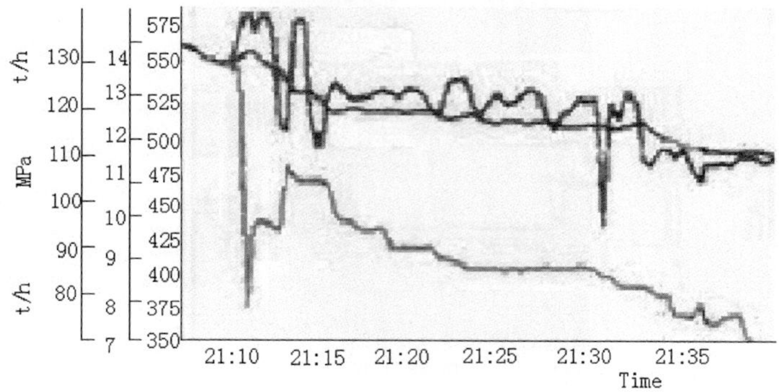

(b) Sliding boiler stop parameters graph

Figure 2. Boiler slip process stops running parametric curve.

Adjacent 8 temperature measuring points in Figure 2 (a) of the process of stopping the operation of the boiler slip at the exit of the vertical temperature profile Wall, 21:41:45 corresponding time temperature values in Table 1, 21:49:25 corresponding time temperature values in Table 2. As can be seen from Figure 2 (a) in a graph: There is a big temperature difference between vertical water wall tube. From the time of 21:41:45 temperature values can be found in Table 1. The maximum temperature difference reaches 75.1 °C. Taking into account that not every root canal temperature measuring points are installed, the temperature deviation portion of the tube may be greater. This leads to a temperature difference between tube telescopic tubes to produce pipe between the different stresses. By comparing Tables 1 and 2 can be found at the exit of vertical water wall tube temperature decreases and the rapid change to a larger, lower rate of up to 79.9 °C, the change rate reached 10.4 °C / min, is much larger than vertical water wall tube temperature is allowed to vary the rate of

2.5 °C / min. Rapid changes in tube temperature, the pipe and header will have a greater thermal shock.

Table 1. Corresponds to low temperature curve value table.

Temperature	1	2	3	4	5	6	7	8
Value	415.6	394.1	390.7	362.1	351.8	347.0	344.6	340.5

Table 2. Corresponds to low temperature curve value table.

Temperature	1	2	3	4	5	6	7	8
Value	335.7	329.8	328.5	329.2	326.6	327.3	327.0	326.6

Figure 2 (b) to the corresponding period of the boiler wall temperature of the water flow, water separator outlet pressure and boiler operating parameters to graph the total amount of coal. By comparing Figures 2 (a) and 2 (b) can be found, there is a corresponding relationship between the changes in the boiler water flow and vertical water wall tube temperature fluctuations.

3.2. *Pipe temperature deviation characteristics*

To find out the relationship between the vertical water wall temperature deviations and load, respectively obtain a full liter boiler load during vertical water wall at the exit of the temperature curve and down load during the vertical temperature profile at the outlet Wall.

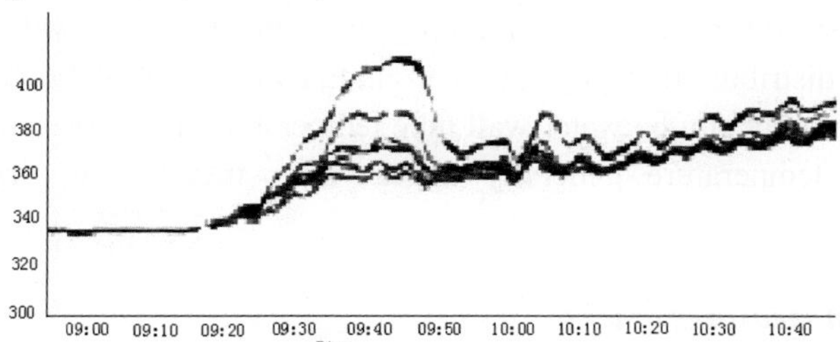

(a) Boiler load temperature rise curve

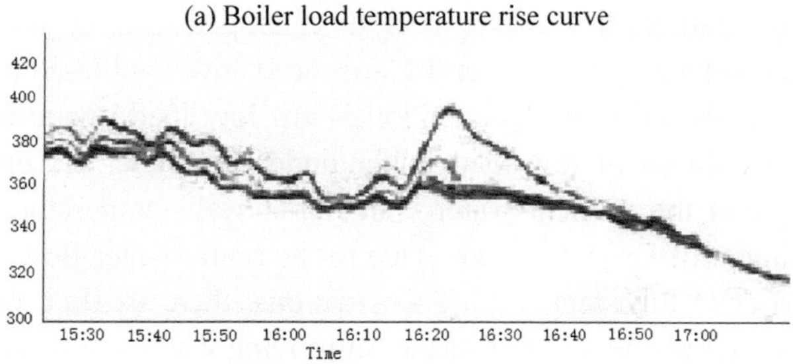

(b) Boiler load drop temperature graph

Figure 3. Boiler raising and lowering the load at the exit of the vertical temperature profile Wall.

Figure 3 is a boiler raising and lowering the load at the exit of the vertical temperature profile Wall, Figure 3 (a) is the boiler water wall at the exit of the vertical temperature rise during the load graph and Figure 3 (b) of the boiler water wall at the exit of the vertical temperature drop during load graph. From Figure 3 can be found that large deviation between the tube temperature and pipe temperature rapid change usually occurs in the boiler turn the dry state of low load conditions. Vertical Water wall at the outlet of which 80% BMCR conditions corresponding to the position of Figure 2 temperature values shown in Table 3, 25% of the vertical water wall at the exit of BMCR condition shown in Figure.2 corresponds to the position of the temperature as shown in Table 4.

Table 3. 80% BMCR conditions of temperature table.

Temperature	1	2	3	4	5	6	7	8
Value	433.6	430.1	430.7	428.9	428.3	427.7	427.2	423.9

Table 4. 25% BMCR conditions of temperature table.

Temperature	1	2	3	4	5	6	7	8
Value	325.6	323.4	322.7	321.8	320.7	320.4	320.2	322.0

From Table 3 can be found at the exit of the vertical water wall temperature deviation is small, a maximum of 9.7 °C. Heavy load operating conditions in the boiler temperature is high, but with little deviation. From Table 4 can be found in the boiler wet operating conditions, the vertical water wall at the exit temperature distribution, the deviation is small, a maximum of 3.6 °C. Integrated 3.2, between vertical tube water wall tube temperature deviation large and rapid changes in temperature generally appear after transduction dry low load conditions.

3.3. *Control measures cracks*

Since the inter-tube vertical water wall tube temperature deviation large and rapid changes in temperature generally appear at low load conditions turn dry after boiler, we should reduce boiler long-term low load operation in the dry state. In the condition of low load boiler under dry state, the operation staff should strengthen the vertical water wall outlet wall temperature monitoring, timely adjustment of large deviation. Due to the boiler water flow variation and vertical water wall tube temperature fluctuations, there exists a corresponding relation, boiler in the operation of further improving water supply automatic coal and to automatically on coal water ratio adjustment ability to prevent in the process of coal water ratio imbalance.

4. Conclusion

Vertical water wall of critical boiler running, in a certain power station, appear several cracks in outlet, causing failure and boiler shutdown. In this paper, the characteristics of the cracks and the boiler operation parameters curve are analyzed, which have found out the reason and the characteristic of the crack, crack control measures are put forward. The following conclusions:

(1) Boiler water wall outlet vertical cracks are made by larger temperature deviation between tube and the thermal stress of tube temperature change rapidly.

(2) Large deviation and the pipe temperature change quickly between vertical boiler water walls produced thermal stress that appearing in the boiler under low load condition after become a cadre state generally.

(3) The change of the boiler feed water flow and vertical there is a corresponding relationship of water wall tube temperature fluctuations.

(4) According to the characteristics and the reason of causing crack and suggested in this paper, the type boiler decrease as far as possible become a cadre state after the long-term low load operation; After become a cadre state in boiler under low load condition, the operation staff should strengthen the vertical water wall outlet wall temperature monitoring, a big deviation adjust; Automatic feed water and for further improvement in the operation of the boiler in automatic coal for coal water ratio adjustment ability, prevent the water coal ratio imbalance in the process of load change.

References

1. Rong Li-Hong, Tong Zhi-Min, Dong De-Fa. Fuzzy Predictive Control on Combustion System of Industrial Boiler. Microcomputer Information. 2007, 23(4):46–47.
2. Xiang-Jie Liu and C.W. Chan. Neuro-fuzzy generalized predicative control of boiler steam temperature. IEEE. Trans. Energy Conversion. Vol. 2l. No. 4. December. pp. 900–908, 2006.
3. Flshwlck P.A. Web-Based simulation: Some personal observations [A]. Proceedings of the 1996 Winter Simulation Conference [C]. San Diego, SCS, 1996, 772–779.

Effect of ecological factors in different latitudes on the quality of 2-year-old *Dendrobium officinale* Kimura et Migo*

S.Y. Teng*, Y.F. He*, J.Q. Shi, C.X. Lei, S.L. Zhong, H. Ma, W. Ren, D.L. Yan and B.S. Zheng†

Nurturing Station for the State Key Laboratory of Subtropical Silviculture, Zhejiang A & F University, Linan, Hangzhou 311300, P. R. China
Suzhou Shenyuan biological science and technology Inc and Suzhou City Dendrobium Institute, Wujiang, Jiangshu 215222, P. R. China
†E-mail: bszheng@zafu.edu.cn
**These authors contributed equally to this work*

The quality of 2-year-old *Dendrobium officinale* Kimura et Migo from four different latitudes Weifang City, Shandong Province (36°78'N, 119°02'E), Wujiang City, Jiangsu Province (31°08'N, 120°66'E), Wenzhou City, Zhejiang Province (28°27'N, 121°05'E) and Menghai City, Yunnan Province (21°59'N, 100°24'E) was analyzed in order to understand the effect of latitude, temperature, humidity, light intensity on its quality. It has important theoretical and practical significance to optimize its cultivation technology and improve its active ingredient content. The results showed that the ecological factors had effects on the quality of 2-year-old *D. officinale*. The quality of 2-year-old *D. officinale* in Weifang City, Shandong Province equals to or is higher than that in source regions. *D. officinale* can be widely planted in high latitude regions such as Weifang City, Shandong Province. The ecological factors for growth *D. officinale* in high latitude regions are as following: temperature three basis points are 4.7-17.2-32.3°C, mean temperature difference is 4.7-20.5°C, effective accumulated temperature is 1831°C, mean humidity is 69.7-89.5% and mean light intensity is 1.9-19.5 Klux.

Keywords: *Dendrobium officinale*; Latitude; Ecological factors; Quality; Correlation analysis.

1. Introduction

Polysaccharide, extract and nutrient content of *Dendrobium officinale* Kimura et Migo are important quality traits (Yang et al., 2004). Plants can adapt to the

* This work is supported by National Natural Science Foundation of China (31070604, 31270716 and 31470683); Zhejiang Provincial Natural Science Foundation for Distinguished Young Scholar (LR13C160001); Suzhou Shenyuan biological science and technology Inc and Suzhou City Dendrobium Institute; Open Foundation of Top Key Discipline of Forestry, Zhejiang Province (KF201314); Pre-research Foundation of Center for Cultivation of Subtropical Forest Resources (CCSFR), Zhejiang A & F University; Key Project of Chinese Ministry of Education (212072).

† Work partially supported by grant Scientific Research Foundation for the Returned Overseas Chinese Scholars, State Education Ministry ([2009]1341).

changes of the environments through changing the traits in vitro or in vivo by different pathways in order to survive in the nature. Therefore, different ecological factors, such as latitude, temperature, humidity and so on, have a certain effect on the quality and physiological index of plant.

Several authors (Zhou et al., 2005, Zheng et al., 2006, Zhang et al., 2008, Chen et al., 2013) have reported the tissue culture, rapid propagation, cultivation method, crude drug identification and chemical composition analysis in the *D. officinale*. Zhou et al. (2005) reported that The B5 medium was the best for bud inducing and proliferating; the highest efficiency of bud proliferation was obtained by the use of B5 medium with the addition of 0.1 mg/L 6-BA, coconut milk; the best effect of protocorm proliferation was obtained by the use of B5 medium with the addition of 1.0 mg/L NAA; the best effect of root inducing was obtained by the use of B5 medium with the addition of 1.0 mg/L NAA. Zheng et al. (2006) reported that the transplantation techniques of tube plantlets of *D. officianle* were developed upon the comprehensive consideration of various factors including media, light intensity, night temperature in winter, culture model and fertilizer. The techniques might be helpful to the protection and utilization of *D. officianle*. Zhang et al. (2008) summarized that the studies on various method of authentication of *D. officinale*, including organoleptic tests, microscopic examination, spectrometry, chromatography and molecular analysis and it can be able to provide some scientific references for the effective authentication of *D. officinale*. Chen et al. (2013) reported that *D. officinale* mainly contains polysaccharides,stilbenes, lignans and phenols. Fingerprint analysis, monosaccharide composition analysis of polysaccharide and multivariate statistical analysis could help in the identification and the quality evaluation of *D. officinale*. It is imperative to determine the low-molecular effective components of *D. officinale*. Multi-component analysis should be beneficial to perfect the quality evaluation system for *D. officinale*.

Hence, the objective of the present study was to examine the effect of ecological factors on the quality for *D. officinale* and its adaptation mechanism by comparative analysis the relationship between ecological factors under different latitude and physiological traits such as polysaccharide, chlorophyll and carotenoid contents.

2. Materials and Methods

2.1. *Plant material and growth condition*

The tests were performed in four different latitudes including Weifang City (WF), Shandong Province (36°78’N, 119°02’E), Wujiang City (WJ), Jiangsu

Province (31°08'N, 120°66'E), Wenzhou City (WZ), Zhejiang Province (28°27'N, 121°05'E) and Menghai City (MH), Yunnan Province (21°59'N, 100°24'E). The variety Shenyuan 2 of *D. officinale* was selected as a test material. The management of water and fertilizer and artificial growth substrates were consistent during growth period. The data of greenhouse temperature, humidity and light intensity was recorded from the ZDR-14 data recorder every hour per day.

2.2. *Methods*

Fifteen months after plant (during harvest period), fresh stems were collected from plant and immediately frozen in icebox. Five gram of frozen stems was used to measure polysaccharide, chlorophyll, and carotenoid contents. The content of polysaccharide was determined as described in the book Chinese Pharmacopoeia (2010) (Yang et al. 2004, Li et al. 2002, Zhang et al. 2004) and according to the phenol – sulfuric acid colorimetric method. The chlorophyll and carotenoids concentration were determined according to the methods of Lichtenthaler (1987) and Cai et al. (2005). 5 g sample of leaves were ground to a powder in liquid nitrogen and repeatedly extracted with 15 ml ice-cold 85 (v/v) and 100 % acetone. The photosynthetic pigments were extracted at 4 °C for 24 h in darkness. The absorbance of the supernatant was measured at 646 and 663 nm after centrifugation at 10,000g for 10 min.

2.3. *Statistical analysis*

The least-square means and analysis of variance (ANOVA, GLM procedure) were performed using SAS Institute Inc. (1999) software.

3. Results

3.1. *The relationship between the latitude and quality of D. officinale*

Table 1 showed that the relationship between the latitude of different experimental sites and the quality of the tested varieties. The quality index differed significantly among different latitudes. With the increasing of latitude, the polysaccharide, chlorophyll a, b and carotenoid increased.

Polysaccharide increased by 1.3%, 33.4% and 27.4% more than that in the lowest latitude MH for WZ, WJ, WF, respectively with the increase of latitude. Chlorophyll a increased by 141.2%, 70.6% and 47.1% more than that in the lowest latitude MH for WZ, WJ, WF, respectively with the increase of latitude. Chlorophyll b increased by 154.5%, 154.5% and 36.4% more than that in the lowest latitude MH for WZ, WJ, WF, respectively with the increase of latitude. Carotenoid increased by 2, 2.3 and 2.4 times more than that in the lowest latitude MH for WF, WJ, WZ, respectively with the increase of latitude.

The correlation coefficient between the latitude and polysaccharide contents was 0.77. The order of polysaccharide contents in different latitudes is WJ, WF, WZ and MH, of which, the polysaccharide content in MH is lowest, reached 37.0% while WJ and WF is the highest, reached 47.2% and 49.4%, respectively. However, the correlation between chlorophyll a, b and latitude respectively was not obvious. The correlation between latitude and carotenoid contents was positive, the correlation coefficient is 0.73.

Table 1. The effect of different latitude on quality of *D. officinale*.

Site	Latitude	Polysaccharide (%)	Chlorophyll a (mg/g)	Chlorophyll b (mg/g)	Carotenoid (mg/g)
WF	36.78°N	47.19	0.25	0.15	0.21
WJ	31.08°N	49.41	0.29	0.28	0.23
WZ	28.32°N	37.53	0.41	0.28	0.24
MH	21.64°N	37.04	0.17	0.11	0.07
Correlation coefficient		0.77	0.28	0.25	0.73

Note: WF, WJ, WZ, and MH means Weifang, Wujiang, Wengzhou, and Menghai, respectively.

3.2. *The relationship between the temperature and quality of D. officinale*

Table 2 showed that the effect of the effective accumulated temperature on the quality of *D. officinale*. The correlation between effective accumulated temperature and polysaccharide, chlorophyll a, b and carotenoid is positive. With the increase of effective accumulated temperature, polysaccharide, chlorophyll a, b and carotenoid increased. The effective accumulated temperature in MH is lowest and induces lack of energy for the growth of *D. officinale*. So it has negative effect on the transformation and accumulation of material. However, in WF, the highest temperature is 24.9°C and difference in temperature is large, so it is benefit to photosynthesis, increases the content of chlorophyll a, b and carotenoid, so as to increase the content of polysaccharide.

Table 2. The effect of effective accumulated temperature on quality of *D. officinale*.

Site	Effective accumulated temperature (°C)	Polysaccharide (%)	Chlorophyll a (mg/g)	Chlorophyll b (mg/g)	Carotenoid (mg/g)
WF	1831	47.19	0.25	0.15	0.21
WJ	2173	49.41	0.29	0.28	0.23
WZ	2218	37.53	0.41	0.28	0.24
MH	968	37.04	0.17	0.11	0.07
Correlation coefficient		0.46	0.86	0.89	0.99

Note: WF, WJ, WZ, and MH means Weifang, Wujiang, Wengzhou, and Menghai, respectively.

3.3. *The relationship between the moisture and quality of D. officinale*

The daily mean moisture and monthly mean moisture in four different latitudes was shown in Table 3. The lowest moisture in the four regions is MH, followed by WF and WJ. The highest temperature is WZ. The relationship between the daily mean moisture, monthly mean moisture and quality of *D. officinale* was evaluated in the growing season (Table 4). The results showed that the correlation coefficient between daily mean moisture and polysaccharide content is negative and reached -0.23. The polysaccharide content decreased with the increase of daily mean moisture and monthly mean moisture except January, February, March, April, July and August. However, the relationship between Chlorophyll a, b, carotenoid and daily mean moisture was positive, and reached 0.98, 0.72 and 0.87, respectively.

Interestingly, the chlorophyll a content increased with the increase of the monthly mean moisture in March, and April and showed a significant positive correlation, the coefficient reached 0.98 and 0.99 respectively. The significant positive correlation also existed between monthly mean moisture and carotenoid in February, April and August, the coefficient reached 0.99, 0.95 and 0.99 respectively.

Table 3. Mean humidity in one year.

	Humidity (%)												
Site	Jan	Feb	Mar	Apr	May	Jun	Jul	Aug	Sep	Oct	Nov	Dec	Mean
WF	89.1	86.2	77.9	78.3	77.3	82.3	86.2	89.5	81.0	69.7	76.1	85.1	81.5
WJ	77.8	88.8	82.8	81.4	75.1	81.9	84.4	88.9	84.8	76.6	83.6	73.6	81.7
WZ	76.0	93.9	87.5	87.1	88.6	89.0	84.0	90.3	88.8	80.7	85.5	83.2	86.2
MH	72.4	69.2	72.1	72.1	75.9	81.7	85.7	86.0	83.6	84.5	86.1	85.9	79.6

Table 4. The relationship between humidity and quality of *D. officinale*.

Month	Polysaccharide (%)	Chlorophyll a (mg/g)	Chlorophyll b (mg/g)	Carotenoid (mg/g)
Jan	0.65	0.02	-0.10	0.38
Feb	0.36	0.91	0.85	0.99**
Mar	0.13	0.98*	0.93	0.94
Apr	0.08	0.99**	0.89	0.95*
May	-0.54	0.84	0.46	0.66
Jun	-0.51	0.87	0.54	0.69
Jul	0.12	-0.80	-0.91	-0.59
Aug	0.34	0.86	0.71	0.99*
Sep	-0.49	0.79	0.70	0.50
Oct	-0.80	-0.08	-0.07	-0.45
Nov	-0.64	0.14	0.24	-0.23
Dec	-0.65	-0.27	-0.74	-0.32
All year	-0.23	0.98*	0.72	0.87

Note: WF, WJ, WZ, and MH means Weifang, Wujiang, Wengzhou, and Menghai, respectively.

3.4. *The relationship between light intensity and quality of D. officinale*

The daily mean light intensity and monthly mean light intensity in four different latitudes was shown in Table 5. The lowest light intensity in the three regions is MH, followed by WF. The highest light intensity is WJ.

The relationship between the daily mean light intensity, monthly mean light intensity and quality of *D. officinale* was evaluated in the growing season (Table 6). The results showed that the correlation coefficient between daily mean light intensity and polysaccharide content is positive and reached 0.88. The polysaccharide content increased with the increase of daily mean light intensity and monthly mean light intensity. The relationship between Chlorophyll a, b, carotenoid and daily mean light intensity was also positive, and reached 0.94, 0.99 and 0.78, respectively.

Interestingly, the chlorophyll a content increased with the increase of the monthly mean light intensity in June and September and showed a significant positive correlation, the coefficient reached 0.99 and 0.99 respectively. The significant positive correlation also existed between monthly mean light intensity and chlorophyll b in November, the coefficient reached 0.99.

Table 5. Mean light intensity in one year.

Site	Mean light intensity (Klux)									
	Mar	Apr	May	Jun	Jul	Aug	Sep	Oct	Nov	Mean
WF	1.93	3.53	6.13	19.46	16.40	17.47	15.92	10.39	8.35	11.07
WJ	12.45	19.02	24.37	29.73	33.56	32.05	22.30	19.25	13.45	22.91
MH	2.69	2.39	4.43	3.71	3.00	5.91	4.69	4.19	6.26	4.14

Note: WF, WJ, WZ, and MH means Weifang, Wujiang, Wengzhou, and Menghai, respectively.

Table 6. The relationship between light intensity and quality of *D. officinale*.

Month	Polysaccharide (%)	Chlorophyll a (mg/g)	Chlorophyll b (mg/g)	Carotenoid (mg/g)
Mar	0.59	0.71	0.96	0.44
Apr	0.68	0.79	0.99	0.55
May	0.70	0.80	0.99	0.57
Jun	0.97	0.99*	0.91	0.92
Jul	0.91	0.97	0.97	0.83
Aug	0.91	0.97	0.97	0.83
Sep	0.98	0.99*	0.90	0.93
Oct	0.90	0.96	0.98	0.81
Nov	0.83	0.91	0.998*	0.72
All year	0.88	0.94	0.99	0.78

4. Discussion

The polysaccharide content increased with the increase of daily mean temperature. Lin et al. (1994) and Wang et al. (1998) found that the respiration rate decreased with the decrease of temperature under the certain temperature period, but as a substrate of respiration, the soluble sugar content increased significantly, which is benefit to resistant to stress and sugar accumulation. This was consistent with the results we reported in *D. officinale*. The polysaccharide content in WF was higher than that in WZ and MH. Ai et al. (2010) also reported that temperature also have a significant effect on the growth and polysaccharide content of *D. officinale*'s stem. The polysaccharide content of *D. officinale* at 20°C significantly higher than that at other temperatures, while the stem length, stem node number, stem fresh weight and stem dry weight were achieved the highest point at 30°C. The results showed that the temperature of 30°C was more appropriate for the photosynthesis of *D. officinale*, while the plants at 20°C have higher polysaccharide content.

There were few reporters about the effect of light on quality of *D. officinale*. Bao et al. (2007) reported that the effect of photoperiod on growth and development of *D. officinale* in vitro under controlled environment in a closed plant factory with artificial lighting. Single explant each with 300 mg fresh mass of *D. officinale* were planted in five treatments with photoperiods of 6, 9, 12, 15, and 18 h/d under controlled environment at air temperature of 22°C, relative humidity of 65%, lighting intensity of 68 umol.m^{-2}.s^{-1}, and CO_2 concentration of 800 umol.mol-1 in photoperiod. The net photosynthetic rate, chlorophyll content, dry mass, and buds in treatment of photoperiod of 12 h/d were higher than that in other treatments, and resulted in greater growth and reproduction for *D. officinale* production in vitro after the explants were cultured 92 days. Therefore, the photoperiod of 12 h.d-1 for *D. officinale* production in vitro is suitable in the closed plant factory with artificial lighting. Based on the former research, this paper studied the relationship between light intensity and quality and found that light intensity promote chlorophyll a, b and carotenoid synthesis, increase photosynthesis of the leaves, then promote organic matter accumulation including polysaccharide. The strong local light intense and long sunshine time can induce high temperature and high humidity in MH. So the open type greenhouse cultivation and management can promote growth in MH faster than that in the other regions, then get higher yield. However, the polysaccharide content in MH is less than that in the other regions and the quality decreased finally.

The cultivation of *D. officinale* is widely cultivated in Yunnan, Guangxi, Zhejiang, Anhui and other regions. Many authors (Lin et al. 2010, Zheng et al. 2012) focused on cultivation, propagation, genetic diversity and so on, but few reporters were found in WF. In this paper, the effect of different latitude on the quality of *D. officinale* was studied and the results showed that the quality of *D. officinale* in WF equal to or higher than that in original regions. The cultivation of *D. officinale* can be widely planted in high latitude regions such as Weifang City, Shandong Province.

References

1. G. F. Li, J. J. Li and J. Y. Xu. Research Summary of *Dendrobium officinale*. *Journal of Chinese Medicinal Materials* 33 (1), 150 (2010).
2. J. H. Zhou, X. F. Zhong and D. W. Cai. Study of tissue culture and rapid propagation in *Dendrobium candidum*. *Journal of ZhongKai university of agriculture and technology* 18 (1), 23 (2005).
3. Y. P. Zheng, C. Wang, J. Y. Yu et al.Techniques on the Transplantation of *Dendrobium officinale* Tube Plantlets. *China forestry science and technology development* 20 (6), 26 (2006).
4. Q. Zhang, C. L Duan. Studies on the Authentication of *Dendrobium officinale*. *Biotechnology bulletin* 6, 69 (2008).
5. X. M. Chen, C. L. Wang, J. S. Yang et al. Research Progress on Chemical Composition and Chemical Analysis of *Dendrobium officinale*. *Chinese pharmaceutical journal* 48 (19), 1634 (2013).
6. Chinese Pharmacopoeia, 1st edn. (2010).
7. H. Yang, S. C. Wang, Z. T. Wang. Structural analysis of polysaccharides from *Dendrobium candidum*. *Chinese pharmaceutical journal* 39 (4), 254 (2004).
8. Y. F Li, X. H. Zhang, G. M. Sun. Determination of Alkaloids and Polysaccharide in *Herba Dendrobii*. *Chinese pharmaceutical affairs* 16 (7), 426 (2002).
9. X. Y. Zhang. Test of polysaccharide content in three genotypes of *Dendrobium officinale* in Sichuan. *Journal of Leshan Teachers College*, 19 (5), 88 (2004).
10. Y. P. Cai, L. Li, H. S. Li. Leaf photosynthetic characteristics and its response for light intensity of three species of *Dendrobium* Sw. in Huoshan County. *Chinese traditional and herbal drugs* 36 (4), 586(2005).

11. G. H. Li, J. S. Liu, Y. B. Wang. Determination of trace elements in edible cactus by atomic absorption spectrometry. *Spectroscopy and spectral analysis* 25 (12), 2079 (2005).
12. Y. Zhu, H. Yuan, G. D. Li. Study on 11 mental element contents in *Dendrobium officinale. China hina journal of chinese materia medica* 36 (3), 356 (2011).
13. M. X. Lin, H. J. Yang. Physiological responses of *Hevea Brasiliensis* during the chilling injury. *Chinese Journal of Tropical Crops* 15 (02), 7 (1994).
14. X. X. Wang, S. D. Li, H. R. Dong et al. The Correlationship of Cold-tolerance with ABA, soluble Sugar and Respiratory Intensity in Tomato. *Acta Horticulturae Sinica* 25 (1), 56 (1998).
15. S. S. Bao, D. X. He and S. X. Guo. Suitable Photoperiod of *Dendrobium Officinale* in vitro under Controlled Environment. *Journal of agricultural science and technology* 9 (6), 90 2007.
16. C. Hong, Z. Zhang, Z. G. Wu. Morphological characters and polysaccharide contents of 4 cultured varieties of *Dendrobium officinale* in Wenzhou Region. *Journal of ZheJiang agricutural sciences* 6, 1261 (2010).
17. P. Zheng, G. L. Zhao, H. J. Ye et al. Main characters and techniques on reproduction and cultivation of *Dendrobium candidum* in Guangdong area. *Guangdong Agricultural Sciences* 39 (22), 25 (2012).
18. H. K. Lichtenthaler. Chlorophylls and carotenoids: pigments of photosynthetic membranes. *Method Enzymol* 148, 350 (1987).
19. J. Ai, N. Yan, H. Hu and S.Y. Li. Effects of temperature on the growth and physiological characteristics of *Dendrobium officinale* (Orchidaceae). *Acta Botanica Yunnanica* 32 (5), 420 (2010).

Effect of different latitudes on the nutrient of 2-year-old *Dendrobium officinale*[*]

W. Ren[*], Y.F. He[*], D.L. Yan[*], S.Y. Teng, J.Q. Shi, C.X. Lei, S.L. Zhong, H. Ma and B.S. Zheng[†]

Nurturing Station for the State Key Laboratory of Subtropical Silviculture, Zhejiang A & F University, Linan, Hangzhou 311300, P. R. China
Suzhou Shenyuan Biological Science and Technology Inc and Suzhou City Dendrobium Institute, Wujiang, Jiangshu 215222, P. R. China
[†]*E-mail: bszheng@zafu.edu.cn*
[*]*These authors contributed equally to this work*

The nutrient of 2-year-old *Dendrobium officinale* Kimura et Migo from four different latitudes Weifang City, Shandong Province (36°78'N, 119°02'E), Wujiang City, Jiangsu Province (31°08'N, 120°66'E), Wenzhou City, Zhejiang Province (28°27'N, 121°05'E) and Menghai City, Yunnan Province (21°59'N, 100°24'E) was analyzed in order to understand the effect of latitude, temperature, humidity, light intensity on its nutrient. The results showed that the ecological factors had effects on the nutrient of 2-year-old *D. officinale*. The nutrient copper and zinc content increased with the increase of latitude, effective accumulated temperature. The magnesium and zinc increased with the increase of humidity. The nutrient of 2-year-old *D. officinale* in WF equals to or is higher than that in the other places.

Keywords: *Dendrobium officinale*; Latitude; Ecological factors; Nutrient; Correlation analysis.

1. Introduction

Dendrobium officinale is perennial epiphytic herb, mainly distributed in the southwest and southern of China. The medicinal part of Dendrobium is fresh or dried stem. In traditional Chinese medicine, it has a long medicinal history, which possesses the function of nourishing the blood and clearing heat, promoting body fluid and benefiting stomach, moistening lung and relieving a

[*] This work is supported by National Natural Science Foundation of China (31070604, 31270716 and 31470683); Zhejiang Provincial Natural Science Foundation for Distinguished Young Scholar (LR13C160001); Suzhou Shenyuan biological science and technology Inc and Suzhou City Dendrobium Institute; Open Foundation of Top Key Discipline of Forestry, Zhejiang Province (KF201314); Pre-research Foundation of Center for Cultivation of Subtropical Forest Resources (CCSFR), Zhejiang A & F University; Key Project of Chinese Ministry of Education (212072).

[‡] Work partially supported by grant Scientific Research Foundation for the Returned Overseas Chinese Scholars, State Education Ministry ([2009]1341).

cough etc. Modern pharmacology research showed that it has anti-tumor, anti-aging, enhances human immunity and dilating blood vessel (Li et al. 2010).

In order to protect this rare plant, since the seventies of the 20th century, scientists from domestic and abroad institutes began to study and development *D. officinale*, especially in tissue culture, rapid propagation and artificial cultivation. It provided strong technical support for its wide cultivation. However, rare deeply studying on nutrition and its active ingredient is conducted, especially lack of system study on its nutritional value.

Several authors (Chen et al. 2005, Jin et al. 2007, Wu et al. 2011, Jin et al. 2009 and Wei et al. 2015) have reported the nutrient components analysis in the *D. officinale*. Chen et al. (2005) reported that C, N, and P absorption were enhanced in inoculated seedlings. Guo et al. (2006) analyzed the nutrients component from cultured *D. officinale* and showed the nutrient is the same with that in natural cultivation. Jin et al. (2007) reported that the effects of mycorrhizal fungi on the growth and mineral nutrition absorption of *D. officinale* and the results showed that the mineral elements contents of the seedlings in six strains of fungi had been higher than those of the control, but the zinc is less. Jin et al. (2009) also reported that the contents of B, Si, Fe, Cu and Mn in inoculated seedlings were increased. Wu et al. (2011) studied that the effect of symbiotic fungi on the physiological characteristics of *D. catenatum* and the results showed that the content of N, P, and K in inoculated seedlings were higher than that in the control. Wei et al. (2015) compared the nutritional composition of between Hainan and Zhejiang. The results showed that *D. officinale* from Hainan and Zhejiang had ample mineral contents and the content of K, Mg, Ca and P was relatively high, and the content of multi-trace elements in *D. officinale* from Hainan was higher than that from Zhejiang.

However, there is no report about the effect of different latitudes on the nutrient content of *D. officinale*. Hence, the objective of the present study was to comparative analyze and evaluate the effect of different latitude on the main nutrient for *D. officinale* in order to lay theoretical and practical foundation for further develop its cultivation and utilization to the north regions of Yangtze River.

2. Materials and Methods

2.1. *Plant Material and growth condition*

The tests were performed in four different latitudes including Weifang City (WF), Shandong Province (36°78’N, 119°02’E), Wujiang City (WJ), Jiangsu Province (31°08’N, 120°66’E), Wenzhou City (WZ), Zhejiang Province

(28°27'N, 121°05'E) and Menghai City (MH), Yunnan Province (21°59'N, 100°24'E). The variety Shenyuan 2 of *D. officinale* was selected as a test material. The management of water and fertilizer and artificial growth substrates were consistent during growth period. The data of greenhouse temperature, humidity and sunshine time was recorded from the ZDR-14 data recorder every hour per day.

2.2. *Methods*

Fifteen months after plant (during harvest period), fresh stems were collected from plant and immediately frozen in icebox. Five gram of frozen stems was used to measure nutrient contents. The content of nutrient elements was determined by flame atomic absorption spectrometry according to Li et al. (2005) and Zhu et al. (2011).

2.3. *Statistical analysis*

The least-square means and analysis of variance (ANOVA, GLM procedure) were performed using SAS Institute Inc. (1999) software.

3. Results

3.1. *The relationship between the latitude and nutrient of D. officinale*

Table 1 showed that the relationship between the latitude of different experimental sites and the nutrient content of the tested varieties. The nutrient content differed significantly among different latitudes. With the increasing of latitude, the Cu, Zn contents increased, while K, Mg and Mn contents decreased. Copper decreased by 34.4% less than that in the lowest latitude MH for WZ while increased by 1.6% and 6.4 times more than that in the lowest latitude for WF and WJ respectively. Magnesium decreased by 13.9% and 63.3% less than that in the lowest latitude MH for WJ and WF, respectively while increased by 21.7% more than that in the lowest latitude for WZ. Zinc increased by 66.5% and 166.2% more than that in the lowest latitude MH for WZ and WF, respectively while decreased by 10.5% less than that in the lowest latitude for WJ.

However, Potassium decreased by 41.7%, 72.6% and 53.2% less than that in the lowest latitude MH for WZ, WJ, WF, respectively with the increase of latitude. Manganese decreased by 46.0%, 63.8% and 62.7% less than that in the lowest latitude MH for WZ, WJ, WF, respectively with the increase of latitude.

The correlation between latitude and nutrient contents was different, among which, the correlation between potassium, manganese, magnesium and latitude was negative, the highest potassium and magnesium content was in MH, while the highest manganese content was in WZ. The correlation between copper, zinc and latitude was positive, the highest copper and zinc was in WF.

Table 1. The effect of different latitude on nutrient of *D. officinale*.

Site	Latitude	Potassium (mg/g)	Magnesium (mg/g)	Copper (mg/kg)	Manganese (mg/kg)	Zinc (mg/kg)
WF	36.78°N	4.54	0.61	6.2	78.2	70.8
WJ	31.08°N	2.66	1.43	45.1	75.9	23.8
WZ	28.32°N	5.65	2.02	4.0	113.2	44.3
MH	21.64°N	9.70	1.66	6.1	209.6	26.6
Correlation coefficient		-0.80	-0.75	0.18	-0.91	0.73

Note: WF, WJ, WZ, and MH means Weifang, Wujiang, Wengzhou, and Menghai, respectively.

3.2. *The relationship between the temperature and nutrient of D. officinale*

Table 2 showed that the effect of the effective accumulated temperature on the nutrient content of *D. officinale*. The correlation between effective accumulated temperature and magnesium, copper, zinc is positive, but potassium and manganese is negative. With the increase of effective accumulated temperature, magnesium, copper and zinc increased. The effective accumulated temperature in MH is lowest and induces lack of energy for the growth of *D. officinale*. So it has negative effect on the transformation and accumulation of material. However, in WF, the highest temperature is 24.9°C and difference in temperature is large, so it is benefit to photosynthesis, so as to increase the content of magnesium, copper and zinc.

Table 2. The effect of effective accumulated temperature on nutrient of *D. officinale*.

Site	Effective Accumulated Temperature (°C)	Potassium (mg/g)	Magnesium (mg/g)	Copper (mg/kg)	Manganese (mg/kg)	Zinc (mg/kg)
WF	1831	4.54	0.61	6.2	78.2	70.8
WJ	2173	2.66	1.43	45.1	75.9	23.8
WZ	2218	5.65	2.02	4.0	113.2	44.3
MH	968	9.70	1.66	6.1	209.6	26.6
Correlation coefficient		-0.87	0.03	0.40	-0.87	0.21

Note: WF, WJ, WZ, and MH means Weifang, Wujiang, Wengzhou, and Menghai, respectively.

3.3. *The relationship between the moisture and nutrient of D. officinale*

The daily mean moisture and monthly mean moisture in four different latitudes was shown in Table 4. The lowest moisture in the four regions is MH, followed by WF and WJ. The highest temperature is WZ.

The relationship between the daily mean moisture, monthly mean moisture and nutrient content of *D. officinale* was evaluated in the growing season (Table 5). The results showed that the relationship between moisture and different nutrient content was different. The correlation coefficient between daily mean moisture and potassium, Copper and manganese content is negative and reached -0.33, -0.18 and -0.40 respectively while the correlation coefficient between daily mean moisture and magnesium and zinc content is positive and reached 0.47 and 0.21 respectively.

Interestingly, the copper content decreased with the increase of moisture in December. The coefficient between them is -0.97. But there was no significant correlation between K, Mg, Mn, Zn content and monthly mean moisture.

Table 3. Mean humidity in one year.

Site	Humidity (%)												
	Jan	Feb	Mar	Apr	May	Jun	Jul	Aug	Sep	Oct	Nov	Dec	Mean
WF	89.1	86.2	77.9	78.3	77.3	82.3	86.2	89.5	81.0	69.7	76.1	85.1	81.5
WJ	77.8	88.8	82.8	81.4	75.1	81.9	84.4	88.9	84.8	76.6	83.6	73.6	81.7
WZ	76.0	93.9	87.5	87.1	88.6	89.0	84.0	90.3	88.8	80.7	85.5	83.2	86.2
MH	72.4	69.2	72.1	72.1	75.9	81.7	85.7	86.0	83.6	84.5	86.1	85.9	79.6

Note: WF, WJ, WZ, and MH means Weifang, Wujiang, Wengzhou, and Menghai, respectively.

Table 4. The relationship between humidity and nutrient of *D. officinale*.

Month	Potassium (mg/g)	Magnesium (mg/g)	Copper (mg/kg)	Manganese (mg/kg)	Zinc (mg/kg)
Jan	-0.54	-0.89	-0.07	-0.70	0.87
Feb	-0.80	0.03	0.23	-0.84	0.33
Mar	-0.65	0.36	0.23	-0.64	0.07
Apr	-0.62	0.34	0.14	-0.64	0.15
May	0.02	0.56	-0.48	-0.08	0.23
Jun	-0.03	0.61	-0.39	-0.11	0.15
Jul	0.36	-0.75	-0.39	0.24	0.48
Aug	-0.75	-0.10	0.06	-0.84	0.52
Sep	-0.03	0.89	0.01	0.03	-0.39
Oct	0.70	0.87	-0.15	0.82	-0.75
Nov	0.45	0.95	0.09	0.60	-0.84
Dec	0.74	-0.09	-0.969*	0.55	0.50
All year	-0.33	0.47	-0.18	-0.40	0.21

3.4. *The relationship between light intensity and nutrient of D. officinale*

The daily mean light intensity and monthly mean light intensity in four different latitudes was shown in Table 6. The lowest light intensity in the three regions is MH, followed by WF. The highest light intensity is WJ.

The relationship between the daily mean light intensity, monthly mean light intensity and nutrient content of *D. officinale* was evaluated in the growing season (Table 7). The results showed that the relationship between light intensity and different nutrient content was different. The correlation coefficient between daily mean light intensity and potassium, magnesium, manganese and zinc content is negative and reached -0.95, -0.14, -0.84 and -0.12 respectively while the correlation coefficient between daily mean light intensity and copper content is positive and reached 0.90.

Interestingly, the copper content increased with the increase of light intensity from March to May. The coefficient between them is 0.998, 0.998 and 0.997 respectively. But there was no significant correlation between K, Mg, Mn, Zn content and monthly mean light intensity.

Table 5. Mean light intensity in one year.

	Mean light intensity (Klux)									
Site	Mar	Apr	May	Jun	Jul	Aug	Sep	Oct	Nov	Mean
WF	1.93	3.53	6.13	19.46	16.40	17.47	15.92	10.39	8.35	11.07
WJ	12.45	19.02	24.37	29.73	33.56	32.05	22.30	19.25	13.45	22.91
MH	2.69	2.39	4.43	3.71	3.00	5.91	4.69	4.19	6.26	4.14

Note: WF, WJ, WZ, and MH means Weifang, Wujiang, Wengzhou, and Menghai, respectively.

Table 6. The relationship between light intensity and nutrient of *D. officinale*.

Month	Potassium (mg/g)	Magnesium (mg/g)	Copper (mg/kg)	Manganese (mg/kg)	Zinc (mg/kg)
Mar	-0.66	0.37	0.998*	-0.46	-0.60
Apr	-0.75	0.25	0.998*	-0.56	-0.49
May	-0.76	0.23	0.997*	-0.58	-0.48
Jun	-0.99	-0.33	0.80	-0.93	0.07
Jul	-0.94	-0.14	0.90	-0.84	-0.12
Aug	-0.95	-0.14	0.90	-0.84	-0.12
Sep	-0.99	-0.36	0.78	-0.94	0.10
Oct	-0.93	-0.11	0.91	-0.82	-0.15
Nov	-0.88	0.03	0.96	-0.73	-0.29
All year	-0.92	-0.06	0.93	-0.79	-0.20

4. Discussion

The nutrient content is one of the most important index in *D. officinale*. Our research reported that the potassium and manganese content in MH is the highest in the four tested regions, but magnesium content in WZ is the highest, copper in WJ is the highest, zinc content in WF is the highest. Wei et al. (2015) compared the nutritional composition of between Hainan and Zhejiang. The results also showed that *D. officinale* from Hainan and Zhejiang had ample mineral contents and the content of K, Mg, Ca and P was relatively high, and the content of multi-trace elements in *D. officinale* from Hainan was higher than that from Zhejiang. This is the same with what we reported about potassium content. However, Guo et al. (2006) reported that the nutrient from cultured *D. officinale* is the same with that in natural cultivation. We reported that the nutrient content is different when *D. officinale* is cultivated in different latitudes. With the increasing of latitude, the Cu, Zn contents increased, while K, Mg and Mn contents decreased. The effective accumulated temperature can promote increase of magnesium, copper, zinc contents, but decrease of potassium and manganese. And daily mean moisture can promote increase of magnesium and zinc content, but decrease of potassium, copper and manganese content. However, daily mean light intensity can promote copper increase and decrease of the other tested nutrient content. This is the first report about the effect of different latitudes on the nutrient contents of *D. officinale*. Different ecological factors can have effect on the nutrient contents in *D. officinale*, which has benefit for its cultivation and utilization in the larger regions, especially in north of China.

References

1. J. Wei, M. T. Gu, Y. Y. Ji et al. Comparison of nutritional composition of *Dendrobium officinale* from Hainan and Zhejiang. *Chinese Journal of Tropical Crops* 36 (6): 1059 (2015).
2. M. B. Guo, L. Y. Feng, M. J. Tian et al. Analysis of nutrients from cultured *Dendrobium officinale*. *Junnan Chemical Technology* 33 (2): 15 (2006).
3. H. Jin, Z. H. Kang, H. Chen et al. Effects of mycorrhizal fungi on the growth and mineral nutrition absorption of *Dendrobium candidum*. *Journal of Fujian College of Forestry* 27 (1): 80 (2007).
4. H. F. Wu, X. Q. Song, F. S. Yang et al. Effects of symbiotic fungi on the physiological characteristics of *Dendrobium catenatum* Lindley (Orchidaceae). *Plant Science Journal* 29 (6): 738 (2011).
5. H. Jin, Z. X. Xu, J. H. Chen et al. Interaction between tissue-cultured seedlings of *Dendrobium officinale* and Mycorrhizal fungus (Epulorhiza

sp.) during symbiotic culture. *Chinese Journal of Plant Ecology* 33 (3): 433 (2009).

6. L. Q. Chen, X. M. Wang and Z. D. Pei. Study on absorbtion and exploitation to P nourishment in the Mycorrhiza of *Dendrobium candidum*. *Forest Research* 18 (2): 163 (2005).
7. L. Q. Chen, X. M. Wang and Z. D. Pei. Absorption of N–element in Dendrobium plants cultivated from tissues and Mycorrhiza. *Scientia Silvae Sinicae* 43 (1): 48 (2007).
8. L. Q. Chen, X. M. Wang and Z. D. Pei. The absorption of C–element by Dendrobium plants cultivated from tissues and mycorrhizae. *Ecology and Environment* 14 (3): 410 (2005).
9. G. F. Li, J. J. Li and J. Y. Xu. Research Summary of *Dendrobium officinale*. *Journal of Chinese Medicinal Materials* 33 (1): 150 (2010).
10. G. H. Li, J. S. Liu, Y. B. Wang. Determination of Trace Elements in Edible Cactus by Atomic Absorption Spectrometry. *Spectroscopy and spectral analysis* 25 (12), 2079 (2005).
11. Y. Zhu, H. Yuan, G. D. Li. Study on 11 mental element contents in *Dendrobium officinale*. *China Journal of Chinese Materia Medica* 36 (3), 356 (2011).

Vibration characteristics of GIS bus based on field measurement

J.J. Huang*, Y.F. Lu, B.K. Yang, D.J. Li, Y.B. Zhang, Q.F. Chen and F.F. Wang
Electric Power Research Institute of Guangxi Power Grid Co., Ltd., Nanning 530023, China
**E-mail: huang_jj.sy@gx.csg.cn*

GIS is widely used in grid for its excellent performance. In the practical operation, besides the accidents caused by discharge faults, there are also a lot of accidents which are induced by mechanical faults. The electrical detection methods are mainly about discharge faults, there's lack of methods about mechanical faults. The field measurement of 110 kV GIS, especially the part of bus, is carried out using the GIS vibration test system in this paper. The results show that the vibration of bus near the part with contact faults is strongest, while the bus vibration is generally weak. The results provide data support for the on-site testing of GIS vibration and fault diagnosis of mechanical faults.

Keywords: GIS; Mechanical faults; GVTS; Vibration characteristics of bus.

1. Introduction

With good operation reliability, GIS is widely used in the power system. In order to ensure its safe operation, the research of internal faults prediction is necessary. At present, there are many ways to detect the discharge failure [1–2]. Starting from practical operation conditions, not only discharge failure but also mechanical faults may cause accidents [3–6]. Research results of China Electric Power Research Institute et al show that vibration of GIS is so harmful for the equipment that damage the insulator and weaken the firmness of enclosure grounding [7–9]. Long-term vibration would cause accidents by loosing screws, making gas leakage and decreasing the gas pressure. Therefore, strengthening the detection of mechanical failure is an important way to ensure the operation safety of GIS. According to this situation, foreign Scholars found that it's possible to detect mechanical faults such as abnormality of electric contact by measuring the vibration of enclosure [10–12]. There are also domestic research about diagnosing GIS faults by detecting vibration [13–14]. Researchers found that vibration signal can be used to detect internal faults, but didn't have systematic study in detail or application of field detection device. In order to further study of vibration signals in GIS enclosure, analyze the characteristic of the signal, solve the field corresponding problem, in this paper, the study is about the field measurement of PT vibration and analysis of results by using the GIS vibration test system.

2. Measuring Datum

2.1. *Situation of substation*

The bus of GIS in this 110kV substation is three-phase community bus. PT, whose upper, middle and lower parts are detected, and the connection are chose to be the measuring points. The wiring diagram and measuring points are shown in Figure 1.

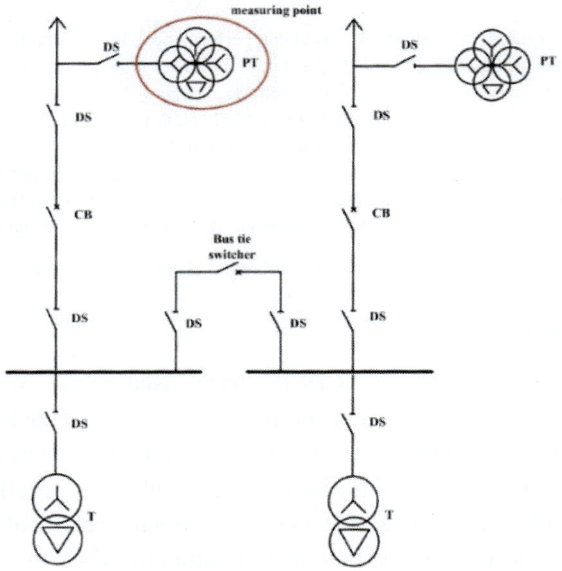

Figure 1. Wiring diagram of substation and measuring points.

2.2. *Vibration of background*

In order to eliminate outside interference from the measured data, the vibration of GIS base is chose to be benchmark. The vibration, measuring by putting probe on the base, is shown in Figure 2.

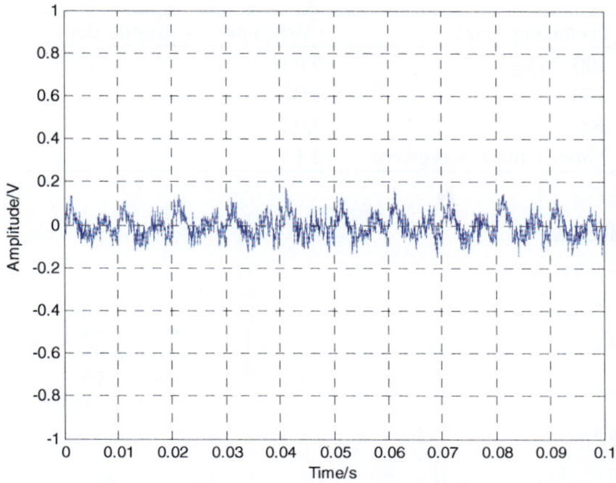

(a) Time-domain waveform

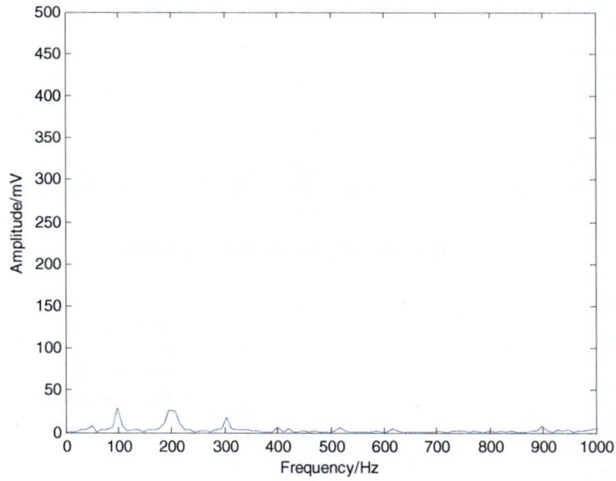

(b) Spectrum

Figure 2. Vibration of background.

As is shown in Figure 2, there is also some vibration in background. The amplitude of time-domain waveform is 0.17V. The vibration dominant frequency is 100Hz, 200Hz and 300Hz. Combined with laboratory and field data analysis, the vibration is caused by electrodynamic force which is made by operation of equipment and surrounding environment. On the other hand, a little bad contact is another reason. But mechanical fastening is good on the whole.

Table 1. Statistics of background vibration.

Frequency (Hz)	Corresponding amplitude(V)
100	0.03
200	0.03
300	0.02
Time-domain waveform	0.17

3. Vibration of Bus

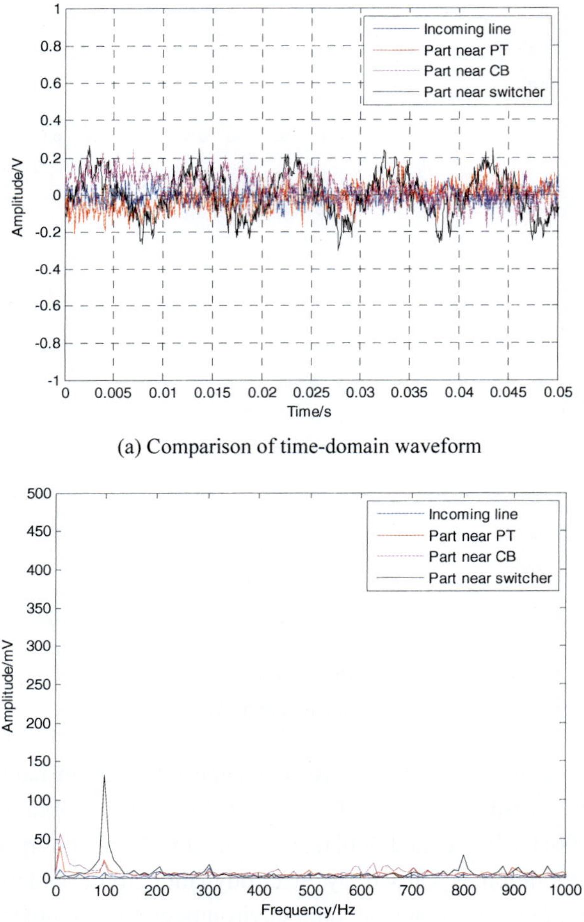

(a) Comparison of time-domain waveform

(b) Comparison of spectrum

Figure 3. Comparison of bus vibration.

Table 2. Comparison of bus vibration.

Location	Amplitude of time–domain waveform (V)	Dominant frequency–Corresponding amplitude
Incoming line	0.14 V	100Hz-0.01V, 300Hz-0.01V
Part near PT	0.20 V	100Hz-0.02V, 300Hz-0.01V
Part near CB	0.24 V	100Hz-0.02V, 300Hz-0.01V, 600Hz-0.01V
Part near switcher	0.29 V	100Hz-0.13V, 300Hz-0.02V, 800Hz-0.01V

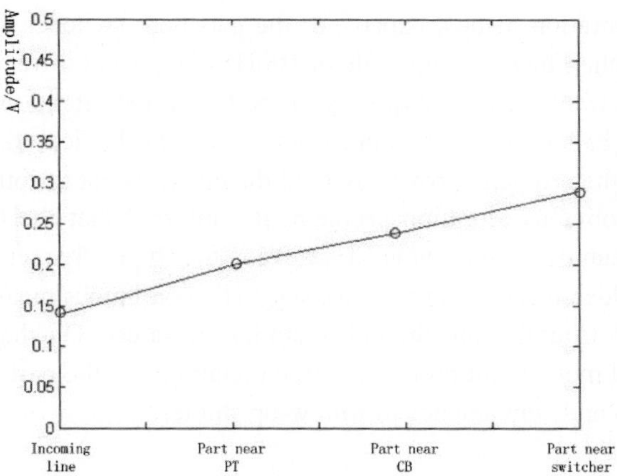

Figure 4. Amplitude trend of different locations in bus.

The vibration, whose amplitude is within 0.3V and lowers than that of double of benchmark, of bus is slight on the whole. Expect for the part near switcher, dominant frequency is 100Hz and amplitude is within 1 time of benchmark. There are no high frequency component or triple-frequency properties in the vibration, which is caused by electrodynamic force. For the part near switcher, amplitude of 100Hz component is 3.3 times high than that of benchmark. It is analyzed that the slight 300Hz and 800Hz components is caused by bad closure in the switcher and little loose mechanical fastening.

4. Summary

In the field measurement of the substation, the operation of GIS is normal, and no fault is found. By analysis of the measured data, it can be concluded as follow:

1) There is a vibration, which is basically, caused by electrodynamic force, in the operation of GIS. The wave with dominant frequency 100Hz is relatively stable, and the amplitude is low. The vibration of base can be chose as benchmark. If the amplitude of measurement point is not higher than 10 times of

benchmark, there is no mechanical fault. The main reason is the aging and normal mechanical loss in the operation of equipment. If there is a strong vibration, whose amplitude is higher than 10 times of benchmark, with no high frequency component, it is caused by poor mechanical fastening such as the loose of grounding screw. If the high frequency component is strong and the triple-frequency properties are outstanding, it can be concluded that there is poor connect such as bad closure of contact.

2) The vibration of bus, expect for the part near switcher, is slight. In the part near switcher, there is amplitude of 100Hz component lower than 5 times of benchmark as well as triple-frequency properties in the vibration. It is analyzed that there is bad closure of contact and poor mechanical fastening in the switcher. But the problem, proved by field diagnosis, is not serious.

With no obvious vibration problem, it's inferred that the GIS is running without mechanical faults now. However, the high frequency component indicates a hidden trouble there. So we suggest a continuous measurement in the part of hidden trouble from the consideration of safety. On the other hand, it should be paid more attention in the initial conditions of the part for verification of conclusions and convenience of follow-up studies.

Acknowledgment

Financial support from the Science and Technology project of Guangxi Power Grid Corp (No. K-GX2014-006) is gratefully acknowledged.

References

1. Liu Xiaohong. Analysis of GIS Partial-Discharge On-line Detection Technology [J]. Sci-Tech Information Development & Economy, 2011, 21(33):184–187. DOI:10.3969/j.issn.1005–6033.2011.33.075.
2. Li Junhao, Che Bin, Lin Min, et al. Partial Discharge Characteristics of Floating Potential Defectin SF6 Gas under Oscillating Lightning Impulse Voltage [J]. Insulating Materials, 2014, (4):62–65. DOI:10.3969/j.issn.1009–239.2014.04.16
3. Sun Jianning, Wu Yanwei. Harm and Solution of GIS Enclosure Abnormal Vibration [J]. Science & Technology Information, 2011, (18):43–43. DOI:10.3969/j.issn.1672–3791.2011.18.031.
4. Yi Xinfang. The Diagnosis of Poor Contact Equipment [J]. High Voltage Apparatus, 1990, (3).
5. Wang Sihua. Analysis and Discussion of GIS Structure [J]. Electrical Manufacturing, 2006, (2):26–28.

6. Gao Linwei, Li Dong. Structural characteristics, installation and debugging of 220kV GIS [J]. Shanghai Dianli, 1996, (6):10–14.
7. Guo Bihong, Zhang Hanhua. Detect Internal Latent Fault by Using GIS Enclosure Vibration Frequency Characteristics [J]. Power System Technology, 1989, (2):44–50.
8. Qian Jiali, Shen Li, Liu Weidong, et al. The Vibration Phenomena and Detection of GIS [J]. High Voltage Apparatus, 1990, (6):3–9.
9. Zhang L, Li J H, Li Y M, et al. The Study of Vibration Diagnosis Method for GIS [J]. Applied Mechanics & Materials, 2013, 295–298:2429–2432.
10. Ohshita Y, Hashimoto A, Kurosawa Y. A Diagnostic Technique to Detect Abnormal Conditions of Contacts Measuring Vibrations in Metal Enclosures of Gas Insulated Switchgear [J]. IEEE Power Engineering Review, 1989, 9(10):51–52.
11. Runde M, Lillevik O, Larsen V, et al. Condition assessment of contacts in gas-insulated substations [J]. IEEE Transactions on Power Delivery, 2004, 19(2):609–617.
12. Charbkeaw N, Suwanasri T, Bunyagul T. Mechanical Defect Detection Of Sf6 High Voltage Circuit Breaker Using Wavelet Based Vibration Signal Analysis [J]. Electrical Engineering/Electronics, Computer, Telecommunications and Information Technology, 2008. E, 2008, 2:901–904.
13. Yang Ruitang. Analysis on the vibration of GIS [J]. HuatongTechnology, 2001, (2):11–14.
14. Cheng Lin. Research on UHV GIS/HGIS Vibration Detection and Diagnosis Method [J]. Electric Power Construction, 2009, 30(7):17–19. DOI:10.3969/j.issn.1000–7229.2009.07.005.

Study on spatial distribution pattern of geological disasters in Wanzhou based on GIS

Z.H.Chen* and H. Xu
Chongqing Key Laboratory of Exogenic Mineralization and Mine Environment
Chongqing Institute of Geology and Mineral Resources, Chongqing, China
**E-mail: chenzhenghua369@126.com*

R. Zhou
Nanjiang Hydrogeological and Engineering Geology Brigade, Chongqing, China

ArcGIS, with excellent characteristics of spatial analysis, has been widely used in various aspects. In this paper, we use DEM as background data, project geological disasters investigation data of Wanzhou area on DEM in ArcGIS, and analyze spatial distribution of different attributes of geological disasters data, explore reasons for distribution patterns, and summarize spatial distribution patterns.

Keywords: Geological Disasters; Spatial Distribution; ArcGIS; Wanzhou.

1. Introduction

With the advent of information society, human society has entered the era of information explosion.

Facing vast information, people has undergone tremendous changes to requirements for information, which are more accurate, rapid and comprehensive. With rapid development of computer technology, geography information system which manages spatial location and other attributes information develops fast as well. In this paper, applications of ArcGIS spatial analysis technique for landslide distribution research in Wanzhou were discussed. With visual display of spatial position of all kinds of geological disasters, distribution pattern of different disaster, we explore causes of disaster spatial distribution in Wanzhou region, then sum up spatial distribution pattern of study area in Wanzhou.

2. Background of Study Area

Wanzhou district is located in the interland of the Three Gorges reservoir, the upper reaches of the Yangtze River, covers an area of 3457 km2, located at the northeast edge of Chongqing City, located at longitude from 107°55′22″ to

108°53'25", latitude from 30°24'25" to 31°14'58". The Yangtze River flows in Shicaoxi river in Nanxi town, flow out from 3.3km site from Huangbai town downstream. After water storage of Three Gorges reservoir, water level in front of dam fluctuated between 145m and 175m (Wu Song Elevation), the highest water level of Wanzhou urban section is 173.31m (Wu Song Elevation is 175.10m).

According to Wanzhou meteorological station data, multi-year average precipitation is 1181.20mm, maximum is 1635.2mm, rainfall concentrated from May to September, which can accounted for 70% of annual rainfall. Rainfall in January, February and December accounted for only 3.72 to 5.37% of average rainfall. Heavy rains and torrential rains occurred in summer, maximum precipitation was 711.8mm (occurred on August 16, 1987), continuous precipitation lasted 16 days at longest.

Low mountain and hills accounted for about a quarter of Wanzhou area, lower-to-middle mountain and plains accounted for about a quarter, Pingba and mesa is rare and scattered. Northwest is Tiefengmountain, Southeast is Qiyao mountains, between two mountains gradually transition to low mountain, elevation above sea level in general is from 800 to 1000m. Landform is mainly strip mountain, terrain slope in general is from 30° to 40°.

3. Data and Methods

3.1. *Data*

This study involves SRTM3 data with accuracy 90m from National Aeronautics and Space Administration (NASA) and State Bureau of Surveying and mapping (NIMA), Wanzhou disaster investigation data (687 in total) in 2011, which includes 553 landslides, 114 dangerous rocks, 20 unstable slopes.

3.2. *Methods*

The development of earth observation and computer technology greatly enhanced the ability of analyzing and processing spatial information. People are eager to exploit spatial information to understand and grasp pattern of earth and society movements to establish theory and methods of spatial information analysis system. ArcGIS is a powerful spatial analysis tool in geographic information system, which has been widely used in various industries.

In this paper we use DEM data to extract terrain data and analyze different attributes of spatial distribution pattern of Wanzhou District disaster via ArcGIS.

4. Analysis

4.1. *Types of geological disasters*

Projecting all kinds of disaster points to DEM, we can easily see spatial positions of all disaster points. Statistical results show that most geological disasters are landslides. Low mountain to hilly transitional zone and thick soil on both sides of valley provide topographic and geomorphologic conditions to formation of landslide and unstable slope. Landslide often occurred in wings of the anticline. Dangerous rock (collapse) occurred mainly in steep slope over 60 degrees, especially in deep cutting areas of low mountains and steep river shore.

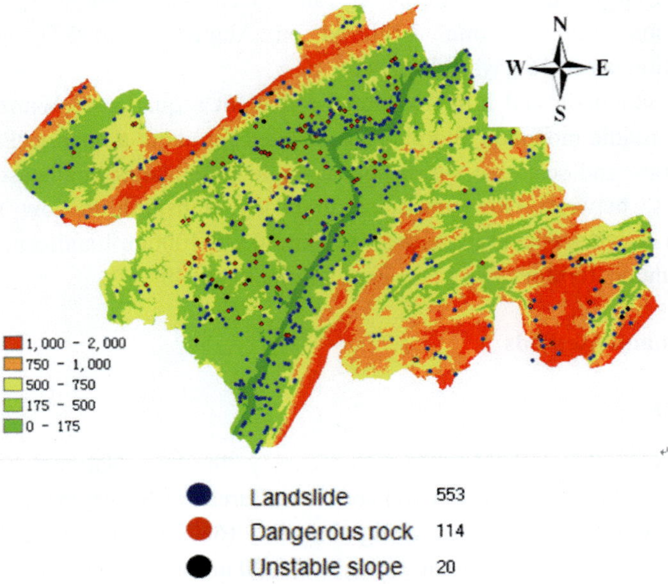

Figure 1. Spatial Distribution of Geological Disasters.

4.2. *Stability*

According to stability classification statistical results, landslides and unstable slope are basically stable type, its development is relatively slow. Landslides in this area are mostly creep deformation, less occurrence of sudden high speed landslide. As for dangerous rock, it's also basically stable type, which is influenced by weathering and groundwater.

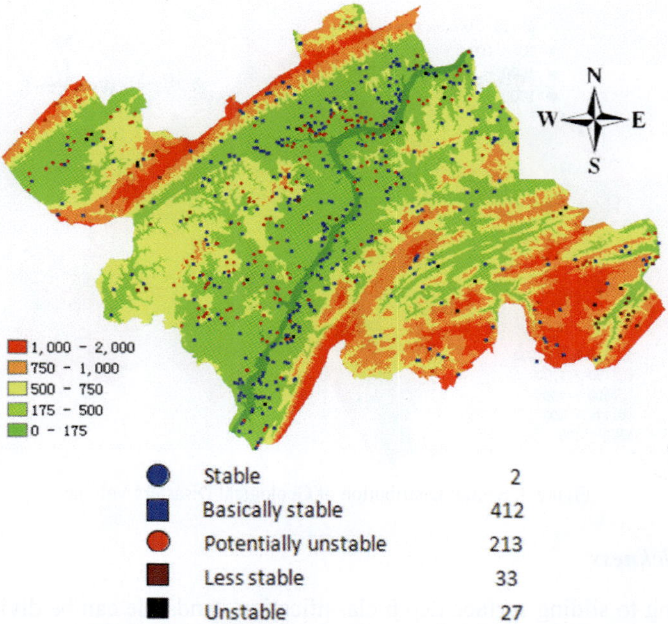

Figure 2. Spatial Distribution of Geological Disasters Stability.

4.3. *Volume*

According to landslide size classification, landslide can be divided into 4 types, including: ① small landslide, V is less than $10^5 m^3$; ② medium-sized landslides, V is between10^5 and $10^6 m^3$; ③ large-scale landslides, V is between10^6 and 10^7 m^3; ④ giant landslide, V is over10^7 m^3.

Landslides and unstable slope are mainly small to medium type, large-scale landslides are less and giant landslides are barely developed. Large-scale landslides are mainly concentrated on the banks of the Yangtze River, Only 2 giant landslides are located on the shore of Yangtze River downstream.

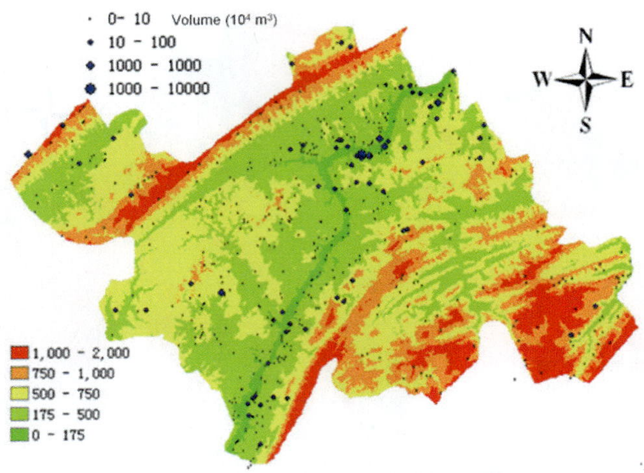

Figure 3. Spatial Distribution of Geological Disasters Volume.

4.4. *Thickness*

According to sliding surface depth classification, landslide can be divided into 4 types,including: ① shallow landslide, maximum depth is less than 6m; ② medium landslide, maximum depth is between 6 and 20m; ③ deep landslide, maximum depth is between 20 and 50m; ④ ultra deep landslide, maximum depth is over 50m.

Shallow landslides and medium landslides are all over the area, deep landslides and ultradeep landslides are 8 in total, of which 4 are located in the Yangtze River shore.

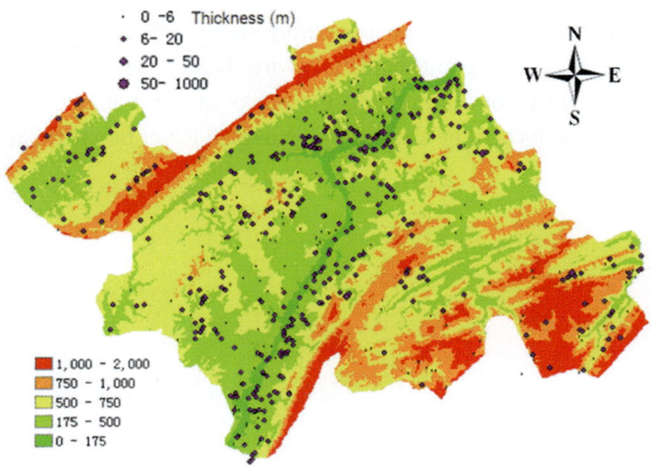

Figure 4. Spatial Distribution of Geological Disasters sliding surface depth.

5. Discussion

According to analysis results, geological disasters mainly distributed along the banks of the Yangtze River, especially in Wanzhou city, which indicated geological disasters are water-related, and human activities are becoming more and more influential on geological disasters.

Statistical data show that most major geological disasters are triggered by rainfall, a few landslides and unstable slope are influenced by river rise or fall. We speculated that rainfall came into overland flow, when overflow segment benefits seepage, safety factor of soil stability is reduced.

6. Conclusions

1. Landslide often occurred in wings of the anticline. Dangerous rock (collapse) occurred mainly in steep slope, especially in deep cutting areas of low mountains and steep river shore. Landslides and dangerous rock in this area are basically stable type, which is mainly affected by weathering and groundwater.

2. Landslides and unstable slope are mainly small to medium type, large-scale landslides are less and giant landslides are barely developed. Large-scale landslides are mainly concentrated on the banks of the Yangtze River; Shallow landslides and medium landslides are all over the area, deep landslides and ultra-deep landslides are 8 in total, half of which are located in the Yangtze River shore.

3. Geological disasters mainly distributed along the banks of the Yangtze River, especially in Wanzhou city. Most major geological disasters are triggered by rainfall, which reduces the stability of landslide.

4. ArcGIS with high performance characteristics makes work less insensitive and more efficient. Vivid visualization of disaster distribution is more intuitive than conventional statistics, and provides convenience for further mining disaster related information.

References

1. Tang G.A. & Yang X. 2006. *Arcgis spatial analysis GIS experimental tutorial*. Beijing: Science Press.
2. Chongqing Geological Prospecting Bureau of Hydrology and Engineering. 2011. *Geology Team of Chongqing geological disaster investigation in Wanzhou District*, Chongqing.
3. Xu. H.L., 2009. *Environmental geology*. Beijing: Geology Press.
4. Fang Y. & Lin T. 2003. *Soil Mechanics*. Wuhan: China University of Geosciences Press.

Application of hour by hour date in drawing galloping distribution map by metrological geographic method

J.H. Li, L.C. Zhang*, J.L. Yang, K.J. Zhu and B. Liu
China Electric Power Research Institute, Beijing 102401, China
E-mail: bkdzlc@126.com

The power grid galloping region distribution map as the basis of the prevention and treatment of galloping for transmission lines which is in the use or under construction, it plays an important role in ensuring the safe and stable operation of power grid. After the frequency method, the meteorological geographical method as a drawing method for galloping region distribution map is used widely in the range of SGCC which does not consider the influence of the actual lines. Based on using daily meteorological data may lead to larger galloping meteorological day calculation error,this paper put forward a method which use hourly meteorological data to draw the galloping region distribution map,and introduce the feasibility of this method by taking the Zhejiang province as an example.

Keywords: Meteorological geographical method; Transmission line galloping; Distribution map; Hourly meteorological data.

1. Introduction

Research on galloping rule of the line and drawing of the galloping distribution map have important reference value for design, construction, operation and maintenance of the line, which are important basis for prevention works of galloping. Some foreign countries with multiple generation of galloping have obtained rich achievement in galloping theory and prevention aspect[1–4], but research on drawing of the galloping distribution map is also short. Galloping disaster is characterized with high generation frequency and scope increasing following enlargement of construction scale of Chinese grid in recent year, galloping prevention conditions is more and more serious, therefore our country has enlarged research strength on galloping, a series of important achievements have been obtained in theory, test and simulation aspects[5–7]. But galloping prevention still faces passive situation on firstly galloping and later treatment. Therefore it is urgent to study generation and distribution rules of galloping and draw scientific and reasonable galloping distribution map on basis of Chinese climate and geographical characteristics, so as to implement targeted galloping prevention measures.

State Grid Corporation draw galloping distribution map based on frequency method in 2010. This version of galloping distribution map takes actual generation times of galloping in the power transmission line in some area to judge galloping strength, it completely follows the galloping operation experience of the line, it can reflect galloping disaster distribution rules of Chinese grid at certain degree, it has certain reference value during galloping prevention works. But for the area with shorted galloping data or less line, it can't reflect true galloping distribution rules, therefore it is necessary to explore a set of galloping area distribution theory which is more in accordance with actual conditions, and instruct to develop drawing works of the galloping distribution map. In 2012, metrological geographical method came out, State Grid Corporation develops trial point drawing works of the galloping distribution map in Hubei, Henan and Liaoning according to this method. Seen from drawing results, its applicable scope is wider than frequency method, classification accuracy is higher because influence of the power transmission line isn't considered in metrological geographical method, this method is gradually popularized and applied in the whole grid scope, 27 companies have drawn new version galloping distribution map on basis of it. The metrological geographical method applies day by day metrological data as basic data for classification of galloping area, but day by day metrological geographical data are not completely applicable to every area, great error may generate in some area, for example above problem is encountered during drawing of the galloping distribution map in Zhejiang province. This paper forwards classification method which takes hour by hour metrological data as base for this issue, result obtained through calculation of this method is more accurate.

2. Limitation of Day By Day Metrological Data in Drawing of Galloping Distribution Map

2.1. *Problem submittal*

There are 68 conventional metrological stations with complete document series in Zhejiang province. Firstly organize metrological documents collected by 68 traditional metrological stations at winter and spring from 2003 to 2012 (November to March of next year), and then select day number of every conventional metrological station which is in accordance with day number of condition combination in Table 1 according to metrological parameter threshold area in Table 1, and 10 years galloping metrological day of every conventional metrological station are converted.

Table 1. Margin settings for A4 size paper and letter size paper.

Parameters	Threshold area
Leading wind direction	North wind
Maximum wind speed at leading wind direction /($m \cdot s^{-1}$)	4–15
Daily average relative humidity /%	70–100
Daily minimum temperature /°C	-5–3

Take DEM with resolution of 90 m×90 m in Zhejiang province as base map, calculate difference between maximum altitude height and lowest altitude height as terrain waviness in this area within area with radius of 2000m, and obtain terrain waviness distribution map of whole province in Zhejiang province.

Obtain terrain waviness of every metrological station from the terrain waviness map according to longitude and latitude of the conventional metrological station. Select the metrological station with waviness greater than 100m, and establish relationship model between terrain waviness and 10 years galloping metrological day number (refer to Figure 1). The grid point with terrain waviness less than 100m is not necessary to be modified; the grid point with terrain waviness greater than 100m shall be modified according to Equation 1 below:

$$Y=-23.8\ln(x)+212.7 \tag{1}$$

where Y is 10 years galloping metrological day number after modification of terrain waviness, x is terrain waviness.

10 years galloping metrological day number after terrain modification is classified into 4 classes according to galloping area classification standard in Table 2. In which, 10 years galloping modified day number in non-galloping area (0 class) is less than 50 day, 10 years galloping modified day number in weak galloping area (1 class) is 50 day–90 day, 10 years galloping modified day number in middle galloping area (2 class) is 90 day–140 day, 10 years galloping modified day number in strong galloping area (3 class) is greater than 140 day.

Table 2. Relationship between classification for galloping region and galloping revising day in 10 years (Daily meteorological data).

Galloping class	10 years galloping modified day number
3 class	≥140
2 class	90–140
1 class	50–90
0 class	0–50

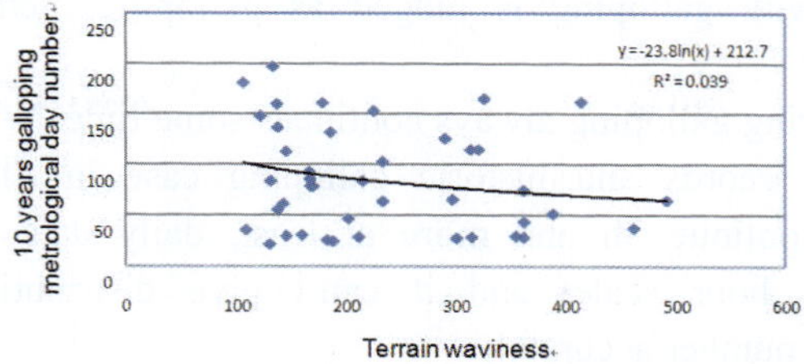

Figure 1. The scatter diagram between topographic prominence and galloping meteorological day in 10 years.

Galloping day number of every calculation grid point in Zhejiang province is obtained after modification of terrain waviness. Most areas in Zhejiang province are light galloping area (1 class galloping area) in Zhejiang province; area around Hangzhou bay is strong galloping area (3 class galloping area); Hangjiahu plain, Ningshao plain and Jinqu base areas are middle galloping areas (2 class galloping area). But seen from historic galloping accident records in Zhejiang grid, galloping accident of the power transmission line never occurs in Hangjiahu plain, in particularly to area around Hangzhou bay, glaze and rime etc weather phenomenon are very seldom. Therefore distribution of 10 years galloping metrological day number on basis of daily metrological data statistical can't be taken as classification basis on galloping distribution map of Zhejiang grid.

2.2. *Reason analysis*

Learn from above analysis, galloping weather day number is calculated according to day by day metrological data in recent 10 years in conventional metrological station in Zhejiang province, the obtained conclusion isn't in accordance with actual operation experience, its statistic result obviously enlarges galloping scope. There are mainly following reasons:

(1) When ice covering galloping occurs, metrological factors must be in accordance with threshold conditions in Table 1. In adverse, metrological factors meet threshold conditions in Table 1, it can't determine whether ice covering galloping occurs, its calculation result always enlarges galloping scope.

(2) When daily statistic data of metrological factors are calculated and analyzed, occurrence time that minimum air temperature and maximum wind speed meet threshold conditions are not synchronous. If one day maximum wind speed may occur in morning, and lowest air temperature may appear at night, there is obvious time difference between them, it isn't in accordance with metrological conditions of ice covering galloping. Metrological day which isn't

in accordance with galloping is judged as galloping metrological day by mistake.

(3) Ice covering galloping always continues some time. When it is analyzed from document records and historic galloping case in Zhejiang province, galloping will continue 3h and more at least, daily data can't distinguish characteristic of hour scale, and it can't give distribution of galloping metrological day number accurately.

In fact, occurrence conditions of galloping in line is very severe, it is analyzed according to galloping metrological conditions of the power transmission line in Quzhou area on January 5, 2010, it is considered that hour by hour metrological data are taken advantage, and wind duration time is added as selection condition, it is better than result of day by day metrological data statistic calculation.

3. Drawing Galloping Distribution Map on Basis of Hour by Hour Metrological Data

3.1. *Calculate metrological weather day number with hour by hour metrological data*

There are about 2000 scale automatic metrological stations in Zhejiang territory, these encrypted metrological observation station network can provide hour by hour metrological observation data, which provides reliable data base for drawing of the galloping distribution map in Zhejiang grid.

The former 10 years galloping metrological day number which is calculated by daily metrological data of the conventional metrological station is interpolated into data of the whole province through simple Cregeen method. This interpolation method doesn't completely consider distribution characteristic of metrological elements, it is very rough. Therefore, different metrological elements of all middle scale metrological stations are firstly interpolated into the space grid point through interpolation method of the common metrological element, and then carry out statistic calculation according to metrological parameter threshold area, and obtain the fine galloping metrological day number in every area of Zhejiang province.

3.2. *Space difference of hour by hour metrological data*

Factors which affect space distribution of the metrological elements are classified into two great categories: one category is latitude, longitudinal, large mountain orientation and large water body etc macroscopic geographic environment factors; another category is altitude height of observation station, slope, slope direction, terrain covering, water body, plantation covering etc microcosmic terrain factors. Therefore metrological element Y of any station point can be expressed as Equation 2:

$$Y = F(\lambda, \phi, h) + \varepsilon \tag{2}$$

Where λ, φ and h are latitude, longitudinal and altitude height. $F(\lambda, \phi, h)$ is metrological element trend value which is determined by microcosmic level distribution function and height distribution function; ε is difference between actual measurement value and trend value, which reflects influence and random error of small terrain.

Trend value $F(\lambda, \phi, h)$ of the metrological element in Equation 2 applies three dimensions secondary trend face of latitude λ, longitudinal φ and altitude height h to simulate as shown in Equation 3:

$$F(\lambda, \phi, h) = b_0 + b_1\lambda + b_2\phi + b_3h + b_4\lambda\phi h + b_5\lambda h + b_6 h\lambda + b_7\lambda^2 + b_8\phi^2 + b_9 h^2 \tag{3}$$

Where $b_{0,} b_1, \ldots, b_9$ are coefficients which are determined later. Actual observation value is deducted with trend value $F(\lambda, \phi, h)$ of the metrological elements simulated in Equation 3, the obtained residual difference ε of every station is waving tope of the metrological elements under action of micro-terrain. Carry out space interpolation of ε of every station, and grid map of residual difference is obtained. Add trend plain of the metrological element on the grid map of residual differential, space distribution of the metrological element is obtained (refer to Figure 2).

In which, minimum air temperature and relative humidity apply interpolation method of "trend plane + residual differential", all data for interpolation include data of the conventional metrological station and the automatic metrological station. Precipitation and wind speed observation data quality of the middle scale automatic metrological station are difficult to guarantee under ice covering weather, therefore only data of the conventional metrological station are applied for space interpolation of precipitation and average wind speed, interpolation applies reverse distance weighting method. The above interpolation method is applied, after hour date at T20:00 on January 5, 2010 are interpolated, space distribution map of hour minimum air temperature, relative humidity, precipitation and average wind speed in Zhejiang province are obtained. Space distribution which various metrological elements are very reasonable and precision are obtained after interpolation.

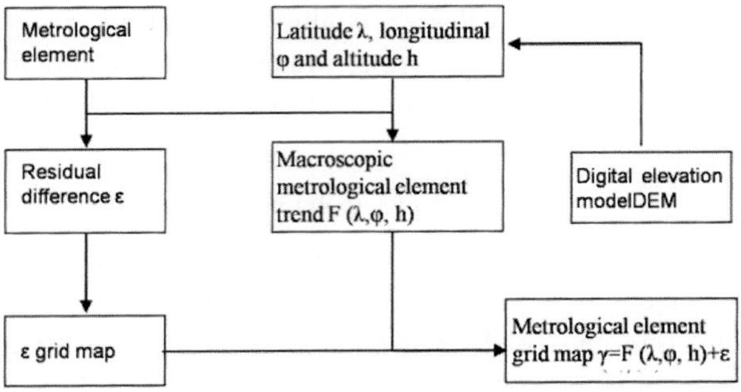

Figure 2. The flow chart of spatial interpolation of meteorological observation data.

3.3. *Determination of hour by hour metrological parameter threshold area*

Galloping in the power transmission line must occur under ice covering conditions of certain lines. Ice covering of the line is generally caused by glaze, rime or wet snow, it is traced with freezing rain or rain with snow when galloping occurs. On another side, galloping occurs after the line is covered by ice, there must be durable and stable wind. Therefore, determine parameter and parameter threshold area for hour by hour metrological data selection through analysis of metrological conditions with generation of glaze and rime in Zhejiang province in combination with analysis of metrological conditions during occurrence period of galloping accident in line, show as Table 3.

Table 3. The threshold value interval of hourly meteorological parameters.

Parameter	Threshold value area
Hours accumulation precipitation /mm	>0.0
Duration time wind speed greater than 4 m/s at stable wind direction /h	≥3
Hour minimum air temperature /°C	-5–1.0
Hour relative humidity /%	≥75

3.4. *Galloping day number calculated by hour by hour metrological data*

Former 10 years galloping metrological day number which is calculated by daily metrological data parameter threshold method can't calculate actual galloping metrological day number accurately. Metrological conditions for galloping generation must be in accordance with parameter threshold combination of daily metrological data. It means actual galloping metrological day number of Zhejiang province is include in metrological day number of former daily metrological data statistic. Therefore select date which is in accordance with metrological data parameter threshold value in recent 10 years, and then take daily hour by hour metrological information for statistic analysis, so as to obtain galloping metrological day number distribution which is in accordance with actual condition in Zhejiang.

This paper samples hour by hour metrological information of 816 days of 2000 metrological stations in Zhejiang and around area, including minimum air temperature, relative humidity, accumulated precipitation, average wind speed and wind direction, and carry out space interpolation according to interpolation method in 3.2, and calculate galloping metrological day number according to parameter threshold combination in Table 3 after interpolation.

Galloping distribution scope is obviously reduced after hour by hour metrological document selection, in particularly to remove Hangjiahu plain area where galloping accident never occurs. Galloping mainly occurs in Jinqu base, Ningbo, Shaoxing, Taizhou and Zhoushan area. Galloping accident occurs in Jinqu area in 2010, galloping accident point is coincided to classified galloping area well. Doubtful galloping accident has occurred in mountain area in Ningbo and Shaoxing, Yuhuan metrological station in Taizhou has observed glaze and rime for several time, the steel tower in Zhoushan has also observed a few ice covering. Therefore 10 years galloping metrological day number distribution calculated by hour by hour metrological data statistic is in accordance with actual conditions of Zhejiang province. But because wind speed interpolation is only applied in data of the conventional metrological station, result of interpolation calculation may enlarge wind speed in mountain area in Ningbo and Shaoxing, which cause galloping scope in this area great. Therefore, it is determined to equip the galloping prevention device or not in combination with micro-terrain characteristics.

4. Drawing Galloping Distribution Map of Zhejiang Grid

Galloping accident which is determined in recent year in Zhejiang is only once, i.e., galloping in the power transmission line which occurred in Quzhou area in January 5–6, 2010, galloping occurred in Ningbo and Shaoxing etc area are doubtful galloping, galloping class in Zhejiang province is determined as 2 class maximum according to current operation experience and analysis of metrological data. Corresponding relationship between 10 years galloping modified days number calculated according to hour by hour metrological data and galloping class is shown as Table 4. The galloping distribution map of Zhejiang grid can be drawn according to galloping class classification in Table 4.

Table 4. Relationship between classification for galloping region and galloping revising day in 10 years (Hourly meteorological data).

Galloping class	10 years galloping modified day number
3 class	≥60
2 class	40–60
1 class	20–40
0 class	0–20

5. Conclusion

This paper forwards drawing method of galloping distribution map on basis of hour by hour metrological data for great galloping metrological daily calculation error caused by day by day metrological data when the galloping distribution map is drawn by the metrological geographical method, and Zhejiang province is taken as sample to introduce feasibility of this method, following conclusions are obtained through calculation and analysis in this paper.

(1) Applying day by day metrological data may enlarge galloping scope; galloping days which are judged by hour by hour metrological data shall meet day by day metrological parameter threshold area. In adverse it isn't certain, galloping day selected by hour by hour metrological data is finer, accuracy is also higher.

(2) Because cost calculate by hour by hour metrological data is higher, the area with result calculated by day by day metrological data and not great operation experience difference shall apply day by day metrological data, vice versa it shall apply hour by hour metrological data.

(3) Propose of hour by hour metrological data will further perfect metrological geographical theory system, relevant influence metrological factor is further optimized following acknowledge improvement of galloping distribution rule and accumulation on drawing experience of the galloping distribution map, and drawing technical level of the galloping distribution map is fatherly improved.

References

1. Rawlins G R. Research on vibration of overhead ground wires [J]. IEEE Trans. on Power Delivery, 1988, 3(2) :769–775.
2. Gupta S, Wipf T. Structural failure analysis of 345 kV transmission line [J]. IEEE Trans. on Power Delivery, 1994, 9(2) :894–903.
3. Tsujimoto K, Iisaka H, Shimojima K, et al. Report on experimental observation of galloping behavior in 8-bundled conductors [J]. IEEE Trans. on Power Apparatus and Systems, 1983, 102(5) :1193–1201.

4. Parkinson G V, Smith J D. The square prism as an aeroelastic non-linear oscillator [J]. The Quarterly Journal of Mechanics and Applied Mathematics, 1964, 17(2) : 225–239.
5. Guo Ying-Long, Li Guo-Xing, You Chuan-Yong. The Galloping for Transmission Line [M]. Beijing: China Electric Power Press, 2003: 1–20.
6. Ma Jian-Guo. Regional Division of Conductor Galloping of Power Networks in Hubei Province [J]. Electric Power Construction, 1999(5) : 39–40.
7. Liu Liang, Liu Hong-Bin, Deng Chun, et al. The Classification Principle of Galloping Probability Distribution Map in North China Power System [C]//Paris, France: CIGRE 2012, 2012.

The effect of crude oil and dispersants on physiological and biochemical characteristics of *Isochrysis galbana*

J.W. Gao[a], S.Y. Wang, R.N. Chen, X.Y. Jia and W.L. Zhou
Tianjin Key Laboratory of Aqua-ecology and Aquaculture, Department of Fisheries Science of Tianjin Agricultural University Tianjin 300384, P.R.China
[a]E-mail: supper_snow@163.com

The effects of oil water accommodated fraction (WAF), emulsion (DWAF) and dispersants on chlorophyll-a content, protein content and enzymatic activity and of *Isochrysis galbana* were studied by experimental ecology method with 96 hours. Thee results showed that content of protein and chlorophyll-a were decreased with the extension of time under the three stress. There were significant effects of three solutions on SOD, GSH-PX and MDA. After 96 hours, the lowest content of protein, chlorophyll-a, SOD and GSH-PX, the highest content of MDA were found in WAF group. Therefore WAF stress had the most influential effects on physiological and biochemical characteristics of *Isochrysis galbana* in the experiment.

Keywords: *I.galbana*; Boxi light crude; Dispersants; Enzymatic activity; Chlorophyll.

1. Introduction

With the rapid development on ocean exploitation in modern society, human factors have caused great damage to the marine environment, among which oil pollution is a very serious type [1]. The acreage and degree of oil pollution in the marine are increasingly expanding and deepening, which not only seriously affect marine ecological environment, but also destroy the structure and the function of marine ecosystem [2]. Marine microalgae are an important part of marine primary production, which plays an indispensable role in the marine food chain [3, 4]. The damage of marine microalgae will have a direct impact on food chains trophic levels of marine ecosystem, destroy phytoplankton and marine bacteria, and reduce the survival rate of zooplankton and larvae, thus affecting the structure and function of the whole marine ecosystem [5–7]. Therefore, take *I.galbana* as study subject, this paper will study Boxi light crude oil and the impact of dispersants to *I.galbana*'s growth and to antioxidant enzymes by the way of experimental ecology, thus providing scientific theoretical evidence of crude oil and the impact of dispersants on the marine ecosystem, as well as the theoretical guidance of the usage of dispersants and marine habitat restoration.

2. Materials and Methods

The *I.galbana* used was provided by aquaculture and ecology key laboratory in Tianjin. Seawater was formulated by biological sea crystal and distilled water, which could be used when cooling to room temperature after sterilization under 121.0°C. Cultured by f/2 nutrient solution, under the temperature of (20 ± 0.1) °C, with the illumination of 40 μmol/ (m2 • s), and the period was 12 L: 12 D. Shaking nutrient flask several times per day to prevent pavementing and sediment [8].

On the basis of preliminary experiments on the selected *I.galbana* influential WAF, dispersants and DWAF concentration as an official test concentrations. The experimental set up three treatment groups and one control group, each group 3 repeats. Treatment group and concentrations were: WAF solution group, the concentration of 2.099mg / L; DWAF solution group, the concentration of 2.949mg / L; dispersant was set at a concentration of 10g / L. The experimental period of 96h, 24h each sampling time, the determination of the protein content, chlorophyll content, MDA content, SOD content and GSH-PX content [9].

In this study, all treatments were repeated three times, the experimental results are Mean ± SD indicated. The results using SPSS 18.0 software One-way ANOVA test for differences between treatment and control groups, LSD method statistically significant differences between groups for multiple comparisons (P <0.05 indicates a significant difference; P <0.01 indicates a significant difference).

Results

Impact of WAF, dispersants, DWAF *I.galbana* protein content shown in Figure 1, along with extension of time, the three treatment groups showed a protein content *I.galbana* first increased and then decreased. Three substances on *I.galbana* protein content had significant effects (p <0.05), which affect the WAF *I.galbana* protein content of the most obvious, DWAF followed dispersants its minimal impact.

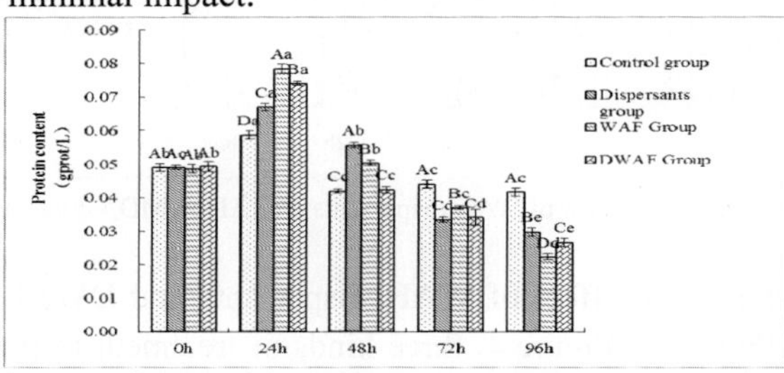

Figure 1. Influence of WAF, dispersants, DWAF on protein content.

Affect the outcome of WAF, dispersants, DWAF *I.galbana* chlorophyll content is shown in Figure 2. As can be seen from Figure 2 there was no significant difference (p> 0.05) compared to 0h and 24h period of three treatment groups *I.galbana* chlorophyll content and the control group. 24h after three treatments have a significant effect (p <0.05) of *I.galbana* chlorophyll content, which WAF influence *I.galbana* chlorophyll content of the most obvious, and with the extension of the three treatment groups *I.galbana* chlorophyll content of the role of time showing a downward trend.

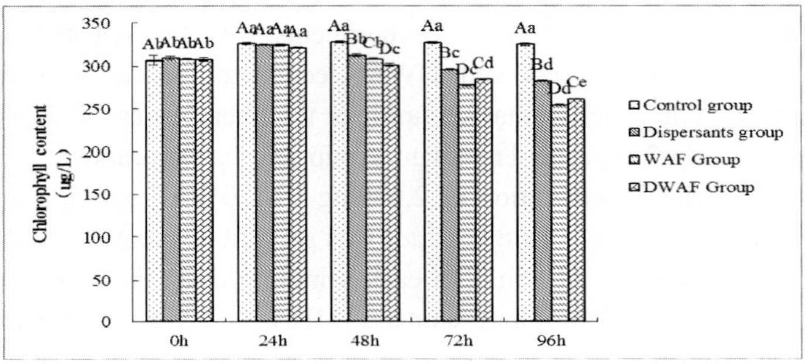

Figure 2. Influence of WAF, dispersants, DWAF on chlorophyll content.

Impact of WAF, dispersants, DWAF *I.galbana* MDA content in Figure 3. As can be seen from Figure 3, each group with the extension of time, the MDA content first and then decreased. 0h, 24h, WAF, dispersants, DWAF impact on the *I.galbana* MDA was not significant (p> 0.05), at 48h, 72h, when 96h, different groups affect the differences significant for *I.galbana* content of MDA (p <0.05), which WAF *I.galbana* MDA content on the most obvious, DWAF its influence, followed by dispersants minimal impact.

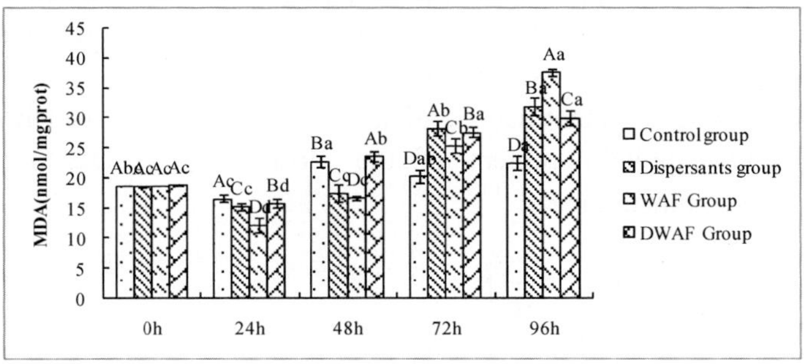

Figure 3. Influence of WAF, dispersants, DWAF on MDA content.

Figure 4 shows the effect of WAF, dispersants and DWAF towards SOD content in *I.galbana*. By Figure 4, three kinds of treatment to the SOD activity of *I.galbana* present the trend of increasing at first then declining, reaching the maximum value at the time of 24 hours. After 48 hours, the activity of SOD in experimental group apparently lower than that in control group. As for control group, three ways already have a significant effect on the activity of *I.galbana* SOD during the experiment (P<0.05), of which WAF has the most obvious influence.

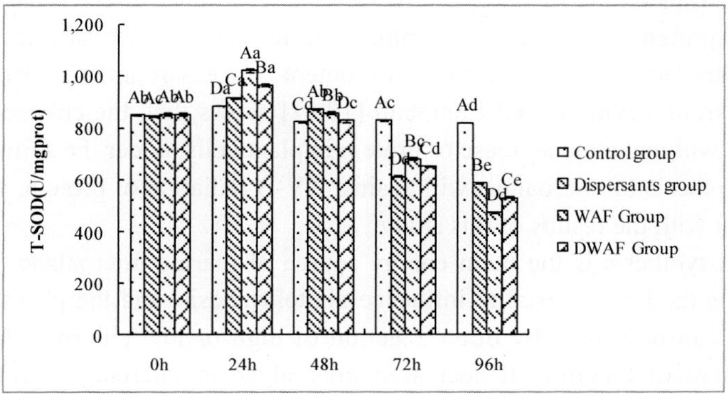

Figure 4. Influence of WAF, dispersants, DWAF on SOD activity.

Figure 5 shows the effect of WAF, dispersants and DWAF toward the activity of *I.galbana* peroxidase. From Figure 5, three kinds of treatment all have a significant effect on the activity of *I.galbana* peroxidase (P<0.05). Meanwhile, the differences among three different groups during the same period are remarkable (P<0.05), of which WAF is the most obvious then DWAF, dispersants is the least one. From the result, it can be concluded that compared with control group, the group of dispersants and DWAF tend to rise before reduce, while the WAF group shows a trend of declining.

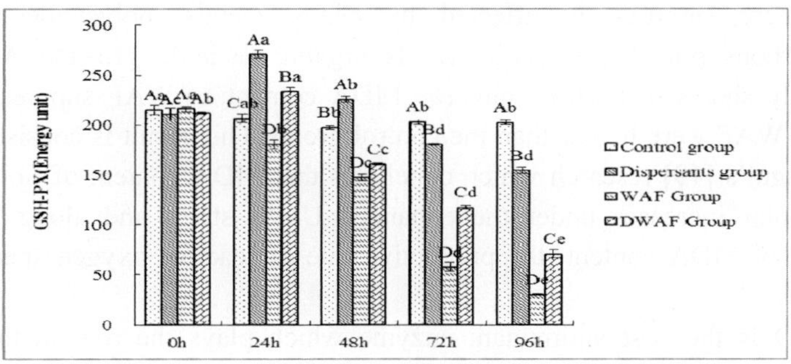

Figure 5. Influence of WAF, dispersants, DWAF on GSH-PX activity.

3. Discussion

Protein is the material basis of life, the basic organic of the cells and is mainly responsible for life activities. With the increase of the incubation time, the *I.galbana* protein content decreased significantly. The reason is probably the three materials on the *I.galbana* cells were induced injury affected the synthesis of body protein, may also be synthesized in the body resistance by other biomacromolecules affected protein content of environmental stress. The research from Tevini M and Campbell [10,11] shows that, the enhanced UV-B radiation will make the increase of protein in plant cells, under the stimulation of UV-B synthesis of microalgae will against UV-B radiation of protein, which are consistent with the results of this study.

Photosynthesis is the main energy source of marine microalgae, light and pigment is the key material to influence photosynthesis, and the photosynthesis intensity can be shown by direct reaction of high or low chlorophyll content. The content of chlorophyll decreased microalgae in microalgae will directly affect the photosynthesis, thereby inhibited the growth of microalgae cells. In this study, the content of chlorophyll in WAF, dispersants and DWAF these three groups of golden algae with the prolongation of time showed a downward trend, this is in accord with the research results of Xu Da [12] and the others. As to the specific mechanism, research has shown that because the surface active component of the WAF and dispersants has strong lipotropism which can enter in the microalgae cells destroy their membrane structures [13,14].

MDA is one of the main products of lipid peroxidation. Its volume can reflect the extent of lipid peroxidation. Research has shown that, MDA content in phytoplankton and the activity of antioxidant enzymes and antioxidant content are significantly affected by heavy metals, polycyclic aromatic hydrocarbons, polychlorinated biphenyls, organic pesticides [15–18]. After 48h, this study shows that Chrysophyceae MDA content of WAF suppression, oil agent, DWAF were higher than the control group. This result is consistent with Xue Lingui's [19] research which concludes that MDA content of Spirulina in intracellular increases under the enhanced UV-B stress and along with the increase of MDA content, the productive rate of reactive oxygen species also increases.

SOD is the first antioxidant enzyme which plays the role in the active oxygen elimination process. It will be capable of superoxide anion free radical (O^{2-}) rapid dismutation to hydrogen peroxide and oxygen. In this study, under these three conditions, the activity of SOD is higher than that of control group at the beginning 24 hours of the experiment, which is a stress reaction of the microalgae to external stimuli. Then the SOD activity decreased with the prolonging of time, which shows that the antioxidative system in microalgae is damaged. This is in accord with the studies of WangYou and Meng Fanping [20,21], that the SOD activity will be increased when the microalgae is on the interaction time under the environmental stress.

4. Conclusions

From the perspective of the protein content, chlorophyll content, the activity of MDA, SOD and GSH-PX, WAF stress had the most influential effects on physiological and biochemical characteristics of *Isochrysis galbana*.

Acknowledgements

This work was supported by Key Projects of Applied Basic and Frontier technology Research of Tianjin (13JCZDJC29300); Scientific research plan program of Tianjin Agricultural University (2014N11).

References

1. Shi Ehou. Sources of petroleum hydrocarbons in the marine environment [J]. Marine Environmental Science, 1988, 7(3):36–47.
2. Wan Banghe. Marine oil pollution and its hazards [J]. Marine Environmental Science, 1986, 5(3):52–63.
3. Qu Liang. Offshore oil spill dispersants and its application in the disposal of contaminated [J]. Marine Information, 2012, 1(2):45–47.
4. Fingas M. Oil spill clean up (second edition) [M]. New York, Lewis Publishers, 2001.
5. Li Chuanhui, Xia Bin, Chen Bijuan. Shengli crude enzyme activity of sulfur transfer of Gunther juvenile liver glutathione peroxidase and glutathione [J]. Marine Environmental Science, 2011, 30(6):827–830.
6. Yang Qingxiao, Gao Gunagzhi, Su Jun etc. Research on Impact of chemical dispersants on the growth of Ulva drugs [J]. Acta Oenological Sinica, 1997, 19(3):45–49.
7. Fisher WS, Foss SS. A simple test for toxicity of number 2 fuel oil and oil dispersants to embryos of grass shrimp, Palaemonetes Pugio [J]. Marine Pollut Bull, 1993, 26(3):85–91.
8. Gulec I, Leonard B, Hold way DA. Oil and dispersed oil toxicity to amphipods and snails [J]. Spill Sci Technol Bull 1997, 4(1):l–6.

9. Chen Yaqu, Rong Pei. Fuel oil, phenol and sodium sulfide on the growth and reproduction of Nitzschia and chlorophyll a content [J]. Oenological Eelimnology Sinic, 1991, 22(l):21–28.
10. Tevini M, Iwanzik W, Thoma U. Some effects of enhanced UV-B irradiation on the growth and co-mposition of plants [J]. Planta, 1981, 15(3):388–394.
11. Campbell D. The cyanobacterium synechococcus resists UV-B by exchanging photosystem reaction center DI protein [J]. Proc Natl Acad USA, 1998, 95(1):364–371.
12. Xu Da, Tang Xuexi, Zhang Peiyu. Physiological effects of UV-B radiation on two species of marine microalgae [J]. Qingdao Ocean University, 2003, 33(20):240–244.
13. Zhou Wenli, Qiao Xiuting, Xiao Hui etc. Effects of antibiotics on chlorophyll-a level in marine microalga [J]. Marine Environmental Science, 2009, 28(3):268–271.
14. Abrahamsson K, Choos K S, Pedersen M, Johansson G, Snoeijs P. Effects of temperature on the production of hydrogen peroxide and volatile halocarbons by brackish-water algae [J]. Phytochemistry, 2003, 64(8):725–734.
15. Singer A C, Thompson I P, Bailey M J. The tritrophic trinity: A source of pollutant-degrading enzymes and its implications for phytoremediation [J]. Current Opinion Microbiology, 2004, 7(5): 239–244.
16. Sabatini S E, Juarez A B, Eppis M R, Bianchi L, Luquet C M, Molina M C R. Oxidative stress and antioxidant defenses in two green microalgae exposed to copper [J]. Ecotoxicology and Environmental Safety, 2009, 7(2): 1200–1206.
17. Yang C Y, Liu S J, Zhou S W, Wu H F, Yu J B, Xia C H. Allelochemical ethyl 2-methyl acetoacetate (EMA) induces oxidative damage and antioxidant responses in Phaeodactylum tricornutum [J]. Pesticide Biochemistry and Physiology, 2011, 100(8): 93–103.
18. Tian Dan, Zhao Wen, Wang Yuan, Xiang Bei. Effects of cadmium stress on marine microalgae growth and antioxidant systems [J]. Journal of DaLian Ocean University, 2010, 25(5):417–421.
19. Xue Lingui, Xu Shijian, Li Shiweng. Spirulina response to enhanced UV-B stress [J]. Acta Hydrobiologica Sinica, 2007, 31(2):201–207.
20. Wang You, Yang Zheng, Tang Xuexi etc. Seven species of marine microalgae for radiosensitivity difference analysis of UV-B [J]. Environmental Science, 2002, 2(22):225–230.
21. Meng Fanping, Li Zhuona, Zhao Shunshun, Liu Jiao. BDE-47 impact on the four kinds of marine microalgae antioxidant enzyme activity [J]. Ecology and Environmental Sciences, 2009, 18(5):1659–1664.

Chapter 9
Environmental Science and Engineering

Evaluation on environment quality of Bai-Yun mountain scenic area[1]

M.Z. Lin*, S.T. Ji, J.L. Zhao and B.Z. Fang
School of Geography Science, Guangzhou University, Guangzhou 510006, China
E-mail: lmzh888@163.com

Data used in the paper were collected by field investigation and relative references in the district of scenic area in Baiyun Mountain. Selected six main scenic areas of Baiyun Mountain as the research object, then built the indicator system and establishes indexes standard to analyze each single index of environmental quality. The analyzed result shows that the air quality we monitor in Bai-yun Mountain scenic area is excellent. Water environment is influenced by the pollution of domestic sewage, some scenic areas' TP numerical value is too high. The voice environment can reach the evaluation standard at night. The whole concentration of the anion in these six scenic areas is higher than that in city center and belongs to moderate coordination. The microbial content is the high coordination. The vegetation coverage is too high to reach 95% at least. The kind of plants is very rich, it reaches the moderation coordination. It shows that the environmental quality indexes in most areas reflect the good environment of Bai-yun Mountain scenic area, except some indexes of water environment and sound environment. The analyzed result can provide the based data and decision support for the comprehensive and quantitative evaluation.

Keywords: Environment capacity; Environmental quality indexes; Evolution; Ecotourism; Bai-Yun Mountain.

1. Introduction

The first concept of environmental capacity was putted forwarded by the Belgian mathematician, biologist F Hoost according to the Malthusian population theory [1], measuring the largest tourist number tourist destination can accommodate [2]. But with the deep research on tourism environmental carrying capacity, more and more scholars try to use various methods to calculate the environmental capacity [3–7]. For example, they used system dynamics [8], economics theory and mathematical model simulation [9] to investigate tourism environment capacity calculation theory [10]. Traditional management of tourism environmental capacity tend to calculate through

[1] Supported by the National Natural Science Foundation of China (Grant No. 40771002) and Science Foundation of Ministry of Education of Guangzhou, China (Grant No.2012A007)

analysis methods and models. For example, they used Butler model [11], the cask theory [12] to calculate ecological capacity of tourism resources, using social and economic capacity formula and standard to estimate quantitatively the capacity of the "threshold value" [13]. However, the traditional understanding and calculation of tourism environmental always fail in practice. The birth of LAC theory [14–17] had realized the change the environmental capacity study from theory research to management tool.

In recent years, the tourism development of Baiyun Mountain is so fast that the rapid increase numbers of visitor put pressure on the ecological environment. Therefore, this paper tries to set up a management framework of environmental capacity of Baiyun Mountain based on LAC theory and PSR model. Providing quantitative analysis for each quality of ecological environment index and clearly understanding tourist area's situation of ecological environment recently, which makes Baiyun Mountain scenic area to have a environment with environmental maintenance, sustainable development and efficient operational control.

2. Overview of Research Area

Baiyun Mountain scenic area is located in the northeast, which is far from 7.5km to the urban. Its geographical coordinates is about 23°09′ to 23°11′N, 113°16′to 113°19′E. Baiyun Mountain is composed by more than 30 peaks, including 21.80km^2 and belongs to Dayu Bridge, history with the title of "the first mountain in Nanyue" [18]. In May 2002, Baiyun Mountain has been examined and approved by the State Council as "national key (national) scenic spot". In March 2012, upgraded into a "national 5A-rated tourist attraction", Baiyun Mountain is also the first 5A-rated tourist attraction among Guangzhou scenic spots. At present, it is the sole scenic spot in Guangzhou that has been granted two honorary titles, including "national 5A-rated tourist attraction" and "national civilized scenery tourist area".

3. Materials and Methods

Data used in the paper were collected by field investigation and relative references in the district of scenic area in Baiyun Mountain. Selected six main scenic areas of Baiyun Mountain as the research object, including Luhu, Santai Ridge, Mingchun Valley, Mount Davis, Mingzhulou, Heyi Ridge scenic area. Collecting materials of scenic area by fieldwork, then built the indicator system and establishes indexes standard to analyze each single index of environmental quality. The evaluation result can lay the foundation for the future of the whole quantification of environmental quality in Baiyun Mountain scenic area.

4. Evaluation Indicator System of Environmental Quality

4.1. *Building system of evaluation indicator*

In Baiyun Mountain, standard system of environment capacity management is composed of three major subsystems and four layers. Three major subsystems include state, pressure and feedback subsystems, while four layers contain objectives, criteria, factors and index. This study focuses on evaluating current quality of ecological environment in scenic spots, so more attention is paid to indices of ecological environment in index selection. Meanwhile, indices of characteristic quality of environmental ecology in this scenic spot shall be highlighted, and great importance shall be attached to air quality. See details from Figure 1:

The detailed calculation of each indicator can be shown below:

(1) Air environmental quality: use the air pollution index API to evaluate.

(2) Water environmental quality: use the classification index method to evaluate.

(3) Sound environmental quality: equivalent consecutive sound level A.

(4) Anion content: the number of anions in per cubic centimeter of air.

(5) Microbe concentration: the number of bacterial colony in per cubic centimeter of air.

(6) Vegetation cover rate: [vegetation are/ (total land area -water area)] ×100.

(7) Vegetation types:the number of total vegetation types in scenic area.

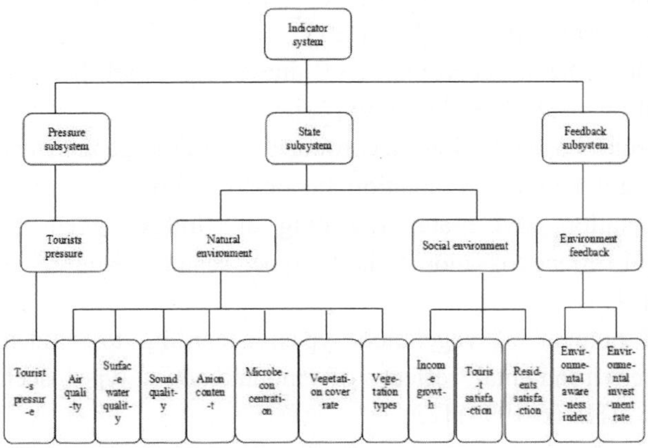

Figure 1. TECC management assess index system of Bai-yun Mountain.

4.2. *Ensuring evaluation standard*

4.2.1. *Ensuring the method of evaluation standard*

Ensuring evaluation standard value of each indicator above should be first in the environmental evaluation work. In current work,the evaluation indexes are so various and complicated. In the past related research had no unified standard value to evaluate the environment. Consequently, we should take different types of evaluation standards when deciding the the standard value. At present, the main evaluation standards include the following three terms:

(1) The international,state or local standards;

(2) Refer to the same type of nature reserves, forest parks and ecological tourist areas' indexes;

(3) The background value(standard value): according to the local standard value before.

4.2.2. *The state index evaluation standard of Baiyun Mountain scenic*

In the specific selection work,using the international, state and local standard, consulting the relevant academic research results and local situation to finally determine the evaluation standard of the indicators.

(1) Air quality: according to the <Ambient air quality standards> (GB3095-1996), adopting the value that atmosphere pollution index (API) sets [19];

(2) Surface water quality: according to the <Quality standard for surface water environment> (GB3838-2002), adopting the background value of evaluation that administration of environmental monitoring sets [20];

(3) Sound environmental quality: according to <Standard of environmental noise of urban area> (GB3096-2008) [21];

(4) Anion content: according to <Chinese forest park landscape resources grade evaluation> (GB/T18005-1999) [22];

(5) Microbe concentration: according to <Microorganism and monitoring technology> and the practice situation the local monitors;

(6) Vegetation cover rate: according to <Index system for evaluating environmental effect of mountainous type scenic resource development> (HJ/T6-1994) [23];

(7) The number of vegetation types: according to <Index system for evaluating environmental effect of mountainous type scenic resource development> (HJ/T6-1994) [23].

The evaluation standard value we analyzed can be seen in the following Table 1.

Table 1. Evaluation standards for state indicators of ecosystem environment.

Evaluation indicators	Different grade evaluation standard				
	I (high coordination)	II (moderate coordination)	III (reluctant coordination)	IV (inharmonious)	V (very inharmonious)
Air quality (API)	0-50	>50-100	>100-150	>150-200	>200
Surface water quality	100-97	<97-90	<90-80	<80-75	<75-10
Sound Environment quality (dB)	daytime ≤45 night ≤40	daytime >45-50 night >40-45	daytime >50-55 night >45-50	daytime >55-60 night >50-55	Daytime >60 night >55
Anion concentration (a/cm^3)	≥2000	<2000-1000	<1000-800	<800-700	<700
Microbe concentration (cfu/cm^2)	<15	≥15-25	>25-45	>45-75	>75
Vegetation cover rate (%)	forest park ≥95 ordinary scenic ≥85	forest park <95-90 ordinary scenic <85-80	forest park <90-85 ordinary scenic <80-75	forest park <85-80 ordinary scenic <75-70	forest park <80 ordinary scenic <70
The number of vegetation types	≥1500	<1500-1000	<1000-600	<600-300	<300

5. Evaluation Content

5.1. *Atmospheric environmental quality*

Atmospheric environmental quality is evaluated by air pollution index (API). To reflect air quality and degree of pollution, air pollution concentration is converted into an index. It is simple and easily understood to clearly compare threshold limit value, air quality standards and impacts of detected air pollution upon human health. In China, current air pollution index mainly include sulphur dioxide, nitrogen dioxide and TSP (total suspended particle).

According to the <Ambient air quality standards> (GB3095-1996), the relationship between the value of API and air quality grades is showed in Table 2. The pollution consistence value grade of API of TSP,SO_2 and NO_2 is shown in Table 3. In Table 3, the data of concentration of SO_2 and NO_2 are the average value per hour, but the PM_{10} are the daily average value.

Table 2. API and air quality level.

API	Classification of air quality	Evaluation of air quality
0–50	I	excellent
51–100	II	fine
101–150	III	slightly polluted
151–200	III	between slightly and moderate polluted
201–250	IV	moderate polluted
251–300	IV	between moderate polluted and heavy polluted
>300	V	heavy polluted

Table 3. Pollution consistence value grade in API.

Classification of air quality	API	Pollutants content		
		SO_2	NO_2	PM_{10}
I	50	0.15	0.12	0.05
II	100	0.50	0.12	0.15
III	200	0.70	0.24	0.25

Dealing with the monitoring data of each scenic area's air quality and calculate the air pollution index (API) to have a comprehensive evaluation of environmental quality in Baiyun Mountain.

See the monitoring results in Table 4, which were collected from the Forestry Bureau in Baiyun Mountain scenic area.

Table 4. Calculate of air quality ($mg\cdot(Nm^3)^{-1}$).

Scenic area	SO_2	NO_2	PM_{10}
Jufang Park (Luhu)	0.069	0.075	0.046
Yuntai Park (Santai Ridge)	0.047	0.079	0.035
Mingchun Valley (Mingchun Valley)	0.040	0.049	0.030
Mount Davis (Mount Davis)	0.035	0.050	0.028
Peach Blossom Stream (Mingzhulou)	0.031	0.034	0.034
Heyi Ridge (Heyi Ridge)	0.043	0.058	0.036

Using interpolation method and piecewise linear equation to calculate the pollution sub-index I_i, so the air pollution index $API = \max\left(I_1, I_2, ..., I_n\right)$.

$$I_i = \frac{C_i - C_{in}}{C_{in+1} - C_{in}} \times \left(I_{in+1} - I_{in}\right) + I_{in} \tag{1}$$

Where, I_i is pollution sub-index of the ith type of pollutant, $I_{i,n}$ is pollution sub-index of the ith type of pollutant at the turn point of n, $I_{i,n+1}$ is pollution sub-index of the ith type of pollutant at the turn point of $n+1$, C_i is actual concentration of the ith type of pollutant, $C_{i,n}$ is concentration limit of the ith type of pollutant at the turn point of n (corresponding to $I_{i,n}$), and $C_{i,n+1}$ is concentration limit of the ith type of pollutant at the turn point of $n+1$ (corresponding to $I_{i,n+1}$).

The result of calculation can be seen in Table 5.

Table 5. Calculate outcome of API.

Scenic area	Sub-index of pollution mg $(Nm^3)^{-1}$			*API=*	Classification	Air quality
	SO_2	NO_2	PM_{10}	*max(I_i)*	of air quality	condition
Luhu	23	31	46	46	I	excellent
Santai Ridge	16	33	35	35	I	excellent
Mingchun Valley	13	20	30	30	I	excellent
Mount Davis	12	21	28	28	I	excellent
Mingzhul-ou	10	14	34	34	I	excellent
Heyi Ridge	14	24	36	36	I	excellent

According to the table, the overall air quality is good in Baiyun Mountain and shall be maintained. Air quality is excellent in all scenic spots investigated. The air quality is the best in the Peach Blossom Stream and the Mount Davis of the Mingzhulou Tourist Area. The air quality is slightly worse in Luhu Tourist Area where the Jufang Park is located than other areas, possibly because Huanhu Road is the major tourist route within the region. In the future, traffic flow shall be reduced by traffic control.

5.2. *Water environmental quality*

Adopting the classification index method that Chinese environmental monitoring station uses, which is match with Chinese surface water quality standard. The evaluation standards have been showed in Table 6 and Table 7.

The monitoring results of water quality in Baiyun Mountain scenic have been showed in Table 8, which can be used to calculate the total evaluation score of water quality of each scenic area. The calculation results shows in Table 9. The data in Table 8 were collected from the Authority in Baiyun Mountain scenic area.

Table 6. Grade of water quality.

Grade of water quality	The score of single factor	Total evaluation score
I	10	97–100
II	9	90–96
III	8	80–89
IV	6	60–79
V	3	30–59

Table 7. Evaluation standard of surface water quality ($mg \cdot L^{-1}$).

Project	Water environmental quality standard of different evaluation grades				
	I	II	III	IV	V
DO	≥7.5	≥6	≥5	3	2
BOD_5	≤2	≤3	≤4	6	10
COD_{cr}	≤15	≤17	≤20	30	40
TP	≤0.01	≤0.025	≤0.05	0.1	0.2
NH_3-N	≤0.15	≤0.5	≤1.0	1.5	2.0
C_6H_5OH	≤0.002	≤0.002	≤0.005	0.01	0.1
Cu	≤0.01	≤1.0	≤1.0	1.0	1.0
Oil	≤0.05	≤0.05	≤0.05	0.5	1.0
C_r^{6+}	≤0.01	≤0.05	≤0.05	0.05	0.1
As	≤0.05	≤0.05	≤0.05	0.1	0.1

Table 8. Monitor data of water quality.

Project	Luhu	Yuntai	Pugu	Jiulong Spring	Huangpodong Reservoir	Shangkeng Reservoir
DO	5.40	3.23	6.70	6.17	5.925	6.675
BOD_5	7.50	8.00	2	2.5	6.00	3.00
COD_{cr}	37.75	29.7	10.6	15.7	28.45	12.7
TP	0.07	0.065	0.135	0.08	0.035	0.11
NH_3-N	0.33	0.28	0.08	0.23	0.55	0.355
C_6H_5OH	0.002	0.002	0.002	0.002	0.002	0.002
Cu	0.00002	0.00002	0.002	0.0038	0.0068	0.0031
Oil	0.075	0.105	0.01	0.045	0.075	0.075
C_r^{6+}	0.004	0.004	0.056	0.029	0.034	0.033
As	0.007	0.007	0.007	0.007	0.004	0.012

Table 9. Evaluation of water quality ($mg \cdot L^{-1}$).

Project	Luhu	Santai Ridge	Mingchun Valley	Mount Davis	Mingzhulou	Heyi Ridge
DO	8	6	9	9	8	9
BOD_5	6	6	10	10	6	10
COD	6	8	10	9	8	10
TP	6	8	3	6	8	3
NH_3-N	9	9	10	9	8	9
C_6H_5OH	10	10	10	10	10	10
Cu	10	10	10	10	10	10
Oil	9	9	10	8	9	9
C_r^{6+}	10	10	10	10	10	10
As	10	10	10	10	10	10
Total evaluation score	84	86	92	91	87	90
Grade of water quality	III	III	II	II	III	II

We can see from the table that the water quality in Baiyun Mountain had obviously polluted which influenced by sewage emission. For example, Pugu and Shangkeng Reservoir have high TP and more than discharge standard the country sets. These fields should be strong managed to control the emissions and increase of pollutants.

5.3. *Sound environmental quality*

Seven areas are designated in the scenic spot known as Baiyun Mountain to measure noises at day and night. Major sound sources differ in all tourist areas. There are mainly tourists' noises in Xinghai Park, Heyi Ridge, Yuntai Garden and Mount Davis, but primarily birdsongs in Plum Blossom Valley, Mingchun Valley and Songtao Bieyuan. Specific data monitored are shown in Table 10, which were collected from the authority in Baiyun Mountain scenic area. Although the sound is louder at daytime than night in Baiyun Mountain, tourists have higher requirements for sound at night. Thus, experts determine weight of noise at daytime and night in Baiyun Mountain, where the acoustic environmental quality average. Xinghai Park, Plum Blossom Valley, Songtao Bieyuan and Mount Davis belong to Level II, while Heyi Ridge, Yuntai Garden and Mingchun Valley are confirmed to be Level III in terms of acoustic environmental quality. At night, acoustic environment may reach evaluation standard, while it doesn't meet the standard in several areas in the daytime. The acoustic environment is acceptable in the daytime in Mount Davis and Songtao Bieyuan. It is relatively noisy in Yuntai Park and Mingchun Valley mainly because of impacts of tourists' travel and motor vehicles.

Table 10. Monitor data of noise (dB).

Scenic area	Ld	Ln
Xinghai Park (Luhu)	48.2	43.4
Heyi Ridge (Heyi Ridge)	52.2	47.7
Mingchun Valley (Mingchun Valley)	53.8	48.3
Yuntai Park (Santai Ridge)	54.6	49.6
Mount Davis (Mount Davis)	43.8	42.6
Plum Blossom (Mingzhulou)	45.3	43.1

5.4. *Anion content*

As the very important tourism resources for ecological tourist area, anion can reduce dustiness and kill bacteria while cleaning the atmosphere,which are helpful for human visceral health. Furthermore, negative ions can make blood pressure drop and keep the heart and lung healthy, but also stimulate generation of several enzymes in humans. As a result, the anion content has been the key to evaluation of normal ecological environment [24]. According to pertinent medical data, human beings' physiological requirement for concentration of negative ion in air isn't lower than $700/cm^3$. It will be effective for preventing and treating diseases once the concentration is up to 10, 000/cm^3. See details in Table 11.

Table 11. Relationship between health and anion concentration (a·cm^{-3}).

Anion content	Relationship with health
<50	Induce physiological disorder
1000-2000	Keep healthy basic requirements and good for health
5000-50000	Strengthen human immunity and resistance
50000-100000	Kill germs and reduce infections
100000-500000	Have natural heal power

See monitoring data in Table 12 which were collected from the authority in Baiyun Mountain scenic area.

Table 12. The anion concentration of Bai-yun Mountain (a·cm^{-3}).

Scenic area	The average of anion	The maximum of anion
Jufang Park (Luhu)	1057	1120
Yuntai Park (Santai Ridge)	820	880
Mingchun Valley (Mingchun Valley)	838	900
Mount Davis(Mount Davis)	912	1100
Peach Blossom Stream (Mingzhulou)	1203	1280
Heyi Ridge (Hey-i Ridge)	642	710

The monitoring results suggest that there are more negative ions generated by buckets effect of sprayed flowing water beside rivers than that on slopes, beside still ponds and in office areas. For instance, there are the most negative ions in Peach Blossom Stream and the fewest in Heyi Ridge. This may have much to do with the areas detected. On the whole, the concentration of negative ions is moderate and much higher than urban areas in Baiyun Mountain. For tourists, negative ions are pretty important tourist resources. They may suppress dust, kill bacteria, clean air, improve human beings' pulmonary functions, lower blood pressure, enhance cardiopulmonary functions and activate several enzymes of human body, so they are helpful for health. Concentration of negative ions may be used as an important index for evaluating regional ecological environment [19].

5.5. *Microbe concentration*

Microbe concentration refers to content of microbe in air. The most fundamental method is to measure number of colonies on each plate by natural sedimentation, which is also an essential standard for evaluating air quality of a forest tourist area. The content of microbe in air is just the opposite to the environmental quality of a scenic spot. See specific data in Table 13 which were collected from the authority in Baiyun Mountain scenic area.

Table 13. Microorganism number of each area ($CFU \cdot cm^{-3}$).

Scenic area	Total bacteria	Grade
Jufang Park (Luhu)	6.7	I
Yuntai Garde (Santai Ridge)	7.9	I
Mingchu-n Valley (Mingc-hun Valley)	8.5	I
Mount Davis (Mount Davis)	2.9	I
Peach Blossom Stream (MIingzhul-ou)	4.6	I
Heyi Ridge (Heyi Ridge)	3.5	I

Seen from the monitoring data, each scenic area of Baiyun Mountain belongs to I highly coordinated. It shows that the air quality in Baiyun Mountain is very good and the bacteria are few, which is profitable to the development of ecological tourism.

5.6. *Vegetation cover rate*

Located in the southern part of China, Baiyun Moutain is an area with subtropical climate. There is an abundant variety of vegetation in Baiyun Mountain, where the forest vegetation mainly includes coniferous forests, broadleaf forests and bamboo forests. There are 29 types altogether, such as Pinus massoniana Lamb, Acronychia pedunculata, Schima superba Gardn. et Champ and Castanopsis chinensis Hance. From 1995 to 1999, a project was carried out in Baiyun Mountain for renovating the form of forest. In other words, original simple form of forest was transformed by creating broadleaf and mixed broadleaf-conifer forest, in order to develop Baiyun Mountain into a forest park with multiple tree species, colors, layers, fragrance and functions. After renovating forest stand for years, luxuriant broadleaf forest with excellent tree species has completely replaced original fragile pinus massoniana forest, so that the forest vegetation landscape may become more complete and reasonable in the Baiyun Mountain. More diversified seasonable changes better reflect local characteristics of the south of the Five Ridges [20].

At present, green coverage is up to more than 95% in Baiyun Mountain. According to statistics, the green area reaches 42,000 acres within the scenic spot. On a daily basis, 2,800t carbon dioxide may be absorbed, and 2,100t oxygen is released, which is enough for normal breath of 3 million people, so Baiyun Mountain is known as “lung of city” in Guangzhou, which means Baiyun Mountain, as a natural solid reservoir with large green area in downtown area of Guangzhou, isn't only a natural barrier for combating natural disaster, but also an air purifier [21].

5.7. *Vegetation type*

There are diversified types of vegetation and a wide range of species that make up a balanced system in Baiyun Mountain. On the whole, this scenic spot is highly resistant to interference and can't be easily destructed. It has become a core area to be monitored in the city of rams for its diversified plants and animals. According to a survey on plant resources in Baiyun Mountain conducted from 1998 to 2001 and reference to pertinent surveys carried out in 1994, there are 180 families, 631 genera and 1038 species of vascular plants. Some of these plants have great potential economic value, including 482 species of plants with medical value, 452 species of plants with ornamental value, 152 tree species used as timber, 75 species of fiber plants, 75 species of plants for extracting oil, 73 species of fruit trees, 70 species of aromatic plants, 53 species of fodder plants, 48 species of tannic plants and 28 species of plants used for heath drinks. In addition, there are 10 species of rare endangered plants and 10 species of national key protected species [11]. There are 1,038 species of plants in Baiyun Mountain altogether, among which a majority of vegetation belongs to tropical and subtropical vegetation. There are numerous subordinate species, including 11 distribution types and 10 mutants. Therefore, it may be concluded according to certain numerical value that there are 1,038 species in each area, which belongs to Level-II moderate harmony.

6. Conclusion and Discussion

The analyzed result shows that the air quality we monitor in Baiyun Mountain scenic area is excellent. Water environment is influenced by the pollution of domestic sewage, some scenic areas' TP numerical value is too high. The voice environment can reach the evaluation standard at night. However, influenced by the tourists and motor vehicles, many scenic areas' TP numerical value is too high, especially the Heyi Ridge, Yuntai Park and Mingchun Valley. The whole concentration of the anion in these six scenic areas is higher than that in city center and belongs to moderate coordination. The microbial content is the high coordination. The vegetation coverage is too high to reach 95% at least. The kind of plants is very rich, it reaches the moderation coordination. It shows that the environmental quality indexes in most areas reflect the good environment of Baiyun Mountain scenic area, except some indexes of water environment and sound environment. The analyzed result can provide the based data and decision support for the comprehensive and quantitative evaluation.

This paper uses the modern manage method of environmental capacity to build the environmental quality evaluate system based on LAC theory. Providing quantitative analysis for each quality of ecological environment index and clearly understanding tourist area's situation of ecological environment recently. The analysis results present the current situation of ecological environment in Baiyun Mountain directly. At present, Guangzhou Baiyun Mountain scenic area lacks research about environmental capacity physical indexes quantitatively. As a consequent, the research can replenish the research data in this field, but also provide data and reference for the analysis of ecological environment quality in Baiyun Mountain more deeply and scientifically. The indicators in the research include the physical indexes and social and cultural indexes which reflect the subjective factors. But the research belongs to the tentative exploration because it only quantitatively evaluate the physical indexes. In the future job, it can be further optimized and improved through the cultural indexes and comprehensive evaluation.

References

1. Stephen F, David W. Tourism carrying capacity: Tempting fantasy or useful reality. *Journal of Sustainable Tourism*, 2001, (9):4–6.
2. Reilly A. Tourism carrying capacity – Concepts and issues. *Tourism Management*, 1987, 7 (3):154–167.
3. When C H, Yang G H, Wang H X. Integrated indicator system of eco-tourism environmental load capacity in nature reserves. *Agro-environmental protection*, 2002, 21(4):365–368.
4. Li T Y. Views on the application of tourism carrying capacity theory. *Nankai business review*, 2001(3):57–60.
5. Feng X G. The theory, method and empirical study of tourism management capacity. [D]. Nanjing*: Nanjing University,* 1999.
6. Lin M Z, Xie S X, Yang M Z et al. Comprehensive assessment of environmental quality of the district of ecological tourism in Nankun Mountain. Acta Ecologica Sinica, 2010, 30(12):3270–3279.
7. Zhang M, Ma S C, Qiong D. Evaluation on eco-tourism environment capacity of the Sejila Mountain national forest park. *Problems of forestry economics,* 2011, 31(1):51–55.
8. Lin M S, Xie H B. On the managing dynamics of tourism environmental carrying capacity. *Journal of Subtropical Resources and Environment*, 2009, 4(2):90–94.

9. Liang Z. An analysis on the models of social-economic carrying capacity of a destination. *Nankai Business Review,* 2002(4):65–68.
10. Dai X J, Ding D S, Lin C. Research on tourist carrying capacity of sustainable tourism. *Human Geography*, 2002, 17(6):32–36.
11. Butler R. The concept of a tourist area cycle of evolution: Implication for management of resources [J]. *Canadian Geographer*, 1980, 14:5–12.
12. Getz D. A Rational and Methodology for Assessing Capacity to Absorb Tourism [J]. *Ontario Geography*, 1982, 19:92–102.
13. Liu N N. The management research of tourism environmental capacity in Wuhan Mulan scenic area. Beijing: *Beijing Forestry University*. 2006.6
14. Cole, David N.; Stankey, George H. 1998. Historical development of Limits Of Acceptable Change: Conceptual clarifications and possible extensions. VS: *Missoula*, MT, 1997.
15. Stankey George H. Limits of Acceptable Change: New Framework for Managing the Bob Marshall Wildness Complex [J]. *Western Wild lands*, 1984, 10(3):3–37.
16. Papageorgiou K, Brotherton I. A management Planning Framework Based on Ecological, Perceptual and Economic Carrying Capacity: The Case Study of Vikos Aoos National Park, Greece [J]. *Journal of Environmental Management,* 1999, 56:271–284.
17. Fernando J, Garrig S Simn. Carrying Capacity in the Tourism Industry: A Case Study of Hengistbury Head [J]. *Tourism Management*, 2004, 25:275–283.
18. Gu Y K. The theories and practice of sustainable development of ecological resources – The key tourist attractions of Baiyun Mountain in Guangzhou. Beijing: *Chinese Forestry Press*, 2005.1.
19. State environmental protection agency. GB3095-1996 Ambient air quality standards. Beijing: *China Environmental Science Press*, 1996.
20. State environmental protection agency. GB3838-2002 Quality standard for surface water environment. Beijing: *China Environmental Science Press*, 2002.
21. State environmental protection agency. GB3096-2008 Standard of environmental noise of urban area. Beijing: *China Environmental Science Press*, 2008.
22. Management office of forestry park. GB/T18005-1999 Chinese forest park landscape resources grade evaluation. Beijing: *China Standards Press*, 1999.
23. State environmental protection agency. HJ/T6-1994 Index system for evaluating environmental effect of mountainous type scenic resource development. Beijing: *China Environmental Science Press*, 1994.

24. Zheng Q M. The application of air negative ions in forestry ecological tourism. Hunan: *Central South Forestry University*. 2000.
25. Zhang W Y, Ye R H, Gu D Q. The utilization and sustainable development research of resource in Guangzhou Baiyun Mountain scenic area – tourists questionnaire analysis. *Guangdong landscape aechitecture*, 2004, 1:15–20.
26. Zeng H L. The development countermeasure research of Guangzhou Baiyun Mountain scenic area. Hubei: *Central China Normal University*, 2008.

CFD simulation of flow and heat transfer in a thermosyphon

K.M. Yang[†,a], Z. Cong[b] and W.F. Wang[c]
[a]*College of Thermal Energy Engineering, Shandong Jianzhu University, Jinan, China*
[b]*Zhongshi Yitong Group, Jinan, Shandong province, China*
[c]*Tengzhou Xinyuan Thermal Power Co. Ltd, China Huadian Corporation, China*
[†]*Email: ykm_168@163.com*

Heat pipes are devices of very high heat transfer capability and have been widely used in many thermal management applications. An experimental investigation and CFD simulation of thermal characteristics of heat pipe was presented in this paper. It can be found that UDF in FLUENT can simulate the evaporation and condensation in heat pipe. Phase change happened in evaporation section of heat pipe is boiling. In steady state, the fluctuation of axial velocity is very small in the most area (without evaporation section) in heat pipe. In general, the magnitudes of velocity vary from 0 to maximum from the end of both evaporation section and condenser, and the maximum value was maintained in the adiabatic section.

Keywords: Flow, Heat transfer, CFD, Thermosyphon.

1. Introduction

Grooved heat pipe is a reliable and efficient heat transport device. The heat pipe has been used in several technologically important processes requiring augmented heat transfer e.g., in the electronic packaging industry, in micro-gravity environments, and spacecraft thermal control because of its high efficiency, reliability and cost effectiveness[1–4].

Cotter[5] first proposed the concept of micro-heat pipe. That is essentially a wickless heat pipe for the uniform temperature distribution in electronic chips. After Cotter's research, 1D model[6], 2D model[7,8] and 3D model[9] for theory analysis of fluid flow and heat transfer in heat pipe were built. Tiselj et al.[10] simulated the single phase flow in a 3D model using the CFD code CFX. Sandra C.K.et al.[11] modeled the phase change and two phase flow in a pipe using FLUENT.

In the present study, 2D model of two phase flow and heat transfer in a thermosyphon was built. The evaporation and condensation were simulated in heat pipe using FLUENT, and the experimental data are compared with the simulating results in order to verifying the simulating model.

2. Experimental Facilities

The experiments were performance using a copper-water heat pipe with a axial length is 250 mm. And the geometric schematic of cross section of heat pipe for this experiment is shown in Figure 2.

The schematic illustration of the experimental apparatus is shown in Figure 1. The heat pipe was divided into three sections representing the evaporator, the adiabatic section, and the condenser.

As shown in Figure 1, the evaporator section was heated by the heater band, and the test error of the input power is within the range of 1 W. The condenser section was cooled by cooling water with constant temperatures. And the glass rotameters having 5% error range are used to measure the water flow rate, which is controlled by the valve. The experiments are carried out at the room temperature, about 20°C, the axial temperature distribution of the heat pipe is measured by thermocouples with an uncertainty of ±0.25 °C, and the working temperature is regarded as the mean value of these measured temperatures. The numbers of thermocouples, which are symmetrically distributed at evaporator sector, adiabatic sector and condenser sector, are 3, 3 and 3 respectively. An intermediate plate having thickness of 0.2 mm is placed between the thermocouples and the resistor. This intermediate plate can protect the thermocouples against a direct contact with resistor and can help to evenly heat the evaporator sector. The electrical power to the band heaters was supplied using a 220 V variable DC power source with uncertainty ±2.5 W.

The experiments were performed in a vertical heat pipe. In the series of tests, the heat input to the evaporator was increased in steps of 20W from 20W to 120W while the mean temperature of cooling water was maintained constant. Tests were performed for cooling water temperatures of 5 and 15°C. The heat pipe was allowed to reach steady state and the data was recorded over a time interval of 3 minutes.

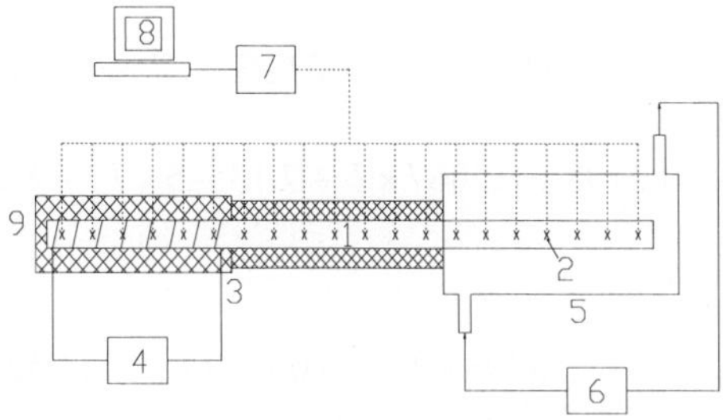

Figure 1. Schematic of experimental set up. 1-heat pipe, 2-thermocouples, 3-electrical heaters, 4-DC electrical source (HY1794-10S), 5-watercooling jacket, 6-thermostatic waterbath (LAUDA RP855), 7-data acquisition system (Fluke 2680A), 8-computer.

3. CFD Modeling

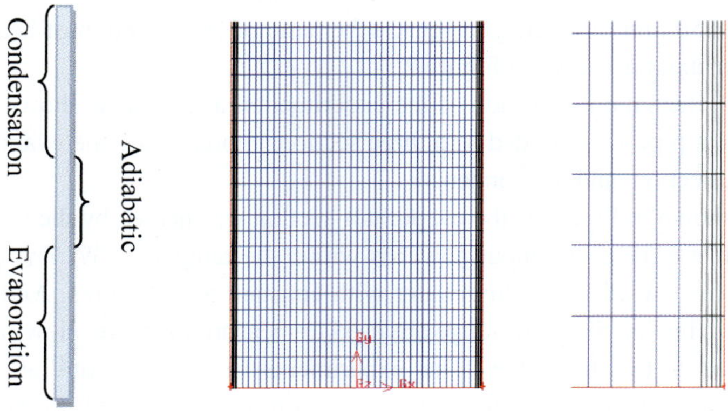

(a) (b) (c)

Figure 2. Schematic of heat pipe for experiment, (a) heat pipe, (b) mesh, (c) prism.

Modeling of gas-liquid interface is quite important in the prediction of interphase heat and mass transfer, and VOF (volume of fluid) model is chosen. The VOF method relies on the fact that two or more phases are not interpenetrating and for each additional phase the volume fraction of the phase must be added in the computation. In the VOF model, the sum of the volume fractions of all phases in each control volume is equal to one. In the CFD modeling, the Navier–Stokes equations are solved simultaneously[11]. The surface tension was implemented using the method of Brackbill et al. User define subroutines (UDF) were developed to employ these equations for modeling of the boiling and condensation phenomena.

The variation of properties of working liquid due to temperature change was considered, and the equations are shown as follows:

density:

$$\rho_{liq} = -0.0022 \times T^2 + 0.9237 \times T + 919.8 \tag{1}$$

$$\rho_{vap} = -20.0198 + 0.2067 \times T - 7.13545 \times T^2 + 8.247 \times T^3 \tag{2}$$

thermal conductivity:

$$\lambda_{liq} = -9.59298 \times T^2 + 0.00739 \times T - 0.7427 \tag{3}$$

$$\lambda_{vap} = 2.86556 \times T^2 - 1.0526 \times T + 0.02445 \tag{4}$$

surface tension:

$$\delta = -2.560 \times 10^{-7} \times T^2 - 2.24 \times 10^{-6} \times T + 0.09537 \quad (5)$$

The calculations were performed by combination of the SIMPLE algorithm for pressure-velocity coupling, first order upwind calculation scheme for the discretization of momentum and energy equations and Geo-Reconstruct for the volume fraction. Because of the dynamic behavior of the two-phase flow, unsteady state calculations with a time step of 0.001 s were carried out for all cases.

4. Validation of CFD Model

Firstly, the grid-independence test had been done to ensure the accuracy. Then a comparison between CFD predicted and measured temperature profile along the thermosyphon wall is given in Figure 3. A reasonable agreement can be observed between measured and predicted data in the evaporation section. This difference can be explained by the fact that in the experiments, there is some energy loses in the adiabatic section and condenser, while it is ignored in the modeling. Therefore, the values of temperature in adiabatic section are over predicted. In the other words, applying the adiabatic condition in the modeling causes the temperature to be constant in this section in the predicted results.

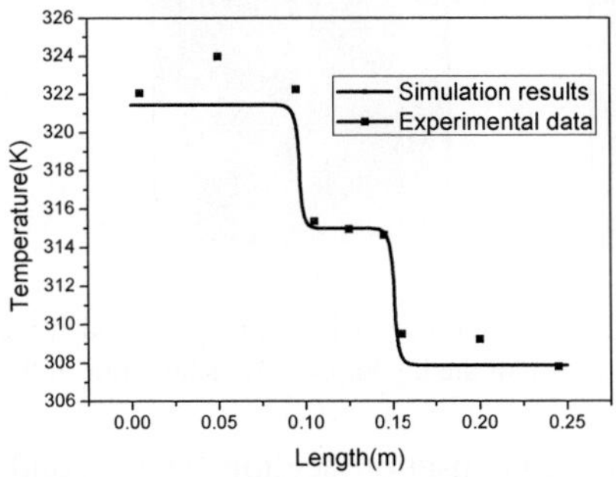

Figure 3. Comparison between simulation results and measured data.

5. Results and Discussion

The contour plots of volume fraction and velocity vectors are presented in Figures 4 and 5. In condenser (Figure 4(a)), when condensation happens, there is a stable liquid film on the solid surface, and it has a tendency to increase the thickness. As liquid moves forwards to the bottom, it will be affected by the countercurrent vapor flow. So it leads to the increase of the film thickness is not obvious and the film is nearly flat. In adiabatic section (Figure 4(b)), here there is no phase change, so the liquid film thickness is constant. The situation in evaporation section (Figure 4(c)) is very complex, there is a liquid pool, in which lots of bubble exist. At the location of surface of the liquid pool, free surface between liquid and vapor fluctuates remarkably. It can be seen from Figures 4(c) and 4(d).

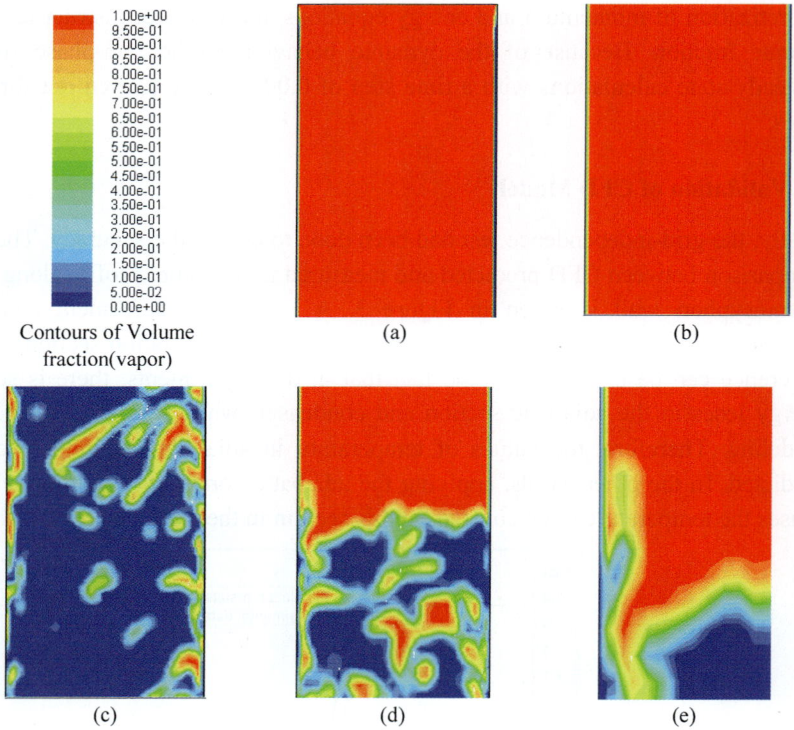

Figure 4. The volume fraction of vapor in heat pipe, (a) condenser, (b) adiabatic section, (c) evaporation section, (d) surface of the liquid pool, (e) detail view of the liquid pool.

As said above, in condensation section where condensation is happening, the fluid flow direction is from the center of pipe to solid surface (Figure 5(a)). Because there is no phase change in the adiabatic section, a sable velocity field is that vapor flows upwards and liquid flows downwards (Figure 5(b)). Similar to the volume fraction contour, the velocity field in the evaporation section is also complex (Figure 5(c)). And it is more complex at the surface of liquid pool (Figure 5(d)).

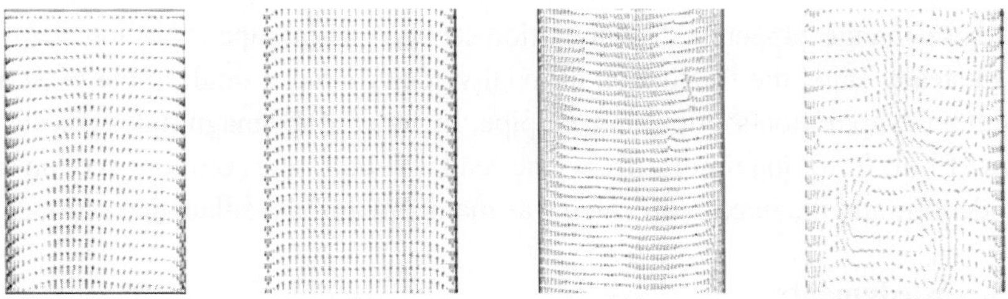

(a) (b) (c) (d)

Figure 5. The velocity vectors in heat pipe, (a) condenser, (b) adiabatic section, (c) evaporation section, (d) surface of the liquid pool.

Vaporization of liquid can be divided into two kinds of evaporation and boiling, the former refers to the evaporation process occurs on liquid surface, and the latter refers to the vaporization with bubbles occurs inside the liquid. From Figure 4(c), it is can be seen that phase change happened in evaporation section is boiling. To improve that further, volume fraction contour in evaporation section near to solid surface with a distance 0.2mm is given in Figure 6, the values of volume fraction change dramatically. Because in this area there are a lot of bubbles, and they are different in size.

The axial distribution of velocity for vapor at the center of the pipe (without evaporation section) is shown in Figure 7. It indicates that the mean velocity at the end of condenser is 0. In evaporation section, the velocity of the vapor vary from 0 to a maximum from the end to adiabatic section due to condensation in the condenser. There are no evaporation and condensation in the adiabatic section, hence the velocity of the vapor is almost a constant.

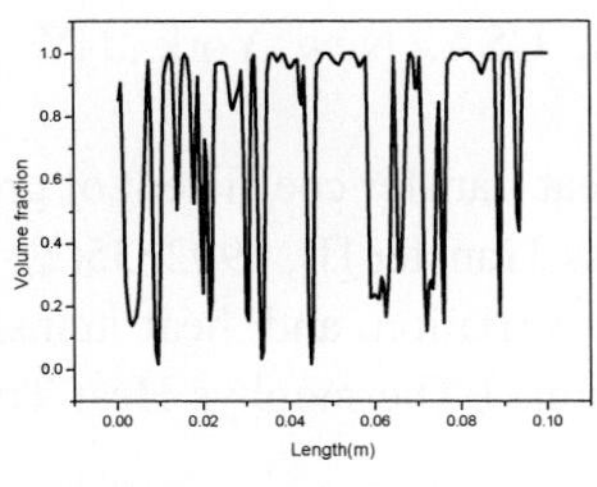

Figure 6. Volume fraction near to solid surface in evaporation section.

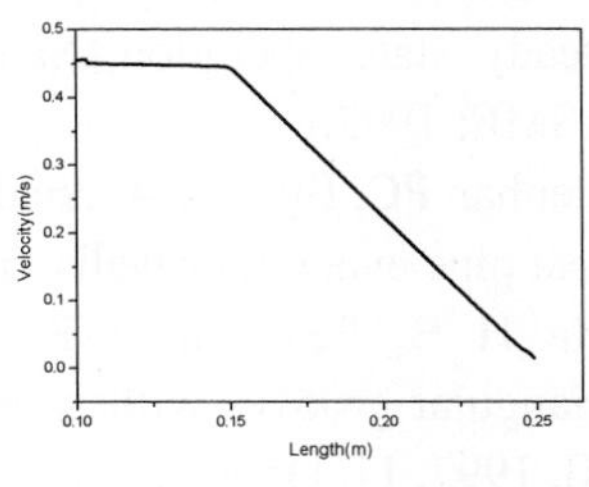

Figure 7. Axial distribution of velocity at the center of the pipe (without evaporation section).

6. Conclusion

1. With UDF in FLUENT, the evaporation and condensation in heat pipe can be simulated.

2. Phase change happened in evaporation section of heat pipe is boiling.
3. In steady state, the fluctuation of axial velocity is very small in the most area (without evaporation section) in heat pipe. In general, the magnitudes of velocity vary from 0 to maximum from the end of both evaporation section and condenser, and the maximum value was maintained in the adiabatic section.

Acknowledgments

This paper is supported by Research Fund for the Doctoral Program of Shandong Jianzhu University, project number is 0000601337.

References

1. Zhuang Jun, Zhang Hong. Heat Pipe Technology and Engineering Application [M]. Beijing: Chemical Industry Press, 2000.
2. Faghri A. Heat Pipe Science and Technology [M]. Washington DC: Taylor & Francis, 1995.
3. Masataka Mochizuki, Thang Nguyen, Koichi Mashiko, et al. Practical application of heat pipe and vapor chamber for cooling high performance personal computer [A]. In: 13th International Heat Pipe Conference [C], Shanghai, 2004.
4. Cotter T P. Principles and prospects for micro heat pipes. In: Fifth international heat pipe conference [C], Tsukuba, Japan; 1984.
5. T P Cotter. Theory of heat pipes [J]. Los Alamos Scientific Lab: Report No., LA-3246-MS, 1965.
6. Longtin J P et al. One-dimensional model of a micro heat pipe during steady state operation, San Diego, CA, USA. New York, NY, USA: ASME; 1992.
7. Stephan PC, Busse CA. Analysis of the heat transfer coefficient of grooved heat pipe evaporator walls. Int J Heat Mass Transfer [J], 1992, 35(2): 383.
8. Ma H B, Peterson G P. Temperature variation and heat transfer in triangular grooves with an evaporating film. J Thermophys Heat Transfer [J], 1997, 11(1): 90.
9. Sartre V et al. Effect of interfacial phenomena on evaporative heat transfer in micro heat pipes. Int J Therm Sci [J], 2000, 39: 498–504.
10. Tiselj I et al. Effect of axial conduction on the heat transfer in micro-channels. Int J Heat Mass Transfer [J], 2004, 47(12–13): 2551.
11. Sandra C.K. De Schepper et al. Modeling the evaporation of a hydrocarbon feedstock in the convection section of a steam cracker. Computers and Chemical Engineering [J], 2009, 33: 122-132.

The environmental Kuznets curve in China – foreign and Chinese research after 2000

E. Nagy

Shanghai University, Sydney Institute for Language and Commerce, Shanghai City, Baoshan District, Shangda Road 99. Shanghai, People's Republic of China

E-mail: 1929643954@qq.com

The purpose of this paper is to give a summary of post-2000 foreign and Chinese research on the relationship of economic development and environmental degradation in the case of China. The Environmental Kuznets Curve theory describes this relationship, although the related research gives different results. Following the initial introduction and a brief review of the theory, the major part introduces the research of certain foreign and Chinese scholars on the subject, focusing on their methods, econometric models and empirical results. After comparing a number of works, we got to the following conclusion: there is no agreement in the scientific community about the EKC theory in China. We believe that this paper can widely introduce some of the latest research results on the topic which haven't crossed the language barrier before and might contribute to the scientific field of sustainable development.

Keywords: EKC; China; Econometric model; Review.

1. Introduction

In recent decades, environmental economics has become a significant and relevant field amongst economic researchers and the theory of the Environmental Kuznets Curve (EKC) has become a popular topic in this field. In the case of the People's Republic of China, this topic is extremely relevant, since the country is facing numerous socio-economic problems brought forth by its rapid economic growth in the past decades, one of these being environmental pollution. Since the People's Republic of China has the world's highest population and it is one of the major actors on the international political and economic field, it is extremely important that the scientific community reaches a common view on the Environmental Kuznets Curve in China and provides a viable route for policy makers. However, at this point there is no general agreement on the topic amongst researchers, especially if we compare the works of foreign and Chinese scholars.

2. Theoretical Overview

In 1991 Grossman and Krueger established a reverse U-shaped relationship between income and environmental pollution in their progressive article, this theory later became the Environmental Kuznets Curve theory. (Grossman, G. and Krueger, A., 1991) After the initial establishment of the EKC hypothesis, numerous scholars have introduced their ideas on the subject. Shafik and Bandyopadhyay's pioneer work in the beginning of the 1990s was particularly influential, since it was used in the 1992 World Development Report (IBRD, 1992). These researchers have come to the same conclusion: as an economy develops, environmental problems will initially get more serious, but will eventually decline.

The above mentioned scholars established the EKC theory in the early 1990s, since then many others conducted research on the matter whose research supports the existence of the EKC (Shafik and Bandyopadhyay, 1992; Panayotou, 1993; Selden and Song, 1994; Cropper and Griffith, 1994; Lopez, 1994; Grossman and Krueger, 1995; de Bruyn et al., 1998; Suri and Chapman, 1998; Schmalensee et al., 1998; Galeotti and Lanze, 1999; Sachs et al., 1999; Andreoni and Levinson, 2001). On the other hand, several researchers believe that the EKC theory is a promising hypothesis, but it doesn't offer any substantial evidence. A number of critical surveys of the EKC literature has been published (e.g., Pearson, 1994; Arrow et al., 1995; Stern et al., 1996; Ekins, 1997; Ansuategi, Barbier and Perrings, 1998; Stern, 1998; Zaim and Taskin, 2000; Dasgupta et al., 2002; Copeland and Taylor, 2004; Stern, 2004).

3. Researchers on the EKC in China

Numerous scholars have conducted their research on the EKC in China, both foreign and Chinese. The research on the Chinese EKC usually falls into two categories: the samples being the entire country or a city/province.

3.1. *Research on the entire country*

De Groot et al. (2001) use province-level panel data to estimate the EKC for waste gas, waste water and solid waste emission (three wastes) between 1982 and 1997 to the 30 provinces and major cities. They find that the EKC exists, although the shape of the relationship heavily depends on the pollutant and on the specification of the dependent variable.

Following the footsteps of de Groot et al. (2001), many others turned their attention towards the PRC. Most of the scholars found the EKC connection in the case of the PRC, but some of them think the theory does not apply here, since China cannot follow the conventional path of developed countries.

Shen Jun Yi and Yoshizo Hashimoto (2004) using provincial level panel data illustrate the EKC connection in the case of some pollutants. Five of the indicators show a reverse U-shaped curve, while two of them show an N-shaped curve.

Chen Wen (2007) conducts the EKC research based on provincial level panel data between 1992 and 2005 (excluding Tibet and Chongqing) for industrial waste water, solid waste, SO2, dust and soot. The results show that the EKC hypothesis is not clear in China, the inverted U-shaped curve cannot be generalized for all emissions. Chen points out that environment cannot be improved at higher income levels automatically in China and all governmental levels should introduce strict environmental policies.

Hu and Pan (2007) point out the fact that there are numerous problems with the EKC hypothesis, as model, data, index and analysis. They believe that the future research on the subject should focus on the relationship between world trade, FDI and environmental degradation.

Jiang et al. (2008) determine the Environmental Kuznets Curve for the coastal, the central and the western region. They believe that by 2015 the whole country can pass the turning point of the EKC curve for waste gas. Every province (except Guangxi) will pass through the waste gas turning point individually. According to their calculations, the central and western region will catch up with the pollution decrease rate of the coastal region.

Song et al. (2008) also use provincial level panel data to research the EKC curve for the whole country of China, their selected indicators are waste gas, waste water and solid waste. Song et al. determine that a reverse U-shaped connection exists.

Llorca and Meunié (2009) conducted research on the SO2 EKC, using provincial level data between 1985 and 2003. Their results support the hypothesis, in the case of the coastal regions, it already reached the decreasing part of the curve. However, when they removed Beijing, Shanghai and Tianjin from the sample, the turning point dropped significantly.

Xu and Song (2011) conducted a research on the carbon emission EKC in central, eastern and western China between 1990 and 2007. According to them, the carbon emission EKC of China in the eastern and central region have inverted U-shaped relationship, while that of the western region is U-shaped.

Gao (2012) carried out an analysis for CO2 and SO2 using panel data between 1995 and 2009; and discovered the existence of a reverse U-shaped EKC for both. However the SO2 emission curve crosses the turning point earlier than the CO_2 curve, due to the significant difference between their emission characteristics. Although in the case of SO_2 both developing and developed countries show the same reverse U-shaped curve, but in the case of CO_2 they haven't been able to find relevant connection.

3.2. *Research on specific areas and cities*

Liu Ying Xi (2005) studies the Guangzhou city economic development and SO_2 emission connection, he discovers that the EKC relationship is not reverse U-shaped, rather M-shaped, with 1996 being the lowest point in the M curve. According to his findings, Guangzhou city has already reached the turning point in 2002.

Li Chun Sheng (2006) also researches the EKC curve in Guangzhou city between 1982 and 2004 for total waste water emission, industrial and city-level household waste water emission, industrial waste gas emission, etc. Li discovered that the per capita GDP and environment (waste water) change corresponds to the reverse U-shaped EKC curve theory.

Wang et al. (2007) determine the existence of the EKC curve in Beijing between 1990 and 2004. Wang et al. chose 10 pollutants as indicators; except for the industrial waste gas emission and the industrial solid waste emission, which have an N-shaped curve, other pollutants produce a reverse U-shaped curve with decreasing tendency. Most of the environmental variables already reached the peak of the reverse U-shaped EKC curve.

Sun Ke (2008) studies the EKC curve in Tianjin city between 1985 and 2006. Sun chose industrial three wastes as indicators. According to the results, waste gas has a normal U-shaped plus a reverse U-shaped connection, whereas waste water and solid waste have an N-shaped curve. Sun points out that the EKC curve only reflects the connection between the environmental indicators and economic growth, but not a cause and effect relationship, therefore it isn't conclusive enough to explain what the most effective measures are to improve the environmental situation in times of economic growth.

Jiang Yi's (2009) analysis shows a well-behaving EKC curve for Shanghai. For industrial waste water emission, the economy is already at the right side of the EKC curve, but household waste water emission sharply rises. The total waste water emission gives us an N-shaped EKC curve.

Long Li Lian (2009) chose data from Shanghai between 1991 and 2007, analyzed the total emission of industrial waste water, waste gas, smoke, dust, and SO_2. The research indicates that most of these indicators show the EKC connection, however in the case of SO_2 emission the regression does not indicate the existence of the EKC.

Ni et al. (2010) research the Shanghai EKC between 1989 and 2004. According to them SO2 has a normal U curve, PM10 has a decreasing tendency, NOx has a reverse U-shaped curve and so has waste water. They point out that the pollutants that actually show the EKC connection with economic development have already crossed the turning point.

Han Jun (2012) thinks that for the two most developed Chinese areas – namely Beijing and Shanghai – regardless of the environmental pollution indicators they use, all the Environmental Kuznets Curves – without exception – will be reverse U-shaped. Compared to Beijing and Shanghai, in other areas with a lower level of development, the shape of the EKC curve depends on the chosen environmental pollution indicator.

Jiang Zai Ping (2012) determines the EKC curve for Beijing city between 2001 and 2010. Jiang uses the following emission numbers as indicators: SO2, NO2, nitric oxide and microscopic air particles. Jiang reaches the conclusion that in the short run Beijing already passed the turning point.

Fan Chuan Qi and Zheng Xiao Jun (2013) do a research on the EKC curve in Sichuan province between 1985 and 2010; their chosen indicators are industrial the three wastes. They discover that the EKC curve does not exist in Sichuan province.

Mao et al. (2013) study the Environmental Kuznets Curve situation between 1990 and 2008 in the Ningxia Yellow River Irrigation Area for the three wastes. They believe that the cultivation industry in the area has already crossed the turning point.

Table 1 and Table 2 summarize the above discussed research methods and results.

4. Conclusion

Table 1. Research results on the entire country.

Researcher	Time Frame	Indicator	Model	Result
De Groot et al. (2001)	1982–1997	three wastes	$ES_{it} = \alpha_{0i} + \beta_1 (GRP\_CAP_{it}) + \beta_2 (GRP\_CAPit)^2 + \beta_3 (GRP\_CAP_{it})^3 + \varepsilon_{it}$	reverse U, no EKC
Shen, Hashimoto (2004)	1990–2001	SO_2, industrial dust, COD, solid waste	$\ln p_{it} = \beta_{0i} + \beta_1 \ln y_{it} + \beta_2 (\ln y_{it})^2 + \beta_3 (\ln y_{it})^3 + \beta_4 ind_{it} + \beta_5 \ln pd + \beta_6 trend + \varepsilon_{it}$	reverse U, N
Chen (2007)	1992–2005	solid waste, waste water, SO_2, soot, smoke	$\ln E_{it} = \alpha_i + \gamma_t + \beta_1 \ln Y_{it} + \beta_2 (\ln Y_{it})^2 + \beta_3 (\ln Y_{it})^3 + \beta_4 \mathit{ind}_{it} + \beta_5 \mathit{ex}_{it} + \beta_6 \mathit{im}_{it} + \beta_7 \ln \mathit{FDI}_{it} + \beta_8 \ln \mathit{POP}_{it} + \beta_9 \ln R_{it} + \varepsilon_{it}$	reverse N, U, rising

Jiang et al. (2008)	1985–2005	three wastes	$E_{ikt} = \Sigma(\beta_{1rk}Q_{it} + \beta_{2rk}Q_{it}^2) + \alpha_{ik} + \mu_{tk} + \varepsilon_{ikt}$	reverse U
Song et al. (2008)	1985–2005	three wastes	$\ln (S/P)_{it} = \alpha_{it} + \theta_i + \beta_1 \ln(GDP/P)_{it} + \beta_2 [\ln(GDP/P)_{it}]^2 + \beta_3 [\ln (GDP/P)_{it}]^3 + u_{it}$	reverse U
Llorca and Meunié (2009)	1985–2003	SO_2	$Yit = \alpha_i + t_j + \beta_1 (GDPc_{it}) + \beta_2 (GDPc_{it})^2 + \beta_3 (GDPc_{it})^3 + \sum\beta_k Zk_{it} + e_{it}$	reverse U
Xu and Song (2011)	1990–2007	CO_2	$lncit = \alpha i + \alpha t + \beta_1 i\ lnyit + \beta_2 i\ (lnyit)^2 + uit$	reverse U, U
Gao (2012)	1995–2009	SO_2, CO_2	$E_{SO2,\ CO2,\ it} = \alpha_i + \beta_{1it}\ GDP_{it} + \beta_{2it} (GDP_{it})^2 + u_{it}$	reverse U
Han (2012)	1991–2009	industrial waste gas, dust, smoke, solid waste	$\ln Y_{it} = \beta_0 + \beta_1 \ln X_{it} + \beta_2 \ln^2 X_{it} + \beta_3 \ln^3 X_{it} + \varepsilon_{it}$	reverse U, N

Table 2. Research results on specific areas.

Researcher	Time Frame	Indicator	Model	Result
Sun (2008)	Tianjin 1985–2006	industrial three wastes	$Y = b_0 + b_1X + b_2X^2 + b_3X^3$	reverse U, N
Jiang (2009)	Shanghai 1990–2007	COD, dust, smoke, SO_2, garbage	$\ln E = \beta_0 + \beta_1 \ln Y + \beta_2(\ln Y)^2 + \beta_3(\ln Y)^3 + \varepsilon$	reverse U, U, M
Long (2009)	Shanghai 1997–2007	industrial three wastes	$E = \beta_0 + \beta_1 Y + \beta_2 Y^2 + \varepsilon$	reverse U
Ni et al. (2010)	Shanghai 1989–2004	SO_2, NO_x, PM_{10}, COD	$I_t = \beta_0 + \beta_1 (GDP/P)_t + \beta_2 [(GDP/P)_t]^2 + e_t$	reverse U, U, declining
Jiang (2012)	Beijing 2001–2010	SO_2, NO_2, NO_x, TSP	$y = \beta_0 + \beta_1 x + \beta_2 x^2 + \beta_3 x^3 + \varepsilon$	reverse U
Fan, Zheng (2013)	Sichuan Province 1985–2010	industrial three wastes	$Yt = \alpha + \beta_1 AGDP_t + \beta_2 AGDP^2_t + \beta_3 AGDP^3_t + \varepsilon_t$	no EKC
Mao et al. (2013)	Ningxia Irrigation Area 1990–2008	agricultural three wastes	$E = \alpha + \beta_1 Y + \beta_2 Y^2 + \beta_3 Y^3 + u$	reverse U

The above mentioned research results show us that in most cases the Environmental Kuznets Curve is an existing relationship, but many of these scholars emphasize the importance of firm environmental policies to ensure the decrease of pollution.

It is also obvious that in some cases (e.g. the existence of sulfur dioxide EKC) the scientific community has not yet reached an agreement. In order to provide assistance to policy makers, we might need to revolutionize the Environmental Kuznets Theory.

References

1. Chen, W. 2007. Economic Growth and the Environment in China – An Empirical Test of the EKC Using Provincial Panel Data. *Annual Conference on Development and Change.* Cape Town.
2. Fan, Ch. & Zheng, X. 2013. An empirical study of the EKC in Sichuan Province, China. *Environment and Pollution*, 2(3).
3. Gao, J. 2012. Steady on SO_2 and CO_2 Emission Path and Environment Quality: Based on the EKC by panel data pf 30 provinces and cities in China. *Economic Research* 8: 120–129.
4. Groot, H.L.F. de, Withagen, C. & Zhou, M. 2001. Dynamics of China's Regional Development and Pollution. *Tinbergen Institute Discussion Paper*, No. 01-036/3.
5. Han, J. 2012. The Stabilization Testing of Regional EKC in China: Based on Panel Data of Provinces. *Statistics and Information Forum*, 27(8): 56–62.
6. Hu, L. & Pan, L. 2007. The Impact of Economic Growth on Environment under Opening Condition. *International Trade*, 10: 101–107.
7. Jiang, Y. 2009. The Empirical Analysis of the EKC in Shanghai, Master's Degree Thesis. Antai College of Economics and Management, *Shanghai Jiao Tong University*.
8. Jiang, Y. L. 2008. EKC curves in the People's Republic of China: turning points and regional differences. *Asian Development Bank No. 141.*
9. Jiang, Z. P. 2012. The EKC test under the long-term influence of economic development on environmental quality in Beijing, Master's Degree Thesis. *Northeastern University of Finance and Economics.*
10. Li, C. S. 2006. Analysis on the EKC in Guangzhou. *Green Economy,* 50–54.
11. Li, X. X. & Pan, X. Q. 2007. Foreign Trade and Environmental Policy Analysis on EKC. *Contemporary Finance and Economics*, 11: 78–84.
12. Liu, Y. 2005. Research on the Relationship Between Discharged SO_2 and GDP in Guangzhou. *Guangzhou Environmental Sciences*, 20(2): 42–44.
13. Llorca, M. & Meunié, A. 2009. SO_2 emissions and the EKC: the case of Chinese provinces. *Journal of Chinese Economic and Business Studies*, 7(1): 1–16.
14. Long, L. 2009. Analysis on EKC in Shanghai. *Environmental Protection Science*, 36(3): 89–92.
15. Mao, Ch., Zhai, N., Yang, J., Feng, Y., Cao, Y., Han, X., Ren, G., Yang, G. & Meng, Q. 2013. EKC Analysis of the Economic Development and

Nonpoint Source Pollution in the Ningxia Yellow River Irrigation Districts in China. *Hindawi Publishing Corporation, BioMed Research International.*

16. Ni, X., Lu, J., Lan, L., Gao, F. Q. & Pan C. L. 2010. Interactions between environmental quality and economic development in Shanghai, China. *International Journal of Engineering, Science and Technology*, 2(7): 56–64.
17. Shen, J. & Hasimoto, Y. 2004. Environmental Kuznets Curve on Country Level: Evidence from China. *Discussion Papers In Economics and Business, OSIPP*, No. 04–09.
18. Shen, J. 2006. A simultaneous estimation of Environmental Kuznets Curve. Evidence from China. *China Economic Review*, 17(4): 383–394.
19. Song, T., Zheng, T. & Tong, L. 2008. An empirical test of the EKC in China: A panel cointegration approach. *China Economic Review*, 19: 381–392.
20. Sun, K. 2008. Research on the EKC in Tianjin, Master's Degree Thesis. *Tianjin Normal University.*
21. Wang, Z., Wen Z.G., Meng, L. & Chen J. N. 2007. Verifying the EKC Hypothesis and its Conditions in Beijing. *China Population, Resources and Environment*, 17(2): 40–47.
22. Xu, G. & Song, D. 2011. An Empirical Study on the EKC for China's Carbon Emissions: Based on Provincial Panel Data. *Chinese Journal of Population Resources and Environment*, 9(3): 66–76.
23. Yang, H., Jia, J., Zhou, Y. & Wang, Sh. 2005. The impact on EKC by Trade and Foreign Direct Investment in China. *Chinese Journal of Population Resources and Environment*, 3(2): 14–19.
24. Yue, L.P. & Bai, Y.X. 2006. Modeling Regional Relationshipbetween Economic Development Quality Based on the EKC. *Journal of Nanjing University of Science and Technology*, 19(4): 41–46.
25. Zhang, H. C. 2012. Urban–rural income disparities and development in a panel data set of China for the period from 1978 to 2006. *Applied Economics*, 44(21): 2717–2728.

Environmental burden shifting consideration within the re-evaluation of environmental impacts

X. Yu
Chengdu University, Computer Science, China
E-mail: oliveryx@163.com

A. Nongaillard
Université de Lille - Sciences et Technologies

A. Sekhari and H.Q. Zhang
University Lyon 2 Lumière, DISP laboratory, France

A. Bouras
Qatar University, Computer Science Department, Doha, Qatar

Life cycle assessment (LCA) can help enterprises evaluate their product's environmental impacts through the entire product life cycle. However, re-evaluatation by using LCA is very time-consuming and needs a lot of up-front work. Actually, the activities taken in a phase may influence the environmental impacts in other phases, so the effects of the activities should be re-evaluated. This paper proposes a novel approach to re-evaluate the environmental impacts of products based on LCA at design phase. The proposed graph-based model is capable to allow users easier to analyze the propagating of the environmental burden shiftingthrough the entire lifecycle. By analyzing the changing trend of materials and energies, the new environmental impacts can be calculated efficiently. A printed circuit board case study is conducted by using the proposed approach. The greenhouse impact of production printed circuit board is calculated. The experiments results reveal that the new proposed approach can efficiently re-evaluate the environmental impacts.

Keywords: Product Life Cycle; Printed Circuit Board; Life Cycle Assessment; Life Cycle Impact Assessment; Material Flow Analysis.

1. Introduction

The pollution emission through the entire life cycle of products should be restricted to protect the environment. Enterprises need to evaluate whether the released products satisfy the environmental performance or not. Many methods have been proposed to evaluate the environmental aspects of products, which can be divided into three categories in terms of qualitative, semi-qualitative, and quantitative (Brezet et al. 1997; Graedel et al. 1996; ISO14040 2006). The qualitative and semi-qualitative methods such as MET Matrix (Material Energy and Toxic), ERPA matrix (environmentally responsible product assessment) can assess the environmental performance quickly, yet the result is not precise enough. The quantitative methods such as LCA can assess the environmental impacts of product precise. Since LCA can provide more accurate information about the environmental impacts of products, it is still the most popular method using in industry so far. However, for the enterprises who already conducted an LCA, they cannotefficiently re-evaluate the improvement of the effects of activities.

The existing methods cannot satisfy the requirements of re-evaluation. Any improvement activity taken in one phase to reduce the environmental impact in a phase may increase the environment impacts in other phases (Lubrizol 2012). Moreover, in the theory of product lifecycle phases' analysis, the strong connection in different phases is that the output of this phase is the input of the other phases (Zhang et al., 2014). This so called environmental burden shifting (EBS) phenomenon forces analysts to collect all the related data in every time when assess the environmental effects of activities. In order to make the re-evaluation more efficient and without influence the validity of the results, this paper proposes a novel approach based on LCA method to re-evaluate environmental impacts after the improvement activities have been taken.A graph-based model is proposed to make the environmental burden shifting analysis more intuitive and comprehensive.

2. Proposed Graph-based Approach by Considering Environmental Burden Shifting

In this section, a graph-based model is proposed to fully consider the burden shifting phenomenon, which leads to a new evaluation model of the environmental performance for product lifecycle. The new environmental release can be computed based on reverse engineering. The reverse engineering using in this paper refers to the environmental release of materials or energies which is calculated based onthe consumption quantities of the materialsand energieswith its corresponding emission factors (Eilam 2011; EPA 1995). Finally, the environmental impacts of each phase can be calculated and aggregated to determine the environmental performance of the whole product lifecycle.

2.1. *Problem statement*

All activities in a product's lifecycle will cause emissions of substances into the natural (Rebitzer 2002). The environmental impacts (EI) are mainly obtained from two aspects that including materials and energies. The consumption quantities of materials and energies are determined by the newly added values in current phase and the inherited values in the other phases. Thus, the EI is calculated based on the following equation:

$$em_i = (\sum_{k=1}^{m} emm_{k,i} + \sum_{k=1}^{n} emm_{k,j}) + (eme_i + eme_j) \qquad (1)$$

$$(\forall i, j, k, m, n \in N)$$

Where, em_i represents the EI of *i* phase. $\sum_{k=1}^{m} emm_{k,i}$ represents the EI generated from all of the materials processed in the current phase. $\sum_{k=1}^{n} emm_{k,j}$ represents the EI generated from all of the materials that inherited from the other phases. eme_i represents the EI brought by energies used in the current phase. eme_j represents the EI brought by energies used in the other phases.

In practice, any activity taken in a lifecycle phase that includes changing the quantityof materials or energies can lead to the materials or energies varied in other lifecycle phases, which in turn influences the EI as well. This phenomenonis called environmental burden shifting (EBS). The EBS force the enterprise must consider the entire lifecycle when conduct an environmental impact assessment. In order to get reliable environmental impact assessment result, the exiting methods have to collect thedata from all phases. However, collecting the complete data is quite time-consuming and requires a lot of up-front work. Therefore, a new approach is proposed to solve these issues.

2.2. *Graph-based Model*

The environmental burden shifting phenomenon can be illustrated in a graph. Each node in the graph represents a lifecycle phase. Each phase is characterized by the consumed materials and energies. $A\delta^{n}_{i \to j}$ represents the transferred environmental impact value caused by the changed quantities of materials or energies in *jth* phase due to the *nth* activity occurred in the *ith* phase. The direct impacts and the indirect impacts are respectively depicted by doted arrows and dashed arrows. The impact Aδ has to be defined specifically between each pair of phases. In practice, many activities may occur at same time, each activity may have impacts on the other phases. Then the graph-based model can be depicted in Figure 1. In Figure 1, the final obtained quantities of materials and energies in the *jth* phase is the sum of the materials and energies caused by different activities occurred in different phases and newly added values in the current phase.

Remark 1 (Sub-graph (1): The transferred EI flow based on the changed impact value of Aδ and the corresponding activity A).

The quantities of materials (math notation is 'quantity(emm)' in Figure 1 (1)) or the amount of energies (math notation is 'quantity(eme)' in Figure 1(1)) used in different phases can be directly acquired by companies. The transferred impact value is determined based on the company experiences and the corresponding activities of used materials and energies. This sub-graph demonstrate the environmental impact transfer from phase i to the other phases once an activity (math notation is 'A_i^n' in Figure 1(1)) is taken in this phase. In practice, more activities have been considered in different phases.

Remark 2 (sub-graph (2): The generated EI from materials in different phases)

This sub-graph discussed the total changed value of materials based on activities. The *n* activity taken in *i* phase can make a material/energy changing in different phases. Therefore, when calculate the variation of a material in the specific phase, all the activities which have influence in the specific phase must be taken into account. The totalchanged value of a material in a specific phase is the accumulation of each variation due to different activities. Similarly, we can obtain the same analysis for energies.

Remark 3 (Environmental emission factor f_{factor})

From the existing databases or statistic data, each material/energy has its own environmental emission factor. Enterprises can estimate the environmental emission factor of each material/energy used in their own phase based on their process and statistic data. For example, 2.56kg carbon dioxide is released by combustion 1kg gasoline. The value of the carbon dioxide emission is called the environmental emission factor of 1kg gasoline. The changed material/energy is multiplied by the environmental emission factor to obtain the final changed environmental emission.

Remark 4 (The final environmental impact from all of the materials and energies)

Based on above analysis, the final environmental emission of materials and energiescan then be acquired by adding the original value and the variation value. The environmental impacts in a specific category of the influenced phase can be calculated by summing all the impacts based on the related materials and energies. Hence, the Equation (1) has a solution.

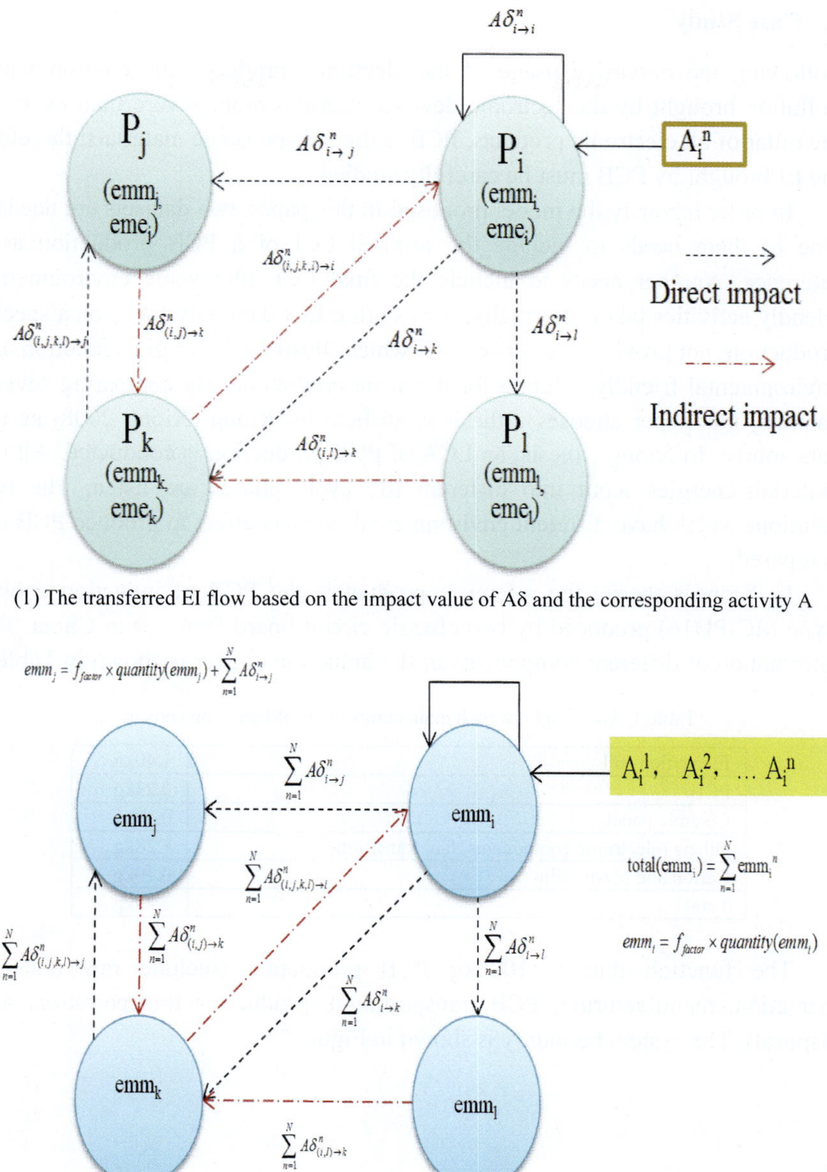

(1) The transferred EI flow based on the impact value of Aδ and the corresponding activity A

(2) The generated EI from materials in different phases

Figure 1. Graph-based model by considering the environmental burden shifting.

3. Case Study

Following the pervasive usage of the electronic products, the environmental pollution brought by theelectronic devices becomes more severe than ever. As the pillar of all electronic products, PCB is the indispensable materials; therefore the EI brought by PCB must be carefully studied.

In order to verify the model proposed in this paper, two datasets are needed. One of them needs to include the original LCI of a PCB production as a reference. Another needs to include the final LCI after some environmental friendly activities taken. Normally, the existing LCI data only focus on a special production, not provide two sets of LCI which illustrate the original solution and environmental friendly solution for the same production. By comparing several datasets, this paper chooses a thesis contribute by Xiong (Xiong 2008) as the data source. In Xiong's thesis, an LCA of PCB production is conducted. All the materials/energies input into different life cycle phases are listed. The two solutions which have different environmental impacts effect to produce PCB are compared.

In Xiong's thesis, the reference product is the PCB of induction cooker (type MC-PD16) produced by two electric circuit board factories in China. The information of different components of the induction cooker is shown in Table 1.

Table 1. The weight of each main components of induction cooker.

Electrified coil	1.40kg
PCB	0.23kg
Ceramic panel	0.17kg
Others (electronic components, fan, transformer)	1.28kg
Carton and plastic film	0.10kg
Total	3.18kg

The function unit is 1000kg PCB production (include raw material extraction, manufacturing, PCB transportation, production transportation, and disposal). The system boundary is shown in Figure 2.

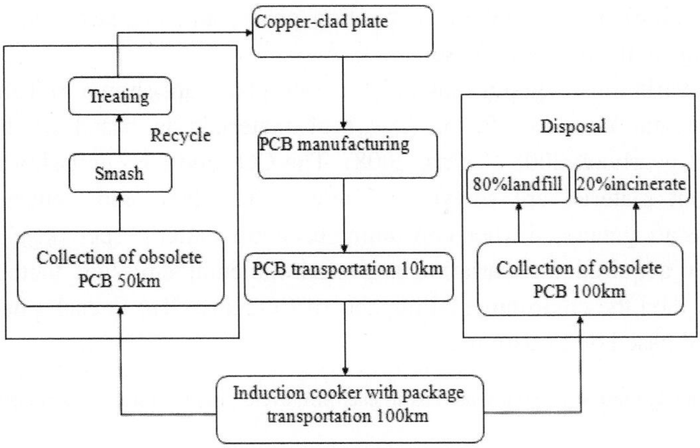

Figure 2. System boundary of the production of PCB life cycle.

The main environmental issues of PCB production are the greenhouse effect in raw material abstraction, manufacturing, and disposal phases. In order to reduce the greenhouse effect, two activities are used. One is to deal with the waste air emission in the manufacturing phase. Another one is to recycle the scrap PCB.

The collected data is analyzed. Since the goal of this case study is to evaluate the greenhouse effect of the two environmental friendly activities, the materials/energies which have contribution to greenhouse effect need to be collected. (Metz *et al.* 2005; Bousquet *et al.* 2006) points out that the main contribution of the greenhouse gases is the fuel consumed in each life cycle phase. Therefore, the data collection can only focus on the mass of the fuel used in each life cycle phase after the two activities have been taken. In practice, the enterprises have no time to collect a set of data after one activity is taken. They always collect a set of data after all activities are taken. The input materials which have main contributions to GHG are: *coal, gas, gasoline and limestone*. The quantities of these materials used and the corresponding GHG emissions in each phase are shown in Table 2 and Table 3.

The graph-based approach is used to analyze the EIs in manufacturing and recycle phases after the activities have been taken. The variation of the quantities of consumed materials in each phase is shown in Figure 3 based on the collected data in Table 2 and Table 3. The relationship between different phases is analyzed based on the effects of the activities. Based on the data collection and activity analysis, the material's variation in each phase can be investigated. For example, the activity taken in manufacturing and recycle phases changed the amount of coal used in the current phase, reduced the used coal (38.08kg) in raw material extraction phase, and increased the used coal (133.94kg) in the disposal phase.

The variations of input materials in each phase are shown in Table 4. The environmental emission factor (f_{factor}) of materials is based on the China statistical yearbook 2008 (China, 2008). The CO_2 and CH_4 emission factors of each input material are shown in Table 5. The table representshow many emissions are generated when consuming each raw material (per 1kg).

After acquired the variation and environmental emission factor of each material, then the environmental impacts of CO2 and CH4 in each phase can be calculated based on Equation 1.

Table 2. The quantities of input materials in each life cycle phase before the conducted activities (unit: kg).

		Raw material extraction	Manufacturing	PCB transport	Product transport	Disposal
Input materials	coal	117.68	305.50	0.00	0.00	-133.80
	gas	961.70	653.70	0.00	0.00	0.88
	gasoline	802.60	217.20	58.40	4.88	60.08
	limestone	1.80	2.58	0.00	0.00	300.00
Emissions	CO2	264.54	264.54	192.82	16.09	658.74
	methane (CH4)	10.20	9.12	0.362	0.03	0.16

Table 3. The quantities of input materials in each life cycle phase after the conducted activities (unit: kg).

		Raw material extraction	Manufacturing	PCB transport	Product transport	Disposal	Recycle
Input materials	coal	79.60	204.80	0.00	0.00	0.14	23.52
	gas	581.89	484.79	0.00	0.00	0.70	73.81
	gasoline	379.01	196.65	107.06	8.95	25.34	68.81
	limestone	203.10	118.44	0.00	0.00	16.71	0.00
Emissions	CO2	193.59	594.70	243.92	20.36	80.50	228.92
	methane (CH4)	5.15	5.83	2.02	0.17	0.11	4.24

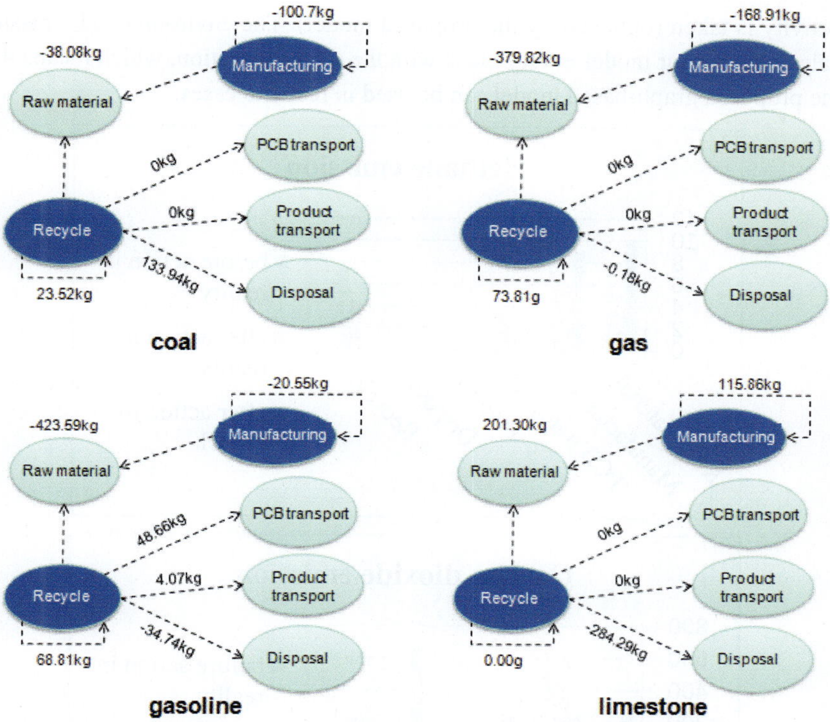

Figure 3. The variation of input material based on the conducted activities.

Table 4. The variation of input materials after the activities are taken.

	Raw material extraction	Manufacturing	PCB transport	Product transport	Disposal	Recycle
coal	-38.08	-100.70	0.00	0.00	133.94	23.52
gas	-379.82	-168.91	0.00	0.00	-0.18	73.81
gasoline	-423.59	-20.55	48.66	4.07	-34.74	68.81
limestone	201.30	115.86	0.00	0.00	-283.29	0.00

Table 5. The environmental emission factor of each material.

	CO2	CH4
coal	2.00	0.15
gas	1.20	0.02
gasoline	1.05	0.03
limestone	4.50	0.12

The comparison analysis of environmental emission from the reality and the proposed approach has been conducted. Three situations have been presented in Figure 4: the emission before the activity is taken (collected in reality), the emission after the activity is taken (collected in reality), and the emission after activity is taken (obtained by the proposed model). The environmental emission calculated by our model is very close with the reality situation, which means that the proposed graph-based model can be used in real-life cases.

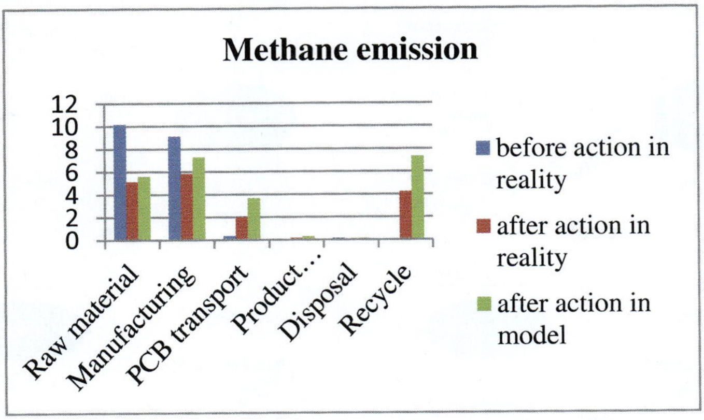

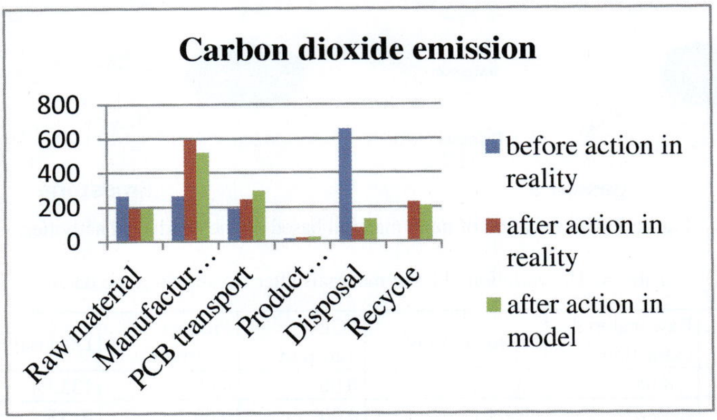

Figure 4. The results comparison between the real-life values and the values obtained from the proposed model

4. Conclusion

Enterprises take the activities in life cycle phases to improve the environmental performance of their products. However,through the entire product life cycle, any activities taken in a phase may bring some influences in other phases, which is the environmental burden shift from one phase to other phases. Therefore, a graph-based approach is proposed to evaluate the environmental effects of the activities, which can help users analyze the interaction among different phases. The variations of the materials or energies can be efficiently calculated based on this model. Based on the results obtained from the case studies, this new approach can provide efficient results of re-evaluation environmental impacts. In addition, the proposed approach can effectively reduce the re-evaluation time by comparing with the LCA approach. However, the proposed approach still some limitations that only the most important materials have been considered in lifecycle phases. Therefore, more materials and the related criteria will be studied in the future work.

References

1. Bousquet, P., P. Ciais, J. Miller, E. Dlugokencky, D. Hauglustaine, C. Prigent, G. Van der Werf, P. Peylin, E.-G. Brunke and C. Carouge. 2006. Contribution of anthropogenic and natural sources to atmospheric methane variability. Nature, 443(7110) 439-443.
2. Brezet., H., C. V. Hemel., U. N. E. Programme., I. a. E. (Paris)., R. Instituut. and D. U. o. T. (Delft). 1997. Ecodesign: a promising approach to sustainable production and consumption, UNEP.
3. Bringezu, S. and Y. Moriguchi. 2002. Material flow analysis. A handbook of industrial ecology, 79.
4. China. 2008. China Statistical Yearbook 2008. China Statistics Press. Retrieved 14 Feb, 2014, from http://www.stats.gov.cn/tjsj/ndsj/2008/indexeh.htm.
5. Eilam, E. 2011. Reversing: secrets of reverse engineering, John Wiley & Sons.
6. EPA. 1995. missions Factors & AP 42, Compilation of Air Pollutant Emission Factors. U.S. Environmental protection agency. Retrieved 2.28, 2014, from http://www.epa.gov/ttn/chief/ap42/index.html.
7. Graedel, T. E., B. R. Allenby, A. Telephone and T. Company. 1996. Design for environment, Prentice Hall Upper Saddle River, NJ.
8. ISO14040. 2006. Environmental management—life cycle assessment—principles and framework. Environmental Management. 3.
9. Lubrizol. 2012 Understanding the environmental life-cycle of a BlazeMaster Fire Sprinkler System.
10. Metz, B., O. Davidson, H. De Coninck, M. Loos and L. Meyer. 2005. Carbon dioxide capture and storage. Intergovernmental Panel on Climate Change (IPCC).
11. Rebitzer, G. 2002. Integrating life cycle costing and life cycle assessment for managing costs and environmental impacts in supply chains. Cost management in supply chains, Springer: 127-146.

12. Xiong, R. 2008. Life cycle assessment methods used to achieve cleaner production of printed circuit boards [M.S. thesis], Guangdong University of Technology, 2008.
13. Zhang, H., Ouzrout, Y., Bouras, A., & Savino, M. M. 2014. Sustainability consideration within product lifecycle management through maturity models analysis. International Journal of Services and Operations Management, 19(2), 151-171.

Research on relation between El Nino climate and summer electricity consumption

J.Y. Lin[a], B. Miao[b], L. Cai[c], C. Liu[d] and B. Li[e]
China Electric Power Research Institute, Beijing, China
E-mail: [a]linjingyi@epri.sgcc.com.cn, [b]miaobo@epri.sgcc.com.cn, [c]cailiang@epri.sgcc.com.cn, [d]liuchang3@epri.sgcc.com.cn, [e]libin5@epri.sgcc.com.cn

2015 is defined as El Nino year. Such phenomena would have influence on climate in China and furthermore impact the electricity condition. This paper is purposed to explore how El Nino phenomena affecting electricity and make prediction on summer electricity consumption. Since meteorological characteristics are complex and multiplex, a variety of meteorological factors should be considered and the paper used Body Feeling Temperature to measure it. Furthermore, to make prediction on summer electricity, the paper used the Pearson Analysis to measure the correlation between weather and electricity and then extracted the weather-used electricity from the whole society electricity using least square method. Finally, the paper built the model on relation between weather-used electricity and body feeling temperature, and took Beijing as an example to make electricity prediction. The prediction idea and model the paper put forward is reliable and practicable.

Keywords: El Niño; Climate; Electricity.

1. Impact of El Nino on the Climate in China

1.1. *Introduction of El Niño*

El Niño is defined by prolonged warming in the Pacific Ocean sea surface temperatures when compared with the average value. The U.S NOAA definition is a 3-month average warming of at least 0.5°C in a specific area of the east-central tropical Pacific Ocean, other organizations defines the term slightly differently. Typically, this anomaly happens at irregular intervals of two to seven years, and lasts nine months to two years. The average period length is five years. When this warming occurs for seven to nine months, it is classified as El Niño "conditions"; when its duration is longer, it is classified as an El Niño "episode".

The first signs of an El Niño are a weakening of the Walker circulation and strengthening of the Hadley circulation and may include:

- Rise in surface pressure over the Indian Ocean, Indonesia, and Australia.
- Fall in air pressure over Tahiti and the rest of the central and eastern Pacific Ocean.
- Trade winds in the south Pacific weaken or head east.
- Warm air rises near Peru, causing rain in the northern Peruvian deserts.

Since 2000, a number of El Niño events have been observed. El Niño events were observed in 1997–98, 2002–03, 2006–07 and 2009–10. A strong El Niño has not occurred since 1997–98.

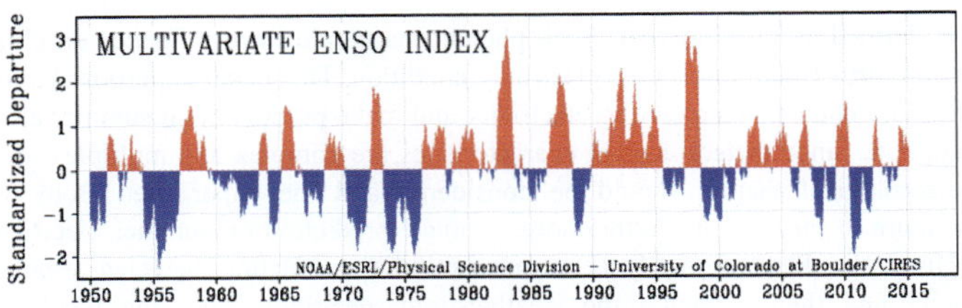

Figure 1. Multivariate ENSO INDEX of El Niño phenomena from 1980–2015.

In December 2014, the Japan Meteorological Agency had declared the onset of El Niño conditions, as warmer than normal sea surface temperatures were measured over the Pacific, which indicates the coming of El Niño phenomena. Months later other meteorological institutions around the world (including China) had declared the arrival of El Niño conditions, in March and May 2015, respectively.

1.2. *Impact on China*

Although this global phenomenon cannot have a direct impact on China, it can indirectly affect the mainland through atmospheric circulation. Due to the sea temperatures in the Eastern Pacific has failed rising to meet expectations in the summer of 2014, its impact on summer climate in China is mainly droughts in the North China, humid and rainy in the South, East and Central China, resulting that the average summer temperature in Central and Eastern China is lower than the average level. Different from the situation in 2014, the strength of El Niño in 2015 is continually improving. It is predicted that the Songhua River basin may suffer flood situation, heavy flood may occur from the middle reaches of the Yangtze River to the western parts of South China. In the Southwest Region, mountain torrents may increase triggered by heavy rainfall and there is a great probability that the North and North-East areas are under a drought climate.

To sum up, summer temperatures in northern China is higher than the same period of normal years, but there would not be large scale and long period extreme high temperature events this summer. While, the cold air activity in Northeastern China is frequent, the temperature in summer could be in an anomaly low condition. Also, continuing precipitation in middle and lower reaches of the Yangtze River will bring down the temperature, and abnormal high temperatures will not be as significant as usual. In addition, the number of typhoon in the Northwest Pacific Ocean will be less than in a normal year.

2. Correlation Analysis between El Nino and Meteorological Elements

El Niño phenomena have an effect on not only temperature, but also humidity and wind load. Maximum electric load weather generally occurs in high temperature, high humidity and breezy weather condition. However, how to correctly reflect the heat degree is a complex question, it can't be described accurately only relying on single meteorological element such as temperature or humidity. In recent years, the Meteorological Department usually defines the level of hot weather using body comfort index. Among these impact factors, temperature, humidity and wind are the most important. Particularly, temperature is the primary factor. Here, the paper measures the body feeling temperature T_e by the following formula:

$$T_e = 37 - \frac{37 - T_a}{0.68 - 0.14R_h + 1/(1.76 + 1.4V^{0.76})} - 0.29T_a(1 - R_h) \qquad (1)$$

T_e=body feeling temperature, T_a=temperature, R_h=relative humidity, V=wind speed.

Since El Niño approach, the value of these factors should be revised. The general climate condition would be drought and high temperature in North China; humid, rainy and relatively hot in South China. According to historical data, in 2015 El Niño condition, the Ta should be multiplied by 1.31 in North and 1.17 in South, the Rh should be multiplied by 0.88 in North and 1.23 in South. So the revised formulas are as below.

In North China,

$$T_e = 37 - \frac{37 - 1.31T_a}{0.68 - 0.123R_h + 1/(1.76 + 1.4V^{0.76})} - 0.38T_a(1 - 0.88R_h) \qquad (2)$$

In South China,

$$T_e = 37 - \frac{37 - 1.17T_a}{0.68 - 0.172R_h + 1/(1.76 + 1.4V^{0.76})} - 0.34T_a(1 - 1.23R_h) \quad (3)$$

3. Correlation Analysis between Climate and Electricity Type

Abnormal weather caused by El Niño phenomenon would have an influence on the electricity consumption of the whole society directly or indirectly. Collect recent 10 years meteorological data and electricity data, we can do correlation analysis on them. Since 2008–2009 and 2014–2015 are judged as El Niño years, here the paper uses the revised formula to describe the Te.

Once get Te and yearly electricity consumption (Abbreviation: E), we can conduct Pearson product-moment correlation analysis which is widely used in statistics and mathematics.

Data analysis display that the correlation coefficient between Te and whole society summer electricity consumption is 0.402, presenting weak correlation. The table 1 below shows the details.

Table 1. Correlation coefficient between electricity and society.

Whole Society	Primary Industry	Secondary Industry	Tertiary Industry	Resident
0.402	0.079	0.367	0.410	0.650

As for the Tertiary Industry, the condition is more complex, and here the Table 2 below shows the details.

Table 2. Correlation coefficient between electricity and different type of tertiary industry.

Transportation Industry	Warehousing Industry	Post Industry	Telecommunication Industry	Software Industry
0.331	0.346	-0.089	0.282	0.375
Wholesale and retail Industry	Lodging and Catering Industry	Financial Industry	Real Estate Industry	Facility Management Industry
0.343	0.438	0.402	0.384	0.302

From the table, we can see that restricted to the huge volume and the economic factors, electricity consumption of primary industry and secondary industry is not sensitive to temperature. Since the tertiary industry include a variety of industries, different item has different correlation coefficient. On the other hand, as temperature sensitive load, resident load display strong correlation with temperature, it can be predicted that when the extremely weather condition happens during the El Niño event, the resident load is mostly influenced as it can be seen in Figure 2 below.

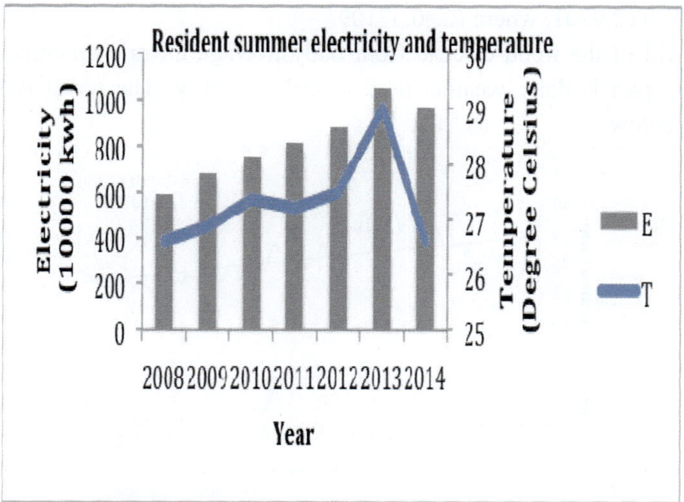

Figure 2. Relationship between resident summer electricity and temperature.

As the main influencing factors of in-used electricity quantity is the development of economy in primary, secondary and tertiary industry. While the electricity quantity used by residents mainly depends on climate. So it is known that the El Niño mostly have influence on resident electricity.

4. Prediction of Summer Electricity Consumption Based on El Nino Phenomena

To accomplish a more accurate prediction on temperature-influenced electricity consumption, the weather electricity quantity should be calculated from the entire social electricity quantity as $P=P_t+P_m+u$.

Among them, p stands for entire social-used electricity quantity; P_t is tendency electricity quantity; P_m is weather-used electricity quantity; while u stands for the contribution from random elements. Among these parameters, u is barely calculated accurately, so the formula above should better be estimated as $P_m=P-P_t$. The trend-electric quantity is influenced by the development of economy and macro-economy. Both of the two factors are stable relatively, so least square method can be used here to accomplish the linear fitting function between monthly average electric quantity which can be written as $P_t=a+bT$.

Take Beijing City as an example, trend-function was created using the data of daily averaged electric quantity in the summer during the EI Niño period: P_t= -0.0053t+2.9201, where R^2=0.37109.

Get rid of the trend-electric from daily average electric quantity, then the remaining part is daily weather-used electric quantity. The result is shown in Figure 3 below.

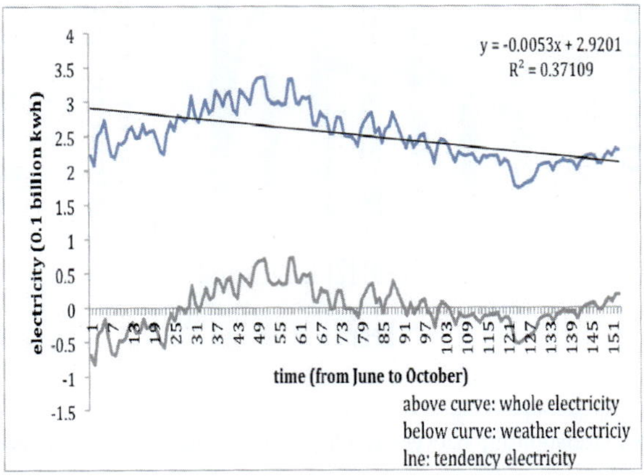

Figure 3. Relationship between whole society electricity, weather electricity and tendency electricity.

Depending on the weather-used electric quantity, according to formula (2) above, the comfortable temperature index has been determined by several kinds of climate factors. And the index is used as climate variables to accomplish tropic model. Considering that the relationship between daily electric quantity and temperature is not perfect linear. Square curve model regression is adopted here, and Figure 4 below shows the relationship.

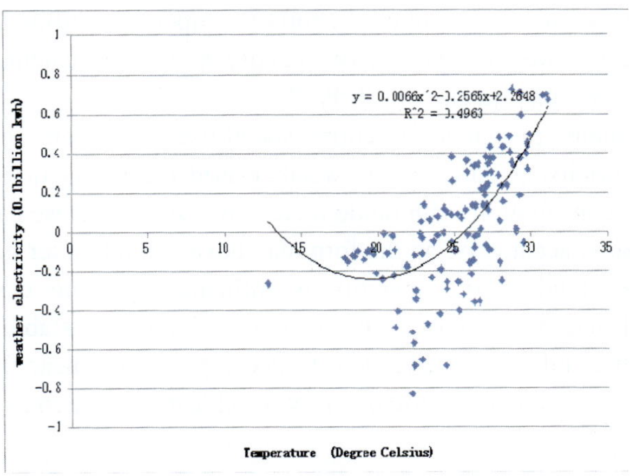

Figure 4. Relationship between weather electricity and temperature.

The regression equation shows that, the change of electricity quantities related to climate ranges along with comfortable index. When the comfortable degree is 19.43, the weather-used electric quantity reaches its lowest spot. As the summer temperature gets higher, 1 comfortable index corresponds to an increase of daily weather-used electricity quantity as 0.0132Te-0.2565.

It can be summarized that, influenced by EI Niño extreme climate, it comes out to be different level of increase for summer daily electricity quantity under different temperature. Thus, the prediction of electricity under extreme high temperature climate is essential. What's more, such method can be references for electricity apartment to arrange work under extreme weather condition.

5. Conclusions

This paper focus on discovering the relationship between climate conditions and electricity consumption when El Niño event happens. By analyzing the climate characteristic, the paper firstly get different climate phenomena in different parts of China, so as to making accurate research on electricity consumption in different districts. Since the whole society electricity is comprised of tendency electricity and weather electricity, filtering the tendency electricity and getting the single weather electricity is the key point, by using basic algorithm, the paper find a method solving the problem and extract the weather electricity suitably. Once getting the weather electricity data and body feeling temperature (which contains a variety of factors such as temperature, humidity, wind speed and so on) data, further electricity consumption prediction can be made.

The prediction method the paper proposes can be applied on practical work, which is also helpful for power grid enterprise on electricity control and prediction.

References

1. Douglas, M.L.C. & Henry, E.W. 1981. Modeling the impact of summer temperature on national electricity consumption. *J. Appl. Meteor.* 20:1415–1419.
2. Quayle. R.G & Diaz. H.F. 1980. Heating degree day data applied to residential heating energy consumption. *J. Appl. Meteor.* 19:241–246.
3. Taylor, J.B & Woodford, M. 1999. *Handbook of macroeconomics*. 1(C):1341–1393.
4. Zhu, Z.W. & Fang, G.F. & Wang, G.F. 2008. Characteristic analysis for summer weather sensitive loads of power grids. *East China Electric Power* (in Chinese). 36(9):53–58.

5. Tan, J.G. & Shao, D.M. & Ma, L.M. 2001. Human body heat balance model with its application in comfort index forecast. *Journal of Nanjing Institute of Meteorology*. 24(3): 384–390.
6. Pu, W.Y. & Wei, M. & Li, H.B. 2004. Application of elementary assumptions of least square method in linear regress model to Z-L relation. *Journal of Meteorological Sciences*. 24(3):644–648.
7. Sheng, Q. & Zhu, X.D. & Luo, L.N. 2011. Characteristics of electricity demand and its relationship with meteorological condition in Huzhou City. *Transaction of Atmosphere Sciences*. 34(1): 122–127.

Study on the disruption management methods of steelmaking and continuous casting process for green manufacturing

H.G. Bo[1], Z.X. Li[1,*], Y. Liu[2], S.H. Liu[2] and Y. Guo[3]
[1]*Institute of Production Operation and Logistics Management, Dalian University of Technology, Dalian 116023, China*
[2]*CSR Qingdao Sifang Co., Ltd. Qingdao 266111, China*
[3]*Dalian Huatie Hai Xing Technology Co., Ltd, Dalian 116023, China*
[*]*E-mail: xiaoyaodae@sina.com*

A novel scheduling method based on disruption management is presented to solve the problem of increased energy consumption with the increase of make span caused by the interference in co-smelting furnace, steelmaking and continuous casting process. The scheduling model is built considering both the targets to minimize total weighted completion time (the original objective) and the target to minimize total weighted delay time (the disruption repairing objective). By combining multi-objective approaching policy with the bi-level particle swarm optimization and stochastic probability multi-neighborhood search mechanism, a heuristic hybrid algorithm is proposed to improve scheduling efficiency while reducing energy consumption indicator so as to achieve green manufacturing goal. The numerical experiments show that the hybrid PSO algorithm is effective.

Keywords: Steelmaking and continuous casting; Production scheduling; Disruption management; PSO algorithm.

1. Introduction

With the development of steel industry, the steel market competition becomes more intense, at the same time, the steel industry has the greatest potential for energy saving and occupies a pivotal position in the energy saving work. On this background, the application of green manufacturing in Steel industry to save energy has been paid more attention by enterprises and countries. Therefore, how to improve the production scheduling level in iron and steel enterprise contributes greatly to the smooth production line and energy consumption.

So far, the scheduling problem of steelmaking and continuous casting process has been extensively investigated from different angles by the research community. The mainstream method is model solutions, and they can be classified into the following categories: (i) Yu Sheng-Ping et al establishment non-linear programming mode and mixed integer linear programming model for various disturbances in steelmaking and continuous casting process, and solved by software package [1, 2]. (ii) Tang et al established steelmaking — continuous casting process scheduling model based on Lagrange relaxation algorithm to optimize make span [3]. (iii) Liu Hang, Missbauer et al, employ heuristic algorithm and heuristic algorithm to assign and sort charges so as to form a rough schedule, and then imply linear programming model to elimination scheduling conflicts in rough schedule [4,5]. But the three methods mentioned have limitations in the application or calculation accuracy.

Steelmaking and continuous casting process is the key step of metallurgy in which scheduling plays an important role in reducing energy consumption and cost saving in iron and steel enterprises. From the point of view of production management, the characteristics of steelmaking and continuous casting process can be classified as following: (i) It requires continuity of batch on the casting stage [6]; (ii) It should take the transport time of charges as well as the continuity between Steelmaking and continuous casting into consideration. (iii) The production rhythm fluctuations should be as small as possible to ensure that steel Maintain the temperature to save energy. Therefore, in theory, steelmaking and continuous casting scheduling is provided with characteristics of Hybrid No-wait Flow shop (HNWFS) problem.

Based on the above research, the following works has done in this paper: solve the problem of steelmaking and continuous casting scheduling by employ disruption management for hybrid no-wait flow shop (HNWFS) with Batch constraints. In this paper, a novel scheduling method based on disruption management is presented. The scheduling model is based on considering both the target to minimize total weighted completion time (the original objective) and the target to minimize total weighted delay time (the disruption repairing objective), then a disruption management for hybrid no-wait flow shop (HNWFS) with Batch constraints is built; Finally, a combining hybrid PSO algorithm which based on particle swarm optimization strategies and random neighborhood search mechanism is proposed.

2. Formulation of the Disruption Management

To make problem more concise and clear, steelmaking, refining and continuous casting are designated by step, and the converter, refining furnace and continuous casting machine are designated by machine. Let charge denote the molten steel produced in converter once, and Let cast denote a set of charges produced in continuous casting machine once. In this way, the steelmaking and continuous casting production process can be formulated as follow. Without loss of generality, we assume that there are a set of n charges $J=\{1,2,\cdots,j,\cdots,n\}$

(n>1) are simultaneously available for being processed, let ω_j denote the processing priority of each charge, and each charge has the same processes, they are arranged to go through $L(L \geq 2)$ steps in HNWFS system and divide into n'_B casts in a step i $(i=1,2,\cdots,L)$, i.e. $J=\{B_1,B_2,\cdots,B_h,\cdots,B_{h'},\cdots,B_{n'_B}\}$ $(B_h=\{j\,|\,j\in J, j\notin B_{h'}\})$. The HNWFS system processing environment: $l_i \geq 1$ parallel identical machines are available in any step, the must be one step which is arranged $l_i \geq 1$ machines. Any machine can only process one charge at a time; each charge is processed in each process only once; each charges must be processed uninterruptedly the HNWFS system. Interrupted charges must re-enter HNWFS system. Charge j is processed on machine $m_{ij}^{(k)}$ $(1 \leq m_{ij}^{(k)} \leq l_i)$ at step i in HNWFS system, let $s_{ij}^{(k)}$ denote the start time, $p_{ij}^{(k)}$ denote the processing time, $C_{ij}^{(k)}$ denote the completion time. The completion time of charge j in HNWFS system is designated by C_j, and the feasible scheduling is designated by π.

2.1. *Initial scheduling*

We make the minimize total weighted completion time of all charges as the optimization objective of original scheduling, and the initial scheduling problem can be formulated as:

$$FF_s\left|nwt,\{l_1,l_2,\cdots,l_m\},\{B_1,B_2,\cdots,B_{n'_B}\}\right|f(\pi)=\sum_{j=1}^{n}\omega_j C_j.$$

It is apparent that such initial scheduling problem is strongly NP- hard problem [7], which can be solved by employing particle swarm algorithm to search global optimization [8], by this way ,we can obtain the optimal processing schedule $\overline{\pi}$ for initial scheduling and the optimal objective evaluation value is then computed by the following formula: $f(\overline{\pi})=\sum_{j=1}^{n}\omega_j\overline{C}_j$.

Since charge change disruption (such as: added charges, canceled charges, etc.) occurs, the original optimal scheduling becomes no longer optimal, even invalid. We assume that at certain time t_0 (we can provide $t_0=0$) while initial optimal scheduling are implemented ,a predictable charge disturbances occurs, i.e. new charge is inserted into the cast B_h, Let $(\pm)\Delta J\,|\,t_0$ denote the disturbances such as added charges, canceled charges, which can be denoted by ΔJ shortly. To reduce the impact which caused by the charge disturbance ΔJ on the processing system and the scheduling indicators, it is necessary to rescheduling the ready charge $J' = \{1, 2, \quad \cdots, j, \cdots, n'\}$ ($n' = n \pm \Delta J$) in HNWFS system to Quickly generate a new scheduling that considering both the original objective and the disruption repairing objective quickly, i.e. disruption repair scheme.

2.2. *Disruption repairing scheme*

Reducing make span and ensuring the continuity and stability in the implement of every steps in steelmaking and continuous casting process can reduce energy consumption and improve product quality, which has great significance for green manufacturing. So, a green manufacturing-oriented disruption repair scheme for steelmaking and continuous casting process is proposed in this paper. Disruption repair problem is a specific manifestation of initial scheduling in the condition that charges disturbances occur. Disruption management scheme should be developed with fully consideration of the optimization objective in initial scheduling scheme. Disruption repair operation usually leads to a bias between the former scheduling and the latter scheduling. To reduce the deviation of the two schedules we ought to ensure the consistency of the two schemes in the scheduling as much as possible.

Thus, two optimization goals interference management scheme are described as:

(1) Initial scheduling objectives: Minimize total weighted completion time, namely $f_I(\pi') = \sum_{j=1}^{n'} \omega_j C_j$.

(2) the disruption repairing objective o: Minimize total weighted delay time, in the Optimal scheduling of initial schedule, charge j is completed at time $\overline{C}_j$ in HNWFS system; In the Optimal scheduling of disruption repairing scheme, charge j is completed at time C_j. The minimization total weighted delay time can be donated by $T_j = max\{0, C_j - \overline{C}_j\}$, thus, the minimized deviation function of the initial scheduling scheme and the disruption management scheduling scheme can be expressed as $f_{II}(\pi') = \sum_{j=1}^{n'} \omega_j T_j$. At the same time we introduce the indicator $C_{ec} = \partial C_{max} | f_I(\pi'), f_{II}(\pi')$ to express the energy consumption in the flow shop under the two aforementioned targets, ∂ is the factor of energy consumption in unit time, and $C_{max} = max(C_1, C_2, C_3 \cdots, C_j)$.

In summary, the HNWFS-dmsp problem for steelmaking and continuous casting process oriented to green manufacturing can be described as: $HNWFS - dmsp : FF_s | nwt, \{l_1, l_2, \cdots, l_m\}, \{B_1, B_2, \cdots, B_{n'_B}\}, \Delta J | f_I(\pi'), f_{II}(\pi')$.

3. Scheduling Model for Disruption Management

In order to facilitate the description of the modeling, we define $\sigma(ijk)$ as the rear charge of charge *j* which processed on machine $m_{ij}^{(k)}$ in step *i*; Let $x_{ij}^{(k)}$ ($x_{ij}^{(k)}=0$ or *1*) denote weather charge *j* is processed on machine $m_{ij}^{(k)}$ in step *i*; Let $x_{ij}^{(k)}$ and $s_{ij}^{(k)}$ which donates the start time that charge *j* is processed on machine $m_{ij}^{(k)}$ in step *i* as decision variables of the mode Thus, the problem of HNWFS-dmsp model can be described as:

$$\min_{i\in L, j\in J'} \{ f_I(\pi')=\sum_{j=1}^{n'} \omega_j C_j \ , f_{II}(\pi')=\sum_{j=1}^{n'} \omega_j T_j \} \tag{1}$$

$$s.t.:\ s_{ij}=s_{ij}^{(k)}\cdot x_{ij}^{(k)}, p_{ij}=p_{ij}^{(k)}\cdot x_{ij}^{(k)} \tag{2}$$

$$C_{ij}=s_{ij}+p_{ij}\ ,\ C_j=s_{1j}+\sum_{i=1}^{L} p_{ij} \tag{3}$$

$$s_{(i+1)j}=s_{ij}+p_{ij}\ ,\ i\in\{1,2,\cdots,(L-1)\} \tag{4}$$

$$s_{i\sigma(ijk)}^{(k)}\cdot x_{ij}^{(k)}\geq s_{ij}^{(k)}\cdot x_{ij}^{(k)}+p_{ij}^{(k)}\cdot x_{ij}^{(k)} \tag{5}$$

$$(s_{ij}^{(k)}\cdot x_{ij}^{(k)}\geq C_{ij'}^{(k)}\cdot x_{ij}^{(k)}\vee s_{ij'}^{(k)}\cdot x_{ij}^{(k)}\geq C_{ij}^{(k)}\cdot x_{ij}^{(k)})$$

$$\vee (m_{ij}^{(k)}\neq m_{ij'}^{(k)})\ ,\ \forall j,j'\in J' \tag{6}$$

$$J'=\{B_1,\cdots,B_h,\cdots,B_{h'},\cdots,B_{n_B'}\}, B_h=\{j\mid j\in J', j\notin B_{h'}\} \tag{7}$$

In HNWFS-dmsp model, equation (1) describes the optimization objective of disruption management scheduling problem, equation (2) describes the start time and processing time that charges processed in each step, equation (3) denotes the completion time that charges processed in each step and in whole HNWFS system; for the heats in the process completion Time and calculation formula in completion time, equation (4) denotes the operating constraint of No - wait flow, namely, the charge must be processed continuously until consummation Once the charge is put in HNWFS system, equation (5) illustrates that one charge can be processed only after the completion of the previous charge on the same machine, equation (6) denotes the disjunctive constraints of hybrid flow shop, namely, it will not be permitted if two charges age arranged on the same machine, and equation (7) denotes batch constraints of charges.

4. Hybrid PSO Algorithms

Particle swarm optimization (PSO) algorithm, first proposed by Kennedy and Eberhart (1995), is an evolutionary metaheuristic based on swarm intelligent theory, which is inspired by bird predation [9].

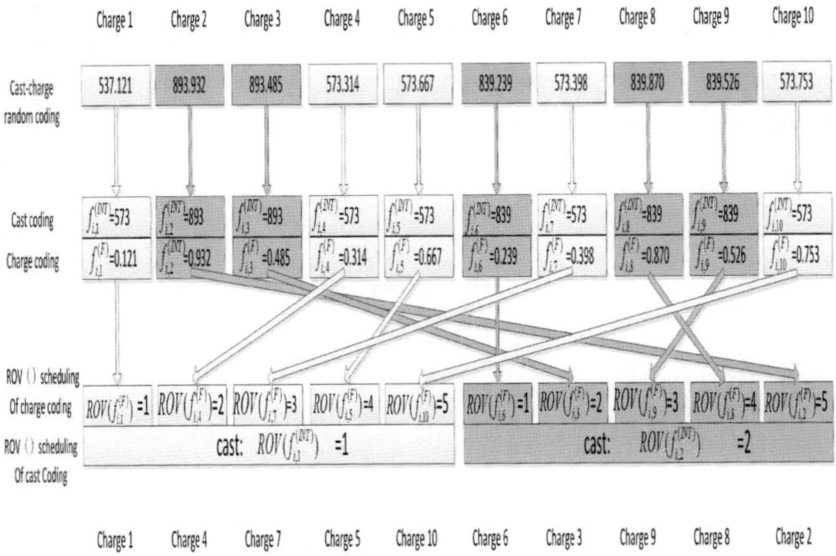

Figure 1. "Batch - charges" random key encoding and bi-level ROV conversion process.

Owing to the feature that PSO algorithm possesses fast global optimization and the advantage that stochastic probability neighborhood local search mechanism can improve the performance of local search algorithm significantly. In this paper, a hybrid PSO algorithm which is based on the tight combination of particle swarm optimization strategy and heuristic neighborhood search mechanism, then hybrid PSO algorithm is employed to solve the HNWFS-dmsp model orientated to green manufacturing.

4.1. *Solution expression and bi-level ROV rules*

To ensure that the coding strategy of HNWFS scheduling scheme will not miss the possible global optimal solution, and the rationality and feasibility of PSO iteration evolutionary operation, firstly, we need to construct appropriate mapping relationship between particle position vector and batch scheduling of charges. next, a bi-level ascending Ranked-Order-Value (ROV) rules based on random key coding mode, which is employed transcode particle position vector (continuous real value) to batch scheduling of charges (continuous real value), the conversion process is shown in Figure 1.

4.2. *Bi-level particle iterative evolution strategy*

Taking the features of HNWFS-dmsp problem into account, in this paper, a bi-level particle iterative evolution strategy is proposed, in which the outer particles are designed to search the scheduling solution of casts, the inner particles are designed to search the scheduling solution of charges in each cast. Scheduling solution obtained by outer search will be delivered to inner search as constraints of inner search. The specific operation of bi-level particle iterative evolution strategy can be described as following:

Outer Search populations: the position vector and velocity vector of the kth particle in n'_B dimension search space are expressed as $F_i^{(n'_B)}=[f_{i,1}^{(INT)},f_{i,2}^{(INT)},\cdots,f_{i,n'_B}^{(INT)}]$ and $V_i^{(n'_B)}=[v_{i,1}^{(INT)},v_{i,2}^{(INT)},\cdots,v_{i,n'_B}^{(INT)}]$.

At the time *t*, the best position that obtained by the a outer particle can be expressed as $P_{i,best}^{(n'_B)}=[p_{i,1}^{(INT)},p_{i,2}^{(INT)},\cdots,p_{i,n'_B}^{(INT)}]$, the best position that obtained by the outer particle Swarm can be expressed as $P_{g,best}^{(n'_s)}=[p_{g,1}^{(INT)},p_{g,2}^{(INT)},\cdots,p_{g,n'_s}^{(INT)}]$. The update equation for the velocity and position of outer particle at time $t+1$ is then: $v_{i,j}^{(INT)}(t+1)=\omega_l v_{i,j}^{(INT)}(t)+c_l r_l[p_{i,j}^{(INT)}$.

$$-f_{i,j}^{(INT)}(t)]+c'_l r'_l[p_{g,j}^{(INT)}-f_{i,j}^{(INT)}(t)] \tag{8}$$

$$f_{i,j}^{(INT)}(t+1)=f_{i,j}^{(INT)}(t)+v_{i,j}^{(INT)}(t+1),\ \ j\in\{1,2,\cdots,n'_B\} \tag{9}$$

The update equation for the velocity and position of inner particle is similar to the outer particle, so it will not be described in detail here.

4.3. *Multi neighborhood structure based on randomization*

Previous studies have shown that the capability of local search depends on the neighborhood structure [10]. For the general optimization problems of permutation and combination, there are common neighborhood structure, they are $InsertChange(\pi,v_1,v_2)$ t (Insert-exchange neighborhood structure), $Insert(\pi,v_1,v_2)$ (insert neighborhood structure) and $Swap(\pi,v)$ (adjacent swap neighborhood structure).

Here, we crystal-clear a combination neighborhood structure $COM<IC,I,S>$ to express the combination state about three neighborhood operators proposed, and we make three kinds of neighborhood operators priority order as follows: first, perform $InsertChange(\pi, v_1, v_2)$, next perform $Insert(\pi, v_1, v_2)$ and last, perform $Swap(\pi, v)$. For the Overlapping probability distribution area, such as $\alpha_2 < \beta_1$, i.e. it is overlapping area $[\alpha_2, \beta_1]$. If the random numbers meets the conditions $\alpha_2 \le rand() \le \beta_1$, the combination neighborhood structure is formed, in which the $InsertChange(\pi, v_1, v_2)$ operator and the $InsertChange(\pi, v_1, v_2)$ operator are performed alternately according to the neighborhood operator priority.

The combination neighborhood structure can be employed both by inner neighborhood structure that is similar to the outer neighborhood structure to improve searching performance.

4.4. *Flow frame of the hybrid PSO algorithm*

Based on the above algorithm design, the specific steps and process framework of hybrid PSO are shown in Figure 2.

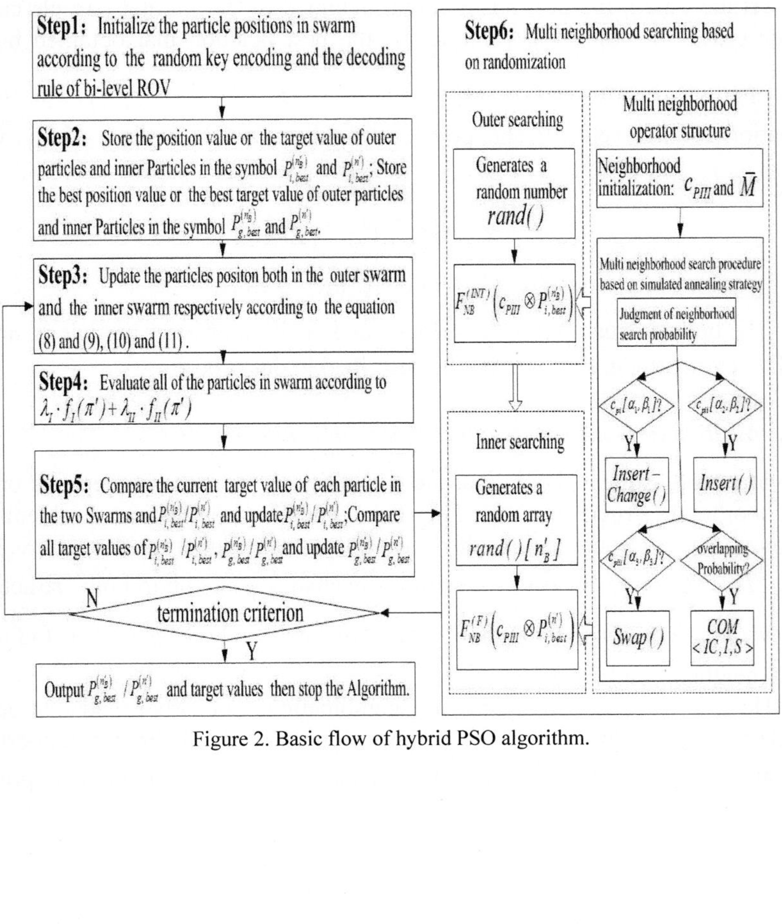

Figure 2. Basic flow of hybrid PSO algorithm.

Owing to the probabilistic jumping ability of simulated annealing (SA) algorithm, which can avoid falling into the local minimum solution effectively in the searching process [9]. As shown in step 3 of 6 basic processes, we use simulated annealing mechanisms to avoid the premature convergence in the multi neighborhood searching process, which is shown at 6th processing step in Figure 2.

5. Implementation and Experimental Results

5.1. *Experiment implementation*

Experiment protocol of HNWFS-dmsp problem modeling oriented to Green Manufacturing for steelmaking and continuous casting process shows as follows: steelmaking and continuous casting process involves three steps, they are steelmaking, refining and continuous casting, m machines. Without loss of generality, we assume there has 2 same electric furnaces for steel-making step, 2 same refining furnaces for refining step and 1 caster for continuous casting step. 58 charge s are be arranged, they are divided 8 casts. Table 1 shows their work number, weighting coefficients, processing time in each step.

Now, we assume that emergency processing tasks are added in, such as the newly added No. 59 and No. 60 charges (they correspond to No. 3 cast and No. 5 cast respectively, as shown in Table 1) they would be inserted into the initial scheduling consisted of 58 previous charges.

We assume that Linear weight parameters of two scheduling objectives for disruption management are: $\lambda_I=1$ and $\lambda_{II}=10$. We take the starting time in the case of charge disturbances as, $t_0=0$ and then teste the performance of the aforementioned hybrid PSO algorithm by several independent experiments.

Parameter setting of Hybrid PSO algorithm: (i) particle iterative optimization: population size: $M = 80$, the maximum generation iteration times: $Max = 100$, the particles acceleration factor $c_1 = c_1' = c_2 = c_2' = 2$, particles inertia weight [11] $\omega_1 = \omega_2 = (0.9-0.4)\times(Max - M_{cur})/Max + 0.4$ (M_{cur} is the current generation iteration times); (ii) Stochastic probability multi-neighborhood search mechanism: to examine the effectiveness of various neighborhood operator in stochastic probability multi-neighborhood search mechanism, three representative algorithms is designed. HPSOI algorithm: $c_{PIII} = \{c_{pi}[0.00,0.85], c_{pii}[0.75,0.95], c_{piii}[0.80, 1.00]\}$; HPSOII algorithm: $c_{PIII} = \{c_{pi}[0.75,0.95], c_{pii}[0.00,0.85], \quad c_{piii}[0.80, 1.00]\}$; HPSOIII algorithm: $c_{PIII} = \{c_{pi}[0.80, 1.00], c_{pii}[0.75,0.95], c_{piii}[0.00,0.85]\}$.

The three proposed hybrid PSO algorithms employ simulated annealing algorithm, which has initial temperature 3.0, annealing rate is 0.9, and maximum iteration of termination criterion is 150.

5.2. *Result analysis*

We conduct 30 times independent numerical experiment on HPSOI algorithm, HPSOII algorithm and HPSOIII algorithm respectively, a summary of the statistical results is displayed in Table 2. It includes the original objective $f_I(\pi')$, the disruption repairing objective $f_{II}(\pi')$, make span and disruption management scheduling objective $\lambda_I \cdot f_I(\pi')+\lambda_{II} \cdot f_{II}(\pi')$ obtained by three proposed algorithms on HNWFS-dmsp problem.

Table 1. Results comparison of 30 times independent numerical experimental on three types hybrid PSO algorithm.

Algorithm		HPSOI	HPSOII	HPSOIII
High probability operator		$Insert\ Change(\pi, v_1 v_2)$	$Insert\ (\pi, v_1 v_2)$	$Swap(\pi, v)$
$f_I(\pi')$	Maximum	2928024	2940565	2970108
	Minimum	2782500	2848149	2792091
	Average	2782500	2896059	2889215
$f_{II}(\pi')$	Maximum	322825	321461	323427
	Minimum	170716	181543	168354
	Average	238632	246945	245206
C_{ec}	Maximum	9045	9042	9042
	Minimum	8860	8907	8907
	Average	8946	8947	8954
$\lambda_I \cdot f_I(\pi') + \lambda_{II} \cdot f_{II}(\pi')$	Maximum	6022478	6117720	6026361
	Minimum	4600704	4710293	4574358
	Average	5261521	5365508	5341272

Statistical results displayed in Table 1 shows that the original objective average obtained by HPSOI algorithm and HPSOII algorithm is better than HPSOIII algorithm; In terms of both disruption repairing objective and disruption management objective, HPSOI algorithm performance better than HPSOII algorithm and HPSOIII algorithm; as for C_{ec}, HPSOI algorithm and HPSOII algorithm obtain better results than HPSOIII algorithm; Preliminary analysis shows that, HPSOI algorithm and HPSOII algorithm are more suitable for disruption management scheduling for energy saving in steelmaking and continuous casting process.

Given the above, we then select the following six indicators: the overall non-dominated vector generation (ONVG) [12], the C-metric (CM) [9], and distance between the non-inferior solutions and Pareto optimal boundary (D_{av} and D_{max}) [13], the maximum spread (MS) [14] and the average quality (AQ) [14], with which, we evaluated the non-inferior solutions quality obtained by

HPSOI algorithm and HPSOII algorithm so as to further research and compare the effectiveness of hybrid PSO algorithms. The results which obtained by 10 independent experiments are displayed in Table 2.

Statistical results shows that both HPSOI algorithm HPSOII algorithms can get approving results for solving the HNWFS-dmsp problem oriented to Green Manufacturing for steelmaking and continuous casting process, there are still some delicate differences. Comparing two algorithms on ONVG value, we found that HPSOI algorithm obtained more non-inferior solutions than HPSOII algorithm; It also shows that HPSOI algorithm outperforms HPSOII algorithm on CM value in most cases; The comparison of two algorithms on D_{av} value and D_{max} value shows that in terms of the approximation degree to optimal Pareto frontier in theory, the non-inferior solutions obtained by HPSOI algorithm are more close to the optimal Pareto frontier in theory in terms of the average distance and the maximize closest distance to optimal Pareto value, i.e. HPSOI algorithm outperforms HPSOII algorithm; Compared on MS value, we found that the non-inferior solutions obtained by Pareto HPSOI algorithm covers theory optimal Pareto front completely, However HPSOII algorithm is not as good HPSOI algorithm; Comparing on AQ value, we conclude that the result of HPSOI algorithm and HPSOII algorithm is similar, HPSOI algorithm slightly is better than HPSOII algorithm, which shows HPSOI algorithm performance better in terms of approximation and discreteness of non-inferior solutions. In summary, HPSOI algorithm is a more effective hybrid algorithm to solve HNWFS-dmsp problem, it is suitable for disruption management scheduling in steelmaking and continuous casting process.

Table 2. Performance comparison of HPSOI and HPSOII.

Index	***ONVG***		***CM***		D_{av}		D_{max}		***MS***		***AQ***	
Algorithm	**HPSOI**	**HPSOII**	**HPSOI**	**HPSOII**	**HPSOI**	**HPSOII**	**HPSOI**	**HPSOII**	**HPSOI**	**HPSOII**	**HPSOI**	**HPSOII**
Test 1	57	38	0.500	0.439	0.009	0.008	0.041	0.034	0.999	0.963	1244935	1244406
Test 2	78	52	0.404	0.423	0.005	0.004	0.050	0.021	1.000	0.984	1245258	1246471
Test 3	51	59	0.288	0.588	0.010	0.004	0.041	0.030	0.986	0.997	1243318	1240422
Test 4	41	50	0.940	0.049	0.000	0.031	0.006	0.130	1.000	0.951	1241698	1248722

AVG	Min	Max	Test 10	Test 9	Test 8	Test 7	Test 6	Test 5
59	41	78	67	47	60	57	67	67
52	38	70	58	39	47	52	70	58
0.464	0.069	0.940	0.069	0.359	0.553	0.692	0.129	0.707
0.454	0.049	0.910	0.910	0.638	0.433	0.211	0.642	0.209
0.009	0.000	0.019	0.018	0.019	0.012	0.003	0.010	0.002
0.010	0.001	0.031	0.001	0.010	0.011	0.014	0.001	0.013
0.041	0.006	0.104	0.044	0.104	0.051	0.020	0.034	0.014
0.048	0.016	0.130	0.016	0.084	0.062	0.044	0.017	0.047
0.992	0.974	1.000	0.996	0.974	0.991	0.989	0.989	1.000
0.986	0.951	1.000	1.000	1.000	0.990	0.997	0.985	0.997
1246055	1240475	1256700	1251100	1256700	1249030	1241499	1246536	1240475
1244693	1240216	1253505	1242512	1253505	1243730	1244170	1240216	1242771

Table 2 shows that both HPSOI algorithm HPSOII algorithms can get approving results for solving the HNWFS-dmsp problem oriented to Green Manufacturing for steelmaking and continuous casting process, there are still some delicate differences.

Comparing two algorithms on ONVG value, we found that HPSOI algorithm obtained more non-inferior solutions than HPSOII algorithm; It also shows that HPSOI algorithm outperforms HPSOII algorithm on CM value in most cases; The comparison of two algorithms on Dav value and Dmax value shows that in terms of the approximation degree to optimal Pareto frontier in theory, the non-inferior solutions obtained by HPSOI algorithm are more close to the optimal Pareto frontier in theory in terms of the average distance and the maximize closest distance to optimal Pareto value, i.e. HPSOI algorithm outperforms HPSOII algorithm; Compared on MS value, we found that the non-inferior solutions obtained by Pareto HPSOI algorithm covers theory optimal Pareto front completely. However HPSOII algorithm is not as good HPSOI algorithm; Comparing on AQ value, we conclude that the result of HPSOI algorithm and HPSOII algorithm is similar, HPSOI algorithm slightly is better than HPSOII algorithm, which shows HPSOI algorithm performance better in terms of approximation and discreteness of non-inferior solutions. In summary,

HPSOI algorithm is a more effective hybrid algorithm to solve HNWFS-dmsp problem, it is suitable for disruption management scheduling in steelmaking and continuous casting process.

In summary, HPSOI algorithm is a more effective hybrid algorithm to solve HNWFS-dmsp problem, it is suitable for disruption management scheduling in steelmaking and continuous casting process for green manufacturing.

6. Conclusion and Future Research

The model and algorithm of steelmaking and continuous casting process oriented to the green manufacturing and the process management in steel enterprises are researched in this paper. The main results obtained are: (i) A novel disruption management scheduling method based on the feature analysis of steelmaking and continuous casting process is presented. The scheduling method is designed by considering both the target to reduce energy consumption indicator and the target to minimize total weighted delay time. We take the continuity of production and the cast constraints into account, built the multi-objective disruption management scheduling model under charge disturbance condition with the optimization goals that minimize total weighted completion time (the original objective) and minimize total weighted delay time (the disruption repairing objective). (ii) Bi-level particle iterative evolution strategy is designed ,in which the outer particles are designed to search the scheduling solution of casts, the inner particles are designed to search the scheduling solution of charges in each cast, and the multi-neighborhood operators construction method based on random strategy is proposed. On those bases, a hybrid PSO algorithm is proposed to solve disruption management scheduling model oriented to green manufacturing in steelmaking and continuous casting process. Last, we test and verify the validity of algorithm proposed.

However, this study is based on a hybrid flow shop scheduling method, and it is necessary to further study the scheduling condition which meets the requirements of steelmaking and continuous casting process route, and with the features of Job shop.

Acknowledgments

This work was supported by the National Science and Technology Support Program (2015BAF08B02) and the National Science and Technology Support Program (2013BAF02B03).

References

1. S.P. Yu & X.F. Pang etc. 'Research on production scheduling for steelmaking and continuous casting with processing time uncertainty', Control and Decision, 2009, 24(10): 1467–1472.
2. A. Bellabdaoui & J. Teghem. 'A mixed-integer linear programming model for the continuous casting planning', International Journal of Production Economics 104.2 (2006): 260–270.
3. H. Xuan & Li-Xin Tang. 'Scheduling a hybrid flowshop with batch production at the last stage', Computers & Operations Research 34.9 (2007): 2718–2733.
4. H. Guang & Tie-Ke Li. 'A steelmaking-continuous casting production scheduling model and its heuristic algorithm', Systems Engineering 6 (2002): 008.
5. H. Missbauer, H. Wolfgang & S. Werner. 'A scheduling system for the steelmaking-continuous casting process. A case study from the steel-making industry', International Journal of Production Research 47.15 (2009): 4147–4172.
6. Tie-Ke Li, J. Zhou & L. Sun. 'Model and algorithm for steelmaking-continuous casting scheduling in DHCR and CCR environment', Systems Engineering-Theory & Practice 26.6 (2006): 117–123.
7. M. Pinedo. Scheduling: 'Theory, algorithms, and systems', Springer Science & Business Media, 2012.
8. Quan-Ke Pan, Hua-Zhao Bao & Yu-Gui Qu. 'Heuristics for the no-wait flow shop problem with makespan criterion', Chinese Journal of Computers 31.07 (2008): 1147–1154.
9. L. Wang & B. Liu. 'Particle swarm optimization and scheduling algorithms.' Beijing: Tsinghua University (2008): 36–56.
10. Yew-Soon Ong et al. 'Classification of adaptive memetic algorithms: a comparative study', Systems, Man, and Cybernetics, Part B: Cybernetics, IEEE Transactions on 36.1 (2006): 141–152.
11. L. Li & B. Niu. 'Particle Swarm Optimization Algorithms', Beijing: Metallurgical Industry Press(2009).
12. Veldhuizen Van & A. David: 'Multiobjective evolutionary algorithms: Classifications, analyses, and new innovations', No. AFIT/DS/ENG/99-01. Air Force Inst of Tech Wright-Pattersonafb Oh School of Engineering, 1999.

13. P. Czyżak & J. Adrezej. 'Pareto simulated annealing—A metaheuristic technique for multiple-objective combinatorial optimization', Journal of Multi-Criteria Decision Analysis 7.1 (1998): 34–47.
14. K.C. Tan et al. 'Evolving better population distribution and exploration in evolutionary multi-objective optimization', European Journal of Operational Research 171.2 (2006): 463–495.

Analysis about the impact of the environment factors on senior intellectual's health and corporeity*

J. Yang
P.E Department Harbin Engineering University Harbin, Heilongjiang Province, China
E-mail: yangjun3289072@163.com

Y.F. Zhang, Z.Z. Li, L.X. Zhu, J. Qi and W. Li
P.E Department Harbin Normal University Harbin, Heilongjiang Province, China

Environmental issue has seriously affected the physical and mental health of many people in our country, and at the same time, it plays an important role in our country's the economic and social development, which has become a major social problem. This article aims to raise the whole society's awareness of environmental protection, strengthen the propaganda of environmental protection and environmental protection consciousness.

Keywords: Health; Environment; Corporeity; Senior intellectual; City.

1. Introduction

Science and technology is the first productive force, an important symbol of human society progress [1]. And intellectuals who has the modern science and technology in the hands are valuable wealth of country and the main power of implementation of the strategy of developing the country. Middle-aged and young teachers in colleges is a very special group in the social life. They not only shoulder the burden of high-level talents but also are the disseminator of modern knowledge as well as the main power of scientific research work. In our country, they play an important role in modernization construction. And its health determine science and technology progress and social development of our country.

2. Definitions of Related Concepts

2.1. *The definition of "Corporeity constitution"*

Corporeity constitution is the material basis of human's living activities and

---

* This work is supported by the Fundamental Research funds for the Central Universities (HEUCF141605).

labor work abilities. Briefly, corporeity constitution is the quality of a person's organism. It is a relatively stable feature of the human body in the form, physiology, biochemistry and behavior on the basis of genetic variation. Corporeity constitution reflects not only the level of the human body's life activity, but also the level of human body movements [2]. Corporeity constitution reflects the unity of opposites between life movements and body movements. Only scientifically grasping and dealing with the contradiction and unification of life movements and body movements, we can achieve the highest achievement of physical development.

The factors that affect the corporeity constitution are various. Genetic traits may provide the possibility or prerequisite for the development of the constitution. The reality of physical strength depends on the influence of the environment, and nutrition, health, education and physical exercise are the most important factors.

2.2. *The definition of "health"*

In 1948, the World Health Organization made the following definition of health in its constitution, " Health is not a state of feeling good. It is a state of complete peace in the body, spirit and society, and it is not a state that only far from disease and illness". Later, this organization adds "moral" to the evaluation of the definition of health, which make the definition of health more comprehensive and more social. We can do the following explanation for these aspects of health:

2.2.1. *Physical health*

Generally refers to the health of human body.

2.2.2. *Mental health*

Generally there are three aspects of the indication: First, personality integrity, feeling good about self; emotional stability, positive emotions are more than negative emotions, having a good self-control ability and can keep the psychological balance, having self-esteem, self-respect, self-confidence and self-knowledge. Second, in his environment, he must have full sense of security and keep moderate anxiety. Third, he should have a clear goal of life in the future, he can meet the reality, he can improve himself continuously, he have the pursuit of ideals and career.

2.2.3. *Good social adaptation*

It means that the individual's social behavior can adapt to the complex environment changes, and everyone can understand and accept him, he can keep normal interpersonal relationship and be welcomed by others.

2.2.4. *Moral health*

Do not meet self needs by damaging the interests of others, he have the ability to distinguish between good and evil, beauty and ugliness, honor, etc. he can control his own behavior according to the social norms, and he can make contribution to the happiness of the people [3].

Health level reflects the level of life movements. The coordination of life movement, exuberant and longevity is a good performance of the health of the state. We must be combined with the World Health Organization Regulations and the requirements for health care for all, recognize it from the height of the international community, pursuit and enjoyment of the highest standards of health is a basic human right; health is a part of social development, a duty to mankind, and everyone has the right to own health and equality.

Definitions of "body constitution" and "health" are not exactly the same, body constitution has long and stable characteristics, but the health has the characteristics of short-term and variable. Basic physical conditions may vary widely for two person who are in a state of health. The same physical condition of the people, in the short term may bring different performance on the health due to the impact of disease. At the same time, the body constitution also includes the evaluation of people's living abilities, labor abilities and sports abilities [4].

3. Urban Environmental Health Problems

The impact of environment on health is multifaceted.Generally there are two aspects. One is the impact of natural environment on health. The other is the impact of social environment on health. Natural environment has the biggest impact on health.

3.1. *The development of modern industry and development and utilization of nature made people destroy environment unconsciously*

Pushing waste water, waste air and waste into natural environment caused serious pollution of water, soil and air. The traditional development mode of consumption of resources-based stressed the protection of ecosystem. Accelerate of the pace of urbanization and the rapid development of the total economy requested the improvement of ecosystem. The natural environment is the basis of human survival and reproduction, but this natural environment for human survival in severe damage has already begun to bring damage to human life and health. Since 1980s, our government has been to "environmental protection" as a basic national policy, but as a populous country, our ecosystem is weak because of geography and climate and other reasons. The extensive mode of production results in excessive resources and energy consumption, which leads to the growing environmental problem. The ecological environment carrying capacity in China has been close to and over the critical path. The environmental problems in the developed countries in stages in the hundreds of years process of industrialization in our country to be displayed in a short period of time. Nowadays, the main characteristics of environmental and health problems in our country is serious compound pollution, a wide range, a large exposed population, long exposure time of the crowd. Significant difference was found in urban and rural. The main pollution in the city area is air pollution and in the rural is water pollution and soil pollution. Traditional environment caused by inadequate sanitation, and health problems have not been properly solved, at the same time, environmental pollution and health risk caused by industrialization and urbanization gradually increased. From the development trend, the problem of the above four aspects is difficult to solve in short term, the health risks of environmental pollution gradually enhanced in the future, the situation of health work and environment is still grim. As the rapid development of industrialization and urbanization, the environmental pollution has been the prominent problem in affecting the health of urban and rural residents. It has been one of the important factors which affects the sustainable development of our country, the construction of well-off society and a harmonious society. protecting the environment, ensuring the health become the most pressing needs in urban and rural residents in our country. Pollution of the environment on the health effects have characteristics of a low exposure levels and long incubation period, influence factors, difficult to determine causal relationships, etc. In the rapid economic development today, our country is facing the deterioration of living environment and working environment, serious food safety problems, health damage problems caused by environmental pollution, which has caused a serious threat to health to the crowd and social stability, also has a bad influence at home and abroad.

3.2. *The impact of social and environmental factors on health*

The city is a composition of non-agricultural population, which occupies a certain space of population centers. Urbanization process is a constantly cohesive population distribution process. Urban and rural distribution of the population is the world's most basic type of distribution of the population. Modern Urban is on behalf of the advanced productive forces, the city not only play a function of cultural and political center, but also the economic center and the center of gravity. Each of a modern city is an important part of the local regional economy, regional ecosystems, and also part of its national and global economies and global ecosystems. With the acceleration of global urbanization process, the world economy has been developing rapidly, but a variety of urban health problems are emerging. World Health Organization (WHO) for the world's urban construction and development developed the new requirements: the city is not just one-sided pursuit of economic growth in the efficiency of economic entities, but should be able to enhance and improve the ideal environment for human health, should be considered a living reality of space,which can breathe, can grow and change.

The factors which have influence on the urban "healthy" are complex and diverse. Many people find the approach to measure urban ecosystem health factors. In analyzing the main dimensions which influence the urban ecosystem health, they tend to focus on human and social capital, including population health, urban community social welfare.

4. Environmental Health View from the Perspective of the Co Students

"Synergetics is proposed by H. [5] Haken, a theoretical physicist of west University in 1977". It outlines the general characteristics of the evolution of the system, points out that the phase transition of a complex system is between subsystems interconnected, the coordinating role of the results, Hawking will this theory further promotion presented slaving (dominate) principle. The theory holds that: "All systems evolve at the critical point, and the fast variable plays a minor role in the evolution of the system, and the slow variable dominates or dominates the evolution of the system, and controls the behavior of the fast variable" [6]. There is a wide range of cooperation and competition among the various subsystems or elements in the system, and this cooperation and competition is the self-organization process of the system.

The environment of human existence is correspond to the system's two conditions: (1) the system is open. The open system is a system that can exchange energy with the outside world. Living environment system is an open system. People need to keep on attaining a variety of material basis and energy substances from the outside world. At the same time, they discharge the end product of their metabolism and invalid energy; people through social activities or exchanges continue to carry out information exchange, and according to their actual state and psychological needs, and constantly improve lifestyle and the way participate in social activities. (2) there is a nonlinear interaction between the various elements of the system [7]". In a certain external conditions, the interaction of the system can not make the system form an ordered structure, only the nonlinear interaction, the system of the state change is possible. The relationship between the system is not simple causal, or linear dependence relationship, but there is a positive feedback of the multiplier effect, there is a limit of growth of the saturation effect, that is, the system of memory in the fluctuations.

The so-called fluctuation refers to a deviation of a variable to the average in the system and it enables the system to deviate from its original state or track. Compared with human beings' living environment itself, in different states, its change is not linear. Even though it is in the same condition,its physical condition is different. In the view of major environment, If the entire environmental system is always in balance, there will be no space for human development and lose its internal impetus in social systems. To survive and develop, its system must break this kind of symmetry then form a broken symmetry. In the process of human social development, in order to achieve their own survival needs, they constantly absorb energy from the environment and a variety of information as well as destroy symmetry, so that they can achieve the internal order of human society, and the orderly behavior.

In this human system, the pathogenic factors and healthy factors interact with each other all the time, which restricts people's physical and mental health. In the critical point between health and sickness, pathogenic factors and health factors interacting with each other, if the pathogenic factors absorb a lot of energy and information, amplified and strengthened in the process of random fluctuation, the next state will be the illness. At that time, the pathogenic factor is the order parameter. On the contrary, in the critical point between health and sickness,if the healthy factors absorb a lot of energy and information, amplified and strengthened in the process of random fluctuation,the next state will be the health. It is worth mentioning that, between health and sickness, people join a "sub-healthy" concept. "Sub-health", if using systematic science to explain it, is a state of "chaos" where the order and disorder are in critical point, which also is the critical state of the fluctuation of system. This state has a great realistic significance for system evolution. If the Evolution is good, the system would be a healthy state; otherwise it would be sick.

5. Conclusion

The behavior depends on the dominance of a certain idea and consciousness. The scientific fitness behavior also needs the guidance of scientific concept of health and environmental consciousness. The poor condition of the urban population living environment, to a certain extent, is due to their lack of environmental awareness and environmental sustainable development consciousness. It is not only the senior intellectuals personal affair, but also the department leadership's to enhance the whole society's awareness of environmental protection, strengthen the propaganda of environmental protection, and improve environmental protection consciousness. Senior intellectuals have relatively heavy task of teaching scientific research. Their job is busy, and the psychological pressure is very big, at the same time they are facing the challenge of the new era of knowledge economy. In addition, with the enrollment expansion of colleges, teachers' workload increased significantly. So, school leaders should attach great importance to nor only the role of the teacher, but also improve working conditions for them and working environment, alleviate psychological pressure and mental burden, increase their leisure time, create a more fitness environment, create a more fitness atmosphere, so they can put themselves into the fitness activities of colleges and universities with exuberant energy and enthusiasm.

References

1. Xinhan Chen. Marxist philosophy tutorial [M]. Shanghai University Publishing House, 171.
2. Zhongwei Yang. Health evaluation and system [M], referring to Lou Yuan town, Social Sports, Higher Education Press, 2003189.
3. Zhongwei Yang. Exercise prescription [M], referring to Lou Yuan town, Social Sports, Higher Education Press, 2003185.
4. H. Hawking. Synergetics, Nature is the secret of success [M]. Dai Mingzhong interpreted, Shanghai popular science publishing house, 1988.
5. Jinchun Zhang. Self-organization theory application in the military academy graduate student education [J], Journal of Harbin Institute of Technology, 2003, (2) 238–241.
6. Junfa Liang. System fluctuation, coordination and evolution [J]. Shanxi Higher School Journal of Social Science, 1995.

The effects of protein level on feather pecking of Hailan layers*

C.P. Wang
College of Life Science, Jiamusi University,
Jiamusi, 154007, China
E-mail: wcp100200@163.com
www. jmsu.org

Y. Wang
College of Food and Biological Engineeriny, Qiqihar University
Qiqihar, China
E-mail: wangyan0468@126.com

C.B. Wei
College of Animal Science and Veterinary Medicine,
Heilongjiang Bayi Agricultural University,
Daqing, China
E-mail: weicb298@163.com

From 9 week to 11 week, 240 Hailan layers were assigned at random to groups of 5 individuals and distributed among 48 pens. They were kept in pens with different protein levels in diet. The three protein levels are 14%, 17% and 19%. We observe the feather pecking of Hailan layers. We study feaher pecking in detail, we observe partitions of body pecked by laying hens, behavior of the pecked laying hens. We find that protein levels had significant effects on the rate of feather pecking. With the increasing of protein levels increasing, the occurrence of feather pecking was decreased. When feather pecking occurred, the main parts of body pecked by laying hens was followed by tail (44.3%), back (26.5%), head (17%). The behavior of the pecked laying hens was standing (45.4%), feeding (23.7%), groveling (23.3%).

Keywords: Protein levels; Feather pecking; Behavior activities.

1. Introduction

Feather pecking is defined as the pecking at and pulling out of feathers of another chicken. Often these feathers are eaten as well. Feather pecking is an indicator of reduced welfare in both victim and performer. Pulling out feather is painful and chickens with feather damage are more susceptible to further feather

---

* This work was funded by Natural science foundation of China (31302008) and Natural science foundation of Heilongjiang province (QC2012C104; QC2013C031).
Corresponding author: C.B. Wei

pecking and injurious pecking (Mcadie and Keeling, 2000). Feather pecking is considered to be a welfare problem due to the pain experienced when feathers are being pulled out and also due to the increased likelihood of cannibalism resulting (McAdie et al., 2000). Apart from reduced animal welfare, feather pecking in commercial farm is an economic problem, even if it does not result in increased mortality or lower production: chickens with feather damage need more feed in order to maintain their body temperature, which can rise up to 27% (Tauson and Svensson, 1980). Feather pecking can be gentle or severe and is distinct from aggressive pecking, which is often aimed at the head.

Feather pecking is influenced by many different environmental factors (Green et al., 2000). It has been shown that many factors such as stocking density, group size, dietary composition, light intensity and genetic background can influence the occurrence of feather pecking (Hughes & Duncan, 1972; Blokhuis, 1989; Vestergarrd, 1994; Savory, 1995).

Feeding high-fibre, low-energy diets or roughage reduces feather pecking (Van Krimpen et al., 2005). Insoluble fibre (non-starch polysaccharides and lignin) affects gut functions and modulates nutrient digestion and there are indications that diets high in insoluble fibre are preventive of cannibalism outbreaks in laying hens (Hetland et al., 2004). In previous studies, low protein diets have been associated with increased feather pecking and mortality attributable to cannibalism (Cain et al., 1984; Ambrosen and Petersen, 1997). But the effects of protein level on feather pecking of Hailan layers is known little. The purpose of this study was to describe and examine the effects of protein level on feather pecking of Hailan layers.

2. Material and Methods

2.1. *Subjects and husbandry*

The experiment was carried out with a total of 240 Hailan layers assigned at random to groups of 5 individuals and distributed among 48 hens. The birds were not beak-trimmed and had been reared in an cage system, the cage floor was covered with nothing. Feed and water were given ad libitum. Incandescent light gave different light intensity. Providing different protein levels in diet.

2.2. *Procedures*

The 240 Hailan layers were assigned at random to groups of 5 individuals in cage. The chick has no visual contact. Also protein level is designed to three levels (14%, 17%, 19%). A scheme of how the dietary of laying hens was designed is depicted in Table 1.

Table 1. The composition and chemical analysis (g kg^{-1} as fed) in each of three diets.

Dietary constituent (g/kg)	Feeding stuff		
	Low protein feeding stuff (14%)	Middle protein feeding stuff (17%)	High protein feeding stuff (19%)
corn meal	71.6	67.2	64.6
soyabean meal	21	22.5	28.3
wheat meal	4	7	4
methionine	0.096	0.096	0.1
lysine	0.2	0.15	---
salt	0.32	0.32	0.32
limestone flour	1.2	1.1	1.1
dicalcium phosphate	1.2	1.2	1.2
choline	0.1	0.1	0.1
vitamin mix	0.025	0.025	0.025
minal mix	0.2	0.2	0.2
enzyme	0.1	0.1	0.1
VC	0.004	0.004	0.01
VE	0.01	0.01	0.01

The 240 Hailan layers were selected at the 9th week balanced for body weight and all birds were healthy. All birds were divided into 3 treatments. Each treatment contained 5 birds covering 16 replicates, each replicate has 5 birds. The chick of different treatment has no visual contact. Instantaneous-sampling method for collecting behavioural perameters was adopted with the time interval of 5 seconds. Feather pecking, the parts of pecked body, the record of the score of damaged feather and the behaviors of the pecked laying hens were recorded.

2.3. *Behavioural observations*

Focal animal observations were carried out to investigate behaviour of feather pecking in more detail. Chosing a focal animal in each group, and the focal animals were marked individually with black paint. Data collection by one persons, ran from week 9 to week 11. The observer sit on the chair, which distance to the ground about 2 m. Thus the observer can scan the focal animal in each group. Hens were observed 96 min every day, in the morning between 09:30 and 11:06. Each group was observed for 2 min, recording one behaviour data in 5 second.

2.4. *Behaviour parameter*

We observed the behaviour of focal animal in each pen, we adopted scan methods to obtain behaviour data. The all occurrences sampling was briefly interrupted every 5 second for a scan sample of the activity of chicks. For each scan we recorded the number of chicks engaged in eight mutually exclusive activities: feather pecking, feeding, groveling, preening, foraging, drinking, standing and walking. They were defined as follows: (1) Feather pecking: feather pecking interaction was attributed to 1 of the following 4 types of behaviour (increasing intensity): (a) pecking a feather without pinching, (b) pinching a feather and pulling weakly, (c) pulling at a feather with a vigorous backward movement of the head and (d) plucking a feather. (2) Feeding: the chick was by the feeder with its head above the food; (3) Groveling: the chick showed no other defined behaviour while its chest feathers touched the ground. (4) Preening: the chick nibbled, stroked or combed its plumage with its beak (Kruijt 1964) or stretched its wings or legs; (5) Foraging: the hen is pecking at the floor or at other parts of the pen (but not at the food) or it is standing/moving with its head in a lower position than the rump. (6) Drinking: the chick's bill was oriented to, and within 5 cm of the cup drinkers; (7) Standing: standing on both feet without showing another defined behaviour; (8) Walking: locomotion without showing another defined behaviour.

2.5. *Partitions of the body pecked by laying hens*

Focal birds from each pen were individually scored at 9 weeks, 10 weeks and 11 weeks. The body was divided (Gunnarsson et al., 1995) into 11 regions: head, upper neck (back side of the neck), back (part between wings), rump, tail, belly (abdomen), breast, under neck (front side of the neck), wings-primary feathers, wings-coverts and legs.

2.6. *Statistical analyses*

The pens were treated as independent units (n=48) in all analyses. Mean values for the behavioural activities were calculated for each pen. The analyses were performed using SPSS and Microsoft Excel. All statistical tests are 2-tailed with an alpha level of 0.05. Data from weeks 9 to 11 were analysed with protein levels. Prior to the ANOVA, square-root and arcsine square-root transformations (Sokal and Rohlf, 1981) were applied to rates and percentages, respectively,to achieve normal distribution of the residuals (Lilliefors, 1967). Untransformed data are given in the tables and the figures.

3. Result

3.1. *Feather pecking behavioural observations*

Protein level had significant effects on rate of feather pecking.Feather pecking was decreased with the increasing of protein levels (P<0.01 F=29.015)(Table 2).

Table 2. The comparison of feather pecking frequency at different protein density (Mean±SE).

P values	Protein level		
	14%	17%	19%
P<0.01	7.41±0.34[a]	5.65±0.22[b]	4.23±0.31[c]

3.2. *Behaviour of the pecked laying hens*

When the behaviour of feather pecking occur, the pecked laying hens were standing (45.4%), feeding (23.7%), groveling (23.3%), preening (4.2%), foraging (1.9%), drinking (0.71%), feather pecking (0.51%), walking (0.28%).

3.3. *Partitions of the body pecked by laying hens*

From the total of 736 feather pecks (gentle and severe) observed throughout the 48 pens, 326 pecks were targeted towards tail (44.3%), 195 on back (26.5%), 125 on head (17.0%), 42 on neck (5.7%), 13 on wings-primary feathers (1.8%), 11 on belly (abdomen) (1.5%), 10 on breast (1.4%), 6 on under neck (front side of the neck) (0.82%), 4 on rump (0.54%), 3 on legs (0.41%), 1 on wings-coverts (0.14%).

4. Discussion

Through the analyse of the experimental data, our experiment finds that with the decreasing of protein levels feather pecking was increased (P<0.01 F=29.015). The results of our experiment demonstrate that protein levels had significant effects on rate of feather pecking.

The feather pecking of Hailan layers is relation to protein levels in detail, thus elevating the protein level of dietary can reduce the occurrence of the feather pecking. It is because that the dietary is not lack of protein. Hailan layers can ingest sufficient protein nutrition, thus the Hailan layers can not offset the deficiency of amino acid through pecking and eating other chick plumage. Because the plumage has the amino acid (the amino acid has sulfur).

The different peck time of different partitions of the body pecked by laying hens is mainly due to the structure of different body part and the degree of easy to be pecked. Wood-Gush & Rowland (1973), in an experiment with pen-housed

Brown Leghorns, reported that most feather pecks were delivered to the rump,followed by the tail and back. Savory & Mann (1997) observed that most feather pecks were deliver on the back in a group of Hisex hens and on the thigh in White Leghorns. Norgaard-Nielsen et al. (1993) found feather pecking on the breast and back of White Leghorns was the most pronounced and developed 1 st. B. Bilcik & L.J. Keeling (1999) find most feather pecks were on the tail, followed by the rump and back. Hierden et al. (2004a) find most feather pecks were on the tail and belly. In our experiment, our experiment find that when feather pecking occurred, the main parts of body pecked by laying hens was followed by tail (44.8%), back (23.6%), head (16%), neck (6.4%). Tail plumage is the eastest to be pecked, which is due to feedstuff is provided adequately, feeding time is short, the space of movement is small. Thus laying hens mainly grovel, stand and feed.thus tail plumage is very easy to expose, it make pecking chick to peck tail plumage easily. Back, head and neck plumage is the easier to be pecked, which is due to laying hens mainly grovel in cage system. Thus back, head and neck plumage is very easy to expose, it make pecking chick to peck back and head plumage easily.

As to the behaviour of the pecked laying hens,when feather pecking occurrences,the most behaviour of the pecked chicks are standing (42%), feeding (29.9%), groveling (19.3%). It is because of when the laying hens are, standing, feeding and groveling. They are easy to be pecked. Also their body is exposed to other chicks.

5. Acknowledgments

This work was funded by Natural Science Foundation of China (31302008) and Natural Science Foundation of Heilongjiang province (QC2012C104).

References

1. Allen, J., and G. C. Perry, 1975. Feather pecking and cannibalism in caged layer flocks. Br. Poult. Sci. 16:441–451.
2. Ambrosen, T, Petersen, VE 1997: The influence of protein level in diet on cannibalism and quality of plumage of layers. Poult Sci 76: 559–563.
3. Appleby, M.C. 1998. Modification of laying hen cages to improve behaviour. Poultry Science 77: 1828–1832.
4. Bilcik, B, Keeling, LJ 1999: Changes in feather condition in relation to feather pecking and aggressivebehaviour in laying hens. Br Poult Sci 40: 444–451.

5. Blokhuis, HJ 1989: The development and causation of feather pecking in the domestic fowl. PhD. Thesis, Agricultural University, Wageningen 109 p.
6. Green, LE, Lewis, K, Kimpton, A, Nicol, CJ 2000: Cross-sectional study of the prevalence of feather pecking in laying hens in alternative systems and its associations with management and disease. Vet Rec 147:233–238.
7. Gunnarsson, S., Odenk., Algers. B., Svedberg, J. & Keeling, L. (1995) Poultry health and behaviour in a tiered system for loose housed layers. Report 35, Swedish University of Agricultural Sciences, Department of Animal Hygiene, Skara, Sweden.
8. Hughes, B.O., Black, A.J., 1974. The effect of environmental factors on activity, selected behaviour patterns and 'fear' of fowls in cages and pens. British Poultry Science, 15, 375–380.
9. Hughes, BO, Duncan, IJH 1972: The influence of strain and environmental factors upon feather pecking and cannibalism in fowls. Br Poult Sci 13: 525–547.
10. Keeling, LJ 1994: Feather pecking – Who in group does it, how often and under what circumstances? Proceedings 9th European Poultry Conference, Glasgow, pp. 288–289.
11. Kjaer, JB, Vestergaard, KS 1999: Development of feather pecking in relation to light intensity. Appl Anim Behav Sci 62: 243–254.
12. Kjaer, JB, Sorensen, P, SU, G 2001: Divergent selection on feather pecking behaviour in laying hens (Gallusgallus domesticus). Appl Anim Behav Sci 71: 229–239.
13. Koelkbeck, KW, Amoss, MS Jr, Cain, JR 1987: Production, physiological, and behavioral responses of laying hens in different management environments. Poult Sci 66: 397–407.
14. Mcadie, T.M., Keeling, LJ., 2000. Effect of manipulating feathers of laying hens on the incidence of feather pecking and cannibalism. Appl. Anim. Behav. Sci. 68:215–229.
15. Nicol, C.J., Gregory, N.J. Knowles, T.G., et al, 1999: Differential effects of increased stocking density, mediated by increased flock size, on feather pecking and aggression in laying hens. Appl. Anim. Behav. Sci. 65: 137–152.
16. Nørgaard-Nielsen, G., J. Kjær, and H. B. Simonsen, 1993. Field test of two alternative egg production systems the Hans Kier System and the BOLEG II Aviary. Research report no. 9, National Institute of Animal Science, Research Centre Foulum, pp. 1–89 (Danish with English summary).
17. Savory, CJ 1995: Feather pecking and cannibalism. World Poultry Sci J 51: 215–219.

18. Savory, CJ, Mann, JS 1997: Behavioural development in groups of pen housed pullets in relation to genetic strain, age and food form. Br Poult Sci 38: 38–47.
19. Savory, CJ, Mann, JS, Macleod, MG 1999: Incidence of pecking damage in growing bantams in relation to food form, group size, stocking density, dietary tryptophan concentration and dietary protein source. Br Poult Sci 40: 579–584.
20. Simonsen, HB, Vestergaard, K, Willeberg 1980: Effect of floor type and density on the integument of egg-layers. Poult Sci 59: 2202–2206.
21. Sokal, R. Lo R. & Rohlf, F. J. 1981. Biometry. 2nd edn. New York: W. H. Freeman.
22. Tauson, R, Svensson, S.A., 1980. Influence of plumage condition on hen feed requirement. Swedish J. Agric. Res. 10, 35–39.
23. Tauson, R., T. Ambrosen, and K. Elwinger, 1984. Evaluation of procedures for scoring the integument of laying hens—Independent scoring of plumage condition. Acta Agric. Scand. 34:400–408.
24. Tullett, S.G., Macleod, M.G. & Jewitt, T.R. (1980) The effects of partial defeathering on energy metabolism in the laying fowl. British Poultry Science, 21: 241–245.
25. Vestergaard, K. S., 1994. Dustbathing and its relation to feather pecking in the fowl: Motivation and development aspects. Dissertation. The Royal Veterinary and Agricultural University, Copenhagen, Denmark.
26. Wood-Gush, DGM, Rowland, CG 1973: Allopreening in domestic fowl. Rev Comp Animal 7: 83–91.

Recent progress in the preparation of aniline copolymer and its properties

X.W. Cai*, L. Lu, H.H. Yan and Y.Y. Zhao
School of Environmental and Chemical Engineering, Jiangsu University of Science and Technology, Zhenjiang, Jiangsu 212003, China
**E-mail: xin883@163.com*
www.just.edu.cn

Polyaniline (PANI) has many advantages, it is of particular interest due to its high conductivity, unique dopability, good redox reversibility, reversibly switchable electrochromic activity and good environmental stability. However, PANI has very low electrochemical activity at pH > 4.0. Aniline copolymers synthesized via the copolymerization of aniline and aniline derivatives with pH functional groups, such as $-SO_3H$, –OH, and –COOH exhibit a good redox activity in a wide pH range. In this paper, it summarizes the use of electrochemical methods in polyaniline to introduce pH functional groups.

Keywords: Aniline copolymer; pH functional groups; Electrochemical copolymerization.

1. Introduction

1.1. *The introduction of polyaniline*

Polyaniline (PANI), also known as aniline black [1–3], it exists in various forms based on its oxidation level: the fully oxidized pernigraniline base, half-oxidized emeraldine base and fully reduced leucoemeraldine base [1,2](Figure 1). Of these, PANI emaraldine, is the most stable and conductive [1,2]. PANI has many advantages, such as ease of synthesis, low cost, good environmental stability, and the ability to be electrically switched between its conductive and resistive states [4–8]. Unfortunately, its use in biological applications is limited by its low processibility, lack of flexibility and non-biodegrad ability, and has been noted to cause chronic inflammation once implanted [5,9,10]. PANI has been investigated for biosensors, neural probes, controlled drug delivery and tissue engineering applications [11,12].

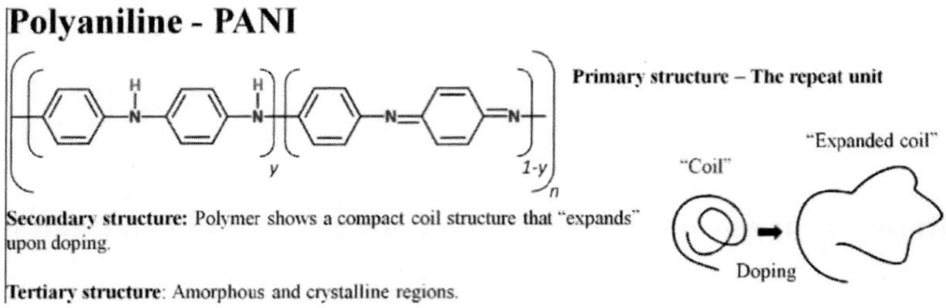

Figure 1. The structure of polyaniline [13,14–17].

At the same time, PANI is unique in the family of conjugated polymers [18,19]. It is of particular interest due to its high conductivity, unique dopability, good redox reversibility, reversibly switchable electrochromic activity and good environmental stability. These properties are favorable to its potential applications in rechargeable batteries [20,21], electrochemical sensors [22–24], biosensors [25], electrochromic devices [26–28], and electrochemical capacitors [29]. Most of these applications are based on its electrochemical activity. However, PANI has very low electrochemical activity at pH > 4.0 [30]. Therefore, it is necessary to improve its activity at high pH values through structural modifications.

Significant progress in the modifications of PANI has already been achieved in enhancing its electrochemical property at high pH values in the past two decades [19]. Basically, there are three commonly used modification approaches [19,31]: homopolymerization of aniline derivatives, modification of PANI by post-polymerization reactions, and copolymerization. Homopolymerization of aniline derivatives provides a simple route to prepare an aniline-based polymer with additional functional groups. However, the synthesis of aniline derivatives often involves a complex multistep synthetic route with a low overall production yield. In addition, this route sometimes fails due to possible side-reactions and interferences of the additional functional groups [32]. Postpolymerization reactions are often limited by the chain stiffness and interchain interactions. Consequently, the resulting polymers may become cross-linked, branched, or electrostatically crosslinked [31,33]. Copolymerization is an effective way to prepare a tailor-made polymer with specially designed physical and chemicalproperties. In recent years, the copolymerization of aniline withother monomer(s) has received an increasing attention [34–36], because it offers a possibility to make new aniline-based polymers that not only retain the good properties of PANI itself but alsopossess new desirable properties.

1.2. *The knowledge of aniline copolymer*

Aniline copolymers synthesized via the copolymerization of aniline and aniline derivatives with pH functional groups, such as –SO_3H [37–39], –OH [19,41,42], and –COOH [38,40] exhibit a good redox activity in a wide pH range. Even though aniline copolymers with one or two pH functional groups still retain the redox activity at higher pH values, their electroactivity decays with increasing pH quickly. Therefore, there are still full of many challenges and opportunities for improving the pH dependence of the electric properties of aniline copolymers.

1.3. *The most common synthesis methods of aniline copolymer*

The copolymerization of aniline and other monomers can be carried out by either chemical or electrochemical methods. Compared with chemical routes, electrochemical methods can obtain aniline-based polymer films on conductive substrates directly. This omits both the extraction of the resulting polymers from the reaction mixtures and the procedure of film coating [31]. Moreover, the electrochemical approach can quickly diagnose the effect of the applied potential and the comonomer ratio on the copolymerization rate and behavior, which is more favorable for optimizing the copolymerization conditions [37]. Therefore, this method is especially suitable for an initial study of a new copolymerization system.

2. The Kinds of the Aniline Copolymer and Its Properties

2.1. *Doping with –SO_3H in polyaniine*

The sulfonic acid ring-substituted polyaniline (self-doped polyaniline), prepared by the reaction of the emeraldine base form of polyaniline with fuming sulfuric acid (EB-SPAN), has been receiving significant attention. This is due to the fact that its conductivity is independent of pH in the aqueous acid solution of pH less than 7 [43,44]. A functional group of SO_3H that was introduced on the polyaniline chain plays an important role in improving the pH dependence of the conductivity. So it is to be expected that this functional group may affect the pH dependence of the electrochemical activity of polyaniline, since the conductivity of the conducting polymer is related to its electrochemical activity. In fact, the pH dependence of the electrochemical activity for EB-SPAN was also improvedpronouncedly, evidence is that there are a sharp redox peaks on the cyclic voltammogram of EB-SPAN in the KCl solution, when the scan rate was set at $50mVs^{-1}$ [45]. This means that EB-SPAN has a good electrochemical activity in the neutral solution, which is better than that of the parent polyaniline. After the first report on the self-doped polyaniline [43,44], several papers with respect to the synthesis, properties and application of the self-doped polyaniline have been reported. Karyakin et al. reported the self-doped polyaniline obtained from the electrochemical copolymerization of aniline with three substituted anilines, i.e. *m*-aminobenzoic acid, anthranilic acid and *m*-aminobenzenesulfonic acid [38,46]. Their cyclic voltammograms show that the copolymers can retain the redox activity in the buffer solution of pH 9 at the scan rate of $25 mVs^{-1}$. It is clear that the pH dependence of the electrochemical activity of the self-doped polyaniline was improved markedly, compared with the parent polyaniline that has a little electrochemical activity at pH >4 [47,48]. Fan et al. reportedthe self-doped polyaniline synthesized via the chemical copolymerization of aniline with *o*-aminobenzenesulfonic acid, in which the conductivity of the copolymer is slightly dependent on the pH in the range of 0–3 [49]. Ohno et al. reported the self-doped sulfonated polyaniline that was preparedby the copolymerization of *p*-aminodiphenylamine and *o*-amino benzene sulfonic acid [50]. The conductivity of this copolymer is independent of pH in the range of 1–7, but its conductivity is only about 4×10^{-5} S cm^{-1}. Lee et al. reported the electrochemical copolymerization of aniline with metanilic acid or orthanilic acid (also known as *meta*- and *ortho*-aminobenzenesulfonic acid) in the solution of $HClO_4$, the copolymer prepared in this way has 40% sulfonation of the aniline ring [51]. Wen et al. reported the in situ UV–vis spectroelectrochemical studies on the copolymerization of diamino benzene sulfonic acid with aniline [52]. Recently, Rahmanifar et al. reported the secondary Zn-polyaniline battery using the self-doped polyaniline as a cathodic material, in which the self-doped polyaniline was synthesized via the electrochemical copolymerization of aniline with *m*aminobenzoic acid [53]. The result indicates that the specific energy of the self-doped polyaniline is $172.8 Whkg^{-1}$, which is higher than that of the parent polyaniline. This is because the self-doped polyaniline has a good electrochemical activity in the wide pH range [38,46]. Lukachova et al. found that the external doping of polyaniline with camphorsulfonic acid extends the electroactivity in neutral and alkaline media [54]. However, a key factor for the electroactive properties and potentialapplications is the degree of sulfonation of polyaniline. As the degree of sulfonation increases, the chemical and physical properties of the self-doped polyaniline were improved significantly. For example, the highly sulfonated polyaniline (LEB-SPAN), prepared using the most reduced form of polyaniline (leucoemeraldine base) as a starting material, has a good pH dependence of the electrochemical activity, and its conductivity (~1 S cm^{-1}) was unaffected by pH over the range $0\leq pH\leq14$ [55,56]. The physical properties of a fully sulfonated polyaniline (N-SPAN, S/N ratio~1.0) were reported [57].

The self-doped polyaniline, prepared by the reaction of the emeraldine base form or the most reduced form of polyaniline with fuming sulfuric acid [43,56] is more difficult to fabricate an electrode for the electrochemical studies. Whereas, the rate of the electrochemical copolymerization of aniline with *m*-aminobenzenesulfonic acid or other monomers is much lower than that of the electrochemical polymerization of aniline. Considering the fact that polyaniline (PANI) can directly react with the concentrated sulfuric acid to form SPAN, and the color of the SPAN film can quickly change with applied potential in 0.3M Na_2SO_4 solution with pH <7 [58]. This electrochromic behavior is much better than that of the parent polyaniline.

Polyaniline reacts with sulfuric acid to form sulfonic acid ring-substituted polyaniline (SPAN), its general form is as follows [43,44]:

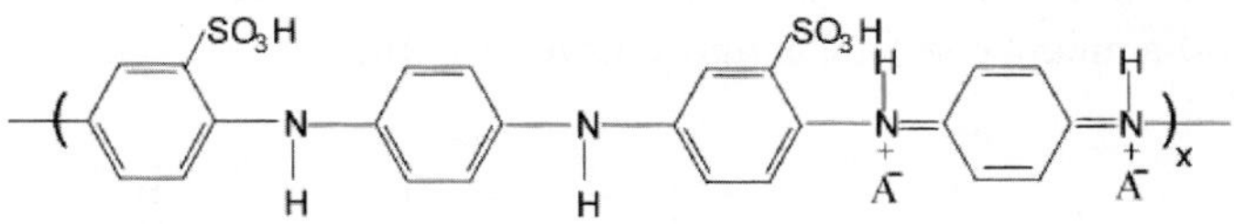

Figure 2. The structure of SPAN [43,44].

2.2. *The introduction of the hydroxyl*

In order to obtain a desirable copolymer whose property is even better than that of the parent polyaniline, the electrochemical copolymerization of aniline and *o*-aminophenol was carried out in the aqueous acid solution in our laboratory. This is because the structure of *o*-aminophenol is similar to that of aniline, and *o*-aminophenol is soluble in aqueous acid solutions and has a functional group –OH.

Hydroxyl group in phenyl ring can be oxidized to quinine and quinone can be reduced. And the hydroxyl group is an electron-donating group. Its electronic effect is different from those of self-doped polyanilines as mentioned above, in which the substituted groups in aniline derivatives are the electron-withdrawing groups, such as $-SO_3H$, –COOH and –CN. The electrochemical copolymerization of aniline and *p*-aminophenol was carried out in the organic electrolyte [59]. This copolymer has potentiometric sensor function forphenol in aqueous solution (phosphate buffer, pH 7) with a detection limit of 2×10^{-5}M. However, its copolymerization behavior is different from that of the electrochemical copolymerization of aniline and *o*-aminophenol.

Mu [60] made the following experiment. Poly (aniline-co-*o*-aminophenol) has been synthesized electrochemically in the solution consisting of 0.2M aniline, 0.6M H_2SO_4 and various concentrations of *o*-aminophenol. The copolymerization rate and the properties of the copolymer are strongly affected by the monomer concentration ratio. The cyclic voltammograms of the copolymer in 0.3M Na_2SO_4 solution of pH value from 5.0 to 9.6 were performed between −0.20 and 0.50V at the scan rate of 60 mV/s. The results indicate that only a 29% decay of the electrochemical activity was observed when the copolymer was moved from pH 5.0 to 9.6. The electrochemical activity of poly (aniline-co-*o*-aminophenol) is about four times as high as that of polyaniline in 0.3M Na_2SO_4solution of pH 5.0. A surprising result shows that two pairs of the redox peaks still appear on the cyclic voltammogram of the copolymer in the solution of pH 9.6 at the scan rate of 60mV/s. In this case, its usable potential range is −0.20 to 0.62V. The copolymer has a good stability and a high reversibility in the above solutions, and has a conductivity of 1.40 S/cm.

The compositions of the reduced and oxidized states of poly (aniline-co-*o*-aminophenol) are assumed as follows [60]:

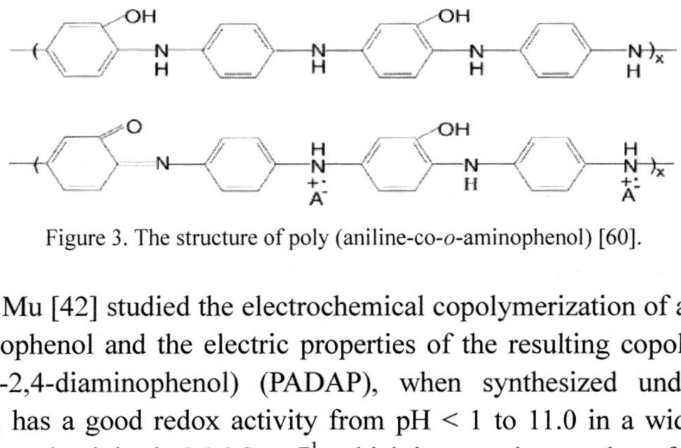

Figure 3. The structure of poly (aniline-co-*o*-aminophenol) [60].

Later, Mu [42] studied the electrochemical copolymerization of aniline with 2,4-diaminophenol and the electric properties of the resulting copolymer. Poly (aniline-*co*-2,4-diaminophenol) (PADAP), when synthesized under optimal conditions, has a good redox activity from pH < 1 to 11.0 in a wide potential region. Its conductivity is 0.26 S cm^{-1}, which is very close to that of polyaniline synthesized under the same conditions in the absence of 2,4-diaminophenol. The pH dependence of PADAP conductivity was found to be better than that of polyaniline.

3. Conclusion

A better route to obtain self-doped PANI derivatives, both soluble and insoluble, presents the electrochemical copolymerization of aniline with its sulfonate- or carboxylate group-pendant comonomers. While homopolymers, derived from sulfonate-containing monomers, possess an excess of negatively charged sulfonate groups, the composition of copolymers can be varied in a broad range by a simple change of comonomer ratio in a feed solution, and electropolymerization conditions as well. By electrochemical polymerization, enhanced the electrochemical activity of aniline copolymers in a wide pH range, which is desirable for some demanding applications such as rechargeable batteries and electrocatalysis used in highly basic environments.

References

1. Ghasemi-Mobarakeh L et al. J Tissue Eng Regen Med 5(2011) 17.
2. Zhou DD, Cui XT, Hines A, Greenberg RJ. Implantable neural prostheses, 2(2010) 217.
3. Blinova NV, Stejskal J, Trchova M, Prokes J. Polym Int. 57(2008) 66–9.
4. Cullen DK, Patel AR, Doorish JF, Smith DH, Pfister BJ. J Neural Eng 5(2008) 374.
5. Borriello A, Guarino V, Schiavo L, Alvarez-Perez MA, Ambrosio L. J Mater Sci: Mater Med 22(2011) 1053.
6. Guo Y et al. 8(2007) 3025.
7. Prabhakaran MP, Ghasemi-Mobarakeh L, Jin G, Ramakrishna S. J Biosci Bioeng 112(2011) 501.
8. Yu QZ, Shi MM, Deng M, Wang M, Chen HZ. Mater Sci Eng BSolid 150 (2008) 70.
9. Huang L et al. Biomacromolecules 9 (2008) 850.
10. Zhang QS, Yan YH, Li SP, Feng T. Biomed Mater 4 (2009) 1.
11. Liu X, Gilmore KJ, Moulton SE, Wallace GG. J Neural Eng 2009;6:1–10.
12. Humpolicek P, Kasparkova V, Saha P, Stejskal J. Synth Met 162(2012) 722.
13. Epstein AJ. Rupprecht L, editor. Norwich, NY: William Andrew Publishing/Plastics Design Library; (1999). 1.
14. MacDiarmid AG, Epstein AJ. Synth Met 65(1995) 85.
15. MacDiarmid AG, Chiang JC, Richter AF. Synth Met 18(1987) 285.
16. Pouget JP, Jdzefowiczt ME, Epstein AJ, Tang X, MacDiarmid AG. Macromolecules 24(1991)779.
17. Tourillon G, Garnier F. Mol Cryst Liq Cryst 118(1985) 221.
18. L.F. Ferreira, L.M. Souza, D.L. Franco, A.C.H. Castro, A.A. Oliveira, J.F.C. Boodts, A.G. Brito-Madurro, J.M. Madurro. Materials Chemistry and Physics 129 (2011) 46.
19. C.X. Chen, C. Sun, Y.H. Gao. Electrochimica Acta 53 (2008) 3021.
20. K. Ghanbari, M.F. Mousavi, M. Sharnsipur, H. Karami. Journal of Power Sources 170 (2007) 513.

21. L. Liu, F.H. Tian, M. Zhou, H.P. Guo, X.Y. Wang. Electrochimica Acta 70 (2012) 360.
22. C.X. Chen, C. Sun, Y.H. Gao. Electrochemistry Communications 11 (2009) 450.
23. C.X. Chen, C. Sun, Y.H. Gao. Electrochimica Acta 54 (2009) 2575.
24. P. Kunzo, P. Lobotka, M. Micusik, E. Kovacova. Sensors and Actuators B: Chemical 171 (2012) 838.
25. M.U.A. Prathap, A.K. Chaurasia, S.N. Sawant, S.K. Apte. Analytical Chemistry 84 (2012) 6672.
26. V.K. Thakur, G.Q. Ding, J. Ma, P.S. Lee, X.H. Lu. Advanced Materials 24 (2012) 4071.
27. J.L. Lu, W.S. Liu,H. Ling, J.H. Kong, G.Q. Ding, D. Zhou, X.H. Lu. RSC Advances 2 (2012) 10537.
28. J. Wei, S.X. Xiong, Y. Bai, P.T. Jia, J. Ma, X.H. Lu. Solar Energy Materials and Solar Cells 99 (2012) 141.
29. X.W. Li, H. Zhang, G.C. Wang, Z.H. Jiang. Journal of Materials Chemistry 20 (2010) 10598.
30. D. Shan, S.L. Mu .Synthetic Metals 126 (2002) 225.
31. T.M. Nguyen, S. Cho, N. Varongchayakul, D. Yoon, J. Seog, K. Zong, S.B. Lee. Chemical Communications 48 (2012) 2725.
32. H. Brisset, A.E. Navarro, C. Moustrou, I.F. Perepichka, J. Roncali. Electrochemistry Communications 6 (2004) 249.
33. Chapter 1 Introduction, in: M.S. Freund, B.A. Deore (Eds.). (2007), p. 23.
34. Y.F. Yang, S.L. Mu. Journal of Physical Chemistry C 115 (2011) 18721.
35. Y.F. Yang, S.L. Mu. Electrochimica Acta 55 (2010) 4706.
36. Y. Kong, X.Y. Li, J.H. Ni, C. Yao, Z.D. Chen. Electrochemistry Communications 14 (2012) 17.
37. Y.F. Yang, S.L. Mu. Electrochimica Acta 54 (2008) 506.
38. A.A. Karyakin, A.K. Strakhova, A.K. Yatsimirsky. Journal of Electroanalytical Chemistry 371 (1994) 259.
39. S.L. Mu. Journal of Physical Chemistry B 112 (2008) 6344.
40. S.L. Mu. Synthetic Metals 161 (2011) 1306.
41. A.A. Shah, R. Holze. Electrochimica Acta 52 (2006) 1374.
42. S.L. Mu, Y. Zhang, J.P. Zhai. Electrochimica Acta 54 (2009) 3923.
43. J. Yue, A.J. Epstein, J. Am. Chem. Soc. 112 (1990) 2800.
44. J. Yue, Z.H. Wang, K.R. Cromack, A.J. Epstein, A.G. MacDiarmid, J. Am. Chem. Soc. 113 (1991) 2665.
45. J. Yue, A.J. Epstein, J. Chem. Soc. Chem. Commun. 21 (1992) 1540.

46. A.A. Karyakin, I.A. Maltsev, L.V. Lukachova, J. Electroanal. Chem. 402 (1996) 217.
47. A.F. Diaz, J.A. Logan, J. Electroanal. Chem. 111 (1980) 111.
48. W.S. Huang, B.D. Humphrey, A.G. MacDiarmid, J. Chem. Soc. Faraday Trans. 1 (82) (1986) 2385.
49. J. Fan, M. Wan, D. Zhu, J. Polym. Sci. Part A: Polym. Chem. 36 (1998) 3013.
50. N. Ohno, H.J. Wang, N. Toshima, Polym. J. 33 (2001) 165.
51. J.Y. Lee, C.Q. Cui, X.H. Su, M.S. Zhou, J. Electroanal. Chem. 360 (1993) 177.
52. T.C. Wen, L.M. Huang, A. Gopalan, Electrochim. Acta 46 (2001) 2463.
53. M.S. Rahmanifar, M.F. Mousavi, M. Shamsipur, J. Power Sources 110 (2002) 229.
54. L.V. Lukachova, E.A. Shkerin, E.A. Puganova, E.E. Karyakina, S.G. Kiseleva, A.V. Orlov, G.P. Karpacheva, A.A. Karyakin, J. Electroanal. Chem. 544 (2003) 59.
55. X. Wei, A.J. Epstein, Synth. Met. 74 (1995) 123.
56. X.L. Wei, Y.Z. Wang, S.M. Long, C. Bobeczko, A.J. Epstein, J. Am. AChem. Soc. 118 (1996) 2545.
57. W. Lee, G. Du, S.M. Long, A.J. Epstein, S. Shimizu, T. Saitoh, M. Uzawa, Synth. Met. 84 (1997) 807.
58. C.M. Li, S.L. Mu, Synth. Met. 144 (2004) 143.
59. I.A. Vinokurov, S.Ya. Khaikin, V.V. Bertsev, T.M. Karkhu, Élektrokhimiya 28 (1992) 181.
60. S.L. Mu. Synth, Met. 143 (2004) 259.

The progress of the functionalized ionic liquids in the application of Knoevenagel condensation reaction

X.W. Cai*, X. Zhang, H.H. Yan and Y.Y. Zhao
School of Environmental and Chemical Engineering, Jiangsu University of Science and Technology, Zhenjiang, Jiangsu 212003, China
**E-mail: xin883@163.com*
www.just.edu.cn

Knoevenagel condensation reaction is an important method for synthesis of carbon carbon double bond. It can directly synthesize a large number of useful compounds in industry, agriculture, pharmaceutical, biological sciences, Pharmaceuticals. The application of functional Ionic liquids provides a high efficient and environmental friendly approach for Knoevenagel condensation reaction. Ionic liquids as catalysts and solvents can promote reaction of a wide range of aldehydes and ketones with methylene compounds. In this paper. the progress in the use of functional ionic liquids for Knoevenagel reactions is reviewed.

Keywords: Knoevenagel condensation reaction; Aldehyde; Ketone; Methylene compounds.

1. Introduction

Knoevenagel condensation which is dehydration condensation of reactioncarbonyl compounds with active methylene compounds, is an important method for synthesis of carbon carbon double bond.[1] It can directly synthesize a large number of useful compounds in industry, agriculture, pharmaceutical, biological sciences, and has been widely used in many other fields. Depending on the type of alkali, Knoevenagel condensation can have two kinds of feasible mechanism.[2] One is to think aldehydes or ketones and amine firstly condense imine formation, and then with carbon anion bonus. Another kind of mechanism is that amine and aldehyde under the effect of alkali aldol become intermediate β-hydroxy dicarbonyl compounds firstly, and then eliminate water to gain condensation products. Knoevenagel condensation usually complete under the action of a catalyst. The traditional Knoevenagel condensation catalyst is mainly ammonia and organic amine, halide, metal oxide and its salt, phosphate, silica gel, alkaline zeolite molecular sieves, clay minerals, organic acids and cation exchange resin catalyst.[3] But in the reaction using a large number of organic solvents used in the reaction has serious harm to the environment, organic chemical reaction proceed under solvent-free conditions has become a research direction of green chemistry.[4]

In recent years, ionic liquids obtained extensive research and application in the field of chemical synthesis. Ionic liquid generally consists of organic cation and inorganic anions.[5] Ionic liquids are meet a series of features of the requirements of the green chemistry such as thermal stability, low volatility, no flammability, good solubility for a variety of organic and inorganic compounds. Therefore, ionic liquids are considered friendly "green" solvent to environmental. For The Past Few Years, ionic liquids as new green solvents replacing volatile solvent induce widely attention, the research about the ionic liquid composition, properties, synthesis and application is booming.

2. Single Site of Functionalized Ionic Liquids

At an early age, someone reported a series of imidazoles ionic liquids continuously, such as in 2004 Xu synthesized a functionalized ionic liquids 1-methyl-3-hydroxyethylimidazole chloride (Figure 1), which can successful catalyze Knoevenagel condensation between aromatic aldehyde with active methylene compounds under solvent-free conditions, product yield is between 82%–97%.[6]

Funcitionalized ionic liquid:

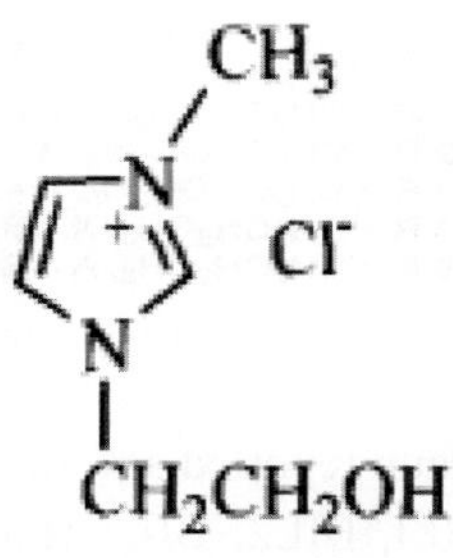

Figure 1. Chemical structure of 1-methyl-3-hydroxyethylimidazole chloride.

In 2006, Cai prepared L-(2-aminoethyl)-3-methylimidazole hexafluorophosphate. The functional ionic liquid catalytic efficiently various Knoevenagel condensation with water as reaction solvent, and the catalyst can be recycled.[7] Except for imidazoles ionic liquid, researchers began to focus on guanidine kind of ionic liquid, quinoline ionic liquids and other ionic liquids. In 2007, Xin synthesized a series of guanidine kind of ionic liquid (Figure 2), All can effectively catalyzed Knoevenagel condensation between aromatic aldehyde with active methylene compounds, short reaction time, obtain excellent product yield, catalyst can be recycled.[8]

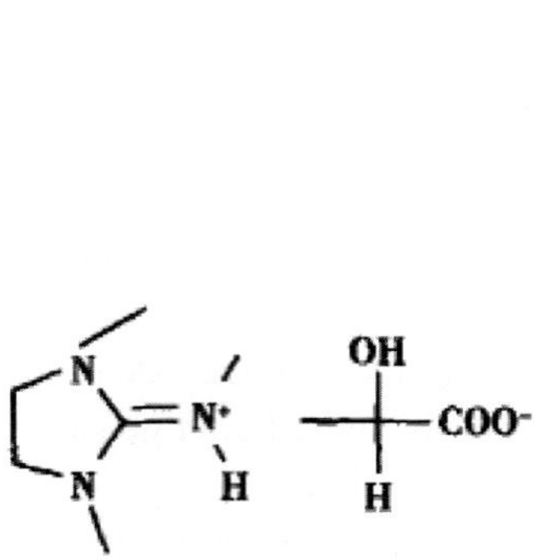

Figure 2. Chemical structure of guanidium ionic liquid.

In 2011, Wu and Li try to sever available quinoline as raw material, under the mild conditions, by quaternary ammonium reaction and anion exchange two-step synthetic quinoline ionic compounds, this series of quinoline ionic compounds used as catalyst for the Knoevenagel condensation reaction (Scheme 1).[9] And under solvent-free conditions with room temperature, the reaction of aromatic aldehyde with malononitrile proceed smoothly, after 30 min, gaining product yield of 80%–95%. Short reaction time, the product purity, simple post-processing reaction, ionic compound synthetic raw material price is low, mild conditions, easy to realize industrialized production.

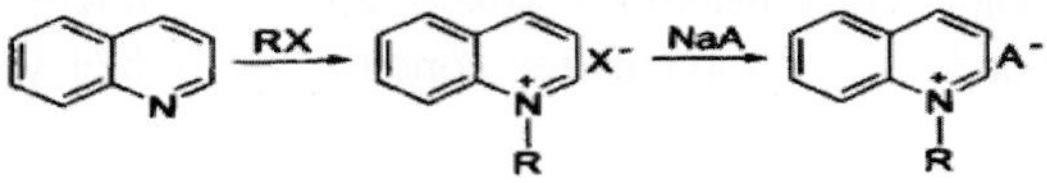

1, 4, 7   2, 3, 5, 6, 8, 9

1 R = $CH_2COOC_2H_5$, X = Cl
2 R = $CH_2COOC_2H_5$, A = BF_4
3 R = $CH_2COOC_2H_5$, A = PF_6
4 R = $CH_2CH_2CH_3$, X = Br
5 R = $CH_2CH_2CH_3$, A = BF_4
6 R = $CH_2CH_2CH_3$, A = PF_6
7 R = $CH_2CH_2CH_2CH_3$, X = Br
8 R = $CH_2CH_2CH_2CH_3$, A = BF_4
9 R = $CH_2CH_2CH_2CH_3$, A = PF_6

Scheme 1. Syntheses of ionic liquids.

In 2008, Paun reported several reusable alkaline functionalized ionic liquids, such as the BILl, BIL2, BIL3 (Figure 3). The condensation reaction between heptaldehyde with ethyl cyanoacetate using these functional ionic liquid as catalyst with room temperature for 1h, could obtain desirable pr oduct of 50%–91% yield.[10]

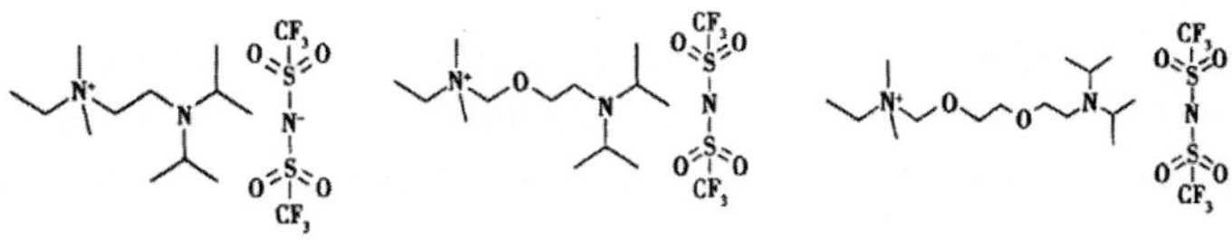

Figure 3. Chemical structure of basic ionic liquids BILl, BIL2, BIL3.

In 2009, Xu studied DABCO-base ionic liquid catalyzed Knoevenagel condensation reaction.[11] It turn out under the condition of room temperature and water as reaction solvents, [C4DABCO]+[BF4]-(R = butyl) ionic liquid can effectively catalyzed series of aromatic aldehyde, aldehyde, ketone and Fatty ketone compounds with malononitrile condensation reaction, the catalyst can be repeatedly used. Then, in 2013, Wang,[12] through the reaction of triethylene diamine (DABCO) and 3-chloro-1,2diol, acquired the intermediate, which was exchanged ion with sodium fluoborate to achieve a new ionic liquid [DABCO][BF4](Scheme 2).

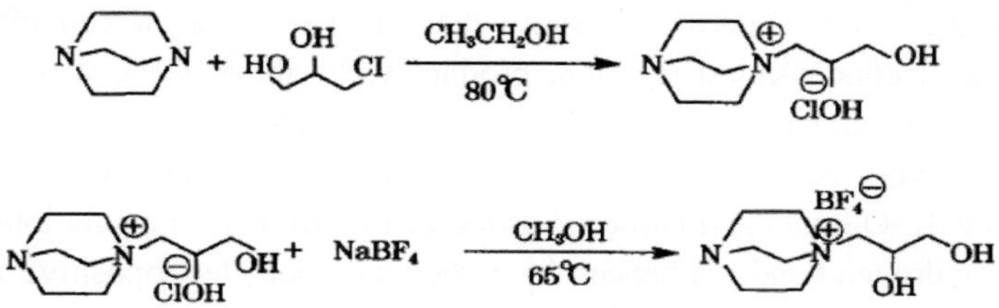

Scheme 2. Syntheses of ionic liquids.

Experiments indicated that under the condition of room temperature, the knoevenagel condensation between different substituent group of aromatic aldehyde with active methylene compounds like malononitrile and ethyl cyanoacetate can proceed in the ionic liquid [DABCO][BF4], under the condition of room temperature.reaction speed, high yield, simple post-processing. The active methylene compounds for malononitrile faster response speed, this is because the cyanoacetic acid ethyl ester of space steric hindrance is larger, and malononitrile is easy to form carbon anion, favors the nucleophilic reaction occurred.

In 2014, Wang[14] prepared the ionic liquid [DABCO-PDO][PF6] through the two-step using triethylene diamine (DABCO) as raw material. Experiments show that this new type of catalytic functional ionic liquid [DABCO-PDO][PF6], used to aromatic aldehydes with various active methylene compounds Knoevenagel condensation reaction, the active methylene compounds into the malononitrile, cyano acetic ester, benzothiazole acetonitrile, phenylacetonitrile, has obtained the good catalytic effect, at the same time, the reaction system without any solvent, reaction at room temperature, only 15% (the amount of material fraction) of catalyst, high yield, simple post-processing reaction, compared with other substrate catalytic system applicability is stronger, is a kind of green feasible solution, has the very good industrial application prospect.

In 2014, Kiumars apply EDAHS-ethane-1,2-diammonium hydrogensulfate as a new Bronsted acidic ionic liquid in aqueous micellar media catalyze Knoevenagel condensation.[15] The reactions were completed at room temperature with excellent yields of desired products in aqueous micellar media.

The advantages of the present reaction are the elimination of the metals and organic solvents, inexpensive reagents, operational simplicity and excellent yields of products.

In 2015, Ban stir triethanolamine, chlorosulfonic acid in methylene for 3h, then add acid (trifluoroacetic acid, nitric acid or methyl sulfonic acid) to exchange ion to gain a new type of acidic ionic liquid [TEOA][CF3COO-] (Scheme 3),[17] the experiments show that the new catalyst can effectively Knoevenagel condensation reaction, product yield above 90%, wide catalytic range, high repeatability. Researchers have also made a study on its catalytic mechanism, verified and put forward the possible reaction mechanism, as shown in Scheme 4. -OSO3H and carbonyl group can be formed in acidic ionic liquid multiple hydrogen bonds, enhances the carbonyl carbon electropositive, makes it vulnerable to attack of the active methylene carbon atoms, accelerate the formation of intermediates A. To further promote the acidic ionic liquid catalyst intermediate B take off A molecule of water to generate the target product. Due to the effect of the huge catalyst steric hindrance of and intermediate B occurred more easily trans dehydration, and acquire E type reaction products, consistent with the reaction results.

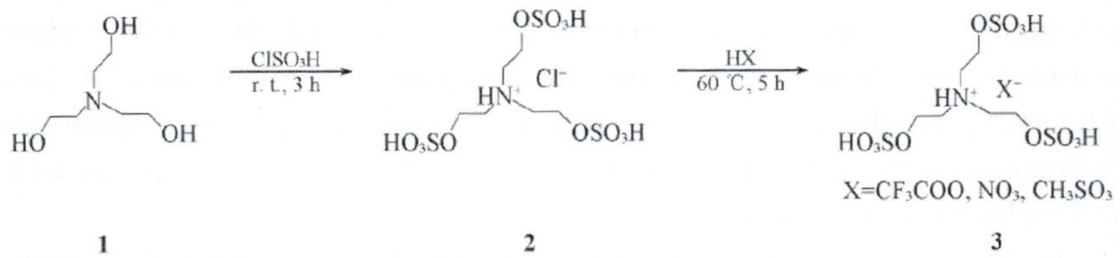

Scheme 3. Syntheses of ionic liquids.

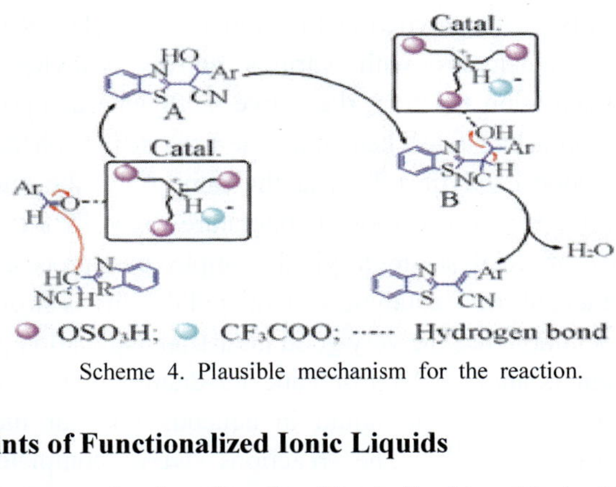

Scheme 4. Plausible mechanism for the reaction.

3. Dual Points of Functionalized Ionic Liquids

In 2008, Wang created a functionalized ionic liquids with double catalytic sites, hydroxide 1-methyl-3-(2-(N,N-diethyl)amino ethylimidazole salt. It contains Lewis alkali cationic and Bronsted base anion.[21] Whereafter, Liu also reported double dot catalytic ionic liquid l-(2-piperidyl-ethyl)-3-methylimidazole morpholine ethyl sulfonate {[Pemlm] Mes}, the anion containing alkaline morpholine, cationic alkaline piperidine group. The two double dot ionic liquids can efficiently catalyze aldehydes or ketones with ethyl cyanoacetate condensation reaction.

In 2011, Shaterian prepared another double dot catalytic ionic liquid [H3N+-CH2-CH2-OH][CH3COO-], which can catalyze aromatic aldehydes and malononitrile or double ketone condensation reaction at room temperature without solvent, and obtain the desirable product of more than 60% yield after a couple of minutes.[22]

In 2012, Dou Reported a double dot catalytic ionic liquid[2-aemim]im catalyze Knoevenagel reaction (Figure 4). The experimental results demonstate that under solvent-free conditions, the ionic liquid has the very high catalytic activity for the Knoevenagel reaction between a series of aromatic aldehydes with the active methylene compounds.[23]

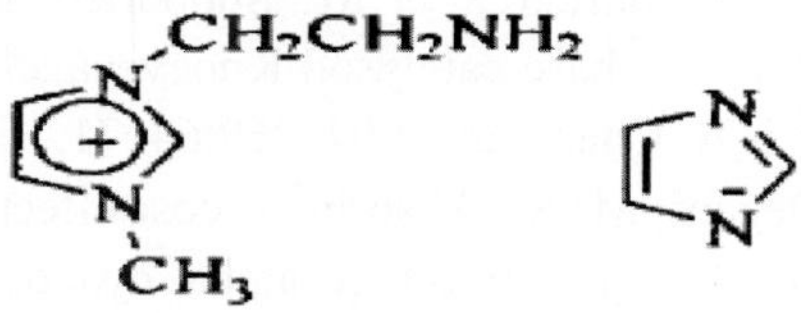

Figure 4. The structure of [2-aemim]im.

In 2012, Xu adopted two steps method to synthesize a series of functional dual-core basic ionic liquids. imidazoles alkaline dual functional ionic liquid has good catalytic performance in the Knoevenagel reaction, likely a double-(3-methyl-l-imidazole) butyl double hydroxide ionic liquids catalyze furfural and propylene cyanide Knoevenagel condensation highly efficiently, the product yield is up 93.2%, and double-(3-methyl-l-imidazole) and hexyl double hydroxide in the study of ionic liquid for Knoevenagel condensation reaction of aldehydes with cyano compounds have good catalytic effect.[24]

In 2013, Ren[25] prepared CaO/HMCM-22 a acid-base bifunctional catalyst which showed obvious acid-base catalysis in the Knoevenagel condensation between benzaldehyde with ethyl cyanoacetate. When CaO load of 5.0%, benzaldehyde conversion rate is as high as 88.9%. It is catalytic more effectively than pure with alkaline earth metal Ca modified MCM-22 in the Knoevenagel condensation, and suggests that the basic sites of the CaO modification and acid HMCM-22 center synergistic effect, promoted the Knoevenagel condensation reaction; At the same time, a small amount of alkaline earth metal modified molecular sieve MCM-22 can get higher catalytic activity.[26] So by the low cost of alkaline earth metal modification of the preparation of new microporous molecular sieve MCM-22 acid-base coordination catalysts, the preparation process is simple, convenient disposition, environment friendly, in line with the development trend of the green catalytic, has a good application prospect.[27–29]

4. Conclusion

Ionic liquids can provide a higher catalytic efficiency in the Knoevenagel condensation reaction, and the performance is better, to avoid the use of additional catalyst. A series of aldehydes and ketones compounds with active methylene compounds of shrinkage reaction can be completed smoothly in ionic liquids. For the Knoevenagel condensation reaction it establish a convenient, economic and environmental protection of green synthetic route.

References

1. Paun C, Barklie J, Goodrich P, et a1. Supported and liquid phase task specific ionic liquids for base catalysod Knoevenagel reactions. Journal of Molecular Catalysis A: Chemical, 2007, 269:64–71.
2. Shanna Y O, Degani M S. Absorbing cost-effective ionic liquid for synthesis of commercially important alpha cyanoacrylic acids: A safe process for activation of cyanoacetic acid. Green Chem, 2009, 11:526–530.
3. Santamarta F, Verdia P, Tojo E. A simple efficient and green procedure for Knoevenagel reaction in [MMIm][MS04] ionic liquid. Catal Commun, 2008, 9:1779–1781.
4. Wang Z, DABCO Ionic liquid catalytic Knoevenagel condensation. Chemical research, 2014 25:353–358.
5. Desai U.V., Pore D.M., Mane R.B., Solabanavar S.B., Wadgaonkar P.P., One Pot Synthesis of Monoalkylated and Mixed, Dialkrlated Meldrum'S Acid derivatives [J]. Synth. Commun., 2004, 34:25–32.
6. Fildes D., Caignaert V., Villemin D., Jaffres P. A., Potassium exchanged zirconium hydrogen phosphate Zr(03POK)2: A heterogeneous basic catalyst for Knoevenagel reaction without solvent [J]. Green Chem. 2001, 3:52–56.
7. Dou H, Luo L. Ionic liquid [2-aemim] im catalytic Knoevenagel and isoxaflutole synthesis. Chemical Research and Application, 2012, 24(3).5–9.
8. Rao E S., Venkataratnam R.V, Indiann Chem. B 1993, 32:484–486.

9. Cui L D, Ma L, Functionalization carbon immobilized ionic liquid catalyst, Zhejiang Chemisical Industry, 43:12–20.
10. Gong K, He Z W, Xu Y, et a1. Green synthesis of 5-benzylidenerhodanine derivatives catalyzed by 1-butyl-3-methyl imidszolium hydroxide in water [J]. Monatsh Chem, 2008, 139:913–915.
11. Hu C G, Chen Z, Functionalization carbon immobilized ionic liquid catalyst, Journal of Xiamen University(Natural Science), 2010, 49:889–892.
12. Cai Y Q, Peng Y Q, Song G H. Amino-functionalized ionic liquid an effitient and recyelable catalyst for Knoevenagel reactions in water [J]. Catalysis Letters, 2006, 109(1–2):61–64.
13. Chen XW, Li XH, Song HB, et a1. Synthesis of a basic imidazelide ionic liquid and Its application in catalyzing knoevenagel condensation [J]. Chin J Catal, 2008, 29:957.
14. Yue C, Mao A, Wei Y, et a1. Knoevenagel condensation reaction catalyzed by task-specific ionic liquid under solvent-free conditions [J]. Catalysis Communications, 2008(9):1571–1574.
15. Shaterian H R Annan M, Rigi F. Domino Knoevenagel condensation, michael addition, and cyclization using ionic liquid, 2-hydroxyethylammonium formate, as a recoverable catalyst. Journal of Molecular Liquids, 2011, 158:145–150.
16. Huang Y B. Research of basic ionic liquid catalytic knoevenagel reaction, Tianjing Chemical Industry, 2010, 24:19–21.
17. Yue C B, Mao A Q, Wei Y Y, et a1. Knoevenagel condensation reaction catalyzed by task-specific ionic liquid under solvent-free conditions [J]. Catal. Commun, 2008, 9(7):1571–1574.
18. Xu X M, Li Y Q, Zhou M Y. Knoevenagel condensation reaction catalyzed by functionalized ionic liquid 1-(2-Hydroxyethyl)-3-methyl Imidazolium Chloride [J]. Chinese Journal of Organic Chemistry, 2004, 24(10):1253–1256(in Chinese).
19. Wang Z, Zeng L K. Ionic liquid [DABCO] [BF4] catalytic Knoevenagel condensation reaction of investigate. Journal of Taizhou University, 2013, 35(6).
20. Chen X W, Song H B. Catalytic Performance of Imidazolide Basic Ionic Liquid in Knoevenagel Reactions in Aqueous Media. Chinese Journal Of Catalysis, 2011, 32(4):693–698.
21. Lai G Q, Peng J J, LU J Y, et a1. Ionic liquid funefionalized silica Gel. Novel catalyst and fixed solvent [J], Tetrahedron Lett, 2006, 47:6951–6953.
22. Liu W, Uu Y. Synergistic effect of an ionic liquid catalyst with two kinds of basic sites on Knoevenagel condensation [J]. Chin J Catal, 2008, 29:771.

23. Balalaie S, Nemati N. One-pot preparation of coumarins by Knoevenagel condensation in solvent-free condition under microwave irradiation [J]. Heterecycl colnfflun, 2001(7):67–72.
24. Rewatkar V B, Joshi J B. Effect of impeller design on liquid phase mixing in mechanically agitated reactors [J]. Chemical Engineering Communication, 1991, 102:1–33.
25. Javed K H, Mahmud T, Zhu J M. Numerical simulation of turbulent batch mixing in a vessel agitared by a Rushton turbine [J]. Chemical Engineering Processing, 2006, 45(2):99–112.
26. Ruszkowski S. A rational method for measuring blending performance and comparison of different impeller types[c]//Proe 8th European Conference on Mixing. Cambridge: Institution of Chemical Engineem, 1994.
27. Tietze LF. Donino reaction in organical synthesis [J]. Chem. Rev. 1996, 96(1):115–136.
28. Li Zhen, Chen Jing, Xia Chungu. Advances in industrial application of ionic liquids [J]. Chemical Industry And Engineering Progress, 2012, 31(10):2113–2122.
29. Zhang X F, Lai E S. M, Martin-Aranda R, et a1. An investigation of Knoevenagel condensation reaction in microreactors using a new ze1ite catalyst [J]. Appl Catal A, 2004, 261(1):109–118.

Quantitative forecast of the inner displaced indexing CAM mechanism impulsive noise*

Y.Q. Li+, C.Y. Li, Y.X. Shang and M.T. Liu
College of Mechanical Engineering, Tianjin University of Science and Technology, Tianjin, China
+E-mail: woshilcylcy@163.com
www.tust.edu.cn

Inner displaced indexing CAM mechanism in operation, due to the number change of meshing tooth needle at the same time, the elastic deformation of needle tooth and tooth caused by load and the manufacture error will cause gear meshing of dynamic meshing force, cause vibration and noise. Within the inner displaced indexing CAM mechanism can be solved by the impulsive noise into two parts: the gear transmission and the CAM & pins meshing. According to the different transmission characteristics of both, respectively quantitative forecast the noise sound pressure and the acoustic radiation power when the mechanism impact. Agency for noise analysis and quantitative calculation, can do noise estimates of institutions in the product design phase, and put forward noise reduction methods. in this paper, deeply discussed the impact of the gear tooth CAM and needle principle, combined the acoustical theory and numerical calculation tool MATLAB to accurately forecast the noise of the institutions, and put forward effective optimization methods.

Keywords: Inner displaced indexing CAM mechanism; Impulsive noise; Noise quantitative forecast; Tip relief.

1. Introduction

Inner displaced indexing CAM mechanism rotate in parallel as the motor drive the input shaft, and via shunt gear the power passed to shunt shaft with eccentric structure, which in the plane drive two CAM phase, have the difference of 180 °, rotate in parallel and engaging with the pin wheel, and drive the output shaft output at a certain motion law, therefore, the vibration and noise of mechanism is mainly caused by gear transmission and the CAM tooth needle transmission.

When the mechanism in operation, due to the number change of meshing tooth needle at the same time, the elastic deformation of needle tooth and tooth caused by load and the manufacture error will cause impact vibration and

* This work is supported by the National Natural Science Foundation of China (50905127) and Tianjin Research Program of Application Foundation and Advanced Technology (15JCZDJC39100).

radiation noise during meshing. Noise has a serious influence to human both on the physiology and psychology, under the strong effect of noise, people's hearing may even permanent loss. Also exposed in the strong noise environment for a long time, people's nervous system, cardiovascular system, digestive system etc. may endure serious damage. In conclusion, controlling noise pollution is extremely urgent[3].

2. Sound Power and Radiation Pressure of Impulsive Noise

The radiation efficiency and sound power of impulsive noise is an important index to show sound radiation ability. Mutual impact between objects can radiate acceleration noise. And impact acceleration noise is related to some parameter such as volume of the impact object, impact speed, impact time and so on[2]. Structural vibration self-ringing noise and radiation efficiency of the object when impact occurs are very different, the impact noise radiation efficiency can be expressed as:

$$\mu_{acc} = \frac{E_{acc}}{\rho_0 (v_s L) v_s^2 / 2} = \frac{(\rho_s \rho_0) E_{acc}}{\rho_s (v_s L) v_s^2 / 2}$$

In the equation: E_{acc} is impact radiation energy, ρ_0 is air density, ρ_s is material density of the impact component, v_s is impact velocity of two components, take the logarithm of it, can have two object's sound power of impact noise radiation:

$$Leq = 10 \lg E_{acc} = 10 \lg \mu_{acc} + 10 \lg[\rho_0 (v_s L) v_s^2 / 2]$$

The impact sound pressure produced by the impact of two objects can be expressed as $P(r,\theta,t) = P_0(r,t)\cos\theta$, $P_0(r,t)$ is sound pressure on the direction of impact, which is a complex function about space distance, direction and time[4-5]. Here, the transmission system between objects shall be seen as acoustic model of two mutual impact variable curvature cylinder. After deduced:

$$\frac{P_{max} r}{\rho_0 a c v_s} = \frac{\cos[(2.3/\delta)^{1/3} - \beta] + 2^{1/2} \times \exp(-(1.5\delta)^{2/3})\cos[(1.5\delta)^{2/3} + \psi]}{2(4\delta^4 + 1)^{1/2}}$$

after simplification: $P_{max}^2 / P_0^2 = \frac{\rho_0^2 c^2}{P_0^2 r^2} 2.56 V^{2/3} v_s^2 [(\frac{c t_0}{V^{1/3}})^4 + 4]^{-1}$

Sound pressure level: $Lp = 10 \lg(P_{max}^2 / P_0^2),\ P_0 = 2 \times 10^{-5}\ N/m^2$

In the formula: V is volume of impact object (active), $V = \pi r_d^2 b$; v_s is impact speed for two objects, r is the distance from a point in space to curvature center of the cylinder.

3. Quantitative Calculation and Structure Improvement for CAM and Tooth Needle Meshing Impulsive Noise

The meshing for needle tooth can be regard as mutual impact between two variable curvature cylinders. During the meshing, the number of tooth, when engaged, is differ, that will produce complex superimposed noise, so calculated the impact noise of 12 section paragraphs respectively, and then superimposed in the interval. As shown in Figure 1, established the coordinate system with the center of the pin wheel o as the origin, and established the coordinate system $x'o'y'$ with CAM center Q as the origin, P is the speed relative instantaneous center of CAM and pin wheel for this moment, *M* is CAM and needle tooth meshing points, Pressure angle αP is the angle between the force direction and speed direction of CAM theoretical profile at the meshing point M, range of values is[0°,90°][6].

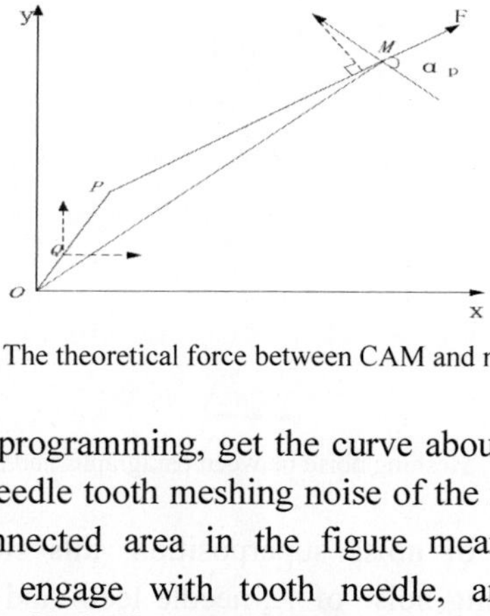

Figure 1. The theoretical force between CAM and needle tooth.

Using MATLAB programming, get the curve about each paragraphs in the indexing period and needle tooth meshing noise of the mechanism, as shown in Figure 3. The disconnected area in the figure means in this time period, paragraphs does not engage with tooth needle, and the noise is gone, dimensionless time $t \in [0,0.5]$ is acceleration section[1].

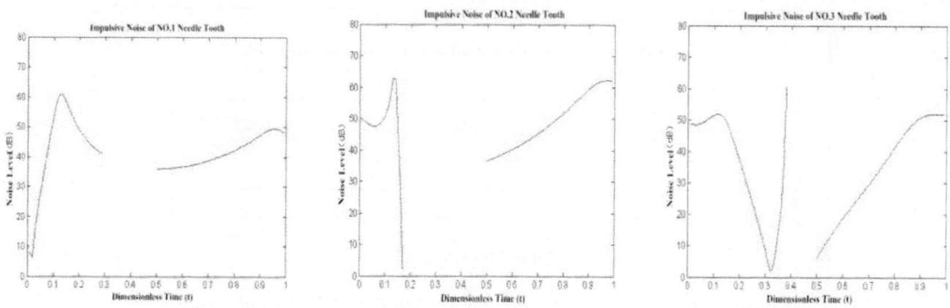

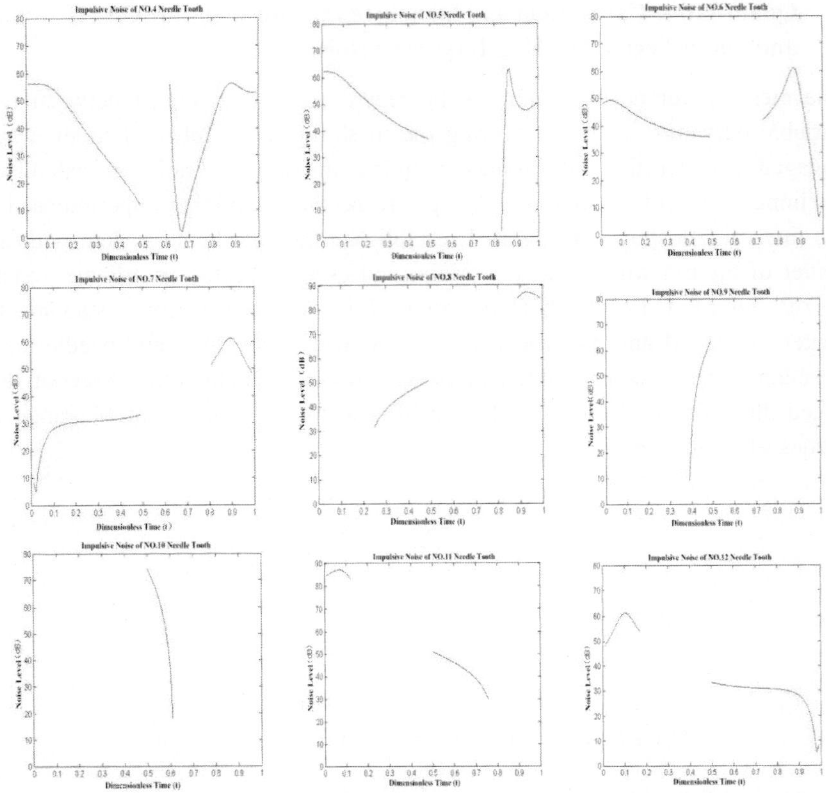

Figure 2. Meshing noise between paragraphs and needle.

With the principle of noise superposition, this study superimposed the calculation of the meshing noise of 12 needle teeth and CAM in the center of gears, and the curve chart of noise after superposition is shown in Figure 3.

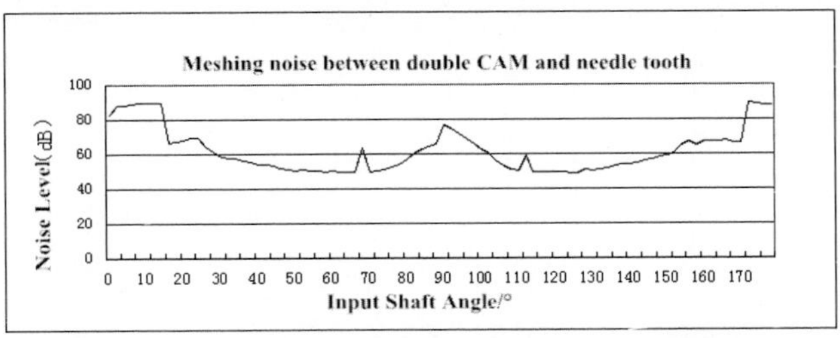

Figure 3. Meshing noise between double CAM and needle tooth.

As shown in Figure 4, at the beginning and the end of the meshing between CAM and needle tooth, the noise level reaches a peak of about 87dB; In the middle-term of the meshing, the meshing noise once increased to 75dB. Although, simultaneously, the size of the noise level and the number of meshing needle teeth is not a simple linear relationship, the trend of the curve is basically the same. After analyzing the value of noise source, it can be found that the 8th, 9th, 10th, 11th sections of paragraphs noise is the main sound source which make the peak of noise, therefore, under the premise of meeting the transmission conditions of the mechanism, get the optimization of the CAM paragraphs line. Figure 4 is a amplifying figure of Transition-curve at crossover point.

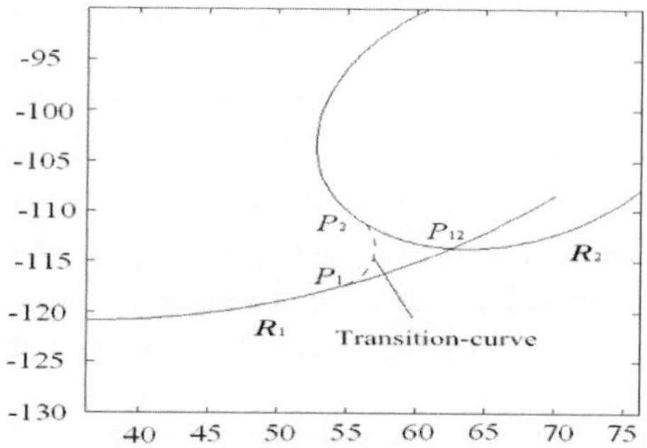

Figure 4. Amplifying figure of smoothed curve at crossover point.

Conditions for Transition-curve at crossover point [7]:

(1) Transition-curve P1P2 must at the inner side of crossover point;

(2) Transition-curve P1P2 must pass through the Point P1and P2;

(3) The first derivative at the Point P1 and P2 of the Transition-curve P1P2 is respectively, equal to the first derivative at the Point P1 and P2 of paragraphs profile equidistant of inside and outside R1, R2;

(4) Transition-curve P1P2 outwardly projecting;

(5) Transition-curve P1P2 not satisfies the conditions of self-locking.

From the above five conditions, the theoretical equation for Transition-curve of paragraphs can be listed, and take it into the original noise equation, get the meshing noise curve for the 8th, 9th, 10th, 11th sections paragraphs after optimization, as Figure 5.

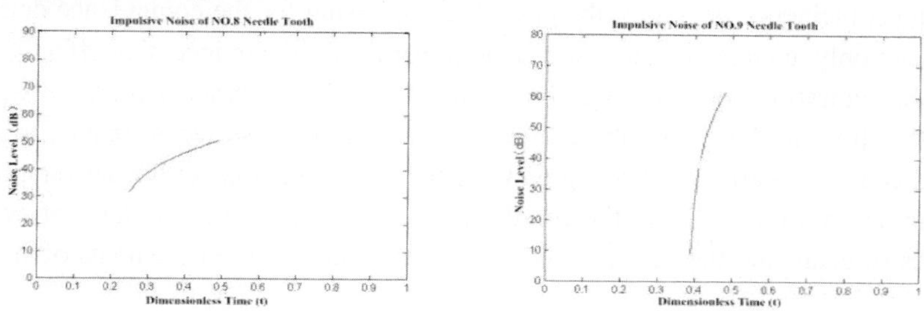

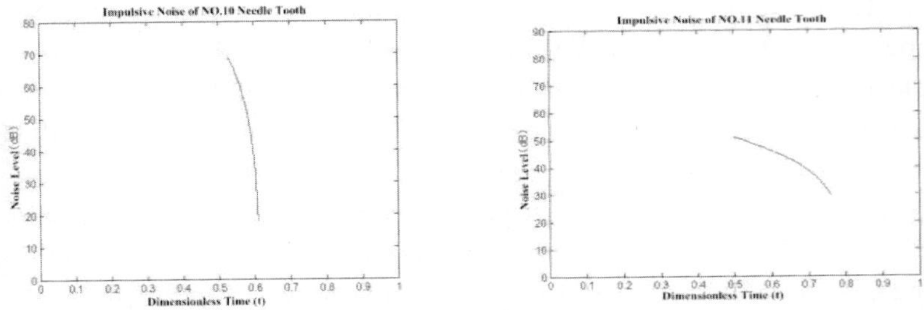

Figure 5. Meshing noise between paragraphs and needle tooth after optimization.

As shown in Figure 6, after optimization of the 8th and 11th sections paragraphs, the high - noise area is successfully avoided, but the meshing area, between the two paragraphs and the needle tooth, is decreased about 25%; after optimization of the 9th and 10th sections of paragraphs, some part of the high – noise reduce 5dB, the meshing area is decreased about 20%. After optimization for paragraphs profile, get meshing noise of CAM in a indexing period and needle tooth as Figure 6.

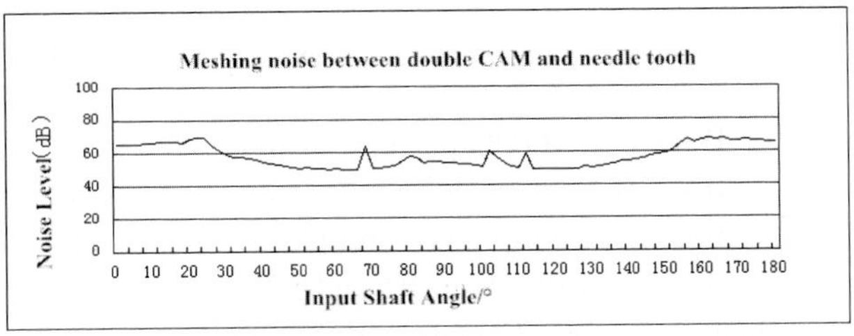

Figure 6. Meshing noise between double CAM and needle tooth.

4. Quantitative Calculation for Gear Meshing Impact Noise and Structure Improvement

Radiation sound power and impact sound pressure is an inversely proportional to the impact time, the shorter impact time, the higher the radiation sound power and impact sound pressure. The gear parameters, when chosen, is mostly related to the coincidence degree of the gear. The increasing for the coincidence degree can not only average transmission error, lower the dynamic incentive effectively, so that vibration and noise can be reduced, but also increasing the number of tooth while meshing, distribute the load to more tooth surfaces, reduce to the load and deformation of the single tooth, and reduce the impact for meshing into tooth and out tooth. Under the condition of comprehensive advantages of every types of gears and the gear parameters and also meet the requirements of speed and power, design a suitable helical gears to replace the original straight gear to the transmission.

Selection for parameters of helical gears:

Table 1. Helical gear parameters.

Modulus m	Teeth Z	Helix angle β	Transfer power (kw)	Accuracy	Tooth width (m)
2	26	15°	1.1	8	0.3

Table 2. Optimization of parameters before and after contrast.

Before and after optimization	Contact ratio β	Radiation power(dB)	The impact pressure(dB)
Spur gear	1.621	47.8	56.9
Helical gear	2.756	42.5	50.5

Compare the optimization of parameters before and after, the gear transmission noise decreased 6.4dB, decreased 11%, and the transmission for helical gear makes the length of meshing line increased, the load on unit length of meshing line decreased, and it is important for improving the stability and carrying ability of the gear transmission.

5. Noise of Entire Mechanism

Noise, which from connection point of Bearing and key etc. is so much lower than the impact noise from mechanism, can be neglect. Before and after the structure improvement, overlay the gear transmission noise and CAM and needle teeth meshing noise at the center of pin wheel, can get the noise curve for entire mechanism, as Figure 7.

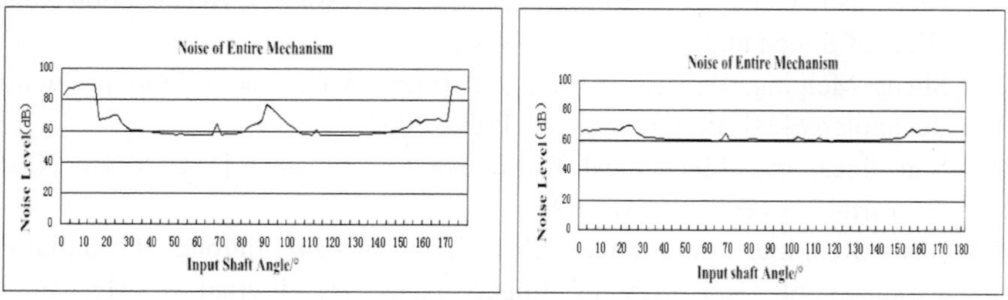

Figure 7. The noise of the machine before and after optimization.

As can be seen from the figure above, after structure improvement, the whole machine noise is significantly decreased: in an indexing period, the peak noise of the initial meshing was decreased by 20dB, decreased 22.2%; the peak noise during the mid-term meshing was decreased by 8dB, decreased 10.2%; the peak noise at the end of meshing was decreased by 20.6dB, decreased 23.1%, above data shows the noise of the mechanism can be greatly decreased after paragraphs optimization.

6. Conclusion

This chapter analyses the gear meshing, caused by machining error and elastic deformation, and the meshing between CAM and needle teeth, and gives equation for impact speed and impact time between two variable curvature cylinders and also the function to calculate the impact radiation noise, through which the impact acceleration noise of mechanism can be calculated, and get the following conclusions:

(1) Meshing impact time and meshing impact noise is shown as inversely proportional relationship, using helical gears to replace the original gear in the mechanism, can increase the impact time and stationary impact force, and has important significance to improving the stability and carrying capacity of gear transmission.

(2) Crossover point of CAM paragraph is the main area of high noise, therefore on the premise of meet the driving conditions, optimized the crossover point of paragraph can avoid the high noise area successfully, so that the noise of mechanism can be decreased sharply.

References

1. Chen Gang. Dynamics and Experimental Investigation of Inner Displaced Indexing Cam Mechanism [D]. Tianjin: Tianjin University of Science and Technology, 2013.
2. Richards E J, et al. On the Prediction of Impact Noise, I: Acceleration Noise [J]. J. of Sound and Vibration, 1979, 62(4).
3. Sheng Meiping, Wang Minqing, Sun Jincai. Noise and Vibration Control Technology [M]. Beijing: Science Press, 2007.
4. Sun Jincai etc. Mechanical Noise Control Principle [M]. Xi'an: Xi'an Industrial University Press, 1993.
5. Shao Renping, Sun Jincai, Shen Yunwen, Jia Jinrong. Quantitative Estimates of Gear Meshing Impact Noise [J]. Journal of Mechanical Science and Technology, 2001, 20(3).
6. Wang Yihui. Academic Design and Experimental Investigation of Inner Displaced Indexing Cam Mechanism [D]. Tianjin: Tianjin University of Science and Technology, 2012.

7. Tang Shanhua. The Intersection of the Parallel Indexing CAM Contour and the Smooth Transition Processing [J]. Mechanical Design, 2008, 25(9).